WORKSHOP ON PHYSICS AT THE FIRST MUON COLLIDER AND AT THE FRONT END OF A MUON COLLIDER

WORKSHOP ON PHYSICS AT THE FIRST MUON COLLIDER AND AT THE FRONT END OF A MUON COLLIDER

Batavia, IL   November 1997

EDITORS
Stephen Geer
Rajendran Raja
Fermi National Accelerator Laboratory

American Institute of Physics

AIP CONFERENCE
PROCEEDINGS 435

Woodbury, New York

Editors:

Stephen Geer and Rajendran Raja
Fermi National Accelerator Laboratory
PO Box 500
Batavia, IL 60510-0500

Email: geer@fnal.gov
       raja@fnal.gov

Authorization to photocopy items for internal or personal use, beyond the free copying permitted under the 1978 U.S. Copyright Law (see statement below), is granted by the American Institute of Physics for users registered with the Copyright Clearance Center (CCC) Transactional Reporting Service, provided that the base fee of $15.00 per copy is paid directly to CCC, 222 Rosewood Drive, Danvers, MA 01923. For those organizations that have been granted a photocopy license by CCC, a separate system of payment has been arranged. The fee code for users of the Transactional Reporting Service is: 1-56396-793-6/ 98 /$15.00.

© 1998 American Institute of Physics

Individual readers of this volume and nonprofit libraries, acting for them, are permitted to make fair use of the material in it, such as copying an article for use in teaching or research. Permission is granted to quote from this volume in scientific work with the customary acknowledgment of the source. To reprint a figure, table, or other excerpt requires the consent of one of the original authors and notification to AIP. Republication or systematic or multiple reproduction of any material in this volume is permitted only under license from AIP. Address inquiries to Office of Rights and Permissions, 500 Sunnyside Boulevard, Woodbury, NY 11797-2999; phone: 516-576-2268; fax: 516-576-2499; e-mail: rights@aip.org.

L.C. Catalog Card No. 98-71902
ISBN 1-56396-793-6
ISSN 0094-243X
DOE CONF- 971194

Printed in the United States of America

CONTENTS

Preface xi
Scientific Advisory Committee xiii
Local Organizing Committee xiii
Plenary Session Chairpersons.. xiii
Parallel Working Group Conveners xiv
Secretariat xiv
Workshop Program xv
Working Groups and Convenors.. xix
Agenda... xxi
The Muon Collider Collaboration xliii

WORKSHOP PARAMETERS

Accelerator Scenario and Parameters for the First Muon Collider
and Front–End of a Muon Collider 3
    C. Ankenbrandt and S. Geer

PLENARY PAPERS

Muon Collider: Introduction and Status................................. 11
    R. B. Palmer for the Muon Collider Collaboration
Physics at a Muon Collider... 37
    J. F. Gunion
Low Energy Physics and the First Muon Collider 58
    W. J. Marciano
Deep Inelastic Scattering and Neutrino Physics........................ 66
    P. Spentzouris
Intermediate Energy Physics in the Strangeness Section with Hadron
Beams... 82
    G. B. Franklin
SUSY Before the Next Lepton Collider................................... 91
    F. E. Paige
Supersymmetry vis-à-vis Muon Colliders.............................. 107
    V. Barger
Physics with Low Energy Hadrons 121
    G. Gutierrez and L. Littenberg
Experimental Neutrino Physics at the Muon Collider Complex 139
    P. Fisher and B. Kayser
Physics with Low Energy Muons at the Front End of the Muon Collider 152
    W. Molzon
Deep Inelastic Scattering at a Muon Collider—Neutrino Physics.............. 166
    H. Schellman

Higgs Boson and Z Physics at the First Muon Collider 177
    M. Demarteau and T. Han
Prospects for SUSY Searches and Measurements at a Muon Collider 193
    M. Carena and S. Protopopescu
Strong Dynamics at the Muon Collider: Working Group Report 208
    P. C. Bhat and E. J. Eichten
Top Quark Physics at Muon and Other Future Colliders 227
    M. S. Berger and B. L. Winer
Physics with a Millimole of Muons 242
    C. Quigg

WORKING GROUP PAPERS

Physics with Low Energy Hadrons

Lepton Flavor Violating Rare Muon Decays and Future Prospects 261
    Y. Kuno
Prospects for Hadronic Physics at a Muon Collider Facility 274
    K. K. Seth
Constraints on Strong Dynamics from Rare B and K Decays 287
    G. Burdman
Can BNL-Style Studies of $K \to {}^-\pi\nu\bar{\nu}$ be Pushed at the FEMC? 299
    L. Littenberg
K^+ Decays in Flight, AGS E865 and E923 308
    H. Ma
Lessons Learned from E871, A Search for Rare Kaon Decays at BNL 318
    M. Bachman

Neutrino Physics

On Future Charged Lepton-Hadron Colliders 327
    S. Ritz
Neutrino Physics at a Muon Collider 334
    B. J. King
Neutrino Oscillation Physics with BooNE 349
    R. Stefanski
Neutrino Physics in a Muon Collider 358
    R. N. Mohapatra
Aspects of Neutrino Reactions at the First Muon Collider 370
    E. A. Paschos
Detectors for Neutrino Physics at the First Muon Collider 376
    D. A. Harris and K. S. McFarland
The Physics Potential of Neutrino Beams from Muon Storage Rings 384
    S. Geer
K2K KEK to Kamioka Long Baseline Neutrino Oscillation Experiment 391
    T. Kobayashi

Measurement of $\sin^2\theta_W$ at the First Muon Collider.................... 398
    J. Yu and A. V. Kotwal

Slow/Stopped Muon Physics

Coherent Muon–Electron Conversion in Muonic Atoms.................... 409
    A. Czarnecki, W. J. Marciano, and K. Melnikov
Muon Capture.................... 419
    P. Kammel
Muonic Processes in Solid Hydrogen.................... 427
    G. M. Marshall, J. M. Bailey, G. A. Beer, J. L. Beveridge, M. C. Fujiwara,
    T. M. Huber, R. Jacot-Guillarmod, P. Kammel, S. K. Kim, P. E. Knowles,
    A. R. Kunselman, M. Maier, G. R. Mason, F. Mulhauser, A. Olin,
    C. Petitjean, T. A. Porcelli, L. A. Schaller, and J. Zmeskal
An Update on MEGA: An Experimental Search for Lepton Number
Non-Conservation.................... 437
    E. V. Hungerford
New Ideas to Improve Searches for $\mu^+ \to e^+ \gamma$.................... 443
    M. D. Cooper
Model for Pion Production in Proton-Nucleus Interactions.................... 453
    N. V. Mokhov and S. I. Striganov
The MECO Muon Beam.................... 460
    M. Bachman
$\mu \to e$ Conversion Status and Prospects.................... 466
    J. Sculli
MECO Physics Backgrounds Studies.................... 475
    T. Liu
MECO Muon Yield Simulation Using Experimental Data.................... 481
    R. M. Djilkibaev
Prospects for High Precision Measurements on Muonic Atoms
at the Front End of a Muon Collider.................... 486
    D. Kawall, M. G. Boshier, V. W. Hughes, K. Jungmann, W. Liu,
    and G. zu Putlitz

Deep Inelastic Scattering Physics

Photon Scattering in Muon Collisions.................... 495
    M. Klasen
A Small Target Neutrino Deep-Inelastic Scattering Experiment
at the First Muon Collider.................... 505
    D. A. Harris and K. S. McFarland
Prospects for a Measurement of $\nu p \to \nu p$ Scattering at the Muon Collider
Complex.................... 511
    E. G. Stern

SUSY Searches and Measurements

SUSY Searches at LEP: Present Status and Future Prospects 519
    J. Nachtman
Search for Supersymmetry at the Tevatron 525
    E. Flattum
Scalar Top Quark Production at $\mu^+\mu^-$ Colliders 531
    A. Bartl, H. Eberl, S. Kraml, W. Majerotto, and W. Porod
Sleptons at a First Muon Collider 537
    F. E. Paige
Precision Measurements of Threshold Chargino Production 543
    M. S. Berger
Supersymmetry at the NLC 549
    D. L. Wagner
SUSY Signatures and Model Discrimination at $\mu^+\mu^-$ Colliders 555
    J. G. Kelly
Flavor and CP Violations from Sleptons at the Muon Collider 561
    H.-C. Cheng
Precision Measurements at The Higgs Resonance: A Probe of Radiative Fermion Masses 567
    F. M. Borzumati, G. R. Farrar, N. Polonsky, and S. Thomas
Lepton Flavor Violation and Fermion Masses 573
    S. Raby

Higgs and Z^0 Physics

Calibrating the Energy of a 50×50 GeV Muon Collider Using Spin Precession 583
    R. Raja and A. Tollestrup
A Muon Collider as Z Factory 597
    A. Blondel
Electroweak and Heavy Flavor Physics at SLD 611
    S. Willocq
Estimates of Vertex Tagging Efficiencies at a Muon Collider Higgs Factory ... 621
    B. J. King
Prospects for Higgs at the Tevatron 627
    J. Womersley
Testing 2HDM at Muon Colliders 635
    M. Krawczyk
Searches for the MSSM Higgs Bosons at LEP 641
    T. Greening
The Search for Higgs Bosons of Minimal Supersymmetry at the LHC 647
    C. Kao
Higgs Resonance Studies at the First Muon Collider 657
    B. Kamal, W. J. Marciano, and Z. Parsa

Precision W—Boson and Higgs Boson Mass Determinations at Muon
Colliders.. 663
    M. S. Berger
Bounds on the Standard Higgs Boson..................................... 669
    J. Erler and P. Langacker
Measuring Trilinear Gauge Boson Couplings at Hadron and Lepton
Colliders.. 675
    U. Baur
Quartic Gauge Boson Couplings 685
    H.-J. He
Probing Anomalous Higgs Coupling through $\mu^+\mu^- \to H\gamma$.................. 701
    A. Abbasabodi, D. Bowser-Chao, D. A. Dicus, and W. W. Repko

Strong Dynamics Physics

Technicolor and the First Muon Collider.................................. 711
    K. Lane
Topcolor and the First Muon Collider 723
    C. T. Hill
Compositeness Test at the FMC with Bhabha Scattering 734
    E. J. Eichten and S. Keller
Constraints on Strong Dynamics from Rare B and K Decays................. 742
    G. Burdman
Technihadron Production at a Muon Collider.............................. 754
    J. Womersley
Testing Technicolor with Scalars at the First Muon Collider 758
    B. A. Dobrescu
Search for Technicolor Particles in W+2jet with b-tag Channel at CDF...... 766
    T. Handa, K. Maeshima, J. Valls, and R. Vilar
A Strong Electroweak Sector at Future $\mu^+\mu^-$ Colliders.................... 772
    R. Casalbuoni, S. De Curtis, D. Dominici, A. Deandrea, R. Gatto,
    and J. F. Gunion

T Tbar Factory

Top Quark at the Upgraded Tevatron to Probe New Physics................. 783
    J. M. Yang, A. Datta, M. Hosch, C. S. Li, R. J. Oakes, K. Whisant,
    B.-L. Young, and X. Zhang
Top Quark Physics at the Next Linear Collider 789
    R. Raja
The Top-Antitop Threshold at Muon Colliders............................ 797
    M. S. Berger
Top Quark Pair Production at Threshold—Uncertainties and Relativistic
Corrections 803
    A. H. Hoang

Flavor Changing Neutral Currents at $\mu^+\mu^-$ Colliders 813
  L. Reina
Gluon Radiation in Top Production and Decay at Lepton Colliders 823
  L. H. Orr
Top Quark Physics at a Polarized Muon Collider 831
  S. Parke

Accelerator

Space-Charge Effects of the Proposed High-Intensity Fermilab Booster 841
  K.-Y. Ng and Z. Qian
High-Intensity Muon Storage Rings for Neutrino Production:
Lattice Design 851
  C. Johnstone

APPENDICES

List of Participants 861
Author Index 867

Preface

The workshop on the muon collider held in November 1997 at Fermilab attracted more than 200 physicists from the high energy community in the U.S, Europe and Japan. The rather unwieldy title "The Workshop on physics at the First Muon Collider and at the front end of a Muon Collider" served to bring together people with a wide variety of interests ranging from atomic physics using slow or stopped muons to Higgs and SUSY physics at the First Muon Collider energy frontier. The workshop addressed accelerator issues associated with the high intensity proton driver needed to produce the muons at high enough rates, the problems associated with cooling and collection of the muons, and the physics capabilities of experiments using muon, neutrino and hadron beams produced at the accelerator complex associated with the First Muon Collider. It became evident that for many neutrino experiments there are significant advantages in using the highly collimated high intensity neutrino beams produced by muons decaying in the straight section of a storage ring. The muons can also be used to collide with proton beams to extend the range of deep inelastic scattering beyond HERA energies and luminosities. The large numbers of slow cold muons available before the acceleration process would enable the search of rare processes such as $\mu \rightarrow e$ conversions to hitherto unprecedented levels of sensitivity.

The neutral Higgs boson may well be discovered by the time the First Muon Collider turns on. The muon collider would then afford a unique opportunity to study the s channel production of Higgs particles. A First Muon Collider, whose energy can be calibrated precisely using $g - 2$ spin precession, can be used to measure the mass and width of the Higgs boson precisely. The beam energy spread is much smaller in a muon collider compared to an electron positron linear collider of comparable energy because both synchrotron radiation and beamsstrahlung occur at much smaller rates. This can be used to good advantage to measure the mass of the W boson precisely using a First Muon Collider and also to explore the threshold of top quark production and obtain the top quark mass precisely. The workshop also showed that detector backgrounds due to muon decay expected at the interaction region are at tolerable levels. Calls were made by various physics groups to the machine physicists to increase the luminosity of the collider by factors of three or more above current predicted levels, since various physics channels would benefit substantially by such an increase.

This workshop was a highly enjoyable event both for the organizers and (we believe) for large numbers of the participants. We would like to thank the local organizing committee, the scientific advisory panel, and the physics conveners for their help with the workshop. We would also like to thank Cynthia Sazama in particular for her untiring efforts, and Maja Christensen and Jody Federwitz for help with the proceedings.

Steve Geer and Rajendran Raja
Fermilab, March 1998

SCIENTIFIC ADVISORY COMMITTEE

Keith R. Ellis — Fermi National Accelerator Laboratory
John Gunion — University of California Davis
Juan C. Gallardo — Brookhaven National Laboratory
Gail G. Hanson — Indiana University
William Marciano — Brookhaven National Laboratory
David Neuffer — Fermi National Accelerator Laboratory
John Peoples — Fermi National Accelerator Laboratory
D. Hywel White — Los Alamos National Laboratory
Lincoln Wolfenstein — Carnegie Mellon University

LOCAL ORGANIZING COMMITTEE

Edmond L. Berger — Argonne National Laboratory
Robert H. Bernstein — Fermi National Accelerator Laboratory
Peter S. Cooper — Fermi National Accelerator Laboratory
Stephen Geer, Co-Chair — Fermi National Accelerator Laboratory
Christopher T. Hill — Fermi National Accelerator Laboratory
Stephen D. Holmes — Fermi National Accelerator Laboratory
Jorge G. Morfin — Fermi National Accelerator Laboratory
Rajendran Raja, Co-Chair — Fermi National Accelerator Laboratory
Cynthia Sazama — Fermi National Accelerator Laboratory
Alvin V. Tollestrup — Fermi National Accelerator Laboratory

PLENARY SESSION CHAIRPERSONS

William A. Bardeen — Fermi National Accelerator Laboratory
Vernon Barger — University of Wisconsin
Gerald T. Garvey — Los Alamos National Laboratory
Gail G. Hanson — Indiana University
Rabi Mohapatra — University of Maryland
John Peoples — Fermi National Accelerator Laboratory
Gordon Thomson — Rutgers University
Bruce Winstein — The University of Chicago
D. Hywel White — Los Alamos National Laboratory

WORKING GROUP CONVENERS

Chuck Ankenbrandt	Fermi National Accelerator Laboratory
Mike Berger	Indiana University
Pushpa Bhat	Fermi National Accelerator Laboratory
Marcela Carena	Fermi National Accelerator Laboratory
Marcel Demarteau	Fermi National Accelerator Laboratory
Tom Diehl	Fermi National Accelerator Laboratory
Estia Eichten	Fermi National Accelerator Laboratory
Peter Fisher	Massachusetts Institute of Technology
Gaston Gutierrez	Fermi National Accelerator Laboratory
Boris Kayser	National Science Foundation
Tao Han	University of Wisconsin
Laurence Littenberg	Brookhaven National Laboratory
William Molzon	University of California, Irvine
Bob Noble	Fermi National Accelerator Laboratory
Serban Protopopescu	Brookhaven National Laboratory
Steve Ritz	Columbia University
Heidi Schellmann	Fermi National Accelerator Laboratory
Brian Winer	Ohio State University

SECRETARIAT

Maja Christensen	Fermi National Accelerator Laboratory
Jody Federwitz	Fermi National Accelerator Laboratory
Carol Picciolo	Fermi National Accelerator Laboratory
Cynthia M. Sazama	Fermi National Accelerator Laboratory
Suzanne Weber	Fermi National Accelerator Laboratory

Workshop on Physics at the First Muon Collider and at the Front End of a Muon Collider

November 6 - 9, 1997
Fermi National Accelerator Laboratory, Batavia, Illinois USA

Workshop Program

Thursday November 6th

Opening Plenary Session (9:00 am - 14:30 pm)

Chair: B. Winstein

9:00-9:15	Welcome	J. Peoples
9:15-9:45	Muon Collider: Introduction and Status	R. Palmer
9:45-10:30	Physics with the Muon Collider	J. Gunion
10:30-11:00	Coffee Break	

Chair: H. White

11:00-11:45	Stopped Muons, Rare Decays, and Low Energy Hadrons	W. Marciano
11:45-12:30	Neutrino Physics and DIS	P. Spentzouris
12:30-13:30	Lunch Break	

Chair: G. Thomson

13:30-14:00	Accelerator Scenario and Parameters for the Workshop	C. Ankenbrandt
14:00-14:30	Working Group Plans	S. Geer/R. Raja

Working Group Sessions (14:30 - 17:30 pm)

14:30-15:30   Working Group Sessions

15:30-16:00   Coffee Break

16:00-17:30   Working Group Sessions

19:00 pm   Reception

Friday November 7th

Plenary Session (9:00 am - 9:45 am)

Chair: G. Garvey

09:00-09:45	Intermediate Energy Physics	G. Franklin
09:45-10:15	Coffee Break	

Working Group Sessions (10:15 am - 17:30 pm)

10:15-12:30   Working Group Sessions

12:30-13:30   Lunch Break

13:30-15:30   Working Group Sessions

15:30-16:00   Coffee Break

16:00-17:30   Working Group Sessions

Muon Collider Collaboration Meeting (19:00 - 22:00 pm)

Saturday November 8th

Plenary Session (9:00 am - 10:30 am)

Chair: R. Mohapatra

09:00-09:40	Introduction to Supersymmetry and SUSY at the LHC	F. Paige

09:40-10:30	Supersymmetry vis-a-vis Muon Colliders	V. Barger
10:30-11:00	Coffee Break	

Working Group Sessions (11:00 am - 12:30 pm)

11:00-12:30	Working Group Sessions	
12:30-13:30	Lunch Break	

Plenary Session (13:30 - 14:30 pm)

Chair: W. Bardeen

13:30-14:30	Recent Developments in String Theory (The Need for Higher Energy Colliders)	E. Witten

Working Group Sessions (14:30 - 17:30 pm)

14:30-15:30	Working Group Sessions	
15:30-16:00	Coffee Break	
15:30-17:30	Working Group Sessions	
20:30 pm	Conference Dinner	

Sunday November 9th

Closing Plenary Session (8:30 am - 12:30 pm)

Chair: G. Hanson

8:30- 9:00	Accelerator Working Group	C. Ankenbrandt B. Noble
9:00- 9:30	Physics with Low Energy Hadrons	L. Littenberg G. Gutierrez
9:30-10:00	Neutrino Physics	B. Kayser P. Fisher
10:00-10:30	Stopped Muon Physics	W. Molzon T. Diehl
10:30-11:00	Coffee Break	

Chair: V. Barger

11:00-11:30	Deep Inelastic Scattering	H. Schellman S. Ritz
11:30-12:00	Higgs and Z0 Physics	T. Han M. Demarteau
12:00-12:30	SUSY Searches and Measurements	M. Carena S. Protopopescu
12:30-13:30	Lunch	

Chair: J. Peoples

13:30-14:00	Strong Dynamics	E. Eichten P. Bhat
14:00-14:30	t tbar Factory	M. Berger B. Winer
14:30-15:30	Workshop Summary	C. Quigg

Workshop on Physics at the First Muon Collider and at the Front End of a Muon Collider

November 6 - 9, 1997
Fermi National Accelerator Laboratory, Batavia, Illinois USA

Working Groups and Convenors

1. Physics with the Front-End

Physics with Low Energy Hadrons (Protons, Pions, Kaons, ..)

Lawrence Littenberg (littenbe@bnl.gov)
Gaston Gutierrez (gaston@fnal.gov)

Neutrino Physics

Boris Kayser (bk@einstein.mps.nsf.gov)
Peter Fisher (fisherp@mail.cern.ch)

Slow/Stopped Muon Physics

William Molzon (molzon@ucihep.ps.uci.edu)
Tom Diehl (diehl@fnal.gov)

Deep Inelastic Scattering (Using Muon and Neutrino Beams and a Muon-Proton Collider)

Heidi Schellman (schellman@fnal.gov)
Steve Ritz (ritz@nevis.columbia.edu)

2. Physics at the FMC

Higgs and Z0 Physics

Tao Han (than@ucdhep.ucdavis.edu)
Marcel Demarteau (demarteau@fnal.gov)

SUSY Searches and Measurements

Marcela Carena (carena@fnal.gov)
Serban Protopopescu (serban@d01.phy.bnl.gov)

Strong Dynamics

Estia Eichten (eichten@fnal.gov)

Pushpa Bhat (bhat@fnal.gov)

t tbar factory

Mike Berger (berger@gluon.physics.indiana.edu)
Brian Winer (winer@fnald.fnal.gov)

3. Accelerator Working Group

Chuck Ankenbrandt (ankenbrandt@fnal.gov)
Bob Noble (noble@fnal.gov)

Agenda of the Physics with Low Energy Hadrons Working Group

Thursday November 6th (14:30 - 17:30 pm)

14:30-14:40	Working Group Plans	L. Littenberg
		G. Gutierrez
14:40-15:20	pi/k/p Yields at 16 GeV	N. Mokov
15:30-16:00	Coffee Break	
16:00-16:25	Physics with a Phase Rotated Pion Beam	Y. Kuno
16:30-17:25	Prospects for Hadronic Physics	K. Seth
19:00 pm	Reception	

Friday November 7th (10:30 - 17:30 pm)

10:30-11:05	Baryon Spectroscopy	H. Spinka
11:05-11:35	Polarized Pbars	H. Spinka
11:35-12:05	CPT	S. Schnetzr
12:10-12:25	CKM	B. Tschirhart
12:30-13:30	Lunch Break	
13:30-14:00	KTeV	B. Hsiung
14:05-14:35	KAMI	K. Arisaka
14:40-15:25	T Violation in Kmu3	H. Ma
15:30-16:00	Coffee Break	
16:00-16:45	Joint Session with the Strong Dynamics Group	
16:00-16:45	Constraints in Strong Dynamics from B and K Decays	G. Burdman

16:45-17:30   Discussion on Physics at the FMC Front End

Saturday November 8th (11:00 - 17:30 pm)

11:00-11:40	Rare K Decays at BNL	L. Littenberg
11:45-12:00	K Decays in BNL-865	H. Ma
12:05-12:30	Discussion on Kaon Physics at the FMC Front End	
12:30-13:30	Lunch Break	
14:30-15:30	Discussion on Kaon Physics at the FMC Front End	
15:30-16:00	Coffee Break	
16:00-17:30	General Working Group Discussion and Summary	
19:30 pm	Conference Dinner	

Sunday November 9th (8:30 - 15:30 pm)

08:30-15:30   Working Group Summaries

Agenda of the Neutrino Physics Working Group

Thursday November 6th (14:30 - 17:40 pm)

14:30-15:30	Joint with DIS Group	
14:30-14:40	Joint Working Group Goals	
14:40-15:05	The Future of Charged Lepton/Proton Scattering	S. Ritz
15:05-15:30	Design for a Muon-Proton Collider	V. Shiltzev
15:30-16:00	Coffee Break	
16:00-17:40	Joint with DIS Group	
16:00-16:30	High Energy Neutrino Physics at a Muon-Collider Complex	B. King
16:30-17:00	Neutrino Oscillation Phyiscs with BooNe	R. Stefanski
17:00-17:30	TBA	
19:00 pm	Reception	

Friday November 7th (10:30 - 17:30 pm)

10:15-12:30	Theoretical Situation	
10:15-11:15	Theoretical Perspectives of Neutrino Mass Phyiscs at the Muon Collider	R. Mohapatra
11:15-11:30	Discussion	
11:30-12:00	Aspects of Neutrino Mass Phyiscs at the Muon Collider	E. Paschos
12:00-12:30	Discussion	
12:30-13:30	Lunch Break	
13:30-15:30	Experimental Environment	

13:30-14:30	Experimental Environment	M. Goodman W. Leeson
14:30-14:50	Neutrino Backgrounds	E. Paschos
14:50-15:10	Non-Neutrino-Oscillation Physics with the BooNe Detector at the Ugraded Booster	R. Tayloe
15:30-16:00	Coffee Break	
16:00-17:30	First Discussion of Experimental Design	

Saturday November 8th (10:30 - 17:30 pm)

10:30-12:30	Experimental Designs - Detection	
10:30-11:00	Sampling Calorimeter	K. MacFarland
11:00-11:30	Emulsions	N. Reay
11:30-12:00	Silicon + Target	S. Mishra M. Cadenas
12:00-12:30	Discussion	
12:30-13:30	Lunch Break	
14:30-15:30	Experimental Designs - Long Baseline	
14:30-15:00	Long Baseline Neutrino Ideas Using an Intense Muon Source	S. Geer
15:00-15:30	K2K	T. Kobayashi
15:30-16:00	Coffee Break	
16:00-17:30	Discussion of Design for Muon Collider	
19:30 pm	Conference Dinner	

Sunday November 9th (8:30 - 15:30 pm)

08:30-15:30	Working Group Summaries

Agenda of the Slow/Stopped Muon Physics Working Group

Thursday November 6th (14:30 - 17:30 pm)

14:30-14:45	Organization and Goals	
14:45-15:10	Polarized Muons for $\mu^+ \to e^+ \gamma$	Y. Kuno
15:50-15:50	Muon EDM Experiment	W. Molzon
15:30-16:00	Coffee Break	
16:00-16:45	Theoretical Issues for Muon Conversion	A. Czarnecki
16:45-17:30	Nuclear Muon Capture	P. Kamel
19:00 pm	Reception	

Friday November 7th (10:15 - 17:30 pm)

10:15-11:00	Muon Catalyzed Fusion Experiments	G. Marshall
11:00-11:45	Status of MEGA	E. Hungerford
11:45-12:15	New Ideas for $\mu^+ \to e^+ \gamma$	M. Cooper
12:30-13:30	Lunch Break	

Joint Session with Accelerator Group

14:00-15:30	Beams Division Seminar:	
	Beam Requirements for MECO	W. Molzon
	AGS Implementation	M. Brennan
15:30-16:00	Coffee Break	

Joint Session with Accelerator Group

16:00-16:20	Preliminary Ideas for Modifications to the FNAL Booster for Muon Experiments	C. Moore
16:20-16:55	A Ring Cooler for Muons	V. Balbekov
16:55-17:30	Plans for the JHF	Y. Kuno

Saturday November 8th (11:00 - 17:30 pm)

11:00-11:20	Models for Low Energy Pion Production	N. Mokhov S. Striganov
11:20-11:50	E910 Data on Low Energy Pion Production	H. Kirk
11:50-12:20	Simulations of Muon Beamlines	M. Bachman
12:20-12:30	Experiments to Measure Pion Yields Discussion	
12:30-13:30	Lunch Break	
14:30-15:30	Muon Conversion Experiment	
14:30-15:00	Present Status and Prospects for Muon Conversion	J. Sculli
15:00-15:30	Physics Backgrounds	T. Liu
15:30-16:00	Coffee Break	
16:00-16:45	Detector Rate Issues	E. Hungerford
16:45-17:30	Triggering and Data Acquisition	D. Koltick
19:30 pm	Conference Dinner	

Sunday November 9th (8:30 - 15:30 pm)

08:30-15:30	Working Group Summaries	

Agenda of the DIS Physics Working Group

Thursday November 6th (14:30 - 17:40 pm)

14:30-15:30	Joint with Neutrino Group	
14:30-14:40	Joint Working Group Goals	
14:40-15:05	The Future of Charged Lepton/Proton Scattering	S. Ritz
15:05-15:30	Design for a Muon-Proton Collider	V. Shiltzev
15:30-16:00	Coffee Break	
16:00-17:40	Joint with Neutrino Group	
16:00-16:30	High Energy Neutrino Physics at a Muon-Collider Complex	B. King
16:30-17:00	Neutrino-Oscillations with BooNe	R. Stefanski
17:00-17:30	TBA	
19:00 pm	Reception	

Friday November 7th (10:30 - 17:30 pm)

10:30-11:00	Prospects for Photo-Production	M. Klasen
11:00-11:30	Neutrino DIS at a Muon Collider: Too Much of a Good Thing?	D. Harris
11:30-12:00	Nu-p Elastic Scattering	P. Spentzouris
12:00-12:30	Discussion	
12:30-13:30	Lunch Break	
13:30-15:30	Discussion	
15:30-16:00	Coffee Break	

16:00-17:00	Neutrino Oscillations "Wine and Cheese"	G. Fuller

---

Saturday November 8th (11:00 - 17:30 pm)

11:00-12:30	Experiments/Beamlines Design Session 1
12:30-13:30	Lunch Break
14:30-15:30	Experiments/Beamlines Design Session 2
15:30-16:00	Coffee Break
16:00-17:30	Experiments/Beamlines Design Session 3
19:30 pm	Conference Dinner

Sunday November 9th (8:30 - 15:30 pm)

08:30-15:30	Working Group Summaries

Agenda of the SUSY Physics Working Group

Thursday November 6th (14:30 - 17:30 pm)

14:30-14:50	Working Group Plan	S. Protopopescu
		M. Carena
14:50-15:25	SUSY General Introduction	C. Wagner
15:30-16:00	Coffee Break	
16:00-16:25	SUSY after LEP	J. Nachtman
16:30-16:55	SUSY after the Tevatron	E. Flattum
17:00-17:25	SUSY after LHC	I. Hinchliffe
19:00 pm	Reception	

Friday November 7th (10:30 - 17:30 pm)

10:15-10:45	NLC	D. Wagner
10:50-11:15	Top Squark Production at Muon Colliders	W. Porod
11:20-11:45	Slepton Production at Muon Collider	F. Paige
11:50-12:30	Discussion Session:	
	- Chargino Production at Muon Collider	M. Berger
		J. Kelly
	- NLC: Similarities with Muon Collider	D. Wagner
13:00-13:30	Lunch Break	

Afternoon Session Joint with Higgs Group

13:30-14:10	Review MSSM Higgs Properties	H. Haber
14:15-14:55	SUSY Higgs at LEP	T. Greening

15:00-15:30	Higgs Physics at the LHC	C. Kao
15:30-16:00	Coffee Break	
16:00-16:50	Strategy for Higgs Discovery and Study at the LHC, NLC and FMC	J. Gunion
16:50-17:10	Higgs Resonance Physics	W. Marciano
17:10-18:00	Discussion Higgs Opportunities at the FMC	

Saturday November 8th (9:00 - 17:30 pm)

11:00-11:25	RPV in the Production of Lepton/ Quark Pairs at Muon Colliders	S. Raychaudhuri
11:30-11:55	RPV and Sneutrino Resonance at Muon Colliders	J. Feng
12:00-12:30	SUSY Signatures at a High Energy Muon Collider	J. Kelly
12:30-13:30	Lunch Break	
13:30-14:30	Witten's talk	
14:40-15:10	Flavour and CP Violation from Sleptons at Muon Colliders	H. Cheng
15:10-15:35	Precision Challenges in SUSY and Higgs Physics	N. Polonsky
15:30-16:00	Coffee Break	
16:00-16:30	Lepton Flavour Violation: Signals of GUT Scale Physics	S. Raby
16:30-17:30	General Working Group Discussion and SUMMARY	
19:30 pm	Conference Dinner	

Agenda of the Higgs and Z-Physics Working Group

Thursday November 6th (14:30 - 17:30 pm)

14:30-14:35	Working Group Goals	M. Demarteau T. Han
14:35-15:10	Energy Calibration	R. Raja
15:10-15:40	Detector and Background Issues	P. Lebrun
15:40-16:00	Coffee Break	
16:00-16:50	Physics at a 10^{33} Z-Pole Factory	A. Blondel
16:55-17:45	Recent SLD Results	S. Willocq
19:00 pm	Reception	

Friday November 7th (10:30 - 17:30 pm)

10:30-11:00	Tau and C-Tagging	B. King
11:00-11:30	SM Higgs Opportunities at the Tevatron	J. Womersley
11:30-11:50	Higgs --> Gamma Gamma at the LHC	D. Rainwater
11:50-12:10	Higgs Opportunities at the NLC	T. Han
12:10-12:30	Higgs Searches in 2-Doublet Models	M. Krawczyk
12:30-12:45	Higgs-Z0 and Higgs-Gamma Production at the FMC	F. Tikhonin
12:45-13:00	Discussion Higgs Opportunities at the FMC	
12:30-13:30	Lunch Break	

Afternoon Session Joint with SUSY

13:30-14:10	Review MSSM Higgs Properties	H. Haber
14:15-14:55	SUSY Higgs at LEP	T. Greening
15:00-15:30	Higgs Physics at the LHC	C. Kao
15:30-16:00	Coffee Break	
16:00-16:50	Strategy for Higgs Discovery and Study at the LHC, NLC and FMC	J. Gunion
16:50-17:10	Higgs Resonance Physics	W. Marciano
17:10-18:00	Discussion Higgs Opportunities at the FMC	

Saturday November 8th (9:00 - 17:30 pm)

11:00-11:30	CP Violation at the FMC	W. Keung
11:30-12:00	Z pole Physics with High Luminosity and Beam Polarization	T. Han
12:00-12:30	Discussion CP Violation and Beam Polarization	
12:30-13:30	Lunch Break	
14:30-15:00	WW Pair Processes & MW Measurements	M. Berger
15:00-15:30	Precision of Electroweak Parameters	J. Erler
15:30-16:00	Coffee Break	
16:00-16:40	Trilinear Gauge Boson Couplings	U. Baur
16:40-17:10	Quartic Gauge Boson Couplings	H. He
17:10-17:25	Unravelling the WWZ Vertex at the FMC	F. Tikhonin
17:25-18:00	Discussion Electroweak Opportunities at the FMC and Summary	
19:30 pm	Conference Dinner	

Sunday November 9th (8:30 - 15:30 pm)

08:30-14:30	Working Group Summaries
14:30-15:30	Workshop Summary

Agenda of the Higgs and Z-Physics Working Group

Thursday November 6th (14:30 - 17:30 pm)

14:30-14:35	Working Group Goals	M. Demarteau T. Han
14:35-15:10	Energy Calibration	R. Raja
15:10-15:40	Detector and Background Issues	P. Lebrun
15:40-16:00	Coffee Break	
16:00-16:50	Physics at a 10^{33} Z-Pole Factory	A. Blondel
16:55-17:45	Recent SLD Results	S. Willocq
19:00 pm	Reception	

Friday November 7th (10:30 - 17:30 pm)

10:30-11:00	Tau and C-Tagging	B. King
11:00-11:30	SM Higgs Opportunities at the Tevatron	J. Womersley
11:30-11:50	Higgs --> Gamma Gamma at the LHC	D. Rainwater
11:50-12:10	Higgs Opportunities at the NLC	T. Han
12:10-12:30	Higgs Searches in 2-Doublet Models	M. Krawczyk
12:30-12:45	Higgs-Z0 and Higgs-Gamma Production at the FMC	F. Tikhonin
12:45-13:00	Discussion Higgs Opportunities at the FMC	
12:30-13:30	Lunch Break	

Afternoon Session Joint with SUSY

13:30-14:10	Review MSSM Higgs Properties	H. Haber
14:15-14:55	SUSY Higgs at LEP	T. Greening
15:00-15:30	Higgs Physics at the LHC	C. Kao
15:30-16:00	Coffee Break	
16:00-16:50	Strategy for Higgs Discovery and Study at the LHC, NLC and FMC	J. Gunion
16:50-17:10	Higgs Resonance Physics	W. Marciano
17:10-18:00	Discussion Higgs Opportunities at the FMC	

Saturday November 8th (9:00 - 17:30 pm)

11:00-11:30	CP Violation at the FMC	W. Keung
11:30-12:00	Z pole Physics with High Luminosity and Beam Polarization	T. Han
12:00-12:30	Discussion CP Violation and Beam Polarization	
12:30-13:30	Lunch Break	
14:30-15:00	WW Pair Processes & MW Measurements	M. Berger
15:00-15:30	Precision of Electroweak Parameters	J. Erler
15:30-16:00	Coffee Break	
16:00-16:40	Trilinear Gauge Boson Couplings	U. Baur
16:40-17:10	Quartic Gauge Boson Couplings	H. He
17:10-17:25	Unravelling the WWZ Vertex at the FMC	F. Tikhonin
17:25-18:00	Discussion Electroweak Opportunities at the FMC and Summary	
19:30 pm	Conference Dinner	

Sunday November 9th (8:30 - 15:30 pm)

    08:30-14:30   Working Group Summaries

    14:30-15:30   Workshop Summary

Agenda of the Strong Dynamics Working Group

Thursday November 6th (14:30 - 17:30 pm)

14:30-14:45	Working Group Plans	P. Bhat E. Eichten
14:45:15:30	Technicolor and Muon Colliders	K. Lane
15:30-16:00	Coffee Break	
16:00-16:45	Top and Strong Dynamics	C. Hill
16:45-17:30	Discussion	
19:00 pm	Reception	

Friday November 7th (10:30 - 17:30 pm)

10:30-11:15	Compositeness Tests at the FMC	E. Eichten S. Keller
11:15-11:45	Discussion	
11:45-12:30	Detector/Background Issues	R. Roser P. Lebrun
12:30-13:30	Lunch Break	
13:30-15:30	Discussion	
15:30-16:00	Coffee Break	
16:00-16:45	Joint Session with Low Energy Hadrons Group	
16:00-16:45	Constraints on Strong Dynamics from B and K Decays	G. Burdman
16:45-17:30	Discussion	

Saturday November 8th (11:00 - 17:30 pm)

11:00-11:45	Detecting Techni-Rho at the First Muon Collider	J. Womersley
11:45-12:30	Testing Technicolor with Scalars at the First Muon Collider	B. Dobrescu
12:30-13:30	Lunch Break	
14:30-15:30	Discussion	
15:30-16:00	Coffee Break	
16:00-16:50	Joint Session with Higgs and Z Physics Working Group	
16:00-16:50	Trilinear Gauge Boson Couplings	U. Baur
16:50-17:30	Discussion	
19:30 pm	Conference Dinner	

Sunday November 9th (8:30 - 15:30 pm)

08:30-14:30	Working Group Reports
14:30-15:30	Workshop Summary

Agenda of the T TBar Factory at the First Muon Collider Working Group

Thursday November 6th (14:30 - 17:30 pm)

14:30-14:40	Working Group's Goals/Overview	M. Berger B. Winer
14:40-15:20	Top Quark Processes at the Upgraded Tevatron to Probe New Physics	J. Yang
15:30-16:00	Coffee Break	
16:00-16:40	Top Measurements at Hadron Colliders	R. Hughes
16:40-17:20	Measurements at the NLC	R. Raja
17:20-17:40	The TTbar Threshold at Muon Colliders	M. Berger
19:00 pm	Reception	

Friday November 7th (10:15 - 17:30 pm)

10:15-10:55	TTbar Production at Threshold - Uncertainties and Relativistic Corrections	A. Hoang
10:55-11:25	The Top Quark Mass and the Threshold	M. Smith
11:25-11:50	Two-Loop Corrections to TTbar Production at Higher Energies	A. Hoang
12:30-13:30	Lunch Break	
13:30-14:20	Flavor Changing Neutral Currents	L. Reina
14:20-15:10	Gluon Radiation in Top Production/Decay	L. Orr
15:10-15:30	Discussion	
15:30-16:00	Coffee Break	
16:00-16:25	Top Quark Physics at Polarized Muon Colliders	S. Parke

16:25-17:10	Detector Issues	R. Roser
17:10-17:30	Discussion	

Saturday November 8th (11:00 - 17:30 pm)

11:00-12:30	Attend other groups
12:30-13:30	Lunch Break
14:30-15:30	Attend Other Groups
15:30-16:00	Coffee Break
16:00-17:30	Attend Other Groups
19:30 pm	Conference Dinner

Sunday November 9th (8:30 - 15:30 pm)

08:30-15:30	Working Group Summaries

Agenda of the Accelerator Working Group

Thursday November 6th (14:30 - 17:30 pm)

14:30-14:45	Organization and Goals	
14:45-15:30	Overview of the Fermilab Proton Source Summer Study	C. Ankenbrandt
15:30-16:00	Coffee Break	
16:00-16:45	RF System Design for a High-Intensity Fermilab Booster; Summary of the Recent Inductive-Insert Experiment at the Los Alamos PSR	J. Griffin
16:45-17:30	Discussion on Generating a Baseline RF System Design for a Fermilab Proton Source	
19:00 pm	Reception	

Friday November 7th (10:30 - 17:30 pm)

10:30-11:30	Longitudinal Beam Simulations in a High-Intensity Proton Synchrotron using ESME	I. Kourbanis Z. Qian
11:30:12:30	Discussion	
12:30-13:30	Lunch Break	
13:30-14:00	Joint Session with Stopped Muon Physics Group	
14:00-15:30	Beams Division Seminar:	
	Beam Requirements for MECO	W. Molzon
	AGS Implementation	M. Brennan
15:30-16:00	Coffee Break	
16:00-17:30	Joint Session with Stopped Muon Physics Group	

	Preliminary Ideas for Modifications to the Fermilab Booster for Muon Experiments	C. Moore
	A Ring Cooler for Muons	V. Balbekov
	Plans for the JHF	Y. Kuno

Saturday November 8th (11:00 - 17:30 pm)

11:00-11:30	Instability Issues for High-Intensity Proton Drivers	K. Ng
11:30-12:30	Discussion	
12:30-13:30	Lunch Break	
14:30-15:30	Review Progress and Discuss Parameters for the Proposed Fermilab Proton Source	
15:30-16:00	Coffee Break	
16:00-17:30	Discussion	
19:30 pm	Conference Dinner	

Sunday November 9th (8:30 - 15:30 pm)

| 08:30-15:30 | Working Group Summeries |

The Muon Collider Collaboration

Charles M. Ankenbrandt[1], Giorgio Apollinari[2], Muzaffer Atac[1], Bruno Autin[3], Valerie I. Balbekov[1], Vernon D. Barger[4], Odette Benary[5], Michael S. Berger[6], S. Alex Bogacz[7], Shlomo Caspi[8], Christine Celata[8], Yong-Chul Chae[9], David B. Cline[10], John Corlett[8], H. Thomas Diehl[1], Alexandr Drozhdin[1], Richard C. Fernow[11], Yasuo Fukui[1], Miguel A. Furman[8], Juan C. Gallardo[11], Alper A. Garren[8], Stephen H. Geer[1], Michael A. Green[8], John F. Gunion[12], Ramesh Gupta [8], Tao Han[12], Katherine C. Harkay[9], Colin Johnson[3], Carol Johnstone[1], Stephen A. Kahn[11], Bruce J. King[11], Harold G. Kirk[11], Masayukiu Kumada[13], Paul LeBrun[1], Kevin Lee[10], Derun Li[8], David Lissauer[11], Laurence S. Littenberg[11], Chang-guo Lu[14], Alfredo Luccio[11], Kirk T. McDonald[14], Alfred D. McInturff[8], Frederick E. Mills[1], Nikolai Mokhov[1], Alfred Moretti[1], David V. Neuffer[1], King-Yuen Ng[1], Robert J. Noble[1], James H. Norem[9,1], Blaine E. Norum[15], Hiromi Okamoto[16], Yasar Onel[17], Robert B. Palmer[11], Zohreh Parsa [11], Jack M. Peterson[8], Yuriy Pischalnikov[1], Milorad Popovic[1], Eric Prebys[14], Zubao Qian[1], Rajendran Raja [1], Pavel Rehak[11], Thomas Roser[11], Robert Rossmanith[18], Jack Sandweiss[19], Ronald M. Scanlan[8], Lindsay Schachinger[8], Andrew M. Sessler[8], Quan-Sheng Shu[7], Gregory I. Silvestrov[20], Alexandr N. Skrinsky[20], Ray Stefanski[1], Sergei Striganov[1], Iuliu Stumer[11], Don Summers[21], Valeri Tcherniatine[11], Lee C. Teng[9], Alvin V. Tollestrup[1], Yăgmur Torun[11], Dejan Trbojevic[11], William C. Turner[8], Andy Van Ginneken[1], Tatiana A. Vsevolozhskaya[20], Masayoshi Wake[22], Robert Weggel[11], Erich H. Willen[11], David R. Winn[23], Jonathan S. Wurtele[24], Yongxiang Zhao[11], Max Zolotorev[8]

[1] *Fermi National Laboratory, P. O. Box 500, Batavia, IL 60510*
[2] *Rockfeller University, New York, NY 10021*
[3] *CERN, 1211 Geneva 23 Switzerland*
[4] *Department of Physics, University of Wisconsin, Madison, WI 53706*
[5] *Tel-Aviv University, Ramat-Aviv, Tel-Aviv 69978, Israel*
[6] *Physics Department, Indiana University, Bloomington, IN 47405*
[7] *Jefferson Laboratory, 12000 Jefferson Ave., Newport News, VA 23606*
[8] *Lawrence Berkeley National Laboratory, 1 Cyclotron Rd., Berkeley, CA 94720*
[9] *Argonne National Laboratory, Argonne, IL 60439*
[10] *Univ. of California Los Angeles, Los Angeles, CA 900095*
[11] *Brookhaven National Laboratory, Upton, NY 11973*
[12] *Physics Department, University of California, Davis, CA 95616*
[13] *National Institute of Radiological Sciences, 4-9-1 Anagawa, Inage, Chiba, Japan*
[14] *Princeton University, P. O. Box 708, Princeton, NJ 08544*
[15] *University of Virginia, 205 McCormick Road, Charlottesville, VA 22901*
[16] *Nuclear Science Research Facility, Institute for Chemical Research, Kyoto University, Gokanoshou, Uji, Kyoto 611, Japan*
[17] *Physics Department, Van Allen Hall, University of Iowa, Iowa City, IA 52242*
[18] *DESY, Hamburg, Germany*
[19] *Physics Department, Yale University, CT 06520*
[20] *Budker Institute of Nuclear Physics, 630090 Novosibirsk, Russia*
[21] *University of Mississippi, Oxford, MS 38677*
[22] *KEK High Energy Accelerator Research Organization, 1-1 Oho, Tsukuba, 305 Japan*
[23] *Fairfield University, Fairfield, CT 06430*
[24] *University of California Berkeley, Berkeley, CA 94720*

WORKSHOP PARAMETERS

Accelerator Scenario and Parameters for the First Muon Collider and Front–End of a Muon Collider

C. Ankenbrandt and S. Geer

Fermi National Accelerator Laboratory, P.O. Box 500, Batavia, Illinois 60510

Abstract. In November 1997 a workshop was held at Fermilab to explore the physics potential of the first muon collider, and the physics potential of the accelerator complex at the "front–end" of the collider. This paper describes the configuration of the muon collider accelerator complex, including the major accelerator parameters, and the particle fluxes and luminosities that would result from such a facility.

INTRODUCTION

The Workshop on Physics at the First Muon Collider and Front–end of a Muon Collider was held at Fermilab from 6–9 November 1997. The goal of the workshop was to explore the physics potential of each of the various options for the first muon collider (FMC), including the physics that could be pursued at the accelerator complex at the "front–end" of the collider. The accelerator parameters assumed for the workshop are based on recent studies of how the facilities at Fermilab might evolve with two goals in mind: to enhance the existing Fermilab physics program based on proton beams, and to provide what is needed for a high-energy muon collider. A summary of these parameters can be found in Tables 1–3. Figure 1 shows in a schematic way how the FMC might fit within the existing accelerator complex at Fermilab.

FRONT–END PARAMETERS

The "Front-End" of a muon collider consists of:

(a) A high-intensity proton source. We will assume that the proton source accelerates protons to 16 GeV/c, is cycling at 15 Hz, and produces 2 proton bunches per cycle, each containing 5×10^{13} particles. These parameters

TABLE 1. Operational parameters of an upgraded Fermilab proton source for a Muon Collider. The right-most column shows parameters for the fully upgraded source, and the other columns for possible intermediate steps in the upgrade.

	Step 1 Scenario 1	Step 1 Scenario 2	Step 2	Step 3
Linac (operating at 15 Hz)				
Kinetic Energy (MeV)	400	1000	1000	1000
Pulse Length (μs)	0.75	0.75	0.75	0.75
H^- per pulse	1×10^{13}	1.5×10^{13}	2.5×10^{13}	1×10^{14}
Pre–Booster (operating at 15 Hz)				
Extraction Kinetic Energy (GeV)				4.5
Momentum Spread (95% FW)				0.5%
Circumference (m)				180.6
Protons per bunch				5×10^{13}
Number of bunches				2
Extracted bunch length (ns)				21
Transverse Emittance (mm–mr)				200π
Longitudinal Emittance (eV-sec)				1.8
Booster (operating at 15 Hz)				
Extraction Kinetic Energy (GeV)	16	8	16	16
Momentum Spread (95% FW)	$< 0.1\%$	$< 0.1\%$	$< 0.1\%$	1.2%
Circumference (m)	474.2	474.2	474.2	474.2
Protons per bunch	1.2×10^{11}	1.8×10^{11}	3×10^{11}	5×10^{13}
Number of bunches	84	84	84	2
Extracted bunch length (ns)	4.9	4.9	4.9	2.3
Transverse Emittance (mm–mr)	50π	30π	50π	240π
Longitudinal Emittance (eV-sec)	2.2	1.8	1.8	4.0

TABLE 2. Parameters of muon bunches downstream of the ionization cooling channel.

	Narrow σ_p	Broad σ_p
muons per bunch	5×10^{12}	5×10^{12}
μ^+ bunches per cycle	1	1
μ^- bunches per cycle	1	1
Momentum (MeV/c)	200	200
σ_p/p	5%	10%
Bunch length (cm)	1.5	10
Normalized $\epsilon_\perp$ (mm-mr)	200π	60π
Repetition rate (Hz)	15	15
μ^+ per year (10^7 secs)	7.5×10^{20}	7.5×10^{20}

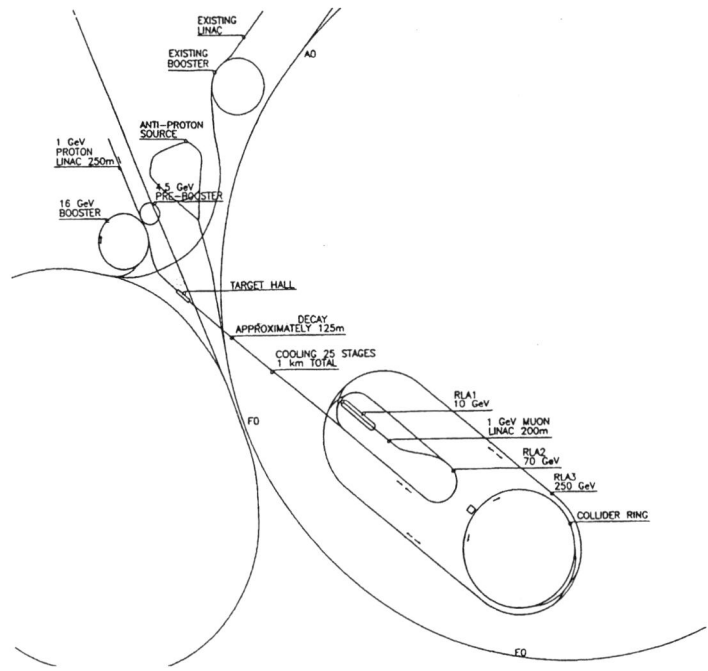

FIGURE 1. Schematic showing a plausible location for the First Muon Collider at Fermilab.

are based on the Fermilab summer study summarized in Ref. [1]. This upgrade to the existing proton source at Fermilab would require upgrading the 400 MeV Linac to a 1 GeV Linac, moving the 8 GeV Booster to a new location to overcome radiation limitations, upgrading the Booster energy to 16 GeV, and finally, adding a 4.5 GeV Pre-Booster to enable the protons to be compressed into short ($\sim$ 2 ns) long bunches. The upgrade can in principle be staged. Plausible staging steps and the associated proton source parameters are summarized in Table 1.

(b) A pion production and collection system, followed by a pion decay channel. Each incident proton bunch interacts in a target to produce $\sim 3 \times 10^{13}$ charged pions of each sign. The $\pi^{\pm}$ are confined within a high field solenoid co-axial with the beam direction. At the end of a 20 m long decay channel consisting of a 7 Tesla solenoid with a radius of 25 cm

TABLE 3. Recirculating linear accelerator parameters.

	RLA 1	RLA 2	RLA 3
Input Energy (GeV)	1.0	9.6	70
Output Energy (GeV)	9.6	70	250
No. of turns	9	11	12
Linac Length (m)	100	300	533.3
Arc Length (cm)	30	175	520
Bunch Length (ps)	158	43	19
Revolution Time (μs)	0.9	3.1	7.0
Decay Losses	9.0%	5.2%	2.4%
Initial muons per bunch	5×10^{12}	4.6×10^{12}	4.3×10^{12}
μ^+ bunches per sec	15	15	15

each incident proton results in about 0.2 muons of each charge. If in each accelerator cycle the first incident proton bunch is used to make and collect μ^+s, and the second bunch used for μ^-s, there will be about 10^{13} muons of each charge available at the end of the decay channel per cycle.

(c) A muon cooling channel. The muons exiting the decay channel populate a very diffuse 6–dimensional phase–space. The diffuse muon cloud must be cooled using a new fast cooling technique to form an intense beam before most of the muons have decayed. The cooling method proposed for the muon collider is ionization cooling [2]. Table 2 summarizes the properties of the muons at the end of the cooling channel. Note that the phase–space occupied by the muons can be optimized either to maximize the luminosity of the collider, or alternatively to minimize the beam energy spread at the expense of luminosity. At the end of the cooling channel each muon bunch is expected to contain about 5×10^{12} muons with a momentum of order 200 MeV/c.

(d) A muon acceleration system. A series of recirculating linear accelerators (RLAs) to accelerate the muons up to the colliding beam energy. Each RLA consists of two Linacs connected together by two arcs. Three RLAs with the operational parameters summarized in Table 3 would be able to accelerate the muons up to 250 GeV.

The front–end accelerator complex could be used for a variety of fixed target type physics experiments. Note first that the new Fermilab Main Injector can probably accept a factor of ~ 5 more protons per cycle than can be provided by the existing Fermilab proton source. Hence, an upgraded proton source of the type required for a muon collider would directly benefit the foreseen FNAL MI program. In addition, a muon collider front–end offers many other possibilities, some of them quite unique. Four working groups (Low Energy Hadron Physics, Neutrino Physics, Deep Inelastic Scattering, and Slow/Stopped Muon Physics) were convened in the workshop to consider the range of possibilities.

TABLE 4. Neutrino beam pulses from the straight sections of the Recirculating Linacs.

	1	2	3	4	5	6	7	8	9	10	11	12
RLA 1												
E_μ(start) (GeV)	1.0	1.96	2.92	3.88	4.84	5.8	6.76	7.72	8.68	9.64		
E_μ(end) (GeV)	1.48	2.44	3.4	4.36	5.32	6.28	7.24	8.2	9.16			
$<E_\mu>$ (GeV)	1.24	2.2	3.16	4.12	5.08	6.04	7.0	7.96	8.92			
$\gamma c\tau$ (km)	7.72	13.7	19.7	25.7	31.7	37.8	43.8	49.6	55.7			
$f_{decay}=100m/\gamma c\tau$(%)	1.3	0.73	0.51	0.39	0.32	0.26	0.23	0.20	0.18			
N_{decay}/bunch ($\times 10^{10}$)	6.5	3.7	2.6	2.0	1.6	1.3	1.2	1.0	0.9			
N_{decay}/year ($\times 10^{18}$)	9.8	5.5	3.8	2.9	2.4	2.0	1.7	1.5	1.4			
RLA 2												
E_μ(start) (GeV)	9.6	15.1	20.6	26.1	31.6	37.1	42.6	48.1	53.6	59.1	64.6	70.1
E_μ(end) (GeV)	12.4	17.9	29.4	28.9	34.4	39.9	45.4	50.9	56.4	61.9	67.4	
$<E_\mu>$ (GeV)	11.0	16.5	22.0	27.5	33.0	38.5	44.0	49.5	55.0	60.5	66.0	
$\gamma c\tau$ (km)	68.7	100	140	170	210	240	270	310	340	380	410	
$\gamma c\tau$ (km)	68.7	100	140	170	210	240	270	310	340	380	410	
$f_{decay}=300m/\gamma c\tau$(%)	0.44	0.30	0.21	0.18	0.14	0.13	0.11	0.097	0.088	0.079	0.073	
N_{decay}/bunch ($\times 10^{10}$)	2.0	1.4	0.97	0.83	0.64	0.60	0.51	0.45	0.40	0.36	0.34	
N_{decay}/year ($\times 10^{18}$)	3.0	2.1	1.5	1.2	0.96	0.90	0.77	0.68	0.60	0.5 4	0.51	
RLA 3												
E_μ(start) (GeV)	70	85	100	115	130	145	160	175	190	205	220	235
E_μ(end) (GeV)	77.5	92.5	108	123	138	153	168	183	198	213	228	243
$<E_\mu>$ (GeV)	73.8	88.8	104	119	134	149	164	179	194	209	224	239
$\gamma c\tau$ (km)	460	550	650	740	840	930	1000	1100	1200	1300	1400	1500
$f_{decay}=533m/\gamma c\tau$(%)	0.12	0.10	0.08	0.07	0.06	0.06	0.05	0.05	0.04	0.04	0.04	0.04
N_{decay}/bunch ($\times 10^{10}$)	0.52	0.42	0.35	0.31	0.27	0.25	0.23	0.21	0.1 9	0.18	0.16	0.15
N_{decay}/year ($\times 10^{18}$)	0.78	0.63	0.53	0.46	0.41	0.37	0.34	0.31	0.28	0.26	0.25	0.23

The neutrino beams would result from decays of circulating muons. A muon collider accelerator complex offers the very attractive possibility of making intense neutrino beams using muon decay channels consisting of straight sections in the RLAs or in the collider itself. The resulting beams would have precisely calculable fluxes [3] and, for μ^- decays, would be a mixture of 50% ν_μ and 50% $\bar\nu_e$. This would provide a uniquely "clean" tool for neutrino physics.

The characteristics of the neutrino pulses downstream of the RLAs are summarized in Table 4.

THE FIRST MUON COLLIDER

The workshop parameters for the First Muon Collider are shown in Table 5. Note that the assumptions that went into computing the luminosities were somewhat conservative. To obtain a more aggressive but still reasonable set of goals for the FMC these luminosities can be multiplied by a factor of three.

TABLE 5. Parameters for (going from left to right) a narrowband low–energy, broadband low–energy, medium–energy, top factory, and higher–energy FMC.

$\sqrt{s}$	100	100	200	350	500
σ_p/p	3×10^{-5}	1×10^{-3}	1×10^{-3}	1×10^{-3}	1×10^{-3}
Muons per bunch	3×10^{12}	3×10^{12}	2×10^{12}	2×10^{12}	2×10^{12}
Number of bunches	1	1	2	2	2
Repetition rate (Hz)	15	15	15	15	15
Norm. $\epsilon_\perp$ (mm-mr)	297π	85π	67π	56π	50π
Collider circum. (m)	380	380	700	864	1000
f_{rev} (Hz)	7.9×10^5	7.9×10^5	4.3×10^5	3.5×10^5	3.0×10^5
turns/lifetime	820	820	890	1260	1560
β^* (cm)	13	4	3	2.6	2.3
σ_z (cm)	13	4	3	2.6	2.3
σ_r (μm)	286	85	47	30	22
L_{peak} ($cm^{-2}s^{-1}$)	6×10^{32}	7×10^{33}	6×10^{33}	1×10^{34}	2×10^{34}
L_{av} ($cm^{-2}s^{-1}$)	5×10^{30}	6×10^{31}	1×10^{32}	3×10^{32}	7×10^{32}

SUMMARY

The tables presented here contain the parameters, fluxes, and luminosities that were used by the working groups as they evaluated the physics potential of a First Muon Collider and of its front end. Summaries of the physics potential of a facility operating with these parameters can be found in references [4] and [5].

Acknowledgments

This work was performed at the Fermi National Accelerator Laboratory, which is operated by Universities Research Association, under contract DE-AC02-76CH03000 with the U.S. Department of Energy.

REFERENCES

1. S. Holmes et al., "A Development Plan for the Fermilab Proton Source", FERMILAB-TM-2021, September 1997, unpublished.
2. A.N. Skrinsky and V.V. Parkhomchuk, Sov. J. Part. Nucl. **12**, 223 (1981).
3. S. Geer, "Neutrino Beams from Muon Storage Rings: Characteristics and Physics Potential", Fermilab-PUB-97/389 (hep-ph/9712290), submitted to Phys. Rev. D.
4. S. Geer, Workshop on Physics at the First Muon Collider and Front-End of a Muon Collider: A Brief Summary", Fermilab-Conf-98/063.
5. C. Quigg, "Physics with a Millimole of Muons", Fermilab-Conf-98/073-T.

Plenary Papers

Session Chairs: William A. Bardeen, Fermilab; Vernon Barger, University of Wisconsin; Gerald T. Garvey, Los Alamos National Laboratory; Gail G. Hanson, Indiana University; Rabi Mohapatra, University of Maryland; John Peoples, Fermilab; Gordon Thomson, Rutgers University; Bruce Winstein, University of Chicago; D. Hywel White, Los Alamos National Laboratory

Muon Collider: Introduction and Status

R. B. Palmer for the Muon Collider Collaboration[1]

*Physics Department
Brookhaven National Laboratory,
Upton, NY 11973-5000, USA*

Abstract. Parameters are given of machines with center-of-mass (CoM) energies of 3 TeV and 400 GeV but, besides a comment on neutrino radiation, the paper concentrates on progress on the design of a machine to operate at a light Higgs mass, assumed, for this study, to be 100 GeV (CoM).

INTRODUCTION

The possibility of muon colliders was introduced by Skrinsky et al. [1] and Neuffer [2]. More recently, a collaboration of over 100 members, lead by BNL, FNAL, LBNL, BNIP, University of Mississippi, Princeton University and UCLA has been formed to coordinate studies on specific designs. Work has been done on designs at a 3-4 TeV, 0.4-0.5 TeV and ≈100 GeV [3-7]. Tb. 1 gives the parameters of such colliders, and Figs. 1 and 2 show possible outlines of the 3 TeV and 100 GeV machines.

The original motive for considering muon colliders was the effective energy advantage of any lepton collider over hadron machines, together with the fact that muons, unlike electrons, generate negligible synchrotron radiation. As a result, a muon collider can be circular and much smaller than the current designs of linear electron colliders, and also much smaller than a hadron machine with the same *effective* energy.

In addition, a $\mu^+\mu^-$ collider would have some unique physics advantages over an e^+e^- collider:

- The direct coupling of a lepton-lepton system to a Higgs boson has a cross section that is proportional to the square of the mass of the lepton. As a result, the cross section for direct Higgs production from the $\mu^+\mu^-$ system is 40,000 times that from an e^+e^- system.

---

[1] Members of the Collaboration can be found at http://www.cap.bnl.gov/mumu/

TABLE 1. Parameters of Collider Rings

(CoM) energy	TeV	3	0.4	0.1		
p energy	GeV	16	16	16		
p's/bunch	10^{13}	2.5	2.5	5		
bunches/fill		4	4	2		
rep rate	Hz	15	15	15		
p power	MW	4	4	4		
μ/bunch	10^{12}	2	2	4		
μ power	MW	28	4	1		
wall power	MW	204	120	81		
collider circ	m	6000	1000	300		
depth	m	500	100	10		
rms $\frac{\Delta p}{p}$	%	.16	.14	.12	.01	.003
6D ϵ_6	10^{-12} $(\pi m)^3$	170	170	170	170	170
rms ϵ_n	π mm mrad	50	50	85	195	280
β^*	cm	0.3	2.3	4	9	13
σ_z	cm	0.3	2.3	4	9	13
σ_r spot	μm	3.2	24	82	187	270
tune shift		0.043	0.043	0.05	0.02	.015
Luminosity	$cm^{-2} sec^{-1}$	$5\ 10^{34}$	10^{33}	$1.2\ 10^{32}$	$2\ 10^{31}$	10^{31}
(CoM) $\frac{\Delta E}{E}$	10^{-5}	80	80	80	7	2
Higgs/year	$10^3\ year^{-1}$			1.6	4	4

- Because of the lack of beamstrahlung, a $\mu^+\mu^-$ collider can be operated with an energy spread of as little as 0.003 %. Furthermore, with the naturally occurring polarization it would be possible, by observing g-2, to determine the absolute energy to an accuracy of 10^{-6} or better [8]. It should thus be possible to use a $\mu^+\mu^-$ collider to make precision measurements of masses and direct measurements of the Higgs width (assumed to be $\approx$ 2 MeV), that would be otherwise impossible, with an e^+e^- collider.

Machines with energies higher than 3-4 TeV, have significant beam current constraints from off site neutrino radiation limits. If the required luminosities are to be reached without unacceptable hazards, then significant improvements in muon emittance over the current base line values are needed. There is however reason to believe such improvements are achievable, and machines with a center of mass energy of 10 TeV and luminosities of 10^{35} $cm^{-2}s^{-1}$ and above may be possible [9]. For energies below 3 TeV, for fixed muon currents, this radiation falls as the energy cubed, and it should be little problem for machines with energies of 1.5 TeV or less.

Recent work in the collaboration has concentrated on the lowest energy machine ($\approx$ 100 GeV), whose energy is taken to be representative of the possible mass of a light Higgs particle. Such a machine would serve as a demonstration of Muon Collider technology, a needed step before high energy machines can be considered, and as a unique physics tool to make and study, if they exist,

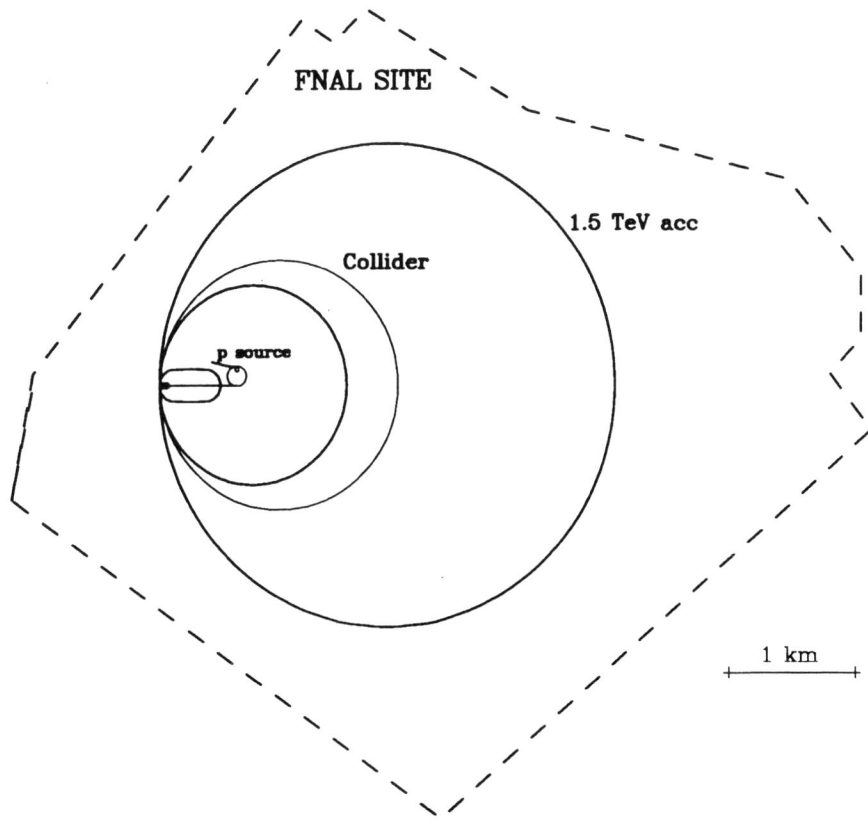

FIGURE 1. Plan of a 3 TeV Muon Collider.

Higgs particles in the S-channel.

PROTON DRIVER

π production rises approximately linearly with proton energy up to about 10 GeV after which it continues rising more slowly, but the requirement of very short bunches sets an effective minimum proton energy of about 16 GeV. The baseline specification used in Tb.2 is for a 16 GeV proton with a repetition rate of 15 Hz, 10^{14} protons per cycle in 2 or 4 bunches (depending on the collider energy), each with an rms bunch length of 1-2 ns. The total beam power is 4 MW. A design worked out at FNAL [10] would involve: a) An upgraded linac (0.4 → 1 GeV); b) higher energy booster (8 → 16 GeV) and c) new pre-booster. Some parameters are given in Tb. 2.

Another study had been done at BNL [11] that, while it did not quite reach the same beam power, involved far less upgrade: a) upgraded linac (0.2 → 0.6

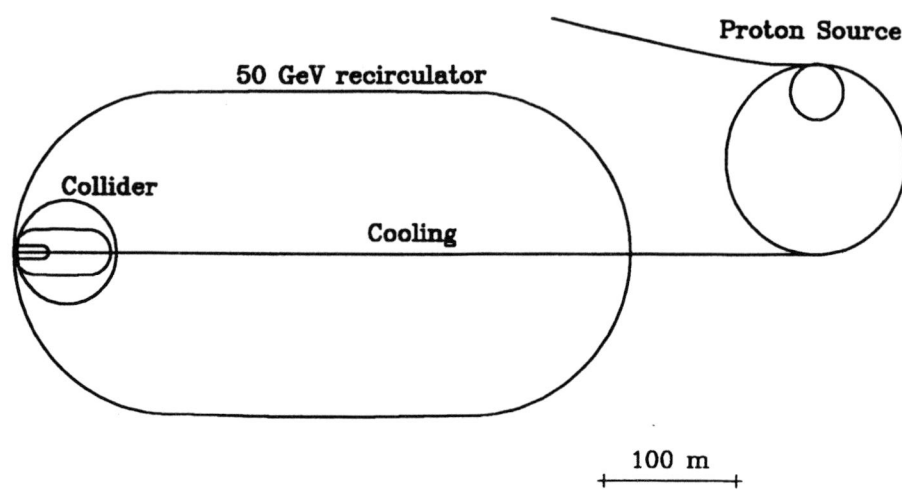

FIGURE 2. Plan of a 100 GeV Muon Collider.

TABLE 2. Proton Driver Specifications

		Linac	Pre-Booster	Booster
Final energy	GeV	1.0	4.5	16
Protons/bunch			$5\,10^{13}$	$5\,10^{13}$
No of bunches			2	2
Rep. freq	Hz	15	15	15
Circumference	m		180.6	474.2
Norm. 95% emit.	$\pi\ mm\ mrad$		200	240
sp ch tune shift			.39	.39
Final field	T		1.3	1.3

GeV); b) increased AGS rep rate: 2.5 Hz.

In order to reduce the cost of the muon phase rotation section and for minimizing the final muon longitudinal phase space, it appears now that the final proton bunch length should be 1-2 ns.

An experiment [12] at the AGS has tested a method to generate such short bunches by rapidly bringing the tune of the machine near transition and allowing a strong phase rotation to occur. Bunches were shortened from 8 ns rms to 2.2 ns with initial longitudinal phase space similar to that specified in the above design. Shorter bunches are expected in later experiments with better control.

Another experiment [13] has used variable inductors to reduce the longitudinal space charge effects.

Target and Pion Capture

π production is maximized by the use a well focused proton beam, small diameter target and a high Z target material. Tungsten, platinum or lead would be good, but the heating could not be easily removed and shock damage could be a problem. The use of a rapidly flowing liquid can solve the heating problem, but the shock could damage the enclosure, if one is used. We are thus considering the use of an open liquid jet. Such a jet has been tested [14] using mercury, although this was never exposed to a beam, and the jet did not move in a strong magnetic field, as required in our case. Theoretical studies of liquid metal flow in magnetic fields are underway [15,16], and the possibilities of using insulating liquids (e.g. PtO_2, Re_2O_3) and slurries (e.g. Pt in water) are being considered.

If the axis of the target is coincident with that of the solenoid field, then there is a relatively high probability that pions produced at the start of the target will reenter, interact again later and be lost. The probability for such interactions is reduced, and the overall production rate increased (by about 60 %) if the target and proton beam are set at an angle (10-15°) with respect to the field axis [17,18].

Three different codes [19–21] have been used to estimate π yields and, despite detailed differences between them, overall μ production was very similar. In addition, the collaboration is involved in an AGS experiment [22] to measure the π yields. The production is peaked at a relatively low pion momentum ($\approx$ 200 MeV/c), but has a very wide distribution: $\frac{\Delta E}{E}$ rms $\approx$ 100 %. The pion multiplicity, per 16 GeV proton, is about 2. At these low energies, the transverse momenta are of the order of 200 MeV/c. If a substantial fraction of these pions are to be captured, a very wide band system is required. A 20 T solenoid, 16 cm inside diameter is found to capture about half of all produced pions, and with target efficiency included, about 0.6 pions per proton emerge from the solenoid end [23]. Such a solenoid is well within the parameters of existing magnets [24]. It would have a superconducting outsert, and an 8 MW water cooled copper insert [25] (see Fig. 3).

After capture, the 20 T solenoid is matched [26] into a decay channel with 5 T fields and diameter of 30 cm.

Phase Rotation Linac

The pions, and the muons into which they decay, have an energy spread with an rms value of approximately 100 %. It would be difficult to handle such a wide spread in any subsequent system. A linac is thus introduced along the decay channel, with frequencies and phases chosen to deaccelerate the fast particles and accelerate the slow ones; i.e. to phase rotate the muon bunch. Tb.3 gives an example of parameters of such a linac. It is seen that the lowest

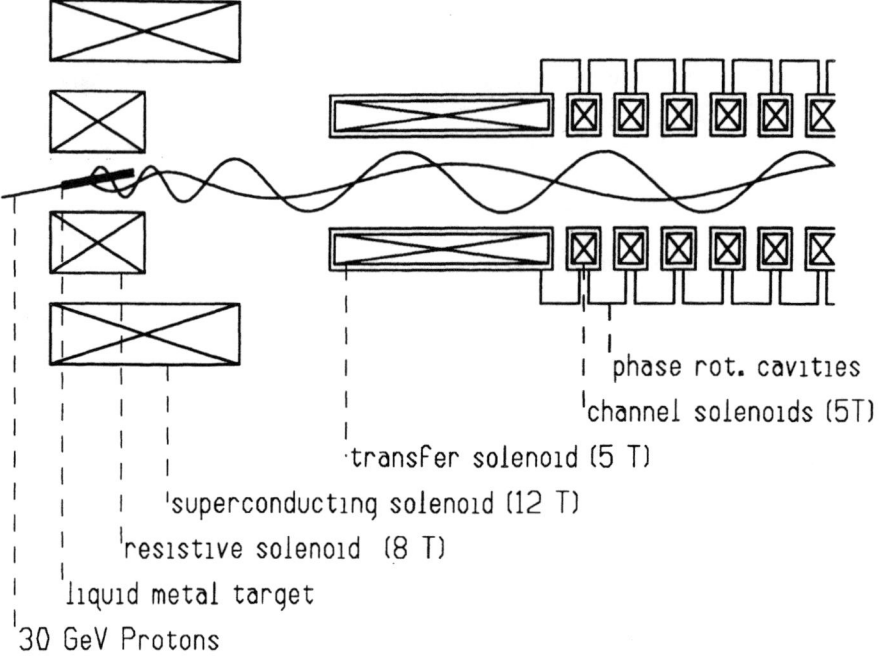

FIGURE 3. Schematics of the front end: skewed target, high field solenoid and decay and phase rotation channel

frequency is 30 MHz, a low but not impossible frequency for a conventional structure.

TABLE 3. Parameters of Phase Rotation Linacs

Linac	Length m	Frequency MHz	Gradient MeV/m
1	3	60	5
2	29	30	4
3	5	60	4
4	5	37	4

Fig. 4 shows the energy vs. c t at the end of the decay and phase rotation channel. A bunch is defined with mean energy 150 MeV, rms bunch length 1.7 m, and rms momentum spread 20 % (95 %, $\epsilon_L = 3.2\,\mathrm{eVs}$) in the Monte Carlo study [6]. The number of muons per initial proton in this selected bunch was 0.38, which can be compared with a value of 0.3 assumed in the baseline parameters.

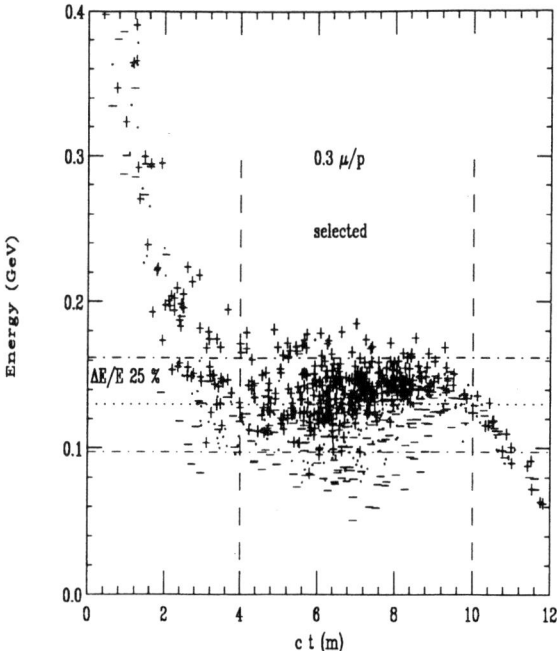

FIGURE 4. Energy vs. ct of μ's at end of decay channel with phase rotation.

Use of Both Signs

Protons on the target produce pions of both signs, and a solenoid will capture both, but the required subsequent phase rotation rf systems will have opposite effects on each. The baseline solution is to use two proton bunches, aim them at the same target one after the other, and adjust the rf phases such as to act correctly on one sign of the first bunch and on the other sign of the second.

A second possibility would be to separate the charges into two channels, and phase rotate them separately. However, the separation, probably using a bent solenoid, is not simple and would not be fully efficient. Whether a gain in overall efficiency could be achieved is not yet known.

Polarization

Polarized Muon Production

In the center of mass of a decaying pion, the outgoing muon is fully polarized (-1 for μ^+ and +1 for μ^-). In the lab system the polarization depends [27] on the decay angle θ_d and initial pion energy. For pion kinetic energy larger than the pion mass, the average is about 20 %, and if nothing else is done, the polarization of the captured muons and phase rotated by the proposed system is approximately this value.

If higher polarization is required, some selection of muons from forward pion decays ($\cos\theta_d \to 1$) is required. Fig. 4, above, showed the polarization of the phase rotated muons. The polarization P$> \frac{1}{3}$, $-\frac{1}{3} < P < \frac{1}{3}$, and P$< -\frac{1}{3}$ is marked by the symbols +, . and − respectively. If a selection is made on the minimum energy of the muons, then greater polarization is obtained. The tighter the cut, the higher the polarization, but the less the fraction F_{loss} of muons that are selected. Fig. 5 gives the results of a Monte Carlo study.

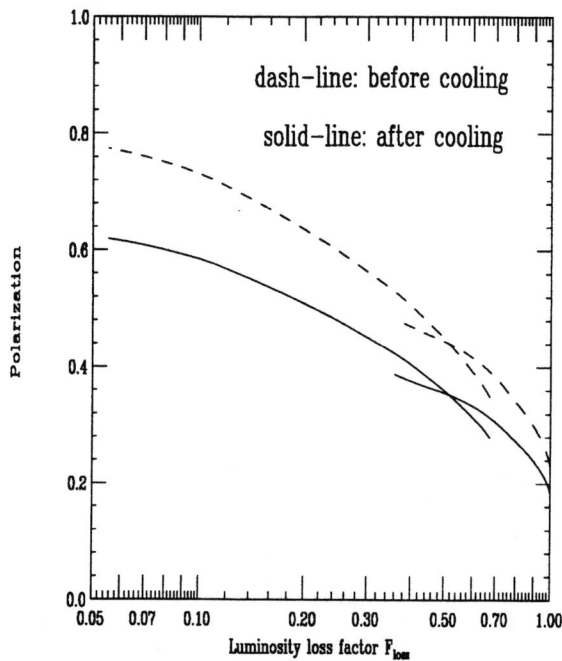

FIGURE 5. Polarization vs F_{loss} of μ's accepted.

If this selection is made on both beams, and if the proton bunch intensity is maintained, then naturally the muon bunch is reduced by the factor F_{loss} and the luminosity would fall by F_{loss}^2. But if, instead, proton bunches are merged so as to obtain half as many bunches with twice the intensity, then the muon bunch intensity is maintained and the luminosity (and repetition rate) falls only as F_{loss}.

One also notes that the luminosity could be maintained at the full unpolarized value if the proton source intensity could be increased. Such an increase in proton source intensity in the unpolarized case would be impractical because of the resultant excessive high energy muon beam power, but this restriction does not apply if the increase is used to offset losses in generating polarization.

Polarization Preservation

A paper [28] has discussed the preservation of muon polarization in some detail. During the ionization cooling process the muons lose energy in material and have a spin flip probability $\mathcal{P}$,

$$\mathcal{P} \approx \int \frac{m_e}{m_\mu} \beta_v^2 \frac{\Delta E}{E} \tag{1}$$

where β_v is the muon velocity divided by c, and $\frac{\Delta E}{E}$ is the fractional loss of energy due to ionization loss. In our case the integrated energy loss is approximately 3 GeV and the typical energy is 150 MeV, so the integrated spin flip probability is close to 10 %. The change in polarization $\frac{\Delta \mathcal{P}}{\mathcal{P}}$ is twice the spin flip probability, so the reduction in polarization is approximately 20 %. This loss is included in Fig. 5.

During circulation in any ring, the muon spins, if initially longitudinal, will precess by (g-2)/2 γ turns per revolution; where (g-2)/2 is 1.166 10^{-3}. A given energy spread $\frac{\Delta \gamma}{\gamma}$ will introduce variations in these precessions and cause dilution of the polarization. But if the particles remain in the ring for an exact integer number of synchrotron oscillations, then their individual average γ's will be the same and no dilution will occur.

In the collider, bending can be performed with the spin orientation in the vertical direction, and the spin rotated into the longitudinal direction only for the interaction region. The design of such spin rotators appears relatively straightforward, but long. This might be a preferred solution at high energies but is not practical for instance, in the 100 GeV machine. An alternative is to use such a small energy spread, as in the Higgs factory, that though the polarization vector precesses, the beam polarization does not become significantly diluted.

COOLING

For a collider, the phase-space volume must be reduced within a time of the order of the μ lifetime. Cooling by synchrotron radiation, conventional stochastic cooling and conventional electron cooling are all too slow. Optical stochastic cooling [29], electron cooling in a plasma discharge [30] and cooling in a crystal lattice [31] are being studied, but appear difficult. Ionization cooling [32] of muons seems relatively straightforward.

Ionization Cooling Theory

In ionization cooling, the beam loses both transverse and longitudinal momentum as it passes through a material medium. Subsequently, the longitudinal momentum can be restored by coherent reacceleration, leaving a net loss of transverse momentum.

The approximate equation for transverse cooling (with energies in GeV) is:

$$\frac{d\epsilon_n}{ds} = -\frac{dE_\mu}{ds}\frac{\epsilon_n}{E_\mu} + \frac{\beta_\perp (0.014)^2}{2 E_\mu m_\mu L_R}, \qquad (2)$$

where ϵ_n is the normalized emittance, $\beta_\perp$ is the betatron function at the absorber, dE_μ/ds is the energy loss, and L_R is the radiation length of the material. The first term in this equation is the coherent cooling term, and the second is the heating due to multiple scattering. This heating term is minimized if $\beta_\perp$ is small (strong-focusing) and L_R is large (a low-Z absorber).

The equation for energy spread (longitudinal emittance) is:

$$\frac{d(\Delta E)^2}{ds} = -2\frac{d\left(\frac{dE_\mu}{ds}\right)}{dE_\mu} <(\Delta E_\mu)^2> + \frac{d(\Delta E_\mu)^2_{\text{straggling}}}{ds} \qquad (3)$$

where the first term is the cooling (or heating) due to energy loss, and the second term is the heating due to straggling.

Energy spread can be reduced by artificially increasing $\frac{d(dE_\mu/ds)}{dE_\mu}$ by placing a transverse variation in absorber density or thickness at a location where position is energy dependent, i.e. where there is dispersion. The use of such wedges can reduce energy spread, but it simultaneously increases transverse emittance in the direction of the dispersion. Six dimensional phase space is not reduced.

Cooling Components

We require a reduction of the normalized transverse emittance by almost three orders of magnitude (from 1×10^{-2} to 5×10^{-5} m-rad), and a reduction of the longitudinal emittance by one order of magnitude. This cooling is obtained in a series of cooling stages. In general, each stage consists of two components:

1. a material in a strong focusing (low$\beta_\perp$) environment alternated with linac accelerators. These components will cool the transverse phase space.

2. a lattice that generates dispersion, with absorbing material wedges introduced to interchange longitudinal and transverse emittance.

Simulations have been performed on examples of each component using the program ICOOL [33] which includes Vavilov distributions (with Landau and Gaussian limits) for dE/dx, and Moliere scattering distributions (with Rutherford limit). The only effects which are not yet included are space-charge and wake-field effects. Analytic vacuum calculations indicate that these effects should be significant, but not overwhelming. A correct simulation must be done before we are assured that no real problems exist.

Transverse Cooling

The baseline solution for the first component involves the use of liquid hydrogen absorbers in strong solenoid focusing fields, interleaved with short linac sections. The solenoidal fields in succesive absorbers must be reversed to avoid build up of the canonical angular momentum. Fig. 6 shows the cross section of one cell of such a system. The top plot in Fig. 7 shows the reduction of transverse emittance in 10 such cells (20 m); the middle one shows the increase in longitudinal emittance induced by straggling and the adverse dependence of dE/dx with energy; while the bottom one shows the overall reduction in 6-dimensional emittance. This simulation has been confirmed, with minor differences by the codes double precision GEANT [34] and PARMELA [35].

Using 30 T solenoids at the end of a cooling sequence can attain a transverse emittance of 190 mm mrad and a six dimensional emittance of 30×10^{-12} m^3 (cf. 280 mm mrad and 170×10^{-12} m^3, respectively required for a Higgs factory).

Other solutions, e.g. rapidly alternating solenoids and LiH absorbers [36] and current carrying Li rods have been and will continue to be studied, but do not appear to be required to meet the baseline parameters (see below).

Linac

The linacs used in the above simulations has a frequency of 805 MHz and required an accelerating gradient (peak phase) of 24 MV/m. The current designs use cavities separated by thin Be foils, $\frac{2\pi}{3}$ or $\frac{2\pi}{4}$ phase advanced per cavity, and powered in 3 of 4 separate interleaved side-coupled standing-wave systems [37,38]. In order to reduce power source requirements the cavities may be operated at liquid nitrogen temperatures.

Longitudinal-Transverse Exchange

The exchange of longitudinal and transverse emittance requires dispersion in a large acceptance channel. One way of achieving this is in a bent solenoid. Fig. 8 shows transverse positions vs. their momenta: a) before the bend, b) after the bend, and c) after hydrogen wedges. The *rms* momentum spread in

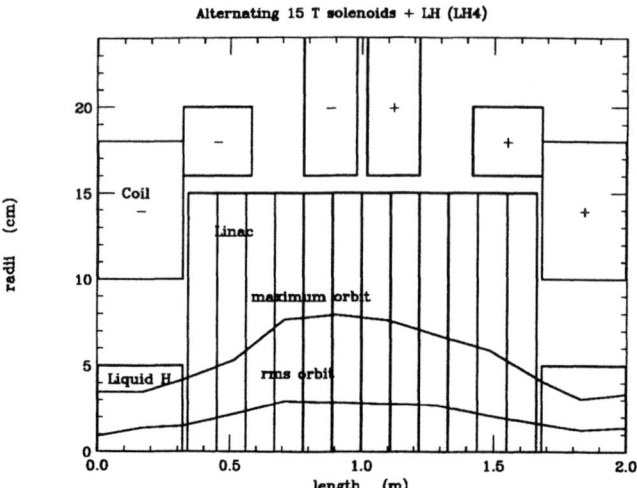

FIGURE 6. The cross section of one cell of an alternating solenoid cooling system

this example is reduced from 8 MeV/c to 4.6 MeV/c with an accompanying approximate equivalent increase in the x-y emittance.

Emittance exchange in solid wedges in the presence of ideal dispersion has also been simulated using SIMUCOOL [39]. Dispersion generation by weak focussing spectrometers [40] and dipoles with solenoids [41] are also studied.

Cooling System

The required total 6 dimensional cooling is about 10^6. Since a single stage, as illustrated above, gives a factor of 2 reduction, about 20 such stages are required. The total length of the system would be of the order of 500 m, and the total acceleration would be of the order of 6 GeV. The fraction of muons remaining at the end of the cooling system is estimated to be $\approx 60\%$.

In a few of the later stages, current carrying lithium rods might replace item (1) above. In this case the rod serves simultaneously to maintain the low $\beta_\perp$, and attenuate the beam momenta. Similar lithium rods, with surface fields of 10 T, were developed at Novosibirsk (BINP) and have been used as focusing elements at FNAL and CERN [42,43]. Cooling in beam recirculators could lead to reduction of costs of the cooling section [40].

ACCELERATION

Following cooling and initial bunch compression, the beams must be rapidly accelerated. A sequence of linacs would work, but would be expensive, some form of circulating acceleration is preferred. At lower energies, the acceleration

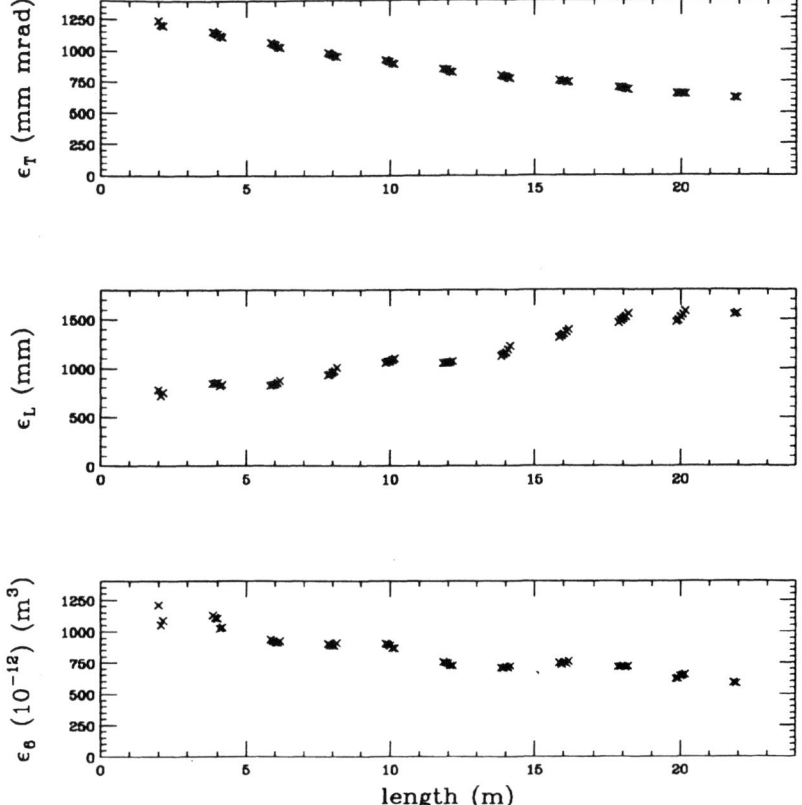

FIGURE 7. Emittance vs. length in 10 alternating solenoid cells; TOP: transverse emittance; MIDDLE: longitudinal emittance; and BOTTOM: 6-dimensional emittance

time is so short that any form of magnet ramping is probably impractical. The conservative option is to use a sequence of recirculating accelerators (similar to that used at TJNL), but fixed frequency alternating gradient acceleration (FFAG) is also being studied [44]. At higher energies, it is probably more economical to use fast rise time pulsed magnets in more conventional synchrotrons [45].

Scenarios

Tbs. 4 and 5 give an example of possible sequences of accelerators for a 100 GeV Higgs Factory and a 3 TeV collider. In both cases, following initial linacs, recirculating accelerators are used. Designs [46] have been made of multiple aperture superconducting magnets for use in recirculating acceleration. The use of such magnets was not assumed in the scenarios, but they would reduce

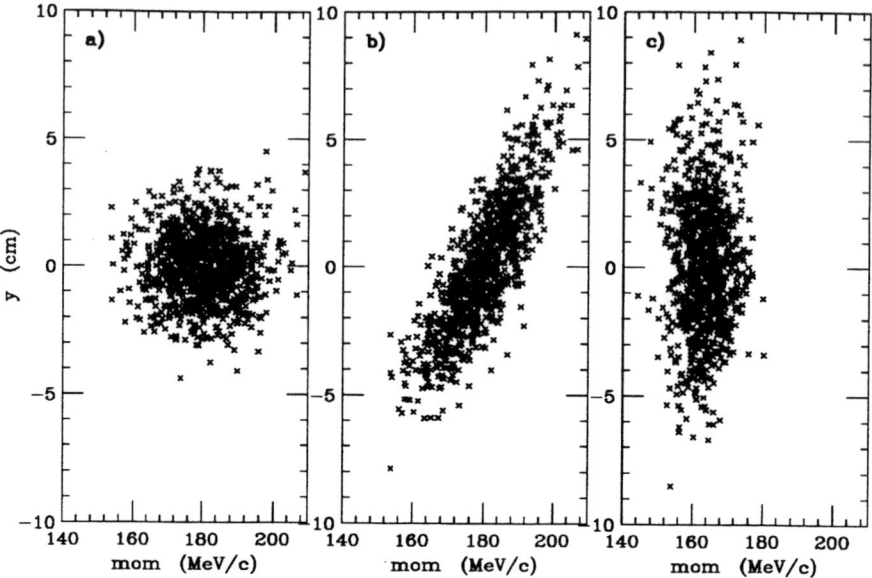

FIGURE 8. Transverse trajectory positions vs. their momenta: a) before the bend, b) after the bend, and c) after hydrogen wedges.

the diameter of the recirculating accelerator ring and lower particle loss from decay.

In the high energy case, the final three stages use pulsed magnet synchrotrons. If only pulsed magnets were used the power consumed by a ring would be high and its circumference large, but a hybrid ring with alternating pulsed warm magnets and fixed superconducting magnets appears practical. In the example, the last two such rings are located in the same tunnel with differing ratios of pulsed to fixed magnets. The fixed magnets are superconducting at 8 T; the pulse d magnets are warm with fields that swing from - 2 T to + 2 T.

In both cases, except for the earliest stages, superconducting rf is employed. The reason for this, in the earlier stage, is that the instantaneous acceleration power requirement is very high, and the use of superconducting cavities allows a longer rf fill time and a reduced rf power source requirement. For the higher energy accelerators the use of superconducting cavities is dictated by the need to achieve high wall to beam efficiency.

A study [47] tracked particles through a similar sequence of recirculating accelerators and found a dilution of longitudinal phase space of the order of 10% and negligible particle loss.

TABLE 4. Parameters Higgs Factory (100 GeV) Accelerators

acc type rf type		linac sledCu	linac sledCu	recirc sledCu	recirc sledCu	recirc SC Nb	sums
E_{init}	(GeV)	0.10	0.20	0.70	2	7	
E_{final}	(GeV)	0.20	0.70	2	7	50	
circ	(km)	0.04	0.07	0.06	0.18	1.21	1.57
turns		1	1	8	10	11	
decay loss	(%)	2.31	3.98	6.74	7.77	9.88	27.29
decay heat	W/m	0.85	1.88	10.50	12.39	12.14	
B_{fixed}	(T)			2	2	2	
pipe width	cm			30.66	21.22	10.44	
pipe ht	cm			10	8	4.30	
rf freq	(MHz)	90	90	120	170	400	
acc/turn	(GeV)	0.20	0.40	0.17	0.50	4	
acc time	(μs)			1	6	43	
acc Grad	(MV/m)	8	8	8	10	15	
grad sag	%			13.08	16.82	27.15	
rf time	ms	0.55	0.56	0.37	0.24	2.04	
peak rf /m	(MW/m)	2.72	2.56	2.21	4.40	0.20	
ave rf power	MW	0.61	1.10	0.28	0.88	1.99	4.88
total wall p	MW	4.71	8.50	1.67	5.20	5.87	25.94
beam power	MW	0.00	0.01	0.03	0.12	0.92	1.08
wall-beam eff	%	0.06	0.15	1.93	2.22	15.62	4.16

COLLIDER STORAGE RING

After acceleration, the μ^+ and μ^- bunches are injected into a separate storage ring. The highest possible average bending field is desirable, to maximize the number of revolutions before decay, and thus maximize the luminosity. Collisions would occur in one, or perhaps two, low-β^* interaction areas. Parameters of the rings were given earlier in Tb. 1.

Lattice Design

In order to maintain the required short bunches, without excessive rf, approximately isochronous Flexible Momentum Compaction lattices [48] would be used.

In the high energy cases, the required betas at the intersection point are very small (e.g. $\beta^* = 3$ mm for 4 TeV), and the quadrupoles needed to generate them are large (20-30 cm diameter). At 100 GeV, the betas are not so small and the quadrupoles are more conventional, but in both cases it has been found that local chromatic correction is essential [49].

Preliminary lattices have been designed for both 4 TeV and 0.5 TeV machines [50], and several designs now exist for the 100 GeV case. Fig. 9 gives the dynamic aperture of one such lattice [51] for the required 1000 turns.

TABLE 5. Parameters 3 TeV Collider Accelerators

acc type magnet type rf type		linac sledCu	recirc warm sledCu	recirc warm sledCu	recirc warm SC Nb	pulsed warm SC Nb	pulsed hybrid SC Nb	pulsed hybrid SC Nb	sums
E_{init}	(GeV)	0.10	0.70	2	7	50	200	1000	
E_{final}	(GeV)	0.70	2	7	50	200	1000	1500	
circ	(km)	0.07	0.12	0.25	1.16	4.65	11.30	11.36	28.93
turns		2	8	10	11	15	27	17	
decay loss	(%)	6.11	12.11	10.38	9.53	10.68	10.07	2.65	47.68
decay heat	W/m	3.46	14.20	16.03	15.49	19.44	30.97	18.09	
pulsed B_{max}	(T)					2	2	2	
B_{fixed}	(T)		0.70	1.20	2	2	8	8	
ramp freq	(kHz)	900	109	40.02	7.99	1.43	0.33	0.53	
sig beam	cm	0.59	0.51	0.39	0.25	0.11	0.08	0.06	
sig width	cm		3.03	3.65	3.85	1.36	0.59	0.18	
mom compactn	%		-1	-2	-2	-1	-1	-1	
pipe width	cm		30.31	36.49	38.53	13.63	5.86	3	
pipe ht	cm		10	8	4.30	3	3	3	
rf freq	(MHz)	90	50	90	200	800	1300	1300	
acc/turn	(GeV)	0.40	0.17	0.50	4	10	30	30	
acc time	(μs)		3	8	41	232	1004	631	
eta	(%)	0.73	0.22	0.33	0.44	10.15	14.37	12.92	
acc Grad	(MV/m)	8	8	10	15	15	25	25	
synch rot's		0.54	0.82	1.91	9.16	27.07	76.78	31.30	
phase slip	deg		6.90	4.62	5.35	1.64			
cavity rad	(cm)	122	220	134	76.52	19.13	11.77	11.77	
loading	%	4.23	6.22	11.98	16.54	210	527	296	
grad sag	%			3.16	6.18	8.65			
rf time	msec	0.56	1.35	0.59	2.04	0.40	1.25	0.96	
peak rf /m	(MW/m)	2.56	3.43	6.05	0.81	0.91	0.56	0.50	
ave rf power	MW	1.11	1.54	2.84	7.20	6.32	21.91	15.07	55.99
rf wall	MW	8.50	9.05	16.69	21.18	18.59	44.72	30.76	149
magnet ps	MJ						34.31	13.19	47.51
magnet wall	MW						3.7	1.4	5.1
total wall	MW	8.50	9.05	16.69	21.18	18.59	48.4	32.2	155
beam power	MW	0.02	0.04	0.15	1.17	3.68	17.54	9.86	32.47
wall-beam eff	%	0.26	0.49	0.91	5.51	19.81	39.23	32.06	21.72

Scraping

Collimation schemes have been designed [52] for colliders at both high and low energies. At low energies, as in the Higgs Factory, tungsten collimators have been shown to be effective. At higher energies, the muons are scattered, but not stopped, by such collimators. For this case it has been shown that electrostatic septa followed by sweeping magnets could effectively extract the tail muons. Lattices [50] have been designed incorporating these systems.

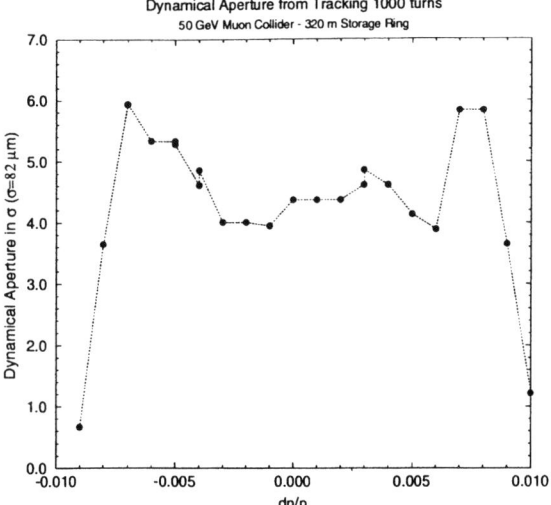

FIGURE 9. Dynamical aperture after 1000 turns.

Instabilities

The studies [53,54] have considered beam emittance growth due to beam-beam tune shift, and both, althought some assumptions were made, predict negligible effects in 1000 tunes at the values shown in Tb. 1.

A study [55] has examined the resistive wall impedance longitudinal instabilities in rings at several energies. At the higher energies and larger momentum spreads, solutions were found with small but finite momentum compaction, and moderate rf. For the special case of the Higgs Factory, with its very low momentum spread, a solution was found with no synchrotron motion, but rf provided to correct the first order impedance generated momentum spread. The remaining off momentum tails, that would not affect the luminosity, but which might generate background, could be removed by a higher harmonic rf correction.

Given the very slow, or nonexistent synchrotron oscillations, the transverse beam breakup instability is significant. But this instability can be stabilized using rf quadrupole [56] induced BNS damping. For instance, in the 3 TeV case, to stabilize the resistive wall instability, the required tune spread, calculated [57] using the two particle model approximation, for a 1 cm radius aluminum pipe, is only 1.58×10^{-4}.

However, this application of the BNS damping to a quasi-isochronous ring, and other head-tail instabilities due to the chromaticities ξ and η_1, needs more careful study.

Bending Magnet Design

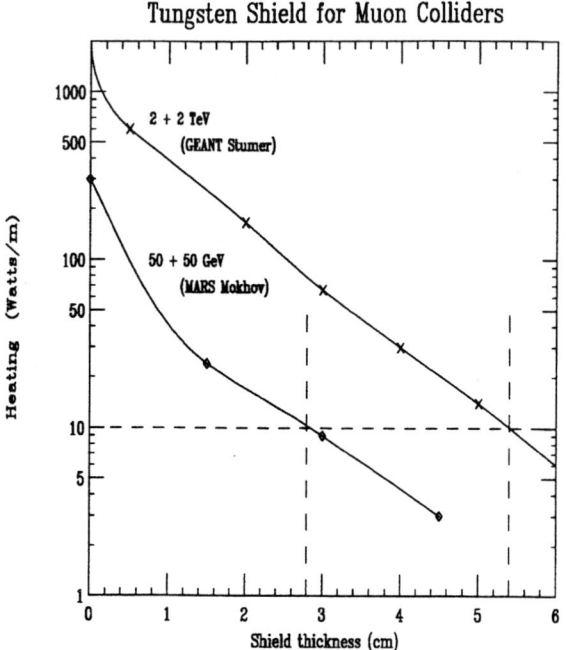

FIGURE 10. Power penetrating tungsten shields vs. their thickness for, a) a 4 TeV, and b) a 100 GeV, Collider.

The magnet design is complicated by the fact that the μ's decay within the rings ($\mu^- \to e^- \overline{\nu_e} \nu_\mu$), producing electrons whose mean energy is approximately 0.35 that of the muons. With no shielding, the average power deposited per unit length would be about 2 kW/m in the 4 TeV machine, and 300 W/m in the 100 GeV Higgs factory. Fig. 10 shows the power penetrating tungsten shields of different thickness. One sees that 3 cm in the low energy case, or 6 cm at high energy would reduce the power to below 10 W/m which can reasonably be taken by superconducting magnets.

The quadrupoles could use warm iron poles placed as close to the beam as practical. The coils could then be either superconducting or warm, placed at a greater distance from the beam and shielded from it by the poles.

NEUTRINO RADIATION

Bruce King [60] has shown that the surface radiation dose $D_B(Sv)$ in a time $t(s)$, in the plane of a bending magnet of field B(T), in a circular collider with beam energy $E(TeV)$, average bending field $< B(T) >$, at a depth $d(m)$ (assuming a spherical earth), with muon current (of each sign) of I(muons/sec/sign) is given by:

$$D_B \approx 4.4 \; 10^{-24} \frac{I_\mu \; E^3 \; t}{d} \frac{<B>}{B} t$$

and that the dose D_S at a location on the surface, in line with a high beta straight section of length ℓ is:

$$D_S \approx 6.7 \; 10^{-24} \frac{I_\mu \; E^3 \; t}{d} \ell <B> t$$

The first formula has been confirmed by a Monte Carlo simulation using the MARS code [61]. In all cases it is assumed that the average divergence angles satisfy the condition: $\sigma_\theta \ll \frac{1}{\gamma}$. This condition is not satisfied in the straight sections approaching the IP, and these regions, despite their length, do not contribute a significant dose.

For the 3 TeV parameters given in Tb. 1 (muon currents $I = 6 \times 10^{20}$ μ^-/yr, $<B> = 6\,T$, $B = 10\,T$, and taking the federal limit on off site radiation Dose/year, D_{Fed} to be 1 mSv/year (100 mrem/year), then the dose D_B per year (defined as 10^7 s), in the plane of a bending dipole is:

$$D_B = 1.07 \; 10^{-5} \; (Sv) \approx 1\% \; D_{Fed}$$

and for a straight section of length 0.6 m is:

$$D_B = 9.7 \; 10^{-5} \; (Sv) \approx 10\% \; D_{Fed}$$

which may be taken to be a reasonable limit.

Special care will be required in the lattice design to assure that no field free region longer than this is present. But it may be noted that the presence of a field of even 1 T over any length, is enough to reduce the dose to the 10 % Federal limit standard. For machines above 3 TeV, the muon current would probably have to be reduced.

For lower energy machines, the requirements get rapidly easier: a 0.5 TeV at 500 m depth could have 130 m straight sections, or if at 100 m depth 25 m lengths, for the same surface dose. For a 100 GeV machine the doses are negligible.

DETECTOR BACKGROUND

There will be backgrounds in the detector from the decay of muons in the ring and approaching the IP. A recent study [58] of electromagnetic, hadronic and muon components of the background has been done using the GEANT codes [59]. This study

TABLE 6. Detector backgrounds from μ decay

Radius	cm	5	10	20	100
Photons hits	cm^{-2}	26	6.6	1.6	.06
Neutrons hits	cm^{-2}	0.06	0.08	0.2	0.04
Charged hits	cm^{-2}	8	1.2	0.2	0.01
Total hits	cm^{-2}	34	8	2	0.12
Pixel size	μm^2	60x150	60x150	300x300	300x300
Occupancy all	%	0.6	0.14	0.4	0.02
Occupancy charged	%	0.14	0.02	0.04	0.002

- followed shower neutrons and photons down to 40 keV and electrons to 25 KeV.

- Used a tungsten shield over the beam, extending in to within 14 cm of the intersection point, and extending outward to an angle of 20 degrees from the axis.

- Inside this shield, between its smallest aperture 1 m from the IP, and its tip, the inner surface is shaped into a series of rising collimating steps and slopes, designed so that, the detector could not *see* any surface directly illuminated by the initial decay electrons, whether seen in the forward or backward (albedo) directions.

- From the aperture point of minimum to a few meters (2.5 m for Higgs Factory) upstream, the inside forms another series of stepped collimators placed at $\pm$ 4 σ_{θ_0} (where σ_{θ_0} is the rms divergence of the beam).

- Further upstream, prior to the first quadrupole (from 2.5 to 4 m in the Higgs case) an 8 T dipole, with collimators inside, is used to sweep decay electrons before the final collimation.

Tb. 6 gives the hit density for the Higgs factory from the various sources and the occupancy of pixels of the given sizes. In all cases the numbers are given per bunch crossing. The hit density for the higher energy machines are found to be somewhat lower than these, due to the small decay angles of the electrons. The radiation damage by the neutrons on a silicon detector has also been estimated. In the Higgs case, at 5 cm from the vertex, the number of hits from neutrons above 100 KeV is found to be $1.8\,10^{13}$ per year ($10^7\,s$). This is an order of magnitude less than that expected at the LHC. The damages for silicon detectors in the higher energy machines are of the same order.

This study also found a significant flux of muons, with quite high energies, from μ pair production in electromagnetic showers (Bethe Heitler). Their most serious effect appears to arise when they make deeply inelastic interactions and deposit spikes of energy in the electromagnetic and hadronic calorimeters. This is not serious in the Higgs case, when the fluxes and cross sections are low, but at higher energies timing and/or longitudinal calorimeter segmentation appears necessary to identify and remove the problem.

An ealier study using the code MARS [62], using less sophisticated shielding, gave results qualitatively in agreement with those from Geant [58].

CONCLUSION

Motive

- Because they can be circular, muon colliders appear to be far smaller than hadron or e^+e^- machines of similar effective energy.

- It is thus hoped that muon colliders could have a lower cost per TeV than other options.

- Their smaller size would allow machines up to 3 TeV effective energy (roughly equivalent to a 30 TeV hadron machine) to fit on existing laboratory sites.

- The low synchrotron and beamstrahlung radiation with muons could allow energy spreads as small as 0.003 % (3×10^{-5}).

- By measuring g-2 of the muon it should be possible to determine the energy to even greater precision.

- The above, plus the large cross section for s-channel production, could make a muon collider into a precision tool to study Higgs particles and their decays.

Progress

- The theory of operation of all components of a muon collider are now well understood.

- Simulations of examples of all components of a baseline design have now been performed. The simulated performances of many of these components has exceeded the baseline specifications. All known effects have been included except space-charge in the cooling, whose effect, calculated analytically, appears not to be too large.

Needed

- More detailed simulations of all components, including space-charge in the cooling.

- Complete scenarios of the cooling stages and acceleration.
- An experimental study of the target.
- The construction and test of one or more of the cooling stages.
- Technical development of components: a large high field solenoid for capture, low frequency rf linacs, multi-beam or pulsed magnets for acceleration, warm bore shielded high field dipoles for the collider, muon collimators and background shields, etc.. But none of these components can be described as *exotic*, and their specifications are not beyond what has been demonstrated.

ACKNOWLEDGMENTS

This research was supported by the U.S. Department of Energy under Contract No. DE-ACO2-76-CH00016 and DE-AC03-76SF00515.

REFERENCES

1. V. V. Parkhomchuk and A. N. Skrinsky, Proc. 12th Int. Conf. on High Energy Accelerators, F. T. Cole and R. Donaldson, Eds., (1983) 485; A. N. Skrinsky and V.V. Parkhomchuk, Sov. J. of Nucl. Physics **12**, (1981) 223; *Early Concepts for $\mu^+\mu^-$ Colliders and High Energy μ Storage Rings, Physics Potential & Development of $\mu^+\mu^-$ Colliders. 2nd Workshop*, Sausalito, CA, Ed. D. Cline, AIP Press, Woodbury, New York, (1995).
2. D. Neuffer, Fermilab Note FN-319, July 1979; Proc. 12th Int. Conf. on High Energy Physics (1983) 481; *Principles and Applications of Muon Cooling*, Part. Acc. **14** 75 (1983)
3. *Proceedings of the Mini-Workshop on $\mu^+\mu^-$ Colliders: Particle Physics and Design*, Napa CA, Nucl Inst. and Meth., **A350** (1994) ; Proceedings of the Muon Collider Workshop, February 22, 1993, Los Alamos National Laboratory Report LA-UR-93-866 (1993) and *Physics Potential & Development of $\mu^+\mu^-$ Colliders 2nd Workshop*, Sausalito, CA, Ed. D. Cline, AIP Press, Woodbury, New York, (1995); Proceedings of the 9th Advanced ICFA Beam Dynamics Workshop, Ed. J. C. Gallardo, AIP Press, Conference Proceedings 372 (1996).
4. R. B. Palmer et al., *Monte Carlo Simulations of Muon Production, Physics Potential & Development of $\mu^+\mu^-$ Colliders 2nd Workshop*, Sausalito, CA, Ed. D. Cline, AIP Press, Woodbury, New York, pp. 108 (1995); R. B. Palmer, et al., *Muon Collider Design*, in Proceedings of the Symposium on Physics Potential & Development of $\mu^+\mu^-$ Colliders, Nucl. Phys B (Proc. Suppl.) **51A** (1996); R. B. Palmer and J. C. Gallardo, *Muon-Muon and other High Energy Colliders, Techniques and Concepts of High Energy Physics IX*, Phys. vol. 365, Ed. T. Ferbel, pp. 183, Plenum Pub. (1997).

5. R. B. Palmer and J. C. Gallardo, *High Energy Colliders*, Proceedings of 250th Anniversary Conference on Critical Problems in Physics, Princeton University, Ed. Fitch, Marlow, Dementi, pp. 247 (1997).
6. R. B. Palmer, *Progress on $\mu^+\mu^-$ Colliders*, submitted to the Proceedings of the PAC97, Vancouver, Canada, June 1997.
7. *$\mu^+\mu^-$ Collider, A Feasibility Study*, BNL-52503, FermiLab-Conf-96/092, LBNL-38946, Proceedings of the 1996 DPF/DPB Summer Study on High-Energy Physics, Snowmass'96. For updated information, see the Muon Collider Collaboration WEB page: http://www.cap.bnl.gov/mumu/.
8. R. Raja and A. Tollestrup, *Calibrating the energy of a 50 x 50 GeV muon collider using spin precession*, LANL preprint archive, hep-ex/9801004; submitted to Phys. Rev. D.
9. B. King, private communication
10. C. Ankenbrandt and B. Noble, *Summary of the Accelerator Working Group*, submitted to the Proceedings of Workshop on Physics at the First Muon Collider and at the Fron End, FNAL, Nov. 1997.
11. T. Roser, *AGS Performance and Upgrades: A Possible Proton Driver for a Muon Collider*, Proceedings of the 9th Advanced ICFA Beam Dynamics Workshop, Ed. J. C. Gallardo, AIP Press, Conference Proceedings 372 (1996).
12. C. Ankenbrandt, K-Y. Ng, J. Norem, M. Popovic, Z. Qian, L. Ahrens, M. Brennan, V. Mane, T. Roser, D. Trbojevic, W. van Asselt, *Bunching Near Transition in the AGS*, Fermilab Pub-98-006, submitted to Phys. Rev. D.
13. J. E. Griffin, K.Y. Ng, Z.B. Qian and D. Wildman, *Experimental Study of Passive Compensation of Space Charge Potential Well Distortion at the Los Alamos National Laboratory Proton Storage Ring*, Fermilab Report, FN-661, Nov. 1997.
14. C. Johnson, *Solid and Liquid Targets Overview*, presentation at the Mini-Wokshop: Target and Muon Collection Magnets and Accelerators, Oxford, MI, Jan. 1997.
15. C. Lu, K. T. McDonald, *Low-Melting-Temperature Metals for possible Use as Primary Targets at a Muon Collider Source*, Princeton/$\mu\mu$/97-3, Revised Dec. 1997, unpublished.
16. R. Weggel, *Deceleration of Conductor by Magnetic Field: 1) Paraxial; 2) Perpendicular*, unpublished
17. M. Green and R. Palmer, *A $\mu - \mu$ collider capture solenoid system for pions froma tilted target*, submitted to the Proceedings of PAC97, May 1997.
18. N.V. Mokhov and A. Van Ginneken, *Pion Production and Targetry at $\mu^+\mu^-$ Colliders*, Fermilab-Conf-98/041 (1998), submitted to Proc. of the 4th Int. Conf. on Physics Potential and Development at $\mu^+\mu^-$ Colliders, San Francisco, CA, December 10-12, 1997
19. D. Kahana, et al., *Proceedings of Heavy Ion Physics at the AGS-HIPAGS '93*, Ed. G. S. Stephans, S. G. Steadman and W. E. Kehoe (1993); D. Kahana and Y. Torun, *Analysis of Pion Production Data from E-802 at 14.6 GeV/c using ARC*, BNL Report # 61983 (1995).
20. N. V. Mokhov, *The MARS Code System User's Guide*, version 13(95),

Fermilab-FN-628 (1995).
21. J. Ranft, DPMJET Code System (1995).
22. Experiment E-910 at AGS, BNL, private communication.
23. N.V. Mokhov and S.I. Striganov, *Towards Reliable Prediction of Particle Production for 6-120 GeV Proton Beams*, presentation at the Workshop on Physics at the First Muon Collider and at the Front End of a Muon Collider, Nov. 1997.
24. J. R. Miller, M. Bird, S. Bole et al., *An Overview of the 45 T Hybrid Magnet System for the National High Field Magnet Laboratory*, IEEE Transactions on Magnetics 30, pp. 1563 (1994).
25. R. Weggel, *4-MW Hollow-Conductor Magnets for 20 T Hybrid Systems to Collect Pions for a Muon Collider*, presentation at the Mini-Wokshop: Target and Muon Collection Magnets and Accelerators, Oxford, MI, Jan. 1997.
26. N. Mokhov, R. Noble and A. Van Ginneken, *Target and Collection Optimization for Muon Colliders*, Proceedings of the 9th Advanced ICFA Beam Dynamics Workshop, Ed. J. C. Gallardo, AIP Press, Conference Proceedings 372 (1996).
27. K. Assamagan, et al., Phys. Lett. **B335**, 231 (1994); E. P. Wigner, Ann. Math. **40**, 194 (1939) and Rev. Mod. Phys., **29**, 255 (1957).
28. B. Norum and R. Rossmanith, *Polarized Beams in a Muon Collider*, in Proceedings of the Symposium on Physics Potential & Development of $\mu^+\mu^-$ Colliders, Nucl. Phys B (Proc. Suppl.) **51A** (1996).
29. A. A. Mikhailichenko and M. S. Zolotorev, Phys. Rev. Lett. **71**, (1993) 4146; M. S. Zolotorev and A. A. Zholents, SLAC-PUB-6476 (1994).
30. A. Hershcovitch, Brookhaven National Report AGS/AD/Tech. Note No. 413 (1995).
31. Z. Huang, P. Chen and R. Ruth, SLAC-PUB-6745, *Proc. Workshop on Advanced Accelerator Concepts*, Lake Geneva, WI , June (1994); P. Sandler, A. Bogacz and D. Cline, *Muon Cooling and Acceleration Experiment Using Muon Sources at Triumf*, Physics Potential & Development of $\mu^+\mu^-$ Colliders 2nd Workshop, Sausalito, CA, Ed. D. Cline, AIP Press, Woodbury, New York, pp. 146 (1995).
32. Initial speculations on ionization cooling have been variously attributed to G. O'Neill and/or G. Budker see D. Neuffer in [3]; D. Neuffer, in Advanced Accelerator Concepts, AIP Conf. Proc. 156, 201 (1987); see also [1-3]; R. C. Fernow and J. C. Gallardo, *Muon Transverse Ionization Cooling: Stochastic Approach*, Phys. Rev. **E52** 1039 (1995).
33. R. Fernow, **ICOOL**, fortran program to simulate muon ionozation cooling.
34. P. Le Brun, *Alternate solenoid in DPGeant*, presented at the Mini-Workshop on Cooling, BNL Jan. 1998, unpublished.
35. H. Kirk, *Parmela modeling of alternating solenoids* presented at the Mini-Workshop on Cooling, BNL Jan. 1998, unpublished.
36. R. C. Fernow, J. C. Gallardo and R. B. Palmer, *Ionization cooling using a FOFO lattice*, BNL Report BNL #64493, submitted to PAC97, Vancouver, Canada, 1997.
37. Y. Zhao, *The preliminary simulation of $\frac{2\pi}{3}$-mode interleaved side coupled standing wave structures*, presented at the Mini-Workshop on Cooling, BNL Jan.

1998, unpublished.
38. A. Moretti, *Rf Update*, presented at the Mini-Workshop on Cooling, BNL Jan. 1998, unpublished.
39. D. Neuffer and A. Van Ginneken, *Recent Cooling Simulation Studies*, presented at the Mini-Workshop on Cooling, BNL Jan. 1998, unpublished.
40. V. Balbekov and A. Van Ginneken, *Ring Cooler for Muon Collider*, presented at the Mini-Workshop on Cooling, Fermilab Oct. 1997, unpublished.
41. D. Neuffer and W. Wan, *COSY transport for μ cooling*, presented at the Mini-Workshop on Cooling, Fermilab Oct. 1997, unpublished.
42. G. Silvestrov, Proceedings of the Muon Collider Workshop, February 22, 1993, Los Alamos National Laboratory Report LA-UR-93-866 (1993); B. Bayanov, J. Petrov, G. Silvestrov, J. MacLachlan, and G. Nicholls, Nucl. Inst. and Meth. **190**, (1981) 9; C. D. Johnson, Hyperfine Interactions, **44** (1988) 21; M. D. Church and J. P. Marriner, Annu. Rev. Nucl. Sci. **43** (1993) 253.
43. G. Silvestrov, *Lithium Lenses for Muon Colliders*, Proceedings of the 9th Advanced ICFA Beam Dynamics Workshop, Ed. J. C. Gallardo, AIP Press, Conference Proceedings 372 (1996).
44. F. Mills and C. Johnstone, presentation at the 4th Int. Conference on Physics Potential and Development of $\mu - \mu$ Colliders, San Francisco, CA, Dec. 1997.
45. D. Summers, presentation at the 9th Advanced ICFA Beam Dynamics Workshop, Montauk 1995, unpublished.
46. G. Morgan, presentation at the 9th Advanced ICFA Beam Dynamics Workshop, Montauk 1995, unpublished.
47. D. Neuffer, *Acceleration to Collisions for the $\mu^+\mu^-$ Collider*, Proceedings of the 9th Advanced ICFA Beam Dynamics Workshop, Ed. J. C. Gallardo, AIP Press, Conference Proceedings 372 (1996).
48. S.Y. Lee, K.-Y. Ng and D. Trbojevic, FNAL Report FN595 (1992); Phys. Rev. **E48**, (1993) 3040; D. Trbojevic, et al., *Design of the Muon Collider Isochronous Storage Ring Lattice, Micro-Bunches Workshop*, BNL Oct. (1995), AIP Press, Conference Proceedings 367 (1996).
49. K. L. Brown and J. Spencer, SLAC-PUB-2678 (1981) presented at the Particle Accelerator Conf., Washington, (1981) and K.L. Brown, SLAC-PUB-4811 (1988), Proc. Capri Workshop, June 1988 and J.J. Murray, K. L. Brown and T.H. Fieguth, Particle Accelerator Conf., Washington, 1987; Bruce Dunham and Olivier Napoly, *FFADA, Final Focus. Automatic Design and Analysis*, CERN Report CLIC Note 222, (1994); Olivier Napoly, it CLIC Final Focus System: Upgraded Version with Increased Bandwidth and Error Analysis, CERN Report CLIC Note 227, (1994).
50. A. Garren, C. Johnstone, *Lattice Design for a 100 GeV Muon Collider*, presentation at the 4th Int. Conference on Physics Potential and Development of $\mu - \mu$ Colliders, San Francisco, CA, Dec. 1997; A. Garren and C. Johnstone, *Progress on a Lattice for a 2 TeV Muon Collider*, submitted to the Proceedings of the PAC97, Vancouver, Canada, June 1997.
51. D. Trbojevic and K.-Y. Ng, submitted to Proc. of the 4th Int. Conf. on Physics Potential and Development at mu+mu- Colliders, San Francisco, CA, Decem-

ber 10-12, 1997

52. A. Drozhdin, C. Johnstone and N. Mokhov, *Muon Collider Beam Collimation System*, unpublished.
53. M. Furman, *The Classical Beam-Beam Interaction for the Muon Collider: A First Look*, BF-19/CBP-Note-169/LBL-38563, April 1996.
54. P. Chen, *Beam-Beam interaction at $\mu^+\mu^-$ Colliders*, in Proceedings of the Symposium on Physics Potential & Development of $\mu^+\mu^-$ Colliders, Nucl. Phys B (Proc. Suppl.) **51A** (1996);
55. W.-H. Cheng, A. M. Sessler and J. Wurtele, *Studies of Collective Instabilities in Muon Collider Rings*, Proceedings of the 9th Advanced ICFA Beam Dynamics Workshop, Ed. J. C. Gallardo, AIP Press, Conference Proceedings 372 (1996).
56. A. Chao, *Physics of Collective Beam Instabilities in High Energy Accelerators*, John Wiley & Sons, Inc, New York (1993).
57. K.Y. Ng, *Beam Stability Issues in a Quasi-Isochronous Muon Collider*, Proceedings of the 9th Advanced ICFA Beam Dynamics Workshop, Ed. J. C. Gallardo, AIP Press, Conference Proceedings 372 (1996).
58. I. Stumer et al., *Study of Detector Backgrounds in a $\mu^+\mu^-$ Collider*, Proceedings of the 1996 DPF/DPB Summer Study on High-Energy Physics, Snowmass'96.
59. *Geant Manual*, Cern Program Library V. 3.21, Geneva, Switzerland, 1993.
60. B. King, presentation at the Muon Collider Mini-Workshop: Lattice and Background, UCLA, Feb. 1997 and private communication.
61. N. V. Mokhov and A. Van Ginneken, *Muon Collider Neutrino Radiation*, presentation at the Muon Collider Collaboration Meeting, Orcas Is., Washington (1997); C.J. Johnstone and N.V. Mokhov, *Shielding the Muon Collider Interaction Region*, presented at the PAC97, Vancouver, Canada, 1997.
62. G. W. Foster and N. V. Mokhov, *Backgrounds and Detector Performance at 2 + 2 TeV $\mu^+\mu^-$ Collider, Physics Potential & Development of $\mu^+\mu^-$ Colliders 2nd Workshop*, Sausalito, CA, Ed. D. Cline, AIP Press, Woodbury, New York, pp. 178 (1995).

Physics at a Muon Collider [1]

John F. Gunion

Davis Institute for High Energy Physics, Department of Physics, University of California, Davis, CA 95616, USA

Abstract. I discuss the exciting prospects for exploring a wide range of new physics at a low-energy muon collider.

The physics possibilities for muon colliders (μC's) are enormous. An incomplete list includes: front-end physics; Z physics; Higgs physics, especially s-channel factory production; precision m_W, m_t measurements; deep-inelastic physics, including lepto-quarks and contact interactions; supersymmetry, including s-channel sneutrino production in R-parity violating models; strong-WW sector physics; light and heavy technicolor resonances; and new Z''s. No matter what physics lies beyond the Standard Model, the muon collider will be a very exciting machine. In this talk, I will emphasize those topics that are relevant to a first 'low'-energy muon collider ($E_{\text{beam}} \sim 50-250$ GeV), paying special attention to s-channel resonance probes of new physics.

The instantaneous luminosity, $\mathcal{L}$, possible for $\mu^+\mu^-$ collisions depends on E_{beam} and the percentage Gaussian spread in the beam energy, denoted by R. The small level of bremsstrahlung and absence of beamstrahlung implies that very small R can be achieved. The (conservative) luminosity assumptions for this workshop were: [2]

- $\mathcal{L} \sim (0.5, 1, 6) \cdot 10^{31} \text{cm}^{-2}\text{s}^{-1}$ for $R = (0.003, 0.01, 0.1)\%$ at $\sqrt{s} \sim 100$ GeV;

- $\mathcal{L} \sim (1, 3, 7) \cdot 10^{32} \text{cm}^{-2}\text{s}^{-1}$, at $\sqrt{s} \sim (200, 350, 400)$ GeV, $R \sim 0.1\%$.

---

[1] Work supported in part by U.S. Department of Energy grant No. DE-FG03-91ER40674.
[2] For yearly integrated luminosities, we use the standard convention of $\mathcal{L} = 10^{32} \text{cm}^{-2}\text{s}^{-1} \Rightarrow L = 1 \text{ fb}^{-1}/\text{yr}$.

With modest success in the collider design, at least a factor of 2 better can be anticipated. Note that for $R \sim 0.003\%$ the Gaussian spread in $\sqrt{s}$, given by $\sigma_{\sqrt{s}} \sim 2$ MeV $\left(\frac{R}{0.003\%}\right)\left(\frac{\sqrt{s}}{100 \text{ GeV}}\right)$, can be comparable to the few MeV widths of very narrow resonances such as a light SM-like Higgs boson, sneutrino resonance, or technicolor boson. This is critical since the effective resonance cross section $\bar{\sigma}$ is obtained by convoluting a Gaussian $\sqrt{s}$ distribution of width $\sigma_{\sqrt{s}}$ with the standard s-channel Breit Wigner resonance cross section $\sigma(\sqrt{\hat{s}}) = 4\pi\Gamma(\mu\mu)\Gamma(X)/([\hat{s}-M^2]^2 + [M\Gamma^{tot}]^2)$. For $\sqrt{s} = M$, the result, [3]

$$\bar{\sigma} \simeq \frac{\pi\sqrt{2\pi}\Gamma(\mu\mu) B(X)}{M^2 \sigma_{\sqrt{s}}} \times \left(1 + \frac{\pi}{8}\left[\frac{\Gamma^{tot}}{\sigma_{\sqrt{s}}}\right]^2\right)^{-1/2}, \qquad (1)$$

will be maximal if Γ^{tot} is small and $\sigma_{\sqrt{s}} \sim \Gamma^{tot}$. [4] Also critical to scanning a narrow resonance and for precision m_W and m_t measurements is the ability [2] to tune the beam energy to one part in 10^6. Finally, by constructing the muon collider at a facility (such as Fermilab) with a high energy proton beam one opens up the possibility of having a μp collider option. The luminosity expected for 200 GeV μ^+ and μ^- beams in collision with the 1 TeV proton beam of the Tevatron (yielding $\sqrt{s} = 894$ GeV) is $\mathcal{L} \sim 1.3 \cdot 10^{33} \text{cm}^{-2}\text{s}^{-1}$.

PHYSICS

• Front-End and μ Beam Physics

A proton driver and intense cooled low-energy muon beam will be the first components of the muon collider to be constructed. These alone will yield a large program of "front-end" physics. In particular, low-energy hadronic physics $(p, \bar{p}, K, \pi)$ [3] and low-energy neutrino physics (analogous to the LSND and BOONE experiments) can be explored with much improved statistics [4]. Great strides in stopped/slow intense muon beam physics (e.g. $g_\mu - 2$, $\mu N \to eN$ conversion, $\mu \to eee$, $\mu \to e\gamma$) will also be possible [5,6]. The search for $\mu N \to eN$ deserves special mention as it would probe for lepton-flavor violation at a level that is generically expected from any one of several sources present in supersymmetric models and other extensions of the SM [5,7].

---

[3] In actual numerical calculations, bremsstrahlung smearing is also included (see Ref. [1]).
[4] Although smaller $\sigma_{\sqrt{s}}$ (i.e. smaller R) implies smaller $\mathcal{L}$, the $\mathcal{L}$'s given earlier are such that when Γ^{tot} is in the MeV range it is best to use the smallest R that can be achieved.

- *Z* Physics

A low-energy muon collider could be run as a *Z* factory that would quickly exceed statistical levels achieved at LEP and SLC/SLD. Using $\sigma(Z)_{\text{peak}} \sim 6 \times 10^7$ fb ($\Gamma_Z^{\text{tot}} \gg \sigma_{\sqrt{s}}$) and $\mathcal{L} \sim 10^{32}$ cm^{-2}s^{-1} for the μC (assuming $R \gtrsim 0.1\%$ as is perfectly acceptable for *Z* physics) leads to $\sim 6 \cdot 10^7$ *Z*'s per year (about four times the best yearly rate achieved at LEP); partial ($\sim 20\%$) polarization for *both* beams would be automatic. [5]

The many important physics topics include the following. (a) $B_s - \overline{B}_s$ mixing. (b) An improved measurement of $\sin^2 \theta_W^{\text{eff}}$, as probed via A_{LR} or A_{FB}, to resolve the LEP/SLD disagreement. (c) Improved α_s determination. (d) CP Violation, as probed *e.g.* by $Z \to B_d \overline{B}_d$ ($B_d, \overline{B}_d \to \psi K_S$) decays. (e) τ Michel parameters using $Z \to \tau^+ \tau^-$ decays. (f) Separation of color-octet from color-singlet J/ψ production; detailed distributions in the final state would allow this, but LEP statistics have proved inadequate. (g) Improved limits (or actual observation) of flavor-changing-neutral-current (FCNC) rare decays; the current limits on $Z \to e\mu$, $Z \to e\tau$, $Z \to \mu\tau$ from the PDG [8] are 1.7×10^{-6}, 9.8×10^{-6}, 1.7×10^{-5}, respectively. Some types of new physics would predict such decays at levels just below this. (h) Improved limits on or observation of $Z \to \gamma X$ decays, which probe many kinds of new physics.

Of these, (a) (b) and (c) received attention during the workshop [9]. With the expected $L \sim 1$ fb^{-1}/yr (20% polarization for the beams being acceptable) one can achieve $\Delta \alpha_s \sim 0.001$ (vs. the current ~ 0.003) and an actual measurement of the x_s parameter of $B_s - \overline{B}_s$ mixing (for which LEP provides only an upper bound). Using $\Delta A_{LR} = (\mathcal{P}\sqrt{N})^{-1}$, where $\mathcal{P} = \frac{P+ - P-}{1 - P+P-}$, and $\Delta \sin^2 \theta_{\text{eff}}^{\text{lept}} \sim \Delta A_{LR}/7.9$, one finds $\Delta \sin^2 \theta_{\text{eff}}^{\text{lept}} \sim 0.0001$ (current error being $\lesssim 0.00025$ from combined LEP data) for a sample of $\sim 10^7$ *Z*'s with $P^\pm \sim \pm 30\%$ polarization for the $\mu^\pm$ beams. This would take at most a few years of operation for current μC designs.

The list of new physics probed by $Z \to \gamma X$ decays is impressive. The factor of ten improvement in sensitivity to such decays, coming from the $\gtrsim 10^8$ *Z*'s produced after a few years at a muon collider *Z* factory, would be very valuable. (i) An anomalous $ZZ\gamma$ CP-conserving and/or CP-violating coupling that might arise beyond the SM would lead to $Z \to \gamma Z^* \to \gamma \nu \overline{\nu}$

---

[5] At the μC, substantial polarization ($\gtrsim 50\%$) for both beams can be achieved only with a significant sacrifice in luminosity.

events; current limits from LEP [10] and D0 [11] are already constraining on SM extensions. (ii) Anomalous trilinear and quartic couplings can lead to $Z \to \gamma\gamma\gamma$ events. The SM prediction is $B(\gamma\gamma\gamma) \sim 10^{-9}$ while the current limit is $\lesssim 10^{-5}$; many SM extensions predict branching ratios of this latter size [12]. (iii) The magnitude of the ν_τ magnetic moment is very relevant to understanding basic neutrino properties and can have a large impact on predictions for this source of dark matter. Non-zero μ_ν leads to γ radiation from the final ν and $\overline{\nu}$ in $Z \to \nu\overline{\nu}$. Current LEP data yields [10] $\mu_{\nu_\tau} \lesssim 3.3 \times 10^{-6} \mu_B$ (90% CL). Limits from elsewhere are competitive. (iv) Improved limits on axions would be possible from searches for $Z \to \gamma A$, where A decays invisibly. Current limits on this branching ratio from LEP are [10] few $\times 10^{-6}$. If axions exist, $Z \to \gamma A$ decays might be observed with improved sensitivity. Stronger limits would significantly constrain many models. (v) Also of interest are decays of the type $Z \to \gamma +$ meson, e.g. $Z \to \gamma\pi^0, \gamma\eta, \gamma J/\psi, \ldots$ Current limits on such branching ratios are $\lesssim$ few $\times 10^{-5}$ [8]. Not only has there been much dispute about the SM predictions, but also new physics could enter. Surprises could emerge with any increase in sensitivity. (vi) A particularly important probe of technicolor theories is the $Z \to \gamma\gamma\gamma, \gamma\ell^+\ell^-, \gamma\not{E}_T, \gamma q\overline{q}, \gamma gg$ class of decays expected from $Z \to \gamma P^0$, where P^0 is an electrically neutral pseudo-Nambu-Goldstone boson (PNGB) that can decay to one or more of the indicated channels [13]. The predicted branching ratio for $Z \to \gamma P^0$ is $B(Z \to \gamma P^0) \sim 10^{-5} \left(\frac{123 \text{ GeV}}{f}\right)^2 (N_{TC} A_{Z\gamma})^2 \beta^3$, where the anomaly factor $A_{Z\gamma}$ is $\mathcal{O}(.05 - 1)$ and f is the technipion decay constant. Improving limits in the above channels by a factor of ten would rule out light P^0's in many technicolor models, whereas currently the light PNGB's of most models would have escaped detection. (vii) Finally, we note that many of the above exotic decays could have large branching ratio if the particles involved are composite.

Overall, a muon collider Z factory would have the luminosity needed to resolve some important outstanding Z physics and would provide increased sensitivity to very important rare processes that probe new physics.

- **Higgs Physics**

The potential of the muon collider for Higgs physics is truly outstanding. First, it should be emphasized that away from the s-channel Higgs pole, $\mu^+\mu^-$ and e^+e^- colliders have similar capabilities for the same $\sqrt{s}$ and $\mathcal{L}$ (barring unexpected detector backgrounds at the muon collider). At $\sqrt{s} = 500$ GeV, the design goal for a e^+e^- linear collider (eC) is $L = 50$ fb^{-1} per year. The conservative $\mathcal{L}$ estimates given earlier suggest that at $\sqrt{s} = 500$ GeV the μC will accumulate *at least* $L = 10$ fb^{-1} per year. If this can be improved somewhat, the μC would be fully competitive with the eC. We will use the notation of ℓC for either a eC or μC operating at moderate to high $\sqrt{s}$.

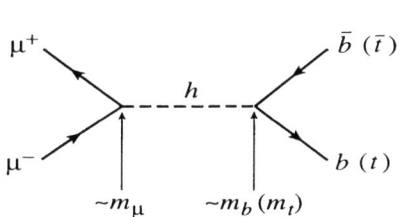 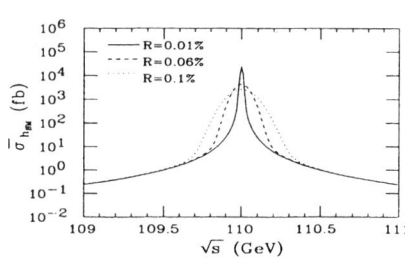

FIGURE 1. Feynman diagram for s-channel production of a Higgs boson.

FIGURE 2. The effective cross section, $\bar{\sigma}_{h_{SM}}$, for $R = 0.01\%$, $R = 0.06\%$, and $R = 0.1\%$ vs. $\sqrt{s}$ for $m_{h_{SM}} = 110$ GeV.

Of course, the totally unique feature of the μC is the very large cross section expected for production of a Higgs boson in the s-channel when $\sqrt{s} = m_h$, see Fig. 1 [1]. Small R is crucial as it leads to dramatically increased peaking of $\bar{\sigma}_h$ [Eq. (1)] at $\sqrt{s} \sim m_h$, as illustrated in Fig. 2 for a SM Higgs (h_{SM}) with $m_{h_{SM}} = 110$ GeV ($\Gamma^{tot}_{h_{SM}} \sim 3$ MeV).

A Standard Model-Like Higgs Boson

For SM-like $h \to WW, ZZ$ couplings, Γ^{tot}_h becomes big if $m_h \gtrsim 2m_W$, and $\bar{\sigma}_h \propto B(h \to \mu^+\mu^-)$ [Eq. (1)] will be small; s-channel production will not be useful. But, as shown in Fig. 2, $\bar{\sigma}_h$ is enormous for small R when the h is light, as is very relevant in supersymmetric models where the light SM-like h^0 has $m_{h^0} \lesssim 150$ GeV. In order to make use of this large cross section, we must first center on $\sqrt{s} \sim m_h$. Once this is done we proceed to the precision measurement of the Higgs boson's properties.

For a SM-like Higgs with $m_h \lesssim 2m_W$ one expects [14] $\Delta m_h \sim 100$ MeV from LHC data ($L = 300$ fb^{-1}) (smaller if ℓC data is available). Thus, a final ring that is fully optimized for $\sqrt{s} \sim m_h$ can be built. Once it is operating, we scan over the appropriate Δm_h interval so as to center on $\sqrt{s} \simeq m_h$ within a

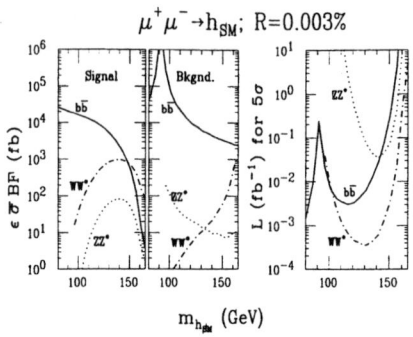
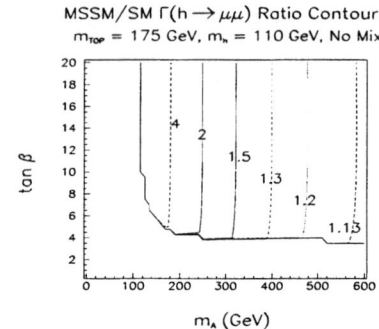

FIGURE 3. SM rates and L required for 5σ observation as a function of $m_{h_{SM}}$, for $R = 0.003\%$.

FIGURE 4. We give $(m_{A^0}, \tan\beta)$ parameter space contours for $\frac{\Gamma(h^0 \to \mu^+\mu^-)}{\Gamma(h_{SM} \to \mu^+\mu^-)}$: no-squark-mixing, $m_{h^0}, m_{h_{SM}} = 110$ GeV.

fraction of $\sigma_{\sqrt{s}}$. For m_h of order 100 GeV, $R = 0.003\%$ implies $\sigma_{\sqrt{s}} \sim 2$ MeV. The luminosity required for a 5σ observation of the SM Higgs boson with $\sqrt{s} = m_{h_{SM}}$ is plotted (along with individual signal and background rates) in Fig. 3. In the "typical" $m_h \sim 110$ GeV case, $\Delta m_h \sim 100$ MeV implies that $\Delta m_h/\sigma_{\sqrt{s}} \sim 50$ points are needed to center within $\lesssim \sigma_{\sqrt{s}}$. From Fig. 3 we find that each point requires $L \sim 0.0015$ fb^{-1} in order to observe or eliminate the h at the 3σ level, implying a total of $L_{\rm tot} \leq 0.075$ fb^{-1} is needed for centering. Thus, for the anticipated $L \sim 0.05 - 0.1$ fb^{-1}/yr, centering would take no more than a year. However, for $m_h \simeq m_Z$ a factor of 50 more $L_{\rm tot}$ is required just for centering because of the large $Z \to b\bar{b}$ background. Thus, for the anticipated $\mathcal{L}$ the μC is not useful if the Higgs boson mass is too close to m_Z.

Once centered, we will wish to measure with precision: (i) the very tiny Higgs width — $\Gamma_h^{\rm tot} = 1 - 10$ MeV for a SM-like Higgs with $m_h \lesssim 140$ GeV; (ii) $\sigma(\mu^+\mu^- \to h \to X)$ for $X = \tau^+\tau^-, b\bar{b}, c\bar{c}, WW^\star, ZZ^\star$. The accuracy achievable was studied in Ref. [1]. The three-point scan of the Higgs resonance described there is the optimal procedure for performing both measurements simultaneously. We summarize the resulting statistical errors in the case of a SM-like h with $m_h = 110$ GeV, assuming $R = 0.003\%$ and an integrated (4 to 5 year) $L_{\rm tot} = 0.4$ fb^{-1}.[6] One finds 1σ errors for $\sigma B(X)$ of $8, 3, 22, 15, 190\%$

---

[6] For σB measurements, $L_{\rm tot}$ devoted to the optimized three-point scan is equivalent to $\sim L_{\rm tot}/2$ at the $\sqrt{s} = m_h$ peak.

for the $X = \tau^+\tau^-, b\bar{b}, c\bar{c}, WW^\star, ZZ^\star$ channels, respectively, and a Γ_h^{tot} error of 16%. These results assume the τ, b, c tagging efficiencies described in Ref. [15]. We now consider how useful measurements at these accuracy levels will be.

If only s-channel Higgs factory μC data are available (*i.e.* no Zh data from an eC or μC), then the σB ratios (equivalently squared-coupling ratios [7]) that will be most effective for discriminating between the SM Higgs boson and a SM-like Higgs boson such as the h^0 of supersymmetry are $\frac{(WW^\star h)^2}{(b\bar{b}h)^2}$, $\frac{(c\bar{c}h)^2}{(b\bar{b}h)^2}$, $\frac{(WW^\star h)^2}{(\tau^+\tau^-h)^2}$, and $\frac{(c\bar{c}h)^2}{(\tau^+\tau^-h)^2}$. The 1σ errors (assuming $L_{tot} = 0.4$ fb^{-1} at $m_h = 110$ GeV) for these four ratios are 15%, 20%, 18% and 22%, respectively. Systematic errors for $(c\bar{c}h)^2$ and $(b\bar{b}h)^2$ of order $5\% - 10\%$ from uncertainty in the c and b quark mass will also enter. In order to interpret these errors one must compute the amount by which the above ratios differ in the minimal supersymmetric model (MSSM) vs. the SM for $m_{h^0} = m_{h_{SM}}$. The percentage difference turns out to be essentially identical for all the above ratios and is a function almost only of the MSSM Higgs sector parameter m_{A^0}, with very little dependence on $\tan\beta$ or top-squark mixing. At $m_{A^0} = 250$ GeV (420 GeV) one finds MSSM/SM ~ 0.5 (~ 0.8). Combining the four independent ratio measurements and including the systematic errors, one concludes that a $> 2\sigma$ deviation from the SM predictions would be found if the observed Higgs is the MSSM h^0 and $m_{A^0} < 400$ GeV. Note that the magnitude of the deviation would provide a determination of m_{A^0}.

If, in addition to the s-channel measurements we also have ℓC $\sqrt{s} = 500$ GeV, $L_{tot} = 200$ fb^{-1} data, it will be possible to discriminate at an even more accurate level between the h^0 and the h_{SM}. The most powerful technique for doing so employs the four determinations of $\Gamma(h \to \mu^+\mu^-)$ below:

$$\frac{[\Gamma(h\to\mu^+\mu^-)B(h\to b\bar{b})]_{\mu C}}{B(h\to b\bar{b})_{\ell C}} ; \qquad \frac{[\Gamma(h\to\mu^+\mu^-)B(h\to WW^\star)]_{\mu C}}{B(h\to WW^\star)_{\ell C}} ;$$
$$\frac{[\Gamma(h\to\mu^+\mu^-)B(h\to ZZ^\star)]_{\mu C}[\Gamma_h^{tot}]_{\mu C+\ell C}}{\Gamma(h\to ZZ^\star)_{\ell C}} ; \qquad \frac{[\Gamma(h\to\mu^+\mu^-)B(h\to WW^\star)\Gamma_h^{tot}]_{\mu C}}{\Gamma(h\to WW^\star)_{\ell C}}. \qquad (2)$$

The resulting 1σ error for $\Gamma(h \to \mu^+\mu^-)$ is $\lesssim 5\%$. Fig. 4, which plots the ratio of the h^0 to h_{SM} partial width in $(m_{A^0}, \tan\beta)$ parameter space for $m_{h^0} = m_{h_{SM}} = 110$ GeV, shows that this level of error allows one to distinguish between the h^0 and h_{SM} at the 3σ level out to $m_{A^0} \gtrsim 600$ GeV. This result

---

[7] From Eq. (1), $\sigma(\mu^+\mu^- \to h \to X)$ provides a determination of $\Gamma(h \to \mu^+\mu^-)B(h \to X)$ (which is proportional to the $(Xh)^2$ squared coupling) when $\sigma_{\sqrt{s}} \gtrsim \Gamma_h^{tot}$, as is the case.

holds for all $m_h \lesssim 2m_W$ ($m_h \neq m_Z$). Additional advantages of a $\Gamma(h \to \mu^+\mu^-)$ measurement are: (i) there are no systematic uncertainties arising from uncertainty in the muon mass; (ii) the error on $\Gamma(h \to \mu^+\mu^-)$ increases only very slowly as the s-channel L_{tot} decreases, [8] in contrast to the errors for the previously discussed ratios of branching ratios from the μC s-channel data which scale as $1/\sqrt{L_{tot}}$. Finally, we note that Γ_h^{tot} alone cannot be used to distinguish between the MSSM and SM in model-independent way. Not only is the error substantial ($\sim 12\%$ if we combine μC, $L = 0.4$ fb^{-1} s-channel data with ℓC, $L = 200$ fb^{-1} data) but also Γ_h^{tot} depends on many things, including (in the MSSM) the squark-mixing model. Still, deviations from SM predictions are generally substantial if $m_{A^0} \lesssim 500$ GeV.

Precise measurements of the couplings of the SM-like Higgs boson could reveal many other types of new physics. For example, if a significant fraction of a fermion's mass is generated radiatively (as opposed to arising at tree-level), then the $hf\bar{f}$ coupling and associated partial width will deviate from SM expectations [16]. Deviations of order 5% to 10% (or more) in $\Gamma(h \to \mu^+\mu^-)$ are quite possible and, as discussed above, potentially detectable.

The MSSM H^0, A^0 and $H^\pm$

We begin by recalling [14] that the possibilities for H^0, A^0 discovery are limited at other machines. (i) Discovery of H^0, A^0 is not possible at LHC for all ($m_{A^0}, \tan\beta$): e.g. if $m_{\tilde{t}} = 1$ TeV, consistency with the observed value of $B(b \to s\gamma)$ requires $m_{A^0} > 350$ GeV, in which case the LHC will not detect the H^0, A^0 if $\tan\beta \gtrsim 3$ (and below a much higher m_{A^0}-dependent value). (ii) At $\sqrt{s} = 500$ GeV, $e^+e^- \to H^0 A^0$ pair production probes only to $m_{A^0} \sim m_{H^0} \lesssim 230 - 240$ GeV. (iii) A $\gamma\gamma$ collider could potentially probe up to $m_{A^0} \sim m_{H^0} \sim 0.8\sqrt{s} \sim 400$ GeV, but only for $L_{tot} \gtrsim 150 - 200$ fb^{-1}.

Thus, it is noteworthy that $\mu^+\mu^- \to H^0, A^0$ in the s-channel potentially allows production and study of the H^0, A^0 up to $m_{A^0} \sim m_{H^0} \lesssim \sqrt{s}$. To assess the potential, let us (optimistically) assume that a total of $L_{tot} = 50$ fb^{-1} (5 yrs running at $<\mathcal{L}> = 1 \times 10^{33}$) can be accumulated for $\sqrt{s}$ in the $250 - 500$ GeV range. (We note that $\Gamma_{A^0}^{tot}$ and $\Gamma_{H^0}^{tot}$, although not big, are of a size such that resolution of $R \gtrsim 0.1\%$ will be adequate to maximize the s-channel cross section, thus allowing for substantial $\mathcal{L}$.)

---

[8] This is because the $\Gamma(h \to \mu^+\mu^-)$ error is dominated by the $\sqrt{s} = 500$ GeV measurement errors.

There are then several possible scenarios. (a) If we have some preknowledge or restrictions on m_{A^0} from LHC discovery or from s-channel measurements of h^0 properties, then $\mu^+\mu^- \to H^0$ and $\mu^+\mu^- \to A^0$ can be studied with precision for all $\tan\beta \gtrsim 1-2$. (b) If we have no knowledge of m_{A^0} other than $m_{A^0} \gtrsim 250-300$ GeV from LHC, then we might wish to search for the A^0, H^0 in $\mu^+\mu^- \to H^0, A^0$ by scanning over $\sqrt{s} = 250-500$ GeV. If their masses lie in this mass range, then their discovery by scanning will be possible for most of $(m_{A^0}, \tan\beta)$ parameter space such that they cannot be discovered at the LHC (in particular, if $m_{A^0} \gtrsim 250$ GeV and $\tan\beta \gtrsim 4-5$). (c) Alternatively, if the μC is simply run at $\sqrt{s} = 500$ GeV and $L_{\rm tot} \sim 50$ fb^{-1} is accumulated, then H^0, A^0 in the $250-500$ GeV mass range can be discovered in the $\sqrt{s}$ bremsstrahlung tail if the $b\bar{b}$ mass resolution (either by direct reconstruction or hard photon recoil) is of order ± 5 GeV and if $\tan\beta \gtrsim 6-7$ (depending on m_{A^0}). Typical peaks are illustrated in Fig. 5. [9]

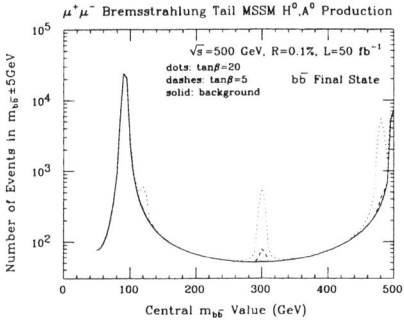

FIGURE 5. $N(b\bar{b})$ in the $m_{b\bar{b}} \pm 5$ GeV interval vs. $m_{b\bar{b}}$ for $\sqrt{s} = 500$ GeV, $L_{\rm tot} = 50$ fb^{-1}, and $R = 0.1\%$: peaks are shown for $m_{A^0} = 120, 300$ or 480 GeV, with $\tan\beta = 5$ and 20 in each case.

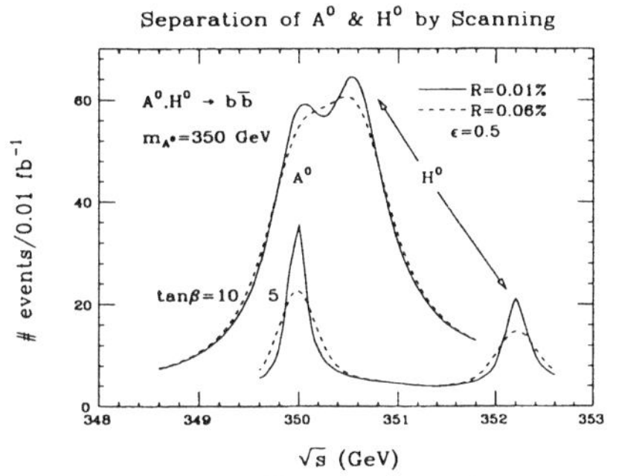

FIGURE 6. $N(b\bar{b})$ (for 0.01 fb^{-1}) vs. $\sqrt{s}$, for $m_{A^0} = 350$ GeV H^0, A^0 resonance (with $\tan\beta = 5$ and 10), including the $b\bar{b}$ continuum background.

Finally, once the closely degenerate A^0, H^0 are discovered, it will be extremely interesting to be able to separate the resonance peaks. This will probably only be possible at a muon collider with small $R \lesssim 0.01\%$ if $\tan\beta$ is large, as illustrated in Fig. 6.

We end with just a few remarks on the possibilities for production of $H^0 A^0$

---

[9] SUSY decays are assumed to be absent in this and the following figure.

and H^+H^- pairs at a high energy μC (or eC). Since $m_{A^0} \gtrsim 1$ TeV cannot be ruled out simply on the basis of hierarchy and naturalness (although fine-tuning is stretched), it is possible that energies of $\sqrt{s} > 2$ TeV could be required for pair production. If available, then it has been shown [17,18] that discovery of $H^0 A^0$ in their $b\bar{b}$ or $t\bar{t}$ decay modes and H^+H^- in their $t\bar{b}$ and $b\bar{t}$ decays will be easy for expected luminosities, even if SUSY decays are present. As a by-product, the masses will be measured with reasonable accuracy.

Regardless of whether we see the H^0, A^0 in s-channel production or via pair production, one can measure branching ratios to other channels, including supersymmetric pair decay channels with good accuracy. In fact, the ratios of branching ratios and the value of $m_{A^0} \sim m_{H^0} \sim m_{H^\pm}$ will be measured with sufficient accuracy that, in combination with one gaugino mass, say the chargino mass (which will also presumably be well-measured) it will be possible [17] to discriminate with incredible statistical significance between different closely similar GUT scenarios for the GUT-scale soft-supersymmetry-breaking masses. Thus, Higgs pair production could be very valuable in the ultimate goal of determining all the soft-SUSY-breaking parameters.

Finally, entirely unexpected decays of the heavy Higgs bosons of SUSY (or other extended Higgs sector) could be present. For example, non-negligible branching ratios for $H^0, A^0 \to t\bar{c} + c\bar{t}$ FCNC decays are not inconsistent with current theoretical model-building ideas and existing constraints [19]. The muon collider s-channel $\mu^+\mu^- \to H^0, A^0$ event rate is sufficient to probe rather small values for such FCNC branching ratios.

Exotic Higgs Bosons

If there are doubly-charged Higgs bosons, $e^-e^- \to \Delta^{--}$ probes λ_{ee} and $\mu^-\mu^- \to \Delta^{--}$ probes $\lambda_{\mu\mu}$, where the λ's are the strengths of the Majorana-like couplings [20–22]. Current $\lambda_{ee,\mu\mu}$ limits are such that factory-like production of a Δ^{--} is possible if $\Gamma^{\rm tot}_{\Delta^{--}}$ is small. Further, a Δ^{--} with $m_{\Delta^{--}} \lesssim 500-1000$ GeV will be seen previously at the LHC (for $m_{\Delta^{--}} \lesssim 200-250$ GeV at TeV33) [23]. For small $\lambda_{ee,\mu\mu,\tau\tau}$ in the range that would be appropriate, for example, for the Δ_L^{--} in the left-right symmetric model see-saw neutrino mass generation context, it may be that $\Gamma^{\rm tot}_{\Delta^{--}} \ll \sigma_{\sqrt{s}}$, [10] leading to

---

[10] For small $\lambda_{ee,\mu\mu,\tau\tau}$, $\Gamma^{\rm tot}_{\Delta^{--}}$ is very small if the $\Delta^{--} \to W^-W^-$ coupling strength is very small or zero, as required to avoid naturalness problems for $\rho = m_W^2/[\cos^2\theta_w m_Z]^2$.

$\overline{\sigma}_{\ell^-\ell^- \to \Delta^{--}} \propto \lambda_{\ell\ell}^2/\sigma_{\sqrt{s}}$. Note that the absolute rate for $\ell^-\ell^- \to \Delta^{--}$ yields a direct determination of $\lambda_{\ell\ell}^2$, which, for a Δ^{--} with very small $\Gamma_{\Delta^{--}}^{tot}$, will be impossible to determine by any other means. The relative branching ratios for $\Delta^{--} \to e^-e^-, \mu^-\mu^-, \tau^-\tau^-$ will then yield values for the remaining $\lambda_{\ell\ell}^2$'s. Because of the very small $R = 0.003\% - 0.01\%$ achievable at a muon collider, $\mu^-\mu^-$ collisions will probe much weaker $\lambda_{\mu\mu}$ coupling than the λ_{ee} coupling that can be probed in e^-e^- collisions. In addition, it is natural to anticipate that $\lambda_{\mu\mu}^2 \gg \lambda_{ee}^2$.

- **Precision Measurements of m_W and m_t**

Let us consider the extent to which the muon collider could contribute to precision measurements of m_W and m_t. Current expectations for the Tevatron, LHC and eC for various benchmark accumulated luminosities appear in Table 1 [24]. Note that more than $L_{tot} = 50$ fb^{-1} is not useful for these measurements at an electron collider since errors become systematics dominated.

At the μC, WW threshold and $t\bar{t}$ threshold measurements are the most accurate ways to determine m_W and m_t. Because of the small R and precise beam energy determination possible at the muon collider, the errors at the μC (given in Table 1) [25,26] are always statistics dominated and the accuracy that can be achieved at the μC is about a factor of two better that at the eC for the same L_{tot}. However, lower yearly L_{tot} is expected at the μC than at the eC. Taking $R \sim 0.1\%$ (better is not useful), the conservative luminosities given earlier imply $L_{tot}(\sqrt{s} = 2m_W) \sim 1$ fb^{-1}/yr and $L_{tot}(\sqrt{s} = 2m_t) \sim 3$ fb^{-1}/yr. Based on these inputs, the conclusion seems to be that systematics from beam energy spread *etc.* are low enough at the μC that accuracies for m_t competitive to 1 year of eC operation can be achieved after 2 years of μC running at a luminosity that is a factor of two better than the conservative assumption of the workshop. The μC measurement of m_W would be competitive only if $\mathcal{L}$ at $\sqrt{s} \sim 2m_W$ can be much larger (a factor of ten or so) than current expectations. Given this, and the fact that the precision electroweak determination of $m_{h_{SM}}$ is optimized for $\Delta m_W/\Delta m_t \sim 0.02$, it would seem best to focus on $t\bar{t}$ threshold measurments at the μC.

Focusing on the $t\bar{t}$ threshold is further motivated by the fact that for small R such measurements are valuable for determining α_s, Γ_t^{tot} and $|V_{tb}|^2$, as well as m_t. (There is also dependence on $m_{h_{SM}}$.) A much more detailed scan is

TABLE 1. Comparison of the achievable precision in m_W and m_t measurements at different future colliders for different L_{tot}.

	Tevatron		LHC	eC		μC			
L_{tot} (fb^{-1})	2	10	10	50	1	3	10	50
Δm_W (MeV)	22–35	11–20	15	15–20	63	36	20	10
Δm_t (GeV)	4	2	2	0.12–0.2	0.63	0.36	0.2	0.1

needed to determine these other quantities than the two- or three-point scan optimized for just the m_t measurement. To give one example, by devoting $L_{tot} = 10$ fb^{-1} to a ten-point scan, one can achieve $\Delta m_t \sim 70$ MeV and $\Delta \alpha_s \sim 0.0015$ [25].

- **Supersymmetry**

The enormous opportunities in this area are detailed in [27]. The program that has been developed for linear e^+e^- colliders is largely applicable also at a $\mu^+\mu^-$ collider. Discovery of pair production of supersymmetric particles with pair mass below $\sqrt{s}$ is generally straightforward, and detailed measurements of their masses and other properties will generally be possible. For example, the lepton and/or jet spectrum end points will typically allow measurement of the LSP mass, the lightest chargino mass, and at least some slepton masses [28–30]. The only drawback at a μC is the loss of luminosity associated with the large beam polarization(s) that would be useful for some SUSY studies. If some of the supersymmetric particles are very heavy, then the fact that a μC may be able to reach to higher energy than an eC could ultimately become crucial. Studies [31,30] suggest that a very high energy μC operating with high luminosity will be able to pin down the GUT-scale boundary conditions of the SUSY model with considerable precision, despite the fact that many different types of SUSY particle pairs will be produced.

- **μp Collisions**

We consider colliding one of the μC beams (μ^+ or μ^-) with whatever proton beam is available, e.g. the 1 TeV (820 GeV) p beam at Fermilab (DESY). Useful benchmark possibilities are $E_\mu \otimes E_p = 30$ GeV $\otimes$ 820 GeV ($\sqrt{s} = 314$ GeV), 50 GeV $\otimes$ 1 TeV ($\sqrt{s} = 447$ GeV), and 200 GeV $\otimes$ 1 TeV ($\sqrt{s} =$

894 GeV). For the $\sqrt{s} = 314$ GeV machine we assume $L = 0.1$ fb^{-1}/yr so as to provide a direct comparison to ep collisions at HERA. For the Tevatron machines, we assume $L = 2$ and 13 fb^{-1}/yr, respectively. [11]

As discussed in [33], the $\sqrt{s} = 894$ GeV machine with $L = 13$ fb^{-1}/yr yields a big increase (compared to HERA) in the kinematic limits accessible, allowing exploration of very large Q^2 values at moderate to high x values. However, event kinematics and detector considerations imply that low-x studies would be very difficult. Backgrounds could be an issue in some kinematic regions.

Contact and/or Lepto-quark Interactions

The potential of the μp collider program is perhaps best illustrated in the context of contact or lepto-quark (L_q) interactions. The relevant contact interactions are denoted $\Lambda^{\mu q}_{LL,LR,RL,RR}$ ($q = u, d, c, s$), where, for example, LR refers to an operator with left-handed μ chirality and right-handed q chirality. Lumping all chiralities together, we [34] can roughly summarize the Λ values that can be probed at 95% CL. For the luminosities quoted above: (a) the $\sqrt{s} = 314$ GeV HERA-analogue machine would be sensitive to $\Lambda^{\mu u} \sim (0.8 - 2.0)$ TeV and $\Lambda^{\mu d} \sim (0.7 - 1.3)$ TeV; (b) the Tevatron machines probe $\Lambda^{\mu u} \sim (7 - 12) \times \sqrt{s}$ and $\Lambda^{\mu d} \sim (4 - 8) \times \sqrt{s}$. In particular, the 200 GeV⊗1 TeV machine probes $\Lambda^{\mu u}$'s $\sim (6-11)$ TeV and $\Lambda^{\mu d} \sim (4-7)$ TeV, *i.e.* far beyond the HERA machine level. However, if Λ^{eu} and/or Λ^{ed} is non-zero (as perhaps indicated by the HERA ep excess), an acceptably small level of FCNC requires $\Lambda^{\mu u,\mu d} \simeq 0$; only $\Lambda^{\mu c,\mu s}$ could be significant in size. Due to the smaller size of the c and s distribution functions in the proton, the $\Lambda^{\mu c}$ ($\Lambda^{\mu s}$) values that can be probed are typically a factor of 3–4 (1.5–2) smaller than the $\Lambda^{\mu u}$ ($\Lambda^{\mu d}$) values quoted above.

We turn next to lepto-quarks. for a first comparison [34] of different colliders we focus on a +2/3 charge spin-0 lepto-quark with $\ell^+ d$ and/or $\ell^+ s$ couplings. The relevant coupling is defined by $\mathcal{L} = \lambda_{\ell q} L_q \bar{q} P_\tau \ell$ ($\tau = L$ or R). We take $B(L_q \to \ell^+ q) = 1$ and require a 95% CL for the signal with respect to predicted background. Table 2 shows that one can probe the same level of $\lambda_{\ell^+ q}$ ($q = s, d$) at much higher M_{L_q} (or increasingly smaller $\lambda_{\ell^+ q}$ at the same M_{L_q}) as the μp collider energy and luminosity increases. Of course, if there is a lepto-quark

---

[11] The former is the result obtained using scaling [32] of $\mathcal{L} \propto E_\mu^{4/3}$ starting with the workshop assumption of $L = 13$ fb^{-1}/yr at the 200 GeV ⊗ 1 TeV machine.

TABLE 2. $\lambda_{\ell+d}$ ($\lambda_{\ell+s}$) values required for a 95% CL signal with respect to background assuming $B(L_q \to \ell^+ j) = 1$.

M_{L_q} (GeV)		200	300	400	500	600	700	800
$\sqrt{s}$ (GeV)	L (fb^{-1})	\multicolumn{7}{c}{$\lambda_{\ell+d} \times 10^3$ ($\lambda_{\ell+s} \times 10^3$)}
314	0.1	14(73)	–	–	–	–	–	–
447	2	4.5(13)	10(53)	55(1130)	–	–	–	–
894	13	2(4)	3(7)	4(12)	6(22)	9(99)	16(140)	45(860)

with $\lambda_{e+d} \neq 0$ (as possibly hinted by HERA data), then FCNC limits require $\lambda_{\mu+d} \simeq 0$; but, $\lambda_{\mu+s}$ for this same lepto-quark could be non-zero. Table 2 shows that at $M_{L_q} = 200$ GeV, the $\lambda_{\mu+s}$ value that can be probed at the $\sqrt{s} = 447$ GeV ($\sqrt{s} = 894$ GeV) Tevatron μp collider is comparable to (much smaller than) the λ_{e+d} value that can be probed at HERA, despite the fact that the distribution function for s quarks in the proton is much smaller than that for d quarks. Further, it is highly possible that the second family $\lambda_{\mu+s}$ coupling would be larger than the first family λ_{e+d} coupling.

We note that if any evidence for contact or lepto-quark interactions is found, it will be very important to look for the corresponding excess events in the $\ell^+\ell^- \to q\bar{q}$ ($\ell = e, \mu$) cross-channels.

Overall, the discovery reach of the μp colliders is quite impressive. Further, it cannot be stressed too strongly that if evidence for contact interactions or lepto-quarks is discovered in ep or e^+e^- collisions, then it will be mandatory to build an analogous muon facility for μp and $\mu^+\mu^-$ interactions so as to explore the lepton flavor dependence of the new physics.

- **Neutrino Beam Physics**

The neutrino beams from an energetic μ beam would be excellent for νN fixed target experiments. One possibility examined at this workshop was a 250 GeV μ beam, which yields $\langle E_\nu \rangle \sim 178$ GeV and a beam of known composition (*e.g.* $\nu_\mu, \bar{\nu}_e$ for a μ^- beam). For νN fixed target experiments, the neutrino flux would be about a thousand times larger than at present machines. The large flux implies that good statistics would be obtained using light targets, allowing a more definitive comparison of $F_2(x, Q^2)$ in charged and neutral current measurements. Improved measurements of xF_3, F_2^{charm}, spin-physics distributions, and $|V_{ub}|^2$ would all be possible as well.

There are also substantial advantages associated with the neutrino beams from an energetic muon beam for long-baseline neutrino oscillation experiments [4,35,36]. For example, by pointing a muon storage ring at an appropriate underground detector (Soudan, Gran Sasso, ...) it would be possible to probe Δm^2 and $\sin^2 2\theta$ neutrino-mixing parameters that are factors of 10 and 100-1000, respectively, smaller than can be probed by the MINOS and MiniBooNE experiments [35]. The known composition of the neutrino beam would again be a big asset.

- **R-parity Violating Scenarios**

If there is $\not{R}$ of form $\lambda_{ijk}\hat{L}_L^i\hat{L}_L^j\overline{\hat{E}_R^k} + \lambda'_{ijk}\hat{L}_L^i\hat{Q}_L^j\overline{\hat{D}_R^k}$ (i.e. baryon number is conserved and lepton number is violated), then many new physics signals arise. (i) $\lambda' \neq 0$ allows an interpretation of the HERA events in which a squark plays the role of a lepto-quark; most likely $e^+d \to \tilde{t}$ or $\tilde{c}$. Even if the HERA excess disappears, the analogous $\mu^+s \to \tilde{c}$ (no family transition) and $\mu^+s \to \tilde{t}$ couplings could be much larger, in analogy to standard Yukawa coupling trends. A μp collider would then be a very exciting machine. (ii) $\lambda \neq 0$ would lead to the possibility of $e^+e^- \to \tilde{\nu}_\tau$ (λ_{131}) and/or $\tilde{\nu}_\mu$ (λ_{121}) and $\mu^+\mu^- \to \tilde{\nu}_\tau$ (λ_{232}) and/or $\tilde{\nu}_e$ (λ_{122}); s-channel e^+e^- and $\mu^+\mu^-$ production of a $\tilde{\nu}$ is an exciting prospect.

Sensitivities to the squark couplings are related to those in the general lepto-quark case (aside from corrections needed for possibly different branching ratios to the final state of interest); for example, we would identify $\lambda_{e+d} \to \lambda'_{1j1}$, $\lambda_{\mu+d} \to \lambda'_{2j1}$, $\lambda_{e+s} \to \lambda'_{1j2}$, $\lambda_{\mu+s} \to \lambda'_{2j2}$, for $L_q = \tilde{t}$ ($j = 3$) or $\tilde{c}$ ($j = 2$). From the lepto-quark discussion it is apparent that fairly small $\lambda'_{2j1,2j2}$ values yield a visible signal for reasonably large squark masses.

Since $\tilde{\nu}_\tau$ is probably the lightest of the sneutrinos and since λ_{232} is probably the largest of the λ's a muon collider looks especially interesting for s-channel sneutrino production. The excellent beam energy resolution of a muon collider would also be a great advantage. To illustrate [37], assume that only λ_{232} and λ'_{333} (surely the largest of the λ''s) are non-zero. For reasonable superpartner masses, appropriate limits are $\lambda_{232} \lesssim 0.06$, $\lambda'_{333} \lesssim 1$, and $\lambda_{232}\lambda'_{333} \lesssim 0.004$. The possible $\tilde{\nu}_\tau$ decays are: (a) $\tilde{\nu}_\tau \to \nu_\tau \tilde{\chi}_1^0$ if $m_{\tilde{\chi}_1^0} < m_{\tilde{\nu}_\tau}$ (with $\tilde{\chi}_1^0$ in turn decaying via $\not{R}$ couplings); (b) $\tilde{\nu}_\tau \to \mu^+\mu^-$ (via λ_{232}); and (c) $\tilde{\nu}_\tau \to b\bar{b}$ (via

λ'_{333}). $\Gamma^{\text{tot}}_{\tilde{\nu}_\tau}$ tends to be large if $m_{\tilde{\chi}^0_1} < m_{\tilde{\nu}_\tau}$, but can be very small if $\tilde{\nu}_\tau \to \nu_\tau \tilde{\chi}^0_1$ is disallowed. Typical decay widths can be found in Ref. [37].

What will be the role of the muon collider? In the most likely case, LHC and/or $\sqrt{s} = 500$ GeV ℓC data will reveal the existence of R-parity violation and yield an approximate determination of $m_{\tilde{\nu}_\tau}$. Expectations for the latter are: (i) $\Delta m_{\tilde{\nu}_\tau} \sim 100$ MeV if $\mu^+\mu^-$ and or $b\bar{b}$ decays are observable; (ii) $\Delta m_{\tilde{\nu}_\tau} \lesssim 2$ GeV if only $\tilde{\nu}_\tau \to \nu_\tau \tilde{\chi}^0_1$ decays give a substantial number of events. However, even though we know $\not{R}$ is present, only in special situations will it be possible to determine the actual magnitude of λ_{232} and λ'_{333} from this LHC or ℓC data. In contrast, unless λ_{232} is very small, the muon collider will allow an accurate measurement of λ_{232} and, possibly, also of λ'_{333}.

The procedure is analogous to that for a Higgs boson. Once the $\tilde{\nu}_\tau$ is observed, we turn to the μC and scan for the precise location. The cross section depends on $\Gamma^{\text{tot}}_{\tilde{\nu}_\tau}$ and $\sigma_{\sqrt{s}}$ as in Eq. (1) leading to the following scenarios: (i) If $\Gamma^{\text{tot}}_{\tilde{\nu}_\tau}$ is as small as is likely if $m_{\tilde{\nu}_\tau} < m_{\tilde{\chi}^0_1}$ ($\tilde{\nu}_\tau \to \nu_\tau \tilde{\chi}^0_1$ forbidden), $\bar{\sigma}_{\tilde{\nu}_\tau}$ is largest if $\sigma_{\sqrt{s}}$ is as small as possible. Thus, it is best to use $R = 0.003\%$ ($L = 0.1$ fb^{-1}/yr) and place scan points using intervals of size $2\sigma_{\sqrt{s}}$. (ii) If $\tilde{\nu}_\tau \to \nu_\tau \tilde{\chi}^0_1$ is observed, $\Gamma^{\text{tot}}_{\tilde{\nu}_\tau}$ will be large and will be dominated by $\Gamma(\tilde{\nu}_\tau \to \nu_\tau \tilde{\chi}^0_1)$ which, in turn, can be computed from known $\tilde{\chi}^0_1$ properties. Then, it is most advantageous to use $R = 0.1\%$ ($L = 1$ fb^{-1}/yr) and use a scan interval of $\max[2\sigma_{\sqrt{s}}, \Gamma^{\text{tot}}_{\tilde{\nu}_\tau}]$. The regions of $(\lambda_{232}, \lambda'_{333})$ parameter space for which the

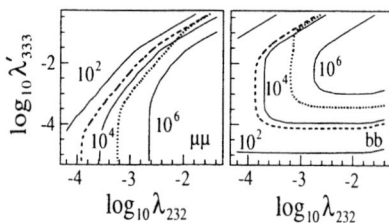 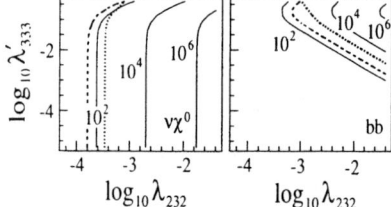

FIGURE 7. We consider $m_{\tilde{\nu}_\tau} < m_{\tilde{\chi}^0_1}$, $m_{\tilde{\nu}_\tau} = 100$ GeV. Solid contours are for $\bar{\sigma}_{\tilde{\nu}_\tau} B(\tilde{\nu}_\tau \to X)$ (in fb), $X = \mu^+\mu^-, b\bar{b}$. The dashed (dotted) contour is the optimistic (pessimistic scan) 3σ discovery boundary for $L_{\text{tot}} = 0.1$ fb^{-1} and $R = 0.003\%$.

FIGURE 8. As in Fig. 7, but for the $X = \nu \tilde{\chi}^0_1$ and $b\bar{b}$ final states in the $m_{\tilde{\nu}_\tau} > m_{\tilde{\chi}^0_1}$ scenario, with $m_{\tilde{\nu}_\tau} = 150$ GeV and a bino-like $\tilde{\chi}^0_1$ with $m_{\tilde{\chi}^0_1} = 100$ GeV. Discovery boundaries assume $L_{\text{tot}} = 1$ fb^{-1} at $R = 0.1\%$.

$\widetilde{\nu}_\tau$ could be discovered by the s-channel scan over the 100 MeV and 2 GeV mass windows in cases (i) and (ii) are illustrated by the dotted contours in Figs. 7 and 8, respectively. The prospects are excellent unless the R-parity violating λ_{232} coupling is really quite tiny.

We have estimated the accuracy with which the R-parity violating couplings can be measured once the $\widetilde{\nu}_\tau$ is discovered, assuming 3 years of operation [$L = 0.3$ fb^{-1} in case (i) and $L = 3$ fb^{-1} in case (ii)] distributed at and/or near $\sqrt{s} = m_{\widetilde{\nu}_\tau}$. For $m_{\widetilde{\chi}_1^0} > m_{\widetilde{\nu}_\tau}$ and $\lambda_{232} = 5 \times 10^{-4}$ (which is on the border of the scan discovery regions and typical of many models), we find 1σ errors of $\Delta\lambda_{232}/\lambda_{232} \sim 2\% - 15\%$; $\Delta\lambda'_{333}/\lambda'_{333} \sim 10\% - 30\%$ is achieved if λ'_{333} is not too small. If $\widetilde{\nu}_\tau \to \nu_\tau \widetilde{\chi}_1^0$, λ'_{333} must be substantial to be measurable. However, even if $\lambda'_{333} \to 0$, the $\nu_\tau \widetilde{\chi}_1^0$ final state will yield $\Delta\lambda_{232}/\lambda_{232} \sim 0.9, 9, 15, 25, 80\%$ for $\lambda_{232} = 10^{-3}, 10^{-4}, 5 \times 10^{-5}, 3 \times 10^{-5}, 10^{-5}$, respectively.

Finally, we note that the ability to achieve $R = 0.003\%$ is a unique muon collider feature that could allow one to resolve the $\mathcal{O}(1$ MeV$)$ splitting between the CP-even and CP-odd $\widetilde{\nu}_\tau$ components that is predicted using generic relationships to neutrino masses and a ν_τ mass in the $\gtrsim 1$ MeV range.

- **Probes of Technicolor and Strong WW Scattering**

The ability of a high energy muon collider ($\sqrt{s} \gtrsim 3$ TeV) with high $\mathcal{L}$ ($\mathcal{L} \gtrsim 10^{33} - 10^{34}$cm^{-2}s^{-1} is anticipated) to search for heavy technicolor or related resonances or explore a strongly interacting WW sector has been well documented [38] and will not be reviewed here. Additional work at this meeting in this area appears in [39]. Here we briefly summarize the ability of a low-energy muon collider to observe the pseudo-Nambu-Goldstone bosons (PNGB's) of an extended technicolor theory. These are narrow states, that, as noted earlier, need not have appeared at an observable level in Z decays at LEP. Some of the PNGB's have substantial $\mu^+\mu^-$ couplings. Thus, a muon collider search for them will bear a close resemblance to the light Higgs and R-parity violating sneutrino cases discussed already. The main difference is that, assuming they have not been detected ahead of time, we must search over the full expected mass range.

The results of PNGB studies at this meeting appear in Refs. [40] and [41]. Here I summarize the results for the lightest P^0 PNGB as given in Ref. [40]. Although the specific P^0 properties employed are those predicted by the extended BESS model [40], they will be representative of what would be found in any extended technicolor model for a strongly interacting electroweak sector. The first point is that m_{P^0} is expected to be small; $m_{P^0} \lesssim 80$ GeV is preferred in the BESS model. Second, the Yukawa couplings and branching ratios of the P^0 are easily determined. In the BESS model, $\mathcal{L}_Y = -i\sum_f \lambda_f \bar{f}\gamma_5 f P^0$ with $\lambda_b = \sqrt{\frac{2}{3}}\frac{m_b}{v}$, $\lambda_\tau = -\sqrt{6}\frac{m_\tau}{v}$, $\lambda_\mu = -\sqrt{6}\frac{m_\mu}{v}$. Note the sizeable $\mu^+\mu^-$ coupling. The P^0 couplings to $\gamma\gamma$ and gg from the ABJ anomaly are also important. Overall, these couplings are not unlike those of a light Higgs boson. Not surprisingly, therefore, $\Gamma^{tot}_{P^0}$ is very tiny: $\Gamma^{tot}_{P^0} = 0.2, 4, 10$ MeV for $m_{P^0} = 10, 80, 150$ GeV, respectively, for $N_{TC} = 4$ technicolor flavors. For such narrow widths, it will be best to use $R = 0.003\%$ beam energy resolution.

For the detailed tagging efficiencies *etc.* described in [40], the L_{tot} required to achieve $\sum_k S_k/\sqrt{\sum_k B_k} = 5$ at $\sqrt{s} = m_{P^0}$, after summing over the optimal selection of the $k = b\bar{b}$, $\tau^+\tau^-$, $c\bar{c}$, and gg channels (as defined after tagging), is plotted in Fig. 9. Very modest L_{tot} is needed unless $m_{P^0} \sim m_Z$. Of course, if we do not have any information regarding the P^0 mass, we must scan for the resonance. The (very conservative, see [40] for details) estimate for the luminosity required for scanning a given 5 GeV interval and either discovering or eliminating the P^0 in that interval at the 3σ level is plotted in Fig. 10. If the P^0 is as light as expected in the extended BESS model, then the prospects

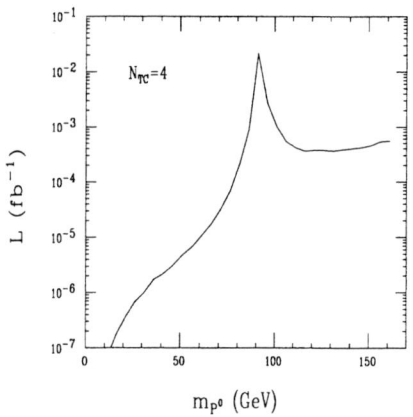
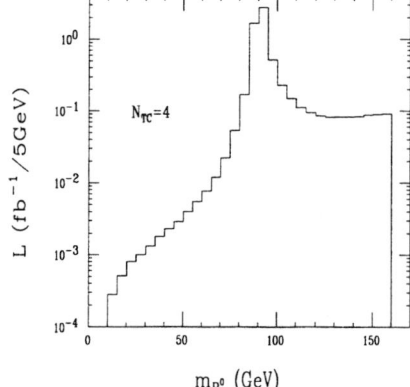

FIGURE 9. L_{tot} required for a 5σ P^0 signal at $\sqrt{s} = m_{P^0}$.

FIGURE 10. L_{tot} required to scan indicated 5 GeV intervals and either discover or eliminate the P^0 at the 3σ level.

for discovery by scanning would be excellent. For example, a P^0 lying in the ~ 10 GeV to ~ 75 GeV mass interval can be either discovered or eliminated at the 3σ level with just 0.11 fb^{-1} of total luminosity, distributed in proportion to the luminosities plotted in Fig. 10. The $\mathcal{L}$ that could be achieved at these low masses is being studied [32]. A P^0 with $m_{P^0} \sim m_Z$ would be much more difficult to discover unless its mass was approximately known. A 3σ scan of the mass interval from ~ 105 GeV to 160 GeV would require about 1 fb^{-1} of integrated luminosity, which is more than could be comfortably achieved for the conservative $R = 0.003\%$ $\mathcal{L}$ values assumed for this workshop.

DISCUSSION AND CONCLUSIONS

There is little doubt that a variety of accelerators will be needed to explore all aspects of the physics that lies beyond the Standard Model and accumulate adequate luminosity for this purpose in a timely fashion. Certainly, a muon collider (preferably in conjunction with a μp option) would make major contributions to understanding any foreseeable type of new physics. It would be of special value in studying narrow resonances with $\mu^+\mu^-$ couplings (such as the SUSY Higgs bosons) and the lepton flavor dependence of many important classes of new physics. The physics motivations for a muon collider are undeniable and we should proceed with the R&D required to assess its viability. Finding designs that yield the highest possible luminosity at low energies, while maintaining excellent beam energy resolution, should be a priority.

REFERENCES

1. V. Barger, M. Berger, J. Gunion and T. Han, Phys. Rev. Lett. **75**, 1462 (1995); *Phys. Rep.* **286**, (1997) 1.

2. R. Raja, these proceedings.

3. G. Gutierrez and L. Littenberg [Physics with Low-Energy Hadrons Working Group], these proceedings.

4. B. Kayser and P. Fisher [Neutrino Physics Working Group], these proceedings.

5. W. Marciano, these proceedings; A. Czarnecki, W. Marciano and K. Melnikov, these proceedings.

6. T. Diehl and W. Molzon [Physics with Slow/Stopped Muons Working Group], these proceedings.

7. S. Raby, these proceedings.

8. Particle Data Group, Phys. Rev. **D54**, 1 (1996).

9. M. Demarteau and T. Han [Higgs, Z Working Group], these proceedings.

10. M. Acciari *et al.* [L3 Collaboration], CERN-PPE/97-82.

11. S. Abachi *et al.* [D0 Collaboration], Phys. Rev. Lett. **78**, 3640 (1997).

12. Seem, for example, F. Boudjema and F.M. Renard, *Z Physics at LEP*, CERN 89-08, Vol. 2, 1989, 185.

13. G. Rupak and E. Simmons, Phys. Lett. **B362**, 155 (1995).

14. J.F. Gunion, L. Poggioli, R. Van Kooten, C. Kao and P. Rowson, in *New Directions for High-Energy Physics*, Proceedings of the 1996 DPF/DPB Summer Study on High-Energy Physics, June 25—July 12, 1996, Snowmass, CO, edited by D.G. Cassel, L.T. Gennari, and R.H. Siemann (Stanford Linear Accelerator Center, 1997) pp. 541–587.

15. B. King, presentation at this conference.

16. F. Borzumati, G. Farrar, N. Polonsky and S. Thomas, these proceedings.

17. J.F. Gunion and J. Kelly, Phys. Rev. **D46**, 1730 (1997.)

18. J. Feng and T. Moroi, *Phys. Rev.* **D56** (1997) 5962.

19. L. Reina, these proceedings.

20. J.F. Gunion, Int. J. Mod. Phys. **A11**, 1551 (1996).

21. P. Frampton, Int. J. Mod. Phys. **A11**, 1621 (1996).

22. F. Cuypers, Nucl. Phys. **B510**, 3 (1997).

23. J.F. Gunion, C. Loomis and K. Pitts, in *New Directions for High-Energy Physics*, Proceedings of the 1996 DPF/DPB Summer Study on High Energy Physics, Snowmass '96, edited by D.G. Cassel, L.T. Gennari and R.H. Siemann (Stanford Linear Accelerator Center, Stanford, CA, 1997) p. 603.

24. H. Haber *et al.*, in *New Directions for High-Energy Physics*, Proceedings of the 1996 DPF/DPB Summer Study on High Energy Physics, Snowmass '96, edited by D.G. Cassel, L.T. Gennari and R.H. Siemann (Stanford Linear Accelerator Center, Stanford, CA, 1997) p. 482; U. Baur *et al.*, *ibid.* p. 499. In the case of the eC, the higher errors are those quoted by U. Baur *et al.*, while the lower come

from J.-F. Grivaz, T. Sjostrand and P.M. Zerwas [Physics Working Group], in Proceedings of the Joint ECFA/DESY Study: *Physics and Detectors of a Linear Collider*, February–November, 1996.

25. V. Barger, M. Berger, J.F. Gunion and T. Han, Phys. Rev. **D56**, 1714 (1997).

26. M. Berger and B. Winer [$t\bar{t}$ Factory Working Group], these proceedings. M. Berger, these proceedings.

27. M. Carena and S. Protopopescu [SUSY Searches and Measurements Working Group], these proceedings.

28. F. Paige, these proceedings.

29. J. Lykken, 4th International Conference on "Physics Potential and Development of Muon Colliders", December, 1997, San Francisco CA.

30. V. Barger, M. Berger, M. Carena, J. Gunion, T. Han, J. Kelly and J. Lykken, in preparation.

31. J. Kelly, these proceedings.

32. R. Palmer, private communication.

33. H. Schellman and S. Ritz [Deep-Inelastic Scattering Working Group], these proceedings.

34. K. Cheung, hep-ph/9802219, presentation at the 4th International Conference on "Physics Potential and Development of Muon Colliders", December, 1997, San Francisco CA; V. Barger, M. Berger, K. Cheung, J.F. Gunion and T. Han, work in progress.

35. S. Geer, these proceedings.

36. R. Mohapatra, these proceedings.

37. J. Feng, J. Gunion and T. Han, UCD-97-25 [hep-ph/9711414].

38. V. Barger, M. Berger, J. Gunion and T. Han, Phys. Rev. **D55**, 142 (1997).

39. P. Bhat and E. Eichten [Strong Dynamics Working Group], these proceedings.

40. R. Casalbuoni, S. De Curtis, D. Dominici, A. Deandrea, R. Gatto and J. F. Gunion, these proceedings.

41. K. Lane, these proceedings.

Low Energy Physics and the First Muon Collider

William J. Marciano

Physics Department, Brookhaven National Laboratory, Upton, New York 11973

Abstract. The utilization of low energy physics studies as a step towards the First Muon Collider is discussed. A program of compelling muon and kaon experiments is outlined.

INTRODUCTORY PERSPECTIVE

The muon collider concept is both stimulating and provocative. It offers the promise of very high energy $\mu^+\mu^-$ collisions with high luminosity. Various possibilities with $\sqrt{s} = 100, 500, 4,000\ldots$ GeV have been studied. They suggest an exciting new accelerator option that could greatly extend the frontiers of high energy physics during the next millennium.

Is the muon collider a "pipe dream" or "an idea whose time has come"? I think the answer still lies somewhere inbetween. After considerable work, a preliminary conceptual design exists [1]. It envisions an accelerator complex based on state-of-the-art technology with varying size storage rings. An attractive feature is the fact that it could fit on an existing laboratory site and take advantage of existing infrastructure.

The many components of a muon collider (muon production, collection, cooling, acceleration, storage, collision, and detection) are all extremely challenging; but so-far, there seem to be no show-stoppers.

How does the "First Muon Collider" (FMC) become a reality? Aggressive R&D is clearly necessary. A consensus on the most compelling physics goals is required. Further accelerator design creativity and reality checks are also needed.

As a concrete first step towards realizing a muon collider, I have previously made the following modest proposal [2]. Why not build an intense low energy muon source? It could be used for collider studies such as muon production, collection, polarization, cooling etc. Just as important, such a facility would support its own world class low energy muon physics program. Furthermore, the demands of real experiments would provide a useful synergism with accelerator R&D.

My specific proposal is based on the possibility of using an existing intense proton synchrotron such as Brookhaven's AGS to copiously produce muons. Indeed, muon collider studies indicate that a proton driver such as the AGS is capable of producing nearly one usable $\mu^\pm$ per proton on target. In principle, the AGS could supply $10^{13}\mu^\pm$/sec. Compare that potential capability with current muon beams at meson facilities such as PSI, TRIUMF, and LAMPF where $10^7 \sim 10^8 \mu$/sec is more typical.

The envisioned muon facility at the AGS would probably have as its realistic goal the collection of about $10^{11}\mu$/sec, about 1% of its maximum potential. (A much less expensive 3 Tesla rather than 30 Tesla (FMC) collection solenoid would be employed.) That relatively modest goal is still nearly four orders of magnitude beyond current state-of-the-art capabilities and could significantly advance low energy muon physics studies. Also, the cost of such a facility and its experiments is of order $30 million, a small part of what one would expect to spend on collider R&D for an eventual multi-billion dollar muon collider.

A more aggressive alternative to the above scenario, being investigated at this workshop, is to build the complete front end of a muon collider and use it to advance all of low energy particle physics, including kaon, antiproton, pion, neutrino, etc. studies. Such a facility would be much more rapid cycling than the AGS and designed to yield $10^{13} \sim 10^{14}\mu^\pm$/sec. If some degree of cooling were available, the muon beams would be extremely clean and could have a very small energy spread. They would be ideal for reducing background effects. In the case of K studies, one could imagine 20 × the intensity of the AGS. Of course, the full front-end of the muon collider along with instrumented beam lines probably represents a cost of order $500 million; so, its main motivation must come from the long term collider potential.

With the two approaches, described above, in mind, I will outline a few low energy muon studies that could be contemplated with $10^{11} \sim 10^{13}\mu$/sec. For the full new front end scenario, I also briefly mention kaon physics goals that could be pursued. Neutrino oscillations, another compelling low energy option, will be covered in other talks [3].

LOW ENERGY MUON PHYSICS

What compelling physics studies might be carried out with $10^{11} \sim 10^{13}\mu^\pm$/sec? Such a 4 ∼ 6 order of magnitude intensity increase could be used for a rich and varied muon physics program. It might, for example, be applied to condensed matter (μSR) or μ^- catalyzed fusion (μ^-dT) research. I will not discuss those interesting options here. Instead, I concentrate on fundamental particle physics issues and mention only a few possibilites.

Precision Measurements

Muon physics has been synonymous with high precision studies. Measurements of the muon lifetime, Michel decay parameters, anomalous magnetic moment, etc. have contributed significantly to the development of the standard model. One might contemplate trying to push those measurements several orders of magnitude further in statistical precision. However, experiments in progress, approved, or being considered are already starting to approach the limitations of theoretical uncertainty in their interpretation. One can, of course, always argue that precision measurements of fundamental quantities should be pushed as far as possible and theory will eventually catch up. Let me discuss two examples.

The muon anomalous magnetic moment, $a_\mu \equiv (g_\mu - 2)/2$, is potentially sensitive to "new physics". That quantity gets contributions from QED, hadronic, and electroweak loops in the standard model. Currently, one finds [4]

$$\begin{aligned} a_\mu^{\text{QED}} &= 116\ 584\ 706(\ 2) \times 10^{-11} \quad &\text{(5 loops)} \\ a_\mu^{\text{Had}} &= 6\ 771(77) \times 10^{-11} \quad &\text{(3 loops)} \\ a_\mu^{\text{EW}} &= 151(\ 4) \times 10^{-11} \quad &\text{(2 loops)} \end{aligned} \quad (1)$$

$$a_\mu^{\text{theory}} = 116\ 591\ 628(77) \times 10^{-11}$$

That prediction is to be compared with a 20 year old CERN result

$$a_\mu^{\text{exp}} = 116\ 592\ 300(840) \times 10^{-11} \quad (2)$$

$$a_\mu^{\text{exp}} - a_\mu^{\text{th}} = 672 \pm 840 \pm 77 \times 10^{-11} \quad (3)$$

Agreement is quite good. An experiment (E821) in progress at Brookhaven aims to reduce the experimental uncertainty by a factor of 20 or more. It will be very sensitive to supersymmetry loops (for large $\tan\beta$)

$$a_\mu^{\text{SUSY}} \sim 140 \times 10^{-11} \tan\beta \left(\frac{100 \text{ GeV}}{m_{\text{SUSY}}}\right)^2 \quad (4)$$

as well as dynamical or loop induced mass generating mechanisms which will be probed at the 5 TeV level and other potential "new physics" effects. At some point, theoretical uncertainties will dominate. How much can they be reduced? I expect the hadronic loop uncertainty to go below $\pm 40 \times 10^{-11}$ particularly when very good $\tau \to \nu_\tau +$ hadrons data becomes available; but, by how much? If one is very optimistic, they may reach $\pm 20 \times 10^{-11}$. It is, therefore, difficult at this time to motivate a future much higher statistics effort beyond the current Brookhaven experiment; but never say never. Theorists may find new techniques for computing low energy hadronic loop effects, for example by using lattice gauge theory approaches. Indeed, theoretical and experimental advances tend to be strongly correlated. So, if a_μ^{exp} can be further improved, it probably should. More interesting however, may be a future follow-up muon electric dipole moment study. I return to that possibility later.

Another important precision measurement is the muon lifetime

$$\tau_\mu = 2.19703(4) \times 10^{-6} \text{ sec} \quad (5)$$

That measurement currently provides the best determination of the Fermi constant

$$G_\mu = 1.16639(1) \times 10^{-5} \text{ GeV}^{-2} \tag{6}$$

It can be used in conjunction with [5]

$$\alpha^{-1} = 137.03599944(57)$$
$$m_Z = 91.186(2) \text{ GeV} \tag{7}$$

and quantum loop calculations (Δr and $\Delta \hat{r}$) which depend on m_t and m_H to predict $\sin^2 \theta_W(m_Z)_{\overline{MS}}$ and m_W (see table 1)

$$\sin^2 2\theta_W(m_Z)_{\overline{MS}} = \frac{A}{1 - \Delta \hat{r}}$$

$$m_W^2 = \frac{1}{2} m_Z^2 \left[1 + \sqrt{1 - \frac{A}{1 - \Delta r}} \right] \tag{8}$$

$$A = \frac{4\pi\alpha}{\sqrt{2} G_\mu m_Z^2} = 0.668590 \pm \underbrace{0.000006}_{\Delta G_\mu} \pm \underbrace{0.000030}_{\Delta m_Z}$$

The uncertainty in those relationships due to the error in G_μ is currently insignificant and cannot at this time be used to strongly motivate a better determination of τ_μ. (An argument to better measure $\tau_\mu(G_\mu)$ would have to be based on the fundamental nature of that quantity. Alternatively, it has been suggested that precisely measuring the lifetime of both stopped μ^+ and μ^- (in hydrogen) could be used to infer the $\mu^- p$ capture rate.) Of course, if we can measure τ_μ 10 to 100 times better, we probably should. Indeed, when the measurement in Eq. (5) was carried out, one probably could not have imagined how it would be used to probe the top quark and Higgs' masses as well as constrain new physics. Perhaps additional uses for G_μ will emerge in the future.

Having given the relationships in Eq. (8), let me take this opportunity to comment on the extraction of the Higgs mass, m_H. For $m_t = 175 \pm 5$ GeV, one finds the results in table 1. Current averages

$$\sin^2 \theta_W(m_Z)_{\overline{MS}} = 0.23122(22)$$
$$m_W = 80.43(8) \text{ GeV} \tag{9}$$

suggest a relatively light Higgs. Note also, that the utility of $\sin^2 \theta_W(m_Z)_{\overline{MS}}$ to constrain m_H is starting to be limited by both Δm_t and hadronic loop uncertainties. However, the use of m_W still has a way to go before those uncertainties set in.

Other interesting muon measurements worth considering are: muonium ($\mu^+ e^-$) H.F.S., muonium-antimuonium oscillations, muon neutrino mass constraints, parity or CP violation in muonic atoms, etc. We are currently investigating those possibilities to see how far they might be pushed with a new intense muon source.

TABLE 1. Standard model predictions and uncertainties due to $\Delta m_t = \pm 5$ GeV, and hadronic loop effects (second error)

m_H (GeV)	m_W (GeV) $\mp 0.034 \pm 0.010$	$\sin^2 \theta_W (m_Z)_{\overline{MS}}$ $\pm 0.0002 \pm 0.0002$
65	80.406	0.23100
100	80.383	0.23121
300	80.309	0.23179
600	80.256	0.23217
1000	80.216	0.23245

Muon-Number Non-Conservation

The search for muon-number violation has played an important role in the standard model's development. In 1959 G. Feinberg computed the loop induced rate for $\mu \to e\gamma$ in an intermediate vector boson model with only one neutrino (coupled to both μ and e). He found $BR(\mu \to e\gamma) \simeq 10^{-4}$ which conflicted with the then experimental bound $BR(\mu \to e\gamma) < 10^{-6}$. That study motivated the two neutrino hypothesis and eventually led to the ν_μ discovery. It triggered the start of accelerator based neutrino physics. So, even null experimental results can lead to revolutionary developments.

Non-zero neutrino masses and mixing could still give rise to $\mu \to e\gamma$; however, the rate would be suppressed by $(m_{\nu_2}^2 - m_{\nu_1}^2)^2/m_W^4$ and thus unobservably small. Hence, the detection of muon-number non-conservation would be an unambiguous signal of "new physics" beyond the standard model.

Current experimental bounds and future goals of various muon-number violating μ and K decays are given in table 2. Of those reactions, coherent muon-electron conversion in the field of a nucleus, $\mu^- N \to e^- N$ offers the best opportunity for significant improvement. The clean signal, a monoenergetic electron with $E_e \simeq 105$ MeV, is not limited by accidentals which beset rare multiparticle decays such as $\mu \to e\gamma$ at about the 10^{-14} level. However, pushing all the reaction sensitivities in Table 2, as far as possible is well motivated.

An approved experiment at the AGS [6] aims for $\lesssim 10^{-16}$ sensitivity in μ^--e^- conversion by employing $10^{11} \mu^-$/sec. It requires a large collection solenoid much like the front-end of the muon collider. Also, very good electron energy resolution and beam pulsing are necessary to reject backgrounds. In principle, using even higher intensity muon beams, one might be able to eventually attain $10^{-18} \sim 10^{-19}$ sensitivity. In my view, muon conversion is one of the best motivated low energy experiments in all of particle physics and should be pushed as far as possible.

The challenge of carrying out the 10^{-16} μ-e conversion experiment will provide a major step towards realizing the first muon collider. It will test the muon production and collection mechanisms while demonstrating their reliability under ex-

TABLE 2. Existing and anticipated bounds (90% CL) on various muon-number violating reactions

Reaction	Current Bound	Ongoing	Future
$B(\mu^- Ti \to e^- Ti)$	$< 7 \times 10^{-13}$	$\sim 2 \times 10^{-14}$	$< 10^{-16}$
$B(\mu^+ \to e^+ e^- e^+)$	$< 1 \times 10^{-12}$	—	—
$B(\mu^+ \to e^+ \gamma)$	$< 4.2 \times 10^{-11}$	$\sim 5 \times 10^{-12}$	10^{-14}
$B(K_L \to \mu e)$	$< 2.4 \times 10^{-11}$	$\sim 8 \times 10^{-13}$	$\sim 10^{-13}$
$B(K^+ \to \pi^+ \mu e)$	$< 2.1 \times 10^{-10}$	$\sim 3 \times 10^{-12}$	$\sim 10^{-13}$
$B(K_L \to \pi^0 \mu e)$	$< 3.2 \times 10^{-9}$	$\sim 10^{-11}$	$\sim 10^{-12}$

TABLE 3. Some examples of physics probed by $B(\mu^- N \to e^- N)$ at the 10^{-16} level.

Heavy Neutrino Mixing	$|V^*_{\mu N} V_{eN}|^2 < 10^{-12}$
Induced $Z\mu e$ Coupling (Equivalent to $B(Z \to \mu e)$)	$g_{Z\mu e} \lesssim 10^{-8}$ $< 10^{-17}$
Induced $H\mu e$ Coupling	$g_{H\mu e} \lesssim 4 \times 10^{-8}$
Compositeness	$\Lambda_c > 3,000$ TeV

perimental demanding conditions. Even more important, it will probe for "new physics" and could make a revolutionary discovery. Some examples are described in table 3. In addition, many supersymmetry scenarios suggest an observable rate for μ-e conversion due to slepton and gaugino mixing [7]. If an effect is seen, it can be thoroughly explored by follow-up target changes and employing polarized muons.

Muon Electric Dipole Moment

An interesting future use for the $g_\mu - 2$ storage has been suggested by Y. Semertzidis et al. in an AGS LOI [8]. They propose increasing the muon storage ring capacity by employing strong focusing. Using an electric field to precess a possible electric dipole moment, they would hope to probe $|d_\mu| \simeq 10^{-24} e$-cm. That would provide about a 6 orders of magnitude improvement beyond the current bound (see table 4). Six orders of magnitude improvement in an amplitude is like twelve orders of magnitude improvement in a rare decay rate bound.

At $|d_\mu| \sim 10^{-24} e$-cm, the muon e.d.m. becomes competitive with the current neutron and electron edm constraints. Indeed, one naively expects.

TABLE 4. Current bounds on some electric dipole moments.

Electron	$\|d_e\| \lesssim 4 \times 10^{-27} e\text{-cm}$
Neutron	$\|d_n\| \lesssim 10^{-25} e\text{-cm}$
Muon	$\|d_\mu\| \lesssim 7 \times 10^{-19} e\text{-cm}$

$$d_e : d_n : d_\mu :: m_e : m_d : m_\mu :: 1 : 10 - 20 : 200 \tag{10}$$

Of course, those quantities could be very independent; so, they should all be pushed as far as possible. The muon has the advantage of a simpler direct interpretation, since the neutron entails strong interaction uncertainties and the electron is indirectly studied via atomic edms.

To reach $|d_\mu| \sim 10^{-24} e$-cm sensitivity requires about $7 \times 10^7 \mu$/sec. If such a measurement proves feasible, one could imagine someday going to even higher intensities. With $10^{10} \sim 10^{12} \mu$/sec, one could (statistically) probe $10^{-25} \sim 10^{-26} e$-cm. Although extremely challenging, such sensitivity would be very exciting and worth pursuing. The standard model predicts unobservably small edms; hence a discovery would uncover a new source of CP violation, perhaps one also responsible for baryogenesis.

KAON PHYSICS GOALS

A new full front end of a muon collider complex would produce about 20 times the kaon flux currently available at the AGS. Can it be fully utilized? For some precision measurements, there is strong motivation to push forward. For example, the rare decays $K \to \pi \nu \bar{\nu}$ ($K = K^+$ or K_L^0) are predicted to be (about)

$$BR(K^+ \to \pi^+ \nu \bar{\nu}) \sim 1 \times 10^{-10}$$
$$BR(K_L \to \pi^0 \nu \bar{\nu}) \sim 3 \times 10^{-11} \tag{11}$$

The theoretical underpinnings of those predictions are very good. Hence, they can be used to extract the CKM element V_{td} with high precision. Indeed, in principle the CP violation parameter η could be determined to $\pm 2\%$ theoretical uncertainty if very high statistics could be obtained. No other known method for measuring η has such a small theoretical uncertainty. It would then set the standard and could be used in comparison with B studies to search for "new physics".

Note that experiment E787 has recently observed its first $K^+ \to \pi^+ \nu \bar{\nu}$ event. The observed rate corresponds to

$$BR(K^+ \to \pi^+ \nu \bar{\nu}) = 4.2^{+9.7}_{-3.5} \times 10^{-10} \tag{12}$$

If the ongoing data analysis points to a larger than expected rate, it would likely suggest "new physics" effects which could be deciphered by much higher statistics.

Other interesting K studies that could potentially be pushed much further or initiated for the first time with increased K flux include: Muon transverse polarization in $K^+ \to \pi^0 \mu^+ \nu_\mu$ or $K^+ \to \mu^+ \nu_\mu \gamma$, spin-spin correlation in $K^+ \to \pi^+ \mu^+ \mu^-$, polarization effects in $K_L \to \mu^+ \mu^-$ etc. Possible future initiatives are under discussion.

CONCLUDING COMMENTS

How do we get the "First Muon Collider"? Serious R&D support is certainly required to demonstrate technical feasibility. In addition, a consensus on its most compelling physics goals and relative worth must be reached.

To start down the road to the FMC and stimulate needed technological advances, a complementary program of low energy physics should prove extremely useful. I have outlined a few possibilities. Currently, I believe that high intensity searches for $\mu^- N \to e^- N$, $\mu^+ \to e^+ \gamma$, $\mu^+ \to e^+ e^- e^+$ are the best bets. They offer robust potential to uncover "new physics". The muon e.d.m. provides another interesting opportunity to unveil a new source of CP violation. Other areas worth pursuing are P and T violation in muonic atoms, rare K decays, neutrino oscillations etc. Collectively, those studies will significantly extend the frontiers of particle physics while advancing the muon collider cause. We can't afford not to pursue them.

REFERENCES

1. Muon Collider Feasibility Study, BNL Report 52503 (1996).
2. W. Marciano, BNL LDRD proposal (1996).
3. See S. Geer, Fermilab-Pub-97/389 (1997).
4. M. Davier and A Höcker, Orsay preprint (1997).
5. G. Degrassi, P. Gambino, and A. Sirlin, Phys. Lett. **B394**, 188 (1997).
6. W. Molzon, in these proceedings.
7. See R. Barbieri, L. Hall, and A. Strumia, Nucl. Phys. **B445**, 219 (1995).
8. Y. Semertzidis *et al.*, BNL Letter of Intent (1997).

Deep Inelastic Scattering and Neutrino Physics

Panagiotis Spentzouris*

*Columbia University, Nevis Labs
136 S. Broadway
Irvington, NY 10533
USA

Abstract. The present status of selected topics in Deep Inelastic Scattering (DIS) with charged lepton and neutrino beams is reviewed. The focus is on experimental results which demonstrate the power of DIS in enhancing our understanding of the hadron structure, and on theoretical issues which are outside of the reach of present experiments. The possibilities to address some of these intriguing issues with the muon collider facility are discussed.

INTRODUCTION

Deep Inelastic Scattering (DIS) with charged lepton and neutrino beams has been the primary laboratory to study hadron structure and quantum chromodynamics (QCD) over the past three decades. The DIS experimental program has completed a very successful cycle, starting with the discovery of the sub-structure of the nucleon and scaling violations in the early fixed target days, and continuing with the high precision perturbative QCD and parton distribution function measurements from the latest fixed target experiments. Currently, the DIS program is again pushing the physics frontier by exploring "extreme" kinematic regions, like very low x and very high Q^2, with the HERA lepton-proton collider. In this presentation I review the status of DIS topics in order to demonstrate the precision and the discovery potential of the DIS process. I begin with a brief introduction to the Deep Inelastic Scattering framework, and follow with a summary of experimental results and theoretical issues which are not yet resolved. The last section is devoted to a discussion of the possibilities to answer some of these open questions with a muon collider DIS program.

Note that there are a lot of other interesting topics related to DIS which are not covered in this presentation, like polarized structure functions and diffractive physics. The contribution of a muon collider facility to these topics could be significant, especially to the spin physics due to the availability of perfectly polarized very intense beams $(\nu, \bar{\nu})$ [1].

LEPTON PROBES ON HADRONS

Cross-Section Formalism and Kinematics

The tree-level diagram for lepton-nucleon DIS corresponds to the exchange of a single virtual boson between the lepton and the target nucleon (see figure 1). This boson could be a γ (electromagnetic scattering), a Z^0 (weak neutral currents) or a W^{+-} (weak charged currents). The kinematic variables which are most commonly used to describe the interaction are: $-Q^2$, the boson's (4-momentum)^2, $\nu = E - E'$, the energy transferred from the leptonic to the hadronic vertex, $y = E/\nu$, and $x = Q^2/2M\nu$, with E, E' the energies of the incoming and the outgoing leptons in the rest frame of the target nucleon, and M the mass of the proton. The differential cross-section is derived from Lorentz invariance and the symmetries of the interaction; for a fixed incoming lepton energy E any two of the above variables are sufficient to describe the interaction. Neglecting the mass of the outgoing lepton, the differential cross-

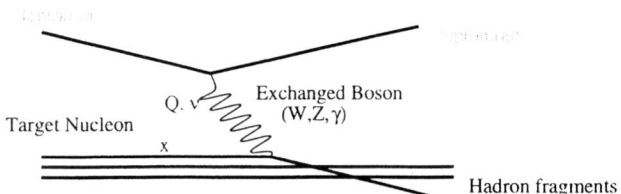

FIGURE 1. Born approximation lepton-nucleon DIS.

section for unpolarized nucleons is expressed in terms of three independent structure functions (SFs): F_2, R, and xF_3 (equation 1 for neutrino DIS and 2 for charged lepton DIS).

$$\frac{d^2\sigma^{\nu(\bar{\nu})N}}{dxdy} = \frac{G_F^2 ME}{\pi (1 + Q^2/M_W^2)^2}$$
$$\left[F_2^{\nu(\bar{\nu})N}(x, Q^2) \left(\frac{y^2 + (2Mxy/Q)^2}{2 + 2R_L^{\nu(\bar{\nu})N}(x, Q^2)} + 1 - y - \frac{Mxy}{2E_\nu} \right) \right.$$
$$\left. \pm xF_3^{\nu(\bar{\nu})N} y \left(1 - \frac{y}{2}\right) \right] + (-) \text{ for } \nu(\bar{\nu}) \quad (1)$$

From equation 1 it is apparent that F_2 can be obtained from the sum of the differential cross-sections for ν and $\bar{\nu}$, while xF_3 may be obtained from the difference.

$$\frac{d^2\sigma^{e(\mu)N}}{dx\,dQ^2} = \frac{2\pi\alpha^2}{xQ^4}\left[2(1-y) + \frac{y^2}{1+R_L^{e(\mu)}(x,Q^2)}\right]F_2^{e(\mu)}(x,Q^2) \qquad (2)$$

R can be obtained from the measurement of the cross-section in the same (x,Q^2) region for several different beam energies.

In the case of fixed target (FT) DIS experiments with charged lepton beams, the kinematics of the interaction could be described using the incoming and outgoing lepton parameters alone: $Q^2 = 2EE'(1-\cos\theta)$ and $\nu = E - E'$.

In the case of charged current neutrino FT DIS, information from the hadronic vertex is required (namely the hadronic energy, E_h), since the incoming neutrino kinematics cannot be determined on an event by event basis: $\nu = E_h - M$, $Q^2 = (E_{had} + E_\mu)E_\mu\theta_\mu^2$. Finally, in the lepton-proton collider case the kinematics could be described using only the scattered lepton and the center-of-mass energy s: $y = 1 - \frac{E_l'}{2E_l}(1-\cos\theta_l)$, $Q^2 = 2E_lE_l'(1+\cos\theta_l)$, $x = Q^2/sy$, and $s = 4E_lE_P$.

Together with the beam energy, the experimental resolution of the above observables defines the kinematic reach for each experiment. Note that at present fixed target energies and with the present beam intensities, the event rates per unit target mass are small in neutrino DIS. Thus, in order to gather high statistics, dense nuclear targets are used. This requirement limits the resolution and so also limits the kinematic reach of neutrino DIS experiments and adds the complication of nuclear effects in the interpretation of the data. On the other hand, the neutrino DIS cross-section has a very small Q^2 dependence, while for charged leptons it falls rapidly with Q^2, making neutrino experiments more suited for QCD studies (see next section).

DIS: a simple picture

The power of DIS measurements is most clearly realized within the parton model and QCD framework, since the experimental observables directly relate to fundamental parameters of the theory. In the quark-parton model x is equal to the fraction of the four-momentum of the nucleon carried by the struck parton, and the SFs are connected to parton distribution functions (PDF) along x: $q_f(x)dx$ is the probability of finding a quark of flavor f carrying a fraction of the four-momentum of the nucleon between x and $x + dx$. The partonic picture is completed with the introduction of QCD effects. The characteristic property of QCD is its asymptotic freedom, which allows the application of perturbation theory in processes which involve short distance scales (large four-momentum transfers). Using the factorization and evolution properties of perturbative QCD (pQCD) the DIS cross-section can be factorized to: $\frac{d\sigma}{dk'd\Omega'} \sim \int_0^x dx \sum_\alpha q_{\alpha/h}(x,\mu)\frac{d\hat{\sigma}_\alpha(\mu,\alpha_s)}{dk'd\Omega'}$, where $\hat{\sigma}_\alpha(\mu,\alpha_s)$ is pQCD calculable, $q_{\alpha/h}(x,\mu)$ is a non-perturbative part, and μ is the factorization scale. If μ is much larger than the scale which characterizes parton-parton interactions

(Λ_{QCD}) the above expression becomes a good representation of the physical cross-section[1]. In order to simplify things, the factorization scale is usually set to be equal to the renormalization scale of the pQCD calculation, and both are set to be $\mu \sim Q$. The requirement that the physical cross-section should be independent of μ leads to the DGLAP evolution equations [2], which govern the Q^2 dependence of the parton distribution functions. The theory cannot predict the x dependence of PDFs so they have to be extracted from the experimental measurements. In this framework, PDFs measured at a given scale μ can be used to predict PDFs at any scale $Q^2 > \mu^2$, and thus allow cross-section calculations. In addition, since in this framework the logarithmic derivatives of the structure functions depend on α_s, $d \ln F(x, Q^2)/d \ln Q^2 \sim \alpha_s(Q^2)$, structure function measurements provide the means for a very precise determination of the strong coupling constant α_s.

In order to quantify the discussion in the following sections, the leading order Structure Function expressions in terms of PDFs are given in equations 3 for ν, $\bar{\nu}$, and μ DIS on an isoscalar target.

$$F_2^{\nu,\bar{\nu}} = xq + x\bar{q}$$
$$F_2^{e(\mu)} = \frac{5}{18} \left(F_2^{\nu,\bar{\nu}}\right) \left[1 - \frac{3}{5}\frac{(s+\bar{s})}{(q+\bar{q})}\right]$$
$$xF_3^{\nu,\bar{\nu}} = xq - x\bar{q} \pm 2(s-c)$$
$$\frac{1}{2}\left(xF_3^{\nu} + xF_3^{\bar{\nu}}\right) = xu_v + xd_v$$
$$R_L^{\nu,\bar{\nu}} = 0 + \mathcal{O}(\alpha_s) \tag{3}$$

It is important to notice that ν DIS has the advantage of flavor discrimination over charged-lepton scattering. In addition, the average xF_3 from ν and $\bar{\nu}$ is purely non-singlet in flavor space[2]. This increases the precision of the α_s determination, since for non-singlet distributions the evolution equations do not involve the gluon distribution while for singlet distributions they do.

To summarize, in this picture the task of a DIS experiment is very well defined: measure the differential cross-sections with high statistics and small systematics. Preferably this is done by concentrating on the scattered lepton reconstruction (to keep things simple), extracting SFs and fitting them to obtain PDFs, and using the logarithmic slopes to measure α_s. After all that is done, the big pay-back is the ability to predict the cross-sections of various processes, and a better understanding of the structure of nuclear matter.

---

[1] Intuitively, we can argue that for high Q^2, the time scale of the interaction is much shorter than the time scale of the interactions among partons which define the PDFs in the nucleon's wave-function, so these dynamics are "frozen" and can be factored out.
[2] The average xF_3 is the sum of the valence quark distributions (equation 3). Valence quarks are the partons which carry the net flavor quantum numbers of the nucleon

Experimental Results

The measurements of modern fixed target DIS experiments reached impressive precision with the final results presented by the NMC [3] muon experiment at CERN, and the CCFR [4] neutrino experiment at Fermilab. A comparison of high-statistics F_2 measurements as a function of Q^2 for selected x bins is shown in figure 2. Note that there is a disagreement ($\sim 20\%$) in the lowest x

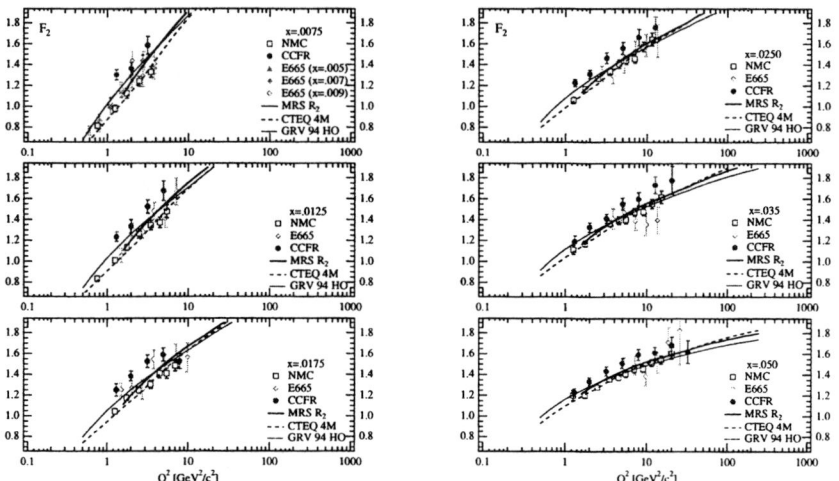

FIGURE 2. F_2 as a function of Q^2 for different x bins. The charged lepton F_2 has been corrected to iron equivalent using nuclear corrections extracted in charged-lepton scattering, and the 5/18ths rule (equation 3) with the strange sea from the direct CCFR measurement.

bin, beyond the tolerance of the experimental errors (more on this issue in the following section). Apart from the low-x discrepancy, these measurements are a valuable source of data for Parton Distribution Function extraction. The kinematic coverage for these experiments in x and Q^2 is shown in figure 3, with CCFR covering $0.0125 < x < 0.65$ and NMC $0.008 < x < 0.6$ in the $Q^2 > 1 \text{GeV}^2$ region. In addition, the QCD Structure Function analysis based on the DGLAP evolution culminates with the high precision CCFR α_s result [4]. This result, $\alpha_s(M_Z^2) = 0.119 \pm 0.002(exp) \pm 0.001(HT) \pm 0.004(scale)$, is of great importance since it is in very good agreement with results from higher energy measurements (see figure 4), and practically eliminates any possibility for contributions from "exotic" physics.

In summary, FT DIS experiments have provided precise information on parton distribution functions and QCD parameters. The sum of the q and $\bar{q}$ distributions is constrained in the $0.001 < x < 0.6, Q^2 > 1$ GeV^2 region and the sum of the valence quark PDFs in the range $0.01 < x < 0.6$, while the α_s determination has reached such high precision that it is now limited by the theoretical uncertainties.

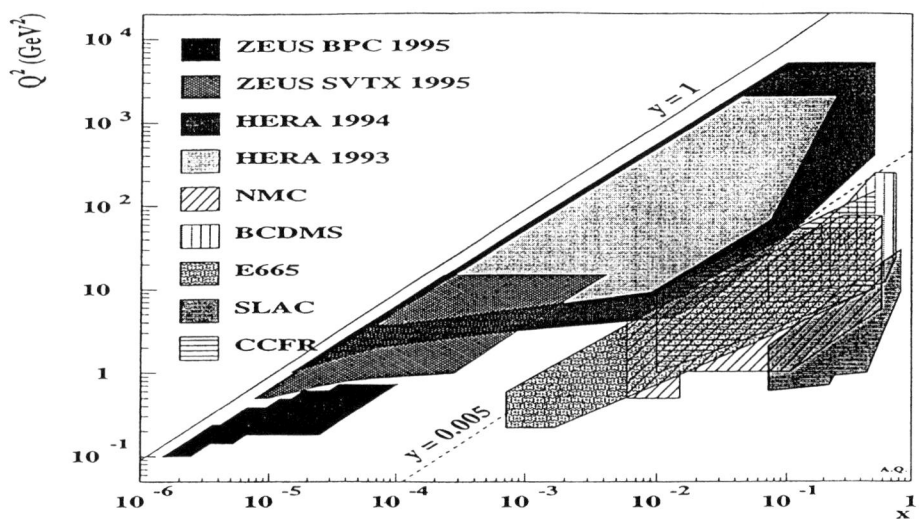

FIGURE 3. Kinematic coverage of high statistics DIS experiments in x and Q^2

DIS: a not so simple picture

While the era of the "traditional" fixed target DIS experiments concluded with the Fermilab 1996 fixed target run and NuTeV, the successor neutrino experiment to CCFR, the emphasis in DIS has been shifting to "extreme" kinematic regions, and to areas where the theoretical understanding is not yet very quantitative. These areas include the very low-x ($x < 0.001$) and very high-x kinematic regions, the very high and very low-Q^2 regions, and heavy quark production, while more conventional topics like nuclear effects are still under investigation. In addition, the precise knowledge of the parton distribution functions is always an issue since they contribute to the accuracy of all measurements of Standard Model parameters which involve hadron interactions (for example the W mass measurement from the Tevatron: $\sim 1/3 \Delta_{syst} M_W$ comes from the PDF uncertainty). Most of the above challenging topics have become the main subject of interest in DIS with the HERA $e^{+-}p$ collider program, since HERA with a center-of-mass energy of 300 GeV has a kinematic reach roughly two orders of magnitude down in x and up in Q^2 compared to FT experiments.

Let us start our discussion of the phenomenology of the "extreme" regions

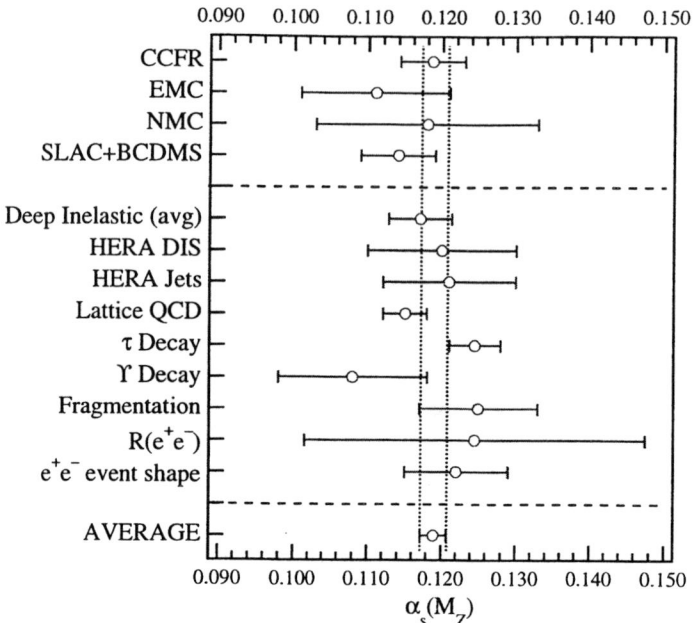

FIGURE 4. A comparison of $\alpha_s(M_Z^2)$ from different processes

with the DGLAP evolution equations. In this picture the corrections which in the conventional kinematic regime are small include power corrections of order $1/Q$, and logarithms of x in the splitting functions. In the first case, as Q^2 decreases, these higher twist corrections become more and more important, up to the point that the pQCD description fails and a smooth transition to the real photo-production limit is expected (F_2 in charged lepton scattering has to go to zero as $Q^2 \to 0$, as a result of charge conservation). The dynamics of this transition are not known. Furthermore, in the neutrino charged current DIS case, F_2 goes to a non-zero constant as $Q^2 \to 0$ (partially conserved axial current), so potential differences in the transition should be interesting to study. In the second case, the logarithmic x corrections become large as x decreases. A different way to describe the low-x dynamics, involves the BFKL evolution equation [6]. BFKL uses a different approach than DGLAP in the factorization of DIS structure functions, by summing logarithms of x at leading power. A schematic of the structure function evolution in $\ln 1/x$ and $\ln Q^2$ is shown in figure 5. The BFKL prediction for the gluon distribution is very singular, as it grows as a power of x. This behavior cannot hold indefinitely, since as the parton distribution functions increase with decreasing x, parton density saturation is expected to occur and factorization should break down. One way or another the increase of PDFs at low-x has to stop, because the total cross-section will violate the unitarity bound. There are two interesting

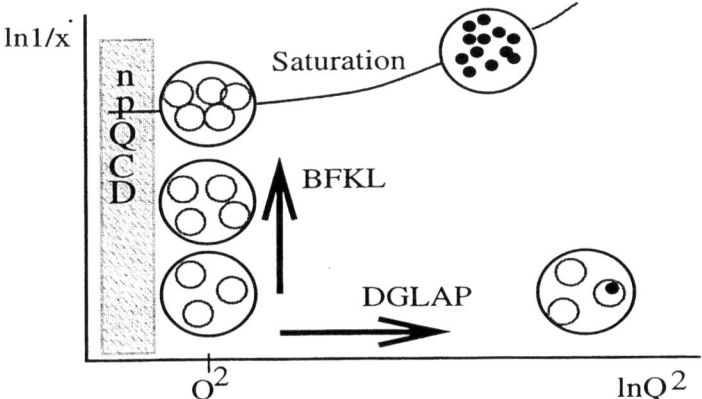

FIGURE 5. Structure Function evolution in x and Q^2.

questions here: at which point does BFKL become more accurate than the DGLAP approach, and at which point does pQCD break down (maybe the latter occurs before the former!).

The experimental data on F_2 as a function of x and for different Q^2 bins are shown in figure 6. The increase of F_2 with decreasing-x is remarkable, but what is even more interesting is that it can be very accurately described by DGLAP (see the curves from a NLO pQCD DGLAP fit over-plotted on the data). There is no need for a BFKL analysis of the data, and there is no indication of parton density saturation effects (the signature would have been a softening in the rise of F_2 with decreasing x). The search for the onset of BFKL dynamics also covers non-inclusive measurements, such as the study of the forward produced DIS jets, where BFKL predicts higher rates than DGLAP [7]. Results from Zeus [8] are shown in figure 7, compared to DGLAP-based and BFKL-based predictions. The latter show a better qualitative agreement with the data. Nevertheless, it seems like neither of the two important questions on low-x dynamics has been definitively answered. Another interesting issue which challenges our simple DIS picture is the heavy quark production. Compared with inclusive DIS, heavy quark production has the additional complication of having two scales; in addition to the Λ_{QCD} scale which separates soft from hard physics, the mass of the heavy quark $(m_H \gg \Lambda_{QCD})$ is also important. It affects the number of active partons as a function of energy (for $\sqrt{s} < m_H$ the production cross-section is zero), and also enters in the calculation of the hard cross-sections and the splitting functions. The problem here is to develop a method which allows the evolution of PDFs through scales $\sim m_H$ without singular behavior. There are currently three

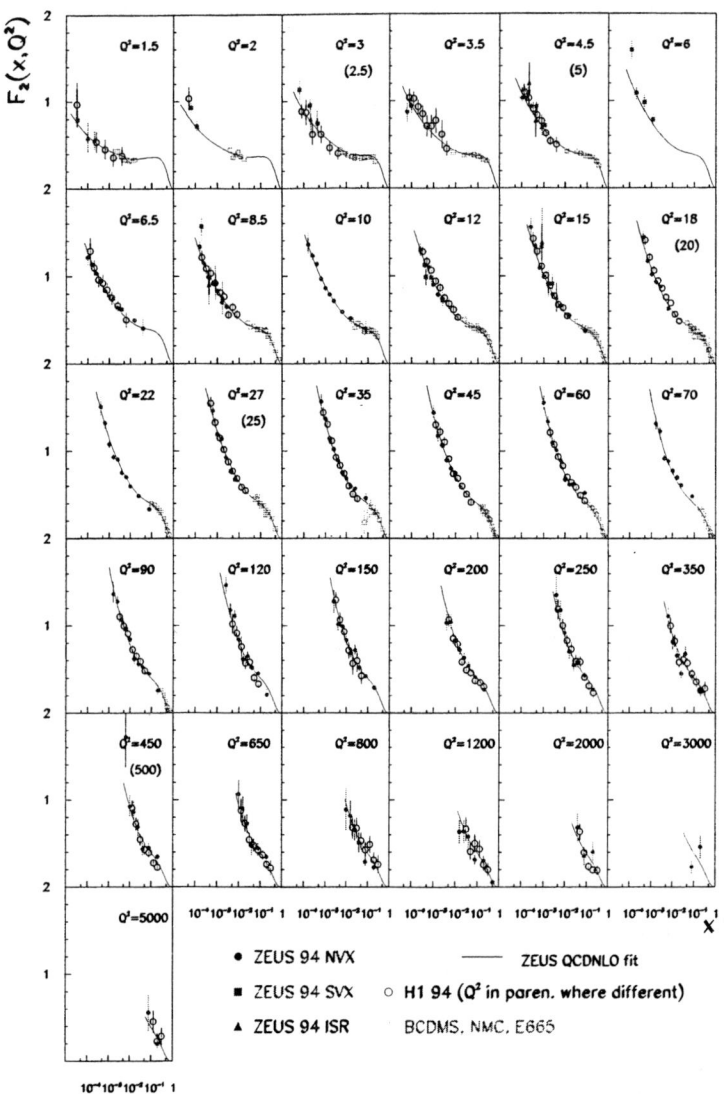

FIGURE 6. F_2 as a function of x in bins of Q^2. The curve shown is a NLO DGLAP fit.

Forward Jet Cross Sections

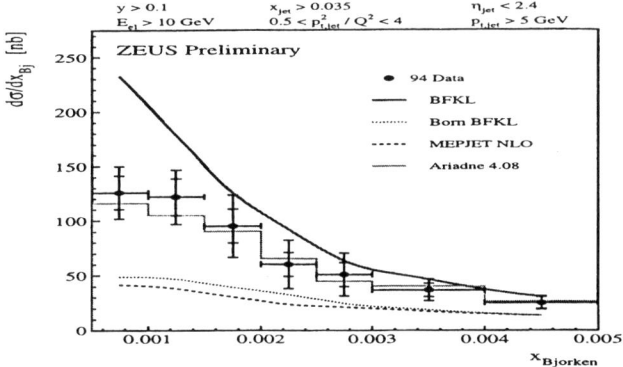

FIGURE 7. Forward jet cross-section from Zeus compared with DGLAP and BFKL calculations.

different factorization prescriptions available in the literature (see [9] for a discussion). Experimental data exist on charged lepton DIS charm production [10] from both HERA and FT, and on neutrino DIS charm production [5]. The experimental data from HERA [11] demonstrate the significance of the charm F_2 at low-x, but they are not precise enough to differentiate between the different factorization prescriptions. The neutrino data have been used within one of the available schemes to determine the NLO strange PDF. The strange PDF parameters are extracted from the experimental data with respect to the non-strange sea PDF parameters for $x < 0.01$, with errors $\sim 10-20\%$. This is the only direct measurement of the strange PDF, and its significance is even more enhanced by the disagreement of inclusive neutrino and charged-lepton scattering F_2 at low-x, since it directly enters the formula which allows the comparison (see equation 3).

The very high Q^2 regime of Deep Inelastic scattering, is a primary candidate for the discovery of new phenomena, complementary or superior to other processes. A powerful demonstration of the physics discovery potential of DIS is provided by the excess of events at high x and Q^2 ($Q^2 > 10000$ GeV^2) over the Standard Model expectation found by both ZEUS and H1 [12]. Possible interpretations of these excess of events include contact interactions and leptoquarks. Although the limits set by other experiments in related searches leave very little parameter space open in the kinematic region probed by HERA, these measurements demonstrate that DIS remains a powerful laboratory to test and expand our physics knowledge.

The HERA high Q^2 results apart from the possibility of discovery in DIS, could be also used in a more "conservative" approach to emphasize the need for precise PDF measurements at the high-x end (x > 0.6) of the kinematic chart. The ratio of the d-quark to u-quark - $d(x)/u(x)$ - is expected to approach 1/5 in the framework of pQCD [13], however current fits are just as consistent with this ratio approaching 0 as the expected 0.2. The limited knowledge of PDFs in this region was made apparent in a recent PDF fit [14], where a toy model which could explain the high-Q^2 HERA events was constructed. In this model, an additional quark contribution, equal to $\sim 1\%$ of the integrated d-quark PDF, was added at x near 1.0 and $q = 2$ GeV without seriously contradicting any data. Through the Q^2-evolution of the PDFs these evolve down to the x of the HERA events at the proper Q^2.

The issue of how a nuclear environment modifies the structure functions measured in DIS is very old, starting with the early SLAC experiments. Ideally, one would like to understand how the parton model and QCD relate to nuclear physics, but even the origin of structure function nuclear effects is not quantitatively understood. These nuclear effects modify the ratio of nuclear target to deuterium structure functions as a function of x without any Q^2 dependence. They are categorized according to the x region examined [15] (see figure 8):

1. Shadowing effect ($x < 0.1$), $\sigma_A < \sigma_D$. There are two kinds of models which try to describe this region: parton recombination models and vector meson dominance models.

2. EMC effect $0.3 < x < 0.6$, $\sigma_A < \sigma_D$. This effect is described using nucleon-nucleon correlation arguments.

3. $0.1 < x < 0.2$, $\sigma_A > \sigma_D$, the transition region. No real explanation.

4. Fermi motion region, $x \geq 1$.

It is interesting that there is no single coherent framework which describes all the effects in a unified way. On the experimental front, there is an abundance of very precise data from charged-lepton scattering [16], but there is very little known experimentally about nuclear effects in neutrino DIS. Since nuclear corrections are needed to extract the nucleon's PDFs from the high precision neutrino data on nuclear targets, the corrections determined in charged lepton DIS are used. This fact is very often used to explain the origin of the low-x discrepancy between charged-lepton and neutrino F_2, so measuring these effects with neutrino DIS is an important measurement. In addition, the measurement of nuclear effects in F_3 would allow the differentiation between the behavior of valence and sea quarks in a nuclear environment.

DIS WITH A MUON COLLIDER FACILITY

A muon collider facility comes along with very intense muon beams not only at the collider level but at the various acceleration stages. These beams pro-

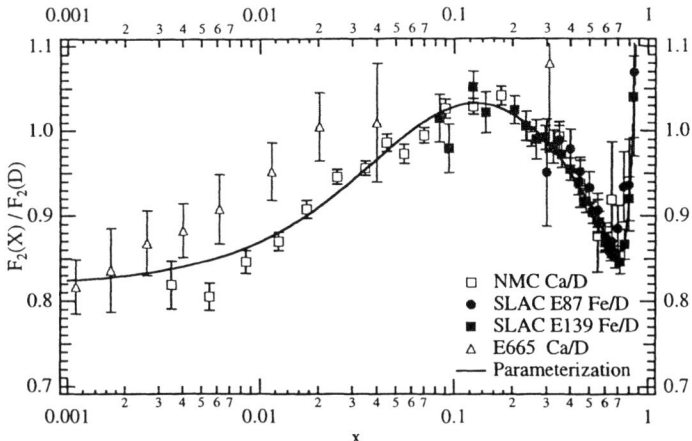

FIGURE 8. A-dependence of structure function data measured in deep inelastic charged-lepton scattering.

vide three different options for Deep Inelastic Scattering experiments: muon fixed target, neutrino fixed target, and muon-proton collider (pairing the muon beam with the Tevatron proton beam). In this section we briefly examine the physics potential of these experiments, using the sensitivity to new physics as a guide, and machine parameters and backgrounds as a constraint.

The possibility of μ^{+-} beams used together with the Tevatron beam is at first glance a very appealing one. From the discussion in the previous section it is obvious that the interesting areas are those of very low-x and very high-Q^2. The key issues from the experimental point of view are the kinematic constraints and the very large beam halo which comes along with the muon beam. The kinematic constraints come from the requirement that in order to be reconstructed, the scattered lepton has to be away from the proton remnant shower, and it has to be away from the beam. Using the HERA experiments as a guide, these requirements translate to cuts on the scattering angle of the muon: $\theta_\mu > 10°$ and $\theta_\mu < 179°$). This two requirements define the optimal configuration of the muon-proton collider for probing either the low-x or the very high-Q^2 regions. In the first case an asymmetric machine configuration is desirable (with the lepton energy being smaller than the proton energy), since the large relative boost of the proton system brings the low-x scatters in the central region of the detector. The kinematic reach for such a configuration is shown in figure 9, together with an estimate of where the gluon density should saturate [17]. There is an improvement over HERA, but in this simple picture there is no consideration of the large background close to the beam inherent in a muon collider which will introduce a harder cut along the beam direction. In order to probe exotic physics at high-Q^2 (leptoquarks, contact interactions,

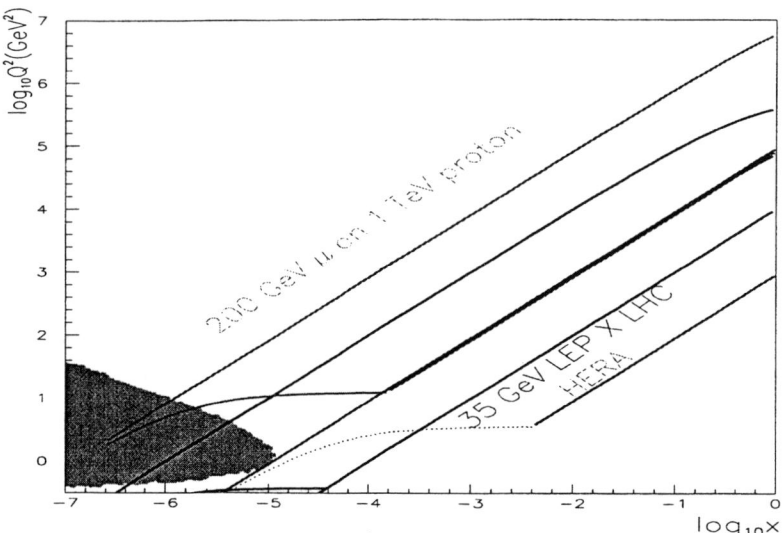

FIGURE 9. The kinematic reach for an asymmetric $\mu - p$ machine. No μ collider specific backgrounds have been considered. The nominal y and θ_μ cuts which correspond to HERA have been applied.

SUSY) a high energy muon beam is desirable, since $s = 4E_\mu E_P$. In order to maximize the acceptance a symmetric configuration is preferred (high-Q^2 events become more central in the detector). Such a configuration (1Tev on 1TeV) will have improved sensitivity with respect to the HERA experiments, with an order of magnitude higher Q^2 and ten times the total HERA integrated luminosity per year (see reference [18] for a detailed discussion).

The second option is that of a fixed target DIS program with either a muon or a neutrino beam. In both cases the key parameters which define the physics output of the experiment are the beam energy and the beam intensity. Given the quality and diversity of existing charged lepton DIS results, a muon fixed target experiment at the muon collider facility seems to have only marginal capabilities for improvements over existing measurements (the high intensity is not an issue, and even in the case of a 4TeV muon beam the kinematic reach of the HERA experiments is unsurpassed). On the other hand, a fixed target neutrino experiment opens up possibilities for precise measurements never done before in $\nu - N$ DIS. The very intense muon beams can provide such high neutrino fluxes that the use of light, thin targets becomes possible. In addition, the beam quality is better than that of conventional neutrino beams from meson decays, with better known energy scale, flavor content and flux. These beams could be either produced by muon decays in the long straight section of the Recirculating Linacs (RLA) or by muon decays in straight sections of the collider ring. Already the physics reach is significant with the RLA3 neutrino beam. This beam, with mean energy $<E_\nu> = 135 \text{GeV}$, will produce an event rate of 1×10^{18} per ton per year, a factor of 1000 higher than the event rates of present neutrino experiments (beam characteristics from the Workshop Parameters for RLA3). This rate allows the use of light thin targets and detectors with good spatial resolution and final state reconstruction ability. The use of light targets will increase the experimental resolution, allowing for high precision neutrino structure function measurements at low-x $(0.001 < x)$. This will be a new kinematic region for xF_3 measurements, with direct impact on the α_s determination from the Gross-Llewellyn Smith Sum Rule (see reference [19]). The use of light targets will also allow the direct comparison between charged lepton and neutrino structure functions without the nuclear correction uncertainty. In addition, the nuclear corrections could be measured, since both thin light and thin nuclear targets can be used in this type of detector. Finally, the flavor dependence of the PDFs could be easily studied using the inclusive structure function measurements from nucleon targets.

The final state reconstruction ability allows for inclusive charm production measurements with an extended kinematic reach and high statistics, compared to the conventional dimuon measurements of present neutrino experiments (an order of magnitude, even with a 1m liquid H_2 target). This measurement can provide information on a wide spectrum of physics topics, in addition to the determination of the strange quark PDF and the charm quark mass. These topics include the direct measurement of $|V_{cd}|$, the study of charmed hadron fragmentation properties, the measurement of branching fractions for D^0, D^+, D_s^+, Λ_c^+, and the measurement of the total charm production cross-section. Assuming good vertexing capabilities, there is also a possibility for a flavor changing neutral current search, using events with a single charm vertex and no primary muon. The sensitivity for this kind of search is high, being proportional to the total number of neutral current interactions. For all of the above topics the existing measurements are not high precision, since the only existing neutrino experiments with final state reconstruction are either emulsion or bubble chamber experiments.

The measurements discussed become even more significant as the neutrino beam energy gets higher. Considering the neutrino beam from a 4TeV μ collider, $E_\nu \sim 1600 GeV$, structure function measurements could be obtained down to $x \sim 10^{-4}$. It will be interesting to study xF_3 to such low-x values and compare its behavior to that of the steeply rising F_2 (the kinematic reach for this case is shown in figure 10). In all of the above measurements the knowledge of the flux and the energy scale will contribute in reducing the systematic errors. One issue that has to be resolved in the design of a neutrino fixed target experiment at the muon collider is that of the muons which will come from the straight section of the machine. Another important issue that

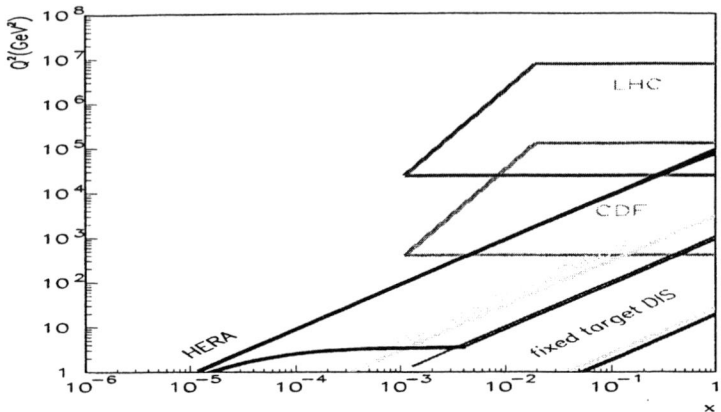

FIGURE 10. Kinematic reach for a high energy neutrino fixed target experiment from a 4TeV μ collider compared to other machines.

should be emphasized is that for both the $\mu-p$ collider and the neutrino fixed target cases the ability to change the beam polarity is necessary.

From the above discussion it is clear that a neutrino DIS experiment at a muon collider facility is a very appealing option, while a successful muon-proton collider program seems to be a much more technically challenging expedition.

CONCLUSIONS

Deep Inelastic Scattering which brought high energy physics to the parton age is still producing exiting results. The high precision QCD measurements are reaching the limit of the theoretical calculations, while new kinematic regions are explored providing more information on the nucleon structure and bringing us to the boundary of perturbative and non-perturbative physics. There are important questions that still need to be answered and a muon accelerator complex with very high intensity beams could provide the tools to investigate. Although they don't bring funding in advance, the most interesting questions are those that haven't been asked yet.

REFERENCES

1. Kevin McFarland and Deborah Harris, DIS Working Group, these Proceedings.

2. G. Altarelli and G. Parisi, *Nucl. Phys.* **B126**, 298 (1977); V. N. Gribov and L. N. Lipatov, *Sov. J. Nucl. Phys.* **15**,438 (1972); Yu. L. Dokshitzer, *Sov. Phys. JETP* **46**, 641 (1977).
3. M. Arneodo et al., *Nucl. Phys.* **B483**, 3 (1997)
4. W.G. Seligman et al., *Phys. Rev. Lett.* **79**, 1213 (1997).
5. A.O. Bazarko et al., *Z. Phys.* **C65**, 189 (1995); S.A. Rabinowitz et al., *Phys.Rev.Lett.* **70**, 134 (1993)
6. E.A. Kuraev, L.N. Lipatov, V.S. Fadin, *Sov. Phys. JETP* **45**,199 (1977); Ya. Ya. Balitskii and L.N. Lipatov *Sov. J. Nucl. Phys.* **28**, 822 (1978).
7. A.H. Mueller, *Nucl. Phys. Proc. Suppl.* **18C**,125 (1991)
8. Rosario Nania, in *Deep Inelastic Scattering and QCD*, J. Repond, D. Krakauer editors, Chicago, AIP Conference Proceedings, 1997, pp. 21-36.
9. Wu-Ki Tung in *Deep Inelastic Scattering and QCD*, J. Repond, D. Krakauer editors, Chicago, AIP Conference Proceedings, 1997, pp. 14-18.
10. J.J. Aubert et al., *Nucl. Phys.* **B213** (1983) 31; C. Adloff et al.,*Z. Phys.* **C72**, 593 (1996).
11. J. Roldan, in *Deep Inelastic Scattering and QCD*, J. Repond, D. Krakauer editors, Chicago, AIP Conference Proceedings, 1997, pp. 366-370.
12. C. Adloff et al., DESY 97-024 (hep-ex/9702012); M. Derrick et al., DESY 97-025 (hep-ex/9702015)
13. G.R. Farrar and D.R. Jackson, *Phys. Rev. Lett* **35**, 1416 (1975).
14. S. Kuhlmann et al., e-print archive: hep-ph/9704338
15. D.F. Geesaman, K. Saito, A.W. Thomas, *Ann. Rev. Nucl. Part. Sci.* **45**, 337 (1995).
16. M. Arneodo et al., *Nucl. Phys.* **B481**, 3 (1996); M. Arneodo et al., *Nucl. Phys.* **B481**,23 (1996).
17. A.H. Mueller, *J. Phys.* **G19**, 1463 (1993).
18. S. Ritz DIS Working Group, these Proceedings.
19. Deborah Harris, DIS Working Group, these Proceedings.
20. M.R. Adams et al., *Nucl. Inst. Methods* **A291**, 533 (1990).

Intermediate Energy Physics in the Strangeness Sector with Hadron Beams

G. B. Franklin

*Department of Physics, Carnegie Mellon University,
Pittsburgh, PA 15213, USA*

Abstract. The field of intermediate energy physics includes the study of systems with one and two s-quarks. These studies allow us to test and expand models of the nonperturbative interactions of light-quark systems. Current efforts in this field are reviewed with emphasis on hadron-beam based experiments serve as examples of programs which could utilize the muon collider front end.

I INTRODUCTION

The field of intermediate energy physics can be described as the study of non-perturbative QCD. These studies encompass topics which range as tests of lattice gauge theory predictions to explorations of quark-model calculations whose form, although motivated by the QCD, cannot be directly connected to the QCD Lagrangian at this time. Rather than attempting an overview of all of intermediate energy physics, I intend to cover the role of strange-sector physics. Systems with non-zero strangeness provide opportunities to combine decades of work in more traditional nuclear physics with flavor symmetry considerations to explore a new regime. The emphasis of this paper will be to review ongoing strange-sector experiments which utilize hadron beams. These experiments can be viewed as examples of the type of intermediate energy studies which could be performed at the collider front-end.

II STRANGE-SECTOR PHYSICS

The strange-sector experiments can be categorized into the following four topics: 1) Production Mechanism Studies, 2) Baryon and Meson Spectroscopy, 3) $NN \rightarrow$ Strange Sector Physics, and 4) Weak Interactions. It is interesting to look at a few examples of work underway in each of these four areas.

A Production Mechanism Studies

These are experiments which measure production cross sections, particularly those involving short range components, which generally explore the effectiveness of one-boson-exchange (OBE) models vs. quark models. For example, reactions of the form $p\bar{p} \to Y\bar{Y}$ have been studied by LEAR PS185 [2]. These experiments take advantage of the self-analyzing properties of the hyperon decays. In the case of $\Lambda\bar{\Lambda}$ production, the $\Lambda \to p\pi^-$ decays can be used to determine the spin-singlet to spin-triplet fractions of the $\Lambda\bar{\Lambda}$ pairs. This quantity is directly related to the singlet to triplet fraction of the $s\bar{s}$ quark pairs. Considerable theoretical effort has been put into understanding these reaction both at a constituent-quark level and in terms of mesons exchange. At this time, for example, these two approaches appear to yield conflicting predictions for the spin depolarization variable, D_{NN} which gives the correlation between the spin of the initial state proton and the final state lambda. Recent measurements by the PS185 collaboration will spur further theoretical work which, in turn, may motivate future experiments. However, with the decomissioning of LEAR and possible end of the AGS slow-extracted beam program there may be no existing facility suitable for these efforts. Additional studies of $p\bar{p} \to Y\bar{Y}^*$ have been discussed by the antiproton working group at the AGS2000 workshop. [3] For example, a measurement of the production rate of the $\Lambda(1405)$ might distinguish between modeling the $\Lambda(1405)$ as a Λ excited state or a bound $K - N$ system. The production of $\Xi\bar{\Xi}$ pairs has a threshold at 2.6 GeV/c $\bar{p}$ momentum and was out of reach of the LEAR beam. At higher momenta, the channel $p\bar{p} \to \Omega\bar{\Omega}$ could be reached. This reaction has the interesting topology that all valence quarks must annihilate and form $s\bar{s}$ pairs as shown in Figure 1.

B Baryon and Meson Spectroscopy

There has been considerable activity in the area of light meson spectroscopy, particularly in the glueball sector. Due in part to the recent work of the Crystal Barrel Collaboration, the surplus of observed mesons which are candidates for the $q\bar{q}$ scalar nonet appears to indicate that states with gluonic admixture have been observed. One possible explanation is that the $f_0(1370)$, the $f_0(1500)$, and $f_0(1750)$ physical states are actual admixtures of $s\bar{s}$, $u\bar{u} + d\bar{d}$, and pure gluonic systems. [4]

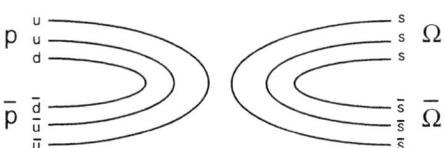

FIGURE 1. Quark-flow diagram for $p\bar{p} \to \Omega\bar{\Omega}$ reaction.

As the theoretical efforts to understand these systems evolve, it is likely that they will motivate additional experimental measurements.

The observation of a meson with exotic quantum numbers may provide a signature of gluonic degrees of freedom requiring less theoretical interpretation. The recent reported observation of AGS experiment E852 indicates that their 18 GeV/c pion beam produced mesons with quantum number $J^{PC} = 1^{-+}$. [5] The experimental difficulties in extracting the signature of this meson are significant. If correct, they indicate the discovery of a meson whose quantum numbers cannot be formed from a simple $q\bar{q}$ valence system and could indicate a quark-gluon hybrid, $q\bar{q}q\bar{q}$, or some other "exotic" form of matter.

There is also activity in the tensor glueball sector. A narrow resonance called the $\xi(2230)$ has been a tensor glueball candidate since it was first observed by the MARK III collaboration. The experimental data have been quite limited until recent work in Beijing by the BES collaboration confirmed the $\xi(2230)$'s existence from J/Ψ decays and extracted several branching ratios. [6,7] The observed rate to $p\bar{p}$ has been particularly puzzling in view of the upper limits for $p\bar{p} \to \xi \to K_s K_s$ from LEAR PS185 and more recent preliminary results from the Crystal Barrel and JetSet Collaborations [8]. A recently approved experiment by Dave Hertzog et. al. (AGS E924) would utilize the antiproton capabilities of the $D6$ line combined with the Crystal Ball to study $p\bar{p} \to \xi \to \eta\eta$. [9] The collaboration is waiting for further clarification of the BES and Crystal Barrel results before proceeding.

There is much work to be done in the field of baryon spectroscopy. David Peaslee has pointed out that the spectrum of 3-quark systems is largely unexplored. [10] In recent years, quark-confinement models have also motivated searches for 6-quark dibaryon systems. Kaon beam lines in the 1 GeV/c to 2 GeV/c momentum range are ideally suited to searches in the S=-1 and S=-2 sectors.

The possible existence of an S=-2, J=0 6-quark system known as the H-Dibaryon is of considerable interest. It was first predicted by Jaffe in 1977. [11] It has been predicted by a variety of confinement models, but has not been experimentally verified. Figure 2 shows the recent AGS E836 90% cl upper limits for H-Dibaryon production through the reaction $K^- + ^3He \to K^+ + H + n$ [12]. These limits are an order of magnitude below the predicted cross sections for H-dibaryon masses more than $30 MeV$ below $2m_\Lambda$. [13] Similar experimental upper limits have been obtained using a carbon target at KEK. [14].

The measurements of the ground state of double-Λ hypernuclei have implications on the possible existence of the H-Dibaryon. There have been three publications of emulsion events which have been interpreted as the formation and decay of double-Λ hypernuclei. The most recent event from KEK is shown schematically in Figure 3 [15]. The event can be interpreted as a Ξ hyperon stopping in the emulsion, forming a $^{10}_{\Lambda\Lambda}Be$ hyperfragment which recoils and undergoes a mesonic decay at point B. This results in a recoiling $^{10}_\Lambda B$ hypernucleus which undergoes a multi-body nonmesonic decay at point C. This interpretation leads to a residual $\Lambda\Lambda$ binding of -4.9 MeV (repulsive). Unfortunately, the emulsion event is also consistent with a second interpretation in which an excited state of $^{14}_{\Lambda\Lambda}C$ is initially formed. This interpretation is consistent with a residual $\Lambda\Lambda$ binding of 4.8 MeV (attractive). There are experimental reasons to favor the former interpretation and theoretical arguments that the latter is most probable. If correctly interpreted, these events can be used to deduce a lower limit on the possible mass of the H-Dibaryon. If one assumes that the $S = -2$ hypernucleus reaches its ground state through strong and electromagnetic transitions before undergoing the weak mesonic decay, the range of the π^- meson and recoiling hyperfragment serve as a direct measure of the ground state of an $S = -2$ system. If the H-Dibaryon exists and is bound by more than 30 MeV, then the transition $\Lambda\Lambda \rightarrow H$ would be kinematically allowed within the nucleus and result in a lighter final-state mass.

While experiments at KEK and the AGS have cast doubt on the possibility of a deeply bound H-dibaryon, the masses at and above $2m_\Lambda$ remains largely unexplored. The preliminary results from a scintillating fiber range stack experiment at KEK [16] show an interesting enhancement in the $\Lambda\Lambda$ invariant mass spectrum from $K^- + ^{12}C \rightarrow K^+ + X$ reaction. One could consider performing this measurement with higher statistics at the AGS or executing a complimentary experiment such as $K^- + ^3He \rightarrow K^+ + \Lambda + \Lambda + n$. Spin-one dibaryons could be created through the process $K^- + d \rightarrow K^+ + X$. Although models with color-magnetic interactions based on quark-gluon perturbative interactions favor the J=0 H-Dibaryon to be the lightest S=-2 dibaryon, recent work indicates that J=1 states may, in fact, be lighter than the H-Dibaryon. [17] Further theoretical developments may increase

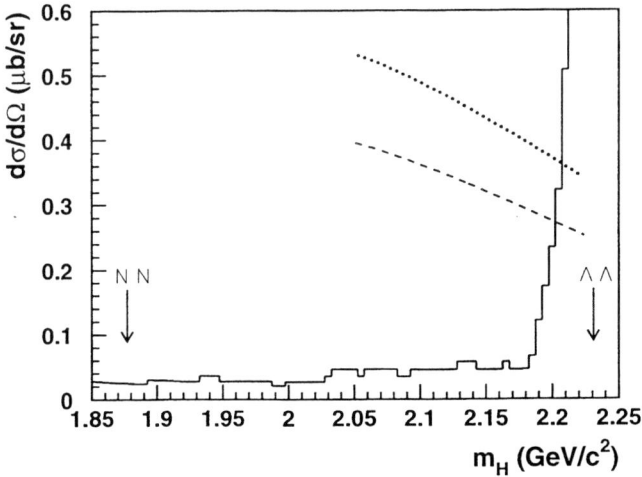

FIGURE 2. Results of AGS E836 search for the H dibaryion through the reaction $K^- + ^3He \rightarrow K^+ + H + n$. Solid line shows the upper limits on H production as a function of the mass m_H. Dotted line shows the predicted production cross section by Aerts and Dover. Dashed line shows the Aerts and Dover prediction modified by recent measurements of the $K^- + p \rightarrow K^+ + \Xi$ elementary production cross section.

the incentive for searching for these exotic systems.

C $NN \to$ Strange Sector

Our understanding of nucleon-nucleon interactions can be tested and expanded by moving to $S = -1$ and $S = -2$ systems. Our understanding of the nucleon-nucleon interaction is largely based on a potential description which is, in turn, motivated from a one-boson-exchange (OBE) description for the long-range portion coupled with additional short-range ingredients. There are many parameters but the wealth of data introduce considerable constraints. To move into the $S = -1$ and $S = -2$ sectors, the OBE terms fit to the NN data are transformed to systems with strangeness using SU(3) flavor symmetry. Although new parameters are introduced in the description of the short range terms, these parameters have only limited influence on the model's ability to describe existing data. The extension of NN models into the strange sector has been studied by the Nijmegen [19] and Jülich [20] groups.

It has long been recognized that competing descriptions which are essentially equivalent in the NN sector give quite differing predictions in the strangeness sector. For example, the well known Nijmegen D and F potentials are largely equivalent in the NN sector but don't even agree on the sign of the interaction for the ΞN case. [22] Thus a limited amount of reliable data in the strangeness sector can have a direct influence in our understanding of the NN interaction.

Much of what we know of the ΛN interaction comes from studies of S=-1 hypernuclei. Excitation spectra from (K^-, π^-) and (π^+, K^+) reactions show clear Λ-shell structure and core excitations. However, these experiments are limited to a few MeV energy resolution and have studied only a limited number of hypernuclear species. Over the last year, it has been shown that the Neutral Meson Spectrometer (formerly used at LAMPF) can be used to measure excitation spectra of hypernuclei created through the (K^-, π^0) reaction. [23] This reaction can

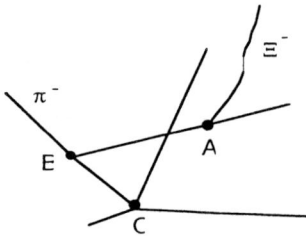

FIGURE 3. Schematic representation of double-lambda hyperfragment formation in emulsion.

create hypernuclear states which cannot be created with the (K^-, π^-) and, for example, provides an opportunity for exploring the large isospin violations which appear to be present in medium-A hypernuclei. Although this program has already been initiated at the AGS, it clearly cannot be completed by the end of the upcoming running period. There is also a need for measurements with greatly improved energy resolution (better than 50 keV) to study the spin-spin and spin-orbit splittings of hypernuclei. Tamura has shown that this can be done using a Ge detector system now under construction in Japan. [18] It will be used to detect γ-rays from hypernuclear created with the (K, π) and (π, K) reactions.

Although a limited number of measurements could be made over the next year at KEK and the AGS, it would be desirable to extend the program. The E929 collaboration also plans to measure γ-rays from hypernuclear decays. [24] They will observe spin-orbit splittings using a NaI array.

Hypernuclear data in the $S = -2$ sector are scarce. They are still limited to a few ambiguous emulsion events identified as $\Lambda\Lambda$ and Ξ hypernuclei. The situation may improve this year with the initial production running of the 'Cylindrical Drift System' (E906) at the AGS. This experiment will attempt to create light hypernuclei through the (K^-, K^+) reaction using the beamline and spectrometer used for the H-Dibaryon searches zE836 and E813 It will attempt to identify and measure the masses of $S = -2$ hyperfragments by measuring the momenta of pions from their sequential mesonic decays. If this is successful, it is natural to assume that follow-up experiments will be proposed. Finally, the construction of a 100 msr superconducting spectrometer designed for (K^-, K^+) studies has been discussed. With such a device installed at the AGS 1.8 GeV/c kaon line one could produce quality Ξ-hypernuclear data despite the prediction of 200 nb/sr cross sections. [25]

Baryon-baryon interactions can also be studied through final-state interactions and scattering experiments. Imai and Ieiri have proposed a study of $\Xi p \to \Xi p$ and $\Xi p \to \Lambda\Lambda$. [26] Ransome has discussed creating Λ hyperons through the reaction $\pi^- p \to K^0 \Lambda$ to study Λp scattering. [27] It could also be interesting to look for final state interactions in the reaction $K^- d \to K^0 \Lambda\Lambda$. These studies could be done using the $D6$ line and the 'Cylindrical Drift System'. Dehnhard et al. have written a Letter-of-Intent [28] to use the Neutral Meson Spectrometer to study final state interactions in the reaction $d(K^-, \pi^0 n)Y^0$. Earlier measurements have shown strong cusp effects in the $d(K^-, \pi^- n)\Lambda$ channel. The existing hyperon-nucleon scattering data are quite poor and, by taking advantage of the self-analyzing properties of the hyperon decays, these experiments could add considerable information in the $S = -1$ and $S = -2$ sectors.

D Weak Interactions

Both the AGS and KEK hyperon program also includes weak decay studies. It is interesting to explore both the mesonic $\Lambda \to N\pi$ rates (which are analogous to free Λ decays and the non-mesonic nucleon stimulated decays of the form $\Lambda N \to NN$.

The mesonic decay rates of hypernuclei are sensitive to the Λ- nucleus potential. However, the mesonic branching ratios are rather small in all but the lightest hypernuclei and thus the data are very limited at this time. Non-mesonic weak decays can be used to explore the empirical $\Delta I = \frac{1}{2}$ rule. This rule is the observation that the effective weak Hamiltonian for $\Delta S = 1$ decays is dominated by $\Delta I = \frac{1}{2}$ isospin transitions. (For an example, consider the mesonic decays of the Λ. One observes the $p\pi^-$ rate to be twice the rate for $n\pi^0$, in agreement with an isospin $\frac{1}{2}$ final state. Since the initial state has isospin zero, this is a nearly pure $\Delta I = \frac{1}{2}$ transition.) First order perturbative w-exchange would predict roughly equal $\Delta I = \frac{1}{2}$ and $\Delta I = \frac{3}{2}$ contributions which would be in large disagreement with the data. There have been various attempts to explain the $\Delta I = \frac{1}{2}$ rule. Some of the models predict large violations in untested non-mesonic decays. Thus, measurements of non-mesonic decay branching ratios provides the means of testing our understanding of the electroweak interaction in the presence of strongly interacting quarks.

A proposal has been recently approved to measure the branching rates of $^4_\Lambda H$ using the NMS spectrometer at the AGS. [29] These data, combined with earlier studies of $^4_\Lambda He$ decays, have been predicted to show large deviations from the $\Delta I = \frac{1}{2}$ rule. Unfortunately, it is not clear if the experiment can be run and completed during the final running of the AGS.

In the longer term, one might hope to study the weak interactions $\Lambda\Lambda \to n\Lambda$ and $\Lambda\Lambda \to p\Sigma$ using $S = -2$ hypernuclei. The antiproton working group [3] has investigated studying CP violation in Λ and $\overline{\Lambda}$ decays. This could be done by extending the $p\bar{p} \to \Lambda\overline{\Lambda}$ studies done by PS185 at LEAR. It is believed that one may be able improve on the CP measurement currently in progress at Fermilab.

III CONCLUSIONS

In view of the wide range of experimental proposals involving hyperons and strange mesons, one might expect to gain considerable insight into strange-sector physics over the next decade. Since the experiments in this field are driven by the state of the theoretical models, it is difficult to predict the experimental priorities years in advance. Given the primitive state of $S = -2$ knowledge and largely unmapped strange-baryon sector, it is probably that there will be considerable work left in these fields even under the best of circumstances. This work would require kaon and antiproton lines in the few GeV/c range.

It should also be noted that little progress may be made over the next few years if the AGS slow-extracted beam program comes to an end. This is particularly true for experiments requiring flux above the rates available at KEK. The situation should improve dramatically in the middle of the next decade as the Japanese Hadron Facility (JHF) comes on-line. Until that time, it is possible that little or no progress will be made in even basic measurements such as hyperon-nucleon scattering.

REFERENCES

1. S. Adler et al., Phys. Rev. Lett. **79**, 2204 (1997).
2. P.D. Barnes et al., Phys Rev. C **54**, 1877 (1996).
3. T.Barnes et al. Summary of the Antiproton Working Group, Proceedings of the Workshop Held at Brookhaven National Laboratory, BNL 521512 Formal Report, 45 (1996).
4. C.A. Meyer, Intersections Between Particle and Nuclear Physics 6th Conf., AIP Conf. Proceedings **412**, 91 (1997).
5. D.R. Thompson et al., Phys. Rev. Lett. **79**, 1630 (1997).
6. Z.Z. Bai et al., Phys. Rev. Lett. **76**, 3503 (1996).
7. Y.C. Zhu, Intersections Between Particle and Nuclear Physics 6th Conf., AIP Conf. Proceedings **412**, 476 (1997).
8. D.W. Hertzog, Intersections Between Particle and Nuclear Physics 6th Conf., AIP Conf. Proceedings **412**, 481 (1997).
9. AGS Proposal E924, D. Hertozg spokesperson.
10. D. Peaslee, unpublished.
11. R.L. Jaffe, Phys. Rev. Lett. **38**, 1995 (1977); 1617(E) (1997).
12. R.W. Stotzer et al., Phys. Ref. Lett. **78**, 3646 (1997).
13. A.T.M. Aerts and C.B. Dover, Phys. Rev. D **28**, 450 (1983); Phys. Rev. Lett. **49**, 1752 (1982).
14. J.K. Ahn et al., Phys. Lett. B **378**, 53 (1996).
15. S. Aoki et al., Prog. Th. Phys. **85**, 1287 (1991).
16. J.K. Ahn et al., Intersections Between Particle and Nuclear Physics 6th Conf., AIP Conf. Proceedings **412**, 923 (1997).
17. S.D. Paginis et al., Los Alamos preprint nucl-th/9706060 (1997).
18. H. Tamura, AGS proposal E930, H. Tamura spokesperson.
19. M.M. Nagels, T.A. Ruken, and J.J. DeSwart, Phys. Rev. **D15** 2547 (1977).
20. A. Reuber et al., AIP Conf. Proc. **338**, 583 (1995). A. Reuber et al., Nucl Phys. **A570** 543 (1994).
21. G. Alexander et al., Phys. Rev. **173**, 1452 (1968); B. Sechi-Zorn et al., Phys. Rev. **175**, 1735 (1968); J.A. Kadyk et al., Nucl Phys. **B27**, 13 (1971).
22. C.B. Dover and A. Gal, *Progress in Particle and Nuclear Physics* **12** (1984).
23. AGS proposal E907, E.V. Hungerford and J.C. Peng spokespersons.
24. AGS proposal E929, T. Kishimoto spokesperson.
25. Y. Yamamoto, T. Motoba, T. Fukuda, M. Takahashi, and K. Ikeda, Prog. Theor. Phys. **S117**, 281 (1994).
26. AGS proposal P928, M. Ieiri and K. Imai spokespersons.
27. R. Ransome, AGS LOI, unpublished.
28. D. Dehnhard, J.M. O'Donnell, and R.E. Chrien, AGS LOI, unpublished.
29. AGS proposal E931, J. Gerald, E. Hungerford, and V. Zeps spokespersons.

SUSY Before the Next Lepton Collider

Frank E. Paige

Physics Department
Brookhaven National Laboratory
Upton, NY 11973

Abstract. After a brief review of the Minimal Supersymmetric Standard Model (MSSM) and specifically the Minimal Supergravity Model (SUGRA), the prospects for discovering and studying SUSY at the CERN Large Hadron Collider are reviewed. The possible role for a future Lepton Collider — whether $\mu^+\mu^-$ or e^+e^- — is also discussed.

I INTRODUCTION

The many attractive features of the Minimal Supersymmetric Standard Model [1] or MSSM have made it a leading candidate for physics beyond the Standard Model. Of course there is no direct evidence for SUSY. The current limits [2] on SUSY masses from LEP are close to its ultimate kinematic reach. LEP will extend the limits on a Higgs boson from the present 77 GeV [2] up to $\gtrsim 95$ GeV [3]. Discovery of a light Higgs would not prove the existence of SUSY but would be a strong hint: the light Higgs boson must have a mass less than 130 GeV in the MSSM and less than 150 GeV in a rather general class of SUSY models [4], while in the Standard Model it must be heavier than about 130 GeV if the theory holds up to a high scale [5]. The next run of the Tevatron will have a better chance to find SUSY particles; the channel $\tilde{\chi}_1^\pm \tilde{\chi}_2^0 \to \ell^+\ell^- \tilde{\chi}_1^0 \ell^\pm \nu \tilde{\chi}_1^0$ can be sensitive to masses up to ~ 200 GeV for some choices of the other parameters [6]. But the reach in this channel is quite model dependent.

The definitive search for weak-scale SUSY, therefore, will have to await the LHC. The LHC with $10\,\text{fb}^{-1}$, 10% of its design luminosity per year, can detect gluinos and squarks up to about 2 TeV in the multi-jet plus missing transverse energy $\not{E}_T$ channels [7–9] compared to an expected mass scale of less than 1 TeV [10]. It is difficult to reconstruct masses directly because every SUSY event contains two missing lightest SUSY particles $\tilde{\chi}_1^0$. It is possible, however, to use endpoints of kinematic distributions to determine combinations of masses. In favorable cases these combinations can be used in a global fit to determine the model parameters. If SUSY is indeed the right answer, there should be a lot known about it before the Next Lepton Collider — whether $\mu^+\mu^-$ or e^+e^- — is built. It is probably difficult to study the whole SUSY spectrum at the LHC, however, so an NLC is also expected to play an important role.

II MINIMAL SUSY STANDARD MODEL

The Minimal Supersymmetric Standard Model [1] (MSSM) has for each Standard Model particle a partner differing in spin by $\Delta J = 1/2$. For each gauge boson there is a $J = 1/2$ gaugino, and for each chiral fermion there is a scalar sfermion. Two Higgs doublets and their corresponding Higgsinos are needed to give masses to all the quarks and to cancel anomalies. The SUSY particles have couplings determined by supersymmetry and are degenerate in mass with their Standard Model partners.

There is at present no experimental evidence for SUSY. There is one possible experimental hint: the renormalization group equations imply that the $SU(3) \times SU(2) \times U(1)$ gauge couplings measured at the Z mass meet in a way consistent with grand unification for the MSSM with $M_{\rm SUSY} \sim 1\,{\rm TeV}$ but not for the Standard Model. [12] The unification is actually not quite perfect, but it is within the range that could be covered by GUT threshold corrections. Other possible hints have been widely discussed but have mostly been discredited.

SUSY must of course be broken, since there is certainly no selectron degenerate with the electron. It is not possible to obtain an acceptable spectrum by breaking SUSY spontaneously using only the MSSM fields. However, mass terms for gauginos, Higgsinos, and sfermions do not break the $SU(3) \times SU(2) \times U(1)$ gauge invariance of the MSSM, so they can be added by hand without spoiling its renormalizability. There are also soft bilinear (B) and trilinear (A_{ijk}) couplings that are also gauge invariant and can be added. Finally, it is necessary to add a SUSY-conserving Higgsino mass (μ). It is generally assumed that SUSY breaking occurs spontaneously in a "hidden sector" and is communicated to the MSSM via some common interaction such as gravity. If SUSY is discovered, understanding the mechanism for its breaking will become an important issue in particle physics.

After SUSY is broken, all the states with the same quantum numbers mix. The $\tilde\gamma, \tilde Z, \tilde H_1, {\rm and}\tilde H_2$ mix to give four neutralinos $\tilde\chi^0_{1,2,3,4}$. The $\tilde W^\pm {\rm and} \tilde H^\pm$ mix to give two charginos $\tilde\chi^\pm_{1,2}$. The left and right squarks and sleptons also mix; this mixing is proportional to the fermion mass and so is significant only for the third generation.

The most general MSSM allows baryon and lepton number violation, giving proton decay at the weak scale. The simplest solution is to impose invariance under a discrete symmetry

$$R = (-1)^{3(B-L)+2S}$$

Note that $R = +1$ for all Standard Model particles, and $R = -1$ for all SUSY particles. Thus R parity invariance implies that SUSY particles are produced in pairs, that they decay to other SUSY particles, and that the lightest SUSY particle (LSP) is absolutely stable. Cosmological constraints then require that the LSP be

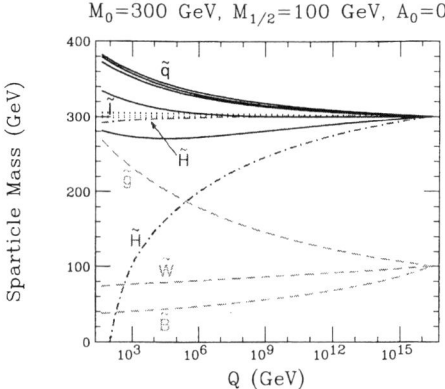

FIGURE 1. Evolution of masses in the SUGRA model, from Ref. 16.

neutral and weakly interacting, so that it escapes from any detector. Thus, the basic signature of (R parity conserving) SUSY is missing transverse energy $\not{E}_T$.

III MINIMAL SUGRA MODEL

The most general MSSM has more than 100 parameters. Many recent phenomenological studies have been based on a more restrictive model, the minimal supergravity (SUGRA) model [13]. This is very similar to the Constrained MSSM [14], although the latter adds some additional constraints. If SUSY breaking is communicated through gravity, then it is plausible that the SUSY breaking terms, like gravity, are universal at the GUT scale. In particular, if all the scalar masses are identical, then electroweak symmetry must be unbroken. It turns out that the Clebsch-Gordon coefficients are such that when the parameters are run from the GUT scale to the weak scale using the renormalization group equations, the large top Yukawa coupling drives the mass-squared of the Higgs negative, breaking electroweak symmetry but not color or charge. A example of this evolution is shown in Fig. 1. It is convenient to eliminate B and μ^2 in favor of M_Z and $\tan\beta$. Then the minimal SUGRA model is characterized by just four parameters and the sign of μ:

- m_0: the common squark, slepton, and Higgs mass at $M_{\rm GUT}$.

- $m_{1/2}$: the common gaugino mass at $M_{\rm GUT}$.

- A_0: the common trilinear coupling at $M_{\rm GUT}$.

- $\tan\beta = v_1/v_2$: the ratio of Higgs vacuum expectation values at M_Z.

- $\mathrm{sgn}\,\mu = \pm 1$.

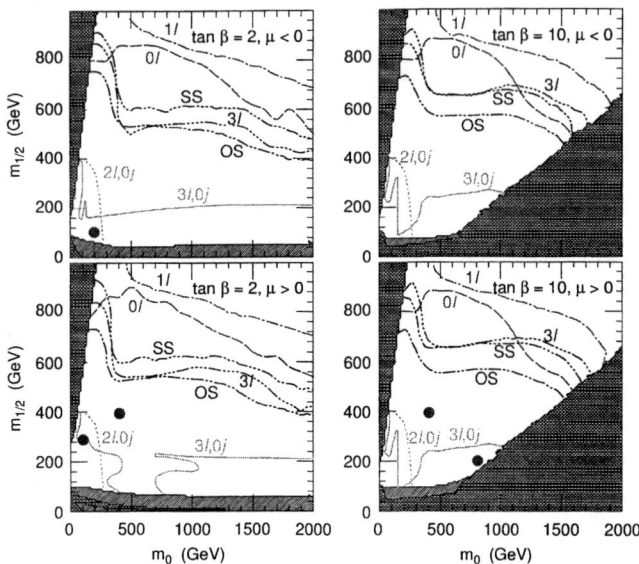

FIGURE 2. SUGRA discovery limits in the SUGRA model at the LHC with $10\,\text{fb}^{-1}$ integrated luminosity. 0ℓ: $\not{E}_T$ + jets + no leptons. 1ℓ: $\not{E}_T$ + jets + one lepton. OS: Opposite sign dileptons. SS: same sign dileptons. 3ℓ: trileptons. $2\ell, 0j$: dileptons with jet veto (from slepton production). $3\ell, 0j$: trileptons with jet veto (from gaugino production). From Ref. 11.

It turns out that A_0 is not very important for phenomenology at the weak scale.

While the SUGRA model provides a much more tractable parameter space than the general MSSM, one should remember that it is only one possible model. It is possible that the assumption of universal masses is not correct. It is also possible that SUSY breaking is communicated at a much lower mass scale, as in gauge mediated models [15].

IV SUSY SIGNATURES AT LHC

The LHC is a pp collider with an energy of 14 TeV, enough to produce $\tilde{g}$ and $\tilde{q}$ pairs with $\lesssim 2\,\text{TeV}$ even with $10\,\text{fb}^{-1}$. These are typically produced with $p_T \sim M$, so they move slowly in the lab frame and their decay products are widely separated. If R parity is conserved, they will decay into the LSP $\tilde{\chi}_1^0$ plus multiple jets and perhaps multiple leptons. A typical decay chain might be:

$$\tilde{g} \to \tilde{q}_L + \bar{q}$$
$$\tilde{q}_L \to \tilde{\chi}_2^0 + q$$
$$\tilde{\chi}_2^0 \to \tilde{\ell}_R + \bar{\ell}$$
$$\tilde{\ell}_R \to \tilde{\chi}_1^0 + \ell$$

Such decay chains produce many possible signatures combining jets, leptons, and $\not{E}_T$ from $\tilde{\chi}_1^0$ and ν's. For example, since the $\tilde{g}$ is self-conjugate, there are isolated same-sign dileptons $\ell^\pm \ell^\pm$.

The 5σ discovery limits at the LHC for $10\,\text{fb}^{-1}$ integrated luminosity are shown in Figure 2. Note that the reach is $> 2\,\text{TeV}$ in the missing energy channels and $> 1\,\text{TeV}$ in the multi-lepton channels. Thus the LHC should find multiple signatures for SUSY with only $10\,\text{fb}^{-1}$ if it exists at the weak scale.

V PRECISION MEASUREMENTS AT LHC

While it is easy to find signals for SUSY at LHC, there are two missing $\tilde{\chi}_1^0$'s in each event, making it difficult to reconstruct masses. However, it is possible [18] to exploit the cascade decays characteristic of the MSSM to determine combinations of masses. The strategy is to start at the bottom of the decay chain and work up, partially reconstruct specific final states and relating precision measurements of endpoints of kinematic distributions to combinations of masses. A global fit to these combinations can then be used to determine the model parameters, at least in favorable cases. It would be better to make the global fit not just to such endpoints but to all distributions, but this is more difficult technically and perhaps premature at this stage.

What combinations of masses can be determined in this way depends on the decay modes and so requires study of specific SUSY models. The CERN LHC Program Committee (LHCC) chose five SUGRA points for detailed study by the ATLAS and CMS Collaborations. The parameters of these points and some representative masses are listed in Table 1. Point 3 is the "comparison point," selected so that every existing or proposed accelerator can discover something. Point 5 was chosen to give the right cold dark matter for cosmology and so is perhaps the most realistic. Points 1 and 2 have gluino and squark masses of order $1\,\text{TeV}$. Point 4 has the squarks much heavier than the gluinos. It is close to the boundary allowed by electroweak symmetry breaking (at least with ISAJET 7.22) and so has a relatively small μ and a large mixing between gauginos and Higgsinos. Studying these specific points has proved surprisingly useful [17–20].

VI EFFECTIVE MASS

Gluinos and squarks are strongly produced at the LHC. But production cross sections fall rapidly with the produced mass, so it is important to find a variable that measures the produced mass for events with missing particles. A variable that works well for SUSY is the effective mass, defined as the sum of the missing energy and the p_T's of the first four jets,

$$M_{\text{eff}} = \not{E}_T + p_{T,1} + p_{T,2} + p_{T,3} + p_{T,4}$$

TABLE 1. Parameters of the LHCC SUGRA points and some representative masses from ISAJET 7.22 [21].

Point	m_0 (GeV)	$m_{1/2}$ (GeV)	A_0 (GeV)	$\tan\beta$	sgn μ	$M_{\tilde{g}}$ (GeV)	$M_{\tilde{u}_R}$ (GeV)	$M_{\tilde{W}_1}$ (GeV)	$M_{\tilde{e}_R}$ (GeV)	M_h (GeV)
1	400	400	0	2.0	+	1004	925	325	430	111
2	400	400	0	10.0	+	1008	933	321	431	125
3	200	100	0	2.0	−	298	313	96	207	68
4	800	200	0	10.0	+	582	910	147	805	117
5	100	300	300	2.1	+	767	664	232	157	104

Samples of signal and Standard Model background events were generated with ISAJET [21]. To separate SUSY from the Standard Model background, events were selected with multiple jets plus missing energy:

- $\not{E}_T > \max(100\,\text{GeV}, 0.2 M_{\text{eff}})$;
- ≥ 4 jets with $p_T > 50\,\text{GeV}$ and $p_{T,1} > 100\,\text{GeV}$;
- Transverse sphericity $S_T > 0.2$;
- No μ or isolated e with $p_T > 20\,\text{GeV}$, $\eta < 2.5$.

Then the SUSY signal emerges from the Standard Model background for large M_{eff}, as can be seen from Figure 3. The signal cross section is of order 1 pb in the region where it dominates, so it could be discovered in about one month at $10^{32}\,\text{cm}^{-2}\text{s}^{-1}$. (Of course, it would take much longer than this to understand the detectors.)

The value of M_{eff} at which the signal emerges from the background scales with the SUSY mass scale. To test this, 100 random SUGRA models were generated, and the peak of the M_{eff} signal was compared with the SUSY mass scale, defined by

$$M_{\text{SUSY}} = \min(M_{\tilde{g}}, M_{\tilde{u}})$$

The scatter plot, shown in Fig. 4, shows a good correlation between the peak and the SUSY mass scale, allowing one to determine the SUSY mass scale to about 10%.

VII RECONSTRUCTION OF SPECIFIC FINAL STATES

The precision measurements of specific combinations of masses are based on the partial reconstruction of the corresponding final states. For each case, SUSY and Standard Model background event samples were generated with ISAJET [21] or PYTHIA [22], a simple particle-level detector simulation incorporating resolutions characteristic of ATLAS and CMS was made, and analysis cuts as described below were applied.

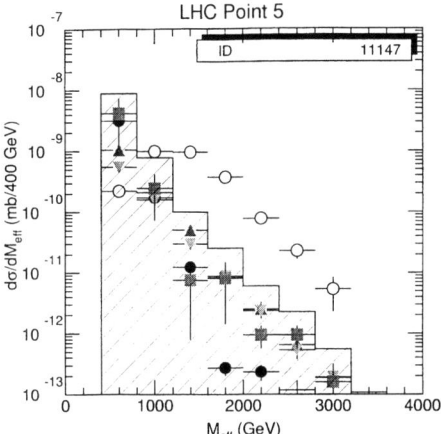

FIGURE 3. M_{eff} distributions after cuts. Open circles: SUSY signal. Solid circles: $t\bar{t}$. Upward triangles: W + jets. Downward triangles: Z + jets. Squares: QCD jets. Shaded histogram: Sum of Standard Model backgrounds. From Ref 18.

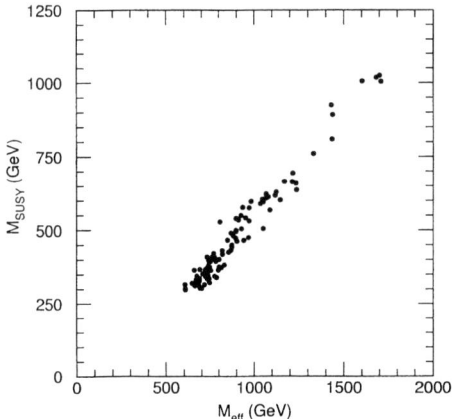

FIGURE 4. Scatter plot of signal peak in M_{eff} vs. M_{SUSY} defined in text. From Ref. 18.

A Measurement of $M(\tilde{\chi}_2^0) - M(\tilde{\chi}_1^0)$

The prototype of all the precision measurements is based on the decay $\tilde{\chi}_2^0 \to \tilde{\chi}_1^0 \ell^+ \ell^-$ at Point 3. Point 3 has unusual branching ratios:

$$B(\tilde{g} \to \tilde{b}_1 \bar{b} + \text{h.c.}) = 89\%$$

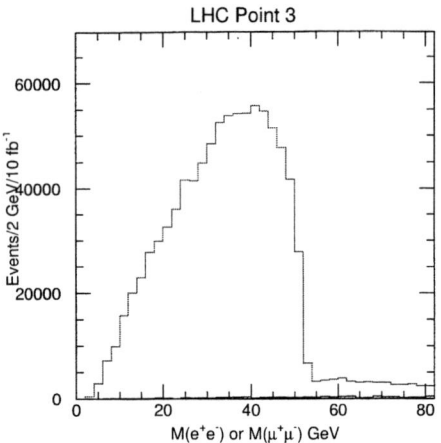

FIGURE 5. $\ell^+\ell^-$ distribution for SUSY events at Point 3 (histogram) and for Standard Model background (shaded) after cuts described in the text. From Ref .18.

$$B(\tilde{b}_1 \to \tilde{\chi}_2^0 b) = 86\%$$
$$B(\tilde{\chi}_2^0 \to \tilde{\chi}_1^0 \ell^+ \ell^-) = 2 \times 17\%$$

The dominant decay of $\tilde{g} \to \tilde{b}_1 \bar{b}$ arises because the $\tilde{b}_1$ is lighter than the $\tilde{g}$ but the other squarks are heavier. Events were selected to have two leptons and two b jets:

- $\ell^+\ell^-$ pair with $p_{T,\ell} > 10\,\text{GeV}$, $\eta < 2.5$.
- ≥ 2 jets tagged as b quarks with $p_T > 15\,\text{GeV}$ and $\eta < 2$.
- No $\not{E}_T$ cut was used.

All distributions shown include a 60% tagging efficiency for b's and 90% efficiency for leptons within the kinematic cuts given above.

The result of this analysis is a spectacular edge at $M(\tilde{\chi}_2^0) - M(\tilde{\chi}_1^0)$ endpoint with almost no Standard Model background, as can be seen in Figure 5. Most of the SUSY background comes from two $\tilde{\chi}_1^\pm$ decays and can be removed by plotting the distribution for

$$e^+e^- + \mu^+\mu^- - 2e^\pm\mu^\mp$$

This analysis clearly would have huge statistics and would be much easier than measuring M_W at Tevatron. Given the current M_W results, it seems conservative to estimate an error

$$\Delta(M(\tilde{\chi}_2^0) - M(\tilde{\chi}_1^0)) = 50\,\text{MeV}$$

for $10\,\text{fb}^{-1}$.

Point 3 is perhaps unusually easy, but there is a similar edge at Point 4 plus Z peak from heavier gauginos, as can be seen from Figure 6. The estimated error in this case is $\pm 1\,\text{GeV}$.

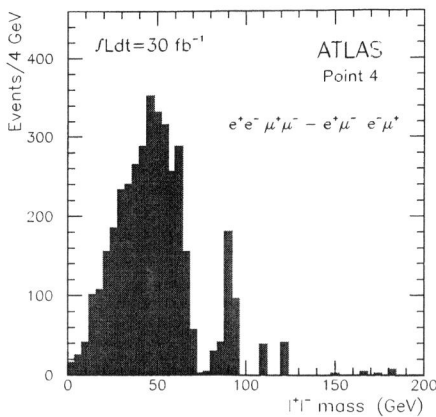

FIGURE 6. $\ell^+\ell^-$ distribution for SUSY events at Point 4. From Ref .23.

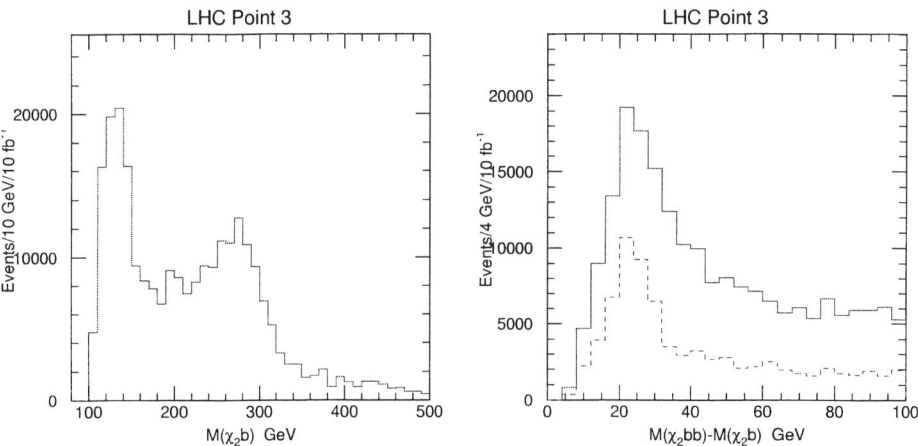

FIGURE 7. Projections of the gluino-sbottom mass scatter plot. From 18.

B $\tilde{g}$ and $\tilde{b}_1$ Reconstruction

The next step at Point 3 is to combine an $\ell^+\ell^-$ pair near the edge with jets to determine the $\tilde{b}_1$ and $\tilde{g}$ masses. Events were selected with

- ≥ 2 jets tagged as b jets with $p_T > 15\,\text{GeV}$, $\eta < 2$;
- $\ell^+\ell^-$ pair with $45 < M(\ell\ell) < 55\,\text{GeV}$.

For an $\ell^+\ell^-$ pair near the endpoint, the $\tilde{\chi}_1^0$ must be soft in the $\tilde{\chi}_2^0$ rest frame, so that

$$\vec{p}(\tilde{\chi}_2^0) \approx \left(1 + \frac{M(\tilde{\chi}_1^0)}{M(\ell\ell)}\right) \vec{p}(\ell\ell)$$

where $M(\tilde{\chi}_1^0)$ must be determined from a global fit. The approximately reconstructed $\tilde{\chi}_2^0$ was combined with one of masses coming from combining the $\tilde{\chi}_2^0$ with one b to make $M(\tilde{b}_1)$ and with a second b to make $M(\tilde{g})$ using the correct $\tilde{\chi}_1^0$ mass. The resulting projections, shown in Figure 7, display good resolution on the mass difference between the $\tilde{g}$ and the $\tilde{b}_1$ masses — just like for $D^* \to D\pi$. By varying the assumed $\tilde{\chi}_1^0$ mass, one finds

$$\Delta M(\tilde{b}_1) = \pm 1.5 \Delta M(\tilde{\chi}_1^0) \pm 3\,\text{GeV}$$

$$\Delta\left(M(\tilde{g}) - M(\tilde{b}_1)\right) = \pm 2\,\text{GeV}$$

C Reconstruction of $h \to b\bar{b}$

For Point 5, the decay $\tilde{\chi}_2^0 \to \tilde{\chi}_1^0 h$ is kinematically allowed and has a branching ratio of 64%. Events were selected with

- ≥ 4 jets with $p_T > 50\,\text{GeV}$, $p_{T,1} > 100\,\text{GeV}$;
- Transverse sphericity $S_T > 0.2$;
- $M_{\text{eff}} = \not{E}_T + \sum_{i=1}^{4} p_{T,i} > 800\,\text{GeV}$;
- $\not{E}_T > \max(100\,\text{GeV}, 0.2 M_{\text{eff}})$.

and the $b\bar{b}$ mass was plotted for all pairs of b jets with $p_{T,b} > 25\,\text{GeV}$ and $\eta_b < 2$. A correction factor was applied to the measured b jet energies to account for neutrinos and energy loss out of the cone, and a 60% b-tagging efficiency was assumed. The resulting distribution, Figure 8, has a peak at the Higgs mass with a substantial SUSY background but very little Standard Model background. This signal is much easier than $h \to \gamma\gamma$ and would be the discovery mode for the Higgs at this point.

The $h \to b\bar{b}$ candidates can be used to reconstruct the decay chain

$$\tilde{g} + \tilde{g} \to \tilde{q}_L q + \tilde{q}_R q$$
$$\tilde{q}_L \to \tilde{\chi}_2^0 q \to \tilde{\chi}_1^0 h q, \qquad \tilde{q}_R \to \tilde{\chi}_1^0 q$$

To select these events exactly two additional jets with $p_T > 75\,\text{GeV}$ were required. Then since one of the two $qb\bar{b}$ combinations comes from the squark decay, the smaller of them must have an endpoint at a function of the squark mass and the other masses in the problem. The squark mass can be measured to about $40\,\text{GeV}$ in this way.

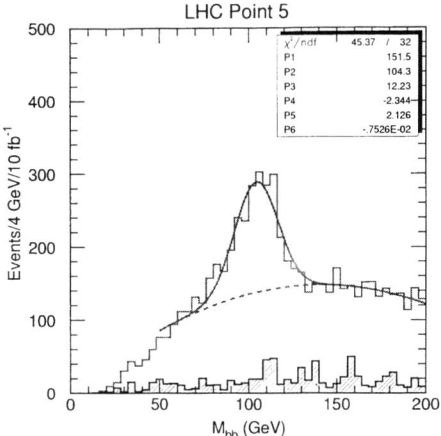

FIGURE 8. $M(b\bar{b})$ for pairs of b jets for the Point 5 signal (open histogram) and for the sum of all backgrounds (shaded histogram). The smooth curve is a Gaussian plus quadratic fit to the signal. The light Higgs mass is 104.15 GeV.

D $\ell^+\ell^-$ Again

Consider dileptons for Point 5. The mass distribution after the by now standard cuts shows a dramatic edge at about 109 GeV. Since this decay must compete with the two-body decay $\tilde{\chi}_2^0 \to \tilde{\chi}_1^0 h$, it cannot be a direct three-body decay $\tilde{\chi}_2^0 \to \tilde{\chi}_1^0 \ell^+ \ell^-$. In fact it comes from two sequential two-body decays, $\tilde{\chi}_2^0 \to \tilde{\ell}^\pm \ell^\mp \to \tilde{\chi}_1^0 \ell^\pm \ell^\mp$, and the edge determines

$$M_{\max}(\ell\ell) = M(\tilde{\chi}_2^0)\sqrt{1 - \frac{M_{\tilde{\ell}}^2}{M_{\tilde{\chi}_2^0}^2}}\sqrt{1 - \frac{M_{\tilde{\chi}_1^0}^2}{M_{\tilde{\ell}}^2}}$$

to about 1 GeV. One should do a complete fit to the Higgs and dilepton events to extract the maximum information on all the masses. This has not yet been done. The variable most sensitive to the slepton mass is $p_{T,2}/p_{T,1}$. This distribution was compared for two different values of m_0, from which it seems that the slepton mass can be estimated to $\Delta M(\tilde{\ell}_R) \sim 3\,\text{GeV}$.

E   Measurement of $M(\tilde{g}) - M(\tilde{\chi}_2^0), M(\tilde{\chi}_1^\pm)$

Gluinos dominate at Point 4 since m_0 is large. This means that there is a lot of combinatorial background from the many jets in the final state. The strategy of this analysis [19,23] is to use trilepton events to select the process

$$\tilde{g} + \tilde{g} \to \tilde{\chi}_2^0 q\bar{q} + \tilde{\chi}_1^\pm q\bar{q}$$

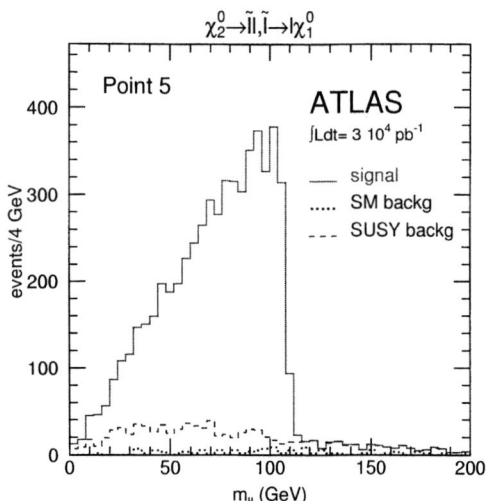

FIGURE 9. $M_{\ell\ell}$ for the Point 5 signal (open histogram) and the sum of all backgrounds (shaded histogram). From Ref. 24.

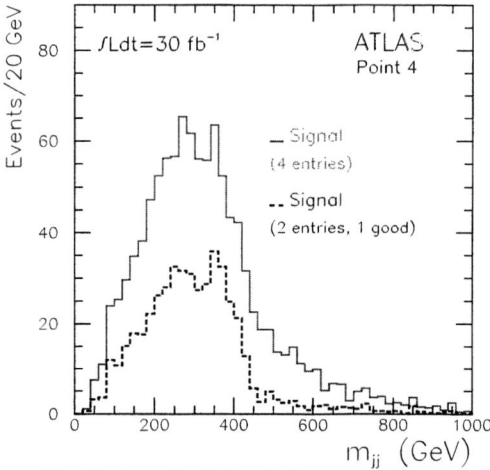

FIGURE 10. Dijet mass distributions for Point 4 after cuts. The dashed curve shows only the right pairing based on generator information. From Ref. 23.

Then the dijet mass distributions for the right jet pairing have a common endpoint since $M(\tilde{\chi}_2^0) \approx M(\tilde{\chi}_1^\pm)$.

Events were selected by requiring three leptons and four jets:

- 3 isolated ℓ: $p_T > 20, 10, 10\,\text{GeV}$, $|\eta| < 2.5$

- One opposite-sign, same-flavor lepton pair with $M_{\ell\ell} < 72\,\text{GeV}$.

- 4 jets; $p_T > 150, 120, 70, 40\,\text{GeV}$, $|\eta| < 3.2$.

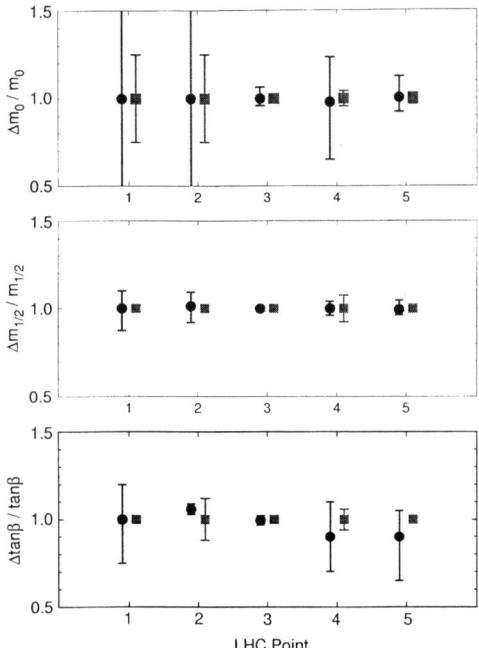

FIGURE 11. Errors from global fits to the SUGRA parameters. Circles: Fit I. Squares: Ultimate Fit II.

- No additional jets with $p_T > 40\,\text{GeV}$ and $|\eta| < 5$ to minimize combinatorics.

No $\not{E}_T$ cut was used. With these cuts there are 250 signal events, 30 $\tilde{g}\tilde{q}$ background events, and 18 other SUSY and Standard Model background events for $30\,\text{fb}^{-1}$. The pairing between the two highest and the two lowest p_T jets is usually wrong and so was eliminated. The dijet mass distribution for the other pairings is shown in Figure 10. There is an endpoint at about the right point.

VIII FITTING SUGRA PARAMETERS

The precision measurements described here are only a fraction of those in Refs. 18, 23, and 24. Ideally one should combine these measurements with a large number of other ones and do a global fit, but this would require generating many signal signal event samples. Instead a much simpler procedure has been adopted. SUGRA parameters were generated at random, the mass spectrum for each SUGRA point was calculated, and these were compared with the precision measurements and their estimated errors.

Two such fits have been made. Fit I [18] uses only the measurements developed in Ref. 18. It assumes that the Higgs mass can be related to the SUGRA parameters

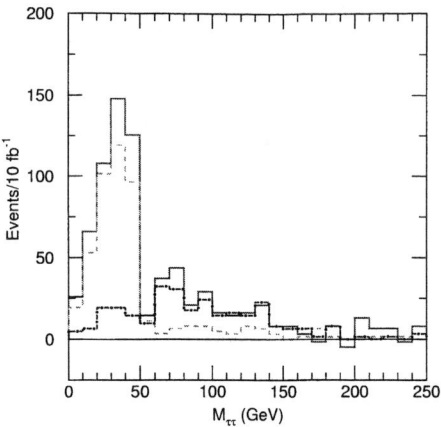

FIGURE 12. Solid: Visible $\tau\tau$ mass distribution for 3-prong τ decays. Dashed: Contribution from $\tilde{\chi}_2^0 \to \tilde{\tau}\tau$. Dash-dotted: Contribution from higher mass gauginos.

with an error $\Delta M_h = 3\,\text{GeV}$, about the current theoretical error. It bases the statistical errors on $10\,\text{fb}^{-1}$. Fit II [25] adds some additional precision measurements developed after Ref. 18. It adds some other data, e.g., from changing the squark masses at Points 1 and 2 and seeing the effect of this on the mean p_T of the hardest jet. This is not fully justified, but it is a plausible way of estimating the improvement from fitting some of the kinematic distributions as well as the precision measurements. The ultimate version of Fit II also assumes a theoretical error on the Higgs mass less than the expected experimental error from $h \to \gamma\gamma$, $\Delta M_h = 0.2\,\text{GeV}$ and scales the statistical errors to $300\,\text{fb}^{-1}$.

Both fits scanned the SUGRA parameter space and determined the 68% confidence interval for each parameter. The results are summarized in Figure 11. No disconnected solutions were found. In particular, $\text{sgn}\,\mu$ was correctly determined, although this required including additional information in the fit in some cases. The gluino and squark masses are insensitive to m_0 at Points 1 and 2, and there is no information available on the slepton masses, accounting for the larger errors on m_0 at these points. Finally, A_0 is poorly constrained in all cases, even for the ultimate version of Fit II. It is possible to determine the weak scale parameters A_t and A_b, but these are insensitive to A_0.

IX τ MODES AT LARGE $\tan\beta$

The five LHCC points do not exhaust the possibilities of even the minimal SUGRA model. For example, the $\tilde{\tau}_1$ is light for large $\tan\beta$, so τ decays can be dominant. Consider the SUGRA point $m_0 = m_{1/2} = 200\,\text{GeV}$, $A_0 = 0$, $\tan\beta = 45$,

$\mu < 0$. Then
$$B(\tilde{\chi}_2^0 \to \tilde{\tau}_1^\pm \tau^\mp) \approx B(\tilde{\chi}_1^\pm \to \tilde{\tau}_1^\pm \nu_\tau) \approx 100\%$$

Discovery of the SUSY signal is still straightforward, but none of the precision measurements discussed above are applicable.

One possible approach is to require two 3-prong hadronic τ's to maximize the visible $\tau\tau$ mass and hence its sensitivity to the endpoint analogous to that discussed in Section VII.D. The difference of $\tau^+\tau^-$ and $\tau^\pm\tau^\pm$ is used to eliminate the contribution from two $\tilde{\chi}_1^\pm$ decays. The resulting distribution is shown in Figure 12. There is clearly an endpoint visible from the contribution of $\tilde{\chi}_2^0 \to \tilde{\tau}\tau$ plus a contribution from heavier gauginos. Signatures like this require more study both for the Tevatron and for the LHC.

X OUTLOOK FOR LEPTON COLLIDERS

If SUSY exists at the electroweak scale, it should be straightforward to find signals for it at the LHC. It is possible in many cases to make precision measurements, and if the SUSY model is relatively simple, these can be used to determine its parameters.

The LHC will mainly produce gluinos and squarks. In SUGRA these tend to decay mainly into the lighter gauginos; the heavier ones are dominantly Higgsino and so are suppressed both by their masses and by their couplings. The direct production of sleptons and sneutrinos is also very small, although they can be produced as decay products of the light gauginos if they are light enough. Finally, the heavy Higgs bosons have small production cross sections, and their dominant decay modes have large backgrounds. One should not underestimate the ingenuity of experimentalists with real data, but it seems likely that the LHC will not be able to study the entire SUSY spectrum.

A Next Lepton Collider with $\sqrt{s} \sim 500\,\text{GeV}$ should be able to detect any SUSY particles except the $\tilde{\chi}_1^0$ that are kinematically accessible. Sleptons probably represent the best opportunity to make significant progress beyond what has been learned from the LHC. One of the attractive features of the SUGRA model is that the $\tilde{\chi}_1^0$ is a good dark matter candidate, and the abundance of cold dark matter favors light sleptons [26]. A lepton collider provides an important additional constraint that the slepton pairs are produced with known energy, and this allows precise measurements to be made [27,28]. But if more than one slepton is being produced, the spectrum can be quite complex, so high luminosity (as well as enough energy) may be essential.

REFERENCES

1. For general reviews of SUSY, see H.P. Nilles, Phys. Rep. **111**, 1 (1984); H.E. Haber and G.L. Kane, Phys. Rep. **117**, 75 (1985).

2. P. Janot, Int. Euro. Conf. on High Energy Physics (Jerusalem, 1997), http://www.cern.ch/hep97/pl17.htm.
3. M. Carena, P. Zerwas, et al., hep-ph/9602250, CERN-96-01 (1996).
4. G.L. Kane, C. Kolda, and J.D. Wells, Phys. Rev. Lett. **70**, 2686 (1993).
5. P.Q. Hung and M. Sher, Phys. Lett. **B374**, 138 (1996).
6. H. Baer, H. Murayama, X. Tata, et al., FSU-HEP-950401, hep-ph/9503479 (1995).
7. H. Baer, C-H Chen, F.E. Paige, and X. Tata, Phys. Rev. **D52**, 2746 (1995).
8. ATLAS Collaboration, *Technical Proposal*, LHCC/P2 (1994).
9. CMS Collaboration, *Technical Proposal*, LHCC/P1 (1994).
10. G.W. Anderson and D.J. Castano, Phys. Lett. **B347**, 300 (1995).
11. H. Baer, C-H Chen, F.E. Paige, and X. Tata, Phys. Rev. **D53**, 6241 (1996).
12. U. Amaldi, A. Bohm, L.S. Durkin, P. Langacker, A.K. Mann, W.J. Marciano, A. Sirlin, H.H. Williams, Phys. Rev. **D36**, 1385 (1987).
13. L. Alvarez-Gaume, J. Polchinski and M.B. Wise, Nucl. Phys. **B221**, 495 (1983);
    L. Ibañez, Phys. Lett. **118B**, 73 (1982);
    J.Ellis, D.V. Nanopolous and K. Tamvakis, Phys. Lett. **121B**, 123 (1983);
    K. Inoue et al. Prog. Theor. Phys. **68**, 927 (1982);
    A.H. Chamseddine, R. Arnowitt, and P. Nath, Phys. Rev. Lett., **49**, 970 (1982).
14. G.L. Kane, C. Kolda, L. Roszkowski, and J.D. Wells, Phys. Rev. **D49**, 6173 (1994).
15. M. Dine, A.E. Nelson, and Y. Shirman, Phys. Rev. **D51**, 1362 (1995).
16. J. Bagger, hep-ph/9508392 (1995).
17. A. Bartl, J. Soderqvist, et al., in *1996 DPF/DPB Summer Study on New Directions for High-Energy Physics (Snowmass 96)*.
18. I. Hinchliffe, F.E. Paige, M.D. Shapiro, J. Söderqvist, and W. Yao, Phys. Rev. **D55**, 5520 (1997).
19. ATLAS Collaboration, SUSY Presentations to the LHCC (October, 1996).
20. CMS Collaboration, SUSY Presentations to the LHCC (October, 1996).
21. H. Baer, F. Paige, S. Protopopescu and X. Tata; in *Physics at Current Accelerators and Supercolliders*, ed. J. Hewett, A. White and D. Zeppenfeld, (Argonne National Laboratory, 1993).
22. T. Sjostrand, LU-TP-95-20, hep-ph/9508391 (1995);
    S. Mrenna, Comput. Phys. Commun. **101**, 232 (1997).
23. F. Gianotti, ATLAS Internal Note PHYS-No-110 (1997).
24. G. Polesello, L. Poggioli, E. Richter-Was, and J. Soderqvist, ATLAS Internal Note PHYS-No-111 (1997).
25. D. Froidevaux, http://atlasinfo.cern.ch/Atlas/GROUPS/PHYSICS/SUSY/lhcc/-daniel.ps.Z.
26. J. Ellis and L. Roszkowski Phys. Lett. **B283**, 252, (1992);
    H. Baer and M. Brhlik, Phys. Rev. **D53**, 597 (1996).
27. T. Tsukamoto, K. Fujii, H. Murayama, M. Yamaguchi, and Y. Okada, Phys. Rev. **D51**, 3153, (1995).
28. M.M. Nojiri, K. Fujii, and T. Tsukamoto Phys. Rev. **D54**, 6756 (1996).

Supersymmetry vis-à-vis Muon Colliders

V. Barger

Physics Department, University of Wisconsin, Madison, WI 53706, USA

Abstract. The potential of muon colliders to study a low-energy supersymmetry is addressed in the framework of the minimal supergravity model, whose predictions are first briefly surveyed. Foremost among the unique features of a muon collider is s-channel production of Higgs bosons, by which Higgs boson masses, widths, and couplings can be precisely measured to test the predictions of supersymmetry. Measurements of the threshold region cross sections of W^+W^-, $t\bar{t}$, Zh, chargino pairs, slepton and sneutrino pairs will precisely determine the corresponding masses and test supersymmetric radiative corrections. At the high-energy frontier a 3–4 TeV muon collider is ideally suited to study heavy scalar supersymmetric particles.

I INTRODUCTION

There are indications that low energy supersymmetry (SUSY) is the right track for physics beyond the Standard Model (SM) [1]. The measurements of gauge coupling strengths $\alpha_1, \alpha_2, \alpha_3$ are consistent with SUSY Grand Unification [2] and global fits to precision electroweak measurements are consistent with SUSY expectations that there is a Higgs boson of mass less than 130 GeV [3]. If nature is indeed supersymmetric, are there compelling arguments why muon colliders should be built? The answer is *YES*, and the reasons why are the subject of this report. The physics at muon colliders discussed herein is largely based on work in collaboration with M. Berger, J.F. Gunion, and T. Han [4,5].

In the minimal supersymmetric model (MSSM) each standard model fermion (boson) has a boson (fermion) superpartner. Two Higgs doublets are required to give masses to the up-type and down-type fermions. SUSY breaking is introduced through all soft masses and couplings that do not introduce quadratic divergences; there are over 100 of these soft SUSY-breaking parameters. In SUSY-breaking models the number of independent soft parameters is greatly reduced. The breaking is transmitted from a hidden sector to the observable sector. There has been an explosive growth in models of SUSY breaking. These models fall into two classes, minimal SuperGravity (mSUGRA) and gauge-mediated symmetry breaking (GMSB). Thus, supersymmetry has many possible faces depending on:

- breaking by general soft parameters or unification of soft parameters;
- whether R-parity is conserved; with R-conservation the lightest supersymmetric particle (LSP) is stable;
- whether the gravitino is heavy (in mSUGRA) or light (in GMSB);
- the nature of the LSP: gaugino ($\tilde{B}$), higgsino ($\tilde{H}$), gravitino ($\tilde{G}$), or gluino ($\tilde{g}$);
- the relative masses of the sparticles.

Fortunately, there are some generic predictions that are not very model dependent. The most important is the guarantee of a light Higgs boson [6], which is accordingly the "jewel in the crown" of supersymmetry. Also a light neutralino, chargino, sleptons and sneutrinos are expected. In this report we concentrate on these lighter SUSY particles.

II RICH SUSY PHENOMENOLOGY

In the mSUGRA model the neutralino $\tilde{\chi}_1^0$ is the LSP and it is a source of missing energy in SUSY events. The signatures of SUSY particle production in the mSUGRA model are leptons + jets + missing E_T (denoted $\not{E}_T$).

In the traditional GMSB models, the gravitino ($\tilde{G}$) is the LSP and it is a source of missing energy. The nature of the next-to-lightest supersymmetric particle (NLSP) in GMSB models determines the phenomenology. The NLSP options and their decays are $\chi_1^0 \to \gamma \tilde{G}$, $\tilde{\ell} \to \ell \tilde{G}$ and $\tilde{\tau}_1 \to \tau \tilde{G}$. Decays of the NLSP may occur within or outside the detector. The signatures of such events are $\gamma\gamma + \not{E}_T$, $\ell\ell + \not{E}_T$, etc. [7]

The mSUGRA model is the usual benchmark for SUSY phenomenology. The renormalization group equations relate masses and couplings at the scale of the Grand Unified Theory (GUT) to their electroweak scale values. The GUT scale parameter set in mSUGRA is $m_{1/2}$, m_0, μ, A_0 and B_0, where $m_{1/2}$ and m_0 are universal gaugino and scalar masses, μ is the Higgs mixing mass, A_0 is the trilinear coupling and B_0 is the bilinear coupling. At the weak scale the phenomenology is determined by the parameters $m_{1/2}$, m_0, A_0, sign(μ) and $\tan\beta$, where $\tan\beta = v_u/v_d$ is the ratio of vacuum expectation values for the two Higgs doublets. A large top quark Yukawa coupling at the GUT scale is necessary to achieve electroweak symmetry breaking as a radiative effect [8]. There are a number of predictions that follow from the large top Yukawa coupling:

1) The Higgs miracle happens at M_Z; electroweak symmetry breaking (EWSB) occurs radiatively.

2) m_b/m_τ is correctly predicted [9-11] from $\lambda_b = \lambda_\tau$ unification [12] at the GUT scale.

3) λ_t has an infrared fixed point [10,11,13,14], predicting (see Fig. 1) $\tan\beta \simeq 1.8$ ($m_t = 200$ GeV $\sin\beta$) or $\tan\beta \simeq 56$.

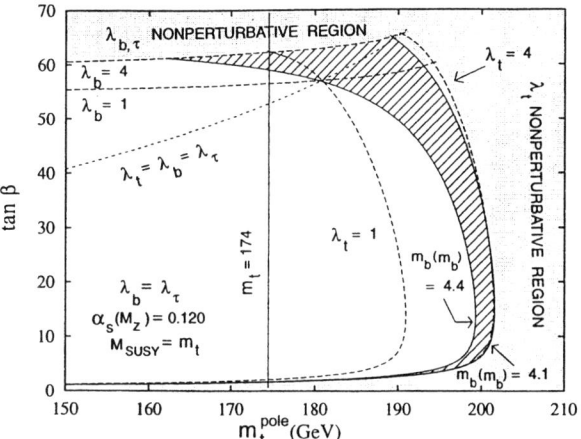

FIGURE 1. Contours of constant $m_b(m_b)$ in the $m_t(m_t)$, $\tan\beta$ plane with contours of constant GUT scale Yukawa couplings. From Ref. [10].

4) Gaugino masses at scale M_Z are given by [15]

$$M_1/\alpha_1 = M_2/\alpha_2 = M_3/\alpha_3 ,$$
$$M_1 = 0.44 m_{1/2}, \quad M_2 = 0.88 m_{1/2}, \quad M_3 = 3.2 m_{1/2} .$$

5) $|\mu|$ is large compared to M_1, M_2 at scale M_Z; this follows from renormalization group evolution from the grand unification scale and minimization of the Higgs potential [16].

6) The chargino mass matrix has the approximate form [16]

$$\mathcal{M} \sim \begin{pmatrix} M_2 & 0 \\ 0 & -\mu \end{pmatrix} \text{ in the } \begin{pmatrix} \tilde{W}^\pm \\ \tilde{H}^\pm \end{pmatrix} \text{ basis}.$$

Thus $\tilde{\chi}_1^\pm \sim \tilde{W}^\pm$ and $\tilde{\chi}_2^\pm \sim \tilde{H}^\pm$.

7) The neutralino mass matrix has the approximate form [16]

$$\mathcal{M} = \begin{pmatrix} M_1 & 0 & & \\ 0 & M_2 & & \\ & & 0 & \mu \\ & & \mu & 0 \end{pmatrix} \text{ in the } \begin{pmatrix} \tilde{B} \\ \tilde{W}^3 \\ \tilde{H}_d^0 \\ \tilde{H}_u^0 \end{pmatrix} \text{ basis}.$$

Thus $\tilde{\chi}_1^0 \sim \tilde{B}^0$ and $\tilde{\chi}_2^0 \sim \tilde{W}^3$.

8) The sparticle mass ratios are approximately in the proportions [15]

$$\tilde{\chi}_1^0 : \tilde{\chi}_2^0 : \tilde{\chi}_1^\pm : g = 1 : 2 : 2 : 7 \,.$$

$\tilde{\chi}_1^0$ is the LSP.

9) The colored particles (squarks and the gluinos) are heavier; the scalar masses depend on m_0 [16]:

$$\tilde{\chi}_1^0, \tilde{\chi}_2^0, \tilde{\chi}_1^\pm, \tilde{\ell}, h \text{ are "light"} \,,$$

$$\tilde{\chi}_3^0, \tilde{\chi}_4^0, \tilde{\chi}_2^\pm, \tilde{g}, \tilde{q} \text{ are "heavy"} \,.$$

10) The $\tilde{\chi}_1^0$ LSP is a natural candidate to explain the dark matter in the universe [17]. The relic density $\Omega_{\tilde{\chi}_1^0}$ is inversely proportional to the thermally averaged $\tilde{\chi}_1^0 \tilde{\chi}_1^0$ cross section, $\Omega_{\tilde{\chi}_1^0} h^2 \propto 1/[\langle \sigma_{\text{ann}} v \rangle]$, where h is the Hubble constant in units of 100 km/s/Mpc. The annihilation diagrams involve sfermion exchanges and Z, h^0, H^0, A^0 s-channel resonances. A cosmologically interesting LSP relic density, $0.1 \lesssim \Omega_{\tilde{\chi}_1^0} h^2 \lesssim 0.5$, singles out the following region of mSUGRA parameters [18,19]:

$$m_0 \lesssim 200 \text{ GeV}, \quad 80 \lesssim m_{1/2} \lesssim 450 \text{ GeV} \quad \text{for } \tan\beta \sim 1.8 \,,$$
$$m_0 \gtrsim 300 \text{ GeV}, \quad 500 \lesssim m_{1/2} \lesssim 800 \text{ GeV} \quad \text{for } \tan\beta \sim 50 \,.$$

The sparticle mass spectra are correspondingly constrained; see Fig. 2.

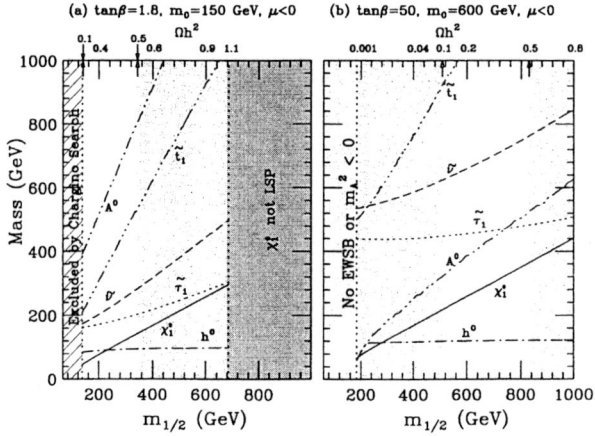

FIGURE 2. The neutralino relic density and SUSY Higgs mass spectrum versus $m_{1/2}$ for $\mu > 0$ with (a) $\tan\beta = 1.8$, $m_0 = 150$ GeV and (b) $\tan\beta = 50$, $m_0 = 600$ GeV. $\tilde{\nu}$ is the lightest scalar neutrino. The shaded regions denote the parts of the parameter space (i) producing $\Omega_{\chi_1^0} h^2 < 0.1$ or $\Omega_{\chi_1^0} h^2 > 0.5$, (ii) excluded by theoretical requirements, or (iii) excluded by the chargino search at LEP-2. From Ref. [18].

11) The mass of the lightest MSSM Higgs boson (h^0) is bounded from above. At tree level $m_h \leq M_Z |\cos 2\beta|$ [20]. Radiative corrections from top and stop loops increase the bound to $m_h \lesssim 130$ GeV [6]. In models with extra Higgs doublets and singlets that remain perturbative to the Planck scale, at least one Higgs boson has mass $m_h \lesssim 150$ GeV [21]. Thus the lightest Higgs boson is a secure target of a low-energy supersymmetry.

12) The other Higgs bosons have masses [18,22]

$$m_A \approx m_H \gg M_Z \quad \text{if} \quad \tan \beta \sim 1.8 \,,$$
$$m_A \sim \mathcal{O}(M_Z) \quad \text{if} \quad \tan \beta \gtrsim m_t/m_b \text{ or } m_0 \sim 50 \text{ GeV} \,.$$

III HIGGS PHYSICS AT A MUON COLLIDER

The production of Higgs bosons in the s-channel with interesting rates is an unique feature of a muon collider [4,23]. The resonance cross section is

$$\sigma_h(\sqrt{s}) = \frac{4\pi \Gamma(h \to \mu\bar{\mu}) \, \Gamma(h \to X)}{(\hat{s} - m_h^2)^2 + m_h^2 \left(\Gamma_{\text{tot}}^h\right)^2} \tag{1}$$

Gaussian beams with root mean square resolution down to $R = 0.003\%$ are realizable. The corresponding root mean square spread $\sigma_{\sqrt{s}}$ in c.m. energy is

$$\sigma_{\sqrt{s}} = (2 \text{ MeV}) \left(\frac{R}{0.003\%}\right) \left(\frac{\sqrt{s}}{100 \text{ GeV}}\right) \,. \tag{2}$$

The effective s-channel Higgs cross section convolved with a Gaussian spread

$$\bar{\sigma}_h(\sqrt{s}) = \frac{1}{\sqrt{2\pi}\,\sigma_{\sqrt{s}}} \int \sigma_h(\sqrt{\hat{s}}) \exp\left[\frac{-\left(\sqrt{\hat{s}} - \sqrt{s}\right)^2}{2\sigma_{\sqrt{s}}^2}\right] d\sqrt{\hat{s}} \tag{3}$$

is illustrated in Fig. 3 for $m_h = 110$ GeV, $\Gamma_h = 2.5$ MeV, and resolutions $R = 0.01\%$, 0.06% and 0.1% [4,23]. A resolution $\sigma_{\sqrt{s}} \sim \Gamma_h$ is needed to be sensitive to the Higgs width. The light Higgs width is predicted to be [23]

$$\begin{aligned}\Gamma \approx 2 \text{ to } 3 \text{ MeV} &\quad \text{if} \quad \tan\beta \sim 1.8 \\ \Gamma \approx 2 \text{ to } 800 \text{ MeV} &\quad \text{if} \quad \tan\beta \sim 20\end{aligned} \tag{4}$$

for $80 \text{ GeV} \lesssim m_h \lesssim 120 \text{ GeV}$.

At $\sqrt{s} = m_h$, the effective s-channel Higgs cross section is [4]

$$\bar{\sigma}_h \simeq \frac{4\pi}{m_h^2} \frac{\text{BF}(h \to \mu\bar{\mu}) \, \text{BF}(h \to X)}{\left[1 + \frac{8}{\pi}\left(\frac{\sigma_{\sqrt{s}}}{\Gamma_{\text{tot}}^h}\right)^2\right]^{1/2}} \,. \tag{5}$$

Note that $\bar{\sigma}_h \propto 1/\sigma_{\sqrt{s}}$ for $\sigma_{\sqrt{s}} > \Gamma_{\text{tot}}^h$. At $\sqrt{s} = m_h \approx 110$ GeV, the $b\bar{b}$ rates are [4,23]

$$\text{signal} \approx 10^4 \text{ events/fb} \qquad (6)$$
$$\text{background} \approx 10^4 \text{ events/fb} \qquad (7)$$

assuming a b-tagging efficiency $\epsilon \sim 0.5$. The effective on-resonance cross sections for other m_h values and other channels (ZZ^*, WW^*) are shown in Fig. 4 for the SM Higgs. The rates for the MSSM Higgs are nearly the same as the SM rates in the decoupling regime [24], which is relevant at $\tan\beta \sim 1.8$ in mSUGRA.

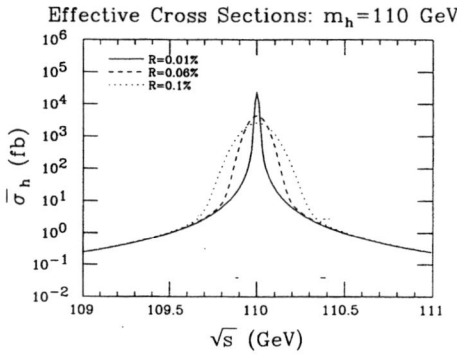

FIGURE 3. Effective s-channel higgs cross section $\bar{\sigma}_h$ obtained by convoluting the Breit-Wigner resonance formula with a Gaussian distribution for resolution R. From Ref. [4].

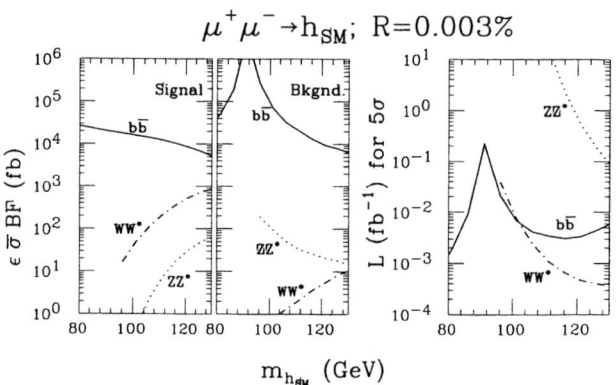

FIGURE 4. The SM Higgs cross sections and backgrounds in $b\bar{b}$, WW^* and ZZ^*. Also shown is the luminosity needed for a 5 standard deviation detection in $b\bar{b}$. From Ref. [4].

The important factors that make s-channel Higgs physics studies possible at a muon collider are energy resolutions $\sigma_{\sqrt{s}}$ of order a few MeV, little bremsstrahlung and no beamstrahlung smearing, and precise tuning of the beam energy to an accuracy $\Delta E \sim 10^{-6} E$ through continuous spin-rotation measurements [25]. As a case study we discuss $m_h \approx 110$ GeV. Prior Higgs discovery is assumed at the Tevatron (in $Wh, t\bar{t}h$ with $h \to b\bar{b}$), at the LHC (in $gg \to h$ with $h \to \gamma\gamma, 4\ell$ with a mass measurement of $\Delta m_h \sim 100$ MeV for an integrated luminosity of $L = 300$ fb^{-1}) or possibly at a NLC (in $Z^* \to Zh, h \to b\bar{b}$ giving $\Delta m_h \sim 50$ MeV for $L = 200$ fb^{-1}). A muon collider ring design would be optimized to run at energy $\sqrt{s} = m_h$. For an initial Higgs mass uncertainty of $\Delta m_h \sim 100$ MeV, the maximum number of scan points required to locate the s-channel resonance peak at the muon collider is

$$n = 2\Delta m / \sigma_{\sqrt{s}} \approx 100 \qquad (8)$$

for a resolution $\sigma_{\sqrt{s}} \approx 2$ MeV. The necessary luminosity per scan point ($L_{\text{s.p.}}$) to observe or eliminate the h-resonance at a significance level $S/\sqrt{B} = 3$ is $L_{\text{s.p.}} \sim 1.5 \times 10^{-3}$ fb^{-1}. (The scan luminosity requirements increase for m_h closer to M_Z; at $m_h \sim M_Z$ the $L_{\text{s.p.}}$ needed is a factor of 50 higher.) The total luminosity then needed to tune to a Higgs boson with $m_h = 110$ GeV is $L_{\text{tot}} = 0.15$ fb^{-1}. If the machine delivers 5×10^{30} cm^{-2} s^{-1} (0.05 fb^{-1}/year), the luminosity criteria specified for this workshop, then 3 years running would be needed. However, luminosities of order 1.5×10^{31} cm^{-2} s^{-1} are currently believed to be realizable [26], in which case only one year of running would suffice to complete the scan and measure the Higgs mass to an accuracy $\Delta m \sim 1$ MeV. Figure 5 illustrates a simulation of such a scan.

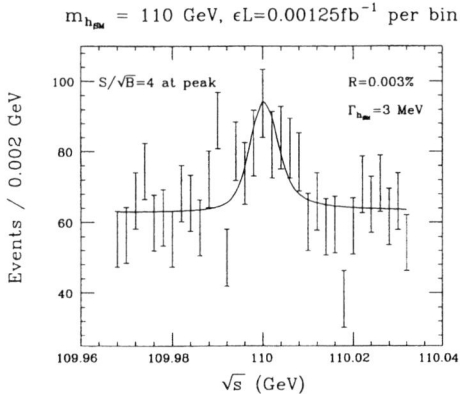

FIGURE 5. Number of events and statistical errors in the $b\bar{b}$ final states as a function of $\sqrt{s}$ in the vicinity of $m_{h_{SM}} = 100$ GeV, assuming $R = 0.03\%$. From Ref. [4].

Once the h-mass is determined to ~ 1 MeV, a 3-point fine scan [4] can be made across the peak with higher luminosity, distributed with L_1 at the observed peak position in $\sqrt{s}$ and $2.5L_1$ at the wings ($\sqrt{s} = \text{peak} \pm 2\sigma_{\sqrt{s}}$). Then with $L_{\text{tot}} = 0.4$ fb^{-1} the following accuracies would be achievable: 16% for Γ^h_{tot}, 1% for σBF($b\bar{b}$) and 5% for σBF(WW^*). The ratio $r = \text{BF}(WW^*)/\text{BF}(b\bar{b})$ is sensitive to m_A for m_A values below 500 GeV. For example, $r_{\text{MSSM}}/r_{\text{SM}} = 0.3, 0.5, 0.8$ for $m_A = 200, 250, 400$ GeV [4]. Thus, it may be possible to infer m_A from s-channel measurements of h.

The study of the other neutral MSSM Higgs bosons at a muon collider via the s-channel is also of major interest. Finding the H^0 and A^0 may not be easy at other colliders. At the LHC the region $m_A > 200$ GeV is deemed to be inaccessible for $3 \lesssim \tan\beta \lesssim 5$-10. At an NLC the $e^+e^- \to H^0A^0$ production process may be kinematically inaccessible if H^0 and A^0 are heavy. At a $\gamma\gamma$ collider, very high luminosity (~ 200 fb^{-1}) would be needed for $\gamma\gamma \to H^0, A^0$ studies. At a muon collider the resolution requirements for s-channel H^0 and A^0 studies are not as demanding as for the h, because the H^0, A^0 widths are broader; typically $\Gamma \sim 30$ MeV for $m_A < 2m_t$ and $\Gamma \sim 3$ GeV for $m_A > 2m_t$. Consequently $R \sim 0.1\%$ ($\sigma_{\sqrt{s}} \sim 70$ MeV) is adequate for a scan. A luminosity per scan point $L_{\text{s.p.}} \sim 0.1$ fb^{-1} probes the parameter space with $\tan\beta > 2$. The $\sqrt{s}$-range over which the scan should be made depends on other information available to indicate the A^0 and H^0 mass ranges of interest.

In mSUGRA with large m_A, $m_{A^0} \approx m_{H^0} \approx m_{H^\pm}$ and the degeneracy in these masses is very close for large $\tan\beta$. In such a circumstance only an s-channel scan with good resolution may allow separation of the A^0 and H^0 states; see Fig. 6.

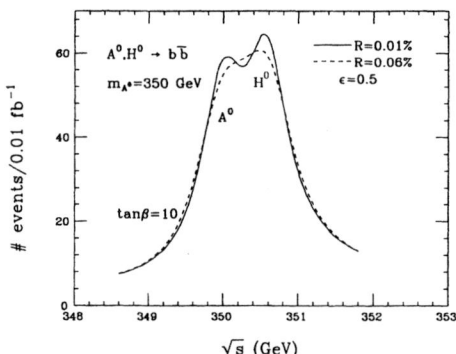

FIGURE 6. Separation of A^0 and H^0 signals for $\tan\beta = 10$. From Ref. [4].

IV THRESHOLD MEASUREMENTS AT A MUON COLLIDER

With 10 fb^{-1} integrated luminosity devoted to a measurement of a threshold cross-section, the following precisions on particle masses may be achievable [27,28]:

$$\begin{aligned}
\mu^+\mu^- &\to W^+W^- & \Delta M_W &= 20 \text{ MeV}, \\
\mu^+\mu^- &\to t\bar{t} & \Delta m_t &= 0.2 \text{ GeV}, \\
\mu^+\mu^- &\to Zh & \Delta m_h &= 140 \text{ MeV} \quad (\text{if } m_h = 100 \text{ GeV}).
\end{aligned} \quad (9)$$

Precision M_W and m_t measurements allow important tests of electroweak radiative corrections through the relation

$$M_W = M_Z \left[1 - \frac{\pi \alpha}{\sqrt{2}\, G_\mu\, M_W^2 (1 - \delta r)} \right]^{1/2}, \qquad (10)$$

where δr represents loop corrections. In the SM, δr depends on m_t^2 and $\log m_h$. The optimal precision for tests of this relation is $\Delta M_W \approx \frac{1}{140} \Delta m_t$, so the uncertainty on M_W is the most critical. With $\Delta M_W = 20$ MeV the SM Higgs mass could be inferred to an accuracy

$$\Delta m_{h_{SM}} = \pm 30 \text{ GeV} \left(\frac{m_h}{100 \text{ GeV}} \right). \qquad (11)$$

Alternatively, once m_h is known from direct measurements, SUSY loop contributions can be tested.

One of the important physics opportunities for the First Muon Collider is the production of the lighter chargino, $\tilde{\chi}_1^+$. Fine tuning arguments in mSUGRA suggest that it should be lighter than 200 GeV [29]. A search at the upgraded Tevatron for the process $q\bar{q} \to \tilde{\chi}_1^+ \tilde{\chi}_2^0$ with $\tilde{\chi}_1^+ \to \tilde{\chi}_1^0 \ell^+ \nu$ and $\tilde{\chi}_2^0 \to \tilde{\chi}_1^0 \ell^+ \ell^-$ decays can reach masses $m_{\tilde{\chi}_1^+} \simeq m_{\tilde{\chi}_2^0} \sim 170$ GeV with 2 fb^{-1} luminosity and ~ 230 GeV with 10 fb^{-1} [30]. The mass difference $M(\tilde{\chi}_2^0) - M(\tilde{\chi}_1^0)$ can be determined from the $\ell^+ \ell^-$ mass distribution.

The two contributing diagrams in the chargino pair production process are shown in Fig. 7; the two amplitudes interfere destructively [31]. The $\tilde{\chi}_1^+$ and $\tilde{\nu}_\mu$ masses can be inferred from the shape of the cross section in the threshold region [28]. The chargino decay is $\tilde{\chi}_1^+ \to f\bar{f}'\tilde{\chi}_1^0$. Selective cuts suppress the background from W^+W^- production and leave $\sim 5\%$ signal efficiency for 4 jets + $\not{E}_T$ events. Measurements at two energies in the threshold region with total luminosity $L = 50$ fb^{-1} and resolution $R = 0.1\%$ can give the accuracies listed in Table 1 on the chargino mass for the specified values of $m_{\tilde{\chi}_1^+}$ and $m_{\tilde{\nu}_\mu}$.

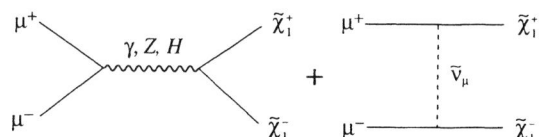

FIGURE 7. Diagrams for production of the lighter chargino.

V SUPERSYMMETRIC RADIATIVE CORRECTIONS

In unbroken supersymmetry, the SUSY gaugino couplings h_i to $\tilde{f}f$ are equal to the SM gauge couplings g_i. In broken SUSY a difference in h_i and g_i couplings

TABLE 1. Achievable uncertainties with 50 fb^{-1} luminosity on the mass of the lighter chargino for representative $m_{\tilde{\chi}_1^+}$ and $m_{\tilde{\nu}_\mu}$ masses. From Ref. [28].

$\Delta m_{\tilde{\chi}_1^+}$ (MeV)	$m_{\tilde{\chi}_1^+}$ (GeV)	$m_{\tilde{\nu}_\mu}$ (GeV)
35	100	500
45	100	300
150	200	500
300	200	300

is induced at the loop level due to different mass scales for squarks and sleptons [32–35]. The differences in the U(1) and SU(2) couplings are [34]

$$\frac{h_1 - g_1}{g_1} \simeq 1.8\% \log_{10}\left(\frac{M_{\tilde{Q}}}{m_{\tilde{\ell}}}\right), \tag{12}$$

$$\frac{h_2 - g_2}{g_2} \simeq 0.7\% \log_{10}\left(\frac{M_{\tilde{Q}}}{m_{\tilde{\ell}}}\right). \tag{13}$$

One-loop amplitudes for SUSY processes are obtained from the tree-level amplitudes by substitution of the modified couplings. The cross-sections of SUSY processes with t-channel exchanges can be enhanced up to $9\% \log_{10}\left(M_{\tilde{Q}}/m_{\tilde{\ell}}\right)$ [33]. Consequently, precision cross-section measurements can be sensitive to squarks of mass $M_{\tilde{Q}} > 1$ TeV. If the first two generations have masses in the 1 to 40 TeV range allowed by naturalness, then precision measurements could provide a way to infer squark masses beyond the kinematic reach of colliders.

Some t-channel exchange processes of interest in this regard at muon colliders are shown in Fig. 8. The technique relies on knowledge of the exchanged particle mass, which must be determined from its production processes. The muon collider advantage in the study of supersymmetric radiative corrections is the accuracy with which mass measurements can be made near thresholds.

VI HEAVY SUSY PARTICLES

Of the many scalar particles of supersymmetry (sleptons, squarks, Higgs) some may have masses of TeV scale. Study of heavy SUSY particles at the LHC will be difficult because of low event rates and high SM backgrounds. At a lepton collider pair production of scalars is p-wave suppressed. Consequently, collider energies well above threshold are necessary to have sufficient production rates; see Fig. 9. A 3 to 4 TeV muon collider offers the promise of high luminosity (~ 1000 fb^{-1}/year) that would allow sufficient event rates to reconstruct heavy sparticles from their complex cascade decay chains.

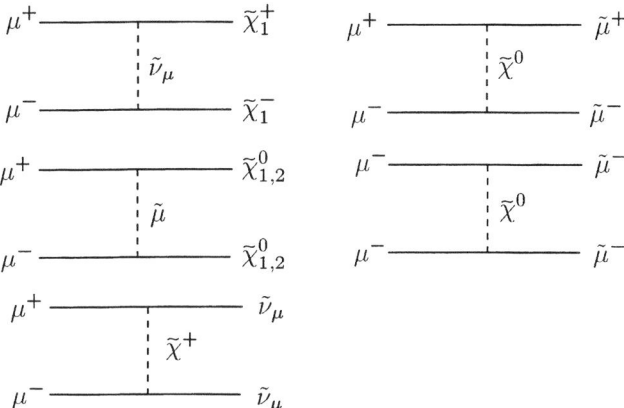

FIGURE 8. *t*-channel exchange diagrams for processes that can be enhanced by SUSY radiative corrections.

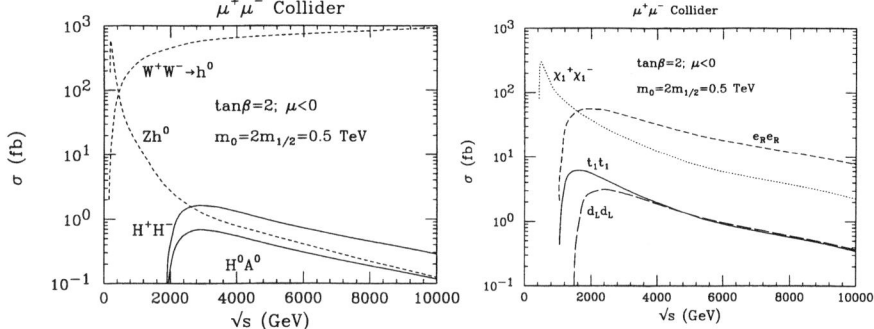

FIGURE 9. Cross sections for pair production of Higgs bosons and scalar particles at a high energy muon collider.

VII CONCLUSIONS

Muon colliders offer unique probes of supersymmetry. The *s*-channel production of Higgs bosons will precisely determine the Higgs mass (to a fraction of an MeV), directly measure the Higgs width, measure the branching fraction ratio BF($h \to WW^*$)/BF($h \to b\bar{b}$) from which m_A can be inferred if $m_A < 500$ GeV, and allow discovery and study of the A^0 and H^0 Higgs bosons. High precision threshold cross-section measurements are possible at a muon collider because of the sharp beam resolution, the suppressed bremsstrahlung and the precise tuning of the beam energy through spin-rotation measurements. Interesting possibilities for precise mass measurements at the First Muon Collider include:

$$\begin{array}{ll} W^+W^-, t\bar{t} & (M_W, m_t) \\ Zh & (m_h) \\ \tilde{\chi}^+\tilde{\chi}^-, \tilde{\chi}^0_{1,2}\tilde{\chi}^0_{1,2} & (m_{\tilde{\chi}^+}, m_{\tilde{\chi}^0_1}, m_{\tilde{\chi}^0_2}) \\ \tilde{\nu}_\mu\tilde{\nu}_\mu & (m_{\tilde{\nu}_\mu}) \\ \tilde{\ell}^+\tilde{\ell}^- & (m_{\tilde{\ell}}) \end{array} \qquad (14)$$

Precision cross-section measurements may allow tests of supersymmetric radiative corrections that may allow us to infer the existence of squarks with mass above 1 TeV. Finally, the next generation muon collider with c.m. energy of 3 to 4 TeV would provide access to heavy SUSY particles.

The bottom line is that muon colliders are a robust option for discovering the nature of supersymmetry.

ACKNOWLEDGMENTS

I would like to thank M. Berger, T. Han and C. Kao for helpful advice in the preparation of this report. This research was supported in part by the U.S. Department of Energy under Grant No. DE-FG02-95ER40896 and in part by the University of Wisconsin Research Committee with funds granted by the Wisconsin Alumni Research Foundation.

REFERENCES

1. For recent reviews of supersymmetry phenomenology, see X. Tata, Proc. of the *IX Jorge A. Swieca Summer School*, Campos do Jordão, Brazil (in press), hep-ph/9706307; R. Arnowitt and P. Nath, *Lectures presented at the VII J. A. Swieca Summer School, Campos do Jordao, Brazil, 1993* CTP-TAMU-52/93; J. Bagger in *QCD and Beyond*, Proceedings of the 1995 TASI, D. Soper, Editor (World Scientific, 1996); M. Drees, KEK-TH-501, hep-ph/9611409 (1996); S. Dawson, hep-ph/9612229 (1996); M. Dine, hep-ph/9612389 (1996); W. de Boer, Prog. Part. & Nucl. Phys. **33**, 201 (1994); V. Barger, to appear in *Proceedings of the FCP97 Workshop on Fundamental Particles and Interactions*, Vanderbilt University, 1997; S. Martin, *A Supersymmetry Primer*, hep-ph/9707356; G.F. Giudice and R. Rattazzi, CERN-TH/97-380 [hep-ph/9801271], to be submitted to Physics Reports.
2. See e.g., R. Barbieri, talk given at 18th International Symposium on Lepton-Photon Interactions (LP 97), Hamburg, Germany, 1997, hep-ph/9711232.
3. P.Langacker and J. Erler, to be published in the Particle Data Book 1998; M. Davier and A. Höcker, LAL 97-85 [hep-ph/9711308].
4. V. Barger, M.S. Berger, J.F. Gunion, and T. Han, Phys. Rep. **286**, no. 1, p. 1 (1997).
5. V. Barger, M.S. Berger, J.F. Gunion, and T. Han, to appear in *Proceedings of the Symposium on New Ideas for Particle Accelerators*, Institute for Theoretical Physics, Santa Barbara, Oct. 1996; V. Barger et al., in *Proceedings of Second Workshop on Physics Potential and Development of $\mu^+\mu^-$ Colliders*, Sausalito, California, 1994,

ed. by D. Cline, American Institute of Physics Conference Proceedings 352, p. 55 (AIP, New York, 1996).
6. M. Carena, J.R. Espinosa, M. Quiros, and C.E.M. Wagner, Phys. Lett. B **355**, 209 (1995); M. Quiros and C.E.M. Wagner, Nucl. Phys. **B461**, 407 (1996); M. Carena et al., hep-ph/9602250, in Vol. 1, Report of the Workshop on Physics at LEP2, ed. by G. Altarelli, T. Sjostrand, and F. Zwirner, CERN-96-01; H.E. Haber, R. Hempfling, and A.H. Hoang, Z. Phyz **C75**, 539 (1997); J.A. Casas, J.R. Espinosa, and H.E. Haber, hep-ph/9801365.
7. See e.g., C. Kolda, *Proceedings of the 5th International Conference on Supersymmetries in Physics* (SUSY 97), Philadelphia, May 1997.
8. K. Inoue, A. Kakuto, H. Komatsu, and S. Takeshita, Prog. Theor. Phys. **689**, 927 (1982); L Alverez-Gaume, J. Polshinski and M.B. Wise, Nucl. Phys. **B221**, 495 (1983); J. Ellis, J.S. Hagelin, D.V. Nanopoulos, and K. Tamvakis, Phys. Lett. **B125**, 275 (1983); L.E. Ibañez and C. Lopez, Nucl. Phys. **B233**, 511 (1984); L.E. Ibañez, C. Lopez, and C. Muñoz, Nucl. Phys. **B256**, 218 (1985).
9. H. Aronson et al., Phys. Rev. Lett. **67**, 2933 (1991); Phys. Rev. **D46**, 3945 (1992).
10. V. Barger, M.S. Berger, and P. Ohmann, Phys. Rev. **D47**, 1093 (1993).
11. V. Barger, M.S. Berger, and P. Ohmann, Phys. Rev. **D49**, 4908 (1994); V. Barger, M.S. Berger, P. Ohmann and R.J.N. Phillips, Phys. Lett. **B314**, 351 (1993).
12. M.S. Chanowitz, J. Ellis and M.K. Gaillard, Nucl. Phys. **B128**, 506 (1997).
13. C.T. Hill, Phys. Rev. **D24**, 691 (1981); B. Pendleton and G.G. Ross, Phys. Lett. **B98**, 291 (1981); C.D. Froggatt, R.G. Moorhouse, Phys. Lett. **B298**, 356 (1993).
14. J. Bagger, S. Dimopoulos, and E. Masso, Phys. Rev. Lett. **55**, 9201 (1985); P. Langacker and N. Polonsky, Phys. Rev. **D50**, 2199 (1994); W.A. Bardeen, M. Carena, S. Pokorski, and C.E.M. Wagner, Phys. Lett. **B320**, 110 (1994); M. Carena, M. Olechowski, S. Pokorski, and C.E.M. Wagner, Nucl. Phys. **B419**, 213 (1994); M. Carena and C.E.M. Wagner, Nucl. Phys. **B452**, 45 (1995); B. Schrempp, Phys. Lett. **B344**, 193 (1995); B Schrempp and W. Wimmer, hep-ph/9606386.
15. S. Martin and P. Ramond, Phys. Rev. **D48**, 5365 (1995) and references therein.
16. J. Ellis and F. Zwirner, Nucl. Phys. **B338**, 317 (1990); G. Ross and R.G. Roberts, Nucl. Phys. **B377**, 571 (1992); R. Arnowitt and P. Nath, Phys. Rev. Lett. **69**, 725 (1992); M. Drees and M. Nojiri, Nucl. Phys. **B369**, 54 (1993); S. Kelley et al., Nucl. Phys. **B398**, 3 (1993); M. Olechowski and S. Pokorski, Nucl. Phys. **B404**, 590 (1993); V. Barger, M.S. Berger, and P. Ohmann, Phys. Rev. **D49**, 4908 (1994); G. Kane, C. Kolda, L. Roszkowski, and J. Wells, Phys. Rev. **D49**, 6173(1994); D.J. Castaño, E. Piard and P. Ramond, Phys. Rev. **D49**, 4882 (1994); W. de Boer, R. Ehret and D. Kazakov, Z. Phys. **67**, 647 (1995); H. Baer, M. Drees, C. Kao, M. Nojiri, and X. Tata, Phys. Rev. **D50**, 2148 (1994); H. Baer, C.-H. Chen, R. Munroe, F. Paige, and X. Tata, Phys. Rev. **D51**, 1046 (1995).
17. See e.g., G. Jungman, M. Kamionkowski and K. Griest, Phys. Rep. **267**, 195 (1996) and references therein.
18. V. Barger and C. Kao, University of Wisconsin-Madison report no. MADPH-97-992 (1997) [hep-ph/9704403], to be published in Phys. Rev. D.
19. H. Baer and M. Brhlik, Phys. Rev. D **53**, 597 (1996); R. Arnowitt and P. Nath, Phys. Lett. B **299**, 58 (1993); B **307**, 403(E) (1993); Phys. Rev. Lett. **70**, 3696

(1993); Phys. Rev. D **54**, 2374 (1996); J.L. Lopez, D.V. Nanopoulos and K. Yuan, Phys. Rev. D **48**, 2766 (1993); M. Drees and A. Yamada, Phys. Rev. D **53**, 1586 (1996); J. Ellis, T. Falk, K.A. Olive and M. Schmitt, Phys. Lett. B **388**, 97 (1996); CERN Report No. CERN-TH-97-105, hep-ph/9705444; M. Drees and M.M. Nojiri, Phys. Rev. D **47**, 376 (1993).
20. J.F. Gunion, H.E. Haber, G.L. Kane, and S. Dawson, *The Higgs Hunters Guide*, (Addison-Wesley, 1990).
21. C. Kolda, G. Kane, and J. Wells, Phys. Rev. Lett. **70**, 2686 (1993).
22. H. Baer, C-h. Chen, M. Drees, F. Paige, and X. Tata, Phys. Rev. Lett. **79**, 986 (1997).
23. V. Barger, M.S. Berger, J.F. Gunion and T. Han, Phys. Rev. Lett. **75**, 1462 (1995).
24. H. Haber and Y. Nir, Nucl. Phys. **B335**, 363 (1990).
25. R. Raja, these proceedings.
26. R. Palmer, talk presented at the 4th International Conference on Physics Potential and Development of $\mu\mu$ Colliders, San Francisco, Dec. 1997.
27. V. Barger, M.S. Berger, J.F. Gunion and T. Han, Phys. Rev. **D56**, 1714 (1997); Phys. Rev. Lett. **78**, 3991 (1997).
28. V. Barger, M.S. Berger, and T. Han, University of Wisconsin-Madison report MADPH-98-1036 [hep-ph/9801410].
29. G.W. Anderson and D.J. Castano, Phys. Rev. **D52**, 1693 (1995); K.L. Chan, U. Chattopadhyay, P. Nath, hep-ph/9710473.
30. H. Baer, Chen, C. Kao, and X. Tata, Phys. Rev. **D52**, 1565 (1995).
31. J.L. Feng and M.J. Strassler, hep-ph/9408359; J.L. Feng, M. Peskin, H. Murayama, and X. Tata, hep-ph/9502260.
32. P.H. Chankowski, Phys. Rev. **D41**, 2877 (1990); P.H. Chankowski and S. Pokorski, hep-ph/9707497.
33. M.M. Nojiri, K. Fujii, and T. Tsukamoto, Phys. Rev. **D54**, 6756 (1996); M.M. Nojiri, D.M. Pierce, and Y. Yamada, hep-ph/9707244.
34. H.-C. Cheng, J.L. Feng, and N. Polonksy, hep-ph/9706476; hep-ph/9706438.
35. M.A. Diaz, S.F. King, and D.A. Ross, CERN-TH-98-26 [hep-ph/9801373].
36. V. Barger, M.S. Berger, J.F. Gunion, and T. Han, in *Proceedings of the Symposium on Physics Potential and Development of $\mu^+\mu^-$ Colliders*, San Francisco, CA (1995), ed. by D. Cline and D. Sanders, Nucl. Phys. B (proc. suppl.) **51A**, 13 (1996).

Physics with Low Energy Hadrons

Gaston Gutierrez[a], Laurence Littenberg[b]

(a) Fermi National Accelerator Laboratory
Batavia, IL 60510

(b) Brookhaven National Laboratory
Physics Department,
Upton NY, 11973, USA

Abstract. The prospects for low energy hadron physics at the front end of a muon collider are discussed.

I INTRODUCTION

The front end of a muon collider as conceived for the purposes of this workshop, is pretty close to the classical idea of a kaon factory. For example, the late lamented KAON [1] was to have been a $30\,GeV$, $100\mu A$ machine. This is to be compared with $16\,GeV$, $60\mu A$ for the machine under discussion. Table 1 shows how this facilities compares with other sources extant, under construction or proposed.

TABLE 1. Front end of the muon collider compared with other multi-GeV fixed target proton sources. 'TP' means trillion protons.

Machine:	AGS	AGS'	FMI	JHF	FMCFE
$p(GeV/c)$:	25	25	120	50	16
Duty factor:	0.33	0.27	0.33	0.16	0.90
TP/sec:	20	30	10	60	400
average forward K_L:	$1.3\,10^9$	$2\,10^9$	10^9	$3.8\,10^9$	$25\,10^9$
2 body acceptance:	.02	.02	.10	.04	.013
"K_L sensitivity":	26	40	100	150	325
K^+ stop:	12	18	8	47	210

AGS' is the expected performance of the AGS in 2000. FMI and JHF indicate the design parameters of the Fermilab Main Injector and the Japan Hadron Facility 50 GeV PS. Most of the entries are obvious, but there are a few slightly obscure measures of usefulness. The average forward K_L are the number of expected K_L per second in a 'typical' modern $0°$ beam. Where possible this is guided by actual experience at similar energy facilities. Similar remarks apply to the two body acceptance entries, although these particular numbers may be slightly unfair to the lower energy accelerators. The K_L sensitivity row is simply the rescaled product of the two rows above it. These numbers give some idea of the relative reach of the accelerators for studying two-body K_L decays. The last row gives a relative measure of the stopping K^+ intensity possible, assuming the use of a $0°$ separated beam. None of the entries in the table have any account of subtleties like background rejection, but they give a rough idea of the situation. The front end of a muon collider has the potential to push certain experiments beyond what can be done, even at the most intense facilities now being planned. Of course one has to do a lot of work to establish whether this is true for any particular experimental target.

II HADRON PHYSICS

'Hadron physics' covers a lot of ground, from subjects deep in the bosom of nuclear physics to ones still generally classified as particle physics. However the line is always shifting monotonically so that more and more of this area is considered nuclear. The fact that it is generally on the border between these two fields has led to problems. Unlike political entities where border territories are jealously competed for, in physics, the border enclaves tend to suffer from neglect. This has led to a lot of people being dispossessed. Gregg Franklin [2] gave an excellent summary of a number of these topics, so we can afford to give most of them short shrift in this report. In our working group we had talks by Kam Seth and Hal Spinka. The former noted that there's about an order of magnitude advantage of the FMC front end over the AGS for K^- and $\bar{p}$ production below about 5 GeV/c. This is quite inspiring to workers in hadron physics. A very interesting use for such enhanced flux was advocated by Hal Spinka.

A Polarized anti-protons

Spinka reported on an idea for making and exploiting a polarized anti-proton beam. It is based on the observation that $\mathcal{O}(1\ GeV/c)$ anti-protons elastically scattered off protons at finite angle are observed to be polarized, at levels up to 50% [3]! However most of the cross section is at small t, where the polarization is rather smaller. Nonetheless quite respectable polarizations can be achieved in this way. Figure 1 shows a conceptual drawing of such a polarized $\bar{p}$ beam.

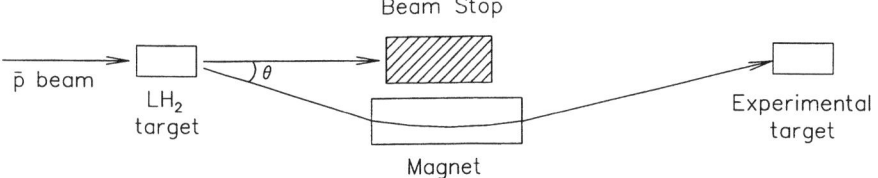

FIGURE 1. Polarized proton beam.

The spin of the $\bar{p}$ will be perpendicular to the scattering plane, and the magnetic field direction is such that it does not precess. To maximize the flux, Spinka envisioned a toroidal geometry with the acceptance centered around $-t \sim 0.12\ GeV/c$. He made a Monte Carlo simulation trying to stick to practical (but not fully optimized) design parameters. For incoming $\bar{p}$'s with $\Delta p/p = \pm 5\%$, spot size = ± 1 cm, and divergence = ± 5 mr, a 10 cm liquid hydrogen target, and defining good events as those with a trajectory passing within ± 2 cm of the beam radius and ± 5 cm in z at the experimental target, he found an $\bar{p}$ intensity of 2×10^{-4} per incident $\bar{p}$. The average polarization was 20%. This is clearly one of the programs that could benefit from the full intensity of the FMC front end. Using $375 TP$ of $16 GeV/c$ protons, one should be able to make on the order of 50,000 polarized $\bar{p}$'s/second in this way.

There is quite an extensive menu of physics that could be done with such a facility. There are five $\bar{N}N$ elastic amplitudes and two isospin states; ≥ 20 spin parameters must be measured at each angle and energy for a full amplitude determination. Using a polarized proton target, one could measure P, C_{NN}, C_{SS}, C_{LL}, and C_{SL} for $\bar{p}p \to \bar{p}p$ and $\bar{p}p \to \bar{n}n$. Using quasielastic scattering from a polarized deuterium target, one could measure these same quantities for $\bar{p}n \to \bar{p}n$. Other measurements that could be made simultaneously are $\Delta\sigma_L(\bar{p}p)$ and $\Delta\sigma_T(\bar{p}p)$ for $\bar{p} \to \pi^+\pi^-, K^+K^-$ and other reactions.

Other areas where high-quality antiproton beams would be welcome are the study of $\bar{p}$ forward scattering parameters, and the time-like form-factor of protons.

B Proton-induced reactions

Kam Seth showed us data from tests of QCD scaling laws, where the ratio of $\frac{d\sigma}{dt}$ was divided by the expected s^{10} factor. The ratio exhibits fascinating oscillations when plotted against $ln(s)$. This is said to be related to the phenomenon of color transparency, another possible target of studies at the FMC front end. Both these kinds of studies were dropped rather than completed by high energy physicists in the past. The problems they addressed were not really solved, but were victims of an insufficiently long attention span.

Another subject discussed by Seth was parity non-conservation in polarized pp interactions. This is allowed by interference between strong and weak interactions, but is predicted to be very small: $|A_L| \equiv |(\vec{\sigma} - \overleftarrow{\sigma})/(\vec{\sigma} + \overleftarrow{\sigma})| \approx 10^{-7}$. This is indeed found to be the case at low energy, but there is one high energy (6 GeV/c) measurement from Argonne [4] which gives $A_L = (26.5 \pm 6.0 \pm 3.6) \times 10^{-7}$. Obviously confirmation is needed, and indeed the entire range above 1 GeV/c should be mapped out. This is an example of a very provocative result that has not been followed up.

C Spectroscopy

Both Seth and Spinka talked about spectroscopy. There has been renewed interest in baryon spectroscopy, mainly because of the advent of new facilities, CLAS at TJNAF and the Crystal Ball at the AGS. The latter program will very probably end in 1999, largely closing the door to the use of hadronic probes in this area. Since the use of hadronic and leptonic probes are complementary, this represents a difficulty for the field, and the loss of a good opportunity. The baryon spectrum needs to be better nailed down. Very basic problems have to be addressed. These include the effective degrees of freedom (3 quarks? quark-diquark?...), how the gluon degrees of freedom are exhibited in the baryon spectrum, the presence or absence of exotic baryons, and the origin of the apparent clustering of baryon states.

Spinka recommended a long program based on two detectors. One would specialize in all-neutral states and the other would have large solid angle acceptance with momentum analysis for charged particles. The latter might include neutrals detection at some level. Ideally, the program would include polarized target measurements. For the most part, this program does not require a large fraction of the flux available at the FMC front end.

Seth discussed meson resonances. He mentioned the recent observation of a candidate for an exotic meson at BNL [5]. There are also of course candidates for glueballs. No type of candidate is exactly where the theorists would like it, but they are probably not out of reach of revisions to the theory. However even assuming theory embraces these objects, there is still a raft of other predicted objects to be found. These include glueballs of higher mass and spin, and strangeonium hybrids in the 2 GeV region that would be relatively narrow. All could profitably be studied at the FMC front end, and for the most part using only a small fraction of the available protons.

D Some General Comments on Hadronic Physics

There's something about this area that makes high energy physicists uncomfortable; maybe it reminds us of unfinished business that we dropped in the rush to the frontier. The more patient intermediate energy types are happy to clean up after us, if only we give them the chance.

Although it is clear there are subjects in this area that require the full intensity of the FMC front end, most can make a lot of progress using only a small fraction of this flux. It's more a matter of having good beamlines and detectors and reliable running time. In other words, they mainly need a home.

III K DECAYS

Certainly one of the most compelling area of physics that could be addressed by a machine with the parameters under discussion is K decay, although this may not be true by the time it is actually built. Most of the discussion in our working group concerned this area.

A $K \to \pi \nu \bar{\nu}$

The most interesting subject in K decays these days is the pursuit of the GIM-suppressed flavor-changing neutral current processes $K \to \pi \nu \bar{\nu}$. In these decays short distance effects are not tiny corrections to a large leading order term, but totally dominate the rate. Long distance contributions are negligible [6], and hadronic matrix elements can be calculated to $\sim 1\%$ accuracy from the rate of the common $Ke3$ decay [7]. In the Standard Model, the amplitudes are dominated by terms proportional to V_{td} [8], a crucial quantity not easy to measure. The charged mode is sensitive to $|V_{td}|$. A next-to-leading-logarithmic order calculation of QCD corrections has been done [9], and it is known that $B(K^+ \to \pi^+ \nu \bar{\nu})$ can give $|V_{td}|$ to 5%, assuming that other SM quantities such as m_t are tied down. Under broad assumptions [10], the neutral mode is essentially a pure CP-violating transition, with a completely negligible indirect (ϵ) component [11]. Unlike the charged mode, there is essentially no charm contribution. A measurement of its rate would yield an unambiguous determination of η, modulo m_t, etc. Combining measurements of the neutral and charged rates determines the unitary angle β, independent of data from the B system [12]. Figure 2 show the relationship between the unitarity triangle and the two kaon FCNC rates. The current ranges of prediction for $B(K^+ \to \pi^+ \nu \bar{\nu})$ and $B(K_L \to \pi^0 \nu \bar{\nu})$ are is $(0.6 - 1.5) \times 10^{-10}$ and $(1 - 3) \times 10^{-11}$ respectively. The uncertainty in each case is given almost entirely by lack of knowledge of the input parameters. These decays compare very well in theoretical cleanliness with those measurements in the B system that have been widely advocated for determining the angles of the unitarity triangle

Besides measuring the magnitude and phase of V_{td} with unique "cleanliness", and with systematics completely different from those of B experiments, it has lately been emphasized that to understand the effects of possible new physics beyond the Standard Model in the B system, it will be essential to measure $K \to \pi \nu \bar{\nu}$ [10,13] as well.

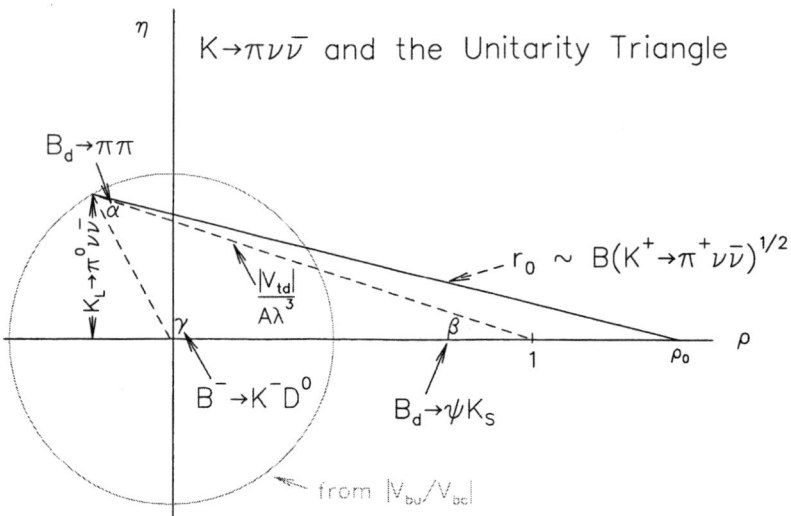

FIGURE 2. Diagram illustrating the relationship of the charged and neutral FCNC kaon decay $K \to \pi\nu\bar{\nu}$ rates to the unitarity triangle. The height of the triangle is proportional to $B(K_L \to \pi^0 \nu\bar{\nu}$.

1 Experimental status and prospects of $K \to \pi\nu\bar{\nu}$

For more than ten years, the E787 collaboration at the AGS has been pursuing $K^+ \to \pi^+ \nu\bar{\nu}$, using a solenoidal magnetic spectrometer in a stopping K^+ beam. This group recently published evidence for the first observation of this decay [14]. The corresponding branching ratio was $(4.2^{+9.7}_{-3.5}) \times 10^{-10}$, consistent with the above-mentioned SM range. E787 has collected data corresponding to about 2.5 times that of the sample containing the first event, and plans to continue to run at least through 1999. This should allow the observation of a few events at the Standard Model level. Beyond this, a proposal for continuing the study of $K^+ \to \pi^+ \nu\bar{\nu}$ into the AGS-2000 era is being prepared [15]. The intention is to collect 15 – 20 events at the SM level. Work is also in progress on a proposal to study this decay in an in-flight geometry at the Fermilab Main Injector [16].

There have as yet been no dedicated searches for $K_L \to \pi^0 \nu\bar{\nu}$, but the KTeV group at Fermilab has recently reported a preliminary result from a special one-day run in a configuration customized for this decay [17]: $B(K_L \to \pi^0 \nu\bar{\nu}) < 1.8 \times 10^{-6}$ at 90% c.l. This group expects to reach the level of a few times 10^{-8} by 1999 [18]. Thereafter, they plan to reconfigure and upgrade their apparatus for working at the Main Injector. They have an Expression of Interest for an experiment aimed at collecting several tens of $K_L \to \pi^0 \nu\bar{\nu}$ events [19]. There is also an approved AGS proposal [20] for an experiment scoped to collect ~ 70 events, which will be discussed below. In addition, there is an approved proposal to search for this decay at KEK [21]

2 $K^+ \to \pi^+ \nu \bar{\nu}$ at the front end of the FMC

Fig 3 shows the apparatus [22] of AGS Experiment 787, a solenoidal spectrometer situated in a ~ 700 MeV/c separated K^+ beam. About 7×10^6 K^+ per AGS spill enter the detector, accompanied by about 2×10^6 pions and muons. The beam strikes a BeO degrader and approximately one quarter of the K^+ penetrate it unscathed and stop in a highly segmented scintillating fiber target. After a $2\,ns$ delay, the detector becomes sensitive to unaccompanied pions exiting the target transversely. These are momentum analyzed by a small, low-mass drift chamber immersed in a $1\,T$ magnetic field, and penetrate a cylindrical array of scintillators and straw chambers ("range stack"), in which they come to rest. The range stack scintillators are read out at both ends by photomultipliers instrumented with 500 MHz, 8-bit transient digitizers. These are used to detect the characteristic $\pi \to \mu \to e$ decay chain. This distinguishes pions very effectively from muons which lack the first step in the chain. An important design principle of the experiment was the minimization of "dead" material, allowing the use of the comparison of range, momentum, and kinetic energy as a powerful means of particle identification. The kinematic and life-cycle methods of particle identification can be used in turn to establish each other's rejection power. Excellent muon rejection power is needed because a major background to $K^+ \to \pi^+ \nu \bar{\nu}$ is $K^+ \to \mu^+ \nu$, whose rate is almost ten orders of magnitude larger than that of the signal.

Surrounding the range stack is a cylindrical array of lead-scintillator shower counters (the "barrel veto") and plugging the upstream and downstream ends of the detector are pure CsI endcap photon vetoes. In addition there are a number of supplementary vetoes in the beam direction. These complete a hermeticity that achieves a 10^6 rejection of π^0's. This is necessary since a second major background to $K^+ \to \pi^+ \nu \bar{\nu}$ is $K^+ \to \pi^+ \pi^0$. The background-rejection power of the experiment has proved quite adequate to reach the Standard Model level of sensitivity.

The main limitation on the experiment is instantaneous detector rate. This leads to both random veto losses and eventually to problems with background rejection. However to the extent that additional protons are available, one can make an immediate gain in sensitivity/hour through increasing the duty factor of the AGS (currently 44%), by extending the flat-top (currently 1.6 seconds every 3.6 seconds). The sensitivity of the experiment increases proportionately, and no improvement in detector performance is required. One can also reduce the momentum of the beam, so that more of the incident K^+ actually decay in the target. This fraction is currently only about 25%. Since the detector rates are proportional to the flux of K^+ impinging on the BeO degrader, but the sensitivity is proportional to the flux of K^+ penetrating it and stopping in the target, this will clearly help. Both increasing the duty factor and reducing the beam momentum require using more of the AGS protons. However, since the experiment uses only about 25% of the presently available proton flux, and the AGS intensity is expected to rise over the next couple of years, significant advances seem quite possible.

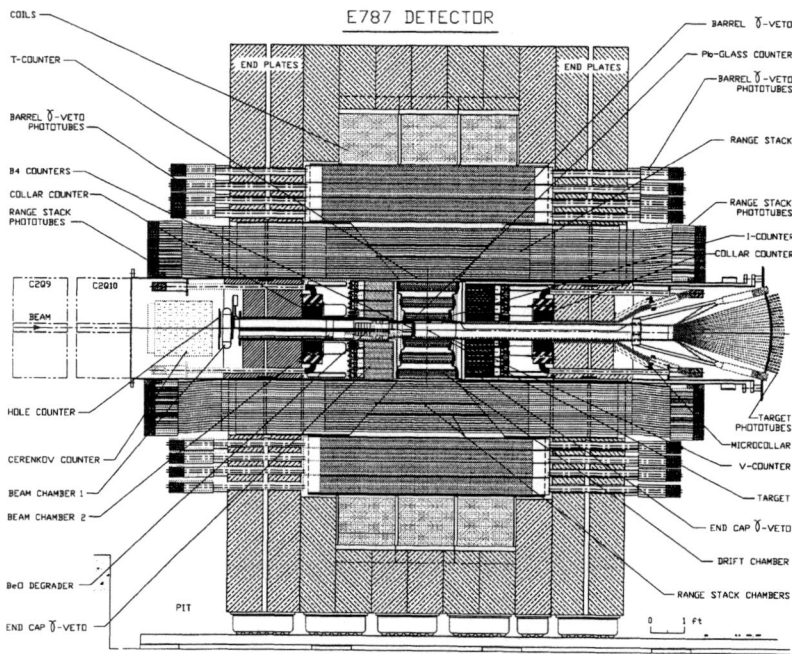

FIGURE 3. E787 detector, mounted in a 1-T solenoid. A $\sim$ 700 MeV/c K^+ beam enters from the left, slows down in a BeO degrader and stops in a highly-segmented scintillating fiber target. Decay π^+ are momentum analyzed by a cylindrical drift chamber and stop in an array of scintillation counters and straw chambers. A barrel lead-scintillator array and CsI (pure) endcaps complete an hermetic photon veto.

Now as mentioned above, there are other improvements under study for the AGS-2000 time scale. All would be applicable to the front end of the First Muon Collider. We should say at the outset that for a low energy forward beam like that of E787, very little K^+ flux is lost in reducing the primary proton energy from the AGS's current $24\,GeV/c$ to the $16\,GeV/c$ of the FMC front end. Table 2 shows a list of possible expedients that could be applied to push the stopping K^+ technique at a higher intensity machine. The units of primary proton intensity shown are TP, *i.e.* trillion protons. The AGS provides a total of about $60\,TP$ per cycle at the moment, we assume that the front end of the First Muon Collider will provide $375\,TP/second$. The potential increase in flux is more than a factor 20, since the AGS pulses only once every 3.6 seconds, whereas the new machine would be practically DC. Note that in Table 2, not quite all the available protons are used.

TABLE 2. $K^+ \to \pi^+ \nu \bar{\nu}$ from E787 to FMC front end

	sensitivity/year	protons required
How we think we're doing lately:	2×10^{-10}	$15\,TP/cycle$
Max spill, double year (to 30wks):	6×10^{-11}	$50\,TP/cycle$
Reduce beam p, use $\pi\nu\bar{\nu}2$:	2×10^{-11}	$100\,TP/cycle$
Go to MCFE (d.f. $0.73 \Rightarrow 0.9$):	1.7×10^{-11}	$25\,TP/sec$
Further reduce beam p:	1.3×10^{-11}	$50\,TP/sec$
Drop e from $\pi \Rightarrow \mu \Rightarrow e$:	9.5×10^{-12}	$50\,TP/sec$
30 weeks $\Rightarrow$ 45 weeks/year:	6.4×10^{-12}	$50\,TP/sec$
Speed up vetoes:	3.2×10^{-12}	$100\,TP/sec$
Reduce Δp, increase geom. acc.:	2.5×10^{-12}	$300\,TP/sec$
Better beam/tgt instrumentation:	1.6×10^{-12}	$300\,TP/sec$
Improved stopping cntr technology:	1.0×10^{-12}	$300\,TP/sec$

Table 2 starts from E787's best guess as to current sensitivity per running year, which is optimistically taken to be 15 weeks long. The second line is the result of running twice as long, and of extending the spill by a large factor (improving the duty factor). The latter costs more than a factor 3 in proton current. The third line assumes that one reduces the beam momentum from the present 700 MeV/c to about 550 MeV/c, and also that one can exploit a large region of phase space that we have not yet accessed. This region corresponds to π^+ with momentum below that of the π^+ from the $K^+ \to \pi^+\pi^0$ background reaction (i.e. $p < 205/ > MeV/c$). This possibility is under study at the moment. If successful, it would allow one to collect about 5 Standard Model events per year, which is the goal of the AGS-2000 initiative. Going to the next line, one enters the world of the front end of the First Muon Collider. One immediately gets a small but significant improvement from the increased duty factor. The availability of so many more protons tempts one to further reduce the beam momentum, to get another small factor. Then, one can try to to drop the electron requirement from the $\pi \to \mu \to e$ decay chain criterion. This reduces the cut and deadtime losses significantly, but it requires a compensating improvement in the kinematic rejection of $K^+ \to \mu^+\nu$ events by about a factor 10. It is thought this can be obtained by upgrading the drift chamber. The next line assumes that one can run for 45 weeks/year at the front end of the First Muon Collider. Why not, since this is a virtual machine? At this point, one is collecting about 15 events/year assuming the central value of the Standard Model predicted range of branching ratio is correct. To make further progress, it is necessary to make major improvements to the detector. Note that one gets pretty far without this!

The next factor of two comes from speeding up the veto counters by a factor two. This would be achieved by replacing the current veto counter technology, and improving the electronics. The time resolution of the present vetoes is not state of the art, so this can certainly be accomplished if the resources are made available. Once the veto gates can be cut in half, one can turn up the wick by a factor two. The next small factor comes from reducing the beam momentum spread by a factor three (one has to compensate for this by increased proton flux), and reconfiguring the apparatus to have better geometrical acceptance. The last two factors come from improving the beam and target instrumentation (whose space and time resolutions could certainly be improved), thus reducing random veto and cut losses, and finally, replacing the present stopping counter technology by something faster, brighter and more granular. This brings one to 10^{-12}/event or $\sim$ 100 SM events/year, which is about as far as any technique so far proposed, and probably about as far as one needs to go until present theoretical uncertainties are reduced.

In our session there was a talk by Bob Tschirhart on the CKM initiative [16]. This is a possible FMI experiment in which $K^+ \to \pi^+ \nu \bar{\nu}$ is studied in flight using a 22.8 GeV/c RF separated beam. This technique turns out to be highly optimized for the high energy regime, and so is not directly adaptable to the FMC front end. However it is quite relevant to the subject at hand because the sensitivity goal of CKM is very similar to that on the bottom line of Table 2. This if CKM is successfully completed in a timely fashion, it may not make sense to pursue $K^+ \to \pi^+ \nu \bar{\nu}$ at the FMC front end by the incremental technique described above. The virtue of that technique is that it is rather well understood. However if the state of the art at the point the FMC front end is ready as moved beyond 10^{-12}, a more aggressive (and imaginative) approach will have to be undertaken. This assumes that advances in theory make higher precision worthwhile.

3 $K_L \to \pi^0 \nu \bar{\nu}$ at the front end of the FMC

Fig. 4 shows a conceptual drawing of a detector [20] proposed to search for $K_L \to \pi^0 \nu \bar{\nu}$ at AGS-2000. It is assumed that when the RHIC collider comes online, the AGS will be free at least 20 hours a day for fixed target experiments. At that point, the available proton flux is expected to be 10^{14} per acceleration cycle. Using about half the available flux, in 80 weeks of running time, on the order of 70 $K^0 \to \pi^0 \nu \bar{\nu}$ events could be recorded with a background contamination of less than 10 events. This would allow a precision on η of < 10% (modulo uncertainty in $|V_{cb}|$).

The principles of the experiment are as follows. First, the neutral beam is extracted at quite a large angle ($\sim 45^o$) so that both the neutron and kaon momentum spectra are quite soft. This minimizes the flux of neutrons that can produce π^0's through interactions with vacuum windows or residual gas. To further suppress background from this source, a vacuum of 10^{-7} Torr must be maintained throughout the beam region. Second, the beam is made highly asymmetric and very carefully collimated. Third, the AGS proton beam is microbunched on extraction with a period of $\sim 40\,ns$. The bunch width is ≤ 200 ps, allowing time-of-flight measurement to determine the neutral kaon's momentum. With this time bunching technique, the massless and other fast debris from the primary target interaction arrive at the detector before the kaons of interest, and so can be vetoed. Fourth, the detector incorporates active pre-radiators that measure the direction of the photons from the $K_L \to \pi^0 \nu \bar{\nu}$ decay. In conjunction with a high resolution calorimeter, this allows one to fully reconstruct the π^0, independent of any assumptions about the beam. Combined with the beam timing information, this allows one to transform the π^0 into the K_L center of mass. Pi-zeros from the major background to $K_L \to \pi^0 \nu \bar{\nu}$, $K_L \to \pi^0 \pi^0$, have a unique energy in this system and so can be recognized. The fifth major requirement is hermetic photon vetoing. Extrapolating from photon vetoing performance achieved in E787, it is estimated that an average single γ rejection of $10^4 : 1$ is possible.

The independent kinematic and photon vetoing of $K_L \to \pi^0 \pi^0$ background allow the power of each technique to be measured. This kind of redundancy is essential in measuring a rare decay mode with such a poor signature. With proper kinematic and vetoing selection, it should be possible to suppress the $K_L \to \pi^0 \pi^0$ background to $\leq 10\%$ of the signal.

Other potential backgrounds are $K_L \to \gamma\gamma$, $K_L \to \pi^- e^+ \nu$, with the e^+ annihilating and the π^- undergoing charge exchange before they are detected, $\Lambda \to \pi^0 n$, and accidentals. These backgrounds have been calculated to contribute to less than 1 event each after 80 weeks of AGS 2000 running time.

Intensive simulation, design, prototype, and beam test work are underway on E926. However since the experiment is not yet built, much less run, any extrapolation to the front end of the First Muon Collider must be far more cautious than in the case of E787. Table 3 shows a possible progression.

TABLE 3. $K_L \to \pi^0 \nu \bar{\nu}$ from E787 to FMC front end

	sensitivity/year	protons required
Nominal estimate of E926:	1.2×10^{-12}	$50\,TP/cycle$
MCFE: Comfort factors/d.f.=0.9:	1×10^{-12}	$50\,TP/sec$
Longer beam		
Filter		
Tune angle/aperture		
Shorter decay volume, smaller beam:	5×10^{-13}	$200\,TP/sec$
Better time response, double rate:	3×10^{-13}	$375\,TP/sec$

There would be an immediate small factor as one exploited the 90% duty factor of the First Muon Collider front end. It would probably be wise to use the next factor of beam on what are labeled "comfort factors" in Table 3. These include a longer beam line for better time resolution and collimation, a

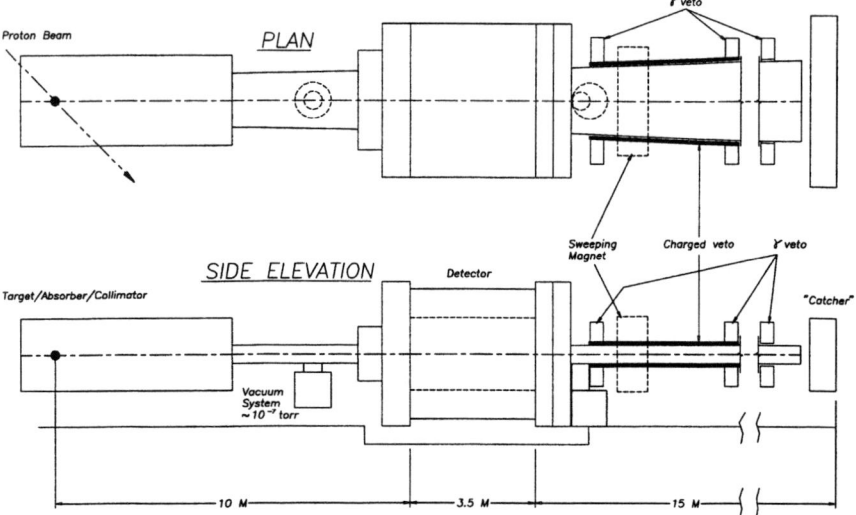

FIGURE 4. Schematic of the proposed 926 detector.

filter to differentially attenuate neutrons and very low energy kaons, and some scope for adjusting the production angle and aperture of the beam. One could then use additional flux by shortening the decay volume, thereby increasing the acceptance of the detector. Finally, if money were no object, faster photon detectors could be deployed so that more beam could be accommodated. This results in a rate of about 70 SM events per year. In a few years of running, in principle η could be determined to about 3%.

B CPT

Another experiment being considered for the Main Injection goes under the acronym 'CPT' [23]. Its primary purpose is to improve the current sensitivity to possible CPT-violation in the K system by a factor large enough to make it sensitive to Planck-scale effects. In particular they seek to measure the phase difference between η_{+-} and ϵ, and evaluate the Bell-Steinberger relation [24]. In addition they will measure CP-violation in $K^0 \to 3\pi$ decay and improve the CP-violation measurements in $K^0 \to \pi^+\pi^-\gamma$. They will also study rare K_S decays. Table 4 is a summary of their goals, compared to current data.

Figure 5 shows the proposed layout. The CPT experiment would share the RF separated K^+ beam with the CKM experiment mentioned above. They would run the beam, set at $25\,GeV/c$ and containing 2×10^8 K^+/pulse, into a W target where K^0's would be produced via charge exchange reactions. The resulting K^0 spectrum peaks at about $15\,GeV/c$. The beam passes through

TABLE 4. Summary of principle measurements of CPT.

	existing data	CPT experiment
ϕ_{+-}	$\pm 1°$	$\pm 0.02°$
$Im x$	$\pm 2.6 \times 10^{-2}$	$\pm 5 \times 10^{-4}$
$Im \eta_{+-0}$	$\pm 1.7 \times 10^{-2}$	$\pm 4 \times 10^{-4}$
$Im \eta_{000}$	$\pm 3 \times 10^{-1}$	$\pm 2 \times 10^{-3}$
$|\eta_{+-}|$	$\pm 1\%$	$\pm 0.1\%$
$|\eta_{+-\gamma}|$	$\pm 3\%$	$\pm 0.1\%$
$B(K_S \to \pi^0 e^+ e^-)$	$< 3.9 \times 10^{-7}$	$\sim 10^{-10}$

a 1.3m long hyperon magnet to remove charged particles and approximately 2000 K_L and 5000 K_S decays/pulse occur in a 14m decay tank. The decays are analyzed in a simple dipole spectrometer augmented by an electromagnetic calorimeter and muon detectors.

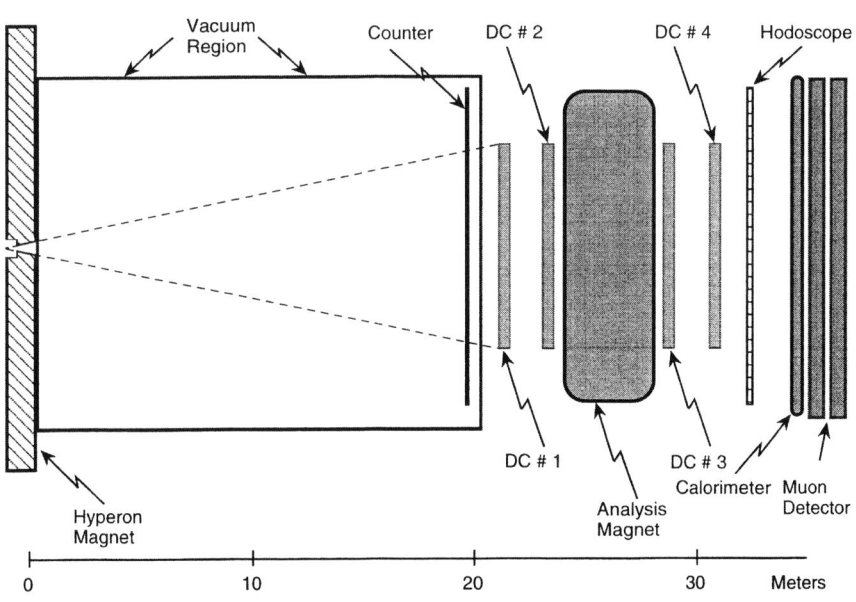

FIGURE 5. Schematic of the CPT detector.

Steve Schnetzer gave a presentation of CPT and discussed the possibility the experiment might be adapted to FMC front-end conditions. Unlike the cases of the other FMI kaon proposals, the answer for CPT is a qualified 'yes'. A certain fraction of the physics targets might remain accessible. Certainly the number of available K^+ is greater at the latter machine. Roughly speaking, the forward cross-section for $16\,GeV/c$ protons to produce K^+ of say $10\,GeV/c$ is about 1/12 of that for $120\,GeV/c$ protons to produce K^+ of $25\,GeV/c$. This is almost completely compensated by the greater charge exchange cross section at the lower energy. However the K_S decay loss is also greater at lower energies. Putting all the factors together, there is an optimum at $p_K \sim 6\,GeV/c$ where the relative number of K_S decays per incident primary protons is $\sim 80\%$ of that at the FMI. Since there are $40\times$ more protons at the FMC front end, a good deal of the physics menu could be further advanced there. There are exceptions, however, such as η_{000}, where the poorer acceptance and photon definition of the lower energy incarnation are bound to hurt.

C Probing symmetry violations through μ polarization in K decay

1 T-violating μ^+ polarization in $K\mu3$ decay

The need for CP-violation in addition to that given by the SM in order to explain the observed baryon asymmetry of the universe [25] motivates investigating low-energy 'windows' where such effects are cleanly identifiable. The CKM model gives virtually no T-violating (out-of-decay-plane) polarization in $K^+ \to \pi^0 \mu^+ \nu$, allowing such a window. Moreover a number of popular attempts to go beyond the Standard Model predict a finite polarization at a level that is experimentally accessible [26].

If T is conserved, the $f_+(q^2)$ and $f_-(q^2)$ form factors that multiply the $(p_K + p_\pi)$ and $(p_K - p_\pi)$ terms respectively in the $K\ell 3$ amplitude are relatively real. Therefore T violation is characterized by the size of the imaginary part of their ratio $Im\xi \equiv Im(f_-/f_+)$. This quantity is in turn approximately proportional to the component of polarization transverse to the $K\mu 3$ decay plane, $\wp_T = (0.2 - 0.3)\,Im\xi$ depending on the phase space sampled.

Experiment 246 at KEK represents a new technique in the study of T-violating μ^+ polarization in $K\mu 3$. It looks promising, but it has not quite proved itself yet. A second approach [27], being advocated for the AGS is to instead optimize the technique of most previous experiments of this type [28]. This was described by Hong Ma in his talk to our session [29]. Fig. 6 shows the layout of the proposed experiment. The source of K^+ is a $2\,GeV/c$ separated beam, a facility quite well suited to the FMC font end. Other improvements with respect to previous experiments include larger acceptance, more nearly complete reconstruction of the decays, finer polarimeter segmentation, and graphite, instead of aluminum, as polarimeter absorbing material. A beam of $\sim 2 \times 10^7$ K^+'s/pulse impinges on a decay tank in which about 5×10^6 decay. π^0 photons are detected in a "shashlyk" calorimeter and μ^+'s penetrate the calorimeter and are tracked into the polarimeter where they stop. When the muons decay, their daughter electrons are tracked through at least two segments of the cylindrically symmetric polarimeter. One is looking for differences in the rates clockwise-going and counter-clockwise-going muon decays. In this case, there are 96 segments as compared to 32 in Ref. [28]. To properly align the the decay plane with the detector, K^+ decays where the π^0 is directed along the beam and the μ^+ approximately perpendicular it in the K^+ center of mass are selected by the trigger. There is no spectrometer magnet, but a ~ 70 G solenoidal field is imposed on the polarimeter to precess the muons. The polarity of this field is reversed every AGS pulse. This technique is very effective in controlling systematic errors. The analyzing power of the polarimeter is calculated to be $\mathcal{O}(30\%)$ which is a large improvement over that of Ref. [28]. The expected statistical sensitivity of the experiment is $\sigma_{\wp_T} = \pm 0.00013$ in about 2000 hours of running. This corresponds to an uncertainty of roughly 7×10^{-4} in $Im\xi$. Systematic errors must be held below this level.

An order of magnitude greater $2\ GeV/c\ K^+$ flux would be available at the FMC front end. About a factor 5 higher singles rates could be accommodated by the proposed apparatus. Perhaps the sensitivity of the experiment could be pushed even further by optimizing the beam. It might be necessary to further segment the polarimeter and make some other apparatus improvements to facilitate tighter control on systematics. Conservatively, one should be able to improve the proposed AGS measurement five-fold, which will yield $\sigma_{\wp_T} = \pm 0.000025$, a very worthwhile level indeed.

2   Polarization effects in $K^+ \to \pi^+ \mu^+ \mu^-$

Top-quark loops very similar to those which make $K \to \pi \nu \bar{\nu}$ sensitive to V_{td} occur in $K^+ \to \pi^+ \mu^+ \mu^-$ as well. However in the latter decay these are overwhelmed by much larger photon exchange effects. The calculation of the branching ratio and decay distribution is an interesting exercise in chiral perturbation theory, but not very revealing of short distance effects. However in the muon polarization such effects are not obscured, and there has been quite a bit of theoretical work on both SM effects and possible non-SM effects in this decay [30].

In the SM there is a parity-violating longetudinal polarization of the μ^+ that is sensitive to the CKM parameter ρ and that can be almost as large as 1% [31]. In principle ρ can be determined to $\sim \pm 0.06$ by such a measurement. This is however, quite an experimental challenge. The reaction $K^+ \to \pi^+ \mu^+ \mu^-$ has only recently been discovered by E787 at BNL [33], with a branching ratio of $(5.0 \pm 0.4 \pm 0.7 \pm 0.6) \times 10^{-8}$. To achieve a $\sim 20\%$ measurement of the μ^+ polarization asymmetry would require at least 8×10^7 events (there are presently about 600 in the world), or a single event sensitivity of about 5×10^{-15}. For an apparatus with 1% acceptance (including K^+ decay probability), which would not be easy, one needs to produce $1.5 \times 10^{17}\ K^+$. If one could

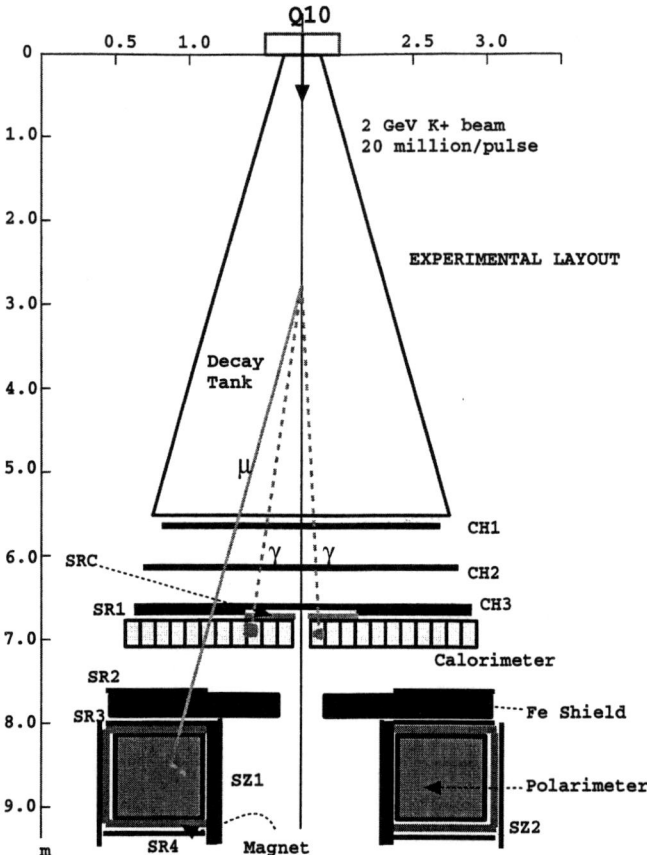

FIGURE 6. Schematic of the proposed 923 detector with a $K\mu 3$ event superimposed.

run for a few years, one would need a beam of $\sim 5 \times 10^9$ K^+/sec. Such a beam is in fact possible, using the entire flux of the FMC front end. This is an experiment that might be a good match for the machine under discussion, given the probably timescale.

If this measurement seems insufficiently difficult, note that a measurement of the two-spin correlation between the μ^+ and the μ^-, is sensitive to CKM η [32], as well as non-SM CP-violating effects.

IV CONCLUSION

There's plenty of potential for interesting physics measurements at the front end of a muon collider. If it were completed tomorrow, there's no question it would be heavily subscribed and produce a raft of important results. However whether it is worth exploiting will be very subject to the vicissitudes of history and politics. Where would one be starting from? What other facilities are available? Also, in order for people to make the large commitment necessary to do these experiments, they would need to have some assurance that the machine would be available for this kind of work for an extended period. One can't expect users to come in, work for two years on extremely complex experimental programs, then pack up and go home because the muon collider needs the protons. Also, any sharing of the protons with the collider would immediately dilute the advantage factors of Table 1.

V ACKNOWLEDGMENTS

We would like to thank K. Arisaka, M. Diwan, S. Kettell, H. Ma, K. Seth, H. Spinka, S. Schnetzer, R. Tschirhart, and H. White for discussions, corrections, plots, access to data and other assistance with this paper. This work was supported by the U.S. Department of Energy under Contract Nos. DE-AC02-76CH03000 and DE-AC02-98CH10886.

REFERENCES

1. TRIUMF, *KAON Factory Proposal*, (1985).
2. G. Franklin, these proceedings.
3. R. Kunne, *et al.*, *Nucl. Phys.* **B323**, 1 (1989).
4. N. Lockyer, *et al.*, *Phys. Rev. Lett.* **45**, 1821 (1980); *ibid. Phys. Rev.* **D30**, 860 (1984).
5. D.R. Thompson, *et al.*, *Phys. Rev. Lett.* **79**, 1630 (1997).
6. J. Hagelin and L. Littenberg, *Prog. Part. Nucl. Phys.* **23**, 1 (1989); D. Rein and L.M. Sehgal, Phys. Rev. D **39**, 3325 (1989); M. Lu and M.B. Wise, Phys. Lett. **B324**, 461 (1994); C.Q. Geng, I.J. Hsu, and Y.C. Lin, Phys. Lett. **B355**, 569 (1995); S. Faijfer, Nuovo. Cim. **110A**, 397 (1997).
7. W.J. Marciano and Z. Parsa, *Phys. Rev.* **D53**, 1R (1996).
8. G. Buchalla and A. J. Buras, *Phys.Rev.* **D54**, 6782 (1996).
9. G. Buchalla and A.J. Buras, *Nucl. Phys.* **B412**, 106 (1994).
10. Y. Nir and M. P. Worah, *Probing the Flavor and CP Structure of Supersymmetric Models with $K \to \pi\nu\bar{\nu}$ Decays*, SLAC-PUB-76010, WIS-97/32/Nov-PH, hep-ph/9711215, Nov. 1997.
11. L. Littenberg, *Phys. Rev.* **D39**, 3322 (1989).
12. G. Buchalla and A. J. Buras, *Phys. Lett.* **B333**, 221 (1994).
13. A. Buras, A. Romanino, and L. Sivestrini, $K \to \pi\nu\bar{\nu}$: *A Model Independent Analysis and Supersymmetry*, TUM-HEP-302-97 (1997); G-C. Cho, *Impacts on Searching for Signatures of New Physics from $K^+ \to \pi^+\nu\bar{\nu}$ Decay*, KEK-TH-554 (1998).
14. S. Adler, *et al.*, *Phys. Rev. Lett.* **79**, 2204 (1997).

15. I-H. Chiang, et al., *An experiment to measure the branching ratio* $K^+ \to \pi^+ \nu \bar{\nu}$ in "AGS-2000 Experiments for the 21st Century", edited by L. Littenberg and J. Sandweiss, **Proc., AGS-2000 Workshop**, Brookhaven National Laboratory, May 13-17, 1996, Formal Report 52512.
16. P.S. Cooper and J. Ritchie, *Summary of the Experiments with Charge Kaons Working Group*, in **Proceedings of the Workshop on Fixed Target Physics at the Main Injector**, G.J. Bock and J.G. Morfin, eds., 113 (1997).
17. R. Ben-David, *XVI International Workshop on Weak Interactions and Neutrinos*, Capri (1997).
18. E. Cheu, et al., *A Letter of Intent to Continue the Study of Direct CP Violation and Rare Processes in Neutral Kaon Decays at KTeV in FY99*, June 1997.
19. E. Chen et al., *An Expression of Intent to Detect and Measure the Direct CP Violating Decay* $K_L \to \pi^0 \nu \bar{\nu}$ *and other Rare Decays at Fermilab Using the Main Injector*, FERMILAB-PUB-97-321-E, hep-ex/9709026 (1997).
20. I-H. Chiang, et al., *Measurement of* $K_L \to \pi^0 \nu \bar{\nu}$, AGS Proposal 926 (1996).
21. T. Inagaki, et al., *Measurement of the* $K_L^0 \to \pi^0 \nu \bar{\nu}$ *decay*, KEK proposal, June 1996.
22. M. S. Atiya, et al., *Nucl. Instrum. Meth.* **A321**, 129 (1992).
23. G. Thomson and H. White, *Summary of the CPT Tests with Kaons Working Group*, in **Proceedings of the Workshop on Fixed Target Physics at the Main Injector**, G.J. Bock and J.G. Morfin, eds., 106 (1997).
24. J.S. Bell and J. Steinberger, **Proc. Oxford Int. Conf. on Elementary Particles, 19-25 September 1965**, T.R. Walsh, et al., eds., 195 (1966).
25. L. McLerran, M. Shaposhnikov, N. Turok, and M. Voloshin, *Phys. Lett.* **B 256**, 451 (1991); N. Turok and M. Voloshin, *Phys. Lett.* **B 256**, 451 (1991); N. Turok and J. Zadrozny, *Nucl. Phys.* **B 358**, 471 (1991); M. Dine, P. Huet, R. Singleton, and L. Susskind, *Phys. Lett.* **B 257**, 351 (1991).
26. G. Bélanger and C.Q. Geng, *Phys. Rev.* **D44**, 2789 (1991); R. Garisto and G. Kane, *Phys. Rev.* **D44**, 2038 (1991); G.-H. Wu and J. N. Ng, *Phys.Lett.* **B392**, 93 (1997); M.Fabbrichesi and F. Vissani, *Phys. Rev.* **D55**, 5334 (1997).
27. M.V. Diwan, et al., "AGS Proposal 923 - Search for T Violating Muon Polarization in $K^+ \to \mu^+ \pi^0 \nu_\mu$ Decay", (1996).
28. S.R.Blatt, et al., *Phys. Rev.* **D27**, 1056 (1983)
29. H. Ma, this workshop.
30. M. Savage and M. Wise, *Phys. Lett.* **250B**, 151 (1990): M. Lu, M. Wise and M. Savage, *Phys. Rev.* **D46**, 5026 (1992); G. Bélanger, C.Q. Geng and P. Turcotte, *Nucl. Phys.* **B390**, 253 (1993).
31. G. Buchalla and A. Buras, *Phys. Lett.* **B336**, 263 (1994).
32. P. Agrawal, et al., *Phys. Rev. Lett.* **67**, 537 (1991); P. Agrawal, et al., *Phys. Rev.* **D45**, 2383 (1992).
33. S. Adler, et al., *Phys. Rev. Lett.* **79**, 4756 (1997).

Experimental neutrino physics at the muon collider complex

Peter Fisher[1] and Boris Kayser[2]

[1] *Laboratory of Nuclear Science, MIT, Cambridge, MA, 02139*
[2] *National Science Foundation, Arlington, VA, 22230*

Abstract. The muon collider complex will provide a large variety of neutrino beams from muon decay in flight. The precisely known flavour composition, intensity and forward collimation of the neutrino beams make a wide variety of experiments possible. In particular, single event long baseline appearance experiments may be feasible. This note summarises the possibilities at the front-end proton source, the recirculating linacs and the collider ring. The possibility of a dedicated muon decay ring is also discussed. Finally, some detector possibilities are presented.

INTRODUCTION

Muon decays in the various elements of the first muon collider complex will provide intense neutrino beams of energies up to 100 GeV [1]. In contrast to neutrinos from pion decay or beam dumps, neutrino beams from the muon collider provide

- high intensities, ranging from fluxes of 10^{19}/yr at the lowest energies to 10^{17}/yr at the highest.

- precisely known neutrino flavor composition, depending on the sign of the decaying muon

$$\mu^- \to \nu_\mu + \bar{\nu}_e + e^-$$
$$\mu^+ \to \bar{\nu}_\mu + \nu_e + e^+$$

Since the energy of the muon at each stage of the system is well known, the neutrino energy spectrum would also be accurately known.

- highly collimated, $\theta_\nu \sim 1/\gamma_\mu$.

These characteristics make possible a number of neutrino oscillation experiments which, in view of present hints of neutrino mass, could prove very useful and interesting.

The present status of the search for neutrino mass was reviewed in our working group from the theoretical perspective by Mohopatra [2]. It was reviewed from the experimental perspective by Goodman [3], Kobayashi [4], Stefanski, White and Tayloe [5]. Prominent among the present hints of neutrino mass is the atmospheric neutrino anomaly, which may be due to $\nu_\mu \to \nu_\tau$ oscillation. The data requires very strong mixing, $\sin^2 2\theta \sim 1$ and mass splitting $\Delta m^2 \sim$ several $\times 10^{-4} eV^2$ to $10^{-2} eV^2$ [6]. An alternative interpretation of the anomaly in terms of $\nu_\mu \to \nu_e$ oscillation is disfavored by a recent bound on $\bar\nu_e \to \bar\nu_x$ from the Chooz reactor experiment [7].

Several long-baseline accelerator neutrino experiments, including K2K and MINOS, will try to confirm or refute the hypothesis that the atmospheric anomaly is due to $\nu_\mu \to \nu_\tau$ oscillation. Unfortunately, these experiments may not be sensitive to the low end of the Δm^2 range favored by the atmospheric data. Now, the probability for neutrinos to oscillate from ν_a to ν_b in a distance L at a neutrino energy E is given by

$$\text{Prob}(\nu_a \to \nu_b) = \sin^2 2\theta \sin^2\left(\frac{\Delta m^2}{4}\frac{L}{E}\right)$$

Thus, the high neutrino flux at a muon collider could extend the Δm^2 sensitivity of the $\nu_\mu \to \nu_\tau$ searches by making possible the use of a longer baseline L, or by increasing the rate at a given L. Since quite low neutrino energies would be available at the collider, the advantage gained by the use of large L would not be canceled by one's being forced to work at a large E. In the section on Neutrino Physics from the Recirculating Linacs, we discuss experiments that could be sensitive to Δm^2 as low as $10^{-4} eV^2$ or below. Such experiments should be able to cover the entire range of neutrino mass splittings which could figure in the atmospheric anomaly, with room to spare.

Owing to the lack of ν_e at accelerators, our exploration of $\nu_e \to \nu_\tau$ oscillation has so far been limited to large mixing angles: $\theta_{e-\tau} > 0.14$ [2]. The neutrino beams at a muon collider, containing as they do a 50% ν_e component, would make possible a much deeper exploration of $\nu_e \to \nu_\tau$.

As already noted, the flavor content of the neutrino beams resulting from muon decay at a muon collider will be quite different from that of the beams at a typical accelerator. We assume that the negative and positive muons at the collider will be sufficiently separated, spatially and in terms of momentum, that the neutrino beams from the decay of muons of one charge will not be contaminated by neutrinos from muons of the opposite charge. Let us then consider the neutrino beam from μ^- decay,

$$\mu^- \to \nu_\mu + \bar\nu_e + e^-$$

This beam contains ν_μ and $\overline{\nu}_e$, but not $\overline{\nu}_\mu$, ν_e, ν_τ or $\overline{\nu}_\tau$ [8]. Its $\nu_\mu, \overline{\nu}_e$ composition, integrated over energy, is obviously 1:1. Since the ν_μ and $\overline{\nu}_e$ have different energy spectra, this 1:1 ratio does not hold at a given neutrino energy. However, for a given muon energy and polarization, the ratio which does hold will be precisely known.

With neutrino detectors that can discriminate between leptons of opposite charge, it will be possible to seek a variety of neutrino oscillations via appearance experiments. One will be able to look for the appearance of ν_e from $\nu_\mu \to \nu_e$, and $\overline{\nu}_\mu$ from $\overline{\nu}_e \to \overline{\nu}_\mu$. With neutrinos above the τ production threshold, one will be able to hunt for the appearance of ν_τ from $\nu_\mu \to \nu_\tau$, and $\overline{\nu}_\tau$ from $\overline{\nu}_e \to \overline{\nu}_\tau$. Obviously, lepton charge discrimination will be crucial for these purposes.

Disappearance experiments will also be possible. At a given neutrino energy, the $\overline{\nu}_e:\nu_\mu$ ratio in the μ^- generated neutrino beam is known, and an anomalous value of this ratio would indicate disappearance of neutrinos from the $\overline{\nu}_e$ and/or ν_μ flavors into flavors not present in the beam. In particular, since CPT invariance requires that $\text{Prob}(\overline{\nu}_e \to \overline{\nu}_\mu) = \text{Prob}(\nu_\mu \to \nu_e)$ [2], an anomalous $\overline{\nu}_e : \nu_\mu$ ratio would indicate the oscillation of beam particles into ν_τ and/or $\overline{\nu}_\tau$.

If the lepton charge discrimination is no possible, one may simply study the ratio $\nu_e + \overline{\nu}_e : \nu_\mu + \overline{\nu}_\mu$ as a function of neutrino energy, much as one studies a similar ratio for the atmospheric neutrinos. An anomalous value of $\nu_e + \overline{\nu}_e : \nu_\mu + \overline{\nu}_\mu$ would indicate the presence of neutrino oscillation. However, it would not be so clear which flavors are involved.

Our working group considered the exotic possibilities of $\nu \to \overline{\nu}$ oscillations. Suppose that in the μ^- generated neutrino beam a $\overline{\nu}_\mu$ component appears. The normal assumption would be that it is from $\overline{\nu}_e \to \overline{\nu}_\mu$. But then since CPT requires that $\text{Prob}(\overline{\nu}_e \to \overline{\nu}_\mu) = \text{Prob}(\nu_\mu \to \nu_e)$, a ν_e component should also appear, at the level demanded by the CPT constraint. If the ν_e component is not present at the required level, we have evidence for $\nu_\mu \to \overline{\nu}_\mu$. Such $\nu \to \overline{\nu}$ oscillations could be induced by right-handed currents or by Lorenz invariance and CPT-violating anomalous neutrino mass terms, as considered by Mohapatra.

As is clear from the discussions, searches for oscillations place two important requirements on neutrino detectors for the muon collider: charge determination and the ability to detect taus. Our study group focused on these two aspects within two general scenarios: intense neutrino beams from the front end proton source, which give neutrino energies of around 1.5 GeV and neutrinos from muon decay in the recirculating linacs are the final collider ring, which give neutrinos of energies from 2-100 GeV. We also considered the use of a dedicated storage ring with straight sections for muon decay. Such a ring could be oriented toward neutrino detectors in Gran Sasso, Soudan or other underground labs.

RLA	E_{min} (GeV)	E_{max} (GeV)	θ_ν at E_{max} mrad	Flux $\times 10^{18} \nu - y$
1	1	10	10	1.4
2	10	70	1.4	0.51
3	70	230	0.4	0.2

TABLE 1. Neutrino flux parameters from the RLAs.

The next section considers neutrino physics from the front end, and the following section considers short and long baseline neutrino physics from the recirculating linacs (RLAs) and the collider ring. The next section describes the basic detector configurations we discussed. Conclusions are given in the final section.

NEUTRINO PHYSICS WITH THE FRONT END

Pions, produced in proton collisions, will decay in flight to produce muons for acceleration in the RLAs and collision in the storage ring. Of necessity, the muon source must be very intense, producing 10^{20} muons per year. As γ_μ is still low, an intense beam of neutrinos will result. For proton energies of 16 GeV, $<E_\nu> \sim 1.5$ GeV with neutrino fluxes 10^{16} $\nu/m^2/yr$. A short baseline of order $L=0.5$-2 km would allow the exploration of the $\Delta m^2 \sim 10^{-2} - 10^{-3}$ eV^2 via the detection of electrons produced by oscillation $\nu_\mu \to \nu_e$, Fig. 1. This covers the region of the reported observation of neutrino oscillations by the LSND collaboration, [9].

Stefanski and Tayloe [5] have proposed the use of the BOONE [10] scintillation detector at the 500 to 2000 meter baseline. The detector consists of a 12 by 12 m tank containing liquid scintillator viewed by 1200 phototubes. The somewhat longer baseline from LSND compensates for the higher neutrino energy. The expected flux gives about $10^7 \nu_\mu$ interactions per year and the increased flux (compared to the Booster) would give roughly a factor of three improved sensitivity to at least $\Delta m^2 \sim$ few$\times 10^{-3}$eV^2.

NEUTRINO PHYSICS FROM THE RECIRCULATING LINACS

The main acceleration takes place in a series of three recirculating linacs (RLAs) which operate in the ranges given in Table 1. The RLAs consist of two straight acceleration sections connected by magnetic channels. Neutrinos resulting from decays in the straight sections will be projected into beams with opening angles $\theta_\nu \sim 1/\gamma_\mu$ and the beam composition will be 50% ν_μ and 50% $\bar{\nu}_e$, for μ^- decay.

As a benchmark, MacFarland [11] considered the response of the NuTeV detector located 600 m from RLA3. Here, 90% of the neutrinos fall in a 40 cm diameter spot and the rates are a factor of 1000 larger than presently available. At this intensity, of order ten neutrino interactions would take place in the NuTeV detector per bunch turn in RLA3. Based on this intensity, King [12] considered the a detector design consisting of a H_2 target of a few kilograms mass surrounded by a magnetic spectrometer consisting of a pixel detector, a RICH, electro-magnetic and hadron calorimeter and muon detector. This concept makes use of the small beam size and is consequently only appropriate for short baseline oscillation experiment. In addition, this experiment would also be able to study proton weak structure functions and the decay of heavy hadrons.

A second short baseline concept, explored by MacFarland [11], consisted of a plastic scintillator target in a magnetic field. The experiment makes use of the precise knowledge of the neutrino beams in a $\nu_e \to \nu_\mu$ (assuming μ^+ decay as the muon source). The muon energy is determined by ranging the muon out in the scintillator and the charge is determined by the curvature of the track in the magnetic field. The experiment could be small (scale size 2m) and, with a baseline of 1 km, could explore the Δm^2 region down to $1 eV^2$. The small size makes it possible to move the experiment over a range of baselines in order to eliminate systematics.

The very high intensity makes long baseline experiments possible as by studied by Geer [13]. Two baselines were considered, Fermilab to Gran Sasso $L=7400$ m and Fermilab to Soudan, $L=732$ m. An obvious problem is that of the angle of inclination (i.e the angle into the earth at which the neutrinos must be aimed) but this may be overcome by the construction of dedicated storage ring constructed at the appropriate angle. The storage ring would be filled from the main acceleration system and would consisted of two parallel transfer lines connected by magnetic channels. With straight sections the same length as the magnetic channel, 25% of the muons would result in neutrinos projected into the forward beam.

For the 20 GeV RLA, the resulting flux in the Gran Sasso would be $10^{10}\nu/m^2/yr$. At the Soudan detector, this would give a flux nearly 200 times that expected for the proposed FNAL long baseline program. The higher flux at still modest energies will allow the exploration of the $\Delta m^2 \sim 10^{-4}$ eV^2 region favored by atmospheric neutrino oscillations. Fig. 2 summarizes the situation. The exclusion area is similar for $\nu_e \to \nu_\tau$ appearance.

The prospect of polarised muons in the RLA give another interesting aspect to the long baseline measurements. A muon beam of polarization P gives a neutrino energy-angle spectrum of $\mu^\pm$ [13]

$$\frac{dn(\nu_\mu)}{dxd\Omega} = \frac{1}{4\pi} = (2x^2(3-2x) \mp 2x^2(1-2x)P\cos\theta$$

$$\frac{dn(\nu_e)}{dx d\Omega} = \frac{1}{4\pi} = (12x^2(1-x) \mp 12x^2(1-x)P\cos\theta$$

where $x = 2E_\nu/m_\mu$ and θ is the angle between the neutrino and muon momenta. Modulation of the muon polarization will modulate the beam energy and divergence. Thus, a polarization modulated beam could be used to beat out beam uncorrelated systematics.

Kobayashi [4] reported on the planned K2K long baseline experiment in which the KEK PS 16 GeV proton beam will be used to produce neutrinos which will illuminate the Super Kamiokande detector with an average energy of 3.5 GeV. The baseline is 250 km which will allow exploration of the $\Delta m^2 \sim$ few $\times 10^{-3}$ eV^2 region searching for $\nu_\mu \to \nu_\tau$ oscillations. In addition, a short baseline monitor detector will also be used to normalize the neutrino flux. This program is expected to begin taking data in late 1998.

The importance of the precise knowledge of the neutrino beam composition must be re-emphasized. In the case of low (zero) background, a simple statistical argument given by Goodman [3] shows the mass sensitivity improves as $\sqrt{N_\nu}$ and the mixing angle sensitivity improves with N_ν. In the case of statistically significant background, the mixing sensitivity improves like $\sqrt{N_\nu}$ while the mass sensitivity plummets to $\sqrt[4]{N_\nu}$. This places a premium on highly efficient charge determination as the only way to fully exploit the well known composition of the neutrino beams.

DETECTORS

Two issues dominated the thinking about detector design: the need to determine whether a neutrino or antineutrino interacted in the detector via measurement of the charge of the final state and the ability to detect τ leptons. A magnetic spectrometer provides and obvious solution to the former while two different solutions obtain for the latter. We will discuss each problem in turn.

Flavor determination

Two designs of magnetic spectrometers were presented by King [12] and MacFarland [11]. In both cases, the implementation is straight forward for for the flavor determination of via the muon final state. Here, the long range of the a muon in absorber make a large spectrometer like NuTeV possible. High energy electrons or positrons, on the other hand, will require a low density tracking system. For example, the sagitta of an electron with $p_t = 50$ GeV over 1 m lever arm in a 1 T magnetic field is of order 0.1 mm. Silicon microstrips interspersed with target material (see below) may provide sufficient spatial resolution to make a charge determination. At lower transverse momenta, the sagitta resolution improves for fixed BL^2, but the effects of multiple scattering begin to become important around a few GeV. The bottom line is that $\nu_\mu \to \nu_e$ may be more difficult in comparison to $\nu_e \to \nu_\mu$ experiments, but the 50-50% flavor mix always present in neutrino beams from muon decay in flight sources constitute a "no-lose" (or equivalently "no-win") theorem for flavour determination.

τ identification

The 300 fs lifetime of the τ lepton provides a unique approach to appearance experiments (see [14]). Here, we summarize the main features of tau decay and discuss two method of detecting tau leptons produced by $\nu_\tau + N \to \tau + X$.

Tau leptons undergo charged current decay to an odd number of charged particles, a neutrino and, possibly, neutral hadrons. A tau lepton with an energy of 25 GeV travels a distance of 1.5 mm before decaying leading to three very sensitive techniques for identifying tau decays:

- impact parameter - the distance of closest approach of the back projection of the final state charged particle to tau production point (the vertex formed by the intersection of tracks from the charged debris of the struck nucleon), of order 90 μm for the example give above.

- lifetime - for cases in which the tau decays to three or more charged particles (15% of tau decays), the decay vertex may be formed from the intersection of the tracks in the final state and the distance between the production and decay points may be determined. This is of order 1.5 mm for the example given above.

- ionization - if the target material is active, the ionization track of the tau may be determined, giving a direct measurement of the lifetime.

These three methods focus on the tau production and decay. The kinematics of the final state particles may also be used since the tau neutrino always present in the final state always results in missing transverse momentum. In the case where the initial neutrino direction is known, the missing p_t provides a powerful discriminant in addition to the vertex reconstruction [14].

Two different schemes are used to for vertex reconstruction: emulsion stacks which are exposed to the neutrino beam and later scanned and silicon sensors sandwiched in between absorber. Emulsion stacks have long been used for vertex identification, most recently by the CHORUS experiment [14] and they are planned for COSMOS [15]. The first silicon target, the NOMAD-STAR prototype, just began operation in the NOMAD detector [14]. We will discuss each in turn.

the NOMAD-STAR prototype is pictured in Fig. 3 and consists of 44 kg of B_4C plates sandwiched between silicon microstrip detectors (total area of silicon is 0.5 m^2). B_4C was chosen to maximize mass per unit radiation length.

Data taking with the prototype in 1997 resulted in 30,000 charged current events and boosted the tau detection efficiency from 3% [1] to 10%. The cost of the prototype was $300 K. A proposal for a full detector with a mass of 990 kg target mass with 2.2 m^2 silicon area has been submitted.

At this point, silicon sensors and associated readout technology is well in hand and large systems could be consider for a long baseline experiments. Using the 10 GeV beam aimed at the Soudan as an example, a target mass of 1 ton giving an τ detection efficiency of 10% would cost $6M and would be able to search $\nu_{\mu,e} \to \nu_\tau$ with a sensitivity of $\Delta m^2 \sim 10^{-4}$ eV^2 .

Reay [17] summarized the current situation with emulsions, which turned out to be somewhat surprising. The basic technique is to expose an emulsion stack to the neutrino beam while also detecting events by means of a precise spectrometer behind the stack. After processing, the emulsion stack is then scanned to search for the characteristic kink signifying a the production of a tau followed by its decay to one or more charged particles. Data from the spectrometer guides the scanning. The single point resolution for emulsion is roughly a micron, making it the most precise track detector available.

Several recent developments have made this technique very attractive. First, the scanning has been automated to a high degree, making it possible to scan the emulsions much more quickly and efficiently. In addition, the track information from the scanning automatically readout from the scanning system and combined with the spectrometer data. The resulting process is much more like a modern high energy experiment in that the complete data is available relatively soon after the data taking has stopped.

A second major improvement has been in the overall quality of the emulsion. The emulsion is supported by a plastic sheet and in the past there have been problems with the emulsion's uniformity, which leads directly to systematic errors in the track reconstruct. In some cases, the emulsion has actually separated from the sheet resulting in lost data. Tests indicate the emulsion planned for MINOS is uniform to better than 1μm for 50μm thick emulsion.

The third major improvement may be the cost. Current emulsion, which is manually produced, costs roughly $1,000 for a 1 m^2 sheet. Recently, high quality X-ray film has become available which sells for $160 per sheet. If such film is usable in an experiment, the contemplation of 100 ton detector masses is not outrageous.

CONCLUSION

The high fluxes and precise beam composition of neutrino beams from muon decay in flight make neutrino physics a major endeavor at the muon collider. However, many questions remain:

---

[1] Previously, only kinematic cuts on the final state particles was used for tau neutrino induced events.

- What is the optimum detection scheme for an experiment at the proton source? Does it make sense to try to design a detector which could be a short baseline experiment first and then serve as a monitor detector for a long baseline experiment?

- What second order effects can degrade the beam purity? What is the optimum design for shielding which reduces "regeneration" via $\nu_\mu + N \to \pi^- + X \to \mu^- + \overline{\nu}_\mu + X$?

- What is the optimum baseline and energy for a dedicated storage ring? How feasible is such a ring in the context of operating cycle of the muon collider? How may the ring best be designed to minimize the overlap of neutrinos from different charged beams?

- What backgrounds will arise from CC induced charm production and decay?

- What is the best way to go about flavor identification via the charge determination of electrons/positrons? Is muon and electron/positron charge identification compatible in the same experiment or is it better to have separate experiments for each channel?

REFERENCES

1. For a complete list of the muon collider parameters, the http://www.fnal.gov/projects/muon_collider/ and these proceedings.
2. Mohapatra, R., in these proceedings
3. Goodman, M., http://www.hep.anl.gov/NDK/Hypertext/nuindustry.html.
4. Kobayashi, T., the proceedings. In introduction to K2K may be found on the wbe at http://pnahp.kek.jp/intro-e.html.
5. Stefanski, R., White, A. and Tayloe, R., these proceedings.
6. Sobel, H., talk presented at the 1998 Aspen Winter Conference on Particle Physics.
7. Apollonio, M. et al., hep-ex/9711002.
8. Contamination of the beam by neutrinos produced in the $\mu^-\mu^+$ collisions in the collider was shown by E. Paschos (thesis proceedings) to be completely neglible.
9. LSND Collaboration (C. Athanassopoulos et al.). UCRHEP-E197, Sep 1997.
   LSND Collaboration (C. Athanassopoulos et al.). UCRHEP-E-191, Jun 1997.
10. Church, E., et al., "A LETTER OF INTENT FOR AN EXPERIMENT TO MEASURE MUON-NEUTRINO TO ELECTRON-NEUTRINO OSCILLATIONS AND MUON-NEUTRINO DISAPPEARANCE AT THE FERMILAB BOOSTER (BOONE)",LA-UR-97-2120, Jun 1997.
11. MacFarland, K. and D. Harris, these proceedings.
12. King, B., these proccedings.
13. Geer, S., Fermilab-PUB/97-389, 1997 and these proceedings.

14. CHORUS Collaboration, Eskut, E., NIM A401(1997)7.
    NOMAD Collaboration, Laveder, M., Nucl. Phys. Proc. Suppl. 48(1996)188.
15. The COSMOS proposal http://pooh.physics.lsa.umich.edu/~e803_disk/documentation/prop/.
16. Ferrere, D., NOMAD Memo 96-031
17. Kodama, K., et al., NIM B93(1994)340.

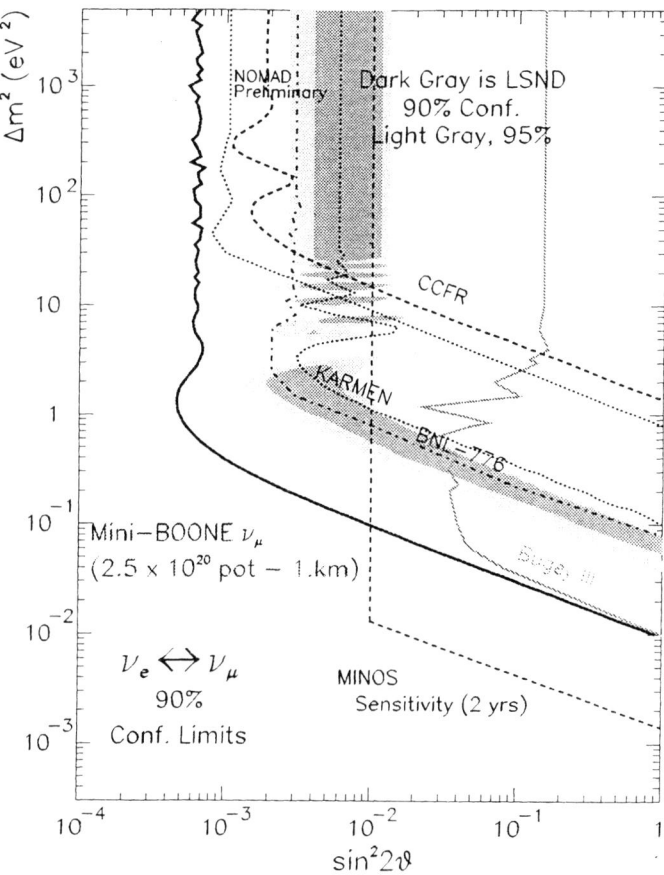

FIGURE 1. Oscillation parameter space showing region of LSND observation and projected sensitivity of BOONE type detector at muon collider proton source.

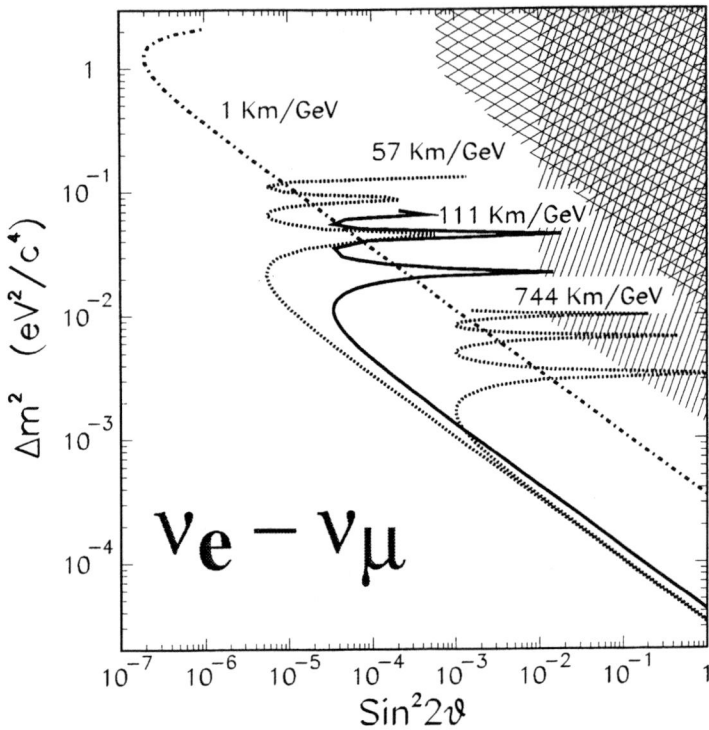

FIGURE 2. Oscillation parameter space accessible by various baseline/energy ratios. Hatched area is the region explored by MINOS, cross hatched is MiniBooNE.

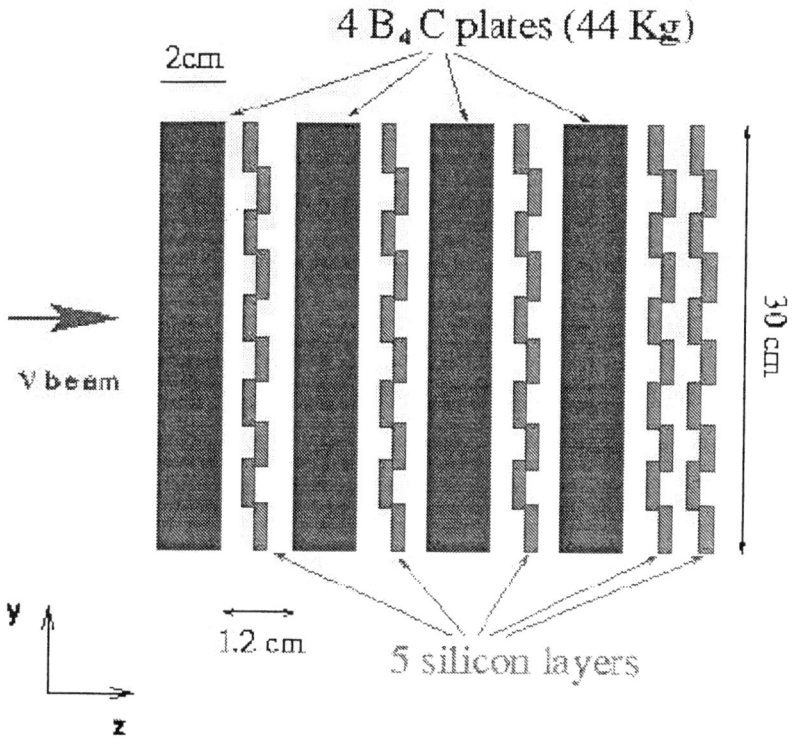

FIGURE 3. Layout of the NOMAD-STAR prototype.

Physics with Low Energy Muons at the Front End of the Muon Collider

William Molzon

University of California, Irvine, California 92697

Abstract. The front end of the muon collider will produce extremely high fluxes of muons, and could provide the opportunity for significantly improved experiments using low energy muons. In this paper, we discuss the physics goals of experiments which could benefit from new, high intensity muon beams. We also discuss the beam requirements, which would require modifications to the source as proposed for the collider.

INTRODUCTION

A tremendous amount has been learned by studying muon decays and interactions, in fields ranging from tests of fundamental symmetries to applications in condensed matter physics. With the commissioning of the front end of the muon collider (FEMC), many experiments could be qualitatively improved. In this workshop, a small group of physicists with diverse interests explored some of the possibilities. Much of the work was on improving tests of muon- and electron-lepton number conservation (LFV). Other topics included exploring the possibility of searching for a muon electric dipole moment, precision measurement of τ_μ, studies of muon nuclear capture, muon catalyzed fusion, etc. Table 1 gives a list of topics which possibly could be improved substantially.

Most of the discussion focused on making substantial improvements using existing facilities, and only peripherally on the usefulness of the FEMC as a muon source for these experiments. This emphasis resulted from the desire and ability to make significant improvements in sensitivity on a time scale shorter than that of the FEMC, and from the realization that some technical aspects of the FEMC are not optimal for the experiments being considered. However, with modifications to the FEMC, even more substantial improvements potentially could be made. In the remainder of this paper, I will discuss the highlights of some of the discussion of experiments that appear to derive real benefit from much cleaner, higher intensity muon beams. Even in most of these cases, the extent to which the experiments can be incrementally im-

TABLE 1. The table lists muon induced processes, the physics addressed, the facilities at which they are being done, and the muon beam requirements. It is adapted from a table by P. Kammel.

PROCESS	THEORETICAL MOTIVATION	MEASUREMENT	EXISTING FACILITY	BEAM
μ- and e number violation	Non-SM physics	$\mu^- N \to e^- N$	PSI, BNL	pulsed
		$\mu^+ \to e^+ \gamma$	LAMPF, PSI	cooled, DC
		$\mu^+ \to e^+ e^+ e^-$		cooled, DC
		$\mu^+ e^- \to \mu^- e^+$	PSI	DC
μ properties	RH couplings	Michel parameters	TRIUMF, PSI	DC
	Fermi constant	μ lifetime	PSI, BNL	pulsed
	Non-SM couplings	μ g-2	BNL	pulsed
	T violation	μ EDM	BNL	pulsed, polarized
Exotic atoms	QED test	Lamb shift	RAL, PSI	pulsed
		hyper-fine splitting	LAMPF,	pulsed
	Parity violation	P-viol. 2S decay	PSI	DC
μ capture	QCD, SM tests	$\mu^3 He$, μp	PSI, LAMPF TRIUMF	pulsed, DC

proved before the FEMC becomes operational is not yet clear. Other topics, such as muon capture studies, muon catalyzed fusion, exotic atoms, etc., while of fundamental interest, do not so clearly benefit from a FEMC class beam. Most are discussed in written contributions to the proceedings, and I will not discuss them here.

MUON- AND ELECTRON-NUMBER VIOLATION

A large fraction of the working group discussion centered on improving searches for LFV using muon induced processes. These processes have been studied theoretically and experimentally since the discovery of the muon. In principle, LFV could be induced by neutrino mixing, but current limits on neutrino mass differences and mixing angles preclude the possibility that observable LFV effects occur via this mechanism. Hence, observation of LFV would indicate the existence of new physics processes, and LFV searches are among the most sensitive means we have to explore physics beyond the Standard Model.

Aside from the motivation to test conservation laws with the best possible sensitivity, these experiments are motivated by the many proposed Standard Model extensions that allow LFV. In many cases, the limits already set restrict the allowed values of parameters within these models. The sensitivity of LFV searches to models for new physics has been discussed, for example by W. Marciano [1] at this conference. The possibilities include contact interactions, heavy neutrinos, lepto-quarks, Z' bosons, supersymmetry, etc. The mass scale which could be reached extends to above 3000 TeV/c^2.

Much interest has occurred recently in grand unified supersymmetric models. It was realized, first by Hall and Barbieri, that LFV will occur at experi-

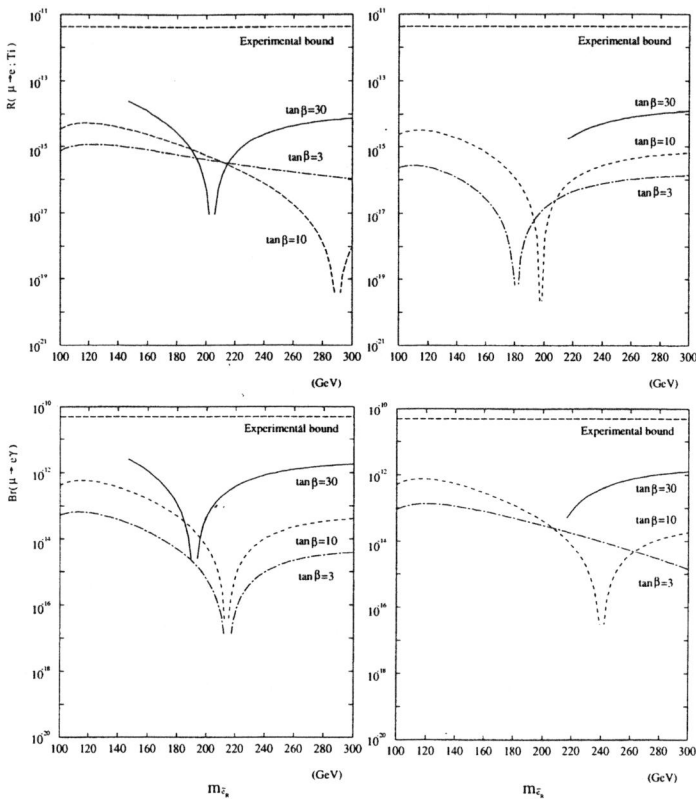

FIGURE 1. Expected rates for $\mu^- N \to e^- N$ and $\mu^+ \to e^+\gamma$ in the model of Hisano, et al., for different values of the ratio of Higgs particle vacuum expectation values, as a function of the right handed selectron mass. The plots are shown for the parameter $\mu > 0$ (left) and $\mu < 0$ (right). The $\mu^- N \to e^- N$ experimental bound has improved to 7.8×10^{-13} since this publication.

mentally accessible levels in a large class of supersymmetric models [2–5]. In some grand unified supersymmetric models, the rate for LFV processes can be related directly to Standard Model parameters, e.g. CKM matrix elements. The predicted rates for $\mu^- N \to e^- N$ and $\mu^+ \to e^+\gamma$ are shown in figure 1 in one such model [5]. These predictions are well within the discovery range of proposed experiments.

The most familiar LFV process is $\mu^+ \to e^+\gamma$ decay. The state of the art is the MEGA experiment [6], which has set a limit $B(\mu^+ \to e^+\gamma) < 3.8 \times 10^{-11}$.

They have data [7] to improve that by a factor of 6-12, and experiments with even better sensitivity [8,9] have been discussed. A closely related decay is $\mu^+ \to e^+e^+e^-$; if this decay occurs by the exchange of a virtual photon, the rate is directly related to that for $\mu^+ \to e^+\gamma$. The current limit [10] is B($\mu^+ \to e^+e^+e^-$) $< 1.0 \times 10^{-12}$ and there are no proposals to do another experiment. A second closely related process is $\mu^-N \to e^-N$. In this working group, Czarnecki, Marciano, and Melnikov [11] reported their new calculation of the relative rate for $\mu^+ \to e^+\gamma$ and $\mu^-N \to e^-N$ for the same underlying physics process. If mediated by a γ, the ratio $R_{\mu e} \equiv \Gamma(\mu^-N \to e^-N)/\Gamma(\mu^-N \to \nu N')$ is about $0.003 \times$ B($\mu^+ \to e^+\gamma$). The SINDRUM2 experiment has set a limit [12] $\Gamma(\mu^-Ti \to e^-Ti)/\Gamma(\mu^-Ti \to \nu Sc) < 7.6 \times 10^{-13}$ and will improve their sensitivity to $\sim 4 \times 10^{-14}$. The MECO experiment [13] will push this to below 10^{-16}.

Experimental Techniques

The $\mu^+ \to e^+\gamma$ experiment is conceptually simple. Temporally coincident e^+ and γ are detected exiting a thin target in which μ^+ beam is stopped. The e^+ and γ each have energy 53 MeV, originate from a common point, and have opposite momenta. At the detection rate required to reach a sensitivity of 10^{-14}, the most important background is events in which the e^+ originates from one μ^+ decay and the γ originates from the radiative decay of a second μ^+. Beam purity is not a significant issue. Depending on the detector acceptance, muon stopping rates of 10^8 to 10^{11} are sufficient to reach the desired sensitivity. The accidental background is reduced with extremely good timing, position, angle, and energy resolution; additionally, ideas to exploit correlations between the direction of the muon polarization and the e^+ and γ to reduce accidental background have been proposed [14–16].

The $\mu^-N \to e^-N$ experiment is in principle even simpler. A μ^- beam is stopped in a thin target, where the μ^- become Coulomb bound to atoms. There, they either capture on the nucleus or decay in orbit. They may convert to electrons; this process is largely coherent, with the nucleus remaining in the ground state, and the signature is an electron with energy equal to $M_\mu c^2$ less the small binding energy. Since the detected final state consists of only one particle, there is no accidental background (unlike the case of $\mu^+ \to e^+\gamma$), and the stopping rate can be extremely high, in excess of $10^{11}\mu^-/s$. There are intrinsic backgrounds from μ^- decay in orbit and radiative μ^- capture, and beam related and cosmic ray induced backgrounds. Previous experiments have reduced beam related backgrounds either by pulsing the μ^- beam [17] or by vetoing events detected in time with a beam particle [12]. At the rates required to reach below 10^{-16}, either a pulsed beam or an extremely pure, low energy μ^- beam is required. In the pulsed beam option, the time structure has short (< 100 ns) pulses separated by a time about equal to the lifetime of μ^- stopped in the target (~900 ns for aluminum). Conversion electrons are detected only after beam associated background has decayed or passed through the detector region, typically 500-700 ns after the main pulse. This motivates the choice of stopping target material, with a lifetime sufficiently long that the loss of sensitivity between the stopping pulse and detection interval is not large.

In the remainder of this section, I discuss the beam requirements for the experiments, and then discuss the status of the lepton flavor violating experiments.

Beam Requirements for New Experiments

Further improvement, by a factor of 100-1000 over the anticipated sensitivities of ongoing experiments, will require significantly higher muon fluxes with less contamination and better background rejection. A number of ideas for new beams have been proposed which should allow significant improvement.

The SINDRUM2 experiment has proposed [12] a novel beam with high purity. A momentum and sign selected pion beam is directed onto a scattering foil on the axis of and at the entrance to a long solenoid. The scattered π^- have helical trajectories, and some decay to μ^-, that have trajectories which do not return to the axis. Undecayed π^- are absorbed in a blocker on the solenoid axis and the μ^- are transported to the stopping target. The SINDRUM2 experimenters calculate that the resulting beam is sufficiently pure that backgrounds from π^- and e^- contamination in the beam is below the 10^{-14} level. It is unlikely that this technique would work at a sensitivity below 10^{-16}.

The MECO experiment to search for $\mu^- N \to e^- N$ has proposed to build a low energy, sign and momentum selected, pulsed muon beam at BNL using some of the ideas of Lobashev and Djilkibaev [18,19]. This type of beam was first proposed to be implemented at the Moscow Meson Factory, a 600 MeV proton accelerator. The BNL implementation was proposed following the observation by the muon collider group that higher μ^- yields could be achieved with a higher energy proton driver, despite the lower proton current. The proposed beam will be produced using a pulsed proton beam of 4×10^{13} p/s, incident on a high Z target in a solenoid with varying axial field. These same ideas have been adopted for the source for the muon collider, with a much higher solenoidal field (~20 T vs. ~3 T).

At BNL, the pulsed proton beam is produced by resonantly extracting a bunched beam from the BNL AGS. M. Brennan has described [20] a modus operandi for the BNL AGS using two filled buckets in the 2.7 μs revolution time of the machine. By operating the AGS below transition ($\Gamma = 8.6$), up to 2×10^{13} protons per bucket could be accelerated and extracted. The level of extinction (ratio of number of protons not in the bunches to that in the bunches) is required to be $< 10^{-9}$ [13]. Preliminary tests of the extinction were made, and an extinction of $< 10^{-6}$ was achieved. Additional improvements are expected, and a pulsed kicker in the extracted beam-line is proposed [13] to achieve the required extinction. This μ^- beam has been simulated extensively by the MECO collaboration [13] and the results were discussed by M. Bachman [21] at this workshop. Figure 2 shows the result of the interaction of a few protons on the production target, from a GEANT simulation of the MECO beam. A possible implementation of such a beam using the Fermilab

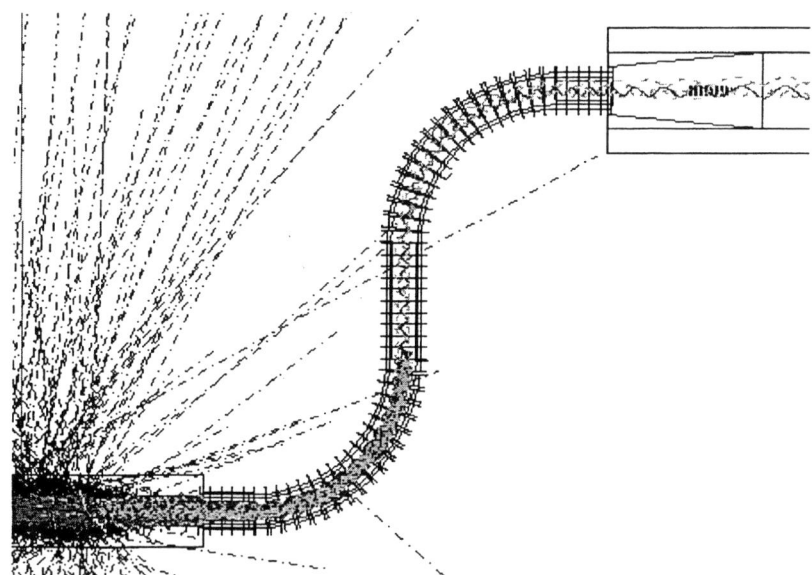

FIGURE 2. An example of a few proton interactions in the production target from the MECO beam simulation. The figure shows the production solenoid at the lower left, the curved transport solenoid, and the detector solenoid at the upper right.

booster as the driver was discussed by C. Moore at the workshop [22]. It now appears that the significant technical difficulties associated with producing a slow spill with the requisite pulse structure preclude this option [23].

In all the new muon beams being discussed, including the muon collider source, a crucial input to the calculation of the beam intensity is the pion yield in low and medium energy pp collisions. An extensive bibliography of pion yield measurements was reported by A. Mokov [24]. E910 at BNL has collected data on pion production over a wide kinematic region, and the status of the analysis of that data was reported by H. Kirk [25]. An experiment at Serpukov has measured pion yields over essentially the full kinematic range, and this has been incorporated into the MECO simulation [26,27].

For MECO, the beam is sign and momentum selected by transporting it in a curved solenoid. Particles drift perpendicular to the plane of the transport and suitable collimators pass only low energy negative particles [13,28]. Current estimates of the yield are ~0.006 μ^- per incident proton, with ~40% stopping in a thin target. The transport in a curved solenoid is crucial in eliminating high energy particles (particularly electrons). The beam has been shown by simulation to have sufficiently small electron and pion contamination that backgrounds from these sources will be below 10^{-18}.

The alternative to using a pulsed beam is to use a very pure beam. The acceptable level of π^- and e^- contamination depends to some extent on the details of the stopping target, but is of order $10^{-12} - 10^{-13}$ if a background below 10^{-18} is to be achieved. The new SINDRUM2 beam is a step in that direction, but will not have sufficient purity.

Y. Kuno has proposed [16] to develop a "phase rotated" muon beam, using the ideas of the muon collider. This beam would have a relatively low momentum bandpass, and hence would be free of high energy electrons. It would be relatively long, and hence the pion contamination would be low by virtue of the long decay time. However, a reliable estimate of the purity which could be achieved in a μ beam has yet to be made.

Two essential differences exist between the muon beam for muon experiments and that for the FEMC. The beam for $\mu^- N \rightarrow e^- N$ or $\mu^+ N \rightarrow e^+ \gamma$ experiments must be low energy (≤ 20 MeV) and must have large duty factor. The phase rotation techniques proposed for the muon collider work only at ~200 MeV, and hence the beam would need to be decelerated. The time structure is problematic. To capture the π's and μ's in the RF structure of the phase rotation, the driver beam must be pulsed, at ~15 Hz in the case of the FEMC. For muon experiments, the beam could be micro-bunched, with a frequency 1 – 10 MHz, while still preserving an essentially uniform μ^+ decay time distribution. However, this would require that the RF be run at high duty cycle, which presents significant difficulties with the power required for conventional RF. Power could be saved with super-conducting RF. However, the phase rotation beam has RF in a solenoidal magnetic field, which appears inconsistent with the use of super-conducting cavities. Further study of this possibility is clearly required and warranted, since, in principle, extremely clean, low emittance beams could be produced.

As discussed by Y. Kuno [14–16], a polarized muon beam would be beneficial both in reducing backgrounds in $\mu^+ \rightarrow e^+ \gamma$ experiments, and in elucidating the mechanism for LFV if a signal is found. The muon collider would also benefit from polarized beams, as would other experiments (the muon EDM measurement, for example). In both cases, substantial polarization is required, beyond what is presently foreseen for the high intensity beams being discussed.

A certain level of synergism exists between R&D for muon collider and that for high intensity muon beams for low energy experiments. For example, both will produce muons in high field solenoids, and will transport them in curved solenoids. To the extent possible, these R&D efforts should be coordinated.

The MECO $\mu^- N \to e^- N$ Experiment

The MECO experiment to search for $\mu^- N \to e^- N$ with sensitivity below 10^{-16} is the next step beyond current experiments. The proposed beam-line and apparatus are shown in figure 3. Briefly, a muon beam pulsed at 0.741

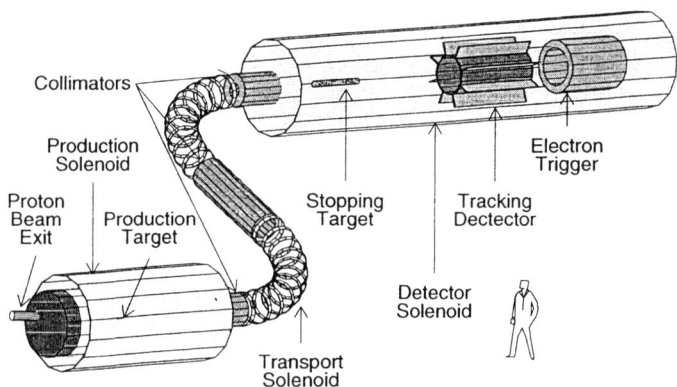

FIGURE 3. Schematic drawing of the MECO production solenoid, transport solenoid, and detector solenoid with the targets, collimators, and detectors.

MHz is transported to the detector solenoid, where a large fraction of it is brought to rest in a thin aluminum target. The field is graded, with an axial field of 2.0 T at the entrance, decreasing linearly to 1.0 T at a point after the target, and constant thereafter. The momenta of conversion electrons are measured with high precision and efficiency in a straw tube spectrometer. The device is triggered with a scintillator calorimeter, which also confirms the electron energy with low precision. The experiment has been described in the proposal [13] and in a paper [29] submitted to this workshop by J. Sculli.

The dominant intrinsic backgrounds are from muon decay in orbit and radiative muon capture. The spectrum falls rapidly near the endpoint, proportional to $(E_{conv} - E_e)^5$, where E_{conv} is the energy of conversion electrons, $\sim$105 MeV. The fraction of all muon decays that produce electrons within 5 MeV of the endpoint is about 2×10^{-12}. Radiative muon capture on aluminum produces electrons from γ conversion, with an endpoint at 102.9 MeV. These backgrounds can be reduced only by improving the energy resolution, and, in the case of radiative decays, by making the target thinner. Reduced energy spread in the beam would allow thinner targets, resulting in better energy resolution due to less straggling in exiting the target, and in fewer conversions of γ's from radiative decay.

Other backgrounds may result from prompt processes, in which an electron is detected close in time to a pion stopping in the target and radiating a photon that converts, or from high energy electrons in the beam scattering in the target, or from muon decay in flight. Some electrons and pions result from the annihilation of anti-protons, and these may have a different time structure than those produced from the target directly. These sources have been eliminated in earlier experiments by vetoing any event which is time coincident with a charged particle in the beam or by making a pulsed beam in which conversion electrons are detected only after all beam particles have decayed or passed through the stopping target. In MECO, a pulsed beam is used.

Additional backgrounds could result from cosmic ray induced events. These are suppressed by active and passive shielding. We note that the cosmic ray background scales with running time and not sensitivity. Improving the sensitivity by a factor of 1000 does not require reducing cosmic ray induced background by a factor of 0.001.

Another concern is the high detector rates resulting from muon capture processes. The MECO detector is arranged so that most electrons from μ^- decay do not strike it. The main sources of rates are protons and photons

TABLE 2. A summary of the expected MECO sensitivity and backgrounds for a one year (10^7 s) run. The sensitivity is one event for $R_{\mu e} = 1.7 \times 10^{-17}$.

Sensitivity	
Running time (sec)	10^7
Proton flux (sec^{-1})	4×10^{13}
μ/p entering solenoid	0.006
Stopping probability	0.370
μ capture probability	0.600
Fraction of μ which capture in time window	0.480
Electron trigger efficiency	0.900
Fitting and selection criteria	0.250
Detected events for $R_{\mu e} = 10^{-16}$	5.800
Backgrounds	
μ decay in orbit	0.29
Radiative μ capture	$\ll 0.05$
μ decay in flight	0.004
Radiative π capture	0.02
π decay in flight	$\ll 0.001$
Beam electrons	< 0.002
Cosmic ray induced	0.004
Total background	0.37

emitted following muon capture. These are minimized by absorbing most of the protons (typically 10-30 MeV energy) and by minimizing material in which photons could convert. The rates are expected to be below 300 kHz per detector element. To set the scale of high rate experiments searching for rare processes, the tracking detector rates in the rare K_L^0 decay experiment E871 at BNL are up to 700 kHz per channel.

The expected sensitivity and background levels of the MECO experiment are shown in table 2. At the proposed sensitivity is not expected to be background limited.

$\mu^+ \to e^+\gamma$ Experiments

Experiments to reach sensitivity of 10^{-14} in $\mu^+ \to e^+\gamma$ will require significant increases in either μ^+ beam intensity or detector acceptance. At the data taking rates required, the experiments will be limited by accidental backgrounds unless significant improvements are made in the ability to detect and reject events in which the e^+ and γ originate from different μ^+ decays. This requirement motivates the use of a continuous beam with very little momentum spread.

The general design considerations to reduce backgrounds and achieve high sensitivity have been discussed by M. Cooper [30] at this workshop. The accidental backgrounds are reduced by minimizing the fraction of electrons and photons from the background processes which are misidentified as signal (i.e. minimizing the energy resolutions), improving the precision of the measurement of the particle positions (to better distinguish events in which the μ^+ and γ are not back to back), improving the timing resolution, and possibly by measuring the γ direction and by detecting the low energy e^+ from the radiative decay. The accidental background depends on these resolutions and on the kinematics of the decay modes. It scales as:

$$B_{acc} \propto (R_\mu/df \bullet \Delta t) \bullet \Delta x \bullet (\Delta y/15)^2 \bullet (\Delta\theta)^2 \qquad (1)$$

where R_μ is the muon stop rate, df is the duty factor, Δt is the timing resolution, Δx and Δy are the e^+ and γ fractional energy resolutions, and $\Delta\theta$ is the e^+ - γ opening angle resolution. A representative example of a $\mu^+ \to e^+\gamma$ experiment which could achieve a background below 10^{-14} was discussed by M. Cooper at this workshop. The typical resolutions required are $\Delta x \simeq 0.2\%$, $\Delta y \simeq 1.7\%$, $\Delta\theta \simeq 400\ ps$, and $\Delta\theta \simeq 4\ mrad$. Achieving these resolutions has not been proved. The particular implementation discussed by Cooper involves a high intensity beam ($> 10^{11}\ \mu^+/s$) and a very thin stopping target to minimize energy loss and scattering as the e^+ exits the target.

Other possible experiments have been discussed [9] at a meeting at PSI about a year ago, and discussed at this meeting by Y. Kuno [16]. Reaching a sensitivity of 10^{-14} with a beam of $10^8\ \mu^+/s$, typical of a surface muon beam produced at PSI, for example, will require relatively large solid angle detectors. The γ detectors under consideration range from pair spectrometers, where the photon is converted and the $e + e^-$ momenta measured, to calorimetric detectors with either crystals or liquid Xenon. Various geometry electron detectors were considered, with different magnetic field geometries and solid angles. All choices are being studied as possibilities for a proposal at PSI.

An idea for reducing background has been discussed by Y. Kuno, A. Maki, and Y. Okada [14–16]. The idea relies on the observation that e^+ are emitted preferentially in the direction of the μ^+ polarization in ordinary μ^+ decay, and that the γ is also emitted preferentially in the direction of the μ^+ polarization in radiative μ^+ decay. Hence, if the μ^+ beam is polarized and maintains its polarization after stopping, the probability of an e^+ from one decay and a γ from another decay being back-to-back is suppressed. Kuno, et al., have shown that a polarization of 90/with for a detector aligned with the polarization and with an opening angle of $\sim$0.3 radian.

The next generation $\mu^+ \to e^+\gamma$ experiment is now being studied intensely, and it is anticipated that a proposal may be produced on a time scale of about a year.

MUON ELECTRIC DIPOLE MOMENT

One experiment which would benefit from increased muon yields is an experiment to measure the electric dipole moment (EDM) of the muon. An electric dipole moment of an elementary particle violates parity and time reversal invariance. For the electron, expectations from Standard Model physics are about 5×10^{-38} and hence this experiment has a large discovery potential between that expectation and the current limit [31,32] of 5.6×10^{-27} ecm (95% CL). For muons, the EDM is enhanced by at least m_μ/m_e. Non-SM CP violation models [33–36] give lepton EDM's in the range 10^{-21} to 10^{-26} ecm. The current limit on the muon EDM is $d < 1.05 \times 10^{-18} e - cm$, derived from the last CERN g-2 experiment [37].

An experiment was described by W. Morse [38] at this workshop, for which a letter of intent has been submitted to BNL [39] proposes to exploit the muon storage ring built for the BNL g-2 experiment to search for a muon EDM with a sensitivity approaching 10^{-24}. Given the enhancement with respect to the electron EDM, this experiment could be competitive in searching for non-SM sources if either there is an enhancement bigger than the ratio of masses, or if the experiment can be done better.

In a storage ring, the frequency of precession of the spin relative to the momentum vector, in the limit that $E/c << B$, is given by:

$$\vec{\omega} = -\frac{e}{m}\left[a\vec{B} + \left(\frac{1}{\gamma^2 - 1} - a\right)\frac{\vec{\beta} \times \vec{E}}{c} + \frac{\eta}{2}(\vec{\beta} \times \vec{B})\right] \quad (2)$$

where $a = (g-2)/2$ and $\eta = 2d\frac{2mc}{\hbar}$. By choosing the momentum such that the first two terms in the expression for $\vec{\omega}$ cancel, the precession frequency is given by $\vec{\omega} = \frac{e\eta}{2m}(\vec{\beta} \times \vec{B})$. Hence, the measured quantity is the asymmetry in the rate of electrons from muon decay detected above and below the plane of the storage ring. Significant experimental challenges exist. The electric field is required to be purely radial to high precision. The field required is determined by the the requirement that the first two terms in the expression for $\vec{\omega}$ cancel, and is 2×10^6 V/m. Any component of $\vec{E}$ in the direction of $\vec{B}$ will fake an EDM; at a sensitivity of 10^{-24} ecm for the EDM, the required alignment is 0.1 μrad.

The proposed BNL experiment would reach its sensitivity goal with a 4800 hour run, with 10^{14} protons of 13.4 GeV/c momentum produced each 1.25 s, resulting in a total of 10^{15} stored muons. The requisite muon polarization is achieved by selecting only those muons which decay in the backward direction with respect to the pion direction.

The proponents of this experiment believe that substantial gains could be achieved with higher muon flux, which could be achieved at a muon collider front end. The sensitivity to a muon EDM is proportional to the square root of the number of detected muon decays and to the muon polarization. The desired muon beam properties are polarization > 50%, a repetition rate of 15 Hz, and 5×10^{11} stored muons per pulse.

SUMMARY

A number of ongoing and proposed experimental efforts have been described. The physics motivation for many of these experiments is extremely strong. These efforts, if they succeed, will push experimental sensitivities to a level which is very interesting from a theoretical perspective.

There are clear benefits to these experiments from improved beams with higher intensity, less contamination, and lower energy spread. In most cases, it appears that experimental sensitivity could be improved and backgrounds could be reduced substantially. However, some technical difficulties are apparent in a straightforward utilization of the FEMC beams. The largest difficulty is with the beam time structure, which, for most experiments, is required to be nearly continuous in order to reduce instantaneous rates. Ways in which the FEMC beam could be made continuous have been discussed, without a resolution yet in hand.

Even if the FEMC beams eventually are not used for these experiments, a great deal of overlapping R&D is required to build high intensity muon beams at existing facilities (i.e. the AGS) and for the muon collider. Among the relevant projects are the target and source region in a high field solenoid, the use of curved solenoids to transport and momentum and sign select the beams, and the eventual use of a phase rotation beam-line to reduce the beam momentum spread. Collaborative efforts on this R&D are beginning and should be encouraged. The experiments using high intensity muon beams that will be done in the coming years may also provide the opportunity for physicists involved in muon collider development to maintain a particle physics research effort during this long effort, and that is also to be encouraged.

ACKNOWLEDGEMENTS

I thank the organizers for providing a productive working environment; the co-convener of this working group, Tom Diehl, for his help in the organization and his contributions; and the members of the working group for their contributions. These include K. Arisaka, M. Bachman, V. Balbekov, A. Blondel, M. Brennan, S. Carabello, M. Cooper, A. Czarnecki, R. Djilkibaev, E. Hungerford, P. Kammel, D. Kawall, H. Kirk, D. Koltic, Y. Kuno, T.J. Liu, W. Marciano, G. Marshall, N. Mokhov, C. Moore, W. Morse, S. Schnetzer, J. Sculli, S. Striganov, E. Swallow.

REFERENCES

1. W. Marciano, plenary talk at the *Workshop on Physics at the First Muon Collider and at the Front end of the Muon Collider*, Nov. 6-9, 1997.
2. R. Barbieri, L. Hall and A. Strumia, *Nucl. Phys.* B **445**, 219 (1995).
3. Arkani-Hamed, *et al.*, "Flavor Mixing Signals for Realistic Supersymmetric Unification", LBL-37343 (1996).
4. Barbieri and Hall, "A Grand Unified Supersymmetric Theory of Flavor", LBL-38381 (1996).
5. J. Hisano, T. Moroi, K. Tobe and M. Yamaguchi, *Phys. Lett.* B **391**, 341 (1997). "Exact Event Rates of Lepton Flavor Violating Processes in Supersymmetric SU(5) Model", LBL-38653 (1996).
6. M.D. Cooper, *et al.*, in *Proceedings of 6th Conference on Intersections Between Particle and Nuclear Physics*, edited by T.W. Donnelly, pp. 34-48.
7. E. Hungerford, talk at the *Workshop on Physics at the First Muon Collider and at the Front end of the Muon Collider*, Nov. 6-9, 1997.
8. Y. Kuno and Y. Okada, LANL hep-ph/9604296 (1996);
   Y. Kuno, A. Maki and Y. Okada, LANL hep-ph/9609307 (1996);
9. C. Walter, in *Proceedings of 6th Conference on Intersections Between Particle and Nuclear Physics*, edited by T.W. Donnelly.
10. Bellgardt *et al.*, *Nucl. Phys.* B **299**, 1 (1988).
11. A. Czarnecki, W. Marciano and K. Melnikov, hep-ph/9801218 (1997).
12. F. Riepenhausen, presented at the *Sixth Conference on the Intersections of Particle and Nuclear Physics*, Big Sky, Montana (1997).
13. M. Bachman, *et al.*, "A Search for $\mu^- N \to e^- N$ with Sensitivity Below 10^{-16}", AGS P940 (1997).
14. Y. Kuno and Y. Okada, *Phys. Rev. Lett* **77**, 434 (1996).
15. Y. Kuno, A. Maki, and Y. Okada, *Phys. Rev.* **D55**, 2517 (1997).
16. Y. Kuno, "Lepton Flavor Violating Rare Muon Decays and Future Prospects", this proceedings (1998).
17. Badert *et al.*, *Nucl. Physics* A**377**, 406 (1979).
18. R.M. Djilkibaev and V.M. Lobashev, *Sov. J. Nucl. Phys.* **49(2)**, 384 (1989).

19. V.S. Abadjev, et al., "MELC Experiment to Search for the $\mu^- A \to e^- A$ Process", INR preprint 786/92, November 1992.
20. M. Brennan, talk at the *Workshop on Physics at the First Muon Collider and at the Front end of the Muon Collider*, Nov. 6-9, 1997.
21. M. Bachman, "The MECO Muon Beam", proceedings of this workshop (1998).
22. C. Moore, talk at the *Workshop on Physics at the First Muon Collider and at the Front end of the Muon Collider*, Nov. 6-9, 1997.
23. C. Moore, private communication.
24. N. Mokhov, talk at the *Workshop on Physics at the First Muon Collider and at the Front end of the Muon Collider*, Nov. 6-9, 1997.
25. H. Kirk, talk at the *Workshop on Physics at the First Muon Collider and at the Front end of the Muon Collider*, Nov. 6-9, 1997.
26. D. Artmutliski, et al., *Sov. J. Nucl. Phys.* **48**, 161 (1988).
27. R. Djilkibaev, "MECO Muon Yield Simulation Using Experimental Data", this proceedings (1998).
28. M. Bachman, "The MECO Muon Beam", this proceedings (1998).
29. J. Sculli, "$\mu \to e$ Conversion Status and Prospects", this proceedings (1998).
30. M.D. Cooper, "New Ideas to Improve Searches for $\mu^+ \to e^+\gamma$, this proceedings (1998).
31. Commins, E., et al., *Phys. Rev.* **A50**, 2960 (1994).
32. Abdullah, K., et al., *Phys. Rev. Lett.* **65**, 2347 (1990).
33. Barger, V., Das, A., and Kao, C., *Phys. Rev.* **D55**, 7099 (1997).
34. Bowser-Chao, D., Chang, D., and Keung, W., *Phys. Rev. Lett.* **79**, 1988 (1997).
35. Bernreuther, W., and Suzuki, M., *Reviews of Modern Physics* **63**, 313 (1991).
36. Geng, C.G., and Ng, J.N., *Phys. Rev.* **D42**, 1509 (1990).
37. Bailey, J., et al., *J. Phys.* **G4**, 345 (1978).
38. W. Morse and Yannis Semertzidis, "Electric Dipole Moment of the Muon", proceedings of this workshop (1998).
39. W. Morse, et al., Letter of Intent to BNL PAC, (1997).

Deep Inelastic Scattering at a Muon Collider - Neutrino Physics

Heidi Schellman
Northwestern University

for the DIS working group:
A. Caldwell, L. de Barbaro, D. Harris, B. King,
M. Klasen, A. Kotwal, D. Krakauer, S. Magill, K. McFarland,
D. Naples, F. Olness, B. Pawlik, S. Ritz*, A. Romosan,
H. Schellman*, P. Spentzouris, E. Stern, J. Yu ...

I INTRODUCTION

Colliding muon beams are not the only source of interesting interactions at a muon-collider complex. A muon-proton collider is a distinct possibility and the muon beams themselves will produce high flux beams of muon and electron neutrinos wherever there is a long straight section. Muon-proton physics would require an additional proton machine joined to the muon collider complex but first rate neutrino physics requires only that a detector be located in the path of the neutrinos coming from the muon-collider complex. The DIS group considered both of these options, our conclusion is that a muon-proton collider is feasible, although not easy or cheap. Neutrino physics at a muon collider appears to be easy, inexpensive and would extend present measurements by several orders of magnitude.

The deep-inelastic-scattering (DIS) group considered two scenarios, a 200 GeVμ × 1000 GeV p collider [1] and neutrino beams generated by the decay of 250 GeV muons. [10] Use of the muon beam in fixed target mode was not considered as most existing experiments are already systematics limited.

II MUON-PROTON COLLIDER

The 200 GeV μ × 1000 GeV pcollider option is described in detail in a paper by V. Shiltzev [1] and a contribution by S. Ritz [3] to these proceedings. Such a collider would reach 3 times the center of mass energy of the current HERA $e-$p collider with much higher luminosity.

A Detector Design

A one TeV proton beam can be considered to consist of quark and gluons beams with typical energies between 0 and 3-400 GeV. As a result, the CM frame for valence quarks colliding with 200 GeV muons would be almost at rest in the lab frame. These kinematics are similar to those at D0 and CDF. This suggests a detector design similar to those at present colliders, with emphasis on the detection (and rejection) of muons in the final state. Due to background from muon halo around the beamline, scattering angles below 10 degrees would be very hard to measure. This essentially rules out measurements at very low quark momentum fraction x.

B Physics Possibilities

Physics possibilities at a μ-p collider were suggested by M. Klasen, S. Magill, A. Caldwell, D. Krakauer, F. Olness and S. Ritz.

An integrated luminosity of 10 fb^{-1}/year was suggested in the preliminary $\mu - p$ collider study by V. Shiltzev. With 10 fb^{-1}, 1 M events/year would be detected with momentum transfer Q^2 greater than 5000, GeV^2. The present Zeus sample from 34 nb^{-1} is 326 events above 5000 GeV^2.

These large data samples, even in the absence of low x acceptance, will allow very accurate measurements of the proton (and photon, see the contribution of M. Klasen) structure functions and their evolution over a Q^2 range where perturbative QCD calculations are reliable. At these Q^2, charged current interactions via W boson exchange also become a significant fraction of the total scattering cross section. This allows measurements of the quark content of the proton and of the electro-weak couplings similar to those presently done in neutrino experiments at $Q^2 << 500$ GeV.

With the same 1 year run, limits on lepto-quarks with standard couplings could be extended up to 800 GeV with coupling limits of 2×10^{-3} at 200 GeV. Limit can also be placed on some Higgs production models for masses less than 120 GeV.

C The bad news

Unfortunately, muon backgrounds due to beam losses elsewhere in the machine are believed to be very large with rates of up to 1 muon/cm^2/crossing. Most μ−p physics depends critically on the detection (or definite non-detection in the case of charged current interactions) of a single muon in the final state. Where $\mu\mu \to H \to bb$ may still be detectable in this environment, studying $\mu p \to \mu X$ will be much more difficult. Our conclusion is that designing a detector able to withstand these background rates would be even more difficult than detectors for the muon-collider itself.

III NEUTRINO BEAMS

The muon-collider complex will produce very intense neutrino beams; so intense that the original calculations of neutrino fluxes at the muon collider were in the context of radiation safety.

The neutrino flux, both ν_μ and $\bar{\nu}_e$, results from decays in flight of muons within the FMC. Intense collimated neutrino beams will be produced from any straight section in the muon-collider complex. As we were interested in high energy neutrino beams, we considered the beams coming from from the last phase of the recirculating linacs (RLA3) and from a 10 m straight section in the muon-collider (FMC) itself. In both cases we considered the fluxes available to a **parasitic** neutrino experiment, one where no modification to the lattice or the number of turns at a given energy is made. Increases of factors of 10 in neutrino fluxes could be achieved by simply lengthening the straight sections or letting the beam coast for a small fraction of a lifetime once it reaches full energy in the RLA3. For more details on the beam flux calculations see the contributions by D. Harris and K.C. McFarland [10] to these proceedings.

The beam line parameters we assumed were:

Source	RLA3	250 GeV Muon Collider (FMC)
E_μ	150 – 250 GeV	250 GeV
turns/pulse	12	1560
decay length	533 m	10 m
$<E_{\nu_\mu}>$	135 GeV	178 GeV
beam size 600 m	50% < 25 cm	50% < 15 cm
Event Rate per 40 tons/year	5×10^9	5×10^9

TABLE 1. Neutrino fluxes and event rates for two parasitic neutrino beams, numbers are for muon neutrinos only. The RLA3 rates assume that the machine is ramping through 12 turns.

Table 1 shows the parameters of the RLA3 and muon-collider interaction regions relevant to neutrino physics. Both will produce energetic neutrino beams, the RLA3 option has lower energy because the muons are at full energy only during the final turn. In both cases, a 40-ton neutrino detector located 600 m away would see 500 neutrino interactions per second of which 50% would be within 15-25 cm of the detector center.

Figure 1 shows the muon neutrino interaction rates as a function of energy expected from the RLA3 and collider straight sections for a 250 GeV muon collider.

In contrast, existing experiments see rates of order one neutrino interaction/second in fiducial volumes ten times larger. This increase of three orders of magnitude in neutrino flux calls for completely new thinking about detector design.

A Neutrino beam characteristics at a muon collider

Neutrino beams from muon decay would be substantially cleaner than those produced by the decay of pions and kaons in present beamlines. Such beamlines use protons and very complex targeting systems to produce pions and kaons with momentum spreads of $\simeq 10\%$. The exact momentum distributions of the decaying mesons are difficult to measure in situ. In addition, other processes involving neutral kaons and charm produced in the target or

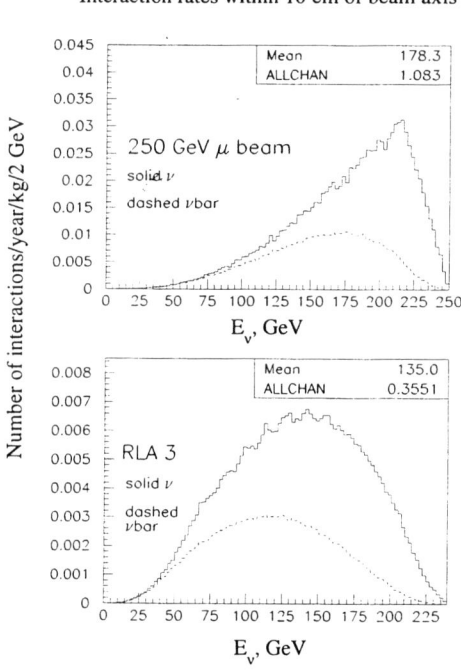

FIGURE 1. Neutrino interaction rates as a function of neutrino energy for ν_μ and $\bar{\nu}_\mu$ beams from a 250 GeV muon-collider complex

dump can produce background neutrinos. Accurate knowledge of the neutrino flux requires careful simulation of the beamline and many underlying physical processes. As a result, the total neutrino-scattering cross section is still not well understood and was last measured in the late 1980's. [5]

In contrast, at the muon collider, the muon beam will be essentially monochromatic and there will be no dumps to produce stray neutrinos. The neutrino flux, for both muons and electrons, will depend only on the beam energy, beam intensity and the muon polarization. Although we did not do detailed studies for this workshop, we believe that the well defined beam spectra at a muon collider will substantially lower the systematic uncertainties in most measurements.

Note: For many neutrino physics topics, comparison of neutrino and antineutrino rates *in the same detector* are needed. Polarity reversal in the beamlines is absolutely necessary for neutrino physics.

B Backgrounds from neutrino-reinteractions

Any neutrino detector must be located far enough away from the accelerator for primary muons to range out. For 250 GeV muons going through concrete this distance is around 600 meters. However, the neutrinos can reinteract in any shielding material and, in the case of a very intense beam, the reinteraction products from the shielding could be significant even at large distances from the accelerator. For example, the studies by Harris and McFarland [10] showed that a 40 ton detector with multiple tons of concrete directly preceding it will be overwhelmed by muons produced in neutrino interactions upstream in the shielding. In present experiments, such upstream interactions already result in 5-10% deadtime. In the high flux environment of the muon-collider complex, vetoing such interactions would be futile, in the RLA3 scenario, this upstream muon rate would be 4/event!

A possible solution is 300 meters of magnetized iron followed by 300 meters of air which would allow muons produced in the iron to sweep away from the detector. A simple calculation indicates that such a system reduces the muon flux from the shielding by more than an order of magnitude. A clean neutrino experiment does appear to be possible in the muon-collider environment.

C Limitations of present Neutrino Experiments

To date, two paths have been taken in studying high energy neutrino interactions. In one, maximum statistical power is gained by the use of massive targets weighing hundreds of tons. This is the path followed by CCFR, CDHSW and CHARM [6] and used in high statistics measurements of structure functions and weak interaction couplings. Such targets provide high statistics with the following compromises:

- High A targets - results from muon and electron scattering [7] indicate that the total cross section can vary by up to 30% as A increases, with diffractive production becoming much more important at high A and low Q^2.

- Coarse grained measurement - the target material itself causes substantial multiple scattering. Typical angular resolutions are 10 mr and momentum resolutions are worse than 10%. The accessible kinematic region is highly restricted and unsmearing the data is very difficult. These measurement difficulties have limited the kinematic regions accessible to high statistics experiment to momentum fraction $x > 0.01$.

The alternatives, fine-grained low mass experiments, (E872, CHORUS, FOCUS, BEBC, LAB C) suffer from low statistics.

Despite these substantial experimental limitations, neutrino experiments are still our only direct source of information on the anti-quark content in the proton and provide competitive measurements of some Kobayashi-Maskawa matrix elements, the strong coupling constant, and the Weinberg Angle.

The factor of a thousand in flux available parasitically at the muon-collider complex changes this picture entirely. For the purposes of this workshop, we followed a suggestion by Bruce King [9] (see his writeup in these proceedings) and considered a radically different experimental design, similar in many ways to the spectrometers used in muon scattering experiments such as SMC and E665. The change in perspective is illustrated in table II where we redefine the figure of merit for a neutrino experiment from interactions/kilo-ton of steel/year to interactions/meter of liquid hydrogen/year.

Table 2 shows the rates expected in a 10 cm diameter ×100 cm long hydrogen target in one year of running. Rates for a larger detector are also shown.

Target radius	RLA3	250 GeV Muon Collider	CCFR
10 cm	750,000	2,400,000	
200 cm	6,500,000	9,400,000	2,000,000

TABLE 2. Muon neutrino interaction rates/year expected in a one meter long liquid hydrogen target located 600 m from the muon-collider complex. The CCFR detector rates, with 680 tons of fiducial volume, are shown for contrast.

A small hydrogen target would see interaction rates/year of 1-2 M, directly comparable to the present NuTeV experiment which uses a 680 ton iron target. The gain in statistical power comes from two factors, first the flux of neutrinos is several thousand times higher, and second, the beam is more collimated, allowing a much smaller and hence less massive target.

Bruce King proposes an experiment with a hydrogen target followed by silicon strip detectors, a TPC, Cerenkov counters for particle ID and electromagnetic and hadronic calorimetry. This detector is quite similar to existing Fermilab and CERN fixed-target spectrometers and would provide very accurate measurements of neutrino events.

IV PHYSICS WITH A LIGHT NEUTRINO DETECTOR

A Weinberg Angle measurements

The ratio of neutral current to charged current interactions observed in muon neutrino interactions provide an interesting measurement of the Weinberg angle $\sin^2 \theta_W$ which is complimentary to direct measurements of the W mass and to indirect measurements made at e^+e^- colliders. The current NuTeV experiment should measure M_W with a precision of 125 MeV using around 1M events. The muon collider neutrino beams would provide both more statistics and eliminate the most important systematic uncertainties in present measurements. This topic is treated in more detail in the contribution

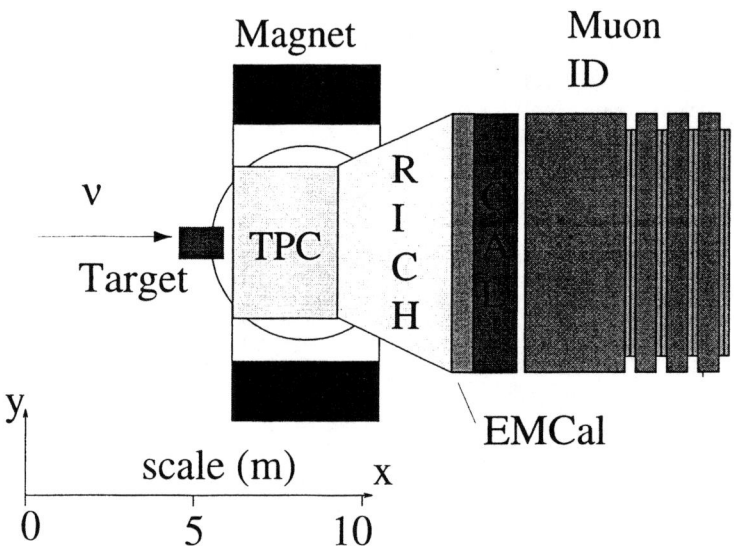

FIGURE 2. A typical fixed target experiment in a neutrino beam. The target is 1m of liquid deuterium followed by a TPC for tracking, a ring imaging Cerenkov chamber for particle ID, electromagnetic and hadronic calorimetry and a muon tagging system.

to these proceedings by J. Yu and A.V. Kotwal [8]. A deuterium target 1 meter in diameter and 1 meter long could detect 20 M ν_μ and 20 M $\bar{\nu}_e$ interactions per year. This would yield a statistical error of 0.0004 on the Weinberg angle, an improvement of a factor of 5 from the current NuTeV measurements. At present, the major sources of systematic errors in this measurement are:

- The kinematic suppression of $s \to c + W^-$ vs $s \to s + Z^0$ which depends on the dynamics of charm production. This suppression artificially raises the ratio of cross sections and introduces an energy dependence which is presently one of the largest errors in neutrino measurements of $\sin^2 \theta_W$.

  Solution - measure charm production directly via a silicon vertex detector.

- $\bar{\nu}_e$ contamination in the beam. In coarse-grained detectors, $\bar{\nu}_e \to eW^+$ looks like $\nu_\mu \to \nu_\mu Z^0$ as final-state electrons are merged with the hadronic shower.

  Solution - a low density detector will allow detection and identification of the final state electron. The anti-neutrino background becomes an advantage, one can use the electron neutrinos from the beam to double the statistical power of the measurement.

A quick estimate for the first year of running a parasitic small experiment is $\Delta \sin^2 \theta_W \leq 0.00010$ which translates into an error on the mass of the W of 30-50 MeV.

B Charm production

Around 5% of charged current neutrino interactions involve charm production. One could expect 100 - 500 K reconstructed charm events per year from a small active silicon target. While these rates are low compared to present Fermilab fixed target experiments and e^+e^- colliders, a charm measurement in a 180 GeV neutrino beam combines the decay length resolution of fixed target experiments with the low backgrounds seen at lepton colliders. In addition, the sign of the charmed quark is tagged by the muon or electron charge, allowing studies of Cabibbo suppressed modes and CP violation in the charm system which are competitive with much higher statistics samples from lepton colliders.

See the contributions by Panagiotis Spentzouris [12] in these proceedings for more detail on charm production.

1 Bubble Chambers?

A 1 meter diameter liquid hydrogen bubble chamber would trigger at 1 Hz, an event rate which has been feasible in such experiments since the 1970's.

With high resolution CCD readout and 2010 pattern recognition algorithms, bubble chamber measurements of short-lived particles similar to those done at SLAC and in the Tohoku bubble chamber at Fermilab may again be feasible.

C Quark content of the proton

Charged-current neutrino scattering is uniquely sensitive to the quark content of the proton as neutrinos are very selective about the helicity and charge of quarks they will scatter from.

- Neutrino scattering from protons is sensitive to d, s and $\bar{u}$ quarks,
- Anti-neutrino scattering is sensitive to $\bar{d}$, $\bar{s}$ and u quarks.
- The relative contributions d and s quarks can be untangled via charm production.
- Data from deuterium targets provides a different mix of u and d quarks in the target and thus allows one to distinguish u from d and $\bar{u}$ from $\bar{d}$.

High statistics neutrino experiments with good charm tagging will be able to give accurate parton distributions for each flavor. Such measurements will be of considerable use to the LHC experiments.

D Spin physics

Neutrino beams, as they are guaranteed to be 100% polarized, are also ideal for spin physics measurements. Neutrinos only scatter from left-handed quarks or right-handed anti-quarks, while anti-neutrinos have the opposite taste. A polarized target allows measurements of the spin asymmetries in parton densities where Δq is defined as the difference between the quark contents parallel and anti-parallel to the proton spin. A large (200 kG) polarized target could see up to 1M events/year from polarized nuclei with charm in the final state. This would yield a 3% measurement of the spin asymmetry in the strange sea. [11]

E B- physics?

Bruce King, Donna Naples and Alex Romosan pointed out that an active silicon target could be used to search for the process $u \to b$. As there is little contribution from $c \to b$ due to the suppression of charm quarks in the proton, this process would provide a very clean measurement of V_{ub}. A small parasitic experiment such as those considered above, would yield 20 such events/year, a bigger target, longer running or longer straight sections could make this a high precision measurement.

F Higher energies

As the neutrino interaction cross section scales with beam energy for center-of-mass energies $\ll M_W$, the event rates shown here will rise linearly with machine energy. Charm and B production rates should rise even faster as they are currently limited by kinematic constraints. Shielding design will become more crucial as the backgrounds, which come from neutrinos, will scale with the event rate.

V CONCLUSIONS

A μ–p collider is very attractive, with very high luminosity and three times the center-of-mass energy currently available at HERA. Unfortunately, backgrounds from muon halo make measuring the final state muon very difficult. A muon collider would be confined to high-x high-Q^2 kinematics.

However, the muon-collider complex, even during intermediate phases of construction, will provide neutrino beams of unprecedented cleanliness and intensity. Measurements which have been done poorly (total cross section) or not at all ($u \to b$) can be done with high precision. Detector designs previously used in fixed target hadron and muon experiments could be directly adapted to neutrino physics. We were unable to find any technical problems which would prevent the muon-collider complex from producing a revolution in neutrino physics at an energy of 250 GeV. Prospects improve even further if the machine energy is raised.

REFERENCES

1. V.D. Shiltzev, An Asymmetric Muon-Proton Collider: Luminosity Considerations, Fermilab-Conf-97/114.
2. D.R. Harris and K.S. McFarland, *Detectors for Neutrino Physics at the First Muon Collider*, these proceedings.
3. S. Ritz, *On Future Charged Lepton-Hadron Colliders*, these proceedings.
4. M. Klasen, these proceeding.
5. P. Berge et al., Z. Phys. **C49**, 187 (1991), R. Oltman et al., Z. Phys. **C53**, 51 (1992).
6. See the *Current Experiments in Elementary Particle Physics* database at: http://www-spires.slac.stanford.edu/find/experiments.
7. M.R. Adams et al., Z. Phys. C67, 403 (1995), P. Amaudruz et al., Nucl. Phys. **B441**, 3 (1995).
8. J. Yu and A.V. Kotwal, *Measurement of* $\sin^2 \theta_W$ *at the First Muon Collider*, these proceedings.
9. B.J. King, these proceedings.

10. D.R. Harris and K.S. McFarland, *Detectors for Neutrino Physics at the First Muon Collider*, these proceedings.
11. D.R. Harris and K.S. McFarland, *A Small Target Neutrino Deep-Inelastic Scattering Experiment at the First Muon Collider*, these proceedings.
12. P.G. Spentzouris, these proceedings.

Higgs Boson and Z Physics at the First Muon Collider

Marcel Demarteau* and Tao Han[†]

*Fermi National Accelerator Laboratory, P.O.Box 500, Batavia, IL 60510
[†]Department of Physics, University of California, Davis, CA 95616
and
Department of Physics, University of Wisconsin, Madison, WI 53706

Abstract. The potential for the Higgs boson and Z-pole physics at the first muon collider is summarized, based on the discussions at the "Workshop on the Physics at the First Muon Collider and at the Front End of a Muon Collider".

INTRODUCTION

Muon colliders offer a wide range of opportunities for exploring the physics within and beyond the Standard Model (SM). Because the muon mass is about 200 times larger than the electron mass, s-channel production of the Higgs boson, and its associated advantages with regard to measurements of Higgs boson properties, is one of the unique features of a muon collider. Since the muons are produced in a decay channel by moving pions, the muons naturally carry a longitudinal polarization of about 20%. A collider allowing for adroit manipulation of the polarization and center of mass energy, combined with the prospect of luminosities in the range of $10^{32} < \mathcal{L} < 10^{33}$cm^{-2}s^{-1} would make for a very powerful probe of the structure of the fundamental forces with unparalleled potential. In this report a summary of the Higgs and Z-pole physics, as well as some other aspects of the electroweak boson physics, as discussed at the workshop on the physics at the First Muon Collider (FMC), is presented [1].

EXPERIMENTAL CONSIDERATIONS

The experiments at the e^+e^--colliders LEP and SLC have shown that the calibration of the luminosity, beam energy and beam polarization is crucial for the physics results obtained. At LEP the luminosity is measured with small angle silicon based calorimeters, counting Bhabha events to a precision of $\frac{\delta \mathcal{L}}{\mathcal{L}} = 10^{-3}$. The Bhabha cross section has been measured down to angles of about 30 mrad with respect to the beam direction. At the FMC, however, it is not clear if the muon Bhabha cross section can be measured down to small angles. The muons, with a lifetime of about 2 µs in their rest frame, can circulate in the machine for only about 800 turns at a center of mass energy $\sqrt{s} = 0.5$ TeV [2]. The electrons from the decay of the muon for a major problem. For $2 \cdot 10^{12}$ muons per bunch there are $3 \cdot 10^5/E_{\text{beam}}$ decays per meter, with E_{beam} in TeV. Because the final focus is tuned to the beam energy of the muon beams, the decay electrons will be sprayed out over the interaction region. Current detector designs [3] include an uninstrumented cone of 10° - 20° with respect to the beamline because of the large direct and induced backgrounds. It is thus unclear if a similar precision on the luminosity measurement can be obtained using muon Bhabha scattering or using an alternative method. For the discussions to follow it is assumed that at the FMC a precision on the luminosity measurement of $\delta \mathcal{L}/\mathcal{L} = 10^{-3}$ is achievable.

The beam energy at LEP is measured most accurately using the technique of resonant depolarization which has an ultimate accuracy of about 200 keV. This calibration, however, cannot be performed very often since it takes a long time for the polarization to build up in the beam. Moreover, it cannot be done during a physics run and has been performed with separated beams only. These and other limitations, resulted in a final uncertainty on $\sqrt{s}$ at LEP of about 1 MeV.

At the FMC the natural polarization of the muons [2] provides a mechanism to measure not only the beam energy but also the polarization itself. The precession of the polarization vector with respect to the momentum vector is governed by the muon spin tune, $\frac{g-2}{2}\gamma$, which corresponds to the number of precessions in one turn around the ring. Coincidentally, for muons at the Z-mass the spin tune is almost exactly 1/2. Thus, the spin flips each turn at $\sqrt{s} \approx M_Z$. The energy spectrum of the decay electrons depends on the muon polarization. By measuring the average energy of the decay electrons each turn at a fixed point along the circumference of the machine a measure of the beam energy and the polarization can be obtained. The measured energy spectrum will exhibit an oscillatory behavior as function of turn number. The frequency and amplitude of the oscillations measure the beam energy and the polarization, respectively. Initial studies for a perfect planar machine geometry show that an absolute energy calibration at the statistical level of $\delta E/E = 10^{-5}$ is easily feasible [4]. Systematic effects will be the dominant sources of uncertainty and are being studied. A clear advantage of this procedure is that the measurements are done concurrently with physics runs, directly sampling the interacting muon bunches.

The beam spread is controlled by the beam optics and a narrow band beam option at lower center of mass energies would provide a beam spread of $R = \sigma(p)/p = 3 \cdot 10^{-5}$.

HIGGS BOSONS

The SM as it stands is incomplete. Many fundamental questions of nature are left unanswered. Among them the question of electroweak symmetry breaking takes a prominent position. In the SM, the electroweak symmetry is broken spontaneously through a fundamental Higgs doublet field, giving rise to a single physical neutral Higgs boson (h^0). In the minimal supersymmetric standard model (MSSM), there are two neutral $\mathcal{CP}$-even Higgs bosons (h^0, H^0), two charged Higgs bosons ($H^{\pm}$), and a $\mathcal{CP}$-odd neutral Higgs boson (A^0). To understand the electroweak symmetry breaking and to explore new physics beyond the SM, the study of the Higgs sector is of highest priority for future collider experiments [5].

Although the Higgs boson masses are largely free parameters, theoretical arguments indicate that there exists an upper limit on the lightest Higgs boson mass, namely, $m_h < 150$ GeV/c^2 [5]. The FMC then is unique in studying light Higgs bosons since they can be produced through s-channel production at resonance, due to the sizeable coupling of a Higgs boson to muons [6–8].

The resonance cross section for Higgs boson production is given by

$$\sigma_h\left(\sqrt{\hat{s}}\right) = \frac{4\pi \Gamma(h \to \mu\mu)\,\Gamma(h \to X)}{(\hat{s} - m_h^2)^2 + m_h^2 \Gamma_h^2}, \tag{1}$$

where $\hat{s}$ is the c.m. energy squared, Γ_h is the total width, and X denotes the final state. The Higgs coupling to fermions is proportional to the fermion mass so the corresponding s-channel process is highly suppressed at e^+e^- colliders with respect to muon colliders. The cross section must be convoluted with the machine energy spectrum, approximated by a Gaussian distribution of width $\sigma_{\sqrt{s}}$,

$$\bar{\sigma}_h\left(\sqrt{s}\right) = \int \sigma_h\left(\sqrt{\hat{s}}\right) \frac{\exp\left[-\frac{\left(\sqrt{\hat{s}} - \sqrt{s}\right)^2}{\left(2\sigma_{\sqrt{s}}^2\right)}\right]}{\sqrt{2\pi}\,\sigma_{\sqrt{s}}} d\sqrt{\hat{s}}. \tag{2}$$

The root mean square spread $\sigma_{\sqrt{s}}$ in c.m. energy is given in terms of the beam resolution R by

$$\sigma_{\sqrt{s}} = (7 \text{ MeV}) \left(\frac{R}{0.01\%}\right)\left(\frac{\sqrt{s}}{100 \text{ GeV}}\right), \tag{3}$$

where a resolution down to $R = 0.003\%$ may be realized at the FMC. In comparison, a value of $R \sim 1\%$ is expected at the Next Linear e^+e^- Collider (NLC). To study a Higgs resonance one wants to be able to tune the machine energy to $\sqrt{s} = m_h$. For this purpose the monochromaticity of the beam energy is vital.

When the resolution is much larger than the Higgs width, $\sigma_{\sqrt{s}} \gg \Gamma_h$, the effective s-channel cross section is

$$\bar{\sigma}_h = \pi\sqrt{2\pi}\,\frac{\text{BF}(h \to \mu\mu)\,\text{BF}(h \to X)}{m_h^2} \cdot \frac{\Gamma_h}{\sigma_{\sqrt{s}}}. \tag{4}$$

It becomes clear that it would be desirable to get as good a beam energy resolution as possible because of the factor $\Gamma_h/\sigma_{\sqrt{s}}$. In the other extreme when the resolution is much smaller than the width, $\sigma_{\sqrt{s}} \ll \Gamma_h$, the effective cross section is

$$\bar{\sigma}_h = 4\pi \frac{\mathrm{BF}(h \to \mu\mu)\,\mathrm{BF}(h \to X)}{m_h^2}. \tag{5}$$

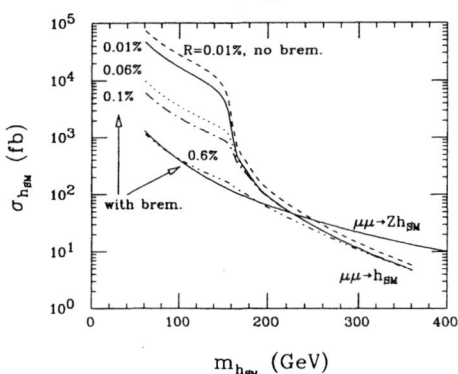

FIGURE 1. The s-channel cross section for $\mu^+\mu^- \to h$ for several choices of the beam resolution R. Also shown is the $\mu^+\mu^- \to Zh$ cross section at $\sqrt{s} = M_Z + \sqrt{2}\,m_h$, from Ref. [8].

Figure 1 illustrates the SM Higgs cross section for several choices of the machine resolution. For $m_h < 2M_W$, Γ_h is very narrow and a better beam resolution can significantly improve the signal rate. On the other hand, for $m_h > 2M_W$ the SM Higgs boson becomes increasingly broad and the effect of $\sigma_{\sqrt{s}}$ is negligible. To effectively explore Higgs physics, the resolution requirements for the machine thus depend on the Higgs width. Figure 2 gives both SM and SUSY Higgs width predictions versus the Higgs mass. A SM Higgs of mass $m_h \sim 100$ GeV/c² has a width of a few MeV. The width of the lightest SUSY Higgs may be comparable to that of the SM Higgs (if $\tan\beta \sim 1.8$) or much larger ($\Gamma_h \sim 0.5$ GeV for $\tan\beta \sim 20$). The width parameter characterizes the fundamental couplings of the Higgs boson to other particles. Figure 3 shows light Higgs resonance profiles versus the c.m. energy $\sqrt{s}$. With a resolution $\sigma_{\sqrt{s}}$ of order Γ_h the Breit-Wigner line shape can be measured and Γ_h determined. For the moment we must plan for a resolution $R \lesssim 0.01\%$ in order to be sensitive to Γ_h of a few MeV.

The prospects for observing the SM Higgs are evaluated in Fig. 4. The first two panels give the signal and background for a resolution $R = 0.003\%$. The third panel gives the luminosity needed for a 5σ detection in the dominant $b\bar{b}$ final state. The luminosity requirements are very reasonable, except for the Z-boson peak region.

It is likely that a SM-like Higgs boson will have been discovered at the LHC or NLC when the FMC starts its mission. The mass of a light Higgs boson will have

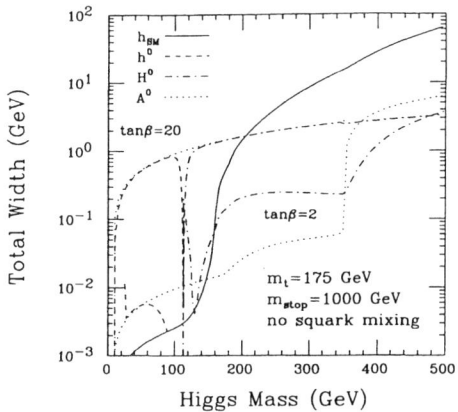

FIGURE 2. Total width of the SM and MSSM Higgs bosons with $\tan\beta = 2$ and 20, from Ref. [8].

been measured to an accuracy of approximately 200 MeV/c^2 [9]. From a rough scan for the s-channel h^0 signal over this 200 MeV range, the mass can be determined to an accuracy $\Delta m_h \sim \sigma_{\sqrt{s}}$. If $S/\sqrt{B} \gtrsim 3$ is required for detection or rejection of a Higgs signal and a resolution $R \sim 0.003\%$ ($\sigma_{\sqrt{s}} \sim 2$ MeV) is employed, then the necessary luminosity per scan point is 0.0015 fb^{-1} for $m_h \lesssim 2M_W$ and m_h not near M_Z. As an example, suppose that the LHC has measured $m_h = 110.0 \pm 0.1$ GeV/c^2. The number of scan points to cover a 200 MeV/c^2 region in $\sqrt{s}$ at the FMC is ~ 200 MeV/2 MeV = 100, and a total luminosity of $100 \times (0.0015$ fb^{-1}/point$) = 0.15$ fb^{-1} is needed to discover the Higgs and reach an accuracy on its mass of

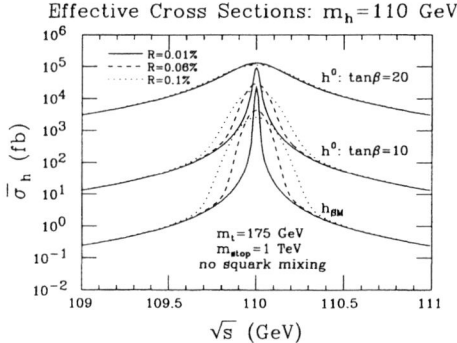

FIGURE 3. Effective s-channel Higgs cross section $\bar{\sigma}_h$ obtained by convoluting the Breit-Wigner resonance formula with a Gaussian distribution for resolution R, from Ref. [8].

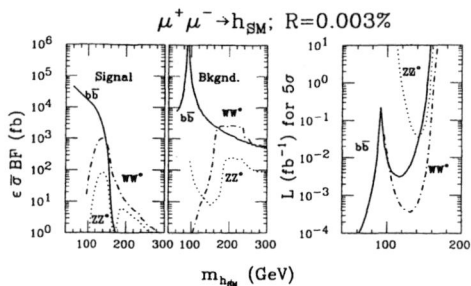

FIGURE 4. The SM Higgs cross sections and backgrounds in $b\bar{b}$, WW^* and ZZ^*. Also shown is the luminosity needed for a 5 standard deviation detection in the $b\bar{b}$ decay mode, from Refs. [7,8].

$$\Delta m_h \simeq \sigma_{\sqrt{s}} \sim 2 \text{ MeV}/c^2. \tag{6}$$

Once m_h is determined to an accuracy $\Delta m_h \sim \mathcal{O}\left(\sigma_{\sqrt{s}}\right)$ a three point fine scan can be made with one setting at the apparent peak and two settings on the wings at $\pm\sigma_{\sqrt{s}}$ from the peak. The ratios of $\sigma(\text{wing}^i)/\sigma(\text{peak}^i)$ determine m_h and Γ_h. With a good energy resolution $R = 0.003\%$ and an integrated luminosity $L_{\text{total}} = 0.4 \text{ fb}^{-1}$, for $m_h = 110 \text{ GeV}/c^2$ and $\Gamma_h = 3$ MeV, accuracies of

$$\Delta\Gamma_h/\Gamma_h = 16\%, \quad \text{and} \quad \Delta m_h \sim 0.1 \text{ MeV}/c^2 \tag{7}$$

could be achieved. At the same time, branching ratios for the dominant decay channels can be also measured to a good precision, i.e., 3% and 15% for $\sigma \cdot \text{BF}(b\bar{b})$ and $\sigma \cdot \text{BF}(W^+W^-)$, respectively. The high precision reached would have significant impact on the electroweak physics within and beyond the SM. For instance, with a measurement of an accuracy about 15% on the ratio $\text{BF}(W^+W^-)/\text{BF}(b\bar{b})$, one should be able to infer A^0 effects up to $M_{A^0} \gtrsim 400 \text{ GeV}/c^2$ by comparing the predictions from the MSSM and the SM [7,8]. However, to reach the necessary precision within a sensible time scale, a machine luminosity of 10^{32} cm^{-2}s^{-1} (or 1 fb^{-1}/yr) and a good energy resolution $R = 0.003\%$ are highly desirable.

The heavier neutral MSSM Higgs bosons are also observable in the s-channel. Figure 5 give the cross sections and significance of the $\mathcal{CP}$-odd state A^0 versus the A^0 mass, assuming $R = 0.1\%$ and $L = 0.01 \text{ fb}^{-1}$. Discovery and study of the A^0 is possible at all m_A if $\tan\beta > 2$ and at $m_A < 2m_t$ if $\tan\beta \lesssim 2$.

The possibility that A^0 and H^0 may be nearly mass degenerate is of particular interest for s-channel Higgs studies. In the large m_A limit, typical of many supergravity models, the masses of A^0, H^0 and $H^\pm$ are similar and h^0 is similar to h_{SM} in its properties. In this situation the A^0 and H^0 contributions can be separated by an s-channel scan; see Fig. 6.

It was reported during the workshop [10] that beam polarization is potentially useful for Higgs resonance studies, but only if the accompanying luminosity reduction is not significant. Large forward-backward asymmetries can also be used to

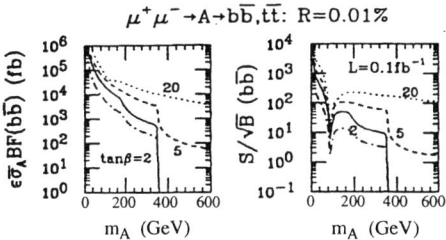

FIGURE 5. Cross sections and significance for detection of the A Higgs boson with an efficiency $\epsilon = 0.5$ and a luminosity $L = 0.1$ fb^{-1}, from Ref. [8].

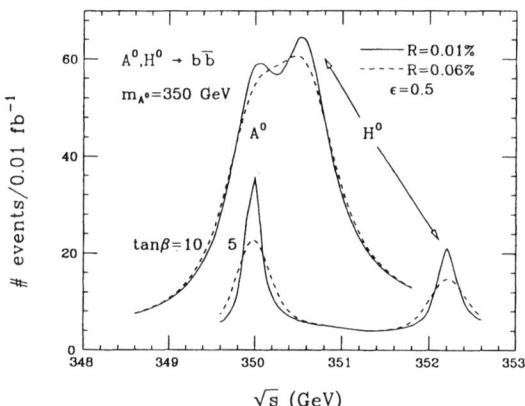

FIGURE 6. Separation of A^0 and H^0 Higgs signals for two values of $\tan\beta$, from Ref. [8].

enhance the Higgs "discovery" signal or improve precision measurements, particularly for the $\tau\bar{\tau}$ final state.

If the muon collider is running at energies above the narrow Higgs resonance, it may still be possible to pick up a signal sample through the photon radiation process $\mu^+\mu^- \to \gamma h$ with a cross section of the order of 0.1 fb. The authors in [11] studied this process and pointed out that the 1-loop contribution is comparable in size to the tree level result, and is especially sensitive to the Higgs coupling to the top quark and to anomalous Higgs couplings to W^+W^- and ZZ. It is also discussed [12] that $\mathcal{CP}$-odd kinematic variables may be constructed for processes like $h^0 \to W^+W^-, ZZ$ and $t\bar{t}$ so that one may be able to probe the $\mathcal{CP}$ properties of the Higgs boson couplings at muon colliders.

During the workshop, many other aspects of Higgs physics were discussed. They include Higgs boson searches at LEP [13,14]; Higgs physics at the Tevatron within the SM [15] and within the MSSM [16], at the LHC [17,18], and at the NLC [19].

Z POLE ELECTROWEAK PHYSICS

The success of the SM has arguably been most beautifully demonstrated by the agreement of the very precise Z pole measurements at the e^+e^- colliders and the direct measurements. It has been unprecedented that an anticipated quark was discovered with a mass exactly within the range predicted from loop corrections within a theoretical framework. This is a remarkable feat for experimentalists and theorists alike and attests to the enormous success of the SM. Even though many measurements are now being carried out with excruciating precision, the SM shows no signs of giving up its claim of being the description of the fundamental interactions as we know them. Despite these enormous successes there are some discrepancies in the data which, given that the LEP Z pole era is over, will most likely stay with us for a long time. The most significant discrepancy from lepton colliders is the measurement of $\sin^2\theta_{\text{eff}}^{\text{lept}}$, defined as

$$\sin^2\theta_{\text{eff}}^{\text{lept}} \equiv \frac{1}{4}\left(1 - \frac{g_V^{f\,2}}{g_A^{f\,2}}\right),$$

where $g_{V(A)}^f$ is the (axial-)vector coupling of the Z boson to fermion f. Currently SLD measures $\sin^2\theta_{\text{eff}}^{\text{lept}} = 0.23055 \pm 0.00041$ [20], derived from the left-right asymmetry of the total cross section and the leptonic forward-backward asymmetries, compared to the LEP average of $\sin^2\theta_{\text{eff}}^{\text{lept}} = 0.23196 \pm 0.00028$, determined from the Z partial decay widths and the forward-backward asymmetries [21]. The discrepancy between the measurements has a significance of 2.8σ. This discrepancy is rather significant. Considering only the A_{LR} measurement, combined with the direct determination of the W and top quark masses, the measurement implies a 95% CL upper bound on the Higgs mass of 77 GeV/c^2, while the direct searches at LEP yield a 95% CL lower limit on the Higgs mass also of 77 GeV/c^2 [22].

The FMC with its anticipated luminosity could contribute significantly to the physics in this sector [23]. Within a one year running period, due to a relaxed requirement on the beam energy resolution R for the broad Z-pole, a sample of 10^8 Z events could be recorded with both beams naturally polarized. To gauge the possible impact of such a large data sample it is instructive to look at the sensitivity of electroweak observables to SM parameters. Table 1 lists electroweak observable $\mathcal{O}$ together with its current measurement and sensitivity, $\Delta\mathcal{O}$, to the top quark mass, $m_t = 175.6 \pm 5.5$ GeV/c^2, the fine structure constant, $\alpha^{-1}(m_Z^2) = 128.896 \pm 0.090$, the strong coupling constant, $\alpha_s(m_Z^2) = 0.121 \pm 0.003$, and the Higgs boson mass, $60 < m_h < 1000$ GeV/c^2 [24]. The observables chosen are R_ℓ, the ratio of the hadronic over leptonic partial decay width of the Z, $R_\ell = \Gamma_{\ell\ell}/\Gamma_{\text{had}}$; $\Gamma_{\ell\ell}$, the Z partial decay width into leptons; $\sin^2\theta_{\text{eff}}^{\text{lept}}$; R_b, the fraction of hadronic Z decays coming from b quarks, $R_b = \Gamma_{b\bar{b}}/\Gamma_{\text{had}}$; and the mass of the W boson, m_W. The choice of these observables is given by their experimental uncertainty compared to their sensitivity to the various parameters in the model.

Observable ($\mathcal{O}$)	Average Value	$\Delta\mathcal{O}(\delta M_t)$	$\Delta\mathcal{O}(\delta\alpha)$	$\Delta\mathcal{O}(\delta\alpha_s)$	$\Delta\mathcal{O}(\delta m_h)$
$R_\ell \cdot 10^3$	20755 ± 27	1.8	4.0	21	28
$\Gamma_{\ell\ell}$ (MeV)	83.91 ± 0.10	0.06	—	0.02	0.25
$\sin^2\theta_{\text{eff}}^{\text{lept}} \cdot 10^4$	2315.2 ± 2.3	2.0	2.3	0.05	15.4
$R_b \cdot 10^4$	2170 ± 9	2.0	0.2	0.05	0.4
m_W (MeV/c^2)	80430 ± 80	37	14	1	200

TABLE 1. Results for various electroweak observables and their sensitivity to the top quark mass, α, α_s and the Higgs boson mass.

From the table one should observe that $\Gamma_{\ell\ell}$ is insensitive to α, whereas R_ℓ is very sensitive to α_s. It is also clear that the constraint on the Higgs mass is dominated by the measurement of $\sin^2\theta_{\text{eff}}^{\text{lept}}$. The measurement of M_W will become as significant as the current measurement of $\sin^2\theta_{\text{eff}}^{\text{lept}}$ when the experimental accuracy reaches a level of 30 MeV/c^2. If at that point, however, the top mass uncertainty has not been reduced, the constraint from the M_W measurement would partially be spoiled by the top mass uncertainty.

Within the framework of the SM the value of $\alpha_s(m_Z^2)$ derived from an analysis of electroweak precision data depends essentially on R_ℓ, Γ_Z and σ_h^0, with σ_h^0 the peak hadronic Z pole cross section. Since R_ℓ is very sensitive to α_s, the strong coupling parameter can be determined from the parameter R_ℓ alone. For $m_Z = 91.1867$ GeV/c^2, and imposing $m_t = 175.6 \pm 5.5$ GeV/c^2 as a constraint, a value of $\alpha_s(m_Z^2) = 0.124 \pm 0.004 \pm 0.002$ is obtained, where the second uncertainty accounts for the change in the result when varying m_h in the range $60 < m_h < 1000$ GeV/c^2 [21]. The experimental uncertainty is dominated by the limited statistics of the leptonic Z decays. With an improvement in statistics of a factor of 10 over the current LEP statistics, an uncertainty on α_s of 0.001 could be obtained [23].

The quantity $\Gamma_{\ell\ell}$ is of particular interest since it is independent of α and is only mildly dependent on α_s. As such, it is a sensitive indicator of possible new physics beyond the SM. A precision measurement of $\Gamma_{\ell\ell}$, however, requires a very accurate absolute luminosity calibration. $\Gamma_{\ell\ell}$ can be determined with better precision indirectly from

$$\Gamma_Z = \Gamma_{\ell\ell}\left(3 + N_\nu \frac{\Gamma_{\nu\nu}}{\Gamma_{\ell\ell}} + R_\ell\right)$$

using the measurement of Γ_Z, the SM prediction for $\frac{\Gamma_{\nu\nu}}{\Gamma_{\ell\ell}}$ and the measurement of R_ℓ. The uncertainty on R_ℓ, as seen above, will be very small and independent of luminosity. The uncertainty is driven by the uncertainty on the Z width, which depends on the luminosity through the point to point errors in the energy scan and on the energy calibration. At the FMC, where a continuous energy calibration should be feasible, the latter can be considerably reduced and an accuracy of $\frac{\delta\Gamma_{\ell\ell}}{\Gamma_{\ell\ell}} = 0.0003$ should be possible, compared to the current measurement of $\Gamma_{\ell\ell} = 83.91 \pm$

0.10.

Currently the most powerful way to determine $\sin^2\theta_{\text{eff}}^{\text{lept}}$ is to measure the left-right asymmetry, A_{LR}, defined as

$$A_{\text{LR}} = -\frac{1}{P}\frac{\sigma_{\text{L}} - \sigma_{\text{R}}}{\sigma_{\text{L}} + \sigma_{\text{R}}} \qquad (8)$$

where $\sigma_{R(L)}$ is the total production cross section for a right (left) handed polarized electron beam with average polarization P. At the Z pole, ignoring photonic corrections, $A_{\text{LR}} = \mathcal{A}_e$, where the asymmetry of couplings, $\mathcal{A}_f$, is given by

$$\mathcal{A}_f \equiv \frac{g_L^{f\,2} - g_R^{f\,2}}{g_L^{f\,2} + g_R^{f\,2}} = \frac{2 g_V^f g_A^f}{g_V^{f\,2} + g_A^{f\,2}}.$$

The measurement of A_{LR} at SLD, using a polarized electron beam with an average polarization during the last run of $\mathcal{P}_e = (76.5 \pm 0.8)\%$, yields a measurement of $\sin^2\theta_{\text{eff}}^{\text{lept}} = 0.23055 \pm 0.00041$ [20], of comparable precision to the combined LEP result, derived from the Z-pole and A_{FB} measurements. The power lies in the availability of polarized beams. The sensitivities of the two measurements are related as $\partial A_{\text{LR}}/\partial \sin^2\theta_{\text{eff}}^{\text{lept}} = \frac{1}{4}\partial A_{\text{FB}}/\partial \sin^2\theta_{\text{eff}}^{\text{lept}}$. Thus, compared to an A_{LR} measurement with fully polarized beams, a sixteen-fold larger data sample is required to achieve a similar accuracy in $\sin^2\theta_{\text{eff}}^{\text{lept}}$ from A_{FB}.

When both beams are polarized, the natural situation for a muon collider, equation 8 generalizes to

$$A_{\text{LR}} = \frac{1}{\mathcal{P}}\frac{\sigma_{\text{L}} - \sigma_{\text{R}}}{\sigma_{\text{L}} + \sigma_{\text{R}}} \quad \text{with} \qquad (9)$$

$$\mathcal{P} = \frac{P^+ - P^-}{1 - P^+ P^-} \qquad (10)$$

where $P^{+(-)}$ refers to the average longitudinal polarization of the positively (negatively) charged incident fermion beam. The total cross section is given by

$$\sigma = \sigma_0 \left\{ (1 - P^+ P^-) + (P^+ - P^-) A_{\text{LR}} \right\}. \qquad (11)$$

From equation 10 it can be seen that the quantity that controls the measurement of A_{LR} ($=\mathcal{A}_e$), and thus the precision of the $\sin^2\theta_{\text{eff}}^{\text{lept}}$ measurement, is $\mathcal{P}$, the effective polarization of the $\mu^+\mu^-$-system, which enhances the sensitivity of the measurement. For a 50% polarization of both beams, $P^\pm = \pm 50\%$, $\mathcal{P} = 80\%$. On top of that the cross section also increases by about 30%. Given the process under study a signal to background enhancement can be obtained by optimizing the polarization [10]. The most clear example is the production of the scalar Higgs boson, where the same sign spin is the favored production mode with a strong suppression of the background.

In the current design of the FMC a big loss in luminosity is incurred for increased polarization [2]. For a polarization of 50% there is a loss in luminosity of about a factor of 4 per beam. The statistical precision with which $A_{\rm LR}$ can be measured is

$$\delta A_{\rm LR} = \frac{1}{\mathcal{P}} \frac{1}{\sqrt{N}}. \tag{12}$$

The relevant quantity therefore to collect the data to measure $A_{\rm LR}$ with a certain precision is $L_P \cdot \mathcal{P}^2$, where L_P is the loss in luminosity for beam polarization P [23]. Given the relation between the loss in luminosity and the beam polarization in the current design of the FMC, this quantity reaches a maximum for a loss in luminosity of about 0.7 corresponding to a beam polarization of about 30%. The advantage of fully polarized beams does not outweigh the loss in luminosity incurred in the current design. The uncertainties on $A_{\rm LR}$ and $\sin^2\theta_{\rm eff}^{\rm lept}$ are related through $\delta A_{\rm LR} = 7.9\,\delta \sin^2\theta_{\rm eff}^{\rm lept}$. A data sample of 10^7 Z events with $\mathcal{P} = 0.5$, assuming equal statistical and systematic uncertainties, should make a measurement of $\delta \sin^2\theta_{\rm eff}^{\rm lept} < 10^{-4}$ easily feasible. The power of a measurement of $\sin^2\theta_{\rm eff}^{\rm lept}$ to such a precision can readily be seen from Fig. 7 which shows the dependence of $\sin^2\theta_{\rm eff}^{\rm lept}$ on m_h taking as input the current measured value for m_Z and assuming $m_t = 173.0\pm0.4$ GeV/c^2 and $\alpha^{-1}(m_Z^2) = 128.923 \pm 0.036$ [25]. Taking the current LEP central value for $\sin^2\theta_{\rm eff}^{\rm lept}$ a precision of 10^{-4} would constrain the Higgs mass to the range $170 < m_h < 370$ GeV/c^2. It should be noted that the uncertainty on the SM prediction is dominated by the uncertainty on α. An error on m_t of 2 GeV/c^2 is equivalent to an uncertainty on α of 0.04. With further improvements to the hadronic contribution of α to be anticipated, a precise measurement of $\sin^2\theta_{\rm eff}^{\rm lept}$ combined with a direct measurement of the Higgs mass, provides a very stringent consistency check of the SM.

It should be noted that relative total cross section measurements with different spin configurations also gives a measure of the polarization. This measures directly the polarization of the interacting muons and no corrections for beam transport to a polarimeter and sampled luminous region need to be applied.

Many more Z pole quantities can be studied at the FMC, notably in the b sector. This, however, has at present not been investigated and should be explored in future studies.

W MASS

Until very recently the mass of the W boson could only be measured directly in $\overline{p}p$ collisions. Precise measurements of the W mass have recently been obtained at LEP2 [21] using the enhanced statistical power of the rapidly varying total W^+W^- cross section at threshold, and the Breit-Wigner peaking behavior of the invariant mass distribution of the W decay products. By measuring the WW threshold cross section at $\sqrt{s} = 161$ GeV, the four LEP experiments have obtained a combined W

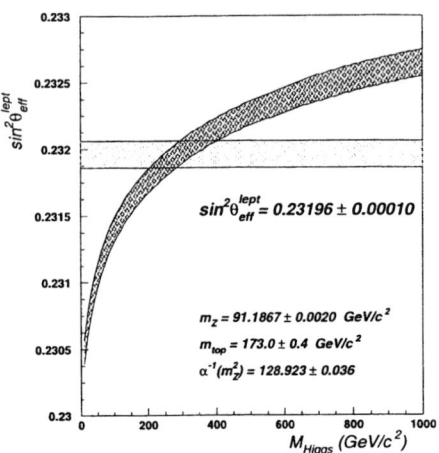

FIGURE 7. Dependence of $\sin^2\theta_{\text{eff}}^{\text{lept}}$ on m_h and the constraint on m_h a hypothetical measurement of $\sin^2\theta_{\text{eff}}^{\text{lept}}$ at the 10^{-4} level would yield, assuming an uncertainty on m_t of 2.0 GeV/c^2.

mass value of $m_W = 80.40^{+0.22}_{-0.21} \pm 0.03$ GeV/c^2. The second error is due to the LEP energy calibration. The direct reconstruction of the W mass from the decay products gives $m_W = 80.37 \pm 0.18 \pm 0.05 \pm 0.03$ GeV/c^2. The second uncertainty is due to effects of color reconnection and the last error is due to the LEP energy calibration. The achievable precision on m_W from LEP for an integrated luminosity of 500 pb^{-1} per experiment is estimated to be $\delta m_W = 35$ MeV/c^2 [26]. For an integrated luminosity of 10 fb^{-1} the Tevatron might be able to constrain m_W to about 20 MeV/c^2 [27]. The FMC is particularly well suited to a threshold measurement because of the very narrow beam spread, the ability to determine the beam energy during a physics run, and the reduced initial state radiation (ISR). Due to the large backgrounds, however, systematic errors arising from uncertainties in both the background level as well as the detection efficiencies will limit the ultimate precision [28,29]. The dominant physics background is mainly due to $Z\gamma$ production, which is almost independent of energy. When detection efficiencies and backgrounds, including the beam induced backgrounds, are largely energy independent, the best precision is obtained through a ratio of cross section measurements at an energy well below the WW threshold and at the threshold. The optimal threshold energy is also for the FMC [26]

$$\sqrt{s} \approx 2\,m_W + 0.5 \text{ GeV} . \tag{13}$$

Figure 8 shows the cross section for $\mu^+\mu^- \to W^+W^-$ in the threshold region for three different W mass values. ISR effects have been included. With an integrated luminosity of $\mathcal{L} = 100$ pb^{-1} a precision of $\delta m_W = 6$ MeV/c^2 from the thresh-

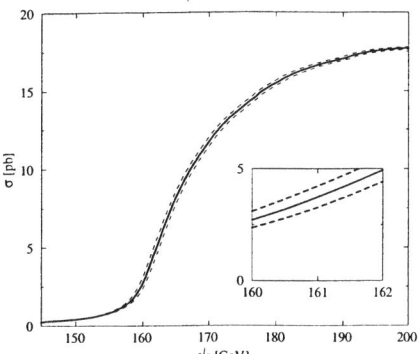

FIGURE 8. The cross section for $\mu^+\mu^- \to W^+W^-$ for $m_W = 80.3$ (± 0.2) GeV/c^2 indicated by the solid (dashed) lines. The inset shows the cross section in the region of maximum statistical sensitivity.

old measurement can be obtained when combining the three decay channels $\bar{q}q\bar{q}q$, $\bar{q}q\ell\nu$ and $\ell\nu\ell\nu$. A 10 MeV/c^2 uncertainty on m_W is equivalent to about a 0.5% uncertainty on the measured cross section.

TRIPLE AND QUARTIC GAUGE BOSON COUPLINGS

The non-Abelian $SU(2) \times U(1)$ gauge symmetry of the SM implies that the gauge bosons self-interact. These self-interactions give rise to very subtle interference effects in the SM. In fact, in the SM the couplings are uniquely determined by the gauge symmetry in order to preserve unitarity. An accurate measurement of the gauge boson self-interactions would constitute a stringent test of the gauge sector of the SM and any observed deviation of the couplings from their SM value would indicate new physics.

The formalism of effective Lagrangians is used to describe gauge boson interactions beyond the SM. The most general effective electroweak Lagrangian contains 2×7 free parameters [30]: $g_1^V, \kappa_V, \lambda_V, g_4^V, g_5^V, \tilde{\kappa}_V, \tilde{\lambda}_V$, with $V = \gamma, Z$. The parameter g_5^V, violates $\mathcal{C}$ and $\mathcal{P}$ but conserves $\mathcal{CP}$; $g_4^V, \tilde{\kappa}_V$ and $\tilde{\lambda}_V$ violate $\mathcal{CP}$. In the SM $g_1^V = 1, \kappa_V = 1$, and all other parameters vanish. For these two parameters one therefore introduces deviations from the SM values, $\Delta\kappa_V = \kappa_V - 1$ and $\Delta g_1^V = g_1^V - 1$.

Gauge boson self-interactions can be studied through di-boson production. The cross sections for di-boson production are generally rather small and a study of the full fourteen-dimensional parameter space is impossible. In general, two approaches are followed to reduce the parameter space. The $\bar{p}p$ experiments generally set all parameters but two to their SM values and concentrate on $\Delta\kappa_V$, λ_V because they have a direct physical connection through the magnetic dipole and electric quadrupole moment of the W boson, $\mu_W = (e/2m_W)(1 + \kappa_\gamma + \lambda_\gamma)$ and $Q_W^e = (-e/m_W^2)(\kappa_\gamma - \lambda_\gamma)$ [31].

The second approach, followed mainly by the LEP experiments, constructs an effective Lagrangian with operators of higher dimension. By imposing some restriction, like retaining only the lowest dimension operators, respecting $\mathcal{C}, \mathcal{P}$ and $\mathcal{CP}$ invariance and requiring the Lagrangian to be invariant under $SU(2) \times U(1)$ and adding a Higgs doublet, the number of free parameters is reduced to just three [32]. With further, rather ad hoc, requirements the parameter space can be reduced to just two free parameters, with definite relations between the different parameters [33].

If in the processes of di-boson production the couplings deviate even modestly from their SM values, the gauge cancellations are destroyed and a large increase of the cross section is observed. Moreover, the differential distributions will be modified. A WWV interaction Lagrangian with constant anomalous couplings would thus violate unitarity at high energies and therefore the coupling parameters are modified to include form factors [34], that is, $\Delta\kappa(\hat{s}) = \Delta\kappa/(1 + \hat{s}/\Lambda^2)^2$ and $\lambda(\hat{s}) = \lambda/(1 + \hat{s}/\Lambda^2)^2$, where $\hat{s}$ is the square of the center of mass energy of the subprocess. Λ is a unitarity preserving form factor scale and indicates the scale at which new physics would manifest itself.

Currently the strongest limits come from the D0 experiment. From a combined fit to the results from the WW, WW/WZ and $W\gamma$ analyses based on the full Run I data, the limits obtained at 95% CL are ($\Lambda = 1.5$ TeV):

$$-0.33 < \Delta\kappa < 0.45 \quad (\lambda = 0)$$
$$-0.2 < \lambda < 0.2 \quad (\Delta\kappa = 0),$$

where it was assumed that the WWZ couplings and the $WW\gamma$ couplings were equal [35]. A relatively small improvement in the limits is anticipated for the combined Tevatron data. With 500 pb^{-1} at $\sqrt{s} = 190$ GeV limits of the order of 0.05 to 0.1 on anomalous couplings are expected from LEP [26]. At the NLC, for a center of mass energy of $\sqrt{s} = 500$ GeV limits of $|\Delta\kappa_\gamma| < 2.4 \cdot 10^{-3}$ and $|\lambda_\gamma| < 1.8 \cdot 10^{-3}$ are expected [36]. The clear advantage of the FMC is the reduced initial state radiation (ISR), which will facilitate the event reconstruction and reduce the systematic uncertainties. No significant improvement in the limits over the constraints from an NLC are expected [37].

On the other hand, it was reported [38] at the workshop that both the NLC and a high energy muon collider with $\sqrt{s} = 0.5 - 1.5$ TeV may have significant sensitivity to perform direct tests of the quartic gauge boson couplings to a precision of theoretical interests. With an integrated luminosity of 200 fb^{-1} limits on the anomalous quartic couplings ℓ_4, ℓ_5, ℓ_6, ℓ_7 and ℓ_{10} of $\mathcal{O}(10^{-1})$ will be possible. Also here, beam polarization significantly improves the sensitivity to anomalous couplings.

CONCLUSIONS

There is a rich physics program available with a muon collider. Among them, s-channel Higgs boson production may prove to be the "Crown jewel": precision measurements on the Higgs mass, width, decay branching fractions and couplings to other particles may provide invaluable information on physics within and beyond the SM. However, extremely good beam energy resolution ($R = 0.003\%$) and a high luminosity ($10^{32} \mathrm{cm}^{-2} \mathrm{s}^{-1}$) are highly desirable to fulfill the physics goal. Z pole physics will yield significant results if a data sample of 10^8 Z events (corresponding to an integrated luminosity of about 2 fb^{-1}) can be recorded with polarized beams. Spin manipulation is extremely powerful if it is built-in in the machine design from the start.

ACKNOWLEDGMENTS

We would like to thank the Workshop organizers for the invitation to the working group, and would like to thank the participants of the working group [1] for their contributions.

REFERENCES

1. The participants of the working group were V. Barger, U. Baur, M. Berger, A. Blondel, D. Bowser-Chao, M. Carena, M. Demarteau, J. Erler, T. Greening, J. Gunion, H. Haber, T. Han, H.-J. He, C. Kao, W.-Y. Keung, B. King, M. Krawczyk, P. LeBrun, W. Marciano, S. Mrenna, R. Raja, D. Rainwater, F. Tikhonin, S. Willocq, J. Womersley.
2. R.B. Palmer, A. Tollestrup, and A. Sessler New Directions for High Energy Physics, Snowmass 1996, eds. D.G. Cassel, L. Trindle Gennari, R.H. Siemann, p. 203.
3. P. LeBrun and R. Roser, *these proceedings*.
4. R. Raja, *these proceedings*.
5. H. Haber, *these proceedings*.
6. V. Barger, *these proceedings*.
7. J.F. Gunion, *these proceedings*.
8. V. Barger, M.S. Berger, J.F. Gunion and T. Han, Phys. Rev. Lett. **75**, 1462 (1995); and Phys. Rept. **286**, 1 (1997).
9. CMS collaboration, the Technical Proposal, CERN/LHCC 94-38.
10. B. Kamal, W. J. Marciano and Z. Parsa, *these proceedings*.
11. D. Bowser-Chao, *these proceedings*; and F. Tikhonin, *these proceedings*.
12. W.-Y. Keung, *these proceedings*.
13. T. Greening, *these proceedings*.
14. M. Krawczyk, *these proceedings*.
15. J. Womersley, *these proceedings*.
16. S. Mrenna, *these proceedings*.

17. C. Kao, *these proceedings*.
18. D. Rainwater, *these proceedings*.
19. T. Han, *these proceedings*.
20. S. Willocq, *these proceedings*.
21. The LEP Electroweak Working Group, LEPEWWG-97-02.
22. M. Chanowitz, LBNL-40877, hep-ph/9710308.
23. A. Blondel, *these proceedings*.
24. Electroweak libraries: D. Bardin et al., Z. Phys. **C44** (1989) 493; Comp. Phys. Comm. **59** (1990) 303; Nucl. Phys. **B351**(1991) 1; Phys. Lett. **B255** (1991) 290 and CERN-TH 6443/92 (May 1992).
25. M. Davier and A. Höcker, LAL 97-85.
26. CERN 96-01, *Physics at LEP2*, eds. G. Altarelli, T. Sjöstrand and F. Zwirner.
27. U. Baur, M. Demarteau, *Proceedings of the 1996 DPF/DPB Summer Study on New Directions for High Energy Physics*, (Snowmass 96), Snowmass, Colorado, June 25 – July 12, 1996.
28. V. Barger, M.S. Berger, J.F. Gunion and T. Han, Phys. Rev. **D56**,1714 (1997).
29. M.S. Berger, *these proceedings*.
30. K. Hagiwara, R.D. Peccei, D. Zeppenfeld, K. Hikasa Nucl. Phys. **B282**, 253 (1987); K. Gaemers, G. Gounaris, Z. Phys. **C1**, 259 (1979).
31. K. Kim and Y-S. Tsai, Phys. Rev. D **7**, 3710 (1973).
32. CERN Report 96-01, *Physics at LEP2*, Vol.I, 525. The three parameters are related to the parameters of the effective Lagrangian through: $\Delta g_1^Z = \frac{1}{\cos^2 \vartheta_W} \alpha_{W\varphi}$, $\Delta \kappa_\gamma = -\cot^2 \vartheta_W (\Delta \kappa_Z - \Delta g_1^Z) = \alpha_{W\varphi} + \alpha_{B\varphi}$, $\lambda_\gamma = \lambda_Z = \alpha_W$.
33. K. Hagiwara, S. Ishihara, R. Szalapski, and D. Zeppenfeld, Phys. Lett. **B283**, 353 (1992), and Phys. Rev. **D48**, 2182 (1993). These so-called HISZ relations are given by $\Delta g_1^Z = \frac{1}{2\cos^2 \vartheta_W} \Delta \kappa_\gamma$, $\Delta \kappa_Z = \frac{1}{2}(1 - \tan^2 \vartheta_W) \Delta \kappa_\gamma$, $\lambda_Z = \lambda_\gamma$.
34. U. Baur and E.L. Berger, Phys. Rev. D **41**, 1476 (1990).
35. M. Demarteau, *Proceedings of the XVIth International Workshop on Weak Interactions and Neutrinos*, Capri, Italy, June 22 – June 28, 1997, Fermilab-Conf-97/296.
36. S. Kuhlman et al., *Physics and Technology of the Next Linear Collider*, SLAC-485.
37. U. Baur, *these proceedings*.
38. H.-J. He, *these proceedings*; and T. Han, H.-J. He and C.-P. Yuan, hep-ph/9711429.

Prospects for SUSY Searches and Measurements at a Muon Collider

Marcela Carena* and Serban Protopopescu[†]

*Fermi National Accelarator Laboratory
P.O Box 500, Batavia, IL 60510, USA
[†]Brookhaven National Laboratory, Upton, NY 11973, USA

Abstract. We summarize the discussions and main results presented at the SUSY Searches and Measurements Working Group of the Workshop on Physics at the First Muon Collider and at the Front End of a Muon Collider.

I INTRODUCTION

Supersymmetry (SUSY) [1–3] provides a well–motivated extension of the Standard Model (SM) with an elegant solution to the naturalness problem [4]. In the Minimal Supersymmetric extension of the SM (MSSM), the particle spectrum is doubled by SUSY. Moreover, to generate masses for up– and down–type fermions while preserving SUSY and gauge invariance, the Higgs sector must contain two doublets [5], yielding five physical Higgs bosons in the spectrum [6]. All the Higgs bosons and other SM particles have superpartners with the same quantum numbers under the SM gauge groups $SU(3)_C \times SU(2)_L \times U(1)_Y$, but with different spin [2]. Given that no superparticles have been observed so far (for present limits on searches at LEP and the Tevatron colliders see [7,8]), it is assumed that SUSY is broken, and that in general the sparticles must be heavier than their partners but not much heavier that a few TeV [9]. The breakdown of supersymmetry can be parametrized by considering all the possible soft SUSY–breaking mass parameters which are compatible with the gauge invariance of the theory. The soft SUSY–breaking parameters are extra mass terms for gauginos and scalar fermions, and trilinear scalar couplings. The exact number of extra free parameters depends on the precise mechanism of SUSY breaking, which is unknown. The supersymmetric spectrum is given as a function of the soft supersymmetry breaking parameters and remains therefore undetermined. The measurement of the Higgs bosons and supersymmetric particles at present and future colliders may provide the only direct way of getting information about the fundamental parameters of the theory and help in understanding the nature of the supersymmetry breaking mechanism. If supersymmetric particles are found at LEP, Tevatron and/or LHC, a higher-energy lepton collider will be necessary to perform precision measurements of particle masses, widths and couplings. It is therefore most important to study the potential of lepton colliders for SUSY searches and to evaluate as well the advantages and drawbacks of different lepton machines. Here we shall concentrate in studying the potential of a $\mu^+\mu^-$ collider in SUSY searches. We should note that for many measurements $\mu^+\mu^-$ and e^+e^- colliders are comparable. The potential of e^+e^- colliders is summarized in [10]. Most contributions to our working group emphasize measurements for which $\mu^+\mu^-$ colliders may have an advantage over a e^+e^- collider or are complementary to it.

In the following we shall assume the MSSM as the general framework to perfom studies of SUSY searches at a muon collider; depending on the actual mechanism of SUSY breaking which determines then the initial boundary conditions for the soft SUSY breaking parameters, relations among supersymmetric particle masses appear that can be useful to explore. The possibility of lepton number violation or lepton flavour violation can also appear naturally within the MSSM and it is of interest to investigate likely signatures at a muon collider.

At present two main mechanism of SUSY breaking are being considered: i) gravity mediated and ii) low energy-gauge mediated supersymmetry breaking scenarios. i)Supergravity (SUGRA) models assume the existence of extra superfields (the so-called "hidden sector") which couple to the MSSM particles only through gravitational-like interactions. When SUSY is spontaneously or dynamically broken in the hidden sector, some of the components of the hidden sector acquire vacuum expectation values. Interaction terms between those components of the hidden sector and the MSSM sparticles give rise to the effective soft SUSY-breaking terms of the MSSM, which are proportional to the hidden sector vacuum expectation values divided by powers of M_{Planck}. In general the number of unknown parameters is very large, but, in the minimal SUGRA scenario, the MSSM sparticles couple universally to the hidden sector, and the number of terms is greatly reduced. Using this guiding principle, at a scale of order M_{Planck} (or, approximately, M_{GUT}, the scale where the gauge couplings unify), all scalars (Higgs bosons, sleptons, and squarks) are assumed to have a common squared-mass m_0^2, all gauginos (Bino, Wino, and gluino) have a common mass $m_{1/2}$, and all trilinear couplings have the value A_0. After specifying $\tan\beta$, the ratio of the two Higgs vacuum expectation values, the model is essentially complete. The value of the supersymmetric Higgs/Higgsino mass parameter μ is determined as a function of m_0^2, $m_{1/2}$ and $\tan\beta$ by requiring proper electroweak symmetry breaking. All that remains to define the sparticle spectrum is to relate the values of the soft SUSY-breaking parameters specified at M_{GUT} to their values at low energies, M_{EW}. This is accomplished using renormalization group equations (RGE's) [11]. On the other hand, it is natural to question the above exact universality of the soft SUSY-breaking parameters [12]. For example, in a $SU(5)$ SUSY GUT model, the left-handed sleptons, $\tilde{e}_L$, $\tilde{\nu}_L$ and right-handed down squark, $\tilde{d}_R$, reside in the same 5-multiplet of $SU(5)$, and naturally have the common mass parameter $m_0^{(5)}$ at the GUT scale. Similarly, $\tilde{u}_L$, $\tilde{d}_L$, $\tilde{u}_R$, and $\tilde{e}_R$, which reside in the same 10–multiplet, have a common mass $m_0^{(10)}$. The two Higgs bosons doublets reside in different 5– and $\bar{5}$–multiplets, with masses $m_0^{(5')}$ and $m_0^{(\bar{5}')}$. There is no symmetry principle that demands that all these mass parameters should be the same, therefore extensions of minimal SUGRA need to be explored as well.

ii) In models in which the soft SUSY–breaking terms are generated through gauge interactions, the mass degeneracies between sfermions with the same quantum numbers (and, hence, the same gauge couplings) occur naturally, suppressing FCNC's. Also, the scale of the SUSY breaking is much smaller than the scale where gravity becomes strongly interacting, so there is no possibility of Planck–scale corrections to these degeneracies (as there can be between the GUT and Planck scales in SUGRA). In most models [13] of gauge–mediated, low–energy supersymmetry breaking, the gaugino and scalar masses are roughly of the same order of magnitude. Even after RG evolution [14] (ignoring the effects of Yukawa couplings), sfermions with the same quantum numbers acquire the same masses yielding a natural mass hierarchy between weakly and strongly interacting sfermions; the mass hierarchy of the gauginos is fixed by the gauge couplings (as in SUGRA models). One distinctive feature of these models is that the spin–3/2 superpartner of the graviton, the so called gravitino $\widetilde{G}$, can be very light and become the LSP. The lightest SM superpartner becomes the next–to–lightest supersymmetric particle (NLSP), which is unstable and decays into its SM partner plus the Goldstino component of the gravitino [15]. Generically the NLSP can be a neutralino, or a slepton (most plausibly a right–handed slepton and, due to the larger Yukawa coupling, a $\tilde{\tau}$). If a gaugino–like neutralino is the NLSP then, signatures of photons plus $\not{E}_T$ can point towards models of low energy SUSY breaking. If the scale of SUSY breaking is not far above the electroweak scale ($\leq$ a few 1000 TeV), the NLSP will decay within the detector, leading to distinctive signatures as displaced vertices or heavy charged sleptons decaying into leptons [16–19].

One simple extension of the MSSM is to break the multiplicative R–parity symmetry [38]. Presently, neither experiment nor any theoretical argument demand R–parity conservation, so it is natural to consider the most general case of R–parity breaking. The R–parity violating (RPV) terms which can contribute to the superpotential are [1].

$$W_{RPV} = \lambda_{ijk} L^i L^j \bar{E}^k + \lambda'_{ijk} L^i Q^j \bar{D}^k + \lambda''_{ijk} \bar{U}^i \bar{D}^j \bar{D}^k \qquad (1)$$

where i, j, k are generation indices (1,2,3), $L_1^i \equiv \nu_L^i$, $L_2^i = \ell_L^i$ and $Q_1^i = u_L^i$, $Q_2^i = d_L^i$ are lepton and quark components of $SU(2)_L$ doublet superfields, and $E^i = e_R^i$, $D^i = d_R^i$ and $U^i = u_R^i$ are lepton, down and up– quark $SU(2)_L$ singlet superfields, respectively. Due to $SU(2)_L$ and $SU(3)_C$ invariance, it follows that $i \neq j$ in $L^i L^j \bar{E}^k$

---

[1] In Eq. (1) bilinear terms are ignored. A discussion of the phenomenological implications of such terms can be found in the literature [21]

and $j \neq k$ in $\bar{U}^i \bar{D}^j \bar{D}^k$, respectively. The coefficients λ_{ijk}, λ'_{ijk} and λ''_{ijk} are Yukawa couplings, and there is no *a priori* generic prediction for their values. W_{RPV} contains 45 extra parameters over the R–parity–conserving MSSM case. The $LL\bar{E}$ and $LQ\bar{D}$ terms above violate lepton number. The $\bar{U}\bar{D}\bar{D}$ term, instead, violates baryon number. In principle, all types of R–parity violating terms may co–exist, but this can lead to a proton with a lifetime shorter than the present experimental limits. The simplest way to avoid this is to allow only operators which conserve baryon–number but violate lepton–number or vice versa. There are several effects on the SUSY phenomenology due to these new couplings: (1) lepton or baryon number violating processes, including the production of single sparticles (instead of pair production), (2) the LSP is no longer stable, but can decay to SM particles within a collider detector, and (3) because it is unstable, the LSP need not be the neutralino or sneutrino, but can be charged or colored. Present data are in remarkable agreement with the SM predictions, and very strong bounds on the R–parity–breaking operators can be derived [22], but there is still room for study.

In most SUSY extensions of the SM, new flavor mixing matrices, analogous to the CKM matrix may appear. If there are universal scalar masses and trilinear couplings at high energies, then the mass matrices of the fermions and their scalar superpartners are diagonalizable in the same basis and lepton flavor is conserved. In this case processes like $\mu \to e\gamma$, $\mu \to 3e$, $\mu \to e$ conversion or $\tau \to \mu\gamma$ are forbbiden. However, quite generically, if the messenger scale for supersymmetry breaking is above the scale generating the fermion mass structure, then new flavor mixing matrices will appear at the gaugino-fermion-sfermion vertices. Although possible flavour lepton violating signatures are highly model dependent, searches for flavor lepton violation are intrinsically of great interest. Any observation of such violation may provide a window to GUT scale physics [23].

The key point of supersymmetric scenarios is the presence of a light Higgs boson in the spectrum. Within the MSSM, a general upper bound on m_h can be determined by a careful evaluation of the one–loop and dominant two–loop radiative corrections [24]. For masses of all SUSY particles and the CP-odd Higgs boson mass (M_A) around 1 TeV, $\tan\beta > 20$, and varying the $\tilde{t}$ mixing parameters to maximize the upper bound on m_h, one finds that $m_h \lesssim 130$ GeV for $m_t = 175$ GeV. For more moderate values of the MSSM parameters, the upper bound on m_h becomes smaller. Qualitatively, most known extensions of the MSSM Higgs sectors, with extra singlets or doublets, preserve this feature and lead to an upper bound of about 150 -160 GeV on the lightest Higgs mass. Thus, by the time we can expect a first muon collider to be operative, either LEP (if $m_h \leq 105$ GeV) or the Tevatron (if $m_h \leq 120$–130 GeV, depending on the final reach in Luminosity) should have seen the Higgs, or the LHC will have settled the issue of whether SUSY is relevant at the electroweak scale.

II MSSM HIGGS SEARCHES

Given that the muon is significantly heavier than the electron, its coupling to the Higgs is significantly enhanced and it becomes possible to consider the Higgs s-channel production at interesting rates [25,26]. For s-channel Higgs searches the crucial features of a muon collider are the energy resolution $\sigma_{\sqrt{s}}$ of a few MeV, the precise tuning of the beam energy [27], and the small bremstrahlung together with no bremstrahlung smearing.

i) For the MSSM lightest Higgs boson h in the low $\tan\beta$ regime, where the width is of a few MeV, the achievable beam resolution ($R \simeq 0.01\%$) enhances the peak cross section for a muon collider ring running at energy $\sqrt{s} = m_h$, and provides an opportunity to measure the Higgs width. Given the strong upper bound on the lightest MSSM Higgs mass, the s-channel measurement of the Higgs boson seems to be a clear first goal of a first muon collider. Recent studies show that for a Higgs mass of about 110 GeV an accuracy $\Delta m \simeq 1 MeV$ may be achieved. After determining m_h with such accuracy, a 3-point fine scan [25] may allow to determine Γ_{tot}^h with 16 % accuracy. Also the branching ratios to the dominant decay channels $\sigma BF(b\bar{b})$, and $\sigma BF(WW^*)$ can be determined with a good precision. To accumulate the necessary integrated luminosity to perform these measurements in a reasonable amount of time, instantaneous luminosities an order of magnitude higher than contemplated during this workshop will be required. A very good energy resolution is also highly desirable [28–30].

ii) For the other MSSM neutral Higgs bosons, the heavy CP-even, H, and the CP-odd Higgs, A, the search at the muon collider via the s-channel production is also of major interest [28–30]. In particular if both mass eigenstates are heavy and, hence, almost degenerate, an s-channel scan with high resolution may be the only way to allow identification of both states. S-channel production of H and A can also provide an interesting mechanism to search for stop production at a muon collider (see section V, this chapter).

iii) It is also possible to produce the lightest Higgs boson via the Higgs-strahlung process $\mu^+\mu^- \to Zh$, by running the machine well above the threshold. By measuring the cross section at three center of mass energies around the Zh threshold one could determine the Higgs mass, the total width, and $g_{ZZh}^2 \times BR(b\bar{b})$ with a reasonable accuracy [31].

For a detail discussion of Higgs searches see the chapters by V. Barger and J. Gunion and the one summarizing the Higgs Working Group activities in these proceedings.

III CHARGINO SEARCHES

After the final run of LEP, chargino masses up to the kinematical limit, $m_{\chi_1^\pm} \leq \sqrt{s}/2$, will have been probed [32], with the exception of the light sneutrino case, for which complementary experimental data may be needed [33]. Indeed, the two processes which contribute to chargino pair production at l^+l^- colliders are s-channel production via interchange of γ and Z (or Higgs at the muon collider) and t-channel production via interchange of the corresponding sneutrino. The two amplitudes interfere destructively and, quite generally, better reach is expected for heavier sneutrinos (the destructive interference gets reduced as the sneutrino mass increases). Charginos can decay as $\chi^\pm \to \chi^0 f \bar{f}'$ and also via the two body decay channel $\chi^\pm \to l^\pm \tilde{\nu}$ or $\tilde{l}^\pm \nu$. At LEP both decay possibilities are being studied. The Tevatron can have very good coverage for lightest chargino mass searches via chargino-neutralino associated production in the trilepton decay mode and via chargino pair production through the dilepton decay mode, with a maximal reach of about $m_{\chi_1^\pm} \simeq 200 - 230$ GeV for an integrated luminosity of about 10 fb^{-1} [34]. These results are, however, model dependent and charginos may escape detection for much smaller masses in some specific cases. Therefore, while there is good coverage for a wide range of chargino masses, it is not possible to set lower mass limits at the Tevatron. At the LHC chargino searches will mainly allow to measure chargino-neutralino mass differences [35]. If no chargino is found by the end of the LEP running, then a muon collider with $\sqrt{s} \geq 200 GeV$, provides a good opportunity for chargino searches measuring the chargino mass via chargino pair production at the threshold. If the chargino is mainly gaugino-like, then a measurement of the production cross section can yield indirectly the chargino and sneutrino masses as this cross section strongly depends on both. A simultaneous measurement of the chargino and sneutrino masss requires to do at least a two point sampling of the cross section as a function of center-of-mass energy. Present studies [36] assume a high integrated luminosity, L $\simeq$ 50-100 fb^{-1}, and a beam energy spread of $R = 1\%$. The main chargino decay mode considered is $\chi^\pm \to \chi^0 f \bar{f}'$, and W exchanges dominate for $m_{\chi_1^\pm} - m_{\chi_1^0} > M_W$. Selective cuts to suppress the main background from W-pair production have been optimized for the chargino threshold energy region[2]. Since the chargino cross section decreases as the mass increases, one gets better precision for smaller chargino masses. Table 1 shows the results for 50 fb^{-1} An approximate re-scaling of the above results will lead to an accuracy of about 300 MeV for chargino masses of about 100 GeV, with sneutrino masses in the 300–500 GeV range and 1-2 fb^{-1}, and of about 300 (700) MeV for chargino masses of about 200 GeV , with sneutrino masses of about 500 (300) GeV and 7 fb^{-1} of luminosity per year. The larger beam energy spread significantly reduces the accuracy of these threshold measurements at e^+e^- colliders, but their

---

[2]) If the chargino were lighter than the sneutrino, and the two body decay modes into lepton-sneutrino or slepton-neutrino open up, the signal to background efficiency would need to be reconsidered.

TABLE 1. Achievable uncertainties in the lighter chargino masses for given $m_{\chi_1^\pm}$ and $m_{\tilde{\nu}_\mu}$ for a total luminosity of 50 fb^{-1}.

$\Delta m_{\chi_1^\pm}[MeV]$	$m_{\chi_1^\pm}[GeV]$	$m_{\tilde{\nu}_\mu}$ [GeV]
35-45	100	500–300
150	200	500
300	200	300

higher luminosity can partially compensate [10].

Chargino mass measurements can also be performed by finding the endpoints in the spectrum of the chargino decay products [37]. Since the endpoint method depend on the kinematics of the decay products, the results in this case will also be sensitive to the neutralino mass. The end point method is less precise but may be useful for determining the approximate chargino mass and so indicate at what energies should the threshold measurements be done.

Finally, beam polarization, depending on the exact trade in luminosity, can provide a useful tool for measuring the gaugino and Higgsino components of the chargino. If the chargino is mainly gaugino-like, this means mainly wino-like, it couples to the left-handed μ^- (and right-handed μ^+). If the chargino is mainly Higgsino-like, a right-handed μ^- polarized beam can be used to turn off the t-channel destructive interference and to reduce the W^+W^- background (although for the latter other methods are also possible, see below). At e^+e^- colliders, large polarization can be achieved together with high luminosities, making them suitable for this particular measurement. At a muon collider, the two muon beams can be used to improve the mass determination provided the luminosity loss can be kept under control.

IV SLEPTON SEARCHES

LEP is well suited for slepton searches, in particular for selectrons and smuons, but high luminosity is needed to approach the kinematical limit [32]. At the LHC some masses may be measured if they are below 300 GeV, but they are hard to study except in the case when the neutralino decays into lepton-slepton [35]. At a muon collider one can produce left-left or right-right sleptons via s-channel interchange of photons and Z's. T-channel neutralino-exchange allows associated production of left-right smuons as well. For left-left or right-right smuon production both amplitudes interfere destructively, but for light gaugino-like neutralinos the t-channel exchange becomes dominant. Slepton mass measurements can be possible at a muon collider using kinematic methods [37–40]. In this workshop, an study using the kinematical endpoint method was presented [40]. Selecting two acollinear leptons from slepton pair production with subsequent decay of the sleptons into lepton-lightest neutralino, one can use the lepton energy distribution, find the endpoints and then determine the slepton and neutralino masses as a function of the lepton energy endpoints. The dominant background in these analyses come from leptonic decays of W^+W^-. Since the W's decay with equal rate into $e\nu_e$ and $\mu\nu_\mu$, a simple way to minimize the SM background up to statistical fluctuations, is to consider the combinations $e^+e^- + \mu^+\mu^- - \mu^+e^- - e^-\mu^+$ or $e^+e^- - \mu^+\mu^-$. This provides a useful tool to detect the signal relaxing the need for a high beam polarization to kill the background [10,40]. However, a small polarization can be useful in interpreting the signature in specific cases. Present studies have been done with some simplifying assumptions, in particular, without taking into account backgrounds from muon decays. An example shows that for a right-handed selectron of mass $m_{\tilde{e}_R} \simeq 150 GeV$ a precision in $\Delta m_{\tilde{e}_R} \simeq 1 GeV$ can be obtained with $L \simeq 10 fb^{-1}$. Although preliminary present studies do not show much difference between the capabilities of an e^+e^- or a $\mu^+\mu^-$ collider, it is clear that any higher–energy lepton collider can complement well the LHC capabilities in slepton physics. In some cases, the small lower limits in the energy distributions ($E_l^{min} \leq 10 GeV$) may make it very difficult to identify and measure the electrons in the presence of the muon backgrounds. Such small lower energy limits are specific of the mass spectrum so, depending on the parameters of the theory, a higher integrated luminosity or a different center of mass energy may be necessary. The ability to measure low energy end points is crucial for these studies. For a recent detail discussion about different kinematic fitting methods see ref. [39].

A New Flavour Mixings

At a muon collider and its front end interesting searches involving sleptons can be done to study new flavour mixing effects both indirectly, through rare flavour changing processes, and directly, through slepton production processes involving flavour violating production or decays [41]. In most SUSY models, the scalar mass matrices which give mass to $\tilde{e}$, $\tilde{\mu}$ and $\tilde{\tau}$ are non-diagonal in the basis in which the fermions, e, μ and τ get masses. Considering then the interactions of lepton and slepton mass eigenstates, flavour mixing matrices appear in the fermion-sfermion-gaugino/Higgsino vertices. Such non-trivial flavour mixing matrices generate contributions to processes such as $\mu \to e\gamma$ which are proportional to $\sin(2\theta_{e\mu})\left(m_{\tilde{e}} - m_{\tilde{\mu}}\right)/m_{\tilde{e}/\tilde{\mu}}$. The current bound on BR($\mu \to e\gamma$) < $4.9 \cdot 10^{-11}$ [42] demands $\sin(2\theta_{e\mu})\left(m_{\tilde{e}} - m_{\tilde{\mu}}\right)/m_{\tilde{e}/\tilde{\mu}} \leq 10^{-2}(10^{-3})$ for small (large) $\tan\beta$. This implies that either the mixing angle is very small or $\tilde{e}$ and $\tilde{\mu}$ are quite degenerate [41]. Most interesting, at a muon collider an extremely intense low energy muon source will be able to do measurements of BR($\mu \to e\gamma$) down to 10^{-14} and even better for $\mu - e$ conversion [43], yielding much more stringent constraints on the $\tilde{e}$ and $\tilde{\mu}$ mass splitting and the corresponding mixing angle. However, since in SUSY theories the BR($\mu \to e\gamma$) depends on many parameters and there may be diagrams which can either add up or cancel, this single measurement is not sufficient to understand the whole flavour mixing matrices. Current bounds on other leptonic rare decays, such as $\tau \to \mu\gamma$, are too weak to provide any constraints. Through products of rare decays contributing to $\mu \to e\gamma$ via loops one can obtain indirectly some additional, but rather weak, constraints for a reduced region of parameter space.

At a muon collider, besides measuring BR($\mu \to e\gamma$), it may also be possible to probe the flavour mixing matrices through flavour changing slepton production and decay processes, One interesting case which has been analyzed is that of on-shell slepton pair production at a muon collider, with a signal of a pair of unlike flavor leptons in the final state $\mu^+\mu^- \to l_i^+ l_j^- \chi_1^0 \chi_1^0$. For simplicity one explicit example considers muon-electron final states and a given assumption for the neutralino and left and right–handed selectron and smuon masses (see ref. [41]). With 20 fb^{-1} and assuming a 3 σ discovery limit one can probe regions of $\sin(2\theta_{e\mu})$ and $(m_{\tilde{e}} - m_{\tilde{\mu}})/m_{\tilde{e}/\tilde{\mu}}$ beyond those probed by the present BR($\mu \to e\gamma$) constraint. However, with the expected reach of the intense muon source at the muon collider front end, the bounds on BR($\mu \to e\gamma$) will be more constraining than those coming from direct lepton production (see Fig 3. in ref. [41], these proceedings). Although the $\mu \to e\gamma$ process is more dependent on the exact SUSY model than the collider signal, it is clear that the front end of the muon collider can give significant information about the SUSY flavour mixing matrices and thus provide important clues to the understanding of the flavour structure of the theory. In addition, a muon collider can probe the muon-electron/tau mixing in a way that is complementary to an electron-collider leading to a more complete understanding of flavour mixing.

B R-Parity Violation

At present there are many ongoing studies considering R-parity violating decays both at LEP and the Tevatron. The muon collider, however offers some unique opportunities which are worth analysing in detail [44–46]. In fact, if R-parity is violated, scalar neutrinos can be produced as s-channel resonance in analogy to the neutral Higgs bosons. The same as in the Higgs case, at a muon collider this resonance production can be used to performed precision measurements of SUSY parameters. Since sneutrinos are likely to be among the lightest supersymmetric particles, and such resonance can probe masses up to $\sqrt{s}$, already a first stage of a muon collider can test most of the typically expected mass range. Given the SU(2) invariance of the $L_i L_j \bar{E}_k$ operators, which demands i $\neq$ j, at a muon collider one can produce $\tilde{\nu}_e$ and $\tilde{\nu}_\tau$ at resonance. If only one R-parity violating coupling at a time is assumed, then the $\tilde{\nu}_{e,\tau}$ can decay back into $\mu^+\mu^-$ or through R-parity conserving decays like $\tilde{\nu}_{e,\tau} \to \nu_{e,\tau} \chi_1^0$, if $m_{\tilde{\nu}} > m_{\chi_1^0}$. Strong bounds on the LLE and LQD R-parity violating couplings are derived from different processes [22]. Products of couplings involving different LLE or LQD type of couplings and products of LLE $\times$ LQD couplings are also constrained in various ways [47]. In the case of sneutrino resonance production, an interesting process will occur if a product of LLE and LQD R-parity violating couplings is allowed, leading to dijet production. In particular the product $\lambda_{2j2} \times \lambda'_{j33}$, with $j \neq 2$, yields $\mu^+\mu^- \to b\bar{b}$ via $\tilde{\nu}$ s-channel exchange.

Since the product of couplings involved are comparable, this example can mimic the Higgs channel exchange at a muon collider. The different signals from s-channel sneutrino production depend on the possible decay patterns and have been analyzed by two groups independently [45,46]. If the R-parity violating decays are dominant the decay width involved is very small because of the smallness of these couplings. In such case the achievable beam energy resolution at a muon collider makes it a unique machine for probing R-parity violating couplings. Studies were done assuming both that the $\tilde{\nu}$ mass is exactly known and that the total luminosity is applied at the resonance peak and also considering the less optimistic situation in which the $\tilde{\nu}$ mass is approximately known from other colliders and a scan over the possible allowed range must be performed [44,46]. For the case of $\tilde{\nu}_\tau$ resonance, considering both R-parity violating decays ($\tilde{\nu}_\tau \to \mu^+\mu^-$, $b\bar{b}$), it is possible to be sensitive to R-parity violating couplings as low as 10^{-3}–10^{-4}. If the R-parity conserving decays are allowed and dominate ($\tilde{\nu}_\tau \to \nu_\tau \chi^0 \to \nu_\tau(\nu_\tau\mu^+\mu^-,\ \nu_\mu\mu^\pm\tau^\mp,\ \nu_\tau b\bar{b})$) the discovery reach for the R parity violating couplings may be greater [44,46]. The above results assumed a total integrated luminosity of 0.1–1 fb^{-1}, depending on the corresponding assumption about the beam energy resolution which maximizes the cross section for a given total width. After detecting a sneutrino resonance one could envisage precise measurement of the corresponding R-parity violating couplings with an accuracy of a few per cent.

Other possible measurements of R-parity violating couplings have also been studied. $\mu^+\mu^- \to b\bar{b}$ via t-channel squark exchange, involving λ'_{2i3} couplings is one interesting example. Assuming a reasonably good b-tagging efficiency for each jet one can measure these couplings if they are greater than about $5 \cdot 10^{-2}$ given a total integrated luminosity of 10 fb^{-1} for $\sqrt{s} = 350 - -500$ GeV [45]. Also other products of (LLE)(LQD) or different (LLE)(LLE) couplings both in s- and t- channel processes (besides the $\mu^+\mu^- \to b\bar{b}$ case already discussed) have been studied with promising results [45]. By analogy with the Yukawa couplings, one would expect R-parity violating couplings involving higher generational indices to be larger, implying that a muon collider will be better than an electron collider for such searches. The analyses performed so far indicate that a muon collider has very good prospects for dicovering s-channel sneutrino resonances and measuring their R-parity violating couplings. Contrary to other colliders which require R-parity violating couplings comparable to the R-parity conserving ones, muon colliders, because of their excellent beam energy resolution are unique in measuring small R-parity violating couplings.

V STOP SEARCHES

The lightest stop can be the lightest supersymmetric charged particle and experiments at both LEP and the Tevatron are vigorously searching for it. At hadron colliders one can search for stop production via strongly interacting processes. At LEP, stop searches are via s-channel production through photon and Z interchange and the results depend on the strength of the stop coupling to the vector bosons, hence left–handed stops are better off due to their enhanced coupling to the Z. At NLC one could perform nice measurements of the stop masses and mixing angles using highly polarized left and right handed electron beams. At a muon collider there is also the possibility of producing stops via s-channel production of Higgs bosons [48–50]. This implies that the CP-even and CP-odd Higgs masses need to be known to sit at the peak of the resonance. Production of stop mass eigenstate pairs, $\tilde{t}_i\tilde{t}_i^*$, will involve the heaviest CP-even Higgs and be maximal at the resonance peak. The associated production of both mass eigenstates will involve the CP-odd Higgs boson as well. In the case of $\tilde{t}_1\tilde{t}_2^* + \tilde{t}_2\tilde{t}_1^*$ production, for m_A above 200 GeV, the peak in the cross section is an overlap of the H and A resonances since their masses are approximately degenerate and their expected widths are of several GeV [51]. A couple of specific examples have been considered [50], and an increase in the $\tilde{t}_1\tilde{t}_1^*$ cross section of 20 % at the resonance has been obtained, while for the associated production $\tilde{t}_1\tilde{t}_2^* + \tilde{t}_2\tilde{t}_1^*$ the increase in the cross section with respect to the vector boson s-channel interchange can be at least of 30 %. Both enhancements at the resonance can reach as much as 100 % depending on the value of the left-right stop mixing parameter A_t. The stops decay mainly into bottom-chargino, or into charm-neutralino via loops if the chargino decay is not allowed by phase space. If they are sufficiently heavy they can also decay into top-neutralino. A detail analysis of the luminosity requirements for the detection of stops via Higgs s-channel production need still to be performed, but it seems clear that a muon collider operating as a heavy Higgs factory may provide an important tool to probe the neutral Higgs bosons-stop-stop couplings and the stop mixing parameter sector.

VI OTHER INTERESTING STUDIES

A generic problem in SUSY "models"[3] is that the richness of the expected spectrum of sparticles can make the data difficult to interpret, if the collider center-of-mass energy is above a large number of sparticle pair threshholds. An interesting question is how to use such data to discriminate between various SUSY models. A study was done on the effectiveness of using a matrix of the number of events observed in a large number of channels as discriminant between models [52]. The results are very encouraging but the study needs further refinement to properly take into account the effect of channels contaminating others because of detector inefficiencies or particle misidentification.

During this workshop it was also presented the possibility of radiative generation of fermion masses, which can occur in certain non-minimal supersymmetric frameworks [53]. Such scenarios imply an enhancement of the Higgs-fermionic coupling, which can be between a few percent and up to an order of magnitude larger than in the MSSM. Therefore, next generation of lepton colliders and, in partic-

---

[3] Here we define models as different sets of low energy MSSM parameters.

ular, a muon collider can provide an excellent opportunity to test fermion masses generated radiatively.

VII CONCLUSIONS

If low energy Supersymmetry is realized in nature, some of the supersymmetric particles will be discovered at LEP, the Tevatron and/or the LHC. However, to do SUSY precision measurements and learn the details of the SUSY spectrum that may help us in understanding the mechanism of supersymmetry breaking, we shall need the Future Lepton Colliders (FLC). The FLC have better chances for measuring the masses of the heavier Higgs bosons, A, H and $H^{\pm}$, particularly if their decays into supersymmetric particles become possible. They obviously have better and more diverse capabilities for slepton mass measurements. Quite generally, one can conclude that the FLC, in any of their forms, can provide higher precision measurements of masses, couplings, spin, decay widths, chargino mixing and lepton flavour mixing than hadron colliders. It is therefore easy to argue that at least one of the higher energy lepton colliders must have a high priority among the High Energy Physics community projects. An exact comparison between the NLC-type and the muon collider will demand a more detail study of the limitations and advantages of both machines, which is beyond the scope of this workshop. However, some special features seem to be rather obvious. An electron collider can have large electron longitudinal polarized beams compatible with a high luminosity environment. This can be most useful to enhance certain signals and minimize backgrounds. Polarization can also be most useful in probing the gaugino and Higgsino components of charginos and neutralinos. On the other hand, a muon collider can extend its energy reach up to 3 TeV, without facing the problems arising in an electron collider, and gradual upgrades can yield many competitive precision measurements studies at the intermediate stages. A higher energy reach, above 1 TeV, will be most important in opening up the production of the heaviest possible supersymmetric particles which are predicted in certain theoretical scenarios. Furthermore, the muon collider can be designed to have superior beam energy resolution, which is crucial to achieve many physics results, particularly those related with Higgs physics. In fact, one of the most exciting possibilities of a muon collider is to operate as an s-channel Higgs factory. In addition, in the presence of R-parity violation it can also provide excellent opportunities as an s-channel sneutrino-factory. Moreover, given the different initial particles involved, electron and muon colliders will be complementary in testing R-parity violating and new lepton flavour mixing couplings. In conclusion, all present studies seem to leave no doubt that a muon collider with design properties similar to those discussed in these proceedings is very likely to contribute substantially to our knowledge of Higgs and Supersymmetric particle physics.

ACKNOWLEDGEMENTS

We would like to thank the organizers of this workshop for their support to the working group activities and the participants of the working group for their contributions through talks and discussions.

REFERENCES

1. Y. Gol'fand and E. Likhtam, JETP Lett. **13**, 323 (1971); P. Ramond, Phys. Rev. **D 3**, 2415 (1971); A. Neveu and J.H. Schwarz, Nucl. Phys. **B 31**, 86 (1971); J.L. Gervais and B. Sakita, Nucl. Phys. **B 34**, 632 (1971); D. Volkov and V. Akulov, Phys. Lett. **B46**, 109 (1973); J. Wess and B. Zumino, Nucl. Phys. **B78**, 39 (1974).
2. J. Wess and J. Bagger, *Supersymmetry and Supergravity* (Princeton University Press, Princeton, N.J., 1983); H.–P. Nilles, Phys. Rep. **110**, 1 (1984); H. E. Haber and G. L. Kane, Phys. Rep. **117**, 75 (1985); R. Barbieri, Riv. Nuovo Cim. **11**, 1 (1988); R. Arnowitt, A Chamseddine and P. Nath, *Applied N=1 Supergravity* (World Scientific, Singapore, 1984); P. West, *Introduction to Supersymmetry and Supergravity* (World Scientific, Singapore, 1986); R. Mohapatra, *Unification and Supersymmetry* (Springer–Verlag, Berlin, 1986).
3. For those experimentalists just beginning on SUSY, we recommend: S. Dawson, talk *SUSY and Such*, given at the NATO Advanced Study Institute on Techniques and Concepts of High Energy Physics, July 11-22, 1996, St. Croix, Virgin Islands, hep–ph/9612229; X. Tata, *QCD and Beyond*, ed. D.E. Soper, 163 (World Scientific, Singapore, 1996), hep–ph/9510287, and Ref. 3 therein.
Intended for theorists: M. Dine, '96 Tasi Lectures, *Supersymmetry Phenomenology (with a Broad Brush)*, hep–ph/9612389; J. Lykken, '96 Tasi Lectures, *Introduction to SUSY*, hep–ph/9612114; J. Bagger, *QCD and Beyond*, ed. D.E. Soper, 109 (World Scientific, Singapore, 1996), hep–ph/9604232.
4. J. Wess and B. Zumino, Phys. Lett. **49B**, 52 (1974; J. Iliopoulos and B. Zumino, Nucl. Phys. **B76**, 310 (1974); S. Ferrara, J. Iliopoulos and B. Zumino, Nucl. Phys. **B77**, 413 (1974); E. Witten, Nucl. Phys **B188** 513 (1981). K. Wilson, as quoted by L. Susskind, Phys. Rev. **D20**, 2619 (1979); G. 't Hooft, in *Recent Developments in Gauge Theories*, eds. G. 't Hooft, *et al.*, (Plenum Press, New York, 1980); L. Maiani, in *Proceedings of the Gif-sur-Yvette Summer School*, edited by R. Barloutand, J.F. Cavaignac, and D. Nanopoulos (Natl. Inst. Nucl. Phys., Paris, 1980); M. Veltman, Acta Phys. Polon. **12B**, 473 (1981).
5. P. Fayet, Nucl. Phys. **B90**, 104 (1975).
6. See J.F. Gunion, H.E. Haber, G.L. Kane and S. Dawson, *The Higgs Hunter's Guide* (Addison–Wesley, Reading, MA, 1990).
7. J. Nachtman, talk presented at this Workshop.
8. E. Flattum, *these proceedings* and references therein.
9. J. Ellis, E. Enqvist, D.V. Nanopoulos and F. Zwirner, Mod. Phys. Lett. **A1**, 57 (1986); R. Barbieri and G.F. Giudice, Nucl. Phys. **B306**, 63 (1988); G. Anderson, D. Castaño and A. Riotto, Phys. Rev. **D55**, 2950 (1997).

10. D. Wagner, *these proceedings* and references therein.
11. See for example, M. Carena, M. Olechowski, S. Pokorski and C.E.M. Wagner, Nucl. Phys. **B426**, 269 (1994) and Nucl. Phys. **B419**, 213 (1994); M. Drees and S. P. Martin, Report of Subgroup 2 of the DPF Working Group on Electroweak Symmetry Breaking and Beyond the Standard Model, hep–ph/9504324; M. Carena, P. Chankowski, M. Olechowski, S. Pokorski and C.E.M. Wagner, Nucl. Phys. **B491**, 103 (1997).
12. M. Olechowski and S. Pokorski, Phys. Lett. **B344**, 201 (1995); D. Matalliotakis and H.P. Nilles, Nucl. Phys. **B435**, 115 (1995); M. Carena and C.E.M. Wagner, Proceedings of the 2nd IFT Workshop on Yukawa Couplings and the Origins of Mass, Gainesville 1994, hep–ph/9407209.
13. M. Dine, A.E. Nelson and Y. Shirman, Phys. Rev. **D51**, 1362 (1995) and Phys. Rev. **D53**, 2658 (1995).
14. C. E. M. Wagner, hep-ph/9801376 and references therein.
15. P. Fayet, Phys. Lett. **B70**, 461 (1977); P. Fayet, Phys. Lett. **B84**, 416 (1979); R. Casalbuoni, S. de Curtis, D. Dominici, F. Feruglio and R. Gatto, Phys. Lett. **B215**, 313 (1988).
16. S. Dimopoulos, M. Dine, S. Raby and S. Thomas, Phys. Rev. Lett. **76**, 3494 (1996) and Phys. Rev. **D55**, 1372 (1997).
17. S. Ambrosanio, G.L. Kane, G.D. Kribs, S.P. Martin and S. Mrenna, Phys. Rev. Lett. **76**, 3498 (1996).
18. S. Dimopoulos, S. Thomas and J. Wells, Phys. Rev. **D54**, 3283 (1996) and Nucl. Phys. **B488**, 39 (1997).
19. S. Dimopoulos, M. Dine, S. Raby, S. Thomas and J. Wells, Nucl. Phys. Proc. Suppl. **A52**, 38 (1997).
20. G.R. Farrar and P. Fayet, Phys. Lett. **B76**, 575 (1978).
21. M. Diaz, talk presented at the International Europhysics Conference on High-Energy Physics (HEP 97), Jerusalem, Israel, hep–ph/9712213; talk given at International Workshop on Quantum Effects in the Minimal Supersymmetric Standard Model, Barcelona, Spain, hep–ph/9711435 and references therein; M. Carena, S. Pokorski and C.E.M. Wagner, hep–ph/9801251, Dec. 1997, to appear in Phys. Lett. B.
22. H. Dreiner, An introduction to Explicit R–parity Violation, to be published in *Perspectives on Supersymmetry*, ed. G.L. Kane, (World Scientific, Singapore, 1997), hep–ph 9707435 and references therein; G. Bhattacharyya, A Brief Review on R–Parity-Violating couplings, invited talk at Beyond the Desert, Castle Ringberg, Tengernsee, Germany, June 1997, hep–ph/9709395.
23. S. Raby, *these proceedings*.
24. M. Carena, J.R. Espinosa, M. Quirós and C.E.M. Wagner, Phys. Lett. **B335**, 209 (1995); M. Carena, M. Quirós and C.E.M. Wagner, Nucl. Phys. **B461**, 407 (1996); H. Haber, R. Hempfling and H. Hoang, Z. Phys. **75**, 539 (1997); M. Carena, P. Zerwas and the Higgs Physics Working Group, *Physics at LEP2*, Vol. 1, eds. G. Altarelli, T. Sjöstrand and F. Zwirner, CERN Report No. 96–01; S. Heinemeyer, W. Hollik, G. Weiglein, hep–ph/9803277.
25. V. Barger, M.S. Berger, J.F. Gunion and T. Han, Phys. Rep. **286** no 1, 1 (1997).
26. V. Barger, M.S. Berger, J.F. Gunion and T. Han, Phys. Rev. Lett. **75**, 1462 (1995).
27. R. Raja and A. Tollestrup, hep-ex/9801004.

28. V. Barger, *Supersymmetry vis-à-vis Muon Colliders*, *these proceedings*.
29. J.F. Gunion, *Physics at a Muon Collider*, *these proceedings*.
30. M. Demarteau and T. Han, Higgs Working Group Summary, *these proceedings*.
31. V. Barger, M.S. Berger, J.F. Gunion and T. Han, Phys. Rev. **D56**, 1714 (1997); Phys. Rev. Lett. **78**, 3999 (1997); M.S. Berger, hep-ph/9802213.
32. Searches for New Physics at LEP 2, hep-ph/9602207.
33. M. Carena, G. F. Giudice and C.E.M. Wagner, Phys. Lett.**B390**, 234(1997) and references therein.
34. H. Baer, C. Chen, C. Kao, X. Tata Phys. Rev. **D52**, 1565 (1995); S. Mrenna, G.L. Kane, G.D. Kribs and J.D. Wells, Phys. Rev. **D53**, 1168 (1996); H. Baer, C. Chen, F. Paige and X. Tata, Phys. Rev. **D54**, 5866 (1996).
35. I. Hinchliffe, talk presented at this Workshop.
36. M. Berger, *these proceedings*; V. Barger, M. Berger and T. Han, hep-ph/9801410 and references therein.
37. T. Tsukamoto, K. Fujii, H. Murayama, M. Yamaguchi and Y. Okada, Phys. Rev. **D51** 3153 (1995); I. Hinchliffe, F. Paige, M. D. Shapiro, J. Soderqvist and W. Yao, Phys. Rev. **D55** 5520 (1997).
38. J. Feng and D. Finnell, Phys. Rev. **D49** 2369 (1994).
39. J. Lykken, *4th Int. Conference on Physics Potential and Developments of a $\mu^+\mu^-$ Colliders*, hep-ph/9803427.
40. F. Paige, *these proceedings*.
41. H-C Cheng, *these proceedings* and references therein.
42. R. D. Boltom, et al., *Phys. Rev.* **D 38**,2077 (1988).
43. W. Marciano, these proceedings.μ^+e^-
44. J. Feng, J.F. Gunion and T. Han, hep-ph/9711414;
45. S. Raychaudhuri, talk presented at this Workshop;
46. J. Feng, *these proceedings* and references therein.
47. G. Bhattacharyya and Amitava Raychaudhuri, hep-ph/9712245.
48. A. Bartl, H. Eberl and W. Majerotto, Nucl Phys. **B472** 481 (1996).
49. A. Bartl, H. Eberl, S. Kraml, W. Majerotto and W. Porod, in preparation.
50. W. Porod, *these proceedings*.
51. See for example, J. Gunion, H. Haber, G. Kane and S. Dawson, *The Higgs Hunter's Guide* and references therein; A. Djouadi, J. Kalinowski, P. Ohmann and P. Zerwas, Zeit. für Phys. **C74** 93 (1997). A. Bartl, H. Eberl, K.. Hikada, T. Kon, W. Majerotto and Y. Yamada, Phys. Lett. **B402** 303 (1997).
52. J. G. Kelly, *these proceedings*.
53. F. Borzumati, G. R. Farrar, N. Polonsky and S. Thomas, *these proceedings*.

Strong Dynamics at the Muon Collider: Working Group Report[*]

Pushpalatha C. Bhat[†] and Estia Eichten[†]

Fermi National Accelerator Laboratory, Batavia, IL 60510

Working group members: Gustavo Burdman, Bogdan Dobrescu,
Daniele Dominici, Takanobu Handa, Christopher Hill, Stephane Keller,
Kenneth Lane, Paul Mackenzie, Kaori Maeshima, Stephen Parke,
Juan Valls, Rocio Vilar and John Womersley

Abstract. New strong dynamics at the energy scale ≈ 1 TeV is an attractive and elegant theoretical *ansatz* for the origin of electroweak symmetry breaking. We review here, the theoretical models for strong dynamics, particularly, technicolor theories and their low energy signatures. We emphasize that the fantastic beam energy resolution ($\sigma_E/E \sim 10^{-4}$) expected at the first muon collider ($\sqrt{s}$ = 100–500 GeV) allows the possibility of resolving some extraordinarily narrow technihadron resonances and, Higgs-like techniscalars produced in the s-channel. Investigating indirect probes for strong dynamics such as search for muon compositeness, we find that the muon colliders provide unparalleled reaches. A big muon collider ($\sqrt{s}$ =3–4 TeV) would be a remarkable facility to study heavy technicolor particles such as the topcolor Z', to probe the dynamics underlying fermion masses and mixings and to fully explore the strongly interacting electroweak sector.

INTRODUCTION

The success of the Standard Model of Particle Physics has been spectacular, thus far! But, new physics beyond the Standard Model (SM) seems inevitable since some critical issues remain unresolved. The cause of electroweak symmetry breaking (EWSB) is not experimentally established, nor is the origin of fermion masses and mixings known. There are two enticing theoretical approaches to understand EWSB – introducing supersymmetry or invoking new

---

[*] Summary talk presented by P.C. Bhat
[†] Co-convenors

strong interactions. Our working group explored the latter scenario, that of a new strong dynamics, and, its search and study at the First Muon Collider.

In this report, we review and summarize the activities of the working group. We first explore the existing theories for new strong dynamics. These models provide an intuitively attractive (though presently disfavored) approach to origin of EW symmetry breaking. A large fraction of this report is devoted to topics related to technicolor, topcolor and their variants. We give an overview of the technicolor models, their low energy signatures, the potential of the First Muon Collider (FMC with $\sqrt{s}=$ 100-500 GeV) for direct searches and detailed measurements of these signatures. Some comments on how these compare with what would be attainable at the Tevatron, LHC and the possible NLC machines are included. We also explore the indirect probes for strong dynamics such as tests for compositeness and comment on constraints from rare B and K decays. We briefly state the long-range opportunities to discover and study new strong dynamics with a big muon collider (BMC with $\sqrt{s} =$ 3-4 TeV).

At the working group meetings, technicolor theories [1] and relevant issues for the FMC were described in detail by Ken Lane [2]. The specific details for topcolor theories [3] and other recent new ideas in strong dynamics were expounded by Chris Hill [4]. Cross sections for production of some low mass technihadrons at a $\mu^+\mu^-$ collider and their signatures were discussed by John Womersley [5]. A model of technicolor with scalars and the prospects for discovering non-standard Higgs-like scalars in the s-channel at the FMC were presented by Bogdan Dobrescu [6]. A study of vector resonances in the framework of BESS (breaking electroweak symmetry strongly) model [7] was presented by Daniele Dominici [8]. A search for technicolor particles using the Tevatron Run I data and the resulting 95% confidence level (C.L.) upper limits on the production cross-section and exclusion of certain mass regions for these particles were reported by the CDF collaboration [9]. Studies of various indirect tests of strong dynamics such as compositeness tests (Eichten and Keller [10]), strong WW scattering (Gunion [11]) and constraints on strong dynamics from rare B and K decays (Burdman [12]), were also presented. During the workshop, prompted by comments from Hill and Mackenzie, stressing the importance of narrow neutral technipion production at a muon collider, Eichten and Lane calculated the cross sections for resonance production of these particles at the FMC. Subsequently, Dominici et al., have also studied the production of such particles (called pseudo-Nambu-Goldstone bosons – PNGBs) in the framework of the BESS model [8].

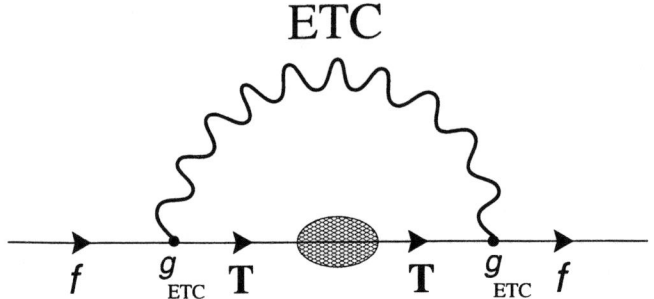

FIGURE 1. Generation of quark and lepton masses via ETC interactions.

TECHNICOLOR AND VARIANTS

Technicolor (TC) is a new strong interaction of fermions and gauge bosons at the scale $\Lambda_{TC} \sim 1$ TeV, which causes dynamical breaking of electroweak symmetry [1]. No elementary scalar bosons (such as the Higgs) are required. Technicolor model, in its simplest form, is a scaled-up version of QCD with massless technifermions that strongly interact at a scale $\Lambda_{TC} \sim 1$ TeV and acquire a dynamical mass $\mathcal{O}(\Lambda_{TC})$. The chiral symmetry is spontaneously broken through technifermion condensation, producing three massless Goldstone bosons. These Goldstone bosons (technipions) have Higgs-like coupling to fermions and correspond to the longitudinal components $W_L^\pm$ and Z_L^0 of the weak gauge bosons. If left- and right-handed technifermions are assigned to weak SU(2) doublets and singlets, respectively, then $M_W = M_Z \cos\theta_W = \frac{1}{2}gF_\pi$ where $F_\pi = 246$ GeV is the technipion decay constant, analogous to $f_\pi = 93$ MeV for the pion. In non-minimal technicolor model, with a large number of technifermion doublets, additional Goldstone bosons arise from technifermion chiral symmetry breaking. The technicolored and the SM fermions however remain massless. They can acquire masses if they couple to technifermions via additional gauge interactions as shown in Fig 1. In this **Extended Technicolor (ETC)** model [13], the quark and lepton masses (m_f) proportional to the dynamical mass of the technifermions (condensate $<T\bar{T}>$) are generated:

$$m_f(M_{ETC}) \approx \frac{g_{ETC}^2}{M_{ETC}^2} <T\bar{T}>_{ETC} \qquad (1)$$

where g_{ETC} is the coupling strength of the fermions to the ETC boson and M_{ETC} is the mass of the ETC boson. The ETC symmetry which pertains to a larger gauge group into which technicolor, color and flavor symmetries are embedded, is broken at a scale $\Lambda_{ETC} = \mathcal{O}(100 \text{ TeV})$.

To avoid large flavor-changing neutral currents and to obtain quark masses of a few GeV, the strong technicolor coupling α_{TC} must run very slowly or "walk", all the way up to the ETC scale of several hundred TeV [14]. **Walking Technicolor** needs a large number of technifermions for α_{TC} to "walk".

Another major turning point in the development of technicolor theories came with the discovery of the top quark and the measurement of its large mass (m_t) [16,17]. The direct measurements from the CDF and DØ collaborations yield a value of $m_t = 175.6\pm 5.5$ GeV/c^2 (current world average) [18,19]. To generate this large m_t, the ETC models would have to violate experimental constraints on the ρ parameter ($\rho = \frac{M_W}{M_Z \cos\theta_W}$) or the $Z \to b\bar{b}$ decay rate. To resolve this problem, a new strong **Topcolor** interaction was introduced by Hill [4] and **Topcolor-assisted Technicolor** ($TC2$) was born [15]. The top quark is very heavy compared to all the other quarks and leptons. This fact suggests that it might be strongly coupled to the mechanism of mass generation and to the dynamics of EWSB itself. It is conceivable that the top quark has unique dynamics. The simplest $TC2$ model has the following group structure:

$$\begin{aligned} G_{TC} \times SU(2)_{EW} \times SU(3)_3 \times SU(3)_{1,2} \times U(1)_3 \times U(1)_{1,2} \\ \longrightarrow G_{TC} \times SU(2)_{EW} \times SU(3)_C \times U(1)_Y \\ \longrightarrow SU(3)_C \times U(1)_{EM} \end{aligned} \quad (2)$$

where G_{TC} and $SU(2)_{EW}$ are the technicolor and electroweak gauge groups; $SU(3)_3$ and $U(1)_3$ are topcolor gauge groups coupled to the third generation fermions (with stronger couplings) while $SU(3)_{1,2}$ and $U(1)_{1,2}$ couple to first and second generations only. Technicolor causes most of the EWSB, while topcolor contributes only feebly with $f_t \approx 60$ GeV. The light quark and lepton masses are generated via ETC dynamics which contributes only a GeV to the third generation masses. The strong topcolor dynamics (top quark pair condensate) generates $m_t \approx 175$ GeV. The $U(1)_3$ provides the difference that causes only top quarks to condense. Thus, the top quark mass may be perceived as being generated by a combination of a dynamical condensate component, $(1-\epsilon)m_t$ (from topcolor dynamics) and a small fundamental component, ϵm_t (from e.g., technicolor) with $\epsilon << 1$. A number of additional particles called "top-pions" π_t and "top-rho" ρ_t, are expected. The small ETC component of the top quark implies that the masses of the top-pions depend on ϵ and Λ. The top-pion mass induced from the fermion loop can be estimated as,

$$M_{\pi_t}^2 = \frac{N\epsilon m_t^2 M_B^2}{8\pi^2 f_\pi^2} = \frac{\epsilon M_B^2}{\log(M_B/m_t)} \quad (3)$$

where the Pagels-Stoker formula is used for f_π^2. For $\epsilon = (0.03, 0.1)$, $M_B \approx (1.5, 1.0)$ TeV, and $m_t = 180$ GeV, this predicts $M_{\pi_t} = (180, 240)$ GeV. The bare values of ϵ generated at Λ_{ETC} is subject to large radiative enhancements (~ 10) by topcolor and $U(1)_3$. Thus, we expect that even a bare value of $\epsilon \sim 0.005$ can produce sizeable M_{π_t} ($> m_t$). The breaking of $U(1)_3 \times U(1)_{1,2} \to U(1)_Y$ in the vicinity of 2 TeV leads to eight color-octet vector bosons V_8 or B (colorons or top-gluons) and an additional Z boson, Z'. The mass of the Z' is expected to be in the range of 1-3 TeV.

Top See-saw Model

The topcolor models have met with some problems in their implications for limits on custodial symmetry violation and other phenomenological constraints. The proximity of the measured top quark mass, m_t, to the electroweak scale, however, suggests that EWSB may have its origin in dynamics associated with the top quark. An explicit realization of this idea is the top condensation mechanism [20], in which the top-antitop quark pair acquires a vacuum expectation value, much like the chiral condensate of QCD or the electron condensate of superconductivity (BCS theory). The EWSB occurs via the condensation of the top quark in the presence of an extra vector-like, weak-isosinglet quark. The mass scale of the condensate is ~0.6 TeV corresponding to the electroweak scale $v \approx 246$ GeV. The vector-like isosinglet then naturally exhibits a see-saw mechanism, yielding the physical top quark mass, which is then adjusted to the experimental value. The choice of ~TeV scale for the topcolor dynamics determines the mass of the weak-isoscalar see-saw partner. The model also implies the existence of PNGBs. The lower bound on the mass of a PNGB that couples to the top quark is less than m_t.

More work is needed to extend the scheme to generate masses and mixing for all quarks and leptons, and to construct attractive schemes for topcolor breaking.

Technicolor Production and Signatures at the FMC

In the minimal technicolor model, with just one technifermion doublet, the only prominent signals at the hadron and lepton colliders would be the enhancements in longitudinally-polarized weak vector boson production. These are due to the s-channel production of color-singlet technirho resonances near 1.5-2.0 TeV and the subsequent decay into vector boson pairs ($\rho_T^0 \to W_L^+ W_L^-$ and $\rho_T^\pm \to W_L^\pm Z^0$). Observing these enhancements would be extremely difficult, since the $\mathcal{O}(\alpha^2)$ production cross sections are small at such high technirho masses and efficiency for reconstructing vector boson pairs low.

The non-minimal technicolor models, however, predict a rich spectrum of light, color-singlet technihadrons—the isotriplet vectors $\rho_T^0, \rho_T^\pm$ and their isoscalar partner ω_T, and pseudoscalars $\pi_T^0, \pi_T^\pm$ and $\pi_T^{0\prime}$—accessible at the Tevatron, LHC and the FMC. (A search at the Tevatron by CDF has been discussed later). Since techni-isospin is likely to be a good approximate symmetry, ρ_T and ω_T are approximately degenerate and so are the technipions. The masses are expected to be: $m_{\pi_T} \approx 100\ GeV$ and $m_{\rho_T} \approx m_{\omega_T} \approx 200$ GeV.

The technipions with Higgs-like ETC couplings to quarks and leptons decay to the heaviest fermion pairs allowed. The isosinglet component of neutral technipions, $\pi_T^{0\prime}$, may decay into a pair of gluons if the constituent technifermions are colored. Thus the predominant decay signatures of the light technipions would be:

$$\begin{aligned} \pi_T^0 &\to b\bar{b},\ c\bar{c},\ \tau^+\tau^- \\ \pi_T^{0\prime} &\to gg,\ b\bar{b},\ c\bar{c},\ \tau^+\tau^- \\ \pi_T^+ &\to c\bar{b},\ c\bar{s},\ \tau^+\nu_\tau. \end{aligned} \quad (4)$$

The signatures for technirhos and techniomegas are as follows:

$$\begin{aligned} \rho_T^\pm &\to W^\pm Z,\ W^\pm \pi_T^0,\ Z\pi_T^\pm,\ \pi_T^\pm \pi_T^0 \\ \rho_T^0 &\to W^+ W^-,\ W^\pm \pi_T^\mp,\ \pi_T^\pm \pi_T^\mp,\ q\bar{q},\ \ell^+\ell^-,\ \nu\bar{\nu} \\ \omega_T &\to \gamma\pi_T^0,\ Z\pi_T^0,\ q\bar{q},\ \ell^+\ell^-,\ \nu\bar{\nu}. \end{aligned} \quad (5)$$

If the large ratio of $\frac{\langle \bar{T}T \rangle_{ETC}}{\langle \bar{T}T \rangle_{TC}}$ significantly enhances technipion masses relative to technivector masses, then $\rho_T \to \pi_T \pi_T$ decay channels may be closed.

If technicolor exists and technihadrons have masses low enough to be produced at the FMC, they will most probably be first discovered at the Tevatron or at the LHC. An interesting aspect of the technihadrons is that several of them, particularly the neutral ones are very narrow. Therefore, a $\mu^+\mu^-$ collider which is expected to have very fine energy resolution is ideally suited for their studies.

Figure 2 shows the production cross section for ρ_T at a muon collider as a function of $\sqrt{s}$, for M_{ρ_T}=210 GeV and M_{π_T}=110 GeV. The peak cross section is ~ 1 nb which translates to 10^6 events/year with $\int \mathcal{L}dt = 10^{32}\ cm^{-2}\ s^{-1}$. Figure 3 shows the cross section for ω_T production (M_{ω_T}=210 GeV). The peak cross section here is even larger, ~ 10 nb, that would provide an yield of 10^7 events/year for the same luminosity. Also, note that the peak is extremely narrow, with a width < 1 GeV. The production cross sections decrease, if ρ_T, ω_T are heavier. For M_{ρ_T}=400 GeV and M_{π_T}=150 GeV, the event rate is still 10^4 events/year.

The neutral technipions, like the SM Higgs boson, are expected to couple to $\mu^+\mu^-$ with a strength proportional to m_μ. In the *ansatz* of the non-minimal technicolor model with N_D technifermion doublets, the coupling is enhanced by a factor of $\sqrt{N_D}$. Therefore, the FMC can serve as a neutral technipion factory with phenomenal rates for production in the s-channel, far exceeding those at any other collider.

Once a neutral technipion has been found in ρ_T or ω_T decays at a hadron collider, it should be relatively easy to locate the precise position of the resonance at the FMC and, take data at the resonance. The cross sections for $f\bar{f}$ and gg final states are isotropic. The $\pi_T^{0\prime}$ production cross sections and the Z^0 backgrounds are shown in Fig. 4 for $M_{\pi_T} = 110\,\text{GeV}$ (for description of other parameters see ref. [2]). The peak signal rates approach 1 nb. The

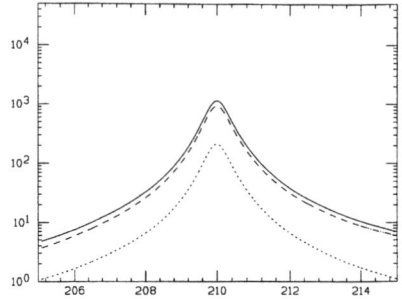

FIGURE 2. Cross section (pb) for technirho production at a muon collider as a function of $\sqrt{s}$(GeV), for $M_{\rho_T}=210$ GeV and $M_{\pi_T}=100$ GeV. The solid curve is the total cross section, the dashed curve is for $\rho_T \to W\pi_T$ and the dotted curve is for $\rho_T \to W^+W^-$.

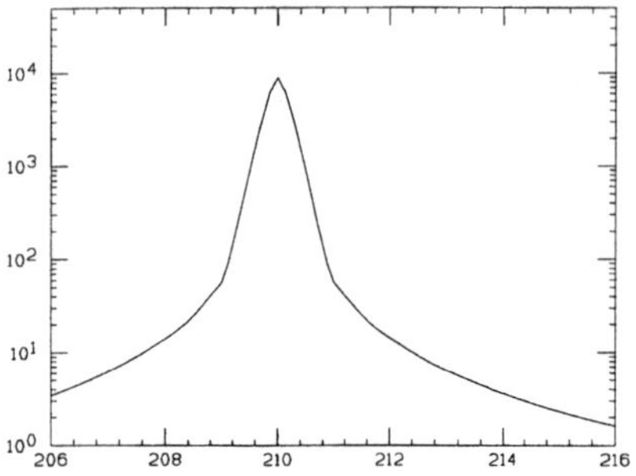

FIGURE 3. Total cross section (pb) for techniomega production at a muon collider as a function of $\sqrt{s}$(GeV), for $M_{\omega_T}=210$ GeV and $M_{\pi_T}=110$ GeV. The dominant decay mode is $\gamma\pi_T^0$.

$b\bar{b}$ dijet rates are much larger than the $Z^0 \to b\bar{b}$ backgrounds, while the gg rate is comparable to $Z^0 \to q\bar{q}$. Details of these and other calculations in this section, including the effects of the finite beam energy resolution, will appear in Ref. [21].

The cross sections for technipion production via the decay of technirho and techniomega s-channel resonances are calculated using vector meson (γ, Z^0) dominance [22–25]. For $M_{\rho_T} = M_{\omega_T} = 210\,\text{GeV}$, $M_{\pi_T} = 110\,\text{GeV}$, and other parameters as above, the total peak cross sections are [21]:

$$\sum_{AB} \sigma(\mu^+\mu^- \to \rho_T^0 \to \pi_A\pi_B) = 1.1\,\text{nb}$$
$$\sigma(\mu^+\mu^- \to \omega_T \to \gamma\pi_T^0) = 8.9\,\text{nb}\,. \qquad (6)$$

The technirho decay rate is 20% W^+W^- and 80% $W^\pm\pi_T^\mp$.

Further, there might be a small nonzero isospin splitting between ρ_T^0 and ω_T. This would appear as a dramatic interference in the $\mu^+\mu^- \to f\bar{f}$ cross section, provided the FMC energy resolution is good enough in the ρ_T-ω_T region. The cross section is most accurately calculated [2] by using the full γ-Z^0-ρ_T-ω_T propagator matrix (Δ).

Figure 5 shows the theoretical ρ_T^0-ω_T interference effect in $\mu^+\mu^- \to e^+e^-$ for input masses $M_{\rho_T} = 210\,\text{GeV}$ and $M_{\omega_T} = 212.5\,\text{GeV}$. The propagator shifts the nominal positions of the resonance peaks by $\mathcal{O}(\alpha/\alpha_{\rho_T})$. The theoretical

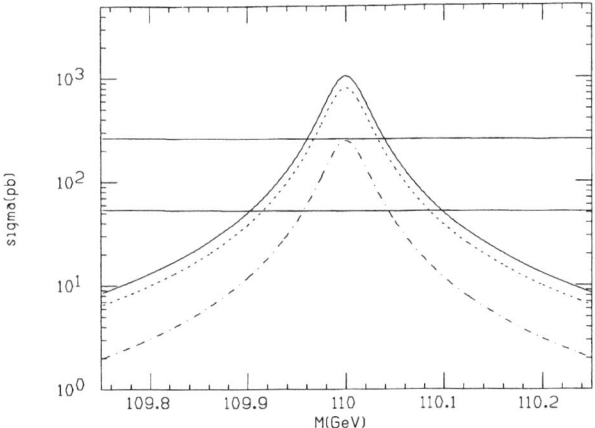

FIGURE 4. Theoretical (unsmeared) cross sections for $\mu^+\mu^- \to \pi_T^{0\prime} \to b\bar{b}$ (dashed), gg (dot-dashed) and total (solid) for $M_{\pi_T} = 110\,\text{GeV}$ and other parameters defined in the text. The solid horizontal lines are the backgrounds from $\gamma, Z^0 \to b\bar{b}$ (lower) and $Z^0 \to q\bar{q}$ (upper). Note the energy scale.

peak cross sections are 5.0 pb at 210.7 GeV and 320 pb at 214.0 GeV. This demonstrates the importance of precision resolution in the 200 GeV FMC.

The detectors [28] at the muon collider should be capable of identifying and measuring electrons, muons, taus, jets and, of tagging b-jets with high efficiency. It would be useful if c-jets could be distinguished from b-jets.

Topcolor Signatures

Topcolor-assisted technicolor introduces additional particles called top-pions (π_t), top-gluons (B or V_8) and topcolor Z', as discussed in the previous section. Top-pions can be as light as $\sim$150 GeV, in which case they would emerge as a detectable branching fraction of top quark decay. However, not to violate constraints on $Z \to b\bar{b}$ rate, $M_{\pi_t} \geq 300$ GeV may be required. Top-gluons are expected to have mass in the range of 0.5-2 TeV and topcolor Z' in the range of 1-3 TeV. The decays are expected to be:

$$\begin{aligned} \pi_t &\to t\bar{b}, \quad \text{or} \quad t \to \pi_t b \\ B &\to b\bar{b}, t\bar{t} \\ Z' &\to t\bar{t}. \end{aligned} \quad (7)$$

Top-pions may be produced copiously at the FMC in the s-channel as previously discussed in the case of technipions. The LHC experiments should be

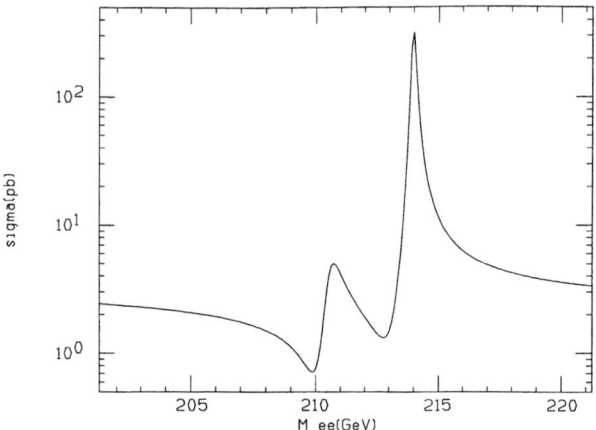

FIGURE 5. Theoretical (unsmeared) cross sections for $\mu^+\mu^- \to \rho_T^0, \omega_T \to e^+e^-$ for input masses $M_{\rho_T} = 210\,\text{GeV}$ and $M_{\omega_T} = 212.5\,\text{GeV}$ and other parameters as defined in the text.

sensitive over the entire range of the expected masses for both top-gluons and topcolor Z'. If topcolor Z' is not found at LHC, it can be discovered at the big muon collider ($\sqrt{s} = 3\text{--}4$ TeV). There are a number of other effects of topcolor that can be observed at the FMC [4]. For example, new effects in Z physics involving the third generation such as $Z \to b\bar{b}$, might be observed. The generational structure of topcolor may induce GIM violation in low energy processes such as $K^+ \to \pi^+ \nu \bar{\nu}$ and lepton family number violation such as $\mu\bar{\mu} \to \tau\bar{\mu}$. There may be induced FCNC interactions giving rise to anomalous $\mu\bar{\mu} \to b\bar{s}$. The FMC and the front-end of the FMC provide great opportunities to study such effects that are enhanced due to topcolor w.r.t. SM.

Technicolor with Scalars

Technicolor models that include scalars are an interesting class of models for dynamical EWSB. In the current model [6], in addition to SM fermions, one doublet of technifermions, P and N, and three scalars, ϕ, χ and Φ are considered. The gauge group considered is $SU(4)_{TC} \times SU(3)_C \times SU(2)_{EW} \times U(1)_Y$. Only the third generation couples to the technicolor fields, and, as in QCD, the $SU(4)_{TC}$ techincolor interactions trigger the formation of technifermion condensates $<P\bar{P}> \simeq <N\bar{N}> \simeq 2\sqrt{3}\pi f^3$, which breaks the electroweak symmetry at a scale f. This also results in the generation of masses for t, b and τ. The masses of the first and second generations are generated by coupling to a scalar Φ that behaves like a Higgs doublet under gauge transformations.

The scalar acquires a small vacuum expectation value (VEV) by coupling to the new strong interactions sector and would have Yukawa couplings to the first and second generations which are larger than in the SM.

If this model is the correct description of physics up to a TeV scale, then the components of the Φ scalar should be accessible at a $\mu^+\mu^-$ collider with $\sqrt{s}$ below the first technihadron resonance. Since the Yukawa couplings are proportional to the fermion mass, the s-channel production is very large at a muon collider. The scalar Φ decomposes into an isosinglet σ and an isotriplet π'^3, a=1,2,3. The neutral real scalar σ and the charged scalars $\pi'^\pm = \frac{(\pi'^1 \mp i\pi'^2)}{\sqrt{2}}$ are almost degenerate, with a mass M_Φ. For $M_\Phi < \sqrt{s} < 2M_\Phi$, only the σ and π'^3, can be produced.

The total decay widths of the σ and π'^3 scalars are equal. The VEV of Φ is taken to be in the range,

$$1 GeV \leq f' \leq 10 GeV, \qquad (8)$$

where the lower bound is chosen to avoid Yukawa coupling constants larger than 1.0, and upper bound is chosen to satisfy condition $f' << f$. The width for decay into pairs of gauge bosons, $\Gamma(W^+W^- + ZZ)$, is at most a few percent of the width for $\sigma, \pi'^3 \to c\bar{c}$, and is neglected here. The widths of the σ and π'^3 scalars are dominated only by the $c\bar{c}$ final state:

$$\Gamma \approx \frac{3 m_c^2 M_\Phi}{8\pi f'^2} \approx 13.2 GeV \left(\frac{3 GeV}{f'}\right)^2 \left(\frac{M_\Phi}{500 GeV}\right)^2. \qquad (9)$$

Given the enhanced couplings to the second generation, the s-channel production of the neutral scalars at a $\mu^+\mu^-$ collider is large. The natural spread in the muon collider beam energy, $\sigma_{\sqrt{s}}$, is rather small, and can be ignored in computing the effective s-channel resonance cross section:

$$\bar{\sigma}(\mu^+\mu^- \to \sigma, \pi'^3 \to X) \approx \frac{4\pi\Gamma^2}{(s - M_\Phi^2)^2 + M_\Phi^2 \Gamma^2} B(\sigma, \pi'^3 \to \mu^+\mu^-) B(\sigma, \pi'^3 \to X). \qquad (10)$$

For the final state is $X \equiv c\bar{c}$, this cross section becomes,

$$\bar{\sigma}(\mu^+\mu^- \to \sigma, \pi'^3 \to c\bar{c}) \approx \frac{4\pi\Gamma^2}{(s - M_\Phi^2)^2 + M_\Phi^2 \Gamma^2} \left(\frac{m_\mu^2}{3 m_c^2}\right). \qquad (11)$$

The main background comes from $\mu^+\mu^- \to \gamma^*, Z^* \to c\bar{c}$, and amounts to

$$\sigma(\mu^+\mu^- \to c\bar{c}) \approx 0.7 pb \frac{(500 GeV)^2}{s}. \qquad (12)$$

The discovery potential of a $\mu^+\mu^-$ collider operating at a maximum center of mass energy of 500 GeV has been studied. Two scan points, at 300 and 500

GeV, are sufficient to find the neutral scalars with masses roughly between 200 and 600 GeV.

Once the resonance is found, the beam energy can be adjusted to the peak (even if this requires a significant reduction in the luminosity) and then the production cross section becomes very large:

$$\bar{\sigma}(\mu^+\mu^- \to \sigma, \pi'^3 \to c\bar{c}) \approx \frac{8\pi}{M_\Phi^2}\left(\frac{m_\mu^2}{3m_c^2}\right) \approx 80 pb \left(\frac{500 GeV}{M_\Phi}\right)^2. \qquad (13)$$

With a luminosity of $2 \times 10^{34} cm^{-2} s^{-1}$ ($7 \times 10^{32} cm^{-2} s^{-1}$), and a c-tagging efficiency of 30%, the observed rate should be 10^7 ($\sim 2 \times 10^5$) events per year.

BESS Model Study of SEWS

The BESS model is an effective Lagrangian parametrization approach to the symmetry breaking mechanism. The symmetry group of the theory is $G' = SU(2)_L \times SU(2)_R \times SU(2)_V$, where $SU(2)_V$ is the hidden symmetry through which new vector particles are introduced. The spontaneous breakdown of the symmetry group $G' \to SU(2)$ gives rise to six Goldstone bosons. Three of these are absorbed by new vector particles while the other three give mass to the SM gauge bosons when gauging of the subgroup $SU(2)_L \times SU(2)_Y \subset G$ is performed.

The parameters of the BESS model are the masses of the new bosons M_V, their self-coupling g'', and a parameter b that characterizes the coupling strengths of V to the fermions. Taking $b \to 0$ and $g'' \to \infty$, the new bosons decouple and the SM is recovered. Bounds on the parameter space obtained by an analysis of $d\sigma(\ell^+\ell^- \to W^+_{L,T}W^-_{L,T})/d\cos\theta$ (θ being the scattering angle of the W in the center of mass), are shown in Fig. 6. The solid lines show the case relevant to an e^+e^- machine with $\sqrt{s}$=500 GeV and $\int Ldt = 20 fb^{-1}$, the dashed lines correspond to a $\mu^+\mu^-$ machine with same $\sqrt{s}$ and luminosity. The $\mu^+\mu^-$ collider provides some improvement in the bounds. The result for LHC with $pp \to W^\pm V^\mp \to W^\pm Z$ is shown by dot-dash curves for comparison.

Partial wave unitarity bounds from WW scattering deduced in the $(M_V, g/g'')$ and (Γ_V, M_V) planes (see Fig. 7) imply that one or more of the heavy vector resonances should be discovered at the LHC, NLC or a $\sqrt{s}$ ~500 GeV muon collider or, for certain, at a 3-4 TeV muon collider, unless g'' is very large and b is very small so that they are largely decoupled.

Since the workshop, the production of the lightest neutral PNGB (P^0) in the s-channel, and the potential for discovering it at the FMC have also been studied using an extension of the model with $SU(8) \times SU(8)$ symmetry [8].

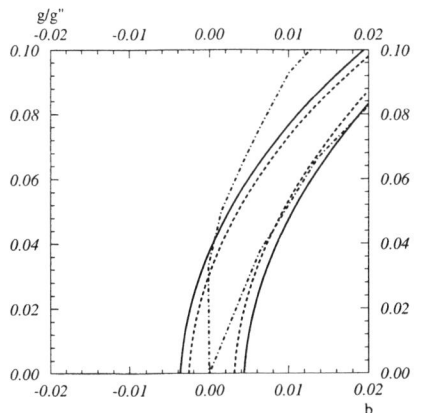
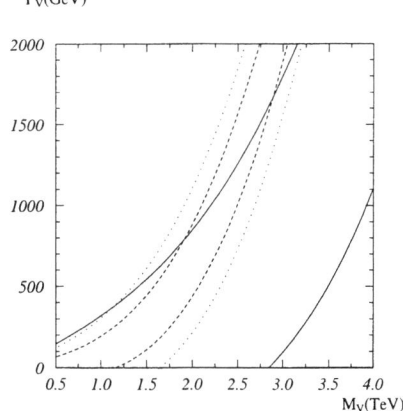

FIGURE 6. 90% C.L. contours from BESS model, for $M_V = 1$ TeV. The allowed region is between two lines. See text for details.

FIGURE 7. Partial wave unitarity bounds in the (M_V, Γ_V) plane for $\Lambda/M_V = 1.5$. The dashed lines corresponds to the partial wave a_{00}, the dotted ones to a_{20} and the solid lines to a_{11}.

Search for Technicolor at CDF

The CDF collaboration reported on their search for technipion and technirho signals in the W + 2 jets + b-tag channel in the Run I data. The signautres sought are for the processes:

$$q\bar{q} \to W^{*\pm} \to \rho_T^{\pm} \to W^{\pm}\pi_T^0$$

and

$$q\bar{q} \to Z^*, \gamma^* \to \rho_T^0 \to W^{\pm}\pi_T^{\mp}$$

with $W^{\pm} \to \ell\nu$ ($\ell = e$ or μ) and $\pi_T^0 \to b\bar{b}$, $\pi_T^{\pm} \to b\bar{c}$, $c\bar{b}$ ($\approx 95\%$) and $\pi_T^{\pm} \to c\bar{s}$, $s\bar{c}$ ($\approx 5\%$).

The candidate event selection requires an isolated electron (muon) with $E_T(p_T) > 20$ GeV within $|\eta| < 1.0$, $\not{E}_T > 20$ GeV and two or more jets with $E_T > 15$ GeV. At least one of the jets is required to be a b-jet, tagged by the silicon vertex detector (SVX). The Z boson candidates are rejected by requiring $|M_{ee} - M_Z| > 15$ GeV/c^2. A total of 42 events are selected while the expected number of background from $Wb\bar{b}$, $Wc\bar{c}$, Wc, top production, mis-tags, Z+heavy flavor amount to 31.6±4.3 events.

The technicolor signal is modeled using PYTHIA MC and GEANT-based detector simulation. Signal MC events are generated at a number of (π_T, ρ_T)

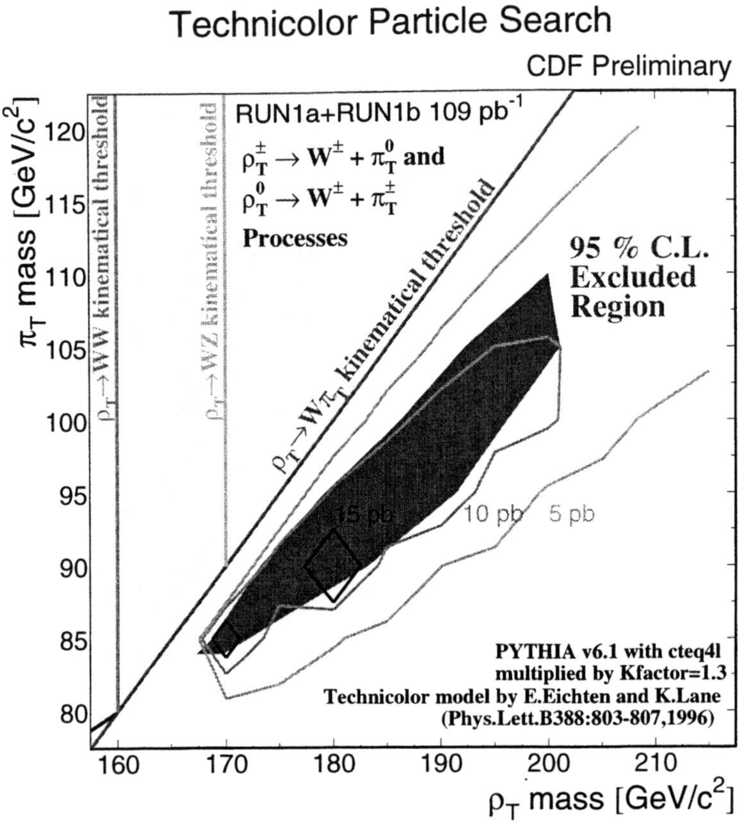

FIGURE 8. The 95% C.L. exclusion region in (M (π_T), M(ρ_T)) plane. Some production cross section contours are also shown.

mass values. The combinations with more than 5 pb cross section are used. The technicolor model parameters used are the ones from ref [24]. Further cuts on kinematic variables $\Delta\phi(jj)$ (the azimuthal angle between two jets) and $p_T(jj)$ (p_T of the dijet system) [5] are employed to enhance the expected signal to background ratio in the selected sample. Finally, M(jj) and M(Wjj) are required to be within $\pm 3\sigma$ of the expected mean values for the signal. No significant excess is seen in the data. The 95% C.L. upper limits on the production cross section then exclude certain region of the (M (π_T), M(ρ_T)) plane as shown in Fig. 8.

PROBING MUON COMPOSITNESS AT THE FMC

The generational pattern of quarks and leptons hints possibly at a substructure (with an associated strong interaction at energy scale Λ) that might manifest at high energies. The existence of such substructure, however, is expected to result in four-fermion "contact" interactions which differ from those arising from the SM, at energies well below Λ. The signals can be sought in a number of ways—inclusive jet production, Drell-Yan production, Bhabha scattering etc. CELLO at the e^+e^- collider PETRA with a $\sqrt{s}=$ 35 GeV and $\int \mathcal{L} dt =$ 86 pb^{-1} was able to set a limit on the electron compositeness scale $\sim$2-4 TeV using Bhabha scattering. These limits are similar to the ones from the Tevatron ($p\bar{p}$, $\sqrt{s} = 1.8$ TeV) [26]. Clearly, the lepton colliders seem to hold great potential for probing lepton compositeness. Probing the muon compositeness using Bhabha scattering measurements and the reach attainable as a function of $\sqrt{s}$ at the muon colliders has been investigated by Eichten and Keller [10].

The four-fermion contact interaction is assumed to be described by the effective Lagrangian proposed by Eichten, Lane and Peskin [27]:

$$\mathcal{L} = \frac{g^2}{2\Lambda^2}[\eta_{LL}j_L j_L + \eta_{RR}j_R j_R + \eta_{LR}j_L j_R] \qquad (14)$$

where j_L and j_R are the left-handed and right-handed currents, respectively; Λ is the compositeness scale and $\frac{g^2}{4\pi}=1$ is assumed (strong coupling). The quantity η is used to set the sign of the coupling i.e., $|\eta| = \pm 1$. Four typical coupling scenarios are considered in the present work : $LL(\eta_{LL} = \pm 1, \eta_{RR} = \eta_{LR} = 0), RR(\eta_{RR} = \pm 1, \eta_{LL} = \eta_{LR} = 0), VV(\eta_{LL} = \eta_{RR} = \eta_{LR} = \pm 1)$ and $AA(\eta_{LL} = \eta_{RR} = -\eta_{LR} = \pm 1)$. The angular distribution of scattered muons (scattering angle θ) are then calculated for each of the models, with and without compositeness hypothesis. The fractional change in the differential cross section due to compositeness,

$$\Delta = \frac{\left(\frac{d\sigma}{d\cos\theta}\right)_{EW+\Lambda} - \left(\frac{d\sigma}{d\cos\theta}\right)_{EW}}{\left(\frac{d\sigma}{d\cos\theta}\right)_{EW}} \qquad (15)$$

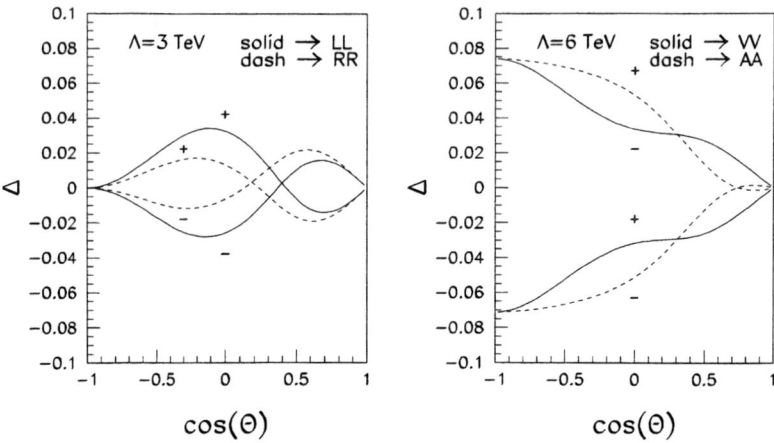

FIGURE 9. The variable Δ versus $\cos\theta$ at $\sqrt{s}=100$ GeV for the four models, LL, RR, VV, and AA, for the two signs of the η's, indicate by + and − on the plot.

is shown in Figures. 9 and 10 for the four different models and for both signs of η's. The plots are made with Λ chosen to provide an average correction of 10% due to compositeness.

The 95% C.L. limits on the compositeness scale Λ are computed by employing an analytical approach that approximates χ^2 fitting of ideal data to theory. The limits extracted for various $\sqrt{s}$ of the muon collider and for various models, together with the expected limits attainable at LEP are shown in Table 1. Since the detectors at a muon collider may not provide coverage down to small angles due to large backgrounds [28], the 95% C.L. limits have also been extracted for different cuts on $\cos\theta$. The results are shown in Table 2. It is seen that the reach only improves by 10% in going from $\cos\theta=$

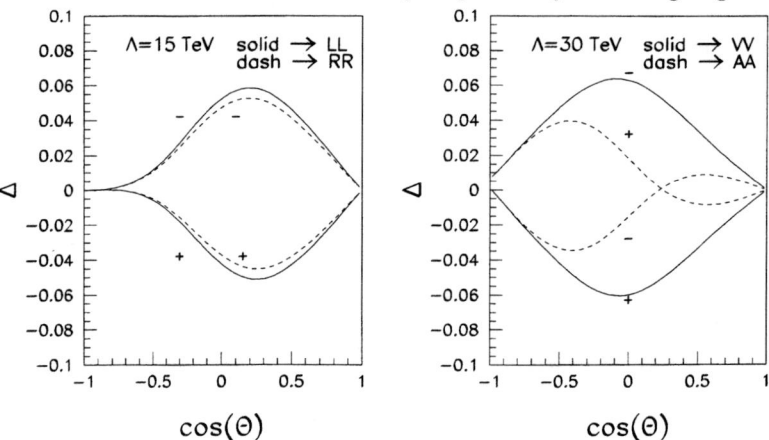

FIGURE 10. The variable Δ versus $\cos\theta$ at $\sqrt{s}=500$ GeV for the four models, LL, RR, VV, and AA, for the two signs of the η's, indicate by + and − on the plot.

TABLE 1. 95% CL limits (in TeV) for various $\sqrt{s}$ (in GeV) of the muon collider ($|\cos\theta| < 0.8$ required). Expected LEP limits ($|\cos\theta| < 0.95$) are also shown.

	LEP(91)	LEP(175)	100	200	350	500	4000
$\mathcal{L}(fb^{-1})$	.15	.1	.6	1.	3.	7.	450.
LL	4.0	5.8	4.8	10	20	29	243
RR	3.8	5.7	4.9	10	19	28	228
VV	6.9	12.	12	21	36	54	435
AA	3.8	7.2	12	13	21	32	263

TABLE 2. 95% CL limits (in TeV) for different on the scattering angle θ cuts ($\sqrt{s} = 500$ GeV, $\mathcal{L} = 7 fb^{-1}$).

$|\cos\theta| <$	.6	.8	.9	.95
LL	26	29	31	32
RR	24	28	30	30
VV	50	54	56	57
AA	28	32	34	35

0.8 to $\cos\theta$=0.95. So, it does not seem necessary to have detector coverage to very small angles.

CONSTRAINTS FROM RARE B AND K DECAYS

The new strong dynamics scenario for EWSB or for the origin of fermion masses can produce sizeable effects in low energy observables at energies much smaller than the scale of new physics. Such effects in rare B and K decays have been studied by Burdman [12], in the framework of an effective Lagrangian Model. These effects in FCNC processes seem to originate from the insertion of anomalous triple gauge boson coupling vertices and four-fermion operators.

In the four-fermion operator scenario, it has been shown that branching ratios for $B \to q\ell^+\ell^-, b \to q\nu\bar{\nu}, b \to q\bar{q}'q'$ can have large deviations (up to a factor of ~ 2) from the SM expectations. However, no significant deviation is expected in $b \to s\gamma$ decay. The effects are very similar in rare K decays such as $K^+ \to \pi^+\nu\bar{\nu}$ and $K_L \to \pi^o\nu\bar{\nu}$. The effective Lagrangian approach for non-SM couplings of fermions to gauge bosons has been examined in the topcolor class of theories. The presence of the relatively light top-pions, and other additional bound states, imposes severe constraints on the topcolor models due to their potential loop effects in low energy observables such as R_b and rare B and K decay rates. These depend not only on f_{π_t} and m_{π_t}, but typically also on one or more elements of the quark rotation matrices necessary to diagonalize the quark Yukawa couplings. So, it can be shown for example, for $f_{\pi_t} \approx 120$ GeV, the effect of a 400 GeV π_t in $b \to s\ell^+\ell^-$ is an enhancement of more than 5% with respect to SM expectations. Similar effects are expected to be present in $K^+ \to \pi^+ \nu \bar{\nu}$. Thus, the measurements of R_b and the rare B and K decay modes can constrain strong dynamics models such as the topcolor model.

SUMMARY

We have reviewed various theories that currently offer to explain the breaking of electroweak symmetry dynamically. In particular, technicolor and related theories have been examined in detail. Direct searches for signals of new strong dynamics as well as indirect tests for existence of strong dynamics, at the FMC, have been studied. Long-range opportunities at the high energy muon colliders (BMC with $\sqrt{s}$=3-4 TeV) are examined. The experimental prospects for strong dynamics at the FMC can be summarized as follows:

- If low energy technicolor signatures (extended walking technicolor, topcolor-assisted technicolor) exist, they would be found at the Tevatron or at the LHC. In this case, the first muon collider will be a remarkable facility to make detailed studies and precision measurements. The narrow neutral technihadrons—π_T, ρ_T and ω_T—would appear as spectacular resonances at the FMC ($\sqrt{s}$=100–200 GeV and energy resolution $\frac{\sigma_E}{E} \leq 10^{-4}$). One can operate on the resonance and study all the decays and branching fractions of the technirhos, techniomegas and technipions. We emphasize that the all-hadronic modes would be very difficult to study at the hadron colliders.

- The good beam resolution achievable at the FMC with $\sqrt{s} \sim 200$ GeV would enable studies of $\rho_T - \omega_T$ interference effects in detail, using fermion-antifermion final states.

- A variety of other models such as technicolor with scalars and top see-saw model predict s-channel resonance production of new particles. Muon collider has a big advantage over other colliders to discover and study these particles.

- Compositeness tests using Bhabha scattering give reaches of several tens of TeV at the FMC. At the big muon collider, the reaches for 95% confidence level limits on compositeness scale are unparelled, far exceeding the reaches possible at any other collider. The reaches would be of the order of 200-300 TeV, at which scale we would be probing the structure of the dynamics that gives rise to fermion masses and mixings.

- Studies of rare B and K decays using the front-end of the FMC can provide tight constraints on strong dynamics and help distinguish between universal (EWSB sector) vs. non-universal (flavor dynamics) scenarios.

These are extremely strong physics motivations to build the first muon collider. A full exploration of the strong electroweak sector can be accomplished at the big muon collider with $\sqrt{s}$=3-4 TeV.

REFERENCES

1. S. Weinberg, *Phys. Rev.* D **19**, 1277 (1979); L. Susskind, *Phys. Rev.* D **20**, 2619 (1979).
2. K. Lane, these proceedings.
3. C. T. Hill, *Phys. Lett.* B **345**, 483 (1995).
4. C. Hill, these proceedings.
5. J. Womersley, these proceedings.
6. B. Dobrescu, these proceedings.
7. R. Casalbuoni, S. De Curtis, D. Dominici, and R. Gatto, *Phys. Lett.* B **155**, 95 (1985); *Nucl. Phys.* B **282**, 235 (1987).
8. D. Dominici *et al.*, these proceedings.
9. T. Handa, K. Maeshima, J. Valls, R. Vilar (for the CDF collaboration), these proceedings.
10. E. Eichten and S. Keller, these proceedings.
11. J. Gunion, these proceedings.
12. G. Burdman, these proceedings.
13. S. Dimopoulos and L. Susskind, *Nucl. Phys.* B **155**, 237 (1979); E. Eichten and K. Lane, *Phys. Lett.* B **90**, 125 (1980).
14. B. Holdom, *Phys. Rev.* D **24**, 1441 (1981); *Phys. Lett.* B **150**, 301 (1985); T. Appelquist, D. Karabali and L. C. R. Wijewardhana, *Phys. Rev. Lett.* **57**, 957 (1986); T. Appelquist and L. C. R. Wijewardhana, *Phys. Rev.* D **36**, 568 (1987); K. Yamawaki, M. Bando and K. Matumoto, *Phys. Rev. Lett.* **56**, 1335 (1986); T. Akiba and T. Yanagida, *Phys. Lett.* B **169**, 432 (1986).
15. K. Lane and E. Eichten, *Phys. Lett.* B **352**, 382 (1995); K. Lane, *Phys. Rev.* D **54**, 2204 (1996).
16. F. Abe, *et al.*, The CDF Collaboration, *Phys. Rev. Lett.* **74**, 2626 (1995).
17. S. Abachi, *et al.*, The DØ Collaboration, *Phys. Rev. Lett.* **74**, 2632 (1995).
18. F. Abe, *et al.*, The CDF Collaboration, *Phys. Rev. Lett.* ??, ???? (1997).
19. S. Abachi, *et al.*, The DØ Collaboration, *Phys. Rev. Lett.* **79**, 1197 (1997).
20. Y. Nambu, in *New Theories in Physics*, Proceedings of the XI International Symposium on Elementary Particle Physics, Kazimierz, Poland, 1988, edited by Z. Adjuk, S. Pokorski and A. Trautmann (World Scientific, Singapore, 1989); Enrico Fermi Institute Report EFI 89-08 (unpublished); V. A. Miransky, M. Tanabashi and K. Yamawaki, *Phys. Lett.* B **221**, 171 (1989); *Mod. Phys. Lett.* **A4**, 1043 (1989); W. A. Bardeen, C. T. Hill and M. Lindner, *Phys. Rev.* D **D41**, 1647 (1990).
21. E. Eichten, K. Lane and J. Womersley, "Narrow Technihadron Production at the First Muon Collider", in preparation.

22. E. Eichten, I. Hinchliffe, K. Lane and C. Quigg, *Rev. Mod. Phys.* **56**, 579 (1984); *Phys. Rev.* D **34**, 1547 (1986).
23. K. Lane and E. Eichten, *Phys. Lett.* B **222**, 274 (1989); K. Lane and M. V. Ramana, *Phys. Rev.* D **44**, 2678 (1991).
24. E. Eichten and K. Lane, *Phys. Lett.* B **388**, 803 (1996).
25. E. Eichten, K. Lane and J. Womersley, *Phys. Lett.* B **405**, 305 (1997).
26. F. Abe, *et al.*, The CDF Collaboration, *Phys. Rev. Lett.* **79**, 2198 (1997).
27. E. Eichten, K. Lane and M. Peskin, *Phys. Rev. Lett.* **50**, 811 (1983).
28. P. Lebrun and R. Roser, these proceedings.

Top Quark Physics at Muon and Other Future Colliders

M. S. Berger* and B. L. Winer[†]

Department of Physics, Indiana University, Bloomington, Indiana 47405
[†]*Department of Physics, Ohio State University, Columbus, Ohio 43210*

Abstract. The top quark will be extensively studied at future muon colliders. The threshold cross section can be measured precisely, and the small beam energy spread is especially effective at making the measurement useful. We report on all the activities of the top quark working group, including talks on top quark physics at other future colliders.

INTRODUCTION

The top quark is expected to be more sensitive than the lighter quarks to new physics effects. It is also the least accessible of the quarks due to its large mass. New colliders are under consideration that could considerably improve our understanding of the top quark.

The fact that the top quark is heavy and there are no heavier quarks (at least probably not) lends some credence to the idea that the top quark is special. Perhaps it is involved in the dynamics of electroweak symmetry breaking, or is subject to some new dynamics. Its Yukawa coupling is comparable in size to the gauge couplings and hence has a significant impact on the evolution of parameters with scale, and is a crucial ingredient in comparisons of weak-scale parameters with possible grand-unified theories. So it is important to measure the mass, couplings, and partial widths of the top quark as well as search for resonances in the $t\bar{t}$ spectrum. Any deviation from SM expectations would be of great interest.

Among the issues of paramount importance are (i) the nature of the absence of flavor changing neutral currents. This can be understood in the Standard Model (SM) with one Higgs boson as arising from the simplicity of the Model. Only one Higgs doublet does not allow tree-level effects (via the GIM mechanism) which are naturally there in almost any extension of the standard model. (ii) The test of QCD in a new regime, and the accurate measurement of its mass and couplings.

We summarize here the activities of the Top Quark Working Group [1]. Discussions highlighted the potential of muon colliders, while there were additional discussions on future electron-positron and hadron colliders. We refer the interested reader to the many recent reviews [2–6] of top quark physics for a more comprehensive treatment.

TOP-QUARK MASS MEASUREMENT AT THE $\mu^+\mu^- \to t\bar{t}$ THRESHOLD

One attractive feature of lepton colliders is the ability to do threshold cross section measurements. The W boson mass has been determined at LEP II by measuring the cross section $e^+e^- \to W^+W^-$ at the center-of-mass energy $\sqrt{s} = 161$ GeV. In general, accurate measurements of particles masses, couplings and widths are possible by measuring production cross sections near threshold. The possibility of measuring the top quark mass as well as other relevant parameters at a Next Linear Collider (NLC) has been under discussion for some years and was nicely review for the group by Raja [7]. This technique has also been investigated more recently in the context of muon colliders. There is very rich physics associated with the $t\bar{t}$ threshold, including the determination of m_t, Γ_t ($|V_{tb}|$), α_s, and possibly m_h [8]. Ref. [7] contains a nice review of the most salient experimental measurements that can be done. This includes not only measurements of the mass and couplings of the top quark from the total cross section, but also extracting information from the various distributions of the detected particles. These issues carry over completely to the muon collider case, with some important differences in the characteristics of the collider beam having some impact on the sensitivities (see below).

Fadin and Khoze first demonstrated that the top-quark threshold cross section is calculable since the large top-quark mass puts one in the perturbative regime of QCD, and the large top-quark width effectively screens nonperturbative effects in the final state [9]. Such studies have since been performed by several groups [10–17]. The phenomenological potential is given at small distance r by two-loop perturbative QCD and for large r by a fit to quarkonia spectra.

The most important parameters affecting the shape of the threshold cross section are the top quark mass m_t and the strong coupling constant α_s. The mass determines at what energy the threshold turns on, while the strong coupling determines the binding between the $t\bar{t}$ pairs and hence causes in principle a resonance structure in the spectrum. However, since the top quark mass has turned out to be so large, only the $1S$ state appears as a structure on the threshold curve. The stronger the strong coupling, the tighter the binding and the lower the $1S$ peak occurs in energy. Weaker coupling also smooths out the threshold peak. These effects are illustrated in Fig. 1. Clearly the effects of varying m_t and α_s are correlated; this fact necessitates that some kind of scan be performed of the threshold cross section.

The scan can be optimized in various ways depending on the parameters one is most interested in measuring. In addition there is information contained in the various distributions of the final state particles. The momentum distribution of the

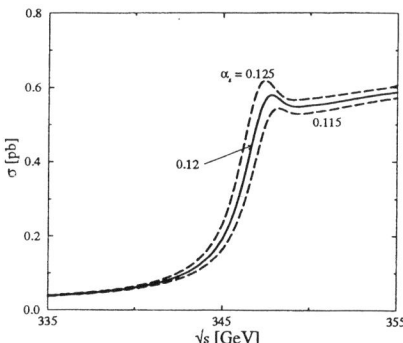

FIGURE 1. The cross section for $\mu^+\mu^- \to t\bar{t}$ production in the threshold region, for $m_t = 175$ GeV and $\alpha_s(M_Z) = 0.12$ (solid) and 0.115, 0.125 (dashes). Effects of ISR and beam smearing are included. This figure is from Ref. [18].

top quark pairs as well as the forward-backward asymmetry are sensitive to the top quark width and α_s.

The presence of the Higgs boson affects the threshold curve. This contribution depends on the Higgs boson mass and the Yukawa coupling with which the Higgs boson couples to the top quark. The Yukawa coupling is fixed in the Standard Model for a given top quark mass, but could be different in extensions to the Standard Model. The Higgs boson contribution mainly affects the overall normalization of the threshold curve. Since the exchange of a light Higgs boson can affect the threshold shape, a scan of the threshold cross section can in principle yield some information about the Higgs mass and its Yukawa coupling to the top quark. Figure 2 shows the dependence of the threshold curve on the Higgs mass, m_h. However, it may be difficult to disentangle such a Higgs effect from two-loop QCD effects, which are not yet fully calculated [19]. Since one does not expect an accurate measurement of the Higgs mass from the $t\bar{t}$ threshold, one should properly think of the Higgs contribution as a systematic uncertainty that can be removed by measuring the Higgs mass and top quark Yukawa coupling elsewhere.

Beamstrahlung is the emission of radiation by one beam due to the action of the effective magnetic field of the other beam. This is expected to be an important issue at electron-positron colliders and clearly depends on the machine design. Muon colliders are expected to naturally have negligible beamstrahlung due to the large mass of the muon.

A more minor difference between the electron colliders and the muon colliders is the difference in the amount of initial state radiation (ISR). The expansion parameter is

$$\beta = \frac{2\alpha}{\pi} \left(\ln(s/m_\ell^2) - 1 \right) , \qquad (1)$$

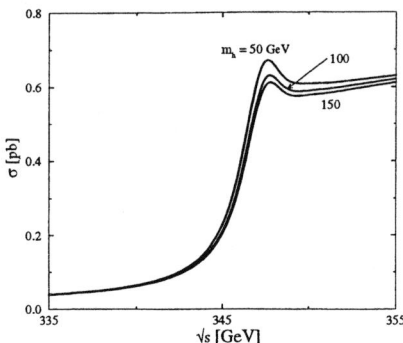

FIGURE 2. The dependence of the threshold region on the Higgs mass, for $m_h = 50, 100, 150$ GeV. Effects of ISR and beam smearing have been included, and we have assumed $m_t = 175$ GeV and $\alpha_s(M_Z) = 0.12$. This figure is from Ref. [18].

where m_ℓ is the mass of the initial state particle (the electron or the muon). The radiator function that must be convoluted with the underlying cross section is

$$\mathcal{D}(x) = 1 + \frac{2\alpha}{\pi}(\pi^2/6 - 1/4)\left[\beta x^{\beta-1}\left(1 + \frac{3}{4}\beta\right) - \beta\left(1 - \frac{x}{2}\right)\right]$$

The ISR is reduced somewhat at a muon collider relative to an electron collider.

Two methods have been used to calculate the threshold cross section. The first (the coordinate-space approach) involves solving a nonrelativistic Schrödinger equation that splices together a QCD potential from perturbative QCD at small distance scales with one that is derived from fits to quarkonia spectra. The other method (the momentum space approach) involves solving a Bethe-Salpeter equation. The construction and the relationship between the QCD potentials used in each case is subject of recent study [20]. At a high luminosity muon collider it is evident that theoretical uncertainties in the threshold cross section might be the limiting factor in the ultimate obtainable precision.

The calculations used in simulations so far have been done mostly at the next-to-leading order (NLO) level. These next-to-next-to-leading order (NNLO) have not been taken into account even though there contributions can be important. For example, the $\mathcal{O}(\alpha_s^2)$ relativistic corrections can shift the location of the 1S peak by $m_t \alpha_s^4 \sim 150$ MeV and introduce a shift in the normalization of the total cross section of order $\alpha_s^2 \sim 3\%$ [21]. Hoang described a procedure to calculate some of the NNLO corrections to the threshold cross section using NRQCD, an effective field theory of QCD for heavy quarks. NRQCD does away with the need for a phenomenological potential, and allows at least in principle the calculation of the cross section and all the distributions from the QCD Lagrangian.

The Abelian part (i.e. those contributions also present in QED) of the NNLO corrections calculated by Hoang are shown in Fig. 3. Notice that the corrections are a few percent and is fairly constant for the part of the cross section including and above the 1S peak, $E > -5$ GeV. (Compare the location of the peak in Fig. 1.)

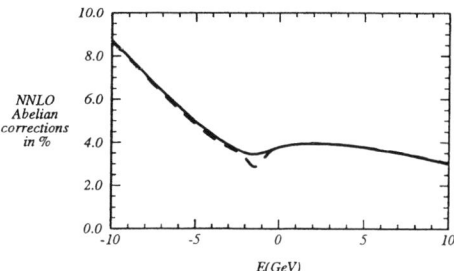

FIGURE 3. The NNLO Abelian corrections to the cross section for $\Gamma_t = 1.56$ GeV (solid line) and 0.80 GeV (dashed line), from Ref. [21].

Muon colliders are expected to naturally have a small spread in beam energy making them an ideal place to study the excitation curve. We present now the parameter determinations that are possible from measuring the total cross section near threshold at a $\mu^+\mu^-$ collider [18,22,23].

The beam energy spread at a $\mu^+\mu^-$ collider is expected to naturally be small. The rms deviation σ in $\sqrt{s}$ is given by [24,25]

$$\sigma = (250 \text{ MeV}) \left(\frac{R}{0.1\%}\right) \left(\frac{\sqrt{s}}{350 \text{ GeV}}\right), \qquad (2)$$

where R is the rms deviation of the Gaussian beam profile. With $R \lesssim 0.1\%$ the resolution σ is of the same order as the measurement one hopes to make in the top mass. For $t\bar{t}$ studies the exact shape of the beam is not important if $R \lesssim 0.1\%$. We take $R = 0.1\%$ here; the results are not improved significantly with better resolution[1].

Suppose one starts with the nominal values of $m_t = 175$ GeV and $\alpha_s(M_Z) = 0.12$. Assuming that 10 fb^{-1} integrated luminosity is used to measure the cross section at each energy in 1 GeV intervals, one can imagine obtaining the hypothetical sample data, shown in Fig. 4. Cuts must be performed to eliminate the backgrounds; following Ref. [17] a 29% detection efficiency has been assumed for the signal where the W's decay hadronically[2] The data points can then be fit to theoretical predictions for different values of m_t and $\alpha_s(M_Z)$; the likelihood fit that is obtained is shown as the $\Delta\chi^2$ contour plot in Fig. 5. The inner and outer curves are the $\Delta\chi^2 = 1.0$ (68.3%) and 4.0 (95.4%) confidence levels respectively for the full 100 fb^{-1} integrated luminosity. Projecting the $\Delta\chi^2 = 1.0$ ellipse on the m_t axis, the top-quark

---

[1] The most recent TESLA design envisions a beam energy spread of $R = 0.2\%$ [26], and a high energy e^+e^- collider in the large VLHC tunnel would have a beam spread of $\sigma_E = 0.26$ GeV [27].
[2] This efficiency includes the decay branching fraction.

mass can be determined to within $\Delta m_t \sim 70$ MeV, provided systematics are under control. With an integrated luminosity of 10 fb^{-1}, the top-quark mass can be measured to 200 MeV.

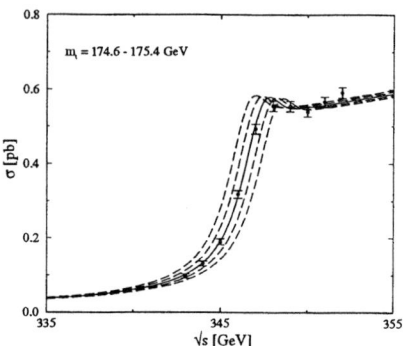

FIGURE 4. Sample data for $\mu^+\mu^- \to t\bar{t}$ obtained assuming a scan over the threshold region devoting 10 fb^{-1} luminosity to each data point. A detection efficiency of 29% has been assumed [17] in obtaining the error bars. The threshold curves correspond to shifts in m_t of 200 MeV increments. Effects of ISR and beam smearing have been included, and the strong coupling $\alpha_s(M_Z)$ is taken to be 0.12. This figure is from Ref. [18].

QCD measurements at future colliders and lattice calculations will presumably determine $\alpha_s(M_Z)$ to 1% accuracy (e.g. ±0.001) [28] by the time muon colliders are constructed so the uncertainty in α_s will likely be similar to the precision obtainable at a $\mu^+\mu^-$ and/or e^+e^- collider with 100 fb^{-1} integrated luminosity. If the luminosity available for the threshold measurement is significantly less than 100 fb^{-1}, one can regard the value of $\alpha_s(M_Z)$ coming from other sources as an input, and thereby improve the top-quark mass determination.

There is some theoretical ambiguity in the mass definition of the top quark. The theoretical uncertainty on the quark pole mass due to QCD confinement effects is of order Λ_{QCD}, i.e., a few hundred MeV [29,30]. For example, this theoretical ambiguity manifests itself in relating quark pole mass to other definitions of the top quark mass (such as the running top quark mass, $\overline{m}_t(\mu)$) that might be relevant as input to radiative correction calculations. So it is not clear that an extraction of the top-quark mass better than Λ_{QCD} is useful, at least at the present time.

Systematic errors in experimental efficiencies are not a significant problem for the $t\bar{t}$ threshold determination of m_t. This can be seen from Fig. 4, which shows that a 200 MeV shift in m_t corresponds to nearly a 10% shift in the cross section on the steeply rising part of the threshold scan, whereas it results in almost no change in σ once $\sqrt{s}$ is above the peak by a few GeV. Not only will efficiencies be known to much better than 10%, but also systematic uncertainties will cancel to a high level of accuracy in the ratio of the cross section measured above the peak to

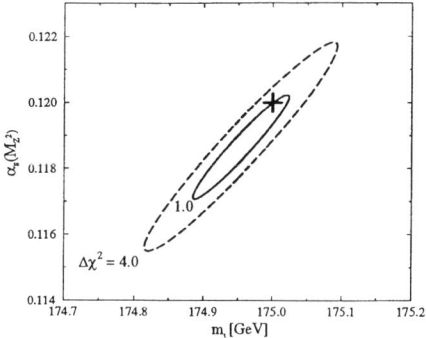

FIGURE 5. The $\Delta\chi^2 = 1.0$ and $\Delta\chi^2 = 4.0$ confidence limits for the sample data shown in Fig. 4. The "+" marks the input values ($m_t = 175$ GeV and $\alpha_s(M_Z) = 0.12$) from which the data were generated. This figure is from Ref. [18].

measurements on the steeply-rising part of the threshold curve.

Differences of cross sections at energies below, at, and above the resonance peak, along with the location of the resonance peak, have different dependencies on the parameters m_t, α_s, m_h and $|V_{tb}|^2$ and should allow their determination. Consequently, the scan procedure described here can be further optimized for extraction of a particular parameter [18].

TOP QUARK PAIRS ABOVE THRESHOLD

The production of top quark pairs at energies above the threshold region at muon colliders will provide a great opportunity for searching for anomalous couplings and rare decays of the top quark. The production of top quarks can be used to test couplings to the neutral gauge bosons γ, Z in the production and to the W in the decays. An important new feature of top quark decays is the fact that the top quark decays before it has a chance to hadronize, so the spin information can be preserved from production to decay. This introduces the possibility that spin correlations between t and $\bar{t}$ might even be measurable [31]. Finally a large sample of top quark pairs allows us to search for possible rare decays, e.g. decays into a charged Higgs boson or flavor changing decays ($t \to c$) which are exceedingly small in the Standard Model.

The off-diagonal basis described by Parke [32] is superior to the standard helicity basis and allows one to describe the $t\bar{t}$ in their simplest possible terms. The basis is characterized by a spin angle ξ between the t spin and the $\bar{t}$ momentum given by

$$\tan\xi = \frac{(f_{LL} + f_{LR})\sqrt{1-\beta^2}\sin\theta^*}{f_{LL}(\cos\theta^* + \beta) + f_{LR}(\cos\theta^* - \beta)}, \qquad (3)$$

where β is the top quark velocity, θ^* is the top quark scattering angle, and f_{IJ} is a combination of muon (or electron) and top quark couplings (see Ref. [31]). This spin basis interpolates between the beam direction at threshold and the top quark direction very far above threshold. Polarization of the incoming beams enhances the sensitivity to the basis choice. The dominant spin component's fraction of the total as function of the polarization of the beams is plotted in Fig. 6 Two different machines are included: the muon collider is assumed to have equal but opposite polarization for the μ^+ and μ^- beams; the NLC has only the electron beam polarized. A muon collider can do as well as an electron-positron machine with relatively less polarization.

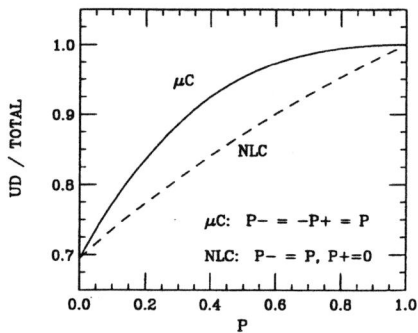

FIGURE 6. Fraction of the total cross section in the off-diagonal basis' Up-Down spin configuration as a function of the polarization. Both beams are assumed to be polarized for the Muon Collider (μC) but only one beam for the NLC. The plot is taken from Ref. [32].

In the future one hopes that detailed studies of the effects of anomalous coupling in the off-diagonal basis will become available. QCD corrections have been calculated and shown to be small [33]. If the muon or electron beams can be polarized, the sensitivity then to anomalous couplings can be enhanced [32].

Hoang described progress in two-loop calculations of the top production cross section in the kinematic region above the threshold [34]. His results for the part of the cross section do not yet include the axial pieces, but the one can see the improvement in the stability under variations of the renormalization scale μ in Fig. 7.

FLAVOR CHANGING NEUTRAL CURRENTS

There are strong phenomenological constraints on flavor changing neutral currents (FCNC) from K physics, for example. However, the possibility of FCNC in top couplings remains to be explored. Flavor changing decays of the top quark (e.g. $t \to c\gamma$) are extremely suppressed in the Standard Model, but new physics

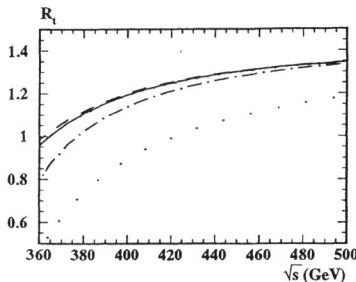

FIGURE 7. The total normalized photon-mediated cross section at the two-loop level versus $\sqrt{s}$ for the renormalization scales $\mu = M_t$ (dashed), $\mu = 2M_t$ (solid) and $\mu = \sqrt{s}$ (dotted line), $M_t = 175$ GeV and $\alpha_s^{(5)}(M_z) = 0.118$. For comparison also the Born (wide dots) and the one-loop cross section for $\mu = 2M_t$ (dashed-dotted line) are displayed. To improve the stability under renormalization scale variations the known three-loop $\mathcal{O}(\alpha_s^3)$ in the large momentum expansion have been added to the two-loop cross section. The plot is taken from Ref. [34,35].

contributions could enhance the rate (see below and Ref. [36]). Another possible source of FCNC is to look for decays of Higgs bosons into single top quarks.

A unique feature of the muon collider is the ability to produce the Higgs boson in the s-channel [24,25]. This possibility arises because the Higgs boson coupling to leptons is proportional to their mass. Since the Higgs boson might be a very narrow object if one can center muon beams with very sharp beam profiles on the resonance energy one could produce a substantial sample of Higgs bosons and study their decays. An interesting set of decays are flavor changing ones, which for the Standard Model Higgs boson are completely absent. Many extensions to the Standard Model have flavor changing processes at the tree-level, and then the question becomes why they are so small in the physics that we see. If the Higgs boson is heavier than the top quark then one can consider the possibility that the Higgs boson decays via $H \to t\bar{c} + c\bar{t}$.

Reina and collaborators [37,38] chose a particular two-Higgs doublet model to provide examples of the kind of effects one might potentially see at a muon collider. The model is given by the Lagrangian

$$\mathcal{L}_Y^{(III)} = \eta_{ij}^U \overline{Q}_{i,L} \tilde{\phi}_1 U_{j,R} + \eta_{ij}^D \overline{Q}_{i,L} \phi_1 D_{j,R} + \xi_{ij}^U \overline{Q}_{i,L} \tilde{\phi}_2 U_{j,R} + \xi_{ij}^D \overline{Q}_{i,L} \phi_2 D_{j,R} + h.c. \,, \quad (4)$$

where η and ξ are non-diagonal Yukawa matrices. Usually at this point one imposes a discrete symmetry to eliminate tree-level FCNCs. Instead a reasonable choice is to assume that the flavor changing couplings adhere to the same hierarchy as the fermion masses [39]

$$\xi_{ij} = \lambda_{ij} \frac{\sqrt{m_i m_j}}{v} \,, \quad (5)$$

and tree-level FC couplings can be substantial for the top quark. This model (like all two Higgs doublet models) is parameterized by a mixing angle α between the two neutral scalars.

An important consideration for producing Higgs bosons in the s-channel, is the relative size of the beam width to the width of the Higgs boson [24,25]. A sufficiently sharp beam, if suitably tuned to the resonance energy, can take full advantage of the resonant cross section. One can define the effective cross section as the convolution of the Breit-Wigner σ_{tc}^{BW} cross section with a gaussian beam energy spread,

$$\sigma_{tc}^{eff} = \int d\sqrt{s'} \frac{\exp[-(\sqrt{s'} - \sqrt{s})^2/2\sigma^2]}{\sqrt{2\pi}\sigma} \sigma_{tc}^{BW}(s') \, , \tag{6}$$

where the rsm of the gaussian distribution is defined in terms of the parameter R.

In the analysis in Ref. [40] the effective cross section after convoluting with the beam width is expressed in units of $R_{\mu\mu}$ (not to be confused with the parameter R describing the beam width) as follows

$$R_{tc} = \frac{\sigma_{tc}^{eff}}{\sigma_0} = R(\mathcal{H})\left(B(\mathcal{H} \to \bar{t}c) + B(\mathcal{H} \to \bar{c}t)\right) \, , \tag{7}$$

where $\sigma_0 = \sigma(\mu^+\mu^- \to \gamma \to e^+e^-)$ and $R(\mathcal{H}) = \sigma_{\mathcal{H}}/\sigma_0$ for $\sigma_{\mathcal{H}}$ the total cross section for producing $\mathcal{H}$.

As described above the results depend on how Γ_{h^0} compares to the resolution parameter R. This is shown in Fig. 8 which shows the results for the pure Breit-Wigner as well as different assumptions for R. Two choices for the mixing angle α shows the kind of variations one can get ($\alpha = 0$ means the s-channel Higgs h^0 does not couple to the gauge bosons which give competing decay channels $h^0 \to W^+W^-, ZZ$.

With 1 fb^{-1} of integrated luminosity and gets around 100 $t\bar{c}+c\bar{t}$ events for $\alpha = 0$ and a few for $\alpha = \pi/4$ [40]. Clearly higher luminosity would be advantageous here.

Flavor changing rare decays of the top occur at an exceeding small rate in the Standard Model and the Minimal Supersymmetric Model (MSSM). However if R-parity violation occurs, then there is at least some hope that rare decays $t \to c$ could be detected at the upgraded Tevatron [41].

R-parity violation also could give an additional source of single top production. For example the lepton-number violating coupling[3] λ' gives rise to the s-channel process

$$u\bar{d} \to \tilde{\ell} \to t\bar{b} \, . \tag{8}$$

The baryon-number violating coupling λ'' gives rise to the s-channel processes

$$c\bar{d} \to \tilde{s} \to t\bar{b} \, , \tag{9}$$

$$c\bar{s} \to \tilde{d} \to t\bar{b} \, , \tag{10}$$

---

[3] For a definition and discussion of the R-parity violating couplings, see Refs. [42,43].

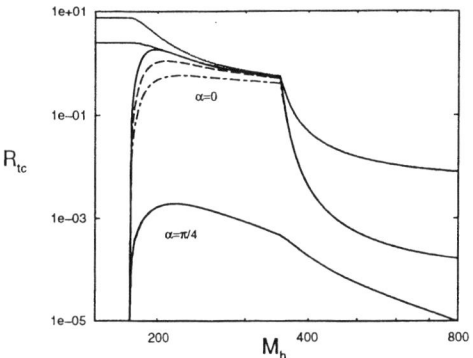

FIGURE 8. The value of $R(h^0)$ is shown as a function of M_{h^0} in a pure Breit-Wigner case (upper dotted line) and when the gaussian width distribution of the beam energy (for $R=0.01$) is taken into account (lower dotted line). The ratio R_{tc} is also shown for different values of the resolution parameter $R=0.001$ (solid), 0.01 (dashed) and 0.03 (dot-dashed), when $\alpha=0$ (upper group of curves) and when $\alpha=\pi/4$ (lower group of curves). This figure is from Ref. [40].

and the t-channel process

$$u\bar{d} \to t\bar{b} \, . \tag{11}$$

The prospects for setting bounds on R-parity violation from these processes is discussed in Refs. [41–43].

GLUON RADIATION

Top quark production involves gluon radiation because the top quark is a colored particle. In a hadron collider the gluon can arise in the initial and final states [44]. As far as gluon radiation at lepton colliders is concerned, there is no significant difference between $\mu^+\mu^-$ and e^+e^- colliders. There is no gluonic ISR, but the radiation must still be divided into production and decay stage radiation, i.e. the gluon can be thought of as originating off the produced t or $\bar{t}$, or it can be thought of as being among the decay products of the t or $\bar{t}$. Gluon radiation needs to be understood if we are able to do precise momentum reconstructions to obtain m_t, and also to identify top events by using mass cuts [45].

What one would really like to do is study the gluon radiation pattern. There are interference effects between the production and decay stage radiation that is potentially sensitive to the top quark width Γ_t. This occurs when the gluon energy is comparable to Γ_t. One such radiation pattern is shown in Fig. 9 for a particular kinematic configuration and a variety of values for Γ_t.

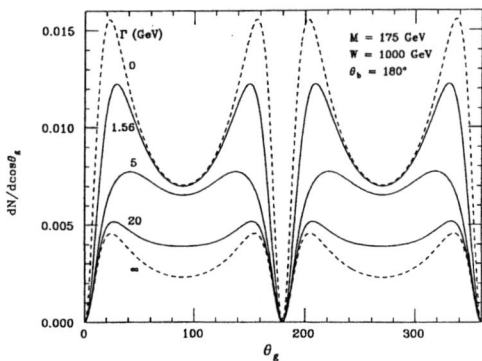

FIGURE 9. Soft gluon distribution in top production and decay at lepton colliders as described in the text, for t's decaying to backward b's and collision energy 1 TeV, from Ref. [45].

Whether one can really extract information about the top quark width by observing the interference patterns remains an open question [45]. The distributions that would be seen at a muon collider would involve all possible kinematic configurations. Moreover, the interference effects are much smaller for a 500 GeV collider which is more likely to be available first. Nevertheless, the interference is an interesting feature of QCD and one would like to see it even if it does not offer a new way to measure Γ_t.

TOP QUARK PHYSICS AT FUTURE HADRON COLLIDERS

Hughes [46] summarized top quark physics at future hadron colliders for the working group. He compared and contrasted the expected measurements at Run I of the Tevatron with what is expected at Run II, Tev33, and the LHC.

A possible manifestation of new physics is resonance structure in the $t\bar{t}$ mass distribution. The Tevatron is sensitive to both color singlet and color octet, while at the LHC should be relatively insensitive to new color singlet gauge bosons (like Z') because gluon fusion is the dominant means of producing $t\bar{t}$ pairs.

Rare top decays ($t \to c$) are expected to be very small in the Standard Model due to the GIM mechanism and the competing (Cabibbo-allowed) decay $t \to b$. Branching fractions are typically of order 10^{-10} in the Standard Model, so the observation of rare decays of this sort indicates new physics. With 2 fb^{-1}, one should be able to get to the 10^{-3} level for $t \to c\gamma$ [46].

Observing single top production will become possible after the Tevatron upgrade. The single top production is proportional to the partial width $\Gamma(t \to Wb)$ and provides an expected precision of 16% for Run II and 9% for Tev33 [46]. Since the final state is a Wjj configuration rather than a $Wjjjj$ one, the background from

QCD processes should be much higher than the usual $t\bar{t}$ case [36]. With 2 fb^{-1}, one should have roughly 100 events in the final sample.

DETECTOR AND BACKGROUND

Roser [47] described progress on a strawman detector for a muon collider and the interplay between the machine designers and detector designers. The detector backgrounds are actually more under control for the high energy (4 TeV) collider where most of the particles continue down the beam pipe without ever entering the dectector.

While the total "background energy" does not depend on the energy of the beam, the "visible" background will depend on details of the lattice and the detector itself. There will be the usual collider backgrounds, such as beam halo, beam-beam, etc. Since backgrounds at a muon collider are expected to be large, one will want to use a large number of detector channels to achieve reasonable occupancies.

The following techniques are being used to study the detector: (1) lattice simulators COSY and MAD, (2) detector simulators MARS and GEANT, and (3) event generators LUND, PYTHIA, etc.

GEANT simulations yield radial particle fluxes per crossing for a layer of silicon at a radius of 10 cm:

$$750 \text{ photons/cm}^2 \to 2.3 \text{ Hits/cm}^2$$
$$110 \text{ neutrons/cm}^2 \to 0.1 \text{ Hits/cm}^2$$
$$1.3 \text{ charged tracks/cm}^2 \to 1.2 \text{ Hits/cm}^2$$
$$\text{Total} \to 3.7 \text{ Hits/cm}^2$$

This translates into a 0.4% occupance in 300×300 μm^2 pixels. The corresponding numbers at a radius of 5 cm are 13.2 Hits/cm^2 for a 1.3% occupancy. The radiation dose in the silicon vertex detector at a 4 TeV muon collider at a radius of 10 cm is comparable to that of the LHC operating at 10^{34}cm^{-2}s^{-1} luminosity.

Backgrounds come from synchrotron radiation from electrons and from electromagnetic showering close to the detector. The lattice focuses the muons at the interaction point, so that the electrons cannot be kept in the beam pipe.

In summary, technology choices and detector optimization will require dedicated work. Because of the large backgrounds arising from the decaying muons, there will be pressure to compromise on 4π coverage.

CONCLUSION

An important issue regarding top physics at a muon collider is the amount of luminosity that would be available. Most of the signals presented here require significantly more than 1 fb^{-1} of integrated luminosity. The years ahead promise to be very exciting as we become able to study the top quark properties in more detail.

ACKNOWLEDGMENTS

We thank all the working group participants [1] for their contributions and discussion, and the conference organizers for an efficient and interesting workshop.

REFERENCES

1. Members of the working group: U. Baur, M. S. Berger, T. Bolton, E. Buckley-Geer, C. Campagnari, D. Chakraborty, D. Dominici, R. Drucker, A. Garren, E. Gottschalk, A. H. Hoang, R. Hughes, C. Kao, S. Keller, G. Landsberg, D. Miller, L. H. Orr, S. Parke, C. Quigg, R. Raja, L. Reina, R. Roser, Y. Shadmi, M. Smith, T. Stelzer, Z. Sullivan, G. Valencia, S. Vejcik, C. White, S. Willenbrock, B. L. Winer, G.-H. Wu, J. M. Yang.
2. *Future Electroweak Physics at the Fermilab Tevatron: Report of the TeV-2000 Study Group*, D. Amidei et al., Fermilab-Pub-96-082.
3. *Top Quark Physics: Future Measurements*, R. Frey et al., hep-ph/9704243.
4. S. Willenbrock, talk presented at 7th International Symposium on Heavy Flavor Physics, Santa Barbara, CA, 7-11 Jul 1997, hep-ph/9709355.
5. A. P. Heinson, talk presented at the 31st Rencontres de Moriond: QCD and High-energy Hadronic Interactions, Les Arcs, France, 23-30 Mar 1996, hep-ex/9605010.
6. T. Liss, talk presented at the International Europhysics Conference on High Energy Physics (HEP 95), Brussels, Belgium, 27 Jul - 2 Aug 1995, hep-ph/9510274.
7. R. Raja, these proceedings.
8. For a review on the top-quark physics near the threshold, see *e.g.*, J.H. Kuhn, TTP-96-18, lectures delivered at SLAC Summer Institute, Stanford, July, 1995.
9. V.S. Fadin and V.A. Khoze, JETP Lett. **46**, 525 (1987); Sov. J. Nucl. Phys. **48**, 309 (1988).
10. J. Feigenbaum, Phys. Rev. **D43**, 264 (1991).
11. W. Kwong, Phys. Rev. **D43**, 1488 (1991).
12. M. Strassler and M. Peskin, Phys. Rev. **D43**, 1500 (1991).
13. M. Jezabek, J.H. Kuhn and T. Teubner, Z. Phys. **C56**, 653 (1992); M. Jezabek and T. Teubner, Z. Phys. **C59**, 669 (1993); M. Jezabek, talk presented at *DESY-Zeuthen Workshop on Elementary Particle Theory: "Physics at LEP200 and Beyond"*, Teupitz, Germany, April 1994 (hep-ph/9406411); M. Jezabek, Acta Phys. Pol. **B26**, 789 (1995); J.H. Kuhn, Acta Phys. Pol. **B26**, 711 (1995).
14. G. Bagliesi, et al., *Workshops on Future e^+e^- Linear Colliders*, Hamburg, Germany and Saariselka, Finland, Sep 2-3 and Sep 9-11, 1991, CERN-PPE/92-05.
15. Y. Sumino, K. Fujii, K. Hagiwara, H. Murayama and C.-K. Ng, Phys. Rev. **D47**, 56 (1993); H. Murayama and Y. Sumino, Phys. Rev. **D47**, 82 (1993); Y. Sumino, Acta Phys. Pol. **B25**, 1837 (1994).

16. P. Igo-Kemenes, M. Martinez, R. Miquel and S. Orteu, CERN-PPE/93-200, Contribution to *the Workshop on Physics with Linear e^+e^- Colliders at 500 GeV*.
17. K. Fujii, T. Matsui and Y. Sumino, Phys. Rev. **D50**, 4341 (1994).
18. V. Barger, M.S. Berger, J.F. Gunion and T. Han, Phys. Rev. **D56**, 1714 (1997).
19. A. H. Hoang, Phys. Rev. **D56**, 5851 (1997); Phys. Rev. **D56**, 7276 (1997); and these proceedings.
20. M. Jezabek, et al., hep-ph/9801419.
21. A. H. Hoang, these proceedings, hep-ph/9712273.
22. M.S. Berger, talk presented at the *Workshop on Particle Theory and Phenomenology: Physics of the Top Quark*, Iowa State University, May 25–26, 1995, hep-ph/9508209.
23. M. S. Berger, these proceedings, hep-ph/9712486.
24. V. Barger, M.S. Berger, J.F. Gunion and T. Han, Phys. Rev. Lett. **75**, 1462 (1995).
25. V. Barger, M.S. Berger, J.F. Gunion and T. Han, Phys. Reports **286**, 1 (1997).
26. D. Miller, private communication.
27. J. Norem, private communication and http://www-ap.fnal.gov/VLHC/electrons/index.html.
28. P. N. Burrows et al., SLAC-PUB-7371, to appear in *Proceedings of 1996 DPF/DPB Summer Study on New Directions for High-Energy Physics* (Snowmass 96), hep-ex/9612012.
29. M. C. Smith and S. Willenbrock, hep-ph/9612329.
30. M. Smith, these proceedings.
31. S. Parke and Y. Shadmi, Phys. Lett. **B387**, 199 (1996).
32. S. Parke, these proceedings, hep-ph/9802279.
33. J. Kodaira, T. Nasuno and S. Parke manuscript in preparation; M. Hori, Y. Kiyo, J. Kodaira, T. Nasuno and S. Parke, hep-ph/9801370.
34. A. H. Hoang, these proceedings, hep-ph/9712275.
35. K. G. Chetyrkin, A. H. Hoang, J. H. Kühn, M. Steinhauser and T. Teubner, hep-ph/9711327, to be published in *Z. Phys.* **C**.
36. T. Han, K. Whisnant, B. L. Young and X. Zhang, Phys. Rev. **D55**, 7241 (1997).
37. D. Atwood, L. Reina and A. Soni, Phys. Rev. **D53**, 1199 (1996); Phys. Rev. **D55**, 3156 (1997).
38. D. Atwood, L. Reina and A. Soni, Phys. Rev. Lett. **75**, 3800 (1995).
39. T.P. Cheng and M. Sher, Phys. Rev. **D35**, 3484 (1987); M. Sher and Y. Yuan, Phys. Rev. **D44**, 1461 (1991); A. Antaramian, L.J. Hall, and A. Rasin, Phys. Lett. **B69**, 1871 (1992); L.J. Hall and S. Weinberg, Phys. Rev. **D48**, R979 (1993).
40. L. Reina, these proceedings, hep-ph/9712426.
41. J. M. Yang, these proceedings.
42. A. Datta, J. M. Yang, B.-L. Young and X. Zhang, Phys. Rev. **D56**, 3107 (1997).
43. R. J. Oakes, et al., Phys. Rev. **D57**, 534 (1998).
44. V. A. Knoze, L. H. Orr, and W. J. Stirling, Nucl. Phys. **B378**, 413 (1992); L. H. Orr, T. Stelzer, and W. J. Stirling, Phys. Rev. **D52**, 124 (1995); Phys. Rev. **D56**, 446 (1997).
45. L. H. Orr, these proceedings, hep-ph/9802215.
46. R. Hughes, these proceedings.
47. R. Roser, these proceedings.

Physics with a Millimole of Muons

Chris Quigg

Fermi National Accelerator Laboratory[1]
P.O. Box 500, Batavia, Illinois 60510 USA
E-mail: quigg@fnal.gov

Abstract. The eventual prospect of muon colliders reaching several TeV encourages us to consider the experimental opportunities presented by very copious stores of muons, approaching 10^{21} per year. I summarize and comment upon some highlights of the *Fermilab Workshop on Physics at the First Muon Collider and at the Front End of a Muon Collider*. Topics include various varieties of $\mu\mu$ colliders, μp colliders, and applications of the intense neutrino beams that can be generated in muon storage rings.

INTRODUCTION

The initial appeal of a $\mu^+\mu^-$ collider is that it may provide a possible path to a few-TeV lepton-lepton collider to address the great issue of our age, the character of the mechanism that breaks electroweak symmetry. It is a commonplace that lepton colliders and hadron colliders offer complementary means to explore the nature of electroweak symmetry breaking [1,2]. It is widely agreed that the rise of synchrotron radiation causes circular electron machines to become impractical for energies above a few hundred GeV. Linear colliders are therefore under development for c.m. energies from a few hundred GeV to about 1.5 TeV. I think it possible that linear-collider technology may only be interesting for about one decade in energy; the growth path beyond 1 to 2 TeV is not clear. But it is a very interesting decade in energy, over which we expect to learn the secrets of electroweak symmetry breaking. That is why there is such intense interest in the linear-collider approach. In contrast, the extrapolation of a $\mu^+\mu^-$ collider to several TeV per beam seems straightforward—if a $\mu^+\mu^-$ collider can be made to work at all [3,4]. If the small size of a $\mu^+\mu^-$ collider is an indication of its cost, which is by no means

---

[1] Fermilab is operated by Universities Research Association Inc. under Contract No. DE-AC02-76CH03000 with the United States Department of Energy.

established, a $\mu^+\mu^-$ collider might even offer a less costly way to a modest-energy lepton collider. Taken together, these two possibilities offer a strong incentive to investigate the technology of a $\mu^+\mu^-$ collider.

Once the technological possibility of a muon collider is raised, there are many interesting possibilities to contemplate [5]. On the way to the ultimate prize of a 2–4-TeV collider, we may consider a high-luminosity Z factory and machines to operate near the W^+W^- and $t\bar{t}$ thresholds, as well as a machine with $\sqrt{s} \approx \frac{1}{2}$ TeV to explore details of a supersymmetric or technicolor world for which the first indications have been found elsewhere. A $\mu^+\mu^-$ collider also offers the unique possibility of a Higgs factory where detailed measurements not possible elsewhere could be undertaken. The front end of a muon collider offers a host of possibilities of its own, including intense low-energy hadron beams, a copious source of low-energy muons, and the neutrino beams of unprecedented intensity and unusual flavor composition that emanate from stored muons. A muon collider in the neighborhood of a hadron storage ring opens the possibility of high-luminosity μp collisions as well.

Many of these possibilities have been explored at this Workshop, which I found notable for the fact that the participants actually did some original work. My first—and most important—conclusion to the Workshop is that there are many *interesting* physics topics to think about.

The Case for Muons

The muon is massive: $m_\mu \approx 106$ MeV/$c^2 \approx 207 m_e$. Compared to electrons in a circular machine of given radius, muons of the same energy lose far less energy to synchrotron radiation, by a factor $(m_e/m_\mu)^4 \approx 5.5 \times 10^{-10}$. A crippling problem for electron machines—and the reason we turn to linear colliders—is of negligible importance for a muon machine.

In common with the electron, the muon is an elementary lepton at our current limits of resolution. Its energy is not shared among many partons, so the muon is a more efficient delivery vehicle for high energies than is the composite proton.

Because the muon is massive, and can be accelerated efficiently in circular machines, and because we can probe the 1-TeV scale with muons of a few TeV, as opposed to protons of several tens of TeV, a muon collider can be small. If a muon collider proves technically feasible, we need to discover whether small translates to inexpensive—both in absolute terms and compared to other paths we might take to high energies.

Beyond the suggestion of these practical advantages, muons offer a possibly decisive physics advantage. The great seduction of a First Muon Collider is that the cross section for the reaction $\mu^+\mu^- \to H$, direct-channel formation of the Higgs boson, is larger than the cross section for $e^+e^- \to H$ by a factor $(m_\mu/m_e)^2 \approx 42,750$. This is a very large factor. The tantalizing question is whether it is large enough to make possible a "Higgs factory" with the luminosities that may be achieved in $\mu^+\mu^-$ colliders. In e^+e^- collisions, of course, the s-channel formation cross section is hopelessly small. That is why the associated-production reaction $e^+e^- \to HZ$ has become the preferred search mode at LEP-2.

The properties of the muon also raise challenges to the construction and exploitation of a $\mu^+\mu^-$ collider. The muon is not free: it doesn't come out of a bottle like the proton or boil off a metal plate like the electron. On the other hand, it is readily produced in the decay $\pi \to \mu\nu$. Still, gathering large numbers of muons in a dense beam is a formidable engineering challenge, and the focus of much of the R&D effort over the next few years. The muon is also not stable. It decays with a lifetime of 2.2 μs into $\mu^- \to e^-\bar{\nu}_e\nu_\mu$. We must act fast to capture, cool, accelerate, and use muons, and must be able to replenish the supply quickly. Multiply 2.2 μs by whatever Lorentz (γ) factor you like for a muon collider, it is still a very short time.

The muon's decay products complicate experimentation as well. Just to indicate the dimensions of the problem, in a $2 \oplus 2$-TeV collider with 2×10^{12} muons/bunch, every meter the bunch travels sees 2×10^5 decays, with an average electron energy of about 700 GeV.

Finally, the neutrinos emitted in μ decay may constitute a radiation hazard. You need not fear the neutrinos themselves. The interaction length of a 100-GeV neutrino is about 25 million kilometers in water, so it has only about 1 chance in 10^{11} of interacting in the column depth of your body. The potential hazard comes from neutrino interactions in the Earth surrounding a $\mu^+\mu^-$ collider, which generate hadronic showers. Estimates suggest that the potential radiation dose from these showers becomes a serious concern for $E_\mu \approx 1 - 2$ TeV.

The Big Questions for $\mu^+\mu^-$ colliders

When we discuss whether there should be muon colliders in our future, we must answer a number of important questions.

- What machines are possible? When? At what cost?
- What are the physics opportunities?
- Can we do physics in the environment? (What does it take?)
- How will these experiments add to existing knowledge not just in the abstract, but *when they are done?*

These questions are not the unique concern of a muon collider, but need to be addressed for any new accelerator we might contemplate. I would like to underscore the importance of the last question: it is crucially important to try to judge what will be known from ongoing experiments and initiatives already launched at the moment that a new experimental tool could be ready. What seems like essential information—if we could have it today—may fade in significance a decade or more hence. Our goal must be to develop the means to do experiments that can change the way we think. It is worth keeping in mind Bob Palmer's estimate that a First Muon Collider might be in operation around the year 2010 [6].

The Focus of This Workshop

The Workshop on Physics at the First Muon Collider and at the Front End of a Muon Collider was organized around nine working groups. One dealt with accelerator issues, concentrating on the design of a proton driver for the Fermilab site. Progress on an RF system, longitudinal space-charge effects, the formation of short bunches a few ns in length, and instability questions was reported by Bob Noble [7]. Four working groups addressed physics prospects for muon colliders. They were organized around Higgs and Z factories [8], top physics [9], supersymmetry [10], and strong dynamics [11]. Four more working groups explored the physics interest of beams associated with the front end of a muon collider. Those groups considered low-energy hadron physics [12], neutrino physics [13], deep inelastic scattering [14], and low-energy muon physics [15].

The Front End of a Muon Collider

The Front End of a Muon Collider consists of four basic elements.

- A high-intensity proton source. An example design developed for the Fermilab site ends in a rapid-cycling synchrotron that delivers 16-GeV protons at 15 Hz [16]. In each cycle, two bunches of 5×10^{13} protons are accelerated, for a total of 1.5×10^{22} protons per year. That is about 10^3 the number of protons delivered at 8 GeV by the Fermilab Booster.

- A system for pion production, collection, and decay. Charged pions created in the collision of the proton beam with a target are confined in a high-field solenoid and guided into a 20-meter-long decay channel within a 7-Tesla solenoid that keeps the muons from escaping. Such a system might yield about 0.2 μ^+ and μ^- per proton, or about 10^{13} μ^+ and μ^- per cycle, for a total of about 1.5×10^{21} μ^+ and μ^- per year.

- A muon cooling channel to concentrate the muons in six-dimensional phase space. It is hoped that an "ionization cooling" system [17] could compress the muons' phase space by a factor of 10^5–10^6, leading to dense

TABLE 1. Recirculating linear accelerator parameters.

	RLA 1	RLA 2	RLA 3
Input energy [GeV]	1.0	9.6	70
Output energy [GeV]	9.6	70	250
Turns	9	11	12
Linac length [m]	100	300	533.3
Arc length [m]	30	175	520
Bunch length [ps]	158	43	19
Revolution time [μs]	0.9	3.1	7.0
Decay losses	9.0%	5.2%	2.4%
Initial muons per bunch	5×10^{12}	4.6×10^{12}	4.3×10^{12}
μ^+ bunches per sec	15	15	15

bunches of 5×10^{12} muons at 200 MeV/c. In the simplest version of ionization cooling, passage through matter degrades a muon's longitudinal and transverse momentum in proportion. An RF cavity adds longitudinal momentum. Iterating these steps cools the beam in the transverse dimensions. An important refinement uses wedge-shaped degraders in a region of high dispersion, so that high-momentum muons see more material than low-momentum muons. By this device one can cool the beam in both longitudinal and transverse dimensions.

- A muon acceleration system to raise the captured muons quickly to the desired energy. An example presented at the Workshop consists of a series of three recirculating linear accelerators (RLAs), whose properties are summarized in Table 1. The muons are raised in steps from 1 to 10 GeV, from 10 to 70 GeV, and from 70 to 250 GeV. Notice that the number of turns in each RLA is quite small: 9, 11, and 12. The decay losses in the RLAs, while not crippling, are noticeable. From the acceleration system, the muons would be passed to a collider ring of quite modest dimensions.

We see that while the front end of a muon collider is small, it is also complex. The important questions to answer are whether the construction and operation of such a device is feasible, and whether the size or the complexity is decisive in determining its cost.

A HIGGS FACTORY

The important possibility that a $\mu^+\mu^-$ collider can operate as a Higgs factory has been studied extensively [18] and received considerable attention at the Workshop [8]. If the Higgs boson is light ($M_H \lesssim 2M_W$), and therefore narrow, then the muon's large mass makes it thinkable that the reactions

$$\mu^+\mu^- \to H \to b\bar{b} \text{ and other modes}$$

will occur with a large rate that will enable a comprehensive study of the properties of the Higgs boson. We assume that a light Higgs boson has been found, and that its mass has been determined with an uncertainty of $\pm(100\text{-}200)$ MeV/c^2 [19]. Then suppose that an optimized machine is built with $\sqrt{s} = M_H$.

The muon's mass confers another important instrumental advantage: the momentum spread of a muon collider is naturally small, and can be made extraordinarily small. The Higgs factory can operate in two modes:

- modest luminosity (0.05 fb^{-1}/year) and high momentum resolution ($\sigma_p/p = 3 \times 10^{-5}$);

- standard luminosity (0.6 fb^{-1}/year) and momentum resolution ($\sigma_p/p = 10^{-3}$).

At high resolution, the spread in c.m. energy is comparable to the natural width of a light Higgs boson: $\sigma_{\sqrt{s}} \approx$ a few MeV $\approx \Gamma(H \to \text{all})$. At normal resolution, $\sigma_{\sqrt{s}} \gg \Gamma(H \to \text{all})$.

Parameters of the Higgs factories are given in Table 2, along with those of other candidates for a First Muon Collider [20]. It is worth remarking that the Higgs factory would be small, with a circumference of just 380 meters, and that the number of turns a muon makes in one lifetime is 820.

The first order of business is to run in the high-resolution mode to determine the Higgs-boson mass with exquisite precision. The procedure contemplated is to scan a large number of points (determined by $2\Delta M_H/\sigma_{\sqrt{s}} \approx 100$), each with enough integrated luminosity to establish a three-standard-deviation excess. If each point requires an integrated luminosity of 0.0015 fb^{-1}, then the scan requires 100×0.0015 fb$^{-1} = 0.15$ fb^{-1}, about three nominal years of running.

TABLE 2. Parameters considered at the Fermilab workshop for narrow-band and broad-band Higgs factories, a LEP2 equivalent, a top factory, and a $\frac{1}{2}$-TeV FMC.

$\sqrt{s}$ [GeV]	100	100	200	350	500
Momentum spread, σ_p/p	3×10^{-5}	1×10^{-3}	1×10^{-3}	1×10^{-3}	1×10^{-3}
Muons per bunch	3×10^{12}	3×10^{12}	2×10^{12}	2×10^{12}	2×10^{12}
Number of bunches	1	1	2	2	2
Repetition rate [Hz]	15	15	15	15	15
$\epsilon_\perp$ [mm·mr]	297π	85π	67π	56π	50π
Circumference [m]	380	380	700	864	1000
f_{rev} [Hz]	7.9×10^5	7.9×10^5	4.3×10^5	3.5×10^5	3.0×10^5
Turns per lifetime	820	820	890	1260	1560
β^* [cm]	13	4	3	2.6	2.3
σ_z [cm]	13	4	3	2.6	2.3
σ_r [μm]	286	85	47	30	22
$\mathcal{L}_{\text{peak}}$ [cm^{-2}s^{-1}]	6×10^{32}	7×10^{33}	6×10^{33}	1×10^{34}	2×10^{34}
$\mathcal{L}_{\text{av}}$ [cm^{-2}s^{-1}]	5×10^{30}	6×10^{31}	1×10^{32}	3×10^{32}	7×10^{32}

The reward is that, after the scan, the Higgs-boson mass will be known with an uncertainty of $\Delta M_H \approx \sigma_{\sqrt{s}} \approx 2$ MeV/c^2, which is quite stunning.

Extended running in the form of a three-point scan of the Higgs-boson line at $\sqrt{s} = M_H, M_H \pm \sigma_{\sqrt{s}}$ would then make possible an unparalleled exploration of Higgs-boson properties. With an integrated luminosity of 0.4 fb^{-1} one may contemplate precisions of $\Delta M_H \approx 0.1$ MeV/c^2, $\Delta \Gamma_H \approx 0.5$ MeV $\approx \frac{1}{6} \Gamma_H$, $\Delta(\sigma \cdot B(H \to b\bar{b})) \approx 3\%$, and $\Delta(\sigma \cdot B(H \to WW^\star)) \approx 15\%$.

These are impressive measurements indeed. The width of the putative Higgs boson is an important discriminant for supersymmetry, for it can range from the standard-model value to considerably larger values. Within the minimal supersymmetric extension of the standard model (MSSM), the ratio of the $b\bar{b}$ and $WW^\star$ yields is essentially determined by M_A, the mass of the CP-odd Higgs boson. In the decoupling limit, $M_A \to \infty$, the MSSM reproduces the standard-model ratio. Deviations indicate that A is light. In the most optimistic scenario, this measurement could determine M_A well enough to guide the development of a second (CP-odd) Higgs factory using the reaction $\mu^+ \mu^- \to A$.

Again, these remarkable measurements exact a high price. At the Workshop luminosity of 0.05 fb^{-1}/year, it takes 8 years to accumulate 0.40 fb^{-1} *after the scan* to determine M_H within machine resolution. It is plain that this program becomes considerably more compelling if the Higgs-factory luminosity can be raised by a factor of 2 or 3—or more!

These projections are based on theorists' simulations; more attention is needed to experimental realities. Precision measurements at LEP and SLC have benefitted from excellent determinations of the luminosity $\mathcal{L}$, the beam energy, and the lepton polarization. For a muon collider, it has been shown that the muon spin tune $\gamma(g_\mu - 2)/2$ offers a means of determining the beam energy to a few parts per million and the lepton polarization in real time [21]. Exploiting the fact that, for a muon collider ring with $\sqrt{s} \approx M_Z$ the muon's spin approximately flips from turn to turn, one measures the decay-electron energy spectrum as a function of turn number. The frequency of the spin oscillations yields the Lorentz factor γ, and hence the beam energy, while the amplitude of the modulations in the energy spectrum is a measure of the beam polarization.

It is less clear how to make a precision determination of the luminosity. An analogue of the standard e^+e^- method of small-angle Bhabha monitors seems ruled out by the high flux of decay electrons. Indeed, the first-pass concepts for muon collider detectors do not instrument a cone of $\pm(10\text{-}20)^\circ$ around the beam line [22]. For now we will assume that $\delta \mathcal{L}/\mathcal{L} = 10^{-3}$, but it is an important exercise to develop robust schemes for making this measurement.

Let us note finally that the flux of decay electrons challenges the operation of silicon detectors close to the interaction point [23].

OTHER OPTIONS FOR THE FMC

Several other candidates for the First Muon Collider have been studied at this Workshop. In order of increasing energy, they are a Z factory, machines to explore the W-pair and top-pair thresholds, and a continuum machine operating at $\sqrt{s} = 500$ GeV. The parameters assumed for these machines are displayed in Table 2. It is worth noting that the average luminosities considered at the Workshop are about an order of magnitude smaller than those projected for e^+e^- linear colliders [24]. Unless there are compensating advantages for a $\mu^+\mu^-$ collider—the superior beam energy resolution, for example—the luminosity that can be achieved will be decisive.

A very-high-luminosity Z factory, say twenty times the luminosity of LEP, would be a superb device for B physics. There is also unfinished business in the precision measurement of electroweak observables, particularly in light of the discrepancy between the value of the weak mixing parameter $\sin^2\theta_W$ inferred from the SLD measurement of A_{LR} and the value determined from a host of measurements at LEP. Alain Blondel [25] emphasized the desirability of controlling independently the polarizations of μ^+ and μ^- for refining our understanding of $\sin^2\theta_W$. Apart from the challenge of attaining adequate luminosity, an open issue for precision electroweak measurements in a $\mu^+\mu^-$ collider is how to monitor the luminosity to high precision.

Although a $\mu^+\mu^-$ collider operating at W^+W^- threshold could make impressive measurements of the W-boson mass, with $\delta M_W \approx 20$ MeV in 10 fb^{-1} [26], it is hard to imagine that this will be an important goal in the year 2010. Experiments at LEP2 and the Tevatron Collider may soon give us a world average uncertainty approaching 50 MeV, and future running at the LEP2, the Tevatron, and the LHC will push the precision further.

It is possible that extensive measurements near top threshold could hold greater interest [27,9]. In principle, such measurements might yield extraordinarily precise measurements of the top-quark mass m_t, and give information on the strong coupling constant α_s and the Higgs-$t\bar{t}$ coupling ζ_t. For those studies, the superb momentum spread of a $\mu^+\mu^-$ collider—about an order of magnitude better that the momentum spread of a linear collider—could be a winning advantage. I have to say that I am not convinced that the advertised determinations of m_t, α_s, and ζ_t are actually attainable. I fear that the statement that the ambiguity in defining m_t is no larger than $\pm\Lambda_{QCD}$ may be too glib. I am also concerned that the theoretical link between the shape of the $t\bar{t}$ excitation curve and m_t, α_s, and ζ_t is more ambiguous than has generally been assumed [28]. It is important to look critically at these questions as we assess the capabilities of both a $\mu^+\mu^-$ collider and a linear collider.

Let us now look briefly at some physics prospects of a 500-GeV $\mu^+\mu^-$ collider. There are rich possibilities for detailed study of the spectrum and properties of superpartners. Strategies for constraining the (many) parameters of supersymmetric models in linear colliders have been documented extensively.

For the most part, the case for the study of supersymmetry in a $\mu^+\mu^-$ collider is quite parallel to that for a linear collider [29,10]. (We have already noted the unique possibility to form the Higgs bosons in the s-channel reactions $\mu^+\mu^- \to h, H, A$.) Linear colliders and $\mu^+\mu^-$ colliders have different possibilities for exploiting beam polarization; how best to use polarization in a muon collider is a good issue for further study. In specific cases considered at the Workshop, luminosity appeared to be a concern. This was especially the case for the discovery and study of sleptons. Since hadron colliders are not well suited to the search for sleptons, it is important that a lepton collider excel in slepton physics.

If evidence for new strong dynamics represented by light-scale technicolor is found elsewhere, a $\mu^+\mu^-$ collider will also have very significant capabilities for following up that discovery [11,30,31]. Technivector mesons with masses in the range 200 - 400 GeV/c^2 would be produced copiously even at a luminosity of 10^{32} cm^{-2}s^{-1} [32]. A linear collider would offer similar possibilities, within the limitations of its $\sim 3\%$ beam energy resolution. It was recognized at this Workshop that a $\mu^+\mu^-$ collider could be an impressive technipion factory, forming $\mu^+\mu^- \to \pi_T^0$ at an appreciable rate [11]. The rate for $e^+e^- \to \pi_T^0$ is, of course, negligible.

A new element in the comparison with a linear collider is the claim by the DESY group [33] that it may be possible to increase the projected luminosity of a 500-GeV linear collider by more than an order of magnitude, perhaps to $\sim 10^{35}$ cm^{-2}s^{-1}. We have an obligation to explore how physics reach depends on luminosity for e^+e^- linear colliders and $\mu^+\mu^-$ colliders alike.

A μp COLLIDER?

If an energetic muon beam is stored in proximity to a high-energy proton beam, it is natural to consider the possibility of bringing them into collision. One concept considered at the Workshop was to collide a 200-GeV muon beam with the Tevatron's 1-TeV proton beam, with a mean luminosity of 1.3×10^{33} cm^{-2}s^{-1}, for an annual integrated luminosity of about 10 fb^{-1} [34]. Such a machine would have an impressive kinematic reach, with $\sqrt{s} \approx 0.9$ TeV and $Q^2_{\max} \approx 8 \times 10^5$ GeV^2. For comparison, the $e^{\pm}p$ collider HERA currently operates with 27.5-GeV electrons on 820-GeV protons, for $\sqrt{s} \approx 0.3$ TeV and $Q^2_{\max} \approx 9 \times 10^4$ GeV^2. The energy of the proton beam will increase over the next two years to 1 TeV, raising the c.m. energy by about 10%. The lifetime integrated luminosity of HERA is projected as 1 fb^{-1}.

Because of the high luminosity and the large kinematic reach, physics at high Q^2 is potentially very rich. In one year of operation (*i.e.*, at 10 fb^{-1}), the μp collider would yield about a million charged-current $\mu^- p \to \nu_\mu +$ anything events with $Q^2 > 5000$ GeV^2. The ZEUS detector at HERA has until now recorded 326 charged-current events in that régime. The search for new phenomena, including leptoquarks and squarks produced in R-parity–violating interactions, would be greatly extended.

On the other hand, the study of low-x collisions appears very difficult because of the asymmetric kinematics and the angular cutoffs foreseen for detectors in the muon-storage-ring setting. A general question is what kind of detectors would survive the harsh environment of the μp collider.

NEUTRINO BEAMS FROM STORED MUONS

The idea of using stored muons to produce neutrino beams of a special character has arisen repeatedly. A neutrino beam derived from the decay

$$\mu^- \to e^- \nu_\mu \bar{\nu}_e$$

is very different from the traditional beams derived from the decays of pions and kaons. The neutrino beam generated in μ^- decay contains ν_μ and $\bar{\nu}_e$, but no $\bar{\nu}_\mu$, ν_e, ν_τ, or $\bar{\nu}_\tau$. It is much richer in electron (anti)neutrinos than a traditional neutrino beam, and muon *neutrinos* are accompanied by electron *antineutrinos*. A neutrino beam derived from muon decay has therefore been seen as a way to remedy the absence of ν_e and $\bar{\nu}_e$ beams at high-energy accelerators. The idea of storing very large quantities of muons—about a millimole per year—adds an important new element to the discussion, for now we can consider muon storage rings as extremely intense neutrino sources.

Neutrino beams generated by the decay of 10^{20} - 10^{21} stored muons per year would make possible investigations of an entirely unprecedented nature: studies of deeply inelastic scattering in thin targets, and neutrino-oscillation studies over a wide range of distance/energy and at very great distances.

In the rest frame of the decaying muon, the distribution of muon-type neutrinos produced in the decays $(\mu^- \to e^- \nu_\mu \bar{\nu}_e, \mu^+ \to e^+ \bar{\nu}_\mu \nu_e)$ is

$$\frac{d^2 N_{(\nu_\mu, \bar{\nu}_\mu)}}{dx d\Omega} = \frac{x^2}{2\pi}[(3 - 2x) \pm (1 - 2x)\cos\theta] ,$$

where θ is the angle between the neutrino momentum and the muon spin and $x = 2E_\nu/m_\mu$ is the scaled energy carried by the neutrino. The distribution favors $x = 1$ with (ν_μ opposite, $\bar{\nu}_\mu$ along) the muon spin direction. The distribution of electron-type neutrinos produced in $\mu^\mp$ decay is somewhat softer; it is given by

$$\frac{d^2 N_{(\bar{\nu}_e, \nu_e)}}{dx d\Omega} = \frac{3x^2}{\pi}[(1 - x) \pm (1 - x)\cos\theta] ,$$

which peaks at $x = \frac{2}{3}$ for ($\bar{\nu}_e$ along, ν_e opposite) the muon spin direction. In a neutrino beam generated by μ^- decay, we would study at the same time, and in approximately equal proportions, the charged-current reactions $\nu_\mu N \to$

μ^- + anything and $\bar{\nu}_e N \to e^+ \to$ anything, along with the corresponding neutral-current reactions in a statistical mixture.

Let us examine the capabilities of a high-energy neutrino beam for deeply inelastic scattering experiments. Two variants were considered at the Workshop [20]. In the first, the 533-m straight section of RLA 3, the final recirculating linear accelerator in the Front End, provides the decay region. Muons enter RLA 3 at 70 GeV and are accelerated in 12 turns to 250 GeV. The muon energy is therefore different on each turn, and increasing along the linac. The mean neutrino energy $\langle E_\nu \rangle \approx 135$ GeV. The resulting neutrino beam is well collimated; at 600 meters downstream, half the neutrinos lie within 25 cm of the linac axis. In the second scheme, a 10-meter straight section in a 250-GeV $\mu^+\mu^-$ collider ring yields neutrinos with $\langle E_\nu \rangle \approx 178$ GeV during 1560 turns. This beam is even better collimated, with about half the neutrinos within 15 cm of the axis 600 meters downstream. The neutrino flux per year is prodigious, about a thousand times the flux the NUTEV detector received in a year of running with a traditional neutrino beam.

The gigantic flux of neutrinos from a millimole of stored muons means that the familiar massive neutrino detectors would be inappropriate devices [35]. Thin targets, instead of extremely massive target calorimeters, become the order of the day. For example, a 1-meter liquid hydrogen target 600 meters downstream of RLA 3 would record 10^7 deeply inelastic events per year. We could therefore measure parton distributions of the proton directly, instead of inferring them from measurements made on heavy (typically, iron) targets. The high rates and light targets should also make it possible to extend measurements of the parton distributions to smaller values of x_{Bjorken} than has been possible before in neutrino scattering. The neutral-current / charged-current ratio could be measured with tiny statistical error, making possible an indirect measurement of the W-boson mass with $\delta M_W = (20$-$50)$ MeV/c^2. By reconstructing 10^5 charmed particles per year, we could make improved measurements of the quark-mixing matrix element $|V_{cd}|$ and significantly advance our knowledge of the strange quark and antiquark distributions within the nucleon.

There are other possibilities as well. Polarized targets might make it possible to probe details of the distribution of spin within the proton, perhaps even to study the polarization of minority components like the s and $\bar{s}$ sea. And we could consider the uses of high-resolution silicon detectors for special studies involving heavy flavors.

Neutrino beams from muon decay offer dramatic new possibilities for the study of neutrino oscillations. The paucity of electron neutrinos and antineutrinos in traditional neutrino beams is the reason why we have limited knowledge of $\nu_e \leftrightarrow \nu_\tau$ oscillations: the $\bar{\nu}_e$ available at reactors are too low in energy to permit τ-lepton appearance experiments. That limitation would be removed with muon-decay neutrino beams. In addition, the intense fluxes will permit flexible experimentation over great distances.

Consider a beam of ν_μ and $\bar{\nu}_e$ produced in μ^- decay. In a detector that can measure the charge of leptons produced in charged-current interactions, it will be possible to distinguish the expected reactions

$$\nu_\mu N \to \mu^- + \text{anything} \text{ and } \bar{\nu}_e N \to e^+ + \text{anything}$$

from the oscillation-induced reactions

$$(\nu_\mu \to \nu_e)N \to e^- + \text{anything} \text{ and } (\bar{\nu}_e \to \bar{\nu}_\mu)N \to \mu^+ + \text{anything}.$$

In addition to these appearance experiments (of a new and interesting kind), we can look for distortions of the charged-lepton energy spectra that might signal oscillations. For beams of sufficiently high energy, it will also be possible to perform appearance experiments in search of

$$(\nu_\mu \to \nu_\tau)N \to \tau^- + \text{anything} \text{ and } (\bar{\nu}_e \to \bar{\nu}_\tau)N \to \tau^+ + \text{anything}.$$

Steve Geer has made a preliminary study of the fluxes and event rates that could be anticipated from a muon storage ring [36]. A rough optimization of a storage ring to maximize the neutrino flux in a given direction results in a ring that consists of two semicircular arcs and two straight sections, with all segments of equal length. In this way, 25% of the muons decay while pointing at the detector. In the conceptual designs under consideration, the typical length of an arc (hence, of a straight section) is about

$$\ell = 75 \text{ m} \times \left(\frac{p_\mu}{40 \text{ GeV}/c}\right),$$

which is short. Accordingly, it is reasonable to consider installing a ring sloped at a steep angle to point to a distant detector [37]. Some interesting possibilities are presented in Table 3. In the case of conventional neutrino beams from meson decay, which require a decay region about a kilometer long, tunneling costs threaten to become prohibitive for dip angles greater than a few degrees.

Not only are the dimensions (including the maximum depth) of the muon storage ring reasonable, the fluxes at distant detectors are impressively large. Geer has estimated that a 20-GeV muon beam would generate a flux of a few$\times 10^{10}$ $\nu/\text{m}^2/\text{year}$ at the Gran Sasso Laboratory, some 7332 km from Fermilab. [A useful comparison may be that the NuTeV detector saw a flux of

TABLE 3. Possible sites for long-baseline neutrino experiments using beams generated in a muon storage ring at Fermilab.

Location	Distance [km]	Dip Angle	Heading
Soudan Mine, Minnesota	729	3°	336°
Gran Sasso, Italy	7332	35°	50°
Kamioka Mine, Japan	9263	47°	325°

about 10^9 $\nu/\text{m}^2/$minute in the 1997 run.] The fluxes at the Soudan Mine in Minnesota would be about a hundred times larger, and ten times the flux planned for the MINOS experiment.

Since an important figure of merit for neutrino-oscillation searches is L/E, the ratio of path length to neutrino energy, it may be advantageous to keep the muon energy low. For 20-GeV muons, about 100 charged-current events would occur per kiloton per year in the Gran Sasso. Both the fluxes and the rates rise with muon-beam energy, but there is a price to pay in L/E.

The properties of neutrino beams produced in the decay of large numbers of muons are altogether very remarkable. The possibilities for experiments are quite astounding. We need to ask what a plausible experimental program might be, and whether the experiments are merely amazing, or truly interesting. We also need to ask the important practical question: can this really be done?

SUMMARY REMARKS

We do not yet know whether a $\mu^+\mu^-$ collider will be a practical tool for particle physics, but the animated discussions at this Workshop and the diversity of ideas reported in this volume are evidence that the prospect of a $\mu^+\mu^-$ collider gives us much to think about. Some of the possibilities I have discussed in this short summary, as well as others to be found elsewhere in these Proceedings, represent opportunities that are both unique and remarkable. This has been an unusually stimulating workshop, for the novelty and reach of the ideas we have discussed. An important conclusion is that the campaign to explore the feasibility and utility of a $\mu^+\mu^-$ collider is serious—and fun.

The original motivation for the $\mu^+\mu^-$ collider remains the central goal: a practical lepton collider with multi-TeV beams.

I would like to conclude with a few general observations inspired by what we have heard during the Workshop.

- The various machines discussed as the First Muon Collider (which some have called the Next Lepton Collider) all are luminosity poor. The interesting—and unique—program that has been outlined for a Higgs factory would be a far more compelling prospect if it could be carried out over a few years, rather than a decade.

- A program that includes many collider rings dedicated to specific studies: a Higgs factory, a top factory, a $\frac{1}{2}$-TeV collider, etc., appears very rich. We have to keep in mind the realities of the muon economy: not all elements of a multiring complex will operate at once. That means that different kinds of experiments will necessarily be sequential or interleaved. We cannot ignore *when* experiments might be done when we try to assess the impact they will have on physics.

- Even modest polarization can be highly useful, especially if it can be controlled flexibly, and separately for μ^+ and μ^-. It is an advantage if polarization can be reversed on demand.

- Single-muon-ring devices do not seem to lack intensity. The capabilities of the intense neutrino beams produced in the decays of stored leptons appear very well matched to the demands of the physics.

It is important for us to learn whether a $\mu^+\mu^-$ collider should be part of our future. I see four important short-term goals. ¶ Determine the overall feasibility of the muon-collider idea, with the goal of a high-performance, low-cost lepton collider that reaches several TeV. ¶ Learn whether it is possible to build a $\mu^+\mu^-$ collider as a Higgs factory, with adequate luminosity to carry out the initial survey in only a few years and growth potential to make it worthwhile to exploit Higgs physics for a decade. ¶ Make serious designs of muon storage rings as neutrino sources and investigate their potential for transforming neutrino physics. It is possible that this approach to neutrino physics might make sense even before we know whether a muon collider is viable. ¶ Develop realistic conceptual designs for muon-collider detectors, paying careful attention to the challenges of the experimental environment, especially for heavy-flavor tagging. Explore adventurous designs for neutrino detectors that would take advantage of the unique character of muon-produced neutrino beams.

In assessing all the possibilities for muon-collider physics and for the physics opportunities that arise from the front end of a muon collider, we must judge as carefully as we can what will be the scientific impact of experiments we could carry out using these adventurous new devices. The idea of a $\mu^+\mu^-$ collider is bold indeed; it calls for bold experiments that can change the way we think.

ACKNOWLEDGEMENTS

It is a pleasure to thank Steve Geer and Rajendran Raja for the stimulating and pleasant atmosphere of the workshop. I am grateful to the workshop staff for providing me with instantaneous copies of transparencies throughout the week. I thank the working-group convenors for advice and assistance in the preparation of this talk.

REFERENCES

1. H. Murayama and M. E. Peskin, *Ann. Rev. Nucl. Part. Sci.* **46**, 533 (1996).
2. J. Ellis, "Physics at Future Colliders," CERN-TH-97-367A (hep-ph/9712444), to appear in the Proceedings of the International Europhysics Conference on High-Energy Physics (HEP 97), Jerusalem.

3. The Muon Collider Collaboration, "$\mu^+\mu^-$ Collider Feasibility Study," BNL-52503 & FERMILAB–CONF–96/092, July 1996, unpublished.
4. R. Palmer, A. Tollestrup, and A. Sessler, "Status Report of a High Luminosity Muon Collider and Future Research and Development Plans," *Snowmass '96*.
5. For a very useful survey of the physics of $\mu^+\mu^-$ colliders, see V. Barger, M. S. Berger, J. F. Gunion, and Tao Han, *Phys. Rep.* **286**, 1 (1997). See also J. F. Gunion, "Physics at a Muon Collider," UCD-98-5 (hep-ph/9802258), *These Proceedings*.
6. R. Palmer for the Muon Collider Collaboration, "Muon Collider: Introduction and Status," *These Proceedings*.
7. Charles Ankenbrandt and Robert J. Noble, "Summary of the Accelerator Working Group," FERMILAB–CONF–98/074, *These Proceedings*.
8. M. Demarteau and T. Han, "Higgs Boson and Z Physics at the First Muon Collider" FERMILAB–CONF–98/030 & MADPH-98-1037 (hep-ph/9801407), *These Proceedings*. See also B. Kamal, W. J. Marciano, and Z. Parsa, "Higgs Resonance Studies at the First Muon Collider" (hep-ph/9712270), *These Proceedings*.
9. M. S. Berger and B. L. Winer, "Top Quark Physics at Muon and Other Future Colliders," IUHET-383 & OHSTPY-HEP-E-98-003 (hep-ph/9802296), *These Proceedings*.
10. M. Carena and S. Protopopescu, "Report of the Supersymmetry Working Group," *These Proceedings*.
11. P. Bhat and E. J. Eichten, "Strong Dynamics at the Muon Collider: Working Group Report," FERMILAB–CONF–98/072, *These Proceedings*.
12. G. R. Gutierrez and L. Littenberg, "Physics with Low Energy Hadrons," *These Proceedings*.
13. P. Fisher and B. Kayser, "Experimental neutrino physics at the muon collider complex," *These Proceedings*.
14. H. Schellman, "Deep Inelastic Scattering at a Muon Collider—Neutrino Physics," *These Proceedings*. See also B. J. King, "Neutrino Physics at a Muon Collider," *These Proceedings*; P. Spentzouris, "Deep Inelastic Scattering and Neutrino Physics," *These Proceedings*.
15. W. Molzon, "Physics with Low Energy Muons at the Front End of the Muon Collider," *These Proceedings*. See also W. J. Marciano, "Low energy physics and the first muon collider," *These Proceedings*.
16. S. D. Holmes, *et al.*, "A Development Plan for the Fermilab Proton Source," FERMILAB–TM–2021, September 1997, unpublished.
17. A. N. Skrinskiĭ and V. V. Parkhomchuk, *Fiz. Elem. Chastits At. Yadra* **12**, 557 (1981) [English transl.: *Sov. J. Part. Nucl.* **12**, 223 (1981)].
18. See, for example, V. Barger, M.S. Berger, J.F. Gunion, and T. Han, *Phys. Rev. Lett.* **75**, 1462 (1995); David B. Cline, *Int. J. Mod. Phys.* **A13**, 183 (1998).
19. For an assessment of the prospects for discovering and determining the mass of a light Higgs boson, see J. F. Gunion, L. Poggioli, R. Van Kooten, C. Kao, P. Rowson, *et al.*, "Higgs Boson Discovery and Properties," *Snowmass '96*.
20. C. Ankenbrandt and S. Geer, "Accelerator Scenario and Parameters for the

Workshop on Physics at the First Muon Collider and Front–End of a Muon Collider," FERMILAB–CONF–98/086, *These Proceedings*.

21. R. Raja and A. Tollestrup, "Calibrating the energy of a 50 × 50 GeV muon collider using spin precession," FERMILAB–PUB–97/402 (hep-ex/9801004).

22. S. Geer, "Backgrounds and detector issues at a muon collider," FERMILAB–CONF–96/313, *Snowmass '96*.

23. The overall radiation environment is similar to that of the Large Hadron Collider at CERN. For an imaginative proposal to deal with the flux of soft photons, see J. Chapman and S. Geer, "The Pixel Microtelescope," FERMILAB–CONF–96/375, *Snowmass '96*.

24. For a convenient tabulation of the expectations for 500-GeV linear colliders, see Table 1 of Reference [1].

25. A. Blondel, "A Z Factory at a Muon Collider," X-LPNHE/98-01, *These Proceedings*.

26. M. S. Berger, "Precision W-Boson and Higgs Boson Mass Determinations at Muon Colliders" (hep-ph/9712474), *These Proceedings*.

27. M. S. Berger, "The Top–Antitop Threshold at Muon Colliders" (hep-ph/9712486), *These Proceedings*.

28. For a discussion of some of the threshold uncertainties, see A. H. Hoang, "Top Quark Pair Production at Threshold: Uncertainties and Relativistic Corrections" (hep-ph/9801273), *These Proceedings*.

29. V. Barger, "Supersymmetry vis-à-vis Muon Colliders," *These Proceedings*.

30. K. Lane, "Technicolor and the first muon collider," BUHEP-98-01 (hep-ph/9801385), *These Proceedings*.

31. For an analysis in the framework of the BESS model, see R. Casalbuoni, *et al.*, "A Strong Electroweak Sector at Future $\mu^+\mu^-$ colliders," DFF-296-01-98 (hep-ph/9801243), *These Proceedings*.

32. J. Womersley, "Technihadron Production at a Muon Collider," FERMILAB–CONF–98/078, *These Proceedings*. See also E. Eichten, K. Lane, and J. Womersley "Narrow Technihadron Production at the First Muon Collider," FERMILAB–PUB–98/065–T & BUHEP-98-02 (hep-ph/9802368).

33. Bjørn Wiik (private communication).

34. V. Shiltsev, "An Asymmetric Muon-Proton Collider: Luminosity Consideration," FERMILAB–CONF–97/114.

35. D. A. Harris and K. S. McFarland, "Detectors for Neutrino Physics at the First Muon Collider," *These Proceedings*; "A Small Target Neutrino Deep-Inelastic Scattering Experiment at the First Muon Collider," *These Proceedings*.

36. S. Geer, "Neutrino Beams from Muon Storage Rings: Characteristics and Physics Potential," FERMILAB–PUB–97/389 (hep-ph/9712290). See also S. Geer, "The Physics Potential of Neutrino Beams from Muon Storage Rings," FERMILAB–CONF–97/417, *These Proceedings*.

37. This is not to say that it is entirely trivial to operate cryogenic systems for the superconducting magnets of the muon storage ring with a large pressure head from the top of the ring to the bottom.

Physics with Low Energy Hadrons

Conveners: Lawrence Littenberg, Brookhaven National Laboratory
Gaston Gutierrez, Fermilab

Lepton Flavor Violating Rare Muon Decays and Future Prospects

Yoshitaka Kuno

Institute of Particle and Nuclear Studies (IPNS),
High Energy Accelerator Research Organization (KEK),
Tsukuba, Ibaraki, Japan 305

Abstract. Future prospects of a search for lepton-flavor-violating rare muon decays, in particular of $\mu^+ \to e^+\gamma$ decay, are presented. At first, the use of muon polarization to suppress backgrounds in a search for $\mu^+ \to e^+\gamma$ is discussed. Second, current efforts and ideas to improve experimental detections of e^+s and photons in $\mu^+ \to e^+\gamma$ decay are shown. Finally, a new intense muon source being considered at Japan Hadron Facility (JHF) is briefly mentioned.

INTRODUCTION

Lepton flavor violation (LFV) has been known for a long time as one of the most critical topics in particle physics. LFV searches were initiated back in the 1940's, as shown in Fig.1. Processes of major interest here are rare muon decays such as $\mu^+ \to e^+\gamma$, $\mu^+ \to e^+e^+e^-$ and $\mu^- N \to e^- N$ conversion in nuclei. Since 1940's, the experimental upper limits have been continuously improved at a rate of about two orders of magnitude per decade ! Originally, physics interest on LFV has been just to explore new exotic particles or interaction at a very high energy scale, like a few 100 TeV for LFV sensitivities of about 10^{-12} in branching ratio. Recently, the physics motivation for LFV becomes even stronger. One of the strongest motivation comes from supersymmetric (SUSY) extension to the standard model, in particular supersymmetric grand unification (SUSY GUT) which predicts large branching ratios for some of the LFV processes [1,2]. In SU(5) SUSY GUT models, for instance, the prediction for $\mu^+ \to e^+\gamma$ can be given approximately by [1],

$$\Gamma(\mu \to e\gamma) = \frac{\alpha}{4}\left|\frac{\alpha(M_Z)}{24\pi \cos^2\theta_W}\frac{(\Delta m_{21}^2)}{m_{\tilde{\mu}}^2}\right|^2 \frac{m_\mu^5}{m_{\tilde{\mu}}^4} \tag{1}$$

$$\Delta m_{21}^2 = -V_{32}^* V_{31} I(y_t) \tag{2}$$

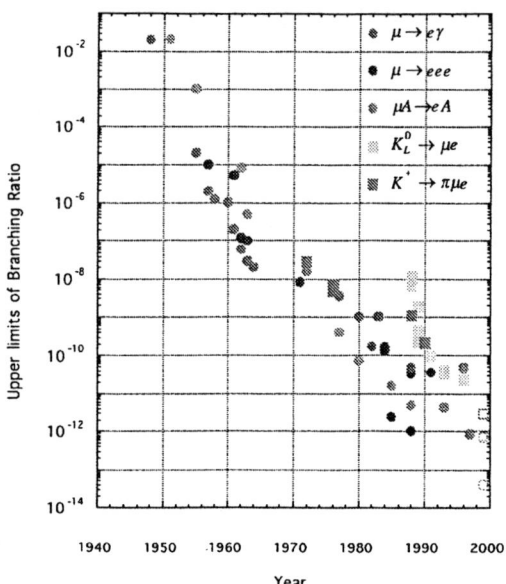

FIGURE 1. A history of Lepton Flavor Violation Searches for various processes of kaon and muon.

where V_{ij} is the Kobayashi-Maskawa quark matrix elements (at a GUT energy scale), $I(y_t)$ is a function of the top Yukawa coupling (y_t). m_μ and $m_{\tilde{\mu}}$ are masses of muons and smuon. The branching ratio thus predicted ranges from 10^{-15} to 10^{-13} for the singlet smuon mass of $m_{\tilde{\mu}_R}$ of 100 to 300 GeV, and the SO(10) SUSY GUT models give an even larger value of 10^{-13} to 10^{-11} by an enhancement of (m_τ^2/m_μ^2) [1]. The $\mu^- - e^-$ conversion processes in nuclei has about 1/200 lower predicted branching ratio than $\mu^+ \to e^+\gamma$ in SUSY GUT models. The theoretical predictions for those processes are only one or two orders of magnitudes smaller than the present experimental upper limits. Therefore, future experimental searches with higher sensitivity are now strongly desired.

The experimental upper limit for $\mu^+ \to e^+\gamma$ compiled in Particle Data Group (PDG) is $B(\mu^+ \to e^+\gamma) \leq 4.9 \times 10^{-11}$ at a 90% confidence level [3]. The latest experiment, MEGA, at LAMPF completed its data-taking in 1995 and reported a preliminary new limit of 3.8×10^{-11} with 16 % of the data analyzed [4]. Regarding $\mu^- - e^-$ conversion in nuclei, a new upper limit from SINDRUM II experiment at Paul Scherer Institute (PSI) of $B(\mu^- + Ti \to e^- + Ti) \leq 7 \times 10^{-13}$ is reported [5]. There is a new experimental proposal approved at BNL-AGS on $\mu^- - e^-$ conversion by the MECO collaboration, which aims at a sensitivity of 10^{-16} level [6]. In this talk, future prospects of a new experimental approach for $\mu^+ \to e^+\gamma$ as well as a new muon source at Japan Hadron Facility (JHF) are presented.

$\mu^+ \to e^+\gamma$ SEARCH WITH POLARIZED MUONS

The event signature of $\mu^+ \to e^+\gamma$ is that a e^+ and a photon are in coincidence, and moving collinear back-to-back with their energies equal to a half of the muon mass ($m_\mu/2 = 52.8$ MeV). There are two major backgrounds to a search for $\mu^+ \to e^+\gamma$. One is a physics (prompt) background from a radiative muon decay $\mu^+ \to e^+\nu\bar{\nu}\gamma$ when e^+ and photon are emitted back-to-back with two neutrinos carrying off little energy. The other background is an accidental coincidence of a e^+ in a normal muon decay, $\mu^+ \to e^+\nu\bar{\nu}$, accompanied by a high energy photon. The sources of a high energy photon might be either that in $\mu^+ \to e^+\nu\bar{\nu}\gamma$ decay, or external bremsstrahlung or annihilation-in-flight of e^+s in the normal muon decay.

In the following, it is discussed how these backgrounds could be suppressed by using polarized muons to aim at a sensitivity of an order of 10^{-14}.

Polarized $\mu^+ \to e^+\gamma$ Signal

First of all, when a muon is spin-polarized, the $\mu^+ \to e^+\gamma$ decay has an asymmetric angular distribution which is different for a left-handed e^+ (e_L^+) and a right-handed e^+ (e_R^+). Namely, as shown in Fig.2, $\mu^+ \to e_L^+\gamma$ decay

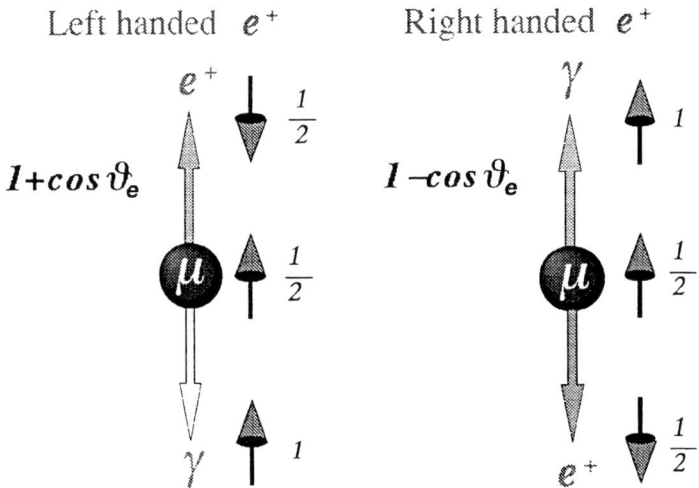

FIGURE 2. Angular distribution of $\mu^+ \to e_L^+\gamma$ and $\mu^+ \to e_R^+\gamma$ for polarized muons.

follows a $(1 + P_\mu \cos\theta_e)$ distribution, whereas $\mu^+ \to e_R^+ \gamma$ does a $(1 - P_\mu \cos\theta_e)$ distribution, where θ_e is an angle of the e^+ emission with respect to the muon polarization (P_μ). Therefore, once the signal is observed, the angular distribution of polarized $\mu^+ \to e^+\gamma$ decay could tell a helicity of e^+s which would be useful to understand the LFV mechanism, as shown later.

Physics Background

The radiative muon decay, $\mu^+ \to e^+\nu\bar{\nu}\gamma$ (branching ratio equal to 1.4% for $E_\gamma > 10$ MeV), would become a background when the e^+ and photon are emitted back to back with two neutrinos carrying off little energy. The differential decay width of this radiative decay was calculated [7]. Given the detector resolution, the sensitivity limitation from this physics background can be estimated by integrating the differential decay width over the kinematic signal box. It is given by [8]

$$dB(\mu \to e\nu\bar{\nu}\gamma) \cong \frac{\alpha}{16\pi}\Big[J_1(1 - P_\mu \cos\theta_e) + J_2(1 + P_\mu \cos\theta_e)\Big]d(\cos\theta), \quad (3)$$

where J_1 and J_2 are given as a function of sizes of the kinematical signal box for the e^+ and photon energies. The sizes of the signal box are determined by the detector resolutions. It is found that the above distribution follows approximately a $(1 + P_\mu \cos\theta_e)$ distribution (i.e. $J_2 \gg J_1$) when the energy resolution of photon detection is worse than that of e^+ [8]. They imply that the selective measurement of e^+s antiparallel to the muon polarization direction would suppress the physics background, improving a sensitivity of the search for $\mu^+ \to e_R^+ \gamma$.

Accidental Background

It turns out that the accidental background becomes more important than the physics background in a new generation experiment with a very high rate of stopped muons. The angular distribution of the sources of accidental background are examined for polarized muons [9]. It is known that e^+s in the normal Michel muon decay follows a $(1 + P_\mu \cos\theta_e)$ distribution. We have studied the inclusive angular distribution of a high energy photon (e.g. ≥ 50 MeV) from $\mu^+ \to e^+\nu\bar{\nu}\gamma$. As shown in Fig.3, it is found to be emitted preferentially along the muon spin direction; namely follows a $(1 + P_\mu \cos\theta_\gamma)$ distribution, where θ_γ is an angle of the photon direction with respect to the muon spin direction. It should be noted that this inclusive angular distribution was obtained after the integration of energy and direction of e^+s, which is in contrast to the case of physics background where only the extreme kinematics of e^+ and photon being back-to-back in $\mu^+ \to e^+\nu\bar{\nu}\gamma$ decay is relevant. It is further noted that the other sources of high energy photons such as external

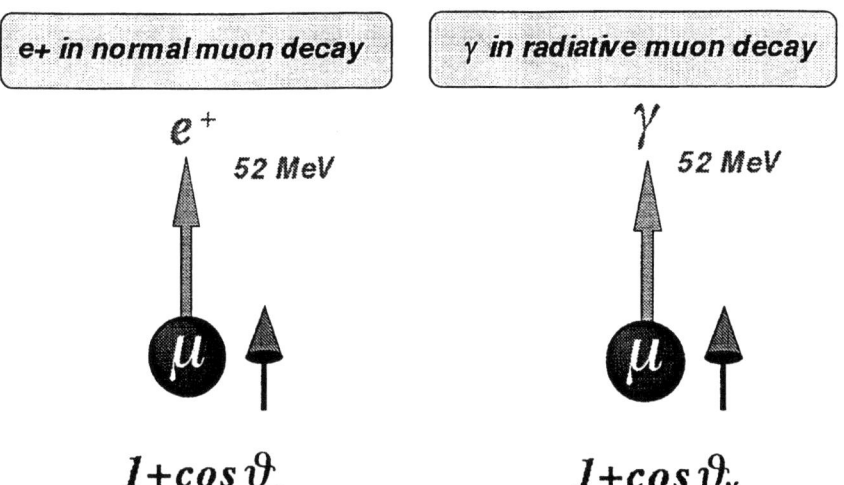

FIGURE 3. Schematic drawing of direction of a 52.8 MeV e^+ in Michel muon decay (left) and that of a 52.8 MeV photon in $\mu^+ \to e^+\nu\bar{\nu}\gamma$ (right). Both are preferentially emitted along the muon spin direction.

bremsstrahlung and annihilation-in-flight of e^+s in the normal muon decay also follow a $(1 + P_\mu \cos\theta_\gamma)$ distribution.

This inclusive angular distribution of a high energy photon in $\mu^+ \to e^+\nu\bar{\nu}\gamma$ implies that the accidental background could be suppressed for $\mu^+ \to e_L^+\gamma$ where high energy photons must be detected at the opposite direction to the muon polarization. A similar suppression mechanism of accidental background can be seen for $\mu^+ \to e_R^+\gamma$ when high energy positrons are detected at the opposite direction to the muon polarization. As a result, the selective measurements of either e^+s or photons antiparallel to the muon spin direction would give the same accidental background suppression for $\mu^+ \to e_R^+\gamma$ and $\mu^+ \to e_L^+\gamma$ decays respectively. This favorable situation comes from the fact that the inclusive distributions of both high energy e^+s and photons, respectively in the normal and radiative muon decays, follow a $(1 + P_\mu \cos\theta)$ distribution, where θ is either θ_e or θ_γ.

The rate of the accidental background (B_{acc}) normalized to the total decay rate can be estimated by

$$B_{\text{acc}} = R_\mu \cdot f_e^0 \cdot f_\gamma^0 \cdot (\Delta t) \cdot \left(\frac{\Delta\omega}{4\pi}\right) \cdot \eta \qquad (4)$$

where R_μ is an instantaneous muon intensity. f_e^0 and f_γ^0 are an integrated fraction of the spectrum (with their branching ratios) of e^+ in the normal muon decay and that of photon in $\mu^+ \to e^+\nu\bar{\nu}\gamma$ decay within the signal region

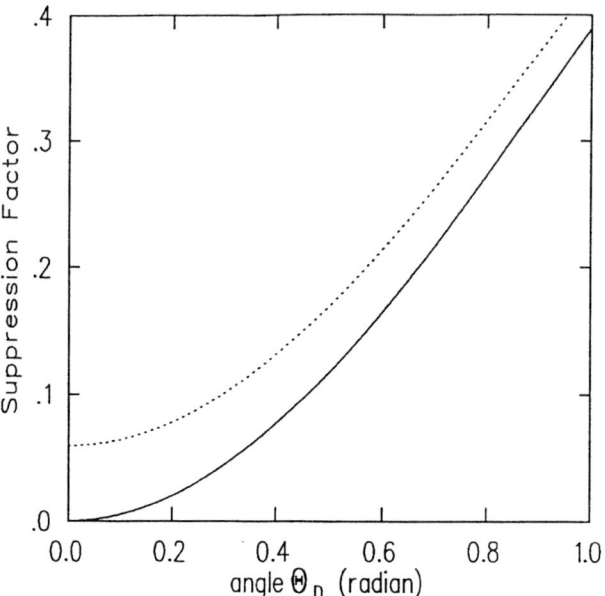

FIGURE 4. Accidental background suppression as a function of (half) detector opening angle.

for unpolarized muons, respectively. Δt and $\Delta \omega$ are respectively a full width of the signal regions for timing coincidence and angular constraint of the back-to-back kinematics. η is a suppression factor of the accidental background. $\eta = 1$ for the case of unpolarized muons. For instance, take some reference numbers such as the e^+ energy resolution (FWHM) of 1%, the photon energy resolution of 6%, $\Delta \omega = 3 \times 10^{-4}$ steradian, $\Delta t = 1$ nsec, and $R_\mu = 3 \times 10^8 \ \mu^+/\text{sec}$, B_{acc} is 3×10^{-13}. Unless there are significant improvements made on the detector resolution, the accidental background might appear at a level of 10^{-13}.

As explained before, the selective measurement of either e^+s or photons antiparallel to the muon spin direction would provide a large background suppression factor. By taking account of the angular distributions of the $\mu^+ \to e^+ \gamma$ signal, the suppression factor η is calculated for polarized muons by

$$\eta \equiv \int_{\cos \theta_D}^1 d(\cos \theta)(1 + P_\mu \cos \theta)(1 - P_\mu \cos \theta) / \int_{\cos \theta_D}^1 d(\cos \theta)$$
$$= (1 - P_\mu^2) + \frac{1}{3} P_\mu^2 (1 - \cos \theta_D)(2 + \cos \theta_D) \quad (5)$$

where θ_D is a half opening angle of detection with respect to the muon polarization direction. For instance, for $\theta_D = 300$ mrad, an accidental background can be suppressed down to about 1/20 (1/10) when P_μ is 100 (97)%. Note that an opening angle of $\theta_D = 300$ mrad give a geometrical solid angle coverage of 2.3%. Therefore, with a reasonable detector acceptance, a suppression of an order of magnitude could be easily achieved. It would be possible to aim for a sensitivity of 10^{-14} level.

Furthermore, for polarized muons as shown in Fig.2, the angular distribution of $\mu^+ \to e^+\gamma$ signal is known to be useful to discriminate different models, since the helicity of e^+ in $\mu^+ \to e^+\gamma$ is sensitive to the mechanism of lepton flavor mixing. For instance, SU(5) SUSY-GUT models predict only $\mu^+ \to e_L^+\gamma$, whereas non-GUT SUSY extension with heavy right-handed neutrino, or left-right symmetric models do $\mu^+ \to e_R^+\gamma$. SO(10) SUSY-GUT models allow both $\mu^+ \to e_R^+\gamma$ and $\mu^+ \to e_L^+\gamma$.

DETECTOR OPTIONS FOR $\mu^+ \to e^+\gamma$

In addition to background suppression by the use of muon polarization mentioned in the previous section, R&D efforts to improve the detector resolutions should be pursued at the same time. Currently, there are some ideas proposed to improve detections of e^+s and photons for a new experiment of $\mu^+ \to e^+\gamma$ decay planned at PSI [10]. Some of them are shown as follows.

First, as for a e^+ spectrometer, a solenoid magnetic field with a large solid angle of 10-20 % of 4π is being studied. Although it is similar to MEGA, possible improvement is to consider a magnetic field gradient which allows Michel e^+s swept out from the fiducial volume of detector quickly. This solenoid geometry is well suited to have an extended slanted muon-stopping target thin enough to minimize multiple scattering of e^+s. The other option is double or triple solenoid magnets to produce sinusoidal variation of a magnetic field along the z axis. This sinusoidal magnetic field creates a ring-shaped quasi-focusing image (at a rotation angle of $\phi = 180°$) where e^+ momentum dispersion is maximum, and the second focus at the detector axis (at a rotation angle of $\phi = 360°$) of zero dispersion. It would yield an excellent "intrinsic" momentum resolution of 10^{-3} when tracking detectors are placed at these focusing positions [5]. This scheme however can not have an extended target, and therefore a real momentum resolution might be deteriorated by multiple scattering in a thicker target.

Regarding a photon detector, two options are being studied. They are liquid Xenon, and high-speed luminous inorganic crystals such as Ce-doped orthosilicates or ortho-aluminates of Yttrium, Lutetium and/or Ytterbium. Liquid Xenon is known to have a large light yield (comparable to NaI(Tl)) with faster timing. The liquid Xenon calorimeter consists of a single volume, the surface of which is densely covered with UV sensitive photomultipliers [11]. Monte Carlo simulation showed that the energy-weighted centroid would give a position resolution of a few mm with good energy resolution (of a few % for 52.8 MeV photon) and fast response. In the crystal option, preconverter which consists of these new scintillation crystals and the back part consists of ordinary crystals, such as CsI-doped NaI. A preconverter crystal plate is viewed from the photon entrance side by photomultipliers or avalanche photodiodes.

At JHF, a $\mu^+ \to e^+\gamma$ search with polarized muons should be pursued. There are at least two different approaches: one is a small solid-angle spectrometer of ultra-high momentum resolution with a high muon beam intensity. A possible spectrometer could be a 180°-bending magnetic dipole spectrometer, to which the muon spin polarization points. At the same time, an ultra-high resolution calorimeter of expensive scintillating crystals, which could be afforded only for a small solid-angle, can be used. A similar design of future experiment is already shown elsewhere [12]. The other is a cylindrical solenoid spectrometer of a large solid angle together with a micro bunched muon beam (which will be mentioned later). In this case, a repetition rate of a bunched beam is adjusted so as to be equal to a cyclotron frequency of muon spin precession under a solenoid magnetic field. A muon decay timing with respect to a beam-incident timing would give information of the muon spin direction. This is called "stroboscopic". Although the timing resolution is smeared by a lifetime of pion decay for a surface μ^+ beam, an appropriate repetition rate of a bunched beam will give sufficient background suppression. Either case had better need a muon-spin rotator consisting of crossing magnetic and electric fields in order to change the muon spin direction transverse to the beam direction.

NEW MUON BEAM WITH PHASE ROTATION AT JHF

The Japan Hadron Facility (JHF) is a next accelerator project at KEK to construct a 50-GeV proton synchrotron with 10 μA and a 3-GeV Booster ring with 200 μA [13]. A high intensity muon source is being considered at JHF to carry out LFV searches. For a conventional beam line design at JHF, it is expected to have an only slightly-higher beam intensity for μ^+ and μ^- than PSI, where a surface μ^+ beam of 3×10^8/sec is currently available. However, to make significant experimental breakthrough, it is definitely needed to build a muon source of much higher intensity like in several orders of magnitude. Some new ideas are coming recently, in conjunction with the $\mu^+\mu^-$ collider R&D [14]. They are, for instance, (a) a high magnetic-field capture of pions at the production target, (b) a phase rotation to make a longitudinal momentum compaction, (c) ionization cooling of muons. Those ideas could be adopted to construct a unique muon source with high intensity and high brightness and low-momentum spread. Such a cooled muon beam of very low energy would be crucial to carry out LFV searches with stopped muons (such as $\mu^+ \to e^+\gamma$, $\mu^+ \to e^+e^+e^-$, and $\mu^- - e^-$ conversion in nuclei), since it allows higher muon stopping efficiency with less backgrounds and a thiner muon-stopping target

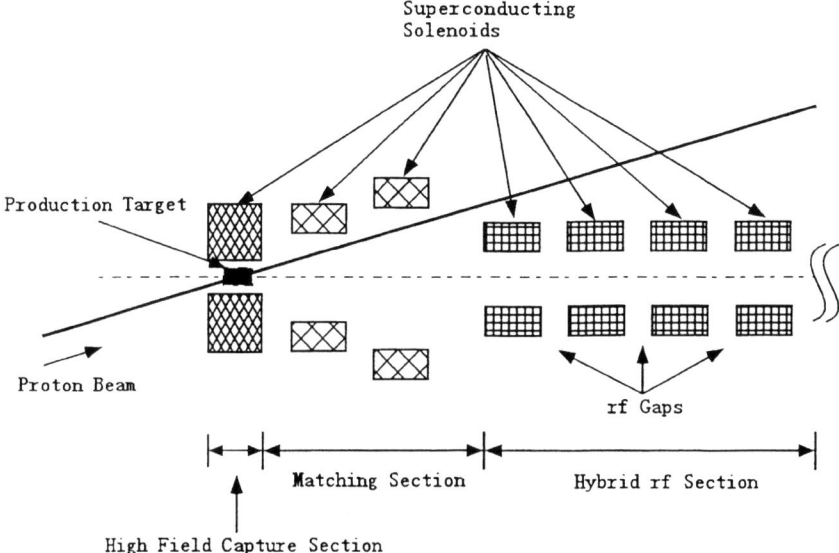

FIGURE 5. Schematic layout of a preliminary muon source design at JHF with a high-field solenoid capture followed by a hybrid rf phase-rotation section.

(or even a gas target).

JHF Design[1]

A schematic layout of a preliminary muon beam design is shown in Fig.5. Protons from a 50-GeV accelerator hit a production target where secondary particles are produced. Secondary particles (such as pions) are collected by a high solenoid magnetic field of 15 T or more. A magnetic field of 15 T is gradually changed into a lower field of 5 T to reduce transverse component of momentum with adiabatic condition of P_T^2/B = constant. And then, they come into an rf phase-rotation section, where an rf field is applied to make longitudinal momentum spread narrower by phase-rotation. In this phase-rotation section, particles with higher energy than reference energy are decelerated, whereas those with lower energy than reference are accelerated in order to yield a high-intensity beam with narrow energy spread. A solenoid filed of about 5 T is given at the phase-rotation section to contain secondary particles inside the rf cavities.

---

[1] The works presented in this section is done by S. Sawada [15].

Time Structure of JHF Beam

In contrast to the case of $\mu^+\mu^-$ collider option which will be running at a few Hz, a time structure of a muon beam at JHF has to be continuous or semi-continuous to carry out coincidence experiments as a secondary beam line. In order to achieve this, special approaches different from the $\mu^+\mu^-$ collider R&D are considered: they are

- micro-bunching extraction of protons from a 50-GeV machine and,
- superconducting rf cavity for phase rotation/ionization cooling.

The phase rotation technique requires an initially narrow-bunched secondary beam. Therefore, it is desired to have a narrow-bunched proton beam. Micro-bunching extraction allows a bunched beam with high repetition in a slow extraction from a proton accelerator [16]. If a repetition is faster than a muon lifetime, a beam is regarded as almost continuous for stopped-muon decay experiments. The time structure can be selected depending on experimental requirements. For instance, it is told that a repetition rate of about 10 MHz with a bunch width of several 100 psec is possible. This technique has been already demonstrated at BNL/AGS.

A superconducting rf cavity might be needed to reduce a required power consumption when rf runs in a DC operation. Furthermore, superconducting rf cavity allows easily a high rf field of about 10–20 MV/m with a few 100

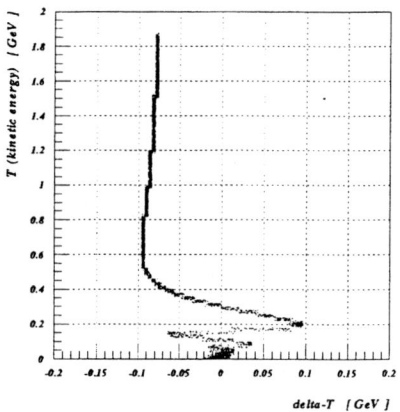

FIGURE 6. Energy gained or lost by phase rotation (delta-T) as a function of initial kinetic energy (T).

TABLE 1. Structure of the rf system simulated.

	rf components	a number of cells	total length
Drift space		1	2.6m
rf system No.1	500MHz, 0.3m, 5MV (15MV/m)	10	3.0m
rf system No.2	250MHz, 0.3m, 5MV (15MV/m)	10	3.0m
a total length			8.6m

MHz, although a leakage of a magnetic field has to be reduced significantly at the cavity locations [17].

The time structure of micro-bunching extraction depends on experimental requirements. For example, $\mu^+ \to e^+\gamma$ search needs a high repetition (of a few 10 MHz to 100 MHz) being shorter than a muon lifetime, whereas $\mu - e$ conversion experiment requires a slower repetition (of a few 100 kHz) in an order of muon lifetime with high beam extinction between beam pulses [6]. It should be also noted that this kind of new muon beam could be useful as a pion and a neutrino source as well. Further studies are underway.

Phase Rotation Simulation

A simple preliminary simulation was done at JHF [15]. One example of a reference kinetic energy of 300 MeV is shown below. The rf parameters used in simulation are listed in Table 1. It is assumed for simplicity that secondary particles have a flat energy distribution.

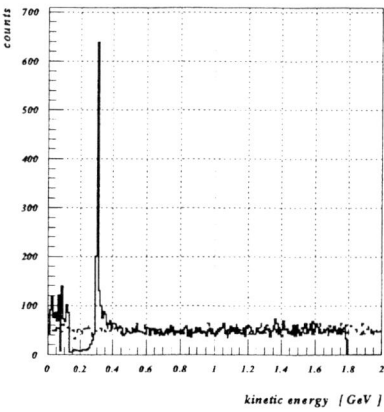

FIGURE 7. Longitudinal kinetic energy distribution after phase rotation.

Fig.6 shows a kinetic energy gained or lost by rf phase-rotation (delta-T) as a function of initial kinetic energy (T) at entrance. It is clearly seen that particles of energy higher than 300 MeV lost their energy and those of energy lower gained after the phase-rotation section. The resultant longitudinal kinetic-energy distribution after the phase rotation is shown in Fig.7, where a strong peak at 300 MeV can be seen. In the same figure, the initial distribution is shown in a dashed line. The intensity of secondary particles is increased by a factor of about 10 in a very narrow energy width of about 10 MeV. The yield of secondary particles is roughly estimated to be about 10^{12}/sec in 300 MeV $\pm$ 5 MeV for 6×10^{13} protons/sec of 50 GeV. To reduce a momentum spread further, muon ionization cooling will be needed. More studies is being undertaken.

SUMMARY

Future prospects of lepton-flavor-violating rare muon decays, in particular of $\mu^+ \to e^+\gamma$, are presented. At first, a new idea to suppress backgrounds by using muon polarization for $\mu^+ \to e^+\gamma$ search is discussed. Secondary, some ideas on possible improvement of detection of e^+s and photons in $\mu^+ \to e^+\gamma$ decay are shown. Finally, a new muon source with high intensity and high brightness at JHF is mentioned. Although more works and efforts should be needed, it is hoped to be able to carry out LFV searches with higher sensitives in future.

ACKNOWLEDGMENTS

The author is grateful to Profs. M. Cooper, T. Doke, A. Maki, S. Noguchi, Y. Mori, W. Molzon, S. Nagamiya, K. Nishikawa, Y. Okada, S. Orito, N. Sasao, S. Sawada, A. van der Schaaf, H.C. Walter (in an alphabet order) for their discussions.

REFERENCES

1. L.J. Hall, V.A. Kostelecky and S. Raby, *Nucl. Phys.* B **267**, 415 (1986);
   R. Barbieri and L. Hall, *Phys. Lett.* B **338**, 212 (1994);
   R. Barbieri, L. Hall and A. Strumia, *Nucl. Phys.* B **445**, 219 (1995).
2. J. Hisano, T. Moroi, K. Tobe and M. Yamaguchi, *Phys. Lett.* B **391**, 341 (1997);
   J. Hisano, D. Nomura, T. Yanagida, KEK Preprint KEK-TH-548, November, 1997 (hep-ph/9711348).
3. R.D. Bolton *et al.*, *Phys. Rev.* D **38**, 2077 (1988).
4. E. Hungerford, presentation in this workshop.
5. A. van der Schaaf, private communication.

6. W. Molzon, presentation in this workshop.
7. C. Fronsdal and H. Überall, *Phys. Rev.* **118**, 654 (1959).
8. Y. Kuno and Y. Okada, *Phys. Rev. Lett.* **77**, 434 (1996).
9. Y. Kuno, A. Maki and Y. Okada, *Phys. Rev.* D **55**, 2517 (1997).
10. H.K.Walter, presentations in International Conference on "The Intersections between Particle and Nuclear Physics", Big Sky, USA, May 27-June 2, 1997, and in 4th International Conference on "Physics Potential and Development of $\mu^+\mu^-$ Colliders", San Francisco, December 10-12, 1997.
11. S. Orito, private communication.
12. M. Cooper, presentation in this workshop, and in the proceedings of the 4th KEK Topical Conference on "Flavor Physics", October 29-31, 1996.
13. Y. Kuno, Proceedings of the 5th Tamura Symposium of International Workshop on "Exciting Physics with New Accelerator Facilities (EXPAF97)", Nishi-Harima, Japan, March 11-13, 1997 and KEK Preprint 97-149, September, 1997; Y. Kuno, Proceedings of the Third Workshop on "Particle Physics Phenomenology", Chin-Shan Center, Taipei, November 14-17, 1996, and KEK Preprint 97-58, June, 1997;
    Y. Kuno, Proceedings of the LAMPF Users Group (LUGI) Symposium on "20 Years of Meson Factory Physics: Accomplishments and Prospects", Los Alamos, U.S.A, October 25-26, 1996, and KEK Preprint 97-5, April, 1997.
14. "$\mu^+\mu^-$ Collider – A Feasibility Study", BNL-52503, Fermi Lab-Conf.-96/092, LBNL-38946, 1996.
15. S. Sawada, private communication, and Proceedings of the KEK workshop on "Kaon, Muon, Neutrino Physics and Future", Tukuba, Japan, October 31-November 1, 1997.
16. Y. Mori, private communication.
17. S. Noguchi, private communication.

Prospects for Hadronic Physics at a Muon Collider Facility

Kamal K. Seth

Northwestern University, Evanston, IL 60208
(kseth@nwu.edu)

Abstract. Some open problems in hadronic physics are discussed, and it is proposed that these could be very effectively addressed at the proton Booster-Stretcher facility proposed for the First Muon Collider.

I DIMENSIONS OF THE DREAM

When I was invited to contribute to this Workshop on the First Muon Collider (FMC), I was told to skip Stage I of the plan, and to catapult to Stage II, which contains a high intensity Booster and Stretcher. I was told that I should base my considerations on having a proton source of 16 GeV with a DC intensity of 4×10^{14} protons/sec, i.e., about 20 times larger than what is the final goal for the AGS at the Brookhaven National Laboratory. There were no restrictions placed on what kind of secondary beams may be made available. The dimensions of this dream situation essentially had no constraints. I was asked to simply discuss what kind of interesting and exciting hadronic physics I could envisage doing at such a facility. So here we go!

II SECONDARY BEAMS

Besides the superhot proton beams, such a facility must certainly provide record intensities of secondary beams. Let us examine the possibilities for kaon and antiproton beams.

In Fig. 1, we show results of calculations made by using the well- known Sanford-Wang formula, [1,2] with parameters taken from BNL 22452 (1977). It is claimed that the essential correctness of the results of this formula have been verified at Brookhaven, CERN and Fermilab.

Fig. 1 shows the double differential yield per interacting proton for K^- and antiprotons per unit solid angle (sr) and momentum bite (GeV/c) for two different incident momenta, 25 GeV/c (AGS) and 16 GeV/c (FMC), of the primary proton beam. The advantage of a 25 GeV/c proton beam over a 16 GeV/c beam are obvious. However, as soon as we put in the factor 20 larger intensity for the 16 GeV/c beam, the advantage

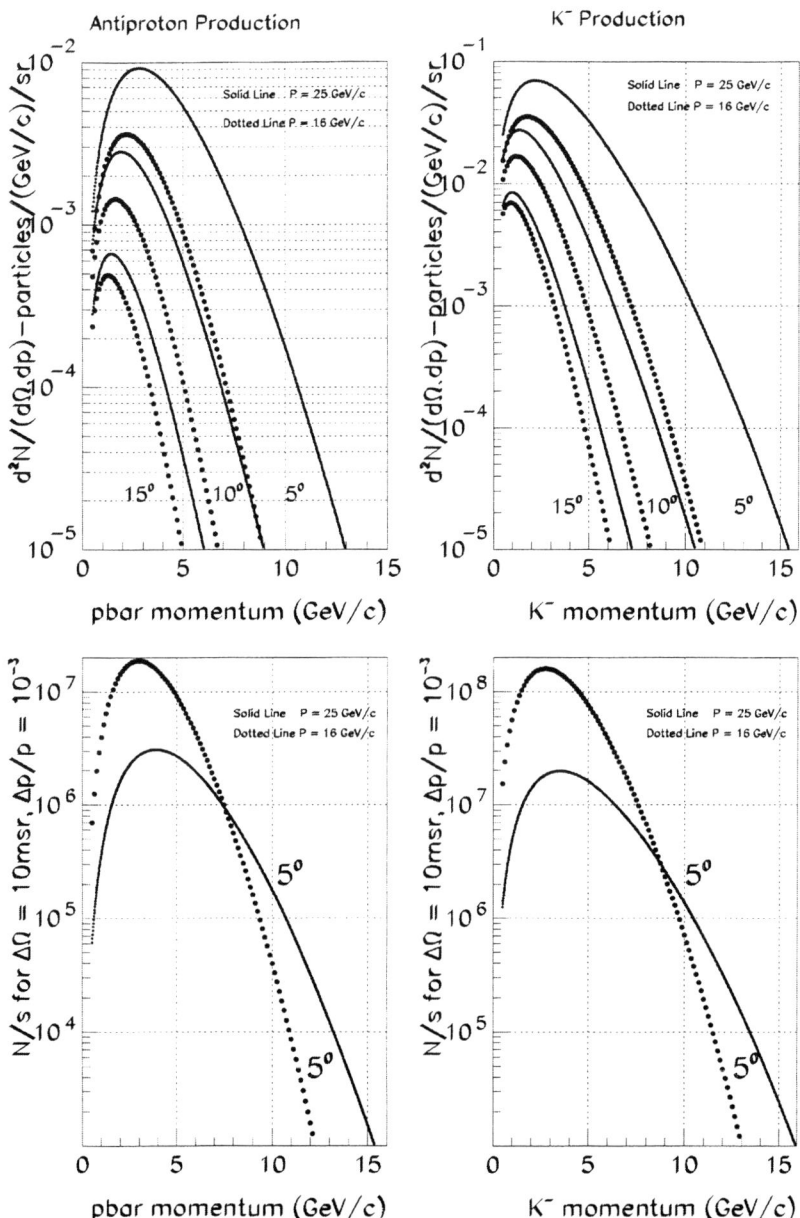

FIGURE 1. Top: K^- and $\bar{p}$ production cross sections per interacting primary proton per steradian per GeV/c momentum bite. Bottom: K^- and $\bar{p}$ intensities in particles/sec in a channel with an acceptance of $\Delta\Omega = 10$ msr and $\Delta p/p = 10^{-3}$.

TABLE 1. Calculated K^- and $\bar{p}$ production characteristics for 25 and 16 GeV/c primary proton beams. In items 5, 6 and 7, primary proton intensities of item 4 have been assumed and K^- and $\bar{p}$ are accepted into a channel at 5° with $\Delta\Omega - 10$ msr, $\Delta p/p = 10^{-3}$.

K^- and $\bar{p}$ production	Proton Momentum = 25 GeV/c		Proton momentum = 16 GeV/c	
	K^-	$\bar{p}$	K^-	$\bar{p}$
1. $F_{max} = \left(\frac{d^2N}{d\Omega dp}\right)$ per proton	70×10^{-3}	9.2×10^{-3}	35×10^{-3}	3.6×10^{-3}
2. p(GeV/c) at max	2.2	2.8	1.8	2.2
3. p(GeV/c) at $F = 10^{-4}$	13.1	10.6	9.2	7.1
4. Proton Intensity (p/s)	2×10^{13}		4×10^{14}	
5. Max intensity (p/s)	20×10^6	3.1×10^6	160×10^6	18.8×10^6
6. p(GeV/c) at max	3.5	3.9	2.8	3.0
7. p range (GeV/c) at $\geq 10^6$/s	0.5 - 10.5	1.6 - 7.4	0.14 - 9.7	0.6 - 7.5

largely disappears in the lower regions of K^- and $\bar{p}$ momenta. In Fig. 1 we also plot the expected intensities with 50% interacting protons into a hypothetical beam line for K^- and $\bar{p}$ at 5°, with a solid angle $\Delta\Omega = 10$ msr and a momentum bite of $\Delta p/p = 10^{-3}$. If we arbitrarily designate K^- or $\bar{p}$ intensities greater than 10^6/s as the useful intensities, the beam line for the 16 GeV/c protons turns out to be superior by larger factors to the beam line for the 25 GeV/c protons for both K^- and $\bar{p}$ over almost the entire range of useful momenta. These results are summarized in Table 1.

III THE OVERLAP

During the last two years several laboratories have developed plans forhigh intensity proton accelerators in the range of 10 GeV/c - 50 GeV/c. Several conferences, workshops, and planning group meetings have been held. The proceedings, numerous and voluminous, attest to the large range of exciting physics that can be addressed by such machines. Notable among these are:

BNL:	30 GeV AGS	- Proc. AGS 2000 (1996) [3]
KEK:	50 GEV JHP	- Proc. Workshop JHP (1997) [4]
GSI:	≤ 50 GeV	- Reports of the Working Group on Hadron Physics [5]
RCNP:		- Proc. QULEN '97 [6]
INDIANA:	≤ 20 GeV	- LISS - Physics Motivation and Design Study '96 [7]

In addition, there is the whole body of somewhat outdated literature associated with the ill-fated LAMPF-II and KAON.

The proposed proton source at the muon collider overlaps in its physics capabilities with some of the above real and imaginary projects. A perusal of these reports is instructive. The hadronic physics topics discussed in these workshops extend over a wide range:

- Proton exclusive scattering and polarization
- Proton-proton and Proton-antiproton forward scattering
- Proton form factors
- Hypernuclear physics
- Baryon spectroscopy
- Rare η decays and symmetry tests
- Light quark spectroscopy
- Charmonium spectroscopy
- CP violation
- Exotic meson and glueball searches

The range is too wide. I will therefore discuss only a select few topics with protons and antiprotons, since physics with kaons has been discussed at length elsewhere in this workshop. The choice is strictly personal.

IV EXPERIMENTS WITH PROTON BEAMS

Here I have three different investigations to suggest.

A PQCD scaling phenomena

Brodsky and Farrar, and others [8] have pointed out that in exclusive reactions

$$A + B \to C + D$$

the differential cross sections for hard scattering (large pt) should scale simply as:

$$d\sigma/dt \propto s^{-n_A+n_B+n_C+n_D-2}$$

where $\sqrt{s}$ is the cm energy, and $n_{A,B,C,D}$ are the number of elementary constituents (quarks, leptons, photons) in A,B,C,D. For pp elastic scattering the predicted scaling law $d\sigma/dt \propto s^{-10}$ holds remarkably well in the range s =15-50 GeV^2 (see Fig. 2). A closer examination by Pire and Ralston [9] has shown more interesting behavior, an oscillatory behavior of $R = (d\sigma/dt)/s^{-10}$ with respect to $\ln s$ (see Fig. 3). This behavior is ascribed to interference between the quark counting amplitude and the Landshoff

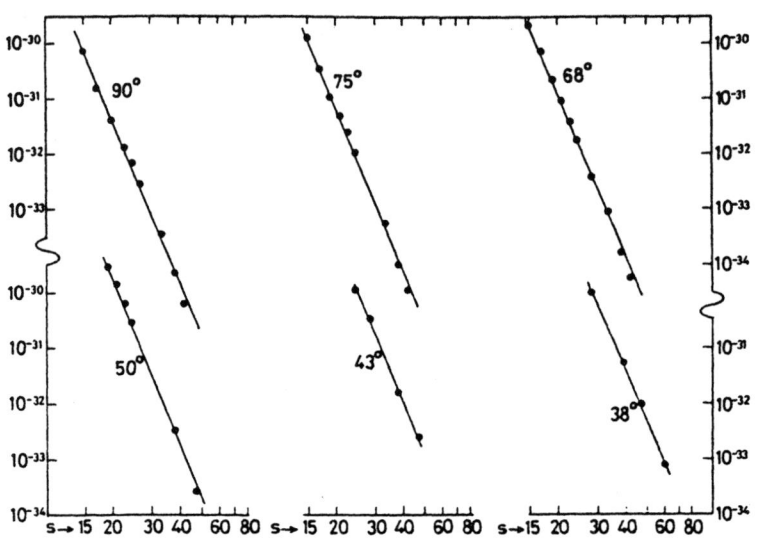

FIGURE 2. Illustrating the s^{-10} dependence of pp elastic scattering cross sections at large angles.

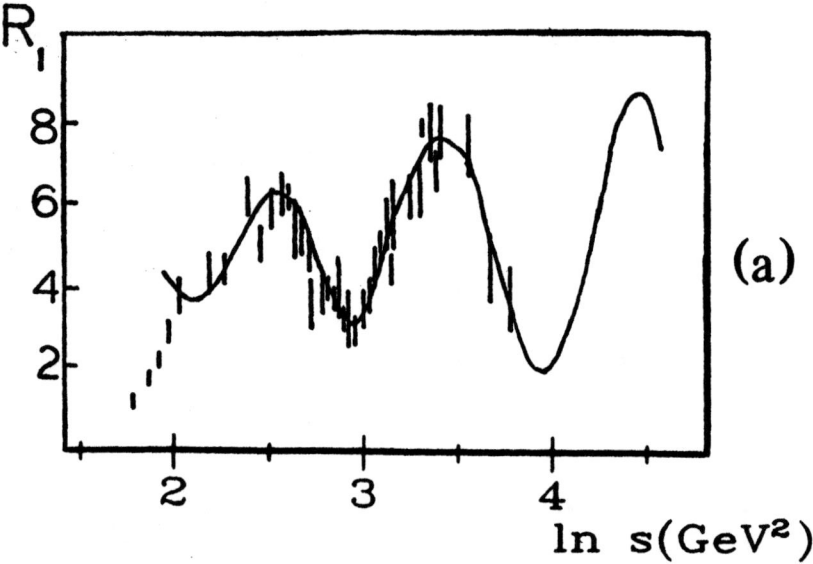

FIGURE 3. Illustrating the oscillations observed in the $(d\sigma(90°)/dt)/s^{-10}$ vs $\ln s$.

amplitude (as corrected for chromo-Coulomb phase). These are extremely interesting phenomena, because they touch on some of the most basic aspects of QCD and nucleon structure. It is very important to establish their validity. Unfortunately, a flag of caution has to be raised about the existing data. The data presented in Figs. 2 and 3 come from an assortment of different measurements, extrapolated and interpolated. This has perhaps little effect on the general idea of a power-law scaling, but detailed things like the Pire-Ralston oscillations and their explanation depend crucially on the quality of data. What is needed is precision measurements of large angle pp elastic scattering in the incident momentum range of about 10 GeV/c to the maximum possible. Measurements should also be made for elastic pd scattering. An s^{-13} scaling is expected in this case, and it will become possible to look into Pire-Ralston oscillations up to s=60 GeV^2.

B Color transparency

The phenomenon of color transparency [10] is related to the ideas underlying scaling. The basic idea is that in hard scattering of a hadron on a nucleon in the nucleus the colored constituents of the nucleon (quarks) must be very close together and therefore color should be greatly screened. This should result in a much reduced cross section for the scattering of the hadron from a nuclear proton compared to that from a free proton, i.e., the hadron should find the nucleus quite transparent. Inspite of the fact that basic ideas of color transparency are firmly rooted in our present understanding of QCD, the phenomenon has not been experimentally confirmed in an unambiguous manner. Only one experiment [11] has ever claimed the observation of color transparency. Several subsequent measurements have failed to do so. What is needed is precision measurements of (p,p), (p,2p) cross section for a variety of nuclei over an extended range of incident momenta and momentum transfers.

C Parity non-conservation

In strong interactions parity non-conservation arises due to the presence of the weak coupling PNC vertex, as shown in Fig. 4. The PNC vertex is expected to give rise to very small asymmetries ($\approx 10^{-7}$) in longitudinal pp cross sections, $A_z = (\vec{\sigma} - \overleftarrow{\sigma})/(\vec{\sigma} + \overleftarrow{\sigma})$.

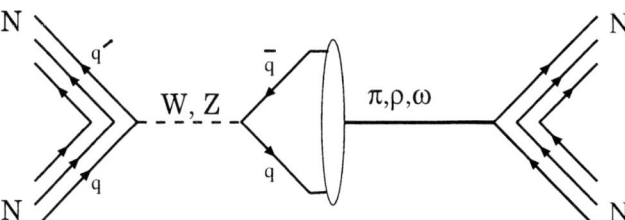

FIGURE 4. Illustrating the parity non-conserving vertex in nucleon-antinucleon scattering.

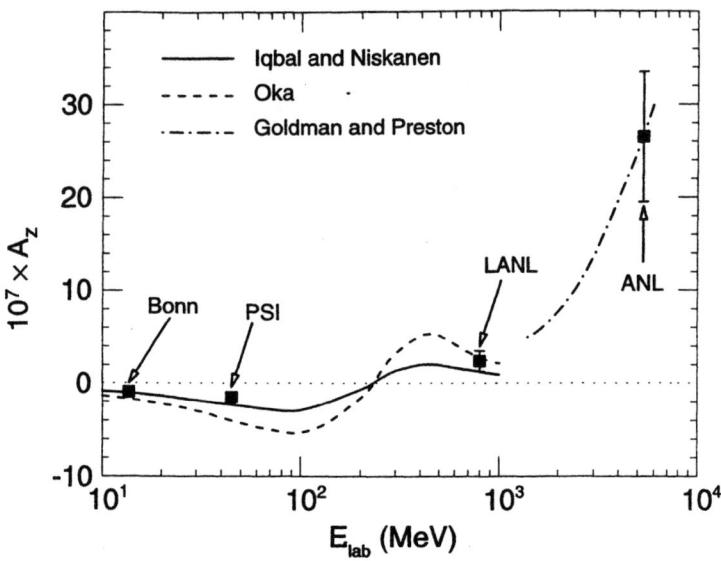

FIGURE 5. Results of the existing measurements of asymmetry in the scattering of longitudinally polarized protons.

All measurements below 1 GeV indeed give $A_z \leq 2 \times 10^{-7}$. However, the only existing measurement above 1 GeV, made at the ZGS at 5 GeV [12], gives $A_z = (26.5 \pm 6.0) \times 10^{-7}$, in severe disagreement with all other measurements, and with conventional theory (see Fig. 5). Of course, exotic theoretical models have been proposed to accommodate this observation, but what is really needed is an independent experimental check of the Argonne result. A program of measurements of A_z in the entire range 1 GeV/c to 16 GeV/c would make extremely valuable contribution to the understanding of the role of weak interactions in the hadronic sector.

V EXPERIMENTS WITH ANTIPROTONS

Antiproton physics at the FMC can be done either with extracted beams or with stored beams. I have already shown that the facility can provide extracted beams of between $10^6 - 10^7 \bar{p}/s$ in the momentum range of 0.2 to 7.0 GeV/c in a rather high resolution channel with an acceptance of $\Delta\Omega = 10$ msr, $\Delta p/p = 10^{-3}$. The extracted beam line will have to have highly efficient RF separators in order to remove π^- and K^- contaminations. The other approach, which is to be much preferred, is to extract $\bar{p}$ at the peak production momentum of ~ 3 GeV/c into a large acceptance debuncher ring, cool them, and store them in an accumulator ring. Such a facility would look very much like the present antiproton source at Fermilab, except that the present facility stores at 9 GeV/c and decelerates to momenta down to 3 GeV/c, whereas the FMC antiproton complex would store at 3 GeV/c and accelerate to momenta up to 9 or 10 GeV/c or decelerate down to very low energies. Such an antiproton facility would be absolutely unique in the world, and it would provide the opportunity to do a broad variety of very interesting physics, assuming, of course that it will be equipped with state of the art detectors.

Among the physics programs that can be pursued with such a facility are antihydrogen physics (e.g., time reversal invariance tests) and charmonium physics, both of which are so well known that I will not dwell on them here. Let me, instead, describe some of the less well known problems of fundamental importance which can be addressed at this facility.

A Nucleon-antinucleon interaction

Understanding of nucleon-nucleon interaction is the paramount objective of nuclear physics. With the advent of QCD it has become obvious that the NN and $N\overline{N}$ interactions have to be studied and understood within a common framework, and over a larger range of energies than before. The realization of this goal has been seriously compromised by the paucity of precision data. The problem is best illustrated by one of the most important parameters of scattering theory, the real part of the forward elastic scattering amplitude, or equivalently the ρ-parameter, defined as $\rho = (Ref(0))/(Imf(0))$. The sparse nature and poor quality of both $\rho(pp)$ and $\rho(\bar{p}p)$, particularly below 10 GeV/c, are illustrated in Fig. 6. Recently, it has been demonstrated that very high precision measurements of ρ can be made by measuring recoils near 90° rather than scattering near 0°. [13,14] See Fig. 7. The technique is particularly well suited for measurements with a gas-jet target in a stored beam. If the accumulator ring described earlier were available both $\rho(pp)$ and $\rho(\bar{p}p)$ could be measured over the entire range of energies with unparalleled precision. Such measurements would make an important contribution to the understanding of the seemingly simple, but in fact most complicated of reactions, elastic scattering.

B Proton form factors

The study of proton form-factors for space-like momentum transfers ($Q^2 \geq 0$) by elastic scattering of electrons, done mainly at SLAC, has revealed a precocious PQCD behavior at momentum transfers as small as $Q^2 \approx 10$ (GeV/c)^2; the quantity $Q^4 G_M/\mu_p$ appears to scale as α_s^2.(see Fig. 8) This has posed the serious question whether these observations herald the inauguration of the PQCD regime, or whether they have some other, more mundane, explanation. One way of shedding new light on this question is to study the behavior of the magnetic form factor GM in a different regime, i.e., for timelike momentum transfers, via the reaction $\bar{p}p \rightarrow e^+ e^-$. Recently the E760/E835 experiment at Fermilab has attempted to do so, with even more intriguing results [15]. Even with measurements of limited precision (the count rates are extremely low) they have shown that from $Q^2 \approx$ 9-13 (GeV/c)^2, $Q^4 G_M/\mu_p$ again scales as α_s^2, but at twice

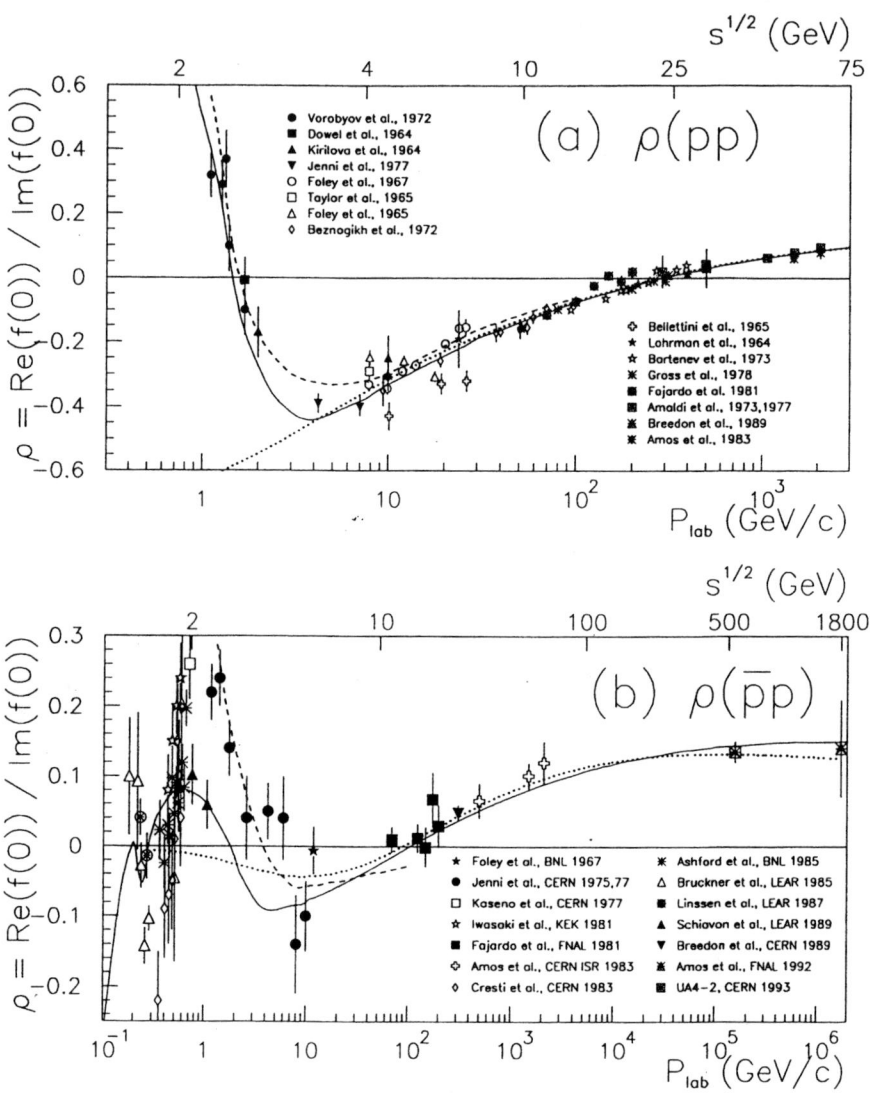

FIGURE 6. Results for the ρ parameter from the literature. (a) pp scattering (b) $\bar{p}p$ scattering. The curves refer to various theoretical attempts to explain the data. (from Ref. [13])

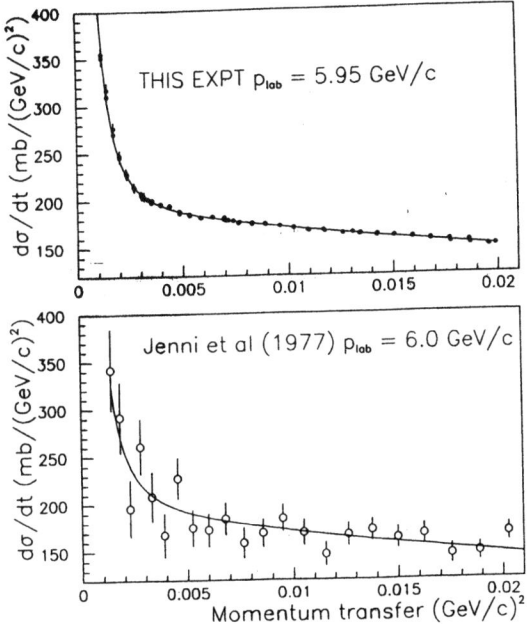

FIGURE 7. $\bar{p}p$ forward scattering cross sections as a function of momentum transfer $|t|$; (top) by measuring proton recoils near 90°, (bottom) by measuring antiprotons scattered near 0°. (t = 0.002 (GeV/c)^2 corresponds to a scattering angle of 0.2°. [13].

the magnitude for spacelike momentum transfers.(Fig. 9) This is difficult to understand. Somewhat exotic diquark-quark models for the structure of the proton have been proposed to explain this observation, but no entirely satisfactory explanation has emerged so far.

In order to really understand the evolution of the timelike form factor we need precision measurements of $\bar{p}p \to e^+ e^-$ from the threshold ($Q^2 = 3.52$ (GeV/c)^2), all the way up. The existing data from 4-5 (GeV/c)^2 are of poor quality, and above 5 (GeV/c)^2 they are essentially non-existent. This situation needs to be rectified. The questions that these data potentially address are at the very heart of QCD - where does the PQCD regime start?

C Light quark structures and searches for exotics

The Crystal Barrel collaboration at LEAR (CERN) has convincingly demonstrated the great versatility of annihilation in populating light quark mesons. Their measurements were primarily devoted to annihilation at rest and therefore limited to mesons of masses less than about 1.5 GeV/c^2. In this domain they made several important discoveries. E760 at Fermilab demonstrated that the success of $\bar{p}p$ annihilations in reaching such structures could be extended to higher masses by annihilations in flight. They identified

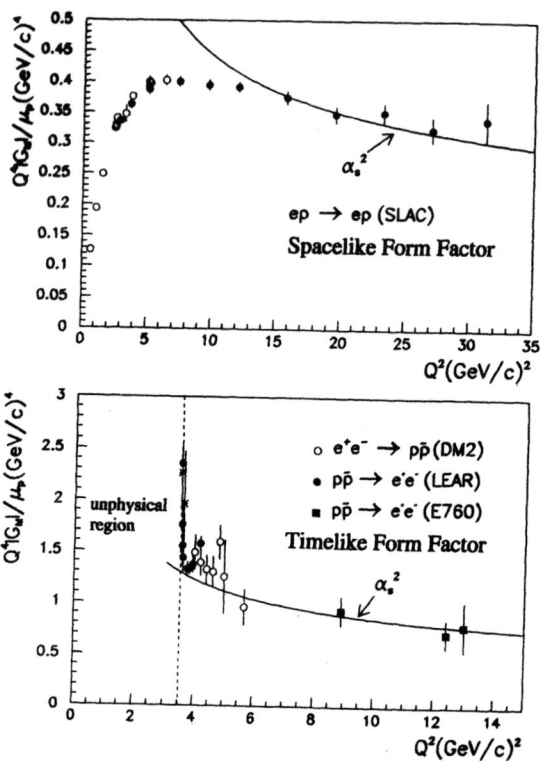

FIGURE 8. Proton form factors: (top) for spacelike momentum transfers, (bottom) for timelike momentum transfers. (based on Ref. [15])

several new mesons but could not determine their spin-parity. [16] This was primarily due to the fact that the E760 detector was far from hermatic. It is clear that with an improved detector at the FMC antiproton facility significant contributions can be made to light quark spectroscopy.

As has been noted before, $\bar{p}p$ annihilations are particularly suited for the production and formation of exotics, glueballs, hybrids, and dimesons (see Fig. 9). These are among the most unique predictions of QCD, and their identification would be a most significant confirmation of QCD. At the $\bar{p}$ facility described here the search for these objects should be a major goal.

1 Glueballs

One of the most important discoveries by Crystal Barrel was $f_0(1500)$ which is currently one of the prime candidates for the scalar glueballs. [17] The alternate candidate, $f_J(1710)$, appears to be out of the reach for annihilation at rest measurements. At the

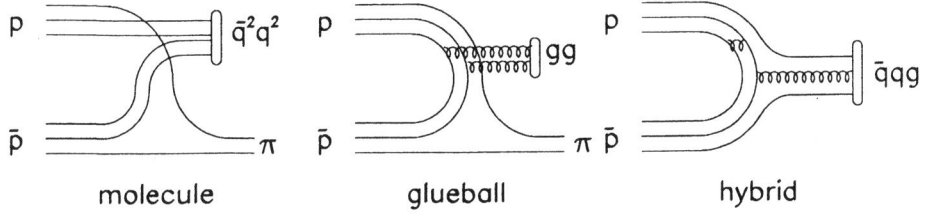

FIGURE 9. Illustrating the versatility of $\bar{p}p$ annihilation in populating exotics.

FMC facility, one can make a serious effort to study the nature of $f_J(1710)$ in a production experiment because of the higher available energy.

The tensor glueball offers even a greater challenge. The extremely narrow ($\Gamma \sim 20$ MeV) $\xi(2230)$, which was recently revived by measurements reported from BEPC (Beijing), [18] appears to have again fallen in disfavor, [19] and a renewed search for the 2^{++} glueball is called for. Lattice calculations predict the mass of the 2^{++} tensor glueball to be ≈ 2300 MeV, and it should be accessible both in formation and production experiments at the FMC.

2 *Hybrids*

These structures with gluonic excitation, , have received great publicity recently, because they can have some very special characteristics which should lead to their unambiguous identification. One of these characteristics is that hybrids can have $J^{PC} = 0^{+-}, 1^{-+}, 2^{+-}$ ..., which are forbidden for $q\bar{q}$ mesons, making them manifestly exotic. Model calculations predict that the lowest u, d $\equiv$ (n) quark hybrids have masses $m(n\bar{n}g) \approx 1.9$ GeV, the strange quark hybrids have $m(s\bar{s}g) \approx 2.2$ GeV and the charm quark hybrids have masses $m(c\bar{c}g) \approx 4.2$ GeV.

Recently the first such exotic meson has been sighted. In a $\pi^- p \to p + (\eta \pi^-)$ measurement done at Brookhaven with 18 GeV/c pions, our group has established [20] the existence of a resonance structure with a mass of 1370 ± 16 MeV, width of 385 ± 40 MeV, and $J^{PC} = 1^{-+}$. This is an extremely exciting discovery; it even met the non-scientific criteria of mention in the NY Times, Scientific American, etc. It gives us hope that the higher mass hybrids may also be found by appropriate measurements.

For the charmed quark hybrids ($c\bar{c}g$) the preferred mode of production has to be $\bar{p}p$ annihilation. In fact it is in order to ensure that we can reach $m = 4.2$ GeV, that I indicated that our $\bar{p}$ accumulator booster should be able to go up to 10 GeV/c.

In order to search for strange quark hybrids ($s\bar{s}g$), the choice reaction would be $K^- p \to \Lambda + K_1 \overline{K}$. I have shown in Sec. 2 that we will indeed have rather copious K^- beams, and the measurements can be successfully done at the FMC secondary beams channel.

REFERENCES

1. J.R. Sanford and C.L. Wang, Brookhaven National Laboratory, AGS Internal Report, 1967 (unpublished)
2. C.L. Wang, Brookhaven National Laboratory Report, BNL 22452 (1977).
3. "AGS 2000: Experiments for the 21st century", ed. by L. Littenberg and J. Sandweiss, BNL 52512, 1996.
4. "Proc. International Workshop on Physics with 50 GeV PS", ed. by T. Fukuda et al., JHP Supp.-18 (1996); "Proc. INS Workshop on Hadron Physics at 50 GeV PS," ed. by H. Hamagaki and K. Imai, JHP Supp.-24 (1997).
5. "Reports of the Working Groups on Hadron Physics at GSI 1997-98" (unpublished).
6. "Proc International Conference on Quark Lepton Nuclear Physics", Osaka, 1997 (in press).
7. "Light-Ion Spin Synchrotron," Indiana University Report (1996) (unpublished).
8. S. Brodsky and G. Farrar, *Phys. Rev. Lett.* **31** (1973) 1153, V. Matueev et al., *Lett. Nuovo Cimento* 7 (1972) 719.
9. B. Pire and J.P. Ralston, *Phys. Lett.* **117B** (1982) 233; J.P. Ralston and B. Pire *Phys. Rev. Lett.* **49** (1982) 1605.
10. S.J. Brodsky and A.H. Mueller, *Phys. Lett.* **B206** (1988) 685.
11. A.S. Carroll et al., *Phys. Rev. Lett.* **61** (1988) 1698.
12. N. Lockyer, *Phys. Rev.* **D30** (1984) 860.
13. S. Trokenheim, Ph.D. dissertation, Northwestern University (1995) unpublished.
14. T. A. Armstrong et al., (E760 Collaboration), *Phys. Lett.* **B385** (1996) 479.
15. T. A. Armstrong et al., (E760 Collaboration), *Phys. Rev. Lett.* **70** (1993) 1212.
16. T. A. Armstrong et al., (E760 Collaboration), *Phys. Lett.* **B307** (1993) 394.
17. C. Amsler and F. E. Close, *Phys. Lett.* **B353** (1995) 385.
18. J. Z. Bai et al., (BES Collaboration), *Phys. Rev. Lett.* **76** (1996) 3502.
19. K.K. Seth (Crystal Barrel Collaboration), "Proc 22nd Rencontres de Moriond '97 QCD and High Energy Hadronic Interactions," ed. by J. Tran Thanh Van, Editions Frontiers (Paris, 1997) p. 599-606.
20. D. R. Thompson et al., (E852 Collaboration), *Phys. Rev. Lett.* **79** (1996) 1630.

Constraints on Strong Dynamics from Rare B and K Decays

Gustavo Burdman

Department of Physics, University of Wisconsin, Madison WI 53706.

Abstract. We discuss the constraints from rare B and K decays on the Electroweak Symmetry Breaking (EWSB) sector, as well as on theories of fermion masses. We focus on models involving new strong dynamics and show that transitions involving Flavor Changing Neutral Currents (FCNC) play an important role in disentangling the physics in these scenarios. In a model-independent approach to the EWSB sector, the information from rare decays is complementary to precision electroweak observables in bounding the contributions to the effective lagrangian. We compare the pattern of deviations from the Standard Model (SM) that results from these sources, with the deviations associated with the mechanism for generating fermion masses.

INTRODUCTION

Two of the most intriguing questions in particle physics are the the EWSB mechanism and the origin of fermion masses. Although the SM remains a successful theory when compared with all the available data, it lacks predictability in the Higgs sector, which determines the masses of gauge bosons, as well as of fermions through *ad hoc* Yukawa couplings. This suggests the possibility that new physics beyond the SM might be associated with either of these questions. In general, the energy scales and dynamics behind the EWSB sector and the fermion masses may be unrelated. In order to avoid fine-tuning, the scale associated with EWSB cannot be much higher than a few TeV, whereas the scales where light fermion masses are generated could be much higher. If the mechanism responsible for the breaking of the electroweak symmetry involves some new strong dynamics, deviations from the SM might be observable in low energy signals even at energies much smaller than the scale of new physics Λ. Reaching this new frontier by direct observation of new physical states or even of tree-level effects in the couplings of SM particles, may require not only very large energies but also some previous knowledge of what (and what not) to expect. Thus, low energy measurements might be of paramount importance in planning experiments and search strategies at high energy machines.

Among these low energy signals are electroweak measurements such as those at LEP and the Tevatron. On the other hand, processes involving Flavor Changing Neutral Currents (FCNC) can play a complementary role, since the fact that these processes are largely suppressed or forbidden in the SM may compensate the suppression by factors of m/Λ (with m the low energy scale, e.g. m_K, m_B, etc.). Here we address the potential of rare B and K decays as a complement to other low energy measurements in constraining models where strong dynamics is associated to either the EWSB sector and/or the origin of fermion masses. In the absence of a completely satisfactory theory of dynamical symmetry breaking and fermion masses, it is convenient to carry out a model-independent analysis that makes maximum use of the known properties of the electroweak interactions. This is the case with the EWSB sector, where an effective lagrangian approach allows us to parameterize the effects of the new strong dynamics in very much the same way chiral perturbation theory parameterizes low energy QCD. On the other hand, the effects from fermion mass generation can also be addressed by a general operator analysis. However, in addition, most theories predict the existence of relatively light states (scalars, pseudo-Goldstone bosons, etc.) which generally couple to mass in one way or another. To exemplify the effects of such states (which cannot be integrated out) we work with a particular set of models known as Topcolor-assisted Technicolor (TaTC). This provides a current example of how strong dynamics model building deals with the large top-quark mass and illustrates the distinct low energy phenomenology emerging from non-standard EWSB scenarios.

LOW ENERGY EFFECTS OF ELECTROWEAK SYMMETRY BREAKING

In the absence of a light Higgs boson the symmetry breaking sector is represented by a non-renormalizable effective lagrangian corresponding to the non-linear realization of the σ model. The essential feature is the spontaneous breaking of the global symmetry $SU(2)_L \times SU(2)_R \to SU(2)_V$. To leading order the interactions involving the Goldstone bosons associated with this mechanism and the gauge fields are described by the effective lagrangian [1]

$$\mathcal{L}_{LO} = -\frac{1}{4} B_{\mu\nu} B^{\mu\nu} - \frac{1}{2} \text{Tr}\left[W_{\mu\nu} W^{\mu\nu}\right] + \frac{v^2}{4} \text{Tr}\left[D_\mu U^\dagger D^\mu U\right], \tag{1}$$

where $B_{\mu\nu}$ and $W_{\mu\nu} = \partial_\mu W_\nu - \partial_\nu W_\mu + ig[W_\mu, W_\nu]$ are the the $U(1)_Y$ and $SU(2)_L$ field strengths respectively, the electroweak scale is $v \simeq 246$ GeV and the Goldstone bosons enter through the matrices $U(x) = e^{i\pi(x)^a \tau_a / v}$. The covariant derivative acting on $U(x)$ is given by $D_\mu U(x) = \partial_\mu U(x) + ig W_\mu(x) U(x) - \frac{i}{2} g' B_\mu(x) U(x) \tau_3$. To this order there are no free parameters once the gauge boson masses are fixed. The dependence on the dynamics underlying the strong symmetry breaking sector appears at next to leading order. To this order, a complete set of operators includes one operator of dimension two and nineteen operators of dimension four [1,2]. The effective lagrangian to next to leading order in the basis of Ref. [1] is given by

$$\mathcal{L}_{\text{eff.}} = \mathcal{L}_{LO} + \sum_{i=0}^{19} \alpha_i \mathcal{O}_i , \qquad (2)$$

where $\mathcal{O}_0$ is a dimension two custodial-symmetry violating term absent in the heavy Higgs limit of the SM. If we restrict ourselves to CP invariant structures, there remain fifteen operators of dimension four. The coefficients of some of these operators are constrained by low energy observables. For instance precision electroweak observables constrain the coefficient of $\mathcal{O}_0$, which gives a contribution to the electroweak parameter T. The 3 σ limit requires

$$\alpha_0 < 6 \times 10^{-3}. \qquad (3)$$

The combinations $(\alpha_1 + \alpha_8)$ and $(\alpha_1 + \alpha_{13})$ contribute to the electroweak parameters S and U. For instance, the constraint on S translates into

$$|\alpha_1 + \alpha_{13}| < 1.5 \times 10^{-2}. \qquad (4)$$

In addition, the coefficients α_2, α_3, α_9 and α_{14} modify the triple gauge-boson couplings (TGC) and will be probed at LEPII and the Tevatron at the few percent level [3].

The remaining operators contribute to oblique corrections only to one loop and, in some cases, only starting at two loops. To the last group belong $\mathcal{O}_{11}$ and $\mathcal{O}_{12}$ given that their contributions to the gauge boson two-point functions only affect the longitudinal piece of the propagators. Of particular interest is the operator $\mathcal{O}_{11}$ defined by [1]

$$\mathcal{O}_{11} = \text{Tr}\left[(\mathcal{D}_\mu V^\mu)^2\right], \qquad (5)$$

with $V_\mu = (D_\mu U)U^\dagger$ and the covariant derivative acting on V_μ defined by $\mathcal{D}_\mu V_\nu = \partial_\mu V_\nu + ig[W_\mu, V_\nu]$. The equations of motion for the $W_{\mu\nu}$ field strength imply [4]

$$\mathcal{D}_\mu V^\mu = \frac{2i}{v^2} \mathcal{D}_\mu J_w^\mu , \qquad (6)$$

where the $SU(2)_L$ current is $J_w^\mu = \sum_\psi \left(\bar{\psi}_L \gamma^\mu \frac{\tau^a}{2} \psi_L\right)\tau^a$, ψ_L denote the left-handed fermion doublets. The dominant effect appears in the quark sector due to the presence of terms proportional to m_t. After the quark fields are rotated to the mass eigenstate basis, the operator $\mathcal{O}_{11}$ can be written as [5]

$$\mathcal{O}_{11} = \frac{m_t^2}{v^4} \left\{ (\bar{t}\gamma_5 t)^2 - 8 \sum_{i,j} V_{ti}^* V_{tj} (\bar{q}_{iL} t_R)(\bar{t}_R q_{jL}) \right\} + \ldots \quad (7)$$

where $i, j = d, s, b$, the V_{ti} are Cabibbo-Kobayashi-Maskawa (CKM) matrix elements and the dots stand for terms suppressed by small fermion masses.

From the above discussion we see that the leading effects of the EWSB sector in FCNC processes are coming from the insertion of anomalous TGC vertices and four-fermion operators like (7). In the rest of this section, we review the status and future impact of these constraints on the symmetry breaking sector.

Four-fermion Operators

The effects of the four-fermion operators in (7) in rare B and K decays were considered in Ref. [6]. The loop insertion will result in contributions to several FCNC processes, that are controlled by both the coefficient α_{11} of the effective lagrangian (2) as well as by the high energy scale Λ. To one loop, only one parameter is needed, namely

$$y = \alpha_{11} \log \frac{\Lambda^2}{m_t^2} . \quad (8)$$

This parameter also governs the contributions of (7) to other neutral processes, both flavor changing and flavor conserving. For instance, the $(\bar{b}_L t_R)(\bar{t}_R b_L)$ term in (7) gives a contribution to $Z \to b\bar{b}$, whereas the terms like $(\bar{b}_L t_R)(\bar{t}_R d_L)$ appear in $B^0 - \bar{B}^0$ mixing [5]. Thus the measurements of R_b and the rate of B mixing (together with all other CKM information) can be used to derive a bound on y. Although the bound carries some uncertainty mainly associated with CKM quantities like f_B and V_{ub}, we will take it to be, approximately [5,6]

$$|y| < 0.50 . \quad (9)$$

Next, we use this as the allowed range for y in order to explore the possible impact of this physics in rare B and K decays. The one-loop insertion of the terms

$$\mathcal{O}_{11} = -\frac{8m_t^2}{v^4} \left\{ V_{ts}^* V_{tb} \bar{s}_L t_R \bar{t}_R b_L + V_{td}^* V_{tb} \bar{d}_L t_R \bar{t}_R b_L + V_{td}^* V_{ts} \bar{d}_L t_R \bar{t}_R s_L \right\} + \ldots \quad (10)$$

induces new contributions to various FCNC vertices in B decays (the first two terms in (10)), as well as in K decays (third term in (10)).

First, let us consider $b \to q\gamma$ processes leading, for instance, to the inclusive $B \to X_s \gamma$, since this rate has been recently measured [7]. The one-loop insertion of the operator $\mathcal{O}_{11}$ does not give a contribution to these processes given

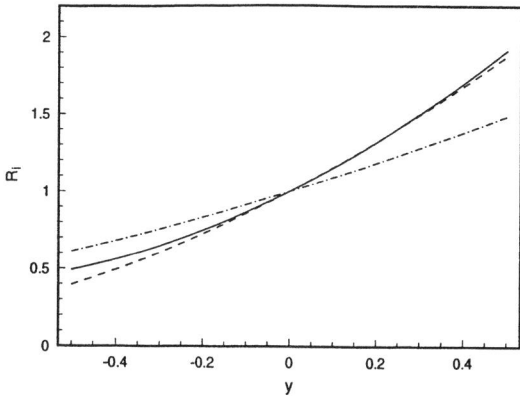

FIGURE 1. Ratio of the modified branching ratio to the standard model expectation as a function of $y = \alpha_{11} \log \frac{\Lambda^2}{m_t^2}$. The solid line corresponds to the ratio R_ℓ for $B \to X_{(s,d)} \ell^+ \ell^-$ inclusive decays, the dashed line to R_ν for $B \to X_{(s,d)} \nu \bar{\nu}$ and the dot-dashed line to R_g for $b \to s\bar{s}s$ decays. From Ref. [6].

that it does not mix with the operator $\bar{s}_L \sigma_{\mu\nu} b_R$ responsible for the on-shell photon amplitude. Mixing only occurs at two loops, when QCD corrections are taken into account. As a result the effect, in all $b \to q\gamma$ transitions is expected to be only a few percent of the SM branching ratios [6].

On the other hand, the off-shell amplitudes for photons, Z's and gluons are non-zero at one loop. They generate contributions to processes such as $b \to q\ell^+\ell^-$, $b \to q\nu\bar{\nu}$, $b \to q\bar{q}'q'$; as well as to similar rare kaon decays like $s \to d\nu\bar{\nu}$, etc. In order to asses the potential effects we define

$$R_\ell \equiv \frac{Br(B \to X_{(s,d)} \ell^+ \ell^-)}{Br(B \to X_{(s,d)} \ell^+ \ell^-)_{\text{SM}}}, \quad (11)$$

which is plotted in Fig. 1 as a function of the parameter y defined in (8), for the allowed range of y (9). Analogously, we can define the ratio R_ν, which tracks the effects in $B \to X_{s,d} \nu \bar{\nu}$ decays; whereas the contribution to gluon penguin processes such as $b \to s\bar{s}s$ is represented by the ratio R_g. As it is clear from Fig. 1, the effects of the operator $\mathcal{O}_{11}$ are very similar in all three types of B decays.

We see that, even with the R_b and $B^0 - \bar{B}^0$ mixing constraints, large deviations from the SM predictions for these modes are possible. The current experimental bounds on these processes are still not binding on y. However, sensitivity to SM branching ratios will be reached in the next round of experiments at the various B factories at Cornell, KEK, SLAC and Fermilab. The distinct feature of this effect is that no significant deviation is expected in $b \to s\gamma$, even when large deviations are observed in all the other modes.

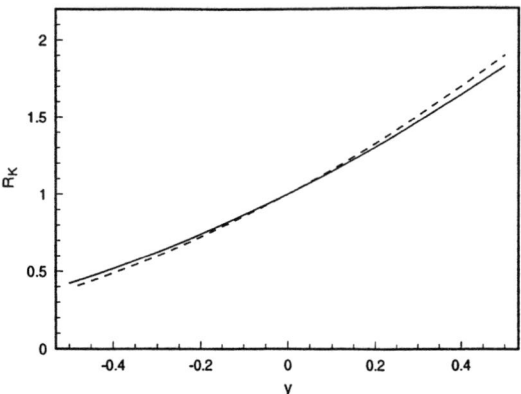

FIGURE 2. Ratio of the modified branching ratio to the standard model expectation for $K^+ \to \pi^+ \nu \bar{\nu}$ (solid line) and $K_L \to \pi^0 \nu \bar{\nu}$ (dashed line). From Ref. [6].

The effects are very similar in rare K decays such as $K^+ \to \pi^+ \nu \bar{\nu}$ and $K_L \to \pi^0 \nu \bar{\nu}$, etc. In Fig. 2 we plot R_K, a quantity analogous to R_ℓ in (11). Again, large effects of up to factors of 2 deviations, are allowed. The recently reported [8] observation of one event in $K^+ \to \pi^+ \nu \bar{\nu}$ roughly translates into $R_K \simeq (0.50 - 5.0)$, which is still not constraining.

Although in this model-independent approach we cannot, as a matter of principle, calculate the size of the coefficients α_i, we can use general arguments to estimate their approximate value. Using naive dimensional analysis [9] we have

$$\alpha_{11} \simeq \mathcal{O}(1) \times \frac{v^2}{\Lambda^2}, \tag{12}$$

with the scale of new physics obeying $\Lambda \lesssim 4\pi v$. For instance, taking $\Lambda = 4\pi v$, one would obtain $y \simeq \mathcal{O}(1) \times 0.04$. On the other hand, if $\Lambda = 2\pi v$, one has $y \simeq \mathcal{O}(1) \times 0.12$. In any case, these are meant to be order of magnitude estimates. Therefore, the experimental relevance of the effect strongly depends on details of the dynamics we are not able to compute in a model-independent fashion.

Triple Gauge-boson Couplings

Imposing C and P conservation, the most general form of the WWV ($V = \gamma, Z$) couplings can be written as [3]

$$\mathcal{L}_{WWV} = g_{WWN} \left\{ i\kappa_N W_\mu^\dagger W_\nu N^{\mu\nu} + ig_1^N \left(W_{\mu\nu}^\dagger W^\mu N^\nu - W_{\mu\nu} W^{\dagger\mu} N^\nu \right) \right. \tag{13}$$

$$\left. + i\frac{\lambda_N}{M_W^2} W_{\mu\nu}^\dagger W_\nu^\mu N^{\nu\lambda} \right\}, \tag{14}$$

with the conventional choices being $g_{WW\gamma} = -e$ and $g_{WWZ} = -g\cos\theta$ [10]. In principle, there are six free parameters. Making contact with the electroweak lagrangian (2), these parameters can be expressed in terms of the next-to-leading order coefficients [4] $\alpha_1, \alpha_2, \alpha_3, \alpha_8, \alpha_9, \alpha_{13}$ and α_{14}. Conservation of the electromagnetic charge implies $g_1^\gamma = 1$. Furthermore, to this order in the energy expansion (2) $\lambda_N = 0$. Then we are left with κ_γ, κ_Z and g_1^Z. Finally, when considering rare B and K decays, we can neglect the contribution of κ_Z since it will be suppressed by powers of the small external momenta over m_Z. Thus, in this simplistic approach, there are only two parameters relevant at very low energies. The SM predicts $\kappa_\gamma = g_1^Z = 1$. The effects of anomalous TGC have been previously studied in the literature [11]. However, this hierarchical approach to the couplings has not been the one used in the various analyses and a more comprehensive study is needed. The experiments at LEP II and the next Tevatron run are going to be sensitive to deviations from the SM prediction at the $(5 - 10)\%$ level [10]. Effects of this size might be also observed in rare B and K decays. For instance, $\delta g_1^Z = g_1^Z - 1 = 0.10$ can produce enhancements in the branching ratios of $b \to s\ell^+\ell^-$ decay modes of up to $(60 - 70)\%$ [11]. In the near future, B factory experiments will have sensitivity to these processes at the SM level, turning these low energy measurements into an excellent complement of direct probes of the TGC.

FERMION MASSES AND ELECTROWEAK DYNAMICS

Up to now, we have only considered the effects of the dynamics associated with the EWSB. These are encoded in the effective lagrangian (2), which only involves the Goldstone boson and gauge fields. Additionally, it is possible that the new strong dynamics may also affect some or all fermions. We first comment on the effective lagrangian approach for non-SM couplings of fermions to gauge bosons, and then examine the effects of a prototypical class of theories (Topcolor) where the dynamical generation of fermion masses imply the existence of relatively light new states.

Anomalous Couplings of Fermions to Gauge Bosons

The effects of new dynamics on the couplings of fermions with the SM gauge bosons can be, in principle, also studied in an effective lagrangian approach. For instance, if in analogy with the situation in QCD, fermion masses are dynamically generated in association with EWSB, residual interactions of fermions with Goldstone bosons could be important [12] if the $m_f \simeq f_\pi \simeq v$. Thus residual, non-universal interactions of the third generation quarks with gauge bosons could carry interesting information about both the origin of the top quark mass and EWSB. In a very general parameterization, the anomalous couplings of third generation quarks can be written as

$$\Delta \mathcal{L} = -\frac{g}{\sqrt{2}} \left\{ C_L \left(\bar{t}_L \gamma_\mu b_L \right) + C_R \left(\bar{t}_R \gamma_\mu b_R \right) \right\} W^{+\mu}$$
$$-\frac{g}{2 c\theta_W} \left\{ N_L^t \left(\bar{t}_L \gamma_\mu t_L \right) + N_R^t \left(\bar{t}_R \gamma_\mu t_R \right) \right.$$
$$\left. + N_L^b \left(\bar{b}_L \gamma_\mu b_L \right) + N_R^b \left(\bar{b}_R \gamma_\mu b_R \right) \right\} Z^\mu , \quad (15)$$

where the parameters $C_{L,R}$, $N_{L,R}^{t,b}$ contain the residual, non-universal effects associated with the new dynamics, perhaps responsible for the large top quark mass. Then, if we assume that the new couplings are CP conserving, there are six new parameters. They are constrained at low energies by a variety of experimental information, mostly from electroweak precision measurements and the rate of $b \to s\gamma$. Several simplifications are usually made in order to reduce the number of free parameters. For instance, in most of the literature, it is assumed that $N_{L,R}^b = 0$ [14]. A stringent bound on the right-handed charged coupling is obtained from $b \to s\gamma$ [15]: $-0.05 < C_R < 0.01$. The bounds obtained on a particular coupling from electroweak observables such as S, T, U and R_b generally strongly depend on assumptions about the other couplings. For example, if $C_L = 0$, then the combination $(N_L^t - N_R^t)$ is strongly constrained since it contributes to T. On the other hand, if $C_L = N_L^t$, then $N_R^t < 0.02$ [12,16] since it is the only (linear) contribution to T. Thus, although in general most parameters are confined to a few percent, some of them are allowed to be as large as 0.30 under certain conditions. This "model-dependent" situation requires more experimental information. A global analysis of the effects of the couplings of eqn. (15) in rare B and K processes such as $b \to s\ell^+\ell^-$, $s \to d\nu\bar{\nu}$, etc. may help disentangle the various possible effects and perhaps will give constraints that may be of importance in interpreting data from higher energy experiments [17].

The effects of light states: the example of Topcolor

The description of the residual effects of strong dynamics at low energies on fermion couplings by using (15) corresponds to cases where the states associated with the new physics are heavy compared to the weak scale. Thus, integrating out the heavy states, leaves us with effective couplings which might be generated at tree level or through loops in the full theory. However, most theories in which electroweak symmetry and/or fermion masses have a dynamical origin also contain states with masses comparable to the weak scale. Such is the case, for instance, in Technicolor models where the breaking of large chiral symmetries imply the presence of pseudo-Goldstone bosons with masses of at most a few hundred GeV. It is also the case in Topcolor-assisted Technicolor (TaTC) models [18], where a top-condensation mechanism generated by the Topcolor interactions is responsible for the large dynamical top quark mass, whereas Technicolor breaks the electroweak symmetry giving (most of) the W and Z masses. The TaTC scenario is designed to relief the problems of Extended Technicolor (ETC) in generating a heavy top [19]. Although the new gauge bosons associated with the TaTC gauge group are heavier than 1 TeV, the presence of several scalar and pseudo-scalar states with masses in the few-hundred GeV range, forces us to take these into account directly in our calculations. From the point of view of their impact in low energy observables, the most important of these states are the top-pions $\vec{\pi}_t$, the triplet of Goldstone bosons associated with the breaking of the top chiral symmetry. Since top condensation does not fully break the electroweak symmetry ($f_{\pi_t} \simeq (60 - 70)$ GeV $< v$), after mixing with the techni-pions, there will be a triplet of physical top-pions in the spectrum, with a coupling to third generation quarks given by

$$i\frac{m_t}{\sqrt{2}f_{\pi_t}} \left\{ \bar{t}\gamma_5 t \pi^0 + \bar{t}_R b_L \pi^+ + \bar{b}_L t_R \pi^- \right\} \ . \tag{16}$$

They acquire masses of a few hundred GeV due to explicit ETC quark mass terms. Additionally, in most models there are scalar and pseudo-scalar bound states due to the strong (although sub-critical) effective coupling of right-handed b-quarks. The closer the effective couplings are from criticality, the lighter these bound states tend to be. The spectrum and properties of these states, unlike those of top-pions, are not determined by model-independent features of the symmetry breaking pattern but depend on details of the model. Finally, in all TaTC model there will be pseudo-Goldstone bosons from the breaking of techni-fermion chiral symmetries. However, their couplings to third generation quarks are reduced with respect to (16) by m_{ETC}/m_t, where m_{ETC} is a small ETC mass of the order of 1 GeV[1]. The presence of the relatively light top-pions, as well as the additional bound states, imposes severe constraints on Topcolor models due to their potential loop effects in low energy observables, most notably R_b and rare B and K decays.

Top-pion Effects in R_b : The one-loop contributions of top-pions to the $Z \to \bar{b}b$ process were studied in Ref. [21]. There it was shown that they shift R_b negatively by an amount controlled by m_{π_t} and f_{π_t}. For instance, for $f_{\pi_t} \simeq 60$ GeV the correction is about -1% for $m_{\pi_t} = 800$ GeV, and top-pions with masses in the expected $(100 - 300)$ GeV range give unacceptably large deviations. This value of the top-pion decay constant is obtained by using the Pagels-Stokar formula, which gives f_{π_t} a logarithmic dependence on the Topcolor energy scale, chosen here to be a few TeV. Potentially cancelling contributions by other states, such as the scalar and pseudo-scalar bound

---

[1] Multi-scale Technicolor models such as the one in Ref. [20], in the absence of Topcolor, have un-suppressed top couplings to pseudo-scalars. This could lead to effects similar to those of top-pions.

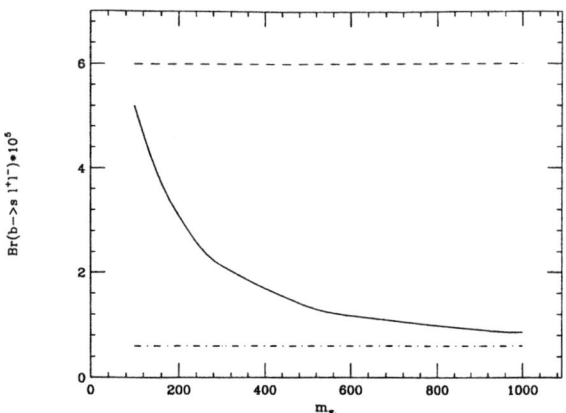

FIGURE 3. Br($b \to s\ell^+\ell^-$) vs. m_{π_t}, for $f_{\pi_t} = 70$ GeV. The dashed horizontal line is the current experimental limit for the inclusive rate [23], whereas the dot-dash line is the SM expectation.

states, Topcolor vector and axial-vector mesons, etc., are either of the wrong sign or not large enough. Possible ways out of this constraint are: larger toppion masses or larger values of f_{π_t}. The larger f_{π_t} is, the smaller the coupling, and the top-pions are more Goldstone-boson-like. For $f_{\pi_t} \simeq 120$ GeV, for instance, the shift of R_b is well within the experimentally allowed region even for $m_{\pi_t} \simeq (200-300)$ GeV. However, in order to obtain such an enhancement in the decay constant we must either assume large corrections to the Pagels-Stokar expression or introduce new and exotic fermion states.

Rare B and K Decays : The top-pions and other scalar states, give one-loop contributions to FCNC processes. These depend not only on f_{π_t} and m_{π_t} but typically also on one or more elements of the quark rotation matrices necessary to diagonalize the quark Yukawa couplings. The contributions of top-pions, as well as "b-pions" (scalar and pseudo-scalar bound states in models where b_R couples to the Topcolor interaction) to $b \to s\gamma$ depend on $D^{bs}_{L,R}$, the $b \to s$ element in the left or right down rotation matrix. Furthermore, the two contributions tend to cancel. Thus, the freedom in this model-dependent aspects of the prediction makes it possible to have quite low masses and still satisfy the bound from the experimental measurement of $B \to X_s \gamma$ [22]. The situation changes drastically in $b \to s\ell^+\ell^-$ processes, where the cancellations are much less efficient. Although experiments have not yet reached sensitivity to SM branching ratios [23], it will be soon achieved at both hadron and lepton B factories. As an example, we plot in Fig. 3 the Br($b \to s\ell^+\ell^-$) as a function of the top-pion mass with no other contributions, for $f_{\pi_t} = 70$ GeV. The $b \to s\gamma$ constraint is in this case (200 GeV $< m_{\pi_t} <$ 800 GeV). However, one can see that, even for heavier top-pions the effect can still be a $(30-60)\%$

enhancement over the SM prediction of 6×10^{-6}. On the other hand, in the presence of a 400 GeV charged b-pion the curve changes little, but the $b \to s\gamma$ bound is now $m_{\pi_t} < 600$ GeV. Finally, to compare the potential of these FCNC transitions with the R_b constraints, let us say that if we take $f_{\pi_t} \simeq 120$ GeV (which avoids conflict with R_b measurements), then the effect of a 400 GeV top-pion in $b \to s\ell^+\ell^-$ is still an enhancement of more than 50% with respect to SM expectations. Thus, the observation of these modes will further constrain Topcolor models beyond the R_b bounds. We expect similar effects due to top-pions and/or b-pions to be present in kaon processes such as $K^+ \to \pi^+ \nu \bar{\nu}$.

CONCLUSIONS

We have seen that a complete, model-independent analysis of the effects of strong dynamics in rare B and K decays could shed light on the nature of the EWSB mechanism and the origin of fermion masses. The signals are also likely to be important in models where relatively light scalars couple strongly to mass, like in the case of TaTC. In most cases, the next round of experiments will have sensitivity to SM branching ratios. This will be the case, for instance, for the Tevatron experiments, as well the KEK and SLAC B factories in the $B \to X_{(s,d)} \ell^+ \ell^-$ modes. It will also be the situation in the next generation of kaon experiments for $K^+ \to \pi^+ \nu \bar{\nu}$ and $K_L \to \pi^0 \nu \bar{\nu}$. The amount and variety of experimental information from these processes is such that suggests a parallel to the role of electroweak measurements at the Z pole as not only a constraint on new physics sources but also as guidance in the searches to be carried out at high energy machines such as the Tevatron in Run II, the LHC and eventually the NLC and/or the muon collider. It is possible to imagine a scenario where deviations from the SM in B and/or K decays point to a particular source, e.g. corrections to Goldstone boson propagators given by $\mathcal{O}_{11}$, anomalous TGC or anomalous couplings of third generation quarks to gauge bosons as in (15). The nature of the deviation might dictate the road to follow at high energies. As an example, if the source of an effect is in one the top quark couplings $N_{L,R}^t$, there would be a strong case for a lepton collider running at $t\bar{t}$ threshold. Other scenarios may not be so clear, and may require a comprehensive and careful analysis of all the data to come (including issues like hadronic uncertainties in B decays). This, however, constitutes a very well defined research program.

REFERENCES

1. A. Longhitano, *Phys. Rev.* **D22**, 1166 (1980), *Nucl. Phys.* **B188**, 118 (1981).
2. T. Appelquist and G. Wu, *Phys. Rev.* **D48**, 3235 (1993).

3. K. Hagiwara, K. Hikasa, R. D. Peccei and D. Zeppenfeld, *Nucl. Phys.* B282, 253 (1987); K. Hagiwara, S. Ishiara, R. Szalapski and D. Zeppenfeld, *Phys. Lett.* B283, 353 (1992) and *Phys. Rev.* D48, 2182 (1993).
4. F. Feruglio, *Int. J. Mod. Phys.* A8, 4937 (1993).
5. J. Bernabéu, D. Comelli, A. Pich and A. Santamaria, *Phys. Rev. Lett.* 78, 2902 (1997).
6. G. Burdman, *Phys. Lett.* B409, 443 (1997).
7. R. Balest et al., the CLEO collaboration, *Phys. Rev. Lett.* 74, 2885 (1995).
8. S. Adler et al., the BNL 787 Collaboration, *Phys. Rev. Lett.* 79, 2204 (1997).
9. A. Manohar and H. Georgi, *Nucl. Phys.* B234, 189 (1984); H. Georgi, "*Weak Interactions and Modern Particle Theory*", Benjamin/Cummings, Menlo Park, California, 1984.
10. T. Barklow et al., SLAC-PUB-7366, In the Proceedings of "New Directions for High-Energy Physics (Snowmass 96), Snowmass, CO, 25 Jun - 12 Jul 1996.
11. S. P. Chia, *Phys. Lett.* B240, 465 (1990); K. A. Peterson, *Phys. Lett.* B282, 207 (1992); G. Baillie, Z. Phys. C61, 667 (1994).
12. R. D. Peccei and X. Zhang, *Nucl. Phys.* B337, 269 (1990); R. D. Peccei, S. Peris and X. Zhang, *Nucl. Phys.* B349, 305 (1991).
13. For a treatment of CP violating effects in B decays from these couplings see A. Abd El-Hady and G. Valencia, *Phys. Lett.* B414, 173 (1997).
14. E. Malkawi and C. P. Yuan, *Phys. Rev.* D50, 4462 (1994).
15. J. Hewett and T. Rizzo, *Phys. Rev.* D49, 319 (1994); K. Fujikawa and A. Yamada, *Phys. Rev.* D49, 5890 (1994);
16. B. Dobrescu and J. Terning, *Phys. Lett.* B416, 129 (1998).
17. G. Burdman, in preparation.
18. C. T. Hill, *Phys. Lett.* B345, 483 (1995).
19. E. Eichten and K. Lane, *Phys. Lett.* B352, 382 (1995).
20. E. Eichten and K. Lane, *Phys. Lett.* B222, 274 (1989); *ibid*, **388**, 803 (1996).
21. G. Burdman and D. Kominis, *Phys. Lett.* B403, 101 (1997).
22. G. Buchalla, G. Burdman, C. T. Hill and D. Kominis, *Phys. Rev.* D53, 5185 (1996).
23. C. Albajar et al., the UA1 collaboration, *Phys. Lett.* B262, 163 (1991); S. Glenn et al., the CLEO collaboration, CLNS-97/1514, hep-ex/9710003; B. Abbot et al., the D0 collaboration, hep-ex/9801027.

Can BNL-Style Studies of $K \to \pi\nu\bar{\nu}$ be pushed at the FEMC?

Laurence Littenberg

*Brookhaven National Laboratory
Physics Department,
Upton NY, 11973, USA*

Abstract. Techniques developed for studying $K^+ \to \pi^+\nu\bar{\nu}$ and $K_L \to \pi^0\nu\bar{\nu}$ at the Brookhaven National Laboratory Alternating Gradient Synchrotron are briefly described. The applicability of these approaches at the front end of the First Muon Collider is assessed.

I THEORETICAL MOTIVATION

It has long been appreciated [1] that the flavor-changing neutral current processes $K \to \pi\nu\bar{\nu}$ are uniquely clean probes of short distance physics in K decays. Long distance contributions are tiny [2], and the hadronic matrix elements are given to $\sim 1\%$ accuracy by the rate of the common $Ke3$ decay [3]. In the Standard Model, the amplitudes are dominated by terms proportional to V_{td} [4], offering unique access to this elusive quantity. The charged mode is sensitive to the modulus of V_{td}. Now that a next-to- leading-logarithmic order calculation of QCD corrections has been done [5], its rate gives $|V_{td}|$ to 5%, modulo knowledge of other SM parameter such as m_t. The neutral mode is essentially a pure CP-violating transition, with only a negligible indirect (ϵ) component [6]. It is even cleaner from a theoretical point of view than the charged mode. A measurement of its rate would yield an unambiguous determination of η, given knowledge of $|V_{cb}|$ (the rate is actually proportional to $[Im(V_{ts}^* V_{td})]^2$). The "true" theoretical uncertainty is only $\sim 1\%$. Combining measurements of the neutral and charged rates determines the unitary angle β, independent of the B system [7]. Figure 1 illustrates the relationship between the unitarity triangle and the two kaon FCNC rates. The current ranges of prediction for $B(K^+ \to \pi^+\nu\bar{\nu})$ and $B(K_L \to \pi^0\nu\bar{\nu})$ are is $(0.6-1.5)\times 10^{-10}$ and $(1 \div 3)\times 10^{-11}$ respectively. The magnitudes of these ranges are almost entirely due to uncertainty in the input parameters such as $|V_{cb}|$ and $|V_{td}|$. These decays compare favorably in theoretical cleanliness with B system measurements that have been proposed for determining the angles of the unitarity triangle

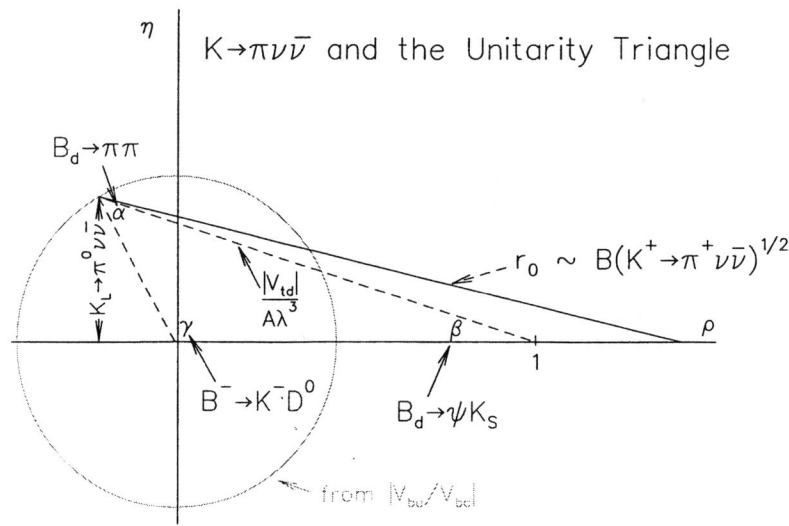

FIGURE 1. Diagram of the relationship of the charged and neutral FCNC kaon decay $K \to \pi \nu \bar{\nu}$ rates to the unitarity triangle.

Beyond this, it has lately become clear that to understand the possible effects of new physics beyond the Standard Model in the B system, it will be essential to have these K measurements [8]

II EXPERIMENTAL STATUS AND PROSPECTS

For more than ten years, the E787 collaboration at the AGS has been pursuing $K^+ \to \pi^+ \nu \bar{\nu}$, using a solenoidal magnetic spectrometer in a stopping K^+ beam. This group recently published evidence for the first observation of this decay [9]. The corresponding branching ratio was $(4.2^{+9.7}_{-3.5}) \times 10^{-10}$, consistent with the above-mentioned SM range. E787 has collected data corresponding to about 2.5 times that of the sample containing the first event, and plans to continue to run at least through 1999. This should allow the observation of a few events at the Standard Model level. Beyond this, a proposal for continuing the study of $K^+ \to \pi^+ \nu \bar{\nu}$ into the AGS-2000 era is being prepared [10]. The intention is to collect 15 ÷ 20 events at the SM level. Work is also in progress on a proposal to study this decay in an in-flight geometry at the Fermilab Main Injector [11].

There have as yet been no dedicated searches for $K_L \to \pi^0 \nu \bar{\nu}$, but the KTeV group at Fermilab has recently reported a preliminary result from a special one-day run in a configuration customized for this decay [12]: $B(K_L \to$

$\pi^0 \nu\bar{\nu}) < 1.8 \times 10^{-6}$ at 90% c.l. This group expects to reach the level of a few times 10^{-8} by 1999 [13]. Thereafter, they plan to reconfigure and upgrade their apparatus for working at the Main Injector. They have an Expression of Interest for an experiment aimed at collecting several tens of $K_L \to \pi^0 \nu\bar{\nu}$ events [14]. There is also an approved AGS proposal [15] for an experiment scoped to collect ~ 70 events, which will be discussed below. In addition, there is an approved proposal to search for this decay at KEK [16]

III $K^+ \to \pi^+ \nu\bar{\nu}$ AT THE FRONT END OF THE FMC

Fig 2 shows the apparatus [17] of AGS Experiment 787, a solenoidal spectrometer situated at the end of a ~ 700 MeV/c separated K^+ beam. About 7×10^6 K^+ per AGS spill enter the detector, accompanied by about 2×10^6 pions and muons. The beam strikes a BeO degrader and approximately one quarter of the K^+ penetrate it unscathed and stop in a highly segmented scintillating fiber target. After a $\sim 2ns$ delay, the detector becomes sensitive to unaccompanied pions exiting the target transversely. These are momentum analyzed by a small, low-mass drift chamber immersed in a $1T$ magnetic field, and penetrate a cylindrical array of scintillators and straw chambers ("range stack"), in which they come to rest. The range stack scintillators are read out at both ends by photomultipliers instrumented with 500 MHz, 8-bit transient digitizers. These are used to detect the characteristic $\pi \to \mu \to e$ decay chain. This distinguishes pions very effectively from muons which lack the first step in the chain. An important design principle of the experiment was the minimization of "dead" material, allowing the use of the comparison of range, momentum, and kinetic energy as a powerful means of particle identification. The kinematic and life-cycle methods of particle identification can be used in turn to establish each other's rejection power. Excellent muon rejection power is needed because a major background to $K^+ \to \pi^+ \nu\bar{\nu}$ is $K^+ \to \mu^+ \nu$.

Surrounding the range stack is a cylindrical array of lead-scintillator shower counters (the "barrel veto") and plugging the upstream and downstream ends of the detector are pure CsI endcap photon vetoes. In addition there are a number of supplementary vetoes in the beam direction. These complete a hermeticity that achieves a 10^6 rejection of π^0's. This is necessary since a second major background to $K^+ \to \pi^+ \nu\bar{\nu}$ is $K^+ \to \pi^+ \pi^0$. The background-rejection power of the experiment has proved quite adequate to reach the Standard Model level of sensitivity.

The main limitation on the experiment is instantaneous detector rate. This leads to both random veto losses and eventually to problems with background rejection. However to the extent that additional protons are available, one can make an immediate gain in sensitivity/hour through increasing the duty factor of the AGS (currently 44%), by extending the flat-top (currently 1.6 seconds every 3.6 seconds). This increases the sensitivity of the experiment proportionately, without requiring any improvement in detector performance. Another expedient is to reduce the momentum of the beam, so that a higher fraction of the incident K^+ actually stop and decay usefully in the target. This fraction is currently only about 25%. Since the detector rates are proportional to the K^+ striking the BeO degrader used to slow the beam, and the sensitivity is proportional to the K^+ penetrating the degrader and stopping in the target, one can clearly win in this way. Increasing the duty factor and reducing the beam momentum both require expending more of the AGS protons. Since the experiment uses only about 25% of the presently available flux, and the total AGS intensity is expected to rise over the next couple of years, significant advances seem quite possible.

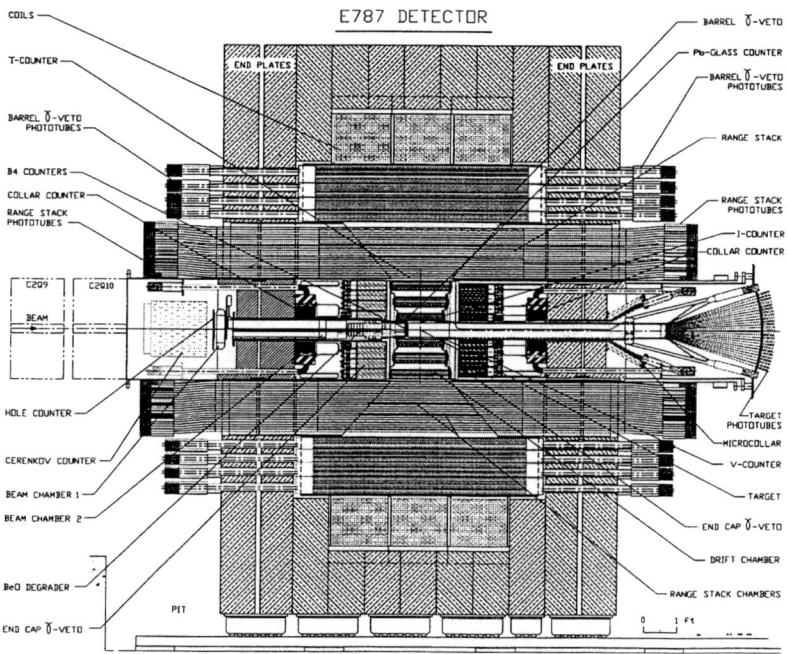

FIGURE 2. E787 detector, built into a 1-T solenoid. A $\sim$ 700 MeV/c K^+ beam enters from the left, slows down in a BeO degrader and stops in a highly-segmented scintillating fiber target. Decay π^+ are momentum analyzed by a cylindrical drift chamber and range out in an array of scintillation counters and straw chambers. A barrel lead-scintillator array and CsI (pure) endcaps complete an hermetic photon veto.

Now as mentioned above, there are other improvements under study for the AGS-2000 time scale. All would be applicable to the front end of the First Muon Collider. I should say at the outset that for a low energy forward beam like that of E787, very little K^+ flux is lost in reducing the primary proton energy from the AGS's current $24\,GeV/c$ to the $16\,GeV/c$ of the FMC front end. Table 1 shows a list of possible expedients that could be applied to push the stopping K^+ technique at a higher intensity machine. The units of primary proton intensity shown are TP, i.e. trillion protons. The AGS provides a total of about $60\,TP$ per cycle at the moment, and the front end of the First Muon Collider is supposed to provide $375\,TP/second$. The potential increase in flux is more than a factor 20, since the AGS pulses only once every 3.6 seconds, whereas the new machine would be practically DC. Note that in Table 1, not quite all the available protons are used.

TABLE 1. $K^+ \to \pi^+ \nu \bar{\nu}$ from E787 to FMC front end

	sensitivity/year	protons required
How we think we're doing lately:	2×10^{-10}	$15\,TP/cycle$
Max spill, double year (to 30wks):	6×10^{-11}	$50\,TP/cycle$
Reduce beam p, use $\pi\nu\bar{\nu}2$:	2×10^{-11}	$100\,TP/cycle$
Go to MCFE (d.f. $0.73 \Rightarrow 0.9$):	1.7×10^{-11}	$25\,TP/sec$
Further reduce beam p:	1.3×10^{-11}	$50\,TP/sec$
Drop e from $\pi \Rightarrow \mu \Rightarrow e$:	9.5×10^{-12}	$50\,TP/sec$
30 weeks $\Rightarrow$ 45 weeks/year:	6.4×10^{-12}	$50\,TP/sec$
Speed up vetoes:	3.2×10^{-12}	$100\,TP/sec$
Reduce Δp, increase geom. acc.:	2.5×10^{-12}	$300\,TP/sec$
Better beam/tgt instrumentation:	1.6×10^{-12}	$300\,TP/sec$
Improved stopping cntr. technology:	1.0×10^{-12}	$300\,TP/sec$

Table 1 starts from E787's best guess as to current sensitivity per running year, which is optimistically taken to be 15 weeks long. The second line is the result of running twice as long, and of extending the spill by a large factor (improving the duty factor). The latter costs more than a factor 3 in proton current. The third line assumes that one reduces the beam momentum from the present $700\,MeV/c$ to about $550\,MeV/c$, and also that one can exploit a large region of phase space that we have not yet accessed. This region corresponds to π^+ with momentum below that of the π^+ from the $K^+ \to \pi^+\pi^0$ background reaction (i.e. $p < 205/ > MeV/c$). This possibility is under study at the moment. If successful, it would allow one to collect about 5 Standard Model events per year, which is the goal of the AGS-2000 initiative. Going to the next line, one enters the world of the front end of the First Muon Collider. One immediately gets a small but significant improvement from the increased duty factor. The availability of so many more protons tempts one to further reduce the beam momentum, to get another small factor. Then, one can try to to drop the electron requirement from the $\pi \to \mu \to e$ decay chain criterion. This reduces the cut and deadtime losses significantly, but it requires a compensating improvement in the kinematic rejection of $K^+ \to \mu^+\nu$ events by about a factor 10. It is thought this can be obtained by relatively minor upgrades to the drift chamber. The next line assumes that one can run for 45 weeks/year at the front end of the First Muon Collider. Why not, since this is a virtual machine? At this point, one is collecting about 15 events/year assuming the central value of the Standard Model predicted range of branching ratio is correct. To make further progress, it is necessary to make major improvements to the detector. Note that one gets pretty far without this!

The next factor of two comes from speeding up the veto counters by a factor two. This would be achieved by a combination of replacement of veto counter technology, and improvements to the electronics. The time resolution of the present vetoes is not state of the art, so this can certainly be accomplished if the resources are made available. Once the veto gates can be cut in half, one can turn up the wick by a factor two. The next small factor comes from reducing the beam momentum spread by a factor three (one has to compensate for this by increased proton flux), and reconfiguring the apparatus to have better geometrical acceptance. The last two factors come from improving the beam and target instrumentation (whose space and time resolutions could certainly be improved), thus reducing random veto and cut losses, and finally, replacing the present stopping counter technology by something faster, brighter and more granular. This brings one to 10^{-12}/event or $\sim$ 100 SM events/year, which is about as far as any technique so far proposed; and possibly about as far as one needs to go until present theoretical uncertainties are reduced.

IV $K_L \to \pi^0 \nu \bar{\nu}$ AT THE FRONT END OF THE FMC

Fig. 3 shows a conceptual drawing of a detector [15] proposed to search for $K_L \to \pi^0 \nu \bar{\nu}$ at AGS-2000. It is assumed that upon the startup of the RHIC collider, the AGS will be free at least 20 hours a day for fixed target proton experiments. At that point, the available flux is expected to be 10^{14} protons per acceleration cycle. Using about half the available flux, it is estimated that in 80 weeks of running time, on the order of 70 $K^0 \to \pi^0 \nu \bar{\nu}$ events could be recorded with a background contamination of less than 10 events. This would yield a precision on η of $< 10\%$ (modulo uncertainty in $|V_{cb}|$). The techniques for obtaining the required sensitivity and background rejection are as follows. First the neutral beam is extracted at a very wide angle ($\sim 45°$) to soften both the neutron and kaon momentum spectra. This minimizes the flux of neutrons above π^0 production threshold that can produce background by interacting with vacuum windows or residual gas. To further suppress background from the latter, a vacuum of 10^{-7} Torr must be maintained throughout the beam region. Second, the proton beam from the AGS is bunched on extraction with a time microstructure of period $\sim 40\,ns$. The rms bunch width is ≤ 200 ps to allow time of flight measurement to determine the neutral kaon's momentum to a few percent. The soft kaon spectrum ($\bar{p} \sim 750\text{MeV}/c$) is necessary for this to work. Also, with this time bunching technique, the massless and other fast debris from the primary target interaction arrive at the detector before the kaons of interest, and so can be distinguished from the decay products of the latter. Third, the detector incorporates active shower pre-converters that allow measurement of the direction of the π^0 photons coming from the $K_L \to \pi^0 \nu \bar{\nu}$ decay. In conjunction with a high resolution scintillating fiber calorimeter, this allows one to fully reconstruct the π^0, independent of any assumptions about the beam. Finally, combining this with the beam timing information, one can transform the π^0 into the K_L center of mass. The last major requirement is hermetic photon vetoing. Extrapolating from photon vetoing performance measured in E787, it is estimated that an average single γ rejection of $10^4 : 1$ can be achieved. The main expected background to $K_L \to \pi^0 \nu \bar{\nu}$ is the 300-million-fold more frequent $K_L \to \pi^0 \pi^0$ decay ($K_{\pi 2}$). These events become background when two of the four final state photons are missed. If the two missed photons are from the decay of the same π^0 ("even" case), then the detected pair will reconstruct properly to a π^0 meson. The energy of this reconstructed π^0, in the rest frame of the K_L, will equal 248.84 MeV/c (modulo the resolution). If the two detected photons each originate from a different π^0 ("odd" case), then they will not, in general, reconstruct to a π^0 mass. In addition, the $K_{\pi 2}$ events which evade the photon veto tend to have rather small values of missing energy and missing mass compared to signal events. Therefore, with proper kinematic cuts, one is able to suppress the $K_{\pi 2}$ background to $\leq .10\%$ of the signal.

Other potential backgrounds sources are $K_L \to \gamma\gamma$, $K_L \to \pi^- e^+ \nu$, with the e^+ annihilating and the π^- undergoing charge exchange before they are detected, $\Lambda \to \pi^0 n$, and accidentals. These backgrounds have been calculated to contribute to less than 1 event each after 80 weeks of AGS 2000 running time.

The E926 proposal received scientific approval by the AGS PAC in October, 1996. Intensive simulation, design, prototype, and beam test work is underway.

Since E926 is not yet built, much less run, any extrapolation to the front end of the First Muon Collider must be far more cautious than in the case of E787. Table2 shows a possible progression.

TABLE 2. $K_L \to \pi^0 \nu \bar{\nu}$ from E787 to FMC front end

	sensitivity/year	protons required
Nominal estimate of E926:	1.2×10^{-12}	$50\,TP/cycle$
MCFE: Comfort factors/d.f.=0.9:	1×10^{-12}	$50\,TP/sec$
Longer beam		
Filter		
Tune angle/aperture		
Shorter decay volume, smaller beam:	5×10^{-13}	$200\,TP/sec$
Better time response, double rate:	3×10^{-13}	$375\,TP/sec$

There would be an immediate small factor as one exploited the 90% duty

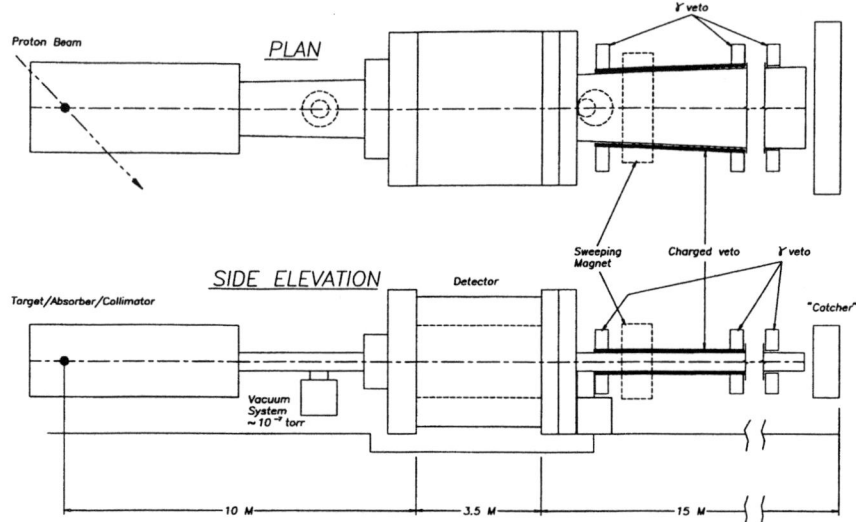
FIGURE 3. Schematic of the proposed 926 detector.

factor of the First Muon Collider front end. It would probably be wise to use the next factor of beam on what are labeled "comfort factors" in Table 2. These include a longer beam line for better time resolution and collimation, a filter to differentially attenuate neutrons and very low energy kaons, and some scope for adjusting the production angle and aperture of the beam. One could then use additional flux by shortening the decay volume, thereby increasing the acceptance of the detector. Finally, if money were no object, faster photon detectors could be deployed so that more beam could be accommodated. This results in a rate of about 70 SM events per year. In a few years of running, in principle η could be determined to about 3%.

V CONCLUSION

Techniques for exploiting high-intensity low energy kaon beams have been developed at Brookhaven for the study of $K \to \pi^+ \nu \bar{\nu}$. These emphasize measurement of all possible kinematic quantities, redundant methods of background rejection, hermetic photon vetoing and the use of deadtimeless electronics. Extending these techniques, the sensitivity achievable for both charged and neutral $K \to \pi \nu \bar{\nu}$ at the front end of the First Muon Collider is excellent. The theoretical motivation for doing such searches is likely to remain very strong. The main open question is whether these branching ratios will not be already well-measured at other facilities, such as Japanese Hadron Facility or the Fermilab Main Injector, by the time the front end of the FMC opens for business.

This work was supported by the U.S. Department of Energy under Contract No. DE-AC02-76CH00016.

REFERENCES

1. M.K. Gaillard and B.W. Lee, Phys. Rev. D **10**, 897 (1974).
2. J. Hagelin and L. Littenberg, *Prog. Part. Nucl. Phys.* **23**, 1 (1989); D. Rein and L.M. Sehgal, Phys. Rev. D **39**, 3325 (1989); M. Lu and M.B. Wise, Phys. Lett. **B324**, 461 (1994); C.Q. Geng, I.J. Hsu, and Y.C. Lin, Phys. Lett. **B355**, 569 (1995); S. Faijfer, Nuovo. Cim. **110A**, 397 (1997).
3. W.J. Marciano and Z. Parsa, *Phys. Rev.* **D53**, 1R (1996).
4. G. Buchalla and A. J. Buras, *Phys.Rev.* **D54**, 6782 (1996).
5. G. Buchalla and A.J. Buras, *Nucl. Phys.* **B412** (106) 1994.
6. L. Littenberg, *Phys. Rev.* **D39**, 3322 (1989).
7. G. Buchalla and A. J. Buras, *Phys. Lett.* **B333**, 221 (1994).
8. Y. Nir and M. P. Worah, *Probing the Flavor and CP Structure of Supersymmetric Models with $K \to \pi \nu \bar{\nu}$ Decays*, SLAC-PUB-76010, WIS-97/32/Nov-PH, hep-ph/9711215, Nov. 1997; A. Buras, A. Romanino, and L. Sivestrini, $K \to \pi \nu \bar{\nu}$: *A Model Independent Analysis and Supersymmetry*, TUM-HEP-302-97 (1997); G-C. Cho, *Impacts on Searching for Signatures of New Physics from $K^+ \to \pi^+ \nu \bar{\nu}$ Decay*, KEK-TH-554 (1998).
9. S. Adler, *et al.*, Phys. Rev. Lett. **79** 2204 (1997).
10. I-H. Chiang, *et al.*, *An experiment to measure the branching ratio $K^+ \to \pi^+ \nu \bar{\nu}$* in "AGS-2000 Experiments for the 21st Century", edited by L. Littenberg and J. Sandweiss, **Proc., AGS-2000 Workshop,** Brookhaven National Laboratory, May 13-17, 1996, Formal Report 52512.
11. R. Tschirhart, *Proceedings of the Workshop on K Physics*, ed. L. Iconomidou-Fayard, 397 (1997).
12. R. Ben-David, *XVI International Workshop on Weak Interactions and Neutrinos*, Capri (1997).
13. E. Cheu, *et al.*, *A Letter of Intent to Continue the Study of Direct CP Violation and Rare Processes in Neutral Kaon Decays at KTeV in FY99*, June 1997.
14. E. Chen *et al.*, *An Expression of Intent to Detect and Measure the Direct CP Violating Decay $K_L \to \pi^0 \nu \bar{\nu}$ and other Rare Decays at Fermilab Using the Main Injector*, FERMILAB-PUB-97-321-E, hep-ex/9709026 (1997).
15. I-H. Chiang, *et al.*, *Measurement of $K_L \to \pi^0 \nu \bar{\nu}$* AGS Proposal 926 (1996).
16. T. Inagaki, *et al.*, *Measurement of the $K_L^0 \to \pi^0 \nu \bar{\nu}$ decay*, KEK proposal, 7 June 1996.
17. M. S. Atiya, *et al.*, *Nucl. Instrum. Meth.* **A321**, 129 (1992).

K^+ decays in flight, AGS E865 and E923

Hong Ma

Brookhaven National Laboratory [1]
Upton, NY 11973

Abstract. The status of the current experiment E865 at the AGS, which is designed to search for the lepton number violating decay ($K^+ \to \pi^+\mu^+e^-$) and study other rare K^+ decays, is summarized. A new experiment at the AGS, E923, designed to search for T-violation in $K^+ \to \pi^0\mu^+\nu$ decays, is also described. The limitation of these experiments and possible improvements at the FEMC are discussed.

INTRODUCTION

Historically, studies of kaon decays played a very important role in modern particle physics. Recently the focus has been on understanding the CP violation, and the very rare kaon decay processes to test the Standard Model and search for new physics beyond the Standard Model. With the large numbers of kaons available at the AGS, and with modern detector techniques, the rare kaon decay experiments are reaching such sensitivities that they effectively explore an energy scale in the 50-100 TeV range, much beyond what is directly accessible in any planned accelerators. Experiment E865 is designed to search for lepton number violating decay, $K^+ \to \pi^+\mu^+e^-$. The observation of such a decay will have a great impact on our understanding of physics beyond the Standard Model, and the absence of such decays in a high sensitivity experiment provides a stringent constraint on theoretical models.

In addition, the measurements of rare decays and precision measurements of the no-so-rare decays provide strong tests of the Standard Model and precise measurements of Standard Model parameters. The E865 detector, as described in the following section, can be used to study most of the K^+ decays that involve three charged particles. The excellent particle identification system provides clear separation of pions, muons and electrons. Decays such as $K^+ \to \pi^+e^+e^-$, $K^+ \to \pi^+\mu^+\mu^-$ and structure dependent part of $K^+ \to \mu^+\nu e^+e^-$ and $K^+ \to e^+\nu e^+e^-$ can be described in the framework of chiral perturbation theory. The precise measurements of the branching ratios as well as the form factors provide a direct test of the

---

[1] This research was supported in part by the U.S. Dept. of Energy under contract DE-AC02-76CH00016 with Brookhaven National Laboratory

chiral perturbation theory. The $\pi\pi$ phase shift near threshold can be extracted from analysis of $K^+ \to \pi^+\pi^-e^+\nu$ decay. A precise measurement of $K^+ \to \pi^0 e^+\nu$ branching ratio can improve the error on CKM matrix element V_{us}. Other rarer decays, such as $K^+ \to \pi^+\pi^0 e^+e^-$ and $K^+ \to \pi^+e^+e^-\gamma$ are also within the reach of the experiment.

Another interesting aspect of kaon decay is CP violation. Experimentally, CP violation has only been observed in the neutral kaon system so far. Although a theoretical description of the CP-violation in the neutral kaon system exists through the complex phase in the Standard Model CKM matrix, part or all of this phase could be consequences of deeper causes that have so far eluded experiments, and CP violation from superweak interaction has yet to be ruled out. Currently ambitious efforts towards understanding CP-violation and the CKM matrix elements are planned with new $\frac{\epsilon'}{\epsilon}$ experiments and B-factories. The importance of these efforts is undeniable, yet it must also be important to investigate the possibility that some or all of the CP-violation comes from effects outside the minimal Standard Model, particularly the CKM matrix. It has been observed that such CP violation, stronger than what is in the Standard Model, may be needed for baryongenesis. Transverse muon polarization in $K^+ \to \pi^0\mu^+\nu$ decays explores a window of such possible new physics [1,2]. The triple product, $\vec{s}_\mu \cdot (\vec{p}_\mu \times \vec{p}_\pi)$, is odd under time reversal transformation. While the transverse muon polarization from the final state interaction is rather small (10^{-6}), the contribution from CKM phase is zero. A scalar particle with a complex phase with respect to the standard weak interaction, such as the charged Higgs particle in multi-Higgs extension of the Standard Model, can induce a sizable transverse muon polarization. The best previous experimental limits were obtained over 15 years ago with both neutral [3] and charged kaons [4] at the BNL-AGS. The K^+ experiment produced a measurement of the transverse polarization, $P_\mu^T = 0.0031 \pm 0.0053$, based on 2.1×10^7 events and was limited by statistics and backgrounds. Currently an experiment is in progress at the KEK-PS, E246 [5], to measure P_μ^T with a new technique of using a stopping K^+ beam and measuring the muon decay direction without spin precession. They expect to reach a sensitivity of $\delta P_\mu^T \sim 9 \times 10^{-4}$ with 1.8×10^7 events.

A non-zero value of P_μ^T of radiative $K_{\mu 2}$ decay, $K^+ \to \mu^+\nu\gamma$, also violates time-reversal invariance [6], although final state interactions could contribute to P_μ^T on the order of 10^{-3}. While P_μ^T in $K_{\mu 3}$ is sensitive to a new scalar interaction, that in $K_{\mu 2\gamma}$ is sensitive to a new pseudoscalar interaction, as well as new vector and axial vector interactions, exploring a broader class of extensions of the Standard Model [8,7]. The structure dependent part of this decay has been measured recently by E787 at the AGS [12], and the related decays with internally converted photons to e^+e^- pairs are being studied in E865 (Fig 3). The muon polarization of this decay, however, has never been measured. The importance of this measurement should not be overlooked.

Measurements of muon polarizations in $K^+ \to \pi^+\mu^+\mu^-$ decays could lead to important constraints on the Standard Model CKM parameters, in particular the

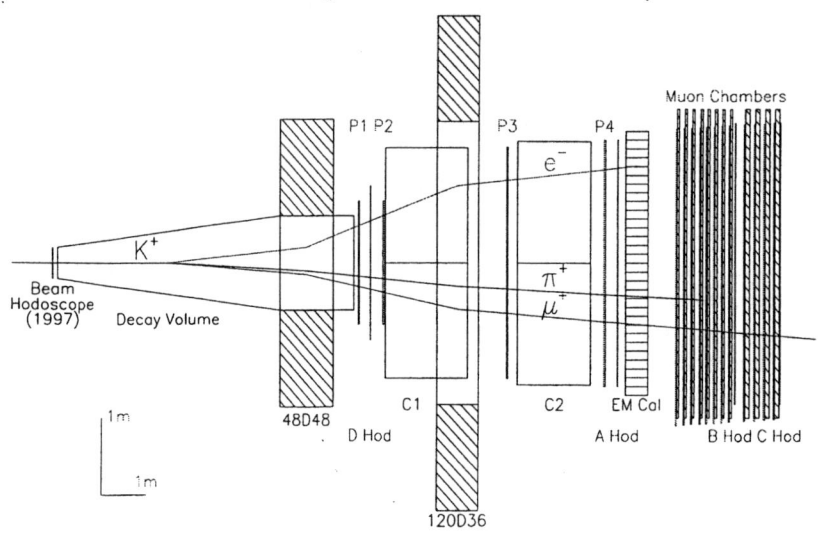

FIGURE 1. Schematic of the E865 detector.

Wolfenstein parameters ρ and η, and on T-violation beyond the Standard Model. Within the context of the Standard Model with CKM matrix, the longitudinal polarization of the muon, a parity-violating quantity, depends on ρ [9,10]. The two spin correlations, such as the product of the longitudinal polarization of muon and the transverse polarization of the anti-muon, is directly T-violating, providing measurement of the parameter η [11]. This can be regarded as complimentary to the other interesting rare kaon decay, $K_L \to \pi^0 \nu \bar{\nu}$, which is a CP-violating process whose branching ratio is directly proportional to the η^2. Recently, E787 reported the observation of 200 $K^+ \to \pi^+ \mu^+ \mu^-$ events [13] and E865 has now 400 events to be analyzed (Fig 2). A few order of magnitude increase in statistics is needed in order to exploit the physics potential of this decay. The Front End of a Muon Collider (FEMC) can provide such an opportunity.

STATUS OF E865

The detector design for E865 was based primarily on its predecessor, AGS E777/E851 [14]. An unseparated 6 GeV/c beam with two stages of collimation is used for this experiment. With nominal intensity of 7×10^{12} proton on target per spill, it yields 5×10^7 K^+'s in the beam, along with 20 times more pions and protons. The detector follows a vacuum decay tank of 5 meters long immediately downstream of the last quadrupole. Approximately 9% of the K^+'s decay in that volume. Figure 1 is the plan view of the detector, showing the π, μ, e tracks schematically.

The detector design is dictated by three basic requirements for background rejection: particle identification to separate pions, muons and electrons, precision tracking to provide kinematic constraints, and the ability to operate in a high rate environment.

The 48D48 dipole magnet separates particles by charge (negative to beam left, positive to beam right) and reduces the low momentum charged particle background that comes from the decay volume. The proportional chambers (P1-P4) along with the 120D36 dipole magnet form the momentum analyzing spectrometer system. Trigger hodoscope D requires that the charged particles come from the decay volume outside the beam region, and the trigger hodoscope A along with the shower counter identifies the charged shower cluster. The hodoscopes B and C in the muon stack are used to identify muons.

There are two gas filled Čerenkov counters (C1 and C2) at atmospheric pressure which are divided by thin membranes into left and right sides. Since the left side of the apparatus must detect electrons from $\pi\mu e$ decay, and since a major source of potential background is misidentification of the π^- from $\pi^+\pi^+\pi^-$ decays, this side of the Čerenkov counters is filled with hydrogen. On the right side the major background comes from $K^+ \to \pi^+\pi^0$ or $K^+ \to \pi^0\mu^+\nu$ followed by $\pi^0 \to e^+e^-\gamma$ decays. Thus the Čerenkov counters on this side are filled with a lower threshold gas (CO_2 or methane) to efficiently identify the positrons.

Downstream of P4 is an electromagnetic calorimeter consisting of 600 $11.4 \times 11.4 cm^2$, 15 radiation lengths long Pb-scintillator sandwich modules, read out by wavelength shifting fibers in a shashlik configuration. Downstream of the calorimeter is a stack of twelve planes of muon detector, each consisting of a wall of proportional tubes with x and y readout, separated by steel absorber plates.

With this apparatus we identify electrons as those particles giving light in the appropriate Čerenkov counters and having pulse height in the calorimeter consistent with the momentum measured in the spectrometer. Muons are seen as particles giving no light in the Čerenkov counters, minimum ionizing pulse height in the calorimeter, and a range in the muon stack consistent with their momentum. Pions should have no light in the Čerenkov counters, and a range shorter than that expected for minimum ionizing particles.

The physics data collection started in the second half of the 1995 run, and continued through the 1997 run. The primary trigger for the 1995-1996 runs are $K^+ \to \pi^+\mu^+e^-$ and final states with high e^+e^- mass. For normalization purposes, we also take prescaled minimum biased three charged track events which is dominantly $K^+ \to \pi^+\pi^+\pi^-$ decays, and prescaled e^+e^- events without the high mass requirement, which is dominantly $K^+ \to \pi^+\pi^0$ followed by $\pi^0 \to e^+e^-\gamma$ decays. The expected single event sensitivity for $\pi\mu e$ search from the combined 1995-96 data set is 1.6×10^{-11}. We expect to increase the sensitivity by a factor of 3 in the 1998-1999 run. This is an improvement of sensitivity by a factor of 15 over the

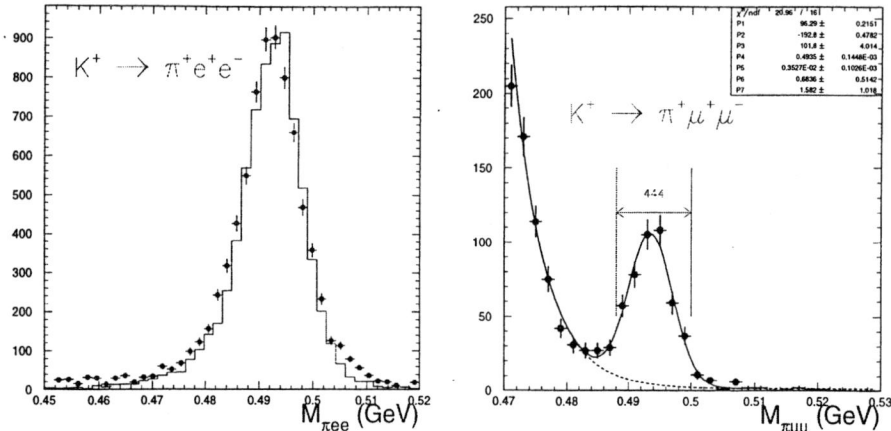

FIGURE 2. Kaon invariant mass distribution of $K^+ \to \pi^+ e^+ e^-$ decay (left) and $K^+ \to \pi^+ \mu^+ \mu^-$ decay (right).

previous experiment, E777.

The 1997 run was relatively short, and devoted to some special triggers, which could not be incorporated into the runs for $K^+ \to \pi^+ \mu^+ e^-$. With methane gas on both sides of the Čerenkov counters, a large sample of $K^+ \to \pi^+ \pi^- e^+ \nu$ events were collected. This will greatly improve the measurement of the $\pi\pi$ scattering phase shift. At 10 times lower intensity, a large sample of events with $e^+ e^-$ (mostly from π^0 dalitz decays) were collected. The ratio of $K^+ \to \pi^+ \pi^0$ to $K^+ \to \pi^0 e^+ \nu$ branching ratios is expected to be measured to better than 1%, reducing the error on the CKM matrix element, V_{us}. The third trigger in the 1997 run required two muons in the final state to study the $K^+ \to \pi^+ \mu^+ \mu^-$ decay.

Table 1 summarizes the number of events in each decay mode, in comparison with the statistics of the previous experiments.

Decay Mode	Branching Ratio	Previous Statistics	E865 Statistics
$\pi^+ e^+ e^-$	2.7×10^{-7}	500	10,000
$\pi^+ \mu^+ \mu^-$	5×10^{-8}	200	400
$\mu^+ \nu e^+ e^-$	$8 \times 10^{-8*}$	14	1,500
$e^+ \nu e^+ e^-$	$3 \times 10^{-8*}$	4	350
$\pi^+ \pi^- e^+ \nu$	4×10^{-5}	30,000	300,000
$\pi^0 e^+ \nu$	4.8×10^{-2}	4,000†	50,000

TABLE 1. Number of events for each decay mode, expected from the 1995-97 data sets, in comparison with the statistics of the previous experiments. *$M_{ee} > 0.14$MeV; † statistics for direct K_{e3} BR measurements.

The high statistics data of $K^+ \to \pi^+ e^+ e^-$, $K^+ \to \pi^+ \mu^+ \mu^-$, $K^+ \to \mu^+ \nu e^+ e^-$

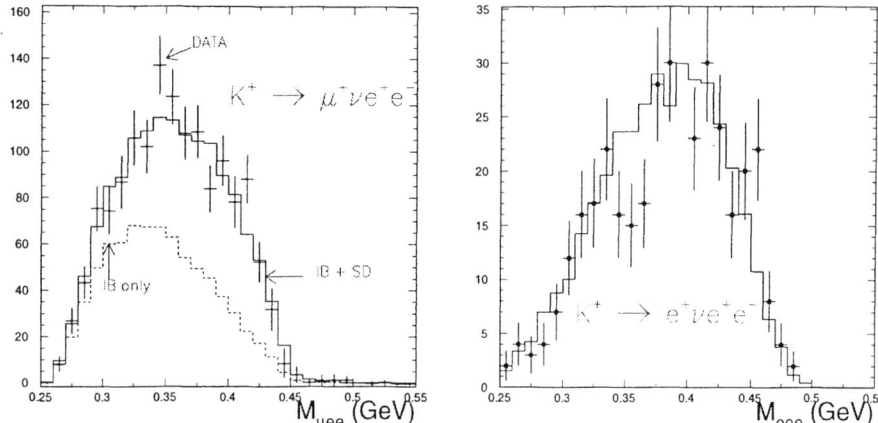

FIGURE 3. Left: μee mass distribution of $K^+ \to \mu^+ \nu e^+ e^-$ decays, in comparison with Monte Carlo with only inner bremsstrahlung, and with bremsstrahlung + structure dependent part; Right: eee mass distribution of $K^+ \to e^+ \nu e^+ e^-$ decays, in comparison with Monte Carlo, which is dominated by the structure dependent part.

and $K^+ \to e^+ \nu e^+ e^-$ are shown in Figure 2 and 3. The analyses are in progress, and the result will provide good measurements of the form factors, as well as checking the consistency of the chiral perturbation theory.

After three years of operation, great experience has been gained in understanding the E865 detector system. Although it has achieved remarkably well in studying these rare decays described above, it also revealed some limitations in the system, which can be improved in future experiments of this kind. Although the beam achieved the desired kaon flux, the beam halo exceeded the naive expectation. Part of the halo can be attributed to the beam pion decays. These high momentum muons not only raised the rates in the primary trigger counters (especially the muon trigger system) and the tracking system, but are also above the threshold of Čerenkov radiation in methane, causing a very high rate in the Čerenkov counters on the right side. This became one of the limiting factors in Čerenkov counter performance. The other damaging effect of the pions and protons in this unseparated beam is that they interact inelastically with the material downstream of the decay volume, producing in-time multiple hits in the trigger system. This effect was reduced by having a set of trigger counters (D-counters) closer to the end of the decay volume, and placing hydrogen and helium in the path of the beam wherever possible. Clearly, in a possible future experiment with higher kaon flux, a separated kaon beam will be essential. This can be made possible by the much larger number of protons available at FEMC.

SEARCH FOR T-VIOLATION IN $K_{\mu 3}$ DECAYS

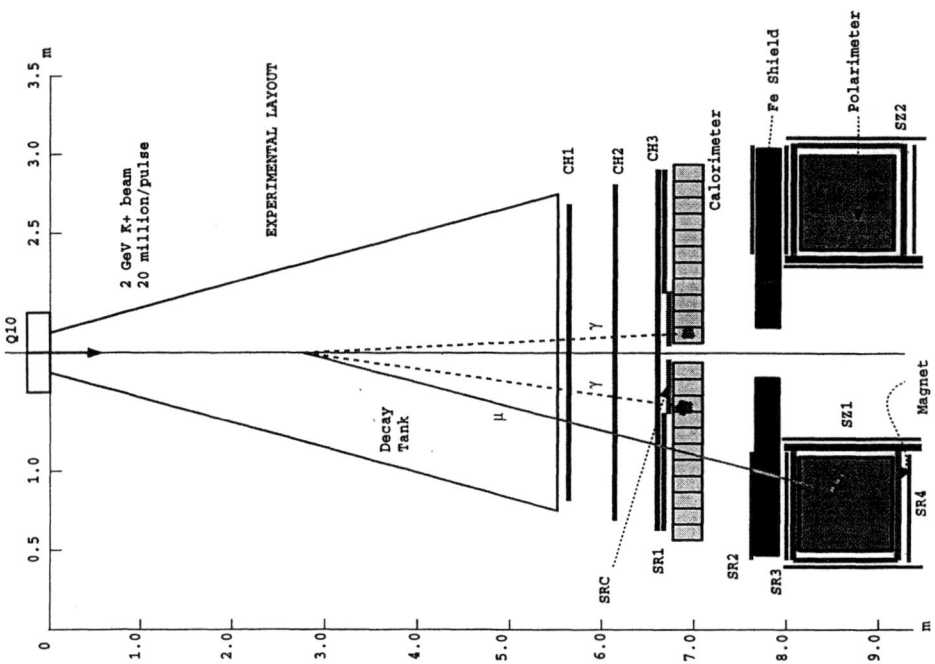

FIGURE 4. Schematic of the detector. A typical $K^+ \to \mu^+ \pi^0 \nu$ events is superimposed.

The E923 experiment [15] will be performed with 2 GeV/c electro-statically separated charged kaons decaying in flight. The beam intensity will be $2 \times 10^7 K^+$'s/spill with 3×10^{13} protons on target every 3.6 sec. Figure 4 shows the plan view of the experiment. The basic workings of the experiment are the same as the experiment in Ref. [4]. The detailed design is, however, optimized for a high intensity 2 GeV/c beam. The cylindrically symmetric detector is centered on the kaon beam. The $K^+_{\mu 3}$ decays of interest occur in the decay tank; the photons from the $\pi^0 \to \gamma\gamma$ decay are detected in the calorimeter; the muon stops in the polarimeter. The decay of the stopped muon is detected in the polarimeter by wire chambers, which are arranged radially with 96 graphite wedges that serve as absorber medium. The hit pattern in the polarimeter identifies the muon stop as well as positron direction relative to the muon stop. By selecting events with π^0 moving along the beam direction and muon moving perpendicular to the beam direction in the K^+ center of mass frame, the decay plane coincides with the radial wedges. A non-zero transverse muon polarization causes an asymmetry between the number of muons that decay clockwise versus the number counter-clockwise. To reduce systematic errors, a weak solenoidal magnetic field along the beam direction (70 gauss, giving a precession period of $\sim 1\mu s$) with polarity reversal every spill is applied to the polarimeter. If there is an initial muon transverse polarization, there will be a small shift in the phase of the sinusoidal oscillation in the measured asymmetry. The difference in the asymmetry for the two polarities is proportional to the T-conserving muon polarization in the decay plane, while the sum is proportional to the T-violating muon polarization normal to the decay plane. Both components will have the same frequency but will be 90 degrees apart in phase.

Compared to the previous in-flight experiment, this experiment has much better background rejection and event reconstruction. The separated K^+ beam should greatly reduce the accidental rate. The polarimeter is more finely segmented and the analyzing power is higher. The positron signature is defined by the coincidence of signals in a pair of neighboring wedges. The larger calorimeter makes it possible to measure both photons and to reconstruct the π^0 momentum for a large fraction of the events. Together with the muon trajectory, the events can be fully reconstructed. The detector acceptance and background rejection is optimized via GEANT simulation.

The experiment will collect approximately 550 events per AGS pulse per 3.6 seconds. Thus the statistical accuracy of the polarization measurement in a 2000 hr (2×10^6 pulses) run will be:

$$\delta P_T \approx \frac{\sqrt{1.20}\sqrt{2}}{0.35\sqrt{2 \times 10^6 \cdot 550}} \approx 1.3 \times 10^{-4}$$

where $\sqrt{1.2}$, $\sqrt{2}$, 0.35, are dilution factors in the analyzing power due to backgrounds, the precession magnetic field, and the muon decay, respectively.

This new experiment is optimized to study muon polarization in $K^+ \to \mu^+\pi^0\nu$ decays. Nevertheless, we have investigated the feasibility of measuring T-violation in $K^+ \to \mu^+\nu\gamma$. The event selection and analysis of $K^+ \to \mu^+\nu\gamma$ will be very similar to $K^+ \to \mu^+\pi^0\nu$ events except that events containing more than 1 photon will be vetoed to reject background from $K^+ \to \mu^+\pi^0\nu$, $K^+ \to \pi^+\pi^0$, and $K^+ \to \pi^+\pi^0\pi^0$ events. Further background rejection will be achieved by matching the measured muon range in the polarimeter with the muon energy from a constrained fit to the photon momentum, the muon direction, and the known kaon momentum. We expect to collect ~ 100 events per AGS pulse. However, the signal to background ratio with out current design will be about 0.3. Two improvements to the detector will reduce the backgrounds further: If the decay volume can be surrounded by photon veto counters with a veto threshold of 10 MeV to detect the low energy photons from π^0 decays, the background level can be reduced to about 10%. Secondly, if the calorimeter resolution can be improved (we have assumed $\sigma(E)/E \sim 8\%/\sqrt{E}$) then the muon range match can be made narrower, thus separating the signal and background better. We are currently calculating the sensitivity that can be gained with these modest improvements for the transverse muon polarization in $K^+ \to \mu^+\nu\gamma$.

The AGS D6 line, to be used by E923, can only be operated with up to 30×10^{12} protons on target per spill at up to 2GeV/c momentum. The lower beam momentum results in a higher π^+ background because less shielding is placed upstream of the polarimeter in order not to raise the muon energy cutoff too high. The tracking system and calorimeter should be able to handle higher rates(maybe a factor of 5 higher), while the polarimeter is designed such that the accidental e^+'s contribute a few percent to the signal sample. To increase the K^+ flux significantly above that of E923, it is essential to keep this accidental rate down, because for each μ^+ stop, any e^+ signal near the muon stop within a few lifetime of muon will be accepted as valid muon decay. Note that the e^+ signal, after vetoing the through tracks, are dominated by random muon decays of stopped muons. A polarimeter with segmentation along the beamline, which can both identify the muon stop and e^+ signal locally, will effectively have a lower e^+ rate for a given μ stop.

The E923 design goal for systematic errors is $\delta P_\mu^T < 10^{-4}$, which calls for mechanical alignment precision of $100 \mu m$ for the polarimeter, detector efficiency uniformity of 4%, beam-detector alignment of 0.5cm and residual magnetic field of 6mG. All these can be achieved at reasonable cost. There could be significant increase in cost if systematic errors have to be reduced by another order of magnitude. It should be noted that there will always be a large sample of muon stop events with no known source of transverse polarization, which can be used to check the systematic errors.

SUMMARY

E865 has acquired abundant physics data of K^+ decays, and will improve the search for $K^+ \to \pi^+ \mu^+ e^-$. Interesting results will be forthcoming in the next a few years. E923 will join the new generation of CP violation study experiments early next century.

At the time the FEMC is built, there may still be physics motivation to study K^+ decays in flight. The experience with the AGS experiments E865 and E923 will be of great value. From theoretical point of view, an experiment exploring $\delta P_T^\mu \sim 10^{-5}$ will be very interesting. It is not inconceivable that such an experiment can be built in the FEMC era, drawing experience from E923.

Another example will be the study of muon polarization of $K^+ \to \pi^+ \mu^+ \mu^-$ decays, which will require a precise tracking system as in E865 to reject the background from $K^+ \to \pi^+ \pi^+ \pi^-$, a polarimeter as in E923 to measure the muon polarization, and a intense proton source as can be provided by the FEMC. Such measurements will be of great interest not only within the Standard Model, but also it may potentially reveal new phenomena beyond the Standard Model. With 5-10 times more intense separated K^+ beam, it is possible to collect about 30 $K^+ \to \pi^+ \mu^+ \mu^-$ events/hour. With a few thousand hours of running, the longitudinal polarization of the μ^+ can be measured to $\sim 1\%$. The real challenge will be to measure the double spin correlation, where the polarization of the μ^- will have to be measured efficiently.

A proton source 100 times more intense than the current leader, the AGS, will provide new opportunities for the field. If the FEMC becomes a reality, there will certainly be ways to take full advantage of the intensity.

REFERENCES

1. R. Garisto and G. Kane, *Phys. Rev.* **D 44**, 2038 (1991).
2. G. Belanger and C. Q. Geng, *Phys. Rev.* **D 44**, 2789 (1991).
3. M. Schmidt, et al., *Phys. Rev. Lett.* **43**, 556 (1979).
   W. Morse, et al., *Phys. Rev.* **D 21**, 1750 (1980).
4. M. Campbell, et al., *Phys. Rev. Lett.* **47**, 1032 (1981).
   S. Blatt, et al., *Phys. Rev.* **D 27**, 1056 (1983).
5. J. Imazato, et al., KEK-PS research proposal Exp-246, June 6, 1991.
6. M. Kobayashi, T.-T. Lin and Y. Okada, *Progress of Theoretical Physics* **95**, 261 (1996).
7. C.H. Chen, C.Q. Geng, C.C. Lih, *Phys.Rev.* **D56** 6856(1997)
8. Guo-Hong Wu, John N. Ng, *Phys.Rev* **D55** 2806(1997)
9. Ming Lu, Mark B. Wise, and Martin J. Savage, *Phys. Rev.* **D46** 5026 (1992).
10. G. Belanger, C.Q. Geng, P. Turcotte, *Nucl. Phys.* **B390** 253 (1993).
11. Pankaj Agrawal, John N. Ng, G. Belanger, C.Q. Geng, *Phys. Rev.* **D45** 2383 (1992).
12. M. Convery, *First measurement of structure dependent $K^+ \to \mu^+ \nu \gamma$* , Ph.D. thesis, Princeton Univ. (1996).
13. S. Adler, et al , *Phys.Rev.Lett.* **79** 4756(1997)
14. A. M. Lee et al, Phys. Rev. Lett. **64**, 165 (1990).
    C. Alliegro et al, Phys. Rev. Lett. **68**, 278 (1992).
    A. Deshpande et al, Phys. Rev. Lett. **71**, 27 (1993).
15. M.V.Diwan, et al, Search for T-violating muon polarization in $K^+ \to \pi^0 \mu^+ \nu$ decay. *AGS proposal 923* (1996).

Lessons Learned from E871, A Search for Rare Kaon Decays at BNL

Mark Bachman*

Department of Physics and Astronomy
University of California at Irvine
Irvine, CA 92697-4575
*Representing the E871 collaboration.

Abstract. E871 at the Brookhaven Alternating Gradient Synchrotron (BNL AGS) searched for the rare decays of the neutral Kaon, $K_L \to \mu\mu, K_L \to ee, K_L \to \mu e$. Data were taken during two run periods in 1995 and 1996 resulting in an expected single event sensitivity of $\sim 10^{-12}$. Over 6,000 $K_L \to \mu\mu$ events have been seen in the data. The E871 collaboration is currently concluding its analysis of these decays and results are expected to be published shortly. Experimental technique, as well as problems associated with the measurement of these rare processes in a high rate environment is discussed.

INTRODUCTION

Experiment 871 at the Brookhaven Alternating Gradient Synchrotron (BNL AGS) searched for the rare decays of the neutral Kaon: $K_L \to \mu\mu, K_L \to ee, K_L \to \mu e$. These decays, summarized in table 1 below, probe the limits of our understanding of the Standard Model. In particular, the $K_L \to \mu e$ decay is forbidden under conservation of lepton flavor, and an observed signal would indicate physics beyond the Standard Model. Indeed, most popular extensions to the Standard Model predict non-conservation of lepton flavor at some level [1,2].

The measurements were made in 1995 an 1996 using the 24 GeV/c proton beam available at the AGS with 15×10^{12} protons per 1.2 second spill. The proton beam was directed on to a platinum target rod, $0.318 \times 0.254 \times 12.45$ cm^3 in

TABLE 1. E871 decay modes

Decay	Branching ratio	Physics probed
$K_L \to \mu\mu$	$(7.2 \pm 0.5) \times 10^{-9}$	Flavor changing neutral currents
$K_L \to ee$	$< 4.1 \times 10^{-11}$	Flavor changing neutral currents
$K_L \to \mu e$	$< 3.3 \times 10^{-11}$	Conservation of lepton number

dimension, and a collimator made of lead and concrete sat at 3.75° to the proton beam. Charged particles were swept away from the collimator by sweeping magnets; photons were blocked by 17 thin lead foils. The resulting neutral beam emerging from the collimator consisted of neutrons and long-lived neutral Kaons (K_L^0), with roughly 20 neutrons per Kaon and 2×10^8 Kaons per second. The beam profile was 5×20 mrad^2. This beam passed through an 11 m vacuum decay tank where most of the Kaons decayed. The Kaons' charged decay products (electrons, muons and pions) were identified and momentum analyzed by the E871 detectors downstream of the decay region. Neutrals remaining after the decay region were absorbed in a large beam stop made of a variety of materials including lead, copper, tungsten, boron and polyethylene. The E871 experiment is shown in figure 1.

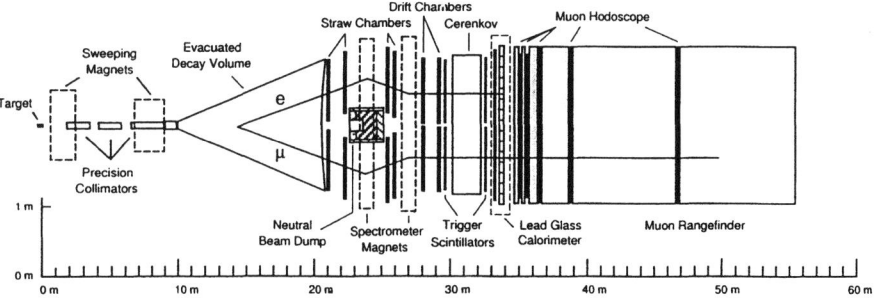

FIGURE 1. Plan view drawing of the E871 experimental apparatus. See text for details.

Several factors allowed E871 to achieve higher sensitivity over the previous best measurement (BNL E791). Primarily, an increased proton yield from the AGS made the experiment possible. Furthermore, E871 adopted several strategies to increase acceptance for the Kaon decays: The collimation angle was moved; the decay tank was lengthened; the spectrometer and detectors were moved closer to the beam.

These improvements to the experiment resulted in a higher Kaon yield and higher acceptance for the Kaon decays. However, they also resulted in a much higher particle flux into the detectors. To accommodate this, the beam stop was placed directly inside the E871 spectrometer. This reduced the hit rates downstream of the stop (where the particle identification and trigger counters were placed), but increased rates in the upstream potion of the spectrometer. Fast, low mass straw detectors were used in this potion of the spectrometer to handle the high flux environment ($\sim 10^4$ cm^2 s^{-1}). In addition, the spectrometer magnets were set to bend the charged products such that a two-body decay results in downstream tracks which are nearly parallel to the beam. This allowed for a trigger which looked for parallel correspondence in the trigger scintillators. Finally, eight parallel VME-based processors performed kinematic reconstruction of the events in real time, reducing the final event rate from ~ 12 kHz to ~ 300 Hz.

To separate signal from background, E871 analysis required good tracks to be reconstructed from the spectrometer. Kinematic signatures of a two-body Kaon decay include an invariant mass equal to the Kaon mass and a reconstructed Kaon momentum vector which points in the known Kaon direction (deduced from the decay vertex). The momentum cut, known as a P_T cut, removes three-body events where the third particle (the neutrino) imparts a "kick" to the remaining two particles. This kinematic requirement removed the dominant backgrounds from $K_L \to \pi\mu\nu$ and $K_L \to \pi e\nu$. Other backgrounds from rarer processes and mistakes in analysis are discussed below.

E871 DETECTORS

The E871 detector system consisted primarily of a large, two-arm spectrometer followed by four particle identification counters. The spectrometer consisted of two bending dipole magnets and twelve tracking stations. The eight upstream stations consisted of low mass straw drift chambers, each straw 5.0 mm in diameter, filled with Ar/CF_4 gas at atmospheric pressure. The four downstream stations were conventional drift chambers filled with Ar/Ethane gas at atmosphere. The straw has a response time of ~ 25 ns; the drift chambers ~ 100 ns. Wires were situated in hexagonally oriented layers, three rows in x-measuring views, two rows in y-measuring views, which greatly facilitated pattern recognition. The magnets were set to produce momentum "kicks" of 0.416 GeV and 0.215 GeV, resulting in nearly parallel tracks for two-body Kaon decays. The beam stop was situated in the center of the upstream magnet.

Particle identification counters sat downstream of the spectrometer. A Cerenkov counter, filled with hydrogen gas at atmospheric pressure, and containing 32 cells performed electron identification based on velocity. An array of 168 lead glass blocks, each 15×15 cm^2, 14 radiation lengths thick, performed electron identification based on electromagnetic shower calorimetry. A muon hodoscope, consisting of 86 scintillating bars sandwiched between iron slabs, provided muon identification. In addition, proportional gas chambers placed in a stack of iron and marble slabs, provided a muon energy measurement based on range.

RATE LIMITATIONS

The E871 system handled the high hit rates (800 kHz/wire) and trigger rates well. In particular, the data acquisition showed no signs of dead-time as beam intensity was increased. Practical limitations during run time were from heating of the platinum production target which typically reached 350° C during each spill, and once reached $> 650°$ C, resulting in an unfortunate melting of the braze between the target and its cooling stem.

Offline analysis, however, reveals that the high rates provide an important potential limitation to the measurement. In particular, multiple hits in the front spectrometer detectors can confuse pattern recognition and event reconstruction. Errors in the front of the spectrometer have a disastrous effect on the invariant mass calculation, which varies roughly with the opening angle. To accommodate high hit rates, pattern recognition employed a maximum likelihood algorithm which incorporated the measured occupancy from noise (1–2% per wire) and inefficiencies. In addition, the two magnets made essentially independent measurement of the momentum, and these were required to agree with each another to within a few percent.

Pattern recognition and event analysis are sometimes fooled by pile-up events, where two decay events overlap in the spectrometer, and the track from one event is paired with the track from another. These background events can usually (though not always) be cut because they fail to reconstruct a good vertex. In addition, E871 analysis checks for relative track timing (to with about one nanosecond) and searches for more than two tracks in the spectrometer in an effort to remove these events.

Due to the beam plug in the spectrometer, rates in the particle ID counters was not a problem.

BACKGROUND LIMITATIONS

Background for the three decays come from different sources. For the $K_L \to \mu\mu$ events, background comes from $K_L \to \pi\mu\nu$ decays where the pion decays into a muon before reaching the muon hodoscope. Many of these events are rejected due to a poorly reconstructed momentum match in the spectrometer (the muon momentum is different from the pion momentum), however, enough remain so that a background subtraction is required. E871 currently counts 6100 $\mu\mu$ events with 650 background events, enough events to reach $\sim 1.3\%$ statistical precision (see figure 2). The largest obstacle to achieving the desired measurement precision of under 2% comes from an incomplete understanding of the relative detection efficiency for $K_L \to \mu\mu$ and $K_L \to \pi\pi$. Pattern recognition in the muon rangefinder is sometimes confused by crossing tracks and the resulting inefficiency is difficult to estimate. To help estimate this effect, we are performing GEANT3 simulations of the muon range stack. In addition, pion interactions in the detector sections affect the efficiency of measuring the $K_L \to \pi\pi$ decays. This effect is estimated by several methods: (1) hand calculation using nuclear tables, (2) Monte Carlo calculation using GEANT3, (3) estimation using measured data taken during special "pion interaction runs". All three methods have the potential of introducing systematic errors and it is not clear at this time which method gives the best estimate.

For the $K_L \to ee$ measurement, the dominant background comes from $K_L \to ee\gamma$ and $K_L \to eeee$. About 75% of the high-mass 4e events (events which reconstruct to an invariant mass of greater than 490 MeV) result in an extra electron entering the spectrometer. These electrons have far too little momentum to make it all the way through the spectrometer; however, they do leave tracks in the front detectors.

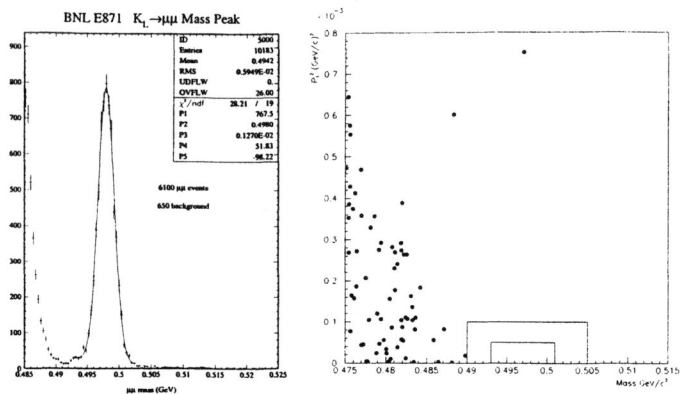

FIGURE 2. Left: Plot of $K_L \to \mu\mu$ signal showing 6100 dimuon events. Right: Plot of background data for $K_L \to ee$. The larger box at the center represents the "blind" area—researchers do not have access to this data. The smaller box represents the probable signal region. Note: *these represent preliminary results only.*

These tracks (called "stubs") are searched for by pattern recognition and the high mass 4e events identified (and cut) with about 65% efficiency. Since the detector system was not built to look for photons, the $K_L \to ee\gamma$ cannot be identified directly. Although both sources of background generally fall short of reproducing Kaon mass, the $K_L \to ee$ events are slightly shifted to lower mass due to inner bremsstrahlung during the decay. An improved measurement for this decay should include a photon detection scheme, as well as the ability to look for extra tracks from the soft electrons. However, if the standard model prediction is correct (9×10^{-12}) [3], then the current signal-to-noise ratio should be a little better than about 5 to 1 with ≈ 4 expected signal events. In this case, one could imagine doing the experiment with higher statistics and using subtraction techniques for handling the background.

For the $K_L \to \mu e$ measurement the principal background comes from $K_L \to \pi e \nu$ where one particle scatters before entering the spectrometer and the pion decays into a muon. To mimic the μe signal, the scattering at the front must be on the order of 10 mrad or more. The dominant process leading to this is Mott scattering from the window of the vacuum decay tank (Mylar and Kevlar) and the material in the first straw tracking chamber. Together, these constitute about 0.003 radiation lengths. Events from this background can fall at high reconstructed mass, but have a tendency to have a non-zero P_T in the plane of the two-body decay. This results in a good vertex, yet a high mass. By cutting explicitly on the P_T component in the scattering plane, E871 can reduce the events from Mott scattered background.

CONCLUSIONS

Experiment 871 took enough data to reach a single event sensitivity of 2×10^{-12}. The experiment appears to have reached its design limit—further sensitivity is limited by background or systematics. In the case of $K_L \to \mu e$, the background from Mott scattering at the front forms the sensitivity limit. For $K_L \to ee$, real physics background competes with signal. For $K_L \to \mu\mu$, systematics in the acceptance calculations form the precision limit. For $K_L \to \mu e$ and $K_L \to ee$, a "blind" analysis is being performed—researchers tune their cuts and algorithms without looking at the signal region. The E871 collaboration expects to "open the box" and present their results by this winter or spring. The $\mu\mu$ analysis should be completed later, probably by summer 1998.

The problems of E871 are not insurmountable, and one could imagine designing a similar experiment for a facility with 20 times the Kaon intensity. Rates appear to be manageable if one rethinks detector locations for a potential factor of 20 increase in flux. Smart triggering and online analysis are a must, but newer computers are already ten times faster than those used in E871. Pattern recognition would require the addition of more tracking stations, particularly at the front, where an accurate track slope measurement is crucial.

Probably the single most important improvement to the experiment would be to add more tracking facilities to the experiment. Backgrounds to all decay modes (especially from $K_L \to \mu e$) come primarily from events with a decayed pion. The E871 tracking system is able to reject these events only if the decay occurs inside the spectrometer where the tracking stations were located. More tracking stations downstream of the spectrometer could (in principle) identify pion decays and reject them.

E871 is poised to present results for the most sensitive measurement of a Kaon decay ever made. The new result, representing an order of magnitude improvement over the previous best measurement, continues an important tradition of pushing our understanding through rare event searches in the Kaon system. Almost all major advances in this field over the past three decades have been driven by the availability of high intensity Kaon sources. Should new sources appear in the future, there is no reason to believe that the tradition can not continue.

REFERENCES

1. Barbieri, R., *Phys. Lett.* **B338**, 212-218 (1994).
2. Lazarides, G., Panagiotakopoulos, C., and Shafi, Q. *Nucl. Phys.* **B278**, 657 (1986).
3. Valencia, G. pre-print, hep-ph/9711377 (1997).

Neutrino Physics

Conveners: Boris Kayser, National Science Foundation
Peter Fisher, MIT

On Future Charged Lepton-Hadron Colliders

S. Ritz[1]

Physics Department, Columbia University
Mail Code 5226, 538 W.120th St., New York, NY 10027

Abstract. Using lessons from HERA, the physics potential of a μp collider facility is explored. The focus here is on kinematics. The physics reach of a μp collider is compared to that of other potential lepton-hadron colliders. The facilities considered are mainly enhancements to other proposed pp or e^+e^- colliders.

I INTRODUCTION AND MOTIVATION

Given the success of past ep experiments and the variety of important results that have already come from HERA, it is worthwhile to review what physics could be explored with a new $l^\pm p$ collider. Realistically, if a new $l^\pm p$ capability will materialize it will most likely come in the form of an add-on to one of the lepton-lepton or pp colliders rather than as a new standalone facility. The discussion here centers on pairing the muon beam from the muon collider with a conventional proton beam. This is an extension of previous work discussed in reference [1].

II KINEMATICS

One of the great advantages of charged lepton-proton collider experiments is that, for the neutral current case at least, the final state is over-constrained. The measurement of the basic kinematic variables, x and Q^2, can be made with just the scattered lepton, or with just the hadrons, or with various combinations of the two. For simplicity, the study here only considers the measurement using the scattered lepton alone. The situation is shown schematically in figure 1 with minimal assumptions about the underlying physical processes. The positive Z direction is taken as the proton beam direction, θ_e is the azimuthal

---

[1] Supported by the National Science Foundation and A.P. Sloan Foundation Fellow

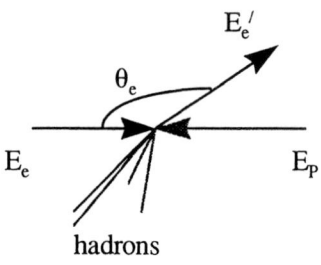

FIGURE 1. Gross features of a NC scatter. The positive, or forward, direction is defined as the direction of the proton beam.

lepton scattering angle, and E_e and E'_e are the incoming and outgoing lepton energies, respectively. The usual DIS kinematic variables are then given by

$$y = 1 - \frac{E'_e}{2E_e}(1 - \cos\theta_e) \tag{1}$$

$$Q^2 = 2E_e E'_e (1 + \cos\theta_e) \tag{2}$$

with

$$x = Q^2/sy \tag{3}$$

where $s = 4E_e E_P$. (At HERA, for example, $E_e = 27.5$ GeV and $E_P = 820$ GeV, so $s \sim 10^5$ GeV^2: a 50 TeV lepton beam would be required to do the equivalent experiment in fixed-target mode.) Combining these, one obtains

$$Q^2 = 4E_e^2 x \frac{1 + \cos\theta_e}{\frac{E_e}{E_P}(1 - \cos\theta_e) + x(1 - \cos\theta_e)} \tag{4}$$

which will be used to study the $\{x, Q^2\}$ range accessible at future machines. Another useful quantity is the virtual photon-proton center-of-mass energy, W, which, at low x is simply $W \sim sy$.

III MESSAGES FROM HERA

A Low x

One of the remarkable HERA results is the enormous rise of F_2 with decreasing x. This translates directly into a virtual photon-proton cross-section that is quickly rising with center-of-mass energy, W (see equation 3). Where and how will this rise slow down? This will likely not be answered at HERA.

A QCD analysis based on DGLAP evolution indicates the rise is driven by a remarkably large gluon density at low x [2]. This evolution must fail as the parton densities saturate, and the familiar QCD evolution will no longer be valid. We can use what we now know about the gluon density at low x to investigate the range of $\{x, Q^2\}$ where this will happen. An estimate of where the gluon density should saturate is given by [3]

$$xg(x, Q^2) \geq \frac{1 fm^2}{1/Q^2} \approx 25 Q^2 (GeV^2). \tag{5}$$

Using a parametrization of the gluon density extracted from present-day data, the region in which this condition is satisfied is shown in figure 2. The experimental signature is a softening in the rise of F_2 with decreasing x in this region. Since F_2 must be rising quickly to begin with (otherwise there would be nothing to soften), the effect will likely be visible only for $Q^2 > 2$ GeV^2, indicated by the solid line.

More generally, HERA has proven that there is a wealth of beautiful physics to explore at low x, and this is clearly an important region for any future lepton-hadron collider facility.

B High x and Q^2

Last year, both HERA experiments ZEUS and H1 reported an excess of events at high x and Q^2. The portion of the $\{x, Q^2\}$ plane containing the possible excess is shown in figure 2.

For the high Q^2 physics potential of a new lepton-hadron collider, we can again look to the experience with HERA. A recent summary is given in reference [4]. For contact interactions, leptoquarks, and R-parity violating SUSY, HERA searches are complementary to those at e^+e^- and pp machines, covering comparable ranges with different assumptions. For many of the new particle searches it is relatively easy to explore up to the kinematic limit (usually $\sqrt{s}$).

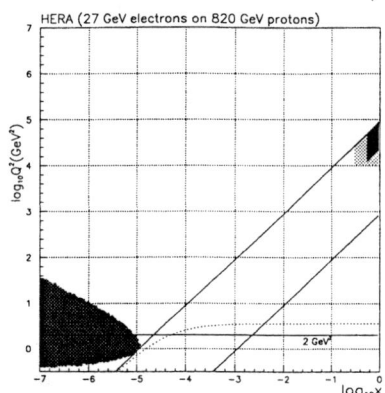

FIGURE 2. Messages from HERA. The region in the $\{x, Q^2\}$ plane in which the gluon density is estimated to show signs of saturation is shown in red at the lower left corner. The effects may only be perceptible for $Q^2 > 2$ GeV^2, indicated by the solid line. The high x, high Q^2 regions that may contain an excess of events beyond the Standard Model prediction are shown in blue and yellow.

IV KINEMATIC COVERAGE OF VARIOUS MACHINES

We use equation 4 to explore the sensitivities of various machines to new physics. The practical range of measurements is limited. We require

1. $y > 0.01$ since the resolution on x degrades unacceptably at low y,

2. $\theta_e > 10°$ since it will be difficult to identify the scattered lepton in the hadronic splash from the proton remnant,

3. $\theta_e < 179°$ since the lepton has to leave the beamline to be measured.

Note that these limitations could be relaxed by an aggessive detector design, but the assumptions become speculative. There is already good experience at HERA doing measurements with cuts like these, so we adopt them for the present study. It is important to state again that we are ignoring the information from the hadrons, which in some cases might expand the kinematic coverage.

The results are shown in figure 3, with the regions of interest overlayed as a reference. The measurable domain, from above requirements, along with the kinematic limit $y < 1$, is contained within the solid lines. Some conclusions are immediately clear:

1. To cover the low x region, the optimal configuration is obtained by pairing a very high energy proton beam with a relatively low energy lepton beam.

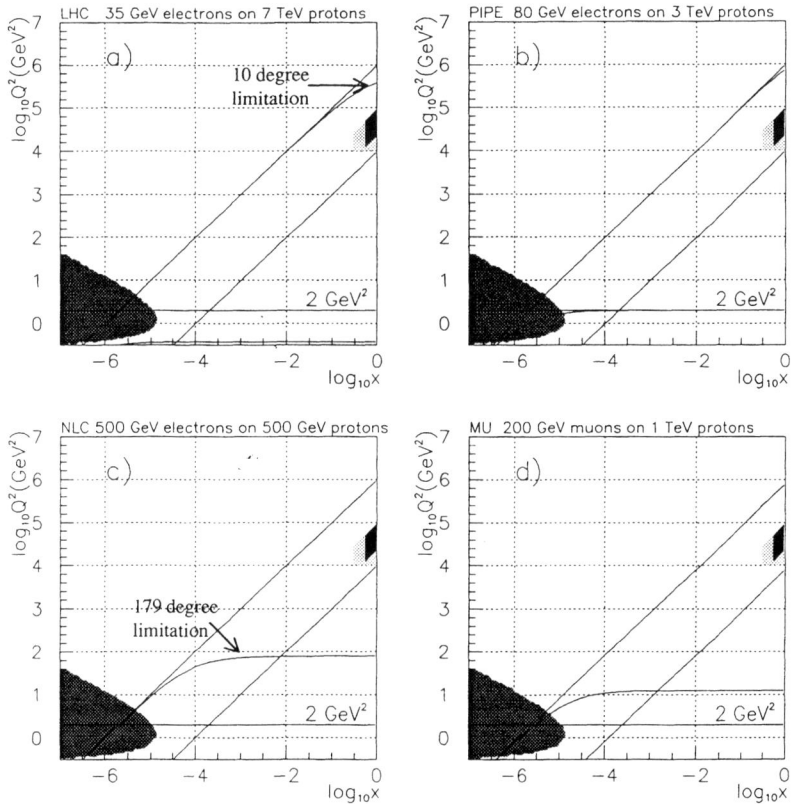

FIGURE 3. The region of the kinematic plane accessible at various lp collider configurations. The diagnal lines form a band corresponding to the requirement $0.01 < y < 1.0$.
(a) Using the LHC proton beam with a conventional 35 GeV electron beam. The limitation at very high Q^2 is due to the $\theta_e > 10°$ requirement. (b) Using a high energy low-field proton beam (3 TeV) with a 80 GeV electron beam. (c) Using a 500 GeV NLC electron beam with a conventional 500 GeV proton beam. The *lower* limit on Q^2 is due to the $\theta_e < 179°$ requirement. Such a configuration is limited to $Q^2 > 100$ GeV^2 and $x > 10^{-4}$, but provides access to very high Q^2. (d) Using a 200 GeV muon collider beam with a 1 TeV proton beam. This configuration accesses the highest Q^2 values, but is limited at low x to $x > 10^{-5}$. The interesting frontier regions based on lessons from HERA are overlayed on each as a reference.

The $\theta_e < 179°$ requirement implies a lower limit on Q^2: for high energy lepton beams, even a 1 degree scatter from the incident direction implies a large Q^2, essentially independent of proton beam energy. The large relative boost of the proton system brings very low x scatters more into the central region of the detector.

2. Conversely, for such an asymmetric configuration, the high Q^2 reach is limited by the $\theta_e > 10°$ requirement. To explore the highest Q^2 physics, a more symmetric pairing of beam energies is desirable.

Thus, a very modest conventional lepton beam paired with a very high energy proton beam is useful for low x physics, while an NLC or muon collider beam would best be paired with a more conventional proton beam to explore high Q^2 physics.

The μp collider option considered here will cover about one decade more in Q^2 than HERA. This is not uninteresting. The unfortunate news, however, is that low x will be *very* difficult to explore with a μp collider. Figure 4 shows contours of constant scatter muon angle. It does not seem practical to measure the momentum and angle of muons so close to the beamline, where backgrounds are likely to be enormous.

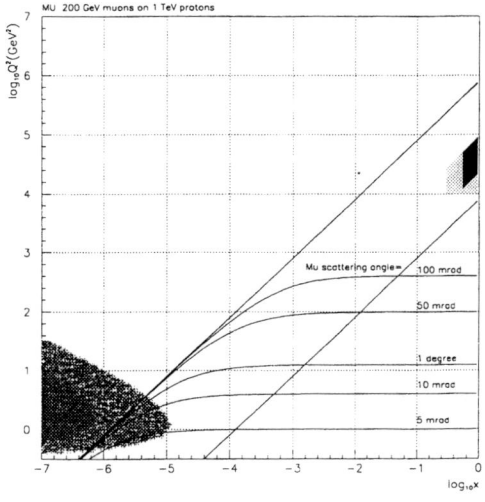

FIGURE 4. Contours of constant muon scattering angle, with respect to the incident muon beam direction, in the x and Q^2 plane.

SUMMARY

We have started to investigate the physics potential of a μp collider. The ep collider approach was a great success, opening up two orders of magnitude at both high Q^2 and low x. HERA is a tough act to follow! A 200 GeV muon beam colliding with a 1 TeV proton beam would open up approximately one decade in Q^2 beyond the reach of HERA. It is not clear when such a collider would be operational, so it is fair to ask whether this regime will still be interesting in the future. Unfortunately, low x physics is best explored with asymmetric beams (very high energy proton beam with a modest lepton beam); in this case, the higher energy muon beam available at a muon accelerator facility is not an asset.

ACKNOWLEDGEMENTS

I thank S. Geer and R. Raja for organizing a successful and informative workshop. I also thank my co-organizer of the DIS group, Heidi Schellman, for doing most of the hard work for the group and for her dry humor.

REFERENCES

1. S. Ritz, 'On Future Lepton-Hadron Colliders', in Proceedings of Future High Energy Colliders, AIP **397**, Z. Parsa, ed. (1997).
2. ZEUS Collaboration, M. Derrick et al., Phys. Lett. B345 (1995)576.,
   H1 Collaboration, S. Aid et al., Phys. Lett. B354(1995)494 .
3. A.H. Mueller, J. Phys G19(1993)1463.
4. Proc. 1995-96 HERA Workshop; Beyond the Standard Model Group Summary, H. Dreiner, H.-U. Martyn, S. Ritz, and D. Wyler; also HEP-PH/9610232.
5. ZEUS Collaboration, M. Derrick et al., Z.Phys.C74(1997)207;
   H1 Collaboration, C. Adloff et al., Z.Phys.C74(1997)191.

Neutrino Physics at a Muon Collider

Bruce J. King

Brookhaven National Laboratory
email: bking@bnl.gov

Abstract. An overview is given of the neutrino physics potential of future muon storage rings that use muon collider technology to produce, accelerate and store large currents of muons.

INTRODUCTION

This paper gives an overview of the neutrino physics possibilities at a future muon storage ring, which can be either a muon collider ring or a ring dedicated to neutrino physics that uses muon collider technology to store large muon currents.

After a general characterization of the neutrino beam and its interactions, some crude quantitative estimates are given for the physics performance of a muon ring neutrino experiment (MURINE) consisting of a high rate, high performance neutrino detector at a 250 GeV muon collider storage ring.

The paper is organized as follows. The next section describes neutrino production from a muon storage rings and gives expressions for event rates in general purpose and long baseline detectors. This is followed by a section outlining a serious design constraint for muon storage rings: the need to limit the radiation levels produced by the neutrino beam. The following two sections describe a general purpose detector and the experimental reconstruction of interactions in the neutrino target then, finally, the physics capabilities of a MURINE are surveyed.

NEUTRINO PRODUCTION AND EVENT RATES

Neutrinos are emitted from the decay of muons in the collider ring:

---

[1] This work was performed under the auspices of the U.S. Department of Energy under contract no. DE-AC02-76CH00016.

$$\mu^- \to \nu_\mu + \overline{\nu}_e + e^-,$$
$$\mu^+ \to \overline{\nu}_\mu + \nu_e + e^+. \quad (1)$$

The thin pencil beams of neutrinos for experiments will be produced from long straight sections in either the collider ring or a ring dedicated to neutrino physics. From relativistic kinematics, the forward hemisphere in the muon rest frame will be boosted, in the lab frame, into a narrow cone with a characteristic opening half-angle, θ_ν, given in obvious notation by

$$\theta_\nu \simeq \sin\theta_\nu = 1/\gamma = \frac{m_\mu}{E_\mu} \simeq \frac{10^{-4}}{E_\mu(\text{TeV})}. \quad (2)$$

For the example of 250 GeV muons, the neutrino beam will have an opening half-angle of approximately 0.4 mrad. The final focus regions around collider experiments are important exceptions to equation 2 since the muon beam itself will have an angular divergence in these regions that is large enough to significantly spread out the neutrino beam.

For TeV-scale neutrinos, the neutrino cross-section is approximately proportional to the neutrino energy, E_ν. The charged current (CC) and neutral current (NC) interaction cross sections for neutrinos and antineutrinos have numerical values of [1]:

$$\sigma_{\nu N} \text{ for } \begin{pmatrix} \nu\_CC \\ \nu\_NC \\ \overline{\nu} - CC \\ \overline{\nu} - NC \end{pmatrix} \simeq \begin{pmatrix} 0.72 \\ 0.23 \\ 0.38 \\ 0.13 \end{pmatrix} \times \frac{E_\nu}{1 \text{ TeV}} \times 10^{-35} \text{ cm}^2. \quad (3)$$

These cross sections are easily converted into approximate experimental event rates for the example of a 250+250 GeV collider with a 200 meter straight section and the example design parameters used for this workshop. For a general purpose detector subtending the boosted forward hemisphere of the neutrino beam:

$$\text{Number of} \begin{pmatrix} \nu_\mu - CC \\ \nu_\mu - NC \\ \overline{\nu}_e - CC \\ \overline{\nu}_e - NC \end{pmatrix} \text{events/yr} \simeq \begin{pmatrix} 2.6 \\ 0.8 \\ 1.4 \\ 0.5 \end{pmatrix} \times 10^7 \times l[\text{g.cm}^{-2}], \quad (4)$$

where l is the detector length. For a long baseline detector in the center of the neutrino beam:

$$\text{Number of} \begin{pmatrix} \nu_\mu - CC \\ \nu_\mu - NC \\ \overline{\nu}_e - CC \\ \overline{\nu}_e - NC \end{pmatrix} \text{events/yr} \simeq \begin{pmatrix} 1.4 \\ 0.4 \\ 0.7 \\ 0.2 \end{pmatrix} \times 10^7 \times \frac{M[\text{kg}]}{(L[\text{km}])^2}, \quad (5)$$

where M is the detector mass and L the distance from the neutrino source.

These event rates are several orders of magnitude higher than in today's neutrino beams from accelerators.

POTENTIAL RADIATION HAZARD

The neutrino fluxes strong enough to constitute a potential off-site neutrino radiation hazard [2]. The problem comes where the neutrinos from the collider ring exit the Earth's surface and their interactions initiate showers of ionizing particles in people and their surroundings. The neutrino interaction cross section is tiny but this is greatly compensated by the huge numbers of neutrinos. A simple but conservative order-of-magnitude calculation [2] predicts the following maximum radiation dose downstream from a straight section of length S:

$$\frac{\text{Radiation dose}}{\text{U.S. Fed. limit}} \simeq 0.3 \times \left(\frac{S}{\text{collider depth}}\right) \times \left(\frac{\text{muon current}}{6 \times 10^{20}\ \mu^-/\text{yr}}\right) \times \left(\frac{E_\mu}{250\ \text{GeV}}\right)^3 \tag{6}$$

The first bracket on the right hand side is valid when the collider ring is not tilted and the land around the collider is flat enough to assume the approximation of a spherical Earth. In practice, a dedicated ring for neutrino physics would almost certainly be tilted downwards towards a long-baseline neutrino experiment of order 1000 km away. Obviously, the radiation hazard would be much reduced for this orientation. The muon current of 6×10^{20} is the example design specification for this workshop.

The cubic dependence on energy means that neutrino radiation is a serious design constraint for colliders at the TeV center-of-mass (CoM) energy scale and above. In practice, this is partially compensated by the increasing luminosity per unit current that can be obtained at higher energies, and high luminosity designs with decreased muon current appear to be feasible to at least the 10 TeV CoM energy scale [3].

A GENERAL PURPOSE NEUTRINO DETECTOR

Figure 1 is an example of the sort of high rate general purpose neutrino detector that would be well matched to the intense neutrino beams. Note the contrast with the kilotonne-scale calorimetric targets used in today's high rate neutrino experiments.

The neutrino target is a stack of CCD tracking planes. A target vessel containing hydrogen could equally well be used, alternating between runs with protium and deuterium. The target is 1 meter long and has a 10 cm radius, which matches the beam radius at approximately 200 meters from production for a 250 GeV muon beam. Besides providing the mass for neutrino interactions, such a detector allows precise reconstruction of the event topologies from charged tracks, including event-by-event vertex tagging of those events containing charm or beauty hadrons (or tau leptons). The vertexing geometry

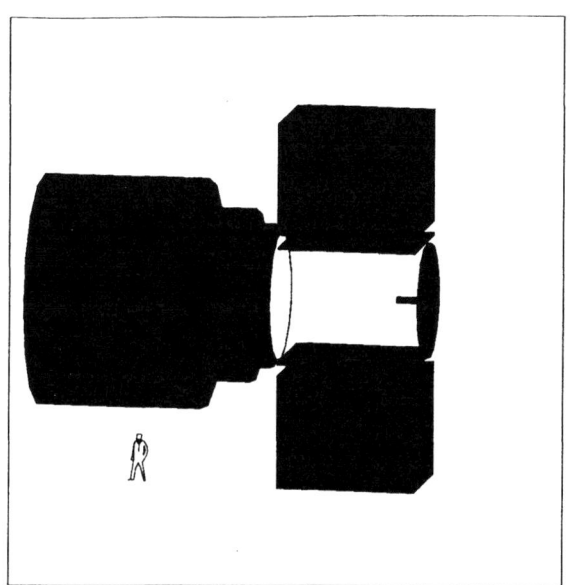

FIGURE 1. Example of a general purpose neutrino detector. A human figure in the lower left corner illustrates its size. The neutrino target is the small horizontal cylinder at mid-height on the right hand side of the detector. Its radial extent corresponds roughly to the radial spread of the neutrino pencil beam, which is incident from the right hand side. Further details are given in the text.

can be made essentially optimal for vertex tagging, with 3.5 micron hit resolution or better (this is the resolution of the CCD's at the SLD detector), shorter extrapolations to secondary vertices than is possible in collider experiments and negligible acceptance inefficiency. It is reasonable to expect almost 100 percent efficiency for b tagging, perhaps 70 to 90 percent efficiency for charm tagging (depending on the tag purity and how 1-prong decays are treated), and excellent discrimination between b and c decays.

The optimization of the number of planes and their thickness is a trade-off between increased detector mass and degradation of the tracker performance due to multiple coulomb scattering and electromagnetic and hadronic secondary interactions in the tracker. For example, 750 planes of 300 micron thick silicon CCD's corresponds to a mass per unit area of approximately 50 g.cm^{-2}, about 2.5 radiation lengths and 0.5 interaction lengths. This seems reasonable for a general purpose detector, while detector masses perhaps up to an order of magnitude larger might be considered for a very high rate experiment dedicated to rare processes. In the neutrino beam assumed for equation 4, such a general purpose detector would record a very healthy 2×10^9 CC interactions per year and even higher statistics are obviously possible using a higher mass target, higher energy neutrino beam and/or a dedicated muon ring.

In figure 1 the target is surrounded by a gas-based tracking subdetector, using a time projection chamber (TPC) geometry with a vertical magnet field provided by a dipole magnet with poles above and below the tracker. The characteristic dE/dx signatures from the tracks would identify each charged particle. Further particle ID is provided by the Cherenkov photons produced in the gas tracker, which are reflected by a spherical mirror at the downstream end of the tracker and focused onto a read-out plane at the upstream end of the target. If desired, a third level of electron ID could be provided by transition radiation detectors (TRD's) behind the mirror. The mirror is backed by electromagnetic and hadronic calorimeters and, lastly, by iron-core toroidal magnets for muon ID.

NEUTRINO INTERACTIONS AND THEIR EXPERIMENTAL INTERPRETATION

The dominant interaction of TeV-scale neutrinos is deep inelastic scattering (DIS) off nucleons (i.e. protons and neutrons). Scattering off atomic electrons also occurs, but at a rate which is lower by three orders of magnitude. There are 2 types of DIS: neutral current (NC) and charged current (CC) scattering, as follows.

In neutral current (NC) scattering, the neutrino is deflected by a nucleon (N) and loses energy with the production of several hadrons (X):

$$\nu + N \to \nu + X, \tag{7}$$

This comprises about 25 percent of the total cross section and is interpreted as elastic scattering off one of the many quarks inside the nucleon through the exchange of a virtual neutral Z boson:

$$\nu + q \to \nu + q. \tag{8}$$

Charged current (CC) scattering is similar to NC scattering except that the neutrino turns into its corresponding charged lepton:

$$\nu + N \to l^- + X,$$
$$\bar{\nu} + N \to l^+ + X, \tag{9}$$

where l is an electron/muon for electron/muon neutrinos. At the more fundamental quark level a charged W boson is exchanged with a quark (q), which is turned into another quark species (q') whose charge differs by one unit.

$$\nu + q \to l^- + q'. \tag{10}$$
$$\bar{\nu} + q' \to l^+ + q. \tag{11}$$

The relativistically invariant quantities that are routinely extracted in DIS experiments are 1) Feynman x, the fraction of the nucleon momentum carried by the struck quark, 2) the inelasticity, $y = E_{\text{had}}/E_\nu$, which is related to the scattering angle of the neutrino in the neutrino-quark CoM frame, and 3) the momentum-transfer-squared, $Q^2 = 2M_{proton}E_\nu xy$. MURINE's will have the further capability of reconstructing the hadronic 4-vector, resulting in a much better characterization of each interaction.

Equation 9 shows that the interactions of muon- and electron-flavor neutrinos are essentially identical for NC interactions but easily distinguishable in CC interactions by the flavor of the final state charged lepton. Similarly, the charge of the final state lepton clearly distinguishes the CC interactions of neutrinos from antineutrinos. Even the NC interactions of neutrinos and antineutrinos are distinguishable on a statistical basis because their y distributions differ for reasons involving the differing helicity states of neutrinos and antineutrinos.

The final state quark always "hadronizes" at the nuclear distance scale, combining with quark-antiquark pairs to produce the several hadrons seen in the detector. It is an important aspect of the MURINE that the hadronic final state will retain some memory of the final state quark flavor since this quark must still exist in one of them. This is a particularly good signature for the neutrino-induced production of the heavy charm and beauty quark flavors since these flavors of quark-antiquark pairs are very rarely produced during hadronization. Hence, the observation of the characteristic displaced vertex of a charm or beauty hadron in an event gives a fairly reliable identification of c or b production from the neutrino interaction.

Some statistically based flavor tagging will be available even when the final state quark is one of the lighter u, d or s quarks. This comes from the so-called "leading particle effect" and is currently used, for example, in analyses at the LEP experiments. The hadron containing the struck quark is known to be more energetic on average than the others, so the quark content of the most energetic, i.e. "leading", hadron provides a potential tag for the final state quark flavor.

A particularly good example is those events where most of the hadronic energy is concentrated in a kaon-anti-kaon pair from the $\phi(1020)$ resonance. The $\phi(1020)$ consists of an s quark and s antiquark, so such events can reliably be interpreted as a final state strange quark (or antiquark) that has subsequently combined with its strange antiparticle during hadronization.

The relatively high purity $\phi(1020)$ tag is one of several possible final state tags. Events with most of the hadronic energy concentrated in a single kaon provide another tag for a strange quark final state which has lower purity but higher efficiency and, clearly, the complementary sample of events with no high energy kaons will be enriched in interactions with u and d final state quarks.

As an important technical detail, it appears that the efficiency and purity of the various u, d and s tags should be measurable with little uncertainty by taking advantage of the various possible calibration event samples. For example, neutrino interactions with the u and d valence quark components of protons and neutrons (which dominate at high x) will produce accurately predictable fractional compositions of u, d and/or s quark final states. Also, as a more general handle, charge conservation dictates that the hadron containing the final state quark from neutrino (anti-neutrino) CC interactions must be either neutral or positively (negatively) charged, so the production of wrong-sign high energy hadrons gives a direct measure of the impurity of each charged hadron tag. To the extent that such calibration analyses can remove systematic uncertainties in the final state tagging of light quarks, the only price to pay for the statistical nature of these tags will be an effective dilution of the event statistics.

PHYSICS OPPORTUNITIES

Neutrino interactions are interesting both in their own right and as probes of the quark content of nucleons, so a MURINE has wide-ranging potential to make advances in many areas of research in elementary particle physics. There is insufficient space to do justice to all the physics possibilities and it actually seems almost easier to list the areas that can't be studied at a MURINE! Significant exceptions are some types of rare decay processes, studies involving the decay of b hadrons and the direct production (as opposed to virtual production) of particles heavier than b hadrons.

The first part of this section gives a discussion of one area the author finds particularly interesting – measurements involving the CKM quark mixing matrix. Briefer overviews are then given of several other areas of physics. These include tests of nucleon structure and QCD, electroweak measurements, neutrino oscillations, the search for exotic particles and, finally, studies of charmed hadrons.

For a benchmark event sample, CC statistics of 10^{10} events are assumed for the general purpose detector, corresponding to 5 years running with a target mass of 50 g.cm^{-2}. This sample size is several orders of magnitude higher than in today's experiments. In all cases, any quantitative predictions for the expected precision of analyses are little more than guesses based on cursory examinations of quark distributions etc. and rough comparisons with analyses of existing neutrino experiments. Clearly, further work in this area is both needed and interesting.

CKM Quark Mixing Matrix

There is considerable theoretical interest in the mixture of final state quarks produced in CC interactions. The struck quark can be converted into any of the three final state quarks that differ by one unit of charge: a down (d), strange (s), or bottom(b) quark can be converted into an up (u), charmed (c), or top (t) quark and vice versa. In practice, production of the heavy top quark is kinematically forbidden at these energies and the production of other quark flavors is influenced by their mass. Beyond this, the Standard Model (SM) predicts the probability for the interaction to be proportional to the absolute square of the appropriate element in the so-called Cabbibo-Kobayashi-Maskawa (CKM) quark mixing matrix. The squared terms are given in table 1 [4].

TABLE 1. Quark mixing probabilities. Threshold suppression due to quark masses has been neglected. In practice, this will reduce the mixing probabilities to the heavier c and b quarks to below the values given in the table and will prevent any mixing to the top quark.

	d	s	b
u	0.95	0.05	1×10^{-5}
c	0.05	0.95	0.002
t	1×10^{-4}	0.001	1

TABLE 2. Percentage uncertainties in quark mixing probabilities. The two terms in brackets have not been measured directly from tree level processes.

	d	s	b
u	0.1%	1.6%	50%
c	15%	35%	15%
t	25%	(40%)	30%

The 4 independent parameters in the underlying 3-by-3 unitary CKM matrix are as fundamental as the masses of elementary particles and, like particle masses, their values are phenomenological parameters to be determined by experiment rather than being predicted by the SM. One of the parameters is a complex phase that is postulated as an explanation for CP violation: the intriguing experimental phenomenon that particles may have tiny deviations from the properties that mirror those of their antiparticles. This phase is poorly constrained by experiment and, in fact, experimental measurements involving the CKM matrix are sufficiently difficult that even the unitary form of the matrix as predicted by the SM is not particularly well established. Improved measurements involving the CKM matrix will test the SM hypotheses and, speculatively, the values and pattern of the parameters might even help provide insights towards some deeper future theory that explains such mysteries of the SM as why nature chooses to have 3 quark generations.

The current percentage uncertainties in the 9 mixing probabilities are given in table 2 [4]. One of the CKM mixing probabilities – that between the d and c quarks – is already best measured in neutrino-nucleon scattering. It appears that the new realm of neutrino experiments at muon colliders could make a dominant contribution to this field, with measurements of perhaps 4 of the 9 mixing probabilities that should be far better than possible in any other type of experiment. The relevant transitions are that between u and b quarks, and those between c quarks and d, s or b quarks.

The d-to-c transition is the cleanest of the 4 measurements, with statistics of order 10^8 events, a well measured distribution of valence d quarks at high x and clean event-by-event vertex tagging of the charm final state. The threshold effect due to the charm quark mass must also be modeled, but this should be little problem with such statistics and the measurement precision should reach the parts-per-mil level.

Measurement of the s-to-c transition will also benefit from of order 10^8 statistics and clean charm tagging. The major difficulty here will be an incomplete knowledge of the initial strange sea distribution, and presumably this will estimated using neutral current interactions with a resonant $\phi(1020)$ final state, as explained in the preceding section. The resulting measurement accuracy might be at the percent level. As in analyses of today's neutrino experiments, the d and s contributions to the charm event sample will in practice be separated from one another in a fit involving the x distributions etc. and with the charm quark mass as a fitted parameter.

A similar analysis will be applied for the two transitions with a b quark in the final state: u-to-b and c-to-b. The statistics for these two processes will be perhaps of order 10^4 events, depending on the energy of the muon storage ring and the consequent threshold suppression due to the b quark mass. (The b quark mass is approximately 5 GeV compared to about 1.5 GeV for the charm quark. These masses would be measured with unique precision in a MURINE.) Besides the threshold suppression, other reasons for the reduced statistics are that the u-to-b transition has a low mixing probability – about 10^{-5} – while the initial charm quark distribution (either an intrinsic charm sea or charm production through higher order Feynman diagrams) will much smaller than those of the lighter quark.

Given the nearly optimal vertexing geometry possible at a MURINE, the vertexing discrimination between b and c hadrons should be able to separate out much of the fractionally small b hadron event sample. (Technically, this can be done by requiring a secondary vertex with a total invariant mass larger than possible for a charm hadron and/or using other "topological vertexing" signatures.) If this is so then the u-to-b analysis should be a relatively straightforward copy of the d-to-c analysis, and the precision of the measurement may approach the statistical limit of around 1 percent. In contrast, the c-to-b transition has the additional challenge of estimating the initial charm quark spectrum using NC charm events and/or CC scattering involving the c-to-s transition, and this may well limit the measurement accuracy to the few percent level.

In summary, a first look indicates the opportunity for tremendous improvements in measuring the quark mixing matrix. More detailed studies are clearly desirable.

Structure Functions and QCD

Another major motivation for MURINE's is the potential for greatly improved measurements of nucleon structure functions (SF) and the consequent tests of quantum chromodynamics (QCD) – the theory of the strong interaction that is widely accepted for its elegance and simplicity but which has not been experimentally verified at the level of the electroweak theory.

Stated simply, quark SF's are the momentum distributions of the quarks inside the nucleon. They are assumed to be universal properties of nucleons which can also be measured and compared at charged lepton scattering experiments, either at a fixed target or a collider (of which HERA is the only operating example). QCD predicts a weak dependence of the SF on the energy transferred in the scattering interaction and the improved verification of this prediction and of various "quark sum rules" at a MURINE will constitute some of the best tests for QCD and also provide one of the best measurements of the coupling strength parameter of the strong interaction, α_s.

Neutrino scattering experiments differ from their charged lepton counterparts in that the scattering proceeds predominantly through the exchange of W and Z bosons, rather than photons. This gives neutrino scattering experiments two important advantages:

1. the heavy mass scale of the W and Z bosons naturally produces hard scattering interactions that can be well treated by assuming scattering off a quasi-free quark with QCD corrections. In contrast, charged lepton scattering suffers from a huge background of soft photon interactions that provide experimental rate problems and are not very amenable to analysis using perturbative QCD.

2. the relatively evenly divided event statistics from W^+, W^- and Z interactions provides much more discrimination between the quark flavors than is possible with a single scattering probe. For example, neutrino experiments have the unique capability to disentangle the valence quark SF without even requiring identification of the final state quarks.

Beyond the intrinsic advantages of neutrino interactions, new capabilities will arise from a MURINE with its precision high rate detector. For the first time, the precisely known beam spectra and detailed reconstruction of the hadronic final state will allow precise SF from NC interactions, while measurement of CC event kinematics will be overconstrained and hence lead to extremely precise CC structure functions with minimal systematic uncertainties. Further, the new capability to identify the final state quark will enable flavor-by-flavor SF measurements. The use of several different target materials will also allow studies of nuclear effects, polarized structure functions and separate SF for protons and neutrons. Finally, the reconstruction of the hadronic final state should, for the first time in lepton scattering experiments, allow high statistics studies of gluon splitting and jet topologies similar to those done, e.g., on the LEP sample of Z decays.

In summary, a MURINE should greatly enrich and extend our knowledge of nucleon structure functions and might well be the best single experiment of any sort for the examination of perturbative QCD.

Electroweak Parameters

Neutrino physics has had an important historical role in measuring the electroweak mixing angle, which is simply related to the mass ratio of the W and Z intermediate vector bosons:

$$\sin^2 \theta_W \equiv 1 - \left(\frac{M_W}{M_Z}\right)^2. \tag{12}$$

(To be precise, this equation is the Sirlin on-shell definition of $\sin^2 \theta_W$ that is conventionally used in neutrino physics. Several other definitions of $\sin^2 \theta_W$ that differ from equation 12 by small amounts are used for convenience in other types of experimental measurement.)

Now that M_Z has been precisely measured at LEP, measurements of $\sin^2 \theta_W$ in neutrino physics can be directly converted to predictions for the W mass. The comparison of this prediction with direct M_W measurements in collider experiments constitutes a precise prediction of the SM and a sensitive test for exotic physics modifications to the SM. Today's neutrino measurements correspond to an uncertainty in M_W of $\Delta M_W = 180$ MeV [5], with contributions from both event statistics and experimental and theoretical systematic uncertainties.

A crude estimate of the predicted uncertainty in M_W from a MURINE analysis was made by reviewing each of the uncertainties in a contemporary $\sin^2 \theta_W$ analysis from the CCFR collaboration [6]. This exercise was made more difficult by the enormously improved experimental conditions and the consequently large extrapolations in experimental accuracy.

The huge statistics at a MURINE should reduce the M_W statistical uncertainty to only a couple of MeV and, given the high quality beam and high performance detector, the experimental uncertainties may well be reducable to a similar level. Further, all of the large sources of large theoretical systematic uncertainties in the CCFR analysis – such as the threshold suppression of charm production, longitudinal SF and the heavy flavor content of the nucleon sea – should be controllable through direct measurements in a MURINE. Radiative corrections, which can't be measured directly, were assigned an uncertainty corresponding to only about 5 MeV. Based on these observations, it appears that MURINE's have the potential for precisions more than an order of magnitude better than today's $\sin^2 \theta_W$ measurements. Thus, W mass predictions with precisions of order 10 MeV seem achievable. This is comparable with the projected best direct measurements from future collider experiments.

Another electroweak measurement of interest for MURINE's is the $\sin^2 \theta_W$ measurement from neutrino scattering off electrons in the target:

$$\nu e^- \to \nu e^- \quad \text{(NC)}$$
$$\nu e^- \to l\, \nu_e \quad \text{(CC)}. \tag{13}$$

This is a theoretically clean process that turns out to give interesting physics information orthogonal to the measurement from neutrino-quark scattering. However, it has been experimentally less accessible because the cross section is lower by three orders of magnitude. Current measurements are limited both by statistics and from the reconstruction limitations of today's high mass neutrino targets, both of which would be enormously improved at a MURINE.

Other new and/or greatly improved electroweak measurements could be performed peripherally to these major analysis topics. These include determinations of the left-handed and right-handed neutrino-quark couplings for each quark flavor, and a precise check of the SM prediction for the inverse muon decay cross section.

A   Neutrino Oscillations

A neutrino property that is currently drawing much interest is the question of whether neutrinos have a non-zero mass. If they do then it is possible that the 3 neutrino flavors mix to produce neutrino oscillations that can be observed using a neutrino beam. The probability for an oscillation between two of the flavors is given by [7]:

$$\text{Oscillation Probability} = \sin^2 \theta \times \sin^2 \left(1.27 \frac{\Delta m^2[\text{eV}^2].L[km]}{E_\nu[GeV]} \right). \tag{14}$$

The first term gives the mixing strength and the second term gives the distance dependence. The discovery potential for a neutrino experiment is conventionally expressed as a characteristically shaped [7] region in an exclusion/discovery plot of Δm^2 vs. $\sin^2\theta$. The projections of this plot on the x and y axes give the limits on mixing strength for most favorable mass difference and on the mass difference for full mixing, respectively. These two limits are now crudely estimated, with both estimates applying generically to all 3 possible mixings between 2 flavors:

$$\nu_e \leftrightarrow \nu_\mu,$$
$$\nu_e \leftrightarrow \nu_\tau,$$
$$\nu_\mu \leftrightarrow \nu_\tau. \qquad (15)$$

The best mass difference limit would come from a long baseline experiment. The background-free 90 % confidence limit of 2.3 oscillated events becomes a reasonable approximation with a low enough event rate and with cuts applied so that only reliably tagged oscillations are accepted. If it is crudely assumed that this corresponds to a 10 % tagging efficiency for oscillated events then 5 years running at a 10 kilotonne long-baseline experiment is easily found to give the following order-of-magnitude mass limit for full mixing:

$$\Delta m^2|_{min} \sim O(10^{-4})\, eV^2, \qquad (16)$$

independent of the distance to the detector. This is more than an order of magnitude better than any proposed accelerator or reactor experiments for $\nu_\mu \leftrightarrow \nu_\tau$ and $\nu_e \leftrightarrow \nu_\tau$, and competitive with the best such proposed experiments for $\nu_e \leftrightarrow \nu_\mu$.

The limit on mixing strength would benefit not only from the neutrino beam properties but also from the high performance of the general purpose neutrino detector. The overconstrained event kinematics for CC events (see the subsection on SF's) should greatly benefit the selection of a clean potential oscillation sample. The search for oscillations to tau neutrinos would also take advantage of the exceptional vertex tagging abilities of such a detector. The huge event samples would allow an analysis strategy of cutting hard on event kinematics to keep only very well reconstructed events with an unambiguous primary lepton. Assuming that 1 in 100 oscillating events would pass the cuts and requiring a signal of 10 such events corresponds to a mixing probability sensitivity for 10^{10} events of:

$$\sin^2\theta|_{min} \sim O(10^{-7}). \qquad (17)$$

This would be a unique sensitivity for each of the three possible oscillations – orders of magnitude better than in any other current or proposed experiment.

Searches for Exotic Particles

It is clear that MURINE's will offer expanded opportunities for searches for new types of exotic particles that couple to neutrinos, such as some hypothesized types of neutral heavy leptons.

B Charm Physics

Interestingly, MURINE's should be rather impressive factories for the study of charm, with a clean, well reconstructed sample of several times 10^8 charmed hadrons produced in 10^{10} neutrino interactions. The particle ID and energy/momentum capabilities of the detector should facilitate the full reconstruction of a good fraction of the decay final states, and the precision vertexing should give accurate event-by-event lifetime information. It is of particular importance to oscillation and CP studies that the production sign of the charm quark is tagged by the final state lepton charge in CC interactions:

$$\nu q \to l^- c$$
$$\overline{\nu} q' \to l^+ \overline{c}. \qquad (18)$$

There are several interesting physics motivations for charm studies at a MURINE [8]. Measurement of charm decay branching ratios and lifetimes are useful for both QCD studies and for the theoretical calibration of the physics analyses on B hadrons. Charm decays also provide a "clean laboratory" to search for exotic physics contributions, with the SM predicting 1) tiny branching ratios for rare decays, 2) small CP asymmetries and 3) slow $D^0 \to \overline{D^0}$ oscillations, with only of order 1 in 10^4 oscillating before decay. In fact, $D^0 - \overline{D^0}$ mixing [9] has yet to be observed, and it is quite plausible that a MURINE would provide the first observation.

SUMMARY

The intense neutrino beams at muon collider complexes should usher in an exciting new era of neutrino physics experiments. It has been shown that great advances are to be expected in traditional areas of neutrino physics and elsewhere.

REFERENCES

1. See, for example, Chris Quigg, *Neutrino Interaction Cross Sections*, FERMILAB-Conf-97/158-T.
2. B.J. King, *Assessment of the prospects for muon colliders*, paper submitted in partial fulfillment of requirements for Ph.D., Columbia University, New York (1994); B.J. King, *A Characterization of the Neutrino-Induced Radiation Hazard at TeV-Scale Muon Colliders*, BNL Center for Accelerator Physics internal report 162-MUON-97R, to be submitted for publication.
3. Private communications within the Muon Collider Collaboration.
4. Values extracted from Andrzej J. Buras, *CKM Matrix: Present and Future*, TUM-HEP-299/97.

5. Janet M. Conrad, Michael H. Shaevitz and Tim Bolton, *Precision Measurements with High Energy Neutrino Beams*, hep-ex/9707015, submitted to Rev. Mod. Phys. (1997)
6. K.S. McFarland *et al.* (CCFR/NuTeV Collaboration) *A Precision Measurement of Electroweak Parameters in Neutrino-Nucleon Scattering*, FNAL-Pub-97/001-E. B.J. King, Columbia University Ph.D. Thesis, 1994; Nevis Report: Nevis-283, CU-390, Nevis Preprint R-1500 (1994).
7. R.M. Barnett et al., Physical Review D54, 1 (1996) and 1997 off-year partial update for the 1998 edition available on the PDG WWW pages (URL: http://pdg.lbl.gov/).
8. I.I Bigi, *Open Questions in Charm Decays Deserving an Answer*, CERN-TH.7370/94, UND-HEP-94-BIG08 (1994). I.I Bigi, *The Expected, The Promised and the Conceivable - on CP Violation in Beauty and Charm Decays.*, UND-HEP-94-BIG11 (1994).
9. Tiehui (Ted) Liu, *The D0-Dobar Mixing Search – Current Status and Future Prospects*, HUTP-94/E021 (1994). Gustavo Burdman, *Charm Mixing and CP Violation in the Standard Model*, FERMILAB-Conf-94/200 (1994).

Neutrino Oscillation Physics with BooNE

Ray Stefanski

Fermi National Accelerator Laboratory[1]
P.O.Box 500
Batavia, IL. 60510

INTRODUCTION

A proposal [1] will be submitted to Fermilab for a Booster Neutrino Experiment (BooNE) to confirm the discovery of neutrino oscillations at LANL using a Liquid Scintillator Neutrino Dtetector (LSND). The location of the experiment at the Fermilab Booster will provide for higher signal rates than were possible at LSND by about an order of magnitude. BooNE will also provide an opportunity for observing the signal under very different conditions and with different systematics than were present at LSND.

The muon collider will provide an opportunity to further explore this region of parameter space with a different set of systematics. Most important will be that the neutrino flux will be accurately known, since the current of the parent muon beam can be measured very precisely. This source will provide a $\nu_\mu(\bar{\nu}_\mu)$ and $\bar{\nu}_e(\nu_e)$ flux equal in magnitude and with easily calculable energy and spatial distributions.

THE BooNE DETECTOR

The proposal to carry out BooNE at Fermilab may have at least two stages, the first stage is MiniBooNE. It will include the construction of a spherical tank, 12 meters in diameter that will be filled with nearly pure mineral oil. The volume of the tank will be observed with an array of about 1250 8″ phototubes. Figure 1 gives a schematic overview of the detector. The detector will be built so that roughly its center will be at grade level, with about a twenty foot earth cover over the top.

The Booster beam will be extracted from the Main Injector tunnel to a target station facing north on the site. An overview of the target station and

---

[1] Operated by Universities Research Association, Inc. Under Contract with the United States Department of Energy.

beam tunnel is given in Figure 2 and shows the location of the detector just southwest of the Lederman Science Center, and 500 meters from the target. Also shown are potential sites for future extensions of BooNE. Locating the extraction point to BooNE at the injection point of the Main Injector, guarantees that beam can be delivered to BooNE under any future scenario for Booster relocation.

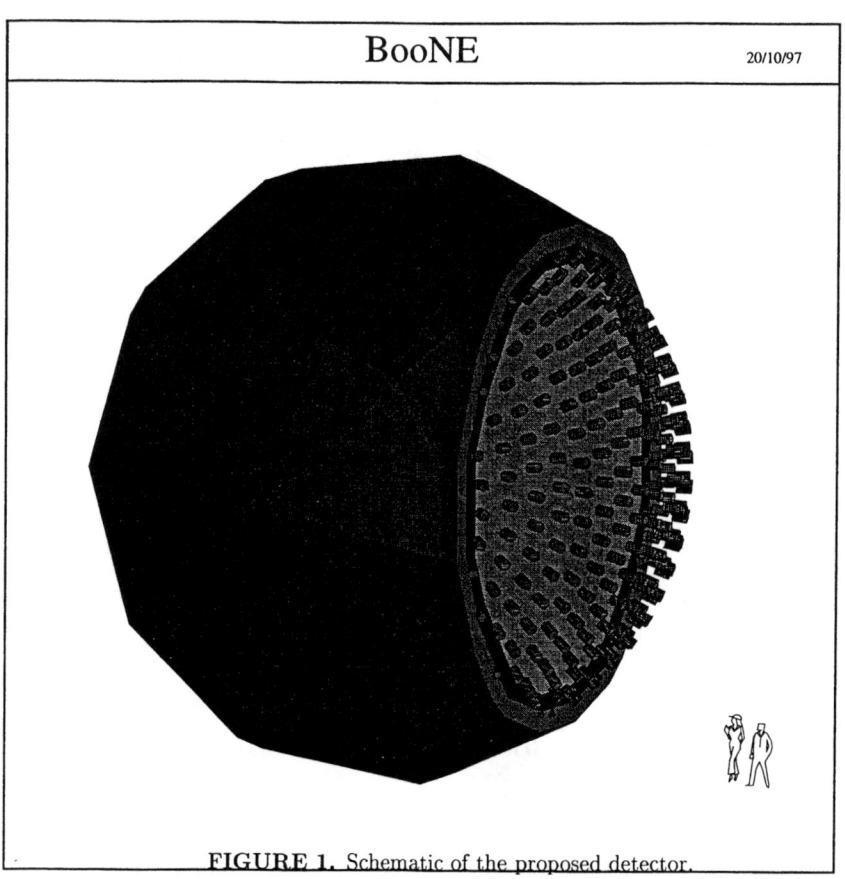

FIGURE 1. Schematic of the proposed detector.

Physics Reach and Neutrino Flux

MiniBooNE will use a two horn beam [2] similar to the BNL design. The ν_μ flux is presented in Figure 3. The flux is well matched to the acceptance of the detector, which is optimal roughly between 1/2 GeV and 2 GeV.

In Figure 4 the parameter space covered by MiniBooNE for the $\nu_\mu \to \nu_e$ is given, including anticipated results from future experiments. The figure

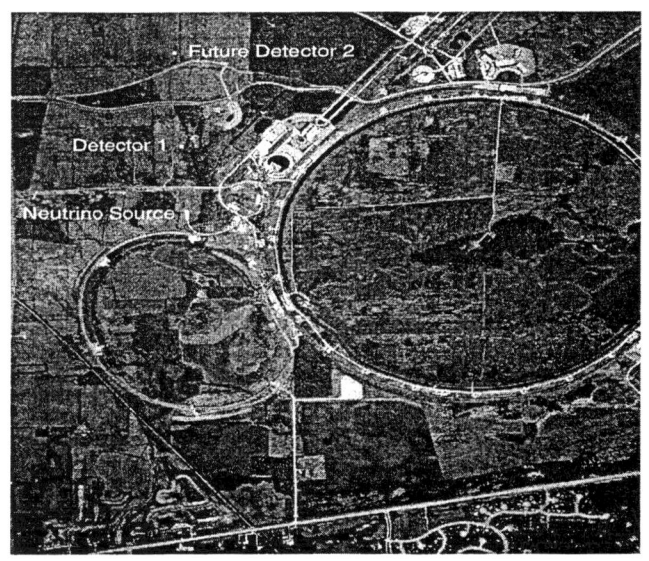

FIGURE 2. An aerial view suggesting the location of MiniBooNE , future extensions of BooNE , and the targeting station located near the Main Injector .

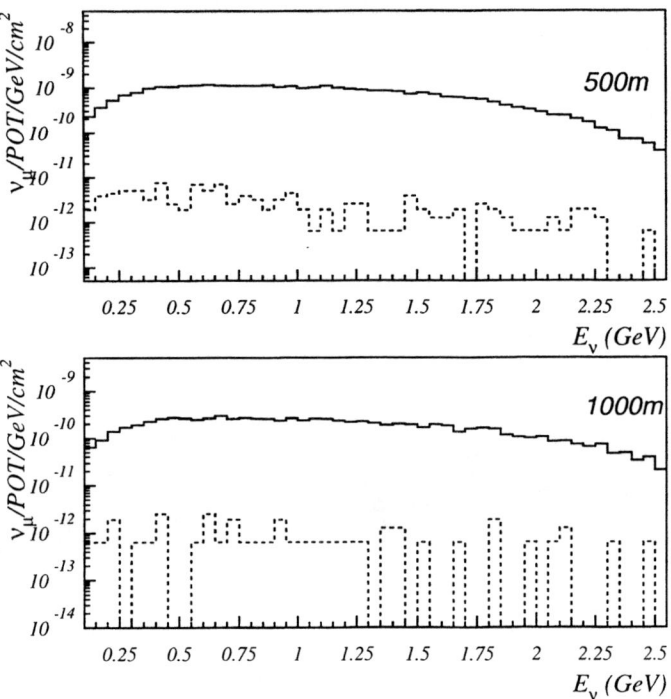

FIGURE 3. Flux of ν_μ (solid histogram) and ν_e (dashed histogram) from a 50m decay length beam line at 500 m and 1000 m from the target.

gives the allowed region for the LSND effect, and the coverage anticipated by MiniBooNE. Figure 5 gives the Δm^2 and $\sin^2 2\theta$ plot for the ν_μ disappearance experiment. It can be seen that a MiniBooNE detector at one km has good reach into the region of the atmospheric neutrino effect.

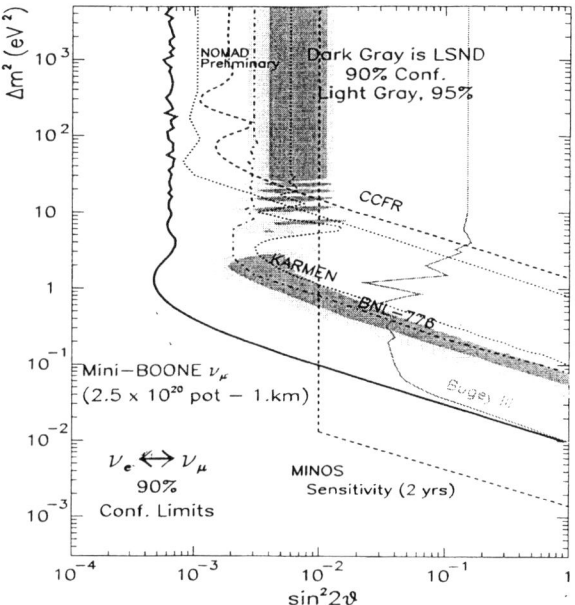

FIGURE 4. 90% C.L. limit expected for MiniBooNE for $\nu_\mu \to \nu_e$ appearance after one year of running, including systematic and statistical error, if LSND signal is not observed (solid line). Summary of results from past experiments and expectations for the future MINOS experiment are also shown.

Future Prospects with the Muon Collider Front End

A first step toward the development of the muon collider might be a low energy facility capable of storing 1.5 GeV muons in a small storage ring. A ten GeV muon collider using superconducting magnets will provide 1.5 GeV in the same ring circumference with conventional magnets [3]. The size of the storage ring can be seen in Figure 6. The facility would fit nicely within the footprint of the Pbar Source. Operating at 1.5 GeV, the ring would provide neutrinos from muon decay for a low energy facility that could include MiniBooNE, and other high resolution detectors.

The neutrino flux and event distributions from decays from muons in a

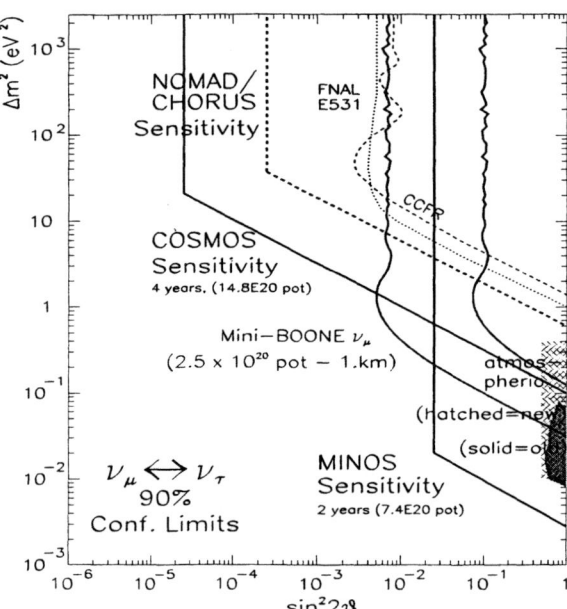

FIGURE 5. Summary of results from past experiments (narrow, dashed and dotted), future approved experiments (wide, dashed) and 90% C.L. limit expected for MiniBooNE (solid) for ν_μ disappearance. Solid region indicates the favored region for the atmospheric neutrino deficit from the Kamioka experiment. No zenith angle dependence would extend the favored region to higher Δm^2 as indicated by the hatched region.

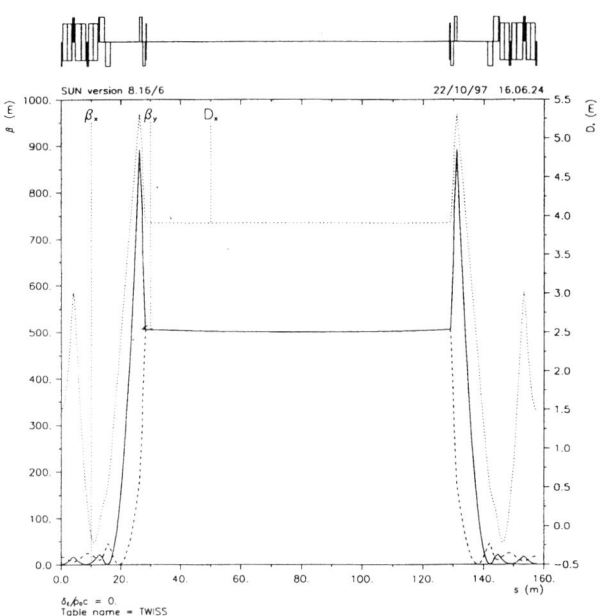

FIGURE 6. Half ring configuration for a Muon Storage Ring. Ring energy is 10 gev if superconducting magnets are used, and about 1.5 gev if conventional magnets are used.

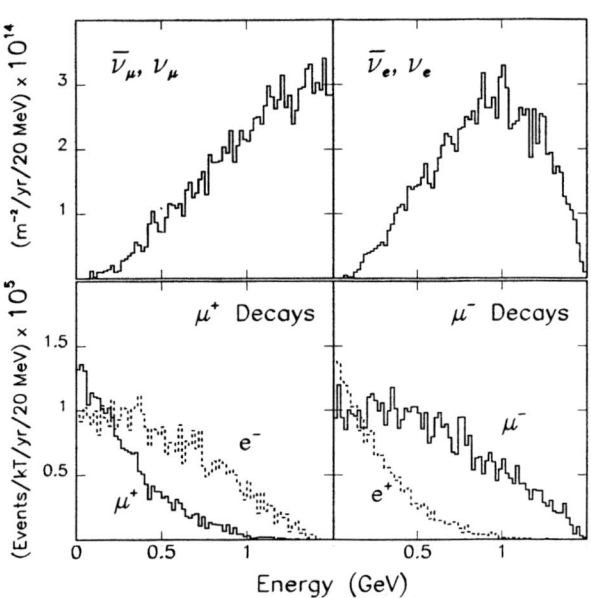

FIGURE 7. Expected neutrino flux and event rates from a 1.5GeV muon storage ring. The half-ring configuration is given in Figure 6.

storage ring of μ^+ is given in Figure 7 [4]. By comparison with Figure 3, and assuming 2×10^{20} protons per year from the Booster accelerator, we can see that the integrated yearly flux is similar between the two beams. More important, however, is that the muon generated beam provides an opportunity for greater control of systematics than the Booster generated beam.

Conclusion

The Muon Collider would benefit from a program that would permit an intensive R&D effort on a small, low energy facility. Low energies can be of physics interest as well. We've described here an approach to developing a low energy (1.5 GeV) muon storage ring that would test most of the concepts associated with capture and storage of a muon beam, with a goal of providing neutrinos for oscillation experiments.

The BooNE program, as outlined in the proposal, will continue into the year 2005. Construction of a small storage ring and muon collection facility on this time scale may be feasible. The facility could become part of a program to verify and extend our understanding of neutrino oscillations within the later phases of the BooNE program.

REFERENCES

1. E. Church *et. al.* , A Proposal for an Experiment to Measure $\nu_\mu \to \nu_e$ Oscillations and ν_μ Disappearance at the Fermilab Booster: BooNE. A proposal submitted to Fermilab, December 15, 1997.
2. L. A. Ahrens *et. al.* , Phys. Rev. D **34**, 75 (1986)
3. The half ring design was provided by Carol Johnstone.
4. Steve Geer, submitted to PRD (Fermilab-pub-97/389).

Neutrino Physics in a Muon Collider

Rabindra N. Mohapatra

Department of Physics, University of Maryland, College Park, MD-20742[1]

Abstract. The muon collider is expected to produce a high intensity neutrino beam which is an admixture of either $\nu_\mu + \bar{\nu}_e$ or $\bar{\nu}_\mu + \nu_e$ which can can be directed to underground detectors far away from the source. It will not only allow a probe of the $\nu_e - \nu_\mu$ as well as $\nu_\mu - \nu_\tau$ oscillations in a range of mixing angle and Δm^2 not probed heretofore but it will also provide information about the mixing angle $\theta_{e\tau}$ for a wide range of $\Delta m^2_{\nu_e \nu_\tau}$ from 10^{-4} eV^2 to 10^{-1} eV^2 which cannot be obtained from any other existing or proposed machine. One can also search for violations of Lorentz invariance and deviations from equivalence principle for neutrinos at a level which is three to four orders of magnitude more sensitive than possible at the moment. This will test for instance some unorthodox suggestions to understand both solar and atmospheric neutrinos using a single mass difference -squared between the ν_e and ν_μ. It can also test various proposed models neutrino masses and mixings to understand existing neutrino data.

I. INTRODUCTION

Neutrino physics is now going through a very exciting period. For the first time in its history, there are several hints for a nonvanishing mass for at least two of three known neutrinos which look very promising and credible. They come (i) from the observations of the solar neutrinos in various experiments and their disagreement with the predictions of the standard solar model [1]: the earlier experiments from Homestake, Kamiokande, SAGE and GALLEX [2] and the most recent high statistics confirmation of these results by the super-Kamiokande experiment [3] and (ii) from the observations of the atmospheric neutrinos by several previous experiments [4,5] and the most recent confirmation of the earlier results by the super-Kamiokande [3] collaboration. Then there is the result from the Los Alamos liquid scintillation neutrino detector (LSND) which gives the first laboratory evidence for the oscillation of both $\bar{\nu}_\mu \rightarrow \bar{\nu}_e$ [6] as well as $\nu_\mu \rightarrow \nu_e$ type [7].

---

[1] Invited talk presented at the *Workshop on Physics at the First Muon Collider and the Front End of a Muon Collider* held in Fermilab, November 6-9, 1997. Work supported by the National Science Foundation grant no. PHY-9421386.

Once the neutrinos have mass, they can mix with each other leading to a rich variety of new physical phenomena, which in turn may lead to insight into the kind of new physics responsible for such mixings and masses.

While at the moment detailed fits to all the above data considerably restrict the nature of the masses and mixings among the three neutrino species, they do not fix the complete neutrino mass texture. On top of this, there are ambiguities that open the neutrino oscillation interpretation of the solar neutrino data to question. Therefore it is crucial that other experiments are performed not only to confirm what is known but also to gain complete knowledge of this basic sector of the standard model.

Presently in planning and constuction stage are several such experiments- MINOS and PALO VERDE to give two examples; several others which are beginning to provide this information are the CHORUS, NOMAD, CHOOZ. These are variously known as long base line and short base line experiments, which will either involve low energy electron neutrino beams from the reactors or high energy ν_μ beams from the accelerators. None of them will have a high energy ν_e beams. While the disappearance of the ν_e in the reactor experiments one can get information about the mixing angles for a certain mass range, a large range of mixing angles and masses of theoretical interest remains unexplored at the moment. Specifically, information on the $\nu_e - \nu_\mu$ mixing angle at the moment is very poor. A similar remark also applies to the ν_μ case where appearance experiments involving the ν_μ beam provide knowledge of the mixing of ν_μ with ν_e or ν_τ (i.e. $\theta_{\mu e}$ or $\theta_{\mu \tau}$) for some mass range while leaving a considerable range of interest unexplored.

The goal of this article is to explore whether the neutrino beams from a muon collider can provide any useful information regarding the neutrino masses and mixings that are not already available (or will not be available once the above mentioned experiments are completed). In other words is there a neutrino physics justification for the muon collider?

The muon collider will be the first place where one can get an energetic beam of electron neutrinos. Therefore with a suitable long baseline experiment, one can probe the mixing angle $\theta_{e-\tau}$ in a completely unexplored domain. Of course needless to say that the muon collider will also provide extensive information on $\theta_{e-\mu}$ or $\theta_{\mu-\tau}$ for very small mass difference squared as well and thus nicely complement the other experiments and may even extend the domain of the search.

Besides with the high flux neutrino beams which are supposed to result in a muon collider, one can also test the validity of several fundamental laws of physics such as Lorentz invariance, CPT theorem as well as equivalence principle and I will show that improvements by several orders of magnitude are possible with the muon colliders in these cases.

This article is organized as follows: in section 2, the implications of the solar, atmospheric and the LSND experiments for neutrino masses and mixings are briefly summarized; in section 3 a summary of the various popular scenarios for neutrino masses and mixings are touched upon; in section 4, the neutrino mixings that can be probed by the muon collider is given using different experimental scenarios and in section 5, we discuss possible tests of the Lorentz invariance, equivalence principle and CPT theorem are considered.

II. INDICATIONS FOR NONZERO NEUTRINO MASS

II.a Solar neutrino deficit

We will assume the explanation of the solar neutrino deficit in terms of the oscillation between the ν_e and ν_x where x is another species of neutrino not necessarily of muon or tau type. The oscillation can be pure vacuum oscillation which requires a mass difference-squared $\Delta m^2 \sim 10^{-10}$ eV^2 and large mixing or it could be matter enhanced MSW [8] type in which case the neutrino mass differences and mixing angles fall into one of the following ranges [1],

$$\begin{aligned}&\text{a) Small} - \text{angle MSW}, \ \Delta m_{ei}^2 \sim 5 \times 10^{-6} - 10^{-5} \text{eV}^2, \ \sin^2 2\theta_{ei} \sim 7 \times 10^{-3},\\ &\text{b)} \qquad \text{Large} - \text{angle MSW}, \ \Delta m_{ei}^2 \sim 9 \times 10^{-6} \text{eV}^2, \ \sin^2 2\theta_{ei} \sim 0.6.\end{aligned} \quad (1)$$

If the solar neutrinos oscillate into sterile neutrinos, the MSW effect is different from the ν_e to ν_μ case and the large angle solution is no more allowed. The above results are based on the approximation that only two of the neutrino species are involved in the oscillation.

II.b Atmospheric Neutrino Deficit

The atmospheric ν_μ's and ν_e's arise from the decays of π's and K's and the subsequent decays of secondary muons produced in the final states of the π and K decays. In the underground experiments the ν_μ and $\bar{\nu}_\mu$ produce muons and the ν_e and $\bar{\nu}_e$ lead to $e^\pm$. Observations of $\mu^\pm$ and $e^\pm$ indicate a far lower value for ν_μ and $\bar{\nu}_\mu$ than suggested by naïve counting arguments which imply that $N(\nu_\mu + \bar{\nu}_\mu) = 2N(\nu_e + \bar{\nu}_e)$ [4]. The assumed oscillation in this case could a priori be between ν_μ to ν_e or ν_μ to ν_τ. However, a recent CHOOZ collaboration result rules out the Kamiokande allowed ν_μ to ν_e mass-squared mixing region [9]. Thus we can assume that the oscillation of ν_μ to ν_τ provides the explanation of the atmospheric neutrino results. Fits to both the sub-GeV and multi-GeV Kamiokande data require that [5]

$$\Delta m_{\mu\tau}^2 \approx 0.025 \text{ to } 0.005 \text{ eV}^2, \ sin^2 2\theta_{\mu\tau} \approx .6 \ to \ 1. \quad (2)$$

The most recent Super-Kamiokande data has confirmed the deficit in both the sub-GeV and the multi-GeV data. Also there is now evidence for zenith angle dependence in the multi-Gev data which according to preliminary analysis [3] would indicate a similar mass range as above for maximal mixing angle.

II.c Results from the LSND experiment

The LSND collaboration first reported seeing indications for $\bar{\nu}_\mu$ to $\bar{\nu}_e$ oscillation using the liquid scintillation detector at Los Alamos in 1996 [6]. Their results in conjunction with the negative results by the E776 group and the Bugey reactor data imply a mass difference squared between the ν_e and the ν_μ lying between

$$0.27 \ eV^2 \leq \Delta m^2 \leq 10 \ eV^2 \tag{3}$$

with a mixing angle $\theta_{e\mu} \sim .05 - .1$. The region for Δm^2 above 10 eV^2 has been ruled out both by the recent CCFR data and the NOMAD data [10]. In a recent paper [7], LSND group has reported preliminary evidence for the $\nu_\mu - \nu_e$ oscillation with mass difference squares and mixings in the similar range as above.

II.d Hot dark matter of the universe

There is increasing evidence that more than 90% of the mass in the universe must be detectable so far only by its gravitational effects. This dark matter is likely to be a mix of $\sim 20\%$ of particles which were relativistic at the time of freeze-out from equilibrium in the early universe (hot dark matter) and $\sim 70\%$ of particles which were non-relativistic (cold dark matter). Such a mixture gives a very good fit to all available cosmological data [11]. This interpretation is however by no means unique and it has been claimed that an equally good fit to the power spectrum can be obtained by a pure CDM model with a tilted spectrum [12].

If however, the mixed dark matter picture is adopted, a very plausible candidate for hot dark matter is one or more species of neutrinos with total mass of $\Sigma_i m_{\nu_i} = 93 h^2 F_H \Omega = 4.8$ eV, if $h = 0.5$ (the Hubble constant in units of 100 km·s^{-1}·Mpc^{-1}), $F_H = 0.2$ (the fraction of dark matter which is hot), and $\Omega = 1$ (the ratio of density of the universe to closure density).

It is usually assumed that the ν_τ would supply the hot dark matter. However, if the atmospheric ν_μ deficit is due to $\nu_\mu \to \nu_\tau$, the ν_τ alone cannot be the hot dark matter, since the ν_μ and ν_τ need to be closer to each other in mass. It is interesting that instead of a single ~ 4.8 eV neutrino, sharing that ~ 4.8 eV between two or among three neutrino species provides a better fit to the universe structure and particularly a better understanding of the variation of matter density with distance scale [13].

II.e: Neutrinoless double beta decay constraints

Finally, let us note the very stringent constraints on neutrino masses now implied by the neutrinoless double beta decay searches. The Heidelberg-Moscow ^{76}Ge experiment [14] has provided the most stringent upper limits on the effective Majorana mass of the neutrino: $< m_\nu > \leq .47$ eV where $< m_{\nu_e} > = \Sigma_i U_{ie}^2 m_{\nu_i}$. This is beginning to put very strong constraints on model building. For instance, it has recently been noted [15] that the CHOOZ and Bugey [16] results already imply that $| < m > | \leq 3 \times 10^{-2}$ eV. Thus any signal for neutrinoless double with $< m >$ above 0.1 eV would be evidence against a hierarchical neutrino mass pattern. Similarly, one can infer from the LSND data that one must have $< m > \geq 4 \times 10^{-3}$ eV assuming that $\theta_{e\tau}$ is small (or at least it does not precisely cancel this contribution). Thus high precision double beta searches are extremely important to a complete understanding of the neutrino masses.

III. NEUTRINO MASS TEXTURES IMPLIED BY DATA

In order to discuss the implications of the above data for the neutrino mass pattern, we will assume that all the neutrinos are Majorana particles, since it is easier to understand the smallness of Majorana masses of neutrinos within the framework of grand unified theories. We will then proceed by assuming that the solar and the atmospheric neutrino data are the two core items that appear as the most secure indications of neutrino oscillation and study their significance for neutrino masses. We will then add the HDM and the LSND results and see their implications.

: Including only solar and the atmospheric data:

Since we only have constraints on the mass difference squares from the solar and the atmospheric data, we can have a "staircase" pattern with $m_{\nu_e} \ll m_{\nu_\mu} \simeq \sqrt{\Delta m^2_{solar}}$ and $m_{\nu_\mu} \ll m_{\nu_\tau} \simeq \sqrt{\Delta m^2_{atmos}}$ or a degenerate pattern [17,18] where all masses are nearly equal with appropriate mass differences. The latter is mandatory if one wants to explain the HDM picture of the universe. As far as the mixing angles go, the $\theta_{\mu\tau}$ is always maximal (i.e. near $\pi/4$) whereas $\theta_{e\mu}$ is either maximal (for vacuum oscillation or large angle MSW) or few percent (for small angle MSW). Several theoretical suggestions are now given.

III.a Model A: Minimal SO(10) inspired

This model is based on a minimal SO(10) model where the Yukawa sector involves so few parameters that all neutrino masses and mixing angles can be predicted [19]. Using the type II see-saw formula in the minimal SUSY

SO(10) model it is possible to get solutions that can accomodate both the solar as well as the atmospheric neutrino puzzles. Specifically, it predicts: $m_{\nu_{mu}} \simeq 1.7 \times 10^{-3}$ eV and $m_{\nu_\tau} \simeq 5.9 \times 10^{-2}$ eV and the neutrino mixing matrix given by:

$$\begin{pmatrix} -0.995 & 0.052 & 0.080 \\ 0.070 & 0.792 & 0.603 \\ 0.032 & -0.608 & 0.793 \end{pmatrix} \quad (4)$$

III.b Model B: Maximal mixing scheme

In this scheme [20], the mixing matrix has the form:

$$U_\nu = \frac{1}{\sqrt{3}} \begin{pmatrix} 1 & 1 & 1 \\ 1 & \omega & \omega^2 \\ 1 & \omega^2 & \omega \end{pmatrix} \quad (5)$$

where $\omega = e^{2\pi i/3}$ This scheme becomes essential if the neutrino masses are degenerate and if the limits on the neutrino mass from neutrinoless double beta decay keps going down [21] since the leading term in the $<m_{\nu_e}>$ cancels for this choice of mixing angles. This mixing angle pattern can be shown to fit both the solar and atmospheric neutrino data [21,22] if one assumes vacuum oscillation solution to the solar neutrino problem.

III.c Model C: Democratic mixing among neutrinos

This model for mixings is based on the idea that the neutrino mass matrix may satisfy an approximate permutation symmetry among the three generations [23] and also can fit the solar and atmospheric neutrino data and has a mixing matrix of the following form:

$$U_\nu = \begin{pmatrix} \frac{1}{\sqrt{2}} & -\frac{1}{\sqrt{2}} & 0 \\ \frac{1}{\sqrt{6}} & \frac{1}{\sqrt{6}} & -\frac{2}{\sqrt{6}} \\ \frac{1}{\sqrt{3}} & \frac{1}{\sqrt{3}} & \frac{1}{\sqrt{3}} \end{pmatrix} \quad (6)$$

This model also can support a degenerate mass pattern consistent with neutrinoless double beta decay. The difference between this and the maximal mixing pattern is that the $\theta_{e\tau}$ values are very different.

There are several other schemes based on attractive theoretical assumptions that lead to three neutrino mixing patterns [24] that can fit both solar and atmospheric data. We do not enter into discussions of those ideas due to lack of space but they are generally similar to the SO(10) case discussed above.

: Accomodating solar, atmospheric and the LSND data:

The LSND result have two important implications for our discussion: first, their result implies oscillation from $\bar{\nu}_\mu$ to $\bar{\nu}_e$ i.e. unlike the solar and atmospheric data, the final state is not a matter of speculation but observation; secondly the $\Delta m^2_{e\mu}$ that fits data is between .2 eV^2 to about 10 eV^2, which is very different from the ranges derived from simple interpretations of the data as noted above. Before the latest results from super-Kamiokande experiment came out, two interesting neutrino mass schemes were proposed which seemed in accord (though rather marginally) with the previous Kamiokande data. The basic idea in these papers was the following: three experiments are sensitive to three mass difference squares; however with three neutrinos there are only two possible Δm^2's. Therefore a three neutrino scheme can only fit data if two of the experimentally determined Δm^2's turn out to be equal. The two models described below essentially exploit these two possibilities.

III.d Model D: Cardall-Fuller scheme

This scheme [25] assumes that $\Delta m^2_{LSND} \simeq \Delta m^2_{atmos}$ and that the $\nu_\mu - \nu_e$ oscillation observed at Los Alamos is an indirect oscillation [19] which proceeds as ν_e to ν_τ to ν_μ. To accomodate the LSND results in this picture assumes the LSND Δm^2 to be around .3 eV^2. Since the solar neutrino puzzle requires that $\Delta m^2_{e-\mu} \simeq 10^{-5}$ eV^2, this scenario implies that we must have $\Delta^2_{\mu-\tau}$ be $\approx .3\ eV^2$. so that the LSND neutrino oscillation frequency is determined by ν_e-ν_τ mass difference. Secondly, for the amplitude of indirect oscillation to be compatible with observations, the $\nu_e - \nu_\tau$ mixing angle should be nonnegligible (say $\sim .1-.2$). The main problem for this scenario comes from the atmospheric neutrino data, since the original analysis of the Kamiokande sub-GeV and the multi-GeV data by the Kamiokande group excludes $\Delta m^2 \geq .1\ eV^2$ at 90% confidence level (c.l.). The analysis of the the atmospheric neutrino data from Super-Kamiokande will therefore provide crucial test of this model. Preliminary analysis of the super-Kamiokande data (which has a clear evidence for zenith angle dependence) appears to contradict this scenario. Furthermore, if we want to fit the HDM picture into this model, one must have $m_{\nu_e} \simeq 1.6$ eV. While at its face this value may be in conflict with the neutrinoless double beta decay results [14], one can hide under the uncertainties of nuclear matrix element calculations which typically could be as much as a factor of 2-3. As the precision in $\beta\beta_{0\nu}$ search improves further (say to the level of 0.1 eV), nuclear matrix element uncertainties cannot be invoked to save the model anymore.

III.e Model E: Acker-Pakvasa scheme

The second three neutrino mass texture [26] also uses indirect oscillation to explain the LSND data but makes the assumption that $\Delta m^2_{solar} \simeq \Delta m^2_{atmos}$ and assumed that the atmospheric neutrino oscillation involves ν_μ to ν_e oscillation, which looks implausible in view of the latest CHOOZ data. In any case, they choose $\Delta m_{e\mu}^2 \simeq 10^{-2}$ eV^2, $\Delta m_{e\tau}^2 \simeq \Delta m_{\mu\tau}^2 \simeq 1-2$ eV^2. It is easy to see that in this case the general three neutrino oscillation formula for P_{ee} becomes energy independent if L is chosen to correspond to the distance of the earth from the Sun. It was shown in Ref. [26] that if one reduces the ^{8}B production in the center of the Sun, one can fit all solar neutrino observations despite the energy independence of the oscillations. It has been pointed out that already in the present data, there is evidence for energy dependence [27] disfavoring this scheme. Therefore this can also be tested by the Super-Kamiokande observations.

To complete this model, we give a typical mixing matrix that characterizes this model:

$$U_\nu = \begin{pmatrix} .700 & .700 & .14 \\ -.714 & .689 & .124 \\ -.010 & -.187 & .982 \end{pmatrix} \quad (7)$$

Note the large value of $\Theta_{e\tau}$.

III.f Model F: The case for a sterile neutrino

The case for a sterile neutrino is made clear by noting the difficulty of fitting the solar, atmospheric and the LSND data with three neutrinos as exemplified by the models D and E. The main obstacle, as we saw, comes from the conflict between the LSND data and the MSW resolution of the solar neutrino data.

The general picture for the case of sterile neutrino is as follows [17,28,29]: the solar neutrino puzzle is explained by the $\nu_e - \nu_s$ oscillation; atmospheric neutrino data would be explained by the $\nu_\mu - \nu_\tau$ oscillation. The LSND data would set the overall scale for the masses of ν_μ and ν_τ (which are nearly degenerate) and if this scale is around 2 to 3 eV as is allowed by the data [6], then the $\nu_{\mu,\tau}$ would constitute the hot dark matter of the universe. The mass matrix in this case would be in the basis $(\nu_s, \nu_e, \nu_\mu, \nu_\tau)$,

$$M = \begin{pmatrix} \mu_1 & \mu_3 & 0 & 0 \\ \mu_3 & \mu_2 & \epsilon & 0 \\ 0 & \epsilon & m & \delta/2 \\ 0 & 0 & \delta/2 & m+\delta \end{pmatrix}. \quad (8)$$

Solar neutrino data requires $\mu_2 \ll \mu_1 \simeq 10^{-3}$ eV and $\mu_3 \simeq .05\mu_1$. The ϵ term is responsible for the $\nu_e - \nu_\mu$ oscillation. Clearly the crucial test of the sterile neutrino scenario will come when SNO collaboration obtains their results for neutral current scattering of solar neutrinos. One would expect that $\Phi_{CC} = \Phi_{NC}$ if the ν_e oscillation to ν_s is responsible for the solar neutrino deficit. There should be no signal in $\beta\beta_{0\nu}$ search. Precision measurement of the energy distribution in charged current scattering of solar neutrinos at Super-Kamiokande can also shed light on this issue.

IV. MUON COLLIDER FOR STUDYING THE NEUTRINO MASSES AND MIXINGS

The muon collider is expected to produce a high luminosity beam of neutrinos which is an admixture of either $\nu_\mu + \bar{\nu}_e$ or their antiparticles. Thus in some sense this is a "controlled atmospheric neutrino" beam. The energy of the neutrinos is expected to range from 10 GeV to 100 GeV. In our discussion we will entertain the possibility of two kinds of long base line experiments [30] with beam directed to either Gran Sasso or Soudan mine with distances respectively of $\sim 10^4$ or 750 kilometers. Recall that the oscillation formula

$$P(\nu_e \to \nu_\mu) = sin^2 2\theta sin^2 \frac{1.27 \Delta m^2 L}{E} \qquad (9)$$

where E is in GeV and L is in kilometers. With 7.5×10^{20} $\mu^\pm$ per year, one can have of the order of 10^{20} neutrinos/year [30]. Geer has calculated the charge current event rate for such particles at a 10 kiloton detector located at Gran Sasso as well as at Soudan. He finds that one can expect thousands of charged current events per year.

To discuss its utility in studying neutrino masses and mixings, note that for a 10 GeV neutrino beam and a distance of 10^4 kilometers, one could probe Δm^2 down to 10^{-5} eV^2 for maximal mixing if we take 10 events per year to get a signal and for $\Delta m^2 \geq 10^{-3}$ eV^2, one could probe mixings $sin^2 2\theta$ down to 10^{-4} or so. Thus, one can not only explore $\theta_{e\tau}$ in a totally unexplored region of parameters but also considerably extend our knowledge of the $\theta_{e\mu}$ as well as $\theta_{\mu\tau}$ into a domain further than what MINOS or COSMOS can accomplish. As a comparision, note that at present, $\theta_{e\tau}$ has an upper bound of about .14 from the Fermilab experiment E531 for $\Delta m^2 \geq 10$ eV^2, which is a very weak bound compared to the other two mixing angles. The main reason being that there does not exist any accelerator source of high energy ν_e's and muon collider will be the first one to provide one such source.

IV. TESTING LORENTZ AND CPT INVARIANCE AND EQUIVALENCE PRINCIPLE

Lorentz invariance, CPT invariance (which is a consequence of Lorentz invariance and locality in Quantum Field theories) and the principle of general covariance are some of the fundamental pillars on which the present day theoretical physics rests. While few would doubt that there is any deviation from these principles, science has to be based on experimentally tested ideas. It is therefore important to look for ways to test the validity of these principles. In order to make the tests quantitative, a framework that has some parameters that characterize the departures from the exactness of these principles is useful. Such frameworks have recently been discussed and I summarize them below and point out how a muon collider can be useful.

IV.a Lorentz invariance

It was pointed out recently by Coleman and Glashow [31] that one way to parameterize a departute from Lorentz invariance for massless particles such is to write

$$E_i = p(c + \delta c_i) \tag{10}$$

Applying this to neutrinos, one gets for the energy difference between two eigenstates into which the weak eigenstate resolves as $E_1 - E_2 = E(\delta c_1 - \delta c_2) \equiv E\delta v$. One can then write the oscillation probability of say ν_e to ν_μ to be

$$P(\nu_e \to \nu_\mu) = sin^2 2\theta_v sin^2 \frac{\delta v E L}{2} \tag{11}$$

The energy dependence of the $P(\nu_e \to \nu_\mu)$ in Eq. 11 is clearly very different from the case of mass oscillation where it goes like L/E. Therefore longer the base line and higher the energy, the more precise the test of Lorentz invariance. Present limits on the δv from various oscillation experiments is $\delta v \leq 10^{-21}$. In this case we must choose the neutrino energy from the muon collider to be as high as possible. Again taking $E \simeq 100$ GeV and $L = 10^4$ Km, this limit can be improved to $\delta v \leq 10^{-26}$ which is an improvement of some five orders of magnitude.

IV.b CPT for neutrinos

A simple CPT violating combination of oscillation probabilities for the $\nu_e - \nu_\mu$ system is given by $P(\nu_\mu \to \nu_e) - P(\bar{\nu}_e \to \bar{\nu}_\mu)$ as is very easily checked. Note that as mentioned before the neutrino beam in a muon collider consists of ν_μ and $\bar{\nu}_e$. Let us assume that $N_\mu = N_{\bar{\nu}_e}$ (although the energy spectra in general will be different) for simplicity. Then without CPT violation but with $\nu_e - \nu_\mu$ oscillation, one would expect in the detector $N_{e^-} = N_{\mu^+}$. Thus any deviation from this equality would be a test of CPT violation. This result is independent of any specific underlying model for CPT violation.

IV.c Violations of equivalence principle

It has been pointed out by Halprin and Leung [32] that violations of equivalence principle can also lead to neutrino oscillation phenomena. To see this, let us parameterize the metric as

$$Metric = g_{\alpha\beta} + 2\gamma_i \phi \delta_{\alpha\beta} \tag{12}$$

where ϕ is the gravitational potential. The second term is absent in Einstein's theory and characterizes the departure from the equivalence principle. The energy-momentum relation now looks as follows:

$$E^2(1 + 2\gamma_j\phi) = p^2(1 - 2\gamma_j\phi) \qquad (13)$$

This can be cast in the language discussed in connection with violation of Lorentz invariance identifying $\delta v \equiv 2(\gamma_1 - \gamma_2)\phi$. Translating our earlier discussion then we can conclude that in a muon collider, one can probe $2\Delta\gamma\phi$ down to the level of 10^{-26} as before. Since this experiment will be done in the solar system, the value of $\phi \simeq 10^{-6}$, it will test for violation of equivalence principle down to the level of 10^{-20}. Note the present long range force experiments test this principle down to the level of 10^{-12} or so.

More importantly, Halprin et al have made the unconventional suggestion that perhaps one could use this phenomenon to explain solar and atmospheric neutrino puzzle by setting $2\Delta\gamma\phi \simeq 10^{-21}$ using only $\nu_\mu - \nu_e$ oscillation. A muon collider could therefore provide a clean test of this hypothesis.

In conclusion, we find that the neutrino beams from the muon collider can provide extremely useful insight into the world of neutrino masses and mixings, specifically it can probe the $\nu_e - \nu_\tau$ mixing angle in a domain of parameters that is beyond the range of any proposed experiment. This will allow us to test several three neutrino mixing schemes such as the maximal mixing scheme and the SO(10) scheme. Muon collider can also extend the domain of validity of some of the fundamental laws govorning the physical phenomena such as the equivalence principle, CPT theorem and Lorentz invariance.

Acknowldgement

I would like to thank Boris Kayser for the invitation to join the muon collider working group on neutrinos and many stimulating discussions on many of the topics discussed in this article. I also like to thank G. Sullivan for discussions on the super-Kamiokande results for atmospheric neutrinos.

REFERENCES

1. J. Bahcall, *Proceedings of Neutrino'96* edited by K. Enquist, K. Huitu and J. Maalampi (World Scientific, Singapore); A. Smirnov, hep-ph/9611465.
2. B. T. Cleveland et al. Nucl. Phys. B(Proc. Suppl.) **38**, 47 (1995); K.S. Hirata et al., Phys. Rev. **44**, 2241 (1991); GALLEX Collaboration, Phys. Lett. **B388**, 384 (1996); J. N. Abdurashitov et al., Phys. Rev. Lett. **77**, 4708 (1996).
3. Y. Suzuki, Invited talk at Erice Neutrino workshop, September 17-22, 1997.
4. K.S. Hirata et al., Phys. Lett. **B280**, 146 (1992); R. Becker-Szendy et al., Phys. Rev. D **46**, 3720 (1992); W. W. M. Allison et al., Phys. Lett. **B 391**, 491 (1997).
5. Y. Fukuda *et al*, Phys. Lett. **B 335**, 237 (1994).

6. C. Athanassopoulos et al. Phys. Rev. Lett. **75**, 2650 (1995).
7. C. Athanassopoulos et al. Nucl-ex/9706006.
8. L. Wolfenstein, Phys. Rev. **D17**, 2369 (1978); S. P. Mikheyev and A. Smirnov, Yad. Fiz. **42**, 1441 (1985); Nuovo Cimento, **9C**, 17 (1986).
9. M. Apollonio et al. hep-ex/9711002.
10. K. Zuber, Invited talk in COSMO'97, Ambleside, England, September 15-19, 1997.
11. For a recent review and references, see J. Primack, astro-ph/9707285.
12. S. Sarkar, astro-ph/9710273.
13. J. Primack, J. Hotzman, A. Klypin and D. Caldwell, Phys. Rev. Lett. **74**, 2160 (1995).
14. H. Klapdor-Kleingrothaus, these proceedings and *Double Beta Decay and Related Topics*, ed. H. Klapdor-Kleingrothaus and S. Stoica, World Scientific, (1995) p. 3; A. Balysh et al., Phys. Lett. **B283**, 32 (1992).
15. S. M. Bilenky, C. Giunti, C. W. Kim and M. Monteno, hep-ph/9711400.
16. B. Achkar et al. Nucl. Phys. **B434**, 503 (1995).
17. D.O. Caldwell and R.N. Mohapatra, Phys. Rev. **D 48**, 3259 (1993).
18. S. T. Petcov and A. Smirnov, Phys. Lett. **B322**, 109 (1994); A. S. Joshipura, Zeit. fur Phys. **C64**, 31 (1994); A. Ionissian and J. W. F. Valle, Phys. Lett. **B332**, 93 (1994); P. Bamert and C. P. Burgess, Phys. Lett. **B329**, 289 (1994); D. G. Lee and R. N. Mohapatra, Phys. Lett. **B329**, 463 (1994).
19. K. S. Babu and R. N. Mohapatra, Phys. Rev. Lett. **70**, 2845 (1993); D. G. Lee and R. N. Mohapatra, Phys. Rev. **D 52**, 4215 (1995); B. Brahmachari and R. N. Mohapatra, hep-ph/9710371.
20. S. Nussinov, Phys. Lett. **B63**, 201 (1976); L. Wolfenstein, Phys. Rev. **D 18**, 958 (1978).
21. R. N. Mohapatra and S. Nussinov, Phys. Lett. **B346**, 75 (1995).
22. P. Harrison, D. Perkins and W. Scott, Phys. Lett. **B 349**, 137 (1995).
23. H. Fritzsch and X. Xing, Phys. Lett. **B 372**, 265 (1996); M. Fukugita, M. Tanimoto and T. Yanagida, hep-ph/9709388.
24. Y. Koide, hep-ph/9707505; Y. Koide and H. Fusaoka, Z. Phys. **C71**, 459 (1966).
25. C. Cardall and G. Fuller, astro-ph/9606024; C. Cardall, D. Cline and G. Fuller, hep-ph/9706426.
26. A. Acker and S. Pakvasa, Phys. Lett. **B397**, 209 (1997).
27. P. Krastev and S. Petcov, Phys. Lett. **B395**, 69 (1997).
28. J. Peltoniemi and J. W. F. Valle, Nucl. Phys. **B 406**,
29. S. M. Bilenky, C. Giunti and W. Grimus, hep-ph/9607372; hep-ph/9711311.
30. S. Geer, talk at this workshop.
31. S. Coleman and S. L. Glashow, Phys. Lett. **B405**, 249 (1997); S. L. Glashow et al. Phys. Rev. **D 56**, 2433 (1997).
32. A. Halprin and C. Leung, Phys. Rev. Lett. **67**, 1833 (1991).

Aspects of Neutrino Reactions at the First Muon Collider

E.A. Paschos

Institut für Physik, Universität Dortmund
D-44221 Dortmund, Germany

Abstract. In this article I describe several topics which are unique for the muon collider. Among them are properties of the neutrinos as beams and background radiation, the reaction $\mu^- \mu^+ \to e \bar{\nu}_e W^+$ and the possible sparkling of the QCD vacuum.

INTRODUCTION

The first muon collider will be an accelerator with several new properties and challenges. It will be the source of intense beams of muons and neutrinos, which will be used for new experiments, provided that we can eliminate the radiation hazards from the penetrating neutrinos. In addition, the interactions of muons and neutrinos introduce new physical phenomena. The width of the muon introduces singularities to be discussed explicitly and the high–energy interactions of leptons with protons and neutrons may cause the sparkling of the QCD vacuum. These are some topics covered in this short talk.

I NEUTRINOS AS PARTICLE BEAMS AND BACKGROUND RADIATION

The intense neutrino beams are a consequence of the high luminosity planned for the collider:

$$\text{for} \quad \sqrt{s} \leq 500 \text{ GeV}, \quad \mathcal{L} = \text{few} \times 10^{33} \text{ cm}^{-2} s - 1 \quad \text{and}$$
$$\text{for} \quad \sqrt{s} \leq 4 \text{ TeV}, \quad \mathcal{L} = \text{few} \times 10^{35} \text{ cm}^{-1} s - 1. \quad (1)$$

As the muons circulate around the ring they decay. The lifetime of a 1 TeV–muon ($\gamma \approx 10^4$) is $\tau = 0.02s$, so that practically all the muons decay after the filling and during the operation of the machine. In the decay each muon produces 2 neutrinos which are uniformly distributed around the ring. Since the neutrinos are of high energy and very penetrating, they form a disc around the ring which extends over long distances. This brings in new questions concerning the location, design and operation of the machine. It becomes necessary to reduce the intensity of the neutrinos when they reach the surface of the earth. The problem is already appreciated and several groups are studying its solution. I will present here a simple (back of the envelope) calculation which demonstrates the interaction rates for neutrinos. The radiation levels created by the beams are discussed in references [1] and [2].

a. <u>Intensity</u>. I consider a collider with 8×10^{19} muon decays per year. This will produce 1.6×10^{20} neutrinos with an average energy $\langle E_\nu \rangle \approx 100$ GeV or higher.

b. <u>Geometrical Attenuation</u>. At a distance R from the machine we consider an object with dimensions 2 m $\times$ 50 cm $\times$ 50 cm and the density of water. At this distance there is an attenuation of the beam because of horizontal and vertical divergences. The horizontal divergence is shown in figure 1 and is $\frac{\Delta\theta}{2\pi} = \frac{L}{R}\frac{1}{2\pi}$.

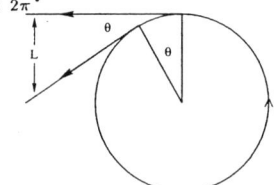

Fig. 1: Horizontal divergence

The vertical divergence is determined by the fact that the neutrinos at decay are within an angle of 80 mrad relative to the muon direction. Thus the number of neutrinos at a target of height h, width L and a distance R from the collider is

$$N_\nu \left(\frac{\Delta\theta}{2\pi}\right) \frac{h}{L_v} = \frac{N_\nu}{2\pi} \frac{L}{R} \frac{h}{L_v}$$

with L_v the vertical spread of the beam ($L_v = 80 \times 10^{-3}$ rad $\times R$). At a distance of 1 km the number of neutrinos through the target is

$$N_\nu \frac{1}{2\pi \cdot 8 \times 10^4} = 2 \times 10^{-6} N_\nu = 3.2 \times 10^{14}.$$

This is the geometric attenuation of the flux.

c. The <u>interaction length</u> of 100 GeV neutrinos in water is

$$L = \frac{\lambda_I}{\rho} \quad \text{with} \quad \lambda_I = \frac{A}{N_A \cdot \sigma}.$$

For the total cross-section we use

$$\sigma_{tot}(\nu p) = (0.682 \pm 0.012) \times 10^{-38} \frac{\text{cm}^2}{\text{GeV}} E_\nu$$
$$= 0.7 \times 10^{-36} \text{cm}^2 \quad \text{for} \quad E_\nu = 100 \text{ GeV}$$
$$\sigma_{tot}(\nu N) = A\,\sigma(\nu p)$$
$$\lambda_I = 2.3 \times 10^{12} \text{g/cm}^2$$
$$L = \frac{\lambda_I}{1\,\text{g/cm}^2} = 2.3 \times 10^{12} \text{ cm} = 2.3 \times 10^{10} \text{m}.$$

d. Thus the interaction on an object $x = 50$ cm thick is

$$N_\nu(\text{at } R)\frac{x}{L} = 3.2 \times 10^{14} \frac{50\,\text{cm}}{2.3 \times 10^{12}\text{cm}} = 6.8 \times 10^2 \text{ neutrino interactions/year}.$$

This estimate is correct provided the space between the collider and the target is free (empty space). If there is material, secondary hadrons (neutrons etc.) will be produced and their interactions are more dangerous. The case of a collider under the soil was studied by Mokhov and Van Ginneken [1,2]. They give the minimum distances where the radiation is below the allowed levels. Their values are summarized in Table 1.

Energy of beams	Decays in a year	Safe distances beyond
250 × 250 GeV	2×10^{20}	1.1 km
500 × 500 GeV	2×10^{20}	3.0 km
1.0 × 1.0 TeV	1.2×10^{21}	20 km
1.5 × 1.5 TeV	1.2×10^{21}	40 km
2.0 × 2.0 TeV	1.2×10^{21}	60 km

Table 1

It is clear that the large neutrino flux will be useful for several studies, however, they bring in a crucial factor in the design of the collider. Special attention must be paid on the design so that the beams are weak when they reach the surface of the earth. Attenuations of the beams can be achieved by building the collider deep underground and modulating the shape and trajectories of the beams.

II INTERACTIONS WITH THE MUONS AS UNSTABLE PARTICLES

It has been pointed out that in the reaction

$$\mu^- + \mu^+ \to e^- \bar{\nu}_e W^+$$

there is a t-channel singularity. Whenever the mass of the $e\bar{\nu}_e$ system is smaller than the mass of the muon, the square of the momentum transfer can be both positive and negative, depending on the scattering angles. Consequently, there is a t-singularity with occurs in the physical region [3]–[5]. This is an artificial singularity because the usual derivation of the Feynman rules for the cross-section assumes that the muons are stable. The singularity is eliminated by introducing

either (1) a width for the muon,

or (2) a finite energy spread of the muon beam.

Both effects have been discussed in articles [4,5]. Thus the singularity is tamed by both effects and presents no problem.

We demonstrate here how the singularity is eliminated in quantum mechanics. In Fermi's Golden Rule the transition amplitude for a state $|i\rangle$ going to state $|f\rangle$ is given by

$$a_{fi}(t) = \frac{-i}{\hbar} \int_0^t \mathcal{H}_{fi}(t') e^{-\frac{i}{\hbar}[(E_f - E_i) - i\gamma_i]t'} dt'$$

with γ_i the width of the initial state and $\mathcal{H}_{fi}(t)$ the matrix element which induces the transition $|i\rangle \to |f\rangle$. We usually assume that $\mathcal{H}_{fi}$ is independent of time. Then integration over t' gives

$$a_{fi}(t) = \mathcal{H}_{fi} \frac{e^{-i/\hbar[E_f - E_i) - i\gamma_i]t} - 1}{(E_f - E_i) - i\gamma_i}.$$

This calculation changes the t-channel pole into a Breit–Wigner resonance and the problem is eliminated [6]. The effect from the width of the muon is always there and it takes effect over time scales of the lifetime of the moving muon. For muon colliders the particles interact when the beams cross, which provides a time–scale much shorter than the lifetime of the muons. This brings in a second width from the energy spread of the beam.

The finite–size effect of the beam was also studied in detail [4,5]. They found that for the parameters of the collider the finite size of the beam is more important. They calculated the cross–section which is now finite and of the order of femto-barns. Thus the singularity for this reaction is safely eliminated, and we expect a few hundred to a thousand events per year. It will be possible to observe this reaction but it is not formidable background.

III SPARKLING OF THE QCD VACUUM

Since the muon collider will have the capability of doing deep inelastic scattering experiments, I will discuss a non-perturbative process : the creation of pairs from the QCD vacuum.

Deep inelastic scattering at large Q^2 involves the correlation of two currents on the light-cone. At small values of the scaling variable $x_{bj} = x$, the distances on the light-cone become very large. In this situation the quark travels long distances and radiates hadrons interacting non-perturbatively with the gluonic field. The quark remains confined by interacting with the gluonic field of the proton. The emission of the pairs has the probability [7]

$$w(E, m^2) = \frac{\alpha_s E^2}{\pi^2} \sum_{n=1}^{\infty} \frac{1}{n^2} e^{-\frac{nm^2\pi}{E}}$$

with E the gluo-electric field seen by the photon, α_s the strong coupling constant and m the mass of the created quark.

At a higher $e - p$ center-of-mass energy, the photon sees a stronger gluo-electric field because of the Lorentz transformation for the field

$$E'_{\parallel} = E_{\parallel} \quad \text{and} \quad E'_{\perp} = \gamma \left[\vec{E} + \vec{v} \times \vec{B} \right]_{\perp} .$$

In this Lorentz frame the relativistic factor $\gamma \propto \frac{1}{\sqrt{x}} = \frac{K(y, Q^2)}{\sqrt{x}}$. In going to higher energy, we rescale the field by γ and obtain

$$\omega(E', m^2) = \frac{\alpha_s}{4\pi} \left(E_{\parallel}^2 + \frac{E_{\perp}^2 K^2}{x} \right) \sum_n \frac{1}{n^2} exp\left\{ -\left[nm^2\pi \left(E_{\parallel}^2 + \frac{E_{\perp}^2 K^2}{x} \right)^{-1/2} \right] \right\} .$$

We see that the production of quark pairs increases as x decreases [8].

This phenomenon can be interpreted as the tunneling of quark-pairs from the QCD vacuum. We consider quark-antiquark pairs in the vacuum, where their energy is zero. When they are subjected to an external field they feel a force and gain energy until they have enough energy to tunnel through the barrier and become free. This phenomenon may be already taking place in hadronic [9] and deep inelastic scattering reactions [8]. What we need are experimental signatures for identifying events. Studies and comparisons in various kinematic regions or among structure functions will very likely reveal this effect.

The measurements at HERA involve electron-proton collisions, where only one structure function $F_2(x)$ plays a significant role. It will be very useful to make comparisons of reactions with different targets and beams. The muon collider may provide the opportunity to compare reactions with the exchange of virtual gammas and $W^{\pm}$. Comparisons between the reactions

$$\mu^- + p \to \mu^- + \text{anything}$$
$$\mu^- + p \to \nu + \text{anything, and}$$
$$\mu^+ + p \to \bar{\nu} + \text{anything}$$

will allow comparison between $F_2^{\gamma p}(x)$, $F_2^{W^\pm p}(x)$ and $F_3^{W^\pm p}(x)$ structure functions. Predictions for all of them are not yet available, but are possible in the various theories. For instance, it is interesting to compare the predictions of perturbative QCD, with the pairs produced from the QCD vacuum.

The tunneling phenomenon, discussed here, should also take place in QED. In fact, the first calculations for sparkling of the vacuum were done in QED [7]. Recently, an experiment at SLAC was interpreted as the creation of electron-positron pairs from the interaction of the intense accelerator beam with the field of a powerful laser [10,11]

The first muon collider will make possible several new and very interesting experiments. Spectacular among the results will be the production of s-channel resonances, like Z^0 and Higgs particles. We expect the physical Higgs of the standard model and additional Higgses from theories beyond the standard model. In case that there are several resonances one can search for the origin of CP-violation through CP-asymmetries induced by widths [12].

I have tried in this talk to touch upon a few new topics which are still under development and should be the subject of future investigations.

REFERENCES

1. N.V. Mokhov and A. Van Ginneken, "Muon Collider Neutrino Radiation", Fermilab–preprint 1997 (to be published).
2. C.J. Johnstone and N.V. Mokhov, "Shielding the Muon Collider Interaction Region", Fermilab–preprint 1997.
3. M. Nowakowski and A. Pilaftsis, *Zeit. f. Phys.* **C60**, 121 (1993).
4. T.F. Ginzburg, *Nucl. Phys. Proc. Suppl.* **51A**, 85 (1996).
5. K. Menlikov and V.G. Serbo, *Phys. Rev. Lett.* **76**, 3263 (1996).
6. H.A. Bethe and R.W. Jackiw, "Intermediate Quantum Mechanics", W.A. Benjamin Inc. (1968), pages 200 and 206.
7. J. Schwinger, *Phys. Rev.* **82**, 664 (1951); **93**, 615 (1954).
8. E.A. Paschos, *Phys. Lett.* **B389**, 383 (1996).
9. A. Casher et al., *Phys. Rev.* **D20**, 179 (1979).
10. D.L. Burke et al., *Phys. Rev. Lett.* **79**, 1626 (1997).
11. 'Search and Discovery' in: *Physics Today*, Feb. 1998, pages 17 - 18.
12. A. Pilaftsis, *Phys. Rev. Lett.* **77**, 4996 (1996).

Detectors for Neutrino Physics at the First Muon Collider

Deborah A. Harris* and Kevin S. McFarland[†]

University of Rochester, Rochester, NY 14627
[†]*Massachusetts Institute of Technology, 77 Massachusetts Ave., Cambridge, MA 02139* [1]
and
Fermi National Accelerator Laboratory, Batavia, IL 60510

Abstract. We consider possible detector designs for short-baseline neutrino experiments using neutrino beams produced at the First Muon Collider complex. The high fluxes available at the muon collider make possible high statistics deep-inelastic scattering neutrino experiments with a low-mass target. A design of a low-energy neutrino oscillation experiment on the "tabletop" scale is also discussed.

INTRODUCTION

This contribution considers the problem of constructing detectors appropriate for doing short-baseline neutrino physics at the First Muon Collider complex. The physics motivations for these detectors are discussed elsewhere in these proceedings [1]. Since the proposed experiments are short-baseline, the physics being considered is primarily the high-energy physics of neutrino-nucleon deep-inelastic scattering; however, the final section of the paper considers an oscillation experiment possible with the lowest energy neutrino beam.

NEUTRINO BEAMS AT THE MUON COLLIDER COMPLEX

The muon collider is expected to use a series of recirculating linacs to accelerate the muons before injection into a collider ring. Any segment along the muon's trajectory that is straight will necessarily create a collimated neutrino beam with an angular divergence of approximately $1/\gamma_\mu$. The recirculating linacs (RLAs) will have $\mathcal{O}(300\text{m})$ in which acceleration takes place, and the interaction points in the collider rings will also have 5-10m straight sections.

---

[1]) permanent address

Experiment:	RLA1	RLA2	RLA3	Med Eng	Top	High Eng
Max Muon Energy (GeV)	10	70	250	100	175	250
Distance (m)	23.3	163	582	233	408	582
50% ν's within radius	25 cm	30 cm	25 cm	15 cm	15 cm	15 cm
90% ν's within radius	75 cm	75 cm	60 cm	40 cm	40 cm	40 cm
ν DIS events,($\times 10^6$), $r < 10cm$	n/a	0.038	0.11	0.13	0.21	0.34
ν DIS events,($\times 10^6$), $r < 150cm$	n/a	0.49	0.92	0.54	0.99	1.34

TABLE 1. For each straight section in the RLAs and for several different collider scenarios, the maximum muon energy, the required shielding distance, the ν beam size and the ν DIS events in Millions per g/cm^2/yr are shown.

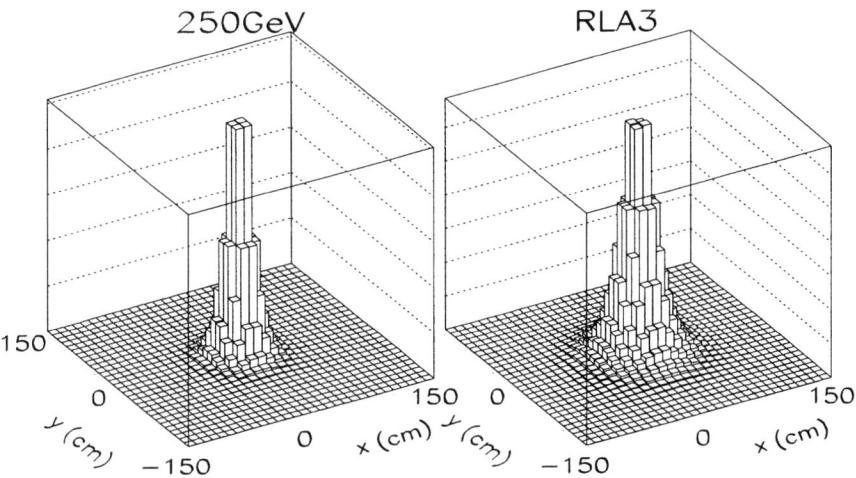

FIGURE 1. Event illuminations in two possible neutrino experiments

We wrote a simple Monte Carlo of the muon collider straight sections to predict fluxes, based on the workshop muon accelerator parameters. We used the μ^- beam, and assumed the beam polarization and divergence were zero. Fluxes were predicted far downstream of each straight section in order to allow for shielding. We considered only a "dumb" shield option in which sufficient concrete shielding to range out the muons from the primary muon beam was used. Table 1 gives a table of the distances of a proposed experiment downstream of each straight section in the RLAs and collider.

Predicted Neutrino Fluxes

As shown in Table 1, the beam is relatively small at all energies (since we have chosen a baseline proposal to energy due to shielding concerns). Figure 1 shows a typical event illumination for two possible neutrino experiments:

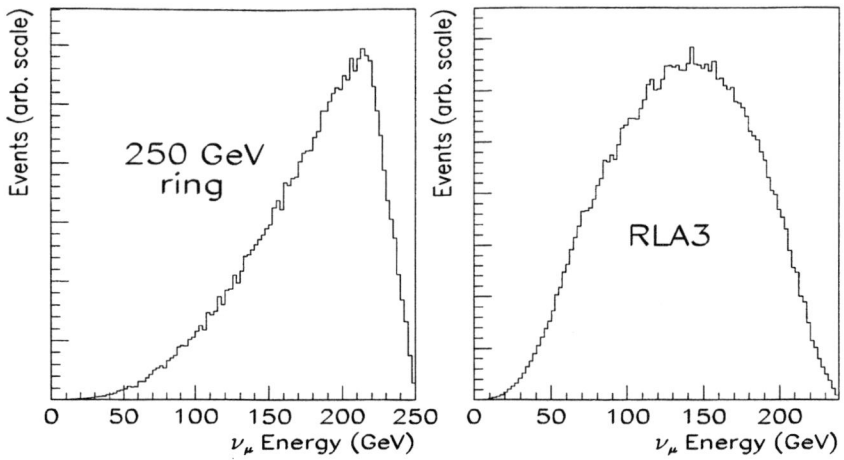

FIGURE 2. Neutrino Energy Spectra for two experiments, each with a target of 10 cm radius: one downstream of the 250 GeV collider ring straight section and one downstream of RLA3. The broader distribution from RLA3 is an artifact of the energy ramp in RLA3; turn-by-turn, the spectra have a similar shape.

one downstream of RLA3, and one downstream of the 250GeV collider ring's straight section.

Figure 2 shows the neutrino event spectra reaching a $10cm$ radius target downstream of RLA3, and also for the same target downstream of a 250 GeV collider ring straight section. Unlike neutrino beams made by decays of hadrons, the energy spectra will be well known from the accelerator parameters at a muon collider. This will make the difficulties of separating flux and cross-section relatively trivial and the flux endpoint will provide an important detector calibration.

Muon Flux at Neutrino Experiments

The shielding necessary to protect a neutrino experiment from primary muons will produce large numbers of muon neutrino DIS interactions some of which will produce muons in the neutrino detector. The Monte Carlo simulation described earlier contained muon production and propagation in shielding to predict the size of this background. Figure 3 shows the surprisingly peaked muon illumination at the detector for experiments downstream of RLA3 and the 250 GeV collider using these assumptions.

The critical number for an experiment is the maximum number of muons arriving at the detector for a single turn in the accelerator. This flux is listed in Table 2. The background for a detector measuring $100cm^2$ is always ≤ 0.06

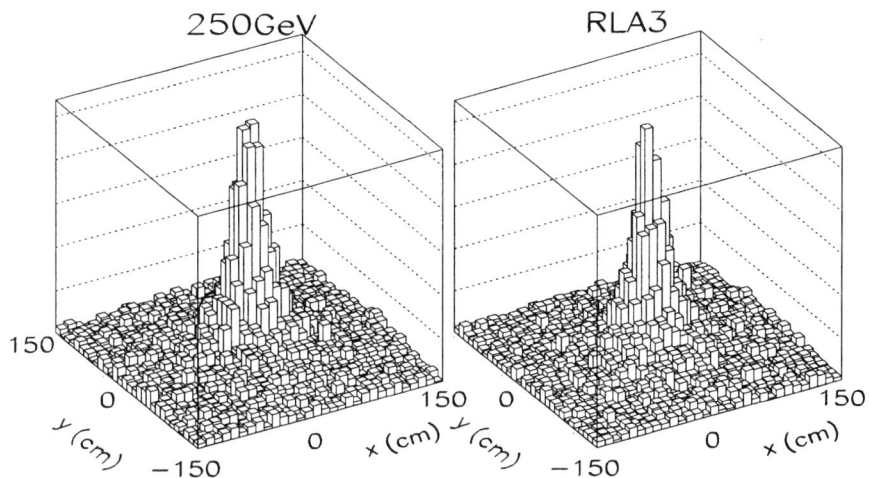

FIGURE 3. Maximum background muon illuminations in one turn for two possible neutrino experiments

Experiment:	RLA1	RLA2	RLA3	Med Eng	Top	High Eng
Muons/Turn	0.015	0.06	0.4	0.0015	0.0025	0.0035
Muons in center $100 cm^2$	0.0028	0.017	0.081	0.00014	0.00024	0.00033
ν_μ DIS events/Turn	0.05	0.07	0.27	0.0015	0.0015	0.002

TABLE 2. Per pulse interaction rates in a 40 cm×40 cm 10 λ_0 steel sampling calorimeter and muon filter located downstream of a "dumb" muon filter which ranges out the direct beam muons.

muons/turn and is therefore not a problem. However, for large detectors downstream of the recirculating linacs more complicated shielding would be necessary. One option would be to have magnetized shielding followed by an empty volume which could bend any produced muons away from the target. Another possibility could be to use shielding heavier than concrete again with a long empty volume just upstream of a large detector to allow the muons to scatter away from the detector acceptance.

HIGH MASS NEUTRINO TARGET

The fluxes detailed above represent improvements by 3 to 5 orders of magnitude over contemporary high energy neutrino beams[2]. It would therefore be

---

[2] The NuTeV SSQT had an interaction rate of $\mathcal{O}(10)$ events/kg/yr with a mean energy of 120 GeV [2]; beams at the CERN PS used by NOMAD and CHORUS had a mean energy of 35 GeV and observed $\mathcal{O}(100)$ events/kg/yr [3].

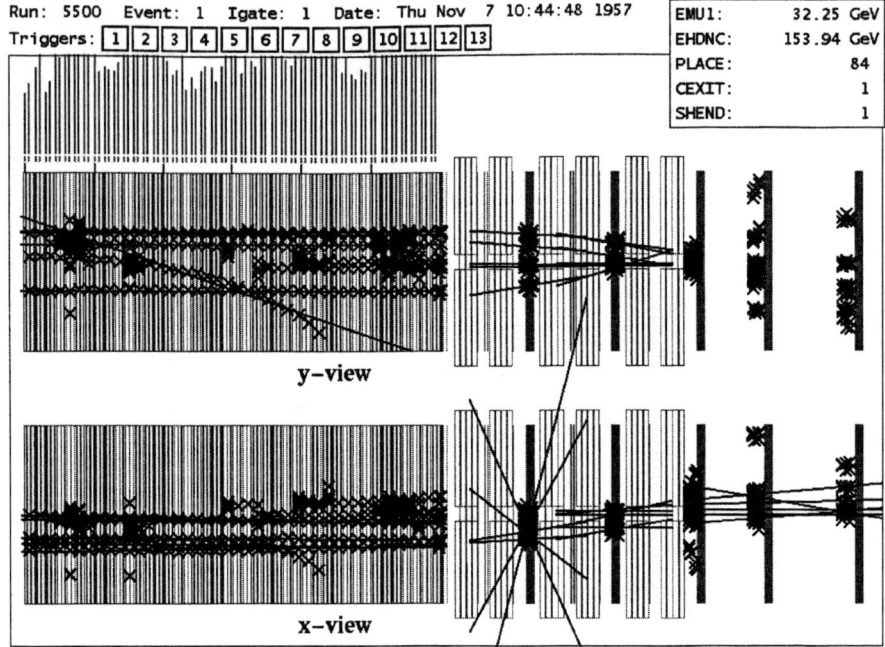

FIGURE 4. An extreme Monte Carlo event in the NuTeV detector, placed 600 meters downstream of RLA3. This event contains six neutrino interactions and three muons produced in upstream shielding from a single turn of the muon beam. This number of muons would be very unlikely, but the number of neutrino interactions is below average in the NuTeV detector

possible to run an experiment with conventional high-mass sampling target-calorimeters and observe perhaps 10^{11} events per year in previously explored energy regimes.

However, sampling calorimeters would have some difficulties in the FMC environment. High mass sampling calorimeters are only designed for the detection of ν_μ or $\overline{\nu_\mu}$ since ν_e and $\overline{\nu_e}$ events are difficult to separate from neutral current interactions of other neutrino species, especially when the final state lepton has a low fraction of the initial neutrino energy. Therefore, unless the muon beam polarity of the FMC were reversible, only neutrinos or antineutrinos could be observed in a single detector. Furthermore, high mass, sparsely instrumented detectors are sensitive to pile-up. As per pulse event rates in Table 2 show, this is a particular concern downstream of the high-energy recirculating linacs where the beam is concentrated in time in a relatively small number of turns. Figure 4 shows an extreme illustration of the difficulties of dealing with pile-up in a sample-calorimeter.

Assuming these difficulties could somehow be overcome, it is also unclear what the novel physics available from 10^{11} neutrino interactions in dense mate-

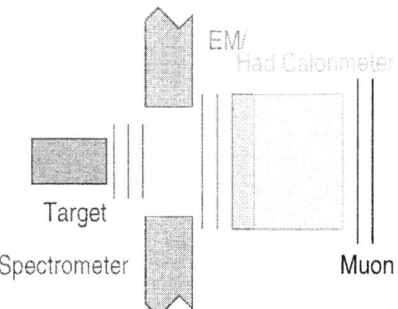

FIGURE 5. A schematic of an FMC neutrino detector in the style of conventional fixed target detectors.

rial would be. In general, precision measurements from neutrinos are systematics limited in such detectors[3] and searches for rare phenomena are background limited.

CONVENTIONAL FIXED-TARGET GEOMETRY

Probably more interesting than the above detector technology would be one where the target is small enough to vary its composition for studies of nuclear effects on nucleon structure or even spin physics in neutrino deep inelastic scattering. This is possible in a geometry more closely resembling traditional fixed-target experiments: target, followed by spectrometry and calorimetry and muon identification as shown in Figure 5. The problems of tracking, particle-ID and calorimetry in such detectors are well-known and will not be discussed here. Such a detector would presumably be able to identify the outgoing lepton in charged-current events, possibly with TRDs in the downstream tracking providing enhanced identification. This detector would be able to tag the production of charmed particles by observing high momentum final state leptons, and with a sufficiently fine-grained upstream tracker, charmed final state particles could be tagged *via* detached vertices [6]. Physics goals of an experiment using this detector are described elsewhere in these proceedings [1].

The most serious technical difficulties of such detectors is pile-up in the dense calorimetry and muon systems. A typical iron-scintillator sampling calorimeter/muon shield of 10 λ_0 (1.6 m of steel or 13 kg/cm^2) would observe the per pulse background rates shown in Table 2. Background rates at the collider are probably manageable, but the background rates at the RLAs, particularly RLA3, could be difficult.

---

[3] For example, the measurements of α_S [4] and $\sin^2\theta_W$ [5] in the CCFR experiment

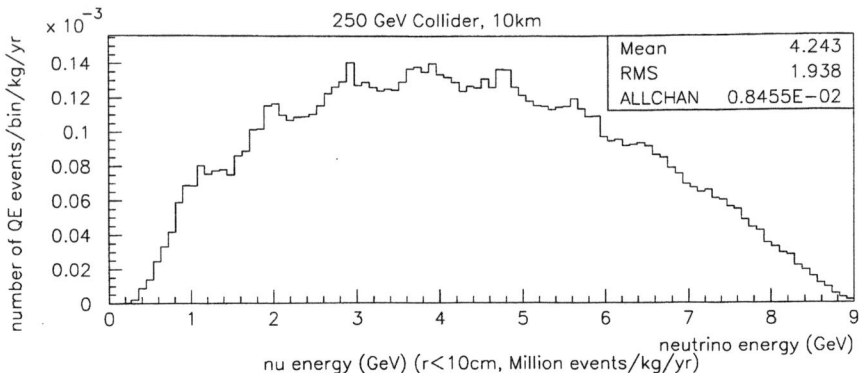

FIGURE 6. Energy spectrum of quasi-elastic ν_μ events downstream of RLA1.

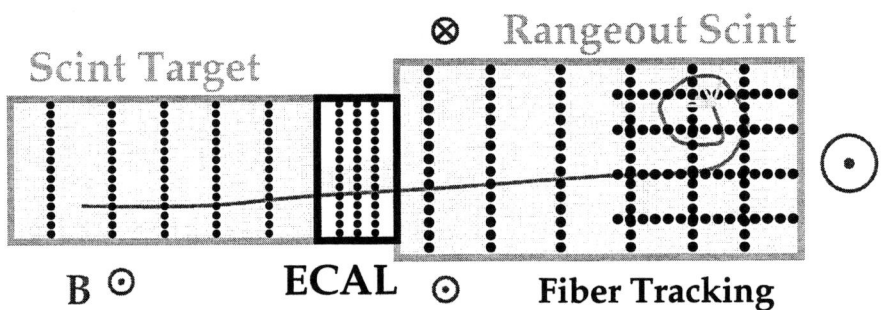

FIGURE 7. A table-top sized neutrino detector for that beam.

A TABLE-TOP LOW ENERGY NEUTRINO DETECTOR

Another avenue that becomes possible with the FMC neutrino beams is searching for neutrino oscillations in a modest-sized experiment. Figure 6 shows the expected energy spectrum of quasi-elastic ν_μ events downstream of RLA1. In a water-density 0.05m^3 target, one would expect to observe approximately 0.4 Million events per year. This sets the scale for a truly "table-top" scale neutrino oscillation experiment.

With a flavor selected beam of $\overline{\nu_\mu}$ and ν_e, the simplest search experimentally[4] is for wrong-sign muon (μ^-) appearance. The primary backgrounds would come from mis-identifying the muon charge or from production of charged pions which could decay or be misidentified as muons. By conclusively iden-

---

[4] e^+ appearance is background-laden at these energies and τ appearance, aside from being heavily suppressed by m_τ, requires a more sophisticated detector.

tifying μ charge in a spectrometer, tracking the particle from the point of appearance and observing its decay with a characteristic muon lifetime, these backgrounds could presumably be kept very low. A naive version of a detector designed for this measurement appears in Figure 7. With proper granularity of tracking and careful optimization of the magnetic field for a broad acceptance in momentum, even such a small scale detector could reach 10^{-5} sensitivity if naive background estimations are correct.

CONCLUSIONS

We have presented outlines of detector designs for short-baseline neutrino experiments at the first muon collider complex. This facility would make available neutrino beams of unprecedented intensity, and we argue that the best way to take advantage of this is to design novel detectors, such as small targets or "table-top" neutrino experiments to explore different aspects of neutrino scattering or neutrino oscillations than those studied in the past.

REFERENCES

1. D.A. Harris and K.S. McFarland, "A Small Target Neutrino Deep Inelastic Scattering Experiment at the First Muon Collider", these proceedings.
2. R. Bernstein *et al.*, "The NuTeV SSQT", FERMILAB-TM-1884, Apr 1994.
3. J. Altegoer *et al.* (NOMAD Collaboration), CERN-PPE-97-059, May 1997.
   E. Eskut *et al.* (CHORUS Collaboration), CERN-PPE-97-033, Mar 1997.
4. W.G. Seligman *et al.*, "A New Determination of α_S from Neutrino-Nucleon Scattering", NEVIS-REPORT-292, Jan 1997, to appear in Phys. Rev. Lett.
5. K.S. McFarland *et al.*, "A Precision Measurement of Electroweak Parameters in Neutrino-Nucleon Scattering", FERMILAB-PUB-97-001-E, Jan 1997, to appear in Z. Phys. **C**.
6. Bruce King, these proceedings.

The Physics Potential of Neutrino Beams from Muon Storage Rings

S. Geer[1]

Fermi National Accelerator Laboratory[1] P.O. Box 500, Batavia, Illinois 60510

Abstract. High-intensity neutrino beams could be produced using a very intense muon source, and allowing the muons to decay in a storage ring containing a long straight section. Taking the parameters of muon source designs that are currently under study for future high luminosity muon colliders, the characteristics of the neutrino beams that could be produced are discussed and some examples of their physics potential given. It is shown that the neutrino and antineutrino beam intensities may be sufficient to produce hundreds of neutrino interactions per year in a detector on the far side of the Earth.

INTRODUCTION

The muon source required for a high-luminosity muon collider would enable intense neutrino beams to be produced from muon decays. In the following, calculated beam fluxes and interaction rates are presented for various long- and short-baseline neutrino configurations that exploit a muon collider muon source. To illustrate the physics potential of these beams some examples are discussed. A more detailed description of the calculation is given in Ref. [1].

The muon lifetime is about 100 times longer than the charged pion lifetime. A linear decay channel of the type used to produce conventional neutrino beams would be too short in practice to use efficiently as a muon decay channel. This problem can be overcome by using a muon storage ring with a straight section pointing in the desired direction. We will consider a storage ring that consists of two parallel straight sections connected together by two arcs. If the straight sections are equal in length to the arcs, 25% of the stored muons will decay whilst they are in the straight section pointing at the experiment.

---

[1] Presented at the Workshop on Physics at the First Muon Collider and Front End of a Muon Collider, Fermilab, November 6-9, 1997. This work was performed at the Fermi National Accelerator Laboratory, which is operated by Universities Research Association, under contract DE-AC02-76CH03000 with the U.S. Department of Energy.

CP435, *Workshop on the Front End of a Muon Collider*
edited by S. Geer and R. Raja
© 1998 The American Institute of Physics 1-56396-793-6/98/$15.00

If the ring lattice is properly designed, a beam divergence $\theta_b \leq O(10^{-4})$ should be achievable [2] in the straight section. Thus, if the circulating muons have momentum $p/m_\mu \ll 10^4$ (corresponding to $p \ll 1000$ GeV/c) the angular divergence of the neutrino beam produced from decays in the straight sections will be dominated by the decay kinematics. In the muon rest-frame the distribution of muon antineutrinos (neutrinos) from the decay $\mu^\pm \to e^\pm + \nu_e\ (\bar{\nu}_e) + \bar{\nu}_\mu\ (\nu_\mu)$ is given by the expression [3]:

$$\frac{d^2 N_{\nu_\mu}}{dx d\Omega} \propto \frac{2x^2}{4\pi}\left[(3-2x) \mp (1-2x)P_\mu \cos\theta\right], \quad (1)$$

where $x \equiv 2E_\nu/m_\mu$, θ is the angle between the neutrino momentum vector and the muon spin direction, m_μ is the muon rest mass, and P_μ is the average muon polarization along the chosen quantization axis. The corresponding expression describing the distribution of electron neutrinos (antineutrinos) is:

$$\frac{d^2 N_{\nu_e}}{dx d\Omega} \propto \frac{12x^2}{4\pi}\left[(1-x) \mp (1-x)P_\mu \cos\theta\right]. \quad (2)$$

The neutrino beam intensity will depend upon the performance of the muon source. In the following we will assume a source with the parameters given for the workshop, which would produce 7.5×10^{20} muons that can be injected into the storage ring each operational year. If 25% of the muons decay in the straight section pointing at the experimental area the resulting neutrino beam will contain 2×10^{20} neutrinos per year and 2×10^{20} antineutrinos per year, with energy- and angular-distributions described by Eqs. (1) and (2).

FLUXES AND INTERACTION RATES

Consider a geometry in which the plane of the storage ring dips at an angle of $\sim 50°$ to the horizon, and the resulting neutrino beam exits the Earth at the "far site" after traversing $O(10^4)$ km. This would correspond to a storage ring sited at the Fermi National Accelerator Laboratory in the United States with the far site at the Kamioka mine in Japan ($L \sim 9300$ km). The calculated neutrino and antineutrino fluxes at the far site are shown in Fig. 1 as a function of the energy and average polarization of the muons in the straight section of the storage ring. The fluxes have been averaged over a 1 km radius "spot" at the far site. The ν_e and $\bar{\nu}_e$ fluxes are very sensitive to the muon spin direction. This can be understood by examining Eq. 2 which shows that for μ^+ (μ^-) decays the ν_e ($\bar{\nu}_e$) flux $\to 0$ for all neutrino energies as $\cos\theta \to +1\ (-1)$. The charged current neutrino and antineutrino rates in a detector at the far site can be calculated using the approximate expressions [4] for the cross-sections: $\sigma_{\nu N} \sim 0.67 \times 10^{-38}\ cm^2 \times E_\nu(GeV)$ and $\sigma_{\bar{\nu} N} \sim 0.34 \times 10^{-38}\ cm^2 \times E_{\bar{\nu}}(GeV)$. The predicted charged current interaction

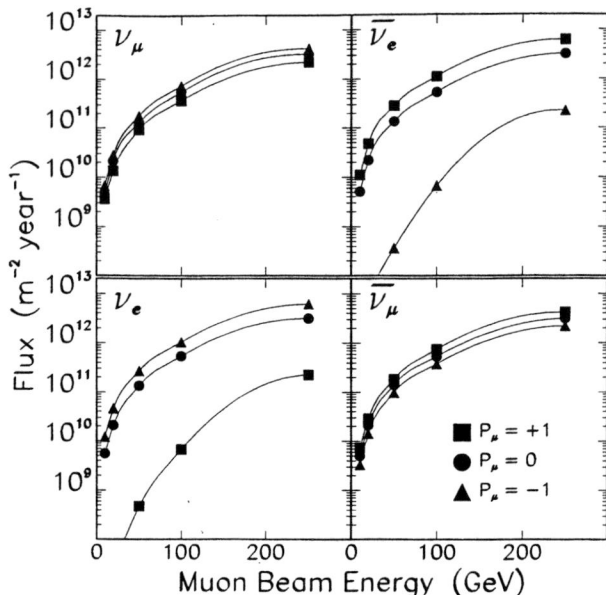

FIGURE 1. Calculated neutrino and antineutrino fluxes at a far site located 10^4 km from a muon storage ring neutrino source. The fluxes are shown as a function of the energy of the stored muons for negative muons (top two plots) and positive muons (bottom two plots), and for three muon polarizations as indicated.

rates are shown in Fig. 2 as a function of the energy and average polarization of the decaying muons. We conclude that, for stored muon momenta greater than ~ 20 GeV/c, neutrino and antineutrino interactions should be readily detectable at the far site.

Consider next a long baseline geometry in which the plane of the muon storage rings tilts at just a few degrees to the horizon and the far site is 732 km from the neutrino source. This would correspond [5] to a storage ring sited at the Fermi National Accelerator Laboratory and a far site at the Soudan underground Laboratory in Minnesota. The neutrino fluxes and charged current interaction rates at the far site can be obtained by scaling the results presented in Figs. 1 and 2 by a factor of 183. Predicted fluxes and spectra corresponding to using 10 GeV/c stored muons are shown in Fig. 3. The calculated neutrino and antineutrino fluxes are both $\sim 1 \times 10^{12}$ m^{-2} year^{-1}, and the corresponding charged current interaction yields are 3.1×10^3 μ^- kT^{-1} year^{-1} and 1.4×10^3 e^+ kT^{-1} year^{-1} when negative muons are stored in the ring, and 1.6×10^3 μ^+ kT^{-1} year^{-1} and 2.9×10^3 e^- kT^{-1} year^{-1} when positive muons are stored in the ring. The mean energies of the charged leptons and

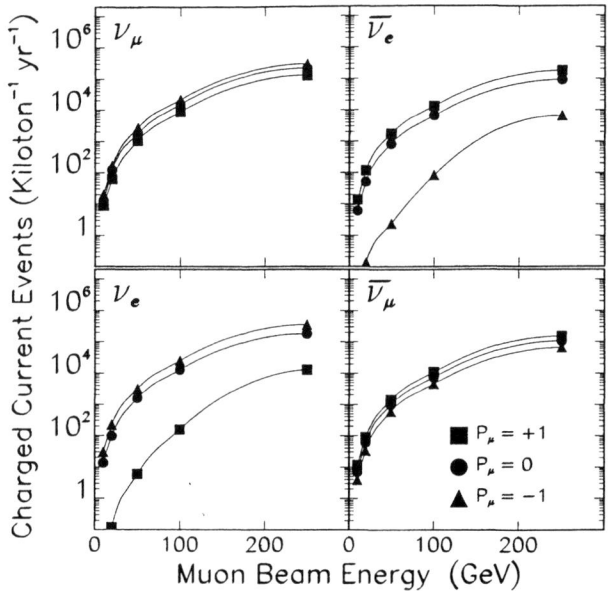

FIGURE 2. Calculated neutrino and antineutrino charged current interaction rates in a detector located 10^4 km from a muon storage ring neutrino source. The rates are shown as a function of the energy of the stored muons for negative muons (top two plots) and positive muons (bottom two plots), and for three muon polarizations as indicated.

antileptons produced in these charged current interactions are respectively ~ 3.5 GeV and ~ 2 GeV. Thus, neutrino and antineutrino interactions should be readily detectable at the far site when the decaying muons in the storage ring have momenta as low as 10 GeV/c. The predicted event rates become very large when high energy muons are stored in the ring (for example, 3×10^7 interactions per KT-year when 250 GeV/c muons are stored).

Finally, consider a short baseline geometry in which the detector is 1 km from a 1.5 GeV/c muon storage ring. Averaging the fluxes at the detector over a "spot" with a radius of 5 m, the predicted neutrino and antineutrino fluxes resulting from unpolarized muon decays are both 1.2×10^{16} m^{-2} per year. The corresponding charged current interaction rates in a 1 kT detector yield 2.9×10^6 μ^+ per year and 5.1×10^6 e^- per year if positive muons are stored in the ring, and 6.0×10^6 μ^- per year and 2.6×10^6 e^+ per year if negative muons are stored in the ring. Hence, a 1 kT detector would record millions of charged current interactions per year with mean charged–lepton energies of ~ 0.6 GeV and mean charged–antilepton energies of ~ 0.3 GeV.

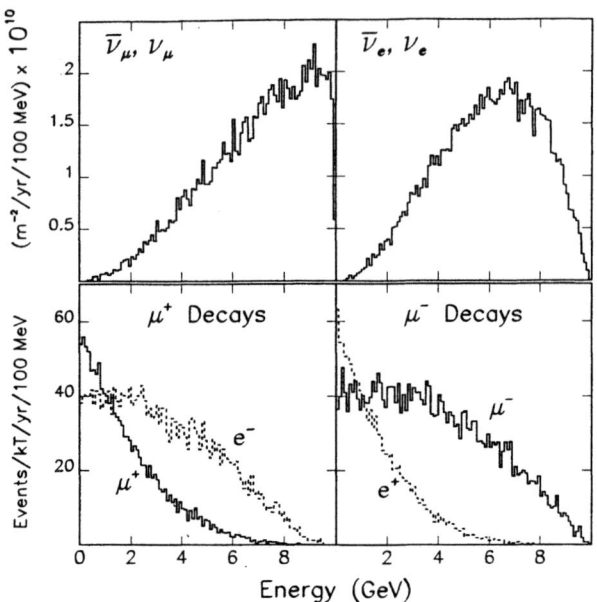

FIGURE 3. Calculated fluxes and spectra in a detector 732 km downstream of a muon storage ring neutrino source in which 10 GeV/c unpolarized muons are circulating. The top two plots show the neutrino and antineutrino spectra, and the bottom two plots show the charged lepton spectra from charged current interactions when positive muons (bottom left) and negative muons (bottom right) are stored in the ring.

PHYSICS POTENTIAL: EXAMPLES

To illustrate the physics potential of the muon storage ring neutrino sources discussed in the previous section, consider the sensitivity of an experiment searching for ν_e-ν_μ or ν_e-ν_τ oscillations performed by searching for charged current interactions producing "wrong-sign" muons. Within the framework of two-flavor vacuum oscillations, the probability that, whilst traversing a distance L, a neutrino of type 1 (mass m_1) oscillates into a neutrino of type 2 (mass m_2) is given by [3]:

$$P(\nu_1 \to \nu_2) = \sin^2(2\theta) \, \sin^2(1.27 \Delta m^2 \, L/E) \,, \tag{3}$$

where θ is the mixing angle, $\Delta m^2 \equiv m_2^2 - m_1^2$ is measured in eV^2/c^4, L in km, and the neutrino energy E is in GeV. In the absence of backgrounds or systematic uncertainties, a neutrino oscillation experiment can be characterized by the total number of neutrino interactions observed (and hence the minimum observable $P(\nu_1 \to \nu_2)$) and the average L/E for the interacting neutrinos. These parameters are summarized in Table 1 for the experimental configurations discussed in the previous section.

Figure 4a compares the ν_e-ν_μ oscillation single–event sensitivity contours in the (Δm^2, $\sin^2(2\theta)$)–plane. The long– and very–long–baseline configurations have similar Δm^2 reaches as $\sin^2(2\theta) \to 1$, approaching single event sensitivities of $\sim 10^{-5}$ eV^2/c^4, more than an order of magnitude better than the expected reaches of the next generation of proposed neutrino experiments. The large charged current event rates expected for the short baseline configuration would enable sensitivities approaching $\sin^2(2\theta) \sim 10^{-7}$ for large Δm^2 ($\Delta m^2 \sim 1$ eV^2/c^4). This $\sin^2(2\theta)$ reach is almost a factor of 1000 better than the expected reaches of presently proposed experiments, but would only be attained in the absence of backgrounds from, for example, secondary production of neutrinos from interactions in the vicinity of the experiment, or a component of neutrinos produced from the decays of "wrong–sign" muons produced upstream of the storage ring. Figure 4b shows the single–event contour in the (Δm^2, $\sin^2(2\theta)$)–plane for a ν_e-ν_τ oscillation search corresponding to the highest energy (250 GeV/c stored muons) configuration in Table 1, where once again the search is based on looking for wrong-sign muons. The pure ν_e component in the neutrino beam from a muon storage ring would enable the sensitivity of ν_e-ν_τ searches to improve beyond the sensitivities of past searches [6,7] by many orders of magnitude.

We conclude that a muon source of the type being developed for future high-luminosity muon colliders could produce very intense neutrino beams. If $O(10^{20})$ muons per year decayed within a 20 GeV/c storage ring with a straight section pointing in the desired direction, the resulting beam would produce hundreds of neutrino interactions per 10 kT-year on the other side of the Earth. High beam intensities, together with the purity of the initial neutrino beam flavor content, would provide unique opportunities for neutrino experiments. The neutrino physics program using muon storage ring neutrino sources could begin with short baseline experiments using low energy muons (O(1 GeV)) as soon as an intense muon source became operational, and be extended to include higher energy storage rings and longer baseline experiments as higher

TABLE 1. Summary of the neutrino oscillation experimental configurations considered in the text.

p (GeV/c)	m_{DET} (kT)	L (km)	$<E_\nu>$ (GeV)	$L/<E_\nu>$ (km/GeV)	ν_e CC interactions/yr
20	10	10^4	13	744	1×10^3
20	10	732	13	57	2×10^5
10	10	732	6.6	111	3×10^4
1.5	1	1	1	1	5×10^6
250	10	732	161	4.5	3×10^8

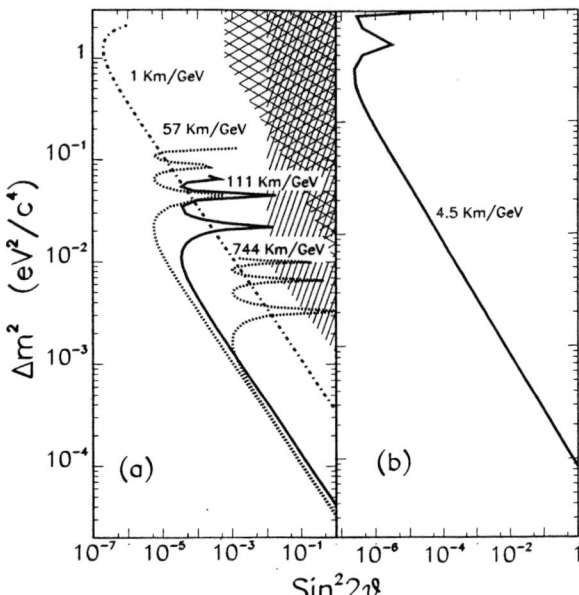

FIGURE 4. Contours of single-event sensitivity for (a) ν_e-ν_μ oscillations and (b) ν_e-ν_τ oscillations. The ν_e-ν_μ contours correspond to 1 year of running with the first four detector configurations summarized in Table 1. The hatched and cross-hatched areas show the expected regions that will be explored by respectively the MINOS experiment [4] after 2 years of running and the MiniBooNe experiment [7] after 1 year of running. The ν_e-ν_τ contour corresponds to the fifth detector configuration in Table 1.

energy muon beams became available.

REFERENCES

1. S. Geer, Neutrino Beams from Muon Storage Rings: Characteristics and Physics Potential, Fermilab-PUB-97/389 (hep-ph/9712290), submitted to Phys. Rev. D.
2. C. Johnstone, private communication.
3. See for example: T. K. Gaisser, "Cosmic Rays and Particle Physics", Cambridge University Press 1990.
4. See for example: F. Boehm and P. Vogel, "Physics of Massive Neutrinos", Cambridge University Press, 1987.
5. E. Ables et al; "P-875: A Long-baseline Neutrino Oscillation Experiment at Fermilab, NuMI-L-63 Minos Proposal", Feb. 1995, unpublished.
6. N. Ushida et al. (The Fermilab E531 Collab.), Phys. Rev. Lett. **57**, 2897 (1986).
7. M. Talebzadeh et al. (BEBC WA66 Collab.), Nucl. Phys. **B291**, 503 (1987).
8. E. Church et al. (BooNE Collab.), "A letter of intent for an experiment to measure $\nu_\mu \to \nu_e$ oscillations and ν_μ disappearance at the Fermilab Booster", May 16, 1997, unpublished.

K2K
KEK to Kamioka Long Baseline Neutrino Oscillation Experiment

Takashi Kobayashi

Institute of Particle and Nuclear Studies,
High Energy Accelerator Research Organization (KEK)
1-1 Oho, Tsukuba 305-0801, Japan
Representing K2K Collaboration[1]

Abstract. We have been constructing a long baseline neutrino oscillation experiment in which a ν_μ beam will be generated at KEK and detected by the Super-Kamiokande (SK) water Čerenkov detector 250 km away from KEK. We will have a front neutrino detector and various beam monitors in KEK in order to predict the flux of the neutrinos at SK as precisely as possible. We will look for the oscillation modes $\nu_\mu \to \nu_e$ appearance and $\nu_\mu \to \nu_x$ disappearance. We can explore the oscillation parameter regions of $\Delta m^2 \gtrsim 2 \times 10^{-3}$ eV^2 and $\sin^2 2\theta \gtrsim 0.5$ for the disapearance and $\sin^2 2\theta \gtrsim 0.1$ for the appearance modes, respectively. The experiment will start at beginning of 1999.

INTRODUCTION

The Standard Model of electroweak interactions is generally presented with the neutrino masses being identically zero. However, experimentally, the mass of the neutrinos have only upper limits. These limits are 15 eV, 170 keV and 24 MeV for ν_e, ν_μ and ν_τ, respectively [2]. Discovery of finite neutrino mass could be a breakthrough in searching for new physics beyond the standard model.

One of the possible ways of probing such small masses is neutrino oscillation [3]. If the neutrinos have mass and the mass and flavor eigenstates are different, a neutrino changes its flavor during propagation. For the case where the third flavor can be neglected, the probability of the oscillation from ν_μ to another flavor eigenstate ν_a is

$$P(\nu_\mu \to \nu_a) = \sin^2 2\theta \sin^2 \left(\frac{1.27 \Delta m^2 [\text{eV}^2] L [\text{km}]}{E_\nu [\text{GeV}]} \right), \quad (1)$$

where θ and Δm^2 are the mixing angle and the difference of mass squared, respectively, between the two mass eigenstates, L is the flight length and E_ν is the energy of the neutrino. The probability has maxima at $\Delta m^2 = \frac{n\pi}{2.54}(E/L)$ for odd integer values of n, where the units are the same as in Eq. (1).

Recently Super-Kamiokande (SK) and Soudan 2 experiments confirmed the previous results that the ratio of the number of ν_μ to that of ν_e in atmospheric neutrinos is significantly smaller than expected [4]. SK also reported that there are fewer upward going muon neutrinos than expected. The results strongly suggest the occurrence of neutrino oscillations of $\nu_\mu \to \nu_e$ and/or $\nu_\mu \to \nu_\tau$ with Δm^2 region of $10^{-3} \sim 10^{-1}$ eV^2. A reactor-neutrino long baseline experiment, CHOOZ, recently published a negative result of a $\bar{\nu}_e$ disappearance search [5]. It excludes the possibility of $\nu_\mu \to \nu_e$ oscillation in the parameter region $\Delta m^2 > 0.9 \times 10^{-3}$ eV^2. The possibility of $\nu_\mu \to \nu_\tau$ oscillation still remains.

The KEK-PS E362, now called K2K (KEK-to-Kamioka), is the first accelerator-neutrino long baseline oscillation experiment to look for neutrino oscillation of the modes $\nu_\mu \to \nu_e$ appearance and $\nu_\mu \to \nu_x$ disappearance [6]. The ν_μ beam is generated at KEK. We have SK at 250 km away from KEK as a far detector and a front neutrino detector on-site at KEK. The sensitive region of the oscillation parameter space is $\Delta m^2 \gtrsim 2 \times 10^{-3}$ eV^2, which covers that suggested by atmospheric neutrino experiments. The experiment is not sensitive to ν_τ appearance because we have very small number of ν_μ which have higher energy than τ production threshold of 3.5 GeV.

NEUTRINO BEAM

A wide-band ν_μ beam is generated at KEK by using the 13 GeV/c proton synchrotron (PS). The PS is operated in the fast extraction mode with a 1.0 μsec spill in every 2.2 sec [7]. The extracted proton beam is transported and bent 94° to the direction of SK, then hits an Al target. The intensity of the proton beam at the target will be 6×10^{12} protons/spill. We will have 10^{20} protons on target (pot) in total during the 3 years of experiment. The neutrino beam line is now being constructed [8].

Positive pions produced at the target are focused by two large horn magnets [9]. Both horns will be operated by a 1-msec width pulsed current of 250 kA. They have been already built and tested with that current. The target-horn system is followed by a decay tunnel 200 m in length in which the π^+'s decay into ν_μ's and muons. The tunnel is terminated by a beam dump. The neutrino beam is directed 1° downward to aim at SK.

In Fig. 1, the expected neutrino flux as a function of the neutrino energy E_ν and the distance from the beam axis at the front neutrino detector (FD) at 300 m from the target and at the SK at 250 km are plotted. They are obtained by GEANT [10] simulation with GCALOR [11]. The average neutrino energy is 1.4 GeV and the contamination of ν_e is estimated to be about 1 %.

BEAM MONITORS

We will have monitors for protons, pions and muons in the neutrino beam line. We will monitor the profile, position and intensity of the proton beam before the target spill by spill by using secondary emission chambers, ionization chambers and current transformers. The muon monitor is a 2 m×2 m segmented ionization chamber and will be installed behind the beam dump. It will monitor the beam position, profile and relative intensity.

The pion monitor is a gas Čerenkov detector which will be installed at just downstream of the 2nd horn [12]. The purpose of it is to make it possible to extrapolate the neutrino flux measured at FD, Φ_{FD} to that at SK without depending on any pion production model by actually measuring the pion distribution.

A gas Čerenkov detector is chosen to reject the primary and secondary protons of which there are many more than pions. Usage of a spherical mirror enables us to measure the momentum p_π and angle θ_π of the pions from the radius and center, respectively, of the Čerenkov rings. The expected number of pions, 10^8 cm^{-2} per spill, at the place where the pion monitor sits is so large that we only observe

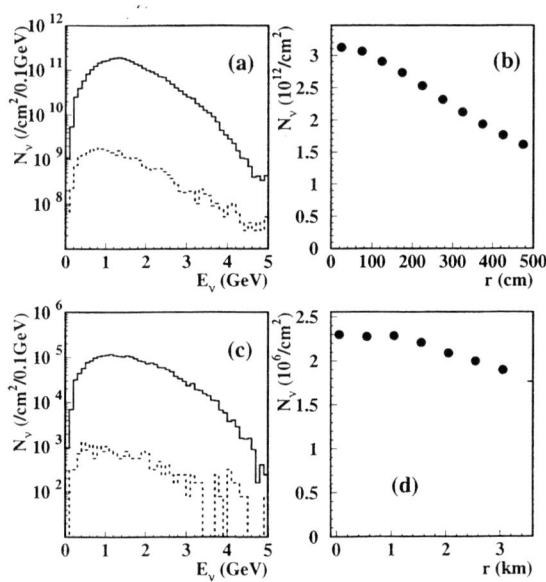

FIGURE 1. Expected spectra and profile of neutrino beam. Plots (a) and (b) are those at FD and (c) and (d) at SK. The solid lines in (a) and (c) are the flux of ν_μ and the dashed lines are of ν_e. All plots are normalized to 10^{20} pot. The horizontal axes of (b) and (d) are the distance from the beam axis.

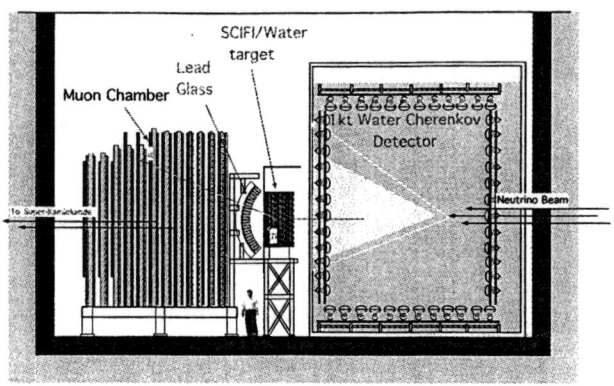

FIGURE 2. Front neutrino detector.

integrated image of the many rings for each spill. From the image, we deduce the pion distribution in the p_π-θ_π plane. Then, from the distribution, we can predict the flux of neutrinos $\Phi(E_\nu)$ at any distance using only the decay kinematics. To eliminate the possible errors in the absolute measurement of the number of pions, we use the flux ratio at SK to FD, $R_\Phi(E_\nu) \equiv \Phi_{\rm SK}(E_\nu)/\Phi_{\rm FD}(E_\nu)$ as a result of the pion monitor. Combining with the absolute $\Phi_{\rm FD}(E_\nu)$ measured at FD, we predict $\Phi_{\rm SK}(E_\nu)$. Our goal is to achieve the accuracy of $R_\Phi(E_\nu)$ better than 10 %.

FRONT NEUTRINO DETECTORS

The front neutrino detector consists of a 1 kton water Čerenkov detector and a fine grained detector (FGD) as shown in Fig. 2. They are installed 17 m underground at 300 m downstream from the target.

The 1 kton detector is a ring imaging Čerenkov detector which works by the same principle as SK. The detector volume is 8.6 m$^\phi \times$ 8.5 m filled with water which is viewed by 860 20-inch PMT's. The fiducial mass directly used for oscillation analysis is 21 ton. The main purpose of the 1 kton detector is to understand the neutrino interactions in water with a well defined neutrino beam in order to reduce systematic errors in the analysis of SK data.

The purpose of the FGD is to measure the neutrino flux $\Phi_{\rm FD}(E_\nu)$, beam profile and ν_e contamination. The strategy of measuring $\Phi_{\rm FD}$ is as follows; measure momentum p_μ and angle θ_μ of a muon from the charged current (CC) quasi elastic (qe) reaction, $\nu_\mu + n \to \mu^- + p$. Then using the two body kinematics, the energy of the ν_μ is

$$E_\nu(p_\mu, \theta_\mu) = \frac{m_n E_\mu - m_\mu^2/2}{m_n - E_\mu - p_\mu \cos\theta_\mu}, \qquad (2)$$

where $m_n(m_\mu)$ is the mass of the neutron(muon), and E_μ is the energy of the muon. Then, the flux is determined from the number of observed CCqe events $N(E_\nu)$:

$$\Phi_{\text{FD}}(E_\nu) = \frac{N(E_\nu) - N_{\text{BG}}(E_\nu)}{\varepsilon(E_\nu) \cdot \sigma_{\text{qe}}(E_\nu) \cdot N_{\text{trgt}}}, \quad (3)$$

where N_{BG} is the number of background events, ε is the detection efficiency, σ_{qe} is the cross section of the process, N_{trgt} is the number of target neutrons in the fiducial volume.

The FGD is composed of a scintillation fiber tracker (SFT), a lead glass (LG) detector and a muon detector. The target is water in Al containers. Its fiducial mass is four tons. The SFT is a stack of the water containers and sheets of staggered fibers. The scintillation light from the fibers are read out by image intensifier tubes and CCD chain. The purpose of the SFT is to detect tracks of charged particles, measure their directions and determine the vertex position of the neutrino reaction precisely. The position resolution of the SFT determines the precision of the fiducial volume, i.e. N_{trgt}. It was measured to be 280 μm by cosmic-ray tests. This corresponds to about 1 % error in N_{trgt}. The purpose of the LG detector is to identify electrons and to measure their energies. The energy resolution is $\sim 10\ \%/\sqrt{E\ [\text{GeV}]}$. The muon detector is a stack of 10- or 20-cm thick iron plates and layers of drift chambers. It will measure the momentum of each muon by its range. The resolution of the muon momentum is $6 \sim 8$ % in the momentum range from 0.5 to 3.5 GeV/c. Our goal is to measure $\Phi_{\text{FD}}(E_\nu)$ with accuracy better than 10 %.

FAR DETECTOR (SUPER-KAMIOKANDE)

SK is a ring imaging water Čerenkov detector located 1000 m underground (2700 m water equivalent) at 250 km away from KEK [4]. The detector volume of 36 m in diameter and 34 m in height is viewed by 11146 20-inch PMT's. The fiducial mass is 22.5 kton. It has been operated since April of 1996 and accumulating data for proton decay, atmospheric, solar neutrino and other particle and astrophysics searches.

For the ν_μ disappearance search, we detect muons from ν_μ CCqe reaction by selecting single ring events. The ν_μ energy is reconstructed from the momentum and angle of the muon by using Eq. (2) to obtain ν_μ energy spectrum. The spectrum will be compared with a spectrum expected from the ν_μ flux extrapolated from FD with the pion monitor data assuming no oscillation. For the ν_e appearance search, we will look for energetic electrons from ν_e CC qe reaction. The e/μ identification capability is important for both appearance and disappearance analysis. The misidentification probability of a water Čerenkov detector was studied by using test beam of particles injected directly to a smaller detector of similar design and found to be less than 2 % [13].

SENSITIVITY

The expected number of events during the experiment (10^{20} pot) in the fiducial volume of the 1 kton, FGD (water target) and SK are summarized in Table 1. In Fig. 3, expected exclusion contours of 90 % C.L. for both ν_e appearance and ν_μ disappearance are shown. In making the contours, we take 10 % overall normalization error in predicting $\Phi_{SK}(E_\nu)$. As shown in the figure, we can explore the Δm^2 region down to about 2×10^{-3} eV^2. The sensitive regions cover the suggested regions by Kamiokande completely for both $\nu_\mu \to \nu_e$ and $\nu_\mu \to \nu_\tau$ modes. Most part of the allowed region by SK for $\nu_\mu \to \nu_\tau$ mode is also within our sensitivity.

TABLE 1. Expected number of neutrino events in the fiducial volume of each detector for 10^{20} pot. The number of events in SK is for the case of no oscillation. The symbols qe, 1π and $m\pi$ mean quasi elastic, single- and multi-pion production, respectively.

Event type	Reaction	1 kton (fiducial mass ~21 ton)	FGD (~4 ton)	SK (~22.5 kton)
ν_μCCqe	$\nu_\mu + n \to \mu^- + p$	142k	44k	120
ν_μCC1π	$\nu_\mu + n \to \mu^- + p + \pi$	130k	40k	110
ν_μCC$m\pi$	$\nu_\mu + n \to \mu^- + p + m\pi$	135k	42k	115
ν_μCC total	$\nu_\mu + n \to \mu^- + X$	408k	127k	350
ν_μNC total	$\nu_\mu + N \to \nu_\mu + X$	144k	45k	120
ν_eCC total	$\nu_e + n \to e^- + X$	4k	1.2k	4

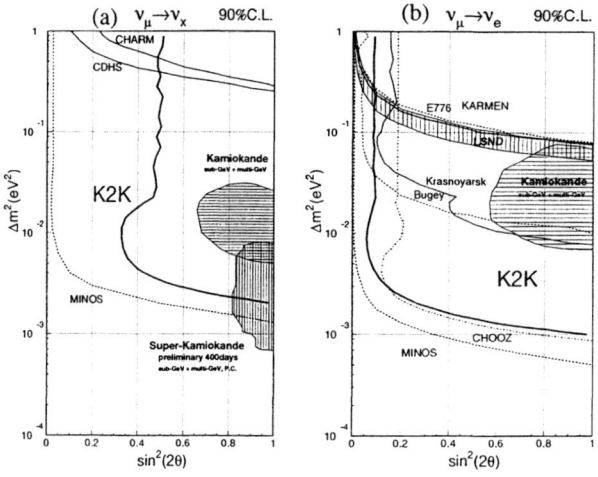

FIGURE 3. Exclusion contours of 90 % C.L. for (a)$\nu_\mu \to \nu_x$ disappearance and (b)$\nu_\mu \to \nu_e$ appearance (thick solid lines). The allowed regions by Kamiokande, LSND and SK and excluded regions by other experiments are also plotted [4,5,14].

SUMMARY

K2K, the first accelerator-neutrino long baseline oscillation experiment, will start from early 1999. In the experiment, the ν_μ flux is measured at the FD and is extrapolated to SK using the pion monitor data. After 3 years we will have about 350 ν_μ CC events at SK in the case of no oscillation. With these statistics, we can explore the oscillation parameter space down to $\Delta m^2 \gtrsim 2 \times 10^{-3}$ eV^2.

I would like to thank the members of the K2K collaboration for their help in preparing my talk and this manuscript.

REFERENCES

1. KEK-PS E362, now called K2K (KEK to Kamioka), is a collaboration of Japan, Korea and the United States. The following institutions are members of the K2K experiment: ICRR, KEK, Kobe, Niigata, Okayama, Tohoku, Tokai, Chonnam, Dongshin, Korea Univ., Seoul Nat. Univ., Boston, Los Alamos, SUNY, Hawaii, UC Irvine, Univ. of Washington.
2. Particle Data Group, *Phys. Rev.* **D54**, "Review of Particle Physics" (1996).
3. See, for example, Klapdor H.V. eds., *"Neutrinos"*, Springer-Verlag, 1988; Fukugita M. and Suzuki A. eds., *"Physics and Astrophysics of Neutrinos"*, Springer-Verlag, 1994.
4. Totsuka Y., in Proceedings of the 18th International Symposium on Lepton Photon Interactions, Hamburg, July 1997, and the references therein.
5. Apollonio M. et al., hep-ex/9711002, Nov. 1997.
6. Nishikawa K. et al., E362 KEK-PS proposal, March, 1995; *Nucl. Phys.* B (Proc. Suppl.) **59**, 289 (1997).
7. Sato H., *"Performance of the KEK 12 GeV-PS and Upgrade"*, in Proceedings of the 16th RCNP Osaka Int. Symp. on Multi-GeV High Performance Accelerators and Related Technology, Osaka, March 1997.
8. Noumi H. et al., *Nucl. Instrum. Meth.* A **398**, 399 (1997).
9. Yamanoi Y. et al., *"Large Horn Magnets at the KEK Neutrino Beam Line"*, KEK Preprint 97-225, November 1997.
10. Brun R. and Carminati F., *GEANT Detector Description and Simulation Tool*, CERN Program Library Long Writeup W5013, September, 1993.
11. Zeitnitz X. and Gabriel T.A., *Nucl. Instrum. Meth* **A349**, 106 (1994).
12. Inagaki T., *"Studies of Secondary Particle Monitors in Neutrino Beam Line"*(in Japanese), Master thesis, Tokyo Univ., Jan. 1998.
13. Kasuga S. et al., *Phys. Lett.* B **374**, 238 (1996).
14. Bergsma F. et al. (CHARM), *Z. Phys.* C **40**, 171 (1988); Dydak F. et al. (CDHS), *Phys. Lett.* B **134**, 281 (1984); MINOS Collab., *"P-875: A long baseline neutrino oscillation experiment at Fermilab"*, NuMI-L-63, Feb. 1995; Borodovsky L. et al. (E776), *Phys. Rev. Lett.* **68**, 274 (1992); Armbruster B. et al. (KARMEN), *Nucl. Phys.* B(Proc. Suppl.) **38**, 235 (1995); Vidyakin G.S. et al. (Krasnoyarsk), *JETP Lett.* **59**, 390 (1994); Achkar B. et al. (Bugey), *Nucl. Phys.* B **434**, 503 (1995).

Measurement of $\sin^2 \theta_W$ at the First Muon Collider

J. Yu* and A. V. Kotwal[†]

*Fermi National Accelerator Laboratory [1]
Batavia, IL 60510
†Columbia University, New York, NY 10027

Abstract. This report summarizes the study of the possibility of measuring $\sin^2 \theta_W$ using the intense neutrino beam expected from the straight sections of the First Muon Collider ring. This study is based on realistic error calculations from the CCFR and the NuTeV experiments. Using a neutrino detector that is capable of identifying and distinguishing electrons and muons, along with a light isoscalar target, it is conceivable to measure $\sin^2 \theta_W$ to the precision equivalent to the W mass uncertainty (experimental) of 30 MeV.

I INTRODUCTION

The weak mixing angle, $\sin^2 \theta_W$, is one of the fundamental parameters in the electro-weak sector of the Standard Model (SM). Neutrino-nucleon deep inelastic scattering experiments provide excellent testing field of the theory due to their wide range of q^2 accessibility. However, since neutrinos interact weakly, the interaction rate is very low, and in the past neutrino fixed target experiments have used dense material as neutrino targets in order to increase the interaction rates per given cycle. While these heavy target detectors increase the interaction rates, the calorimetric nature of the targets did not allow one to distinguish electron neutrino induced charged current interactions (CC) from neutral current (NC) interactions.

II CURRENT ν-N EXPERIMENTS

The CCFR experiment used ratios of the cross sections of NC to CC interactions, expressed in the following Llewellyn-Smith formula [1] :

[1)] Sponsored by the US Department of Energy.

$$R^{\nu(\bar{\nu})} = \frac{\sigma_{NC}^{\nu(\bar{\nu})}}{\sigma_{CC}^{\nu(\bar{\nu})}} = \rho^2 \left(\frac{1}{2} - \sin^2 \theta_W + \frac{5}{9} \sin^4 \theta_W \left(1 + \frac{\sigma_{CC}^{\bar{\nu}(\nu)}}{\sigma_{CC}^{\nu(\bar{\nu})}} \right) \right). \quad (1)$$

to extract $\sin^2 \theta_W$, assuming the SM expectation for ρ [2].

The NuTeV (E815) experiment just finished taking its data, totalling $\sim 3 \times 10^{18}$ protons on target. During the run, the experiment used the Sign-Selected-Quadrupole-Train (SSQT) [3] to run separately with neutrinos or anti-neutrinos at a given running period by reversing the magnet polarities. The beam line optics was set so that for the given mode (ν or $\bar{\nu}$) the polarities of all secondary magnets can be reversed, selecting only the secondary particles with desired charge. This capability of sign selection was necessary for the experiment to utilize the Paschos-Wolfenstein relationship [4]:

$$R^- = \frac{\sigma_{NC}^{\nu} - \sigma_{NC}^{\bar{\nu}}}{\sigma_{CC}^{\nu} - \sigma_{CC}^{\bar{\nu}}} = \frac{R_\nu - r R_{\bar{\nu}}}{1 - r} = \rho^2 \left(\frac{1}{2} - \sin^2 \theta_W \right), \quad (2)$$

where $r = \frac{\sigma(\bar{\nu}, CC)}{\sigma(\nu, CC)}$, to minimize the measurement uncertainty due to the mass threshold effect in CC charm production. The main reason for sign-selection was the inability of the NuTeV detector to distinguish ν_μ induced NC events from the $\bar{\nu}_\mu$ induced NC events.

Table 1 compares various uncertainties on the $\sin^2 \theta_W$ measurements from CCFR and NuTeV, along with the expected First Muon Collider (FMC) uncertainties. There were two dominant systematic uncertainties in the CCFR experiment: 1) ν_e flux and 2) CC charm production. These two major systematic uncertainties in CCFR have been reduced in the NuTeV experiment, utilizing the SSQT, so that the remaining dominant uncertainty is the statistical uncertainty.

Using the FMC beam parameters for the mean muon energy of 200GeV to give a mean neutrino energy of 178GeV [5] and a 10m long straight section. we expect approximately 94k neutrino events per year per cm of H_2 target with a radius of 150 cm. This results in $\sim$ 20million ν_μ induced events per year, for a 1 m long D_2 target with 1 m radius located $\sim$ 500 m away from the end of the straight section of the FMC, because 90% of the beam is contained within 60 cm radius for this energy [6]. With a detector that can distinguish electrons and muons resulting from CC interactions, the effective neutrino interaction statistics double, resulting in a total of 40 million events per year. Since the number of neutrino events from the NuTeV experiment is of the order of 1 million events for ν_μ, the expected 40 million events per year from the FMC would cause a reduction in the statistical uncertainty by a factor of 6. This enormous increase in statistical power also helps to minimize many of the systematic uncertainties in table 1 dramatically.

The remaining error of 0.0004 due to ν_e flux no longer exists for a detector that is capable of distinguishing CC interactions of ν_e from hadronic showers resulting from NC interactions by identifying the outgoing electrons. The

TABLE 1. $sin^2\theta_W$ uncertainties (not all the errors from NuTeV are available).

SOURCE OF UNCERTAINTY	CCFR	NuTeV	μ-Col (20M)
data statistics	0.0019	0.0019	0.0004
Monte Carlo statistics	0.0004		
TOTAL STATISTICS	0.0019	0.0019	0.0004
ν_e flux	0.0015	0.0006	$\ll 0.0004$
Cosmic Ray Background	<0.0001	<0.0001	
Transverse Vertex	0.0004	0.0004	~ 0
Energy Measurement			
Hadron Energy Scale (1%)	0.0004	0.0004	0.0004
Muon Energy Loss in Shower	0.0003	0.0002	~ 0.0001
Muon Energy Scale (1%)	0.0002	0.0002	~ 0.0002
Hadron Energy Resolution	0.0001		<0.0001
NC/CC E_{had} Difference	0.0001		<0.0001
e/π ratio	<0.0001		
Event Length	.		Irrelevant
Hadron Shower Length	0.0007	0.0001	
Counter Fiducial Size	0.0005	0.0004	
Counter Efficiency	0.0004	0.0001	
Counter Noise	0.0001	0.0002	
Vertex Determination	0.0003	0.0007	
TOTAL EXP. SYST.	0.0019	0.0012	< 0.0004
Charm Production, $\bar{s}$			
($m_c = 1.31 \pm 0.24$ GeV)	0.0027	~ 0	~ 0?
Higher Twist	0.0010	0.0006	Need to be controlled
Longitudinal Cross-Section	0.0008	N/A	Need to be measured
Charm Sea, ($\pm 100\%$)	0.0006	0.0004	Need to be measured
Non-Isoscalar Target	0.0004	0.0004	~ 0 for D_2
Structure Functions	0.0002	0.0001	0.0001
Rad. Corrections	0.0001		
$\sigma^{\bar{\nu}}/\sigma^{\nu}$	<0.0001	N/A	
TOTAL PHYSICS MODEL	0.0030	$\sim$0.0008	$\ll 0.0008$?
TOTAL UNCERTAINTY	0.0041	0.0024	< 0.0010
ΔM_W	0.21 GeV/c^2	0.11 GeV/c^2	< 0.050 GeV/c^2
			0.030 GeV/c^2 (EXP)

uncertainties resulting from energy scale can also be minimized by carefully planned calibration runs. The errors in event length, in principle, do not exist, because with the detector described in the following section, one can distinguish CC from NC interactions on an event-by-event basis. The above expectations will enable the experiment to reduce the statistical and experimental systematic uncertainties on $\sin^2\theta_W$ to the equivalent M_W uncertainty of 30 MeV.

III DETECTOR AND BEAM REQUIREMENTS

It is crucial to be able to reverse the polarity of the ring so that one accepts ν_μ and $\bar{\nu}_e$ or $\bar{\nu}_\mu$ and ν_e at a given time. This capability also provides the opportunity for studying possible systematics coming from the beam of a given sign.

The beamline also needs sweeping magnets or sufficient thickness of shielding to filter out electrons or electromagnetic (EM) shower particles resulting from the muon decay in the FMC ring. Due to the intensity of neutrino beam, some neutrino interactions would occur in the shielding. Thus, one requires adequate veto counters before the target in order to flag the charged particles that come from upstream neutrino interactions.

It is extremely important for the detector to distinguish electrons from muons that are coming from neutrino CC interactions. The detector should also be capable of distinguishing the combination of electron and hadron induced showers from purely hadronic ones. Traditional heavy target neutrino detectors could not distinguish ν_e induced CC interactions from NC interactions.

It also is necessary to use light isoscalar targets for the charm and strange sea measurements from the same experiment as this will be very useful in reducing the remaining systematic uncertainties.

In order to satisfy the above requirements, the detector needs to have:

- Good EM and hadronic shower identification.

- High electron detection efficiency along with high efficiency particle identification.

- Good charged particle momentum measurement.

- Good EM and hadronic shower energy containment and measurement.

- Muon identification and momentum measurement.

Figure 1 shows a conceptual design of a neutrino detector [7]. The components of the detector in the figure are discussed in the following section.

Neutrino Detector

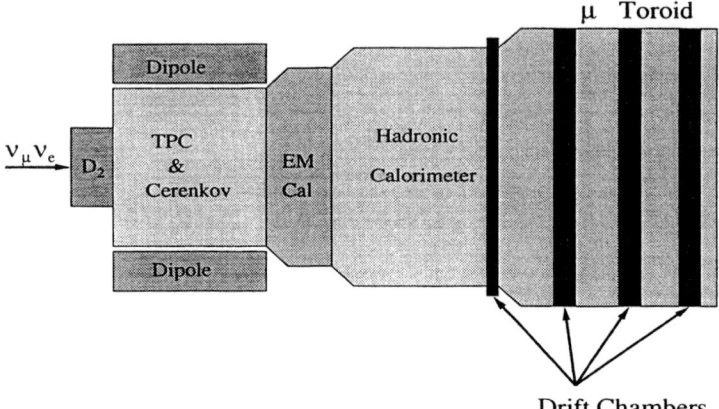

FIGURE 1. A schematic drawing of a conceptual detector design for the $\sin^2\theta_W$ measurement. This detector is essentially the same as B.J. King's conceptual design.

A Detector Components

This section summarizes the suggested components of the detector to meet the requirements discussed in the previous section.

- Target : D_2, $r = 1$ m, $l = 1$ m (light isoscalar)
- Vertex chamber for the interaction vertex determination.
- TPC and Čerenkov counter : particle-ID and momentum measurement.
- EM calorimeter :
    - Good longitudinal segmentation and multiple depth readout
    - Fine transverse granularity (enhanced at the shower max) for shower shape analysis.
    - More than 21 X_0 depth for the full EM shower containment. (The depth has to be optimized so that the hadronic showers do not deposit too much energy in the EM section.)
    - Energy resolution : $\frac{\sigma}{E} = \frac{15\%}{\sqrt{E}}$
- Hadron calorimeter :
    - More than 3 $\sim$ 4 layer longitudinal readout.
    - Some transverse granularity in order to determine shower direction.
    - 10 $\sim$ 20 λ_0 deep (168 cm to 333 cm Fe equivalent)

- Energy resolution : $\frac{\sigma}{E} = \frac{50\%}{\sqrt{E}}$

• μ Toroid magnet interspersed with drift chambers for muon momentum measurements with its field magnitude to be determined.

It is premature at this point to discuss the detailed technologies for the various elements of the detector. However, the above functionality is necessary to identify electrons and muons in CC interactions.

IV BACKGROUNDS AND DIFFICULTIES

We discuss expected backgrounds and some difficulties to be overcome for the precise measurement of $\sin^2 \theta_W$, using the conceptual detector discussed in the previous section.

- $\pi \& K$ from hadron shower decaying in-flight in the particle-ID system, faking CC events.

- π^0 conversions resulting in electrons.

- Electrons from upstream ν_e interactions.

All of the above backgrounds have to do with identification of CC interactions. One may be able to enhance the detector to overcome these backgrounds.

We also expect the following minor difficulties from the CCFR or NuTeV type measurements of $\sin^2 \theta_W$.

- CC-charm error $\Rightarrow 0$ (This error can be reduced by measuring the CC production of charm directly from the same experiment, using oppositely charged dimuon or di-electron final states.)

- Is it straightforward to measure R^-?

$$R_\nu = \frac{\sigma(\nu_\mu, NC) + \sigma(\bar{\nu}_e, NC)}{\sigma(\nu_\mu, CC) + \sigma(\bar{\nu}_e, CC)}$$

$$R_{\bar{\nu}} = \frac{\sigma(\bar{\nu}_\mu, NC) + \sigma(\nu_e, NC)}{\sigma(\bar{\nu}_\mu, CC) + \sigma(\nu_e, CC)}$$

- Are the higher twist effects going to be under better control? (It is extremely important to reduce this error, because this error will be the dominant uncertainty.)

- Is the CC to NC identification error close to 0?

- Are there any other theoretical effects we need to worry about?

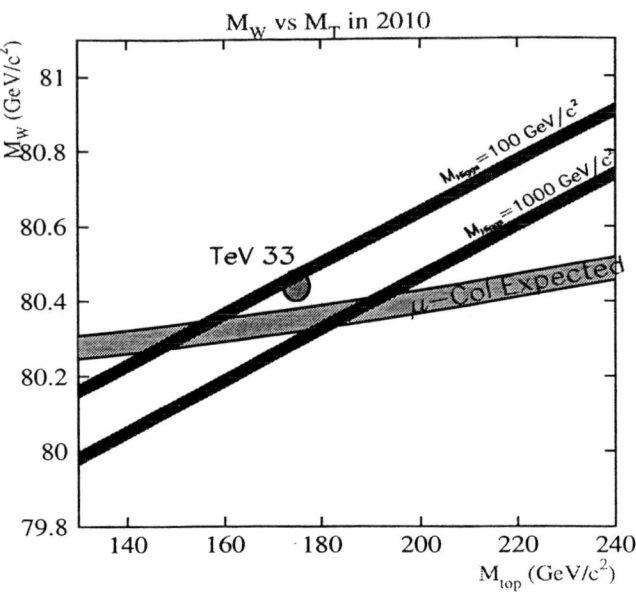

FIGURE 2. Expected uncertainties of TeV 33 and FMC M_W vs M_T in the year 2010.

V EXPECTED M_W STATUS IN 2010

Figure 2 shows the expected status of various W mass measurements in the year 2010. The most precise measurement is expected from direct measurements of the TeV33 project. The contour represents the 68% confidence level from the TeV33 expectations of $\delta M_W = 30$ MeV and $\delta M_t = 2$ GeV, with $\int \mathcal{L}dt = 10 fb^{-1}$ using the traditional M_T method [8]. The FMC measurements of the W mass would be of similar precision after one year of running with the beam parameters provided to us. As can be seen in figure 2, since the errors from both the direct measurements and the FMC are going to be extremely small, and the FMC measurement provides the SM based band in M_W-M_T plane, the measurements would be complementary to each other in testing the SM and nailing down the SM Higgs mass.

VI CONCLUSIONS

We have investigated the possibility of measuring $\sin^2\theta_W$ in an FMC neutrino experiment. With a suitable detector that uses a light isoscalar target along with excellent electron and muon identification, one can expect collecting 40 million events per year, due to the utilization of both ν_μ and ν_e induced events resulting from the FMC beam. Using this high intensity neutrino beam, along with theoretical help in reducing higher twist effects, we expect the precision of $\sin^2 \theta_W$ to be equivalent to 30 MeV on the W mass in the on-shell scheme.

Since the statistical and experimental systematic uncertainties will reduce dramatically using the proper apparatus, it becomes crucial to reduce the remaining theoretical uncertainties. Calculations to minimize the higher twist effects and the uncertainties from longitudinal structure function, R_L, need to be dealt with.

REFERENCES

1. C.H.Llewellyn Smith, Nucl. Phys. **B228**, 205 (1983)
2. C.Arroyo, B.J.King *et. al.*, Phys. Rev. Lett. **72**, 3452 (1994).
3. R.Bernstein *et. al.*, NuTeV Collaboration, Fermilab-TM-1088 (1994).
4. E.A.Paschos and L. Wolfenstein, Phys. Rev. **D7**, 91 (1973)
5. H. Schellman, in this workshop.
6. D.A. Harris and K.S.McFarland, in this workshop.
7. B.J.King, in this workshop.
8. U. Baur and M. Demarteau, "Precision Electroweak Physics at Future Collider Experiments," Fermilab-Conf-96/423 (1996).

Slow/Stopped Muon Physics

Conveners: William Molzon, University of California, Irvine
Tom Diehl, Fermilab

Coherent muon–electron conversion in muonic atoms

Andrzej Czarnecki and William J. Marciano

Brookhaven National Laboratory, Upton, New York 11973

Kirill Melnikov

*Institut für Theoretische Teilchenphysik,
Universität Karlsruhe, D-76128 Karlsruhe, Germany*

Abstract. Transition rates for coherent muon-electron conversion in muonic atoms, $\mu N \to eN$, are computed for various types of muon number violating interactions. Attention is paid to relativistic atomic effects, Coulomb distortion, finite nuclear size, and nucleon distributions. Discrepancies with previously published results are pointed out and explained. Results are presented for several elements of current and future experimental interest.

I INTRODUCTION

In the standard model (SM) of electroweak interactions, leptons of different flavors do not mix because of vanishing neutrino masses. On the other hand, lepton flavor violation is predicted in various extensions of the SM. Hence, its observation would be direct evidence of physics beyond the SM.

Because the muon is a relatively stable particle which can be abundantly produced, muon–number violating processes are of particular interest. Most recently, the following processes have been studied experimentally (for a review see e.g. [1,2]): $\mu \to e\gamma$, muonium-antimuonium oscillations, and, the subject of the present work, muon–electron conversion in the field of a nucleus. Muon number violation has also been searched for in kaon, tau, and Z decays.

A particularly sensitive search for muon number violation is the coherent conversion in a muonic atom, formed by muon and a nucleus. Here by coherent conversion we understand a process $\mu^- N \to e^- N$ in which the nucleus N remains in its initial state (up to recoil effects). The rate of the coherent conversion is enhanced with respect to processes with a nuclear excitation by a factor of the order of the number of nucleons.

Since the original pioneering papers on the conversion theory [3–5], there have been several theoretical efforts intended to provide an accurate description of coherent muon conversion. There are two kinds of theoretical issues: first, the short distance effects which are responsible for the muon number violation (presumably caused by some "new physics"); second, the long distance atomic physics of the muonic atom, which also determines the transition rate.

The first group of problems has been studied in many extensions of the standard model (for a review see [6–8]). In particular, in ref. [9] the rate of the coherent conversion $\mu^- N \to e^- N$ was calculated in a variety of gauge models. It was pointed out in that paper that in a large class of models the conversion can be much more probable than the decay $\mu \to e\gamma$. For example, there can be logarithmic enhancements of the form factors leading to conversion which are absent in the decay rate. Such logarithmic effects were also recently discussed in [10].

Weinberg and Feinberg [3] focused on an electromagnetic mechanism of transferring the energy yield to the nucleus. The structure of this electromagnetic interaction is richer than for the $\mu \to e\gamma$ decay since the photon need not be on mass shell. Therefore, it is possible that conversion can occur even if $\mu \to e\gamma$ is forbidden for some reason. The matrix element for conversion contains monopole terms which do not contribute to the decay $\mu \to e\gamma$, because the longitudinal polarization states are possible only for virtual photons. In addition, there may be conversion amplitudes other than photon mediated processes. This makes the conversion on nuclei a particularly interesting process to study.

The early theoretical studies of muon conversion into electrons on nuclei performed in [3–5] are valid mainly for conversion on light nuclei. The developments before the year 1978 have been summarized in [6]. In heavier atoms new effects become important: relativistic components of the muon wave function, Coulomb distortion of the outgoing electron, and the finite nuclear size. These were first addressed in [11]. Nuclear effects were also analyzed, albeit in a non-relativistic approximation, in [12] and, more recently, in [13,14].

Recently, we have undertaken a new calculation of the full relativistic atomic physics aspect of the conversion. In this talk our main results are summarized. The details of the calculation and a more extensive analysis will be given in a forthcoming paper [15].

II GENERAL DESCRIPTION

The muon–electron transition can occur via various mechanisms (see e.g. [6,16,17]). To keep the description fairly general, it is convenient to write down a low energy effective Hamiltonian for the $\mu \to e$ transition:

$$\mathcal{H} = -\bar{e}\hat{O}\mu + h.c., \tag{1}$$

$$\hat{O} = -\sqrt{4\pi\alpha} \left[\gamma_\alpha \left(f_{E0} - f_{M0}\gamma_5 \right) \frac{q^2}{m^2} + i\sigma_{\alpha\beta} \frac{q^\beta}{m} \left(f_{M1} + f_{E1}\gamma_5 \right) \right] A^\alpha(q) \quad (2)$$

$$+ \frac{G_F}{\sqrt{2}} \gamma_\alpha (a - b\gamma_5) J^\alpha,$$

$$J^\alpha = \bar{u}\gamma^\alpha u + c_d \bar{d}\gamma^\alpha d. \quad (3)$$

In this equation $m = m_\mu$ is muon mass and $a, b, f_{E0,1}, f_{M0,1}$ are dimensionless coupling constants.

The part of the Hamiltonian containing the photon field A_α describes the transition of a muon into an electron and an on–shell photon, $\mu \to e\gamma$, whose rate is $\Gamma(\mu \to e\gamma) = \alpha m \left(|f_{M1}|^2 + |f_{E1}|^2 \right)/2$, which gives the branching ratio $Br(\mu \to e\gamma) = \Gamma(\mu \to e\gamma)/\Gamma(\mu \to e\bar{\nu}_e\nu_\mu) = 96\pi^3\alpha \left(|f_{M1}|^2 + |f_{E1}|^2 \right)/(G_F^2 m^4)$. Due to gauge invariance, the transition vector and axial currents (terms proportional to f_{E0}, f_{M0}) cannot contribute to the transition to an on–shell photon [3]. We explicitly account for this by factoring out q^2/m^2. We further assume that all effective coupling constants introduced in the Hamiltonian are slowly varying functions of all external parameters (photon off–shellness, muon off–shellness etc.) and therefore can be considered as constants in the course of the calculation.

To calculate the transition rate for the coherent conversion $\mu N \to eN$ we first average the effective Hamiltonian over the nucleus. The result of this average depends on muon and electron wave functions and the field of the nucleus:

$$H_{\text{int}} = \langle N|\mathcal{H}|N\rangle, \quad \langle N|N\rangle = 1. \quad (4)$$

To calculate a matrix element of an arbitrary operator $\hat{Q}$ between two nuclei, we use the following approximation:

$$\langle N|\hat{Q}|N\rangle = \int d^3r \left(Z\rho_p(r)\langle p|Q(r)|p\rangle + (A-Z)\rho_n(r)\langle n|Q(r)|n\rangle \right).$$

In the above equation Z is the number of protons in the nucleus N and $(A-Z)$ is the number of neutrons. Also, $|p\rangle$ and $|n\rangle$ denote states of a single proton and neutron respectively, with densities normalized as follows:

$$\int d^3r \rho_p(r) = 1, \quad \int d^3r \rho_n(r) = 1. \quad (5)$$

To obtain matrix elements of the Hamiltonian, one should calculate the matrix element of the quark currents between two nucleons. A typical momentum transfered by the current is of the order of the muon mass, a small quantity compared to the mass of the nucleus. The time component of the soft current counts the number of constituent quarks in the nucleon. We therefore obtain

$$\langle p|\bar{u}\gamma^0 u + c_d \bar{d}\gamma^0 d|p\rangle = 2 + c_d,$$
$$\langle n|\bar{u}\gamma^0 u + c_d \bar{d}\gamma^0 d|n\rangle = 1 + 2c_d. \qquad (6)$$

The matrix element of spatial components of the current is proportional to the velocities of the constituents and is negligible in the present problem.

Calculating the matrix element of the Hamiltonian with respect to the nucleus states, we arrive at an effective Hamiltonian for the coherent $\mu \to e$ conversion in the field of the nucleus:

$$H_{\text{int}} = H_1 + H_2 + H_3,$$
$$H_1 = e \int d^3r \overline{\Psi}_e(r)\gamma_0 \left(f_{E0} - f_{M0}\gamma_5\right) \Psi_\mu(r) 4\pi Z \rho_p(r)$$
$$H_2 = e \int d^3r \overline{\Psi}_e(r) i\sigma_{\alpha\beta} \left(f_{M1} + f_{E1}\gamma_5\right) \Psi_\mu(r) F_{\alpha\beta}(r)$$
$$H_3 = \frac{G_F}{\sqrt{2}} \int d^3r \overline{\Psi}_e(r)\gamma_0 (a - b\gamma_5) \Psi_\mu(r)$$
$$\times \left(Z(2 + c_d)\rho_p(r) + (A - Z)(1 + 2c_d)\rho_n(r) \right) \qquad (7)$$

In the above equations the fields $\Psi_e(r)$ and $\Psi_\mu(r)$ stand for the second quantized operators, and the electromagnetic tensor $F_{\alpha\beta}$ is understood as a classical electric field produced by the nucleus. The matrix element of H_{int} taken between appropriate initial and final states will give the amplitude for the coherent conversion.

In our case the initial state is the muon in the $1S$ orbit around the nucleus; the final state is an electron with the energy equal to the energy of the initial muon. The corrections due to the nucleus recoil are small and we do not consider them here.

In their pioneering work on muon–electron conversion [3] Weinberg and Feinberg performed an approximate calculation of the coherent conversion rate. Using simple estimates, we would like to show when complete treatment of the problem requires going beyond the approximations used in Ref. [3]. For this purpose we describe the scales relevant for the problem. The wave function of the muon bound in the lowest orbit is characterized by the Bohr radius $a_B = (\alpha Z m)^{-1}$. The radius of the nucleus R_N scales like $m_\mu R_N \sim (Z/4)^{1/3}$. Evidently, the nucleus can be considered as point-like only if the Bohr radius is much larger than the radius of the nucleus $a_B \gg R_N$. This implies $Z \ll 60$. On the other hand, the relativistic corrections are governed by the parameter $Z\alpha$. Therefore, for high Z elements such as Pb, it is not clear a priori if the non-relativistic treatment of the muon bound state is sufficient. Moreover, another physical effect is governed by the same parameter $Z\alpha$. Consider the case when $\mu \to e$ conversion occurs due to an exchange of the photon with the nucleus. In this case the kinematics of the decay dictates that the virtuality of the photon is determined by the mass of the decaying muon. The process can be considered point–like, if this scale is much less than the Bohr radius. This implies that the photon–mediated conversion can be considered as a point–like process for $Z\alpha \ll 1$. Therefore, for heavy nuclei such as Pb the consideration of the conversion process as point like is no longer valid. The appearance of the part H_2 in the effective Hamiltonian H_{int} reflects this observation: one notes that the $\mu \to e$ transition current couples to the electric field of the nucleus, not to the proton density directly.

Clearly, for light nuclei one can rely on the hierarchy of the scales and perform an approximate calculation of the rate. In this case the conversion rate will be proportional to the square of the muon wave function at the origin and the square of the nucleus form-factor. Such an approximation was used in Ref. [3]. When the charge of the nucleus grows, all scales relevant to the problem become comparable and the above approximations cannot be trusted. In this case a correct treatment of the problem requires solving the Dirac equation for both muon and electron wave functions in the field of the nucleus.

To describe proton and neutron distributions in the nucleus we use two-parametric Fermi functions:

$$\rho_{p(n)}(r) = \frac{\rho_0}{1 + e^{(r-r_{p(n)})/a_{p(n)}}}, \qquad \int d^3r \, \rho_{p(n)}(r) = 1. \tag{8}$$

In the above equation $r_{p(n)} \sim A^{1/3}$ is the radius of the proton (neutron) fraction of the nucleus and $a_{p(n)}$ is the thickness of the boundary of the proton (neutron) fraction. Precise values of these parameters depend on the nucleus and can be found in tables [18,19].

The transition rate for the coherent conversion is given by:

$$\omega(\mu^- N \to e^- N) = \omega_{conv} = \sum_{\lambda_f} |H^{fi}_{\text{int}}|^2. \tag{9}$$

Here the sum goes over all quantum numbers which the electron in the final state can have in addition to energy. The muon bound state wave function is normalized to unity. The electron wave function of the continuous spectrum is normalized such that:

$$\int d^3r \, \psi^*_{E',\lambda'}(r) \psi_{E,\lambda}(r) = \delta_{\lambda,\lambda'} 2\pi \delta(E' - E) \tag{10}$$

where λ denotes a set of all discrete quantum numbers which electron obeys in addition to energy.

It is convenient to use the expression for the wave functions as suggested in [20]:

$$\psi^m_k = \begin{pmatrix} g^{(k)}(r) \chi^m_k \\ i f^{(k)}(r) \chi^m_{-k} \end{pmatrix}. \tag{11}$$

In this equation m is the eigenvalue of the operator J_z, $J = L + S$; $-k$ is the eigenvalue of the operator $L \cdot \sigma + 1$ (see e.g. [20]); χ_k^m is an orthonormalized spinor. For the muon wave function we use the lowest energy bound state wave function; this suggests that $j = 1/2$, $m = \pm 1/2$, $k = -1$. We will distinguish muon and electron radial wave functions by a subscript.

A calculation of the explicit expression for the transition rate ω_{conv} is now straightforward. We find

$$\omega_{conv} = \left| \int dr\, r^2 \left[\tilde{f}_{E0}(r)(g_e^- g_\mu^- + f_e^- f_\mu^-) + \frac{f_{M1}}{m}(g_e^- f_\mu^- + f_e^- g_\mu^-)\frac{dV}{dr} \right] \right|^2$$

$$+ \left| \int dr\, r^2 \left[\tilde{f}_{M0}(r)(g_e^- g_\mu^- + f_e^- f_\mu^-) + \frac{f_{E1}}{m}(g_e^- f_\mu^- + f_e^- g_\mu^-)\frac{dV}{dr} \right] \right|^2,$$

$$\tilde{f}_{E0}(r) = -f_{E0}\frac{4\pi Z\alpha}{m^2}\rho_p(r) + \frac{G_F}{\sqrt{2}}a\left[Z(2 + c_d)\rho_p(r) + N(1 + 2c_d)\rho_n(r)\right],$$

$$\tilde{f}_{M0}(r) = -f_{M0}\frac{4\pi Z\alpha}{m^2}\rho_p(r) + \frac{G_F}{\sqrt{2}}b\left[Z(2 + c_d)\rho_p(r) + N(1 + 2c_d)\rho_n(r)\right], \quad (12)$$

where $V(r) = -eA^0(r)$ is the muon potential energy in the field of the nucleus.

The above equation provides a general expression for the rate of the reaction $\mu^- N \to e^- N$ which we will use for numerical analysis in the next section. The radial wave functions $f_{e,\mu}^-$ and $g_{e,\mu}^-$ are obtained by solving the Dirac equation. Inside the nucleus and close to it these solution are found numerically. At large distances we match the numerical solutions to the exact Coulomb wave functions.

III NUMERICAL ANALYSIS

In general, the following integrals are needed for the description of the transition rate:

$$I_1^p = -\frac{4\pi Z\alpha}{m^2}\int dr\, r^2 \rho_p(r) g_\mu^- g_e^-, \qquad I_2^p = -\frac{4\pi Z\alpha}{m^2}\int dr\, r^2 \rho_p(r) f_\mu^- f_e^-$$

$$I_1^n = -\frac{4\pi Z\alpha}{m^2}\int dr\, r^2 \rho_n(r) g_\mu^- g_e^-, \qquad I_2^n = -\frac{4\pi Z\alpha}{m^2}\int dr\, r^2 \rho_n(r) f_\mu^- f_e^-$$

$$I_3 = \frac{1}{m}\int dr\, r^2 \frac{dV(r)}{dr} g_\mu^- f_e^-, \qquad I_4 = \frac{1}{m}\int dr\, r^2 \frac{dV(r)}{dr} f_\mu^- g_e^-. \quad (13)$$

Tables with the values of these integrals for various elements will be given in [15]. Using those values, the conversion rate is

$$\omega_{conv} = 3 \cdot 10^{23}(\omega_{conv}^{(1)} + \omega_{conv}^{(2)})\, \text{sec}^{-1} \quad (14)$$

$$\omega_{conv}^{(1)} = \left| f_{E0}I_p - \frac{G_F}{\sqrt{2}}\frac{m^2}{4\pi Z\alpha}a\left(Z(2 + c_d)I_p + N(1 + 2c_d)I_n\right) + f_{M1}I_{34} \right|^2,$$

$$\omega_{conv}^{(2)} = \left| f_{M0} I_p - \frac{G_F}{\sqrt{2}} \frac{m^2}{4\pi Z \alpha} b \Big(Z(2+c_d) I_p + N(1+2c_d) I_n \Big) + f_{E1} I_{34} \right|^2, \quad (15)$$

where

$$I_p = -\Big(I_1^p + I_2^p\Big), \quad I_n = -\Big(I_1^n + I_2^n\Big), \quad I_{34} = I_3 + I_4. \quad (16)$$

All dimensional parameters in the above equations are expressed in fermi. For proton and neutron distributions we use the two-parameter Fermi distribution (8) with parameters taken from Ref. [18,19].

For the present application we neglect the effects of the vacuum polarization; such approximation is justified in view of much bigger errors in wave function integrals induced directly by the nuclear distribution uncertainties.

IV BRANCHING RATIO FOR μ-E CONVERSION IN A SPECIFIC MODEL

We consider here one specific model which predicts lepton flavor violation, a supersymmetric grand-unified theory discussed in Ref. [17].

The analysis of Ref. [17] implies that magnetic couplings f_{M1} and f_{E1} (cf. Eq.(2)) are significantly enhanced in a large region of the parameter space of those models in comparison with other couplings in the effective Hamiltonian. Therefore, it is reasonable to neglect all other couplings in the effective Hamiltonian and to analyze the dependence of $\omega_{conv}/\omega_{capt}$ on the choice of target.

With these approximations the ratio of the conversion rate to the capture rate becomes

$$\frac{\omega_{conv}}{\omega_{capt}} = 3 \cdot 10^{12} \Big(|f_{E1}|^2 + |f_{M1}|^2 \Big) B(A,Z), \quad (17)$$

$$\text{with} \quad B(A,Z) = 10^{11} \frac{I_{34}^2}{\omega_{capt}/\sec^{-1}}. \quad (18)$$

We compare the above formula with the Weinberg–Feinberg approximation, in which the rate is obtained by replacing $B(A,Z)$ in (17) by $B_{WF}(A,Z)$:

$$B_{WF}(A,Z) = 8\alpha^5 m Z_{\text{eff}}^4 Z F_p^2 \frac{1}{\omega_{capt}} \frac{1}{3 \cdot 10^{12}} \quad (19)$$

In this formula Z_{eff} denotes the "effective Z," obtained by averaging the muon wave function over the nuclear density (see e.g. [12]), and F_p is the formfactor describing the charge distribution, given by

$$F_p = \frac{\int d^3 r \rho(r) \frac{\sin mr}{mr}}{\int d^3 r \rho(r)}. \quad (20)$$

In table 1 we show the B and B_{WF} coefficients for three elements. We conclude that in this particular model the ratio of the coherent conversion rate to the capture rate, as described by our $B(A, Z)$, does not change significantly with changing the target. The WF approximation tends to slightly overestimate the conversion rate. In ref. [11] various corrections to the WF approximations have been studied; in that approach B_{WF} should be replaced by $B_S \equiv C_1 C_2 C_3 B_{WF}$, where the correction factors $C_{1,2,3}$ are listed in table III in [11]. We agree with these results for non-photonic mechanisms of conversion. However, for the photonic case we are considering here, our results differ from [11] even stronger than from the WF approximation, as can be seen in table 1. This is because the effect of the non-locality in the interaction with the electric field was not taken into account in [11]. It is especially important for heavy elements.

TABLE 1. Comparison of our results for the coefficient $B(A, Z)$ in the rate formula (17) with the Weinberg-Feinberg approximation $B_{WF}(A, Z)$ and with $B_S(A, Z)$ obtained from ref. [11]. Z_{eff} is taken from [12] and the capture rates ω_{capt} from [21]; F_p is computed using (20).

Element	$B(A,Z)$	$B_{WF}(A,Z)$	$B_S(A,Z)$	Z_{eff}	F_p	ω_{capt}, $[10^6/\text{sec}]$
Al	1.1(1)	1.2	1.3	11.62	0.63	0.7
Ti	1.8	2.0	2.2	17.61	0.53	2.6
Pb	1.25(15)	1.6	2.2	33.81	0.15	13.0

It is instructive to compare the conversion rate to the branching ratio for $\mu \to e\gamma$ in this model:

$$\frac{Br(\mu \to e\gamma)}{\omega_{conv}/\omega_{capt}} = \frac{96\pi^3 \alpha}{G_F^2 m^4} \frac{1}{3 \cdot 10^{12} B(A,Z)} \approx \frac{428}{B(A,Z)}. \quad (21)$$

This ratio varies from 389 for ^{27}Al to 238 for ^{48}Ti, and increases again for heavy elements to 342 for ^{208}Pb.

V BOUNDS ON THE MUON NUMBER VIOLATING COUPLINGS FROM SINDRUM

Using updated results for the atomic part of the theoretical description of the coherent muon electron conversion, we reconsider the analysis performed in Ref. [22]. In that paper the upper limit for the ratio $\omega_{conv}/\omega_{capt}$ has been reported based on the measurement using ^{208}Pb. These results were combined with previous measurement on ^{48}Ti target to obtain the bounds on muon–number violating coupling constants. In that analysis it was assumed that the effective Hamiltonian does not contain a piece which corresponds to the photon mediated conversion.

As a first step we have to switch to the model and notations of Ref. [22]. The model, considered in Ref. [22], corresponds to the exchange of a heavy vector or scalar boson which mediates $\mu \to e$ transition. For this reason, we should equate $f_{E0,1}$ and $f_{M0,1}$ to zero in our general expression for conversion rate, eq. (12). Also, to switch to their notations, one should substitute $b = a$, $ac_d \to (g_V^0 - g_V^1)/2$, and $a \to (g_V^0 + g_V^1)/2$ in that equation. Constants g_V^0, g_V^1 then parameterize the coupling of the heavy gauge boson to isoscalar and isovector parts of the quark current respectively.

Then, one gets for the conversion rate:

$$\omega_{conv} \sim I_p^2 \left[g_V^0 \left(1 - \frac{N}{A}\xi\right) + g_V^1 \left(\frac{(Z-N)}{3A} + \frac{N}{3A}\xi\right)\right]^2. \tag{22}$$

In this equation we use the notation (cf. Eq. (16)): $\xi = I_p/I_n$. Using the bounds on the $\mu \to e$ branching ratios measured with Ti and Pb targets, we derive the new bounds for the coupling constants g_V^0, g_V^1:

$$|g_V^0| < 8 \pm 2 \cdot 10^{-7}, \qquad |g_V^1| < 40 \pm 13 \cdot 10^{-6}. \tag{23}$$

The variation in the boundaries shown above is the estimate of the uncertainty of the result induced by uncertainties in the input nuclear parameters. These new bounds should be compared with the values

$$|g_V^0| < 3.9 \cdot 10^{-7}, \qquad |g_V^1| < 9.7 \cdot 10^{-6}. \tag{24}$$

which were given in Ref. [22] using the results of Ref. [11].

There are two reasons for such large differences. As we have mentioned already, in Ref. [22] the results of Ref. [11] have been used to interpret experimental data. We note in this respect, that Table I of ref. [11] does not provide correct results – the results for $\omega_{conv}/\omega_{capture}$ quoted there are about a factor of 2 too large. Second, in the case of a Pb target the difference in proton and nucleon distributions becomes quite noticeable. However, the numbers in Table I of [11] were obtained using identical proton and neutron distributions. These two reasons conspire to give quite a large discrepancy.

We note, however, that the values for the bounds quoted above (especially for g_V^1) are very sensitive to the input parameters. This happens, because the bounds are derived from a system of linear equations in which the coefficients of g_V^1 there are much smaller than all other parameters. Therefore even a small variation in the input parameters can generate quite substantial change in the resulting value for g_V^1.

VI ACKNOWLEDGEMENT

We are grateful to Dr. P. Goudsmit for providing us with ref. [23] and to Dr. Roland Rosenfelder for discussions and providing his Fortran codes for comparisons. A.C. thanks Dr. T. Kozłowski for explaining some details of the Sindrum II experiment. This research was supported by BMBF-057KA92P; by "Graduiertenkolleg Elementarteilchenphysik" at the University of Karlsruhe; and by U.S. Department of Energy under contract number DE-AC02-76CH00016.

REFERENCES

1. A. van der Schaff, Prog. Part. Nucl. Phys. **31**, 1 (1993).
2. A. Czarnecki, hep-ph/9710425. Talk given at the 5th Intl. Conference *Beyond the Standard Model*, Balholm, Norway, May 1997. (unpublished).
3. S. Weinberg and G. Feinberg, Phys. Rev. Lett. **3**, 111 (1959), erratum: ibid., p. 244.
4. N. Cabibbo and R. Gatto, Phys. Rev. **116**, 1334 (1959).
5. S. P. Rosen, Nuovo Cim. **15**, 7 (1960).
6. W. J. Marciano, in *New frontiers in High Energy Physics*, edited by A. Perlmutter and L. F. Scott (Plenum, New York, 1978), proc. of Orbis Scientiae, Coral Gables, 1978.
7. T. S. Kosmas and J. D. Vergados, Phys. Rept. **264**, 251 (1996).
8. J. D. Vergados, Phys. Rept. **133**, 1 (1986).
9. W. J. Marciano and A. I. Sanda, Phys. Rev. Lett. **38**, 1512 (1977).
10. M. Raidal and A. Santamaria, hep-ph/9710389 (unpublished).
11. O. Shanker, Phys. Rev. **D20**, 1608 (1979).
12. H. C. Chiang *et al.*, Nucl. Phys. **A559**, 526 (1993).
13. T. S. Kosmas, A. Faessler, F. Simkovic, and J. D. Vergados, Phys. Rev. **C56**, 526 (1997).
14. T. S. Kosmas, A. Faessler, and J. D. Vergados, J. Phys. **G23**, 693 (1997).
15. A. Czarnecki, W. Marciano, and K. Melnikov, in preparation.
16. J. Bernabéu, E. Nardi, and D. Tommasini, Nucl. Phys. **B409**, 69 (1993).
17. R. Barbieri, L. Hall, and A. Strumia, Nucl. Phys. **B445**, 219 (1995).
18. H. de Vries *et al.*, At. Data and Nucl. Data Tables **36**, 495 (1987).
19. C. Garcia-Recio, J. Nieves, and E. Oset, Nucl. Phys. **A547**, 473 (1992).
20. M. E. Rose, *Relativistic Electron Theory* (John Wiley, New York, 1961).
21. M. Eckhause *et al.*, Nucl. Phys. **81**, 575 (1966).
22. W. Honecker *et al.*, Phys. Rev. Lett. **76**, 200 (1996).
23. R. Engfer *et al.*, At. Data and Nucl. Data Tables **14**, 509 (1974).

Muon Capture

Peter Kammel

*University of California at Berkeley and
Lawrence Berkeley National Laboratory
Berkeley, CA 94720*

Abstract. Recent results and future research prospects in muon capture are reviewed concentrating on studies to determine the nucleon/nuclear form factors and to test the structure of charged current electroweak interactions.

INTRODUCTION

Muon capture is a basic electroweak charged current reaction involving first generation quarks and second generation leptons. At the quark level the reaction proceeds according to the simple diagram of fig.1a. At the nucleon level the basic reaction is

$$\mu + p \to n + \nu_\mu \tag{1}$$

As indicated schematically in fig.1b the hadronic vertex gets dressed by strong interactions, which can be parameterized by six q^2 dependent form factors $g_i(q^2)$ in the charged current of the nucleon [1] (vector g_V, magnetic g_M, scalar g_S form factors for the vector current and axialvector g_A, pseudoscalar g_P and tensor g_T form factors for the axialvector current, respectively).

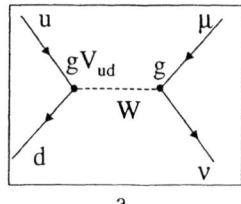 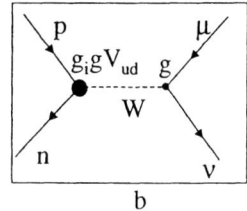

FIGURE 1. Muon capture on (a) quark and (b) nucleon level

Physics within the Standard model. These form factors are constraint by hadronic symmetries embedded in the Standard Model. g_V and g_M are well constraint by CVC and g_A is related to $g_A(0)$ as determined by beta decay. g_P is only poorly known experimentally and muon capture provides a unique probe of this form factor. The value of g_P has been a historic prediction of PCAC and by now can be calculated with impressive accuracy of 1-3% by heavy baryon chiral perturbation theory. It is a challenge to design experiments with similar precision for stringent tests of modern effective field theories of low energy QCD. The second class form factors g_S and g_T vanish in the limit of exact G-parity invariance.

Physics beyond the Standard Model. If the hadronic vertex (i.e. the form factors g_i) is determined with sufficient precision (either from experiments or from theory), muon capture can also test physics beyond the SM, e.g. muon-electron universality as well as additional interactions at the quark-lepton level.

Proceeding to the nuclear level there are two complementary descriptions of muon capture. (a) In the elementary particle model [2] the properties of the nuclear system are parameterized by effective form factors F_i, resembling the more familiar form factor description of the weak nucleon current. As with the nucleon, the form factors have to be determined empirically by relating experimental observables of the nuclear system. (b) In the microscopic model the basic nucleon amplitudes are summed over the nuclear wave function (impulse approximation) and two body currents are taken into account by explicit calculation of meson exchange currents. Accurate calculations of this kind exist for very light nuclei only.

We will review the recent developments and future prospects in muon capture. The discussion will concentrate on ^{1}H and ^{3}He, which are simple enough to study the particle physics aspects of this process. A recent experiment at PSI measured the capture rate in ^{3}He with a precision of 0.3%, by far exceeding the typical precision achieved in this field thus far(5-10%). As regards hydrogen, a TRIUMF experiment succeeded in the first observation of the rare radiative muon capture process. A new effort for a precision measurement of ordinary muon capture in hydrogen gas is under preparation at PSI.

MUON CAPTURE ON ^{3}HE

The exclusive muon capture reaction

$$\mu^- + {}^3\text{He} \longrightarrow {}^3\text{H} + \nu_\mu \qquad (2)$$

can be rather accurately calculated by theory [4]. It connects the two states of the A=3 isodoublet, which are well studied experimentally and theoretically. Thus both the elementary particle model as well as microscopic calculations based on sophisticated 3-body wave functions can be applied. Experimentally this reaction has significant advantages over reaction 1. The final state contains a charged particle and the muon induced kinetics before capture is better understood.

Precision measurement of capture rate

Very recently a high precision experiment [3] of the capture rate λ_{stat} for process 2 was performed at PSI. The index indicates that the capture takes place from the 1S ground state of the muonic He atom in a statistical mixture of its hyperfine states. The accurate knowledge of the initial atomic state is essential for an unambiguous theoretical interpretation. In the new experiment these initial conditions could be verified experimentally by observing the time distribution of capture events.

The experiment used a new technique based on a gridded ionisation chamber filled with isotopically ultrapure ^3He at 120 bar. Muons and tritons were detected in the active target. The drift times and shapes of the tracks in the ionization chamber were recorded by FADC's. This information combined with the segmentation of the chamber anodes allowed the definition of a fiducial volume where the 1.9 MeV tritons from capture were detected with 100% efficiency. The main background was a continuum from the breakup channels in $\mu^3 He$ capture which was highly suppressed by the good energy resolution ($\sigma = 30 keV$) of the chamber.

Several different analysis methods were used leading to a consistent final result of

$$\lambda_{stat} = (1496.0 \pm 4.0) s^{-1}, \qquad (3)$$

which provides a stringent test of theory.

In the elementary particle model the A=3 form factors F_V and F_M at the relevant four-momentum transfer $q_0^2 = -0.954 m_\mu^2$ are derived from elastic electron scattering on ^3He and ^3H. Result Eq. 3 then defines an allowed region in the $F_P(q_0^2)$ versus $F_A(q_0^2)$ plane shown in fig 2. By using $F_A(q_0^2) = 1.052 \pm 0.010$ (extrapolated from tritium beta decay to q_0^2) a value of pseudoscalar form factor $F_P(q_0^2) = 20.8 \pm 2.8$ is obtained [3]. The error in this experimental determination of F_P is dominated by the uncertainty in $F_A(q_0^2)$, while the experiment error only contributes ± 0.5. The result agrees very well with the PCAC prediction $F_P(q_0^2) = 20.7 \pm 0.2$ and constitutes the most precise test of nuclear PCAC.

The microscopic model gives $\lambda_{stat} = 1304 s^{-1}$ in the impulse approximation [4] and $\lambda_{stat} = 1502 \pm 32 s^{-1}$ in the complete calculation including meson exchange currents [5]. By quantitatively comparing this calculation with Eq. 3, a value for the nucleon pseudoscalar form factor $g_P = 8.53 \pm 1.54$ is derived [3], in agreement with recent results from heavy baryon chiral perturbation theory.

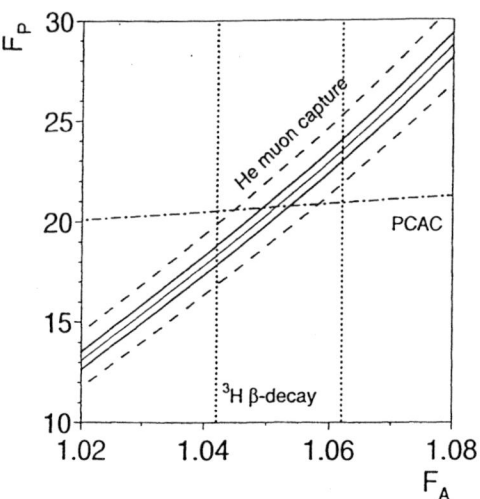

FIGURE 2. Allowed region of axial form factor F_A and pseudoscalar form factor F_P from Ref. [3]. Constraint derived from Eq 3 (solid lines) and by adding errors of F_V and F_M (dashed lines). The vertical dotted lines are the constraints for F_A extrapolated from tritium beta decay and the dot-dashed line is the PCAC prediction.

The precision of the new experiment also allows tests of physics beyond the Standard Model. In the analysis of Refs. [6] the general muon capture Lagrangian on the quark-lepton level is parameterized in terms of effective couplings using a chirality projector formalism, similar to the standard analysis of muon decay parameters. Several limits on couplings and masses for interactions between the first two generations can be improved.

Polarization observables

It is evident from fig. 2 that at least two form factors of the A=3 system (F_A and F_P) are not known sufficiently, while only one experimental observable (λ_{stat}) has been measured with precision. An independent constraint from a second observable is required for a model independent determination of both form factors. Possible observables depending on the polarization and hyperfine population of the initial $\mu^3 He$ atom were investigated theoretically in ref. [4]. The vector analyzing power A_V was found to be ~ 3 times more sensitive to F_P than λ_{stat} and only slightly sensitive to F_A. The measurement of this quantity appears very promising. Recently a new technique has been devised to repolarize μ^3He atoms by neutralization and spin exchange with laser polarized Rubidium atoms [7]. Average polarizations of $(26.8 \pm 2.3)\%$ have been achieved and A_V has been measured for the first time [8], though with an error $> 10\%$. A dedicated effort is necessary to push the experimental precision to 1%, which is required to fully exploit the 0.3% λ_{stat} measurement. Nevertheless, this experiment seems feasible and highly attractive, as it would improve the experimental sensitivity on weak nuclear form factors, muon-electron universality and non Standard Model interactions [6] by typically a factor of 3.

MUON CAPTURE ON THE PROTON

current situation

In spite of considerable effort, the experimental situation concerning muon capture on the nucleon is rather confused at present. Experiments on ordinary muon capture (OMC) Eq. 1 suffer from limited precision and uncertainties in their interpretation. The latter remark applies specifically to measurements with liquid targets where capture proceeds from the ortho and para state of the $pp\mu$ molecules with quite different theoretical rates 506 s^{-1} and 200 s^{-1}, respectively. The interpretation of the observed capture rates relies heavily on the transition rate λ_{OP} between the two $pp\mu$ states. Unfortunately, experiment [9] and theory [10] disagree on the value of λ_{OP}.

A significant achievement in this field was the first measurement of the rare radiative muon capture (RMC) reaction

$$\mu + p \rightarrow n + \gamma + \nu_\mu \qquad (4)$$

in an experiment [11] at TRIUMF. The branching ratio for this reaction is of the order of 10^{-8}, while ordinary muon capture Eq. 1 occurs at the $\sim 10^{-3}$ level. On the theoretical side the 3-body final state in radiative muon capture is very attractive. It allows kinematic regions with momentum transfer close to the pion pole, where RMC is more than 3 times more sensitive to g_P than OMC. After cuts and background subtraction 279 ± 26 RMC photons with energy above 60 MeV are observed, yielding a branching ratio of $(2.10 \pm 0.22) \times 10^{-8}$ for this energy cut. Using the theoretical photon spectrum from Ref. [12] this result corresponds to a value of g_P 1.47 ± 0.12 times higher than the theoretical predictions. Currently several theoretical groups [13] are investigating whether this discrepancy can be resolved by a consistent treatment of radiative muon capture within chiral perturbation theory which includes additional diagrams not present in the earlier calculations.

A summary of the constraints on g_p derived from the present experimental results and their dependence on the ortho-para transition rate is presented in fig. 3 [11,14]. The large uncertainty and inconsistency in the experimental results is contrasted by the precision of the latest chiral perturbation theory results $g_p = 8.44 \pm 0.23$ [15] and 8.21 ± 0.09 [16]. Fig. 3 also shows the substantial improvement in precision planned in a new experiment [14] at PSI.

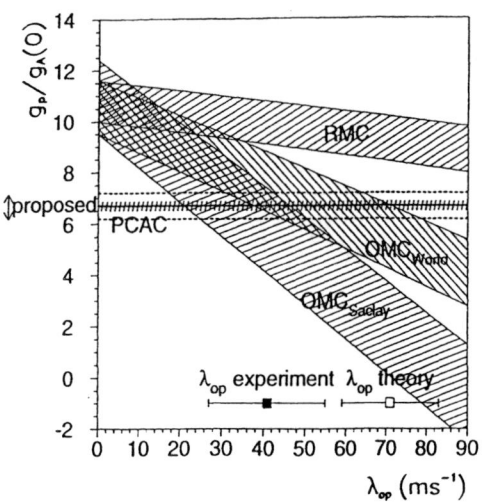

FIGURE 3. Current constraints on g_p as function of the ortho-para transition rate λ_{OP}. Experimental results from ordinary muon capture (OMC), radiative muon capture (RMC) and theory (PCAC). "Proposed" denotes sensitivity of new PSI experiment discussed below.

μ^-/μ^+ lifetime experiment

Reaction 1 will be observed [14] by measuring the μ^- lifetime in hydrogen to better than 10 ppm and comparing it with the μ^+ lifetime. In the first stage of the experiment the world data of τ_{μ^+} (20 ppm error) will be used, in a second stage the precision of τ_{μ^+} will be improved by a factor of 2, yielding the best determination of the Fermi coupling constant. Muon fluxes of $\sim 5 \times 10^4 s^{-1}$ will be stopped in a TPC filled with ultrapure hydrogen of 10 bar. The experimental conditions are chosen such that uncertainties due to $pp\mu$ formation are reduced to a negligible level. Capture takes place predominantly from the F=0 hyperfine state of muonic hydrogen (rate Λ_0). The TPC and surrounding tracking chambers allow the operation at the high rates of modern meson facilities, since pileup problems can be sufficiently reduced by identifying the muon-electron pair originating from a common vertex. The analysis of some 10^{10} events is necessary to achieve the proposed precision. This requires fast tracking in parallel online CPU's. Contiguous time regions will be recorded which are large compared to the muon lifetime. Thus the complete information about all neighboring events will be available for every analyzed $\mu - e$ pair, allowing careful optimization of offline cuts and detailed analyses of systematic effects. The basic principles of the TPC operation with hydrogen have been successfully demonstrated in prototypes at Gatchina and a first test run is foreseen for 1998.

SUMMARY AND OUTLOOK

In this paper a brief overview on muon capture in the lightest nuclei is given. Recent experimental and theoretical work demonstrates that stringent tests of low energy QCD and the structure of charged current electroweak interactions can be derived from precision experiments on semileptonic reactions.

TABLE 1. selected present and future research topics in muon capture

topic	physics	observable		facility/status
μ capture	low energy QCD	$\mu^3 He$	λ_{stat}	PSI
	chiral symmetry		A_V	LAMPF/TRIUMF
	SM tests		RMC	TRIUMF
		μp	Λ_0	PSI
			A_V	future
			RMC	TRIUMF, PSI
T-odd correlations in μ capture	CP violation	triple correlations		future
ν_μ mass	SM test	$\mu + {}^3He \to d + n + \nu$		future
		$\mu + Li \to t + t + \nu$		

Some present and future research topics, which might benefit from higher luminosity muon beams, are compiled in table 1. As regards muon capture in hydrogen the vector analyzing power (as well as the small capture rate from the F=1 μp hyperfine state) are uniquely sensitive to g_P. These elusive observables might be investigated by laser pumping the population of the F=1 state in a pulsed beam. Also mentioned are some topics of fundamental interest, which have been discussed in the past, but not been demonstrated yet. T-violation in muon capture can be searched for in triple correlations between the initial muon spin and the spin and momenta of the reaction products [17]. High intensity beams might allow improved ν_μ mass measurements by high statistics investigations of the sensitive kinematic region in 3 body final states.

Overall, the field will benefit from beams of higher intensity, higher quality and suitable time structure (pulsed and DC). At the muon collider beam intensities exceeding present facilities by 5-6 orders of magnitude are anticipated. The impact of such fantastic beams on muon capture is less clearcut than for rare decays and has to be studied case by case. Very likely completely new ambitious experimental ideas will be stimulated by this quantum jump in muon intensity.

ACKNOWLEDGEMENT

This work was supported by the U.S. Department of Energy (contract No. DE-FG03-87ER40323). It is a pleasure to thank the members of our international collaborations at PSI for many years of exciting and enjoyable work which forms a basis for this survey.

REFERENCES

1. For reviews and references, see for example,
   E. Zavattini, in *Muon Physics*, eds. V.W. Hughes and C.S. Wu (Academic Press, New York, 1975), Vol. II, pp. 219-261;
   N.C. Mukhopadhyay, *Physics Reports* **30**, 1 (1977).
2. C.W. Kim and H. Primakoff, *Phys. Rev.* **140 B**, 566 (1965).
3. P. Ackerbauer et al., hep-ph/9708487, to be published in *Phys. Lett. B.* and references given therein.
4. J.G. Congleton and H.W. Fearing, *Nucl. Phys.* A **552**, 534 (1993); J.G. Congleton, *Nucl. Phys.* A **570**, 511 (1994).
5. J.G. Congleton and E. Truhlík, *Phys. Rev.* C **53**, 956 (1996).
6. J. Govaerts, hep-ph/9701385, to be published in *Nucl. Instr. Meth.*; hep-ph/9711496, to be published in Proceedings of The International Europhysics Conference on High Energy Physics, Jerusalem (1997).
7. A.S. Barton et al., *Phys. Rev. Lett.* **70**, 758 (1993).
8. W.J. Cummings et al., Proc. IV Int. Symposium on WEIN'95, Osaka, 12-16 June 1995, eds. H. Ejiri, T. Kishimoto and T. Sato (World Scientific, Singapore, 1995), pp. 381-385.
9. G. Bardin et al., *Phys. Lett.* **104 B**, 320 (1981).
10. D.D. Bakalov et al., *Nucl. Phys.* A **384**, 302 (1982).
11. G. Jonkmans et al., *Phys. Rev. Lett.* **77**, 4512 (1996).
12. D.S. Beder and H.W. Fearing, *Phys. Rev.* D **39**, 3493 (1989).
13. H.W. Fearing et al., hep-ph/9709459 (1997); T. Meissner, F. Myhrer, K. Kubodera, nucl-th/9707019 (1997); S. Ando and D.P. Min, hep-ph/9707504 (1997).
14. D.V. Balin et al., PSI proposal R-97-05 (1996).
15. V. Bernard, N. Kaiser and U.G. Meissner, *Phys. Rev.* D **50**, 6899 (1994).
16. H.W. Fearing et al., *Phys. Rev.* D **56**, 1783 (1997).
17. J. Deutsch, *Z.Phys.* C **56**, 77 (1992); A.L. Barabanov, nucl-th/9704028 (1997).

Muonic Processes in Solid Hydrogen

G.M. Marshall,* J.M. Bailey,† G.A. Beer,‡ J.L. Beveridge,*[1]
M.C. Fujiwara,§ T.M. Huber,** R. Jacot-Guillarmod,†† P. Kammel,‡‡
S.K. Kim,§§ P.E. Knowles,‡[2] A.R. Kunselman,*** M. Maier,‡
G.R. Mason,‡ F. Mulhauser,†† A. Olin,‡ C. Petitjean,†††
T.A. Porcelli,‡ L.A. Schaller,†† and J. Zmeskal‡‡‡

*TRIUMF, 4004 Wesbrook Mall, Vancouver, BC, Canada
†Chester Technology, Chester, UK
‡University of Victoria, Victoria, BC, Canada
§University of British Columbia, Vancouver, BC, Canada
**Gustavus Adolphus College, St. Peter, MN, USA
††University of Fribourg, Fribourg, Switzerland
‡‡Lawrence Berkeley Laboratory, Berkeley, CA, USA
§§Jeonbuk National University, Jeonju City, S. Korea
***University of Wyoming, Laramie, WY, USA
†††PSI, Villigen, Switzerland
‡‡‡IMEP, Vienna, Austria

Abstract.
    Muonic hydrogen participates in many different interactions, including muon induced fusion of hydrogen nuclei. Conventional experimental techniques cannot always unravel and separate the processes of interest. Some of the most important measurements may be more reliably accomplished with the use of a unique and versatile target consisting of layers of different solid hydrogen isotope mixtures.

INTRODUCTION

A negative muon readily forms a muonic atom in the surrounding medium when it comes to rest. In an environment consisting of hydrogen isotopes, many interactions occur which are analogous to those in normal electronic atoms, but with scales different due to the reduced muonic hydrogen mass; energies are larger, while times and lengths (or sizes) are smaller. There are also interactions which do not readily occur in normal atoms, such as fusion of the nuclei bound into a very small muonic

---

[1] Now at Oncometrics, Vancouver, BC.
[2] Now at University of Fribourg.

molecular ion. Because the muon accelerates the rate at which fusion occurs but is not consumed by fusion, this process is known as muon catalyzed fusion (μCF).

The catalysis of fusion in muonic hydrogen was at one time considered as a candidate for clean and inexpensive energy production, but it was recognized that two major limitations existed. First, the rate-limiting process of muonic molecular ion formation is not fast enough; whereas the fusion rate following molecular ion formation is $\sim 10^{12}$ s^{-1} (for $dt\mu$), formation itself via Auger processes plods along at only $\sim 10^5$ s^{-1}. Once the ion is formed, the separation of the hydrogen nuclei is such that, in the case of $dt\mu$, fusion proceeds rapidly. When experiment [1] and theory [2] indicated that in some cases a fascinating resonant molecular interaction speeds molecular ion formation, interest in the process was rekindled. However the second limitation, the sticking of the muon to the charged fusion product (an alpha particle in the case of dt fusion), removes the possibility for the muon to participate in further cycles. On average, one muon can induce less than 200 fusions if the probability of sticking is 0.5%. Although estimates vary widely, something greater than an order of magnitude more fusions per muon is deemed necessary to break even with practical energy production schemes.

Despite this disappointing outlook for energy production, research in muon catalyzed fusion remains challenging. Detailed theoretical calculations can be tested, and real applications exist. Our understanding is still incomplete, as demonstrated by persistent disagreement between theory and experiment in several important cases. These could be resolved via important measurements which would benefit enormously from more intense muon beams.

Experiments with solid hydrogen targets are among those which would benefit, and they have advantages which enable or simplify certain measurements. One can create "hot" beams of muonic deuterium ($d\mu$) or muonic tritium ($t\mu$) with energies of up to tens of eV, a range practically inaccessible with target techniques relying on high temperature. Solid layers of different isotopic composition allow separation in space and time of different muonic processes, and make time of flight a practical tool for energy measurement. In addition, detection of low energy charged fusion products (such as protons or alpha particles) is simplified because a detector can be placed close to the layer structure without any intervening window; energy loss is minimized in this geometry, aiding in the definition of a distinctive, low-background fusion signature.

There are also disadvantages which must be recognized. Due to the very different nature of the thermalization mechanism at energies close to the Debye temperature, muonic hydrogen in the solid does not lose energy as it does in liquid or gas. Kinetics models, which rely on rates independent of time, are not always applicable when the rate is energy dependent and the energy is changing with time. Analysis therefore relies on simulations which require a detailed and correct description of numerous rates and cross sections as a function of energy.

IMPORTANT PROCESSES IN MUON CATALYZED FUSION

The muonic interactions which take place in hydrogen can be categorized as muonic atom formation, elastic scattering, inelastic scattering (including hyperfine transitions, isotopic transfer, and chemical transfer to heavier nuclei), molecular ion formation, fusion, and sticking. In the following discussion, the case of dt fusion is emphasized because of its high efficiency. Several reviews [3–5] describe this and other systems in great detail.

Muons are initially captured into excited atomic states of muonic hydrogen; while deexcitation to the ground state occurs in some 10^{-12} s in condensed hydrogen, excited state transfer to heavier isotopes occurs as well [6]. There also exists a mechanism by which the muonic atom is accelerated, if deexcitation occurs in the field of a neighboring nucleus [7]. When the ground state is reached, the inelastic process of isotopic transfer results in the muon rapidly seeking out the heavier isotopes. For example, with only 0.1% concentration of deuterium or tritium in protium ($^1H_2$), the mean time for transfer to form $d\mu$ or $t\mu$ is of order 10^{-7} s [8]. The resulting heavier muonic atoms are produced with approximately 40 eV kinetic energy because of the difference in reduced mass. This is the source of the "hot" or energetic muonic atom beams in experiments with solid hydrogen.

Elastic collisions moderate the energy of muonic atoms until thermal equilibrium is established with the surrounding medium. Although this equilibrium is usually established very rapidly (in times of order 10^{-9} s), there are exceptions. For $t\mu$ or $d\mu$ scattering via protons, the cross section in the kinetic energy range of 1 to 100 eV is significantly reduced due to the Ramsauer-Townsend mechanism; thermalization is much slower, and the muonic atoms travel appreciable distances. In a thin solid layer, they may escape to an adjacent layer or into the vacuum environment [9]. In *all* solid muonic hydrogen systems, thermalization at very low energies is inhibited because the muonic atom is interacting with a lattice structure, which cannot absorb the kinetic energy efficiently [10].

Molecular ion formation occurs when the muonic atom combines with one nucleus of a hydrogen molecule. The rate sets the limitation on how fast the μCF cycle can proceed. One mechanism is the Auger process, where a molecular electron is ejected, as in the reaction:

$$t\mu + D_2 \rightarrow [(dt\mu)^+ de] + e \tag{1}$$

The Auger rate is relatively slow ($\sim 10^6$ s^{-1}), but molecular resonance mechanisms exist which proceed much more quickly at certain energies, as in:

$$t\mu + D_2 \rightarrow [(dt\mu)^+ dee]^* \tag{2}$$

In this case, the kinetic energy of the projectile ($t\mu$) is absorbed in the excitation of internal degrees of freedom, namely rotational and vibrational modes of the muonic molecular ion and the complex molecule. The discovery of this resonance mechanism [2] provided motivation to reconsider muon catalyzed fusion as a potentially efficient energy generation system. Recent calculations [11] for targets of D_2 and HD show that the resonances lead to very high molecular formation rates of $\sim 10^{10}$ s^{-1}, for $t\mu$ kinetic energies near 0.5 eV. Verification of the huge rates therefore requires muonic atoms in this energy range, corresponding to temperatures near 3000 K.

Fusion proceeds rapidly with a rate of order 10^{12} s^{-1} from the $dt\mu$ molecular ion, releasing a neutron and an alpha particle which share the fusion energy of 17.6 MeV, while the muon is typically left with some 20 keV [12]. In a small fraction ω_s^0, the muon "sticks" to the alpha. As the resulting $(\alpha\mu)^+$ slows down, the muon may be stripped or reactivated with probability R, so that final sticking is given by

$$\omega_s = \omega_s^0(1 - R). \qquad (3)$$

For $dt\mu$, several different theoretical approaches [12,13] predict $\omega_s^0 \approx 0.91\%$, while other calculations show $R = 0.34$ [15–17] at high density.

EXPERIMENTS WITH SOLID TARGETS

Although many different muonic hydrogen interactions have been investigated with different solid target compositions and geometries, only two important examples will be described. The first is the energy dependence of resonant muonic molecular ion formation, and the second is a proposal for a direct measurement of sticking and stripping in muon catalyzed fusion. Both measurements utilize a 3 K gold foil of thickness 0.05 mm, one surface of which is exposed to cooled hydrogen gas [18] at approximately 10^{-9} bar introduced into the ultra-high vacuum target apparatus through a diffusing plate. A layer is thus formed of desired thickness [19], typically in the range of $1 - 500$ μm, of the chosen isotopic mixture. Other layers of a different isotopic mixtures can be applied, depending on the measurement.

The TRIUMF proton beam (500 MeV, 140 μA) produces pions in a 10 cm Be production target. Low energy π^- decaying near the target are the source of a beam of 5×10^3 s^{-1} muons of 26 MeV/c (kinetic energy 3.1 MeV), which is transported to the hydrogen target system by conventional dipoles and quadrupoles and a dc crossed-field velocity separator. The beam has a momentum spread (fwhm) of $\sim 4\%$ which combines with range straggling to give a range spread of ~ 7 mg cm^{-2} in hydrogen, so that a substantial fraction of the beam can be stopped in layer of less than 1 mm thickness. Muons pass through a thin beam-defining scintillator, and then the gold foil support, before stopping in the thin layer of solid hydrogen; the momentum of the muon beam is tuned for maximum stop rate in hydrogen. The cryogenic target is protected by a thermal shield in which silicon detectors are mounted. It is contained within an ultra-high vacuum cube surrounded with detectors for fusion neutrons, muon decay electrons, and muonic x-rays, as shown in Figure 1. An order of magnitude increase in beam intensity would be easily accommodated by this arrangement, keeping in mind the limitation of one muon at a time in the target region. With minor design changes, a lower momentum beam with less range straggling would also allow a better definition of the muon's environment.

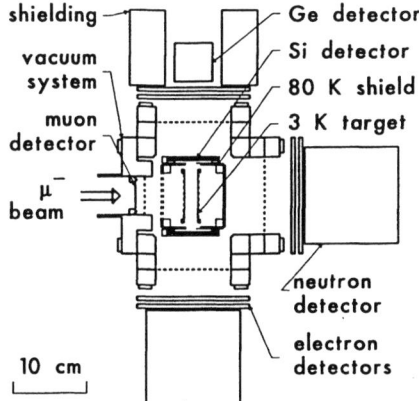

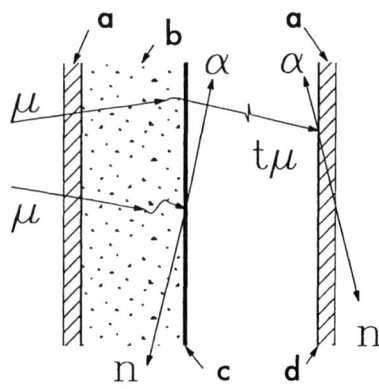

FIGURE 1. Arrangement for time-of-flight experiments, showing the cryogenic solid hydrogen target and vacuum system, with detectors for incident muons, muon decay electrons, muonic x-rays, and fusion products (α, p, n).

FIGURE 2. Diagram of the target layers for TOF experiments, horizontally expanded to show relative layer thicknesses: (a) gold foils, (b) upstream formation and emission layer, (c) moderation layer, (d) downstream reaction layer.

The time-of-flight technique is applied to test predictions for the energy dependence of resonant muonic molecule formation. In this case, a second support foil is used, as shown in Figure 2. The time for a muonic atom to travel between the hydrogen isotope layers on the first foil to the adjacent surface of the second foil is compared with a simulation in order to confirm the predicted resonance structure. Most muons stop in a production layer of protium ($^1H_2$) containing 0.1% tritium, of thickness 3.5 $mg\ cm^{-2}$, almost always forming muonic protium ($p\mu$). Transfer to the small tritium concentration occurs with a characteristic time of $\sim 10^{-7}$ s, and the increase in binding energy corresponding to the greater reduced mass results in a typical $t\mu$ energy of 45 eV. The Ramsauer-Townsend scattering cross section of $t\mu$ on H_2 is low enough that the muonic atom can escape the layer before thermalization. In fact, a moderation layer of D_2 (0.1 $mg\ cm^{-2}$), in which $t\mu$ energy loss is much greater, is added to the surface of the production layer to optimize the energy distribution. After traversal of the 18 mm separation between the first and second foils, the $t\mu$ atoms which possess an energy corresponding to the high resonant rate of equation (2) may form muonic molecular ions in the reaction layer of thickness 0.02 $mg\ cm^{-2}$. If so, fusion follows almost immediately, in about 10^{-12} s.

It is also possible that the $t\mu$ atoms scatter and lose energy before formation and fusion, so that the time of flight is not correlated with the energy of molecular ion formation. This indirect process leads to an unavoidable background. In either case, a fusion product such as an alpha particle at 3.5 MeV can be detected, while other muonic atoms pass through the reaction layer into the gold substrate, giving no fusion signal. The interval between muon arrival and fusion is dominated by the time of flight between the foils. Lower energy resonances near 0.5 eV appear with times greater than 3 μs, while higher energy resonances are in the range $2-3$ μs. Emission from the moderation layer at different angles implies different lengths of flight path and smears the resonant time-of-flight structures, necessitating analysis via comparison with simulations.

Fusion also occurs in the D_2 moderation layer. Since there is no delay due to time of flight, these events show up with times less than about 1.5 μs and are easily separated from reaction layer events. These more prompt events can be used to confirm important assumptions in the simulation and analysis. For example, the time distribution has been used [8] to extract effective rates $\lambda_{pp\mu}$ and λ_{pt} for non-resonant $pp\mu$ molecular ion formation and muon transfer from protons to tritons, respectively.

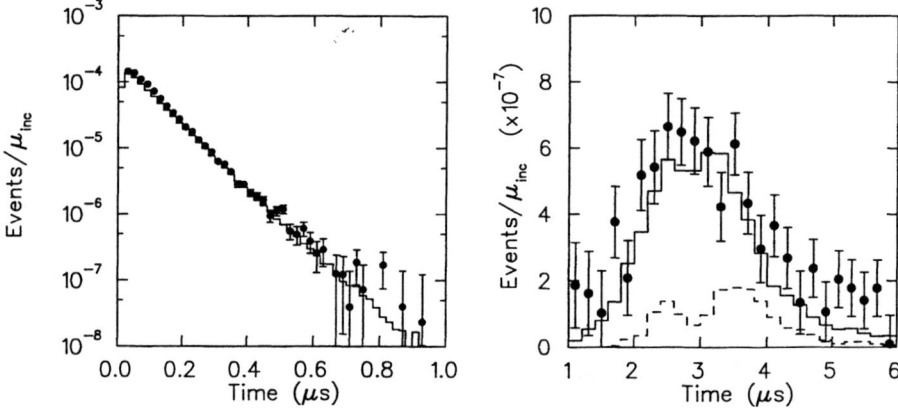

FIGURE 3. Fusion product (α) time distributions following muon arrival, normalized to muons incident on the upstream gold foil support. Time ranges correspond to fusion in moderation layer (left) and reaction layer (right), both of which were D_2. Histograms in solid lines are from a simulation using theoretical and experimental rates and cross sections, estimated detector efficiency and muon stopping fraction, and are not otherwise normalized. The dashed line corresponds to direct molecular ion formation in the reaction layer. These data are preliminary.

The energy distributions of alpha particles from fusion in the moderation and reaction layers are different, because the monoenergetic alphas lose more energy in the moderation layer which is some five times thicker. However, in both cases, efficiency for alpha detection is essentially determined by the solid angle. The analysis of the absolute rate is therefore simplified compared to neutron detection, which is one significant advantage of using solid hydrogen layers. After making appropriate energy cuts and accounting for the solid angle and the fraction of muons stopping in the production layer, both time spectra can be compared to the simulation. Figure 3 [20] shows the results. In the moderation layer, the agreement is within the systematic uncertainty in the solid angle and stopping fraction estimates. In the reaction layer, the agreement is also reasonable although the statistical uncertainties are large. For comparison, the relative strength of the direct formation is indicated by the dashed line where structures reflecting different formation resonances can be seen. This preliminary result confirms approximate agreement of the data with the calculations of [11].

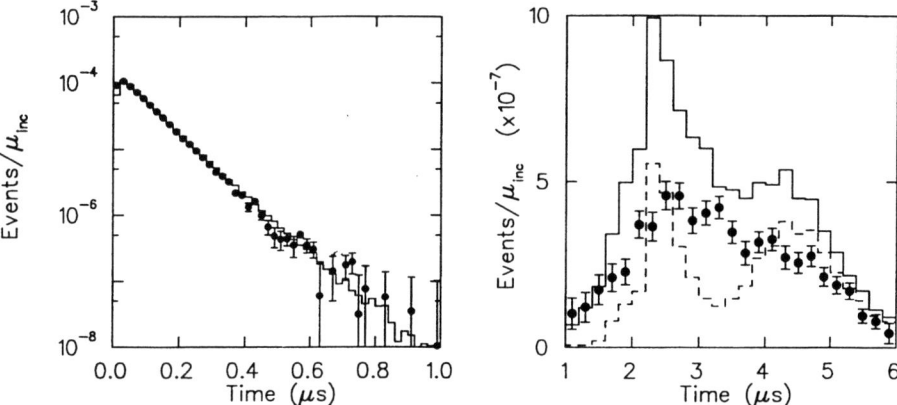

FIGURE 4. As in Figure 3, but with a reaction layer of HD rather than D_2. The thickness of the moderation layer was also slightly less (0.07 $mg\ cm^{-2}$ rather than 0.1 $mg\ cm^{-2}$). These data are preliminary.

A similar experiment was performed in which the reaction layer was composed of HD rather than D_2. The calculations for HD suggest resonances with a maximum rate some four times higher than for D_2, and occurring at a slightly lower energy. In this case, the data disagrees substantially with the simulation based on the theory in [11], being inconsistent with the higher predicted rate (see Figure 4 [21]). Potential sources of this apparent disagreement are being studied.

Whereas the measurements of muonic molecular ion resonant formation were performed with two separated parallel foils supporting hydrogen layers perpendicular to the incident beam (Figures 1 and 2), the proposed experiment for a direct measurement of sticking and stripping uses only one support foil at an angle to the beam as in Figure 5. The angle reduces the spread in energy loss of charged fusion products as they exit the hydrogen layers toward the silicon detector. It also enables a coincidence measurement of the charged products, both α and $(\mu\alpha)^+$, with a neutron in collinear geometry. Due to backgrounds, it is not now possible to operate the silicon detector in the beam direction; a higher quality low-momentum muon beam might make it so.

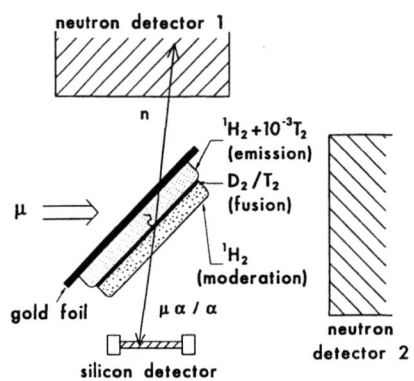

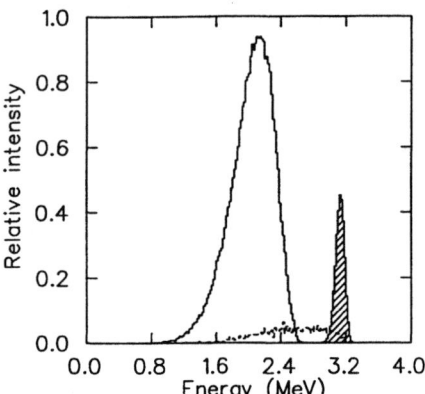

FIGURE 5. Arrangement of solid target layers and detectors to measure initial sticking and stripping.

FIGURE 6. Simulation of energy distribution in Si detector. The hatched region shows $\mu\alpha$ events from sticking, while the solid line shows the α spectrum, which has been divided by ten for easier comparison. The dashed line shows events for which the μ has been stripped, multiplied by 10.

The principle of the experiment is simple, the extraction of the sticking coefficient ω_s is direct, and the correction for stripping (R in equation (3)) is small and measurable. The target consists of three separate layers of different isotopic mixtures of solid hydrogen. Muons which stop in an emission layer of protium, with tritium concentration $\sim 10^{-3}$, are converted into $t\mu$. They are subsequently emitted with high probability from the surface of the layer, exactly as in the resonant molecular formation measurement, and enter the adjacent and comparatively thin source or fusion layer in which molecular formation and fusion take place. The doubly-charged $^4He^{++}$ (α) ions from fusion as well as $(\mu^4He)^+$ or $(\mu\alpha)^+$ ions from the sticking process are produced at nearly identical energies of 3.5 MeV; if sticking does not occur, the muon is a spectator and retains only a small momentum. Both ions escape through the third degrader or moderation layer, consisting of pure protium, and can be detected by the ion-implanted silicon charged particle detector. On passage through the moderation layer, the α and the $\mu\alpha$ lose different amounts of energy, since dE/dx is roughly proportional to Z^2 of the projectile. Detection of the collinear fusion neutron coincident with either ion is used for background rejection as well as normalization. A more complete description can be found in [22].

Figure 6 is a simulation of the Si detector energy spectrum. Three types of events are plotted; α events which form the majority, $\mu\alpha$ events from sticking, and stripping events where the muon has been stripped during passage through the layers. To simplify comparison, the solid line (α) peak is divided by ten, and the dashed line (stripping) is magnified by ten. The simulation included detector energy resolution as determined in previous experiments (65 keV at 5.5 MeV), and differences in energy loss due to path length differences (due to different ion flight angles, different production depths in the fusion layer, and measured layer nonuniformities). The fusion layer thickness was taken to be 0.1 $mg\ cm^{-2}$ and the moderation layer thickness was 0.2 $mg\ cm^{-2}$. The values N_α and $N_{\mu\alpha}$, derived from an analysis of the energy spectrum, give the effective sticking ω_s^{eff} for the $\mu\alpha$ equivalent path length l in the target medium:

$$\omega_s^{eff}(l) \equiv \omega_s^0[1 - R^{eff}(l)] \tag{4}$$

$$= \frac{N_{\mu\alpha}}{N_{\mu\alpha} + N_\alpha} \tag{5}$$

where

$$R^{eff}(l) \equiv \int_0^l \frac{dR}{dl} dl, \tag{6}$$

i.e., the probability that stripping occurs by the time $\mu\alpha$ travels a distance l in the target medium. With a thin degrader and source, hence a small l, the method is sensitive to initial sticking, since only a small correction R^{eff} is required. By increasing the thickness of the degrader, even to the point where the α peak is ranged out and monitoring of fusion is accomplished via neutron counting, a systematic measurement of stripping as a function of $\mu\alpha$ energy can be performed. Thus, experimental separation of initial sticking and stripping is possible, with precision expected to be limited by statistics.

SUMMARY

Some unique results on the processes related to muon catalyzed fusion have been obtained by the use of solid hydrogen targets. The potential exists for more results of much higher precision. One limitation is the quality of low energy muon beams, which do not have adequate intensity and stopping rate to best take advantage of the new techniques. A source of cooled, low energy muons, preferably with a high duty factor, would ameliorate the experimental difficulties and provide more stringent tests of theoretical predictions.

The authors would like to thank Dr. M.P. Faifman for providing rate calculations for molecular formation at 3 K, similar to those of [11]. The research described was supported by the Natural Sciences and Engineering Research Council (Canada), the Swiss National Science Foundation, the United States National Science Foundation, a Cotrell Grant from Research Corporation, and by NATO grant LG-930162.

REFERENCES

1. Dzhelepov, V.P., et al., *Zh. Eksp. Teor. Fiz.* **50**, 1235 (1966) [*Sov. Phys. JETP* **23**, 820 (1966)]
2. Vesman, E.A., *Pis'ma Zh. Eksp. Teor. Fiz.* **5**, 113 (1967) [*JETP Lett.* **5**, 91 (1967)]
3. Petitjean, C., et al., *Hyp. Int.* **82**, 273 (1993).
4. Ponomarev, L.I., *Contemporary Physics* **31**, 219 (1991).
5. Breunlich, W.H., et al., *Ann. Rev. Nucl. Part. Sci.* **39**, 311 (1989).
6. Lauss, B., et al., *Phys. Rev. Lett.* **76**, 4693 (1996).
7. Markushin, V.E., *Phys. Rev. A* **50**, 1137 (1994).
8. Mulhauser, F., et al., *Phys. Rev. A* **53**, 3069 (1996).
9. Forster, B.M., et al., *Hyp. Int.* **65**, 1007 (1990); Fujiwara, M.C., et al., *Hyp. Int.* **106**, 257 (1997).
10. Adamczak, A., *Hyp. Int.* **101/102**, 113 (1996).
11. Faifman, M., and Ponomarev, L., *Phys. Lett. B* **265**, 201 (1991); a subsequent calculation, including effects of quadrupole interactions, can be found in Faifman, M., et al., *Hyp. Int.* **101/102**, 179 (1996).
12. Melezhik, V.S., *Hyp. Int.* **101/102**, 365 (1996).
13. Cohen, J.S., *Hyp. Int.* **101/102**, 349 (1996).
14. Stodden, C.D., et al., *Phys. Rev. A* **41**, 1281 (1990).
15. Markushin, V.E., *Muon Catalyzed Fusion* **3**, 395 (1988).
16. Cohen, J.S., *Phys. Rev. Lett.* **58**, 1407 (1987).
17. Knowles, P.E., et al., *Nucl. Instrum. Methods* **A368**, 604 (1996).
18. Fujiwara, M.C., et al., *Nucl. Instrum. Methods* **A395**, 159 (1997).
19. Fujiwara, M.C., Ph.D. thesis, University of British Columbia (in preparation).
20. Porcelli, T.A., Ph.D. thesis, University of Victoria (in preparation).
21. Fujiwara, M.C., et al., *Hyp. Int.* **101-102**, 613 (1996).

An Update on MEGA: An Experimental Search for Lepton Number Non-conservation

Ed V. Hungerford[1]

Department of Physics
University of Houston
Houston, TX 77204

Abstract.
Mega is an experiment to search for the decay $\mu \to e\gamma$. It is designed to test lepton-number conservation in muon decay by a background free measurement in this branching ratio at a sensitivity of a few $\times 10^{-13}$. This paper reports particularly on the design and performance of the MEGA detector, as the lessons learned previously are important when new experiments are proposed which would increase the sensitivity of such a measurement down to a branching ratio on the order of 10^{-14}.

I INTRODUCTION

MEGA is a search for the decay of a stopped, positive muon into an positron and a photon. This decay is allowed by energy, momentum, and charge conservation, but violates lepton-number conservation. Lepton-number conservation is simply assumed by the standard model, although many extensions to the model [2] [3] require the violation of this symmetry. Some examples of lepton-number non-conserving processes are given in Table 1. Experimentally the process, $\mu \to e\gamma$, is well defined kinematically, so that a redundant measurement of the kinematic parameters of the particles emitted in the decay is sufficient to reduce the probability of background events to a negligable level. Thus MEGA was originally designed to be a background free measurement at a sensitivity equivalent to a branching ratio of a few $\times 10^{-13}$.

MEGA measures the energy and vector momentum of coincident electrons and photons (both temorally and spatially) from stopped positive muons. A true $\mu \to e\gamma$ event would have essentially equal energy sharing between these coincident particles, and equal but oppositely directed momenta. The allowed decay, $\mu \to e\nu\bar{\nu}\gamma$ has very low probability of simultaneously producing a photon and an electron in the upper momentum corner of the phase space, so for reasonable

TABLE 1. Limits on Lepton Number Non-conservation

Process	Limit	Δ Generation Number	Reference
$K_L^0 \to \mu e$	2.4×10^{-11}	0	[4]
$K_L^0 \to \pi \mu e$	3.2×10^{-10}	0	[5]
$K^+ \to \pi^+ \mu e$	2.1×10^{-10}	0	[6]
$\mu N \to eN$	7.0×10^{-13}	1	[7]
$\mu \to eee$	1.0×10^{-12}	1	[8]
$\mu \to e\gamma$	3.8×10^{-11}	1	[9]
$\mu \to \gamma\gamma\gamma$	7.2×10^{-11}	1	[10]

resolution this process is not a significant source of background. Indeed, the dominant background source is random coincidences between high energy photons and electrons produced in different decay events.

The presently published limit on the branching ratio for $\mu \to e\gamma$ was established by the Crystal Box experiment at the Los Alamos National Laboratory, LAMPF [10]. In this experiment, the decay positrons were tracked by a volume of drift chambers inside an array of segmented NaI crystals. The direction of the decay positrons was tracked by the drift chambers, and the energies of the positrons and photons determined by the NaI crystals. The system energy resolution was 8% at 50 MeV. The single-particle solid angle acceptance was approximately 45% of 4π, and detection efficiencies for both positrons and photons was essentially 100%.

In order to reduce the branching ratio sensitivity below that of the Crystal Box, MEGA was designed to trade detection efficiency for better energy resolution and rate capability. Table 2 compares various performance factors between the MEGA and Crystal Box detectors. Items in this table will be discussed in more detail later, but current performance figures of the MEGA detector are still improving as the data analysis proceeds.

II DETECTOR

In the MEGA experiment, positive muons from the surface muon beam at LAMPF are stopped in a thin mylar foil, positioned at the center of a large, solenoidal, superconducting magnet, Fig. 1. These muons decay, producing positrons, photons, and neutrinos. As the positrons spiral in the magnetic field, their momentum is detemined by a set of high-rate, cylindrical MWPCs which have wire anode and foil cathode-strip readouts. Coincident photons are analyzed in one of three independent, cylindrical pair spectrometers. The stopping foil is sloped with respect to the axis of the cylindrical field to increase the stopping power of the target while reducing the material which contributes to the multiple scattering of the spiraling positrons. A sloped target also enhances the ability to determine the stopping position, which is the intersection of the positron trajectory with the target. The continuation of the tangent to the positron trajectory must

TABLE 2. Comparison between MEGA and the Crystal Box

Parameter	MEGA		Comments	Crystal Box
	Design	Measured		
ΔE_γ	1.7 MeV	2.6 ± 0.5 MeV	All Events	4 MeV
		2.0 ± 0.5 MeV	Outer Conversion	
		3.6 ± 0.5 MeV	Inner Conversion	
ΔE_e	0.3 MeV	0.41 ± 0.01 MeV	Average	4 MeV
$\Delta t_{e\gamma}$	0.8 ns	1.7 ns	IB Data	1.2 ns
$\Delta \theta_{e\gamma}$	0.6°	1.3°		5.0°

also strike a pair spectrometer at a photon conversion point. In addition, though not yet used in the data analysis, the photon direction as determined from the pair spectrometer provides a constraint on the events which can be used to further reduce background. The positron detector [11] consists of a set of cylindrical

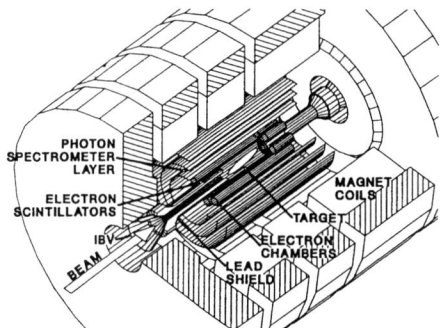

FIGURE 1. The MEGA Detector showing the major detector components including the positron MWPCs, photon pair spectrometers, positron scintillators, and the magnet.

MWPCs which surround the stopping target. These detectors are constructed of 15μm Au coated W wire centered between mirror-imaged, sterio-striped, 25μm, cathode foils. The sterio wrap is 2π. Each hit in an MWPC is characterized by a coincidence between an anode wire and stripes on both inner and outer cathode foils. This triple coincidence defines a 3 dimensional point in space, and each reconstructable trajectory will have at least 6 hits per loop in the set of MWPCs. A typical positron event forms a helical trajectory, spiraling from the decay point and striking one of a set of sintillators placed in a barrel arrangement around both upstream and downstream nose-cones of the magnet. These scintillators provide the positron event time, and also the starting point for the reconstruction of the positron trajectory. Thus starting at the scintillator, hits along the trajectory are "unwound" by tracing the event backward with the aid of the time between the electon and photon event. Typically the positron hit pattern is complicated, but few, if any, patterns reconstruct to a possible electron-photon coincidence event.

A photon may interact in one of three independent pair spectrometers [12], where by measuring the orbits of the conversion pair, the perpendicular component of the photon momentum, and the dip angle of the photon direction can be determined. After one-half loop, the conversion pair passes through scintillators which provide timing for the photon event.

A typical photon event consists of an electron and positron conversion pair which spiral in the axial magnetic field of the magnet. The photon energy is determined by fitting the particle trajectories from the conversion vertex to the edges of the shower. The position of the vertex relative to the edges, the diameter of the circles, and the shape of the hit pattern determine if an acceptable event has occured. Hits within the interior of a lepton loop are due to subsequent passes of this particle through the system. A "good" event is produced by pair production within one of the two lead converters in a spectrometer layer, has a total energy above a given threshold, and has sufficient energy sharing between the pairs to allow the orbits to be reconstructed. The total photon energy is obtained from a fit to the projected circular trajectory, corrected by the dip angle due to the photon direction. The dip angle is determined by the displacements of the hits in the axial direction.

The energy resolution of the positron detector is determined by measuring the Michel edge of the positron spectrum at 52.8Mev. This energy resolution is determined to be about 0.4 MeV. The energy resolution of the photon is determined by measuring the coincident photons from π^0 decay, where the π^0 is produced via the reaction, $\pi^- p \to \pi^0 n$. The energy resolution depends on whether the photon conversion occured in the inner or outer lead conversion layer, but but the total energy resolution is primarily constrained by the measurement of the photon dip angle. The presently accepted value of this resolution is about 2.6 MeV for all events.

III TRIGGER

A sophisticated hardware and software trigger system reduced the event rate to a level which allowed data storage on magnetic tape for subsequent off-line analysis [13]. There were two stages of hardware trigger, and a workstation farm to implement a software filter before the data was committed to permanent storage. Detailed Monte Carlo studies are used to identify event patterns with which to encode the hardware and software triggers, so that unwanted background events can be rejected. The efficiency of these triggers and the data cuts are measured using photons produced from $\pi p \to \pi^0 n \to \gamma\gamma n$, and $\mu \to e\nu\bar{\nu}\gamma$.

Since high energy photons are rare, and the MEGA photon pair spectrometers are shielded by the magnetic field from charged particles occuring in muon decay, the MEGA trigger is generated from photon shower patterns in the photon arm.

TABLE 3. Trigger Rates

	Per μ Stop	Instantaneous(Hz)	Average(Hz)
μ Stop	1	2.6×10^8	1.3×10^7
1st	1.2×10^{-4}	3.1×10^4	1.6×10^3
2nd	6.4×10^{-5}	1.7×10^4	8.3×10^2
Reconstructed Photons> 47 MeV	8.5×10^{-6}	2.2×103	1.1×10^2

The first-level trigger thus uses the sum of the measured orbital diameters of a conversion pair to roughly determine the momentum of the photon perpendicular to the magnetic field. The system is triggered if this momentum lies above some threshold value, typically 40 MeV. Each pair spectrometer operates independently, and generates its own trigger. All triggers are coordinated with each other and the system live time by an electronic controller. After the system has been triggered, a fast clear may be issued if the shower hit pattern does not correspond to a reconstructable event. However, even after the events have been filtered by this second-level trigger the data rate is still too high to commit it to permanent storage. Therefore the events are entered into a computer workstation farm which further filters the data. Calculations in-this farm test for hits in the positron sperctrometers in windows along a 3-dimensional road based on the photon conversion point and on the relative timing of the positron to photon scintillators. Data is stored in a Fast-Bus interface, and read-out during the beam off-gate in the LAMPF beam duty cycle. Table 3 gives data rates at various trigger stages.

Measurement of photons from the IB decay, $\mu \to e\gamma\nu\bar{\nu}$, can provide information on photon angular resolution, trigger efficiency, and the relative timing between a photon and a positron. An IB photon energy spectrum compared to a Monte Carlo simulation indicates that the fraction of IB events passing the energy cut in the signal region is well reproduced, showing that the hardware and software triggers, and the cut efficiencies are well understood.

IV RESULTS

Approximately 16% of the data have been analyzed. This represents about 2.3×10^{13} muon stops. The error in the final number of the effective muon stops is dominated by the uncertainty in the muon stopping rate, which is known to about 10%. A box analysis of the two dimensional energy spectra, E_e vs E_γ, has no events within the energy cuts. Thus there is approximately a 90% confidence level that the number of true candidates is below 2.3. From this the measured branching ratio becomes;

$$\frac{\Gamma(\mu \to e\gamma)}{\Gamma(\mu \to e\nu\bar{\nu})} \leq \frac{2.3}{N_\mu} = 3.8 \times 10^{-11} (90\% CL)$$

This result is somewhat better than that of the Crystal Box, but will improve as more data is analyzed. In addition it is expected that further improvements in detector performance, and the application of additional cuts, such as the photon pointing cut mentioned above, will allow an increase in acceptance, pushing the limit on the branching ratio even lower. However the final result will be something like a factor of 5 to 10 poorer than the MEGA design limit. There is no one reason for this degradation, as a number of small factors in each component of the system conspired to reduce the experimental sensitivity. MEGA was quite an ambitious experiment, pushing technological boundries in many areas. That it did not quite reach its goals in all areas before the experiment concluded is not so suprising. However it is important to remember that small changes in the performance of several components can make a significant change in the final result. New proposals should expect similar problems.

REFERENCES

1. For the MEGA collaboration. Institutions in MEGA are University of California Los Angeles, University of Chicago, Fermilab, Hampton University, University of Houston, Indiana University, Los Alamos National Laboratory, Texas A&M University, Valpariso University, University of Virginia, and Virginia Tech University.
2. Riccardo Barbieri, Lawrence Hall, and Alessandro Strumia, preprint Department of Physics, University of California at Berkeley, CA 94720 UBC-PTH-94/29.
3. Lincoln Wolfenstein and Yue-Liang Wu, preprint Department of Physics, Carnegie Mellon University, Pittsburgh, PA CMU-HEP94-26.
4. K. Ariska, et. al., Phys. Rev. Let. **71**,3910(1993).
5. P. Krolak, et. al., Phys. lett. **B320**,407(1994).
6. A. M. Lee, et. al., Phys. Rev. Let. **64**,165(1990).
7. P. Wintz, PhD thesis RWTH Aachen(1995); S. Eggli, PhD thesis, University of Zurich(1995).
8. U. Bellgardth, et. al., Nucl. Phys. **B229**,1(1988).
9. M. Cooper, Conference on Intersections of Particle and Nuclear Physics, Big Sky, Montana(1997).
10. R. D. Bolton, et. al., Phys. Rev. **D38**, 2077(1988)
11. V. Armijo, et. al., Nucl. Inst. and Meth., to be published.
12. M. Barakat, et. al., Nucl. Inst. and Meth. **A349**, 118(1994).
13. Y. K. Chen, et. al., Nucl. Inst. and Meth. **A373**,195(1996).

New Ideas to Improve Searches for $\mu^+ \to e^+\gamma$

M. D. Cooper

Los Alamos National Laboratory, Los Alamos, NM 87545

Abstract. Lessons are drawn from the experience of the MEGA experiment in searching for $\mu^+ \to e^+\gamma$. In light of that experience, some ideas are evaluated regarding new searches that might take place in the era of a source of low-energy muons associated with a muon collider.

INTRODUCTION

Searching for decays that change total lepton family number is an excellent method to explore potential physics beyond the Standard Model because those processes are predicted to be zero except when new physics is present. Essentially all extensions of the Standard Model that introduce new, heavy particles predict the existence of these rare decays, though the most probable channel is highly model dependent. Recently, the prejudice has grown within the physics community that supersymmetry is an extension that is likely to be related to nature. Barbieri, Hall, and Strumia (1) show that rare decays are signatures for grand unified supersymmetry and calculate the rates for $\mu^+ \to e^+\gamma$ and related processes for a wide range of parameters of these models. They conclude that $\mu^+ \to e^+\gamma$ has the largest rate by more than two orders of magnitude, and it ranges between the current experimental limit and 10^{-14}. Hence, there is continuing interest in the community for an experiment that could have a sensitivity near 10^{-14}.

The experimental signature for an at-rest $\mu^+ \to e^+\gamma$ is a 52.8-MeV positron that is back-to-back and in time coincidence with a 52.8-MeV photon. The MEGA experiment, designed to search for it, has been described several times (2). Briefly, it consists of a magnetic spectrometer for the positron and three pair spectrometers for the photon. The apparatus has been optimized for high rates and for good resolution to suppress backgrounds; the principal background is random coincidences. In high-rate experiments, any design that has sufficient resolution to suppress the random-coincidence background will easily eliminate any prompt background from $\mu^+ \to e^+\gamma\nu\nu$; in fact, it may be a problem to observe the $\mu^+ \to e^+\gamma\nu\nu$ process, which would provide nice confirmation of the proper operation of the detector, with the same experimental settings as used in the search. The experiences of the MEGA collaboration provide a useful guide in designing future experiments.

DESIGN CONSIDERATIONS

In preparing any design for a $\mu^+ \to e^+ \gamma$ search, one must maximize the sensitivity while minimizing the backgrounds and rate parameters. The sensitivity of an apparatus to $\mu^+ \to e^+ \gamma$ depends on the product of the solid angle and the rate:

$$S\,(90\%\ \text{C.L}) = 2.3/M, \tag{1}$$

where

$$M = (\Omega_0/4\pi) \bullet \varepsilon_\gamma \bullet \varepsilon_P \bullet E_c \bullet R_\mu \bullet T \tag{2}$$

and Ω_0 is the overlap solid angle, ε_γ is the gamma-ray detection efficiency, ε_P is the positron detection efficiency, E_c is the cut efficiency, R_μ is the average stop rate, and T is the live time.

The branching ratio where the accidental backgrounds set in is

$$B_{\text{acc}} = \left(R_\mu/d \bullet \Delta t\right) \bullet \Delta x \bullet (\Delta y/15)^2 \bullet (\Delta\theta)^2 \bullet f(\theta_\gamma) \bullet f(IB) \bullet f(P_\mu), \tag{3}$$

where d is the duty factor, Δt is the positron-photon time resolution, Δx and Δy are the fractional positron and photon resolutions, and $\Delta\theta$ is the positron-photon angular resolutions. Typically, Δ should be taken to be about ±1 full-width at half maximum (FWHM) of the detector response to a variable for clean background separation. The set of variables $\{\Delta t, \Delta x, \Delta y, \Delta\theta\}$ are the classic variables for identifying the $\mu^+ \to e^+ \gamma$ process. In addition, some configurations can get additional separation of signal from background by measuring the angle of the photons θ_γ, the low-energy positron associated with high-energy photons IB, or from the correlation of the positron or photon decay direction with the muon spin P_μ. The corresponding additional suppression factors f are indicated in Eq. (3). The validity of Eq. (3) depends on the photon spectrum being dominated by bremsstrahlung processes; care should be taken to evaluate sources of photons from positron annihilation in flight to see that they do not dominate. In MEGA, the central region of the detector was free enough of mass that the cross-over point was at $y = 0.99$ and the photon energy resolution Δy was 0.03 FWHM, so Eq. (3) was applicable.

Figure 1 gives representative plots for Eqs. (1)–(3) for the 90% confidence limits obtainable as a function of run time. The solid curve is for the MEGA detector at LAMPF with its 6% duty factor. The stopping rate is $5 \infty 10^8$ Hz, and the solid angle is greater than actually achieved. The curve shows the two components: the $1/T$ dependence in the background free region and the $1/\sqrt{T}$ in the background limiting region. The short dashed piece connecting the two sections is a likelihood representation of the transition between the two regions. Nevertheless, the point of intersection is roughly the practical experimental reach of this design. The long dashed curves illustrate one reason why the reach is not just linear in the duty factor, because they are all

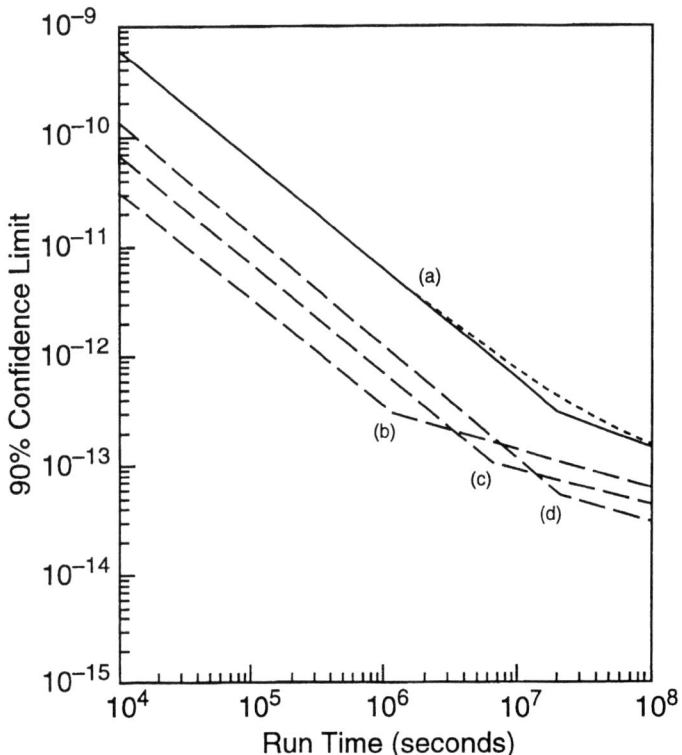

FIGURE 1. A typical characterization of a $\mu^+ \to e^+ \gamma$ design. The solid curve (a) is for the MEGA detector for a higher acceptance than actually achieved and an instantaneous rate of $5 \infty\ 10^8$ Hz. Curve (b) is for the same instantaneous rate and a duty factor of 1. Curves (c) and (d) are at instantaneous rates of $2.5 \infty\ 10^8$ Hz and $1.25 \infty\ 10^8$ Hz with a duty factor of 1.

calculated for a duty factor of 1.0. In curve (b), the instantaneous rate is the same as at LAMPF, and the result is the same except that it is achieved in a shorter clock time. As the rate is decreased, the ultimate limit gets better. In a practical sense, curve (d) is about as well as could be done with unit duty factor, i.e., a factor of 6 for a factor of 16 improvement in duty. Such curves are very useful in characterizing the capabilities of a design.

The rate parameters that need minimizing are

$$R_{E\gamma > \sigma\gamma} \cdot t_{\text{col-}\gamma} , \tag{4a}$$

$$R_{\gamma\text{-reg}} \bullet \tau_{\text{col-}\gamma}, \qquad (4b)$$

$$R_e \bullet \tau_{\text{col-}e}, \qquad (4c)$$

$$R_{\text{trig}} \bullet \tau_{\text{readout}}. \qquad (4d)$$

Each of these should be kept as small as possible compared to one for easy data analysis. Equation (4a) is the requirement that the product of the rate of low-energy photons times the collection time of the photons be small to avoid pileup. Equation (4b) states that the rate of high energy photons in a regions of the detector times the collection time of the photons needs to be small to have well separated events. Equation (4c) states that the electron rate times their collection time should be small for well-separated events. In MEGA, Eq. (4c) was not small, and a difficult pattern recognition problem had to be overcome. Equation (4d) is the usual readout dead-time effect.

Equation (4d) can really limit the ability to take data. It is not often appreciated that the MEGA experiment was 90% dead. Using electronic manipulations, this dead time was arranged to coincide with the periods between beam spills. The actual dead time for data during the beam was only about 6%. Trying to achieve the factor of 6 improvement at a continuous accelerator would require some new triggering scheme for reducing the data flow, and such a technique is not known because MEGA was near several bandwidth limits.

It is useful to catalog in one place the areas that must be studied for a successful $\mu^+ \to e^+\gamma$ design. Some have been discussed above, and some of the remainder will be considered below. The list includes (1) having high sensitivity, (2) being free from background, (3) being well-matched beam properties, (4) having a large product of decaying particles and solid angle, (5) having a high efficiency for detecting the signal, (6) achieving good detector rate capability, (7) achieving good detector resolutions, (8) having a plan to prevent the stopping positrons from making high energy photons, (9) finding a method to measure the resolutions, (10) measuring enough of the background to know its level in the signal region, (11) designing an adequate trigger, (12) measuring a known process like $\mu^+ \to e^+ \gamma \nu \nu$, (13) calibrating the elements, (14) stabilizing the detector elements against temporal drifts, (15) developing a good Monte Carlo simulation, and (16) controlling the cost.

Figure 2 is a display of the four classic variables on expanded scales from the MEGA data analysis and illustrates several points. The solid curves are from the data and the dotted curves are the response functions of the detector from a Monte Carlo simulation. A useful rule-of-thumb is that one should measure about 5 FWHM around the signal region. In the upper left panel, the positron spectrum shows the built-in calibration from the edge of the Michel spectrum. Also, the high-energy tail is small and of little consequence since high-energy positrons are highly probable. The plot of the photon energy shows a high-energy tail which would be quite damaging to the experiment if later stages of the analysis had not been successful in eliminating these poorly reconstructed events. The panel containing the relative time would be expected to have a flat distribution for random events; the distortions are due to the on-line trigger. Finally, the data have a relative angle that dies away at 180°. The important point is that only one of the panels contains a feature that is useful for calibration. In the case of the photon energy, MEGA resorted to measuring the decays of π^0's. For the relative timing, MEGA used low-rate $\mu^+ \to e^+ \gamma \nu \nu$ events. MEGA never found an experiment method for measuring the relative angle but relied on measuring quantities that checked the Monte Carlo simulation and used it to predict the response function.

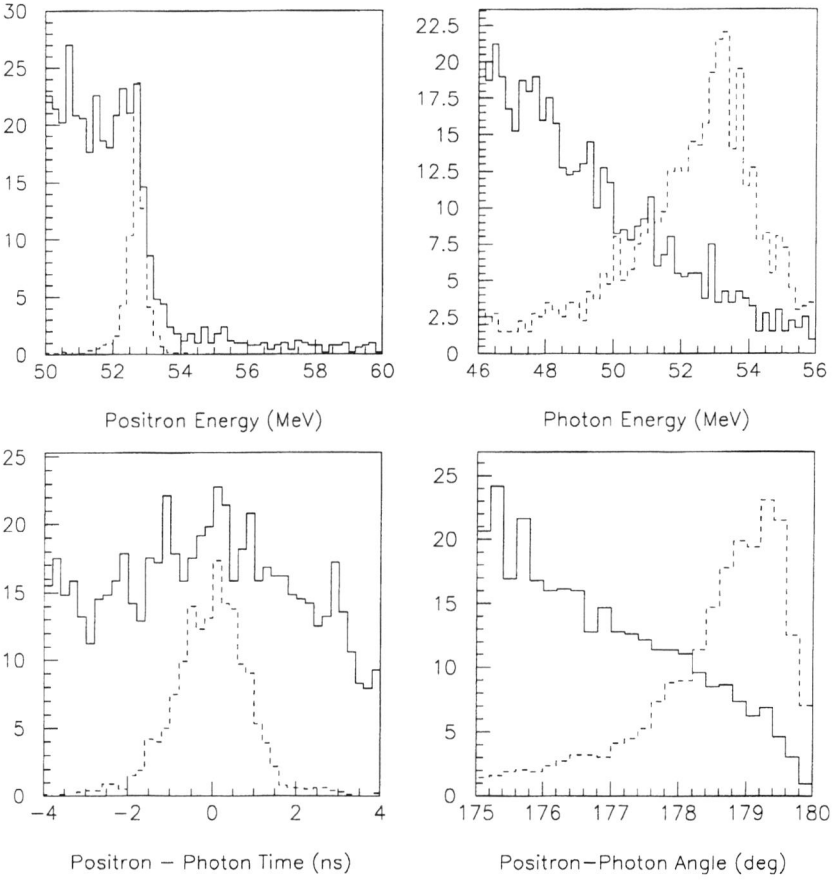

FIGURE 2. Solid curves are data for random events near the signal region for the positron and photon energies as well as the relative timing and angle. Dashed curves are the Monte Carlo simulated events for the $\mu^+ \to e^+ \gamma$ signal.

The primary beam conditions with stopping muons do not contain any sharp photon lines. In order to get a sharp photon line, negative pions are stopped in polyethylene. Some pions charge exchange and produce a slowly moving π^0 that, in turn, decays into two photons. By selecting the photons that happen to be nearly back-to-back, one gets a narrow line at 55 MeV from the low-energy photon, quite near any possible photon from $\mu^+ \to e^+ \gamma$. The spectrum of such events is shown in Fig. 3. The energy resolution is near that predicted (1.7 MeV FWHM).

The relative time resolution can be measured by looking for the allowed process $\mu^+ \to e^+ \gamma \nu \nu$. This internal bremsstrahlung correction to ordinary muon decay can only be seen easily at low rates where the random backgrounds are greatly reduced. The timing spectrum is shown in Fig. 4. Improvements in the calibration constants are

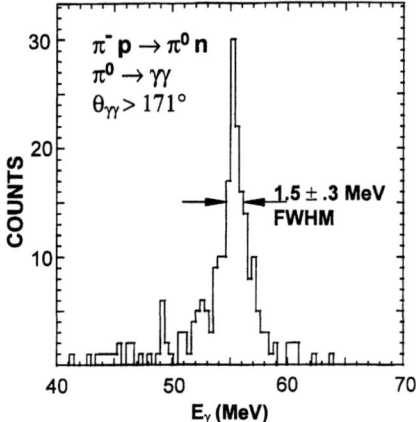

FIGURE 3. Photon energy response for 55-MeV gamma rays from π^0 decays.

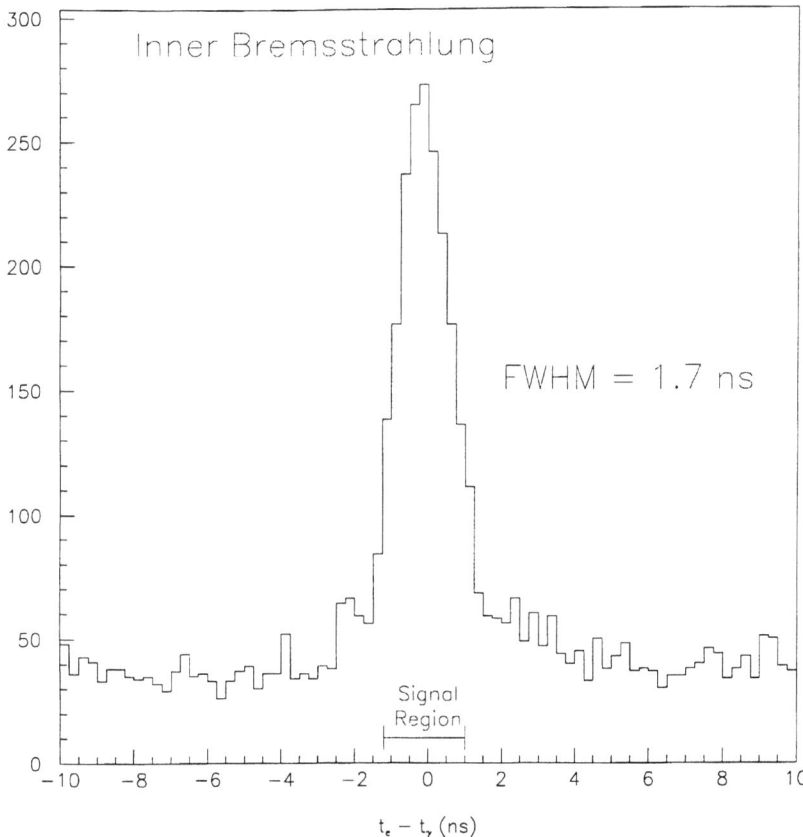

FIGURE 4. Positron-photon timing for the process $\mu^+ \to e^+ \gamma \nu \nu$ process at low rates.

expected to improve the timing to be nearly 1 ns FWHM. Observation of this decay is reassuring because it is the proof that the detector sees some events that it should. As new designs restrict the acceptance more in order to search with more sensitivity, observation of this peak will become increasingly more problematic because the branching ratio falls very rapidly as the energy thresholds on the two particles are increased; if it did not fall so rapidly, the prompt backgrounds would be a real problem. Finding this peak should be an early goal for any new experiments; for example, the data (~1%) from the 1992 engineering run had to be discarded because it was discovered that the peak was outside the coincidence window. Even getting the peak in the rough acceptance of the MEGA apparatus required the addition of additional scintillators whose sole purpose was to calibrate the positron and photon arms to one another and watch for temporal drifts.

The back-to-back angle is taken from the dot product of the positron direction after decay in the target and the vector that points from the electron's origin to the photon conversion point into charge particles. Hence, there are several contributions to this resolution: the resolution on the point of photon conversion, the knowledge of the positron's point of origin, and the uncertainty in the direction of the positron. In addition, in MEGA, where there was a magnetic field surrounding the decay point, the positrons propagate in circles that require a precise knowledge of the absolute location of the target to obtain the correct direction. Getting reliable measurements to test the simulation against is a challenge for each of these components. Even if all these resolutions are quite good, there are some decays that come nearly parallel to the target and cannot be included in the acceptance.

The other response functions beyond the classic four can also be quite useful. For example, in MEGA, the use of pair spectrometers allowed crude (~10°) reconstruction of the photon direction at the conversion point to see if the origin of the photon seemed to be the same as that of the positron. This knowledge completely eliminated certain backgrounds not associated with the target and suppressed about two thirds of the random coincidences from the target.

The conclusion of this section is that the design of a new experiment that has a chance of achieving its goals requires examining a substantial number of elements and proving that all the critical ones have been optimized. Such a design usually requires a careful simulation.

IDEAS FOR NEW EXPERIMENTS

In March of 1997, there was an informal workshop held at the Paul Scherrer Institute in Switzerland on "A New $\mu^+ \to e^+ \gamma$ Experiment." Many configurations were studied, but a final design is unsettled. In general, the problem is to keep the acceptance high while suppressing accidental coincidences. However, the conclusion of the workshop was that a 10^{-14} experiment looks feasible. However, no design was pushed sufficiently to be evaluated at the level suggested in the previous section.

A promising idea from the workshop was developed by A. Van der Schaaf for a magnetic positron detector that uses two bends. The idea is that the first bend limits the acceptance and thereby shields the active elements from the high rates near the target. Magnetic analysis with position sensitive detectors is done in the second bend to get the energy, position, and direction of the positron. If the optics is arranged correctly, the tracking to the target region should have adequate precision to maintain the back-to-back angle precision.

One interesting idea for suppressing accidental backgrounds has been developed for stopped, polarized muons (3). If the muons are polarized along the beam direction, then the angular distribution of the positrons and photons is given by

$\mu^+ \to e^+ \nu\nu$ $d\Gamma/d6_e \sim 1 + P_\mu \cdot k_e$ for $E_e \sim 53$ MeV,   (5a)

$\mu^+ \to e^+ \gamma\nu\nu$ $d\Gamma/d6_\gamma \sim 1 + P_\mu \cdot k_\gamma$ for $E_\gamma \sim 53$ MeV,   (5b)

$\mu^+ \to e^+ \gamma$ $d\Gamma/d6_e \sim$ unknown   for $E_e = 53$ MeV.   (5c)

As the angular correlation of the $\mu^+ \to e^+\gamma$ process is unknown, it is necessary to search in both the forward and backward hemispheres. At backward angles, either the high-energy positron or photon is suppressed. The suppression factor $f(P_\mu)$ in Eq. (3) can be large and is crudely $(1-\cos\theta)/(1+\cos\theta) \sim 0.05$ for $\theta \sim 25°$. To realize this factor, two back-to-back apparatuses are needed, one with the photon detector at back angles and the other with the electron detector at back angles. With a large solid angle detector, the suppression factor is considerably worse but still worth incorporating into a design.

To reach a sensitivity of 10^{-14}, a large solid-angle detector is needed for beam intensities of 10^8/s. However, for intensities of 10^{10}/s that might be available in association with the source of a muon collider, a small solid-angle detector would be practical. Additionally, if a technique can be developed that would make the muons polarized, the detector would be ideal for taking advantage of the suppression mechanism described above. One possibility would be to use a beam similar to that planned for MECO (4), though the muons would not be polarized; as all the beam potentially available is not required, perhaps a small fraction can be separated and polarized. The idea is based on the fact the result in Eq. (2) depends on the product of the solid angle and the rate. If the rate is as high as suggested above, then the solid angle can be small. Hence, small solid-angle, special-purpose spectrometers can be used that solve the problems of high singles rates and costs. In particular, of the four classic resolutions, this geometry looks to make major improvements in the positron resolution to stay background free. The specification for this idea were given previously by M. Cooper (5). The sensitivity is estimated to be 10^{-14}, and the result would be free of background. If the suppression factor from using polarized muons were practical, then even better limits could be achieved. A real simulation of this geometry remains a necessity.

SUMMARY

The MEGA detector has provided much useful information on the experimental difficulties involved in very sensitive searches for the process $\mu^+ \to e^+\gamma$. Many design considerations must be optimized simultaneously. Any new design will have to have some of its specifications significantly better than MEGA if a sensitivity of 10^{-14} is to be achieved. The intense beams associated with the source of a muon collider open the possibilities for new designs with this potential reach.

REFERENCES

1. Barbieri, R., Hall, L., and Strumia, A., *Nucl. Phys.* **B449**, 437 (1995).
2. Hogan, G. E., et al., International Conference on High Energy Physics, Warsaw (1996); Los Alamos National Laboratory document LA-UR-96-3749.
3. Kuno, Y., et al., *Phys. Rev.* **D55**, 2517 (1997); Kuno, Y., and Okada, Y., *Phys. Rev. Lett.* **77**, 434 (1996).

4. Bachmen, M., et al., University of California at Irvine Phys. Tech. Report 96-30.
5. Cooper, M., Fourth KEK Topical Conference on Flavor Physics, Tsukuba, Japan, 1996, eds. Kuno, Y., and Nojiri, M. M., *Nucl. Phys.* **B59**, 209 (1997).

Model for Pion Production in Proton-Nucleus Interactions

N. V. Mokhov and S. I. Striganov[+]

Fermi National Accelerator Laboratory, Batavia, IL 60510[1]
[+] *Now at Institute for High-Energy Physics, Protvino, Moscow region, Russia*[2]

Abstract. A new phenomenological model has been developed to describe pion production in high-energy proton-nucleus interactions. Special attention is paid to low-momentum pions (0.1< p <2 GeV/c) for intermediate proton momenta 5< p_0 <30 GeV/c. It is shown that the model predictions are in an excellent agreement with data in the entire kinematic region. Comparisons to other models are also presented. The model is embedded into the MARS13 code.

INTRODUCTION

Reliable prediction of pion yield in hadron-nucleus (hA) collisions is vital in numerous applications, particularly in the planning of future experiments and accelerators. The newest examples include a $\mu^+\mu^-$ collider project [1] and neutrino experiments at Fermilab Main Injector [2] and Booster [3]. There are a few models capable of generating pions in $pA \to \pi^\pm X$ reactions, e. g., [4-9]. Theoretical calculations based on the intranuclear cascade model are reliable at proton momenta p_0 <5 GeV/c, but drastically overestimate hadron yield at higher energies. Microscopic models, such as DPMJET [5] (based on the dual topological unitarization approach) and FRITIOF [6] (based on the LUND model) were developed mainly for high energies $\gtrsim$50 GeV/c. As it is shown in [1], there is an uncertainty up to a factor of 5 in the pion yield at p <1 GeV/c on heavy nuclei for proton momenta 5< p_0 <30 GeV/c – the region that is especially interesting for the $\mu^+\mu^-$ collider project. On the other hand, there are many data on inclusive charged pion production in hA collisions obtained over the last three decades. Based on those data and our original model [4,9], we develop a phenomenological model for a reliable description of inclusive pion production in the entire kinematic range for pA collisions at 5 GeV/c< p_0 <10 TeV/c.

---

[1] Work supported by the Universities Research Association, Inc., under contract DE-AC02-76CH00300 with the U. S. Department of Energy.
[2] Supported by Russian Foundation for Basic Research, under contract RFBR-96-07-89230.

PHENOMENOLOGICAL MODEL

Many reliable data and parameterizations exist on pion yield in pp-collisions. We can compensate for the lack of data for pA reactions by using the following form (see, e. g., [4,9]) for the double differential cross section of the $pA \to \pi^{\pm} X$ reaction:

$$\frac{d^2\sigma^{pA\to\pi^{\pm}X}}{dpd\Omega} = R^{pA\to\pi^{\pm}X}(A,p_0,p,p_\perp)\frac{d^2\sigma^{pp\to\pi^{\pm}X}}{dpd\Omega}, \quad (1)$$

where p and $p_\perp$ are the total and transverse momenta of $\pi^{\pm}$, and A is an atomic mass of the target nucleus. The function $R^{pA\to\pi^{\pm}X}$, measured with much higher precision than the absolute yields, is almost independent of $p_\perp$ and its dependence on p_0 and p is much weaker than for the differential cross-section itself. Because of rather different properties of pion production on nuclei in the forward ($x_F \gtrsim 0$) and backward ($x_F \lesssim 0$) hemispheres, where x_F is the Feynman's longitudinal variable, we treat these two regions differently.

R at $x_F \gtrsim 0.05$. In this region we assume $R^{pA\to\pi^{\pm}X} \sim A^\alpha$. The power α is almost independent of the pion sign. The following parameterization was proposed in [10] for $p_0 \geq 70$ GeV/c:

$$\alpha_g = 0.8 - 0.75 \cdot x_F + 0.45 \cdot x_F^3/|x_F| + 0.1 \cdot p_\perp^2. \quad (2)$$

Fig. 1(a) shows our compilation of data [11–15] on α for π^--production. It turns out that (2) describes data very well at $p_0 \geq 24$ GeV/c and can be successfully used at lower momenta ($5 \leq p_0 \leq 24$ GeV/c) if it is replaced with (see Fig. 1(a)):

$$\alpha = \alpha_g - 0.0087 \cdot (24 - p_0). \quad (3)$$

The $R^{pA\to\pi^{\pm}X} \sim A^\alpha$ form doesn't extrapolate well to $A=1$ because of the difference in the π-yield in proton-proton and proton-neutron collisions. This difference can be taken into account if one uses the following form for $R^{pA\to\pi^{\pm}X}$ [10]:

$$R^{pA\to\pi^{\pm}X} = \left(\frac{A}{2}\right)^\alpha \cdot f(p_0,Y), \quad (4)$$

where $f(p_0,Y) = \frac{d\sigma}{dp}(pd \to \pi^{\pm})/\frac{d\sigma}{dp}(pp \to \pi^{\pm})$. It turns out that pion yields in pd and pp collisions are not very different, i. e. $f(p_0,Y) \approx 1$. Using FRITIOF results, we found that $f(p_0,Y)_{\pi^-} = 1 + 0.225/N_{\pi^-} - a_{\pi^-} \cdot Y_{cms}$, where N_{π^-} is mean π^- multiplicity in pp collisions and Y_{cms} is pion rapidity in the center-of-mass system (CMS). Data [16] show linear dependence of N_{π^-} on *free energy* $W = \frac{(\sqrt{s}-2\cdot m_p)^{0.75}}{s^{0.25}}$, where $\sqrt{s}$ is the CMS collision energy. Our fit to the data gives $N_{\pi^-} = 0.81 \cdot (W - 0.6)$. The other parameter $a_{\pi^-} = 0.16$ for $p_0 \leq 20$ GeV/c, and depends on energy for higher momenta as $a_{\pi^-} = -0.055 + 0.747/log(s)$. $f(p_0,Y)_{\pi^-}$ is forced to be 1 if it becomes less than 1. For π^+ production the approximation is much simpler $f(p_0,Y)_{\pi^+} = 0.85 + 0.005 \cdot p_0$ for $p_0 \leq 30$ GeV/c and $f(p_0,Y)_{\pi^+} = 1$ for higher momenta.

R at $x_F \lesssim 0.05$. In this region, due to the lack of experimental data on α, we use the following expression for the function R in (1):

$$R^{pA \to \pi^{\pm} X} = \frac{dN/dY(pA)}{dN/dY(pp)} \quad (5)$$

The following scaling law was proposed in [17] for charged shower particle ($\beta > 0.7$) production in pA collisions at 20 - 400 GeV/c:

$$\frac{Y_0}{<N_s>} \cdot \frac{dN}{d\eta} = f(A, \frac{\eta}{Y_0}), \quad (6)$$

where $<N_s>$ is a mean multiplicity of shower particles, Y_0 is rapidity of primary proton and $\eta = -log(tan(\theta/2))$ is pseudorapidity of a secondary particle. We found that this approximation is in a reasonable agreement with data at $p_0 > 7.5$ GeV/c [18]. Unfortunately, η is not a convenient variable to describe forward pion production ($\theta \approx 0$). Our analysis of the $pA \to \pi^- X$ data [19,20] at $10 < p_0 < 100$ GeV/c shows that replacing η in (6) with rapidity Y

$$\frac{dN}{dY} = \frac{<N_\pi>}{Y_0} \cdot F(A, \frac{Y}{Y_0}) \quad (7)$$

provides better description of the pion yield in the entire kinematic range. Here $<N_\pi>$ is mean pion multiplicity, $Y_0 = log(\frac{E_0 + p_0}{m_p})$ is rapidity of incident proton and $Y = log(\frac{E_\pi + p_z}{m_\perp})$ is π rapidity, $m_\perp = \sqrt{p_\perp^2 + m_\pi^2}$. We choose the Gaussian form for the scaling function:

$$F(A, \frac{Y}{Y_0}) = c_1 \cdot exp(-(\frac{Y}{Y_0} - c_2)^2/c_3), \quad (8)$$

where for π^-: $c_1 = 1.149 \cdot A^{0.0479}$, $c_2 = 0.492 \cdot A^{-0.0565}$, and $c_3 = 0.214 \cdot A^{-0.121}$. Reliable rapidity distributions for π^+ at $x_F < 0$ are measured only for $p_0 \geq 100$ GeV/c. Assuming that the scaling (7)-(8) is valid for π^+ also, we found the following parameters from data [20]: $c_1 = 1.6$, $c_2 = 0.521 \cdot A^{-0.0416}$, and $c_3 = 0.12$. The data on dN/dY for $pp \to \pi^{\pm} X$ reaction is well described by a Gaussian [21]: $dN/dY = C_{pp} \cdot exp(-Y_{cm}^2/2\sigma^2)$, where $\sigma_{\pi^+} = 0.402 + 0.198 \cdot log(p_0)$ and $\sigma_{\pi^-} = 0.465 + 0.157 \cdot log(p_0)$. The normalization parameter in (5), combined of $<N_\pi>$, C_{pp} etc, is chosen to match the functions (4) and (5) at $x_F = 0.05$.

$pp \to \pi^{\pm} X$. To describe the invariant cross section of charged pion production in pp collisions we use the form proposed in [22] that we modified at low [13,23] and high [4] $p_\perp$:

$$E\frac{d^3\sigma^{pp \to \pi^{\pm} X}}{dp^3} = A(1 - \frac{p^*}{p^*_{max}})^B exp(-\frac{p^*}{C\sqrt{s}})V_1(p_\perp)V_2(p_\perp), \quad (9)$$

TABLE 1. Parameters in formula (9).

	A	B	C	D	E	F
π^+	60.1	1.9	0.18	0.3	12	2.7
π^-	51.2	2.6	0.17	0.3	12	2.7

where p^* and p^*_{max} are pion momentum and maximum momentum transfer in CMS and parameters are given in Table 1. The best description of the $p_\perp$ dependence is obtained with:

$$V_1(p_\perp) = \begin{cases} (1-D)exp(-Ep_\perp^2) + Dexp(-Fp_\perp^2), & p_\perp \leq 0.933 \text{ GeV/c}, \\ 0.2625/(p_\perp^2 + 0.87)^4, & p_\perp > 0.933 \text{ GeV/c}, \end{cases}$$

$$V_2(p_\perp) = \begin{cases} 0.7363\, exp(0.875 p_\perp), & p_\perp \leq 0.35 \text{ GeV/c}, \\ 1, & p_\perp > 0.35 \text{ GeV/c}. \end{cases}$$

COMPARISON TO DATA

The model developed agrees very well with available data and reliable DPMJET-II [5] predictions at proton momenta 50 GeV/c$< p_0 <$10 TeV/c (as used in the original model [4,9]). In this section we compare pion yield predicted by the new model and other models with data available at lower proton momenta down to $p_0 \approx$5 GeV/c on thin and thick nuclear targets. Fig. 1(b) shows comparison of calculated π^- rapidity distributions in pC and pTa collisions at p_0=10 GeV/c with

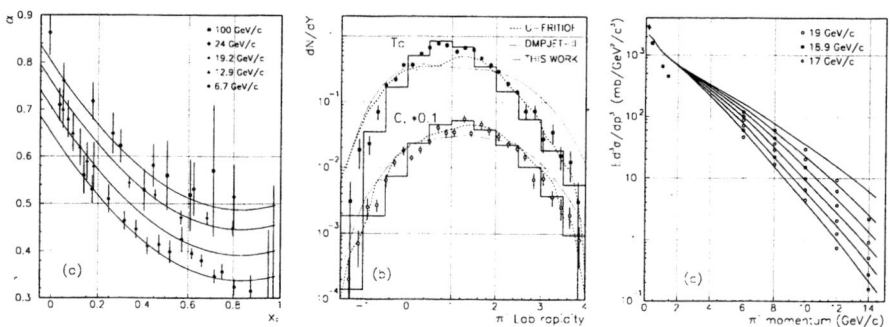

FIGURE 1. (a) Parameter α (2)-(3) calculated for p_0=6.7, 12.9, 19.2 and >24 GeV/c (from bottom up) in comparison with data [11-15]; (b) rapidity distributions of π^- in pC and pTa interactions at 10 GeV/c as calculated with FRITIOF, DPMJET and the model developed in this paper in comparison with data [19]; (c) $pCu \to \pi^- X$ at p_0=17 - 19 GeV/c for θ=12.5, 30, 40, 50, 60 and 70 mrad (from top down), data from [13,25,26].

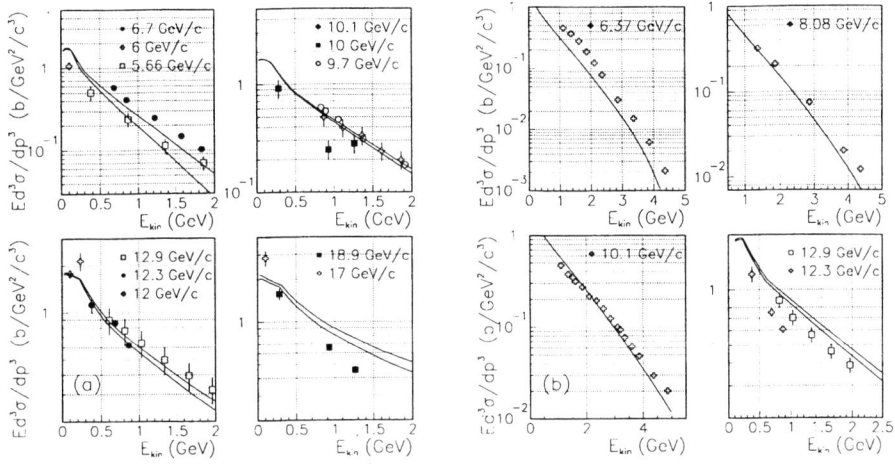

FIGURE 2. (a) $pCu \to \pi^- X$ ($\theta \approx 0$), data from [11,12,24–28]; (b) $pCu \to \pi^+ X$ ($\theta \approx 0$), data from [12,27–29]. Two curves presented in some plots correspond to higher (top) and lower (bottom) p_0 shown in those plots.

data [19]. One sees that our model gives much better results than the DPMJET and FRITIOF codes for soft pions at $0 < Y < 2$. Production of energetic pions is also nicely described by our model (Fig. 1(b),(c)). Fig. 2 shows that the developed model gives a reliable description of pions generated at $\theta \approx 0$ in the very 'difficult' region of intermediate momenta, $5 < p_0 < 19$ GeV/c, for other models.

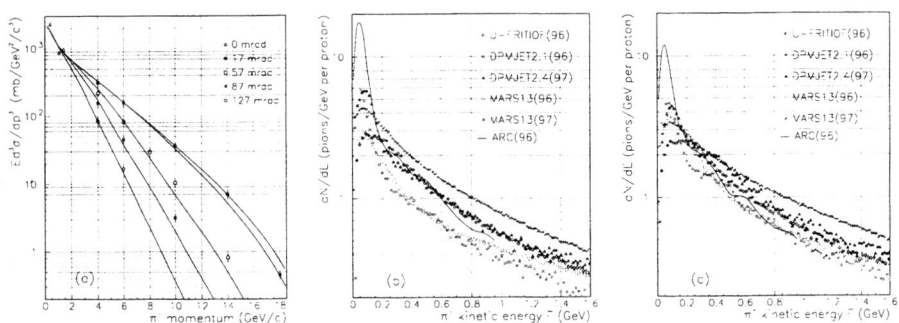

FIGURE 3. Pion spectra at $p_0 = 24$ GeV/c: (a) $pCu \to \pi^- X$ at 5 angles, data from [14,26]; (b) $pHg \to \pi^- X$ and (c) $pHg \to \pi^+ X$ according to several codes (MARS13(97) with the new model).

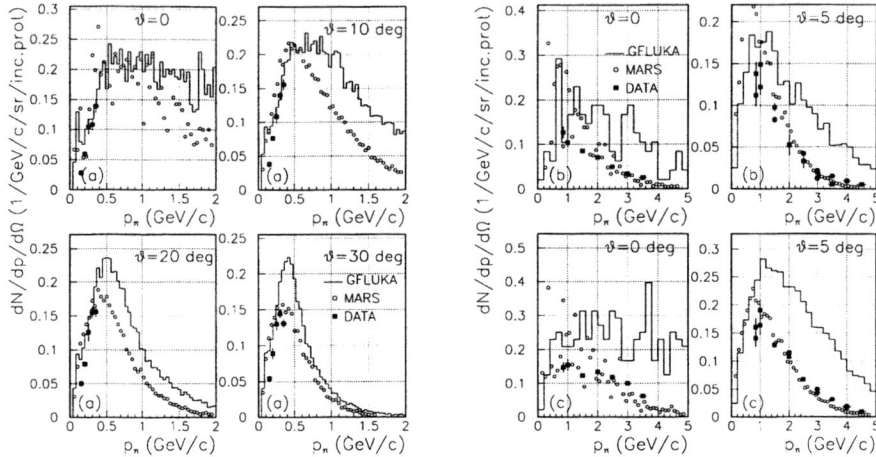

FIGURE 4. Double differential spectra from thick targets as calculated with GEANT-FLUKA and MARS13(97) and measured: (a) π^- from 6-in copper target at p_0=6 GeV/c, data [25]; (b) π^- and (c) π^+ from 10-cm lead target at p_0=8 GeV/c, data [30].

The 'mystery' with the soft pion production began with the analysis of the pHg reaction at 24 GeV/c [1]. As Fig. 3(a) shows, the model developed gives a good agreement with data (calculations for nuclei other than copper nuclei at this energy are not shown but agree very well with data), and gives guidance on the other codes in this considered case (see Fig. 3(b),(c)). Calculations with the MARS13(97) code (with the new model) of the pion double differential spectra with cascading in the thick copper and lead targets at p_0=6 and 8 GeV/c agree nicely with data [25,30] in the momentum region 0.1< p <5 GeV/c (so crucial for $\mu^+\mu^-$ collider applications) (see Fig. 4). At the same time, GEANT, even in the most appropriate FLUKA mode, certainly has some problem here.

CONCLUSION

This model development began while one of us (NM) pointed out that there is a great degree of uncertainty in central ($x_F \approx 0$) pion production on medium and heavy nuclei in the medium proton momentum range 5< p_0 <30 GeV/c [1]. Successful benchmarking, performed with the MARS13(97) code with the new model embedded, against data in the wide kinematic range for nuclei ranging from hydrogen to lead, assure that we have now a tool for reliable prediction of pion yield for the proton momentum range estimated as 5 GeV/c< p_0 <10 TeV/c.

REFERENCES

1. $\mu^+\mu^-$ Collider: A Feasibility Study, The $\mu^+\mu^-$Collider Collaboration, BNL–52503; Fermilab–Conf–96/092; LBNL–38946, July 1996.
2. http://www-numi.fnal.gov:8875/.
3. Church, E., et al., 'A proposal for BooNE Neutrino Experiment at the Fermilab Booster', Fermilab-P-898 (1997).
4. Kalinovskii, A. N., Mokhov, N. V., and Nikitin, Yu. P., 'Passage of High-Energy Particles through Matter', AIP, New York (1989).
5. Ranft, J., Phys. Rev., **D51**, p. 64 (1995); Gran Sasso report INFN/AE-97/45 (1997).
6. Uzhinskii, V.V., JINR-E2-96-192, Dubna (1996)
7. Mokhov, N. V., 'The MARS code system Users Guide, version 13(95)', Fermilab–FN–628 (1995).
8. Kahana, D., and Torun, Y., BNL-61983, Brookhaven (1995).
9. Mokhov, N. V., Striganov, S. I., and Uzunian, A. V., IHEP-87-59, Serpukhov (1987).
10. Geist, W. M., Nucl. Phys., **A525**, p. 149 (1991).
11. Bayukov, Y. D., et al., Yad. Fiz., **29**, p. 947 (1979).
12. Yamamoto, A., KEK 81-13, Tsukuba (1981)
13. Allaby, J. V., et al., CERN 70-12 (1970).
14. Eichten, T., et al., Nucl. Phys., **B44**, p. 333 (1972).
15. Barton, D. S., et al., Phys.Rev., **D27**, p. 2580 (1983).
16. Gazdzicki, M., and Rohrich, D., Z. Phys., **C65**, p. 215 (1995).
17. Stenlund, E., and Otterlund, I., CERN-EP/82-42 (1982).
18. Bayukov, Y. D., et al., Yad. Fiz., **42**, p. 1414 (1985).
19. Armutliysky, D., et al., Yad. Fiz., **48**, p. 161 (1988).
20. Whitmore, J. J., et al., Z. Phys., **C62**, p. 199 (1994).
21. Uvarov, V. A., and Shlyapnikov, P. V., Sov. J. Nucl. Phys., **50**, p. 1048 (1982).
22. Folomeshkin, V. N., IHEP-71-22, Serpukhov (1971).
23. Smith, D. B., et al., Phys. Rev. Lett., **23**, p. 1064 (1969).
24. Papp, J., LBL-3633, Berkeley (1975).
25. Berley, D., et al., IEEE Trans. Nucl. Sc., **23**, p. 997 (1973).
26. Amman, J.F., et al., LA-9486-MS, Los Alamos (1982).
27. Vorontsov, I. A., et al., ITEP-11, Moscow (1988).
28. Marmer, G. J., et al., Phys. Rev., **179**, p. 1294 (1969).
29. Arefiev, A. S., et al., ITEP-25, Moscow (1985).
30. Audus, M. F. et al., Nuovo Cimento, **A46**, p. 502 (1966)

The MECO Muon Beam

Mark Bachman*

Department of Physics and Astronomy
University of California at Irvine
Irvine, CA 92697-4575
*Representing the MECO collaboration.

Abstract. The muon beam required by MECO, a muon conversion experiment, has been studied and simulated using GEANT3. The beam selects low energy, negative muons sufficient to perform a measurement of muon conversion to a level of 10^{-16} by relying on a graded magnetic solenoidal field and a curved transport solenoid equipped with collimators. Some details of the beam and simulation are discussed.

INTRODUCTION

The MECO experiment (Muon to Electron COnversion) proposes to measure muon conversion to a level of 10^{-16} performing the most stringent test on lepton flavor violation ever performed [1]. Muon conversion, where a muon (through interaction with a nucleus) converts directly into an electron, is predicted by several extensions to the Standard Model and forms a stringent test of supersymmetry [2-5].

To perform such a search, MECO requires an intense source ($\sim 10^{11}$/sec) of low energy, negative muons to stop on a conversion target. To achieve this, MECO proposes to use two unique solenoids. The first, called the "production solenoid" employs a 3.3 T field which adiabatically reduces to a 2.0 T field throughout its volume. An 8 GeV proton beam, which enters from the low field side, hits a tungsten target producing pions which then decay into muons. The muons (and pions) spiral in the magnetic field. Those particles continuing in the same general direction as the incident proton face increasing magnetic field, causing them to tighten their spirals and eventually reverse their drift trajectory. Thus the produiction solenoid forms a "half-magnetic bottle" which bounces the forward moving particles back.

Following the production solenoid at the low field end is a long 13.1 m transport solenoid which curves in two 90° bends. This solenoid delivers the muons to the conversion target and following detectors. The curved sections of the solenoid produce two toroids which cause charged particles to drift in a direction perpendicular to the plane of the curve, and proportional to the sign of their charge. The transport solenoid is equipped with collimators which absorb particles that drift in the wrong direction (positives) or particles with too large a helical radius (large momentum particles). The resulting beam consists of low momentum, negatively charged electrons and muons. In addition to having low momentum, the electrons arrive quickly at the conversion target, while the muons are delayed by several hundred nanoseconds. Thus, the low momentum muons are time separated and a clean beam is accomplished.

To study the MECO muon beam, simulations were performed using GEANT3 with realistic magnetic fields and all physics processes included. Results presented are from those studies.

PRODUCTION DETAILS

The layout of the MECO apparatus is described in other articles from these proceedings. The production region and beamline are shown in figure 1, along with many sample events. The nominal design calls for $\sim 10^{13}$ 8 GeV protons per second to be directed at 10° on a tungsten production tasrget which sits in the center of the production solenoid. The target is radiatively cooled and should maintain a temperature of 2430 K. The solenoid is built from a copper and tungsten cylinder, 30 cm inner radius, with various configurations for outer radius ranging from 60 cm to 75 cm. These configurations keep the heat load on the 6 cm aluminum superconducting coils to under 30 Watts.

Figure 2 shows how the magnetic field varies in the production solenoid. In general, for a graded magnetic solenoid,

$$\frac{P_T^2}{B} \approx \text{constant},$$

where P_T is the component of momentum perpendicular to the field. As the field increases, the P_T increases and the longitudinal component (P_L) necessarily decreases until it reaches zero and the trajectory can continue no further into the increasing magnetic field. At this point, the trajectory reverses. This magnetic bottle idea is used to capture the pions and muons which are produced in the direction away from the transport solenoid, thus increasing the acceptance for muons. This technique also allows the proton beam to be directed in a direction *away* from the collection region, thus minimizing beam contamination from unwanted high energy particles.

Only muons of 50 MeV/c or less are likely to be absorbed in the conversion target, so it is these muons which are of interest. High energy electrons (above 100 MeV) which exit the transport are also of interest, as they can mimic the signal from a muon conversion. Figure 3 shows the distribution of muons and pions which are created in the production solenoid, and also the distribution of those which successfully exit the production solenoid. Most low energy muons come from low energy pions, many of which are absorbed in the tungsten target before they have a chance to exit.

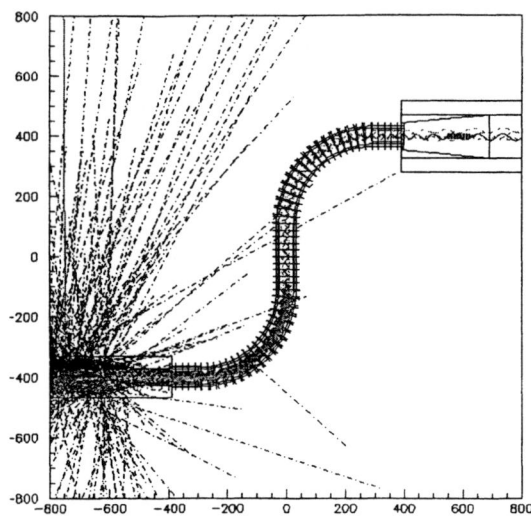

FIGURE 1. Diagram of production and transport solenoids showing typical events. The effects of the collimators is clearly seen. The proton beam enters the production solenoid from the right.] Dashed lines are neutrinos.

An accurate estimate of low energy pion yield (and thus, muon yield) is difficult to establish owing to the lack of data for low energy of pion production off of high energy beams. Production models are unreliable and vary considerably in this region, as shown in figure 4. For our simulations, we assumed the GHEISHA model which is the default production model used by GEANT3.

TRANSPORT DETAILS

The transport solenoid is built from 54 coil packs which are pieced together to form a "snake" solenoid maintaining a magnetic field of 2 T throughout. The packs have coils at 35 cm, and are made of copper rings, nominal inner radius 30 cm, outer radius 45 cm. Particles which enter the transport solenoid are directed by the magnetic field along the length of the solenoid until they reach the conversion target at the end. Along the path, however, are collimators designed to absorb trajectories with too large a radius (thus, large momenta) or particles with positive charge.

Collimators at the beginning and end of the transport region are 1 meter long and have inner radii of 20 cm. These absorb high momentum particles and collimate the beam for transport.

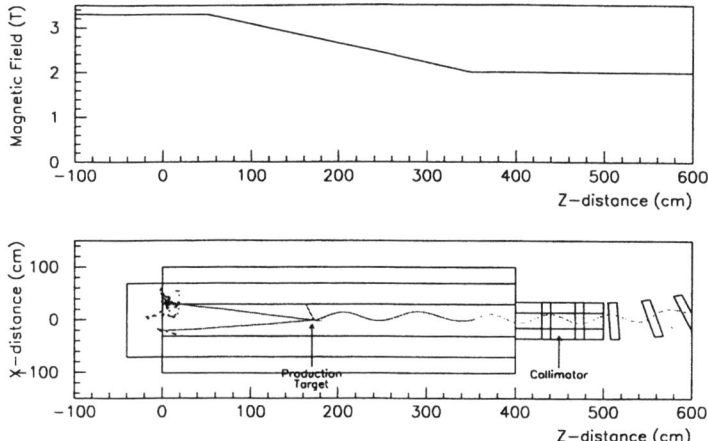

FIGURE 2. Schematic drawing of the production solenoid with a proton interaction producing a muon superimposed. The incident proton beam enters from the right. Above the drawing is a plot of the axial component of the magnetic field in this region as a function of z. The horizontal scales of the two drawings are the same.

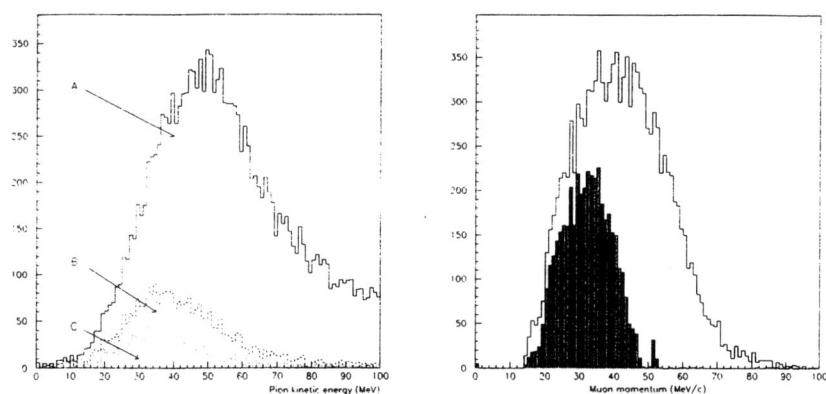

FIGURE 3. Left: Energy distributions for pions which create muons for (a) all muons, (b) muons exiting the transport solenoid, (c) muons with energy below 50 MeV/c. Right: Momentum distribution for muons exiting collimator and for muons which are stopped by the conversion target.

Charged particles moving in a toroidal field will drift in a direction perpendicular

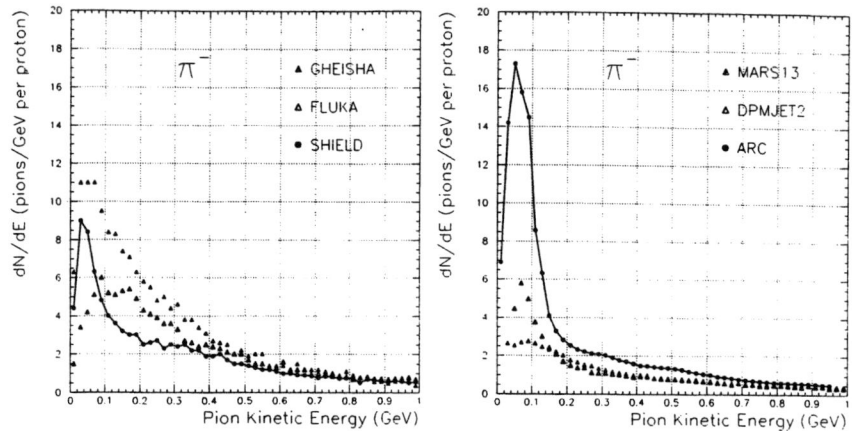

FIGURE 4. Predicted π yields for 24 GeV protons on mercury from six hadronic codes: GHEISHA, FLUKA, SHIELD (left) and MARS, DPMJET2, ARC (right). At low kinetic energies (of most interest to MECO), the predictions vary dramatically. Similar differences exist for 8 GeV protons on tungsten.

to the bend plane according to the formula,

$$D = \frac{Q}{0.3B} \frac{S}{R} \frac{(P_L^2 + \frac{1}{2}P_T^2)}{P_L},$$

where D is the total drift displacement (m), Q is the particle charge (in units of e), R is the bend radius of the toroid, B is the magnetic field strength (T), S is the distance traversed along the field, P_L is the component of particle momentum along the field (GeV/c), P_T is the component of momentum perpendicular to the field.

Positive particles which enter the transport solenoid drift in the positive vertical direction, while negative particle drift down. The central collimator is 2.18 meters long and has asymmetric absorbers placed at +5 cm and −19 cm to create a small window through which particles may pass. Positively charged particles and high momentum electrons have a small probability of passing this collimator, whereas negative, low momentum particles (electrons and muons) have higher acceptance.

CONCLUSIONS

The muon beam for the MECO experiment is created using a graded solenoidal field coupled with a curved solenoid. This configuration allows MECO to collect a high yield of muons, then collimate the unwanted particles out of the beam. The resulting yield is approximately 1.2% muons per incident proton, and about 0.44%

muons stopped in the conversion target per proton. For 2×10^{13} protons/sec, this corresponds to a muons stopping rate of 0.9×10^{11} muons/sec. To the limit of our simulation statistics (10^{-7}), no electrons with energy above 100 MeV pass the collimators. This, combined with experimental acceptance, allows one to perform a search for muon conversion at the level of 4.6 events for a branching fraction of 10^{-16}. Uncertainties in these calculations arise due to the lack of reliable production models. However, even if one scales back these results to the model with the lowest yield, the resultant yields are sufficiently high to perform the experiment with the desired sensitivity of 10^{-16}.

REFERENCES

1. Bachman, *et al.*, "A Search for $\mu^- N \to e^- N$ with Sensitivity Below 10^{-16}", Proposal to BNL, E940, October 1997.
2. Barbieri, *et al.*, "Violations of Lepton Flavour and CP in Supersymmetric Unified Theories", IFUP-TH 72/94, January 1995.
3. Arkani-Hamed, *et al.*, "Flavor Mixing Signals for Realistic Supersymmetric Unification", LBL-37343 (1996).
4. Barbieri and Hall, "A Grand Unified Supersymmetric Theory of Flavor", LBL-38381 (1996).
5. Hisano *et al.*, "Exact Event Rates of Lepton Flavor Violating Processes in Supersymmetric SU(5) Model", LBL-38653 (1996).

$\mu \to e$ Conversion
Status and Prospects

John Sculli*

*New York University

INTRODUCTION

When a negatively charged muon comes to rest in matter, it is captured in an atomic orbit, cascades down to the 1s state and disappears, either by decaying to an electron and two neutrinos or by muon capture on the nucleus with the emission of a neutrino. The search for the neutrinoless process,

$$\mu^- + (Z, A) \longrightarrow e^- + (Z, A), \tag{1}$$

began in the early fifties, not long after a decade (1937-1947) of cosmic ray experiments had established the existence of the pion and muon. The first accelerator search, that of Steinberger and Wolfe [1], took place in 1955. The phenomenology of the $\mu \to e$ transition, for the reaction in equation 1 and for the decays $\mu \to e + \gamma$ and $\mu \to e + e + e$, was developed in 1959 by Weinberg and Feinberg [2] and, independently, by Cabbibo and Gatto [3]. Feinberg, a year earlier, had estimated the rate for $\mu \to e + \gamma$ decay at 10^{-4} that of the decay $\mu \to e + \nu + \bar{\nu}$, assuming the decays were mediated by an intermediate boson. In the 1959 paper, Weinberg and Feinberg began with the observation:

> *The existence of the ordinary μ decay, $\mu \to e + \nu + \bar{\nu}$, seems to prove that the muon and the electron do not differ in any quantum numbers.*

There were no subscripts on the neutrinos in 1959, three years before the two neutrino experiment of Danby et al. [4]. It is now known that there are three families of leptons and quarks, and the conservation of lepton family number is empirically well established. In the Standard Model, this conservation law is not associated with a global symmetry. Fig. 1 shows one Feynman graph, written down originally by Feinberg, that would produce the $\mu \to e$ transition in the Standard Model and the corresponding graph in the quark sector responsible for the transition, $b \to s + \gamma$, which occurs through quark mixing with a BR of 4×10^{-5}. The smallness, or absence, of such mixing in the lepton sector and the electron volt, or less, mass of the neutrino effectively forbid the $\mu \to e$ transition in the Standard Model.

This is no longer so in extensions of the Standard Model. In supersymmetric grand unified theories with weak-scale symmetry breaking, these lepton number changing processes are expected. Very roughly, supersymmetry provides the massive particles, the scalar partners of the leptons, to mediate the leptonic reaction in fig. 1, unification, which puts leptons and quarks in the same multiplets, forces the mixing, and weak-scale breaking makes the rate observable, suppressed by powers of order $(G_F)^{1/2}$, rather than the inverse of the mass scale that characterizes the grand unified theory [5]. The calculated ratio

$$R = \frac{\mu^- + (Z,A) \to e^- + (Z,A)}{\mu^- + (Z,A) \to \nu + (Z-1,A)} \quad (2)$$

is model dependent– it varies with $\tan \beta$, the ratio of the vacuum expectation values of the two Higgs doublets, the masses of the scalar leptons and other parameters as well– but falls generally in the range 10^{-14} to 10^{-17}. Remarkably, these ratios appear to be within the reach of current and planned experiments, as we now discuss.

EXPERIMENTAL STATUS

Fig. 2 shows the fifty year history of accelerator searches for $\mu \to e$ conversion, beginning with the 1955 experiment of Steinberger and Wolfe ($R < 5 \times 10^{-4}$) and including the most recent results from TRIUMF ($R < 4.6 \times 10^{-12}$ 1988) and the PSI ($R < 4.3 \times 10^{-12}$ 1993, and $R < 8.4 \times 10^{-14}$ 1997). Shown also are the limits expected from experiments planned for the PSI and BNL.

The features of the most recent experiments are listed in table 1. The backgrounds for this measurement come principally from four sources: muon decay in orbit, radiative muon capture, prompt processes coincident, or nearly so, with the muon stop, and cosmic rays. A high resolution momentum measurement is required to reject the highest energy electrons coming from muon decay in orbit and

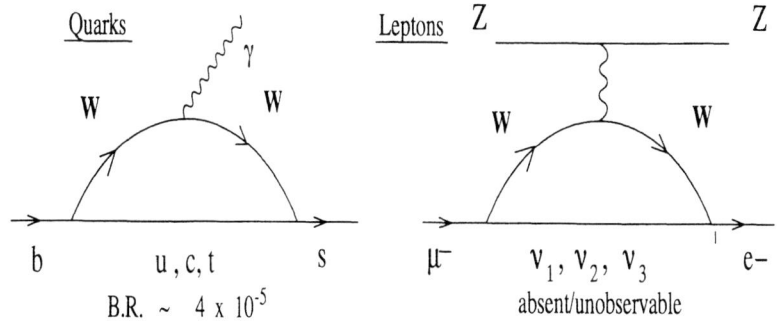

FIGURE 1. Feynman diagrams for $b \to s + \gamma$ and $\mu \to e$ conversion.

radiative muon capture. Prompt background comes mostly from pions in the beam and can be rejected using beam counters if the beam intensity is not too large. An active cosmic ray shield is required to reject electrons from cosmic rays. The cosmic ray background depends only on the exposure time.

The $\mu \to e$ conversion, equation 1, occurs coherently in the Coulomb field of the nucleus, the electron recoiling against the nucleus with energy $\approx m_\mu c^2$,

$$E_e = E_\mu - \frac{E_\mu^2}{2M_A}, \qquad (3)$$

where E_μ is the muon energy, mass plus binding energy, before capture. An electron of this energy detected in a window delayed with respect to the muon stop signals the conversion. While a free muon decaying at rest can produce an electron whose energy is at most $m_\mu c^2/2$, the decay of a bound muon can result in an electron with energy approaching that given in equation 3. At the kinematic limit in bound decay, the two neutrinos carry away little momenta and the electron recoils against the nucleus, simulating the two-body final state of $\mu \to e$ conversion. The spectrum falls rapidly near the endpoint, proportional to $(E_e - E_e(max))^5$. In titanium, a popular choice for the target material, the fraction of all muon decays that produce electrons within 5 MeV of the endpoint is about 2×10^{-12}.

Radiative muon capture will sometimes produce photons with energy approaching that given in equation 3, but falling short because of the difference in mass of the initial and final nuclear states. For muons stopping in titanium, the conversion electron is expected to have energy 104.3 MeV. The maximum photon energy is 100 MeV. The photon can convert in the target to an asymmetric electron-positron pair, resulting in an electron within 5 MeV of the $\mu \to e$ conversion energy.

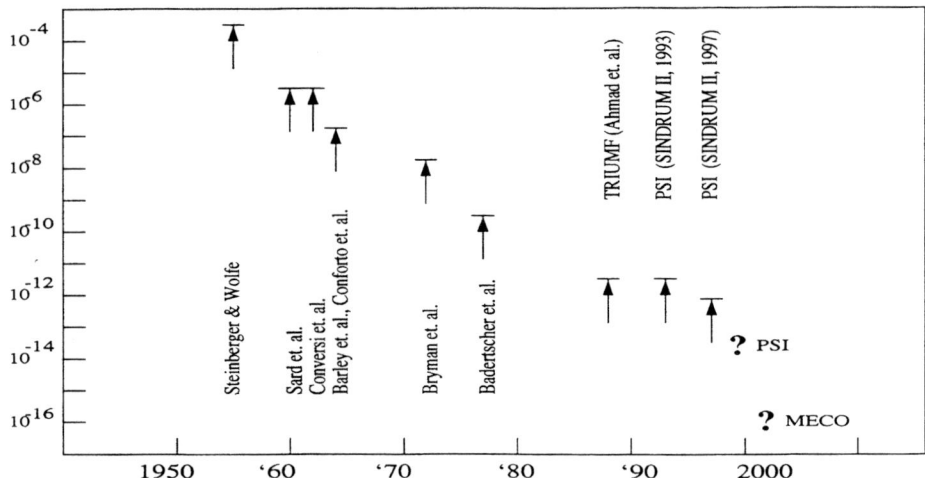

FIGURE 2. Fifty years of $\mu \to e$ accelerator experiments. See references [1,6].

Features	TRIUMF Ahmad et al. 1988	SINDRUM II (PSI) Dohmen et al. 1993	MECO BNL Proposal 940
Principal Detector	TPC, 0.9 T field	Drift Chamber, 1.2 T field	Straw tubes, 1.0 T field
Target Material	Titanium	Titanium	Aluminum
Muons In/Stopped	$1.3/1.0 \times 10^6 sec^{-1}$	$12/3.3 \times 10^6 sec^{-1}$	$2.5/1 \times 10^{11} sec^{-1}$
π/μ stops	10^{-4}	10^{-7}	10^{-4}
Reject Prompts	beam ctrs	beam ctrs	pulsed beam
Resolution FWHM @ 100 MeV	$4.5\ MeV$	$2.3\ MeV$	$750\ KeV$
Exposure time	100 days	25 days	120 days
Cosmic Ray Bkgrnd	$\sim 0.15/MeV$	Negligible	Negligible
90 % Limit	4.6×10^{-12}	4.3×10^{-12} (1993)	10^{-16}

TABLE 1. Main features of the two most recent $\mu \to e$ searches, columns 2 and 3, and the MECO experiment proposed for BNL, column 4.

These will be the dominant physics backgrounds provided prompt processes can be rejected. Pions stopping in the target are the major source of this background, and can produce photons with energy up to 140 MeV. Electrons in the beam that scatter in the target is another such prompt background, as is muon decay in flight.

Both the 1988 TRIUMF experiment and the 1993 SINDRUM II experiment used beams of high purity, with small π/μ ratio. In these experiments, the beam intensities were low enough that it was possible to use scintillation counters in the beam to eliminate prompt background. The SINDRUM II experiment also had no background from cosmic rays during the 25 day exposure. Fig. 3 shows graphically the events in the region from $85 - 120$ MeV in this experiment. The plot shows the data (i) before suppression of prompts and cosmic rays, (ii) after suppression of cosmics and (iii) after suppression of all backgrounds. The remaining events are consistent with coming entirely from muon decay in orbit. The highest energy electron detected had an energy of 100.6 MeV. The data are shown in fig. 4. In the earlier TRIUMF experiment, there were no events in the window from $96.5 \leq P_e \leq 106\ MeV/c$, where 85 % of all $\mu - e$ conversion electrons were expected, but nine events with momenta $> 106\ MeV/c$ were observed (see fig. 5). The source of most of these events was thought to be cosmic rays. This leakage of cosmic rays through the shield was confirmed in a separate experiment in which the cosmic ray induced background was measured with the beam off. The two experiments described achieved similar sensitivities, $R < 4 \times 10^{-12}$.

PROPOSED EXPERIMENTS

It is interesting that except for the cosmic ray induced events in the TRIUMF experiment, neither of the experiments described was background limited. The

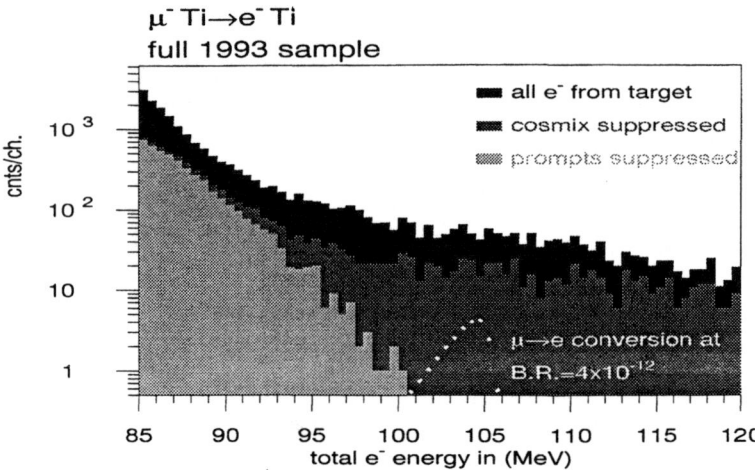

FIGURE 3. Electron energy spectrum from SINDRUM II experiment. There is no background above 101 MeV after suppression of cosmics and prompts.

limit from the SINDRUM II experiment at PSI has since been lowered a factor of five, to 8.4×10^{-13}, in a 60 day exposure completed this year and should reach $\sim 2 \times 10^{-14}$ in the 1998/1999 period [7]. To get to this level, the beam intensity will be raised an order of magnitude or more from the value given in table 1. At this intensity, beam counters can no longer be used to reject prompts. A new high flux beam line and a pion to muon converter situated inside an 8.5 meter long superconducting solenoid will be used to produce a muon beam of high intensity and purity.

The design of the MECO experiment at BNL is based in large part on the the 1992 MELC proposal [8] intended for the Moscow Meson Factory. The parameters of this experiment, which expects to reach a sensitivity of 10^{-16} or less, are listed in column four of table 1. The prescription for achieving the considerable improvement in sensitivity is as follows.

- Increase the muon beam intensity to 10^{11} sec^{-1}. The high intensity is achieved in much the same way it would be in a muon collider. A graded solenoidal field is used, but here varying only up to 3 Tesla. The single particle final state is exploited fully; there is no physics background associated with pileup, as there is for multiparticle final states, .e.g., $\mu \to e\gamma$, where the signal event topology can be built up from interactions of different beam particles. In a pure muon beam, there are three potential sources of electrons with $E_e > 100$ MeV and $P_T^e > 90$ MeV/c: (1) muon decay in orbit, (2) radiative muon capture and (3) muon decay in flight. This last is a prompt background (the muon must decay in the few meters before the detector). Requiring $E_e > 85 - 90$ MeV in an electromagnetic calorimeter not directly in the beam will result in a

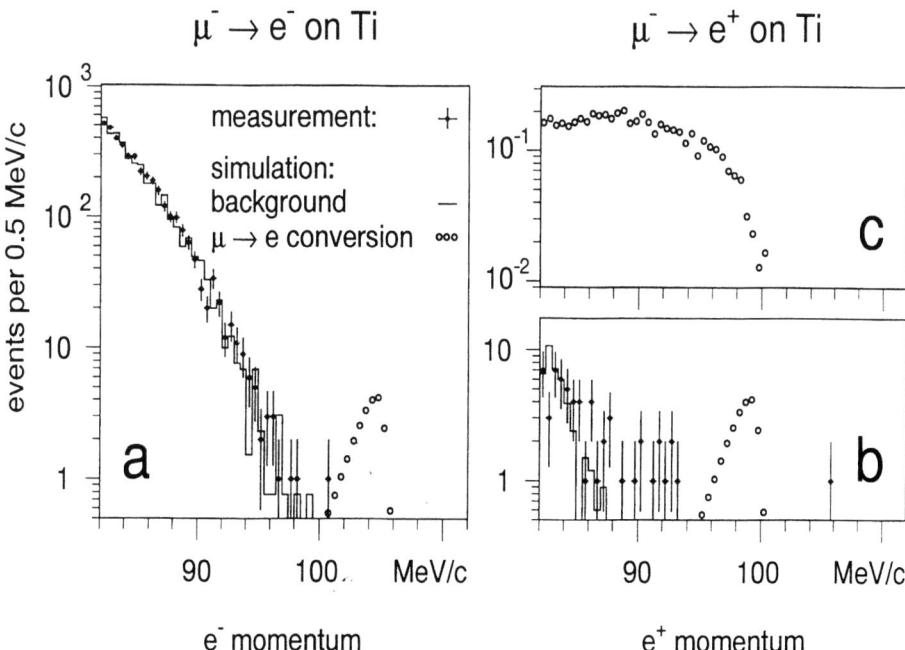

FIGURE 4. Results from SINDRUM II experiment. Muon decay in orbit is the only remaining background after suppression of prompts and cosmic rays and is described well by the theoretical curve. The open circles show the signal expected for $R = 5 \times 10^{-11}$. Figures (b) and (c) are the results of their search for $\mu^- \to e^+$ conversion.

manageable trigger rate. Of course, extra, misidentified, hits can spoil the resolution or lead to reconstruction errors that fake a conversion electron.

- Use a pulsed beam, one bunch approximately every microsecond, to reject prompt background. The conversion electron is detected in a ~ 400 nsec time window between bunches when, ideally, there is no beam in the detector region.

- Situate the target in which the muons are stopped in a graded solenoidal field and displace the detector several meters downstream of the target to a region of uniform field. The graded field reflects electrons produced in the backward direction resulting in large acceptance, $\Delta\Omega/4\pi = 1/2$. Conversion electrons in a $\pm 30°$ cone about $90°$ ($P_T^e > 90~MeV/c$) are projected forward in helical trajectories of large radii that intercept the cylindrical detector. Beam particles and decay electrons at smaller P_T^e pass undisturbed down the center

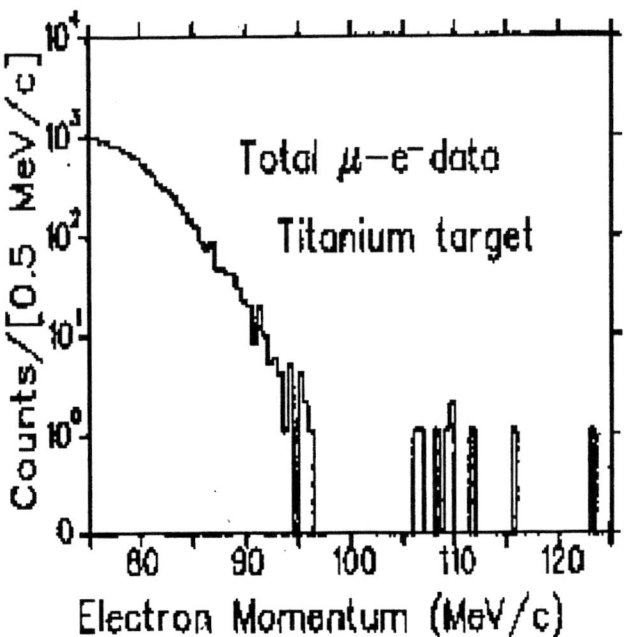

FIGURE 5. TRIUMF data. The events above 106 MeV come mostly from cosmic rays.

of the solenoid.

Fig. 6 is a schematic drawing of the proposed MECO experiment, showing the detectors and the production, transport and detector solenoids. The S-shaped transport solenoid transmits low energy negatively charged muons from the production solenoid to the detector solenoid. High energy negatively charged muons and muons of positive charge are absorbed in the collimators. The proposed tracking detector would be made from straw tubes oriented along the axis of the solenoid. A cylindrical detector with 8 vanes extending radially outward has been studied using GEANT and appears to provide good acceptance. The electron momentum resolution determined from the same GEANT simulation is 750 KeV/c (fwhm), the uncertainty coming largely from fluctuations in the energy lost in the target and from multiple scattering. The simulation of the signal shape and the background from muon decay in orbit are shown in fig. 7 for $R = 1 \times 10^{-16}$.

CONCLUSION

The search for $\mu \rightarrow e$ conversion began over forty years ago and continues with ever increasing sensitivity. A sensitivity of $R < 8.4 \times 10^{-13}$ was reached in an

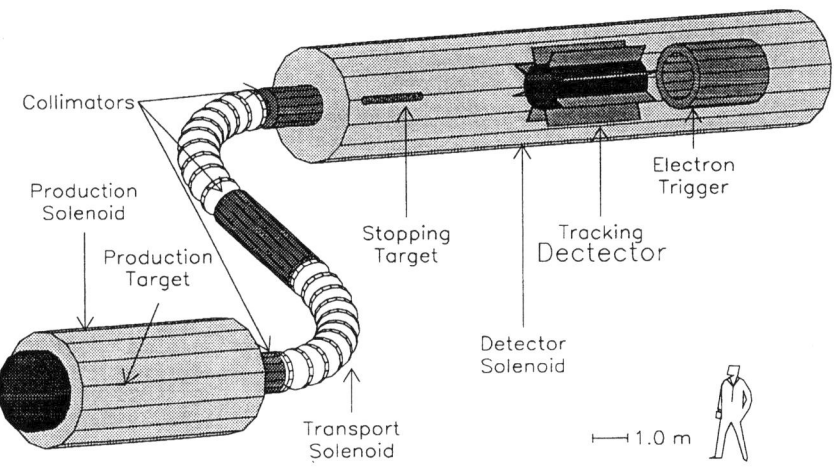

FIGURE 6. MECO Detector.

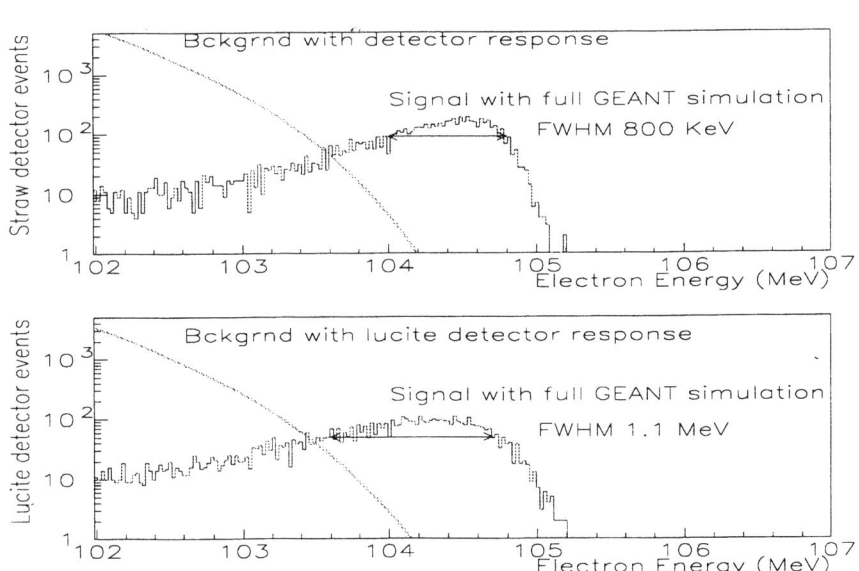

FIGURE 7. GEANT simulation of signal and muon decay in orbit in the MECO detector for $R = 10^{-16}$. At this value of R, ~ 15 events are expected above $104\ MeV$ in one year of data taking.

experiment at the PSI completed just this year, and experiments in the planning stage may well reach sensitivities of 10^{-16} or lower. Some of the issues involved in reaching these levels have been examined briefly in this presentation. The limit $R < 2 \times 10^{-14}$ expected at the PSI in 1998/1999 seems achievable, given the absence of background in the last two experiments. The MELC/MECO proposal is a prescription for going well beyond this limit. Interest in the experiment remains high because it may well serve as a window to physics beyond the Standard Model. From the authors of reference [5]:

... *the study of the corresponding experimental signals [$\mu \to e$, and $\mu \to e + \gamma$] provides a test of supersymmetric unification at least as significant as the one that can be obtained from either proton decay or neutrino masses.*

REFERENCES

1. J. Steinberger and H. B. Wolfe, *Phys. Rev.* **100**, 1490 (1955).
2. S. Weinberg and G. Feinberg, *Phys. Rev. Lett.* **3**, 111 (1959).
3. N. Cabbibo and R. Gatto, *Phys. Rev.* **116**, 1334 (1959).
4. G. Danby et al., *Phys. Rev. Lett.* **9**, 36 (1962).
5. R. Barbieri, L. Hall and A. Strumia, *Nucl. Phys.* **B445**, 219 (1995).
6. M. Conversi, L. Di Lella, A. Egidi, C. Rubbia and M. Toller, *Phys. Rev.* **D122**, 687 (1961).
   R. Sard, K. Crowe, H. Kruger, *Phys. Rev.* **121**, 619 (1961).
   G. Conforto, M. Conversi, L. Di Lella, G. Penso, C. Rubbia and M. Toller, *Nuovo Cimento* **26**, 261 (1962).
   J. Bartley, H. Davies, H. Muirhead and T. Woodhead, *Phys. Lett.* **13**, 258 (1964).
   D. Bryman, M. Blecher, K. Gotow and R. Powers, *Phys. Rev. Lett.* **28**, 1469 (1972).
   A. Badertscher et al., *Phys. Rev. Lett.* **39**, 1385 (1977).
   S. Ahmad et al., TRIUMF, *Phys. Rev.* **D38**, 2102 (1988).
   SINDRUM II Collaboration, C. Dohmen et al., *Phys. Lett.* **B317**, 631 (1993).
7. A. van der Schaaf, private communication.
8. V.S. Abadjev et al., "MELC Proposal to Search for $\mu^- A \to e^- A$ Process," *INR Preprint 786/92*, (1992). R. Djilkibaev and V. Lobashev, *Sov. J. Nucl. Phys.* **49**, 384 (1989).
9. The MECO Proposal, W. Molzon et al., BNL Proposal 940.

MECO Physics Backgrounds Studies

Tingjun Liu[1]

Dept. of Physics & Astronomy, University of California, Irvine, Ca 92697

Abstract. MECO(Muon – Electron COnversion) is a proposed experiment to search for $\mu^- N \to e^- N$ with sensitivity below 10^{-16}. The physics backgrounds for this experiment is discussed here.

The MECO(Muon – Electron COnversion) collaboration proposes to search for $\mu^- N \to e^- N$ with sensitivity below 10^{-16}. In this process, the signature is mono-energetic electrons emitted at almost the energy of the muon mass at about 105 MeV. At such a high sensitivity, the suppression of backgrounds becomes a crucial issue. In the following sections, we discuss 8 primary sources of background to the $\mu^- N \to e^- N$ signal, which motivates many of the basic ideas of our experiment. The first two sources are intrinsic to μ's stopped in the target; they can be minimized by improving the measurement of the electron energy. Sources 3-6 derive from prompt processes, with the electron detected close in time to the arrival of a beam particle in the detector. Because of MECO's unprecedented high sensitivity, these sources are very serious so that we conclude that a pulsed beam is necessary. In this study, a beam extinction (defined to be the ratio of the number of protons between pulses to that during pulses) of $\sim 10^{-10}$ is assumed.

We next discuss the results of our calculations of each source of background in more detail.

I μ DECAY IN ORBIT

The rate for production of electrons from μ decay in Coulomb bound orbit is approximately proportional to $(E_{max} - E_e)^5$ near the endpoint [1]. Hence the signal/background ratio is extremely sensitive to resolution. Removing high energy tails in the energy resolution function is of critical importance to reducing the background. Figure 1 shows the signal and background for $R_{\mu e} = 10^{-16}$ for two different designs of the detector, with effects of target energy

---

[1] On behalf of the MECO collaboration.

loss and detector spatial resolutions taken into account. These distributions were calculated in a full GEANT simulation [2].

In figure 1, the detected energy is shifted below 105 MeV due to energy loss in the target and in the detector flanges. By accepting events between 103.9 MeV and 105.4 MeV, the noise to signal ratio is below 0.05 with large acceptance for signal events. The FWHM for both detectors is 1 MeV or below, with very little high energy tail in the resolution function.

II RADIATIVE μ CAPTURE

Radiative μ capture background results from the process $\mu^- Al \to \gamma \nu_\mu Mg$. The photon endpoint energy is 102.5 MeV and the probability (per μ capture) of producing a photon with energy exceeding 100.5 MeV is $\sim 4 \times 10^{-9}$ [3]. The conversion probability in the target is ~ 0.005, and the probability of the electron energy exceeding 100 MeV is ~ 0.005. The probability of producing an electron above 100 MeV is then $[4 \times 10^{-9}] \times [5 \times 10^{-3}] \times [5 \times 10^{-3}] = 10^{-13}$. Further, these electrons are all less than 102 MeV (most are near 100 MeV),

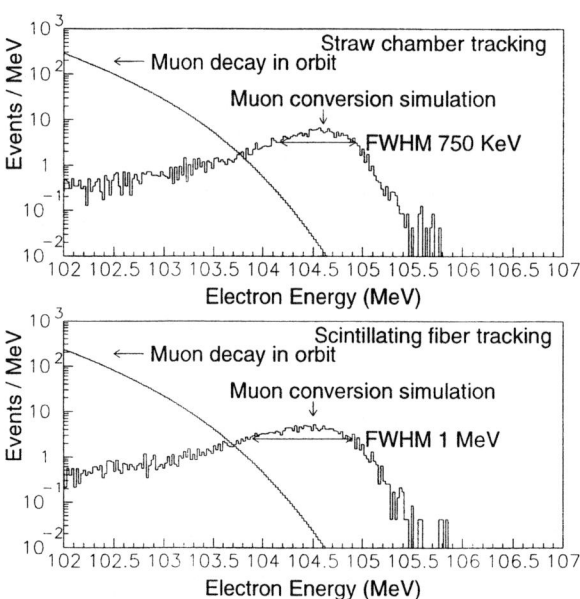

FIGURE 1. A simulation of the expected signal and background for $R_{\mu e} = 10^{-16}$ for two different proposed detectors. The vertical scale is arbitrary; the relative normalization of the signal and background curves is for $R_{\mu e} = 10^{-16}$ and we would see ~ 5 events above 103.9 MeV at this value for the nominal running time of 10^7 seconds.

and for an electron to be considered as signal, its measured energy must exceed 103.9 MeV. The integral of the high energy tail in the resolution function above 1.9 MeV is less than 10^{-6} (limited by statistics). Hence, the probability of getting an electron above 103.9 from radiative μ capture is less than 10^{-19} or a signal/noise ratio of greater than 1000 for $R_{\mu e} = 10^{-16}$.

III BEAM ELECTRONS

Beam electrons which cause background must be produced in the production or transport solenoid region and then scatter in the stopping target in order to simulate signal events.

The rate for electrons scattering at $\sim$100 MeV is defined by the Mott cross section multiplied by a nuclear form factor for the target material [4].

The collimator system is designed to suppress high energy electrons. A GEANT simulation of the production of electrons and their transport to the detector solenoid yielded no transmitted electrons above 100 MeV for 10^7 incident protons. We take the transverse momentum distribution to be that of electrons of 70–90 MeV and use that distribution to calculate the probability of scattering in the target to a transverse momentum sufficient to fake a signal. This probability is about 3×10^{-7}, taking into account of the fact that most of the outgoing high momentum electrons miss the target hence no scattering occurs for them. With a run time of 10^7 sec, a proton intensity of 2×10^{13} p/s, and a beam extinction of 10^{-10}, the total number of background events is less than:

$$[10^7] \times [2 \times 10^{13}] \times [10^{-10}] \times [10^{-7}] \times [3 \times 10^{-7}] = 0.0006$$

IV μ DECAY IN FLIGHT

Muon decay in flight can result in energetic electrons with sufficient transverse momentum to fake the signal directly or that reach the required p_T by scattering in the stopping target. In order for the electron to have energy above 102 MeV, the μ momentum must exceed 77 MeV. Electrons produced by μ decays before and within the transport solenoid are included in the beam electron background calculation. Background from decays in the detector solenoid are calculated using a GEANT beam simulation, yielding less than 0.003 in the total running time. A second background source is electrons from μ decay which scatter in the stopping target. This background is estimated to be 0.004.

V π DECAY IN FLIGHT

Beam π's decaying to electrons with $E_e > 102$ MeV and $p_T > 90$ MeV/c are also a potential source of background. The π momentum must exceed 60 MeV for this background process. A GEANT simulation was used to calculate the probability of a proton producing a beam π with $p_\pi > 54$ MeV/c passing the transport solenoid; it is 4×10^{-6}. The probability for a π to decay into an electron after the transport solenoid and before the tracking detector is 1×10^{-4} and the probability of the decay electron to have $E_e > 102$ MeV and $p_t > 90$ MeV is 5×10^{-6}. The background from this source is: $[10^7] \times [2 \times 10^{13}] \times [10^{-10}] \times [4 \times 10^{-6}] \times [1 \times 10^{-4}] \times [5 \times 10^{-6}] = 0.00004$
A second background mechanism is π decay electrons which scatter in the stopping target. This background was calculated in much the same way as the similar process for μ decay. The number of electrons from π decay with 103 MeV $< E_e <$ 105 MeV per proton is 1.6×10^{-11} and the probability of scattering to $p_t > 90$ MeV/c is 4×10^{-5}, resulting in an expected number of background events of: $[10^7] \times [2 \times 10^{13}] \times [10^{-10}] \times [1.6 \times 10^{-11}] \times [4 \times 10^{-5}] = 0.00001$.

VI RADIATIVE π CAPTURE

Stopped π^-s are immediately captured by a nucleus after they stop in the target; about 2% of the captures result in the emission of a photon [5] without significant nuclear excitation. The photon energy spectrum has a peak at 110 MeV and endpoint at 140 MeV. The probability of photon conversion in the Al target, with a conversion electron in a 1.5 MeV energy interval around 104 MeV is 3.5×10^{-5}, as calculated in a GEANT simulation. The acceptance for electrons from photon conversion is large (~ 0.8), since the path length for conversion is largest for photons emitted at 90°. The yield of π's which pass the transport solenoid and stop in the target is $\sim 6 \times 10^{-7}$ per proton. Accounting for the beam extinction of 10^{-10}, the background is: $[10^7] \times [2 \times 10^{13}] \times [10^{-10}] \times [6 \times 10^{-7}] \times [0.02] \times [3.5 \times 10^{-5}] \times [0.8] = 0.007$

A contribution which is more difficult to calculate is that due to π's which take a very long time to traverse the production and transport solenoid and arrive at the stopping target. For these events, the suppression factor of 10^{-10} from the beam extinction is absent. However, since our detection window starts ~ 600 ns after the proton pulse, the π's must live approximately 600 ns and must follow a trajectory in the transport solenoid which results in a flight time of 600 ns in order to be a source of background. This background is estimated to be 2.2.

Two means exist to further suppress these events. First, the late arriving π's are very low energy, and can be absorbed with high efficiency in a very thin window. The second means of suppression is simply to delay the start of the timing window from the nominal 600 ns, so that the late arriving π's are further suppressed due to decays. We've estimated that, by using an aluminum absorber of 100 μm thickness and delaying the start of the timing window to 700 ns, the background is reduced to 0.014.

VII ANTI-PROTONS

The anti-protons are a potentially serious background. They can annihilate in the transport system or the stopping target and hence produce signals directly or indirectly through creation of secondary particles. Since the tranport system does not allow transmission of negatively charged particles with momentum over 80 MeV/c, these anti-protons which got through are arriving, hence not suppressd by pulsed beam. We have not done a full scale study on them, partially because the production cross section near threshold is not well measured. However, we note that, The threshold for the incoming proton momentum is 6.5 GeV/c in pp collisions, and it is reduced to 5 GeV/c taking into account of the Fermi momentum of 300 MeV/c. Of curse, the nucleon momentum distribution has talis beyond the Fermi momentum, but that is a tiny fraction. Furthermore, the anti-protons produced at threshold are boosted forward in the lab frame, and do not reflect in the production field hence will not get transported. At this stage, we do not know whether or not the anti-prorons will be a problem. If they do turn out to be so, we can work below the threshold, which translates to a 40-50% loss in the μ stop rate.

VIII COSMIC RAYS

Cosmic ray induced electrons are potentially a limiting background and we have studied it using a GEANT simulation [6] of the detector and shielding. The details of the simulation and the shielding required to reduce the background to a negligible level are discussed in a later section. The conclusion of these studies is that cosmic ray background can be reduced to a negligible level with a combination of active and passive shielding and detection of extra particles in the tracking detector. These consist of the following:

- A passive shield of modest thickness (2 m of concrete and 0.5 m of steel).

- Two layers of scintillator veto counter surrounding the detector, with a combined efficiency for charged particles of 99.99%.

- Selection criteria which eliminate events having significant evidence of extra particles in the detector in time with the electron candidate.

With this suppression, the expected background from cosmic rays in a 10^7 second run is estimated to be $\sim$ 0.0035 events.

IX SUMMARY

Table 1 Summarizes the background levels except for the anti-protons. The dominate background source is from μ decay in orbit, representing a 5% noise to signal ratio. This ratio can be reduced significantly by moving the signal window up in energy, without much loss of acceptance. Of the backgrounds which scale with the beam extinction rate, the primary source is radiative π capture representing a noise to signal ratio of 0.14%. One sees that there is enough room to relax the beam extinction rate by a factor of 10. Overall, in the current configuration, the total noise to signal ratio is 8%. As we have discussed, this number can be substancially reduced without much of loss of acceptance.

TABLE 1. A summary of the level of background from various sources, calculated for the sensitivity given in the previous table, and with scaling as discussed in the text. The numbers given are for a sensitivity of 5 events for $R_{\mu E} = 10^{-16}$.

Source	Events	Comment
μ decay in orbit	0.19-.33	signal/noise = 20 for $R_{\mu e} = 10^{-16}$
Radiative μ capture	< 0.005	
μ decay in flight	< 0.003	without scatter in target
μ decay in flight	0.004	with scatter in target
Radiative π capture	0.007	from proton during detection time
Radiative π capture	0.014	from late arriving π
π decay in flight	<< 0.001	
Beam electrons	< 0.0006	
Cosmic ray induced	0.004	assuming 10^{-4} CR veto inefficiency
Total background	0.29-.41	

REFERENCES

1. O. Shankar, Phys. Rev. **D25**, 1847 (1982).
2. T. Liu, meco-004, "Initial Resolution Studies" (1997).
3. A. Frischknecht, et al., Phys. Rev. **C2**, 1506 (1985).
4. T. Stoudl, et al., Nucl. Phys. **A91**, 520 (1967).
5. R.A. Eramzhyan, et al., Nucl. Phys. **A290**, 294 (1977).
6. M. Overlin, meco-014.

MECO Muon Yield Simulation Using Experimental Data

Rashid M. Djilkibaev[1]

Department of Physics and Astronomy
University of California, Irvine, CA 92697-4575

Abstract. A comparison of experimental low energy pion spectra with GHEISHA and FLUKA hadron code simulation for Tantalum target at 10 GeV/c has been done. Muon flux simulations for MECO setup using experimental data and the hadron codes for different proton energy are considered.

INTRODUCTION

The MECO muon beam [1] is based on a solenoid muon capture scheme [2] proposed for the MELC experiment using a pulsed proton beam.

The basic idea of the MELC scheme is to place proton target into a gradually decreasing magnetic field on the solenoid axis with a small tilt angle relative to the axis. The gradually decreasing magnetic field is arranged to reflect some part of the backwards muon flux and hence increase the forward muon flux. The backwards direction for the proton beam with respect to the muon beam is selected to have more convenient conditions for utilizing the proton beam and to reduce muon beam contamination.

Because of the high melting point temperature of tungsten (3653 $°K$), this material W(A=183.85, Z=74) was selected for proton target. In this case we can use the simplest technique for target cooling of radiation cooling.

Pions with a transverse momentum below $P_t(\text{MeV}/c) = 1.5 \times B(T) \times R_{bore}(cm)$, where R_{bore} is a radius of production solenoid bore, will be captured. In our case $R_{bore} = 30$ cm, B field in the proton target region $\simeq 2.6$ T, so $P_t \simeq 120$ MeV/c. We can expect that the main contribution to the muon flux will give pions with momentum $\simeq 100$ MeV/c. A detailed MC simulation indicates that the main contribution to the muon flux is pions with kinetic energy from 20 to 60 MeV, which correspond to pion momentum 77 and 143 MeV/c, respectively.

---

[1] Permanent address: Institute for Nuclear Research, 60-th Oct. pr. 7a, Moscow 117312, Russia

LOW ENERGY PION SPECTRA

Invariant cross sections of low energy pion(π^-) production as a function of kinetic energy (T) at fixed angles within an interval from 0 up to 180° for Tantalum (A=180.95, Z=73) in the reaction p + Ta $\to \pi^-$ + X have been measured [3]. Thin Ta plates (1 mm) with spacing of 93 mm were placed in a 2 m propane bubble chamber which was operated under a magnetic field of 1.5 Tesla. Pion trajectories were confidently identified with minimum momentum of 80 MeV/c (T = 21 MeV). Measured average π^- multiplicity at 10 GeV/c is 1.51 ± 0.03. The experimental π^- inclusive differential cross section and fit results in different angle range are shown in Figure 1.

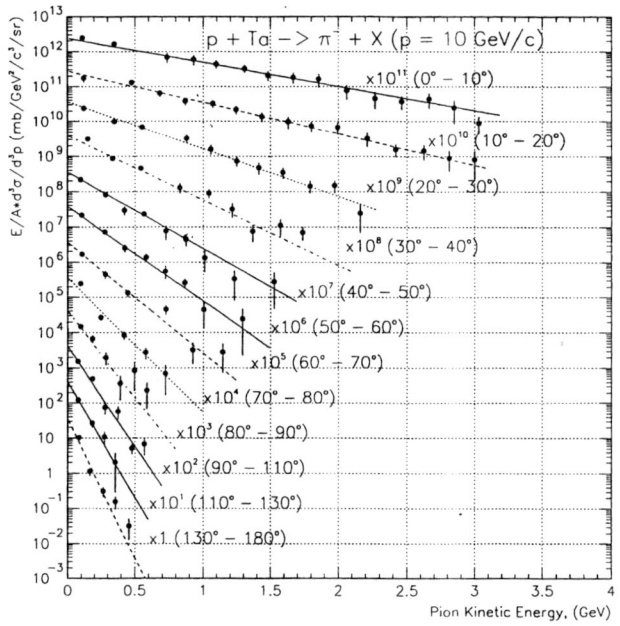

FIGURE 1. The π^- inclusive differential cross section for Ta at 10 GeV/c in different angle intervals. Full curves are fitted result by exponential function $f = C \cdot \exp(-T/T_0)$.

The dependence of invariant cross sections versus pion kinetic energy are well approximated by exponential function $f = C \cdot \exp(-T/T_0)$. This means that the kinetic energy of the hadron is a better parameter for describing cross sections at low energies than total energy. Unfortunately the article [3] presented the results of fits for T_0 parameters only. So we used scaned data from the original plots to fit C constants using fixed parameter T_0 in each angle interval.

In Table 1 the resulting fitfor normalized invariant cross-sections $E/A \cdot d^3\sigma/d^3p$ by function $f = C \cdot \exp(-T/T_0)$ are presented.

TABLE 1. The fit for normalized invariant cross-sections $E/A \cdot d^3\sigma/d^3p$ by function $f = C \cdot \exp(-T/T_0)$.

Angle Range (deg.)	Kinetic Energy Range T (GeV)	Parameter T_0 (GeV)	Parameter C (mb/GeV^2/c^3/sr)
0 - 10	0.02 - 3.0	0.632 ± 0.048	23.9 ± 1.8
10 - 20	0.02 - 3.0	0.484 ± 0.022	27.7 ± 2.2
20 - 30	0.02 - 3.0	0.321 ± 0.013	37.1 ± 1.7
30 - 40	0.02 - 2.6	0.235 ± 0.012	40.0 ± 2.5
40 - 50	0.02 - 1.65	0.200 ± 0.012	36.1 ± 2.3
50 - 60	0.15 - 1.5	0.162 ± 0.013	37.9 ± 2.6
60 - 70	0.02 - 1.2	0.138 ± 0.015	36.7 ± 2.8
70 - 80	0.15 - 0.75	0.114 ± 0.020	35.8 ± 3.6
80 - 90	0.02 - 0.8	0.088 ± 0.012	41.2 ± 5.3
90 - 110	0.02 - 0.6	0.076 ± 0.007	41.5 ± 3.2
110 - 130	0.02 - 0.6	0.066 ± 0.009	40.6 ± 3.1
130 - 180	0.02 - 0.5	0.055 ± 0.006	34.6 ± 4.2

The slope parameter T_0 was approximated [3] by the function $T_0 = \frac{T_0^\pi}{1-\beta \cdot cos(\Theta)}$ where $\beta = 0.78 \pm 0.03$ and $T_0^\pi = 86 \pm 5$ MeV.

We can calculte the total pion production cross section for Ta at 10 GeV/c, using determined C and T_0 parameters. The total cross section is 2.36 barn. Nuclear inelastic cross section for Ta is 1.56 barn. So the total pion production cross section is in a good agreement with the measured pion multiplicity of 1.51.

We can compare hadron simulation results GHEISHA [4] and FLUKA [5] with the experimantal data for Ta at 10 GeV/c, comparing normalized energy spectra of π^- for 10 GeV/c protons on Ta nuclei. The energy spectra of π^- as calculated with GHEISHA, FLUKA and the experimental data integrated over angle are shown in Figure 2.

MUON YIELD SIMULATION

To compare MECO muon flux simulation, using the hadron codes and the experimental data, Ta proton target (ρ=16.6 g/cm^3) with length 19.34 cm (1.67 nuclear lengths) and radius 0.4 cm was selected. For muon flux calculations GEANT 3.21 code version (1997) [6] with GHEISHA and FLUKA interface have been used.

The proton beam has Gaussian form with $\sigma_x = \sigma_y = 0.2$ cm. Figure 3 shows the simulation results for muon yield per primary proton for different beam momenta. The muon yield is defined to be the relative number of muons which stop in the detector target of MECO setup per primary proton.

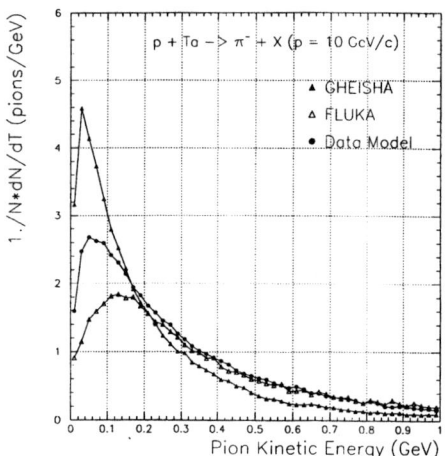

FIGURE 2. Energy spectra of π^- for 10 GeV/c protons on Ta nuclei as calculated with GHEISHA and FLUKA and the results derived from experimental data.

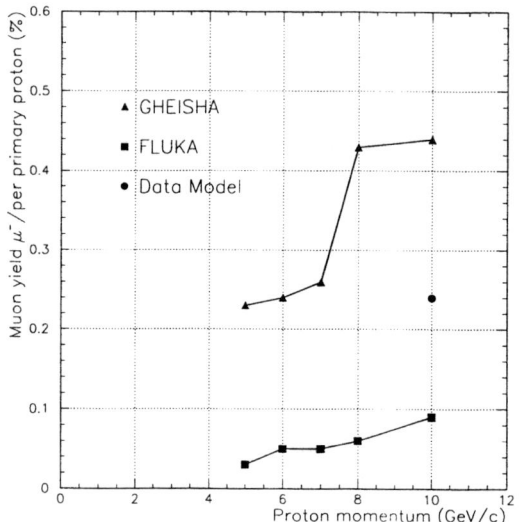

FIGURE 3. Muon yield simulations for Ta target using hadron codes GHEISHA and FLUKA and the results derived from experimental data for muons stopping in the detector target of MECO setup.

ACKNOWLEDGEMENTS

We would like to thank N. Mokhov and S. Striganov for the helpful discussion and remarks and to M. Bachman and W. Molzon for their helpful advice.

REFERENCES

1. M. Bachman et al. *MECO proposal P-940 to BNL AGS* Sep. (1997).
2. R.M. Djilkibaev and V.M. Lobashev, *Sov. J. Nucl. Phys.* **49** (2), 384 (1989). V. Abadjev et al., MELC proposal to search $\mu^- N \to e^- N$, Prep. INR-786/92 (1992).
3. D. Artmutliski et al., *Sov. J. Nucl. Phys.* **48**, 161 (1988), Prep. JINR P1-91-191 (1991).
4. H.C. Fesefeldt 'Simulation of Hadronic Showeres, Physics and Applications.' Report PITHIA 85-02, Achen, 1985.
5. P.A. Aarnio, J. Lindgren, J. Ranft, et al., CERN/TIS-RP/190, (1987).
6. R.Brun et al. Program GEANT3, DD/EE/84-1, CERN (1984).

Prospects for High Precision Measurements on Muonic Atoms at the Front End of a Muon Collider

D. Kawall[†], M. G. Boshier[§], V. W. Hughes[†], K. Jungmann[‡],
W. Liu[†], and G. zu Putlitz.[‡]

[†] *Yale University, New Haven, CT 06520, USA*
[§] *University of Sussex, Brighton, BN1 9RH, UK*
[‡] *Universität Heidelberg, D-69120, Heidelberg, Germany* [1]

Abstract. Low energy pulsed muon beams at the front end of a muon collider may provide some 5 orders of magnitude greater intensity than present sources. This suggests dramatic possibilities for high precision spectroscopic and other measurements on muonic atoms, which could importantly extend our tests of QED, measurements of fundamental constants, and tests of physics beyond the standard model.

INTRODUCTION

Much of the early development and testing of quantum electrodynamics was founded on measurements of energy levels in electronic atoms, as was our knowledge of many fundamental constants. In the recent past, the precision of these measurements has reached astounding levels; for example, the ground state hyperfine interval in hydrogen, $\Delta\nu(H)$, is known to better than a part in 10^{12} [1]. Our enthusiasm is tempered when we note that the theoretical prediction for $\Delta\nu(H)$ is six orders of magnitude less precise (at the 1 ppm level), due to the difficulties in calculating strong interaction effects which lead to structure and polarizability of the proton [2]. Similarly, another important test of low energy QED, the $1S$ ground state Lamb shift in hydrogen has been measured to about 3.5 ppm, while the theoretical prediction is limited to 5 ppm due to the uncertainty in the root mean square radius of the proton, $\langle r_p^2 \rangle^{1/2}$ [3,2].

Higher precision measurements on the purely leptonic atom muonium (M) (μ^+e^-) [4], where hadronic structure effects do not appear in leading order and

---

[1]) Supported by U.S. DOE., NSF, NATO and BMBF(Germany).

where theoretical values of intervals can be carried much further, would increase significantly the test of QED and fundamental lepton interactions. And higher precision measurements on simple muonic atoms such as μ^-p^+, taken together with the measurements on electronic atoms, will provide hadronic structure data as well as tests of the QED and weak interaction terms.

MEASUREMENTS ON MUONIC ATOMS

In addition to the Lamb shift, the other classic test of QED has been the determination of the electron anomalous g-factor, $a_e \equiv (g_e - 2)/2$. Here the measurements have been carried out to a precision of 3.4 ppb [5], while the QED theoretical predictions are limited by the $\sim$ 20 ppb uncertainty in independent measurements of the fine structure constant, α [6]. Even with this success, tests of QED in simple atoms are not redundant since the physics involved of the relativistic two-body bound state is quite different. In contrast to the physics of a_e, the bound state is essentially nonperturbative.

Understanding the Lamb shift, hyperfine structure interval, or other bound state properties, has come about by performing expansions in the small parameters α and m_e/m_N [2]. For positronium, where $m_{e^-}/m_{e^+} = 1$, the recoil correction expansion terms are difficult to calculate. For muonium, the parameter is $\sim$ 8.9 times that in hydrogen but still much less than unity. Without the hadronic structure contributions, the calculations for muonium become more tractable than those for hydrogen, and because of the larger mass ratio, recoil and other interesting higher-order bound state effects become larger [2], making this system an ideal testbed to probe our understanding.

It is interesting to consider if one could ever achieve the experimental precision of hydrogen energy level measurements in muonium; a difference currently of about four orders in magnitude. One difficulty to be overcome is the small average rate of muonium atom formation, $\leq 10^7$ M/s in a 1000 cm^3 gas target using existing sources (Rutherford Appleton Laboratories (RAL), Paul Scherrer Institut (PSI) and the Los Alamos Meson Physics Facility (LAMPF)) where average muon fluxes of a few$\times 10^5$ μ/s to a few$\times 10^7$ μ/s are obtained. In contrast, fluxes of 10^{13} H atoms s^{-1} are produced in atomic beams. The 5×10^{12} muons per bunch after the ionization cooling channel at a muon collider [7] goes a long way towards addressing this imbalance, though the need for high muon polarization in many experiments might reduce this flux appreciably. Still, increases in statistical precision over current measurements of two or more orders of magnitude are conceivable, and some rate-limited experiments become feasible. In addition, new approaches to existing experiments may be possible which take advantage of the high intensities to reduce dominant systematic errors.

A second difficulty occurs as the muonium lifetime of 2.2 μs implies a large natural width to spectral lines of 145 kHz, and also limits the density of M atoms present at any time. This limitation can be overcome by observing muonium atoms which have lived and interacted coherently with an electromagnetic field for periods longer than the 2.2 μs mean lifetime [8]. The bunch lengths anticipated of 1.5 to 10 cm in the pulsed beam of a muon collider, are short compared to the muon lifetime and naturally suited to this line-narrowing technique, and would also allow for an important reduction in systematic errors which often scale with the resonance linewidth.

An advantage of muonium spectroscopy over hydrogen is the high efficiency of detection through the energetic decay positron, which also carries information on the spin state of the muon.

Laser Spectroscopy of Muonium and Muonic Atoms

Many important energy intervals should be measurable in muonium, muonic hydrogen (μ^-p), muonic deuterium (μ^-d) and the muonic helium ions (^{4}Heμ^-)$^+$ and (^{3}Heμ^-)$^+$ using the techniques of laser spectroscopy [9,10]. Measurements of the $1S - 2S$ interval, fine and hyperfine structure intervals, and Lamb shift intervals are sensitive to fundamental QED processes, and can be used to extract precise values of particle masses, magnetic moments and nuclear radii of p, d, ^{3}He and ^{4}He, as well as nuclear polarizabilities. At a more sensitive level, electroweak effects appear as well.

Of particular interest is the use of laser spectroscopy to determine the $1S - 2S$ transition frequency in muonium. First performed at KEK [11], then improved at Rutherford Appleton Laboratory (RAL) [12], the transition is induced using a two-photon, Doppler-free process, and detected through the photoionization of the $2S$ state in the same laser field, which releases a slow μ^+. The latter experiment achieved a precision of 10^{-8} in the $1S-2S$ interval, and determined the $1S - 2S$ Lamb shift to 0.8%. The muonium-hydrogen isotope shift in the transition (which is due primarily to reduced mass) was used to make a 5 ppm determination of the μ^+ mass. Alternatively, the experiment may be interpreted as the best test of charge equality between μ^+ and e^-. The possibilities for this experiment and others are given in Table 1.

Microwave Spectroscopy of Muonium

Microwave spectroscopy has been used to measure the $2^2S_{1/2} - 2^2P_{1/2}$ Lamb shift [13,14] and the $2^2S_{1/2} - 2^2P_{3/2}$ fine structure interval [15] in muonium. In the theoretical interpretation of the results, the uncertainty due to $\langle r_p^2 \rangle^{1/2}$ is not present, though for the 1% precision on the Lamb shift of these experiments this is not a serious concern. Since the errors in the experiments were almost exclusively statistical (due to the very low production rates into vacuum of muonium in the $2S$ state), one could expect significant improvements simply from the higher incident muon flux at the front end of a muon collider.

TABLE 1. Some possibilities for laser spectroscopy of muonic atoms. The potential is given for line centers determined to 1 part in 10^3. (From [10]).

System & Transition	Laser System & Frequency	Physics Interest and Potential
Muonium $1S - 2S$	CW dye/diode laser with an enhancement cavity 2×1228.5 THz	Measures Lamb shift (without nuclear structure effects), QED recoil, μ^+ mass. Improve current 0.8 % Lamb shift and 5 ppm μ^+ mass determinations by several orders of magnitude.
Muonic ^{4}He ion $2S - 2P$	Dye or Ti:Sapphire 369.6, 334.2 THz	Test QED vacuum polarization. Improve current 0.2% α-particle charge radius measurement by two orders of magnitude.
Muonic ^{4}He ion $3D - 3P$	Carbon dioxide 30.4 THz	Probes QED vacuum polarization (insensitive to nuclear structure). Sensitivity of 1/5 of linewidth would provide new test of vacuum polarization. Potential for factor 200 more.
Muonic Hydrogen $2S - 2P$	Carbon monoxide or nonlinear mixing 48.4 THz	Measures proton charge radius, polarizability, QED vacuum polarization. PSI aims at a 500 MHz measurement, giving $\langle r_p^2 \rangle^{1/2}$ to a few parts in 10^4.
Muonic Hydrogen $1S, F = 0, F = 1$	Difference frequency generation 43.9 THz	Measures proton charge radius and polarizability. A 10 GHz uncertainty determines $\langle r_p^2 \rangle^{1/2}$ to about 2%. Potential for five more orders of magnitude.
Muonic Hydrogen $3D - 3P$	Free electron laser 1.6 THz	Probes QED vacuum polarization (insensitive to nuclear structure). Potential for 6 ppm test of vacuum polarization, an improvement of 1000 over current tests.

Fertile ground for precision tests of QED and the determination of fundamental constants has also been found in microwave magnetic resonance spectroscopy of muonium. An experiment at LAMPF determined the muonium ground state hyperfine structure interval and the muon to proton magnetic moment ratio [16]:

$$\Delta\nu_{\text{exp}}(M) = 4\,463\,302.88(16) \text{ kHz } (36 \text{ ppb})$$
$$\mu_\mu/\mu_p = 3.183\,346\,1(11) \; (360 \text{ ppb}).$$

The present theoretical value for $\Delta\nu(M)$ is known to 0.3 ppm [6,17,18], better by almost a factor of four than the equivalent value for hydrogen :

$$\Delta\nu_{\text{th}}(M) = 4\,463\,303.04(1.34)(0.04)(0.16)(0.06) \text{ kHz } (0.3 \text{ ppm}).$$

The principal error of 1.34 kHz arises from the uncertainty in the muon mass, m_μ, which is itself derived from μ_μ/μ_p determined from the Zeeman effect in muonium [16] and from μSR in liquid bromine [19]. The second error comes from the uncertainty in α determined from the electron g-2 experiment, and the other two from numerical uncertainties in the calculations of radiative correction terms and uncalculated higher order terms respectively. The experimental and theoretical values agree well, $(\Delta\nu_{th}(M) - \Delta\nu_{exp}(M))/\Delta\nu_{exp}(M) = 0.04 \pm 0.3$ ppm, and together comprise the most sensitive test of the relativistic two-body bound state in QED.

Data-taking has been completed on a new experiment at LAMPF which used a chopped muon beam to obtain linewidths narrower than the natural linewidth by up to a factor three [8], and which should reduce the uncertainties in $\Delta\nu_{exp}(M)$ and μ_μ/μ_p by about a factor of four. This experiment is still statistics limited, and could benefit if high fluxes of polarized muons were available at a muon collider. In addition to the line narrowing technique, one might hope to implement cooling, trapping and phase space compression of the muon beam [20], so the muons will stop in low pressure gases of smaller volume. This would reduce by up to a factor of three the dominant systematic uncertainty of knowing the absolute magnetic field strength over the volume in which muonium is formed. New field measuring techniques would be required for further improvement.

With a factor of a hundred reduction in the uncertainty in m_μ, a factor of ten in $\Delta\nu(M)$, and some progress in the theory of the hyperfine structure interval in muonium, one would have a 25% measurement of the parity conserving weak interaction contribution to $\Delta\nu^M$ of -0.065 kHz [6,17]. This would also allow a determination from $\Delta\nu(M)$ of α to about 2 ppb, double the precision of the value extracted from a_e.

Other Tests

Apart from tests of QED and the determination of fundamental constants, muonium has been used in searches for physics beyond the standard model. The experimental search for spontaneous muonium-antimuonium conversion [21] has been used to set limits on couplings in models which allow the violation of lepton family number by two units. Since the conversion probability increases with the square of the evolution time, and the major backgrounds decay exponentially [22], using the intense low energy beams at a muon collider could allow searches with much higher signal/background probability and more stringent limits to be set. The high muon flux would also be an advantage in tests of T-invariance in nuclear muon capture [23], and for the broad field of low energy muon science [24].

Finally we note that measurements of parity violation in muonic atoms [25] (as yet unobserved) would be extremely sensitive to physics beyond the standard model, such as extra Z bosons or leptoquarks [26]. The possibility of observing parity-violation in muonic boron has been investigated at PSI, and appears extremely difficult because the required $2S$ metastability is achieved only at low pressures, restricting the stopping rate to 10^5 μ^-/s, while 10^{13} stopped muons are required for a measurement [27]. It is unclear if the high intensity at a muon collider could make such an experiment feasible.

CONCLUSIONS

The high intensity, low energy muon beams at the front end of a muon collider contain the possibility for orders of magnitude increase in precision in many fundamental measurements involving muonium and muonic atoms. The sensitivity to physics beyond the standard model using these systems also stands to benefit enormously.

REFERENCES

1. N. F. Ramsey, in *Quantum Electrodynamics*, ed. T. Kinoshita (World Scientific, Singapore 1990), p. 673.
2. J.R. Sapirstein and D.R. Yennie, *ibid*, p. 560.
3. Th. Udem *et al.*, *Phys. Rev. Lett.* **79**, 2646 (1997); S. Bourzeix *et al.*, *Phys. Rev. Lett.* **76**, 384 (1996); D.J. Berkeland, E.A. Hinds and M.G. Boshier, *Phys. Rev. Lett.* **75**, 2470 (1995).
4. V.W. Hughes, *Ann. Rev. Nucl. Sci.* **16**, 445 (1966),
5. R.S. Van Dyck Jr. in *Quantum Electrodynamics*, ed. T. Kinoshita (World Scientific, Singapore 1990), p. 322.
6. T. Kinoshita, *Rept. Prog. Phys.* **59**, 1459 (1996).
7. Parameters for the Workshop on Physics at the First Muon Collider and at the Front End of a Muon Collider, November 6-9, 1997, Fermi National Accelerator Laboratory, Batavia, Illinois, USA.
8. M.G. Boshier *et al.*, *Phys. Rev.* **A52**, 1948 (1995).
9. K. Jungmann, *Z. Phys.* **C56**, S59 (1992).
10. M.G. Boshier *et al.*, *Comments At. Mol. Phys.* **33**, 17 (1996).
11. S. Chu *et al.*, *Phys. Rev. Lett.* **60**, 101 (1988).
12. F.E. Maas *et al.*, *Phys. Lett.* **A187**, 247 (1994).
13. K.A. Woodle *et al.*, *Phys. Rev.* **A41**, 93 (1990).
14. C.J. Oram *et al.*, *Phys. Rev. Lett.* **52**, 910 (1984).
15. V.W. Hughes and G. zu Putlitz, in *Quantum Electrodynamics*, ed. T. Kinoshita (World Scientific, Singapore 1990), p. 822.
16. F.G. Mariam *et al.*, *Phys. Rev. Lett.* **49**, 993 (1982).
17. M. Nio and T. Kinoshita, *Phys. Rev.* **D55**, 7267 (1997).
18. S.A. Blundell, K.T. Cheng and J.R. Sapirstein, *Phys. Rev. Lett.* **78**, 4914 (1997).
19. E. Klempt *et al.*, *Phys. Rev.* **D25**, 652 (1982).
20. L.M. Simons, *Phys. Scr.* **T22**, 96 (1988); D. Taqqu, *Nucl. Instr. and Meth.* **A247**, 288 (1986).
21. R. Abela *et al.*, *Phys. Rev. Lett.* **77**, 1950 (1996).

22. L. Willmann and K. Jungmann, in *Atomic Physics Methods in Modern Research*, Lecture Notes in Physics, Vol. 499, (Springer, Heidelberg, 1997), p. 43.
23. S. Ciechanowicz, *Z. Phys.* **A337**, 97 (1990).
24. *Proceedings of the International Workshop on Low Energy Muon Science LEMS '93*, compiled by M. Leon, Los Alamos, LA-12698-C (1994).
25. G. Feinberg and M.Y. Chen, *Phys. Rev.* **D10**, 190 (1974).
26. P. Langacker, *Phys. Lett.* **B256**, 277 (1991).
27. K. Kirch et al., *Phys. Rev. Lett* **78**, 4363 (1997); J. Missimer and L.M. Simons, *Phys. Rep.* **118**, 179 (1985).

Deep Inelastic Scattering

Conveners: Heidi Schellman, Fermilab
 Steven Ritz, Columbia University

Photon Scattering in Muon Collisions[1]

Michael Klasen[2]

Argonne National Laboratory[3]
High Energy Physics Division
Argonne, Illinois 60439

Abstract. We estimate the benefit of muon colliders for photon physics. We calculate the rate at which photons are emitted from muon beams in different production mechanisms. Bremsstrahlung is reduced, beamstrahlung disappears, and laser backscattering suffers from a bad conversion of the incoming to the outgoing photon beam in addition to requiring very short wavelengths. As a consequence, the cross sections for jet photoproduction in μp and $\mu^+\mu^-$ collisions are reduced by factors of 2.2 and 5 compared to ep and e^+e^- machines. However, the cross sections remain sizable and measurable giving access to the photon and proton parton densities down to x values of 10^{-3} to 10^{-4}.

INTRODUCTION

Muon colliders offer an interesting alternative and complement to future e^+e^- linear colliders. If the considerable design difficulties, in particular in the multistage cooling, the neutrino radiation problems arising from decaying muons, and the management of detector backgrounds, can be solved, they can be precisely tuned to a Higgs or $t\bar{t}$ factory, be ramped in several stages to much higher energies up to 3 or 4 TeV, and, due to a smaller total size, eventually be built at lower cost. In addition, an existing high energy proton beam might be collided with one of the muon beams to give a high energy lepton-proton collider. The physics interest focuses on measurements of the Higgs and weak gauge boson properties, top quark physics, supersymmetry with and without R-parity violation, and other phenomena beyond the Standard Model [1].

Little attention has been paid in this context to photon initial states and QCD measurements. This is surprising given the many studies at present and future e^+e^- colliders, the success of HERA in determining the proton structure and hard QCD processes, and the significance of QCD background for many

---

[1] Talk given at the Workshop on Physics at the First Muon Collider, Fermilab, Nov. 1997.
[2] E-mail: klasen@hep.anl.gov.
[3] Work supported by the U.S. Department of Energy, Division of High Energy Physics, Contracts W-31-109-ENG-38 and DEFG05-86-ER-40272.

of the new physics processes under consideration. At LEP, ALEPH, DELPHI, and OPAL have studied the photon structure function $F_2^\gamma(x, Q^2)$ in electron-photon scattering [2], and L3 and OPAL have obtained first measurements of the energy dependence of the total $\gamma\gamma$ cross section for hadron production [3]. Whereas LEP1 was dominated by e^+e^- annihilation at the Z pole, $\gamma\gamma$ scattering is important at LEP2 and will be dominant at a future linear e^+e^- collider. This is due to the fact that the annihilation cross section drops like $\sigma_{l^+l^-} \propto \frac{1}{s}$ or at best like $\frac{\log s}{s}$ ($l = e, \mu$), whereas the photon-photon cross section rises like $\sigma_{\gamma\gamma} \propto \log^3 s$ or even $\propto s$ taking into account the hadronic structure of the photon. This leads us naturally to the presumption that photoproduction might not be as negligible in muon collisions as is widely believed.

Literature on muon colliders in general and on photon radiation by muons in particular is very restricted if not absent. However, two detailed studies on two-photon physics at e^+e^- linear colliders exist [4,5], where the authors discuss photon emission and distribution functions, soft and hard two-photon reactions, and total cross sections. This study follows the lines given in these references, looks for changes from e^+e^- to $\mu^+\mu^-$, extends them to μp collisions and compares these to ep collisions. In the following section, we estimate the production rate of photons from muon beams in three different mechanisms. The results are used in the third section to calculate the cross section for dijet production as a function of the transverse energy of the jets and the center-of-mass energy of the muon pair. Finally, we determine the ranges in x in which the photon and proton structure functions could be measured at a $\mu^+\mu^-$ and μp collider.

(QUASI-)REAL PHOTON EMISSION

Three mechanisms can contribute to the emission of photons from leptons: bremsstrahlung, beamstrahlung, and laser backscattering. To begin with, bremsstrahlung is an approximation of the complete two-photon process $l^+l^- \to l^+l^- X$, where l can be an electron or a muon, and X is a hadronic final state in our case. The photons are radiated from the leptons in the scattering process, and their spectrum can be expressed through the usual Weizsäcker-Williams or Equivalent Photon Approximation [6]

$$f_{\gamma/l}^{\text{brems}}(x) = \frac{\alpha}{2\pi} \left[\frac{1 + (1-x)^2}{x} \log \frac{Q_{\max}^2}{Q_{\min}^2} + 2m_l^2 x \left(\frac{1}{Q_{\max}^2} - \frac{1}{Q_{\min}^2} \right) \right], \quad (1)$$

where $x = \frac{E_\gamma}{E}$ is the fraction of the photon energy E_γ of the lepton beam energy E, and where the virtuality of the photon

$$Q^2 = \frac{m_l^2 x^2}{1-x} + E^2(1-x)\theta^2 + \mathcal{O}\left(E^2\theta^2, m_l^2\theta^2, \frac{m_l^2}{E^2}\right) \quad (2)$$

is always smaller than

$$Q^2 = -(p_l - p_{l'})^2 = 2EE'(1 - \cos\theta) < 4EE' = 4E^2(1-x). \tag{3}$$

This upper limit on Q^2 is used if no information on the scattered lepton is available. For a safe factorization of the lepton tensor and phase space, it is preferred to anti-tag the outgoing lepton and use the maximum scattering angle θ in Eq. (2). Consequently, the photon density will depend on this maximum scattering angle. This is shown in Table 1 for a $\sqrt{s} = 500$ GeV e^+e^- collider. The variation in this case can reach up to 40%. Which changes

TABLE 1. Dependence of the photon density on the maximum scattering angle.

θ_{max}	$f_{\gamma/e}^{brems}(x=0.5)$
—	0.078
10 mrad	0.047
20 mrad	0.051
30 mrad	0.053

occur in the Weizsäcker-Williams spectrum for muon beams? Eq. (1) depends explicitly on the lepton mass m_l in its non-logarithmic contribution as well as implicitly through the minimal photon virtuality $Q^2_{min} = \frac{m_l^2 x^2}{1-x}$. Taking into account only the leading logarithmic contribution, we expect a reduction by a factor of

$$\frac{\log \frac{Q^2_{max}(1-x)}{m_\mu^2 x^2}}{\log \frac{Q^2_{max}(1-x)}{m_e^2 x^2}} = \frac{-\log m_\mu^2}{-\log m_e^2} \simeq \frac{1}{3.38}, \tag{4}$$

where we have taken $Q^2_{max} = 1$ GeV^2 and $1-x = x^2$ or $x = 0.618$ for simplicity. If one considers a muon collider with a photon spectrum on either side, one expects the cross section to be reduced by a factor of $\left(\frac{1}{3.38}\right)^2 \simeq \frac{1}{11.4}$. In Figure 1, we compare the Weizsäcker-Williams (WW) spectra of electrons and muons as a function of x. We have not assumed any anti-tagging conditions on the scattered lepton. The shape of the muon spectrum is basically unchanged with respect to the electron spectrum, and the normalization drops by a factor of two. This is in qualitative agreement with our naïve estimate above.

A second contribution to photon emission comes from beamstrahlung. The particles in one bunch experience rapid acceleration when they enter the electromagnetic field of the opposite bunch. Then the scattering amplitudes between particles within the characteristic length add coherently. This can involve up to 10^6 particles. The intensity and spectrum of the beamstrahlung depend sensitively on the size and shape of the bunches and thus on the machine parameters. It is known that for particular e^+e^- linear collider designs

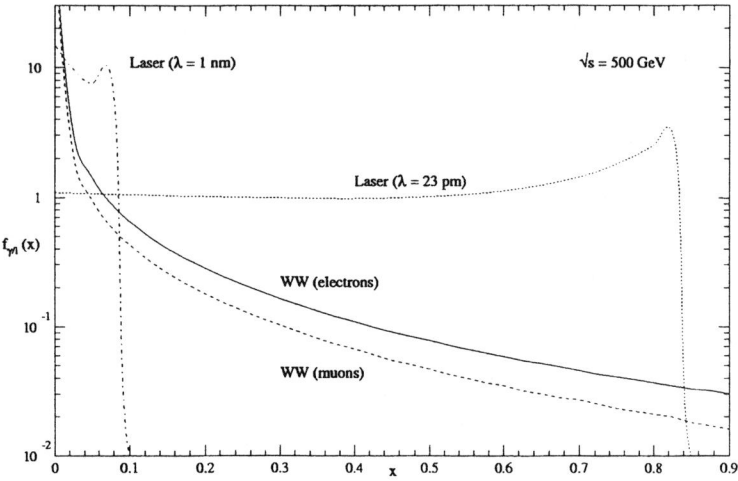

FIGURE 1. *(Quasi-)real photon emission at a $\sqrt{s} = 500$ GeV $\mu^+\mu^-$ collider. We show the Weizsäcker-Williams (WW) spectra for electrons and muons and the spectra for laser backscattering off muons for incident photons with wavelengths of 23 pm and 1 nm.*

beamstrahlung can be more important than bremsstrahlung over a wide range in x [4]. Within a semiclassical calculation and for a Gaussian longitudinal bunch profile, Chen et al. find

$$f_{\gamma/l}^{\text{beam}}(x) = \frac{1}{\Gamma\left(\frac{1}{3}\right)} \left[\frac{2}{3\mathcal{Y}}\right]^{1/3} x^{-2/3}(1-x)^{-1/3} \exp\left[-\frac{2x}{3\mathcal{Y}(1-x)}\right] G(x), \quad (5)$$

where the function $G(x)$ is of order 1 and can found in [5]. The beamstrahlung intensity is controlled by the effective beamstrahlung parameter

$$\mathcal{Y} = \frac{5r_l^2 EN}{6\alpha\sigma_z(\sigma_x + \sigma_y)m_l} \quad (6)$$

mostly through the exponential function.

$$r_l = \frac{e^2}{4\pi\epsilon_0 m_l c^2} \quad (7)$$

is the "classical radius" of each of the N leptons with mass m_l in a bunch of length σ_z and transverse dimensions σ_x and σ_y. E is the beam energy, and $\alpha = \frac{e^2}{4\pi}$ is the fine structure constant. The beamstrahlung parameter $\mathcal{Y}$ depends explicitly and implicitly through the classical lepton radius on the lepton mass m_l. We therefore expect the beamstrahlung parameter for muon beams to be reduced by

$$\frac{\mathcal{Y}_\mu}{\mathcal{Y}_e} = \frac{m_e^3}{m_\mu^3} \simeq 8 * 10^{-6}, \qquad (8)$$

i.e. by three powers of the mass ratio of electrons to muons. Since the beamstrahlung intensity will be suppressed by this factor exponentially, we do not expect any beamstrahlung for muon beams and in fact do not see any in a numerical simulation using the formula of Eq. (5).

e^+e^- colliders can also be run in a mode dedicated to $\gamma\gamma$ collisions using backscattering of laser light off the incident electron and positron beams. One considers even dedicated interaction regions. For unpolarized lepton beams, the spectrum of the backscattered photons depends only on the beam energy, on the laser wavelength, and — this is crucial for muon beams — on the beam particle mass. Ginzburg et al. determined this spectrum to be [7]

$$f_{\gamma/l}^{\text{laser}}(x) = \frac{1}{N}\left[1 - x + \frac{1}{1-x} - \frac{4x}{X(1-x)} + \frac{4x^2}{X^2(1-x)^2}\right], \qquad (9)$$

where the normalization is

$$N = \left[1 - \frac{4}{X} - \frac{8}{X^2}\right]\log(1+X) + \frac{1}{2} + \frac{8}{X} - \frac{1}{2(1+X)^2}. \qquad (10)$$

As already stated above, the crucial parameter

$$X = \frac{4EE_\gamma^{\text{in}}}{m_l^2} \qquad (11)$$

in this case depends on the beam energy E, on the incident photon energy $E_\gamma^{\text{in}} = \frac{hc}{\lambda}$, and on the lepton mass m_l. Telnov estimates the optimal value of X from the threshold of the e^+e^- pair creation in photon collisions being

$$E_\gamma^{\text{out}} E_\gamma^{\text{in}} = \frac{X^2 m_l^2}{4(X+1)} > m_e^2. \qquad (12)$$

For electron beams, $m_l = m_e$ cancels in the equation above leading to an optimal value of $X_e^{\text{opt}} = 2(1+\sqrt{2})$. For muon beams, we find

$$X_\mu^{\text{opt}} = 2\frac{m_e^2}{m_\mu^2}\left(1 + \sqrt{1 + \frac{m_\mu^2}{m_e^2}}\right) \simeq 9.72 * 10^{-3}. \qquad (13)$$

In Table 2 we calculate the optimal laser wavelength as a function of the center-of-mass energy $\sqrt{s}$ of the lepton collider or equivalently the lepton beam energy $E_{e,\mu}$. For electron beams, the optimal wavelengths lie in the region of

TABLE 2. Dependence of the laser wavelength on the center-of-mass energy.

$\sqrt{s}$/GeV	$E_{e,\mu}$/GeV	λ_γ^e/nm	$\lambda_\gamma^{\mu,\text{equiv}}$/pm	$\lambda_\gamma^{\mu,\text{opt}}$/nm
100	50	197	4.6	2.29
200	100	393	9.2	4.57
350	175	688	16.1	8.00
500	250	983	23.0	11.4

visible light (197 - 983 nm) and can be provided by current laser technology. To obtain the same backscattered photon spectrum as with electron beams, much shorter wavelengths are needed at a muon collider (4.6 - 23.0 pm). The spectrum for $\lambda = 23$ pm is shown in Figure 1. This spectrum does, however, not take into account the bad conversion of the incoming to the outgoing photon beam due to enhanced e^+e^- pair creation. If one chooses the optimal value of $X_\mu^{\text{opt}} = 9.72 * 10^{-3}$ for 100 % conversion and no e^+e^- pair production, one finds incident laser wavelengths of 2.29 - 11.4 nm. However, a laser beam of 11.4 nm produces backscattered photons at much too low energies which do not appear on the spectrum of Figure 1 at all. For illustration, we show the spectrum for $\lambda = 1$ nm, where the spectrum is still concentrated at low energies, but at least visible. However, the conversion will be bad, and these short wavelengths can only be obtained with new laser technology like free electron lasers.

QCD AND $\gamma\gamma$ SCATTERING

In this section we will restrict ourselves to the reduced bremsstrahlung spectrum of muon beams and apply it to QCD and $\gamma\gamma$ scattering. As an example, we compare the differential dijet cross section $d\sigma/dE_T^{2-\text{jet}}$ at LEP2 and a muon collider of the same center-of-mass energy $\sqrt{s} = 166.5$ GeV as a function of the transverse energy E_T in Figure 2. This cross section has recently been measured by OPAL for 3 GeV $< E_T <$ 20 GeV, $-2 < \eta_{1,2} < 2$, and a maximum electron scattering angle of $\theta = 33$ mrad using an integrated luminosity of about 20 pb^{-1} [3]. It drops from 133 to 0.16 pb/GeV from the first to the last E_T-bin. The prediction for a muon collider is reduced by factors of 4.64 to 6.47 in this E_T-range. The cross section is still large enough to be measured with good precision.

It is also interesting to look at the dependence of the total dijet cross section on the center-of-mass energy. The result is shown in Figure 3 for two cuts on the transverse energy at 10 and 20 GeV. We observe a linear rise of the cross section as we expect when taking into account the hadronic structure of the photon. Photoproduction of jets will therefore increase linearly in importance with the energy at which a muon collider is operated.

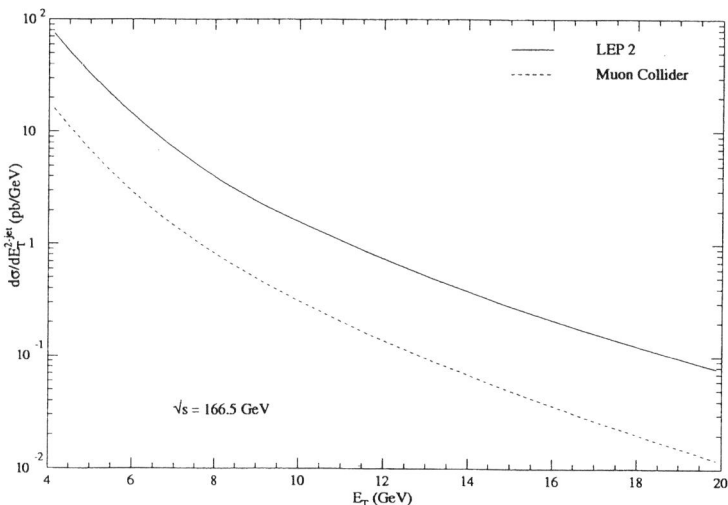

FIGURE 2. *Comparison of the differential dijet cross section at LEP2 and a muon collider of the same center-of-mass energy as a function of the transverse energy E_T.*

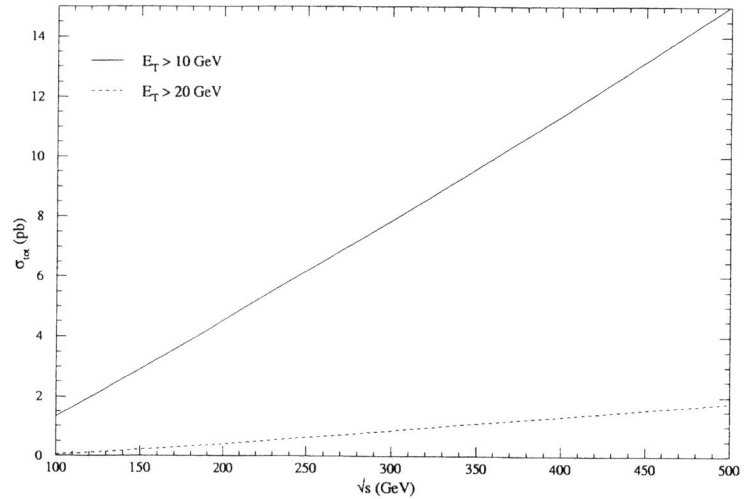

FIGURE 3. *Center-of-mass energy dependence of the total dijet cross section for two thresholds of the transverse energy E_T.*

PARTON DENSITY MEASUREMENTS

We will now investigate the prospects for measuring the photon and proton parton densities at $\mu^+\mu^-$ and μp colliders in photoproduction processes. The x-ranges that are accessible can be calculated from four-momentum conservation and therefore depend completely on the kinematics of the initial and final states. The momentum fraction x_b of a parton b in the correspondent parent particle (muon or proton) is given by

$$x_b = \frac{x_a E_a E_T e^{\eta_1}}{2 x_a E_a E_b - E_b E_T e^{-\eta_1}}. \tag{14}$$

It depends on the momentum fraction x_a of parton a in its parent particle, on the beam energies $E_{a,b}$, on the transverse momentum of the outgoing particles E_T, and on the rapidity η_1 of one of these. The second rapidity η_2 is linearly dependent on these variables. We have plotted this function in Figure 4 for a muon collider with $\sqrt{s} = 500$ GeV and different rapidities $\eta_1 \in [-2; 2]$. This range corresponds to the fairly limited rapidity coverage of a detector at muon colliders due to massive shielding of decay electrons from the muon beams. The transverse energy $E_T = 10$ GeV has been chosen in such a way that one may hope for a sufficient suppression of non-perturbative effects and a good determination of the energy scale which are indispensable for a safe extraction of the parton densities. Figure 4 also shows the envelope

$$x_{b,\text{env}} = \frac{E_T^2}{x_a E_a E_b}, \tag{15}$$

which gives the absolute limits on $x_{a,b}$ independent of the rapidity η_1. We find that for the kinematical conditions described above, rather low values of $x_{a,b} > 1.6 * 10^{-3}$ can be reached where little is known about the parton densities (in particular of the gluon) in the photon.

Since it is not clear at which energy a muon collider will eventually operate, we also consider the center-of-mass energy dependence of the minimal x value that can be reached in Figure 5. For $E_T = 10$ (20) GeV, one can only go down to 0.04 (0.16) for a 100 GeV collider. Higher energies are therefore a clear advantage for photoproduction both for higher cross sections and for larger x-ranges.

Finally, we repeat the above analysis for a 200 GeV x 1000 GeV muon-proton collider. The result is shown in Figure 6. We find that the parton densities in the photon and proton could be measured down to values of $\mathcal{O}(5 * 10^{-4})$.

CONCLUSIONS

We have studied several aspects of photon physics at muon colliders. Photon emission due to bremsstrahlung is found to be reduced, there is no beamstrahlung, and laser backscattering appears to be difficult. However, QCD

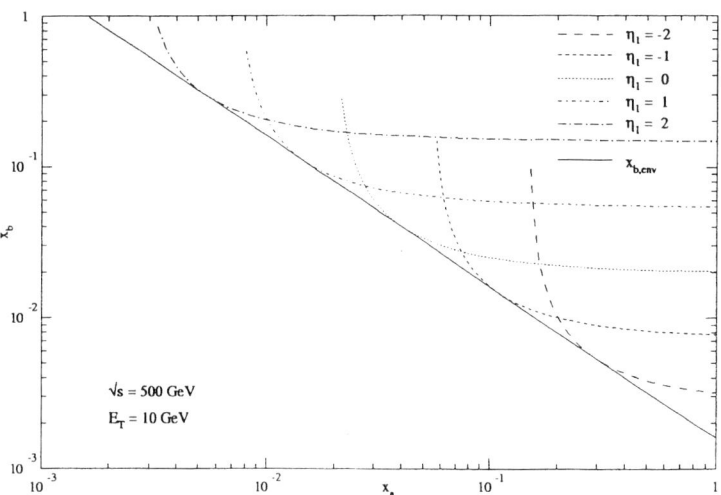

FIGURE 4. Parton density regions in the $x_a - x_b$ plane that can be accessed at a $\sqrt{s} = 500$ GeV muon collider for different rapidities η_1 and $E_T = 10$ GeV.

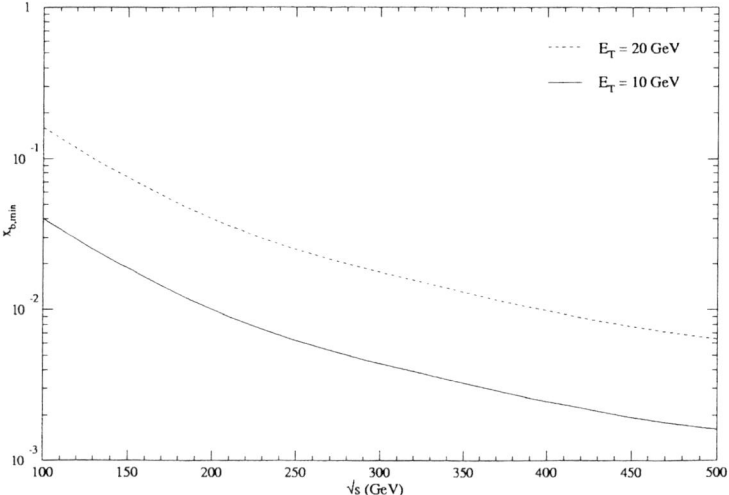

FIGURE 5. Center-of-mass energy dependence of the minimal x_b value for two different transverse energies $E_T = 10$ and 20 GeV.

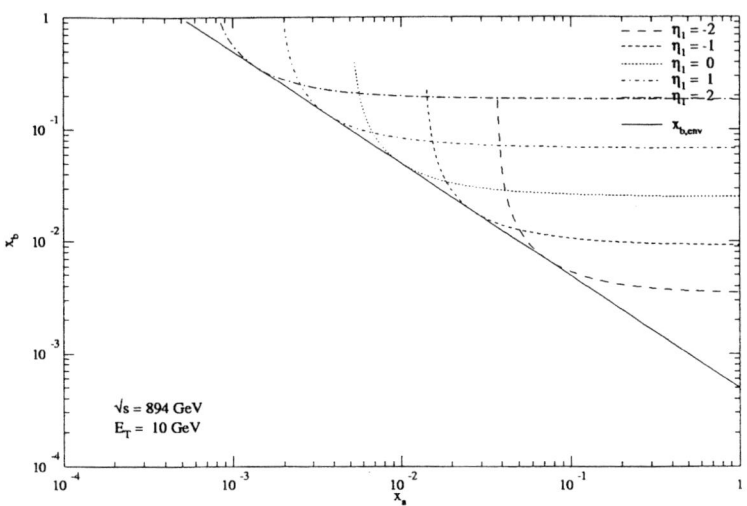

FIGURE 6. *Parton density regions in the $x_a - x_b$ plane that can be accessed at a $\sqrt{s} = 894$ GeV muon-proton collider for different rapidities η_1 and $E_T = 10$ GeV.*

cross sections like photoproduction of dijets remain sizable and increase linearly with the center-of-mass energy. Photon and proton parton densities could be measured down to x values of 10^{-3} to 10^{-4}.

REFERENCES

1. Gunion J.F., in: Proc. of *Beyond the Standard Model V*, Balholm, Norway, May 1997, hep-ph/9707379.
2. ALEPH Coll., paper LP-315 submitted to LP 97, Hamburg, Germany; DELPHI Coll., *Z. Phys.* **C69** (1996) 223; DELPHI Coll., paper 416 submitted to EPS 97, Jerusalem, Israel; OPAL Coll., *Z. Phys.* **C74** (1997) 33; OPAL Coll., CERN-PPE/97-87, hep-ex/9708019; OPAL Coll., CERN-PPE/97-103, hep-ex/9708028.
3. Söldner-Rembold S., plenary talk P01 at LP 97, Hamburg, Germany, hep-ex/9711005.
4. Drees M., and Godbole R.M., *Z. Phys.* **C59** (1993) 591.
5. Chen P., Barklow T.L., and Peskin M.E., *Phys. Rev.* **D49** (1994) 3209.
6. Frixione S., Mangano M.L., Nason P., and Ridolfi G., *Phys. Lett.* **B319** (1993) 339.
7. Ginzburg I.F., Kotkin G.L., Serbo V.G., and Telnov V.I., *Nucl. Instrum. Methods* **205** (1983) 47; Ginzburg I.F., Kotkin G.L., Panfil S.L., Serbo V.G., and Telnov V.I., *Nucl. Instrum. Methods* **219** (1984) 5; Telnov V.I., *Nucl. Instrum. Methods* **294** (1990) 72.

A Small Target Neutrino Deep-Inelastic Scattering Experiment at the First Muon Collider

Deborah A. Harris* and Kevin S. McFarland[†]

*University of Rochester, Rochester, NY 14627
[†]Massachusetts Institute of Technology, 77 Massachusetts Ave., Cambridge, MA 02139

Abstract. Several different scenarios for neutrino scattering experiments using a neutrino beam from the muon collider complex are discussed. The physics reach of a neutrino experiment at the front end of a muon collider is shown to extend far beyond that of current neutrino experiments, since the high intensity neutrino beams one would see at the muon collider allow for a large flexibility in choosing neutrino targets. Measurements of quark spin, A-dependence of the structure function xF_3 and neutral current chiral couplings to quarks are outlined.

INTRODUCTION

Neutrino deep-inelastic scattering has proven an invaluable tool to study hadron structure, QCD, and the electroweak force. Neutrinos give the only clean measurement of the valence quark distribution inside a nucleon, and have a very distinct signature for scattering off a strange quark inside the nucleon sea. Neutrino-nucleon scattering also provides two important tests of QCD, through structure function evolution and the Gross Llewellyn Smith Sum Rule. Both tests can also provide precise measurements of the strong coupling constant, α_s [1]. Neutrino scattering experiments which can reconstruct neutral current interactions also provide a stringent test of electroweak symmetry breaking.

In the past, neutrino experiments have been shaped by the low neutrino scattering cross section. To get ample statistics experiments have had to use massive active targets. For example, the CCFR/NuTeV experiments used a target of 700 tons of steel instrumented with scintillators [2], and the NOMAD experiment used a target of 2.7 tons of drift chambers [3]. In reference [4] the results from a simulation of the neutrino beams resulting from high energy muon beams were given, and event samples of comparable statistics are attainable using targets of only several g/cm^2 thickness. In this paper we consider targets that have previously been used for muon scattering, and discuss the novel physics that suddenly becomes feasible when placing these targets in extremely intense neutrino beams.

NEUTRINO SCATTERING KINEMATICS

By including known ν-nucleon differential cross sections in the Monte Carlo described in reference [4], one can predict the kinematic reach of a year's worth of data per g/cm^2 of target. It is important to note that since the neutrino beams from a muon collider are not more energetic than conventional beams, the reach in momentum transfer squared (Q^2) of these experiments is comparable to what has already been measured. However, since α_s is large and changing rapidly in the region $1 < Q^2 < 100\,GeV^2$, this is precisely where one wants to make perturbative QCD measurements. The kinematic variable one could extend in reach is x, or the struck quark's fractional momentum. To probe smaller and larger x regions one needs both high statistics and good detector acceptance and resolution. To maximize statistics one could use a target which would result in a optimal number of interactions (for example, .05) per turn. This target would still be rather small, and would not degrade the event reconstruction, compared to previous neutrino experiments. The remainder of the experiment could then be a low-mass spectrometer with some particle identification, modeled on non-neutrino fixed target experiments (see reference [4]). A target of the size described above with a low mass detector to reconstruct the scatters would result in about 2×10^7 events per year for one of the recirculating linacs, and a factor of 100 more than that for an experiment downstream of a collider ring's straight section. Figure 1 shows the number of events expected per g/cm^2 as a function of x and Q^2, assuming a 20 cm radius target. The reach in x for a "high statistics target" in a given Q^2 region is an order of magnitude beyond what is currently measured. As there have been surprises in the low x region, for example, the fast rise of F_2 as seen by HERA [5], the question arises, is the valence quark distribution well-behaved at low x, or does something unexpected happen there too?

Strong Coupling Constant Determinations

Neutrino deep-inelastic scattering provides two clean measurements of the strong coupling constant α_s: one from the Q^2 dependence of the non-singlet structure function xF_3, and one from the Gross Llewellyn Smith Sum Rule. Both of these measurements are independent of the gluon distribution, and the latter has very small perturbative QCD uncertainties. Currently, however, both measurements are limited by experimental systematic uncertainties: the xF_3 evolution is limited by the uncertainty in the energy calibration [2], and the GLS Sum Rule is limited by the uncertainty in both the overall neutrino cross section, and the ratio of neutrino to anti-neutrino cross sections [6]. Furthermore, both cross section measurements are themselves limited by systematic uncertainties [7]. A neutrino experiment at a muon collider could address all three of these issues in fundamentally different ways, because of the way the beam is formed.

Neutrinos from a monochromatic beam of muons have a very distinct spectrum

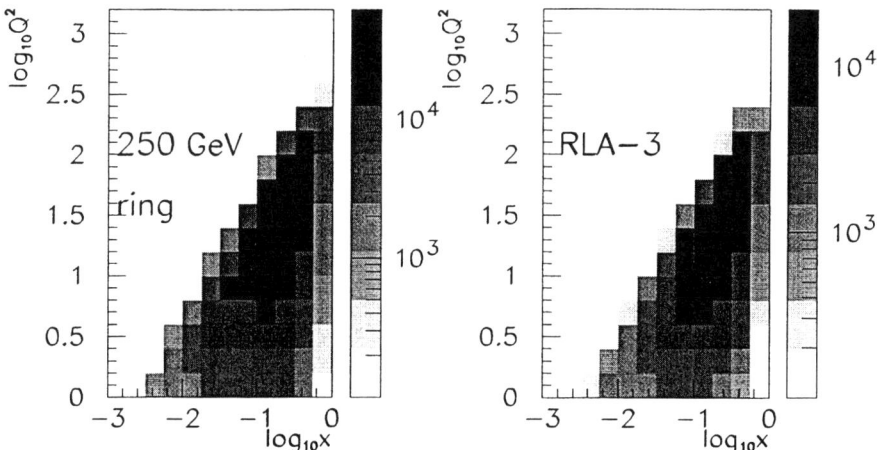

FIGURE 1. x versus Q^2 distributions for two possible neutrino experiments. Units are in events per year per g/cm^2 of target for a $20\,cm$ radius target.

peaked towards the high end of the available energy from the muon. Muon decay itself is very well understood, and the muon energies at the collider are predicted to be measurable to a few parts per million [8]. By calibrating the detector to the end point of the neutrino spectrum one should be able to achieve much better than the current energy scale uncertainty in neutrino experiments, which is about 1% [2].

Similarly, the muon currents in the beam-lines can be measured, and given the energy of the muons and the currents and the geometry of the beam-line, the neutrino flux at a given detector can be accurately predicted. The availability of an independent flux measurement for a neutrino experiment will be an important improvement and will dramatically change the nature of absolute neutrino cross section measurements, as well as the scale of the uncertainties in those measurements.

LIGHT TARGETS

Large low-Z targets have already been used successfully by muon scattering experiments. The sizes of liquid hydrogen targets are in general limited by safety considerations, and a liquid hydrogen target of $1m$ thickness and radius approximately $10\,cm$ has been used by experiment E665 at Fermilab [9]. A target of this volume of hydrogen with a $20\,cm$ radius would have a thickness of $1.75g/cm^2$, resulting in data samples of slightly over a million ν charged current events per year. The per turn event rate on this target would still be small enough so that pile-up in the target itself would not be a problem.

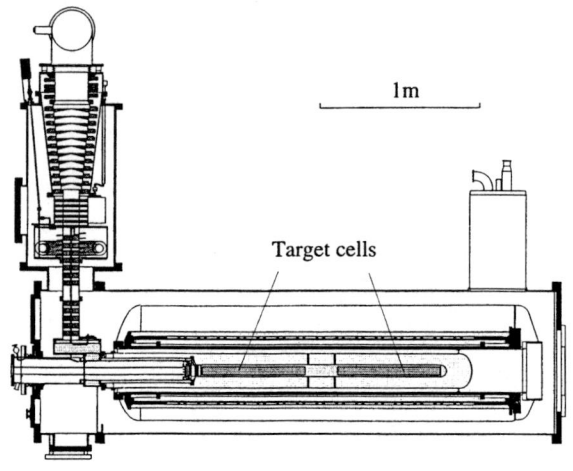

FIGURE 2. Polarized butanol target used in the SMC experiment (see text). The total target mass was 3 kg, and the polarized fraction of the target was approximately 10%.

NUCLEAR EFFECTS

Charged lepton experiments have studied nuclear dependences of the structure function F_2 and have seen a sizeable effect [10] [9], yet no nuclear dependences have been seen in xF_3, and for now theory must only assume that the integral of the valence quark distribution is independent of nuclear corrections. By scattering neutrinos off several light targets one could finally determine whether or not valence quarks know anything about what kind of nucleus they are in. If one had both a hydrogen and a deuterium target, one could measure if $u(x)$ in the proton is indeed equal to $d(x)$ in the neutron. Such a measurement would either confirm or challenge our assumptions about isospin symmetry. The effects seen in charged lepton scattering are large; for example, $2F_2^d/F_2^p - 1$ is $0.935 \pm 0.008(\text{stat}) \pm 0.034(\text{syst})$ at x below 0.01 [9], and at higher x this ratio is even farther away from unity. With hydrogen and deuterium targets of the sizes suggested above one would have enough statistics to measure the quantity $2xF_3^d/xF_3^p - 1$ as a function of x to better than a per cent, over a reasonably large x range.

POLARIZED TARGETS

One exciting possibility in such a light-target neutrino experiment is measuring neutrino scattering from polarized targets. The neutrino beams from the muon collider can produce on the order of 5×10^5 events/kg/year in a target; therefore, a target with approximately 20 kg of polarized protons would produce excellent statistics.

An example of a large polarized target used in the past is the SMC solid butanol target [11] (Figure 2), which was a 2.5 cm radius cylinder containing 0.3 kg of polarized protons. If such a target could be scaled up to the size of the neutrino beam, approximately 20 cm in radius, this would provide sufficient mass for a high statistics neutrino experiment.

SPIN PHYSICS

If the polarized target described above were feasible, the physics motivations would be many-fold. Deep-inelastic scattering from hydrogen targets would probe the following processes

$$\nu u \to \mu^- d \quad \nu \bar{d} \to \mu^- \bar{u}$$
$$\bar{\nu} d \to \mu^+ u \quad \bar{\nu} \bar{u} \to \mu^+ \bar{d}$$
$$\bar{\nu} s \to \mu^+ c \quad \nu \bar{s} \to \mu^- \bar{c},$$

and their Cabbibo-suppressed analogs. The beauty of probing polarized targets with neutrinos is that these six processes are clearly separable. y distributions can be used to separate quark and anti-quark targets; neutrinos and anti-neutrinos each pick one flavor of a quark doublet; strange quarks can be tagged by final state charm in its decay to high momentum leptons or *via* its finite lifetime.

This leads to the possibility of a measurement of the polarization of the proton's quarks by flavor, with sea and valence contributions separated. This measurement would be relatively independent of details of final state hadronization in contrast with current experimental programs of measuring the polarization of the strange sea which rely on final state strange mesons or baryons.

The statistical challenge for such an experiment would be the measurement of the strange polarization from final state di-lepton events, where one lepton comes from the decay of charm. The total di-lepton cross-section is approximately 2%, therefore, on a 200 kg butanol target, approximately 1×10^6 neutrino and 5×10^5 anti-neutrino di-lepton events per year would be observed, about half of which come from scattering off of strange quarks. With these statistics, a raw asymmetry could be measured with a 1.5×10^{-3} precision. Assuming the d quark asymmetry would be better known from the more common processes listed above, this would translate into a precision in the strange sea polarization of 3% in a one year run.

This precision is reasonable for testing models for the spin of the strange sea. A recent prediction for the strange sea polarization in an effective chiral quark model [12] estimates a -8% polarization.

NEUTRAL CURRENT QUARK COUPLINGS

Neutrino deep-inelastic scattering from light, non-isoscalar nuclei also gives the possibility of doing precision measurements of the left- and right-handed couplings of the neutral current to quarks. To date, precision neutrino and atomic physics experiments have probed primarily heavy nuclei and thus could only measure isoscalar combinations of couplings. Such observations could determine whether the lingering evidence for some enhanced coupling of the neutral current to b-type quarks [13] is only observed in the b system or whether it is perhaps the first evidence of some generation-independent phenomenon.

CONCLUSIONS

We have presented several different physics motivations for a deep-inelastic scattering neutrino experiment at the front end of the muon collider complex. The possibility of scattering neutrinos off either light elements or polarized targets will address a wide range of issues in our understanding of hadron structure. The experience from polarized targets used in muon scattering experiments will be particularly useful, as it would enable measurements of the individual quark contributions to the spin of the proton. In order to best take advantage of the neutrino beams at the muon collider complex a run of two or more years using several different cryogenic targets would be optimal. Furthermore, the characteristics of this new neutrino beam would be ideal for new precision measurements of both the strong coupling constant, and the neutral current quark couplings. In summary, deep-inelastic scattering neutrino experiments at the muon collider have the potential to fundamentally change our understanding of the baryons that make up most of the observed universe.

REFERENCES

1. D.J. Gross and C.H. Llewellyn Smith, Nucl. Phys. **B14** 337 (1969), and G. Altarelli and G. Parisi, Nucl. Phys. **B126** 298 (1977).
2. W.G. Seligman et al., Phys. Rev. Lett. **79**, 1213 (1997).
3. M. Laveder, Nucl. Phys. Proc. Suppl. **48**, 188 (1996).
4. D.A. Harris and K.S. McFarland, "Detectors for Neutrino Physics at the First Muon Collider", these proceedings.
5. J. Breitweg et al., Phys. Lett. **B407**, 432 (1997).
6. J. Yu et al., proceedings to XXXI Rencontres de Moriond, QCD and Hadronic Interactions (1997).
7. W.G. Seligman, Ph.D. Thesis, NEVIS-292 (1997).
8. R. Raja, A. Tollestrup, FERMILAB-Pub-97/402-E, submitted to Phys. Rev. **D**.
9. M.R. Adams et al.,Phys. Rev. Lett. 75,1466 (1995).
10. EMC, J.J. Aubert et al., Phys. Lett. **123B**, 275 (1983)
11. Adams, D., et al., Phys. Rev. **D56**, (1997)
12. E. Eichten, I. Hinchliffe, C. Quigg, Phys. Rev. **D45**, 2269 (1992).
13. Particle Data Group, Phys. Ref. **D54**, (1996).

Prospects for a measurement of $\nu p \to \nu p$ scattering at the muon collider complex

Eric G. Stern

Nevis Laboratories
Columbia University
P.O. Box 137
Irvington, NY 10533

Abstract.
There has been much interest [1] in the nuclear and spin physics community in the measurement of νp elastic scattering as a tool to understanding the contribution of the strange quark to the proton spin. Previous [2] measurements have had difficulties due to large errors from non-elastic contamination and smearing from nuclear effects due to their heavy targets. The high intensity and well collimated neutrino beams from the recirculating linacs in the proposed muon collider complex offer the possibility of a measurement of this process with a well understood beam at low energies using a liquid hydrogen target eliminating many of the systematic effects of the previous measurements.

There has been much interest [1] in the nuclear and spin physics community in the measurement of νp elastic scattering as a tool to understanding the contribution of the strange quark to the proton spin. The spin of the proton may receive a contribution from the strange quark through its axial coupling to the weak hadronic current $\langle p|\bar{s}s|p\rangle$. Charged lepton scattering is not sensitive to the axial contribution of this term; it requires a parity violating weak component which appears in neutrino elastic scattering form factors.

I EXPERIMENT ISSUES

The most recent experiment [2] to measure νp elastic scattering was experiment E734 at Brookhaven National Laboratories. This experiment used a 170 metric ton fine grained calorimeter as a neutrino target. Each module consisted of 112 modules with a plane of sixteen liquid scintillator cells for measuring energy and timing, and a plane of X and Y proportional drift tubes for measuring position to within 1.3 mm in each view. This was followed by a dense section of scintillator planes with lead absorber for shower containment and a magnetic spectrometer.

The neutrino beam [3] was a wide band horn based beam with a mean energy of 1.2 GeV. The experiment had a sample of 1686 νp elastic and 1821 $\bar{\nu} p$ elastic candidates after final cuts. Of these events, 40% of the νp sample, and 52% of the $\bar{\nu} p$ sample were estimated to be due to contamination by background sources reducing the sensitivity of this measurement to form factor parameters.

Elastic events are only distinguished by the absence of other particles in the event. In general, the kinematics of the elastic scatter at these energies dictate that the angle of the outgoing proton is quite large. Figure 1 (Left) shows the measured proton angle for the sample of νp and $\bar{\nu} p$ events. Figure 1 (Right) shows an example event display of a typical νp elastic event illustrating several of the problems with extracting a clean sample of these events. In a neutrino-proton elastic scattering event, the only visible particle will be an outgoing proton and the only observables are the proton energy and angle. The proton (kinetic) energy T is related to the invariant variable Q^2 by the relation $Q^2 = 2M_p T$. At low Q^2 where the sensitivity to strange quark content is most evident, the energies are so low and the angles are large so that the proton track does not cross enough modules to determine its parameters.

The detector was primarily composed of carbon and aluminum so over 80% of target protons are bound. As a result, nuclear effects complicate matters in several ways. At low Q^2 especially, the cross section is suppressed by the Pauli exclusion principle. The apparent proton energy is reduced by the energy with which it is bound in the nucleus. Fermi momentum causes smearing of the apparent proton energy and angle once it escapes the nucleus. In addition, as the proton exits the nucleus it can scatter off of the spectator nucleons ejecting additional particles which will deposit energy and obscure the selection of events without additional particles.

The nuclear composition of the detector contributes to the mismeasurement of the proton energy and path length. A medium Q^2 elastic scatter event with a $Q^2 = 0.6 \,\text{GeV}/c^2$ has a proton energy of about 320 MeV. At this energy, the normal range of the proton [4] is $80 \,\text{g/cm}^2$, approximately equal to the nuclear interaction length in scintillator oil which makes up the bulk of the detector material. So in this detector, most events that can be measured have a high probability to undergo some sort of possibly catastrophic nuclear interaction during their passage through the detector. This effect has to be modeled in a Monte Carlo resulting in poor resolution for event parameters and difficulties in extracting a clean event sample.

II NEUTRINO BEAM FROM RECIRCULATING LINAC

Neutrino beams produced as part of the process of accelerating muons for a muon collider have characteristics markedly different than those produced in a conventional neutrino beam. As the muons travel in the beam pipe, they decay producing neutrinos that are close in direction to the parent muon. The energy of the parent muon is well determined so there is a close correlation between the neutrino direction and its energy.

We will concentrate on the first recirculating linac (RLA1) in the proposed muon collider complex. Since the cross section for νp and $\bar{\nu} p$ elastic scattering is constant above a neutrino energy of 1.5 GeV while the total cross section for neutrino interactions is rising linearly, it is evident that running with higher energy neutrinos just contributes non-elastic contamination events that must be rejected to arrive at an elastic scattering signal. Muons in RLA1 [5] are accelerated in 9 turns beginning at an energy of 1 GeV and ending at 9.64 GeV. It would make sense to use just those neutrinos produced during the first turn when the muon energy is less than 1.5 GeV. This rules out a bubble chamber as the detector technology since the formation time for bubbles is on the order of milliseconds while a single turn takes place in 1 microsecond. An electronic counting experiment should be able to take advantage of this however and only accept triggers during the first turn.

To determine what sort of neutrino rate is available, we have performed a simulation of expected neutrino events from the first turn of RLA1. From the workshop parameters, we have that there are $9.8 \cdot 10^{18}$ muon decays in the straight section of RLA1 in the first turn when the energy goes from 1.0 GeV to 1.5 GeV. We allow these muons to decay along the length of the 100 m straight section. We use the full dynamics of muon decays to determine the distribution in neutrino energy positions and angles. We assume a liquid hydrogen target located at the end of the straight section. We use the calculated cross section for $\nu p \to \nu p$ elastic scattering to determine the event rate per gram of liquid hydrogen in such a target. The results are shown in figure 2 in units of millions of $\nu p \to \nu p$ elastic vents/gram/year as a function of the neutrino energy producing the event.

From the total area of this plot, we can calculate that for a target that is a 50 cm radius cylinder, 100 cm long, there will be 2488 events $\nu p \to \nu p$ events per year. Muon decays also produce electron antineutrinos. There will be 975 elastic scatters from these antineutrinos per year. To do an experiment in this configuration, one would use a liquid hydrogen target followed by traditional fixed target tracking and particle identification methods. Since the target is plain hydrogen and a detector itself can be low mass, this gives the possibility of a measurement of *nuptonup* elastic scatterint process with low systematics. Rozhdestvensky and Sapozhnikov [6] have calculated that with a low systematics experiment, a very large event sample is not necessary, one on the order of 1000 events is enough to make a useful measurement. Of course we have not included any effects having to do with low energy protons leaving the target or anything about the actual detector. A true estimate of the feasibility of such an experiment would need a more complete design and Monte Carlo simulation.

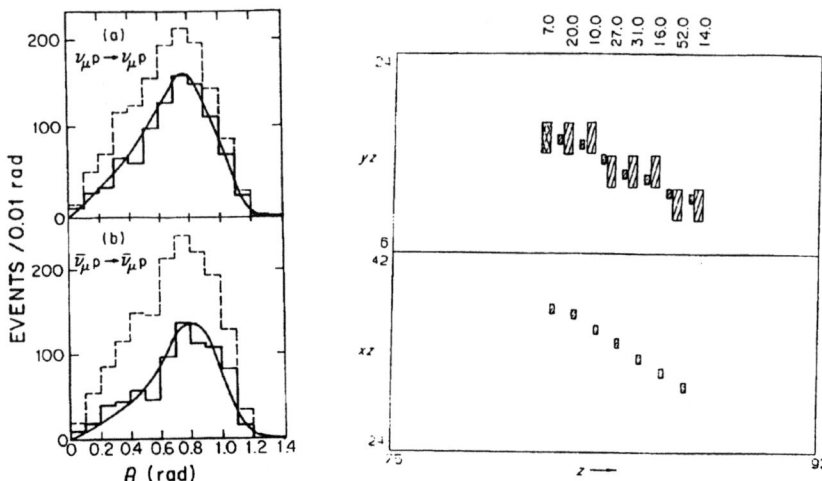

FIGURE 1. (Left) Distributions of the angle θ_p of the proton candidates for (a) $\nu p \to \nu p$ and (b) $\bar{\nu} p \to \bar{\nu} p$. The dashed lines represent the data after particle identification and cuts which eliminate events with obvious extra energy deposits. The solid histograms are the observed distributions after Monte Carlo calculated background subtraction while the curves are the Monte Carlo calculated elastic scattering distributions. (Right) Event display of a typical $\nu p \to \nu p$ candidate event. The two views are xz and yz projections. The large boxes in the yz view show scintillator cells with the numbers above the display indicating the deposited energy in MeV for each cell. The small boxes are proportional drift hits indicating particle position.

ACKNOWLEDGMENTS

The author gratefully acknowledges useful discussions with P. Spentzouris.

REFERENCES

1. Kaplan and Manohar, Strange Matric Elements in the Proton From Neutral Current Experiments, preprint HUTP-88/A024, Harvard University (1988), Alberico, Bilenky, Giunti, Maieron, Elastic νN and $\bar{\nu} N$ Scattering and Strange Form Factors of the Nucleons, preprint hep-ph/9508277 (1995).
2. L.A. Ahrens et al., Phys. Rev. **D35**, 785 (1987).
3. L.A. Ahrens et al., Phys. Rev. **D34** 75 (1986).
4. R.M. Barnett et al., Extracted from the Review of Particle Properties, Phys. Rev. **D54** 1 (1996).
5. Workshop on Physics at the Front End of the Muon Collider Neutrino Beam Parameters from
   http://fnphyx-www.fnal.gov/conferences/femcpw97/neutrino_params.html.
6. A.M.Rozhdestvensky and M.G.Sapozhnikov, Polarized instrinsic nucleon strangeness

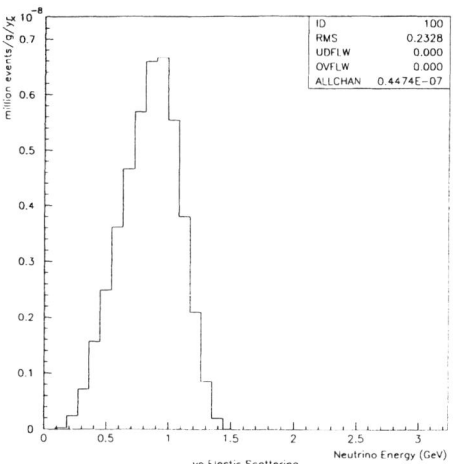

FIGURE 2. Event rate for neutrinos producing νp elastic scattering events in a liquid hydrogen target in units of million events/year/gram of liquid hydrogen as a function of the neutrino energy.

and elastic neutrino scattering on nucleons, presented at the 1997 Main Injector Stationary Target workshop, Fermilab, Batavia, Illinois (1997).

SUSY Searches and Measurements

Conveners: Marcela Carena, Fermilab
           Serban Protopoposcu, Brookhaven National Laboratory

SUSY Searches at LEP: Present Status and Future Prospects

Jane Nachtman

University of California, Los Angeles
Los Angeles CA 90095

Abstract. This talk presents the latest results as of the summer of 1997 of direct searches for supersymmetry at LEP and extrapolates to the discovery potential with the full LEP2 data set.

INTRODUCTION

Direct searches for supersymmetric particles (charginos, neutralinos, sleptons and squarks) have been performed by LEP experiments (ALEPH, DELPHI, OPAL, and L3) in many possible decay modes and extended to experimentally difficult cases. Extra efforts have been made to reduce the model dependence of the results, and to combine the results of all four experiments [1]. Alternate scenarios of allowing R parity violation and non-MSSM models have been considered.

The current LEP2 data set consists of 6 pb^{-1} of data at $\sqrt{s} = 130$ and 136 GeV taken in November, 1995, 20 pb^{-1} at $\sqrt{s} = 161$ and 172 GeV taken in 1996, and ~ 60 pb^{-1} at $\sqrt{s} = 183$ GeV taken in 1997. The results presented here include only ~ 7 pb^{-1} at $\sqrt{s} = 183$ GeV. The full LEP2 data set will include at least an additional 200 pb^{-1} at energies of up to 200 GeV.

MSSM-BASED SEARCHES

When R parity conservation is assumed, sparticles are pair-produced and decay to the lightest SUSY particle (LSP) usually assumed to be the lightest neutralino (denoted χ), which is stable and undetected, leading to the classic missing energy signature. In this scenario, one of the most important experimental aspects of searches for $X \to Y_{stable} + Standard\ Model\ particles$ is $\Delta M = M_X - M_Y$. This quantity determines the visible mass and energy of the detected event, which determines the most relevant backgrounds. Very low ΔM (< 10 GeV/c^2) leads to events with low visible mass and energy, where the most difficult background is the two-photon process, $e^+e^- \to f\bar{f}$. Intermediate ΔM (≈ 40 GeV/c^2) signals have more background from $e^+e^- \to q\bar{q}\gamma$, WW, ZZ (where one Z can be a Z* or γ^*). High ΔM (> 50 GeV/c^2) signals have large background from $e^+e^- \to$ WW.

Searches for Charginos and Neutralinos

The expected signature for pair production of charginos assumes that each chargino decays to the lightest neutralino and a virtual W, leading to three possible topologies based on the possible decays of the W's. The topologies are leptons, leptons and hadrons, and hadrons. The branching fractions into each topology are those of the W, except in the case of light sleptons, which causes an enhancement of decays through virtual sleptons, leading to more events in the purely leptonic topology. In order to set limits which are independent of the details of the model, searches are developed for all possible decay topologies and for a wide range of masses and mass differences.

No evidence of chargino production has been found by the LEP experiments [2], and limits are set. The general procedure is to derive the basic experimental result, the limit on the production cross section. The result is then interpreted in the context of a model. The masses and cross section depend on the parameters of the model, and so the experimental constraints can be applied to constrain those parameters. With the latest data at $\sqrt{s} = 183$ GeV, charginos with mass less than 90.4 GeV/c^2 are excluded for gaugino-like charginos ($mu = -500$ GeV/c^2) for heavy sneutrinos ($m_0 = 200$ GeV/c^2) and $\tan\beta = \sqrt{2}$ [2]. Typically, some parameters are fixed to representative values, and others are scanned to find the most conservative limits. More restrictive limits are placed when more model assumptions are used, and the goal is to try to remove as many assumptions as possible in order to obtain the most conservative and widely applicable limits.

An example of a model assumption commonly applied is the gaugino mass unification condition, $M_1 = \alpha \frac{5}{3} \tan^2 \theta_W M_2$, where α is taken to be 1. This relation fixes the relation between $M_{\chi^\pm}$ and M_χ. However, the searches are still viable without this condition – varying α allows almost any M_χ to be obtained for a given $M_{\chi^\pm}$, and the searches have high efficiency for a large range of ΔM. The ALEPH limit on the chargino mass as a function of neutralino mass is shown in Fig. 1 [2], for several values of the sneutrino mass, and $\tan\beta = \sqrt{2}, \mu = -500$ GeV/c^2.

The simplest experimental scenario for neutralino production is associated production of the lightest and next-to-lightest neutralinos, χ and χ', with the subsequent decay chain, $\chi' \to f\bar{f}\chi$. This leads to a signature of acoplanar jets and missing energy, when the fermions in the final state are quarks. No evidence for supersymmetry was found in this channel by the LEP experiments [2].

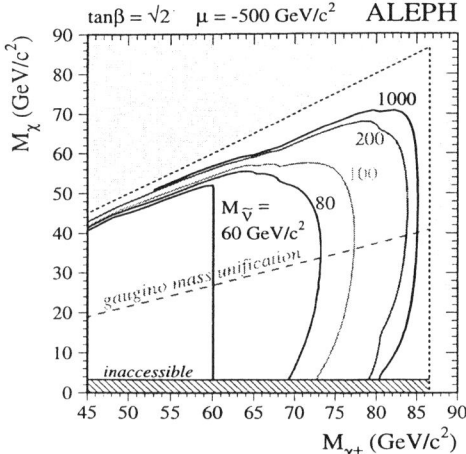

FIGURE 1. Excluded chargino and neutralino masses from ALEPH searches at $\sqrt{s} \leq 172$ GeV, for $\tan\beta = \sqrt{2}$ and $\mu = -500$ GeV/c^2, from Ref. [2].

The search for neutralino production can be extended to consider heavier neutralinos, some of which are accessible at LEP energies, and lead to cascade decay topologies [2]. In addition, radiative decays ($\chi' \to \gamma\chi$) have been considered.

Searches for Squarks and Sleptons

Due to large mixing in the third generation, the stop (and in some cases, the sbottom) can be much lighter than the other squarks. The stop can decay as $\tilde{t} \to c\chi$ or, if sneutrinos are light, $\tilde{t} \to b\ell\tilde{\nu}$. The signature for sbottom production is $\tilde{b} \to b\chi$, leading to a similar topology as for $\tilde{t} \to c\chi$. The cross sections for stop and sbottom pair production depend on the mixing angle, and limits are presented in terms of mixing angle and sparticle masses. Limits have been combined for the LEP experiments [3]. In the $\tilde{t} \to c\chi$ channel, $\tilde{t}$ with mass less than 78 (75) GeV/c^2 are excluded for minimal (maximal) mixing, for $\Delta M > 10$ GeV/c^2. In the $\tilde{t} \to b\ell\tilde{\nu}$ channel, $\tilde{t}$ with mass less than 70 (64) GeV/c^2 are excluded for minimal (maximal) mixing, for $\Delta M > 7$ GeV/c^2. From the $\tilde{b}$ searches, the limits are 71 (51) GeV/c^2 for minimal (maximal) mixing, for $\Delta M > 7(10)$ GeV/c^2.

Searches for selectrons, smuons, and staus look for acoplanar leptons or jets, with missing energy. Mixing is relevant in the stau sector. The combination of all LEP results in the selectron and smuon searches leads to a potential gain of ≤ 5 GeV/c^2 in the excluded range. For $\Delta M > 15$ GeV/c^2, the limit on the production cross section of selectrons (smuons) is 0.2 (0.13) pb [4].

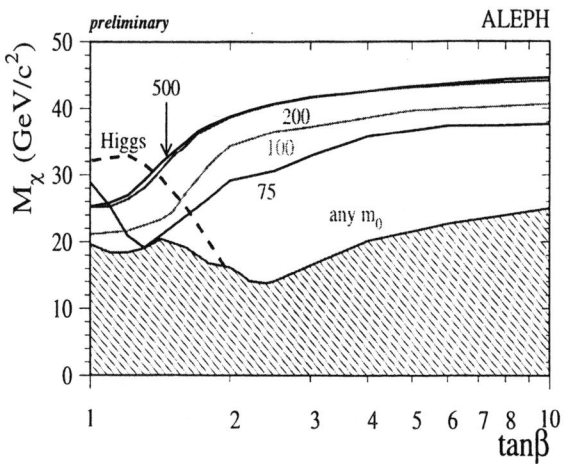

FIGURE 2. Lower limit on the mass of the lightest neutralino derived from data taken at $\sqrt{s} \leq 172$ GeV, for several m_0 and the value of m_0 giving the most conservative result ("any m_0") from Ref. [4].

Combination of Results

The complementarity of searches for different sparticles can be exploited to derive limits on SUSY parameter space. For example, for low m_0, which corresponds to low slepton masses, the chargino results are weak due to a reduction in cross section In this region the slepton limits exclude the region not covered by the chargino searches. Comprehensive analyses [2,5] have been performed with a resulting limit on the mass of the lightest neutralino of 14 GeV/c^2 for any values of m_0 and $\tan\beta$ [5], as shown in Fig. 2.

ALTERNATE SUSY SCENARIOS

R Parity Violation

If R parity is not conserved, sparticles can decay into Standard Model Particles, and additionally, single sparticle production is allowed. The typical missing energy signature of supersymmetry is no longer, and decays such as $\chi^{\pm} \to W^*\chi, \chi \to \ell\ell\nu$ (indirect decay, LLE coupling dominant) and $\tilde{t} \to \ell d$ (direct decay, LQD coupling) lead to a variety of signatures with jets, leptons, and little missing energy. Searches have been performed by the LEP experiments [6], finding no evidence of supersymmetry in this scenario, and limits have been set on sparticle production. For example, constraints are placed in the ($\mu - M_2$) plane which are comparable to those of R-parity conserving searches.

Gauge-Mediated Supersymmetry Breaking

In Gauge-Mediated SUSY Breaking models, the gravitino ($\widetilde{G}$) is the LSP, and typically the χ or $\tilde{\tau}$ is the NLSP. Signatures include decays in which the final step is the process $\chi \to \gamma \widetilde{G}$. Searches have been performed for $\chi\chi$ production, giving a signature of two photons and missing energy; LEP combined results [7] exclude cross sections of greater than 0.35 pb for $M_\chi < 85.5$ GeV/c^2. Cascade decays of charginos and heavier neutralinos lead to signatures that resemble the MSSM-inspired searches, with additional photons [8]. For example, DELPHI sets limits at $\sqrt{s} = 183$ GeV of 88.4 GeV/c^2 on the chargino mass, for heavy sneutrinos ($M_{\tilde{\nu}} > 300$ GeV/c^2).

Sparticles with Lifetime

SUSY particles can have lifetime if ΔM is very small (< 1 GeV/c^2) [9] or for example, in Gauge-Mediated SUSY Breaking models where the process $\tilde{\tau} \to \tau \widetilde{G}$ [10] can occur. Experimental signatures depend on the sparticle lifetime. Short lifetimes make no noticeable difference experimentally, whereas intermediate lifetimes allow the particle to decay in the detector, and a search for kinks is performed. Long-lived sparticles do not decay within the detector, giving a signature of high-mass charged particles which can be detected through anomalous ionization in time projection chambers or absence of light in Cerenkov detectors. No evidence of any of these scenarios was found at LEP. The combined results of LEP searches exclude long-lived left(right) smuons or staus in the mass range of 45 to 76.5(75.6) GeV/c^2 [9].

FUTURE PROSPECTS

The LEP energy and luminosity will continue to increase, up to energies of 200 GeV and at least 300 pb^{-1} of integrated luminosity for each experiment. The discovery potential was studied in the LEP2 workshop [11], where the input parameters of $\sqrt{s} = 192$ GeV and $\mathcal{L} = 300$ pb^{-1} were assumed.

Charginos are the best candidate for discovery of supersymmetry due to the high production cross section. The reach is limited only by the center-of-mass energy in most cases, and charginos with mass of up to ($\sqrt{s} - 2$ GeV/c^2) could be discovered. Special cases such as light sneutrinos and very small ΔM weaken the discovery range. The sensitivity to all possible topologies, as presented previously, allows discovery for numerous possible models.

Scalars have a lower production cross section which is much more dependent on the model parameters. Due to kinematic suppression (as β^3), reaching the kinematic limit in discovery potential is difficult. For the LEP2 workshop parameters, right-handed sleptons could be discovered up to masses of $83 - 88$ GeV/c^2. Stops could be discovered up to masses of 75-90 GeV/c^2, depending on the model parameters. Higher potential for discovery exists if the findings of all four LEP experiments are combined, as has been demonstrated by the current combination of exclusion limits.

Studies have been performed by Baer et al. [12] to determine to potential exclusion in minimal SUGRA models for LEP2 and the upcoming Tevatron Run 2. LEP2 can exclude up to $m_{1/2} \simeq 90, 150$ GeV/c^2 ($\mu > 0, < 0$) for all m_0, with higher limits for lower m_0.

CONCLUSIONS

A wide variety of SUSY signatures can be probed at LEP, and successive increases of energy and luminosity allows for great discovery (or exclusion) potential in a less model-dependent way.

REFERENCES

1. See http://www.cern.ch/LEPSUSY/
2. OPAL Coll., CERN-PPE/97-083;
   DELPHI Coll.,CERN-PPE/97-107;
   DELPHI Coll., cont. to IECHEP, Jerusalem, Israel, Aug. 1997, Ref. no. 427;
   DELPHI Coll., cont. to IECHEP, Jerusalem, Israel, Aug. 1997, Ref. no. 858;
   L3 Coll., cont. to IECHEP, Jerusalem, Israel, Aug. 1997, Ref. no. 859;
   ALEPH Coll., CERN-PPE/97-128;
   L3 Coll., CERN-PPE/97-130.
3. The LEP SUSY Working Group, LEPSUSYWG/97-02.
4. The LEP SUSY Working Group, LEPSUSYWG/97-03
5. ALEPH Coll., cont. to IECHEP, Jerusalem, Israel, Aug. 1997, Ref. no. 594.
6. OPAL Coll.,cont. to IECHEP, Jerusalem, Israel, Aug. 1997, Ref. no. 213;
   L3 Coll.,cont. to IECHEP, Jerusalem, Israel, Aug. 1997, Ref. no. 524;
   DELPHI Coll.,cont. to IECHEP, Jerusalem, Israel, Aug. 1997, Ref. no. 589;
   ALEPH Coll.,cont. to IECHEP, Jerusalem, Israel, Aug. 1997, Ref. no. 621.
7. The LEP SUSY Working Group, LEPSUSYWG/97-04.
8. DELPHI Coll.,cont. to IECHEP, Jerusalem, Israel, Aug. 1997, Ref. no. 855.
9. The LEP SUSY Working Group, LEPSUSYWG/97-01.
10. DELPHI Coll.,cont. to IECHEP, Jerusalem, Israel, Aug. 1997, Ref. no. 350.
11. *Physics at LEP2, CERN Report 96-01, Volume 2 (1996)*.
12. *H. Baer et. al., FSU-HEP-940625, hep-ph/9408265*

Search for Supersymmetry at the Tevatron

Eric Flattum

Fermi National Accelerator Laboratory
Batavia, Il 60510-0500

(for the DØ and CDF collaborations)

Abstract. We discuss is the search for supersymmetry at the Fermilab Tevatron by the DØ and CDF collaborations in $p\bar{p}$ collisions at $\sqrt{s} = 1.8$ TeV. The searches are performed in the jets plus $\not{E}_T$, leptons plus $\not{E}_T$, and photons plus $\not{E}_T$ channels. In these channels there is no excess of events over the expected backgrounds.

INTRODUCTION

The Standard Model (SM) has been very successful in describing our current understanding of high energy physics. However, the SM is not without its defects, and there are strong reasons to believe that new physics, and particles, exist beyond what have been observed. One extension of the SM is called Supersymmetry [1] (SUSY). Supersymmetry states that every SM particle has a supersymmetric partner with the same quantum numbers but differing by 1/2 unit of spin. This paper is a description of *some* of the searches for SUSY at the DØ and CDF collaborations.

THE DETECTORS

The Tevatron is currently the highest energy accelerator in the world, colliding protons and antiprotons with a center of mass energy of 1.8 TeV. The high energy allows a unique opportunity to search for new particles with large masses. The two collider detectors at Fermilab are DØ and CDF. Each is a large multipurpose detector used to measure charged leptons, photons, and jets. Moving radially from the beamline the DØ detector consists of a non-magnetic central tracking system, a compact uranium-liquid argon calorimeter, and a muon spectrometer. The CDF detector has a central tracking system, immersed in a 1.4T magnetic field, which is surrounded by a calorimeter and a muon system.

SQUARK AND GLUINO SEARCHES

At the Tevatron squarks ($\tilde{q}$) and gluinos ($\tilde{g}$) would be produced through the strong interactions since they carry color. If R parity is conserved SUSY particles are produced in pairs. When squarks and gluinos decay they may cascade decay into quarks, gluons, charginos ($\tilde{\chi}^{\pm}$), and neutralinos ($\tilde{\chi}^0$). The charginos and neutralinos may decay into leptons plus the lightest stable supersymmetric particle (LSP). The LSP is typically taken to be $\tilde{\chi}_1^0$ and is not observed in the detector. Therefore SUSY signatures arise from the jets from the quarks and gluons, the leptons from the charginos and neutralinos, and $\not{E}_T$ from the LSPs.

Dileptons and $\not{E}_T$

The DØ collaboration searches for squarks and gluinos using the dielectron plus $\not{E}_T$ channel in 93 pb^{-1} of data [2]. The search requires two electrons with $E_T > 15$ GeV and two jets with $E_T > 20$ GeV. The $\not{E}_T$ is required to be greater than 25 GeV, or 40 GeV if the dielectron mass is within 12 GeV of the Z mass. After these cuts only two events are left with an expected background of 3.0 ± 1.3. The backgrounds are mainly due to $t\bar{t}$ production and $Z \to \tau\bar{\tau}$. Since there is no excess of events over background the result is presented, in the minimal supergravity (SUGRA) framework, as a limit in the m_0 and $m_{1/2}$ plane with $A_0 = 0$, $\tan\beta = 2$, and $\mu < 0$ (see Fig. 1). For equal $\tilde{q}$ and $\tilde{g}$ masses a limit of 267 GeV is obtained.

The CDF collaboration searches for gluinos using same sign dileptons [3]. Since the gluino is a Majorana particle the two gluinos decay into charginos with the same sign 50% of the time. In 90 pb^{-1} of data the analysis requires an electron or muon with $E_T > 11$ GeV, a second with $E_T > 5$ GeV, two jets with $E_T > 15$ GeV, and $\not{E}_T > 25$ GeV. The requirement of same-sign leptons leaves two events. The background from $t\bar{t}$ and Drell-Yan events is estimated to be $1.28 \pm 0.61(stat) \pm 0.35(syst)$. The limit is presented in the $M_{\tilde{q}}$-$M_{\tilde{g}}$ plane for the grand unified theory (GUT) inspired minimal supersymmetric standard model (MSSM) parameters (see Fig. 1).

Jets and $\not{E}_T$

A search in the jets plus $\not{E}_T$ channel complements the dilepton modes. This is because the decay of the SUSY particles into leptons can be very sensitive to the choice of model parameters, e.g. $\tan\beta$. Therefore, a search which does not depend on leptons explores a region of parameter space not accessible to the lepton modes. Each experiment has searched for squarks and gluinos in the jets and $\not{E}_T$ channel.

The most recent DØ search [4] in the jets plus $\not{E}_T$ channel is based on 72 pb^{-1} of data. The analysis requires 3 jets with $E_T > 25$ GeV, with the leading jet having $E_T > 115$ GeV and $|\eta| < 1.1$. The leading jet is used to confirm the primary vertex because a mismeasured vertex will lead to significant spurious $\not{E}_T$. Also, the event $\not{E}_T$ is required to be uncorrelated with the jets, and any isolated electrons and muons are vetoed. The backgrounds are due to vector boson production in association with jets and $t\bar{t}$ production. The vector boson backgrounds are simulated with VECBOS [5] and the $t\bar{t}$ background with HERWIG [6]. The multijet background is determined from a data set taken without $\not{E}_T$ in the trigger. The $\not{E}_T$ distribution is fit and extrapolated into the region of interest for this analysis. The cuts on $\not{E}_T$ and H_T, where H_T is the scalar sum of the non-leading jets, are optimized at each point in parameter space. The number of events observed is 15 with an expected background of 9.3 ± 3.4. Since there is no significant excess over background this result is interpreted as an exclusion contour in the m_0 and $m_{1/2}$ plane for $\tan\beta = 2$, $A_0 = 0$, and $\mu < 0$ (see Fig. 1).

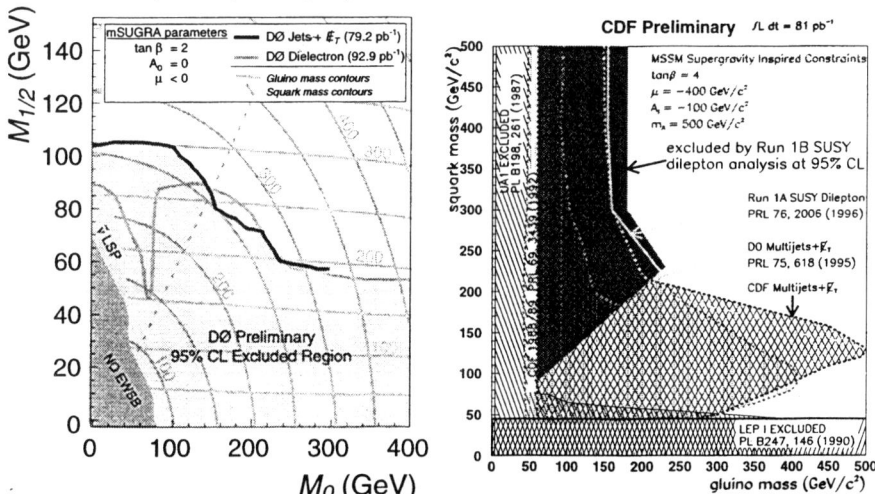

FIGURE 1. The DØ dielectron plus $\not{E}_T$ and jets plus $\not{E}_T$ limits in the m_0 and $m_{1/2}$ plane (left). The CDF dilepton plus $\not{E}_T$ and jets plus $\not{E}_T$ limit in the $M_{\tilde{g}}$ and $M_{\tilde{q}}$ plane (right).

The CDF jets plus $\not{E}_T$ search is based on 19 pb^{-1} of data where 3 or 4 jets are required with $E_T > 15$ GeV and $|\eta| < 2.4$ [7]. The $\not{E}_T$ is required to be greater than 60 GeV and $S > 2.2$ GeV$^{1/2}$, where $S = \not{E}_T / \sqrt{\sum_i E_T^i}$. The vector boson background is taken from the VECBOS generator normalized to the Wjj data. The $t\bar{t}$ backgrounds are generated with the ISAJET [8] generator and normalized to the CDF measured cross section. In the 3(4) jets channel there are 24(6) events observed with an expected background of 33.5±19.4(8.0±5.7). This limit is expressed in terms of the $\tilde{q}$ and $\tilde{g}$ masses in Fig. 1.

CHARGINO AND NEUTRALINO SEARCHES

Charginos and neutralinos may be produced directly at the Tevatron through their electroweak couplings to squarks and vector bosons. Though, the decay of these particles is highly sensitive to the choice of model parameters, the chargino and neutralino may be light and therefore accessible to Tevatron energies.

Trileptons and $\not{E}_T$

The search for events with trileptons and $\not{E}_T$ is considered to be one of the golden channels for discovering new phenomena since the SM backgrounds are very small. Both DØ and CDF have searched for events with three leptons (electrons or muons) and $\not{E}_T$.

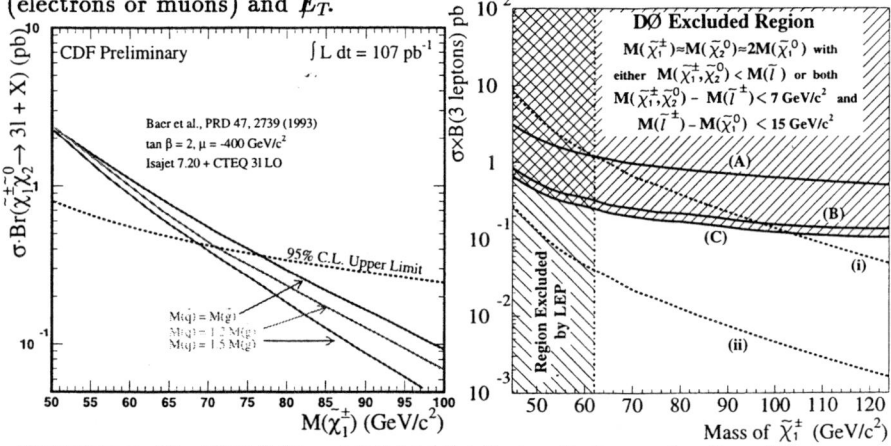

FIGURE 2. The CDF (left) and DØ (right) trilepton limits as a function of the mass.

The CDF search [3] is based on 107 pb^{-1} of data and requires three electrons, muons, or a combination of both, with the leading lepton $E_T>11$ GeV and the trailing two with $E_T>5(4)$ GeV for electrons(muons). Invariant mass cuts are applied on same flavor leptons to remove the J/ψ, Υ, and Z resonances. Also, the leptons are required to be isolated. After the cuts 6 events remain. Adding a $\not{E}_T>15$ GeV cut removes all of these. The expected background of 1.5 events is dominated by Drell-Yan events with a fake lepton. The limits are expressed in terms of the branching ratio into three leptons as a function of the chargino mass and are shown in Fig. 2.

The DØ search [9] is similar to the CDF search in that the analysis requires three leptons with $E_T>5$ GeV, though the leading two leptons may be higher due to trigger requirements. The leptons are required to be isolated and mass cuts are used to remove the resonances. After the $\not{E}_T$>15 GeV cut is applied to the 95 pb^{-1} of data no events remain. The main background is Drell-Yan plus a fake lepton, and total of 1.3 events are expected from all backgrounds. The limit as a function of the chargino mass is shown in Fig. 2.

PHOTONS AND $\not{E}_T$

Even though there has been no excess of signal from SUSY in the data there are events that appear inconsistent with SM description. One event in particular, observed at CDF, contains two high E_T electrons, two high E_T photons, and large $\not{E}_T$ [10]. The observation of this event has led to a number of papers with possible SUSY interpretations [11]. One explanation is from gauge-mediated SUSY breaking models where the neutralino decays into a photon (γ) and gravitino ($\tilde{G}$); the gravitino is the LSP. In this model the event is selectron production where $\tilde{e} \to e\tilde{\chi}_1^0 \to e\gamma\tilde{G}$. A second model is a region of MSSM space where $\tilde{\chi}_2^0$ decays to $\gamma\tilde{\chi}_1^0$. Again the $ee\gamma\gamma\not{E}_T$ event is interpreted as the production of selectrons with the decay ($\tilde{e} \to e\tilde{\chi}_2^0 \to e\gamma\tilde{\chi}_1^0$).

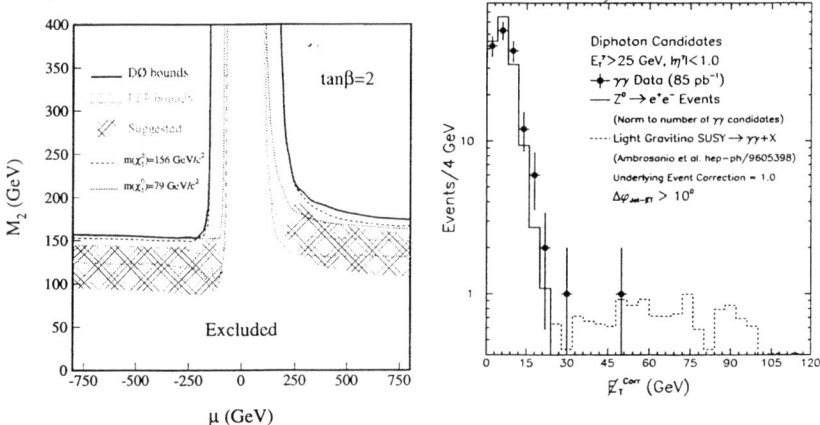

FIGURE 3. The DØ diphoton limit (left) as a function of M_2 and μ, and the $\not{E}_T$ distribution from the CDF analysis (right).

Each experiment has performed a SUSY search of inclusive diphoton events with $\not{E}_T$. The DØ search is based on 106 pb^{-1} of data and requires two photons with E_T>20 and 12 GeV and $\not{E}_T$>25 GeV [12]. With these cuts two events remain with an expected background of 2.3 ± 0.9. In the light gravitino LSP model with $M_2 \sim 2M_1$, and for a heavy $M_{\tilde{q}}$, the limits can be presented as a function of M_2, μ, and $\tan\beta$. The SPYTHIA [13] Monte Carlo is used to generate events as a function of these parameters with the limits shown in Fig. 3. The hatched area in Fig. 3 is the region proposed to explain the $ee\gamma\gamma\not{E}_T$ event and is clearly ruled out by this measurement.

The CDF search is based on 85 pb^{-1} of data and requires two photons with $E_T>25$ GeV and $\not{E}_T>35$ GeV [3]. The photons are required to have $|\eta|<1.0$, and the $\not{E}_T$ cannot be along a jet direction. The distribution of events, as a function of $\not{E}_T$, is shown in Fig. 3 and the determination of limits is underway.

CONCLUSIONS

At the Tevatron the search for SUSY so far has not observed any excess of events over the expected SM backgrounds, though interesting events do exist. The DØ and CDF collaborations continue to search for new SUSY decay channels in the existing data and with improved techniques gained from experience in previous studies. A future run of the Tevatron, to begin in 1999, anticipates a twenty fold increase in luminosity and an increase in the center-of-mass energy to 2 TeV. This will allow the DØ and CDF collaborations with upgraded detectors to stay at the forefront of the search for new phenomena into the next decade.

REFERENCES

1. For a review see H.E. Haber and G.L. Kane, Phys. Rep. **117** 75 (1985).
2. S. Abachi et al. (DØ Collaboration) Proceedings of the 28th International Conference on High Energy Physics, Warsaw, Poland, 1996.
3. J. Done, (CDF Collaboration) Proceedings of the American Physical Society Division of Particles and Fields 1996 Divisional Meeting, Minneapolis, MN, 1996; FERMILAB-Conf-96/372-E.
4. S. Abachi et al. (DØ Collaboration) Proceedings of the XVIII International Symposium on Lepton Photon Interactions, Hamburg, Germany, 1997.
5. F.A Berends, H. Kuijf, B. Tausk, and W.T. Giele, Nucl. Phys. B **357** 32 (1992).
6. G. Marchesini et al. Comput. Phys. Commun. **67** 465 (1992).
7. F. Abe et al. (CDF Collaboration) Phys. Rev. D **56** 1357 (1995).
8. F. Paige and S. Protopopescu, BNL Report, **38034** (unpublished) (1986).
9. B. Abbott et al. (DØ Collaboration) FERMILAB-Pub-97/153-E, submitted to Phys. Rev. Lett.
10. S. Park, (CDF Collaboration) Proceedings of the 10th Topical Workshop on Proton-Antiproton Collider Physics, Batavia, IL, 1996.
11. S. Dimopoulos et al. Phys. Rev. Lett. **76** 3494 (1996); S. Ambrosanio et al. Phys. Rev. Lett. **76** 3498 (1996); J. Ellis, J. Lopez, and D.V. Nanopoulos, Phys. Lett. B **394** 354 (1997).
12. B. Abbott et al. (DØ Collaboration) FERMILAB-Pub-97/273-E, submitted to Phys. Rev. Lett.
13. S. Mrenna, Comput. Phys. Commun. **101** 232 (1997).

Scalar Top Quark Production at $\mu^+\mu^-$ Colliders

A. Bartl*, H. Eberl†, S. Kraml†, W. Majerotto†, W. Porod*,1)

*Institut für Theoretische Physik, Universität Wien, A-1090 Vienna, Austria
†Institut für Hochenergiephysik der Österreichischen Akademie der Wissenschaften, A-1050 Vienna, Austria

Abstract

We discuss the production of stops at a $\mu^+\mu^-$ collider. We present numerical predictions for the cross sections within the Minimal Supersymmetric Standard Model. In particular we consider stop production near $\sqrt{s} = m_{H^0}$ and $\sqrt{s} = m_{A^0}$.

Introduction

The search for supersymmetric particles plays an important rôle at LEP2 and TEVATRON. It will play an even more important rôle at the future colliders LHC, an e^+e^- linear collider with an energy range up to 2 TeV, and a $\mu^+\mu^-$ collider with an energy range up to 4 TeV. We assume that at the time when a $\mu^+\mu^-$ collider starts operation, supersymmetry will have been discovered at TEVATRON or LHC. While proton colliders are good discovery machines [1, 2, 3], one can do precision measurements at $\mu^+\mu^-$ colliders [4]. Another exciting feature of a $\mu^+\mu^-$ collider is the possibility of producing Higgs bosons in the s–channel [4, 5]. This allows one to measure various Higgs couplings at the Higgs resonances.

In the following our framework is the Minimal Supersymmetric Standard Model (MSSM) [6]. The MSSM implies the existence of five physical Higgs bosons: two scalars h^0, H^0, one pseudoscalar A^0, and two charged ones $H^\pm$ [7]. The top quark has two supersymmetric partners, the lighter stop $\tilde{t}_1$ and the heavier stop $\tilde{t}_2$. The top quark and the stops give important contributions to Higgs masses due to radiative corrections (see e.g. [8]). Moreover, their contributions to the renormalization group equations can lead to electroweak symmetry breaking when the Higgs parameters evolve from the GUT scale to the electroweak scale [9]. Therefore, the couplings of

---

[1]Talk presented at the *Workshop on Physics at the First Muon Collider and the Front End of a Muon Collider*, November 6 – 9, 1997, FNAL, Batavia, Illinois, USA.

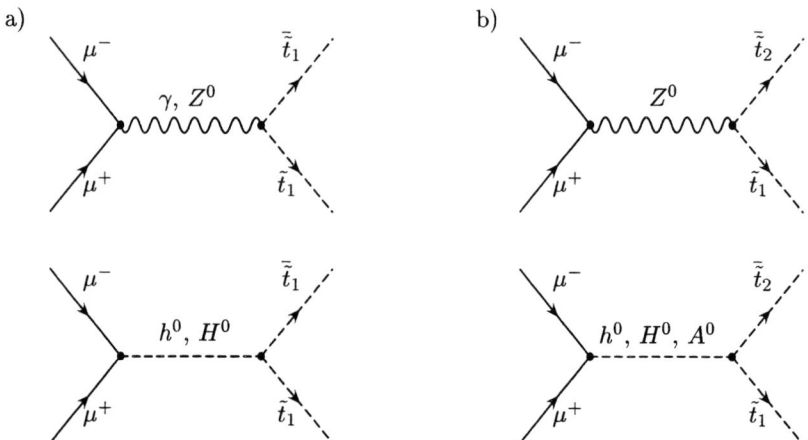

Figure 1: Feynman–graphs for scalar top quark production in $\mu^+\mu^-$ annihilation: a) for $\tilde{t}_1\bar{\tilde{t}}_1$, b) for $\tilde{t}_1\bar{\tilde{t}}_2$.

the stops to the neutral Higgs bosons are of special interest. In this contribution we study stop production in $\mu^+\mu^-$ collisions paying particular attention to the energy range near the Higgs resonances.

Production of Stops

The mass terms of the stops is given by a 2×2 mass matrix. The diagonal elements are $M^2_{\tilde{t}_L} = M^2_{\tilde{Q}} + m^2_Z \cos 2\beta(\frac{1}{2} - \frac{2}{3}\sin^2\theta_W) + m^2_t$ and $M^2_{\tilde{t}_R} = M^2_{\tilde{U}} + \frac{2}{3}m^2_Z \cos 2\beta \sin^2\theta_W + m^2_t$, and the off–diagonal element is given by $m_t(A_t - \mu \cot \beta)$. The physical states are characterized by their mass eigenvalues $m_{\tilde{t}_1}, m_{\tilde{t}_2}$ and the mixing angle $\cos \theta_{\tilde{t}}$.

Figure 1 shows the Feynman–graphs for the processes $\mu^+\mu^- \to \tilde{t}_i\bar{\tilde{t}}_j$ $(i,j=1,2)$. The differential cross section reads

$$\frac{d\sigma}{d\cos\vartheta} = C_{ij}\left(\frac{\kappa^2_{ij}}{s^2}T_{VV}\sin^2\vartheta + \frac{\kappa_{ij}}{s}T^a_{VH}\cos\vartheta + \frac{m^2_i-m^2_j}{s}T^b_{VH} + T_{HH}\right) \quad (1)$$

with s the center–of–mass energy squared, $\kappa^2_{ij} = (s-m^2_i-m^2_j)^2 - 4m^2_im^2_j$, and ϑ the angle between μ^- and $\tilde{t}_i$. T_{VV} denotes the contribution from γ and Z^0 exchange, $T^{a,b}_{VH}$ the interference terms between gauge and Higgs bosons, and T_{HH} the contribution stemming from the exchange of Higgs bosons. The pure gauge boson contribution, the first part of Equation 1, is the same as for $e^+e^- \to \tilde{t}_i\bar{\tilde{t}}_j$ and given in [10]. The explicit form of $T^{a,b}_{VH}$ and T_{HH} will be given in a forthcoming paper [11]. Notice

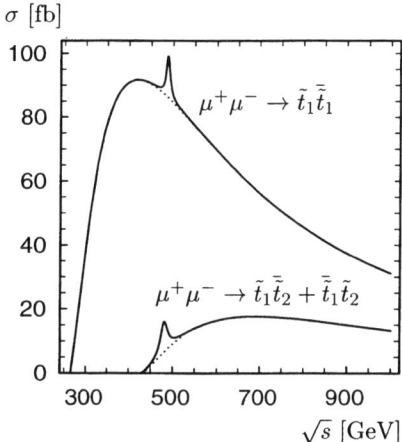

Figure 2: Production cross section for $\mu^+\mu^- \to \tilde{t}_1\bar{\tilde{t}}_1$ and $\mu^+\mu^- \to \tilde{t}_1\bar{\tilde{t}}_2 + \bar{\tilde{t}}_1\tilde{t}_2$ as a function of $\sqrt{s}$. The parameters are: $M_{\tilde{Q}} = 160$ GeV, $M_{\tilde{U}} = 145$ GeV, $M_{\tilde{D}} = 175$ GeV, $A_t = A_b = 350$ GeV, $\mu = 300$ GeV, $M = 140$ GeV, $\tan\beta = 2$ and $m_{A^0} = 480$ GeV. The graphs correspond to: total cross section (full line) and gauge boson contribution (dotted line).

that the gauge boson term has a $\sin^2\vartheta$ dependence whereas T_{HH} and T^b_{VH} are independent of ϑ. The T^a_{VH} term is proportional to $\cos\vartheta$ giving rise to a forward backward asymmetry. However, this asymmetry is proportional to m_μ and of the order $\lesssim 10^{-4}$ [11]. The following parameters enter the couplings of the stops to the Higgs bosons: A_t, μ, $\cos\theta_{\tilde{t}}$, $\cos\alpha$, and $\tan\beta$.

In Figure 2 we show the total cross section as a function of $\sqrt{s}$ for $M_{\tilde{Q}} = 160$ GeV, $M_{\tilde{U}} = 145$ GeV, $M_{\tilde{D}} = 175$ GeV, $A_t = A_b = 350$ GeV, $\mu = 300$ GeV, $M = 140$ GeV, $\tan\beta = 2$, and $m_{A^0} = 480$ GeV. The full lines show the total cross sections and the dotted lines show the gauge boson contributions. The latter ones are identical with the cross sections of $e^+e^- \to \tilde{t}_i\bar{\tilde{t}}_j$. For $\tilde{t}_1\bar{\tilde{t}}_1$ production the peak results from the H^0 exchange leading to an enhancement of ~ 20 fb compared to the gauge boson contribution. For $\tilde{t}_1\bar{\tilde{t}}_2$ production the peak is an overlap of the H^0 and A^0 resonances because $m_{A^0} \simeq m_{H^0}$ and the widths of A^0 and H^0 are of the order of several GeV (see e.g. [7, 12, 13]).

In Figure 3 we show the total cross section near the Higgs resonances for various values of A_t and the other parameters as above. For $A_t = -50\,(350)$ GeV one has $m_{\tilde{t}_1} = 133$ GeV, $m_{\tilde{t}_2} = 296$ GeV, and $\cos\theta_{\tilde{t}} = 0.69\,(-0.69)$. $A_t = 50\,(250)$ GeV gives $m_{\tilde{t}_1} = 187$ GeV, $m_{\tilde{t}_2} = 265$ GeV, and $\cos\theta_{\tilde{t}} = 0.67\,(-0.67)$. For $A_t = 150$ GeV one gets $m_{\tilde{t}_1} = 226$ GeV, $m_{\tilde{t}_2} = 233$ GeV, and $\cos\theta_{\tilde{t}} = 0$. The shifts of the peaks are due to radiative corrections to m_{H^0}. One can clearly see that the widths of the peaks depend on A_t and therefore also on the sign of $\cos\theta_{\tilde{t}}$. Note that for $\cos\theta_{\tilde{t}} = 0$ the $H^0\tilde{t}_1\bar{\tilde{t}}_1$ coupling is rather small and, therefore, the peak nearly vanishes. However, at the same time the $A^0\tilde{t}_1\bar{\tilde{t}}_2$ coupling is large leading to the enhancement and to the shift of the corresponding peak compared to the other A_t values. Note that the decay widths of A^0 and H^0 into stops are an essential part of the total widths. Therefore, when the peaks are narrower for $\tilde{t}_1\bar{\tilde{t}}_1$ production then they are broader for $\tilde{t}_1\bar{\tilde{t}}_2$ production and vice versa.

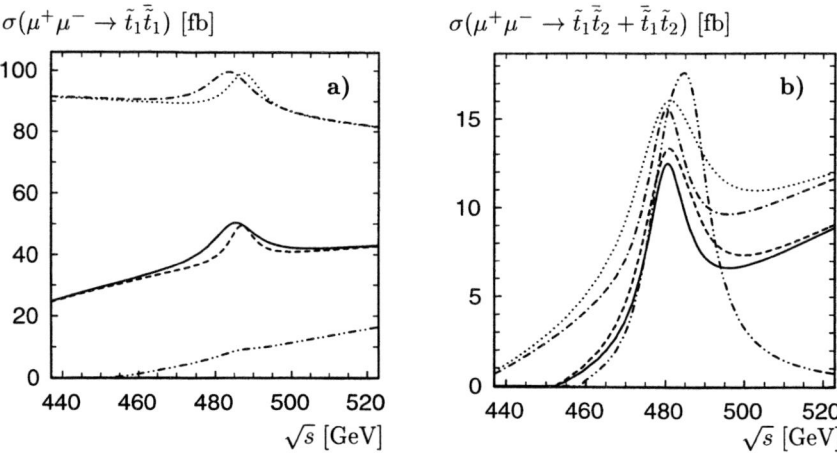

Figure 3: Production cross section (in fb) for a) $\mu^+\mu^- \to \tilde{t}_1\bar{\tilde{t}}_1$ and b) $\mu^+\mu^- \to \tilde{t}_1\bar{\tilde{t}}_2 + \bar{\tilde{t}}_1\tilde{t}_2$ as a function of $\sqrt{s}$. The parameters are: $M_{\tilde{Q}} = 160$ GeV, $M_{\tilde{U}} = 145$ GeV, $M_{\tilde{D}} = 175$ GeV, $\mu = 300$ GeV, $M = 140$ GeV, $\tan\beta = 2$ and $m_{A^0} = 480$ GeV. The graphs correspond to $(A_t = A_b)$: $A_t = -50$ GeV (dash–dotted line), $A_t = 50$ GeV (full line), $A_t = 150$ GeV (dash–dot–dotted line), $A_t = 250$ GeV (dashed line) and $A_t = 350$ GeV (dotted line).

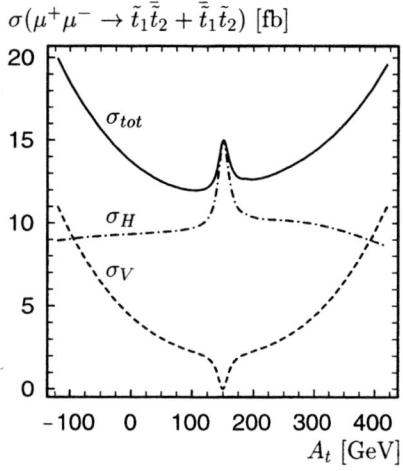

Figure 4: Production cross section (in fb) for $\mu^+\mu^- \to \tilde{t}_1\bar{\tilde{t}}_2 + \bar{\tilde{t}}_1\tilde{t}_2$ as a function of A_t. The parameters are: $M_{\tilde{Q}} = 160$ GeV, $M_{\tilde{U}} = 145$ GeV, $M_{\tilde{D}} = 175$ GeV, $A_b = A_t$, $\mu = 300$ GeV, $M = 140$ GeV, $\tan\beta = 2$ and $\sqrt{s} = m_{A^0} = 480$ GeV. The graphs correspond to: total cross section σ_{tot} (full line), Higgs boson contribution σ_H (dash–dotted line), and gauge boson contribution σ_V (dashed line).

Figure 4 shows the A_t dependence of the cross section $\mu^+\mu^- \to \tilde{t}_1\bar{\tilde{t}}_2 + \bar{\tilde{t}}_1\tilde{t}_2$ for $\sqrt{s} = m_{A^0} = 480$ GeV and the other parameters as above. The Higgs exchange contributes to the total cross section at least 30%. For $\cos\theta_{\tilde{t}} = 0$ ($A_t = 150$ GeV) this contribution reaches 100%. The smaller peak near $A_t = 200$ GeV is due to the

H^0 resonance.

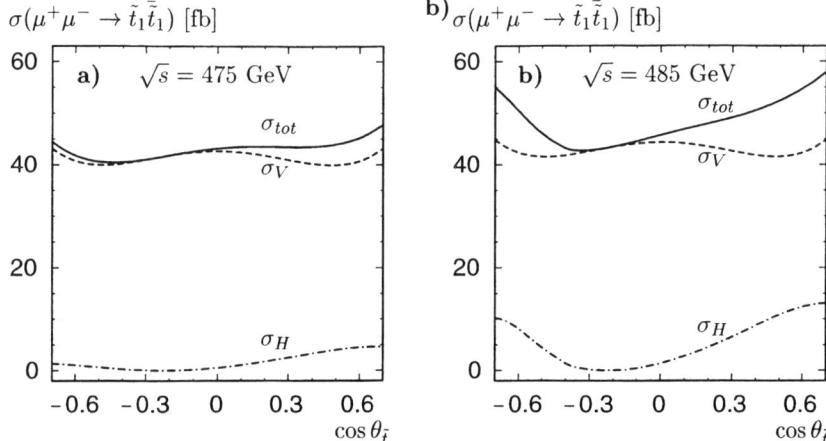

Figure 5: Production cross section for the process $\mu^+\mu^- \to \tilde{t}_1\bar{\tilde{t}}_1$ (in fb) as a function of $\cos\theta_{\tilde{t}}$ for a) $\sqrt{s} = 475$ GeV and b) $\sqrt{s} = 485$ GeV. The parameters are: $m_{\tilde{t}_1} = 180$ GeV, $M_{\tilde{Q}} = 160$ GeV, $M_{\tilde{D}} = 175$ GeV, $\mu = 300$ GeV, $M = 140$ GeV, $\tan\beta = 2$ and $m_{A^0} = 480$ GeV. The graphs correspond to: total cross section σ_{tot} (full line), gauge boson contribution σ_V (dashed line), and Higgs boson contribution (dash-dot-dotted line).

Figure 5 shows the $\cos\theta_{\tilde{t}}$ dependence of the $\mu^+\mu^- \to \tilde{t}_1\bar{\tilde{t}}_1$ cross section in the energy range close to the H^0 resonance. The parameters are $m_{\tilde{t}_1} = 180$ GeV, $M_{\tilde{Q}} = 160$ GeV, $M_{\tilde{D}} = 175$ GeV, $\mu = 300$ GeV, $M = 140$ GeV, $\tan\beta = 2$, and $m_{A^0} = 480$ GeV. Notice that the Higgs contribution depends on the sign of $\cos\theta_{\tilde{t}}$.

Conclusions

We have studied the production of $\tilde{t}_1\bar{\tilde{t}}_1$ and $\tilde{t}_1\bar{\tilde{t}}_2$ in $\mu^+\mu^-$ annihilation focusing on the impact of the Higgs resonances in these processes. In particular we have found that one gets important information on the $H^0\tilde{t}_1\bar{\tilde{t}}_1$, $H^0\tilde{t}_1\bar{\tilde{t}}_2$ and $A^0\tilde{t}_1\bar{\tilde{t}}_2$ couplings.

Acknowledgments

We are very grateful to M. Carena and S. Protopopescu for their kind invitation to this interesting and inspiring workshop. This work was supported by the "Fonds zur Förderung der wissenschaftlichen Forschung" of Austria, project no. P10843-PHY.

References

[1] A. Bartl, J. Soderqvist et al., Proc. of the Workshop on *New Directions for High Energy Physics*, Snowmass, Colorado, 1996, and references therein.

[2] F. Paige, *these Proceedings*.

[3] I. Hinchliffe, *these Proceedings*.

[4] J. Gunion, *these Proceedings*.

[5] H. E. Haber, *these Proceedings*.

[6] H. E. Haber and G. L. Kane, *Phys. Rep.* **117**, 75 (1985).

[7] J. F. Gunion, H. E. Haber, G. L. Kane, and S. Dawson, *The Higgs Hunter's Guide*, Addison-Wesley, 1990, and references therein.

[8] J. Ellis, G. Ridolfi, and F. Zwirner, *Phys. Lett. B* **257**, 83 (1991); *Phys. Lett. B* **262**, 477 (1991).

[9] L. Ibáñez and G. G. Ross, *Phys. Lett. B* **110**, 215 (1982).

[10] H. Eberl, A. Bartl, W. Majerotto, *Nucl. Phys. B* **472**, 481 (1996).

[11] A. Bartl, H. Eberl, S. Kraml, W. Majerotto, and W. Porod, *in preparation*.

[12] A. Bartl, H. Eberl, K. Hidaka, T. Kon, W. Majerotto, and Y. Yamada, *Phys. Lett. B*, **402**, 303 (1997)

[13] A. Djouadi, J. Kalinowski, P. Ohmann, and P.M. Zerwas, *Zeit. f. Phys. C* **74**, 93 (1997)

Sleptons at a First Muon Collider

Frank E. Paige

Physics Department
Brookhaven National Laboratory
Upton, NY 11973

Abstract. Signatures for sleptons, which have been extensively studied for the Next Linear Collider, are reexamined taking into account some of the different features of a First Muon Collider.

Supersymmetry (SUSY) signatures have been extensively studied for the Next Linear Collider (NLC) [1]. The basic strategy [2] [3] uses the fact that SUSY particles are produced in pairs and decay into Standard Model (SM) particles plus an invisible lightest SUSY particle $\tilde{\chi}_1^0$. Hence the maximum and minimum energies of the visible SM particles determine the initial SUSY and LSP mass. However, the NLC studies have used properties of the NLC such as easily variable energy and high electron polarization. This study makes assumptions appropriate for a First Muon Collider (FMC), namely operation at a single energy with a 20° hole for shielding and no polarization. It does not, however, take into account backgrounds from muon decays.

I SLEPTONS AT LHC POINT 5

This analysis is carried out for LHC Point 5, a minimal supergravity (SUGRA) point with $m_0 = 100\,\text{GeV}$, $m_{1/2} = 300\,\text{GeV}$, $A_0 = 300\,\text{GeV}$, $\tan\beta = 2.1$, and $\text{sgn}\,\mu = +1$. For this point, $M(\tilde{\chi}_1^0) = 121.66\,\text{GeV}$, $M(\tilde{\ell}_R) = 157.20\,\text{GeV}$, $M(\tilde{\ell}_L) = 238.82\,\text{GeV}$, and $M(\tilde{\chi}1\pm) = 232.05\,\text{GeV}$. The LHC can trivially discover SUSY at this point and can use precision measurements of combination of masses to determine the SUGRA parameters. Using only such precision measurements, the estimated errors for $10\,\text{fb}^{-1}$ of luminosity are [4]

- $m_0 = 100.5^{+12}_{-5}\,\text{GeV}$,

- $m_{1/2} = 298^{+16}_{-9}\,\text{GeV}$,

- $\tan\beta = 1.8^{+0.3}_{-0.5}$,

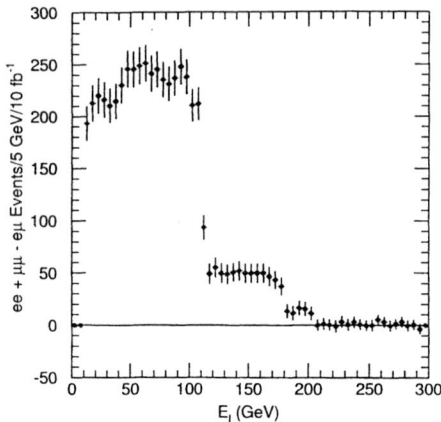

FIGURE 1. The sum of the signal and background for $\mu^+\mu^- + e^+e^- - \mu^\pm e^\mp$ at Point 5 with statistical errors appropriate for $10\,\text{fb}^{-1}$.

- $\text{sgn}\,\mu = +1$.

A_0 is poorly determined because the weak-scale phenomenology is insensitive to it. The ultimate LHC precision is considerably better [5]. Given these results, one would presumably choose the energy of the FMC to be $\sqrt{s} = 600\,\text{GeV}$, the value assumed here.

SUSY and Standard Model events were generated as e^+e^- events with ISAJET 7.31 [6]; e's and μ's were interchanged in the analysis. The toy detector simulation is the same as that used for the LHC studies [4] except that the η range is limited to $|\eta| < 1.8$, approximately equivalent to $\theta > 20°$. Jets were found with a fixed cone algorithm, and leptons were taken from the generator. Slepton candidates are selected by requiring two μ or e leptons with $|\eta| < 1.3$ and $E_\ell > 10\,\text{GeV}$ and no other leptons or jets. The two leptons are required to satisfy

- $E_\ell > 10\,\text{GeV}$, $|\eta_\ell| < 1.3$ to select two identified leptons in the detector,
- $|\vec{p}_1 + \vec{p}_2| < 0.9\sqrt{s}$ to reject lepton pair background,
- $|\vec{p}_{T,1} + \vec{p}_{T,2}| > 10\,\text{GeV}$ to reject lepton pair and $\gamma\gamma$ background,
- $\Delta\phi_{1,2} < 0.95\pi$ to reject lepton pair and $\gamma\gamma$ background.

These cuts eliminate the $\ell^+\ell^-$, $\gamma\gamma \to \ell^+\ell^-$, and $ZZ \to \ell^+\ell^-\nu\bar{\nu}$ backgrounds. A Z mass cut was found to distort the E_ℓ distributions and was replaced by the cut on $|\vec{p}_{T,1} + \vec{p}_{T,2}|$.

After these cuts, the dominant background comes from leptonic decays of WW pairs. Since W's decay equally into $e\nu$ and $\mu\nu$, the SM background vanishes up

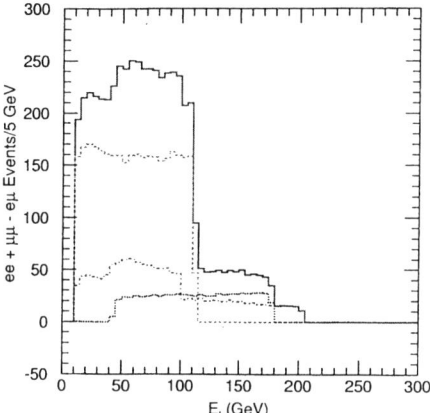

FIGURE 2. Composition of E_ℓ distribution for $\mu^+\mu^- + e^+e^- - e^+\mu^- - e^-\mu^+$ at Point 5. The dashed curve is from $\tilde{\ell}_R\tilde{\ell}_R$; the dotted curve is from $\tilde{\ell}_L\tilde{\ell}_L$; and the dashed-dotted curve is from $\tilde{\ell}_R\tilde{\ell}_L$.

to statistical fluctuations in the combinations $\mu^+\mu^- + e^+e^- - e^+\mu^- - e^-\mu^+$ (and also for $\mu^+\mu^- - e^+e^-$). This distribution is shown in Figure 1 with error bars appropriate for $10\,\text{fb}^{-1}$ but with larger Monte Carlo statistics.

The origins of the signal are shown in Figure 2. There are contributions from $\tilde{\ell}_R\tilde{\ell}_R$, $\tilde{\ell}_L\tilde{\ell}_L$, and $\tilde{\mu}+_R\tilde{\mu}_L$, the last coming from gaugino exchange in the t-channel. Two-body kinematics implies that for production of $\tilde{\ell}_i\tilde{\ell}_j$, $i,j = R, L$, the maximum and minimum energies are

$$E^\pm_{i\to\ell} = \frac{M_i^2 - M_{\tilde{\chi}_1^0}^2}{4M_i^2}\left[\frac{s + M_i^2 - M_j^2 \pm \sqrt{(s - M_i^2 - M_j^2)^2 - 4M_i^2 M_j^2}}{2\sqrt{s}}\right]$$

There are four distinct ranges for the lepton energy, one each for RR and LL events and two for LR events:

$$E_{RR} = (111.4\,\text{GeV}, 11.2\,\text{GeV})$$
$$E_{LL} = (178.3\,\text{GeV}, 33.0\,\text{GeV})$$
$$E_{LR} = (99.5\,\text{GeV}, 10.0\,\text{GeV})$$
$$= (101.9\,\text{GeV}, 19.2\,\text{GeV})$$

If no other cuts had been made, one would obtain a sum of four square distributions with these limits. All of these limits can be seen as edges in Figure 2.

The small lower limits for some of these distributions may make it difficult to identify and measure the electrons in the presence of the muon decay background. These limits are an "accidental" consequence of the masses at this point; they decrease slowly as $\sqrt{s}$ is increased.

II ERROR ANALYSIS

The easiest edge to detect in Figure 1 is the one at 111.4 GeV. Statistically, one could detect this with bins of 0.5 GeV, ten times smaller, but detector resolution and possible confusion from the two edges at about 100 GeV must be included. We assume an error $\sigma_E = 1\,\text{GeV}$. This edge is associated with the ℓ_R sleptons, and its position is given by

$$E_\ell^{\max} = \frac{M_{\tilde{\ell}_R}^2 - M_{\tilde{\chi}_1^0}^2}{4 M_{\tilde{\ell}_R}^2}[\sqrt{s} + \sqrt{s - 4 M_{\tilde{\ell}_R}^2}]$$

The other RR edge is at such low energy, 11.2 GeV, that it will probably be difficult to measure. The LHC can measure the endpoint of the $\ell^+\ell^-$ mass spectrum and so determine

$$M_{\ell\ell}^{\max} = M_{\tilde{\chi}20}\sqrt{1 - \frac{M_{\tilde{\ell}_R}^2}{M_{\tilde{\chi}20}^2}}\sqrt{1 - \frac{M_{\tilde{\chi}_1^0}^2}{M_{\tilde{\ell}_R}^2}}$$

with an estimated error $\sigma_M = 1\,\text{GeV}$ for $10\,\text{fb}^{-1}$. [4] While one can estimate the masses independently, the results are not so accurate, so we assume $M_{\tilde{\chi}20} = 2 M_{\tilde{\chi}_1^0}$. Then the χ^2 error ellipse in the $(M_{\tilde{\chi}_1^0}, M_{\tilde{\ell}_R})$ plane is given by

$$\chi^2 = \sum_{ij}[\frac{1}{\sigma_E^2}\frac{\partial E_\ell^{\max}}{\partial M_i \partial M_j} \frac{1}{\sigma_M^2}\frac{\partial M_{\ell\ell}^{\max}}{\partial M_i \partial M_j}]\Delta M_i \Delta M_j$$

The resulting error matrix is shown in Figure 3 and represents a significant improvement over the LHC alone.

We next attempt to determine the $\tilde{\ell}_L$ mass by measuring the edge at 178.3 GeV in Figure 1 from $\tilde{\ell}_L^+ \tilde{\ell}_L^-$ production. The error is probably about one bin width, 5 GeV. Unfortunately, the position of this edge,

$$E_\ell^{\max} = \frac{M_{\tilde{\ell}_L}^2 - M_{\tilde{\chi}_1^0}^2}{4 M_{\tilde{\ell}_L}^2}[\sqrt{s} + \sqrt{s - 4 M_{\tilde{\ell}_L}^2}]$$

turns out to be very insensitive to $M_{\tilde{\ell}_L}$ for these values of the parameters; the derivative of the $\tilde{\ell}_L \tilde{\ell}_L$ endpoint has a zero as a function of $\sqrt{s}$ that happens to occur very close to 600 GeV for these masses. Numerically,

$$\frac{dE_\ell^{\max}}{M_{\tilde{\ell}_L}} \approx 0.036\,.$$

As a result the sensitivity is accidentally very poor. For "typical" values the error on $M_{\tilde{\ell}_L}$ would be about three times that on the edge.

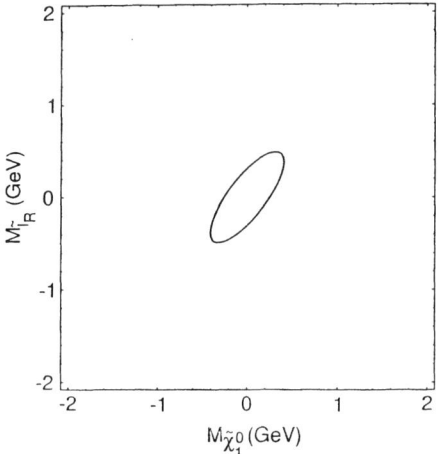

FIGURE 3. Error ellipse in the $(M_{\tilde{\chi}_1^0}, M_{\tilde{\ell}_R})$ plane from measurements at the LHC and FMC at Point 5.

III SLEPTONS AT LHC POINT 3

The second LHC point at which the FMC could contribute is LHC Point 3, a SUGRA point with $m_0 = 200\,\text{GeV}$, $m_{1/2} = 100\,\text{GeV}$, $A_0 = 0$, $\tan\beta = 2$, and $\text{sgn}\,\mu = -1$. These parameters were chosen so that every accelerator could find something; in particular, LEP recently announced discovery of the light Higgs at 68 GeV. The LHC can make many precise measurements at this point. However, since sleptons do not occur in the cascade decays of gluinos and squarks, they are not directly constrained.

Figure 4 shows the lepton energy distribution for Point 3 and its sources; compare with Figure 2 for Point 5. The sleptons are nearly degenerate at this point: $M_{\tilde{\ell}_L} = 206.5\,\text{GeV}$ and $M_{\tilde{\ell}_R} = 215.7\,\text{GeV}$. Hence all the slepton edges are nearly degenerate; the $\tilde{\ell}_R\tilde{\ell}_R$ contribution dominates because the branching ratio for $\tilde{\ell}_R \to \tilde{\chi}_1^0 \ell$ is nearly 100%. There is also a contribution to like-flavor dileptons from $\tilde{\chi}_2^0$ at this point, the long-dashed curve in Figure 4. While the various contributions in probably cannot be resolved, both the upper and the lower edges should be measurable, allowing one to determine $M_{\tilde{\ell}}$ and $M_{\tilde{\chi}_1^0}$ without additional assumptions. [2] [3] The resulting error ellipse is shown assuming a measurement error of 1 GeV on each edge. Clearly this case is much more favorable than that for Point 5, although in part the difference is due to the fact that there is less information on sleptons from the LHC.

While polarization is not essential to detect the signal, it would help to interpret it. In particular, the dominance of the $\tilde{\ell}_R\tilde{\ell}_R$ contribution seen in Figure 4 is due to $\tilde{\chi}_1^0$ exchange in the t-channel. Even rather modest beam polarization would show that this contribution was dominant and so provide another test of the SUSY model. The degree of polarization required needs study but is probably much smaller than

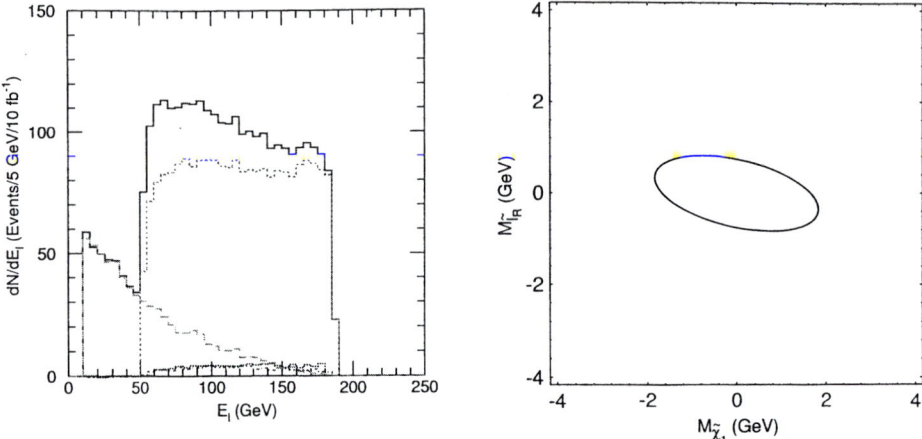

FIGURE 4. Composition of E_ℓ distribution for $\mu^+\mu^- + e^+e^- - e^+\mu^- - e^-\mu^+$ at Point 3. The dashed curve is from $\tilde{\ell}_R\tilde{\ell}_R$; the dotted curve is from $\tilde{\ell}_L\tilde{\ell}_L$; the dashed-dotted curve is from $\tilde{\ell}_R\tilde{\ell}_L$; and the long-dashed curve is from all events with a $\tilde{\chi}20$. The error ellipse is also shown.

that needed to suppress Standard Model backgrounds.

This work was supported in part by the United States Department of Energy under Contract DE-AC02-76CH00016.

REFERENCES

1. S. Kuhlman, et al., *Physics and Technology of the Next Linear Collider*, BNL-52502 (1996)
2. T. Tsukamoto, K. Fujii, H. Murayama, M. Yamaguchi, and Y. Okada, Phys. Rev. **D51**, 3153, (1995)
3. M.M. Nojiri, K. Fujii, and T. Tsukamoto Phys. Rev. **D54**, 6756 (1996)
4. I. Hinchliffe, M.D. Shapiro, F.E. Paige, J. Soderqvist, W. Yao, Phys. Rev. **D55**, 5520 (1997)
5. D. Froidevaux, http://atlasinfo.cern.ch/Atlas/GROUPS/PHYSICS/SUSY/susy.html
6. H. Baer, F. Paige, S. Protopopescu and X. Tata; in *Physics at Current Accelerators and Supercolliders*, ed. J. Hewett, A. White and D. Zeppenfeld, (Argonne National Laboratory, 1993).

Precision Measurements of Threshold Chargino Production

M. S. Berger

Indiana University
Bloomington Indiana 47405

Abstract. We analyze the prospects at a muon collider for measuring chargino masses in the $\mu^+\mu^- \to \tilde{\chi}^+\tilde{\chi}^-$ processes in the threshold region. We find that a measurement of the lightest chargino mass to better than 200 MeV is possible with 100 fb^{-1} luminosity. The muon sneutrino mass can also be simultaneously measured to a few GeV.

INTRODUCTION

Particle masses can be measured quite accurately by producing them near threshold. This has been demonstrated recently at LEP II where W pairs were produced at $\sqrt{s} = 161$ GeV, and a precise measurement of the W mass has been obtained. We have recently shown that future high-luminosity $\mu^+\mu^-$ colliders can measure the W boson, top quark and Higgs boson masses with high precision in the processes $\ell^+\ell^- \to WW, t\bar{t}, ZH$ [1-3]. Threshold production of chargino pairs at a muon collider offers a possible way of measuring the chargino mass and also the muon sneutrino mass that is involved in the production process [4].

Muon colliders [5-9] could be especially useful tools in precision measurements of particle masses, widths, and couplings. Initial state radiation from muons is reduced compared to electrons, and muon colliders have negligible beamstrahlung. The threshold regions for particle production depend on the particle widths. In fact, one can in principle measure the widths of the W boson, the top quark, and the Higgs boson width by performing the appropriate measurements of the production cross sections near threshold. When the lightest chargino is dominantly gaugino, its width is usually negligibly small and the threshold cross section is controlled only by angular momentum considerations and the characteristics of the colliding beam. The measurement of the chargino mass via the threshold cross section has been considered previously for electron-positron machines in Ref. [10,11]. We consider the measurement at a muon collider with high luminosity, carefully taking into account the beam effects and reoptimizing cuts to eliminate the background in the threshold region. We assume here that the muon collider has a relatively modest beam energy spread of $R = 0.1\%$, where R is the rms spread of the energy of a muon beam[1]. We assume that 100 fb^{-1} integrated luminosity is available; high luminosity is necessary if the threshold measurement is to prove interesting.

A precision measurement of the chargino mass will be highly desirable to test patterns of supersymmetry breaking. For example the relationship between the lightest neutralino and the lightest chargino masses can be used to test the existance of a universal soft SUSY-breking parameter. The chargino pair production process has been investigated beyond the tree-level recently [14]. A precision measurement of the cross section can test radiative corrections coming from heavy squarks.

SIGNAL AND BACKGROUND

A simultaneous measurement of the chargino and sneutrino masses requires a sampling of the cross section at at least two points. As in other threshold measurements, the statistical precision on the chargino mass is maximized just above $2m_{\tilde{\chi}^\pm}$. However as is evident from Fig. 1, a change in the cross section at $\sqrt{s} = 2m_{\tilde{\chi}^\pm} + 1$ GeV can be due to a variation in the sneutrino mass, so a second measurement of the cross section must be taken at a higher $\sqrt{s}$ where the dependence of the cross section on the chargino mass and the slepton mass is different. It turns out to be advantageous for the chargino mass measurement to choose this higher energy measurement at a point where the chargino cross section is not flat.

The precision that can be obtained in the chargino mass depends substantially on the chargino mass itself: the heavier the chargino the smaller the production cross section. The cross section also depends on the mass of the sneutrino which appears in the t-channel. The contribution from the sneutrino graph interferes destructively with the s-channel graphs. If the lightest chargino is gaugino-dominated, then changing the parameters of the chargino mass matrix essentially changes the mass but not the chargino couplings significantly. Therefore one can envision a measurement of the cross section that depends on just two parameters: the chargino mass $m_{\tilde{\chi}^\pm}$ and the sneutrino mass $m_{\tilde{\nu}}$.

The chargino decay mode is $\tilde{\chi}^\pm \to \tilde{\chi}^0 f \overline{f'}$. If $m_{\tilde{\chi}^\pm} - m_{\tilde{\chi}^0} > M_W$ then W exchange dominates and the final state is comprised of 49% purely hadronic events, 42% mixed hadronic-leptonic events, and 9% purely leptonic events (these ratios are determined by the W branching fractions). The width of the chargino has a negligible impact on the threshold cross section even the two body decay is possible ($m_{\tilde{\chi}^\pm} - m_{\tilde{\chi}^0} > M_W$) if the lightest chargino is gaugino-dominated. There are several backgrounds to the chargino pair signal, the largest being $\mu^+\mu^- \to W^+W^-$. The cross section is reduced near threshold, so the cuts to reduce this background need to be reoptimized.

---

[1]) The most recent TESLA design envisions a beam energy spread of $R = 0.2\%$ [12] while the NLC design expects a beam energy spread of $R = 1.0\%$. A high energy e^+e^- collider in the large VLHC tunnel would have a beam spread of $\sigma_E = 0.26$ GeV [13] which should give numbers precisions comparable to those considered here.

One must worry about the level of the backgrounds, and the systematic error that the residual background presents for the cross section measurement. Figure 1 indicates that with the amount of integrated luminosity we are assuming will be available for the measurement, the cross section is being measured to the few percent level, so an understanding of the background to at least this level is necessary. The backgrounds to chargino pair-production have been investigated in Refs. [15,16] where the signal efficiencies have been obtained for the various final states when the center-of-mass energy is $\sqrt{s} = 500$ GeV. The primary background is W pair production which is very large, but can be effectively eliminated because the W's are produced in the very-forward direction. However, if the energy is reduced so that the collider is operating in the chargino threshold region, then the effectiveness of these cuts might be reduced (the signal events might be expected to be more spherical as well). Therefore the efficiencies should be reinvestigated for the threshold measurement. The overall signal efficiency of our cuts in Ref. [4] is about 10% for the fully hadronic decays.

We have assumed here that the chargino is lighter than the muon sneutrino. If that is not the case, the chargino has a new decay mode: $\tilde{\chi}^\pm \to \ell^\pm \tilde{\nu}$. The efficiency of the cuts against background would need to be reconsidered if this mode is kinematically allowed.

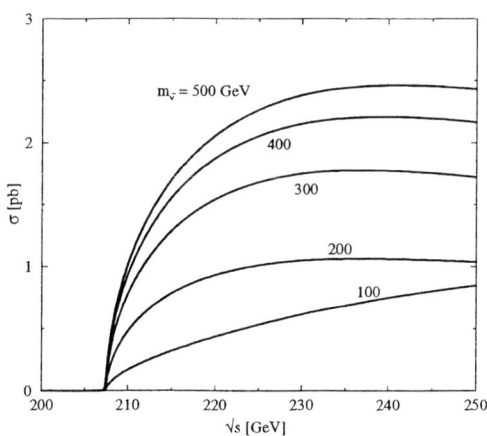

FIGURE 1. The threshold region of $\mu^+\mu^- \to \tilde{\chi}^+\tilde{\chi}^-$ for various sneutrino masses, taking $M_2 = 100$ GeV and $\tan\beta = 4$. The rapid rise of the cross section is due to the pair production of spin-1/2 particles with small decay widths. The muon collider is assumed to have a beam energy spread of $R = 0.1\%$.

A further advantage of the threshold measurement is that the chargino mass measurement is somewhat isolated from its subsequent decays. Distributions in the final state observables, say e.g. E_{jj} from the decay $\tilde{\chi}^\pm \to \tilde{\chi}^0 jj$ [15], depend on the neutralino mass. The cross section for chargino pair production, on the other hand, is independent of the final state particles, and only the branching fractions and detector efficiencies for the various final states impact this measurement (as indicated above, if $m_{\tilde{\chi}^\pm} - m_{\tilde{\chi}^0} > M_W$ the branching fractions of chargino decay is given essentially in terms of the W branching fractions).

The chargino production cross section decreases with increasing chargino mass. Therefore the precision with which the mass can be measured is better at smaller values of the mass. Figure 2 shows the expect precision with 100 fb^{-1} integrated luminosity for sneutrino masses of 300 and 500 GeV. For a lighter sneutrino, for which the destructive interference between the s-channel and t-channel graphs is more severe, the precision obtained is reduced. Furthermore inspection of Fig. 1 demonstrates the variability of the cross section is reduced for heavier sneutrino masses leading to a reduced precision measurement. The sneutrino mass can be measured to about 6 GeV accuracy for $m_{\tilde{\nu}} = 300$ GeV and to about 20 GeV accuracy for $m_{\tilde{\nu}} = 500$ GeV. This provides an indirect method of measuring the sneutrino mass (the sneutrino might be too heavy to produce directly).

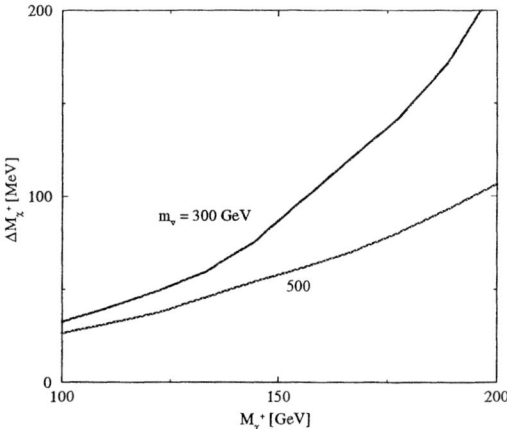

FIGURE 2. The 1σ precision obtainable in the chargino mass taking $m_{\tilde{\nu}} = 300$ and 500 GeV. The precision is better for *larger* sneutrino mass because the contribution from the t-channel sneutrino exchange diagram destructively interferes with the s-channel diagrams.

The result of a fit to the chargino cross section is shown in Fig. 3, taking $M_2 = 120$ GeV and $\tan\beta = 4$ and assuming an integrated luminosity of 100 fb^{-1}. The cross section is measured just above the threshold $\sqrt{s} = 2m_{\tilde{\chi}^\pm} + 1$ GeV, and at a point well above the threshold, $\sqrt{s} = 2m_{\tilde{\chi}^\pm} + 20$ GeV.

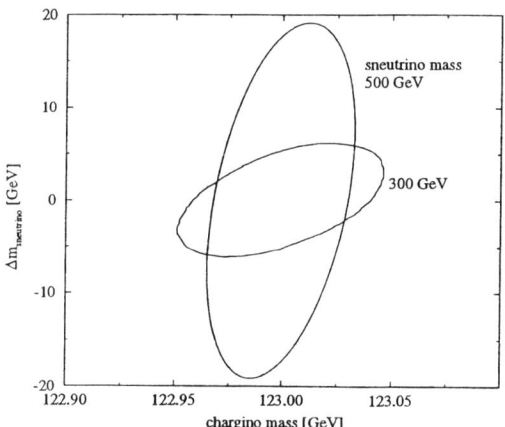

FIGURE 3. The $\Delta\chi^2 = 1$ contours in the chargino mass - sneutrino mass plane, taking $M_2 = 120$ GeV, $\tan\beta = 4$, and $m_{\tilde{\nu}} = 300$ and 500 GeV. This gives a chargino mass $m_{\tilde{\chi}^\pm} \approx 123$ GeV. The curves assume 50 fb^{-1} of integrated luminosity is devoted to $\sqrt{s} = 2m_{\tilde{\chi}^\pm} + 1$ GeV, and 50 fb^{-1} is applied at $\sqrt{s} = 2m_{\tilde{\chi}^\pm} + 20$ GeV. The chargino mass determination is better for higher sneutrino mass since the sneutrino exchange diagrams interferes destructively with the s-channel diagrams.

POLARIZATION

It is expected that the both beams of a muon collider can be partially polarized, although with some loss of luminosity [9]. This could prove a useful tool for measuring the gaugino and Higgsino components of the chargino. When the chargino is gaugino-dominated, it couples to the left-handed μ^- because the chargino is then dominantly the partner to the W. Since the WW background can be reduced by having substantial right-handed polarization, some improvement can be expected in the chargino mass precision, especially in the case where the chargino is higgsino-dominated. Only the gaugino component of the chargino couples to the t-channel sneutrino exchange, and this can be turned off by operating with polarized μ beams.

For the gaugino-dominated chargino considered here, both the signal and background are approximately proportional to $(1 - P)^2$ where P is the polarization of the two muon beams. So for fully polarized μ^+ and μ^- beams one can improve the mass determination by a factor of two.

CONCLUSIONS

We have shown that a measurement of the lightest chargino mass to better than 200 MeV is possible by measuring the pair production cross section near threshold at a muon collider with 100 fb^{-1} luminosity. This is much better than other techniques. The muon sneutrino mass can also be simultaneously measured to a few GeV.

ACKNOWLEDGMENTS

I thank V. Barger and T. Han for a pleasant collaboration on the issues reported here. This work was supported in part by the U.S. Department of Energy under Grant No. DE-FG02-91ER40661.

REFERENCES

1. V. Barger, M.S. Berger, J.F. Gunion and T. Han, Phys. Rev. **D56**, 1714 (1997).
2. V. Barger, M.S. Berger, J.F. Gunion and T. Han, Phys. Rev. Lett. **78**, 3991 (1997).
3. M.S. Berger, talk presented at the *Workshop on Particle Theory and Phenomenology: Physics of the Top Quark*, Iowa State University, May 25–26, 1995, hep-ph/9508209.
4. V. Barger, M.S. Berger and T. Han, hep-ph/9801410.
5. *Proceedings of the First Workshop on the Physics Potential and Development of $\mu^+\mu^-$ Colliders*, Napa, California (1992), Nucl. Instru. and Meth. **A350**, 24 (1994).
6. *Proceedings of the Second Workshop on the Physics Potential and Development of $\mu^+\mu^-$ Colliders*, Sausalito, California (1994), ed. by D. Cline, American Institute of Physics Conference Proceedings 352.
7. *Proceedings of the 9th Advanced ICFA Beam Dynamics Workshop: Beam Dynamics and Technology Issues for $\mu^+\mu^-$ Colliders*, Montauk, Long Island, (1995), to be bibitem.
8. *Proceedings of the Symposium on Physics Potential and Development of $\mu^+\mu^-$ Colliders*, San Francisco, California, December 13-15, 1995.
9. *$\mu^+\mu^-$ Collider: A Feasibility Study*, Snowmass, Colorado, July, 1996.
10. A. Leike, Int. J. Mod. Phys. **A3**, 2895 (1988).
11. *Physics with e^+e^- Linear Colliders*, by ECFA/DESY LC Physics Working Group (E. Accomando et al.), DESY-97-100, May 1997, hep-ph/9705442.
12. D. Miller, private communication.
13. J. Norem, private communication and http://www-ap.fnal.gov/VLHC/electrons/index.html.
14. P. Chankowski, Phys. Rev. **D41**, 2877 (1990); M. M. Nojiri, K. Fujii and T. Tsukamoto, Phys. Rev. **D54**, 6756 (1996); H.-C. Cheng, J. L. Feng and N. Polonsky, hep-ph/9706476; hep-ph/9706438; M. A. Diaz, S. F. King and D. A. Ross, hep-ph/9711307.
15. T. Tsukamoto, K. Fujii, H. Murayama, M. Yamaguchi and Y. Okada, Phys. Rev. **D51**, 3153 (1995).
16. J.-F. Grivaz, preprint LAL 91-63, Talk at the Workshop on Physics and Experiments with Linear Colliders, Saariselka, Finland, 9-14 September 1991.

Supersymmetry at the NLC

David L. Wagner

University of Colorado

Abstract. This document describes how supersymmetry can be discovered and precisely measured, and provides a comparison between the NLC and the FMC.

INTRODUCTION

Since the NLC and the muon collider are both lepton colliders, there are many similarities between the two. It is therefore useful to examine the results from NLC supersymmetry studies to get an idea of the physics possibilities at the muon collider. This paper will cover the results from the Snowmass '96 conference [1], some results since the conference, and results were shown at the SLAC/SLUO supersymmetry workshop.

Since there are also significant differences between the NLC and the muon collider, this paper concludes with a brief summary of these differences, and how they might affect the physics capabilities of the muon collider.

FEATURES OF THE NLC

The NLC will have a number of features that will be very useful for the discovery of supersymmetry and the precision measurements of the supersymmetry parameters. These features include:

- Large electron longitudinal polarization

- Clean beams which allow complete detector coverage

- Beam energy which can be tuned to optimize analysis

- A high signal-to-background ratio, which allows measurements of differential cross sections and relative branching ratios

- Simple kinematics which allow complete mass reconstruction

Electron Polarization

The SLC has obtained electron polarizations up to 80%. Discussions for the NLC have suggested that an electron polarization of up to 90% and a positron polarization of up to 65% might be possible; however, for all of our studies, we have conservatively assumed that no improvements will be made over the SLC number. (Note that we can select any polarization between 80% left-handed to 80% right handed.)

The electron polarization is useful for suppressing standard model backgrounds and for enhancing certain SUSY signals. Figure 1 shows the cross sections versus polarization for SM events. Note that changing the electron polarization from 80% right-handed to 80% left-handed decreases the W-pair background by about an order of magnitude.

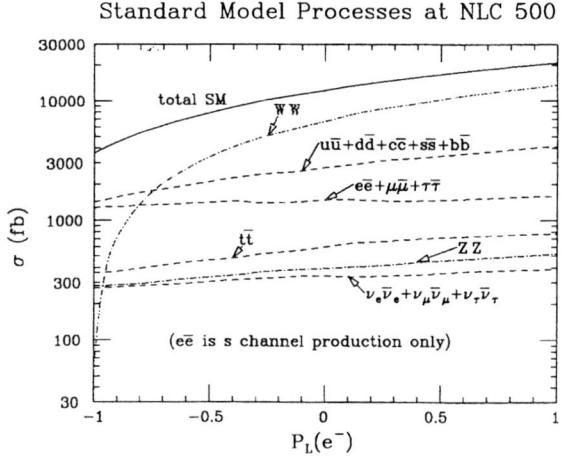

FIGURE 1. The cross-sections versus polarization for various standard model processes [2,3]. The polarization, $P_L(e^-)$ is such that the fraction $f_L = (1 + P_L(e^-))/2$; thus, 80% left-handed polarization corresponds to -0.60 on the plots.

Figure 2 shows the cross sections versus polarization for SUSY signals (for one particular choice of a SUSY model). Note that changing the polarization from left-handed to right-handed enhances the cross section for the production of $\tilde{e}_R^+ \tilde{e}_R^-$ events while reducing backgrounds from $\tilde{e}_L^+ \tilde{e}_L^-$ events (and conversely, using left-polarized electrons will enhance the production of $\tilde{e}_L^+ \tilde{e}_L^-$ events over $\tilde{e}_R^+ \tilde{e}_R^-$ events).

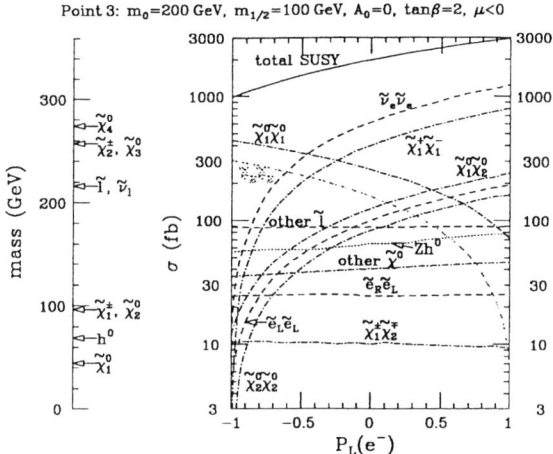

FIGURE 2. The cross-sections versus polarization for various SUSY processes, for one particular choice for a SUSY model [2,3].

Detector Coverage

It is important that the detector be as hermetic as possible. The main source of background from standard model events is $e^+e^- \to W^+W^-$ events, where some of the W decay products escape down the beam pipe, thus mimicking the SUSY signal. This background is particularly acute, since the cross-section for W-pair events is greatly enhanced in the forward direction.

Studies have shown that the NLC will need to have coverage down to at least $|\cos\theta| < 0.95$ so that the signal will not be swamped by the background. The current NLC detector design calls for tracking down to $|\cos\theta| < 0.90$ and calorimetry down to $|\cos\theta| < 0.995$.

The complete coverage will also aid the measurements of the differential cross-sections (and thus the measurements of the particle spins). For pure s-channel processes, spin-0 particles will be produced with a $\sin^2\theta$ distribution while spin-1/2 particles will be produced with a $(1+\cos^2\theta)$ distribution. The endcap is the region where these cross sections are most distinctive.

Mass Reconstruction

In this section, we present an example analysis of simulated NLC data, to serve as an indication of how well the NLC could perform. In this example we generate 50 fb^{-1} (one year) of data at 500 GeV and 80% right-handed electron polarization. We specifically want to look at $e^+e^- \to \tilde{e}_R^+\tilde{e}_R^-$ events where the $\tilde{e}_R^\pm$ decays to a $\tilde{\chi}_1^0 e^\pm$. We select those events with exactly one electron and one positron; we require that each be in the barrel region ($|\cos\theta| < 0.80$), in order to reduce backgrounds from W-pairs; and we require that each electron have less than 200 GeV of energy, to reduce backgrounds from radiative Bhabhas and two-photon events.

If we plot the energy spectrum of the electron and positron, we find a flat distribution, since the electrons come from two-body decays (see Figure 3). By measuring the endpoints of this spectrum, we can determine the masses of the selectron and the neutralino to within a couple of percent. Figure 3 also shows another benefit of a polarized beam. By varying the electron polarization, we can enhance or attenuate the $\tilde{e}_R^+\tilde{e}_R^-$ production, which helps us to distinguish these events from $\tilde{e}_R^\pm \tilde{e}_L^\mp$ events.

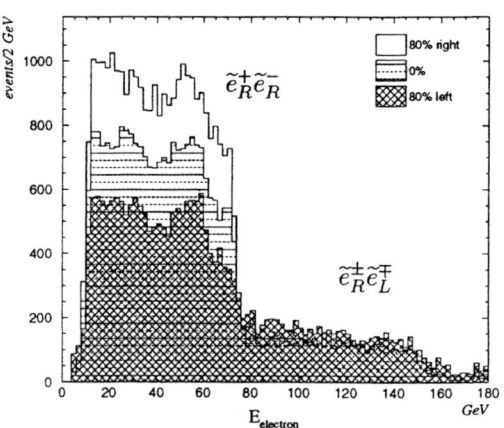

FIGURE 3. Energy spectrum of electrons and positrons from $\tilde{e}_R^\pm \to \tilde{\chi}_1^0 e^\pm$ decays. From the endpoints of this spectrum we can calculate the $\tilde{\chi}_1^0$ and $\tilde{e}_R^\pm$ masses.

We can also use a different method which was developed by Feng and Finnell [5]. This technique correlates the electrons on the two sides of the event. By employing knowledge of the electron energies, the beam energy (which is also the selectron energy) and previous measurements of the neutralino mass, we can construct the minimum kinematically allowed selectron mass, $M_{\tilde{e}_R^\pm}^{min}$. Figure 4 shows a plot of this quantity event by event; there is a clear peak in the $M_{\tilde{e}_R^\pm}^{min}$ distribution at the true selectron mass. Taking the width of the peak as an indication of the measurement uncertainty, this constitutes a 0.5% measurement of the selectron mass.

Beam Energy

At the Snowmass conference, there were extensive discussions on how the tunable energy of the NLC could help optimize the analysis. The details can

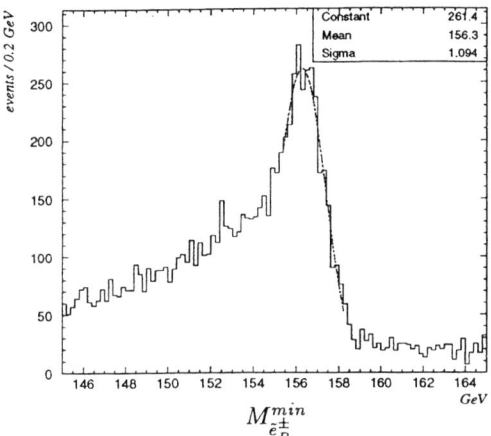

FIGURE 4. Distribution of $M^{min}_{\tilde{e}^{\pm}_R}$; this quantity peaks at the true $\tilde{e}^{\pm}_R$ mass.

be found in [1], but the basic idea is that we can tune the beam to the optimum energy for a particular channel, and then use the results of this analysis to choose the optimal energy for the next analysis. For example, the sequence of steps (assuming point 3) might look like:

- Run at 350 GeV; this is the top-pair threshold. With a couple of inverse femtobarns of data, we would see the $\tilde{\chi}^+_1$ and the $\tilde{\chi}^0_1$ and determine their masses well enough to know that we should run at 250 GeV to enhance the production of $\tilde{\chi}^+_1 \tilde{\chi}^-_1$ events.

- Run at 250 GeV; we will determine the masses of the $\tilde{\chi}^+_1$ and the $\tilde{\chi}^0_1$ to a few percent. By changing the beam polarization, we will be able to determine the cross-sections $\sigma_{L,R}(\tilde{\chi}^+_1 \tilde{\chi}^-_1)$—these are dependent on the amount of Higgsino-wino mixing in the chargino. Assuming the MSSM, we can predict $M_{\tilde{\nu}} < 250$ GeV, so our next step is to run at 500 GeV to pick up the sleptons.

- Run at 500 GeV and measure the slepton and sneutrino masses. With this additional information, we can determine the supersymmetry parameters such as μ, M_2 and $\tan\beta$ to within a few percent, which will be sufficient to predict that the squark masses are below 400 GeV. We then run the machine at 800 GeV to measure the squark properties.

COMPARISON WITH A MUON COLLIDER

The NLC and the muon collider are both lepton colliders, and therefore have many properties in common. But because of their vastly different designs, they have distinct sets of advantages. Some of the advantages of the NLC include:

- tunable beam energy,
- polarized electrons,
- hermetic detector coverage, and
- lower machine backgrounds.

Some of the advantages of a muon collider include:

- higher beam energy, and
- smaller beam energy spread.

The feature of the FMC which is the most worrying is the high background, both due to the machine and due to standard model events that point into the large dead region in the endcap. Careful Monte Carlo studies will need to be performed in order to determine whether these backgrounds will overwhelm the supersymmetry signal.

REFERENCES

1. Danielson, M. N. et al., "Supersymmetry at the NLC," Proceedings of the Snowmass '96 conference, 1996.
2. Baer H., Dubois R., Fahey S., Manly S., Munroe R., Nauenberg U., Tata X., Wagner D. L., "Determination of Supersymmetric Particle Production Cross Sections and Angular Distributions at a High Energy Linear Collider," contributed paper to *Physics and Technology of the Next Linear Collider*, BNL 52-502, Fermilab-PUB-96/112, LBNL-PUB-5425, SLAC Report 485, UCRL-ID-124160, submitted to the Snowmass '96 conference.
3. Baer H., Munroe R., Tata X., "Supersymmetry Studies at Future Linear e^+e^- Colliders," FSU-HEP-960601 (1996) (hep-ph/0906325).
4. Paige F., Protopopescu S., in *Supercollider Physics*, p. 41, edited by D. Soper (World Scientific, 1986); Baer H., Paige F., Protopopescu S., and Tata X., in *Proceedings of the Workshop on Physics at Current Accelerators and Supercolliders*, edited by J. Hewett, A. White, and D. Zeppenfeld, (Argonne National Laboratory, 1993), hep-ph 9305342.
5. Feng J. L., and Finnell D. E., "Squark mass determination at the next generation of linear e^+e^- colliders", Phys. Rev. D **49**, 2369 (1994).

SUSY Signatures and Model Discrimination at $\mu^+\mu^-$ Colliders

James G. Kelly

*Phenomenology Inst., Dept. of Physics, Univ. of Wisc.,
1150 University Ave., Madison, WI 53706*

Abstract. I give a brief sample of what can can be said regarding the MSSM model parameters and possible signatures of MSSM partner pair production for the case of a high energy (3.6 TeV) muon collider. A method for estimating the errors on a large set of MSSM parameters is described and exemplary results are displayed.

INTRODUCTION

I wish to address the issue of what can be said about the Minimal Supersymmetric Standard Model (MSSM) if one turns on a high energy ($\sim$ 3.6 TeV) $\mu^+\mu^-$ collider above a considerable number of sparticle pair thresholds. This clearly is a worst case scenario as most likely various incarnations of a muon collider of increasingly higher energy would cross various thresholds and provide bits of model information along the way. However, there are MSSM models in which the thresholds might be closely spaced, and the prospect raises a few interesting issues regarding how one discriminates between models and signals when presented with potentially confusing experimental data:

- Cascade decays can obscure the relation between the produced sparticle pair and what is seen in the detector. For example, with the production of a heavy squark pair in a model with many lower-lying slepton and gaugino states, a cascade of decays occurs before a signal is seen in the detector. In this case, the usual procedure of assuming a simple decay scheme, running a monte carlo simulation, and choosing appropriate cuts is made ineffective by the number of unknown masses of the lower mass states.

- Cascade decays may also help in sorting out signals and models in the cases where there are significant (yet complicated) correlations between a produced pair and the final detected state. Typically the cascades lead to final states with small or nonexistent Standard model backgrounds.

- Final state signatures resulting from cascades are surprisingly sensitive to location within a large MSSM parameter space [1], and can give rather good bounds on a large number of MSSM parameters.

GENERAL PROCEDURE

We characterize a given final state by the number of various particle types observable in the detector. Our final states include numbers of:

t	(reconstructible hadronic decays only)
b	(no tagging efficiency included here)
other jets	(u,d,c,s,g)
e^+	
e^-	
μ^+	
μ^-	
τ^+	(not decayed)
τ^-	(not decayed)
h^0	(assume all reconstructible $h^0 \to b\bar{b}$)
Z^0	(reconstructible only)
W^+	(reconstructible only)
W^-	(reconstructible only)

For each sparticle $\tilde{X}$, we decay the pair-produced $\tilde{X}\bar{\tilde{X}}$ in all possible ways. The rate (including associated branching fractions) associated with each decay chain is placed in a bin corresponding to the various numbers of final state particles appearing in the list above. At present no Standard Model backgrounds are considered. In this preliminary study we have included the following sparticle pairs $\tilde{X}\bar{\tilde{X}}$ and Higgs pairs:

$$\mu^+\mu^- \to \tilde{e}_L^+\tilde{e}_L^-, \tilde{e}_R^+\tilde{e}_R^-, \tilde{\nu}_e\bar{\tilde{\nu}}_e, \tilde{u}_L\bar{\tilde{u}}_L, \tilde{u}_R\bar{\tilde{u}}_R, \tilde{d}_L\bar{\tilde{d}}_L, \tilde{d}_R\bar{\tilde{d}}_R,$$
$$\tilde{\mu}_L^+\tilde{\mu}_L^-, \tilde{\mu}_L^+\tilde{\mu}_R^-, \tilde{\mu}_R^+\tilde{\mu}_R^-, \tilde{\nu}_\mu\bar{\tilde{\nu}}_\mu, \tilde{c}_L\bar{\tilde{c}}_L, \tilde{c}_R\bar{\tilde{c}}_R, \tilde{s}_L\bar{\tilde{s}}_L, \tilde{s}_R\bar{\tilde{s}}_R,$$
$$\tilde{\tau}_1^+\tilde{\tau}_1^-, \tilde{\tau}_1^+\tilde{\tau}_2^-, \tilde{\tau}_2^+\tilde{\tau}_2^-, \tilde{\nu}_\tau\bar{\tilde{\nu}}_\tau, \tilde{t}_1\bar{\tilde{t}}_1, \tilde{t}_1\bar{\tilde{t}}_2, \tilde{t}_2\bar{\tilde{t}}_2, \tilde{b}_1\bar{\tilde{b}}_1, \tilde{b}_1\bar{\tilde{b}}_2, \tilde{b}_2\bar{\tilde{b}}_2$$
$$h^0A^0, H^0A^0, H^+H^-.$$

Such a binning scheme allows one to make statements about the possible correlations between a particular sparticle pair $\tilde{X}\bar{\tilde{X}}$ and associated rates in the final state bins, all without considering the kinematic variables of the final state particles. An example scenario will be considered in the next section.

It is desirable not only to be able to associate complicated final states with particular sparicle pairs *within* a given MSSM model, but also to have a way to discriminate between such models in the presence of complicated cascade decays. Here I outline a method for constraining the values of 23 important low energy MSSM parameters:

$$\mu, m_A, \tan\beta$$
$$m_{\tilde{e}_L}, m_{\tilde{e}_R}, m_{\tilde{u}_L}, m_{\tilde{u}_R}, m_{\tilde{d}_R} \ (\times \text{ 3 generations})$$
$$A_t, A_b, A_\tau, M_1, M_2.$$

Let N_i denote the number of events in final state bin i (*e.g.* $i \leftrightarrow 2j1e^+1e^- + E_{\text{miss}}$), and let P_j denote one of the 23 MSSM parameters given above (*e.g.* $j \leftrightarrow \tan\beta$). Then we calculate the standard χ^2 statistic for two neighboring points in this 23-dimensional parameter space for any bin i:

$$\Delta\chi_i^2 = \frac{(N_i(P) - N_i(P+\Delta P))^2}{N_i(P)} \simeq \frac{(\sum_j \frac{\partial N_i}{\partial P_j} \Delta P_j)^2}{N_i(P)}, \qquad (1)$$

where P represents the point in parameter space, and I have approximated the difference by derivatives. Since all of the bins are independent, we can sum over i and then perform the square in the numerator to obtain

$$\Delta\chi^2 = \sum_i \Delta\chi_i^2 \simeq \sum_{j,k} M_{jk} \Delta P_j \Delta P_k, \qquad (2)$$

where

$$M_{jk} = \sum_i \frac{1}{N_i} \frac{\partial N_i}{\partial P_j} \frac{\partial N_i}{\partial P_k}. \qquad (3)$$

Equation (2) has the form of an ellipsiod in the 23-dimensional parameter space. The inverse of the matrix of coefficients $\sigma_{kj}^2 = (M_{jk})^{-1}$ is usually called the error matrix. Its diagonal entries give the squares of the "1σ" errors on our free parameters [2]:

$$\Delta P_{k,1\sigma} = \sqrt{\sigma_{kk}^2} = \sqrt{(M^{-1})_{kk}}. \qquad (4)$$

The interpretation of $\Delta P_{k,1\sigma}$ for the simple case of two free parameters is the half-widths of the smallest rectangle with sides parallel to the parameter axes which contains the skewed ellipse described by eq. (2).

RESULTS

To present some results of the methods outlined in the previous section I adopt an mSUGRA scenario with $m_0 = 700$ GeV, $m_{1/2} = 700$ GeV, and $A_0 = 0$ as the GUT-scale scalar mass, gaugino mass, and soft Yukawa, respectively, and with $\tan\beta = 2$ and $\text{sign}(\mu) = +$. I consider a 3.6 TeV machine with 1000 fb^{-1} of integrated luminosity, with the above parameter set chosen to allow for all pairs listed in the second section to be produced.

As an example, consider the pair $\tilde{\nu}_\mu \bar{\tilde{\nu}}_\mu$. The raw number of pairs produced is 45228. After decaying down the pair in all possible ways, binning the final state particle signatures as detailed above, and comparing with associated "backgrounds" from the other pairs listed above (not Standard Model sources) which have the same binning signatures, we find (keeping only bins with ≥ 5 events) the following bins and associated signals and MSSM backgrounds:

$$\text{Signal} = \tilde{\nu}_\mu \bar{\tilde{\nu}}_\mu$$

t	b	j	μ^-	μ^+	e^-	e^+	τ^-	τ^+	h^0	Z^0	W^+W^-	
0	0	0	0	0	0	0	0	0	0	0	0	

Signal= 2374.46, Backgrnds (sum)= 74.49
$\tilde{\nu}_e \bar{\tilde{\nu}}_e$ 34.207
$\tilde{\nu}_\tau \bar{\tilde{\nu}}_\tau$ 34.321
$\tilde{\chi}_1^0 \tilde{\chi}_2^0$ 5.965

| 0 | 0 | 0 | 1 | 1 | 0 | 0 | 0 | 0 | 0 | 1 | 1 |

Signal= 5579.38, Backgrnds (sum)= 27.89
$\tilde{\chi}_3^0 \tilde{\chi}_4^0$ 27.890

0	0	0	1	1	0	0	0	1	0	0	0	1	Signal= 898.72
0	0	0	1	1	0	0	1	0	0	0	1	0	Signal= 898.72
0	0	0	1	1	0	0	1	1	0	0	0	0	Signal= 145.31
0	0	0	1	1	0	1	0	0	0	0	0	1	Signal= 857.64
0	0	0	1	1	0	1	1	0	0	0	0	0	Signal= 138.15
0	0	0	1	1	1	0	0	0	0	0	1	0	Signal= 857.64
0	0	0	1	1	1	0	0	1	0	0	0	0	Signal= 138.15
0	0	0	1	1	1	1	0	0	0	0	0	0	Signal= 132.38
0	0	0	1	2	0	0	0	0	0	0	0	1	Signal= 892.22
0	0	0	1	2	0	0	1	0	0	0	0	0	Signal= 143.72
0	0	0	1	2	1	0	0	0	0	0	0	0	Signal= 137.15
0	0	0	2	1	0	0	0	0	0	0	1	0	Signal= 892.22
0	0	0	2	1	0	0	1	0	0	0	0	0	Signal= 143.72
0	0	0	2	1	0	1	0	0	0	0	0	0	Signal= 137.15
0	0	0	2	2	0	0	0	0	0	0	0	0	Signal= 142.68,

where the backgrounds include only the SUSY and Higgs pairs listed above and the latter bins with no backgrounds shown are free of SUSY and Higgs backgrounds.

The large number of bins with little or no background clearly indicate that distinguishing the $\tilde{\nu}_\mu \bar{\tilde{\nu}}_\mu$ channel from other SUSY pairs should not be a problem given that we can narrow down what model we are in. The first channel represents a pure missing energy channel, the isolation of which would probably involve more complicated experimental issues, so I do not consider it as usable here.

The larger issue of narrowing down the menacingly large number of MSSM models is perhaps more important. If there is sufficient variation in the bin rates as a function of position in parameter space, then this should translate into some expected bounds on the the values of the MSSM free parameters as a function of parameter space position. The method for obtaining approximate bounds on 23 parameters was explained in the preceding section. Here I tabulate the parameters P, their "1σ" errors $\Delta P_{1\sigma}$, and for convenience the relative error $\Delta P/P$, for the same machine and model parameters used above:

	μ	m_A	$\tan\beta$
P	1264	1703	2
$\Delta P_{1\sigma}$	4.8	7.5	0.018
$\Delta P/P$	0.004	0.004	0.009

	$\tilde{m}_{\tilde{Q}1_L}$	$m_{\tilde{d}_R}$	$m_{\tilde{u}_R}$	$m_{\tilde{\ell}1_L}$	$m_{\tilde{e}_R}$
P	1602	1542	1550	852	750
$\Delta P_{1\sigma}$	6.2	1100	92000	7.6	9.9
$\Delta P/P$	0.004	0.7	60	0.009	0.01

	$\tilde{m}_{\tilde{Q}2_L}$	$m_{\tilde{s}_R}$	$m_{\tilde{c}_R}$	$m_{\tilde{\ell}2_L}$	$m_{\tilde{\mu}_R}$
P	1602	1542	1550	852	750
$\Delta P_{1\sigma}$	12	1100	87000	21	14
$\Delta P/P$	0.007	0.7	60	0.02	0.02

	$\tilde{m}_{\tilde{Q}3_L}$	$m_{\tilde{b}_R}$	$m_{\tilde{t}_R}$	$m_{\tilde{\ell}3_L}$	$m_{\tilde{\tau}_R}$
P	1431	1542	1152	851	749
$\Delta P_{1\sigma}$	7.6	8.2	6.2	3.1	17
$\Delta P/P$	0.005	0.005	0.005	0.004	0.02

	A_t	A_b	A_τ	M_1	M_2
P	-1107	-1808	-450	305	596
$\Delta P_{1\sigma}$	11	18	2000	2.1	0.18
$\Delta P/P$	0.01	0.01	4	0.007	0.0003

Some of the bounds are surprisingly good. The reader is reminded that the results are preliminary, and that various suggestions for making the bounds better and also more realistic came up through fruitful discussions during the working group (see next section). It is not yet known why the bounds on the fist and second generation squark masses should be so much worse the than bounds on the other parameters, given that my parameter choices allow for the production of such squark pairs.

FUTURE DIRECTIONS

Clearly the method of binning as outlined here allows a great deal of model information to be extracted from complicated experimental signals in the presence of several generations of cascade decays. Such a method does not rely on the use of kinematic information of the observed particles, and yet one can still pin down model parameters very well for the few test cases we have looked at. For most of the bins the signatures are unique enough to allow one to ignore Standard Model backgrounds completely.

One interesting issue that arose during the working group discussions is the possibility of bin crossover diluting the results. Crossover refers to final state particles being misidentified or missed, in which case a sensitive dependence on parameter space location might me averaged out, leaving substantially larger bounds on the parameters. Such misidentification probabilities are being included in the analysis at present.

Another avenue being explored is the use of a generalized χ^2 statistics which is valid for Poisson distributed data [3], rather than the standard χ^2 statistic for which precise inferences can be made only for normally distributed data. Thus, we could remove the restictions that channels used have ≥ 5 events and see if there is any useful model discriminating power in the smaller rate channels, of which there are many.

REFERENCES

1. Gunion, J. F., and Kelly, J. G., *Phys. Rev.* **D51**, 2101 (1994). This reference looks at the article looks at the similar problem of constraining mSUGRA parameter space using Higgs pair signals.
2. Eadie, W. T., et. al., *Statistical Methods in Experimental Physics*, Amsterdam, North-Holland Publushing, 1971, pp. 192-197.
3. Baker S., and Cousins R. D., *Nucl. Inst. and Meth.* **221**, 437 (1984).

Flavor and CP Violations from Sleptons at the Muon Collider

Hsin-Chia Cheng

Fermi National Accelerator Labortory
P.O. Box 500, Batavia, IL 60510

Abstract.
Supersymmetric theories generally have new flavor and CP violation sources in the squark and slepton mass matrices. They will contribute to the lepton flavor violation processes, such as $\mu \to e\gamma$, which can be probed far below the current bound with an intense muon source at the front end of the muon collider. In addition, if sleptons can be produced at the muon collider, the flavor violation can occur at their production and decay, allowing us to probe the flavor mixing structure directly. Asymmetry between numbers of $\mu^+ e^-$ and $e^+ \mu^-$ events will be a sign for CP violation in supersymmetric flavor mixing.

Weak scale supersymmetry (SUSY) is one of the most attractive candidates for physics beyond the standard model (SM). The discovery of superpartners of the SM particles is promising at the planned future colliders. If SUSY is discovered, measuring the masses and couplings of the superpartners will become the focus of study. Measurements of the superparticle couplings is essential for verifying supersymmetry [1], and the superpartner masses provide information of the origin of SUSY breaking and unification at high scales [2]. In most SUSY extension of the standard model, mass matrices of the fermions and their scalar superpartners are, however, not diagonal in the same basis. New flavor mixing matrices W, analogous to the CKM matrix, will appear at the gaugino-fermion-sfermion vertices. These new flavor mixing matrices may provide clues to the puzzle of the flavor structure, and therefore should also be important to study. At the muon collider and its front end, these new flavor mixing effects can be studied both indirectly through the rare flavor-changing process and directly by the slepton production if they are accessible. In this talk, we will discuss the power in probing the SUSY flavor mixings of the muon collider and compare the indirect and the direct probes. This work grew out from the studies done with N. Arkani-Hamed, J. L. Feng, and L. J. Hall [3,4].

Lepton flavor, although conserved in the SM, is typically violated in most SUSY extension of the SM, since the scalar partners of the fermions must be given mass, and the scalar mass matrices are generally not diagonal in the same basis as the

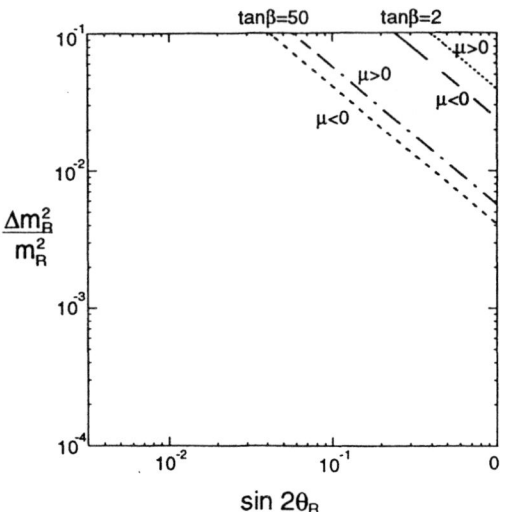

FIGURE 1. Constant contours of current bound on $B(\mu \to e\gamma) = 4.9 \times 10^{-11}$ for the parameters given in the text and different $\tan\beta$ and signs of μ.

fermion masses. When we work in the mass eigenstates of both leptons and sleptons, the flavor mixing matrices W will appear in gaugino/Higgsino vertices,

$$\tilde{e}_{Li}W^*_{Li\alpha}\overline{e_{L\alpha}}\tilde{\chi}^0 + \tilde{e}^*_{Li}W_{Li\alpha}\overline{\tilde{\chi}^0}e_{L\alpha}$$
$$+\tilde{e}_{Ri}W^*_{Ri\alpha}\overline{e_{R\alpha}}\tilde{\chi}^0 + \tilde{e}^*_{Ri}W_{Ri\alpha}\overline{\tilde{\chi}^0}e_{R\alpha}, \quad (1)$$

where the Latin and Greek subscripts are generational indices for scalars and fermions, respectively. Nontrivial W matrices generate contributions to the rare flavor-changing processes, such as $\mu \to e\gamma$. The $\mu \to e\gamma$ rate is proportional to (simplifying to 2 generation mixing, $W_{11} = \cos\theta_{12}$, $W_{12} = \sin\theta_{12}$)

$$\left(\frac{\Delta m^2_{12}}{\bar{m}^2_{12}}\sin 2\theta_{12}\right)^2, \quad (2)$$

where $\bar{m}^2_{ij} = (m^2_i + m^2_j)/2$ is the average squared slepton mass, and $\Delta m^2_{ij} = (m^2_i - m^2_j)/2 \approx 2m\Delta m_{ij}$, with $\Delta m_{ij} = m_i - m_j$. It also depends on other SUSY parameters such as $m_{\tilde{\chi}^0}$, $\bar{m}_{L,R12}$, μ, $\tan\beta$, and so on. The current bound $B(\mu \to e\gamma) < 4.9 \times 10^{-11}$ [5] puts strong constraints on $\frac{\Delta m^2_{12}}{\bar{m}^2_{12}}$ and $\sin 2\theta_{12}$. The constraints on $\frac{\Delta m^2_{R12}}{\bar{m}^2_{R12}}$ and $\sin 2\theta_{R12}$ (assuming right mixing only) for $M_1 = M_2/2 = 130\text{GeV}$, $m_{\tilde{l}_R} = 200$ GeV, $m_{\tilde{l}_L} = 350$ GeV, and $|\mu| = 400$ GeV are shown in Fig. 1. Assuming no accidental cancellation among different diagrams, the $\mu \to e\gamma$ constraint requires $\frac{\Delta m^2_{12}}{\bar{m}^2_{12}}\sin 2\theta_{12} \lesssim 10^{-2}$ for small $\tan\beta$, and about one order of magnitude stronger

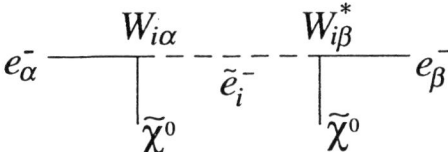

FIGURE 2. The flavor-violating process involving a single slepton production.

for large $\tan\beta$ (~ 50). Therefore, if the mixing angle θ_{12} is not very small, $\tilde{e}$ and $\tilde{\mu}$ have to be quite degenerate in order to suppress the contribution to $\mu \to e\gamma$ by the superGIM mechanism.

Similarly, nonzero W_{32} and W_{31} can contribute to the rare decays $\tau \to \mu\gamma$ and $\tau \to e\gamma$. The current bounds $B(\tau \to \mu\gamma) < 2.9 \times 10^{-6}$, $B(\tau \to e\gamma) < 2.7 \times 10^{-6}$ [7], however, do not constrain the corresponding mixing angles and mass splittings for small $\tan\beta$, and constrain them only weakly for large $\tan\beta$. While W_{32} and W_{31} are not constrained individually, their product is constrained by $\mu \to e\gamma$ if large splitting between $\tilde{\tau}$ and $\tilde{\mu}$, $\tilde{e}$ masses exist, because $\mu \to e\gamma$ can occur through the $\tilde{\tau}$ loop.

At the muon collider, an extremely intense very low energy muon source will be available for greatly improving the search for rare muon decays. It is estimated that the bound on $B(\mu \to e\gamma)$ may be pushed down to $\sim 10^{-14}$ and even better for $\mu \to e$ conversion [6]. This will dramatically improve the probing range of $\frac{\Delta m_{12}^2}{\tilde{m}_{12}^2}$ and $\sin 2\theta_{12}$. The discovery of $\mu \to e\gamma$ will have important implications on the flavor structure and hence will be extremely exciting. However, the prediction of $B(\mu \to e\gamma)$ in SUSY theories still depends on many parameters and there are many diagrams which may add up or cancel each other. A single number $B(\mu \to e\gamma)$ is not enough for us to understand the whole flavor mixing matrices. We would like to get more handles on the W matrices. At the muon collider, in addition to measuring $B(\mu \to e\gamma)$, if sleptons can be produced, we can probe these flavor mixing matrices directly from the flavor-changing slepton production and decay. They have simpler dependence on the SUSY parameters and there can be different modes to be measured. Therefore, they may provide more and clearer information for the flavor mixing matrices

We now consider the flavor-violating signals from on-shell slepton production at the muon collider. The signals we look for consist of a pair of unlike flavor leptons in the final state, $\mu^+\mu^- \to e_\alpha^+ e_\beta^- \tilde{\chi}^0 \tilde{\chi}^0$. For simplicity, let us consider the flavor-violating processes involving a single slepton[1] as shown in Fig. 2. Summing over the amplitudes of different flavor sleptons, the cross section is proportional to

$$\sum_{ij} W_{i\alpha} W_{i\beta}^* W_{j\alpha}^* W_{j\beta} \frac{1}{1 + ix_{ij}}, \qquad (3)$$

---

[1] The formalism for correlated slepton pair production is more complicated and is presented in [4]. However, the essential results remain unchanged.

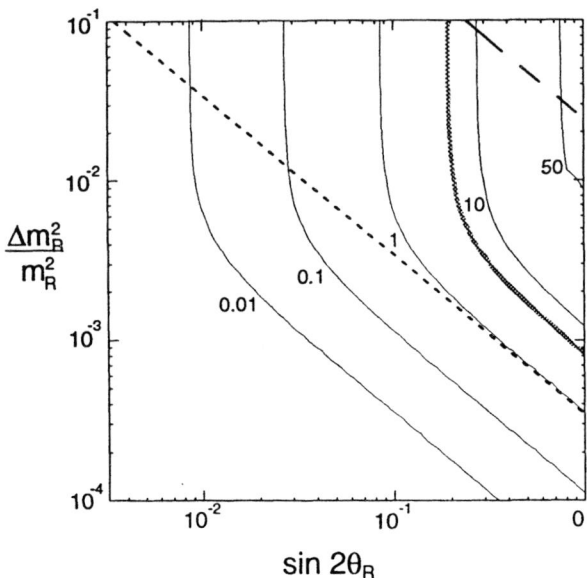

FIGURE 3. Contours of constant $\sigma(\mu^+\mu^- \to e^\pm\mu^\mp\tilde{\chi}^0\tilde{\chi}^0)$ (solid) in fb for $\sqrt{s} = 500$ GeV, $m_{\tilde{e}_R}, m_{\tilde{\mu}_R} \approx 200$ GeV, and $M_1 = 130$ GeV. The thick contour represents the experimental reach with integrated luminosity 20 fb^{-1}. Constant contours of $B(\mu \to e\gamma) = 4.9 \times 10^{-11}$ (dashed) and 10^{-14} (dotted) are also plotted for $m_{\tilde{l}_L} \approx 350$ GeV.

where $x_{ij} \equiv \Delta m_{ij}/\Gamma$, and Γ is the slepton decay width. For two generation mixing, it reduces to (1-2 mixing only)

$$\sin^2 2\theta_{12} \frac{x_{12}^2}{2(1 + x_{12}^2)}. \tag{4}$$

As with the low-energy signal, the flavor-violating collider signal vanishes in the limit of degenerate sleptons. However, the collider signal is suppressed only for $\Delta m < \Gamma$ where the quantum interference between different flavor sleptons becomes important, in constrast with the low energy signal which is suppressed by $\Delta m/\bar{m}$.

Having discussed the flavor-violating cross sections, we now consider the reach in the flavor mixing parameter space at the muon collider. We consider a case with the following SUSY parameters: $m_{\tilde{e}_R}, m_{\tilde{\mu}_R} \approx 200$ GeV, $M_1 = M_2/2 = 130$ GeV, $\mu = -400$ GeV, and $\tan\beta = 2$. The LSP is almost pure bino and the cross section has little dependence on μ, M_2, and $\tan\beta$ in this region. For center of mass energy $\sqrt{s} = 500$GeV, we calculate the cross section of the flavor-violating signal $\mu^\pm e^\mp \tilde{\chi}^0\tilde{\chi}^0$ as a function of $\sin 2\theta_{12}$ and Δm^2, and the result is shown in Fig. 3. The major backgrounds come from WW, $W\nu\mu$, and $\tau\tau$ events. Assuming the similar cuts in the Next Linear Collider studies can be applied, the backgrounds can be reduced to ~ 5 fb while keeping about 30% of the signals. With the integrated luminosity 20 fb^{-1}, the 3σ discovery limit is then $\sigma \sim 5$ fb, which is shown by the thick contour in Fig. 3. We can see that $\sin\theta_{12}$ may be probed to ~ 0.1 for $\Delta m > \Gamma$. For comparison, we also superimpose the contours of the current $B(\mu \to e\gamma)$ constraint and the expected reach by the intense muon source at the muon collider front end. The collider probe reach far below the current constraint for small Δm, but not as far as the expected $\mu \to e\gamma$ reach. However, $\mu \to e\gamma$ could receive contributions from many diagrams and has complicated dependence on many SUSY parameters. Therefore, it is more difficult to be disentangled to give precise information of flavor mixings than the collider signal.

It is also possible to probe 23 and 13 mixings by looking at final states of various flavor leptons. For the 23 mixing, the analysis is similar except that we look for final states with a τ instead of an e. Only hadronic decays of τ can be used as signals so the reach of the 23 mixing is a little worse than the 12 mixing. Nevertheless, it is still very interesting compared with no constraint from the current $\tau \to \mu\gamma$ bound. For the 13 mixing, because the dominant t-channel contribution to the slepton production always involves the initial state muons, we can only probed it (if possible) through the smaller s-channel contribution. Therefore, the probing power is not promising. In constrast, the electron collider can probe 13 mixing quite well, but not the 23 mixing.

The flavor mixing matrices may also contain CP-violating phases. In the presence of CP violation, the cross sections $\sigma_{e_\alpha^+ e_\beta^-}$ and $\sigma_{e_\beta^+ e_\alpha^-}$ are no longer equal [4,9], and the difference is proportional to the SUSY analogue to the Jarlskog invariant, $\tilde{J}$, which is defined by

$$Im\left[W_{i\alpha}W_{i\beta}^*W_{j\alpha}^*W_{j\beta}\right] = \tilde{J}\sum_{k\gamma}\varepsilon_{ijk}\varepsilon_{\alpha\beta\gamma}. \tag{5}$$

Therefore, the asymmetry between the numbers of μ^+e^- and $e^+\mu^-$ events provides a signal for the CP violation in the slepton flavor mixing matrices. This CP-violating asymmetry can only be significant for large 3 generation mixings and the mass splittings among different generations comparable to Γ. Therefore, if it is discovered, it points toward a very specific flavor structure. Fig. 4 shows the 3σ discovery limit of the CP asymmetry for the set of SUSY parameters: $m_{\tilde{l}_R} = 150$ GeV, $M_1 = M_2/2 = 100$ GeV, $\mu = -400$ GeV, $\tan\beta = 2$, and assuming $\Delta m_{12} = \Delta m_{23} \equiv \Delta m$, $\theta_{12} = \theta_{23} = \theta_{13} \equiv \theta$, $\sin\delta = 1$ in the standard parametrization [10].

In conclusion, if supersymmetry is discoverd, there will be a long and exciting road to measure the SUSY parameters in order to understand the underlain theory in nature. The SUSY flavor mixing matrices may provide important clues to the flavor structure and hence should be an important subject to study. At the muon collider, the flavor violation can be probed both by the low energy processes at the front end and direct slepton production at the collider. They can provide complementary information on the flavor mixings of the underlain SUSY theory.

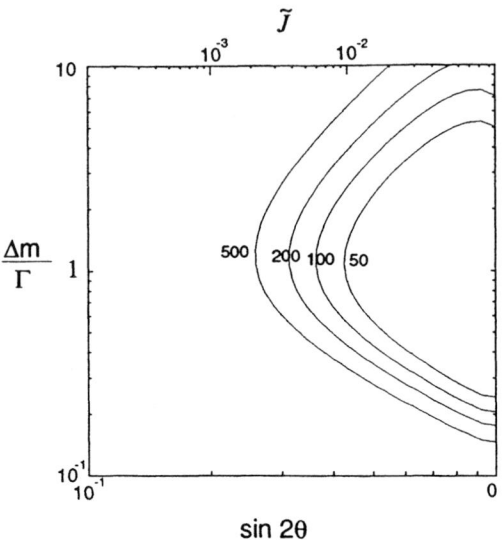

FIGURE 4. 3σ slepton CP violation discovery contours for the integrated luminosity given (in fb^{-1}). The CP-violating phase is fixed to $\sin\delta = 1$. The SUSY parameters are as given in the text.

REFERENCES

1. J. L. Feng, H. Murayama, M. E. Peskin and X. Tata, *Phys. Rev.* **D52**, 1418 (1995); M. M. Nojiri, K. Fujii and T. Tsukamoto, *Phys. Rev.* **D54**, 6756 (1996); H.-C. Cheng, J. L. Feng and N. Polonsky, *Phys. Rev.* **D56**, 6875 (1997), and Fermilab-PUB-97/205-T, hep-ph/9706476, to be published in *Phys. Rev.* **D**.
2. S. Martin and P. Ramond, *Phys. Rev.* **D48**, 5365 (1993); Y. Kawamura, H. Murayama and M. Yamaguchi, *Phys. Lett.* **B324**, 52 (1994); H.-C. Cheng and L. J. Hall, *Phys. Rev.* **D51**, 5289 (1995); S. Dimopoulos, S. Thomas and J. D. Wells, *Nucl. Phys.* **B488**, 39 (1997).
3. N. Arkani-Hamed, H.-C. Cheng, J.L. Feng and L.J. Hall, *Phys. Rev. Lett.* **77**, 1937 (1996).
4. N. Arkani-Hamed, H.-C. Cheng, J.L. Feng and L.J. Hall, *Nucl. Phys.* **B505**, 3 (1997).
5. R.D. Bolton, *et al.*, *Phys. Rev.* **D38**, 2077 (1988).
6. W. Marciano, in these proceedings.
7. CLEO Collaboration, K. W. Edwards *et al.*, *Phys. Rev.* **D55**, 3919 (1997).
8. R. Becker and C. Vander Velde, in *Proceedings of the European Meeting of the Working Groups on Physics and Experiments at Linear e^+e^- Colliders,* ed. P.M. Zerwas, Report No. DESY-93-123C, p. 457.
9. D. Bowser-Chao and W.-Y. Keung, *Phys. Rev.* **D56**, 3924 (1997).
10. L.-L. Chau and W.-Y. Keung, *Phys. Rev. Lett.* **53**, 1802 (1984).

Precision Measurements at The Higgs Resonance: A Probe of Radiative Fermion Masses[1]

F. M. Borzumati[a], G. R. Farrar[b], N. Polonsky[b], and S. Thomas[c]

[a] *Institut für Theoretische Physik, Universität Zürich, CH - 8057 Zürich, Switzerland*
[b] *Department of Physics and Astronomy, Rutgers University, Piscataway, NJ 08854, USA*
[c] *Physics Department, Stanford University, Stanford, CA 94305, USA*

Abstract. The possibility of radiative generation of fermion masses from soft supersymmetry breaking chiral flavor violation is explored. Consistent models are identified and classified. Phenomenological implications for electric dipole moments and magnetic moments, as well as collider probes – in particular those relevant at the Higgs resonance – are discussed. It is shown that partial widths $\Gamma_{h^0 \to ff}$ are enhanced compared with the minimal supersymmetric standard model.

INTRODUCTION

The main motivations for proposing and building the next generation(s) of hadron and lepton colliders are the discovery of the Higgs boson and the search for signals of physics beyond the Standard Model (SM). Most of the related theoretical work has focused so far on discovery strategies and on the extraction of various parameters within a given theoretical framework. These issues have been and are still extensively addressed in the framework of supersymmetry, in general, and the minimal supersymmetric standard model (MSSM), in particular [1]. Recently, it has been shown that electron colliders could efficiently probe various quantum corrections in supersymmetric models. One-loop corrections to gaugino couplings, that grow logarithmically with the scale of heavy (and kinematically unaccessible) sparticles, modify observables involving light sparticles and may be measured [2]. Hence, if supersymmetry is present at low-energies one may gain a more complete understanding of its realization in a similar manner to the knowledge gained from precision probes of quantum corrections in the electroweak theory.

---

[1] Talk presented by N. Polonsky. Work supported by Schweizerischer Nationalfonds and by the US National Science Foundation grant No. PHY-94-23002.

Complementary measurements (in lepton flavor and energy range) of the corrections discussed above are also possible in a muon collider [3]. Here, however, we would like to consider a unique feature of the muon collider – its beam energy resolution which enables it to explore narrow resonances. In particular, an s-channel Higgs boson [4] could be produced and its properties studied. The resonant Higgs production offers an opportunity to explore quantum effects in the Higgs sector of supersymmetric models. Most studies of the Higgs resonance have focused so far (i) on distinguishing the SM Higgs boson from the light supersymmetric Higgs boson, and (ii) on extracting the properties of the heavier Higgs bosons in supersymmetric and two Higgs doublet models [5]. Here, we propose that precision measurements of a partial width $\Gamma_{H \to ff}$, along with other low-energy observables discussed below, can reveal whether the respective fermion mass is radiative [6,7] with a corresponding "soft" (rather than tree-level) Yukawa coupling. If the Yukawa couplings are induced by supersymmetry breaking quantum effects, they would enable one to indirectly probe the supersymmetry breaking sector. We note in passing that even if there are tree-level Yukawa couplings, the quantum corrections in supersymmetric models can be possibly measurable.

We will demonstrate how radiative scenarios may arise in supersymmetric frameworks and explore their implications for fermion masses and couplings. We will show that a soft Yukawa coupling arising from supersymmetry breaking chiral flavor violation is always enhanced in comparison with the case of a tree-level coupling. In the case of the muon, however, the enhancement is constrained by the upper bound on the muon anomalous magnetic moment. Before concluding, we will suggest possible avenues for more detailed collider studies. Here, we will concentrate on the phenomenology of the light supersymmetric Higgs boson, h^0, which may be produced in a low-energy machine ($\sqrt{s} \leq 200 - 500$ GeV). Details, as well as applications involving the heavy Higgs bosons and the Higgsinos, can be found in Ref. [7].

I MODEL BUILDING

In the absence of tree-level Yukawa couplings, chiral flavor symmetries can still be broken by trilinear terms in the scalar potential,

$$V = \sum m_i^2 \phi_i^2 + [B_{ij}\phi_i\phi_j + A_{ijk}\phi_i\phi_j\phi_k + A'_{ijk}\phi_i^*\phi_j\phi_k + h.c.] + \lambda_{ij}\phi_i^2\phi_j^2. \quad (1)$$

In this case, the chiral flavor symmetries in the fermion sector are broken at the quantum level. Gauge loops $\propto A$ or $\propto A'$ which dress the fermion propagator generate the fermion mass and its effective coupling to the Higgs bosons. We will return to the generation of masses and couplings below. First, however, we would like to elaborate on the possible realizations of such a framework.

The flavor symmetries of the high-energy theory can forbid certain fundamental Yukawa couplings but allow for either (i) $ZH\Phi_L\Phi_R/M$ or (ii)

$ZZ^\dagger H^\dagger \Phi_L \Phi_R/M^3$ (superfield) operators in the superpotential or the Kahler potential, respectively. The chiral superfield $Z = z + \theta^2 F_Z$ parameterizes here the supersymmetry breaking sector, and $\langle F_Z \rangle = M_{SUSY}^2$ signals supersymmetry breaking at a scale M_{SUSY}. If the scalar component $\langle z \rangle$ vanishes and the auxiliary component $\langle F_Z \rangle$ does not, then no Yukawa couplings arise but only soft supersymmetry breaking trilinear terms $\propto \langle F_Z \rangle^n$ in the scalar potential. The operators (i) lead to A-type terms, $AH\phi_L\phi_R$, while the operators (ii) lead to A'-type terms, $A'H^*\phi_L\phi_R$. The trilinear terms are not proportional to any Yukawa couplings. The symmetries of the models typically allow for only one type of operators for a given flavor, as we will assume. Note that a sufficiently large $A' \sim M_{SUSY}^4/M^3$ requires that the supersymmetry breaking scale, M_{SUSY}, and the scale that governs the dynamics in the Kahler potential, M, are both relatively low-energy scales. Such a situation could arise, for example, if there is strong dynamics at the scale M.

Our results below imply $A/m \sim m_q(M_{\text{weak}})/(1.5-3 \text{ GeV})$, $m_l(M_{\text{weak}})/(50-100 \text{ MeV})$ for correct quark and lepton mass generation, respectively, assuming a typical sfermion mass scale m (and similarly for A'/m). Hence, the maximal magnitude of the trilinear parameters that can be realized consistently determines which fermion masses can be generated radiatively. Their magnitude is constrained most significantly by the requirement of a stable color and charge conserving minimum. One has the sufficient (but not necessary) constraint (for any flavor indices) $|A_{ijk}/m| \lesssim \sqrt{3\lambda}$, where $\lambda = \lambda_{ij} + \lambda_{ik} + \lambda_{jk}$, and similarly for A'/m. Models with A-type trilinear operators and with minimal (MSSM) matter content have $\lambda = 0$ at tree level and $\lambda < 0$ at one loop. Such models are inconsistent with the stability constraint. On the other hand, there are other viable possibilities: (a) For minimal matter content but with A'-type operators one has $\lambda = g'^2/2 \sim 0.06$ (from the D-term potential), where g' is the hypercharge coupling. For the b-quark one has in addition (from the F-term potential) $\lambda = g'^2/2 + y_t^2 \sim 1$, but only for a large t-quark Yukawa coupling, y_t, at tree-level. (b) Mirror matter flavor breaking: An exotic multi-TeV sector with vector-like matter, which is allowed by the symmetries to mix with the SM matter with a typical Yukawa coupling y, gives $\lambda \sim y^2[(m_{SB}^2 + m_S^2)/m_{SB}^2] \sim$ a few. The exotic matter is assumed to have supersymmetry conserving, m_S, and breaking, m_{SB}, masses of the same order of magnitude. Integrating out the exotic matter generates the quartic scalar terms. One concludes that e, d, s, b, u, c masses can all be realized in either case (a) or (b). The muon mass (and perhaps the τ mass, where the sfermion mixing constraint $A/m \lesssim m/\langle H \rangle$ is also relevant) can be realized only in case (b). Realization of radiative fermion masses from trilinear terms requires either an unconventional scenario of supersymmetry breaking, non-minimal matter content, or both, offering indirect probes of such scenarios.

II THE PHENOMENOLOGY

The one-loop sfermion-gaugino exchange which dresses the fermion propagator generates a finite contribution to the fermion mass. It is given by

$$m_f = -m_{LR}^2 \left\{ \frac{\alpha_s}{2\pi} C_f m_{\tilde{g}} I(m_{\tilde{f}_1}^2, m_{\tilde{f}_2}^2, m_{\tilde{g}}^2) + \frac{\alpha'}{2\pi} m_{\tilde{B}} I(m_{\tilde{f}_1}^2, m_{\tilde{f}_2}^2, m_{\tilde{B}}^2) \right\}, \qquad (2)$$

where $C_f = 4/3, 0$ for quarks and leptons, respectively, and $m_{LR}^2 = A\langle H \rangle$ or $A'\langle H^* \rangle$. The first and second terms correspond to the QCD (gluino) and hypercharge (bino) contributions, respectively. (Corrections due to possible neutralino mixing are omitted here). The function $I(m_{\tilde{f}_1}^2, m_{\tilde{f}_2}^2, m_\lambda^2)$ can be typically approximated $I(m_{\tilde{f}_1}^2, m_{\tilde{f}_2}^2, m_\lambda^2) \times \max(m_{\tilde{f}_1}^2, m_{\tilde{f}_2}^2, m_\lambda^2) \simeq \mathcal{O}(1)$, where λ ($\tilde{f}$) denotes a gaugino (sfermion). This leads to the numerical results given in the previous section. Note that the radiatively generated fermion mass does not vanish for large sparticle masses, provided that A (or A'), $m_{\tilde{f}_i}$, and m_λ, are all of the same order of magnitude. An effective Yukawa coupling $\bar{y}_f H f f$, which is momentum dependent, is generated by the corresponding loop diagrams.[2] When applied to the decay $H \to f f$, it depends only on internal and external masses. Furthermore, it is simplified for a light Higgs boson[3] $(m_{h^0}/m_{\tilde{f}}, m_{h^0}/m_\lambda \to 0)$ [7],

$$\bar{y}_f = \frac{m_f}{\langle H \rangle} \left\{ \sin^2 2\theta_{\tilde{f}} \left[\frac{1}{2} \frac{\sum_i I(m_{\tilde{f}_i}^2, m_{\tilde{f}_i}^2, m_\lambda^2)}{I(m_{\tilde{f}_1}^2, m_{\tilde{f}_2}^2, m_\lambda^2)} - 1 \right] + 1 \right\}. \qquad (3)$$

The ratio $r_f \equiv \bar{y}_f/(m_f/\langle H \rangle)$ is illustrated in Fig. 1 for sfermion mixing angle $\sin 2\theta_{\tilde{f}} = 1$, which maximizes the effect. One observes that the radiative Yukawa coupling could be enhanced by a significant percentage in comparison to the case of a tree-level fermion mass. The enhancement increases with the mass splitting between the sfermion eigenstates. Most importantly, we would like to stress that the soft coupling is always enhanced. Note that the projecting factors between the physical and interaction Higgs eigenstates were omitted above. In the case of A'-type operators these factors are different than in the usual case of tree-level couplings. For the light Higgs boson this is irrelevant in the limit in which the heavy Higgs bosons decouple, and which applies to most of the parameter space.

The radiative fermion mass also has strong implications for low-energy phenomena. The mass and one-loop electric dipole moment arise from closely related diagrams, implying that the phases are aligned, $\text{Arg}(m_e) = \text{Arg}(d_e^{SUSY})$. The one-loop electric dipole moment therefore automatically vanishes. Similarly, if the mass matrix of a whole sector (i.e.. lepton, down, or up) is

---

[2] It differs numerically from the Higgsino-sfermion-fermion coupling [7].
[3] In the case of a massive Higgs boson the vertex is described by a C function. One also finds enhancements in this case [7].

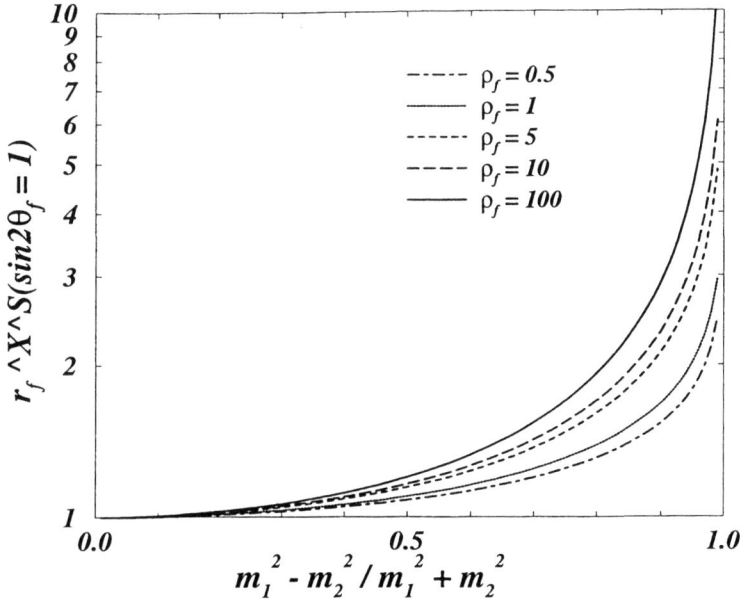

FIGURE 1. The ratio r_f of the radiative soft Yukawa coupling to a tree level Yukawa coupling for a given fermion flavor as a function of the mass splitting $(m_{f_1}^2 - m_{f_2}^2)/(m_{f_1}^2 + m_{f_2}^2)$. The different curves correspond to different values of $\rho_f = (m_{f_1}^2 + m_{f_2}^2)/2m_\lambda^2$.

generated radiatively and the soft supersymmetry breaking masses m_i^2 are flavor independent, then the trilinear parameters and the corresponding fermion masses are diagonalized simultaneously, suppressing potential contributions to flavor changing neutral currents. Such a scenario is possible at least in the case of a radiatively generated down-quark sector [7].

The magnetic moment operator is also given by a loop diagram which is similar to the mass diagram. Hence, it is not suppressed by a loop factor compared to the radiatively generated mass as is the case for a tree-level mass, leading to $g_f - 2 \sim m_f^2/m_\lambda^2$ [7]. If m_μ is radiative one predicts a muon anomalous magnetic moment of the order of current limits ($\mathcal{O}(10^{-(9-8)})$) for $m_{\tilde{B}} \lesssim 300$ GeV. Therefore, an observation at the Brookhaven experiment of a deviation from the SM prediction for $g_\mu - 2$ could signal a radiative muon mass. An s-channel resonant Higgs production could test the radiative muon mass interpretation of a deviation in $g_\mu - 2$, as a radiative m_μ also implies that the Higgs production rate is enhanced. However, because of the constraint $m_{\tilde{B}} \gtrsim 300$ GeV and the smallness of the muon mass, the mixing $\sin 2\theta_{\tilde{\mu}}$ is bounded from above [7]. As a result, the enhancement of $\bar{y}_\mu$ cannot be maximal (see eq. (3)) and is typically only a few percent. Nevertheless, efforts to precisely determine r_μ are strongly motivated in such a situation.

III PROSPECTS

It has been shown that soft Yukawa couplings, *i.e.*, radiative generation of fermion masses, is a logical possibility in certain (non-minimal) supersymmetric frameworks. Aside from low-energy implications for CP violation and magnetic moments, such scenarios imply an enhancement of the Higgs - fermion couplings ranging from a few percent up to an order of magnitude in comparison with the MSSM. Hence, the next generation of lepton colliders, and in particular, the muon collider, offer an opportunity to determine experimentally whether fermion masses are generated radiatively.

The production rate on the s-channel Higgs resonance of a fermion whose mass is generated radiatively is enhanced in comparison with the MSSM by r_f^2, or if m_μ is also radiative, by $r_\mu^2 r_f^2$. The partial width $\Gamma_{h^0 \to ff}$ is enhanced by r_f^2. The enhancement can partially cancel out in branching ratios, so partial widths or the ratios $\Gamma_{h^0 \to ff}/\Gamma_{h^0 \to WW^*}$ are more sensitive observables. However, an enhancement of the total width can also indicate radiative fermion masses. The projected errors of $\sim 5\%$, $10-20\%$, 10% in the determination of $\Gamma^2_{h^0 \to \mu\mu}$, $\Gamma^2_{h^0 \to cc, bb}$ and the total width [5], respectively, suggest that such tests are feasible. However, detailed studies are needed. Decay modes of the heavy Higgs bosons, the Higgsinos, and the sfermions offer additional probes of such scenarios in high energy machines [7]. For example, if the b mass is radiative then one expects enhancement of $\tilde{b}_1 \to H\tilde{b}_2$. Similarly, in the case of, *e.g.*, the c-quark, one would also observe mixing (which vanishes in the MSSM) between the scalar charms.

We would like to acknowledge discussions with V. Berger, J. F. Gunion, H. E. Haber, and W. J. Marciano.

REFERENCES

1. See, for example, contributions by Hinchliffe, Kelly, Paige, and Porod.
2. Cheng C.-H., Feng J.L., and Polonsky N., hep-ph/9706438; *ibid.*, hep-ph/9706476; Pierce D., Nojiri M., and Yamada Y., hep-ph/9707244; Pierce D. and Thomas S., SU-ITP 97-24.
3. Cheng *et al.* [2]; Barger V., in these proceedings.
4. Barger V., Berger M.S., Gunion J.F., and Han T., *Phys. Rev. Lett.* **75**, 1462 (1995); *ibid.*, *Phys. Rept.* **286**, 1 (1997).
5. See Gunion's contribution and a summary talk by Han.
6. del Aguila F., Dugan M., Grinstein B., Hall L.J., Ross G.G., and West P., *Nucl. Phys.* **B250**, 225 (1985); Banks T., *Nucl. Phys.* **B303**, 172 (1988); Ma E., *Phys. Rev. D* **39**, 1922 (1989); Krasnikov V.A., *Phys. Lett. B* **302**, 59 (1993); Arkani-Hamed N., Cheng C.-H., and Hall L.J., *Phys. Rev. D* **54**, 2242 (1996).
7. Borzumati F.M., Farrar G.R., Polonsky N., and Thomas S., *Soft Yukawa couplings in supersymmetric theories*, ZU-TH 23/97, RU-97-56, SU-ITP 97-45.

Lepton Flavor Violation and Fermion Masses

Stuart Raby*[1]

*The Ohio State University
174 W. 18th Ave
Columbus OH 43210
raby@mps.ohio-state.edu

Abstract. Lepton flavor violation may be a signature of "GUT scale" physics, if the messenger scale for SUSY breaking is above the "GUT scale." We elaborate on the details of this simple statement in the following talk.

INTRODUCTION

The minimal supersymmetric standard model [MSSM] is defined by its spectrum and interactions, i.e. the minimal particle spectrum necessary for a self-consistent extension of the standard model, along with R parity so that the only interactions are those in the standard model or supersymmetric extensions thereof. Even with these constraints the theory in principle has many unknown parameters. These are associated with soft SUSY breaking parameters defined at a messenger scale, M. In minimal supergravity [1] the messenger scale $M = M_{Pl} \sim 10^{18}$ GeV. In this case SUSY breaking occurs in a hidden sector and is transmitted to the visible sector via gravitational interactions. It results in 5 soft SUSY breaking parameters, a *universal scalar mass* m_0, a universal gaugino mass $M_{1/2}$, a supersymmetric Higgs mass parameter μ (which in some theories is only generated once SUSY is broken), the Higgs scalar mass $B\mu$ and a *universal soft trilinear interaction parameter* A. In gauge-mediated SUSY breaking [2-4], on the otherhand, the messenger scale is typically much less than M_{Pl}. In this case, scalars with common gauge charges are degenerate and A vanishes at tree level. In this talk we consider minimal supergravity SUSY breaking, unless otherwise stated. Finally in the MSSM, as in the standard model, *neutrinos are massless*.

---

[1] Talk presented at the Workshop on Physics at the First Muon Collider and at the Front End of a Muon Collider, November 6-9,1997, FNAL.

The MSSM as defined above is a symmetry limit. Individual lepton numbers, L_e, L_μ, L_τ, are conserved. Thus processes such as $\mu \to e\gamma$, $\mu \to 3e$, $\mu \to e$ conversion or $\tau \to \mu\gamma$ are forbidden. The experimental branching ratios for these processes are bounded by [5] $B(\mu \to e\gamma) \leq 5 \times 10^{-11}$, $B(\mu \to 3e) \leq 1 \times 10^{-12}$, $B(\mu\,{}^{48}_{22}T_i \to e\,{}^{48}_{22}T_i) \leq 4.3 \times 10^{-12}$ and $B(\tau \to \mu\gamma) \leq 4.2 \times 10^{-6}$. These strong constraints have two significant consequences.

1. **Possible non-universal scalar masses or soft trilinear parameters are severely constrained [6,7].** For example, define

$$\delta^{\bar{e}}_{ij} \equiv \frac{\Delta^{\bar{e}}_{ij}}{\tilde{m}^2}$$
$$\delta^{LR}_{ij} \equiv \frac{\Delta^{LR}_{ij}}{\tilde{m}^2} \quad (1)$$

where $\Delta^{\bar{e}}_{ij}$ (Δ^{LR}_{ij}) is the off-diagonal mass squared term for right-handed (left-to-right handed) scalar leptons in a superbasis where lepton masses are diagonal (i,j are flavor indices). Then typical constraints [7] are,

$$\delta^{\bar{e}}_{12} < 4.3 \times 10^{-3} \left(\frac{\tilde{m}_e(GeV)}{100}\right)^2 \quad (2)$$

(from $\mu \to e\gamma$ with $\left(\frac{m_{\tilde{\gamma}}}{\tilde{m}_e}\right)^2 = 0.3$), or

$$\delta^{LR}_{ij} < 1.5 \times 10^{-6} \left(\frac{\tilde{m}(GeV)}{100}\right)^2 \quad (3)$$

2. **Lepton flavor violation[LFV] is sensitive to "GUT scale" physics [8,9].** Although "GUT scale" physics could easily violate flavor symmetries, one might suspect these flavor violations to be suppressed by powers of $1/M_G$. This is not the case however. As shown by Hall et al. [8] flavor violation in the lepton sector can be induced at the GUT scale due to RG running from M to M_G. Moreover, this flavor violation enters as a boundary condition in the slepton mass matrices; hence it is not suppressed by inverse powers of M_G.

As an illustration of this phenomenon, consider a generic GUT-like theory with heavy states X, Y, Z with mass $M_I \sim M_G$ and the standard model states $F_i = \{Q_i, \bar{u}_i, \bar{d}_i, L_i, \bar{e}_i\}$. Assume some new interactions between the scales M_I and M given by

$$\lambda_{ij}\,F_i\,F_j\,X + k_i\,F_i\,Y\,Z \quad (4)$$

As a consequence of renormalization group running, we find at M_I.[2]

$$\delta^F_{ij} \sim -((\lambda^\dagger \lambda)_{ij} + k^\dagger_i\,k_j)\,ln\frac{M}{M_I} \quad (5)$$

---

[2] Universal scalar masses at M are assumed in all analyses.

Of course, the numerical value depends on the scale M_I and the magnitude of the Yukawa couplings λ_{ij}, k_i. In the rest of this talk we consider three different possible contributions to lepton flavor violating interactions emanating from "GUT scale" physics. In all three cases, M_I is the scale where the structure of the fermion mass hierarchy is generated.

A.   Adding neutrino masses to the MSSM
B.   GUTs and the Third family yukawa couplings
C.   Family mass hierarchy and the FN mechanism

A   Adding neutrino masses to the MSSM

Consider adding to the MSSM some right-handed neutrinos; one for each family. The most general renormalizable superspace potential including the right-handed neutrinos is given by

$$W = \lambda_{ij}^\nu \, \bar{\nu}_i \, L_j \, h \;+\; M_{ij} \, \bar{\nu}_i \, \bar{\nu}_j \tag{6}$$

where $M_{ij} = \delta_{ij} M_i$ (δ_{ij} is a Kronecker delta) and we work in a basis where charged lepton yukawa couplings are diagonal. In this case, the scale M_I is given by $M_I = \min\{M_i\} \gg M_Z$. As long as $\det M \neq 0$, this theory results in 3 light majorana neutrinos (predominantly left-handed) and 3 superheavy majorana neutrinos (predominantly right-handed). RG running leads to radiative mass corrections of the form [9,10]

$$\delta_{ij}^L \sim -(\lambda^{\nu\,\dagger} \lambda^\nu)_{ij} \, \ln(\tfrac{M}{M_I}) \tag{7}$$

A recent analysis by Hisano et al. [10] takes $\lambda_{ij}^\nu = \lambda_i^u \, V_{ij}^{CKM}$ where λ_i^u are the diagonal up quark yukawa couplings and V^{CKM} is the CKM matrix. This form for λ_{ij}^ν is suggested by SO(10) GUTs. A value for M_I of order 10^{12} GeV was also assumed. With this value, the tau neutrino has mass of a few eV and thus it makes a good hot dark matter candidate in a universe with hot + cold dark matter. Branching ratios for LFV processes are obtained which are below the experimental bounds but close enough to be observable in future LFV experiments at Los Alamos or PSI.

B   GUTs and the Third family yukawa couplings

Quark flavor is not conserved; this is the essense of CKM mixing. Since GUTs relate quarks and leptons, it is not surprising that GUT interactions also violate lepton flavor.

For example, consider a simple SUSY SU(5) model with quarks and leptons in the $10_i \supset \{Q_i, \bar{u}_i, \bar{e}_i\}$, and $\bar{5}_i \supset \{\bar{d}_i, L_i\}$. For $\tan\beta \sim 1$, the top quark yukawa coupling is the largest yukawa coupling in the theory. It enters the superspace potential in the expression

$$W \supset \lambda_i^u \, (10_i \, 10_i \, H) \tag{8}$$

where H is a 5 of SU(5) containing the Higgs doublets as well as their color triplet partners and we work in a basis where the up quark yukawa coupling is diagonal. (Note, this simple SU(5) model with Higgs in the 5 and $\bar{5}$ representation and only dimension 4 fermion mass operators cannot fit the known fermion masses. Nevertheless, this is a useful exercise, since in any more realistic theory, the top quark yukawa coupling must still be large. [11]

RG running from M to M_G induces lepton flavor violating masses for right-handed sleptons given by [8,11,12]

$$\delta_{ij}^{\bar{e}} \sim -\lambda_t^2 \, \delta_{3i} \, \delta_{3j} \, \ln(\tfrac{M}{M_G}) \tag{9}$$

In the effective theory below M_G, in a basis where lepton masses are now diagonal, we have

$$\delta_{ij}^{\bar{e}} \sim -\lambda_t^2 \, V_{3i}^* \, V_{3j} \, \ln(\tfrac{M}{M_G}) \tag{10}$$

where $V = V^{CKM}$. Note, in SU(5), only right-handed sleptons are affected.

The CKM elements mixing the first two families with the third are small. In addition, with only $\delta_{ij}^{\bar{e}} \neq 0$, LFV is further suppressed due to a subtle cancellation between neutralino and higgsino contributions [12]. As a result lepton flavor violating processes are well within experimental bounds and possibly beyond the reach of future experiments.

In SO(10), on the otherhand, both $\delta_{ij}^{\bar{e}}$ and δ_{ij}^{L} are non-zero. Non-vanishing contributions to δ_{ij}^{L} occur, even in the limit of small $\tan\beta$, because both L and $\bar{e}$ are contained in a $16 \supset \{Q, \bar{u}, \bar{d}, L, \bar{e}\}$ of SO(10). The combination of both these terms avoids the accidental cancellation discussed previously when only $\delta_{ij}^{\bar{e}} \neq 0$ [12,11]. In this case, observable flavor violating effects may be expected in future experiments. Moreover, certain regions of parameter space are already ruled out. A study of the large $\tan\beta$ regime has also been carried out, see Ciafaloni et al. [13], with results similar to those at low $\tan\beta$.

C Family mass hierarchy and the Froggatt-Nielsen mechanism

The problem with the specific GUT models discussed above is that they give unrealistic fermion masses and mixing angles. In order to improve upon this situation within the context of GUTs one needs to either add several Higgs multiplets (with some in higher dimensional representations of the GUT symmetry) or consider the possibility of a simple Higgs sector but with higher dimension effective fermion mass operators. The latter case can provide effective higher dimensional Higgs representations by incorporating direct products such as $(5 * 24 \supset 45 + \cdots)$ in SU(5). Moreover, it was shown by Froggatt and Nielsen [14] that these effective fermion mass operators are "natural" in theories with heavy intermediate states and softly broken flavor symmetries.

As an example, consider the renormalizable superspace potential given by

$$W = \psi_3 \psi_3 H + \psi_2 \chi H + \bar{\chi}(M_{FN} \chi + \phi \psi_3) \tag{11}$$

where ψ_2 (ψ_3) represent the second (third) generation of quarks or leptons, H is the electroweak Higgs, $(\chi, \bar{\chi})$ are heavy Froggatt-Nielsen states with mass M_{FN} and ϕ contains a scalar whose vev breaks the FN flavor symmetry at a scale below M_{FN}, so that $\epsilon \equiv <\phi>/M_{FN} << 1$. In the effective theory below M_{FN}, the FN states $(\chi, \bar{\chi})$ are integrated out; giving the effective superspace potential

$$W = \psi_3 \psi_3 H + \epsilon \psi_2 \psi_3 H \tag{12}$$

plus calculable corrections of order ϵ^2. Thus we have a 2×2 yukawa matrix of the form

$$\lambda = \begin{pmatrix} 0 & \epsilon \\ \epsilon & 1 \end{pmatrix} \tag{13}$$

In these theories, the scale $M_I = M_{FN}$.

It has been shown by Dimopoulos and Pomarol [15] that the FN mechanism can lead to enhanced flavor violation due large yukawa couplings (of order one) as well as to the mixing of heavy FN scalar states with light squarks and sleptons. Consider the scalar masses

$$\mathcal{L}_{soft} \supset \tilde{m}^2_{\psi_3} |\psi_3|^2 + \tilde{m}^2_\chi |\chi|^2 + \tilde{m}^2_{\psi_2} |\psi_2|^2 + \cdots \tag{14}$$

Assume that at the scale M we have universal boundary conditions

$$\tilde{m}^2_{\psi_3}(M) = \tilde{m}^2_{\psi_2}(M) = \tilde{m}^2_\chi(M) = m_0^2 \tag{15}$$

After RG running from M to M_I and integrating out $(\chi, \bar{\chi})$ we obtain the scalar mass matrix for these two families given by

$$\tilde{m}^2 \sim \begin{pmatrix} \tilde{m}^2_{\psi_2}(M_I) & 0 \\ 0 & \tilde{m}^2_{\psi_3}(M_I) + \epsilon \tilde{m}^2_\chi(M_I) \end{pmatrix} = \begin{pmatrix} \tilde{m}^2_2 & 0 \\ 0 & \tilde{m}^2_3 \end{pmatrix} \tag{16}$$

Since the yukawa couplings in the renormalizable theory above M_I are assumed to be of order one, we have $(\tilde{m}^2_2 - \tilde{m}^2_3)/\tilde{m}^2_3 \sim 1$. When extended to three families, order one flavor splittings between all three families of sleptons are induced. Such large splittings between the third and the first two families has already been discussed in the previous section. It leads to acceptable LFV rates due to the small mixing angles between the third and first two families. Order one splittings between the first and second family, on the otherhand, gives unacceptable LFV rates, since Cabibbo like mixing between the first two families is not small. Recently Lucas [16] has calculated the LFV rates in an SO(10) SUSY GUT with realistic fermion masses and mixing angles [17,18]. LFV interactions place severe constraints on this model.[3] Consistency with present data is only obtained with sufficiently heavy scalars; in particular, sneutrinos can be as light as 800 GeV, but only in a very restricted region of parameter space.

U(2) family symmetry

When using the Froggatt-Nielsen mechanism to generate a fermion mass hierarchy, one may also need to suppress large slepton mass mixing between the first and second families. This can be accomplished by a non-abelian family symmetry which is only broken below the FN scale or by lowering the messenger scale below M_I ($= M_{FN}$), such as in gauge-mediated SUSY breaking models [2,4].[4] Several such symmetries have been considered in the literature. These include: SU(2), SU(3), S_3, U(2), $\Delta(3n^2)$ with $n = 4, 5$.

As an example consider the family symmetry group U(2) [19]. Extensions to include an SU(5) [20] (or SO(10) [21,22]) GUT have also been considered. In the $SO(10) \times U(2)$ model, the first two families transform as a (16,2) and the third family transforms as a (16,1) (represented by the fields 16_a, $a = 1, 2$ and 16_3).

The superspace potential for the fermion mass sector is given by

$$W = 16_3 \, 16_3 \, 10 + 16_a \, \chi^a \, 10$$
$$+ \bar{\chi}_a \, (M_{FN} \, \chi^a + S^{ab} \, 16_b + A^{ab} \, 16_b + \phi^a \, 16_3) \tag{17}$$

where 10 contains the electroweak Higgs doublets and their color triplet partners, $(\bar{\chi}_a, \chi^a)$ are the massive FN states, and $(S^{ab} = S^{ba}, A^{ab} = -A^{ba}, \phi^a)$ contain the scalars which spontaneously break the FN U(2) symmetry.[5] The vacuum expectation values of the latter fields determine the small parameters

$$\epsilon = \frac{<\phi^2>}{M_{FN}} \approx \frac{<S^{22}>}{M_{FN}}$$

---

[3] This analysis assumed that the messenger scale is the Planck scale with universal boundary conditions for soft SUSY breaking parameters at M_{Pl}.

[4] In gauge-mediated SUSY breaking models, the messenger scale may be as small as O(10^5) GeV. The suppression of flavor violating effects is one of the main motivations for these models.

[5] Note, in order to fit fermion masses and mixing angles, S^{ab} transforms as a 45 while M_{FN} transforms as a direct sum of 1 + 45.

$$\epsilon' = \frac{<A^{12}>}{M_{FN}}$$
$$\epsilon' < \epsilon \qquad (18)$$

This theory results in fermion yukawa matrices schematically given by

$$\lambda \sim \begin{pmatrix} 0 & \epsilon' & 0 \\ -\epsilon' & \epsilon & \epsilon \\ 0 & \epsilon & 1 \end{pmatrix} \qquad (19)$$

with $\epsilon \sim V_{cb} \sim 0.03$. For more details, see refs. [21,22].

Using a simple operator analysis, the scalar mass are given by [22]

$$\tilde{m}^2 \sim \begin{pmatrix} m_1^2 & 0 & \epsilon\epsilon' m_5^2 \\ 0 & m_1^2(1+\epsilon^2) & \epsilon m_4^2 \\ \epsilon\epsilon' m_5^2 & \epsilon m_4^2 & m_3^2 \end{pmatrix} \qquad (20)$$

Hence $\delta_{12}^{\bar{e}} \sim \delta_{12}^{L} \sim \epsilon^2 \sim 10^{-3}$; consistent with experimental bounds (see eqn. 2).

THE BOTTOM LINE

If the messenger scale **M** for soft SUSY breaking is above the "structure scale" $\mathbf{M_I}$ for fermion mass hierarchies, then observable *lepton flavor violation* is predicted due to the RG running of slepton masses from **M** to $\mathbf{M_I}$. In the examples discussed in this talk, the messenger scale was assumed to be the Planck scale.

- *For the "structure scale" of fermion masses three cases were considered:*

A. $M_I \sim 10^{12}$ GeV — Neutrino masses in the MSSM, consistent with $m_{\nu_\tau} \sim$ few eV;

B. $M_I = M_G \sim 10^{16}$ GeV — GUTs and the third family yukawa couplings;

C. $M_I = M_{FN} \geq 10^{12}$ GeV — Family hierarchy described by Froggatt-Nielsen mechanism.

- *The results are strongly model dependent, but in most cases LFV effects should be observed in the next generation experiments.*

Lepton flavor violation may be a rich goldmine of "GUT scale" physics.

OR

If $M << M_I$, then lepton flavor violation may be suppressed. This would be the case in gauge-mediated SUSY breaking models where $M << M_{Pl}$.

ACKNOWLEDGEMENT

I would like to thank K. Tobe for interesting discussions. This work is partially supported by DOE contract DOE/ER/01545-731.

REFERENCES

1. A.H. Chamseddine, R. Arnowitt and P. Nath, *Phys. Rev. Lett.* **29**, 970 (1982); R. Barbieri, S. Ferrara and C.A. Savoy, *Phys. Lett.* **B119**, 343 (1983); L.J. Hall, J. Lykken and S. Weinberg, *Phys. Rev.* **D22**, 2359 (1983); P. Nath, R. Arnowitt and A.H. Chamseddine, *Nucl. Phys.* **B322**, 121 (1983).
2. S. Dimopoulos and S. Raby, *Nucl. Phys.* **B192**, 353 (1981); M. Dine, W. Fischler, and M. Srednicki, *Nucl. Phys.* **B189**, 575 (1981) ; M. Dine and W. Fischler, *Phys. Lett.* **B110**, 227 (1982); M. Dine and M. Srednicki, *Nucl. Phys.* **B202**, 238 (1982); L. Alvarez-Gaumé, M. Claudson, and M. Wise, *Nucl. Phys.* **B207**, 96 (1982); C. Nappi and B. Ovrut, *Phys. Lett.* **B113**, 175 (1982).
3. M. Dine and W. Fischler, *Nucl. Phys.* **B204**, 346 (1982); S. Dimopoulos and S. Raby, *Nucl. Phys.* **B219**, 479 (1983); J. Polchinski and S. Susskind, *Phys. Rev.* **D26**, 3661 (1982); T. Banks and V. Kaplunovsky, *Nucl. Phys.* **B211**, 529 (1983).
4. M. Dine, A.E. Nelson and Y. Shirman, *Phys. Rev.* **D51**, 1362 (1995); M. Dine, A.E. Nelson, Y. Nir and Y. Shirman, *Phys. Rev.* **D53**, 2658 (1996).
5. Particle Data Group, *Phys. Rev.* **D54**, 1 (1996).
6. S. Dimopoulos and H. Georgi, *Nucl. Phys.* **B193**, 150 (1981).
7. F. Gabbiani, E. Gabrielli, A. Masiero and L. Silvestrini, *Nucl. Phys.* **B477**, 321 (1996).
8. L.J. Hall, V.A. Kostelecky and S. Raby, *Nucl. Phys.* **B267**, 415 (1986).
9. F. Borzumati and A. Masiero, *Phys. Rev. Lett.* **57**, 961 (1986).
10. J. Hisano, T. Moroi, K. Tobe, M. Yamaguchi and T. Yanagida, *Phys. Lett.* **B357**, 579 (1995).
11. R. Barbieri and L.J. Hall, *Phys. Lett.* **B338**, 212 (1994); R. Barbieri, L.J. Hall and A. Strumia, *Nucl. Phys.* **B445**, 219 (1995).
12. J. Hisano, T. Moroi, K. Tobe and M. Yamaguchi, *Phys. Rev.* **D53**, 2442 (1996); ibid., *Phys. Lett.* **B391**, 341 (1997); K. Tobe, *Nucl. Phys.* (Proc. Suppl.) **B59**, 223 (1997).
13. P. Ciafaloni, A. Romanino and A. Strumia, *Nucl. Phys.* **B458**, 3 (1996).
14. C. Froggatt and H.B. Nielsen, *Nucl. Phys.* **B147**, 277 (1979).
15. S. Dimopoulos and A. Pomarol, *Phys. Lett.* **B353**, 222 (1995).
16. V. Lucas, OSU Thesis, (1997), unpublished; see also, M.E. Gomez and H. Goldberg, *Phys. Rev.* **D53**, 5244 (1996).
17. V. Lucas and S. Raby, *Phys. Rev.* **D54**, 2261 (1996).
18. T. Blazek, M. Carena, S. Raby and C.E.M. Wagner, *Phys. Rev.* **D56**, 6919 (1997).
19. R. Barbieri, G. Dvali and L.J. Hall, *Phys. Lett.* **B377**, 76 (1996).
20. R. Barbieri and L.J. Hall, *Nuovo Cim.* **110A**, 1 (1997).
21. R. Barbieri, L.J. Hall, S. Raby and A. Romanino, *Nucl. Phys.* **B493**, 3 (1997).
22. R. Barbieri, L.J. Hall and A. Romanino, *Phys. Lett.* **B401**, 47 (1997).

Higgs and Z⁰ Physics

Conveners: Tao Han, University of Wisconsin
 Marcel Demarteau, Fermilab

Calibrating the Energy of a 50 x 50 GeV Muon Collider Using Spin Precession

Rajendran Raja
&
Alvin Tollestrup

Fermi National Accelerator Laboratory
P.O. Box 500
Batavia, IL 60510

Abstract. The neutral Higgs boson is expected to have a mass in the region 90-150 GeV/c² in various schemes within the Minimal Supersymmetric extension to the Standard Model. A first generation Muon Collider is uniquely suited to investigate the mass, width and decay modes of the Higgs boson, since the coupling of the Higgs to muons is expected to be strong enough for it to be produced in the s channel mode in the muon collider. Due to the narrow width of the Higgs, it is necessary to measure and control the energy of the individual muon bunches to a precision of a few parts in a million. We investigate the feasibility of determining the energy scale of a muon collider ring with circulating muon beams of 50 GeV energy by measuring the turn by turn variation of the energy deposited by electrons produced by the decay of the muons. This variation is caused by the existence of an average initial polarization of the muon beam and a non-zero value of $g-2$ for the muon. We demonstrate that it is feasible to determine the energy scale of the machine with this method to a few parts per million using data collected during 1000 turns.

THE METHOD

The spin vector $\vec{S}$ of a muon in the muon collider will precess according to the following equation, first derived by Bargmann, Michel and Telegdi [1]

$$\frac{d\vec{S}}{dt} = \vec{\Omega} \times \vec{S} \qquad (1)$$

$$\vec{\Omega} = -\frac{e}{\gamma m_\mu} \left((1+a\gamma)\vec{B}_\perp + (1+a)\vec{B}_\parallel - (a\gamma + \frac{\gamma}{1+\gamma})\vec{\beta} \times \frac{\vec{E}}{c} \right) \qquad (2)$$

where $\vec{B}_\perp$ and $\vec{B}_\parallel$ are the transverse and parallel components of the magnetic field with respect to the muon's velocity $\vec{\beta}c$, e is the electric charge, m_μ the mass of the muon, $a \equiv \frac{g-2}{2}$ is the magnetic moment anomaly of the muon and γ and g are the Lorentz factor and the gyromagnetic ratio of the muon. The value of $a \equiv \frac{g-2}{2}$ for the muon is 1.165924E-3 [2]. In what follows, we will consider the ideal planar collider ring case where $\vec{B}_\parallel$ and $\vec{E}$ are zero. For such a collider ring, $\vec{\Omega}$ is given by

$$\vec{\Omega} = \vec{\Omega}_{cyc}(1 + a\gamma) \tag{3}$$

where $\vec{\Omega}_{cyc}$ is the angular velocity of the circulating beam. From this, it follows that when the beam completes one turn, the spin will rotate by a further $a\gamma \times 2\pi$ radians. We will compute the precision with which γ can be determined by measuring the energy of the electrons produced by muon decay in this ideal case. We will examine the effects of departures from the ideal case in the last section.

It can be shown that the angular distribution of the decay electrons in the muon center of mass is given by the relation [3]

$$\frac{d^2N}{dxdcos\theta} = N(x^2(3-2x) - \hat{P}x^2(1-2x)cos\theta) \tag{4}$$

where N denotes the number of muon decays, $x \equiv 2E/m_\mu$ is the electron energy E in the muon rest frame expressed as a fraction of the maximum possible energy ($\approx 0.5m_\mu$), $cos\theta$ is the angle of the electron in the muon rest frame with respect to the z axis which is the direction of motion of the muon in the laboratory and $\hat{P}$ is the product of the muon charge and the z component of the muon polarization. The muon polarization is defined as the average of the individual muon unit spin vectors over the ensemble of muons considered. We note that the distribution is linear in $\hat{P}$.

The average energy $<E>$ and longitudinal momentum $<P_L>$ of the electron in the muon rest frame can be obtained using equation 4 as follows.

$$<E> = \frac{m_\mu}{2} \int\int x \frac{d^2N}{dxdcos\theta} dxdcos\theta = \frac{7}{10}\frac{m_\mu}{2} \tag{5}$$

$$<P_L> = \frac{m_\mu}{2} \int\int xcos\theta \frac{d^2N}{dxdcos\theta} dxdcos\theta = \frac{\hat{P}}{10}\frac{m_\mu}{2} \tag{6}$$

These two quantities form the components of a 4-vector, whose transverse components are zero, which may be transformed to the laboratory frame to yield the average electron energy $<E_{lab}>$.

$$<E_{lab}> = \frac{7}{20}E_\mu(1 + \frac{\beta}{7}\hat{P}) \tag{7}$$

where E_μ is the energy of the muon beam. Since the polarization $\hat{P}$ precesses from turn to turn by the amount $\omega = \gamma(g-2)/2 \times 2\pi$ radians, and the number of muons decrease turn by turn due to decay and losses, the total energy $E(t)$ due to decay electrons observed during turn t in an electromagnetic calorimeter will have the following expression

$$E(t) = Ne^{(-\alpha t)}(\frac{7}{20}E_\mu(1 + \frac{\beta}{7}(\hat{P}\cos\omega t + \phi))) \qquad (8)$$

where N is the number of muon decays sampled in turn 0, ϕ is an arbitrary phase containing information on the initial direction of polarization and α is the turn by turn decay constant of the muon intensity which in the absence of losses other than decay is given by

$$\alpha = \frac{t_{circ}}{\gamma t_{life}} \qquad (9)$$

where t_{circ} is the time taken to circulate around the storage ring and t_{life} is the muon life time.

For a 100% polarized beam, the amplitude of the oscillations is only 1/7 that of the non-oscillating background. It can be seen from equation 4 that the sensitivity to $\hat{P}$ is enhanced by selecting larger values of $\cos\theta$. This implies selecting electrons with higher laboratory energy. Figures 1(a-c) show the deposited electron energy as a function of turn number for polarization $\hat{P} = 1.0$ for individual electron energy ranges of 0-10 GeV, 10-25 GeV and 25-50 GeV respectively as a function of turn number. Figure 2(b) shows very little oscillatory signal, since the electrons in that energy range have small values of $\cos\theta$. Figure 2(d) shows the deposited electron energy with no electron energy cuts. Superimposed is the predicted behavior according equation 8. This serves as a consistency check for our routines. The signal to background ratio increases as we demand electrons with higher value of $\cos\theta$. In what follows, we use electrons with energy greater than 25 GeV during the investigative phase of this analysis and will later optimize this cut. In practice, we can select electrons with energies above a value by momentum analyzing them with a dipole field before they enter the calorimeter.

The method to determine the energy scale of the collider would then entail fitting a functional form of the type

$$f(t) = Ae^{-Bt}(C\cos(D + Et) + F) \qquad (10)$$

to the energy observed in the calorimeter. The variables A, B, C, D, E, F are parameters to be fitted. The information on the energy scale is contained in the parameter E.

Parameters of a 50 GeV idealized muon storage ring

In order to arrive at reasonable numbers for α and ω, we consider a storage ring of 50 GeV muons with a uniform bending field of 4 Tesla. This would produce a circular ring with the parameters given in table 1.

It should be noted that for an idealized storage ring with constant B field considered here, α does not depend on γ, since

$$t_{circ} = \frac{m_\mu \gamma}{0.3 B c} \tag{11}$$

$$\alpha = \frac{2\pi m_\mu}{0.3 B c t_{life}} \tag{12}$$

where m_μ is the muon rest mass, B is the bending field of the storage ring and c is the velocity of light. A 100 GeV collider ring will have the same α as a 50 GeV collider ring or a 25 GeV collider ring in this idealized case. As γ changes slightly, t_{circ} changes in proportion, α being the constant used to convert measurements of t_{circ} to γ. Measuring the decay rate of muons also affords a second method to determine γ. The beam circulation time t_{circ} can be measured to precisions of the order of a part in 10^6 and the fractional error in muon lifetime is 1.82E-5 [2]. The fractional error in γ obtainable by observing the rate of decay of the muons will then be dominated by the precision that one can measure α, namely $\delta\gamma/\gamma = \delta\alpha/\alpha$.

Generation of events and fitting for γ

Since equation 4 is linear in $\hat{P}$, the decay distribution of an ensemble of muons depends only on $\hat{P}$, the ensemble average of the z component of the individual muon spin vectors. However, because of the momentum spread of the muons, each individual particle will have a γ slightly different from the average and hence the precession of the spin vector around the ring will be different, leading to a slightly different value of $\hat{P}$ for the next turn. We model the beam by generating an ensemble of 100,000 muons each having its own spin vector and momentum. In an actual collider, it will be possible to sample significantly more decays than this. During each turn, we decay all the beam particles once and record the number and total energy deposited by electrons with individual energies above 25 GeV. Approximately 27% of the decay electrons pass this cut, on average. We decrease the number of decays by the appropriate number expected by muon decay alone for the next turn. At this stage we do not introduce fluctuations in the number of decays from turn to turn, since the 100,000 muons are meant to be representative of a much larger number in the actual ring. We precess the 100,000 spin vectors by their individual precession rates and make them decay again. We repeat this for 1000 turns. We re-use the muons after each turn since the 100,000 muons represent our model of the muon ensemble in the collider.

Generation of muon spin vectors

We generate 4 different samples of events with different ensembles of spin vectors. The z component of the unit spin vector of a muon S_z is allowed to vary from -1 to 1, using a binomial distribution of specified mean. The average value of the distributions are 0.9, 0.74, 0.5 and 0.26 respectively. We study negatively charged muons resulting an initial value of $\hat{P}$ of -0.9,-0.74,-0.5 and -0.26 respectively for these samples. In the absence of momentum spread, the decay distributions would only depend on $\hat{P}$ and not on the details of the distribution of S_z. The angles of the spin vectors are precessed by the individual γ dependent precession rate from turn to turn. In what follows, we assume a beam energy spread of 0.03% for the muons for all samples unless otherwise specified.

Fitting procedure and generation of errors

The energy deposited every turn is fitted to the functional form given by equation 10 using the CERN program MINUIT [4]. In order to study the variation of the fractional error $\delta\gamma/\gamma$ with the number of electrons sampled, we fluctuate the energy observed in the calorimeter E_m by

$$\frac{\sigma_{E_m}^2}{<E_m>^2} \approx \frac{1}{N}(1.03153) \qquad (13)$$

where N is the number of electrons sampled. We analyze the case for 41261, 10315, 2579 and 1146 electrons sampled which corresponds to a fractional error in the measured total energy of PERR $\equiv \frac{\sigma_{E_m}}{E_m}$ of .5E-2,1.0E-2,2.0E-2 and 3.0E-2 respectively.

RESULTS

We simulate the muon collider spin precession for a grid of values of $\hat{P}$ =-0.9,-0.74,-0.5 and -0.26 and fractional measurement error for the first turn (PERR) of 0.5E-2, 1.0E-2, 2.0E-2 and 3.0E-2. Figure 2(a) shows the result of the MINUIT fit plotted for 50 turns for $\hat{P}$=-0.26 and PERR=0.5E-2. Figure 2(b) shows the same plot but with the function being plotted only at integer values of the turn number t. A beat is evident in both the theoretical curve and the simulated measurements as a result of sampling the oscillation function at fixed intervals, not connected with the oscillation frequency. The origin of the beat is stroboscopic. Figure 2(c) shows the pulls, defined as $(data - fit)/error$ at each measurement as a function of turn number for 1000 turns. There are no major turn dependent variations in this quantity indicating that the fit converged satisfactorily. Figure 2(d) shows the histogram of the pulls, which approximates a unit Gaussian as desired. Table 2 shows the results of the fit for the grid of values of $\hat{P}$ and PERR. The results presented in table 2 are shown graphically in Figure 3. As an example, for an average polarization $\hat{P} = -0.26$, the fractional error in $\delta\gamma/\gamma$ varies from 5.1E-6 to 1.9E-5 as the fractional error in the electron energy sampled varies from 0.5E-2 to 3.0E-2, corresponding to the number of electrons sampled during the first turn varying from 41261 to 1146. The average number of decays in the muon collider is expected to be 3.2E6 decays per meter for a beam intensity of 10^{12} muons. The error in determining γ is thus going to be dominated by the fluctuations in the number of electrons sampled turn by turn, rather than sampling fluctuations in the calorimeter. We have simulated conditions involving $\approx$ 40,000 decays. It should be possible to go to higher statistical precision than computed here by sampling larger number of electrons.

The results for $\delta\gamma/\gamma$ obtained from the measurement of the turn by turn rate of decay of the electron energy are not competetive with the precession method primarily because of the small value of α (0.8399E-3). This leads to larger fractional errors for γ from this method (which also assumes that the loss of intensity is entirely due to the decay process) by almost three orders of magnitude than from the precession method.

Variation of $\delta\gamma/\gamma$ as a function of muon energy

The spin precession per turn equals 2π for a γ value of 857.689, which corresponds to a muon beam momentum of 90.622 GeV/c. This is the first spin resonance for muons. At this point, the fitting method loses sensitivity completely, since there will be no spin oscillations turn by turn. We now study the error $\delta\gamma/\gamma$ as a function of beam energy for $\hat{P}$=-0.26 and PERR=0.5E-2 (keeping the magnetic field in the idealized storage ring to be 4.0 Tesla) as a function of muon beam energy that straddles the spin resonance. For initial muon collider physics, the interesting beam energies are 45.5 GeV (half the Z mass), 80.3 GeV

Parameter	Value	Parameter	Value
Muon Energy	50 GeV	γ	473.22
spin precession in one turn	3.4667 radians	Magnetic field	4.0 Tesla
radius of ring	41.66666 meters	beam circulation time	0.87327E-06 sec
dilated muon life time	0.10397E-02 sec	turn by turn decay constant	0.8399E-03

TABLE 1. Parameters of an idealized muon storage ring

(W threshold), 175 GeV (top threshold) as well as half the neutral Higgs mass, which could be as low as 55 GeV in some SUSY scenarios. We sample all electrons that have energies greater than half the muon energy. Figure 4

$\hat{P}$	PERR	Number of electrons sampled	$\delta\gamma/\gamma oscillations$	$\delta\gamma/\gamma decay$	χ^2 for NDF=1000
-0.90	0.50E-02	41261	0.14568E-05	0.13227E-02	824.
-0.90	0.10E-01	10315	0.22147E-05	0.20124E-02	936.
-0.90	0.20E-01	2579	0.39999E-05	0.36398E-02	1009.
-0.90	0.30E-01	1146	0.58659E-05	0.53457E-02	1030.
-0.74	0.50E-02	41261	0.17418E-05	0.13019E-02	843.
-0.74	0.10E-01	10315	0.26183E-05	0.19591E-02	954.
-0.74	0.20E-01	2579	0.46981E-05	0.35229E-02	1021.
-0.74	0.30E-01	1146	0.68765E-05	0.51672E-02	1039.
-0.50	0.50E-02	41261	0.25903E-05	0.12813E-02	888.
-0.50	0.10E-01	10315	0.38407E-05	0.19029E-02	973.
-0.50	0.20E-01	2579	0.68338E-05	0.33972E-02	1026.
-0.50	0.30E-01	1146	0.99744E-05	0.49749E-02	1041.
-0.26	0.50E-02	41261	0.51242E-05	0.12688E-02	898.
-0.26	0.10E-01	10315	0.75317E-05	0.18791E-02	1004.
-0.26	0.20E-01	2579	0.13324E-04	0.33447E-02	1053.
-0.26	0.30E-01	1146	0.19380E-04	0.48950E-02	1061.

TABLE 2. Results of fits for $\delta\gamma/\gamma$ as a function of polarization $\hat{P}$ and noise PERR. Also shown is the χ^2 of the fit for 1000 turns.

shows the variation of $\delta\gamma/\gamma$ as a function of muon beam energies that straddle these values. It can be seen that $\delta\gamma/\gamma$ first decreases as one gets close to the resonance and then blows up on the spin resonance. As one approaches the spin resonance, the oscillations slow down. It is nevertheless possible to fit the slowed down oscillations by a rapidly oscillating theoretical function to high accuracy on either side of the resonance. At the resonance, the oscillations die completely, which results in a large value of $\delta\gamma/\gamma$. It may be possible to use this blow-up in $\delta\gamma/\gamma$ to find the spin resonance accurately and (paradoxically) determine γ at resonance accurately. This would depend on the width of the spin resonance, an analysis of which would take us beyond the scope of this paper.

Variation of $\delta\gamma/\gamma$ as a function of beam energy spread

We now calculate the variation of polarization as a function of turn number for an ensemble of muons with initial value of polarization $\hat{P}$ = -0.26 and values of momentum spread $\delta p/p$ varying from 0.02E-2 to 0.00125E-2. This variation is plotted in figure 5. For the larger values of momentum spread, there is a significant degradation of polarization as a function of turn number, due to differential spin precession of the individual beam particles. We note that when the beam energy is at 175 GeV, the spin tune is significantly higher

Machine	Spin tune ν_0	Quadrupoles	RMS Kl_Q meters^{-1}	σ_y meters	$\delta\nu$	$\sigma_{\delta\nu}$
46 GeV LEP	100.47	≈ 600	0.032	0.5E-3	5.7E-6 $\equiv$ 3KeV	6.1E-5 $\equiv$ 30KeV
50 GeV MC	0.5517	70	0.274	0.5E-3	-0.26E-8 $\equiv$ -0.24KeV	1.66E-8 $\equiv$ 1.46KeV

TABLE 3. Predictions for spin tune shift $\delta\nu$ and spread in spin tune shift $\sigma_{\delta\nu}$ caused by quadrupoles for LEP compared to the 50 GeV Muon Collider (MC) ring

and the depolarization is more rapid. Despite this depolarization, there is enough information from the first few hundred turns to extract the excellent value of $\delta\gamma/\gamma$ for 175 GeV beam energy as shown in figure 4.

Figure 6 shows the variation of the fractional energy resolution, $\delta\gamma/\gamma$ as a function of fractional beam energy spread for a muon beam with $\hat{P} = -0.26$, with 41261 electrons sampled. There is little dependence of $\delta\gamma/\gamma$ on the momentum spread. This is due to the fact that the momentum spread is determined from the spin tune and not from the spin oscillation amplitude and the fact that the depolarization is not significant for the first few hundred turns for any of the beam momentum spreads considered here.

Optimization of the electron energy cut

We now vary the cut on electron energy and study the dependence on $\delta\gamma/\gamma$ on the cut. Figure 7 shows the variation of $\delta\gamma/\gamma$ with the cut on individual electron energies for $\hat{P} = -0.26$ for 41261 and 1146 electrons sampled. The fractional error on the average energy of electrons is much smaller than the fractional error on the total energy of electrons. It is possible to measure the average electron energy by counting the number of electrons going into the calorimeter with a scintillator array. However, the precession information is contained increasingly in the number of electrons rather than their average energy as we increase the electron energy cut. Figure 7 shows the variation of $\delta\gamma/\gamma$ calculated from average as well as total electron energy as a function of the electron energy cut. For smaller values of the electron energy cut, the average method produces superior errors than the total energy method. However, with 40,000 electrons or more sampled a total energy method with a cut of 25 GeV or higher seems optimal. It should however be pointed out that the average energy method does not require a model for the rate of decay of muon intensity in the machine, which in practice could be a complicated function of turn number. As such the systematics associated with this would not be present in the average energy method.

EFFECTS DUE TO DEPARTURES FROM THE IDEAL CASE

So far we have considered a planar collider ring with uniform vertical magnetic field and no electric fields. The actual collider ring will depart from the ideal in three respects; a)It will have RF electric fields to keep the muons bunched, b) it will have radial horizontal magnetic fields experienced by partcles in an off-center trajectory at quadrupoles and at vertical correction dipoles, and c) it will have longitudinal magnetic fields due to solenoidal magnets in the interaction region(s). We now consider the effect due to each of these departures from the ideal.

Electric fields

Equation 2 implies that there is no spin precession due to longitudinal electric fields ($\vec{\beta} \times \vec{E} = 0$). RF electric fields are longitudinal, so there will be no precession due to the RF electric fields. At present there are no plans to install electrostatic separators to separate the beams. If and when this happens, one should consider the effect due to the transverse electric fields thus introduced.

Effect of radial magnetic fields

Particles which are off-axis at quadrupoles will experience radial as well as vertical magnetic fields. Even though the net integral of these off-axis fields around the ring is zero, the spin rotation along a horizontal axis followed by spin rotation about a vertical axis (caused by a bend dipole) followed by a reverse rotation in the horizontal direction still produces a net effect since the rotations about the horizontal and vertical axes do not commute. The effects have been analyzed by Assmann and Koutchouk [5] who show that this results in both a net spin tune shift $< \delta\nu >$ as well as a spread in tune $\sigma_{\delta\nu}$.

$$< \delta\nu > = \frac{\cot\pi\nu_0}{8\pi} \nu_0^2 \left(n_Q (Kl_Q)^2 \sigma_y^2 + n_{CV} \sigma_{\theta CV}^2 \right) \quad (14)$$

where $\nu_0 \equiv a\gamma$ is the spin tune of the collider ring, n_Q are the number of quadrupoles with integrated gradient Kl_Q, σ_y is the misalignment spread of the closed orbit at the quadrupoles, n_{CV} is the number of vertical correction dipoles and $\sigma_{\theta CV}$ is the rms beand angle in the vertical correctors. The spread in tune is given by,

$$\sigma_{\delta\nu} = \frac{< \delta\nu >}{\cos\pi\nu_0} \quad (15)$$

Table 3 shows the values for $< \delta\nu >$ and $\sigma_{\delta\nu}$ obtained by Assman and Koutchuk [5] for LEP. We compare this with to the current design for the

50 GeV muon collider ring [6]. Including the low beta section, there are 70 quadrupoles with an RMS value of $Kl_Q = 0.27$ m^{-1}. The effects due to correction dipoles may be neglected in both the LEP and the muon collider cases. We assume a beam misalignment of 5mm at the quadrupoles, which is the same value used in the LEP calculation. This is probably being conservative. The tune shift for LEP corresponds to a shift in beam energy calibration of 3.0 KeV. The tune spread for LEP corresponds to a spread in beam energy calibration of 30 KeV. For the muon collider, the tune shift corresponds to a shift in beam energy calibration of -0.24 KeV and a spread of 1.46 KeV, both of which are negligible. The reason for the smallness of this effect for the muon collider is twofold. Since the circumference of the muon collider is smaller than LEP, there are fewer quadrupoles. Secondly, the muon is two hundred times more massive than the electron and has has a spin tune $a\gamma$ that is smaller by the same factor. The spin tune shift depends on the the square of the spin tune. It should be noted that the above formulae are not valid for a fractional spin tune of 0.5.

Solenoidal magnetic fields

The experimental region will in all likelihood contain a solenoidal magnet. This solenoidal field, if uncorrected, will rotate the spin vector of the muons about the beam direction by a constant amount θ_s per turn, which can be derived using equation 2.

$$\theta_s = -\frac{e}{\gamma m_\mu}(1+a)B_s = -(1+a)\frac{B_s l}{B\rho} \tag{16}$$

where B_s is the field due to the solenoid of length l, B is the dipole bending field of the ring of radius ρ. For a solenoid of 1.5 Tesla and length 6 meters, $\theta_s = 3.09$ degrees for the planar storage ring parameters of table 1. It can be shown analytically [8] that this produces a spin tune shift $\delta\nu$ given by

$$\nu + \delta\nu = \frac{1}{\pi}arccos\left(cos(\pi\nu)cos(\frac{\theta}{2})\right) \tag{17}$$

yielding a spin tune shift $\delta\nu = -1.901$E-5, or a fractional spin tune shift of $\delta\nu/\nu$ = -3.45E-5. For a 50 GeV muon beam, this is a shift in energy calibration of -1.72 MeV. In LEP, a similar solenoid will have a much smaller fractional tune shift [8], since the tune is 200 times larger for electrons. It is important to correct the effect due to the solenoids, since this is cumulative turn by turn. At LEP this is done by a series of vertical orbit correctors [9] followed by normal lattice followed by vertical orbit correctors of reverse polarity, which has the effect of rotating the spin by half the amount produced by the solenoid. A similar set of corrections is inserted after the solenoid to complete the correction. This method depends on a non-zero value of $g - 2$ and as such will be 200 times less effective for muons than for electrons, for any given magnet strength. The most effective method to correct for the solenoid is to surround it on either side by compensating solenoids of minimal radius large enough to allow the beam to go through.

COSY studies

We have studied the effects due to non-linearities in the aperture using COSY [7], a beam optics program based on differential algebra techniques. Figure 8 shows the polarization as a function of turn number for three different cases of emittance. We have tracked 1000 muons from the interaction point with an initial polarization of 0.25. The three different cases considered are

- emittance = 297πmm-mr and $\delta p/p = 0.0025$E-2
- emittance=85π and $\delta p/p = 0.02$E-2 and
- emittance=40π and $\delta p/p = 0.02$E-2.

We have also studied other cases of $\delta p/p$. The general conclusion is that the main depolarization effect seems to be the non-linearities sampled by the larger emittance and not the beam momentum spread, as can be evidenced by the fact the depolarization effects are smaller when one goes from an emittance of 297π to 85π despite the fact the momentum spread is worse for the latter case. However, we have shown that depolarization effects of the type exhibited here can still be tolerated provided that there is initial polarization of the order of 0.25 which can be maintained for a few hundred turns. Cosy gives the spin tune as a function of the position, angle and energy variables of the beam as

$$\nu = 0.5517 + 0.5915\kappa - 64.61x^2 - 0.1017x'^2 - 69.81y^2 - 0.1088y'^2 - 8.341x\kappa - 0.3921\kappa^2 \tag{18}$$

to second order in the variables $\kappa \equiv$ change in kinetic energy/kinetic energy of a 50 GeV/c momentum muon, x, y are deviations from the closed orbit in centimeters and x' and y' are defined as p_x/p and p_y/p where p_x and p_y are the momentum components of a muon of beam momentum $p \equiv 50$ GeV/c.

CONCLUSIONS

We have demonstrated that it is feasible to measure the energy of a 50 GeV muon collider to a few parts per million using the $g - 2$ spin precession technique, provided it is feasible to maintain a muon polarization of the order of $\hat{P}$=0.25 in the ring for a thousand turns. In order to explore the Higgs resonance, it is necessary to measure the bunch by bunch variation in energy to a few parts per million. We have demonstrated that the $g-2$ technique is capable of doing so. It is still possible to tolerate a spin tune shift in the overall energy scale of a few percent, which will act only as a systematic error on the Higgs mass and width.

We would also like to note in passing that polarization information from a calorimeter of the type proposed here can be used in conjunction with a neutrino detector placed along the line of the neutrinos produced in association with the electrons to estimate the variation in the energy spectrum of the muon neutrinos and electron antineutrinos in the beam. Such information can be a valuable tool in untangling various possible neutrino oscillation scenarios.

The authors would like to thank Martin Berz, Weishi Wan and Carol Johnstone for help with COSY calculations and would like to acknowledge useful conversations with Alain Blondel and Robert Rossmanith.

REFERENCES

1. V.Bargmann, L. Michel and V.L. Telegdi, Phys. Rev. Lett. 2, 10 (1958) 435.
2. Particle Data Group, R.M.Barnett et al., Physical Review D54,1 (1996).
3. G.Barr, T.K. Gaisser and T. Stanev, Phys. ReV. D 39(1989) 3532.
4. MINUIT is a CERNLIB program written by F. James.
5. R.Assmann and J.P.Koutchouk, "Spin tune shifts due to optics imperfections", Cern SL/94-13. See also L. Arnaudon et al, "Accurate determination of the LEP beam energy by resonant depolarization ", Z.Phys.C66: 45-62,1995.
6. Carol Johnstone, Private Communication.
7. COSY INFINITY Version 7 User's guide and reference manual, M.Berz, Michigan State University Preprint MSUCL-977.
8. J.P.Kouthcouk, "Spin tune shift due to solenoids", CERN SL-note/93-26 (AP) (1993).
9. R.Rossmanith, LEP Note 525 (1985). A. Blondel, LEP note 629, (1990).

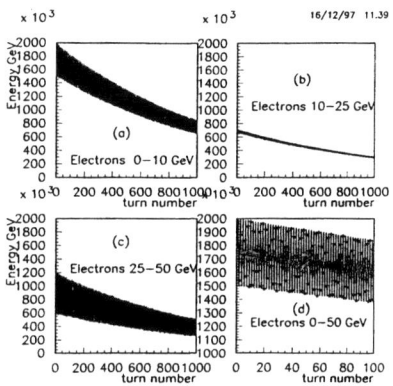

FIGURE 1. (a)Total energy observed as a function of turn number for $\hat{P} = -1.0$ with individual electron energies in the range 0-10 GeV for 100,000 muon decays. (b) Electron energies in the range 10-25 GeV (c) 25-50 GeV (d) All electrons included. Superimposed is a functional form defined by equation 8

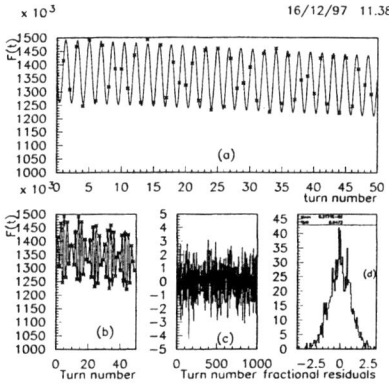

FIGURE 2. (a)Energy detected in the calorimeter during the first 50 turns in a 50 GeV muon storage ring (points). An average value of $\hat{P}=-0.26$ is assumed and a fractional fluctuation of 0.5E-2 per point. The curve is the result of a MINUIT fit to the functional form in equation 10. (b) The same fit, with the function being plotted only at integer turn values. A beat is evident. (c) Pulls as a function of turn number (d)Histogram of pulls.

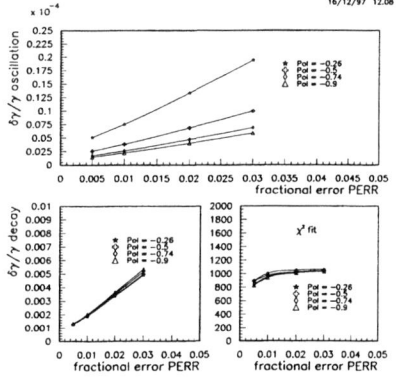

FIGURE 3. (a)Fractional error in $\delta\gamma/\gamma$ obtained from the oscillations as a function of polarization $\hat{P}$ and the fractional error in the measurements PERR (b) Fractional error in $\delta\gamma/\gamma$ obtained from the decay term as a function of polarization $\hat{P}$ and the fractional error in the measurements PERR (c) The total χ^2 of the fits for 1000 degrees of freedom

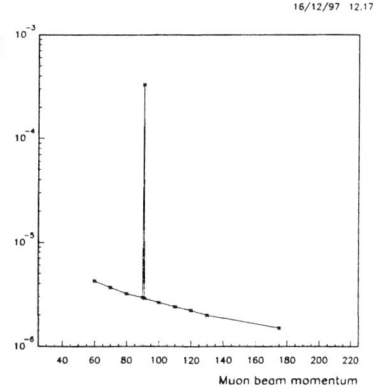

FIGURE 4. Fractional error in $\delta\gamma/\gamma$ obtained from the oscillations as a function of muon beam momentum

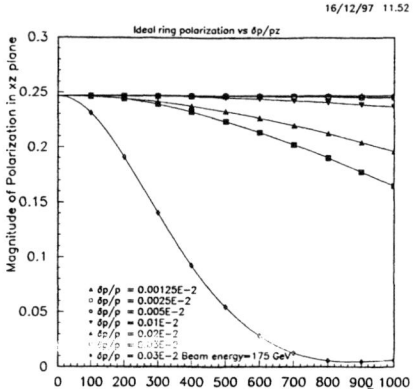

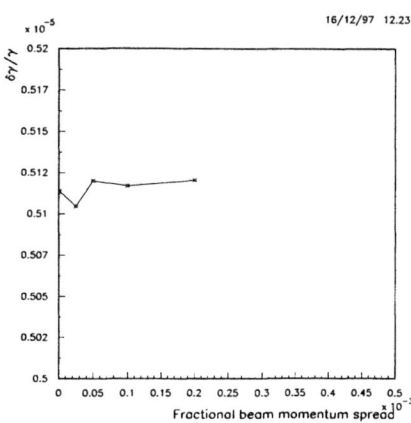

FIGURE 5. Variation of polarization as a function of turn number for 50 GeV muons with initial $\hat{P} = -0.26$ and various values of $\delta p/p$ in an ideal collider ring. The bottom curve is for 175 GeV muons and shows a more rapid depolarization due to the higher spin tune.

FIGURE 6. $\delta\gamma/\gamma$ versus fractional beam energy spread for 50 GeV muons with PERR=.5E-2 and $\hat{P} = -0.26$

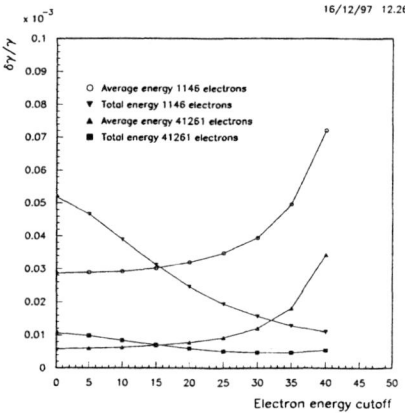

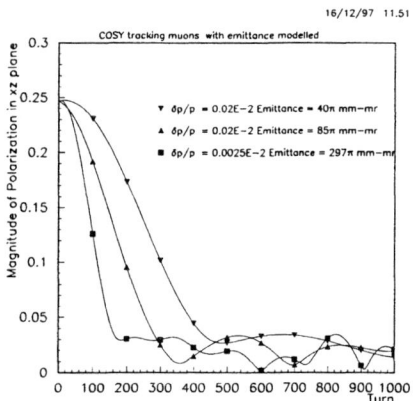

FIGURE 7. The variation of $\delta\gamma/\gamma$ as a function of the electron energy cut for 41261 and 1146 electrons $\hat{P} = -0.26$. We fit the total energy in the calorimeter as well as the average energy per electron

FIGURE 8. Polarization as a function of turn for 1000 muons in COSY. Three separate cases of beam emittance and beam energy spread are considered.

A Muon Collider as Z factory

Alain Blondel

LPNHE Ecole Polytechnique
France 91128 Palaiseau cédex

Abstract. A muon collider running at the Z peak could make important contributions to electroweak precision measurements. From this very incomplete and brief study, the following ones could be identified. The measurement of the hadronic to leptonic Z widths ratio, R_ℓ, could lead to a unique precision on α_s of better than ± 0.001. The high degree of accuracy in the beam energy determination offered by spin precession of the muons would allow very accurate determination of the Z width. Finally, the possibility of longitudinal polarization for both beams, combined with high luminosity, would allow a measurement of A_{LR} with very high precision and reliability. Experimental difficulties could come from machine-related backgrounds which render delicate the measurement of luminosity and possibly high precision tracking. High luminosity (100 million Z a year) and beam polarization would make this option highly worthwhile.

INTRODUCTION

Precision measurements at the Z peak have had a major impact on our understanding of electroweak and strong interactions. Before one envisages further measurements it is worthwhile analyzing the reasons of success so far. At LEP, the assets were: i) high statistics, four million hadronic Z decays recorded by each of four experiments; ii) precise energy calibration, about 1 MeV beam energy resolution; iii) precise measurement of luminosity, to $\pm 10^{-3}$; iv) essentially 4π detectors with high granularity and precision secondary vertexing capabilities, allowing for instance b-tagging efficiencies of 30% for a purity of 98%. At SLC, assets are quite complementary, a high level, 80%, of longitudinal beam polarization known with a relative precision of $\pm 1\%$, and even better secondary vertexing capabilities compensating statistics typically a factor 40 lower than at LEP. All this has allowed experiments at LEP and SLC to measure most electroweak observables with a relative precision of 10^{-3}.

To make it worthwhile, the performance of a Z factory must exceed these figures by far, in more than one way. On the other hand, the interest of a muon collider in this energy range could well become overwhelming if a Higgs scalar lies close by in energy, offering the unique opportunity of (relatively) abundant s-channel Higgs production. In that case an investigation of the additionnal bonus provided by the Z peak experiments is quite natural.

PARAMETER LIST

A parameter list extracted from that of the workshop [1], and very similar to that of [2], is given in Table 1. The narrow momentum spread option is not useful at the Z peak, which has an intrisic width of 2.5%, so that $\sigma_p/p = 10^{-3}$ is good enough. One notices first that the average luminosity according to this table is not so much larger than that of LEP which, with a peak luminosity of $2.5 \; 10^{31}/\text{cm}^2/\text{s}$, an effective duty cycle of almost 0.2 compensated by the presence of four experiments, corresponds to an average luminosity of $2 \; 10^{31}/\text{cm}^2/\text{s}$. A factor of three better is not enough in itself to justify a new machine more than ten years later.

Muon Collider Parameters

parameter	low Energy	medium energy	high energy
$\sqrt{s}$ (GeV)	91.2	500	4000
spin tune	0.503	2.76	22.06
momentum spread σ_p/p	10^{-3}	10^{-3}	10^{-3}
μ/bunch	3.10^{12}	2.10^{12}	2.10^{12}
bunches	1	2	2
rep. rate (Hz)	15	15	15
τ_μ lab (ms)	0.94	5.15	41.2
$1/\tau_\mu$ (Hz)	1066	194	24
lifetimes between injections	71	13	1.6
circumference (m)	380	1000	8000
average luminosity(/cm^2/s)	6.10^{31}	7.10^{32}	10^{35}

TABLE 1. *Parameter list of the "low energy" muon collider.*

One has to hope that further improvement by a factor of ten can be found. Some potential can be seen in the row "lifetimes between injections" of table 1: the low energy machine is empty most of the time. A higher repetition rate would certainly be welcome.

Lets assume in the following that 100 Million Z can be produced each year. In three years, one obtains a factor 20 improvement in Z decay statistics over LEP. What can one learn from that? This first glance at the physics programme only considers standard electroweak precision measurements. A high luminosity Z factory could certainly also be useful for further investigations in b and τ physics, detailed studies of fragmentation and the many physics topics offered by very high statistics in a clean environment. Detailed studies would be necessary however to assess the value of these measurements at an epoch where B factories and LHC will have already taken data.

PRECISION MEASUREMENTS AND RADIATIVE EFFECTS

The major interest of LEP observables is their sensitivity to electroweak (propagator or vertex) radiative effects. QED radiative effects (emission of real or virtual photons) are conceptually straightforward, although uncertainties in their evaluation can contribute to the experimental uncertainties. It should be noted in passing that the QED corrections for $\mu^+\mu^-$ collisions have not been calculated. In particular the Z line shape with muons is not known. It is not impossible that the muon being heavier, related uncertainties become smaller.

Electroweak corrections are sensitive [3] to heavy particles, such as the top quark or the Higgs boson, in an inclusive way. There are four [6–9] main radiative effects at the Z pole:

- The running of the QED coupling constant $\alpha(q^2)$ from $q^2 = 0$ to $q^2 = M_Z^2$. This is not sensitive to heavy physics but, because it is estimated from $e^+e^- \to$ hadrons cross-sections, it is a source of uncertainty. This has recently been reevaluated by Davier and Höcker [4] using QCD predictions when they are arguably more reliable that the data themselves.

- The isospin-breaking loop corrections to the W and Z propagators. They are absorbed conveniently in the ρ parameter, $\rho = 1 + \Delta\rho$. This is where the quadratic dependence in the top quark mass arise, as would the dependence on isotopic mass splitting of other new particles.

- The running of the Z self-energy, absorbed in the parameter Δ_{3Q} or ϵ_3.

- The $Z \to b\bar{b}$ vertex correction.

- The difference in running of the W self energy with that of the Z, which enters in the W mass via the parameter ϵ_2.

The building blocks of electroweak physics at the Z are measured cross sections for various final states, forward-backward and polarization asymmetries. Assuming that Z and photon exchange are the only processes that occur, they can all be expressed in terms of the chiral couplings,

$$g_{Vf} = (g_{Lf} + g_{Rf}) = I_{Lf}^3 - 2Q_f \sin^2\theta_w$$
$$g_{Af} = (g_{Lf} - g_{Rf}) = I_{Lf}^3. \qquad (1)$$

The $Z \to f\bar{f}$ partial width is given by:

Quantity	Main Technologies	Physics Outputs	present relative Precision
line shape			
M_Z	**Absolute energy scale** relative cross sections line shape fit (QED rad. corr.)	input	$2 \cdot 10^{-5}$
Γ_Z	**Relative energy scale** relative cross sections line shape fit (QED rad. corr.)	$\Delta\rho$	10^{-3}
$\sigma_{\text{had}}^{\text{peak},0}$	**Absolute cross sections**	$N_\nu \cdot \frac{\Gamma_{\text{inv}}}{\Gamma_{\ell\ell}}$ universality	$1.2 \cdot 10^{-3}$
$R_\ell \equiv \frac{\Gamma_{\text{had}}}{\Gamma_{\ell\ell}}$	lepton, hadron event selection	universality $f(\alpha_s, \sin^2\theta_w^{\text{eff}}, \delta_{vb})$	10^{-3}
M_W	**absolute energy scale** in pp collisions: structure functions	Δr^{ew}	10^{-3}
$R_b \equiv \frac{\Gamma_b}{\Gamma_{\text{had}}}$	b-tagging	δ_{vb}	$5 \cdot 10^{-3}$
$R_c \equiv \frac{\Gamma_c}{\Gamma_{\text{had}}}$	c-tagging	universality	3%
A_{LR}	Beam polarization	$\sin^2\theta_w^{\text{eff}}$	10^{-3}

TABLE 2. Synopsis of Precision Electroweak Measurements at the Z energy scale. R_ℓ is defined as $R_\ell \equiv \Gamma_{\text{had}}/\Gamma_{\ell\ell}$, where $\Gamma_{\ell\ell}$ refers to the partial width into a pair of massless charged leptons.

$$\Gamma_f = \frac{G_F M_Z^3}{24\sqrt{2}\pi}(g_{Vf}^2 + g_{Af}^2)(1 + \frac{3}{4}Q_f^2\frac{\alpha}{\pi})[N_c(1 + \alpha_s/\pi + ...)] \qquad (2)$$

The last parentheses describe the corrections for final state radiation of photons [and gluons, for quarks]. The total width is the sum over all open channels.

Cross-section formulae will be obtained by analogy with e^+e^- collisions with the simple replacement of initial state e by initial state μ. Around the Z pole, the photon exchange is only a correction to the Z-channel, which dominates the cross section and can then be written as:

$$\sigma_f = \frac{12\pi(\hbar c)^2}{M_Z^2} \frac{s\Gamma_\mu \Gamma_f}{(s - M_Z^2)^2 + s^2 \frac{\Gamma_Z^2}{M_Z^2}}. \qquad (3)$$

Forward-backward asymmetries or polarization asymmetries are sensitive to the chiral coupling asymmetry:

$$\mathcal{A}_f \equiv \frac{g_{Lf}^2 - g_{Rf}^2}{g_{Lf}^2 + g_{Rf}^2} = \frac{2g_{Vf}g_{Af}}{g_{Vf}^2 + g_{Af}^2}. \qquad (4)$$

For unpolarized beams the forward-backward asymmetry is:

$$A_{FB}^{(f)} \simeq \frac{3}{4}\mathcal{A}_\mu\mathcal{A}_f. \tag{5}$$

For the tau lepton, the polarization of the final state fermion is measurable, as a function of polar angle. For unpolarized beams:

$$\mathcal{P}_\tau(cos\theta) \simeq -\frac{\mathcal{A}_\tau + \frac{2cos\theta}{1+cos^2\theta}\mathcal{A}_\mu}{1 + \frac{2cos\theta}{1+cos^2\theta}\mathcal{A}_\mu\mathcal{A}_\tau} \tag{6}$$

from which one can derive both $\mathcal{A}_\mu$ and $\mathcal{A}_\tau$.

Interesting observables are obtainable if longitudinal beam polarization is available: the Left-Right asymmetry of Z production

$$A_{LR} = \frac{\sigma_L - \sigma_R}{\sigma_L + \sigma_R} \simeq \mathcal{A}_\mu, \tag{7}$$

which has the beautiful property of being the same for all final states, thus a very large statistical power; and the forward-backward polarized asymmetry

$$A_{FB}^{pol(f)} = \frac{(\sigma_{L,F} - \sigma_{R,F}) - (\sigma_{L,B} - \sigma_{R,B})}{(\sigma_{L,F} + \sigma_{R,F}) + (\sigma_{L,B} + \sigma_{R,B})} \simeq \frac{3}{4}\mathcal{A}_f. \tag{8}$$

The propagator corrections modify the Neutral current couplings, eq. 1, by an overall scaling factor $\sqrt{\rho}$ and a global change of $\sin^2\theta_w$, in an universal way. Non-universal corrections are small and – with the notable exception of the $Z \to b\bar{b}$ vertex – insensitive to heavy physics. Furthermore, all Z-pole asymmetries with unpolarized beams and the most precise asymmetry with polarized beams are proportional to the muon coupling $\mathcal{A}_\mu$, while the sensitivity to $\sin^2\theta_w$ of unpolarized hadronic asymmetries is contained in the $\mathcal{A}_\mu$ term. It is therefore convenient to express all asymmetry measurements at the Z pole in terms of the effective weak mixing angle [10] defined as:

$$\sin^2\theta_w^{eff} \equiv \frac{1}{4}(1 - \frac{g_{V\mu}}{g_{A\mu}}) \tag{9}$$

where the ratio $\frac{g_{V\mu}}{g_{A\mu}}$ is extracted from pole asymmetries.

As can be seen from Table 3, of all electroweak observables, the effective mixing angle is by far the most sensitive to the Higgs boson mass. Of all asymmetries, the left-right asymmetry is statistically the most powerful and systematically the cleanest one. Since polarization should be available at a muon collider, the understanding of how well A_{LR} could be measured will be investigated here in detail.

The relations between Z peak observables and m_W, the Fermi constant G_F and the QED running constant can be written in terms of these universal electroweak corrections $\Delta\rho, \Delta_{3Q}, \Delta r^{ew}$ and δ_{vb} as [9,8,11,12]:

$$M_Z^2 = \frac{\pi\alpha(M_Z^2)}{\sqrt{2}G_F(1+\Delta\rho)(1+\Delta_{3Q})\sin^2\theta_w^{\text{eff}}\cos^2\theta_w^{\text{eff}}};$$

$$\Gamma_{\ell\ell} = \frac{G_F M_Z^3}{24\sqrt{2}\pi}[1+\Delta\rho]\left[1+\left(\frac{g_{V\ell}}{g_{A\ell}}\right)^2\right]\left(1+\frac{3}{4}\frac{\alpha}{\pi}\right);$$

$$\Gamma_b = \Gamma_d(1+\delta_{vb});$$

$$M_W^2 = \frac{\pi\alpha(M_Z^2)}{\sqrt{2}G_F(1-\Delta r^{\text{ew}})(1-\frac{M_W^2}{M_Z^2})}. \qquad (10)$$

In the SM, δ_{vb} depends on M_t only (quadratically), $\Delta\rho$ depends on M_t (quadratically) and on M_H (logarithmically), while Δ_{3Q} has a logarithmic dependence on both M_t and M_H, and thus is relatively more sensitive to M_H. The leading terms are:

$$\Delta\rho \simeq \frac{\alpha}{\pi}\frac{M_t^2}{M_Z^2} - \frac{\alpha}{4\pi}\ln\frac{M_H^2}{M_Z^2}$$

$$\Delta_{3Q} \simeq \frac{\alpha}{9\pi}\ln\frac{M_H^2}{M_Z^2}$$

$$\delta_{vb} \simeq -\frac{20}{13}\frac{\alpha}{\pi}\left(\frac{M_t^2}{M_Z^2} + \frac{13}{6}\ln\frac{M_t^2}{M_Z^2}\right) \qquad (11)$$

More complete expressions have been calculated and implemented in computer codes. A very common and similar parametrisation is that of Altarelli et al [8], which can be obtained from the equations:

$$\epsilon_1 = \Delta\rho \qquad (12)$$
$$\epsilon_3 = -\cos^2\theta_w \Delta_{3Q} \qquad (13)$$
$$\epsilon_2 = \frac{\sin^2\theta_w}{\cos^2\theta_w \sin^2\theta_w}(\Delta r^{\text{ew}} + \frac{\cos^2\theta_w}{\sin^2\theta_w}\epsilon_1 - 2.\epsilon_3) \qquad (14)$$

Table 2 summarises the main observables, their physics output and the most critical technique involved.

EXPERIMENTAL CONDITIONS

A Measurement of luminosity

One of the strong points of the LEP experiements was the possibility of very precise luminosity measurements. If one considers a proposed muon collider set-up, it jumps to your eye that the detector would consist in a lot of shielding with a little bit of detector inbetween. In particular, the angular cone from the beam line up to typically 20 degrees would be used for shielding purposes exclusively [13].

The measurement of luminosity could be based on the equivalent of Bhabha scattering, elastic scattering $\mu^+\mu^- \to \mu^+\mu^-$ at small angles. Lets call this "Mhamha" scattering. In order to accumulate statistics well in excess of those of hadronic Z decays at the top of the resonance, an angular coverage down to typically 25 mrad is required. This is deep into the region foreseen for shielding! Muons are easier than electrons in that they can go through shielding. However, a precise knowledge of the acceptance boundaries is necessary, as the systematic error related to that goes as:

$$\frac{\Delta\sigma_{\text{Mhamha}}}{\sigma_{\text{Mhamha}}} = 2 \times \frac{\Delta R}{R}$$

For the typical value of R (5 cm at 2 meters) the precision required is 20 μm for a precision well below 10^{-3}. Could one instrumenting the shielding with devices providing both a good signal-to-backgroumd rejection and this mechanical precision? This a problem that has not found a solution yet.

Although it could be thought that this problem is specific of a Z factory, this is not quite true. For energies above the Z peak, radiative returns onto the Z resonance $\mu^+\mu^- \to \gamma + Z$ could provide the necessary normalisation. This works only up to energies where the boost of the Z is so high that one of the two fermions from Z decays would be emmitted below the limit of the shielding. For an angular range of 20°, acceptance for these events vanishes for a center-of-mass energy of about 500 GeV, and the acceptance corrections are already severe at 200 GeV. This underlines the importance of the problem of luminosity measurement at muon colliders.

B Beam energy calibration

One of the strength of LEP over the SLC, is the possibility of knowing the beam energy by measurement of the spin precession frequency. This is in fact a general feature of circular machines with respect to linear ones, and the muon collider has the same advantage. A further advantage is that polarimetry is provided for free by muon decays, the electron energy spectrum being a measure of the muon longitudinal polarization. Longitudinal polarization of the muons at the level of 20% is hardly avoidable, since it is naturally provided by parity violation in pion decays. Statistics is not a problem since all muons eventually decay. The number of spin precessions per turn, or spin tune ν, is proportional to the beam energy:

$$\nu = a_\mu \gamma = \frac{g_\mu - 2}{2} \frac{E_{\text{beam}}}{m_\mu} = \frac{E_{\text{beam}}(\text{GeV})}{90.6223(6)} \ . \tag{15}$$

Its value is 0.503 at the energy corresponding to the Z peak. Energy spread is not a real problem. Since the muons go around only about thousand turns, beam energy spread leads to a gradual, but not complete, depolarisation. Raja [14] has studied the possibility of measuring the decay electron momentum spectrum turn after turn and extracting the spin tune by a fourier transform. This leads to an astonishing relative precision of between 10^{-5} and 10^{-6} for each muon fill! The limit is given by the present relative precision of the muon $g_\mu - 2$ (7.10^{-6}), but this is presently being remeasured and should come well below 10^{-6}. This uncertainty would affect the measurement of the Z mass but not that of the Z width.

These promising numbers tell us that precision energy measurement should be a strong point of the muon collider. At the Z peak this opens the possibility of more precise measurements of the Z mass and width. Given the assumed performance, a precision of 0.4 MeV on both quantities should be achievable.

C Detector acceptance and tracking

This subject has been already alluded to in section A. Fierce backgrounds from muon decays leads to large amounts of shielding in the forward cones, and makes it difficult to bring in tracking devices closer than about 5 cm from the beam line, as studied by Bruce King [15]. Although these numbers are highly preliminary, they are telling us that the muon collider experiments would have a hard time doing much better than the LEP detectors in terms of heavy flavour tagging.

For this first evaluation of performance, it seems reasonable to assume that the muon collider detector will be completely blind below 20°, and that its otherwise behaves as well as a typical LEP detector, say, ALEPH. The systematic uncertainty on most observables is determined by statistically limited cross-checks on the data, and have been assumed here to scale as $1/\sqrt{N}$. The only place hwere this seems hard to accept is for the measurement of R_b, a dedicated study would be necessary to ascertain this point. For R_b, the present limit of correlated systematic errors among LEP experiments has been assumed as limiting precision.

D Beam Polarization

Muons from pion decay are naturally polarized, with a typical level of 28%. Some of this polarization is expectedly lost in the cooling and acceleration process, so that the muon polarization is anticipated to be at the level of 19%.The polarization can be improved by momentum selection of the muons at the expense of flux, as shown on the dashed curve in figure 1.

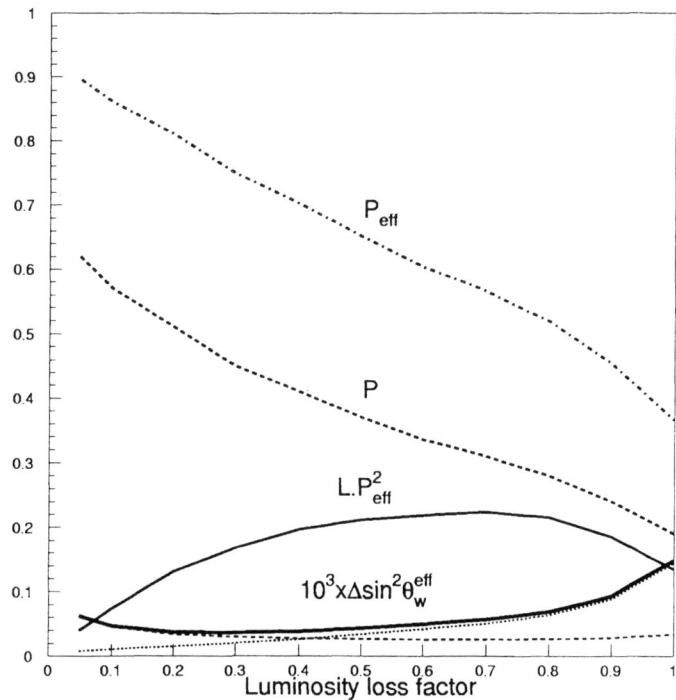

FIGURE 1. Polarization predictions for the Z factory muon collider.
Dashed line labelled "P": single beam longitudinal polarisation estimate from [2].
Dash-dotted line labelled "P_{eff}": polarization of the $\mu^+\mu^-$ system for spins pointing in the same direction.
Full line labelled "$L.P_{eff}^2$": polarization figure of merit for polarization asymmetry measurement.
Full line labelled $\Delta\sin^2\theta_w^{eff}$: resulting error on $\sin^2\theta_w^{eff}$ for a total number of 100 million Z. Below this are the component of the errror from event statistics (dashed) and from the polarization normalization (dotted).

One of the nice features of the muon colliders is that both beams are naturally polarized. If one manages to collide beams of longitudinal polarizations $\mathcal{P}_{\mu^+}, \mathcal{P}_{\mu^-}$, the total coss-section becomes:

$$\sigma(\mathcal{P}_{\mu^+}, \mathcal{P}_{\mu^-}) = \sigma_u(1 - \mathcal{P}_{\mu^+}\mathcal{P}_{\mu^-} + (\mathcal{P}_{\mu^+} - \mathcal{P}_{\mu^-})\mathrm{A_{LR}}), \quad (16)$$

σ_u is the unpolarised cross-section. To maximize the sensitivity to the parity-violating term $\mathrm{A_{LR}}$, it is best to have large and opposite polarizations of the two beams, the longitudinal polarization sign convention being such that positive helicity corresponds to positive polarization for both beams. Comparison of the cross-sections obtained with opposite polarizations gives:

$$\frac{\sigma(\mathcal{P}_{\mu^+}, \mathcal{P}_{\mu^-}) - \sigma(-\mathcal{P}_{\mu^+}, -\mathcal{P}_{\mu^-})}{\sigma(\mathcal{P}_{\mu^+}, \mathcal{P}_{\mu^-}) + \sigma(-\mathcal{P}_{\mu^+}, -\mathcal{P}_{\mu^-})} = \mathcal{P}_{\mathrm{eff}}.\mathrm{A_{LR}} \quad (17)$$

where we have introduced the *effective* polarization:

$$\mathcal{P}_{\mathrm{eff}} = \frac{(\mathcal{P}_{\mu^+} - \mathcal{P}_{\mu^-})}{1 - \mathcal{P}_{\mu^+}\mathcal{P}_{\mu^-}}. \quad (18)$$

The effective polarization is almost twice as large as the single beam polarization, as drawn on figure 1. This property allows rather large levels of polarisation to be obtained. The statistical error on $\mathrm{A_{LR}}$ is given by:

$$\Delta \mathrm{A_{LR}} = \frac{1}{\mathcal{P}_{\mathrm{eff}}\sqrt{N}} \quad (19)$$

So that the figure of merit for the measurement of the polarisation asymmetry is the product $\mathcal{L}\mathcal{P}_{\mathrm{eff}}^2$. This quantity is shown on figure 1 and can be used to optimize the running strategy.

A second nice feature of having two beams polarised is that the comparison between the sum of the cross-sections with opposite helicities and the unpolarised cross-section gives an important cross-check on the beam polarization itself. (One could also compare with the cross-section for like sign helicities). The quality of the cross-check can be expressed in an equivalent beam polarization uncertainty by assuming that the polarizations of the two beams have the same absolute value $\mathcal{P}$.

$$\sigma(\mathcal{P}, -\mathcal{P}) + \sigma(-\mathcal{P}, \mathcal{P}) = 2\sigma_u(1 + \mathcal{P}^2) \quad (20)$$

from which one obtains:

$$\frac{\Delta \mathcal{P}_{\mathrm{eff}}}{\mathcal{P}_{\mathrm{eff}}} = \frac{1 + \mathcal{P}^2}{\mathcal{P}^2} \frac{\Delta \sigma}{\sigma} \quad (21)$$

where $\frac{\Delta \sigma}{\sigma}$ represents the relative precision in the comparison of the polarized and unpolarized cross-sections. Because it is obtained from measured cross-sections, this cross-check applies to the polarization of interacting muons, avoiding difficulties related to the extrapolation from the full beam phase space – the polarization of which would be measured by the asymmetry in decay electron energy – to the population of muons which is sampled by the other beam at the interation point.

A fraction of the statistics could be devoted for this cross-check. For an exposure yielding 100 million Z without optimization of the beam polarization, 10 million Z could be put aside for this, allowing a cross-check at a level of precision given by $\frac{\Delta\sigma}{\sigma} \simeq 3 10^{-4}$ The resulting error on A_{LR} is $\Delta A_{LR} = A_{LR} \frac{\Delta \mathcal{P}_{\text{eff}}}{\mathcal{P}_{\text{eff}}}$.

The resulting uncertainty on $\sin^2 \theta_w^{\text{eff}}$, $\Delta \sin^2 \theta_w^{\text{eff}} = 1/7.9 \Delta A_{LR}$, is shown in figure 1, with the decomposition between the statistical error and that from the polarization measurement. With an exposure of one year an experimental error of better than $\Delta \sin^2 \theta_w^{\text{eff}} \simeq 0.4 10^{-4}$ should be reachable.

All of the above assumes quite a bit of flexibility in the spin manipulation. Because of the smaller spin tune, muon spin polarization is much harder to modify at will than that of electrons – and that is already a hard problem! The amplification factor available for electrons in transverse fields is reduced for muons. This has been discussed in [16]. For spin tune less than 1, i.e. for beam energies below 90 GeV, the simple-minded spin rotator would be a solenoid. Unfortunately, a spin rotation of 180 degrees for a beam energy of 45.6 GeV requires a solenoid of 477.5 T.m! It is very clear that spin manipulation in the muon collider has to be studied a little bit carefully and very early in time, so that it is included in the design of the accelerator. On figure 2 one can see a (im)possible scenario of spin manipulation, that includes two 477.5 T.m solenoids.... Performing spin manipulation at lower energies would probably be more economical.

One of the best surprises of this study was the realization that spin tune at the Z pole for muons is very close to 0.5. In fact the center-of-mass energy for which spin-tune is exactly 0.5 is 90.6223 GeV, energy for which the cross-section is more than 80% of that at the maximum. For this energy, if the first collision occurs for a particular configuration of helicities, the configuration will be exactly opposite for the next turn, as shown on figure 2. This automatically provides the two spin configurations necessary for the measurement of the polarization asymmetry: one on odd turns, the other one on even turns. Magic!

SUMMARY

Having rapidly – and superficially – evaluated the experimental environment at a muon collider Z factory, and in particular the magic bonus of a spin tune of 0.5, it is straightforward to guess the improvement in precision on the major electroweak observables. This is done on table 3, and compared with their sensitivity to the physics output. Of course a global improvement of the tests by a factor of typically 5 would have to be backed up by improved

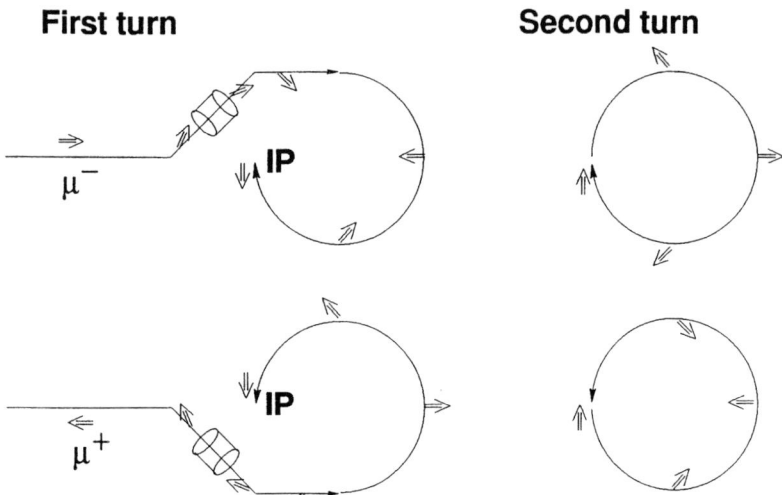

FIGURE 2. A (im)possible scheme assuring longitudinal polarization at the interaction point in a muon collider at a spin tune $\nu = 0.5$. The beam line of the μ^- and μ^+ beams have been shown separately, they should be superimposed, so that the two beams collide at the IP.

calculations at the same degree of precision. A step in this direction has recently been taken by the new calculation of higher orders [5]. In addition, the top quark mass should by then be known with a precision of 1-2 GeV, allowing the corresponding uncertainties to match the precision that could be obtained at a Z factory. Finally, the – important – error on the calculation of $\alpha(M_Z^2)$ from the $e^+e^- \to$ hadrons cross-sections has recently been reduced by use of the QCD prediction with the appropriate systematic error in the energy regions were it is more precise and reliable than existing data.

Increasing the precision by such a factor could lead to a significant discrepancy with the Standard Model. If this would still not happen, these data would still produce:
– a very much improved determination of the strong coupling constant α_s, with a precision of the order of $\Delta\alpha_s = \pm 0.001$ or below. This constitutes the most precise that one could presently imagine.
– a determination of $\Delta\rho$, ϵ_3 with a precision of $2\ 10^{-4}$, from a combination of the improved Z width and A_{LR}. These radiative corrections would offer sensitivity to the Higgs mass with a precision of about $\Delta log_{10} M_H = 0.05$, i.e. ± 10 GeV for a Higgs mass of 100 GeV.

Even in case where the Higgs boson, or some scenario of new physics, were found by direct observation, these data would still provide very strong constraints and would certainly play an essential role in our understanding.

All of this is beautiful, but quite a number of problems were found along the way:
- First, the average luminosity must be about $10^{33}/cm^2/s$;
- A solution must be found for the luminosity measurement;
- Detector backgrounds....
- Spin manipulation must be studied and foreseen at the design stage;

None is a a killer, but all exist, and all must be solved.

Physics sensitivity of Electroweak observables

Observable	Winter'98 value (error)	SM	M_t(GeV) 174 ± 5	M_H(GeV) 60 1000	α_s ±0.003	$\alpha(M_Z^2)^{-1}$ 128.923(36)	(m_b) 4.7 ± 0.3	Higher orders	Possible accuracy at μC Z factory
Γ_Z(MeV)	2494.8(2.5)	2493.3	+1.2	+4.2 -5.3	1.7	0.3	0.2	0.6	0.4
Γ_ℓ(MeV)	83.91(0.10)	83.93	+0.05	+0.11 -0.14	.02	–	–	0.02	0.03
$R_\ell \times 10^3$	20775(27)	20732	-1.5	+15 -13	21	1.5	2	5.	6.
$\sin^2\theta_w^{eff} \times 10^4$	2314.9(2.1)	2320.6	-1.6	-8.5 +6.9	0.05	0.9	–	1.	0.4
$R_b \times 10^4$	2171(9)	2158	-1.6	-0.4 0.	0.05	0.1	0.8	1.	5
M_W(MeV)	80375(65)	80314	+31	+103 -97	1	6	–	7.	20(*)

TABLE 3. Summary of the present experimental errors, physics sensitivity and theoretical uncertainties in SM predictions for the main Electroweak observables measurable at a muon collider Z factory. The value of $\alpha(M_Z^2)^{-1}$ is from the new evaluation by Davier and Höcker [4]. The SM value is for $M_H = 300$ GeV. (*)The W mass precision given here is what is expected from Tevatron run II, for comparison.

REFERENCES

1. General information and parameter list of the Workshop on Physics at the First Muon Collider and at the Front End of a Muon Collider can be found at: http://fnphyx-www.fnal.gov/conferences/femcpw97/workshop.html
2. R. B. Palmer, A. Tollestrup, A. Sessler, "Status Report of a High Luminosity Muon Collider and Future Research and Development Plans", Snowmass, 1996.
3. D. A. Ross and M. Veltman, Nucl. Phys. B95 (1975) 135, M. Veltman, Nucl. Phys. B123 (1977) 89.
4. M. Davier and A. Hocker, "Improved Determination of $\alpha(M_Z^2)$ and the anomalous magnetic moment of the muon" hep-ph/9711308.
5. G. Degrassi, P. Gambino, A. Sirlin, hep-ph/9611363 (1996).

6. B.W.Lynn, M.E.Peskin, R.G.Stuart, "Physics at LEP" CERN 86-02, (1986) 90.
7. M.E.Peskin and T.Takeuchi, Phys. Rev. Lett. 65 (1990) 964; V.A.Novikov, L.B.Okun, M.I.Visotsky, Nucl. Phys. B397 (1993) 35.
8. G.Altarelli and R.Barbieri, Phys. Lett. B253 (1991) 161; G.Altarelli, R.Barbieri, S.Jadach, Nucl. Phys. B369 (1992) 3; Err. Nucl. Phys. B376 (1992); G.Altarelli et al, Nucl. Phys. B405 (1993) 3.
9. A. Blondel, TASI 1991, Ellis, Hill and Lykken eds., world scientific (1992) 283; A.Blondel and C.Verzegnassi, Phys. Lett. B311 (1993) 346.
10. The LEP collaborations, Phys. Lett. B276 (1992) 247.
11. A.Blondel, F.M.Renard and C.Verzegnassi, Phys. Lett. B269 (1991) 419.
12. A.Blondel, A.Djouadi, C.Verzegnassi, Phys. Lett. B 293 (1992) 253.
13. P. Lebrun, "Detector Simulation and Backgrounds", Workshop on Physics at the First Muon Collider and at the Front End of a Muon Collider, Fermilab, November 1997.
14. R. Raja and A. Tollestrup, "Calibrating the energy of a 50×50 GeV muon collider using spin precession", these proceedings.
15. B. King, "Tau and c- tagging", these proceedings.
16. B. Norum and R. Rossmanith, Nucl Phys. B (proc. suppl.) 51A (1996) p. 191.

Electroweak and Heavy Flavor Physics at SLD

Stéphane Willocq*

*Stanford Linear Accelerator Center[1]
Stanford University, Stanford, CA 94309

Abstract. We review recent electroweak and B physics results obtained in polarized e^+e^- interactions at the SLC by the SLD experiment. Unique and precise measurements of the electroweak parameters A_e, A_b, A_c, R_b and R_c provide powerful constraints on the Standard Model. The excellent 3-D vertexing capabilities of SLD are further exploited to extract precise B^+ and B_d^0 lifetimes, as well as measurements of the time evolution of $B^0 - \overline{B^0}$ mixing.

INTRODUCTION

The various measurements presented below rely on the strengths of the SLC/SLD environment. Most important is the fact that the electrons are longitudinally polarized at the interaction point. Average polarizations of $(63.0 \pm 1.1)\%$, $(77.2 \pm 0.5)\%$ and $(76.5 \pm 0.8)\%$ were measured during the 1993, 1994–95, and 1996 data taking periods with a Compton Polarimeter [1]. The numbers of hadronic Z^0 decays collected during these periods are approximately 50K, 100K, and 50K, respectively. A description of the detector can be found in Ref. [2] and references therein.

ELECTROWEAK PHYSICS

In the Standard Model (SM), the tree-level differential cross section for $e^+e^- \to Z^0 \to f\bar{f}$ is expressed by

$$\frac{d\sigma}{d\cos\theta} = (1 - P_e A_e)\left(1 + \cos^2\theta\right) + 2\cos\theta \left(A_e - P_e\right) A_f, \quad (1)$$

where $\cos\theta$ is the cosine of the angle between the final state fermion f and the incident electron directions, P_e is the electron beam longitudinal polarization, and A_e and A_f are the asymmetry parameters for the initial and final

---

[1] Work supported in part by DOE Contract DE-AC03-76SF00515(SLAC).

state fermions, respectively. The parameter A_f represents the extent of parity violation at the $Z^0 \to f\bar{f}$ vertex and is defined as $A_f = \frac{g_L^2 - g_R^2}{g_L^2 + g_R^2}$, with the left- and right-handed coupling constants g_L and g_R.

The existence of parity violation introduces a forward-backward asymmetry $A_f^{FB} = (\sigma_f^F - \sigma_f^B)/(\sigma_f^F + \sigma_f^B)$ which is equal to $\frac{3}{4}A_e A_f$. At the SLC, the electron beam polarization allows a left-right forward-backward asymmetry to be measured

$$\tilde{A}_f^{FB} = \frac{\left[\sigma_f^F - \sigma_f^B\right]^{\text{left}} - \left[\sigma_f^F - \sigma_f^B\right]^{\text{right}}}{\left[\sigma_f^F + \sigma_f^B\right]^{\text{left}} + \left[\sigma_f^F + \sigma_f^B\right]^{\text{right}}} = \frac{3}{4}P_e A_f. \quad (2)$$

The latter asymmetry provides a *direct* measurement of A_f and yields a statistical enhancement factor of $(P_e/A_e)^2 \simeq 25$ over the unpolarized forward-backward asymmetry.

A particularly powerful yet straightforward asymmetry is the left-right cross-section asymmetry $A_{LR}^0 = \frac{\sigma_L - \sigma_R}{\sigma_L + \sigma_R} = A_e$, which yields a direct measurement of the coupling between the Z^0 and the e^+e^- initial state.

Measurements of the ratio between the partial Z^0 decay width into $f\bar{f}$ and that into any hadron are also sensitive to the $Z^0 \to f\bar{f}$ coupling constants $R_f = \frac{\Gamma(Z \to f\bar{f})}{\Gamma(Z \to \text{hadrons})} \propto g_L^2 + g_R^2$.

Precise measurements of A_f and R_f probe the effect of radiative corrections to the Z^0 propagator and the $Zf\bar{f}$ vertex. Since the radiative corrections depend on the top and Higgs masses, precise measurements can measure or constrain these quantities. Furthermore, such measurements also are sensitive to physics beyond the SM. Vacuum polarization corrections affect the value of $\sin^2 \theta_W^{eff}$ which is most precisely measured by the left-right asymmetry $A_{LR}^0 = \frac{2\left[1 - 4\sin^2 \theta_W^{eff}\right]}{1 + \left[1 - 4\sin^2 \theta_W^{eff}\right]^2}$. Heavy quark partial widths and asymmetry parameters are most sensitive to vertex corrections but with different sensitivity to left- and right-handed coupling constants, e.g., R_b (A_b) is more sensitive to deviations in the left- (right-) handed coupling.

Electroweak measurements at SLD are only summarized here, for a complete review of recent measurements see Ref. [3].

Left-Right Cross-Section Asymmetry

The measurement of A_{LR}^0 is a simple counting experiment. One needs to count the number of Z^0 produced with left- and right-handed electron beams, and measure the average electron beam polarization. The event sample is selected using tracking and calorimetry information and mostly consists of hadronic Z^0 decays (99.9%). In the 1996 data, the number of selected events is 28,713 and 22,662 for left- and right-handed electrons, respectively. The resulting measured asymmetry is $A_m = (N_L - N_R)/(N_L + N_R) = 0.1178 \pm 0.0044(\text{stat})$. A very small correction $\delta = (0.08 \pm 0.08)\%(\text{syst})$ is applied to take into account residual contamination in the event sample and slight beam asymmetries. Finally, the result is corrected for the electron beam polarization, the initial and final state radiation as well as for scaling the result to the Z^0 pole energy: $A_{LR}^0 = 0.1570 \pm 0.0057(\text{stat}) \pm 0.0017(\text{syst})$, giving $\sin^2 \theta_W^{eff} = 0.23025 \pm 0.00073(\text{stat}) \pm 0.00021(\text{syst})$. This preliminary measurement can be combined with previous measurements from the 1992, 1993, and 1994–95 running periods [4] to yield $A_{LR}^0 = 0.1550 \pm 0.0034$ and $\sin^2 \theta_W^{eff} = 0.23051 \pm 0.00043$. This represents the most precise measurement of $\sin^2 \theta_W^{eff}$ by a single experiment and its uncertainty is completely dominated by the statistical error.

Further information about $\sin^2 \theta_W^{eff}$ can be obtained from Z^0 decays into a pair of charged leptons (μ or τ). If lepton universality is assumed, the measurements may be combined to yield

$$\sin^2 \theta_W^{eff} = 0.23055 \pm 0.00041. \quad (3)$$

This value can be compared with the LEP average of 0.23162 ± 0.00041 obtained from lepton forward-backward asymmetries and τ polarization measurements. The SLD and LEP lepton averages agree within $1.9\,\sigma$. Fig. 1 compares the SLD measurement with all others obtained at LEP.

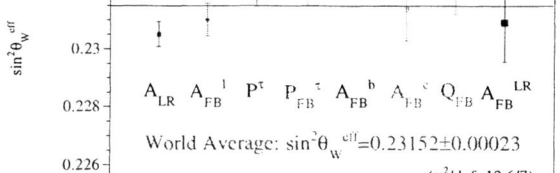

FIGURE 1. Measurements of $\sin^2 \theta_W^{eff}$ grouped by technique.

Z Decay Partial Widths

Another observable providing strong constraints on the SM is $R_b = \Gamma(Z^0 \to b\bar{b})/\Gamma(Z^0 \to \text{hadrons})$. The measurement proceeds by selecting hadronic Z^0 decays and tagging each event hemisphere independently for the presence of a B hadron decay. By computing both the rate for tagging a hemisphere and the rate for tagging both hemispheres in an event, one can extract the value of R_b and the hemisphere tag efficiency ϵ_b from the data. The Monte Carlo (MC) is used to estimate the charm and uds efficiencies as well as the small correlation between hemispheres.

SLD has designed a new approach to achieve high-efficiency and high-purity b tagging. The excellent 3-D vertexing capabilities of the vertex detector allow B decays to be reconstructed with an inclusive topological technique [5]. This technique relies on the precise knowledge of the e^+e^- interaction point (IP), 7 μm (15-35 μm) perpendicular to (along) the beam direction, and the precise tracking achieved with the pixel-based vertex detectors: VXD2 for 1993–95 and VXD3 for 1996 and beyond. The upgraded detector (VXD3) provides much improved polar angle coverage ($|\cos\theta| < 0.90$), lever arm and self-tracking capabilities, as well as a significantly reduced amount of material. Secondary vertices are found in 50% (65%) of b hemispheres, 15% (20%) of c hemispheres but in less than 1% of uds hemispheres for VXD2 (VXD3). The b hemisphere vertex finding efficiency increases with the decay length D to attain a constant level of 80% for $D > 3$ mm. Due to the typical $B \to D$ cascade structure of the decays, not all tracks originate from a single space point. Therefore, isolated tracks are attached to the vertex if they extrapolate close to the vertex line-of-flight and are sufficiently displaced from the IP.

The mass of the reconstructed vertex is used to tag b hemispheres. A clear separation between b and light-flavor hemispheres can be observed in Fig. 2(a). In particular, the charm contribution vanishes above the natural cutoff mass of ~ 2 GeV. Improved tagging efficiency is achieved by constructing the p_T-added mass: $M = \sqrt{M_{raw}^2 + p_T^2} + |p_T|$, where p_T is the total momentum of all tracks in the vertex in the plane perpendicular to the vertex axis. The value of p_T is minimized taking into account the uncertainties in the vertex and IP positions. As a result, the b-tag efficiency is enhanced by approximately 40% without significant degradation in purity [see Fig. 2(b)].

Requiring $M > 2$ GeV yields b-tag efficiencies of (35.3 ± 0.6)% and (47.9 ± 0.8)% for the 1993–95 and 1996 data, respectively. The corresponding samples are 98% pure in B hadrons. Combining all measurements yields $R_b = 0.2124 \pm 0.0024(\text{stat}) \pm 0.0017(\text{syst})$, in agreement with the SM value of 0.2158.

The above mass tag can be expanded to provide both a b tag for $M > 2$ GeV and a c tag for $0.6 < M < 2.0$ GeV. The c tag is improved by utilizing the total momentum of tracks attached to the vertex, exploiting the fact that charm hadrons in $c\bar{c}$ events have a harder momentum spectrum than in $b\bar{b}$ events. Similarly as was done for R_b, a system of five equations can be solved for the various b- and c-tagging fractions. Here, the values of R_b, R_c, as well as the b- and c-tag efficiencies are extracted from the data. The c-tag efficiencies are measured to be (11.4 ± 1.0)% and (14.0 ± 1.3)% for 1993–95 and 1996 data, respectively, with charm purities of 68%. Combining all data samples

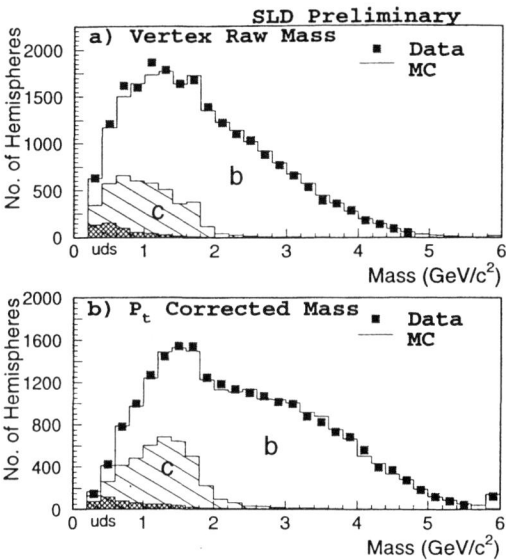

FIGURE 2. Distributions of (a) raw mass and (b) p_T-added mass for topological vertices in the 1993–95 data.

yields $R_c = 0.181 \pm 0.012(\text{stat}) \pm 0.008(\text{syst})$, in agreement with the SM value of 0.172. This double-tag technique has the advantage of having the lowest systematic uncertainty of all current R_c measurements.

Heavy Flavor Asymmetries

Different techniques have been used to measure the asymmetry parameters A_b and A_c. These differ mostly by the method used to determine which hemisphere contains the primary b or c quark: jet charge, vertex charge, kaon charge, $D^{(*)}$ charge or lepton charge. The first step (after hadronic event selection) in each analysis is to tag $Z^0 \to b\bar{b}$ ($Z^0 \to c\bar{c}$) events either by applying the mass tag utilized for the R_b (R_c) analysis, or by utilizing the fact that leptons from B hadron decays have harder p and p_T (with respect to the jet axis) distributions than those from other processes. A_b and/or A_c are determined with a fit to the $\cos\theta$-dependent differential cross section [Eq. (1)]. The thrust axis direction is used to provide the primary quark axis, except for the lepton analysis which uses the jet axis.

The first A_b measurement tags $b/\bar{b}$ with a momentum-weighted track charge defined as $Q = -\sum Q_{track} |\vec{p}\cdot\hat{T}|^\kappa \text{sign}(\vec{p}\cdot\hat{T})$, where $\vec{p}$ is the three-momentum of each track and Q_{track} its charge, $\hat{T}$ is the direction of the event thrust axis, signed such as to make $Q > 0$, and the coefficient $\kappa = 0.5$. The 100%

TABLE 1. Measurements of the $Z^0 \to b\bar{b}$ asymmetry parameter A_b.

Jet Charge	Kaon Charge	Lepton Charge	SLD average
$0.911 \pm 0.045 \pm 0.045$	$0.877 \pm 0.068 \pm 0.047$	$0.891 \pm 0.083 \pm 0.113$	0.898 ± 0.052

tag efficiency allows the analyzing power of the tag to be calibrated directly from the data. The probability to correctly tag the primary $b/\bar{b}$ quark is $P_{correct} = (1+e^{-\alpha_b|Q|})^{-1}$, where the constant α_b is measured to be 0.253 ± 0.013. The average correct tag probability is thus $\langle P_{correct} \rangle = 68\%$.

Clear forward-backward asymmetries are then observed for left- and right-handed electron beams, see Fig. 3.

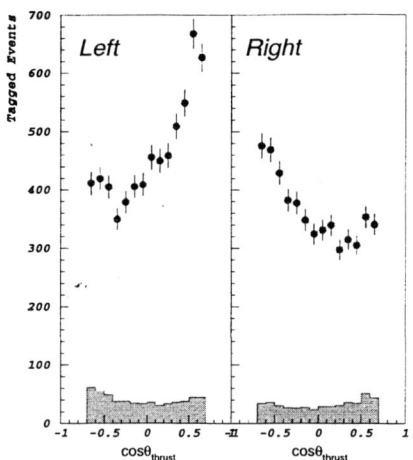

FIGURE 3. Distributions of the thrust axis $\cos\theta$ for left- and right-handed electrons.

The second A_b measurement tags $b/\bar{b}$ by exploiting the dominant $b \to c \to s$ transition in B decays. Therefore, detection of K^- (K^+) mesons tags b ($\bar{b}$) quarks. Right-sign kaon production has been measured by ARGUS [6] in B^+ and B_d^0 decays to be $(85 \pm 5)\%$ and $(82 \pm 5)\%$, respectively. The analysis presented here is the first application of kaon tagging for a heavy quark asymmetry measurement. Charged kaons are identified with the Cherenkov Ring Imaging Detector and the rate of pion misidentification is carefully calibrated from K_s^0 and τ decay data samples.

Another approach to heavy quark selection and flavor tagging is to select leptons from semileptonic decays. The distinctive lepton total and transverse momenta are exploited as well as the charge of the lepton.

The measured values of A_b obtained with the three different tagging techniques are presented in Table 1. The SLD average agrees well with the SM prediction of 0.935.

We have also performed measurements of the asymmetry parameter A_c

TABLE 2. Measurements of the $Z^0 \to c\bar{c}$ asymmetry parameter A_c.

Kaon & Vtx Charge	Lepton Charge	$D^{(*)}$ Charge	SLD average
$0.66 \pm 0.07 \pm 0.04$	$0.61 \pm 0.10 \pm 0.07$	$0.64 \pm 0.11 \pm 0.06$	0.647 ± 0.060

in $Z^0 \to c\bar{c}$ decays. The most powerful measurement relies on combined kaon charge and vertex charge tags. This particular measurement based on only 150K hadronic Z^0 decays is already among the best and should become dominant with a modest increase in statistics. A_c is also measured with two more traditional techniques, as summarized in Table 2. The SLD average agrees well with the SM prediction of 0.67.

B PHYSICS

Several aspects of the weak interaction can be probed by studying the weak decays of B hadrons. First, by measuring lifetimes, we can test our understanding of B hadron decay dynamics. Second, we can test the Cabibbo-Kobayashi-Maskawa (CKM) quark mixing matrix description within the SM.

B^+ and B_d^0 Lifetimes

A strong lifetime hierarchy is observed in the case of charm hadrons: $\tau(D^+) \simeq 2.3\ \tau(D_s) \simeq 2.5\ \tau(D^0) \simeq 5\ \tau(\Lambda_c^+)$. This hierarchy is predicted to scale with the inverse of the heavy quark mass squared and the B hadron lifetimes are expected to differ by only 10-20% [7,8].

The technique used by SLD takes advantage of the excellent 3-D vertexing capabilities of the VXD to reconstruct the decays inclusively. The goal is to reconstruct and identify all the tracks originating from the B decay chain. This then allows charged and neutral B mesons to be separated by simply measuring the total charge of tracks associated with the B decay.

The analysis [9] uses the inclusive topological vertexing technique described earlier. In the hadronic Z^0 event sample, we select 20,783 B decay candidates by requiring $M > 2$ GeV and $D > 1$ mm. The sample is divided into 7942 neutral and 12841 charged vertices corresponding to reconstructed decays with total charge $Q = 0$ and $Q = \pm 1, 2, 3$, respectively, where Q is the charge sum of all tracks associated with the vertex. MC studies show that the ratio between B^+ and B_d^0 decays in the charged sample is 1.55 (1.72) for VXD2 (VXD3), and the ratio between B_d^0 and B^+ decays in the neutral sample is 1.96 (2.24) for VXD2 (VXD3)[2].

The B^+ and B_d^0 lifetimes are extracted with a simultaneous binned maximum likelihood fit to the decay length distributions of the charged and neutral

---

[2] Reference to a specific state (e.g., B^+) implicitly includes its charge conjugate (i.e., B^-).

samples (see Fig. 4). The maximum likelihood fit yields lifetimes of $\tau_{B^+} = 1.698 \pm 0.040(\text{stat}) \pm 0.046(\text{syst})$ ps, $\tau_{B_d^0} = 1.581 \pm 0.043(\text{stat}) \pm 0.061(\text{syst})$ ps, with a lifetime ratio of $\tau_{B^+}/\tau_{B_d^0} = 1.072^{+0.052}_{-0.049}(\text{stat}) \pm 0.038(\text{syst})$. The main

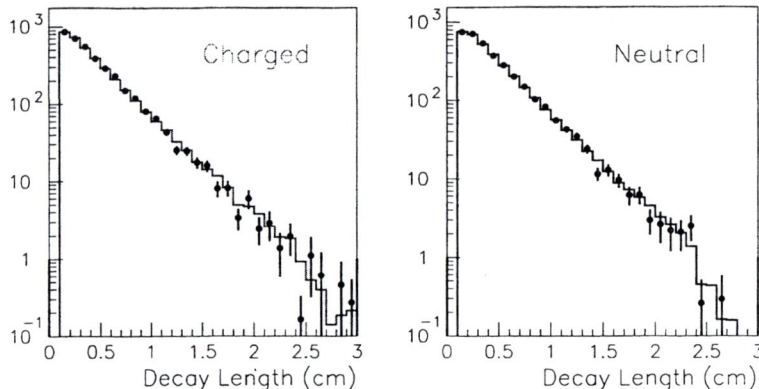

FIGURE 4. Decay length distributions for data (points) and best fit Monte Carlo (histogram).

contributions to the systematic error come from uncertainties in the detector modeling, B_s^0 lifetime, b-baryon fraction, fit systematics, and MC statistics.

These measurements are among the best currently available and confirm the expectation that the B^+ and B_d^0 lifetimes are nearly equal.

B^0-$\overline{B^0}$ Mixing

Transitions between flavor states $B^0 \leftrightarrow \overline{B^0}$ take place via second order weak interactions "box diagrams." The oscillation frequency Δm_d for B_d^0-$\overline{B_d^0}$ mixing depends on the CKM matrix element $|V_{td}|$ for which little is known experimentally. Theoretical uncertainties are significantly reduced for the ratio between Δm_d and Δm_s. Thus, combining measurements of the oscillation frequency of both B_d^0-$\overline{B_d^0}$ and B_s^0-$\overline{B_s^0}$ mixing translates into a measurement of the ratio $|V_{td}|/|V_{ts}|$.

Experimentally, a measurement of the time dependence of B^0-$\overline{B^0}$ mixing requires three ingredients: (i) the B decay proper time has to be reconstructed, (ii) the B flavor at production (initial state $t = 0$) needs to be determined, as well as (iii) the B flavor at decay (final state $t = t_{\text{decay}}$). At SLD, the time dependence of B_d^0-$\overline{B_d^0}$ mixing has been measured using four different methods. All four use the same initial state tagging but differ by the method used to either reconstruct the B decay or tag its final state.

Initial state tagging takes advantage of the large polarization-dependent forward-backward asymmetry in $Z^0 \to b\bar{b}$ decays as described above. For left-(right-) handed electrons and forward (backward) B decay vertices, the initial quark is tagged as a b quark; otherwise, it is tagged as a $\bar{b}$ quark. The initial state tag can be augmented by using a momentum-weighted track charge (see A_b measurement above) in the hemisphere opposite that of the reconstructed B vertex. These two tags are combined to yield an initial state tag with 100% efficiency and effective average right-tag probability of 82% (for $\langle P_e \rangle = 77\%$).

Two B_d^0-$\overline{B_d^0}$ mixing analyses use topological vertexing (see R_b measurement) to reconstruct B decays. The $b/\bar{b}$ flavor tag is performed by using either a kaon-charge tag (as in the A_b analysis) or by exploiting the $B \to D$ cascade charge structure. This latter tag is a novel technique developed by SLD which relies on a "vertex charge dipole" defined as $\delta q = (\sum^+ w_i L_i)/(\sum^+ w_i) - (\sum^- w_i L_i)/(\sum^- w_i)$, where the first (second) term is a sum over all positive (negative) tracks in the vertex and the quantity L_i corresponds to the longitudinal separation between the IP and the point of closest approach of track i to the vertex line-of-flight. The weight w_i is inversely proportional to the uncertainty in L_i. Two other analyses select semileptonic decays and thus use the lepton charge for the $b/\bar{b}$ flavor tag.

The time dependence of B_d^0-$\overline{B_d^0}$ mixing is measured from the fraction of decays tagged as mixed as a function of decay length or proper time. The four measurements are combined to produce the following SLD average: $\Delta m_d = 0.525 \pm 0.043(\text{stat}) \pm 0.037(\text{syst})$ ps^{-1}, consistent with the world average value of 0.472 ± 0.018 ps^{-1}. Further details about the above measurements may be found in Ref. [10] and references therein.

The above techniques (except for the kaon tag) can be extended to study B_s^0-$\overline{B_s^0}$ mixing. Recent studies at LEP indicate that the oscillation frequency is very large: $\Delta m_s > 10.2$ ps^{-1} at the 95% C.L. Therefore, excellent proper time, and thus decay length, resolution is required to improve that limit or observe oscillations. With VXD3, SLD expects to achieve decay length resolutions of 80-100 μm, i.e. a factor of ~ 3 better than those obtained at LEP. Given a sample of 500K hadronic Z^0 decays with VXD3, a limit of $\Delta m_s > 16$ ps^{-1} can be set at the 95% C.L., see Fig. 5.

SUMMARY AND PROSPECTS

Using samples of 150K and 50K hadronic Z^0 decays collected in 1993-95 and 1996, the SLD Collaboration has produced precise and/or unique tests of the Standard Model. These analyses take advantage of the large longitudinal electron beam polarization, the small and stable SLC beam spot, the high-resolution 3-D pixel vertex detector, and the particle identification capabilities of the Cherenkov Ring Imaging Detector.

Many measurements relying on precise tracking will benefit greatly from the increased resolution and coverage of the upgraded vertex detector (VXD3) installed before the 1996 run. SLD expects to collect another 300K to 400K Z^0 by the end of the 1997-98 run. With that sample, SLD will surpass the

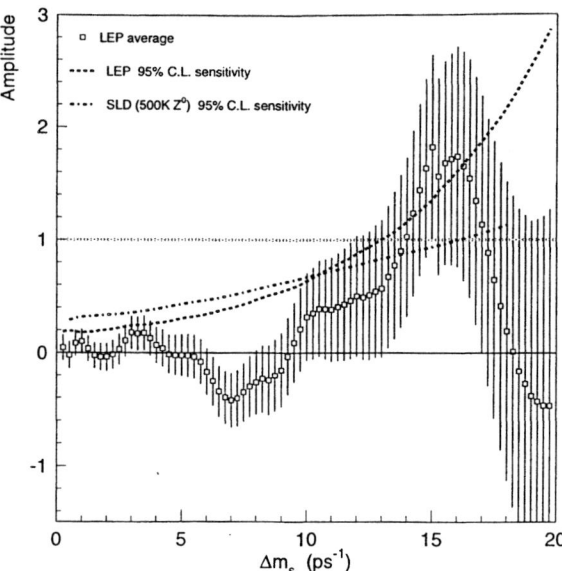

FIGURE 5. Comparison of sensitivity to B_s^0–$\overline{B_s^0}$ mixing as a function of Δm_s for the current LEP average (dashed curve) and the projected SLD average for 500K hadronic Z^0 (dash-dotted curve). Also shown are the current LEP measurements (data points).

precision achieved by the combined LEP measurements of $\sin^2 \theta_W^{eff}$ and A_b, and approach the same precision for R_b. Furthermore, the superior resolution will enable SLD to significantly increase the sensitivity to B_s^0–$\overline{B_s^0}$ mixing beyond that currently attained by the LEP experiments.

REFERENCES

1. M. Woods, in *Proceedings of the AIP Conference*, Vol. 343 (1995), p. 230.
2. SLD Collab., K. Abe et al., Phys. Rev. D **53**, 1023 (1996).
3. B. Schumm, *Electroweak Results from the SLD*, SLAC-PUB-7697, Nov. 1997.
4. SLD Collab., K. Abe et al., Phys. Rev. Lett. **78**, 2075 (1997).
5. D. Jackson, Nucl. Instrum. Methods A **388**, 247 (1997).
6. ARGUS Collab., H. Albrecht et al., Z. Phys. C **62**, 371 (1994).
7. I. I. Bigi et al., in *B Decays*, ed. S. Stone (World Scientific, New York, 1994), p. 132.
8. M. Neubert and C.T. Sachrajda, Nucl. Phys. B483, 339 (1997).
9. K. Abe et al., *Measurement of the B^+ and B^0 Lifetimes using Topological Vertexing at SLD*, SLAC-PUB-7635, August 1997.
10. S. Willocq, *B Physics at SLD*, SLAC-PUB-7567, June 1997.

Estimates of Vertex Tagging Efficiencies at a Muon Collider Higgs Factory

Bruce J. King

Brookhaven National Laboratory
email: bking@bnl.gov

Abstract. Tagging efficiencies and purities are estimated for the decay modes $H \to b\bar{b}$, $H \to \tau\bar{\tau}$ and $H \to c\bar{c}$ of Higgs bosons produced at an s-channel muon collider Higgs factory.

INTRODUCTION

Measurement of the branching ratios (BR) of the Higgs boson would be an important goal of a future s-channel muon collider Higgs factory. This paper derives quick, heuristic estimates of the expected tagging efficiencies and purities for the 3 main decay modes of a light, Standard Model Higgs: $H \to b\bar{b}$, $H \to \tau\bar{\tau}$ and $H \to c\bar{c}$. These estimates should be useful as input to theoretical assessments of the physics capabilities of Higgs factory muon colliders.

The vertex tagging methods are similar for $b\bar{b}$ events and $c\bar{c}$ events and these two modes are treated together in the next section. The $\tau\bar{\tau}$ mode is then discussed in a separate section, before ending with a short conclusion section.

I CHARM AND BEAUTY TAGGING

A Tagging Signatures

Higgs decays to c or b quark-antiquark pairs produce back-to-back 2-jet events with a displaced vertex in each jet from the decay of the c or b hadron.

---

[1] This work was performed under the auspices of the U.S. Department of Energy under contract no. DE-AC02-76CH00016.

The goal of a vertex tagging algorithm is to find the displaced vertices and also to distinguish the $b\bar{b}$ events from $c\bar{c}$ events. The figures of merit for the tagging algorithm are the overall tagging efficiency for each of the 2 event types and the rejection factor against the other, wrong event type. In principle, one should also consider the rejection factor for hadronic events that are neither $b\bar{b}$ nor $c\bar{c}$ but, in practice, this is a much simpler task than distinguishing the 2 types with displaced vertices.

For b jets, the charged tracks in the displaced vertex will have an invariant mass up to the mass of b hadrons – about 5 GeV. There will usually be several tracks. B hadrons almost always decay into a charm hadron plus additional hadrons, so some of the charged tracks will likely form a tertiary vertex downstream from the B decay vertex. If a "topological vertexing" algorithm such as ZVTOP [1] is used then this tertiary vertex may be reconstructed. The characteristic displacement length of the vertex is $\beta\gamma.c\tau$, where $c\tau$ is 450 microns. $\beta\gamma$ would typically be about 7 for a 100 GeV Higgs decay at rest – corresponding to about 70% of the quark energy or 35% of the Higgs mass, as is seen in Z decays to b quarks – so the characteristic displacement length is about 3 mm.

The displaced vertices from charm typically have a lower multiplicity and, more significantly, a lower invariant mass than B decays since the mass of all weakly decaying charm hadrons is less than 2 GeV. In analogy to Z decays, the charm hadrons should typically carry about 50% of the c quark energy. Combining this with $c\tau = 317$ (124) microns for charged (neutral) D mesons gives characteristic decay lengths of 4 mm (1.7 mm). The lower multiplicity and lower invariant mass of charm decays makes them more difficult to tag than B decays, even for the longer lived charged D mesons.

B Jet Tagging Efficiencies

An estimate for jet tagging efficiencies and purities was made based on studies done for the proposed DESY 500 GeV center-of-mass electron-positron linear and presented in the Conceptual Design Report (CDR) [2]. This study was performed for the following scenario:

1. a 3 layer barrel vertex detector with its innermost layer a cylinder at either 1.0 cm or 2.2 cm from the interaction point (ip)

2. pixel tracking elements. Studies were performed for both charge coupled devices (CCD's) and active pixel sensors (APS's)

3. a 50–50 mix of b and c jets

4. vertex reconstruction using the SLD topological vertexing algorithm ZVTOP [1]

5. the central region of the detector was considered.

A muon collider Higgs factory will have worse backgrounds than a linear electron-positron collider and two adjustments were made to allow for this. Firstly, APS's were assumed to be the tracking technology because CCD's are presumably too susceptible to radiation damage for use in the muon collider tracking environment and, secondly, the radius of the innermost tracking layer was assumed to be 5 cm, consistent with background studies for muon colliders [3]. This requires an extrapolation of the DESY studies.

With these adjustments, it is assumed that the Higgs factory jet tagging performance can be estimated using the DESY studies. This implicitly assumes that 2 further differences are not important in the study. Firstly, it is assumed that the effect of the higher uncorrelated "fake hit" density in a muon collider can be minimized in a well designed vertex detector using the redundancy afforded by the several vertexing layers. Secondly, the Higgs factory muon collider will have a beam spot size in the hundreds of microns which, in contrast to the smaller spot at a linear collider, will give little useful vertex constraint to assist the vertexing algorithm. The lack of a vertex constraint is important today's collider detectors that use 2-D vertexing with silicon microstrips and only 2 or 3 layers of vertexing e.g. the LEP and Tevatron detectors. For these geometries the additional constraint would be very helpful. However, it is probably reasonable to assume that the loss would be much less in a multi-layer 3-D pixel vertex detector such as one would expect in a future muon collider.

The purity vs. efficiency curves from the DESY study are shown in figure 2.2.2 of the DESY CDR [2]. All of the curves have a fairly flat purity out to a certain efficiency then dive quite steeply, so little is lost in either purity or efficiency by choosing the efficiency and purity values at this lip of the curve. These values for 1.0 and 2.2 cm inner radii and b– and c–tagging are given in table 1. The table also gives the assumed extrapolation to the 5.0 cm radius at a muon collider.

The extrapolated values were estimated by noting that the extrapolation ratio 2.2:5.0 cm is similar to the ratio between the 2 DESY values, 1.0:2.2 cm. The extrapolation to the muon collider scenario is rather modest since the observed changes between 1.0 cm and 2.2 cm are relatively small. Thus, it is considered that this specific choice of extrapolation introduces little uncertainty and should be entirely adequate for estimates of the physics potential of the Higgs factory muon collider – certainly until a detailed detector design and simulation is arrived at.

The purities in table 1 refer to an equal mixture of c and b jets. For dealing with an arbitrary mixture of b and c jets it is useful to convert these purities into efficiencies for tagging the wrong type of jet. For an equal sample of b's and c's and subscript i (j) denoting the intended (wrong) tag it is clear that:

$$p_i = \frac{e_i}{e_i + e_j}, \qquad (1)$$

TABLE 1. Jet tagging efficiencies and purities for an equal sample of b and c jets in the central region of the vertex detector. The values for 1.0 cm and 2.2 cm were read off from the curves for APS's in figure 2.2.2 of the DESY CDR and these values have been extrapolated to give efficiencies and purities at 5.0 cm.

tag	radius	efficiency (e)	purity (p)
b-tag	1.0 cm	63 %	97 %
b-tag	2.2 cm	59 %	97 %
b-tag	5.0 cm	55 %	97 %
c-tag	1.0 cm	46 %	74 %
c-tag	2.2 cm	42 %	71 %
c-tag	5.0 cm	38 %	68 %

so solving for e_j gives:

$$e_j = e_i \times (\frac{1}{p_i} - 1) \qquad (2)$$

This gives the jet tagging efficiencies of table 2 for the central region of the detector.

TABLE 2. Jet tagging efficiencies for b– and c–tagging in the central region of the vertex detector for a Higgs factory muon collider. The innermost tracking layer of the vertex detector is assumed to be at 5.0 cm from the ip.

tag	eff. for b jets	eff. for c jets
b-tag	55 %	2 %
c-tag	18 %	38 %

C Event Tagging Efficiencies

The jet tagging efficiency must be converted to an event tagging efficiency. Each Higgs decay to $b\bar{b}$ or $c\bar{c}$ will produce an event with two approximately back-to-back jets with the same quark flavor. In this subsection, the event tagging efficiency and rejection factor against wrong flavor events is calculated from the assumed jet tagging efficiencies using a simple algorithm for combining the jet tagging information from the two jets. Several simplifying approximations are used.

Since the jets are fairly back-to-back it will be assumed that either both jets are central or neither jet is. Therefore, the calculation is done assuming central events and then the efficiency is multiplied by an overall geometrical acceptance factor of $\cos(\pi/4) = 1/\sqrt{2}$. Since the Higgs decay is isotropic this is roughly equivalent to assuming that jets more than 45 degrees from the beam direction can be considered central enough for reliable vertex tagging.

TABLE 3. Probabilities for tagging 0,1 or 2 jets if the jet tagging efficiency is e.

number of tagged jets	probability
0	$(1-e)^2$
1	$2e(1-e)$
2	e^2

The simple event tagging strategy is to require either one or both of the 2 jets to be tagged correctly and also to require that neither was tagged as the incorrect flavor. From table 3, it is seen that the probability for the first condition is $1 - (1 - e_i)^2$ and for the second condition is $(1 - e_j)^2$. (As before, i denotes the correct jet tag and j the incorrect tag.) If the simplifying assumption is made that the two probabilities are independent and the $1/\sqrt{2}$ geometrical acceptance is included then the efficiency, E_{ii}, for correctly tagging the event is:

$$E_{ii} = \frac{1}{2} \times [1 - (1 - e_i)^2] \times (1 - e_j)^2. \tag{3}$$

Clearly, the probability, E_{jj} for tagging the wrong event type is obtained by simply swapping the i's and j's:

$$E_{jj} = \frac{1}{2} \times [1 - (1 - e_j)^2] \times (1 - e_i)^2, \tag{4}$$

and the rejection factor R against the wrong flavor event is defined naturally to be

$$R = \frac{E_{ii}}{E_{jj}}. \tag{5}$$

The $b\bar{b}$ and $c\bar{c}$ tagging efficiencies and rejection factor against the wrong flavor are given in table 4.

II TAU EVENT TAGGING

The tagging of $h \to \tau\bar{\tau}$ events appears to be much easier than the tagging of quark jets. An s-channel Higgs factory will produce Higgs at rest and

TABLE 4. Event tagging efficiencies and wrong-flavor rejection factors for b$\bar{\text{b}}$ and c$\bar{\text{c}}$ events at a Higgs factory muon collider. A geometrical acceptance factor of $1/\sqrt{2}$ is included in the efficiencies.

event type	efficiency, E_{ii}	rejection factor, R
b$\bar{\text{b}}$	54 %	50
c$\bar{\text{c}}$	42 %	8

not in association with other particles, so the geometry should be identical to the $Z \rightarrow \tau\bar{\tau}$ events seen in the LEP and SLD detectors operating at the Z pole energy. These distinctive events consist of almost back-to-back high energy tracks emanating from the ip. Each side is usually a single track – i.e. a "1-prong", with an 85.5 % probability per side – with almost all of the remainder being tightly collimated "3-prong" jets. The tau lifetime is long enough ($c\tau = 88$ microns) that the slight offset of the prongs from the ip should be observable in a precise vertex detector, but vertexing information should not generally be required to identify this event sample. Therefore, the purity of the sample should be close to 100 % and the efficiency should be dominated by the geometrical acceptance of the central tracker. For all practical purposes, physics studies could reasonably assume a purity of 100% and a conservative efficiency of $\cos(\pi/4) = 0.71$.

III CONCLUSIONS

Heuristic estimates have been made of tagging efficiencies and purities in an s-channel muon collider Higgs factory for the 3 main decay modes of a light Higgs boson. The estimates for the b$\bar{\text{b}}$ and c$\bar{\text{c}}$ modes are given in table 4, while the $\tau\bar{\tau}$ mode is assumed to have approximately a 71 percent efficiency with essentially 100 percent purity.

REFERENCES

1. C.J.S. Damerell and D.J. Jackson, SLAC-PUB-7215 (1996), to be published in Nucl. Instrum. and Meth.
2. Conceptual Design of a 500 GeV Electron-positron Linear Collider with Integrated X-ray Laser Facility, Editors: R. Brinkmann, G. Materlik, J. Rossbach, A. Wagner, DESY 1997-048, ECFA 1997-182
3. The Muon Collider Collaboration, Status of Muon Colliders and Future Research and Development Plans, to be published in Nucl. Instrum. and Meth.

Prospects for Higgs at the Tevatron

John Womersley

Fermi National Accelerator Laboratory[1]
Batavia, Illinois 60510

Abstract. The current status of simulation studies for the observation of a standard-model or lightest supersymmetric Higgs boson at TeV33 are reviewed. Latest studies indicate that the mass range $60 < m_H \lesssim 130$ GeV can be covered at the 5-standard-deviation level with 30 fb^{-1}, using the WH and ZH channels. This is the full allowed mass range for the lightest Higgs h of minimal supersymmetry.

INTRODUCTION

The observation of a Higgs boson is possibly the most exciting physics prospect for the Tevatron. Understanding the mechanism of electroweak symmetry breaking and the scalar sector of the standard model is the most pressing issue confronting high energy physics today.

The current best limit on the mass of a light Higgs boson H decaying to $b\bar{b}$ is $m_H > 77$ GeV (from LEP2) [1].

Fits to electroweak data from LEP and SLD, using the Tevatron top mass, show some (weak) sensitivity to m_H; the latest results [1] give $m_H = 115^{+116}_{-66}$ GeV and exclude $m_H > 420$ GeV (95% C.L.). Additionally, in the minimal supersymmetric standard model (MSSM), the lightest Higgs h has

---

[1] operated by the Universities Research Association for the U.S. Department of Energy

a mass $m_h < 125$ GeV (< 112 GeV in the absence of stop-quark mixing); even with a non-minimal Higgs sector the lightest neutral Higgs has a mass $m_h \lesssim 150$ GeV [2]. Thus both the available experimental evidence, and theoretical prejudice, point to the existence of a light Higgs. In the following we shall not distinguish between a standard model and MSSM Higgs since both are expected to have similar couplings and decays (predominantly to $b\bar{b}$) in the mass range of interest.

The first accelerator to explore this mass region is likely to be LEP2. Its mass sensitivity should extend to $m_H < \sqrt{s} - m_Z - (5 - 10)$ GeV; with $\sqrt{s} = 200$ GeV, masses up to ~ 105 GeV will be probed by 1999. This covers much, but not all, of the light Higgs mass range. We may therefore set a goal for TeV33: to discover (or exclude) a light Higgs over the whole range $60 < m_H < 130$ GeV before significant data from the LHC becomes available. We shall conclude that this goal is attainable, and that 30 fb^{-1} of integrated luminosity is required to meet it.

HIGGS DISCOVERY MODES

The most promising modes for Higgs discovery are those where the H is produced in association with a vector boson, as proposed by Stange, Marciano and Willenbrock [3].

$p\bar{p} \to WH$ with $W \to \ell\nu$ and $H \to b\bar{b}$

This channel was investigated in some detail for the TeV2000 report [4]. The signal is a $W \to \ell\nu$ decay together with two jets; the invariant mass of the jets reconstructs to the Higgs mass. The dominant background, $W + 2$jet production, can be reduced by requiring that both jets be b-tagged; in the TeV2000 study the tagging efficiency was assumed to be 0.5 and the mistag rate 0.005 for both jets. After double-tagging, the remaining backgrounds are $W + b\bar{b}$ and $W + c\bar{c}$, WZ with $Z \to b\bar{b}$, $t\bar{t}$ and single top production (both $W^* \to t\bar{b}$ and Wg fusion $\to tqb$ processes). Background estimates were verified against CDF data wherever possible.

The study concluded that Higgs signals would be observable up to $m_H = 120$ GeV with 25 fb^{-1} of luminosity. However, it made two assertions. Firstly a factor of two improvement in signal-to-background ratio was put in "by hand" to model the expected improvement from optimized selection cuts [5]. This resulted in an artificial suppression of the irreducible $WZ(b\bar{b})$ background. Secondly, it was assumed that dijet resolutions of order $100\%/\sqrt{m_{jj}}$ were attainable. This gives $\Delta m_{jj} = 11$ GeV at $m_{jj} = 80$ GeV compared with 16 GeV from the full simulation.

A Snowmass study [6] aimed to quantify possible improvements in significance. Events were selected having:

- $p_T(\ell) > 20$ GeV/c;
- $\not{E}_T > 25$ GeV;
- At least two jets with $E_T > 15$ GeV and $|\eta| < 2.5$;
- No more than two jets with $E_T > 30$ GeV and $|\eta| < 2.5$;
- Two b-tags.

Based on CDF Run I results, an improvement in signal efficiency of a factor 1.8 was obtained by adding leptonic b-tags and using a looser silicon vertex tag for the second of the two jets. Background rejection remains adequate. Possible cuts on various center-of-mass angles were also investigated. It was found that requiring $|\cos\theta_H| < 0.8$ (where θ_H is the Higgs scattering angle in the WH center of mass) improves signal-to-background by 50%. Cutting on θ_{bb}, the center of mass angle between the b and $\bar{b}$, or on θ_b, the b decay angle in the Higgs rest frame, did not improve the signal significance. This study did not assume any improvement in dijet mass resolution. A mass window of $84 < m_{jj} < 117$ GeV/c^2 was used for $m_H = 100$ GeV/c^2. Higgs discovery at the 5 standard deviation level is demonstrated up to $m_H \lesssim 125$ GeV in 30 fb^{-1}.

$p\bar{p} \to ZH$ with $Z \to \ell\ell$ or $\nu\nu$ and $H \to b\bar{b}$

This channel was not investigated for TeV2000 but a Snowmass study [7] shows it to be very promising. The signal is either a leptonic Z or missing E_T, together with two jets. Once again, the second jet is allowed to be more loosely-tagged than the first, and Run I CDF tagging efficiencies are applied. The backgrounds are ZZ, $Z + b\bar{b}$, $Z + c\bar{c}$, and for the $\not{E}_T$ signal, QCD $b\bar{b}$ production, $W + b\bar{b}$ with the lepton lost, and $t\bar{t}$. These backgrounds were all estimated from CDF data. Requiring $\not{E}_T > 35$ GeV and $\Delta\phi > 0.5$ between the $\not{E}_T$ and any jet, together with a third-jet veto ($E_T > 8$ GeV in $|\eta| < 2.4$)

and a veto on isolated tracks with $p_T > 10$ GeV/c, reduces the backgrounds to a tolerable level for the $\not{E}_T + 2$jets final state.

For the $\ell\ell + 2$jets final state, events are required to have two leptons with $p_{T1,2} > 20, 10$ GeV and $|\eta_{1,2}| < 1.0, 2.0$. The mass must be within 15 GeV of m_Z. Both final states are required to have two b-tagged jets with $E_T > 15$ GeV and $|\eta| < 2.0$.

The overall ZH acceptance is found to be comparable to that for WH and a Higgs could be observed as a 3 standard deviation effect up to $m_H = 110$ GeV with 30 fb^{-1}.

OTHER MODES STUDIED

$p\bar{p} \to (W, Z)H$ with $(W, Z) \to jj$ and $H \to \tau^+\tau^-$

This channel was suggested by Mrenna and Kane [8]. The signal is two tau-jets together with two hadronic jets having a mass consistent with a W or Z boson. The TeV2000 study [9] used only one-prong tau decays and assumed that the neutrino direction was parallel to the hadronic track; in this case, the tau-tau invariant mass can be reconstructed (provided the opening angle is not 180°). The dominant background is Z+jets with $Z \to \tau^+\tau^-$. Suitable selections can improve the signal to background ratio for Higgs from $1/2 \times 10^4$ at the cross section level to 1/60; in a mass window for a 120 GeV Higgs the signal to background is 1/10. One would therefore need to know the shape of the $Z \to \tau^+\tau^-$ mass peak to better than a few percent, off-resonance, to claim any discovery. This does not seem very credible given the difficulties understanding $\not{E}_T$ response. This channel is not regarded as very promising.

An alternative method of tau-tau reconstruction was also investigated [10]. Here the $\tau\tau \to \ell j \not{E}_T$ final state is used and a three-body transverse mass constructed. A Higgs signal for $m_H = 130$ GeV was simulated but again there does not appear to be sufficient signal-to-background to distinguish it from the high-side tail of the $Z \to \tau^+\tau^-$ transverse mass peak.

Inclusive $H \to b\bar{b}$

At Snowmass [13] the possibility of observing inclusive $H \to b\bar{b}$ production in double-tagged jet final states was investigated. The inclusive Higgs cross section is relatively large (~ 1 pb) but there is an overwhelming background from QCD $b\bar{b}$ production. The first difficulty is triggering; in this study it was assumed that a semileptonic b-decay would be required for a lepton plus jets trigger, and the resulting overall efficiency is ~ 0.05. In 30 fb^{-1} there would be 2.5×10^6 QCD background events in the Higgs mass window and a signal cross section of > 10 pb would then be required for a 3 standard deviation effect, making this impossible as a Higgs observation (unless the efficiency can be raised tenfold). However, it is worth noting that a clear $Z \to b\bar{b}$ signal should be observable even in Run II (with 50,000 signal events). This will provide an important event sample for development of the b-tagging and dijet mass reconstruction algorithms needed for Higgs discovery at TeV33.

Four-b Final States

Gunion and collaborators [11] have studied $4b$ final states as signals of Higgs production. Only parton-level simulations have been performed. In the MSSM, four b-jet final states could offer signals for:

- $gg \to b\bar{b}H \to 4b$ ($H = h, H, A$),
- $p\bar{p} \to H \to hh \to 4b$,
- $p\bar{p} \to H \to AA \to 4b$.

Backgrounds arise from combinatorics, from $p\bar{p} \to b\bar{b}g$, $c\bar{c}b\bar{b}$, $t\bar{t}$ and $t\bar{t}g$ (reducible); and from $p\bar{p} \to 4b$ and $b\bar{b}Z(Z \to b\bar{b})$ (irreducible). They concluded that, with 30 fb^{-1} at TeV33, signals would be observable for $gg \to b\bar{b}H$ for large $\tan\beta$ or small m_A (roughly $\tan\beta > 0.2 m_A(\text{GeV})$), and for $H \to hh/AA$ for $\tan\beta \gtrsim 2$ or $m_A \lesssim 60$ GeV.

Shortcomings with this analysis are that the dijet resolution assumed was probably over-optimistic, and there are no clear peaks above background (making observation of a signal a matter of confidence in the background estimation).

The authors were concerned about the ability to trigger on such $4b$ final states. At Snowmass it was estimated [12] that an efficiency as high as 60%

could be obtained using a lepton plus jets trigger. Also, both CDF and DØ are investigating the use of a displaced vertex trigger to select multi-jet final states rich in b's, which may allow these processes to be selected.

DIJET MASS RESOLUTION

The mass resolution attainable on the $b\bar{b}$ dijet system was identified as a critical issue in the TeV2000 report. It directly feeds into the signal-to-background ratio which can be achieved in the Higgs search. A number of studies have been conducted [13–15], all of which reach broadly similar conclusions.

In realistic simulations, both the CDF and DØ detectors appear capable of a di-b-jet resolution of about $\Delta m_{jj}/m_{jj} \sim 15 - 20\%$ at $m_{jj} \sim 100$ GeV. This contains contributions of about 6% from the intrinsic calorimeter resolution, about 10% from neutrinos in the b-decays, and about 10% from gluon radiation effects. Since these three effects are comparable in magnitude, it seems hard to dramatically improve the resolution. In a DØ study, jets found using cones with radii $R = 0.3, 0.5$ and 0.7 and "k_T" (successive recombination) jets with separation parameter $d = 0.4$ and 1.0 were compared; the k_T jets give slightly (about 20%) better resolution but their use is clearly not the panacea that had been hoped by some. Cutting on jet widths, or vetoing events with nearby third jets, can also improve $\Delta m_{jj}/m_{jj}$ somewhat (10–20%) but at the cost of losing up to one-third of the signal events. Energy deposition from pileup events does not seem to significantly degrade the mass resolution [4].

In summary, then, one may hope to reduce dijet resolutions by perhaps 20% compared with those obtained from the current CDF and DØ detector simulations. More dramatic improvements would be welcome but should not be regarded as particularly likely. The conclusions of this review do not depend on any improvements in dijet resolution being obtained.

DETERMINATION OF HIGGS PROPERTIES

In 30 fb^{-1} of data we expect about 200 Higgs events at $m_H = 100$ GeV. The statistical error on the mass would then be as small as ~ 1 GeV. Systematic effects would probably dominate but with a large sample of $Z \to b\bar{b}$ events available for calibration these should be controllable.

Observation of both $WH \to \ell\nu b\bar{b}$ and $ZH \to (\nu\nu, \ell\ell)b\bar{b}$ could be used to

TABLE 1. Number of signal and background events expected in 30 fb^{-1} for WH and ZH processes, and signal significance, as a function of Higgs mass.

m_H (GeV/c^2)	60	80	90	100	110	120
WH signal (S)	681	420		228		117
Background (B)	2085	1260		789		456
S/B	0.33	0.33		0.29		0.26
$S/\sqrt{B}$	14.9	11.8		8.1		5.5
ZH signal (S)			108	92	82	51
Background (B)			533	495	462	378
S/B			0.20	0.19	0.18	0.13
$S/\sqrt{B}$			4.7	4.1	3.8	2.6

fix the ratio of couplings $(WWH)^2/(ZZH)^2$ to $\pm 15\%$ [16]. If $m_H < 95$ GeV this ratio could be combined with the LEP determination of $(ZZH)^2$ and $B(H \to b\bar{b})$ in order to fix $(WWH)^2$ to about $\pm 20\%$ [16].

Unfortunately it is unlikely that a light standard model Higgs could be distinguished from the lightest SUSY Higgs on the basis of Tevatron measurements.

SUMMARY AND CONCLUSIONS

Table 1 summarizes the number of signal events S, background events B, and signal significance $S/\sqrt{B}$, attainable in the $p\bar{p} \to WH \to \ell\nu b\bar{b}$ and $p\bar{p} \to ZH \to (\ell\ell, \nu\nu)b\bar{b}$ channels. Numbers are taken from Refs. [6] and [7] and scaled to 30 fb^{-1}. By combining these two final states it appears that the whole mass range $60 < m_H \lesssim 130$ GeV can be covered at TeV33.

REFERENCES

1. LEP Electroweak Working Group, as reported at the Europhysics Conference on High Energy Physics, Jerusalem, August 1997.
2. H. Haber et al., Precision Electroweak Physics and Weakly-Coupled Higgs Bosons Working Group Report, in the proceedings of the 1996 DPF-DPB Summer Study on New Directions for High Energy Physics (Snowmass 1996).
3. A. Stange, W. Marciano and S. Willenbrock, Phys. Rev. D **49** 1354 (1994); Phys. Rev. D **50** 4491 (1994).
4. S. Kuhlmann, in TeV2000 report.
5. P. Agrawal, D. Bowser-Chao and K. Chung, Phys. Rev. D **51** 6114 (1995).
6. S. Kim, S. Kuhlmann and W.M. Yao, "Improvement of signal significance in $WH \to \ell + \nu + b + \bar{b}$ search at TeV33," in the proceedings of the 1996 DPF-DPB Summer Study on New Directions for High Energy Physics (Snowmass 1996).
7. W.M. Yao, "Prospects for Observing Higgs in $ZH \to (\nu\bar{\nu}, \ell^+\ell^-)b\bar{b}$ Channel at TeV33," in the proceedings of the 1996 DPF-DPB Summer Study on New Directions for High Energy Physics (Snowmass 1996).
8. S. Mrenna and G. Kane, CALT-68-1938 (1994).
9. U. Heintz, in TeV2000 report.
10. M. Kelly, Prospects for a $Z \to \tau\tau$ Analysis, DØ Note, unpublished.
11. J. Dai, J. Gunion and R. Vega, Phys. Lett. **B371** 71 (1996); Phys. Lett. **B387** 801 (1996).
12. R. Jesik, presentation to the Light Higgs Working Group, Snowmass 1996.
13. D. Hedin, presentation to the Light Higgs Working Group, Snowmass 1996.
14. S. Kuhlmann, presentation to the Light Higgs Working Group, Snowmass 1996.
15. B. Abbott, presentation to the Higgs Working Group, Fermilab TeV33 Workshop, 1996.
16. J. Gunion et al., "Higgs Boson Discovery and Properties," in the proceedings of the 1996 DPF-DPB Summer Study on New Directions for High Energy Physics (Snowmass 1996).

Testing 2HDM at Muon Colliders

Maria Krawczyk [1]

Institute of Theoretical Physics, Warsaw University, Warsaw, Poland

Abstract. One very light neutral Higgs scalar with mass, say, below 40-50 GeV is still allowed in the non-supersymmetric 2HDM (Model II), while the remaining particles of the Higgs sector in this model have to be heavier. The possibility of testing such a scenario at Muon Colliders is discussed.

INTRODUCTION

The Two Higgs Doublet extension of the Standard Model (SM) contains five Higgs particles: two neutral scalars h, H ($M_H > M_h$), one pseudoscalar A and two charged particles $H_\pm$, and is characterized by two parameters α and $\tan\beta$ (for the CP conserved approach), see [1]. Within this extension we study the non-supersymmetric version of the Model II, called here 2HDM. In this framework, unlike in the MSSM, [2] all the masses and α and $\tan\beta$ are independent parameters, which should be constrained one by one.

The present mass limit on the Higgs scalar in the SM is 77 GeV [2]. The non-minimal MSSM neutral Higgs bosons h and A, for the considered (as a) typical supersymmetric particle spectra, individually should be heavier than ~ 60 GeV (for $\tan\beta > 1$) [3]. In contrast to the above limits the low mass range of the neutral Higgs sector is still allowed in the non-supersymmetric case, *i.e.* in 2HDM. In particular, one very light neutral Higgs particle h or A may exist, even as light as 10 to 20 GeV yet with the $\tan\beta$ quite large, $\tan\beta \sim 25$ to 45 [4] [3]. Such an object should in principle lead to visible effects. In fact, however, such a scenario is not yet excluded, since some of the effects are surprisingly weak.

One should keep in mind that in 2HDM the mass limits for the lightest neutral Higgs bosons are of the form: $M_h + M_A \gtrsim 90$-110 GeV, therefore there is still a possibility that both of these particles are relatively light, with mass well below M_Z. Also it is worth noticing here that a very large mass gap between the neutral

---

[1] Supported in part by US-Poland Maria Sklodowska-Curie Joint Fund II (MEN/DOE-96-264).
[2] Note that the Higgs sector of the MSSM also belongs to the Model II.
[3] It may be even lighter for the appropriate limit on the allowed range of $\tan\beta$ [5,6,4].

Higgs bosons may naturally occur in 2HDM [4].

As far as the charged Higgs boson is concerned it should be at least as heavy as 330 GeV [7,8], leading in most cases to negligible contributions to the observables.

The detailed study on the Higgs boson search at Muon Colliders both for the SM and MSSM, can be found in proceedings of the Muon Colliders workshops and of this workshop, devoted to the physics at the First Muon Collider [9]. The aim of my talk is to present unique features of the 2HDM with a possibly one (very) light neutral Higgs boson.

STATUS OF THE 2HDM WITH A LIGHT HIGGS BOSON

We start the discussion of the 2HDM with the presentation of widths expected for the lightest neutral Higgs bosons h and A. These widths depend crucially on the value of $\tan\beta$, as h and A couple to fermions by the coupling $\sim (\cos(\alpha)/\sin(\beta))^{\pm}$ and $\tan\beta^{\pm}$ for the states with $I_3=(1/2)^{\mp}$. In the Fig.1 the LO results are presented for the mass range below 40 GeV, where for simplicity we assume $\alpha = \beta$. The width

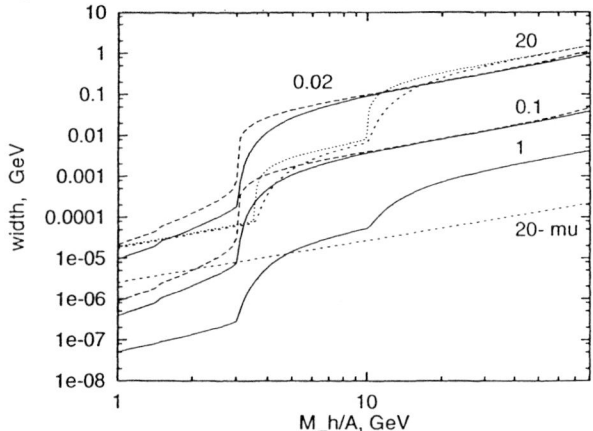

FIGURE 1. The width of the scalar Higgs boson h with $\alpha = \beta$ (solid line) and of the pseudoscalar Higgs boson A (dashed line) for $\tan\beta = 0.1$, and 0.02. The results for $\tan\beta=20$ are denoted by short dashed and dotted lines for h and A, respectively. In addition the SM prediction ($\tan\beta = 1$) and the contribution from the muonic channel for $\tan\beta = 20$ (one line for both h and A) are given.

of the SM Higgs scalar (*i.e.* for $\tan\beta=1$) is small in the whole considered mass region. For a large $\tan\beta$ the width is considerably larger, reaching for $\tan\beta=20$ and Higgs boson mass 40 GeV the value of 1 GeV. It does not mean that for small $\tan\beta$

---

[4]) In MSSM the degeneracy in M_h and M_A is expected for large $\tan\beta$.

the width is always small, see Fig.1 for results for $\tan\beta = 0.1$ and 0.02. Note that, for the mass range 3-10 GeV, the widths obtained for $\tan\beta = 20$ and 0.1 coincide. Similar effect can be found for the mass above 10 GeV, where the predictions for $\tan\beta = 20$ and 0.02 are close to each other.

Similarly to the widths also the preferred decay modes for the studied by us light neutral Higgs bosons strongly depend on the parameters of the model, mainly on $\tan\beta$. Branching ratios of h and of A can be found in Figs. 2a and 2b, where decays of the scalar and pseudoscalar are presented for two choices of $\tan\beta$, $\tan\beta = 0.1$ and $\tan\beta = 20$, respectively (also here for the simplicity we assume that $\alpha = \beta$).

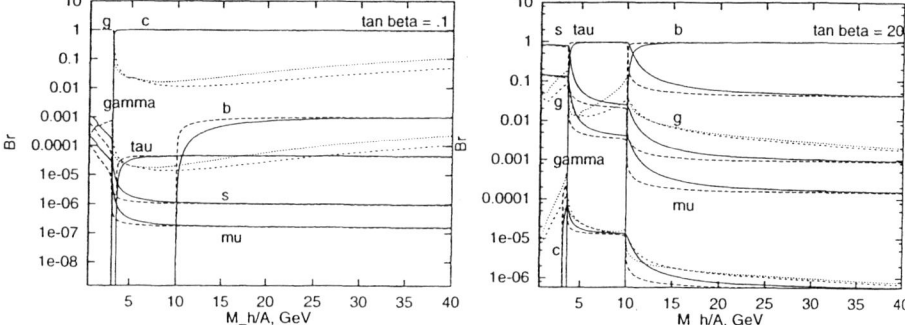

FIGURE 2. The branching ratios for the scalar h with $\alpha = \beta$ (solid and short dashed line) and for the pseudoscalar A (dashed and dotted line) for a) $\tan\beta = 0.1$, b) $\tan\beta = 20$.

The basic LEP limits on the parameters of the considered 2HDM model may be listed as follows:

- the 95% C.L. exclusion derived from the Bjorken process, $e^+e^- \to Zh$, on $\sin^2(\alpha - \beta)$ for M_h smaller than 70 GeV [10], indicates the small value of $\sin^2(\alpha - \beta)$, e.g. below 0.1 for $M_h \leq 50$ GeV.

- one neutral Higgs boson can be very light, as the excluded region from the combined results on $\sin^2(\alpha - \beta)$ and $\cos^2(\alpha - \beta)$ (from the Bjorken process and the Higgs boson pair production $e^+e^- \to hA$, respectively) has a form $M_h + M_A < 90\text{-}110$ GeV [11].

- the 95% C.L limit for $\tan\beta$ for the M_A smallar then 40 GeV [6], obtained from the Yukawa process $e^+e^- \to f\bar{f}h/A$, allows for $\tan\beta$ up to 20-25 for $M_A \sim 5\text{-}10$ GeV and up to 100 for $M_A \sim 40$ GeV.

- measurements of the process $e^+e^- \xrightarrow{Z} h/A\gamma$ for $\tau\tau$, light quarks, b-quarks decay channels performed recently at LEP collider by all experimental groups [12] lead to relatively weak limits on $\tan\beta$ in 2HDM: $\tan\beta \gtrsim 0.12$ for $M_A \sim 10$ GeV, as discussed in Ref. [13].

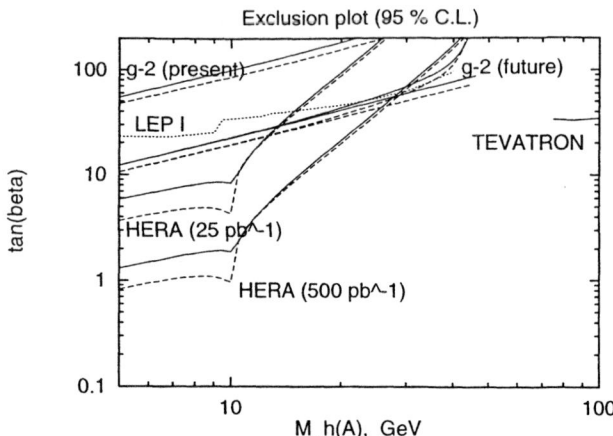

FIGURE 3. *The exclusion plot based on $g-2$ measurements, LEP data on the Yukawa process and TEVATRON data on $bb\tau\tau$, and on possible results from the ep collider HERA and from a low energy NL collider. Solid lines correspond to h ($\alpha = \beta$), dashed and dotted lines to A.*

In addition, as it was mentioned above, $M_{H_\pm}$ should be larger than 330-350 GeV as follows from the $b \to s\gamma$ data [7] and the newest NLO calculation [8]. Note that the above limits from LEP, especially from the measurement of the Yukawa process and the $Z \to h/A\gamma$ decay, may be much more stringent if the luminosity will be higher by factor 10 or so.

The present limits on $\tan\beta$ in 2HDM from LEP and TEVATRON [14], $g-2$ measurement for the muon, together with the potential of the improved measurement of $g-2$ for the muon in the E821 experiment at BNL, and of the HERA collider as well as the NL collider [4] can be found in Fig.3.

SEARCH FOR A LIGHT NEUTRAL HIGGS BOSON AT MUON COLLIDERS

The present constraints on Higgs boson masses (and other parameters) of the general 2HDM may be used to study the potential of the Muon Colliders for searching for light neutral Higgs bosons, if the tight constraints will not appear earlier.

We will concentrate below on the potential of a direct search of neutral Higgs bosons at the First Muon Collider (FMC). The energy of a collision at FMC is planned to be around the mass of the Z resonance, and in the next stage around 300-500 GeV, with the expected luminosity 1-10 fb^{-1}/yr. The end-front search will not be considered here although it may be very useful in the search of a light Higgs particle in 2HDM.

From the point of view of the potential of First Muon Collider one has to distinguish two different Higgs mass ranges allowed in 2HDM:

- *the mass of the lightest Higgs boson is below M_Z*
  This mass range is unique for the considered by us 2HDM, being excluded in the SM and in the standard scenarios discussed in the MSSM (SUSY) approaches. In such case the study of the Higgs sector similar as at LEP I collider can be performed, so obviously there is a need of a higher luminosity than the one obtained at LEP I.

  Let me first concentrate on the mass of the lightest boson *below 40-50 GeV*. In this case the analogous production mechanisms as discussed above for LEP I, with a dominating Z- boson intermediate state, are possible - in particular the Bjorken process, the Higgs pair production and the Yukawa process.

  In the case of the production of the h/A + *photon* final state an important difference is expected in comparison to the LEP I measurements. In $\mu^+\mu^-$ collision not only the loop but also the tree diagram contributes, being negligible for e^+e^- collider. Therefore in addition to the Z pole processes the resonance production of the light neutral Higgs h/A is also possible due the *return to the pole* (tree) process: $\mu^+\mu^- \to h/A\gamma$ [15,16].

  Note that in principle even *two* light Higgs bosons h and A with masses *around 40-50 GeV* may be produced at First Muon Collider.

  If the mass of h or A is *close to M_Z* the scan in the energy (the mass) planned at FMC will allow to discover the resonant object. For large $\tan\beta$ case the width of a light neutral Higgs boson is not small, also a small $\tan\beta$ (as small as 0.1) may lead to relatively large width (see Fig.1). So no extra requirement referring to the beam energy resolution is needed for such a search.

  The interesting option with two lightest neutral Higgs bosons h and A in the direct reach of the FMC, still being very different in masses (*a large mass gap*), is also open in the framework of 2HDM. In this case the corresponding processes as listed in the previous section for the LEP I collider are possible, with the Z to be interchanged by the heavier of two bosons. A very attractive possibility of having as a resonance the heavier Higgs boson with the lighter one in the final state may occur, *e.g.* for $M_A \geq M_h$ the process $\mu^+\mu^- \xrightarrow{A} f\bar{f}h$.

- *the mass of the lightest Higgs boson is above M_Z but below the energy of the collider*
  This case may have a similar signature to the MSSM Higgs bosons production. The basic difference to MSSM, beside the fact that α and β parameters are not correlated with the masses of Higgs bosons, is the possibility to have *a large mass gap* between two lightest neutral Higgs particles h and A. Both particles may in principle be found as resonances, the difference to the SM Higgs scalar case is that their widths may be large.

To summarize, the general 2HDM may lead to unique signature both in the resonant production of neutral Higgs bosons, and in the *higgsstrahlung* processes at low energy First Muon Collider. Useful option with $\sqrt{s} \sim M_Z$ requires the luminosity at least by the factor 10 higher than the one achieved at LEP.

Acknowledgments

The author is very grateful to the Organizers of the Workshop for the invitation to this interesting workshop and many important discussions. She also likes to thank for the financial support.

REFERENCES

1. J. F. Gunion et al., Higgs Hunter Guide (Addison-Wesley Publ. Company, 1990); V. Berger et al., *Phys. Rev.* D41 (1990) 3421; Y. Grossman and Z. Ligeti, *Nucl. Phys.* B426 (1994) 355
2. W. Murray, talk at the Int. Europhysics Conference on High Energy Physics, Jerusalem, August 1997
3. Y. Pan, talk at the Int. Europhysics Conference on High Energy Physics, Jerusalem, August 1997
4. For a review of recent results see M. Krawczyk, talk at the ICHEP'96, Proceedings ed. by Z. Ajduk, A. K. Wróblewski, World Scientific, p. 1460; and in HERA Workshop 1995-96, in proc. p.244
5. P. Lee-Franzini, talk at ICHEP'88, Munich; M. Narain, Ph.D Thesis, Inclusive photon spectra from Υ decays, State Univ. of New York at Stony Brook, 1991
6. ALEPH Coll., submitted to ICHEP'96, Warsaw, PA13-027
7. CLEO Coll.,S. Alam, *Phys. Rev. Lett.*74 (1995) 2885; ALEPH Coll., submitted to the Int. Europhysics Conference on High Energy Physics, Jerusalem, August 1997
8. M. Misiak, S. Pokorski and J. Rosiek, (hep-ph/9703442) M. Ciuchini et al., CERN-TH-97-279 (hep-ph/9710335v2)
9. these proceedings
10. ALEPH Coll., R. Barate et al., EPS-HEP97-748, D. Buskulic et al., *Phys. Lett.* B384 (1996) 427; DELPHI Coll., P. Abreu et al., *Nucl. Phys.* B421 (1994) 3; L3 Coll., M. Acciarri et al., *Z. Phys.* C62 (1994) 551; ICHEP' 96 PA11-016; OPAL Coll., G. Alexander et al., ICHEP'96 PA13-004
11. ALEPH Coll., EPS-0415 (1995); DELPHI Coll., submitted to the Lepton-Photon Conference, Hamburg, July-August 1997 (DELPHI 97-82 CONF 68); L3 Coll., ICHEP'96 (PA11-017); OPAL Coll., CERN-EP/98-029
12. ALEPH Coll., R.Barate et al., CERN-EP/98-022; DELPHI Coll., J.A.Barrio et al., DELPHI 95-73 PHYS 508, submitted to the EPS-HEP Conference '95; L3 Coll., M.Acciarri et al., *Phys. Lett.* B388 (1996) 409 OPAL Coll., G.Alexander et al., *Z. Phys.* C71 (1997) 1.
13. M. Krawczyk, P. Mättig and J.Żochowski - in preparation
14. in preparation, based on M. Drees et al., *Phys. Rev. Lett.* 80 (1998) 2047
15. M. Demarteau and T. Han, in these proceedings
16. D. Bowser-Chao, these proceedings; and F. Tikhonin, these proceedings

Searches for the MSSM Higgs Bosons at LEP

Thomas Greening

University of Wisconsin
Madison, WI 53706 USA

Abstract. With data recorded at 161 and 172 GeV at LEP, the limit on the mass of the lightest neutral MSSM Higgs boson is 62.5 GeV/c^2 while the limit on the charged Higgs boson is 54.5 GeV/c^2. When LEP finishes in the year 2000, each experiment expects to collect 200 pb^{-1} of data at 200 GeV. These data will improve the limit on the lightest neutral and charged Higgs bosons in the MSSM to 95 GeV/c^2 and 75 GeV/c^2 respectively, assuming that no new evidence for the MSSM Higgs bosons is found.

INTRODUCTION

The objectives of my talk are to present current LEP limits on the masses of the Minimally Supersymmetric Standard Model (MSSM) Higgs bosons using data from 1996 taken with center-of-mass energies of 161 and 172 GeV, and to predict final mass limits using all of the expected data taken by LEP through the year 2000 assuming no evidence for the MSSM Higgs bosons is found.

In the MSSM, two Higgs doublets are introduced to separately give masses to the up-type and down-type quarks. The Higgs doublets give rise to five physical states consisting of three neutral bosons — two CP-even h and H, and one CP-odd A — and a pair of charged bosons H$^\pm$. At tree level, only two parameters are necessary to determine the masses and couplings of the Higgs sector. These two parameters are chosen to be the ratio of the vacuum expectation values $\tan\beta$ and the mass of the lightest CP-even Higgs boson m_h. Radiative corrections introduce three other parameters, namely M_{SUSY}, A_t, and μ. In the stop sector, the mass scale is determined by M_{SUSY} while mixing is controlled by A_t and μ. Results are presented for $M_{SUSY} = 1\,\text{TeV}/c^2$ and for extreme configurations of stop mixing: no mixing ($A_t, \mu \ll M_{SUSY}$) and maximal mixing ($A_t - \mu/\tan\beta = \sqrt{6}M_{SUSY}$).

TABLE 1. Efficiencies, expected number of background events, number of observed candidates, and the 95% C.L. lower limits for the SM Higgs boson mass obtained by each experiment.

Experiment	Efficiency	Background	N_{obs}	m_{95}^{lim} (GeV/c^2)
ALEPH	29%	0.8	0	69.6
DELPHI	29%	4.2	2	65.9
L3	40%	7.2	6	69.3
OPAL	31%	4.1	2	68.9
Combination		16.3	10	77.5

NEUTRAL MSSM HIGGS BOSONS

The neutral MSSM Higgs Bosons can be produced at LEP via two complementary processes, the Standard Model-like Higgs-strahlung process $e^+e^- \to hZ$ and the associated pair production process $e^+e^- \to hA$, where the cross sections are proportional to $\sin^2(\beta - \alpha)$ and $\cos^2(\beta - \alpha)$ respectively. (The parameter α is the mixing angle in the CP-even sector.)

Higgs-strahlung Production: $e^+e^- \to hZ$

The four LEP collaborations have searched for the Standard Model-like Higgs-strahlung process in all of the possible decays of the Higgs boson ($b\bar{b}$, $\tau^+\tau^-$) and the Z ($\ell^+\ell^-$, $\nu\bar{\nu}$, $q\bar{q}$). Using the data recorded with center-of-mass energies of 161 and 172 GeV, no significant excess of events was observed above the background expectation by any of the four LEP experiments as can be seen in Table 1. Table 1 also shows the 95% confidence level lower limits on the Standard Model Higgs boson for each experiment, as well as the lower mass limit of 77.5 GeV/c^2 after combining the results of all four experiments [1]. Each experiment reinterpreted the results from the Standard Model Higgs search analyses in the MSSM context by reducing the number of signal events expected by the factor $\sin^2(\beta - \alpha)$.

An MSSM-specific decay is the decay of the Higgs boson into two invisible neutralinos. ALEPH and L3 have set limits of 71.2 and 69.6 GeV/c^2 respectively assuming $\sin^2(\beta - \alpha) = 1$ and that the invisible decay rate R_{inv} is 1 [2,3].

Associated Pair Production: $e^+e^- \to hA$

The four LEP experiments have also searched for hA production in the $\tau^+\tau^- b\bar{b}$ and $b\bar{b}b\bar{b}$ final states. As seen in Table 2, no excess beyond expectation was recorded by any of the experiments in these topologies. Table 2

TABLE 2. Efficiencies, expected number of background events, number of observed candidates, and the 95% C.L. lower limits for m_h obtained by each experiment.

Experiment	Efficiency	Background	N_{obs}	m_h Limit (GeV/c^2)
ALEPH	54%	0.8	0	62.5
DELPHI	36%	2.4	0	59.5
L3	75%	142.5	135	58.4
OPAL	38%	7.4	8	56.0

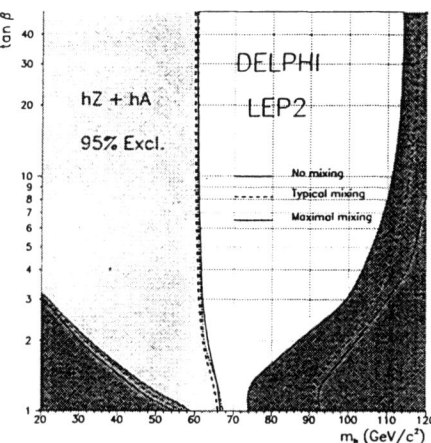

FIGURE 1. Regions in the (m_h, $\tan\beta$) plane excluded at 95% CL by DELPHI. The regions not allowed by the MSSM model are in dark grey. Limits are set for three different sets of mixing parameters.

also indicates the limit on m_h and m_A when $\cos^2(\beta - \alpha) \approx 1$ (the large $\tan\beta$ region) [3,4].

As an illustration, the excluded regions in the m_h, $\tan\beta$ plane by the DELPHI experiment are shown in Figure 1 [5]. The figure indicates three regions corresponding to different choices in the soft SUSY breaking MSSM parameters (so called minimal, maximal, and typical mixing) responsible for radiative corrections to m_h. All values of m_h below 59.5 GeV/c^2 are excluded independently of the chosen parameters. Similarly, all values of m_A are also excluded below 59.5 GeV/c^2 if $\tan\beta \geq 1$. No absolute limit, however, can be set on m_A if the $\tan\beta$ constraint is relaxed.

In addition, by systematically scanning the MSSM parameter space (squark masses and mixings), OPAL found a few unexcluded points with m_h or m_A smaller than the limit indicated by maximal mixing [3].

TABLE 3. Efficiencies, expected numbers of background events, the numbers of observed candidates, and the 95% C.L. lower limits for $m_{H^\pm}$ obtained by each experiment.

Experiment	Efficiency	Background	N_{obs}	$m_{H^\pm}$ Limit (GeV/c^2)
ALEPH	39%	23.3	14	52.0
DELPHI	31%	44.9	37	54.5
OPAL	43%	22.8	15	52.0

CHARGED HIGGS PRODUCTION: $e^+e^- \to H^+H^-$

Results of the search for pair produced charged Higgs bosons were available from ALEPH, DELPHI, and OPAL. The topologies considered were given by the charged Higgs decay into $c\bar{s}$ and/or $\tau^+\nu_\tau$. Again, no excesses were found by the three experiments, as shown in Table 3 [3]. The lower limits given in Table 3 are independent of $\tan\beta$ which is a function of the branching fraction into $c\bar{s}$ or $\tau^+\nu_\tau$. Figure 2 from ALEPH shows the charged Higgs limit as a function of the branching ratio into $\tau^+\nu_\tau$ giving the branching ratio independent limit of 52.0 GeV/c^2 [6].

In the MSSM, however, the charged Higgs boson masses are related at tree level to the W and A masses according to:

$$m_{H^\pm}^2 = m_W^2 + m_A^2.$$

Consequently, the charged Higgs mass is expected to be heavier the m_W and the current LEP limits cannot yet probe into the MSSM region of interest.

END OF LEP EXPECTATIONS

By the end of the LEP program in the year 2000, each experiment is expected to have received about 200 pb^{-1} of data with a center-of-mass energy of 200 GeV. Figure 3 shows the expected limit for the MSSM Higgs bosons in the m_h, $\tan\beta$ plane [7]. The figure indicates that a $\tan\beta$ independent limit of 95 GeV/c^2 can be achieved on m_h, while a discovery is expected if m_h is less than 90 GeV/c^2. The plot also reveals that with 200 pb^{-1} of data at 200 GeV the region near $\tan\beta \approx 1$ can be excluded, although this exclusion depends strongly upon the top mass.

A search for the charged Higgs bosons is limited by the large irreducible background from W^+W^- production coupled with the small signal cross-section. Consequently, even with more data at higher energies, it will be difficult to study the kinematic region near and above the W mass. The expected charged Higgs mass limit, therefore, at 95& C.L. at the end of LEP will be about 75 GeV/c^2.

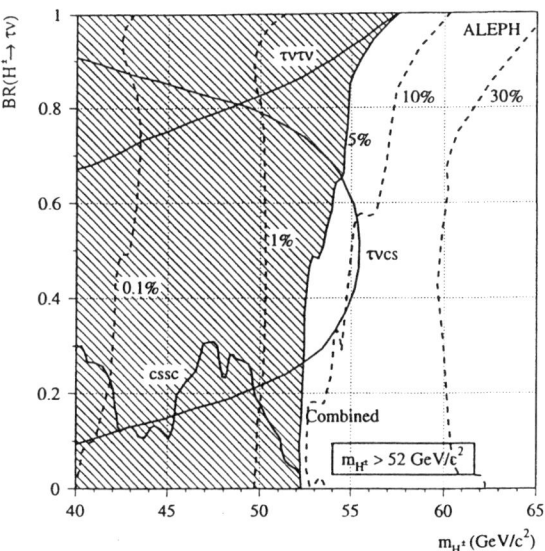

FIGURE 2. The ALEPH limit on the mass of the charged Higgs boson as a function of $\mathrm{Br}(H^+ \to \tau^+\nu_\tau)$. The three solid curves indicate the excluded domains by three separate analyses, whereas the hatched region is the excluded region for the combination.

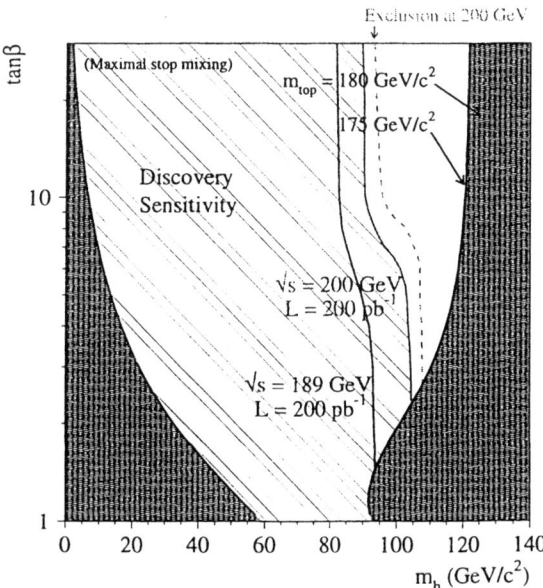

FIGURE 3. Excluded region and discovery potential in the $(m_h, \tan\beta)$ plane for the expected combined LEP search for the MSSM Higgs boson.

SUMMARY

Using the data collected at center-of-mass energies of 161 and 172 GeV, no evidence for the neutral or charged MSSM Higgs bosons was found. The best LEP limit for the neutral MSSM Higgs boson is from ALEPH, which excludes m_h below 62.5 GeV/c^2, while the best limit for the charged Higgs boson from DELPHI excludes $m_{H^\pm}$ below 54.5 GeV/c^2. Using the final LEP data estimation of 200 pb^{-1} of data per experiment at a center-of-mass energy of 200 GeV, LEP can exclude m_h below 95 GeV/c^2 and $m_{H^\pm}$ below 75 GeV/c^2. In addition, the final LEP limit would be able to exclude all neutral Higgs bosons in the low tan β region.

REFERENCES

1. LEP working group for Higgs boson searches, *Lower bound for the SM Higgs boson mass: combined result from the four LEP experiments*, CERN-LEPC/97-11.
2. L3 Collaboration, M. Acciarri et al., *Missing mass spectra in hadronic events from e^+e^- collisions at $\sqrt{s}$ = 161-172 GeV and limits on invisible Higgs decays*, CERN-PPE/97-97, submitted to Physics Letters B.
3. Janot, P., *Searches for new particles at present colliders*, EPS Conference in Jerusalem, August 1997 and references therein.
4. Private communication with M. Felcini.
5. DELPHI Collaboration, P. Abreu et al., *Search for neutral and charged Higgs bosons in e^+e^- collisions at $\sqrt{s}$ = 161 GeV and 172 GeV*, CERN-PPE/97-85, submitted to Zeit. Phys. C.
6. ALEPH Collaboration, R. Barate et al., *Search for charged Higgs bosons in e^+e^- collisions at centre-of-mass energies from 130 to 172 GeV*, CERN-PPE/97-129, submitted to Physics Letters B.
7. Private communication with P. Janot.

The Search for Higgs Bosons of Minimal Supersymmetry at the LHC

Chung Kao

Department of Physics, University of Wisconsin
Madison, Wisconsin 53706

Abstract.
The prospects for discovering neutral Higgs bosons in the minimal supersymmetric model (MSSM) and in the minimal supergravity model at the LHC are investigated. Two special discovery channels are discussed: (i) the decay mode of the MSSM CP-odd Higgs boson into photon pairs, and (ii) the decays of neutral Higgs bosons into muon pairs in the MSSM as well as in the minimal supergravity model.

INTRODUCTION

In the minimal supersymmetric model (MSSM) [1], there are two Higgs doublets ϕ_1 and ϕ_2 coupling to fermions with $t_3 = -1/2$ and $t_3 = +1/2$ respectively [2]. After spontaneous symmetry breaking, there remain five physical Higgs bosons: a pair of singly charged $H^\pm$, two neutral CP-even H^0 (heavier) and h^0 (lighter), and a neutral CP-odd A^0. The Higgs potential is constrained by supersymmetry such that all tree-level Higgs boson masses and couplings are determined by two independent parameters, commonly chosen to be mass of the CP-odd pseudoscalar (m_A) and ratio of the vacuum expectation values (VEVs) of Higgs fields ($\tan\beta \equiv v_2/v_1$).

Extensive studies have been made for the detection of MSSM Higgs bosons at the CERN LHC [3–10]. Most studies have focused on the SM decay modes $\phi \to \gamma\gamma$ ($\phi = H^0, h^0$ or A^0) and $\phi \to ZZ$ or $ZZ^* \to 4l$ ($\phi = H^0$ or h^0). For $\tan\beta$ close to one, the detection modes $A^0 \to Zh^0 \to l^+l^-b\bar{b}$ or $l^+l^-\tau\bar{\tau}$ [11] and $H^0 \to h^0h^0 \to \gamma\gamma b\bar{b}$ [9] may provide channels to simultaneously discover two Higgs bosons of the MSSM. For large $\tan\beta$, the $\tau\bar{\tau}$ decay mode [5,7–9] is a promising discovery channel for the A^0 and the H^0; neutral Higgs bosons might be observable via their $b\bar{b}$ decays [12,13]. In some regions of parameter space, the rates for Higgs boson decays to SUSY particles are dominant. While these decays reduce rates for the standard modes, they might also open up new promising modes for Higgs detection [6]. Recently, the muon pair decay mode was proposed [14,7,9] to be a promising discovery channel for neutral Higgs bosons. For large $\tan\beta$, the muon pair discovery mode might be the only channel at the LHC that allows precise reconstruction of the A^0 and the H^0 masses.

In this article, the prospects for discovering neutral Higgs bosons in the MSSM and in the minimal supergravity model (MSUGRA) at the LHC are investigated. Two special discovery channels are discussed: (i) the search for the MSSM CP-odd Higgs boson via its photon pair decay [15], and (ii) the dectection of neutral Higgs bosons via their muon pair decays in the MSSM [14] and in the MSUGRA [16].

THE PHOTON PAIR DISCOVERY CHANNEL

In this section, we present a realistic study for the observability of the MSSM CP-odd Higgs boson (A^0) via its photon pair decay mode[1] $(A^0 \to \gamma\gamma)$ with the CMS detector performance [15]. The cross section for the process of $pp \to A^0 \to \gamma\gamma + X$ is evaluated from the cross section $\sigma(pp \to A^0 + X)$ multiplied with the branching fractions of $A^0 \to \gamma\gamma$. We take $m_{\tilde{q}} = m_{\tilde{g}} = \mu = 1000$ GeV. The irreducible backgrounds considered are, (i) $q\bar{q} \to \gamma\gamma$ and (ii) $gg \to \gamma\gamma$ (Box). In addition, we consider reducible backgrounds with at least one γ in the final state, (i) $q\bar{q} \to g\gamma$, (ii) $qg \to q\gamma$, and (iii) $gg \to g\gamma$ (Box). In Figure 1, we present number of events for the signal and the background at the LHC versus $M_{\gamma\gamma}$.

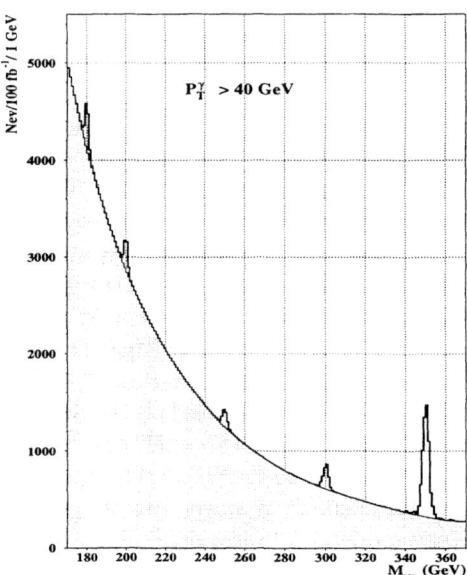

FIGURE 1. Number of events versus $M_{\gamma\gamma}$, generated from a simulation with CMS performance, for the signal and the background at $\sqrt{s} = 14$ TeV with $L = 100$ fb^{-1} and $\tan\beta = 1$.

---

[1] This important channel was not included in the CMS and the ATLAS technical proposals [7,8].

We use PYTHIA 5.7 and JETSET 7.4 generators [17] to simulate events at the particle level. The PYTHIA/JETSET outputs are processed with the CMSJET program [18]. The resolution effects are taken into account by using the parameterizations obtained from the detailed GEANT [19] simulations. The ECAL resolution is assumed to be $\sigma(E)/E = 5\%/\sqrt{E} + 0.5\%$ (CMS high luminosity regime). We require that every photon should have a transverse momentum (p_T) larger than 40 GeV and $|\eta| < 2.5$, and both photons must be isolated, i.e., (i) there is no charged particle with $p_T > 2$ GeV in the cone $R = 0.3$; and (ii) the total transverse energy $\sum E_T^{cell}$ is taken to be less than 5 GeV in the cone ring $0.1 < R < 0.3$. To be conservative, we assume no rejection power against π^0's with high p_T, i.e., all π^0's surviving the cuts (p_T, isolation, etc.) are considered as γ's. [2]

For each m_A and $\tan\beta$, the values of mass window around the peak (within the range 2-6 GeV) and p_T cut (50-100 GeV) were chosen to provide the best value of $N_S = S/\sqrt{B}$. For example, the best values of the mass window and p_T cut for $m_A = 200$ GeV are 2 GeV and 60 GeV respectively, whereas these values equal to 4 GeV and 100 GeV for $m_A = 350$ GeV. Figures 2 shows the discovery contour for $pp \to A^0 \to \gamma\gamma$ at $\sqrt{s} = 14$ TeV, in the (m_A,$\tan\beta$) plane, with an integrated luminosity (L) of 100 fb^{-1} and 300 fb^{-1}.

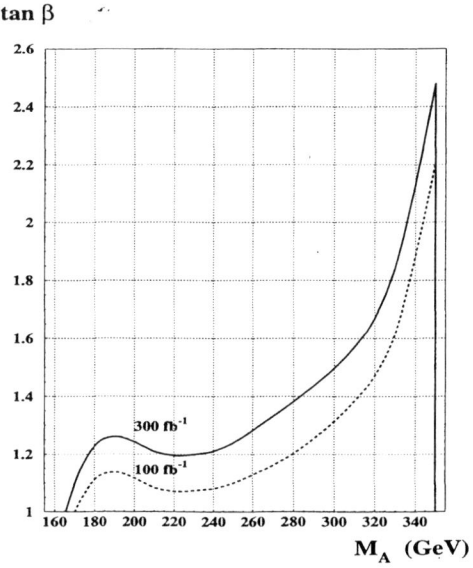

FIGURE 2. The 5σ contour in the (m_A,$\tan\beta$) plane, generated from a simulation with CMS performance, for $pp \to A^0 \to \gamma\gamma + X$ at the LHC with $L = 100$ fb^{-1} and 300 fb^{-1}.

---

[2] The background from the π^0 is overestimated, especially in the low mass $M_{\gamma\gamma}$ region.

THE MUON PAIR DISCOVERY CHANNEL

The cross section of $pp \to \phi \to \mu\bar{\mu}+X$ ($\phi = A^0, H^0$, or h^0) is evaluated from the Higgs boson cross section $\sigma(pp \to \phi+X)$ multiplied with the branching fraction of the Higgs decay into muon pairs $B(\phi \to \mu\bar{\mu})$. The Higgs masses and couplings are evaluated with one loop corrections from the top and the bottom Yukawa interactions in the one-loop effective potential [20].

In the MSSM, gluon fusion ($gg \to \phi$) is the major source of neutral Higgs bosons for $\tan\beta \lesssim 4$. If $\tan\beta$ is larger than about 10, neutral Higgs bosons are dominantly produced from b-quark fusion ($b\bar{b} \to \phi$) [21] because the $\phi b\bar{b}$ couplings are enhanced by $1/\cos\beta$. We have evaluated the cross section of Higgs bosons in pp collisions $\sigma(pp \to \phi+X)$, with two dominant subprocesses: $gg \to \phi$ and $gg \to \phi b\bar{b}$. For $m_A \gtrsim 150$ GeV, the couplings of the lighter scalar h^0 to gauge bosons and fermions become close to those of the SM Higgs boson, therefore, gluon fusion is the major source of the h^0 even if $\tan\beta$ is large.

The QCD radiative corrections to $gg \to \phi$ was found to be large [22], the same corrections to $gg \to \phi b\bar{b}$ are still to be evaluated. To be conservative, we take a K-factor of 1.5 and 1.0 for the contributions from $gg \to \phi$ and $gg \to \phi b\bar{b}$ respectively, to evaluate the cross section of $pp \to \phi+X$. For the dominant Drell-Yan background [14,7,9], we have adopted the well known K-factor from reference [23].

If the $b\bar{b}$ mode dominates Higgs decays, the branching fraction of $\phi \to \mu\bar{\mu}$ is about $m_\mu^2/3m_b^2$, where 3 is the color factor of the quarks. The QCD radiative corrections greatly reduce the decay width of $\phi \to b\bar{b}$ [24]. For $\tan\beta \gtrsim 10$, the $b\bar{b}$ decay mode dominates, and the branching fraction of $B(\phi \to \mu\bar{\mu})$ ($\phi = A^0, H^0$, or h^0) is about 2×10^{-4}. For m_A less than about 80 GeV, the H^0 decays dominantly into $h^0 h^0$, $A^0 A^0$ and ZA^0.

Higgs Bosons of Minimal Supersymmetry

In Figs. 3(a) and 3(b), we present the cross section of the MSSM Higgs bosons at the LHC, $pp \to \phi \to \mu\bar{\mu}+X$, as a function of m_A for $\tan\beta = 15$ and $\tan\beta = 40$. As $\tan\beta$ increases, the cross section is enhanced because for $\tan\beta \gtrsim 10$, it is dominated by $gg \to \phi b\bar{b}$ and enhanced by the $\phi b\bar{b}$ Yukawa coupling. Also shown is the same cross section for the SM Higgs boson h^0_{SM} with $m_{h_{SM}} = m_A$. For $m_{h_{SM}} > 140$ GeV, the SM h^0_{SM} mainly decays into gauge bosons; therefore, the branching fraction $B(h^0_{SM} \to \mu\bar{\mu})$ drops sharply.

To study the observability for the muon pair decay mode, the dominant background from the Drell-Yan (DY) process, $q\bar{q} \to Z, \gamma \to \mu\bar{\mu}$ is considered. We take $\Delta M_{\mu\bar{\mu}}$ to be the larger of the ATLAS muon mass resolution (about 2% of the Higgs bosons mass) [8,9] or the Higgs boson width.[3] The minimal cuts applied are (1) $p_T(\mu) > 20$ GeV and (2) $|\eta(\mu)| < 2.5$ for both the signal and background.

---

[3] The CMS mass resolution will be better than 2% of m_ϕ for $m_\phi \lesssim 500$ GeV [14,7].

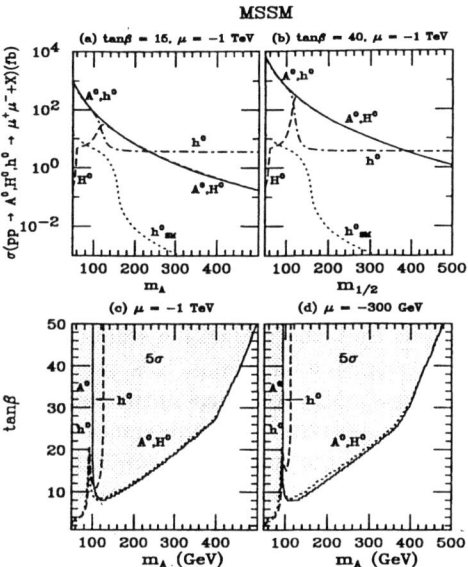

FIGURE 3. The cross sections of $pp \to A^0, H^0, h^0 \to \mu\bar{\mu} + X$ in fb at $\sqrt{s} = 14$ TeV, versus m_A for $m_{\tilde{g}} = m_{\tilde{q}} = -\mu = 1$ TeV, (a) $\tan\beta = 15$ and (b) $\tan\beta = 40$. Also shown is the cross section for the SM Higgs boson with $m_{h_{SM}} = m_A$. The 5σ contours at the LHC with $L = 300$ fb^{-1} are shown for (c) $m_{\tilde{g}} = m_{\tilde{q}} = -\mu = 1$ TeV, and (d) $m_{\tilde{g}} = m_{\tilde{q}} = -\mu = 300$ GeV.

For $m_A \gtrsim 130$ GeV, m_A and m_H are almost degenerate while for $m_A \lesssim 100$ GeV, m_A and m_h0 are very close to each other [14,7]. Therefore, we sum up the cross sections of the A^0 and the h^0 for $m_A \leq 100$ GeV and those of the A^0 and the H^0 for $m_A > 100$ GeV,

We define the signal to be observable if the 99% confidence level upper limit on the background is smaller than the corresponding lower limit on the signal plus background [3,25], namely,

$$L(\sigma_s + \sigma_b) - N\sqrt{L(\sigma_s + \sigma_b)} > L\sigma_b + N\sqrt{L\sigma_b}$$

$$\sigma_s > \frac{N^2}{L}[1 + 2\sqrt{L\sigma_b/N}] \quad (1)$$

where L is the integrated luminosity, and σ_b is the background cross section within a bin of width $\pm \Delta M_{\mu\bar{\mu}}$ centered at M_ϕ; $N = 2.32$ corresponds to a 99% confidence level and $N = 2.5$ corresponds to a 5σ signal.

The 5σ discovery contours at $\sqrt{s} = 14$ TeV and $L = 300$ fb^{-1} are shown in Figs. 3(c) and 3(d) for $m_{\tilde{q}} = m_{\tilde{g}} = -\mu = 1$ TeV and $m_{\tilde{q}} = m_{\tilde{g}} = -\mu = 300$ GeV. The discovery region of $H^0 \to \mu\bar{\mu}$ is slightly enlarged for a smaller μ, but the observable region of $h^0 \to \mu\bar{\mu}$ is slightly reduced because the lighter top squarks make the H^0 and the h^0 lighter and enhance the $H^0 b\bar{b}$ coupling while reduce the $h^0 b\bar{b}$ coupling.

Higgs Bosons of Minimal Supergravity

In the minimal supergravity model (MSUGRA) [26], it is assumed that SUSY is broken in a hidden sector with SUSY breaking communicated to the observable sector through gravitational interactions, leading naturally to a common scalar mass (m_0), a common gaugino mass ($m_{1/2}$), a common trilinear coupling (A_0) and a common bilinear coupling (B_0) at the GUT scale. Through minimization of the Higgs potential, the B parameter and magnitude of the superpotential Higgs mixing parameter μ are related to $\tan\beta$ and M_Z.

The SUSY particle masses and couplings at the weak scale can be predicted by the evolution of RGEs [27] from the unification scale [28,29]. Since A_0 mainly affects the masses of third generation sfermions, it is taken to be zero in most of our analysis. We calculate masses and couplings in the Higgs sector with one loop corrections from the top and the bottom Yukawa interactions in the RGE-improved one-loop effective potential [20] at the scale $Q = \sqrt{m_{\tilde{t}_L} m_{\tilde{t}_R}}$ [30,31]. At this scale, the RGE improved one-loop corrections approximately reproduce the dominant two loop corrections [35] to the mass of the lighter CP-even scalar (m_h).

The mass matrix of the charginos in the weak eigenstates ($\tilde{W}^\pm$, $\tilde{H}^\pm$) has the following form [28]

$$M_C = \begin{pmatrix} M_2 & \sqrt{2} M_W \sin\beta \\ \sqrt{2} M_W \cos\beta & -\mu \end{pmatrix}. \qquad (2)$$

The form of Eq. (2) establishes our sign convention for μ. Recent measurements of the $b \to s\gamma$ decay rate by the CLEO [32] and the LEP collaborations [33] excludes most of the MSUGRA parameter space for $\mu > 0$ with a large $\tan\beta$ [34]. Although we choose $\mu < 0$ in our analysis, our results and conclusions are almost independent of the sign of μ.

Figure 4 shows masses, in the case of $\mu < 0$, for neutral Higgs bosons: the lighter CP-even (h^0), the heavier CP-even (H^0) and the CP-odd (A^0). Also shown are the regions that do not satisfy the following theoretical requirements: electroweak symmetry breaking (EWSB), tachyon free, and the lightest neutralino (χ_1^0) as the lightest supersymmetric particle (LSP). The region excluded by the $m_{\chi_1^+} > 85$ GeV limit from the chargino search [36] at LEP 2 is indicated. There are a couple of interesting aspects to note: (i) an increase in $\tan\beta$ leads to a larger m_h but a reduction in m_A and m_H; (ii) increasing m_0 raises m_A, m_H and masses of the other scalars significantly.

The LHC discovery contours in the minimal supergravity model are presented in Figure 5 for (a) the $m_{1/2}$ versus $\tan\beta$ plane with $m_0 = 150$ GeV, (b) the $m_{1/2}$ versus $\tan\beta$ plane with $m_0 = 500$ GeV, (c) the $m_{1/2}$ versus m_0 plane with $\tan\beta = 15$, and (d) the $m_{1/2}$ versus m_0 plane with $\tan\beta = 40$. The discovery region is the part of the parameter space between the curve of square symbol and the dash line. The QCD radiative corrections to background from the Drell-Yan process are included.

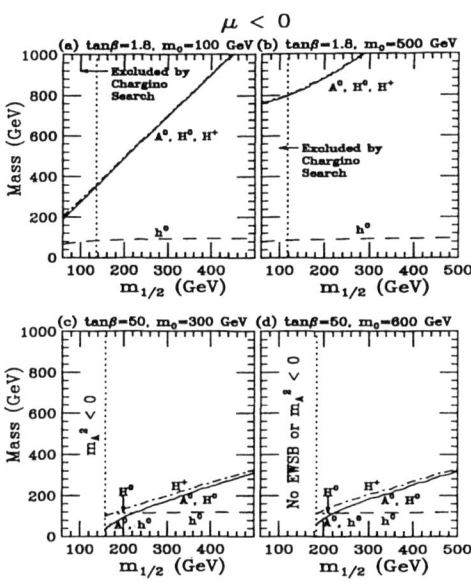

FIGURE 4. Masses of H^0, h^0, and A^0 at the mass scale $Q = \sqrt{m_{\tilde{t}_L} m_{\tilde{t}_R}}$, versus $m_{1/2}$.

CONCLUSIONS

The discovery channel of $A^0 \to \gamma\gamma$ might provide a good opportunity to precisely reconstruct the CP-odd Higgs boson mass (m_A) for 170 GeV $< m_A < 2m_t$ if the decays of the A^0 into SUSY particles are forbidden and $\tan\beta$ is close to one. The impact of SUSY decays on this discovery channel might be significant [6] and it is under investigation with realistic simulations.

The muon pair decay mode can be a very promising channel to discover the neutral Higgs bosons of minimal supersymmetry and minimal supergravity, and this mode will provide a good channel to precisely reconstruct Higgs boson masses. The A^0 and H^0 might be observable in a large region of parameter space with $\tan\beta \gtrsim 10$. The h^0 might be observable in a region with $m_A < 120$ GeV and $\tan\beta \gtrsim 5$. For $m_A \gtrsim 200$ GeV and $\tan\beta > 25$, $L = 10$ fb^{-1} would be enough to obtain Higgs boson signals with a statistical significance larger than 7 [14].

In the MSUGRA, the observable regions of the parameter space are found to be

$$m_0 = 150 \text{ GeV}: \quad m_{1/2} \lesssim 400 \text{ GeV and } \tan\beta \gtrsim 12$$
$$m_0 = 500 \text{ GeV}: \quad m_{1/2} \lesssim 1 \text{ TeV and } \tan\beta \gtrsim 28 \qquad (3)$$

For two specific choices of large $\tan\beta$, the observable regions are

$$\tan\beta = 15: \quad m_{1/2} \lesssim 200 \text{ GeV and } m_0 \lesssim 200 \text{ GeV}$$
$$\tan\beta = 40: \quad m_{1/2} \lesssim 600 \text{ GeV and } m_0 \lesssim 800 \text{ GeV}. \qquad (4)$$

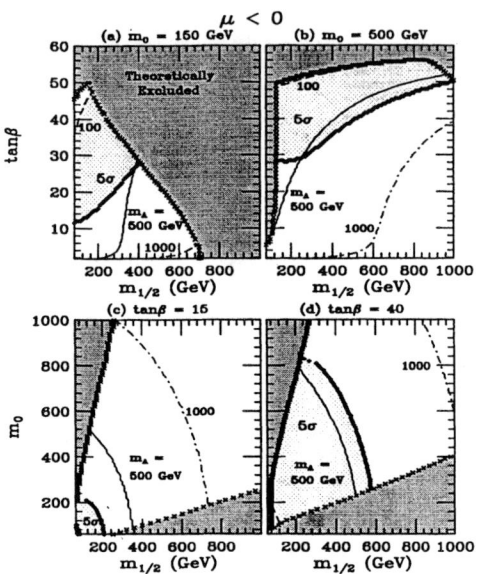

FIGURE 5. The 5σ contours for detecting Higgs bosons of MSUGRA at the LHC with $L = 300$ fb^{-1}. Also shown are (i) the mass contours for $m_A = 100$ GeV, 500 GeV and 1000 GeV, (ii) the parts of the parameter space excluded by theoretical requirements (dark shading), and (iii) the region excluded by the $m_{\chi_1^+} > 85$ GeV limit from the chargino search at LEP 2.

ACKNOWLEDGMENTS

I am grateful to Salavat Abdullin, Vernon Barger and Nikita Stepanov for enjoyable and inspiring collaborations. This research was supported in part by the U.S. Department of Energy under Grant No. DE-FG02-95ER40896, and in part by the University of Wisconsin Research Committee with funds granted by the Wisconsin Alumni Research Foundation.

REFERENCES

1. H.P. Nilles, Phys. Rep. **110** (1984) 1; H. Haber and G. Kane, Phys. Rep. **117** (1985) 75.
2. J. Gunion, H. Haber, G. Kane and S. Dawson, *The Higgs Hunter's Guide* (Addison-Wesley, Redwood City, CA, 1990).
3. H. Baer, M. Bisset, C. Kao and X. Tata, Phys. Rev. **D46** (1992) 1067.
4. V. Barger, M. Berger, A. Stange and R. Phillips, Phys. Rev. **D45** (1992) 4128; J. Gunion, R. Bork, H. Haber and A. Seiden, Phys. Rev. **D46**, 2040 (1992); J. Gunion, H. Haber and C. Kao, Phys. Rev. **D46**, 2907 (1992); J.F. Gunion and L. Orr, Phys. Rev. **D46** (1992) 2052.

5. Z. Kunszt and F. Zwirner, Nucl. Phys. **B385** (1992) 3.
6. H. Baer, M. Bisset, D. Dicus, C. Kao and X. Tata, Phys. Rev. **D47** (1993) 1062; H. Baer, M. Bisset, C. Kao and X. Tata, Phys. Rev. **D50** (1994) 316.
7. CMS Technical Proposal, CERN/LHCC 94-38 (1994).
8. Atlas Technical Proposal, CERN/LHCC 94-43 (1994).
9. E. Richter-Was, D. Froidevaux, F. Gianotti, L. Poggioli, D. Cavalli, and S. Resconi, CERN report CERN-TH-96-111, (1996).
10. Recent reviews can be found in: H. Haber, T. Han, F.S. Merritt, J. Womersley et al., in Proceedings of the 1996 DPF/DPB Summer Study on New Directions for High Energy Physics, Snowmass, CO, hep-ph/9703391; J.F. Gunion, L. Poggioli, R. Van Kooten, C. Kao, P. Rowson et al., ibid., hep-ph/9703330; H. Haber, to appear in Proceedings of the Ringberg Workshop on the Higgs Puzzle, Ringberg Castle, Germany (1996), hep-ph/9703381; J.F. Gunion, to appear in Perspectives on Higgs Physics, ed. G. Kane, 2nd edition (World Scientific Publishing), hep-ph/9705282; V. Barger, to appear in Proceedings of 5th International Conference on Supersymmetries in Physics (SUSY 97), Philadelphia, PA, 27-31 May 1997, hep-ph/9708442; and references therein.
11. S. Abdullin, H. Baer, C. Kao, N. Stepanov and X. Tata, Phys. Rev. **D54** (1996) 6728; H. Baer, C. Kao and X. Tata, Phys. Lett. **B303** (1993) 284.
12. J. Dai, J.F. Gunion and R. Vega, Phys. Lett. **B315** (1993) 355; Phys. Lett. **B345** (1995) 29; **B387** (1996) 801.
13. E. Richter-Was and D. Froidevaux, CERN report CERN-TH-97-210, (1997), hep-ph/9708455.
14. C. Kao and N. Stepanov, Phys. Rev. D **52** (1995) 5025.
15. S. Abdullin, C. Kao, and N. Stepanov, research performed for the 1996 DPF/DPB Summer Study on New Directions for High-energy Physics (Snowmass 96), Snowmass, CO, 25 Jun - 12 Jul 1996, CMS Technical Notes CMS TN/96-102, University of Wisconsin Report MADPH–96–976.
16. V. Barger and C. Kao, University of Wisconsin report MADPH-97-1020, hep-ph/9711328.
17. T. Sjöstrand, Computer Physics Commun. 39 (1986) 347; CERN-TH.7112/93; T. Sjöstrand and M. Bengtsson, Computer Physics Commun. 43 (1987) 367; H. U. Bengtsson and T. Sjöstrand, Computer Physics Commun. 46 (1987) 43.
18. S. Abdullin, A. Khanov, N. Stepanov, CMS TN/94-180 (1994).
19. R. Brun et al., GEANT3, CERN DD/EE/84-1 (1986).
20. H. Haber and R. Hempfling, Phys. Rev. Lett. **66** (1991) 1815; J. Ellis, G. Ridolfi and F. Zwirner, Phys. Lett. **B257** (1991) 83; Y. Okada, H. Yamaguchi and T. Tanagida, Prog. Theor. Phys. Lett. **85** (1991) 1; We use the calculations of M. Bisset, Ph.D. thesis, University of Hawaii (1994).
21. D. Dicus and S. Willenbrock, Phys. Rev. **D39** (1989) 751.
22. S. Dawson, Nucl. Phys. **B359** (1991) 283; A. Djouadi, M. Spira and P.M. Zerwas, Phys. Lett. **B264** (1991) 440; D. Graudenz, M. Spira and P.M. Zerwas, Phys. Rev. Lett. **70** (1993) 1372; M. Spira, A. Djouadi, D. Graudenz and P.M. Zerwas, Nucl. Phys. **B453** (1995) 17; S. Dawson, A. Djouadi and M. Spira, Phys. Rev. Lett. **77** (1996) 16.

23. V. Barger and R. Phillips, *Collider Physics, updated edition*, (Addison-Wesley Publishing Company, Redwood City, CA, 1997).
24. E. Braaten, J.P. Leveille, Phys. Rev. **D22** (1980) 715; M. Drees and K. Hikasa, Phys. Lett. **B240** (1990) 455; (E)-*ibid.* **B262** (1991) 497.
25. N. Brown, Z. Phys. **C49** (1991) 657.
26. A. Chamseddine, R. Arnowitt and P. Nath, Phys. Rev. Lett. **49**, 970 (1982); L. Ibañez and G. Ross, Phys. Lett. **B110** (1982) 215; R. Barbieri, S. Ferrara and C. Savoy, Phys. Lett. **B119**, 343 (1982); L.J. Hall, J. Lykken and S. Weinberg, Phys. Rev. **D27**, 2359 (1983); L. Alvarez-Gaumé, J. Polchinski and M. Wise, Nucl. Phys. **B121** (1983) 495.
27. K. Inoue, A. Kakuto, H. Komatsu and H. Takeshita, Prog. Theor. Phys. **68**, 927 (1982) and **71**, 413 (1984).
28. V. Barger, M.S. Berger, P. Ohmann, Phys. Rev. **D47** (1993) 1093; **D49** (1994) 4908; V. Barger, M.S. Berger, P. Ohmann and R.J.N. Phillips, Phys. Lett. **B314** (1993) 351.
29. J. Ellis and F. Zwirner, Nucl. Phys. **B338** (1990) 317; G. Ross and R.G. Roberts, Nucl. Phys. **B377** (1992) 571; R. Arnowitt and P. Nath, Phys. Rev. Lett. **69** (1992) 725; M. Drees and M.M. Nojiri, Nucl. Phys. **B369** (1993) 54; S. Kelley *et. al.*, Nucl. Phys. **B398** (1993) 3; M. Olechowski and S. Pokorski, Nucl. Phys. **B404** (1993) 590; G. Kane, C. Kolda, L. Roszkowski and J. Wells, Phys. Rev. **D49** (1994) 6173; D.J. Castaño, E. Piard and P. Ramond, Phys. Rev. **D49** (1994) 4882; W. de Boer, R. Ehret and D. Kazakov, Z. Phys. **67** (1995) 647; H. Baer, M. Drees, C. Kao, M. Nojiri and X. Tata, Phys. Rev. D **50** (1994) 2148; H. Baer, C.-H. Chen, R. Munroe, F. Paige and X. Tata, Phys. Rev. D **51** (1995) 1046.
30. H. Baer, C.-H. Chen, M. Drees, F. Paige and X. Tata, Phys. Rev. Lett. **79** (1997) 986.
31. V. Barger and C. Kao, University of Wisconsin report, MADPH-97-992, (1997), hep-ph/9704403, to be published in Phys. Rev. D.
32. M.S. Alam et al., (CLEO Collaboration), Phys. Rev. Lett. **74** (1995) 2885.
33. P.G. Colrain and M.I. Williams, talk presented at the International Europhysics Conference on High Energy Physics, Jerusalem, Israel, August 1997.
34. P. Nath and R. Arnowitt, Phys. Lett. **B336** (1994) 395; Phys. Rev. Lett. **74** (1995) 4592; Phys. Rev. **D54** (1996) 2374; F. Borzumati, M. Drees and M. Nojiri, Phys. Rev. **D51** (1995) 341; H. Baer and M. Brhlik, Phys. Rev. **D55** (1997) 3201.
35. M. Carena, J.R. Espinosa, M. Quiros, and C.E.M. Wagner, Phys. Lett. **B355** (1995) 209; M. Carena, M. Quiros, C.E.M. Wagner, Nucl. Phys. **B461** (1996) 407; H. Haber, R. Hempfling and A. Hoang, CERN-TH/95-216 (1996), hep-ph/9609331.
36. ALEPH collaboration, talk presented at CERN by G. Cowan, February, 1997.

Higgs Resonance Studies at the First Muon Collider*

Basim Kamal, William J. Marciano and Zohreh Parsa

Physics Department, Brookhaven National Laboratory, Upton, New York 11973

Abstract. Higgs resonance signals and backgrounds at the First Muon Collider are discussed. Effects due to beam polarization and background angular distributions (forward-backward charge asymmetries) are examined. The utility of those features for improving precision measurements and narrow resonance "discovery" scans is described.

If the standard model Higgs boson has a mass $\lesssim 160$ GeV (i.e. below the W^+W^- decay threshold), it will have a very narrow width and can be resonantly studied in the s-channel via $\mu^-\mu^+ \to H$ production at the First Muon Collider (FMC). Within the framework of supersymmetry or more general two Higgs doublet scenarios, there can be several neutral spin zero bosons; h, H, and A, all of which might be resonantly produced. The lightest scalar, h, of supersymmetry is expected to be $\lesssim 150$ GeV (and narrow), with the range 80–130 GeV favored. Precision electroweak measurements also tend to suggest, via quantum loop sensitivity, a relatively light Higgs. Hence, there are strong motivations to examine the capabilities of the FMC for producing and studying relatively light scalar resonances [1,2].

A strategy for "light" Higgs physics studies would be to first discover the Higgs particle at LEPII, the Tevatron, or the LHC and then thoroughly scrutinize its properties on resonance at the FMC. There, one would hope to precisely determine the Higgs mass, width, and primary decay rates [3]. Besides those interesting physics studies, such an initiative would provide a nice testing ground for muon collider technology and lay the foundation for future much higher energy facilities.

The FMC Higgs resonance program would entail two stages: 1) "Discovery" via an energy scan which pinpoints the precise resonance position and (perhaps) determines its width. Since pre-FMC efforts may only determine the Higgs mass to $\mathcal{O}(200 \text{ MeV})$ or worse and its width is expected to be narrow $\mathcal{O}(1 \sim 30 \text{ MeV})$ for $m_H \lesssim 160$ GeV, the resonance scan may be very time consuming [3]. 2) Precision measurements of the primary Higgs decay modes. Deviations from standard model

---

*) Supported by U.S. Department of Energy contract number DE-AC02-76CH00016.

TABLE 1. Expected signals and backgrounds (fully integrated) for a standard model Higgs with $m_H = 110$ GeV, $\Gamma_H \simeq 3$ MeV. Muon collider resonance conditions with no polarization, $\Delta E/E \simeq 3 \times 10^{-5}$, and $L = 0.05$ fb^{-1} are assumed. The total number of Higgs scalars produced is ~ 3000. Realistic efficiency and acceptance cuts are likely to dilute signal and backgrounds for $b\bar{b}$ and $c\bar{c}$ by a 0.5 factor.

$H \to$	$b\bar{b}$	$c\bar{c}$	$\tau\bar{\tau}$
N_S (events)	2400	210	270
N_B (events)	2520	2416	945
$\pm\sqrt{N_S + N_B}/N_S$	±0.03	±0.24	±0.13

expectations could point to additional Higgs structure or elucidate the framework of supersymmetry [3]. (Expectations for $m_H = 110$ GeV are illustrated in Table 1.)

The Higgs resonance "discovery" capability and scan time will depend on $N_S/\sqrt{N_B}$, where N_S is the Higgs signal and N_B is the expected background. The precision measurement sensitivity will be determined by $N_S/\sqrt{N_B + N_S}$. For both, it will be extremely important to enhance the signal and suppress backgrounds as much as possible. To that end, one should employ highly resolved $\mu^+\mu^-$ beams with a very small energy spread. The proposed $\Delta E/E \simeq 3 \times 10^{-5}$ is well matched to the narrow Higgs width. It allows $N_S/N_B \sim \mathcal{O}(1)$ for the primary $H \to b\bar{b}$ mode (see Table 1). Unfortunately, high resolution is accompanied by luminosity loss. The current goal of $\mathcal{L}_{\text{ave}} \simeq 5 \times 10^{30}$cm^{-2}s^{-1} on resonance is probably not ambitious enough. One should strive for another order of magnitude in luminosity while maintaining the outstanding beam energy resolution.

In this paper, we examine two additional ways of enhancing the Higgs signal to background ratio: beam polarization and final state angular distributions. The Higgs signal $\mu^-\mu^+ \to H \to f\bar{f}$ results from left-left (LL) or right-right (RR) beam polarizations and leads to an isotropic (i.e. constant) $f\bar{f}$ signal in $\cos\theta$ (the angle between the μ^- and f). Standard model backgrounds $\mu^-\mu^+ \to \gamma^*$ or $Z^* \to f\bar{f}$ result from LR or RL initial state polarizations and give rise to $(1 + \cos^2\theta + \frac{8}{3}A_{FB}\cos\theta)$ angular distributions. Similar statements apply to WW^* and ZZ^* final states, but those modes will not be discussed here.

To illustrate the difference between signal, $\mu^-\mu^+ \to H \to f\bar{f}$, and background, $\mu^-\mu^+ \to \gamma^*$ or $Z^* \to f\bar{f}$, we give the combined differential production rate with respect to $x \equiv \cos\theta = 4\mathbf{p}_{\mu^-} \cdot \mathbf{p}_f/s$ for polarized muon beams and fixed luminosity

$$\frac{dN(\mu^-\mu^+ \to f\bar{f})}{dx} = \frac{1}{2}N_S(1 + P_+P_-) \qquad (1)$$
$$+ \frac{3}{8}N_B[1 - P_+P_- + (P_+ - P_-)A_{LR}](1 + x^2 + \frac{8}{3}xA_{eff}).$$

$P_+(P_-)$ is the $\mu^+(\mu^-)$ polarization with $P = -1$ pure left-handed, $P = +1$ pure right handed, and $P = 0$ unpolarized. N_S is the fully integrated $(-1 < x \leq 1)$

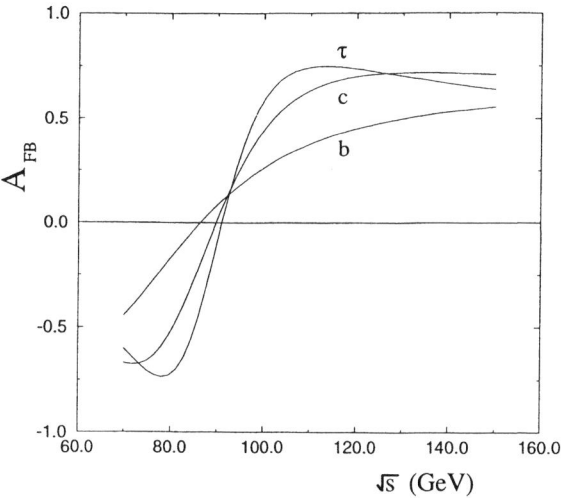

FIGURE 1. Forward-backward asymmetry for $\mu^-\mu^+ \to f\bar{f}$.

Higgs signal and N_B the integrated background for the case of unpolarized beams, $P_+ = P_- = 0$. In that general expression,

$$A_{LR} \equiv \frac{\sigma_{LR \to LR} + \sigma_{LR \to RL} - \sigma_{RL \to RL} - \sigma_{RL \to LR}}{\sigma_{LR \to LR} + \sigma_{LR \to RL} + \sigma_{RL \to RL} + \sigma_{RL \to LR}}, \tag{2}$$

where, for example, $LR \to LR$ stands for $\mu_L^- \mu_R^+ \to f_L \bar{f}_R$. The effective forward-backward asymmetry is given by

$$A_{eff} = \frac{A_{FB} + P_{eff} A_{LR}^{FB}}{1 + P_{eff} A_{LR}}, \tag{3}$$

with

$$P_{eff} = \frac{P_+ - P_-}{1 - P_+ P_-}, \tag{4}$$

$$A_{FB} = \frac{3}{4} \frac{\sigma_{LR \to LR} + \sigma_{RL \to RL} - \sigma_{RL \to LR} - \sigma_{LR \to RL}}{\sigma_{LR \to LR} + \sigma_{RL \to RL} + \sigma_{RL \to LR} + \sigma_{LR \to RL}}, \tag{5}$$

$$A_{LR}^{FB} = \frac{3}{4} \frac{\sigma_{LR \to LR} + \sigma_{RL \to LR} - \sigma_{LR \to RL} - \sigma_{RL \to RL}}{\sigma_{LR \to LR} + \sigma_{RL \to LR} + \sigma_{LR \to RL} + \sigma_{RL \to RL}}. \tag{6}$$

and the $\mu_i^- \mu_j^+ \to f_{i'} \bar{f}_{j'}$, cross sections ($i \neq j$) are to lowest order

$$\sigma_{ij \to i'j'} = (N_C)\sigma_0 \left[1 - \frac{s}{m_W^2} \left(1 + T_{3\mu_i} - \frac{T_{3f_{i'}}}{Q_f} \left(1 + \frac{T_{3\mu_i}}{\sin^2\theta_W} \right) \right) \right]^2, \tag{7}$$

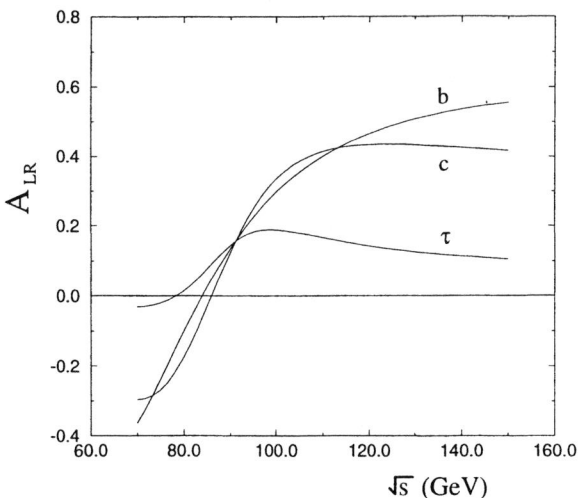

FIGURE 2. Left-right asymmetry for $\mu^-\mu^+ \to f\bar{f}$.

$$T_{3\mu_L} = T_{3\tau_L} = T_{3b_L} = -T_{3c_L} = -1/2,$$
$$T_{3f_R} = 0, \quad Q_\tau = 3Q_b = -\frac{3}{2}Q_c = -1 \qquad (N_C = 3 \text{ for } f = b, c).$$

Realistic cuts, efficiencies, systematic errors etc, will not be considered. They are likely to dilute the $b\bar{b}$ and $c\bar{c}$ event rates by a factor of 0.5. In addition, we ignore the radiative Z production tail under the assumption such events are vetoed.

The (unpolarized) forward-backward asymmetries are illustrated in Fig. 1. Note that A_{FB} is large (near maximal) for $\tau\bar{\tau}$ and $c\bar{c}$ in the region of interest. As we shall see, that feature can help in discriminating signal from background.

In principle, large polarization can be important for enhancing "discovery" and precision measurement sensitivity for the Higgs. From Eq. (1), we find that $N_S/\sqrt{N_B}$ is enhanced (for integrated signal and background) by the factor

$$\kappa_{\text{pol}} = \frac{1 + P_+P_-}{\sqrt{1 - P_+P_- + (P_+ - P_-)A_{LR}}}, \tag{8}$$

where the A_{LR} are shown in Fig. 2. That result generalizes the $P_+ = P_-$ case [4]. For natural beam polarization [1], $P_+ = P_- = 0.2$ (assuming spin rotation of one beam), the enhancement factor is only 1.06. For larger polarization, $P_+ = P_- = 0.5$, one obtains a 1.44 enhancement factor (statistically equivalent to about a factor of 2 luminosity increase). Unfortunately, obtaining 0.5 polarization simply by muon energy cuts reduces each beam intensity [1] by a factor of 1/4, resulting in a

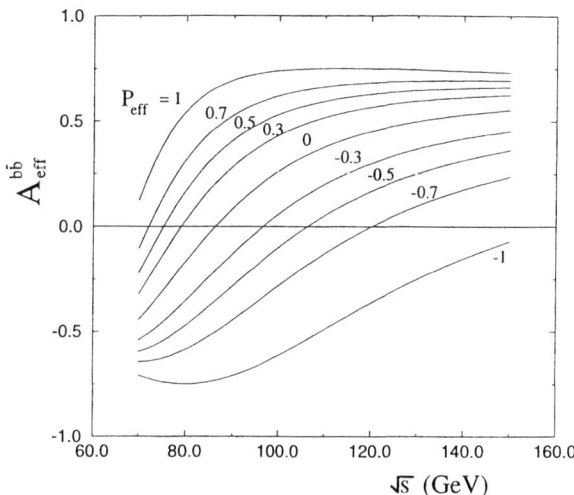

FIGURE 3. Effective forward-backward asymmetry for $\mu^-\mu^+ \to b\bar{b}$.

luminosity reduction by 1/16. Such a tradeoff is clearly unacceptable. Polarization will be a useful tool in Higgs resonance studies only if high polarization is achievable with little luminosity loss. Tau final state polarizations can also be used to help improve the $H \to \tau\bar{\tau}$ measurement, but will not be discussed here.

Some "discovery" or sensitivity enhancement can also be obtained from angular discrimination. A proper study would include detector acceptance cuts and maximum likelihood fits. Here, we wish to only crudely approximate the gain. For that purpose, we assume perfect (infinitesimal) binning and obtain the measurement sensitivity enhancement factor

$$\frac{1}{2}(1 + P_+ P_-)\sqrt{N_S + N_B}\left[\int \frac{dx}{dN/dx}\right]^{1/2}, \qquad (9)$$

which becomes, from Eq. (1),

$$\kappa_{\text{pol}}\sqrt{\frac{2}{3}}\sqrt{\frac{N_S + N_B}{N_B}}\left(\frac{\tan^{-1}\left(\frac{2}{\zeta}\sqrt{1 - \frac{16}{9}A_{eff}^2} + \zeta\right)}{\sqrt{1 - \frac{16}{9}A_{eff}^2} + \zeta}\right)^{1/2}, \quad \zeta \equiv \frac{4}{3}\frac{N_S}{N_B}\frac{\kappa_{\text{pol}}^2}{1 + P_+ P_-}. \quad (10)$$

For $A_{eff} \simeq 3/4$, $\zeta \simeq 0.38$ (which roughly applies to $\tau\bar{\tau}$) and $P_+ = P_- = 0$, one finds a sensitivity enhancement of 1.33. That means the $\pm 13\%$ statistical error in Table 1 would be reduced to $\pm 10\%$. Similar sensitivity enhancements apply to $c\bar{c}$. In the case of $H \to b\bar{b}$, the primary discovery mode, $A_{eff} \simeq 0.4$ and one finds only a 3%

enhancement. One can increase the effective $b\bar{b}$ forward-backward asymmetry via polarization (see Fig. 3). However, one must again confront the issue of luminosity loss.

In the case of "discovery", a very large forward-backward asymmetry (near maximal) can, in principle, significantly reduce the scan time. For the highly idealized coverage and binning assumed above, the time is reduced by the factor

$$\frac{1}{\kappa_{pol}^2}\frac{3}{\pi}\sqrt{1-\frac{16}{9}A_{eff}^2}\;. \tag{11}$$

Of course, that naive formula must be corrected for realistic acceptances, efficiencies, etc.; so, it should not be taken too literally (particularly for $A_{eff} \simeq 3/4$). Nevertheless, applying it to the $b\bar{b}$ discovery mode with "natural" $P_+ = P_- = 0.2$ and $A_{eff} \simeq 0.37$ gives a scan reduction time factor of 0.74.

The $H \to \tau\bar{\tau}$ "discovery" time is about 15 times longer than that of the $b\bar{b}$ (with efficiencies) for fully integrated signals. Employing $A_{FB} \simeq 0.743$ and assuming tau detection down to about $15°$ from the beams, reduces that time by about a factor of $6 \sim 7$, making it somewhat less than 1/2 as effective as $b\bar{b}$. Using both together along with all background angular information should, therefore, reduce the scan time by almost a factor of 2 compared to using the integrated $b\bar{b}$ signal alone. Such a reduction would be extremely welcome, particularly if the luminosity is less than expected.

In conclusion, we have shown that polarization is potentially useful for Higgs resonance studies, but only if the accompanying luminosity reduction is not significant. Large forward-backward asymmetries can also be used to enhance the Higgs "discovery" signal or improve precision measurements, particularly for $\tau\bar{\tau}$. However, to make the s-channel Higgs "factory" a compelling facility, one must focus on attaining the outstanding beam resolution assumed here and maintaining the highest luminosity possible.

REFERENCES

1. Muon Collider Feasibility Study, BNL Report 52503 (1996).
2. Cline, D., "The Problems and Physics Prospects for a $\mu^+\mu^-$ Collider", in *Future High Energy Colliders*, edited by Z. Parsa, AIP Conference Proceedings **397**, 1997, pp. 203–218.
3. Barger, V., Berger, M.S., Gunion, J.F., and Han, T., "The Physics Capabilities of $\mu^+\mu^-$ Colliders", in *Future High Energy Colliders*, edited by Z. Parsa, AIP Conference Proceedings **397**, 1997, pp. 219–233; *Phys. Rep.* **286**, 1–51 (1997); *Phys. Rev. Lett.* **75**, 1462–1465 (1995).
4. Parsa, Z., (unpublished).

Precision W-Boson and Higgs Boson Mass Determinations at Muon Colliders

M. S. Berger

Indiana University
Bloomington Indiana 47405

Abstract. Precise determinations of the masses of the W boson and of the top quark could stringently test the radiative structure of the Standard Model (SM) or provide evidence for new physics. We analyze the excellent prospects at a muon collider for measuring M_W and m_t in the W^+W^- and ZH threshold regions. With an integrated luminosity of 10 (100) fb^{-1}, the W-boson mass could be measured to a precision of 20 (6) MeV, provided that theoretical and experimental systematics are understood. A measurement of $\Delta M_W = 6$ MeV would constrain the mass of a ~ 100 GeV Higgs to about ± 10 GeV. We demonstrate that a measurement at future colliders of the Bjorken process With an integrated luminosity of 100 fb^{-1} it is possible to measure the Standard Model Higgs mass to within 45 MeV at a $\mu^+\mu^-$ collider for $m_h = 100$ GeV.

INTRODUCTION

Muon colliders offer a wide range of opportunities for exploring physics within and beyond the Standard Model (SM). An important potential application of these machines is the precision measurement of particle masses, widths and couplings. We estimate here the accuracy with which the W and Higgs boson masses can be determined from W^+W^- and ZH threshold measurements at a muon collider [1,2].

M_W MEASUREMENT AT THE $\mu^+\mu^- \to W^+W^-$ THRESHOLD

A muon collider is particularly well suited to the threshold measurement because the energy of the beam has a very narrow spread. The threshold cross section [3,4] is most sensitive to M_W just above $\sqrt{s} = 2M_W$, but a tradeoff exists between maximizing the signal rate and the sensitivity of the cross section to M_W. Detailed analysis [5] shows that if the background level is small and systematic uncertainties in efficiencies are not important, then the optimal measurement of M_W is obtained by collecting data at a single energy

$$\sqrt{s} \sim 2M_W + 0.5 \text{ GeV} \sim 161 \text{ GeV},$$

where the threshold cross section is sharply rising.

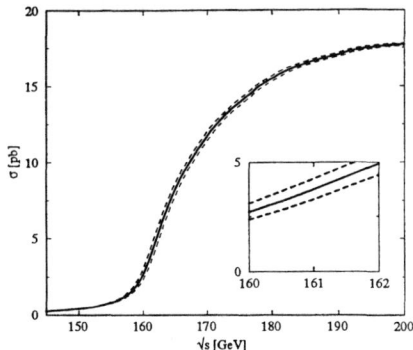

FIGURE 1. The cross section for $\mu^+\mu^- \to W^+W^-$ in the threshold region for $M_W = 80.3$ GeV (solid) and $M_W = 80.1, 80.5$ GeV (dashed). The inlaid graph shows the region of the threshold curve where the statistical sensitivity to M_W is maximized. Effects of ISR have been included.

For a LEP2 measurement with 100 pb^{-1} of integrated luminosity the background and systematic uncertainties are, in fact, sufficiently small that the error for M_W will be limited by the statistical uncertainty of the measurement at $\sqrt{s} = 161$ GeV. But, at a muon collider at high luminosity, systematic errors arising from uncertainties in the background level and the detection/triggering efficiencies will be dominant unless some of the luminosity is devoted to measuring the level of the background (which automatically includes somewhat similar efficiencies) at an energy below the W^+W^- threshold. Then, assuming that efficiencies for the background and W^+W^- signal are sufficiently well understood that systematic uncertainties effectively cancel in the ratio of the above-threshold to the below-threshold rates, a very accurate M_W determination becomes possible.

The dominant background derives from $e^+e^- \to (Z/\gamma)(Z/\gamma)$ which is essentially energy independent [5] below 180 GeV. For our present analysis we model the background as energy independent, and accordingly assume that one measurement at an energy in the range 140 to 150 GeV suffices to determine the background.

We analyze our ability to determine the W mass via just two measurements: one at center of mass energy $\sqrt{s} = 161$ GeV, just above threshold, and one at $\sqrt{s} = 150$ GeV. The optimal M_W measurement is obtained by expending about two-thirds of the luminosity at $\sqrt{s} = 161$ GeV and one-third at $\sqrt{s} = 150$ GeV. Combining the three modes, an overall precision of

$$\Delta M_W = 6 \text{ MeV} \tag{1}$$

should be achievable.

The combination of the measurements of the masses M_Z, M_W and m_t to such high precision has dramatic implications for the indirect prediction of the mass of the Higgs boson and for other sources of physics beyond the Standard Model. This is illustrated in Fig. 2. Assuming the current central values of M_W, m_t, $\alpha(M_Z)$ and $\alpha_s(M_Z)$, and that $L = 10$ fb^{-1} (100 fb^{-1}) is devoted to the measurement of m_t (M_W), the mass of the SM Higgs boson would be determined to be 260 GeV with an error of about ± 5 GeV from $\Delta m_t = 200$ MeV at a fixed M_W, and about ± 20 GeV from $\Delta M_W = 6$ MeV at a fixed m_t. For $m_h = 100$ GeV, the corresponding values would be ± 2 GeV and ± 10 GeV, respectively. More generally, the Δm_h value scales roughly like m_h.

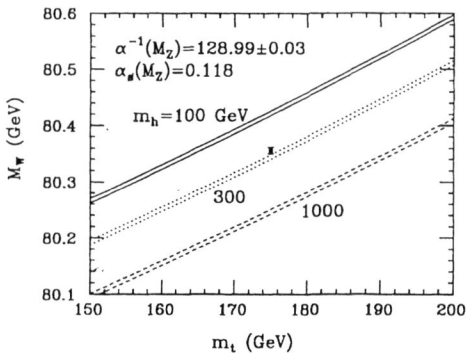

FIGURE 2. Correlation between M_W and m_t in the SM with QCD and electroweak corrections for $m_h = 100, 300$ and 1000 GeV. The data point and error bars illustrate the possible accuracy for the indirect m_h determination assuming $M_W = 80.356 \pm 0.006$ GeV and $m_t = 175 \pm 0.2$ GeV. The widths of the bands indicate the uncertainty in $\alpha(M_Z)$.

HIGGS BOSON MEASUREMENT AT THE $\mu^+\mu^- \to Zh$ THRESHOLD

A very accurate determination of m_h is obtained by measuring the threshold cross section for the Bjorken Higgs-strahlung process [6] $\ell^+\ell^- \to Zh$. With integrated luminosity $L = 100$ fb^{-1}, a 1σ precision of order 45 MeV is possible for a SM Higgs $m_h = 100$ GeV at a $\mu^+\mu^-$ collider. This error in m_h is smaller than that achievable via final state mass reconstruction for a typical detector, and would then be the most accurate determination of m_h at an e^+e^- collider.

The SM Higgs boson is easily discovered in the Zh production mode by running the machine well above threshold, e.g. at $\sqrt{s} = 500$ GeV. For $m_h \lesssim 2M_W$ the dominant Higgs boson decay is to $b\bar{b}$ and most backgrounds can be eliminated by b-tagging. The best means for measuring m_h will be to first determine m_h to within a few hundred MeV in $\sqrt{s} = 500$ GeV running, which will also yield a precise measurement of $\sigma(Zh)$, and then reconfigure the collider for maximal luminosity in the threshold energy region $\sqrt{s} \approx M_Z + m_h$.

In Fig. 3 we show the cross section for the Bjorken process $\ell^+\ell^- \to Zh$ for Higgs masses from 50 to 150 GeV. Since the threshold behavior is S-wave, the rise in the cross section (which is a few tenths of a pb) in the threshold region is rapid.

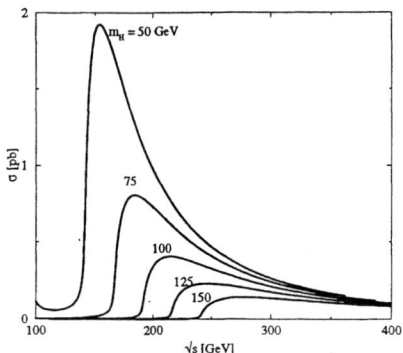

FIGURE 3. The cross section vs. $\sqrt{s}$ for the process $\mu^+\mu^- \to Z^\star h \to f\bar{f}h$ for a range of Higgs masses.

In the ideal case that the normalization of the measured Zh cross section as a function of $\sqrt{s}$ can be precisely predicted, including efficiencies and systematic effects, sensitivity to the SM Higgs boson mass is maximized by a single measurement of the cross section at $\sqrt{s} = M_Z + m_h + 0.5$ GeV, just above the real particle threshold. As an example of the precision that might be achieved, suppose $m_h = 100$ GeV and backgrounds are neglected. The Zh cross section is 120 fb and is rising at a rate of 0.05 fb/MeV. With $L = 50$ fb^{-1} and including an overall (b-tagging, geometric and event identification) efficiency of 40%, this yields 2.4×10^3 events, or a measurement of the cross section to about 2%. From the slope of the cross section one concludes that a m_h measurement with accuracy of roughly 50 MeV is possible.

For a more precise estimate of the accuracy with which m_h can be measured, we employ b-tagging and cuts in order to reduce the background to a very low level. These cuts and other systematic uncertainties are discussed in more detail in Ref. [2]. The background is very much smaller than the signal unless m_h is close to M_Z. We note that electroweak radiative corrections to the cross section are estimated to be less than 1% for $m_H \sim 100$ GeV [10,11]. A precision of the SM Higgs mass determination to within 45 MeV for $m_h = 100$ GeV may be achievable at a muon collider.

Outside the Standard Model the cross section generally depends on the Higgs mass, the ZZH coupling (g_{ZZh}) and the total Higgs width (Γ_H). In order to simultaneously determine m_h, g_{ZZh} and Γ_H, measurements could be made at the three c.m. energies $\sqrt{s} = m_h + M_Z + 20$ GeV, $\sqrt{s} = m_h + M_Z + 0.5$ GeV, and $\sqrt{s} = m_h + M_Z - 2$ GeV. The solid curves in Fig. 4 show the statistical precision that can be obtained for $m_h = 100$ GeV in a three-parameter fit to m_h, $g^2_{ZZh}B(h \to b\bar{b})$ and Γ_H, including smearing effects from bremsstrahlung, beamstrahlung and beam energy spread at a muon collider, using an integrated luminosity of $100/3$ fb^{-1} at each of the above three values of $\sqrt{s}$ in the threshold region. The crosses in the center of the ellipses indicate the input values. With a three-parameter fit, the attainable error in m_h is about 110 MeV at the 1σ level. The second panel in Fig. 4 shows that there is significant sensitivity to the Higgs width Γ_H if it is of order 100 MeV. If Γ_H is very narrow (~ 3 MeV is predicted in the SM) then $\Delta m_h \sim \pm 80$ MeV is possible (see the dashed ellipse) from a fit to m_h and $\sigma(Zh)B(h \to b\bar{b})$ by devoting 50 fb^{-1} at each of two c.m. energies, $\sqrt{s} = m_H + M_Z + 20$ GeV and $\sqrt{s} = m_H + M_Z + 0.5$ GeV. Measurements that would simultaneously determine m_h, $\sigma(Zh)B(h \to b\bar{b})$ and Γ_H could be done at a level of accuracy that could distinguish a Standard Model Higgs boson from its many possible (e.g. supersymmetric) extensions.

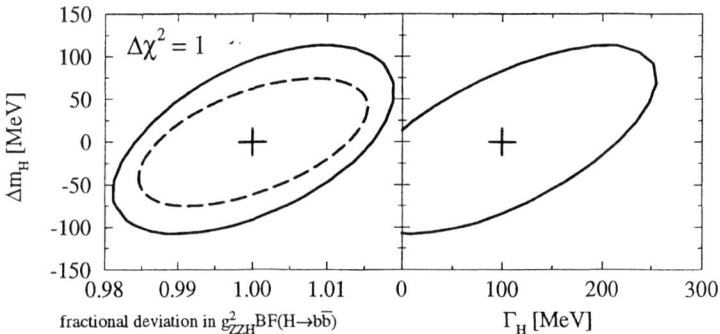

FIGURE 4. Solid curves show the $\Delta\chi^2 = 1$ contours for determining the Higgs mass versus $g^2_{ZZh}B(h \to b\bar{b})$, or versus Γ_H, by devoting $100/3$ fb^{-1} to each of the c.m. energies $\sqrt{s} = M_Z + m_h + 0.5$ GeV, $\sqrt{s} = M_Z + m_h + 20$ GeV and $\sqrt{s} = M_Z + m_h - 2$ GeV at a muon collider; b-tagging and cuts 1)–4) are imposed and initial state radiation and beam energy smearing are included. A Higgs mass $m_H = 100$ GeV is assumed. The dashed curve shows the $\Delta\chi^2 = 1$ contour that results when Γ_H is negligibly small and 50 fb^{-1} is devoted to each of the c.m. energies $\sqrt{s} = m_H + M_Z + 0.5$ GeV and $\sqrt{s} = m_H + M_Z + 20$ GeV.

CONCLUSION

A muon collider offers an unparalleled opportunity for precision W and Higgs mass measurements in the respective threshold regions. These measurements, however, require a collider that can deliver substantial luminosity.

ACKNOWLEDGMENTS

I thank V. Barger, J. F. Gunion and T. Han for a pleasant collaboration on the issues reported here. This work was supported in part by the U.S. Department of Energy under Grant No. DE-FG02-91ER40661.

REFERENCES

1. V. Barger, M.S. Berger, J.F. Gunion and T. Han, Phys. Rev. **D56**, 1714 (1997).
2. V. Barger, M. S. Berger, J. F. Gunion and T. Han, Phys. Rev. Lett. **78**, 3991 (1997).
3. T. Muta, R. Najima and S. Wakaizumi, Mod. Phys. Lett. **A1**, 203 (1986).
4. V. S. Fadin, V. A. Khoze and A. D. Martin, Phys. Lett. **B311**, 311 (1993); V. S. Fadin, V. A. Khoze, A. D. Martin and W. J. Stirling, Phys. Lett. **B363**, 112 (1995); V. S. Fadin, V. A. Khoze, A. D. Martin and A. Chapovsky, Phys. Rev. **D52**, 1377 (1995).
5. Z. Kunszt and W.J. Stirling et al., hep-ph/9602352, in *Proceedings of the Workshop on Physics at LEP2*, eds. G. Alterelli, T. Sjostrand and F. Zwirner, CERN Yellow Report CERN-96-01 (1996), Vol. 1, p. 141; W.J. Stirling, Nucl. Phys. **B456**, 3 (1995).
6. J.D. Bjorken, *Proceedings of the Summer Institute on Particle Physics*, ed. M. Zipf (Stanford, 1976).
7. V. Barger, M.S. Berger, J.F. Gunion and T. Han, Phys. Rev. Lett. **75**, 1462 (1995); Phys. Rept. **286**, 1 (1997).
8. E. Boos, M. Sachwitz, H. J. Schreiber and S. Shichanin, Int. J. Mod. Phys. **A10**, 2067 (1995).
9. *Higgs Boson Discovery and Properties*, J.F. Gunion et al., *Proceedings of the 1996 Snowmass Workshop*.
10. B.A. Kniehl, Z. Phys. **C55**, 605 (1992).
11. R. Hempfling and B. Kniehl, Z. Phys. **C59**, 263 (1993).

Bounds on the Standard Higgs Boson

Jens Erler and Paul Langacker

*Department of Physics and Astronomy,
University of Pennsylvania,
Philadelphia, PA 19104-6396, USA*

Abstract. We review the status of precision electroweak physics with particular emphasis on the extraction of the Higgs boson mass. Global fit results depend strongly on the used value for the hadronic contribution to $\alpha(M_Z)$. We emphasize, however, that the general tendency for a light Higgs persists when using any of the recently obtained values for $\alpha(M_Z)$, and is also less dependent on deviating observables such as A_{LR} than in the past.

Before the discovery of the top quark, precision analyses of the Standard Model (SM) were mainly focussed on constraining its mass, m_t, while the Higgs boson mass, M_H, was fixed to a set of reference values between its direct lower limit and typically 1 TeV. After the top quark was discovered [1] and its mass found to be in perfect agreement with the predictions of precision measurements at LEP and elsewhere, the interest shifted towards finding similar constraints for M_H.

With a first precise measurement of the left-right asymmetry, A_{LR}, at the SLC [2] came also for the first time a preference for a light Higgs boson from precision tests. Indeed, by changing M_H from 1 TeV to 60 GeV the minimum χ^2 decreased by 4.4 units. However, this observation depended entirely on the A_{LR} and R_b measurements, both of which deviated by more than 2 σ from their SM predictions. Removing them resulted in a virtually flat $\chi^2(M_H)$ function [3]. R_b itself is independent of M_H, but it favors a smaller m_t and through the strong m_t–M_H correlation in the ρ parameter, M_H is also driven to smaller values.

Subsequently the A_{LR} and R_b measurements moved closer to the SM, but with their smaller errors the deviations remained at the 2 σ level, and as a result the sensitivity to M_H was enhanced. The direct top mass determinations by CDF and DØ increased the sensitivity further and the minimum χ^2 value now increased by more than 10 when M_H was increased to 1 TeV [4]. Yet, most of the sensitvity was lost upon removing A_{LR} and R_b, and both, central values and upper limits for M_H depended strongly on only 2 input quantities, both in conflict with the prediction.

Extraction of information on M_H is also hampered by the uncertainties in the hadronic contribution to the vacuum polarization, $\Delta\alpha_{\text{had}}^{(5)}(M_Z)$. There is a strong

(70%) anticorrelation between $\Delta\alpha^{(5)}_{\text{had}}(M_Z)$ and M_H.

With the increased precision of the measurements at LEP 1 [5,6] and the SLC [7], a better agreement of R_b with the SM prediction (1.3 σ), accurate measurements of the W boson mass, M_W, at LEP 2 [5] and the Tevatron [8,9], and interesting new developments regarding the determination of $\Delta\alpha^{(5)}_{\text{had}}(M_Z)$ as we will discuss later, the tendency for a light Higgs became stronger. The minimum χ^2 for a 1 TeV Higgs boson is now 16.6 larger than at its direct lower limit [10], and is less dependent on conflicting observations, although A_{LR} continues to play an important role.

We will now discuss the current status on electroweak precision tests within the SM. Most of the results presented here are from the December 1997 off-year partial update of the Particle Data Group (PDG) [10] where more details, an extended list of references, and constraints on parameters describing physics beyond the SM can be found. For implications of electroweak precision studies for supersymmetric extensions of the Standard Model see Ref. [12]. We will conclude with a discussion of the current limits on M_H, its central fit values for a variety of fits, and the impact of $\Delta\alpha^{(5)}_{\text{had}}(M_Z)$ and some very recent developments in its determination.

In Table 1 we give a list of observables used in the fits. The value of $m_t = 175 \pm 5$ GeV includes results from the dilepton, lepton plus jet, and all hadronic channels [13,14]. Γ_Z is the total width of the Z boson, σ_{had} its hadronic peak cross section, and the R_f and $A^{(0,f)}_{FB} = \frac{3}{4}A_e A_f$ are branching ratios (normalized w.r.t. the hadronic width) and forward-backward asymmetries on the Z pole, respectively [5,6]. A_f is a function of the effective weak mixing angle, $\bar{s}_f^2$, appearing in the Zff coupling. The two values of s_W^2 from deep-inelastic neutrino scattering are from CCFR [15] and the global average, respectively. Similarly, the $g^{\nu e}_{V,A}$ are from CHARM II [16] and from the νe scattering world average. The second errors in the weak charges, Q_W, of atomic parity violation in Cs [17] and Tl [18] are theoretical [19,20]. The value of α_s [in brackets] from non-lineshape determinations [21] is for comparison only, and is not used as a fit constraint.

If we include M_H as a fit parameter we find

$$M_H = 69^{+85}_{-43} \text{ GeV}, \quad (1)$$

with the central value slightly below the direct lower limit of 77 GeV (95% CL) [22]. The central value in Eq. (1) is 46 GeV smaller than the best fit value obtained by the LEP Electroweak Working Group (LEPEWWG) [5]. We trace the differences to a different treatment of radiative corrections and to a slightly different and more recent data set. Most importantly, inclusion of $\mathcal{O}(\alpha^2 m_t^2)$ corrections [23] shift the extracted M_H by -17 GeV. Also, we use the recent update for $\Delta\alpha^{(5)}_{\text{had}}(M_Z)$ from Alemany, Davier, and Höcker [24], which drives M_H smaller by another 10 GeV compared to the use of $\Delta\alpha^{(5)}_{\text{had}}(M_Z)$ from Eidelman and Jegerlehner [25]. Our result,

$$\alpha_s = 0.1214 \pm 0.0031 \ (+0.0018),$$

is higher than the one in Ref. [5]. This is mainly due to $\mathcal{O}(\alpha\alpha_s)$ vertex corrections [26] which increase the extracted α_s by 0.001. Taking these and other smaller

TABLE 1. Principal LEP and other recent observables compared with the Standard Model predictions for $M_H = M_Z$. The first value for M_W is from $p\bar{p}$ colliders [8,9], while the second includes the measurements at LEP [5]. The four values of A_ℓ are (i) from $A_{LR} = A_e$, the left-right asymmetry for hadronic final states [7]; (ii) the combined value from SLD including leptonic asymmetries and assuming univerality; (iii) A_τ from the total τ polarization; and (iv) A_e from the angular distribution of the τ polarization. The other A_f are mixed forward-backward left-right asymmetries from SLD [5]. $\bar{s}_\ell^2(A_{FB}^{(0,q)})$ is extracted from the hadronic charge asymmetry. The uncertainties in the SM predictions are from the fit parameters. The SM errors in Γ_Z, R_ℓ, and σ_{had} are completely dominated by the uncertainty in α_s. In parentheses we show the shift in the predictions when M_H is changed to 300 GeV. Older low-energy results are not listed but are included in the fits.

Observable	Value	Standard Model
m_t [GeV]	175 ± 5	173 ± 4 (+5)
M_W [GeV]	80.405 ± 0.089	80.377 ± 0.023 (-0.036)
	80.427 ± 0.075	
M_Z [GeV]	91.1867 ± 0.0020	91.1867 ± 0.0020 (+0.0001)
Γ_Z [GeV]	2.4948 ± 0.0025	2.4968 ± 0.0017 (-0.0007)
σ_{had} [nb]	41.486 ± 0.053	41.469 ± 0.016 (-0.005)
R_ℓ	20.775 ± 0.027	20.754 ± 0.020 (+0.003)
R_b	0.2170 ± 0.0009	0.2158 ± 0.0001 (-0.0002)
R_c	0.1734 ± 0.0048	0.1723 ± 0.0001 (+0.0001)
$A_{FB}^{(0,\ell)}$	0.0171 ± 0.0010	0.0162 ± 0.0003 (-0.0004)
$A_{FB}^{(0,b)}$	0.0984 ± 0.0024	0.1030 ± 0.0009 (-0.0013)
$A_{FB}^{(0,c)}$	0.0741 ± 0.0048	0.0736 ± 0.0007 (-0.0010)
$A_{FB}^{(0,s)}$	0.118 ± 0.018	0.1031 ± 0.0009 (-0.0013)
$\bar{s}_\ell^2(A_{FB}^{(0,q)})$	0.2322 ± 0.0010	0.2315 ± 0.0002 (+0.0002)
A_ℓ	0.1550 ± 0.0034	0.1469 ± 0.0013 (-0.0018)
	0.1547 ± 0.0032	
	0.1411 ± 0.0064	
	0.1399 ± 0.0073	
A_b	0.900 ± 0.050	0.9347 ± 0.0001 (-0.0002)
A_c	0.650 ± 0.058	0.6678 ± 0.0006 (-0.0008)
$s_W^2(\nu N) = 1 - M_W^2/M_Z^2$	0.2236 ± 0.0041	0.2230 ± 0.0004 (+0.0007)
	0.2260 ± 0.0039	
$g_V^{\nu e}$	-0.035 ± 0.017	-0.0395 ± 0.0005 (+0.0002)
	-0.041 ± 0.015	
$g_A^{\nu e}$	-0.503 ± 0.017	-0.5064 ± 0.0002 (+0.0002)
	-0.507 ± 0.014	
$Q_W(\text{Cs})$	$-72.41 \pm 0.25 \pm 0.80$	-73.12 ± 0.06 (+0.01)
$Q_W(\text{Tl})$	$-114.8 \pm 1.2 \pm 3.4$	-116.7 ± 0.1
$\Delta\alpha_{\text{had}}^{(5)}(M_Z)$	0.02817 ± 0.00062	0.02802 ± 0.00049 (-0.00066)
$\sin^2\hat{\theta}_{\overline{MS}}$	—	0.23124 ± 0.00017 (+0.00024)
α_s	$[0.1178 \pm 0.0023]$	0.1214 ± 0.0031 (+0.0018)

671

differences, which are well understood, into account, the agreement with the results of the LEPEWWG is excellent. We would like to stress that this agreement is quite remarkable as the electroweak library ZFITTER [27] is based on the on-shell renormalization scheme, while we use the $\overline{\rm MS}$ scheme throughout. It also demonstrates that once the most recent theoretical calculations, in particular Refs. [23,26] are taken into account, the theoretical uncertainty becomes quite small and is in fact presently negligible compared to the experimental errors. The relatively large theoretical uncertainties obtained in the Electroweak Working Group Report [28] were estimated using different electroweak libraries, which did not include the full range of higher order contributions available now.

The agreement between theory and experiment is excellent. Even the largest discrepancies in A_{LR}^0, $A_{FB}^{(0,b)}$, and $A_{FB}^{(0,\tau)}$, deviate by only 2.4 σ, 1.9 σ and 1.7 σ, respectively. There is an experimental discrepancy of 1.9 σ between A_ℓ from LEP and the SLC,

$$A_\ell(\text{LEP}) = 0.1461 \pm 0.0033,$$
$$A_\ell(\text{SLD}) = 0.1547 \pm 0.0032,$$
(2)

where the LEP value is from leptonic forward-backward asymmetries and τ polarization measurements assuming lepton universality. If one considers this discrepancy as a fluctuation, one can use the average value from Eqs. (2) to extract A_b from $A_{FB}^{(0,b)} = \frac{3}{4} A_e A_b$ and combine it with A_b from SLD to obtain $A_b = 0.877 \pm 0.023$, which is 2.5 σ or 6% below the SM prediction. That means a 30% radiative correction to $\hat{\kappa}_b$ defined through $\sin^2 \hat{\theta}_b^{\text{eff}} = \hat{\kappa}_b \sin^2 \hat{\theta}_{\overline{\rm MS}}$ would be needed to explain the discrepancy in terms of new physics in loops. Only a new type of physics which couples at the tree level preferentially to the third generation, and which does not contradict R_b (including the off-peak R_b measurements by DELPHI [29]), can conceivably account for a low A_b [30].

Let us now return to the implication for the Higgs mass. Results depend strongly on the used input parameter $\Delta \alpha_{\text{had}}^{(5)}(M_Z)$. There has been a lot of activity in the recent past on this subject, and initially not all the obtained results were in agreement with each other. This is due to the difficulty of extracting phenomenologically the function $R(s)$ describing the cross section for e^+e^- annihilation into hadrons from low and intermediate energy collider data. Now, the results obtained from this type of analysis are in reasonable agreement. Alternatively, one may try to employ perturbative QCD (PQCD) down to smaller energies, $\sqrt{s} \sim m_\tau$, and compute the continuum contribution to $R(s)$ theoretically. This approach was advocated by Martin and Zeppenfeld [31], and yields both smaller central values and errors for $\Delta \alpha_{\text{had}}^{(5)}(M_Z)$. The main reason is that some of the measured cross sections lie systematically higher than the theoretical predictions in a regime where PQCD should be reliable. Very recently, Davier and Höcker [32] improved this approach by performing a spectral moment analysis of $R(s)$ and showing that the non-perturbative terms are under control (and very small). Hence this approach appears to be quite reliable. Moreover, a similar technique [33] applied to τ decays yields consistent results [34]. Therefore, it was concluded in Ref. [32] that PQCD can be applied down to $\sqrt{s} = m_\tau$. If we use the resulting $\Delta\alpha_{\text{had}}^{(5)}(M_Z) = 0.02784 \pm 0.00026$, (with the top quark contribution removed) for our fit, we find

$$M_H = 93^{+76}_{-46} \text{ GeV}. \qquad (3)$$

Here the central value is above the direct lower limit. It should be stressed however, that a precise prediction for M_H is impossible to obtain due to the large error, the SLD discrepancy, and the complications from $\Delta\alpha_{\text{had}}^{(5)}(M_Z)$. On the other hand, upper limits and the tendency for a light Higgs are more robust. The 90 (95)% upper limits on M_H from the more experimental [24] and the more theoretical approach [32] are $M_H < 236$ (287) GeV and $M_H < 224$ (266) GeV, respectively, fortuitously in very good agreement. In order to obtain these upper limits, we have taken the Higgs exclusion curve from LEP [22] carefully into account. Since this curve extends above the quoted lower limit of 77 GeV, this results in slightly higher (more conservative) upper limits.

As a demonstration that the tendency for a light Higgs is not entirely due to the high A_{LR} we remove it from the data and the result (1) changes to

$$M_H = 154^{+140}_{-82} \text{ GeV}. \qquad (4)$$

Clearly, the central value and the errors are much larger, but this result is still compatible with the supersymmetric Higgs mass range $M_H < 150$ GeV. A less radical way to deal with deviating data is the use of PDG scale factors. Using them results in an increase of upper limits by $\mathcal{O}(100)$ GeV [10].

In conclusion, the SM of electroweak interactions is in excellent agreement with observations, with only a few deviations in some asymmetries. There is a much stronger tendency for a light Higgs boson than in the past, independently of whether one wishes to rely on PQCD or not. On the other hand, best fit values for M_H are rather volatile and depend more sensitively on input parameters and details of the analysis.

Acknowledgement: We would like to thank D. Zeppenfeld for useful discussions and the Aspen Center for Physics for its hospitality.

REFERENCES

1. CDF Collaboration: Abe, F., et al., *Phys. Rev. Lett.* **74**, 2626 (1995);
   DØ Collaboration: Abachi, S., et al., *Phys. Rev. Lett.* **74**, 2632 (1995).
2. SLD Collaboration: Abe, K., et al., *Phys. Rev. Lett.* **73**, 25 (1994).
3. Erler, J., and Langacker, P., *Phys. Rev.* **D52**, 441 (1995).
4. Langacker, P., and Erler, J., e-print hep-ph/9703428.
5. ALEPH, DELPHI, L3, OPAL, the LEP Electroweak Working Group, and the SLD Heavy Flavour Group: Abbaneo, D., et al., Report LEPEWWG/97–02.
6. DELPHI Collaboration: Abreu, P., et al., *Z. Phys.* **C67**, 1 (1995), and Boudinov, E., et al., submitted to HEP 97, Jerusalem (1997).

7. SLD Collaboration: Abe, K., et al., *Phys. Rev. Lett.* **78**, 2075 (1997), and Rowson, P.C., Talk presented at the 32nd Rencontres de Moriond: Electroweak Interactions and Unified Theories, Les Arcs (1997).
8. CDF Collaboration: Abe, F., et al., *Phys. Rev. Lett.* **75**, 11 (1995), *Phys. Rev.* **D52**, 4784 (1995), and Wagner, R.G., Talk presented at the 5th International Conference on Physics Beyond the Standard Model, Balholm (1997).
9. DØ Collaboration: Abachi, S., et al., *Phys. Rev. Lett.* **77**, 3309 (1996), and Abbott, B., et al., Report Fermilab–Conf–97–354–E, submitted to LP 97, Hamburg (1997).
10. Erler, J., and Langacker, P., *Electroweak Model and Constraints on New Physics*, in Ref. [11].
11. Barnett, R.M., et al., *Phys. Rev.* **D54**, 1 (1996) and 1997 off-year partial update for the 1998 edition available on the PDG WWW pages (URL: http://pdg.lbl.gov/).
12. Erler, J., and Pierce, D.M., e-print hep-ph/9801238.
13. CDF Collaboration: Abe, F., et al., *Phys. Rev. Lett.* **79**, 1992 (1997), and Leone, S., Talk presented at QCD 97, Montpellier (1997).
14. DØ Collaboration: Abachi, S., et al., *Phys. Rev. Lett.* **79**, 1197 (1997), and Abbott, B., et al., e-print hep-ex/9706014.
15. CCFR Collaboration: McFarland, K.S., et al., e-print hep-ex/9701010.
16. CHARM II Collaboration: Vilain, P., et al., *Phys. Lett.* **335B**, 246 (1994).
17. Boulder: Wood, C.S., et al., *Science* **275**, 1759 (1997).
18. Oxford: Edwards, N.H., et al., *Phys. Rev. Lett.* **74**, 2654 (1995); Seattle: Vetter, P.A., et al., *Phys. Rev. Lett.* **74**, 2658 (1995).
19. Dzuba, V.A., Flambaum, V.V., and Sushkov, O.P., e-print hep-ph/9709251.
20. Dzuba, V.A., et al., *J. Phys.* **B20**, 3297 (1987).
21. Hinchliffe, I., *Quantum Chromodynamics*, in Ref. [11].
22. ALEPH, DELPHI, L3, and OPAL: Murray, W., Talk presented at HEP 97, Jerusalem (1997).
23. Degrassi, G., Gambino, P., and Vicini, A., *Phys. Lett.* **383B**, 219 (1996).
24. Alemany, R., Davier, M., and Höcker, A., e-print hep-ph/9703220.
25. Eidelman, S., and Jegerlehner, F., *Z. Phys.* **C67**, 585 (1995).
26. Czarnecki, A., and Kühn, J.H., *Phys. Rev. Lett.* **77**, 3955 (1996).
27. ZFITTER: Bardin, B., et al., Report CERN–TH.6443/92.
28. Bardin, B., et al., e-print hep-ph/9709229.
29. DELPHI Collaboration: Abreu, P. et al., *Z. Phys.* **C70**, 531 (1996) and contribution submitted to HEP 97, Jerusalem (1997).
30. Erler, J., *Phys. Rev.* **D52**, 28 (1995); Erler, J., Feng, J.L., and Polonsky, N., *Phys. Rev. Lett.* **78**, 3063 (1997).
31. Martin, A.D., and Zeppenfeld, D., *Phys. Lett.* **345B**, 558 (1995).
32. Davier, M., and Höcker, A., e-print hep-ph/9711308.
33. Le Diberder, F., and, Pich A., *Phys. Lett.* **289B**, 165 (1992).
34. Höcker, A., e-print hep-ex/9703004.

Measuring Trilinear Gauge Boson Couplings at Hadron and Lepton Colliders

U. Baur

Department of Physics
State University of New York at Buffalo
Buffalo, NY 14260

Abstract. We discuss the measurement of the WWV ($V = \gamma, Z$) gauge boson couplings in present and future collider experiments. The major goals of such experiments will be the confirmation of the Standard Model (SM) predictions and the search for signals of new physics. The present limits on these couplings from Tevatron and LEP2 experiments as well as the expectations from future hadron and lepton collider experiments are summarized. We also study the impact of initial state radiation on the sensitivity limits which can be achieved at the NLC and a $\mu^+\mu^-$ collider.

INTRODUCTION

Over the last seven years e^+e^- collision experiments at LEP and at the SLAC linear collider have beautifully confirmed the predictions of the Standard Model (SM). At present experiment and theory agree at the 0.1 – 1% level in the determination of the vector boson couplings to the various fermions [1], which may rightly be considered a confirmation of the gauge boson nature of the W and the Z. On the other hand, the most direct consequences of the $SU(2)_L \times U(1)_Y$ gauge symmetry, the non-abelian self-couplings of the W, Z, and photon, are known with much less experimental precision.

A direct measurement of these vector boson couplings is possible in hadron and lepton collider experiments, in particular via pair production processes like $e^+e^- \to W^+W^-$ and $q\bar{q} \to W^+W^-$, $W\gamma$, WZ. The first and major goal of such experiments will be a confirmation of the SM predictions. A precise and direct measurement of the trilinear couplings of the electroweak vector bosons and the demonstration that they agree with the SM would beautifully corroborate spontaneously broken, non-abelian gauge theories as the basic theoretical structure describing the fundamental interactions of nature. At the same time, such measurements may be used to probe for new physics. However, if the energy scale of the new physics responsible for the non-standard gauge boson couplings is ~ 1 TeV, these anomalous couplings are expected to be no larger than $\mathcal{O}(10^{-2})$ [2].

In the following, we present an overview of how trilinear gauge boson couplings are measured in collider experiments. For simplicity, we shall restrict our discussion to the WWV ($V = \gamma, Z$) couplings; $Z\gamma V$ couplings are not discussed. Analogous to the introduction of arbitrary vector and axial vector couplings g_V and g_A for the coupling of gauge bosons to fermions, the measurement of the WWV couplings can be made quantitative by introducing a more general WWV vertex. For our discussion of experimental sensitivities we shall use a parameterization in terms of the phenomenological effective Lagrangian [3]

$$i\mathcal{L}_{eff}^{WWV} = g_{WWV}\left[g_1^V\left(W_{\mu\nu}^\dagger W^\mu - W^{\dagger\mu}W_{\mu\nu}\right)V^\nu + \kappa_V W_\mu^\dagger W_\nu V^{\mu\nu} + \frac{\lambda_V}{m_W^2} W_{\rho\mu}^\dagger W^\mu{}_\nu V^{\nu\rho}\right]. \quad (1)$$

Here the overall couplings are defined as $g_{WW\gamma} = e$ and $g_{WWZ} = e\cot\theta_W$, $W_{\mu\nu} = \partial_\mu W_\nu - \partial_\nu W_\mu$, and $V_{\mu\nu} = \partial_\mu V_\nu - \partial_\nu V_\mu$. Within the SM, at tree level, the couplings are given by $g_1^Z = g_1^\gamma = \kappa_Z = \kappa_\gamma = 1$, $\lambda_Z = \lambda_\gamma = 0$. For on-shell photons, $g_1^\gamma = 1$ is fixed by electromagnetic gauge invariance; g_1^Z may, however, differ from its SM value. Deviations are given by the anomalous couplings $\Delta g_1^Z \equiv (g_1^Z - 1)$, $\Delta\kappa_\gamma \equiv (\kappa_\gamma - 1)$, $\Delta\kappa_Z \equiv (\kappa_Z - 1)$, λ_γ, and λ_Z.

The effective Lagrangian of Eq. (1) parameterizes the most general Lorentz invariant and C and P conserving WWV vertex which can be observed in processes where the vector bosons couple to effectively massless fermions. If C or P violating couplings are allowed, four additional couplings, g_4^V, g_5^V, $\tilde{\kappa}_V$ and $\tilde{\lambda}_V$, appear in the effective Lagrangian [3] and they all vanish in the SM, at tree level. For simplicity, these couplings are not considered in this report.

The terms in $\mathcal{L}_{eff}^{WW\gamma}$ correspond to the lowest order terms in a multipole expansion of the W−photon interactions, the charge Q_W, the magnetic dipole moment μ_W, and the electric quadrupole moment q_W of the W^+ [4]:

$$Q_W = eg_1^\gamma, \quad (2)$$

$$\mu_W = \frac{e}{2m_W}\left(g_1^\gamma + \kappa_\gamma + \lambda_\gamma\right), \quad (3)$$

$$q_W = -\frac{e}{m_W^2}\left(\kappa_\gamma - \lambda_\gamma\right). \quad (4)$$

Terms with higher derivatives are equivalent to a dependence of the couplings on the vector boson momenta and thus merely lead to a form-factor behaviour of these couplings. Since a constant anomalous coupling would lead to unitarity violation at high energies [5] such a form factor behaviour is a feature of any model of anomalous couplings. When studying W^+W^- production at a lepton collider at fixed $q^2 = s$ this form factor behaviour is of no consequence. Weak boson pair production at hadron colliders, however, probes the gauge boson couplings over a large q^2 range and is very sensitive to the fall-off of anomalous couplings which necessarily happens once the threshold of new physics is crossed. Not taking this cutoff into account results in unphysically large cross sections at high energy (which violate unitarity) and thus leads to a substantial overestimate of experimental sensitivities. In the following we will assume a simple dipole behaviour, e.g.

$$\Delta\kappa_V(q^2) = \frac{\Delta\kappa_V^0}{(1 + q^2/\Lambda_{FF}^2)^2}, \qquad (5)$$

and similarly for the other couplings. Here, Λ_{FF} is the form factor scale which is a function of the scale of new physics, Λ. Due to the form factor behaviour of the anomalous couplings, the experimental limits extracted from hadron collider experiments explicitly depend on Λ_{FF}.

From the phenomenological effective Lagrangian [see Eq. (1)] it is straightforward to derive cross section formulas for the di-boson production processes, $q\bar{q}' \to W^\pm\gamma$, $W^\pm Z$, and W^+W^- production in $q\bar{q}$, e^+e^- and $\mu^+\mu^-$ annihilation. While the SM contribution to the di-boson amplitudes is bounded from above for fixed scattering angle Θ, the anomalous contributions rise without limit as $\hat{s}$ increases, eventually violating unitarity. This is the reason the anomalous couplings must show a form factor behavior at very high energies. Anomalous gauge boson couplings also affect the angular distributions of the produced vector bosons in a characteristic way (see Ref. [3]). In hadronic collisions, the transverse momentum distribution of the vector boson should be particularly sensitive to non-standard WWV couplings. At lepton colliders, on the other hand, where the center of mass energy is fixed, angular distributions are more useful. Hadron and lepton collider data thus thus yield complementary information on the nature of the WWV couplings [2].

PRESENT LIMITS ON WWV COUPLINGS

Presently, the most stringent limits on anomalous WWV couplings come from the DØ and CDF experiments at the Tevatron, and from the four LEP experiments. While the di-boson analysis of DØ is fairly complete, the CDF collaboration has not presented final results from their analysis of run 1b data yet.

Tevatron Results

The Tevatron experiments obtained information on the structure of the WWV vertices from $W\gamma$ production with subsequent $W \to \ell\nu$ ($\ell = e, \mu$)

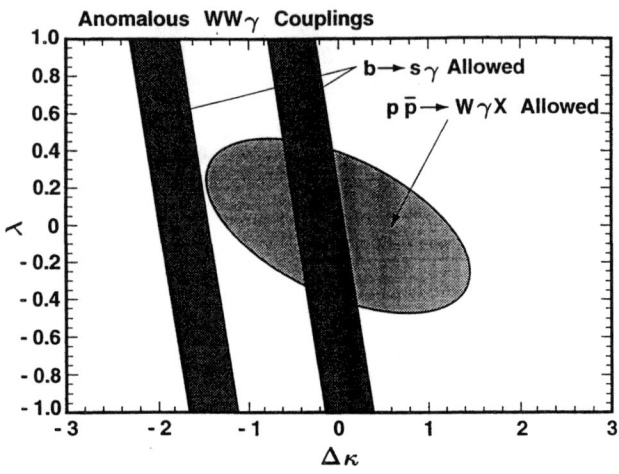

FIGURE 1. Present 95% CL limits on anomalous $WW\gamma$ couplings from $W\gamma$ and $b \to s\gamma$ data.

TABLE 1. 95% CL limits on $WW\gamma$ anomalous couplings from CDF and DØ.

	$\lambda^0_\gamma = 0$	$\Delta\kappa^0_\gamma = 0$
CDF (67 pb^{-1}, preliminary)	$-1.8 < \Delta\kappa^0_\gamma < 2.0$	$-0.7 < \lambda^0_\gamma < 0.6$
DØ (93 pb^{-1})	$-0.93 < \Delta\kappa^0_\gamma < 0.94$	$-0.31 < \lambda^0_\gamma < 0.29$

decay, $p\bar{p} \to W^+W^- \to \ell\bar{\nu}\ell'\nu$ ($\ell, \ell' = e, \mu$), and WW/WZ production with one of the vector bosons decaying leptonically and the other gauge boson decaying into two jets. Limits were also obtained from a combined fit to the three processes.

$W(\to \ell\nu)\gamma$ candidates were selected from the inclusive e/μ channel W samples by requiring an isolated photon with high transverse energy (E_T). The main background sources for $W\gamma$ production are $W+$ jets production where one of the jets "fakes" a photon, and $Z\gamma$ production with one of the leptons from the Z decay being undetected. The signal to background ratio is about 1 to 0.2 – 0.3.

A detailed discussion of the CDF and DØ $W\gamma$ event selection can be found in Refs. [6] and [7]. To set limits on the $WW\gamma$ couplings $\Delta\kappa_\gamma$ and λ_γ, a binned maximum likelihood fit to the photon E_T spectrum was performed, using a Monte Carlo program based on the calculation of Ref. [8], and a form factor scale of $\Lambda_{FF} = 1.5$ TeV. The DØ 95% CL limit contour is shown in Fig. 1, together with the bands allowed by the CLEO [9] and ALEPH [10] $b \to s\gamma$ data. The CDF and DØ 95% CL limits on for $\Delta\kappa_\gamma$ and λ_γ are listed in Table 1.

Candidates for $WW \to$ dilepton production were selected by searching for

TABLE 2. 95% CL limits on WWV anomalous couplings from $WW/WZ \to \ell\nu jj, \ell\ell jj$ production at the Tevatron, assuming $\Delta g_1^Z = 0$, $\Delta \kappa^0 = \Delta \kappa_\gamma^0 = \Delta \kappa_Z^0$, and $\lambda^0 = \lambda_\gamma^0 = \lambda_Z^0$.

	$\lambda^0 = 0$	$\Delta \kappa^0 = 0$
CDF (preliminary)	$-0.49 < \Delta \kappa^0 < 0.54$	$-0.35 < \lambda^0 < 0.32$
DØ	$-0.43 < \Delta \kappa^0 < 0.59$	$-0.33 < \lambda^0 < 0.36$

events with two isolated, high E_T charged leptons (e or μ) and large missing transverse energy, $\not{E}_T$. The main non-instrumental background in this process is $\bar{t}t$ production which can be suppressed by applying a cut on the hadronic energy in the event. The analysis is described in detail in Ref. [11]. From a fit to the electron E_T and muon p_T distribution, the DØ Collaboration obtains the following preliminary 95% CL limits ($\int \mathcal{L} dt = 97$ pb^{-1}) [12]:

$$-0.62 < \Delta \kappa^0 < 0.75 \quad (\lambda^0 = 0), \qquad -0.50 < \lambda^0 < 0.56 \quad (\Delta \kappa^0 = 0), \quad (6)$$

where we have assumed $\Lambda_{FF} = 1.5$ TeV, $\Delta \kappa^0 = \Delta \kappa_\gamma^0 = \Delta \kappa_Z^0$ and $\lambda^0 = \lambda_\gamma^0 = \lambda_Z^0$.

The $WW/WZ \to \ell\nu jj, \ell\ell jj$ data samples are extracted from inclusive e/μ W/Z data, requiring two high E_T jets in addition to the $\ell\nu$ and $\ell^+\ell^-$ system. To reduce the enormous background of QCD multijet and $W/Z + 2$ jet production, a cut on the di-jet invariant mass of 60 GeV $< m(jj) <$ 110 GeV (50 GeV $< m(jj) <$ 110 GeV) is imposed by CDF (DØ). More details of the experimental analysis can be found in Refs. [13] and [14]. The 95% CL limits for $\Lambda_{FF} = 2$ TeV obtained from the W transverse momentum distribution are listed in Table 2. A vanishing WWZ vertex ($\Delta g_1^{Z0} = \kappa_Z^0 = \lambda_Z^0 = 0$) is excluded at the 99% CL level by the experimental data.

The DØ Collaboration has also obtained preliminary limits on $\Delta \kappa^0 = \Delta \kappa_\gamma^0 = \Delta \kappa_Z^0$ and $\lambda^0 = \lambda_\gamma^0 = \lambda_Z^0$ from a combined fit to the $W\gamma$, $WW \to$ dilepton and $WW/WZ \to \ell\nu jj, \ell\ell jj$ data collected in run 1 [12] ($\Lambda_{FF} = 1.5$ TeV):

$$-0.33 < \Delta \kappa^0 < 0.45 \quad (\lambda^0 = 0), \qquad -0.20 < \lambda^0 < 0.20 \quad (\Delta \kappa^0 = 0). \quad (7)$$

LEP2 Results

The four LEP experiments have recently presented [1] measurements of anomalous WWV couplings parameters in $e^+e^- \to W^+W^-$ using the 1996 data set. A total integrated luminosity of approximately 10 pb^{-1} was recorded at each, $\sqrt{s} = 161$ GeV and $\sqrt{s} = 172$ GeV, per experiment. The LEP experiments extract limits on the following combination of anomalous WWV couplings:

TABLE 3. Comparison of the OPAL ($e^+e^- \to W^+W^-$, 1996 data) and DØ ($WW/WZ \to \ell\nu jj, \ell\ell jj$) 95% CL limits on WWV anomalous couplings. Only one of the independent couplings is assumed to deviate from the SM at a time. The OPAL limits have been corrected for form factor effects ($\Lambda_{FF} = 1.5$ TeV).

OPAL	DØ
$-0.77 < \Delta g_1^{Z0} < 0.79$	$-0.64 < \Delta g_1^{Z0} < 0.89$
$-0.92 < \Delta\kappa_\gamma^0 = \Delta\kappa_Z^0 < 1.15$	$-0.48 < \Delta\kappa_\gamma^0 = \Delta\kappa_Z^0 < 0.65$
$-0.80 < \lambda_\gamma^0 = \lambda_Z^0 < 1.22$	$-0.36 < \lambda_\gamma^0 = \lambda_Z^0 < 0.39$

$$\alpha_{W\phi} = \Delta g_1^Z \cos^2\theta_W, \tag{8}$$
$$\alpha_{B\phi} = \Delta\kappa_\gamma - \Delta g_1^Z \cos^2\theta_W, \tag{9}$$
$$\alpha_W = \lambda_\gamma = \lambda_Z \tag{10}$$

with the constraint $\Delta\kappa_Z = \Delta g_1^Z - \Delta\kappa_\gamma \tan^2\theta_W$. Combined results from the four experiments have been obtained by adding likelihood curves from each experiment, taking both cross section and information on the angular distributions of the final state fermions into account. The combined 95% CL limits are [1]

$$\begin{aligned}
-0.28 < \alpha_{W\phi} < 0.33 \quad &(\alpha_{B\phi} = \alpha_W = 0), \\
-0.81 < \alpha_{B\phi} < 1.50 \quad &(\alpha_{W\phi} = \alpha_W = 0), \\
-0.37 < \alpha_W < 0.68 \quad &(\alpha_{W\phi} = \alpha_{B\phi} = 0).
\end{aligned} \tag{11}$$

Because of the different parameterization used by the LEP experiments, it is difficult to compare these limits with the bounds extracted at the Tevatron, except for $\alpha_W = \lambda_\gamma = \lambda_Z$.

The OPAL Collaboration recently has published [15] a measurement of the WWV couplings from the 1996 data which employs the parameterization of Eq. (1). In Table 3, the OPAL results are compared with the limits of the DØ $WW/WZ \to \ell\nu jj, \ell\ell jj$ analysis ($\Lambda_{FF} = 1.5$ TeV) [14]. The limits obtained by the two experiments for Δg_1^{Z0} are similar. The DØ bounds for $\Delta\kappa_\gamma^0$ and λ_γ^0 are about a factor 2 to 3 better than the current OPAL limits.

FUTURE LIMITS: LEP2, TEVATRON AND LHC

The bounds on anomalous gauge boson couplings from LEP2 are expected to improve rapidly. In 1997, about 55 pb^{-1} were recorded by each of the four experiments at a center of mass energy of $\sqrt{s} = 183$ GeV. This should result in bounds on non-standard WWV couplings which are at least a factor 2 to 3 better than the present LEP2 limits. Ultimately one hopes to collect 500 pb^{-1} per experiment and achieve a precision of 0.02 – 0.1 [16].

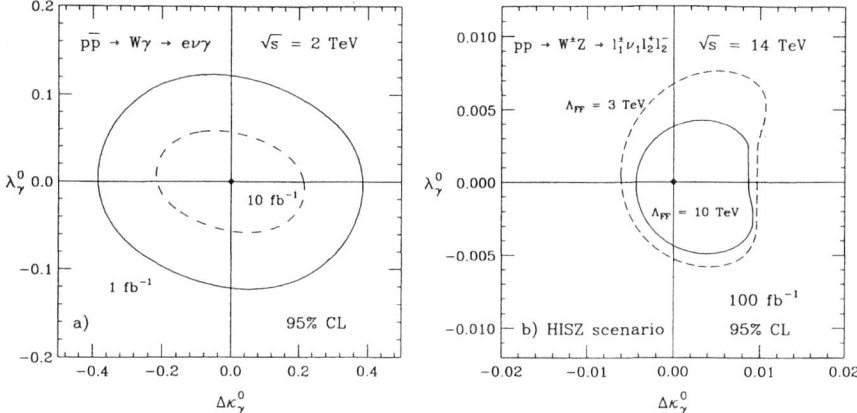

FIGURE 2. Expected 95% CL limits on non-standard WWV couplings from future hadron collider experiments. Part a) shows the projected sensitivity for $W\gamma$ production at the Tevatron. Part b) displays the limits one expects from WZ production at the LHC in the HISZ scenario [17], where $\Delta g_1^Z = \Delta\kappa_\gamma/2\cos^2\theta_W$, $\Delta\kappa_Z = (1 - \tan^2\theta_W)\Delta\kappa_\gamma/2$, and $\lambda_Z = \lambda_\gamma$.

In run 2 at the Tevatron, integrated luminosities of at least 1 fb^{-1} per experiment are foreseen. Through further upgrades in the Tevatron accelerator complex, an additional factor 10 in integrated luminosity may be gained (TeV33). With 1 fb^{-1} (10 fb^{-1}) one expects to improve the present limits on anomalous couplings by about a factor 3 (5). As an example, Fig. 2a shows the 95% CL limits expected from $W(\to e\nu)\gamma$ production for 1 fb^{-1} and 10 fb^{-1} [2].

At the LHC, sensitivities of $\mathcal{O}(10^{-2})$ can be reached [2] with an integrated luminosity of 100 fb^{-1}. This is illustrated in Fig. 2b where we show the limits which can be reached in $pp \to W^\pm Z \to \ell_1^\pm \nu_1 \ell_2^+ \ell_2^-$ (ℓ_1, $\ell_2 = e, \mu$). The sensitivity bounds which can be obtained at the LHC depend significantly on the form factor scale.

MEASURING WWV COUPLINGS AT THE NLC AND A $\mu^+\mu^-$ COLLIDER

Since the LEP2 center of mass energy is only slightly above the W pair threshold, the SM gauge cancellations are not fully operative, and the sensitivity to anomalous gauge boson couplings is limited. Much better limits on WWV couplings will be possible at a linear e^+e^- collider (NLC), or a $\mu^+\mu^-$ collider (FMC), operating in the several hundred GeV range or above.

A variety of processes can be used to constrain the vector boson self-interactions at the NLC or FMC. Because the limits obtained from W pair production [18] are comparable or better than those obtained from other processes, we restrict ourselves to this process in the following.

A study [19] based on ideal reconstruction of W daughter pairs in $e^+e^- \to W^+W^- \to \ell\nu jj$ and ignoring initial state radiation (ISR), found the following 95% CL limits in the HISZ scenario [17] ($\sqrt{s} = 500$ GeV, $\int \mathcal{L}dt = 80$ fb^{-1}):

$$|\Delta\kappa_\gamma| < 0.0024 \quad (\lambda_\gamma = 0), \qquad |\lambda_\gamma| < 0.0018 \quad (\Delta\kappa_\gamma = 0). \tag{12}$$

To ensure that events are well within the detector volume, and to suppress the contribution from the t-channel ν-exchange diagram, a $|\cos\Theta_W| < 0.8$ cut on the W production angle Θ_W has been imposed.

No detailed simulations have yet been carried out for the FMC. Nevertheless, a few general conclusions can be drawn. The main differences between W pair production in e^+e^- and $\mu^+\mu^-$ collisions are

- the background in the detector caused by the decay of the muons [20], and

- the reduced level of initial state radiation due to the larger mass of the muons.

In the following we concentrate on the impact of initial state radiation on the measurement of the WWV-couplings at a lepton collider.

For the NLC, these effects were studied in Refs. [21] and [22]. Initial state radiation is strongly peaked in the beam direction and at zero photon energy. Many photons originating from ISR therefore are not detected. The emission of an undetected photon along the beam direction affects the kinematics of the visible decay products, and a kinematic fit assuming full energy momentum conservation results in incorrect production and decay angles. In addition, photon emission leads to a reduced effective center of mass energy, and thus to distorted angular distributions.

In $WW \to jjjj$ events, the photon momentum vector can in principle be fully reconstructed. In the semileptonic case, $WW \to \ell\nu jj$, the presence of the neutrino makes this more difficult. Assuming that the photon is emitted in beam direction, a kinematic fit to the event can be performed. ISR effects can be reduced by requiring that the fitted photon energy is less than a fraction x_γ^{max} of the nominal center of mass energy. Figure 3a shows the efficiencies obtained for $\sqrt{s} = 500$ GeV and $\int \mathcal{L}dt = 10$ fb^{-1} as a function of x_γ^{max} [21]. The presence of the undetected neutrino leads to a degradation in the photon energy resolution, resulting in a large loss in efficiency for small values of x_γ^{max}. Figure 3b displays the corresponding biases in fitted $\Delta\kappa_\gamma$ values for the HISZ scenario as a function of x_γ^{max}. The bias is small compared to the statistical error only for $x_\gamma^{max} < 0.02$. Due to the loss in efficiency at small x_γ^{max}, the error on $\Delta\kappa_\gamma$ increases significantly with decreasing x_γ^{max}.

For $WW \to \ell\nu\ell'\nu'$ events the momenta of the two neutrinos are unknown. If ISR is ignored, the neutrino momenta can be reconstructed up to a two-fold

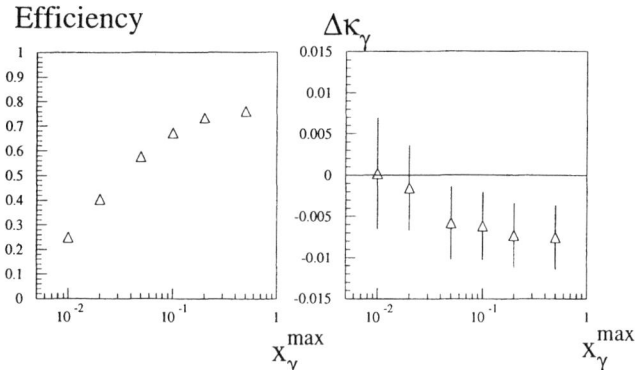

FIGURE 3. Efficiency and bias for $\Delta\kappa_\gamma$ in the HISZ scenario as a function of the maximum allowed fraction of center of mass energy for the fitted energy of a photon undetected along the beam pipe in $e^+e^- \to W^+W^- \to \ell\nu jj$ ($\sqrt{s} = 500$ GeV; $\int \mathcal{L} dt = 10$ fb^{-1}). The error bars indicate the statistical error in $\Delta\kappa_\gamma$.

ambiguity. Initial state radiation influences the existence of solutions, and introduces a large bias in the extracted values of anomalous couplings [22].

At the FMC, one expects the effects caused by ISR to be substantially smaller than at the NLC. It should therefore be possible to measure anomalous WWV couplings at a $\mu^+\mu^-$ collider with similar or better precision than at a e^+e^- machine operating at the same center of mass energy and luminosity, unless backgrounds from muon decay in the detector play an important role. Clearly, more detailed work is needed before definite conclusions can be drawn.

CONCLUSIONS

Within the past few years, our experimental knowledge of the gauge boson self-interactions has grown very rapidly. The WWV coupling parameters have been measured with an accuracy of 20 – 45% at the Tevatron. Within the next few years the limits on anomalous couplings are expected to improve by a factor 4 to 5 by experiments at LEP2 and the Tevatron. At the LHC, one hopes to probe non-standard WWV couplings with a precision of $\mathcal{O}(10^{-2})$. At the NLC, anomalous couplings can be tested at the $\mathcal{O}(10^{-3})$ level. A similar or better precision is expected for a $\mu^+\mu^-$ collider operating at the same energy and luminosity.

REFERENCES

1. D. Ward, hep-ph/9711515, to appear in the Proceedings of the *"International Europhysics Conference on High Energy Physics"*, Jerusalem, August 1997.

2. for a review see: H. Aihara et al., in "Electroweak Symmetry Breaking and New Physics at the TeV Scale", eds. T.L. Barklow, S. Dawson, H.E. Haber and J.L. Siegrist, World Scientific 1996, p. 488.
3. K. Hagiwara, K. Hikasa, R. D. Peccei, D. Zeppenfeld, Nucl. Phys. **B282**, 253 (1987); K. Gaemers and G. Gounaris, Z. Phys. **C1**, 259 (1979).
4. H. Aronson, Phys. Rev. **186**, 1434 (1969); K. J. Kim and Y.-S. Tsai, Phys. Rev. **D7**, 3710 (1973).
5. J. M. Cornwall, D. N. Levin, and G. Tiktopoulos, Phys. Rev. Lett. **30**, 1268 (1973), Phys. Rev. **D10**, 1145 (1974); C. H. Llewellyn Smith, Phys. Lett. **B46**, 233 (1973); S. D. Joglekar, Ann. Phys. **83**, 427 (1974).
6. D. Benjamin, Proceedings of the "10th Topical Workshop on Proton-Antiproton Collider Physics" eds. R. Raja and J. Yoh, AIP Press, p. 370.
7. S. Abachi et al. (DØ Collaboration) Phys. Rev. Lett. **78**, 4536 (1997).
8. U. Baur and E.L. Berger, Phys. Rev. **D41**, 1476 (1990).
9. M.S. Alam et al. (CLEO Collaboration) Phys. Rev. Lett. **74**, 2885 (1995).
10. P.G. Colrain and M.I. Williams, (ALEPH Collaboration), contributed paper #587, "International Europhysics Conference on High Energy Physics", Jerusalem, 19 – 26 August 1997.
11. S. Abachi et al. (DØ Collaboration) Phys. Rev. Lett. **75**, 1023 (1995); F. Abe et al. (CDF Collaboration) Phys. Rev. Lett. **78**, 4536 (1997).
12. T. Yasuda, FERMILAB-Conf-97/206-E, to appear in the Proceedings of the "Hadron Collider Physics XII" Conference, Stony Brook, NY, June 1997.
13. L. Nodulman, in "Proceedings of the 28th International Conference on High Energy Physics", Warsaw, Poland, 1996.
14. S. Abachi et al. (DØ Collaboration) Phys. Rev. Lett. **79**, 1441 (1997).
15. K. Ackerstaff et al., (OPAL Collaboration), CERN-PPE/97-125 (September 1997), submitted to Z. Phys. C.
16. Z. Ajaltouni et al., in "Physics at LEP2", eds. G. Altarelli, T. Sjöstrand and F. Zwirner, CERN 96-01, Vol. 1, p. 525.
17. K. Hagiwara et al., Phys. Lett. **B283**, 353 (1992); Phys. Rev. **D48**, 2182 (1993).
18. P. Mättig et al., Proceedings of the Workshop "e^+e^- Collisions at 500 GeV: The Physics Potential", Munich, Annecy, Hamburg, 1991, Vol. A, p. 223.
19. T. Barklow, in "Proceedings of the International Symposium on Vector Boson Self-Interactions", Los Angeles, 1995, eds. U. Baur, S. Errede and T. Müller, AIP Press, p. 307.
20. S. Geer, Proceedings of the Snowmass96 Workshop "New Directions for High Energy Physics", Snowmass, CO, June – July 1996, eds. D.G. Cassel, L. Trindle Gennari and R.H. Siemann, Vol. 1, p. 453.
21. K. Riles, in "Physics and Technology of the Next Linear Collider", BNL 52-502 (1996).
22. J.B. Hansen and J.D. Hansen, Proceedings of the Workshop "e^+e^- Collisions at 500 GeV: The Physics Potential", Annecy, Gran Sasso, Hamburg, 1995, Vol. D, p. 283.

Quartic Gauge Boson Couplings *

HONG-JIAN HE

*Department of Physics and Astronomy, Michigan State University
East Lansing, Michigan 48824, USA*

Abstract. We review the recent progress in studying the anomalous electroweak quartic gauge boson couplings (QGBCs) at the LHC and the next generation high energy $e^{\pm}e^{-}$ linear colliders (LCs). The main focus is put onto the strong electroweak symmetry breaking scenario in which the non-decoupling guarantees sizable new physics effects for the QGBCs. After commenting upon the current low energy indirect bounds and summarizing the theoretical patterns of QGBCs predicted by the typical resonance/non-resonance models, we review our systematic model-independent analysis on bounding them via WW-fusion and WWZ/ZZZ-production. The interplay of the two production mechanisms and the important role of the beam-polarization at the LCs are emphasized. The same physics may be similarly and better studied at a multi-TeV muon collider with high luminosity.

1. Introduction

The non-Abelian gauge structure of the standard model (SM) predicts the presence of electroweak quartic gauge boson couplings (QGBCs) besides the couplings of triple gauge bosons. The electroweak symmetry breaking (EWSB) sector involves the would-be Goldstone boson dynamics which generates the longitudinal components for $W^{\pm}, Z^0$ so that they acquire the observed masses. Despite the astonishing success of the SM at scales up to $O(100\text{GeV})$ [1,2], this EWSB sector remains yet unverified [3]. Any new physics in the underlying Goldstone boson dynamics will cause the gauge boson self-interactions to deviate from that of the SM. The quartic gauge boson interactions are particularly interesting because they can involve four longitudinal components which, according to the equivalence theorem [4,5], manifest at high energies as pure Goldstone boson interactions (that is independent of the SM gauge couplings). To unambiguously test the couplings of the quartic gauge boson interactions (QGBCs), the high energy WW-fusion and triple gauge boson production

---
*⁾ Invited talk presented at the Workshop on " *Physics at the First Muon Collider and at the Front End of a Muon Collider* ", November 6-9, 1997, at Fermi National Accelerator Laboratory, Batavia, IL, USA.

processes have to be used, where the QGBCs directly appear at the tree level. It is therefore important to study how the future high energy colliders (such as the CERN LHC and $e^{\pm}e^-$ linear colliders [6]) can sensitively probe the QGBCs for unveiling the mystery of the EWSB mechanism.

The EWSB sector can interact weakly or strongly. The weakly coupled case (such as supersymmetric models [7]) ensures the new physics at higher scales to have *decoupling* property [8] at low scales, while in the strongly interacting scenario [9] the nondecoupling guarantees the new physics scale to lie below or at $4\pi v \sim 3$ TeV [10]. In the former case the light Higgs boson(s) plus superpartners have to be first discovered, while for the latter we expect sizable new physics deviations showing up in the quartic (and triple) gauge boson couplings, which is the focus of this review. Below the new physics scale Λ, all the new physics effects in the EWSB sector can be parametrized by a complete set of the next-to-leading order (NLO) effective operators of the electroweak chiral Lagrangian (EWCL) [11], in which the $SU(2)_L \otimes U(1)_Y$ gauge symmetry is nonlinearly realized Without experimental observation on any new light resonance [1,2], this effective field theory approach [12,10] provides the most economic description of the possible new physics effects. Among the complete set of the fifteen NLO operators, five of them characterize only the quartic gauge interactions [11]:

$$\begin{cases} \mathcal{L}_4 = \ell_4 \left(\frac{v}{\Lambda}\right)^2 [\text{Tr}(\mathcal{V}_\mu \mathcal{V}_\nu)]^2 , \\ \mathcal{L}_5 = \ell_5 \left(\frac{v}{\Lambda}\right)^2 [\text{Tr}(\mathcal{V}_\mu \mathcal{V}^\mu)]^2 ; \end{cases} \quad (SU(2)_c : \checkmark)$$

$$\begin{cases} \mathcal{L}_6 = \ell_6 \left(\frac{v}{\Lambda}\right)^2 [\text{Tr}(\mathcal{V}_\mu \mathcal{V}_\nu)]\text{Tr}(\mathcal{TV}^\mu)\text{Tr}(\mathcal{TV}^\nu) , \\ \mathcal{L}_7 = \ell_7 \left(\frac{v}{\Lambda}\right)^2 [\text{Tr}(\mathcal{V}_\mu \mathcal{V}^\mu)]\text{Tr}(\mathcal{TV}_\nu)\text{Tr}(\mathcal{TV}^\nu) , \\ \mathcal{L}_{10} = \ell_{10} \left(\frac{v}{\Lambda}\right)^2 \frac{1}{2}[\text{Tr}(\mathcal{TV}^\mu)\text{Tr}(\mathcal{TV}^\nu)]^2 . \end{cases} \quad (SU(2)_c : \times) \quad (1)$$

In (1), $\mathcal{V}_\mu \equiv (D_\mu U)U^\dagger$, $D_\mu U = \partial_\mu U + \mathbf{W}_\mu U - U\mathbf{B}_\mu$, $\mathbf{W}_\mu \equiv igW_\mu^a \tau^a/2$, $\mathbf{B}_\mu \equiv ig'B_\mu \tau^3/2$, $U = \exp[i\tau^a \pi^a/v]$ (with π^a the would-be Goldstone boson field), and $\mathcal{T} \equiv U\tau_3 U^\dagger$ is the custodial $SU(2)_c$-violation operator. Here, the operators $\mathcal{L}_{4,5}$ conserve $SU(2)_c$ while $\mathcal{L}_{6,7,10}$ violate $SU(2)_c$. The dependence on v and Λ is factorized out so that the dimensionless coefficient ℓ_n of the operator $\mathcal{L}_n$ is naturally of $O(1)$ [10]. Because they contain *only* QGBCs these five operators cannot be directly tested via their tree-level contributions at low energies and are therefore least constrained from the current data. So far, only some rough estimates have been made by inserting them into the one-loop corrections and keeping the log-terms only. Here is an updated estimate at 90% C.L. by choosing $\Lambda = 2$ TeV and setting only one parameter nonzero at a time [25,13]:

$$\begin{array}{cc} -4 \leq \ell_4 \leq 20 , & -10 \leq \ell_5 \leq 50 ; \\ -0.7 \leq \ell_6 \leq 4 , & -5 \leq \ell_7 \leq 26 , \quad -0.7 \leq \ell_{10} \leq 3 . \end{array} \quad (2)$$

(2) shows that the bounds on the $SU(2)_c$ symmetric parameters $\ell_{4,5}$ are about an order of magnitude above their natural size; while the allowed range for the $SU(2)_c$-breaking parameters ℓ_{6-10} is about a factor of $O(10-100)$ larger than that for $\ell_0 =$

$\frac{\Lambda^2}{2v^2}\Delta\rho \left(=\frac{\Lambda^2}{2v^2}\alpha T\right)$ derived from the ρ (or T) parameter: $0.052 \leq \ell_0 \leq 0.12$ [13], for the same Λ and confidence level. To directly test the EWSB dynamics, it is crucial to probe these QGBCs at future high energy scattering processes where their contributions can be greatly enhanced due to the sensitive power-dependence on the scattering energy [13].

2. Quartic Gauge Boson Interactions and Underlying Models

So far the full theory underlying this effective EWCL is not determined, it is thus important to analyze how the typical resonance/non-resonance models contribute to these EWSB parameters. Knowing the theoretical sizes and patterns of these parameters tells how to use the phenomenological bounds derived in following sections for discriminating different new physics models. We mainly focus on the quartic gauge boson interactions (1) and consider typical models such as a heavy scalar (S), a vector (V_μ^a) and an axial vector (A_μ^a) for the resonance scenario, and the new heavy doublet fermions for the non-resonance scenario.

• **A Non-SM Singlet Scalar**   Up to dimension-4 and including both $SU(2)_c$ conserving and breaking effects, we write down the most general Lagrangian for a singlet scalar which is invariant under the SM gauge group $SU(2)_L \otimes U(1)_Y$:

$$\mathcal{L}_{\text{eff}}^S = \frac{1}{2}\left[\partial^\mu S \partial_\mu S - M_S^2 S^2\right] - V(S) \\ - \left[\frac{\kappa_s}{2}vS + \frac{\kappa_s'}{4}S^2\right]\text{Tr}[\mathcal{V}_\mu \mathcal{V}^\mu] - \left[\frac{\tilde{\kappa}_s}{2}vS + \frac{\tilde{\kappa}_s'}{4}S^2\right][\text{Tr}\mathcal{T}\mathcal{V}_\mu]^2 \quad (3)$$

in which $V(S)$ only contains Higgs self-interactions. The SM Higgs boson corresponds to a special parameter choice: $\kappa_s = \kappa_s' = 1$, $\tilde{\kappa}_s = \tilde{\kappa}_s' = 0$ and $V(S) = V(S)_{\text{SM}}$. A heavy scalar can be integrated out from low energy spectrum and the corresponding contributions to (1) are derived as:

$$\ell_4^s = 0, \quad \ell_5^s = \frac{\kappa_s^2}{8} \geq 0; \quad \ell_6^s = 0, \quad \ell_7^s = \frac{\kappa_s \tilde{\kappa}_s}{4}, \quad \ell_{10}^s = \frac{\tilde{\kappa}_s^2}{8} \geq 0. \quad (4)$$

In (4), the deviation from $\kappa_s = 1$ and $\tilde{\kappa}_s = 0$ signals a *non-SM Higgs boson*.

• **Vector and Axial-Vector Bosons**   The S-parameter measurement at LEP disfavors the naive QCD-like dynamics for the EWSB [9], where the vector ρ_{TC} is the lowest new resonance in the TeV regime. This suggests a necessity of including the axial-vector boson [14] in a general formalism for modeling the non-QCD-like dynamics. We consider the vector V_μ^a and axial-vector A_μ^a fields as the weak isospin triplets of (custodial) $SU(2)_c$. $\{V, A\}$ transform under the SM global $SU(2)_c$ as $\hat{V}_\mu \Rightarrow \hat{V}_\mu' = \Sigma_v \hat{V}_\mu \Sigma_v^\dagger$, $\hat{A}_\mu \Rightarrow \hat{A}_\mu' = \Sigma_v \hat{A}_\mu \Sigma_v^\dagger$, where $\hat{V}_\mu \equiv V_\mu^a \tau^a/2$, $\hat{A}_\mu \equiv A_\mu^a \tau^a/2$, and $\Sigma_v \in SU(2)_c$. If $\{V, A\}$ are further regarded as gauge fields of a new local hidden symmetry group $\mathcal{H} = SU(2)_L' \otimes SU(2)_R'$ (with a discrete left-right parity) [14], we can write down the following general Lagrangian (up to two derivatives), in the *unitary gauge* of the group $\mathcal{H}$ [1] and with both $SU(2)_c$-conserving and -breaking effects included,

$$\mathcal{L}_{\text{eff}}^{VA} = \mathcal{L}_{\text{kinetic}}^{VA} - v^2 \left[\kappa_0 \text{Tr}\overline{\mathcal{V}}_\mu^2 + \kappa_1 \text{Tr}\left(J_\mu^V - 2V_\mu\right)^2 + \kappa_2 \text{Tr}\left(J_\mu^A + 2A_\mu\right)^2 + \kappa_3 \text{Tr}A_\mu^2 \right. \tag{5}$$
$$\left. + \tilde{\kappa}_0 \left[\text{Tr}\widetilde{\mathcal{T}}\overline{\mathcal{V}}_\mu\right]^2 + \tilde{\kappa}_1 \left[\text{Tr}\widetilde{\mathcal{T}}(J_\mu^V - 2V_\mu)\right]^2 + \tilde{\kappa}_2 \left[\text{Tr}\widetilde{\mathcal{T}}(J_\mu^A + 2A_\mu)\right]^2 + \tilde{\kappa}_3 \left[\text{Tr}\widetilde{\mathcal{T}}A\right]^2 \right]$$

where
$$\begin{cases} J_\mu^V = J_\mu^L + J_\mu^R \\ J_\mu^A = J_\mu^L - J_\mu^R \end{cases} \quad \begin{cases} J_\mu^L = \xi^\dagger D_\mu^L \xi = \xi^\dagger \left(\partial_\mu \xi + W_\mu \xi\right) \\ J_\mu^R = \xi D_\mu^R \xi^\dagger = \xi \left(\partial_\mu \xi^\dagger + B_\mu \xi^\dagger\right) \end{cases}$$

and, by definition, $V_\mu \equiv i\tilde{g}\widehat{V}_\mu = i\tilde{g}V_\mu^a \tau^a/2$, $A_\mu \equiv i\tilde{g}\widehat{A}_\mu = i\tilde{g}A_\mu^a \tau^a/2$, $\overline{\mathcal{V}}_\mu \equiv U^\dagger D_\mu U = U^\dagger \mathcal{V}_\mu U$, $\widetilde{\mathcal{T}} = \tau^3 = U^\dagger \mathcal{T} U$, and $U \equiv \xi^2$. ($\tilde{g}$ is the gauge coupling of the group $\mathcal{H}$.) Among the above four new $SU(2)_c$-conserving parameters κ_n's, κ_0 is determined by normalizing the Goldstone kinematic term: $\kappa_0 = -4\kappa_2\kappa_3/(4\kappa_2 + \kappa_3)$. After eliminating the V and A fields in the heavy mass expansion, we derive ℓ_n's below:

$$\begin{cases} \ell_4 = \ell_4^v + \ell_4^a \\ \ell_5 = \ell_5^v + \ell_5^a \\ \ell_6 = \ell_6^v + \ell_6^a \\ \ell_7 = \ell_7^v + \ell_7^a \\ \ell_{10} = \ell_{10}^v + \ell_{10}^a \end{cases} \quad \begin{cases} \ell_4^v = -\ell_5^v = 1/[2\sqrt{2}\tilde{g}v\Lambda^{-1}]^2 > 0 \\ \ell_4^a = -\ell_5^a = \left[\eta^2(\eta^2 - 2) + 16\tilde{\eta}^2\right]/[2\sqrt{2}\tilde{g}v\Lambda^{-1}]^2 \\ \ell_6^v = \ell_7^v = 0 \\ \ell_6^a = -\ell_7^a = -\tilde{\eta}\left[4(3 - \eta^2)\tilde{\eta} + (1 - \eta^2)\eta\right]/[2\sqrt{2}\tilde{g}v\Lambda^{-1}]^2 \\ \ell_{10}^v = \ell_{10}^a = 0 \end{cases} \tag{6}$$

in which

$$\eta = \frac{4\kappa_2}{4\kappa_2 + \kappa_3}, \quad \tilde{\eta} = \frac{2\kappa_2 + 4\tilde{\kappa}_2}{(4\kappa_2 + \kappa_3) + 2(4\tilde{\kappa}_2 + \tilde{\kappa}_3)} - \frac{2\kappa_2}{4\kappa_2 + \kappa_3}, \tag{7}$$

and $\Lambda = \min\{M_V, M_A\}$. After ignoring the SM gauge couplings g and g', $\{M_V, M_A\} \simeq \{\tilde{g}v\sqrt{\kappa_1}, \tilde{g}v\sqrt{\kappa_2 + \kappa_3/4}\}$, at the leading order. In (6), the factor $1/[\tilde{g}v\Lambda^{-1}]^2 \simeq \kappa_1(\Lambda/M_V)^2 = O(\kappa_1)$ and all $SU(2)_c$-breaking terms depend on $\tilde{\eta}$. Note that the $SU(2)_c$-symmetric contribution from the axial-vector boson interactions to $\ell_4^a = -\ell_5^a$ becomes negative for $|\eta| < \sqrt{2}$, while the summed contribution $\ell_4 = -\ell_5 = [(\eta^2 - 1)^2 + 16\tilde{\eta}^2]/[2\sqrt{2}\tilde{g}v\Lambda^{-1}]^2 \geq 0$. The deviation of η and/or $\tilde{\eta}$ from $\eta(\tilde{\eta}) = 0$ represents the *non-QCD-like* EWSB dynamics.

• **Heavy Doublet Fermions**   Take for instance a model of one flavor heavy chiral fermions which form a left-handed weak doublet $(U_L, D_L)^T$ and right-handed

---

[1] By "unitary gauge" we mean a gauge containing no new Goldstone boson other than the three ones for generating the longitudinal components of the *known* W, Z. In fact, it is not essentially necessary to introduce such a new local symmetry $\mathcal{H}$ for $\{V, A\}$ [15] since $\mathcal{H}$ has to be broken anyway and $\{V, A\}$ can be treated as matter fields [16]. The hidden local symmetry formalism is more restrictive on the allowed free-parameters (κ_n's etc) due to the additional assumption about that new local group $\mathcal{H}$.

singlets $\{U_R, D_R\}$, and joins a new strong $SU(N)$ gauge group in its fundamental representation. Their small mass-splitting breaks the $SU(2)_c$ and is characterized by the parameter $\omega = 1 - [M_U/M_D]^2$. The anomaly-cancellation is ensured by assigning the $\{U, D\}$ electric charges as $\{+\frac{1}{2}, -\frac{1}{2}\}$. By taking $\{U, D\}$ as the source of the EWSB, the W, Z masses can be generated by heavy fermion loops. The new contributions to the quartic gauge couplings of W/Z come from the *non-resonant* $\{U, D\}$ box-diagrams. The leading results in the $1/M_{U,D}$ and ω expansions are summarized below:

$$\ell_4^f = -2\ell_5^f = \left[\frac{\Lambda}{4\pi v}\right]^2 \frac{N}{12} > 0 \; ; \quad \ell_6^f = -\ell_7^f = -\left[\frac{\Lambda}{4\pi v}\right]^2 \frac{7N}{240} \omega^2 \; , \quad \ell_{10} = 0 \; ; \qquad (8)$$

in which $\Lambda = \min\{M_U, M_D\}$.

3. A Global Analysis on Probing QGBCs versus TGBCs at the LHC

The general EWCL formalism [11] contains in total 15 NLO new operators whose coefficients (ℓ_n's) depend on the details of the underlying dynamics as exemplified in the previous section. It is shown [13] that, except for $\ell_{0,1,8}$ (S, T, U), the current data only bound a few triple gauge boson couplings (TGBCs) to $O(10)$ at the 1σ-level and give no direct tree-level bound on QGBCs. The rough estimates of the bounds from 1-loop corrections still allow QGBCs to be of $O(5-50)$. For a *complete* test of the EWSB sector in discriminating different dynamical models, all these TGBCs and QGBCs (ℓ_n's) have to be measured through various high energy VV-fusion and $f\bar{f}^{(\prime)}$-annihilation processes. ($V^a = W^\pm, Z^0$.) For this purpose, a systematic global analysis [13] has been carried out which reveals the important overall physical pictures and guides us for further elaborate precise numerical studies (cf. Secs. 4-5). In performing such a global analysis we developed a precise electroweak power counting rule (à la Weinberg) for conveniently estimating *all* high energy scattering amplitudes and formulated the equivalence theorem (ET) [5] as a *necessary* physical criterion for sensitively probing the EWSB dynamics. Applying this counting method, we have carried out a systematic analysis for all $V^a V^b \to V^c V^d$ and $f\bar{f}^{(\prime)} \to V^a V^b, V^a V^b V^c$ processes by estimating the contributions to their S-matrix elements from both the leading order operator up to one-loop and the other 15 NLO operators at the tree-level. Based upon the basic features of the chiral perturbation expansion, we further build the following electroweak power counting hierarchy for the S-matrix elements,[2]

$$\frac{E^2}{f_\pi^2} \gg \left[\frac{E^2}{f_\pi^2}\frac{E^2}{\Lambda^2}, g\frac{E}{f_\pi}\right] \gg \left[g\frac{E}{f_\pi}\frac{E^2}{\Lambda^2}, g^2\right] \gg \left[g^2\frac{E^2}{\Lambda^2}, g^3\frac{f_\pi}{E}\right] \gg \left[g^3\frac{Ef_\pi}{\Lambda^2}, g^4\frac{f_\pi^2}{E^2}\right] \gg g^4\frac{f_\pi^2}{\Lambda^2} \; .$$

(9)

---

[2] For $f\bar{f}^{(\prime)} \to VVV$ amplitudes, there is an additional factor $1/f_\pi$ by dimentional counting.

In the typical TeV region, for $E \in (750\,\text{GeV},\ 1.5\,\text{TeV})$, this gives:

$$(9.3, 37) \gg [(0.55, 8.8), (2.0, 4.0)] \gg [(0.12, 0.93), (0.42, 0.42)] \gg$$
$$[(0.025, 0.099), (0.089, 0.045)] \gg [(5.3, 10.5), (19.0, 4.7)] \times 10^{-3} \gg (1.1, 1.1) \times 10^{-3},$$

where E is taken to be the invariant mass of the VV pair. This power counting hierarchy can be nicely understood. In (9), from left to right, the hierarchy is built up by increasing either the number of derivatives (i.e. power of E/Λ) or the number of external transverse gauge boson V_T's (i.e. the power of gauge couplings). This power counting hierarchy provides us a theoretical base to classify all the relevant scattering amplitudes in terms of the three essential parameters E, f_π and Λ plus possible gauge/Yukawa coupling constants.

At the event-rate-level, we have adopted the usual leading-log effective vector boson method (LL-EVBM) [19] to reasonably and conveniently estimate the VV-luminosities. In Fig. 1, the rate $|R_B|$ denotes an intrinsic background defined via the formulation of the ET as a necessary criterion for the sensitivities to the EWSB [5,13]. Fig. 1 shows that, at the 14TeV LHC with $\int \mathcal{L} = 100\text{fb}^{-1}$ Luminosity and for $\Lambda = 2\text{TeV}$, the W^+W^+-fusion is most sensitive to $\ell_{4,5}$ (QGBCs) and marginally sensitive to $\ell_{3,9,11,12}$; while the $q\bar{q}' \to W^+Z$ annihilation can best probe $\ell_{3,11,12}$ and marginally test $\ell_{8,9,14}$. Hence, the VV-fusions and $f\bar{f}^{(\prime)}$-annihilations are *complementary* in probing the different sets of these NLO parameters (the QGBCs and TGBCs) at the LHC.

4. Measuring the QGBCs via WW-Fusion Processes

Though the LHC will give the first direct test on these new quartic gauge boson couplings (QGBCs), the large backgrounds limit its sensitivity and cutting off the backgrounds significantly reduces the event rate. As shown in Ref. [17], for the non-resonance $W^\pm W^\pm$ production channels in the TeV regime only around 10 signal events were predicted at the LHC with a 100 fb^{-1} annual luminosity after imposing necessary cuts in the gold-plated modes (by pure leptonic decays). The corresponding study at the TeV $e^\pm e^-$ LCs opens a more exciting possibility [18].

In this and next sections we review how to make further precision constraints for the QGBCs via the WW-fusion [20,21], WWZ/ZZZ-production [24][3], and their interplay at LCs [24], which is much cleaner than the LHC so that the final state W/Z's can be detected via the dijet mode and with large branching ratios. Due to the limited calorimeter energy resolution, the misidentification probability of W versus Z and the rejection of certain fraction of diboson events should be considered [18]. Inclusion of the leptonic decay of Z to e^-e^+ and $\mu^-\mu^+$ is also useful. The detection efficiencies for WW, ZZ and WZ final states are estimated to be about 34%, respectively.

---

[3] The WWZ/ZZZ-production in the SM was first studied in Ref. [22], and later some analyses on including the anomalous couplings have also appeared [23] for the case of unpolarized $e^\mp$ beams. For a very recent study similar to Ref. [24] for WWZ/ZZZ-production, see Ref. [25].

To completely determine all the QGBCs, we need at least five independent processes. From WW-fusions alone, we can have

$$
\begin{array}{lll}
\text{Full process}: & \text{Sub} - \text{process}: & \text{Relevant parameter}: \\
e^-e^+ \to \nu\bar{\nu}W^-W^+, & (W^-W^+ \to W^-W^+), & (\ell_{4,5}), \\
e^-e^- \to \nu\bar{\nu}W^-W^-, & (W^-W^- \to W^-W^-), & (\ell_{4,5}); \\
e^-e^+ \to \nu\bar{\nu}ZZ, & (W^-W^+ \to ZZ), & (\ell_{4,5}; \ell_{6,7}), \\
e^-e^+ \to e^{\pm}\nu W^{\mp}Z, & (W^{\mp}Z \to W^{\mp}Z), & (\ell_{4,5}; \ell_{6,7}), \\
e^-e^+ \to e^-e^+ZZ, & (ZZ \to ZZ), & ([\ell_4+\ell_5]+2[\ell_6+\ell_7+\ell_{10}]);
\end{array}
\qquad (10)
$$

where in the round backets the corresponding fusion (signal) sub-process. We see that $\ell_{4,5}$ can be cleanly tested via the first two processes in (10), as shown by Fig. 2. (All our plots have chosen the new physics cutoff as $\Lambda = 2$ TeV and the numerical results for other values of Λ can be obtained via re-scaling.) By including the third and fourth reactions $\ell_{6,7}$ can be further disentangled. Finally the fifth channel provides a unique probe on ℓ_{10}. Though this scheme is complete in principle, the realistic situation is more involved. Note that the rate of the last reaction in (10) is significantly lower than all others due to the double suppressions of the e-e-Z couplings while the fourth channel has huge backgrounds which are uneasy to overcome [18,21]. But $ZZ \to ZZ$ also has an advantage due to the absence fusion-type backgrounds and the triple gauge boson couplings have no contribution either. This makes it relatively cleaner than others. Since the parameter ℓ_{10} appears only in $4Z$ vertex, the above last channel has to be used anyway when only the fusion mechanism is made use of. (For the process $e^-e^+ \to ZZZ$ on ℓ_{10}, see Sec. 5.) Since the large backgrounds make the WZ-channel less useful (see Fig. 4a below), we propose to use the production $e^-e^+ \to WWZ$ (cf. Sec. 5) to complete this five parameter determination.

Before concluding this section, we summarize below the 90% C.L. one-parameter fusion-bounds for $\Lambda = 2$ TeV at a later stage of the LC with $\sqrt{s} = 1.6$ TeV and $\int \mathcal{L} = 200$ fb^{-1}:

$$
\begin{array}{lll}
-0.13 \leq \ell_4 \leq 0.10, & -0.08 \leq \ell_5 \leq 0.06; \\
-0.22 \leq \ell_6 \leq 0.22, & -0.12 \leq \ell_7 \leq 0.10, & -0.21 \leq \ell_{10} \leq 0.21;
\end{array}
\qquad (11)
$$

which are very stringent. Here we have used a 90% (65%) polarization for the $e^-(e^+)$ beam.

5. WWZ/ZZZ-Production and its Interplay with WW-Fusion

To probe the QGBCs (1), we know [13] that the WW-fusion amplitudes have the highest E-power dependence in the TeV regime while the s-channel signals of the WWZ/ZZZ-production lose an enhancement factor of $(E/v)^2$ relative to that of the fusion processes. When the collider energy is reduced by half (from 1.6 TeV down to 800 GeV), the sensitivity of the WW-fusion decreases by about a factor of 20 or more [20,21]. We thus expect that $ee \to WWZ, ZZZ$ become more important at the earlier phase of the LCs and will be competitive with and complementary to fusions for the later stages of the LCs around $0.8 \sim 1$ TeV [24]. In fact, it was revealed that even at the 1.5/1.6 TeV, $e^+e^- \to WWZ$ plays a crucial role in achieving a clean five-parameter analysis [24].

To avoid the potential fusion backgrounds from $e^-e^+ \to eeZZ, eeWW$, we now only add the $Z \to \mu^-\mu^+$ decay besides the dijet-decay mode. The detection efficiencies for ZZZ and WWZ final states are thus estimated to be about 16.8% and 18.4%, respectively. It turns out that $e^-e^+ \to WWZ$ has huge backgrounds due to the t-channel ν_e or e-ν_e exchange, and the kinematic cuts alone help very little. However, we find that such type of backgrounds involve the left-handed W-e-ν coupling and thus can be effectively suppressed by using the right(left)-hand polarized $e^-(e^+)$ beam. The highest sensitivity is reached by maximally polarizing *both* e^- and e^+ beams. The crucial roles of the beam polarization and the higher collider energy for the WWZ-production are demonstrated in Fig. 3a, where $\pm 1\sigma$ exclusion contours for ℓ_4-ℓ_5 are displayed at $\sqrt{s} = 0.5, 0.8, 1.0$ and 1.6 TeV, respectively. The beam polarization has much less impact on the ZZZ mode, due to the almost axial-vector type e-Z-e coupling. Including the same polarizations as in the case of the WWZ mode, we find about $10-20\%$ improvements on the bounds from the ZZZ-production. Assuming the two beam polarizations (90% e^- and 65% e^+), we summarize the final $\pm 1\sigma$ bounds for both ZZZ and WWZ channels and their combined 90% C.L. contours for 0.5 TeV with $\int \mathcal{L} = 50$ fb^{-1} in Fig. 3b (representing the *first direct probe* at the LC) and for 1.6 TeV with $\int \mathcal{L} = 200$ fb^{-1} in Fig. 3c (representing the *best* sensitivity gained from the final stage of the LC with energy around 1.5/1.6 TeV). Note that, the 90% C.L. level bounds on ℓ_4-ℓ_5 at 0.5 TeV are within $O(10-20)$, while at 1.6 TeV they sensitively reach $O(1)$. The WWZ channel gives the same bounds for ℓ_4-ℓ_5 and ℓ_6-ℓ_7, while the ZZZ channel imposes stronger bound on ℓ_6-ℓ_7 due to a factor of 2 enhancement from the $4Z$-vertex. ℓ_{10} only contributes to ZZZ final state and can be probed at the similar level.

For comparison, a parallel analysis to Fig. 3b-c is further performed for the case without e^+-beam polarization (but with e^- polarization the same as before). For a two-parameter ($\ell_{4,5}$) study, the results are listed below at 90% C.L.:

at 0.5 TeV : $\quad -12\,(-18) \leq \ell_4 \leq 21\,(27), \quad\quad -17\,(-22) \leq \ell_5 \leq 9.5\,(15);$

at 1.6 TeV : $\quad -0.50\,(-0.67) \leq \ell_4 \leq 1.5\,(1.7), \quad\quad -1.3\,(-1.5) \leq \ell_5 \leq 0.36\,(0.58);$

(12)

where the numbers in the parentheses denote the bounds from polarizing the e^--beam alone. The comparison in (12) shows that without e^+-beam polarization, the sensitivity will decrease by about $15\% - 60\%$. Therefore, making use of the possible e^+-beam polarization with a degree around 65% is clearly helpful. In the above, the total rates are used to derive the numerical bounds. We have further studied the possible improvements by including different characteristic distributions, but no significant increase of the sensitivity is found.

Now, we are ready to analyze the interplay with WW-fusion processes. As noted in Sec. 4, the WZ-channel in (10) has large γ-induced $eeWW$ background in which one e is lost in the beam-pipe and one W misidentified as Z. A cut on the missing $p_\perp(\nu)$ is imposed to specially suppress this background. Even though, the final sensitivity still turns out to be less useful in constraining the ℓ_6-ℓ_7 plane (cf. Fig. 4a) [21]. To sensitively bound $\{\ell_6, \ell_7\}$ (especially ℓ_6) well below $O(1)$, we propose to use the production $e^-e^+ \to WWZ$. Fig. 4a demonstrates the interplay of WW-fusion and WWZ-production for discriminating the $SU(2)_c$-breaking QGBCs ℓ_6-ℓ_7 at $\sqrt{s} = 1.6$ TeV. The ZZZ-production can provide bound on ℓ_{10}, in addition to the $eeZZ$ fusion-channel in (10). Assuming that $\ell_{4,5;6,7}$ are constrained by the processes mentioned above, we set their values to the reference point (zero) for simplicity and define the statistic significance $S = |\mathcal{N} - \mathcal{N}_0|/\sqrt{\mathcal{N}_0}$ which is a function of ℓ_{10}. (Here $\mathcal{N}$ is the total event-number while $\mathcal{N}_0$ is the number at $\ell_{10} = 0$.) As shown in Fig. 4b, at 1.6 TeV, the sensitivity of $e^-e^+ \to eeZZ$ for probing ℓ_{10} is better than that of $e^-e^+ \to ZZZ$.

In summary, the first direct probe on these QGBCs will come from the early phase of the LC at 500 GeV, where the WW-fusion processes are not useful. The two mechanisms become more competitive and complementary at energies $\sqrt{s} \sim 0.8 - 1$ TeV. From (11) and Table 1, we see that at a later stage of the LC with $\sqrt{s} = 1.6$ TeV, the 90% C.L. one-parameter bounds on $\ell_{4,5}$ from WWZ/ZZZ-modes are about a factor of $3 \sim 6$ weaker than that from WW-fusions; while the bounds on $\ell_{6,7,10}$ are comparable. In a complete multi-parameter analysis, the WWZ-channel is crucial for determining ℓ_6-ℓ_7 even at a 1.6 TeV LC (cf. Fig. 4a).

Table 1: Combined 90% C.L. bounds on ℓ_{4-10} from WWZ/ZZZ-production. For simplicity, we set one parameter to be nonzero at a time. The bound on ℓ_{10} comes from ZZZ-channel alone.

$\sqrt{s}$ (TeV)	0.5	0.8	1.0	1.6
$\int \mathcal{L}$ (fb^{-1})	50	100	100	200
WWZ/ZZZ Bounds (at 90%C.L.)	$-9.5 \leq \ell_4 \leq 11.7$	$-2.7 \leq \ell_4 \leq 3.2$	$-1.7 \leq \ell_4 \leq 2.0$	$-0.50 \leq \ell_4 \leq 0.58$
	$-9.8 \leq \ell_5 \leq 8.9$	$-3.1 \leq \ell_5 \leq 2.3$	$-1.9 \leq \ell_5 \leq 1.4$	$-0.54 \leq \ell_5 \leq 0.36$
	$-5.0 \leq \ell_6 \leq 5.8$	$-1.5 \leq \ell_6 \leq 1.6$	$-0.95 \leq \ell_6 \leq 1.0$	$-0.28 \leq \ell_6 \leq 0.28$
	$-5.0 \leq \ell_7 \leq 5.7$	$-1.5 \leq \ell_7 \leq 1.5$	$-0.95 \leq \ell_7 \leq 0.92$	$-0.28 \leq \ell_7 \leq 0.26$
	$-4.3 \leq \ell_{10} \leq 5.2$	$-1.4 \leq \ell_{10} \leq 1.4$	$-0.83 \leq \ell_{10} \leq 0.88$	$-0.26 \leq \ell_{10} \leq 0.26$
Range of $|\ell_n|$	$\leq O(4 \sim 10)$	$\leq O(1 \sim 3)$	$\leq O(0.8 \sim 2)$	$\leq O(0.3 \sim 0.6)$

6. Concluding Remarks

Despite the constantly increasing evidence in supporting the Standard Model (SM) over the past 30 years, we particle physicists have been struggling in search for " *New Physics Beyond the SM* " so far [1,2]. Among the numerous ways for going beyond the SM, the Higgs boson hypothesis [26] stands out. The updated direct Higgs search at LEP [2] puts a 95%C.L. lower bound $m_H \geq 89.3$ GeV. Due to the discrepancy between the precision Z-decay asymmetry measurement and the direct Higgs search limit, the combined 95%C.L. upper Higgs mass bound from the global fit has been shown to significantly increase toward the TeV regime [3]. However, the unitarity and triviality theoretically forbid the SM Higgs mass to go beyond the TeV scale, at which we are facing an exciting strong electroweak symmetry breaking (EWSB) dynamics. Below the new heavy resonance, we have to first probe the EWSB parameters formulated by means of the electroweak chiral Lagrangian (EWCL), among which the quartic gauge boson interactions penetrate the pure Goldstone dynamics. After commenting upon the low energy indirect bounds and analyzing the different patterns of these quartic couplings predicted by the typical resonance/non-resonance models, we estimate the sensitivity of the LHC to probing these couplings, and then analyze the constraints on them via WWZ/ZZZ-production and WW-fusion at the next generation $e^{\pm}e^-$ linear colliders (LCs). The interplay of the two production mechanisms and the important role of the beam-polarization at the LCs are stressed.

Finally, we remark that the same physics may be similarly and better studied at a multi-TeV muon collider (MTMC) ($\sqrt{s} \simeq 3 - 4$ TeV) with high luminosity ($\sim 500 - 1000$ fb^{-1}/year) [27,28]. Due to the higher center mass energy of the MTMC, certain unitarization on the EWCL is needed for studying the WW-fusions. The other muon collider options, like $\mu^-\mu^-$ and $\mu^+\mu^+$ are likely to be as easily achieved as $\mu^-\mu^+$ mode. Furthermore, the large muon mass relative to the electron mass makes the initial state photon-radiation of the muon collider much less severe than that of the electron collider. The two drawbacks of a muon collider in comparison with an electron linear collider are [28]: (i) substantial beam polarization ($\geq 50\%$) can be achieved only with a significant sacrifice in luminosity; (ii) the $\gamma\gamma$ and $\mu\gamma$ options are probably not very feasible.

Acknowledgments I am grateful to the working group convener Tao Han for invitation and FermiLab for hospitality. Special thanks go to Tao Han and C.-P. Yuan for carefully reading the draft and providing many useful suggestions. I thank them and all other collaborators, E. Boos, W. Kilian, Y.-P. Kuang, A. Pukhov and P.M. Zerwas, for productive collaborations [13,21,24] upon which this review is based. I am also indebted to K. Floettmann and R. Frey for discussing the $e^{\mp}$-beam polarizations, and many other colleagues such as R. Casalbuoni, D. Dominici, G. Jikia, I. Kuss, A. Likhoded, C.R. Schmidt and G. Valencia for useful discussions. This work is supported by the U.S. Natural Science Foundation.

References

1. C. Rembser, *Recent Results on LEP2*, plenary talk at International Symposium on " Frontiers of Phenomenology from Non-perturbative QCD to New Physics ", March 23-26, 1998, Madison, Wisconsin.
2. B.C. Allanach et al, *Report of the Working Group on 'Searches'*, hep-ph/9708250; A. Sopczak, *Searches for Higgs Bosons at LEP2*, hep-ph/9712283; and references therein.
3. M.S. Chanowitz, Phys. Rev. Lett. (1998), in press, and hep-ph/9710308.
4. J.M. Cornwall, D.N. Levin, and G. Tiktopoulos, Phys. Rev. **D10** (1974) 1145; B.W. Lee, C. Quigg, and H. Thacker, Phys. Rev. **D16** (1977) 1519; M.S. Chanowitz and M.K. Gaillard, Nucl. Phys. **B261** (1985) 379. G.J. Gounaris, R. Kögerler, and H. Neufeld, Phys. Rev. **D34** (1986) 3257.
5. H.-J. He, Y.-P. Kuang, C.-P. Yuan, Phys. Rev. **D51** (1995) 6463; H.-J. He and W.B. Kilgore, Phys. Rev. **D55** (1997) 1515; H.-J. He, Y.-P. Kuang, and X. Li, Phys. Rev. Lett. **69** (1992) 2619; Phys. Rev. **D49** (1994) 4842; Phys. Lett. **B329** (1994) 278; and references therein.
6. E. Accomando et al, Phys. Rep. (1998), in press and DESY-97-100 (hep-ph/9705442); H. Murayama and M.E. Peskin, Ann. Rev. Nucl. & Part. Sci. **46** (1996) 533.
7. M.E. Peskin, Prog. Theor. Phys. Suppl. **123** (1996) 507; H.E. Haber, hep-ph/9703381; G.L. Kane, hep-ph/9709318.
8. T. Appelquist and J. Carazzone, Phys. Rev. **D11** (1975) 2856.
9. M. Peskin, Talk at the Snowmass Conference (June, 1996), and the working group summary report, hep-ph/9704217.
10. For a nice review, H. Georgi, Ann. Rev. Nucl. & Part. Sci. **43** (1994) 209.
11. T. Appelquist and C. Bernard, Phys. Rev. **D22** (1980) 200; A.C. Longhitano, Nucl. Phys. **B188** (1981) 118; T. Appelquist and G.-H. Wu, Phys. Rev. **D48** (1993) 3235; J. Bagger, S. Dawson, and G. Valencia, Nucl. Phys. **B399** (1993) 364; and references therein.
12. S. Weinberg, Physica **96A** (1979) 327.
13. H.-J. He, Y.-P. Kuang, and C.-P. Yuan, hep-ph/9708402; Phys. Rev. **D55** (1997) 3038, Mod. Phys. Lett. **A11** (1996) 3061; Phys. Lett. **B382** (1996) 149; for an updated comprehensive review, hep-ph/9704276 and DESY-97-056, Lectures in the Proceedings of the CCAST (World Laboratory) Workshop on *Physics at TeV Energy Scale*, Vol. 72, pp.119-234.
14. R. Casalbuoni, et al, Phys. Rev. **D53** (1996) 5201; D. Dominici, hep-ph/9711385; and references therein.
15. H. Georgi, Nucl. Phys. **B331** (1990) 311.
16. C.G. Callan, S. Coleman, J. Wess, B. Zumino, Phys. Rev. **177** (1969) 2247.
17. J. Bagger, V. Barger, K. Cheung, J. Gunion, T. Han, G.A. Ladinsky, R. Rosenfeld, and C.-P. Yuan, Phys. Rev. **D49** (1994) 1246; **D52** (1995) 3878.
18. V. Barger, K. Cheung, T. Han, R.J.N. Phillips, Phys. Rev. **D52** (1995) 3815.
19. M.S. Chanowitz and M.K. Gaillard, *Phys. Lett.* **B142**, 85 (1984); G.L. Kane, W.W. Repko and W.R. Rolnick, *ibid.* **B148**, 367 (1984); S. Dawson, *Nucl. Phys.*

B249, 427 (1985).
20. H.-J. He, DESY-97-037, in the proceedings of " *The Higgs Puzzle* ", pp.207-217, Ringberg, Munich, Germany, December 8-13, 1996, Ed. B. Kniehl, (World Scientific Pub).
21. E. Boos, H.-J. He, W. Kilian, A. Pukhov, C.-P. Yuan, and P.M. Zerwas, Phys. Rev. **D57** (1998) 1553 and hep-ph/9708310.
22. V. Barger and T. Han, Phys. Lett. **B212** (1988) 117;
    V. Barger, T. Han, and R.J.N. Phillips, Phys. Rev. **D39** (1989) 146.
23. G. Belanger and F. Boudjema, Phys. Lett. **B288** (1992) 201;
    S. Dawson, A. Likhoded, G. Valencia, and O. Yuschenko, hep-ph/9610299.
24. T. Han, H.-J. He, and C.-P. Yuan, Phys. Lett. **B422** (1998) 294 and hep-ph/9711429.
25. O.J.P. Eboli, M.C. Gonzalez-Garcia, and J.K. Mizukoshi, hep-ph/9711499.
26. P.W. Higgs, Phys. Lett. **12** (1964) 132; Phys. Rev. Lett. **13** (1964) 508; F. Englert and R. Brout, Phys. Rev. Lett. **13** (1964) 321; G.S. Guralnik, C.R. Hagen, and T.W.B. Kibble, Phys. Rev. Lett. **13** (1964) 585.
27. V. Barger, M. Berger, J. Gunion and T. Han, Phys. Rev. **D55** (1997) 142.
28. J. Gunion, in these proceedings and hep-ph/9802258;
    V. Barger, M. Berger, J. Gunion and T. Han, hep-ph/9604334.

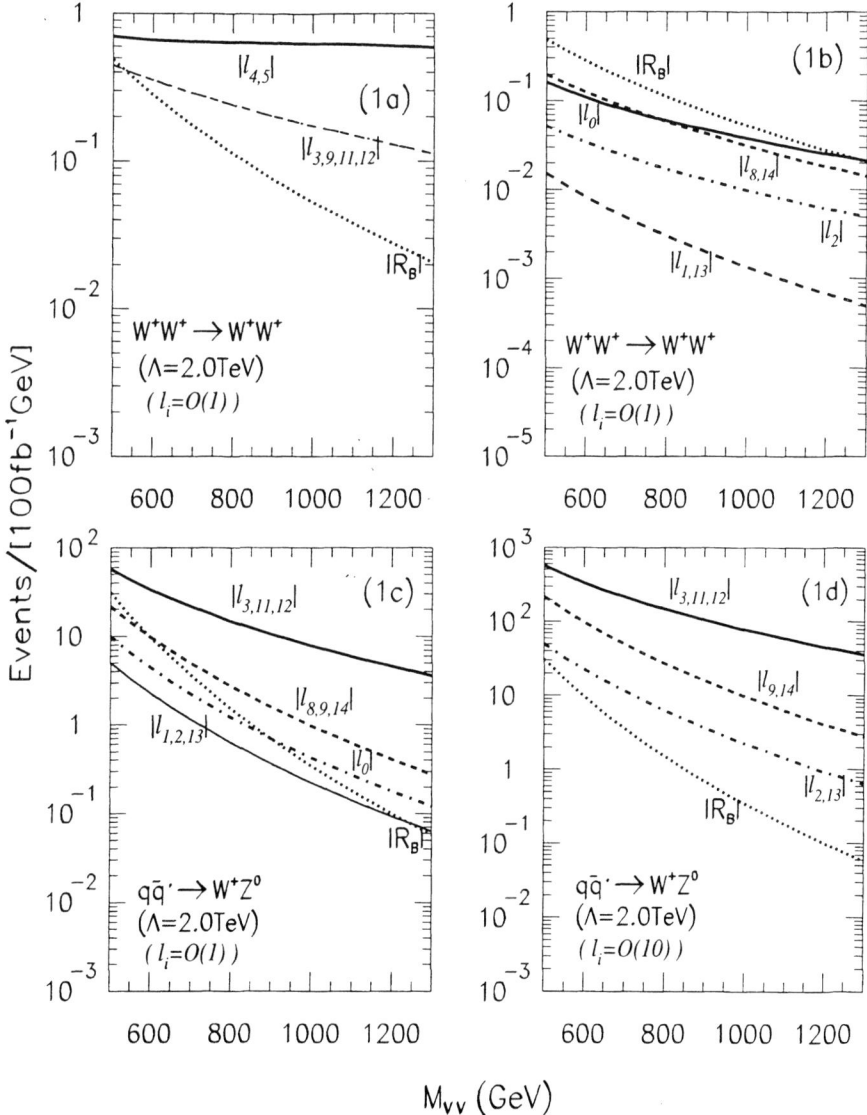

FIGURE 1. A classification of the contributions from all 15 next-to-leading order operators at the 14 TeV LHC (with 100 fb^{-1} annual luminosity) for $\Lambda = 2$ TeV.

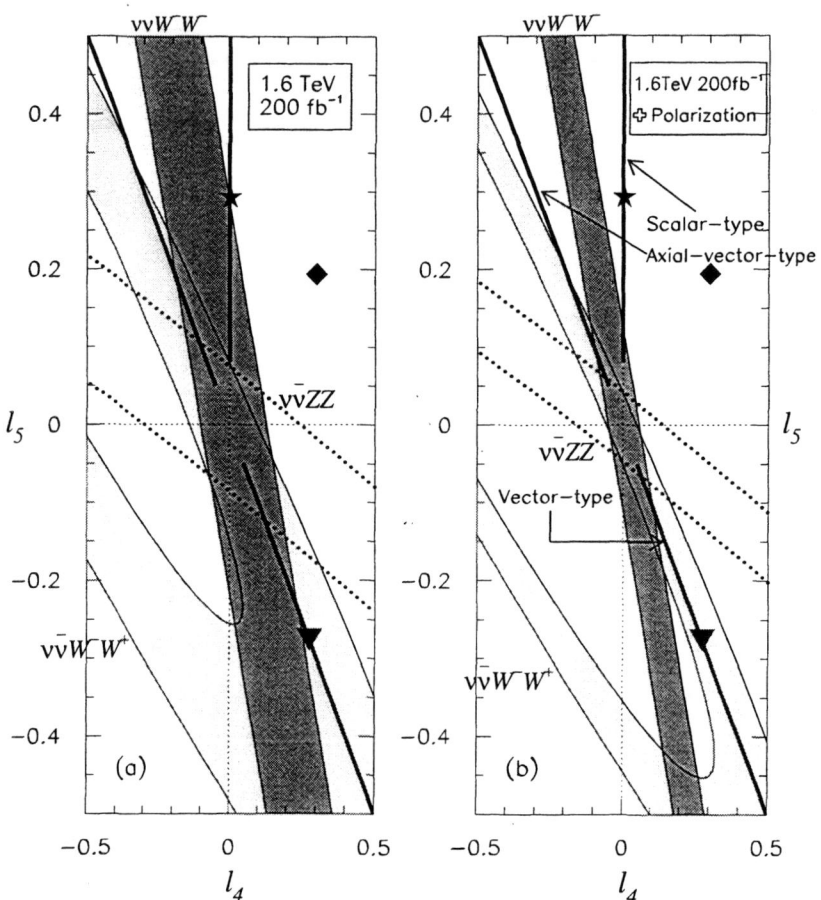

FIGURE 2. Determining the $SU(2)_c$-symmetric parameters ℓ_4-ℓ_5 at 1.6 TeV e^-e^+/e^-e^- LCs. Here the $\pm 1\sigma$ exclusion contours are displayed. (a). unpolarized case; (b). the case with 90%(65%) polarized $e^-(e^+)$ beam. Contributions from three types of resonance models (scalar, vector and axial-vector) to (ℓ_4, ℓ_5) are shown by the thick solid lines. The different points on these solid lines correspond to different values of their couplings to the weak gauge bosons. Note that for axial-vector-type, it is also possible to have $\ell_4 + \ell_5 = 0$ with $\ell_4 \geq 0$, i.e., similar to the vector-type case. This makes the discrimination more involved. Big-star: from a scalar; black-triangle-down: from a vector; black-lozenge: from mixed contributions of a heavy scalar and vector. (Here we typically set these heavy resonances around 2 TeV.)

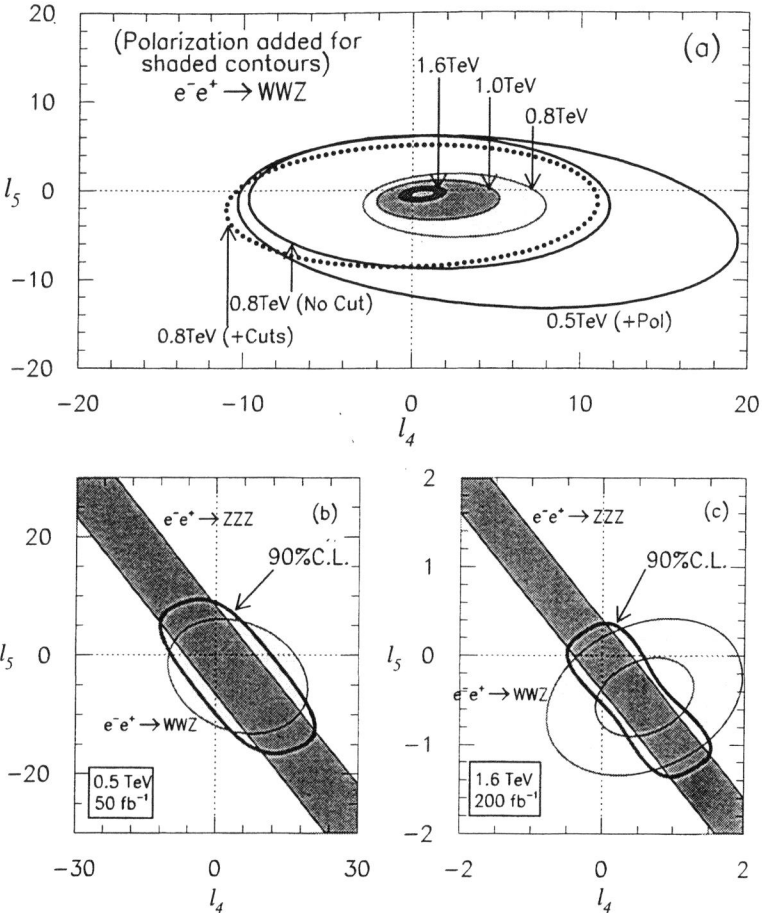

FIGURE 3. Probing ℓ_4-ℓ_5 via WWZ and ZZZ production processes. The roles of the polarization and the higher collider energy for $e^-e^+ \to WWZ$ are shown by the $\pm 1\sigma$ exclusion contours in (a). The integrated luminosities used here are 50 fb^{-1} (at 500 GeV), 100 fb^{-1} (at 800 GeV) and 200 fb^{-1} (at 1.0 and 1.6 TeV). In (b) and (c), the $\pm 1\sigma$ contours are displayed for ZZZ/WWZ final states at $\sqrt{s}$ =0.5 and 1.6 TeV respectively, with two beam polarizations (90% e^- and 65% e^+); the thick solid lines present the combined bounds at 90% C.L.

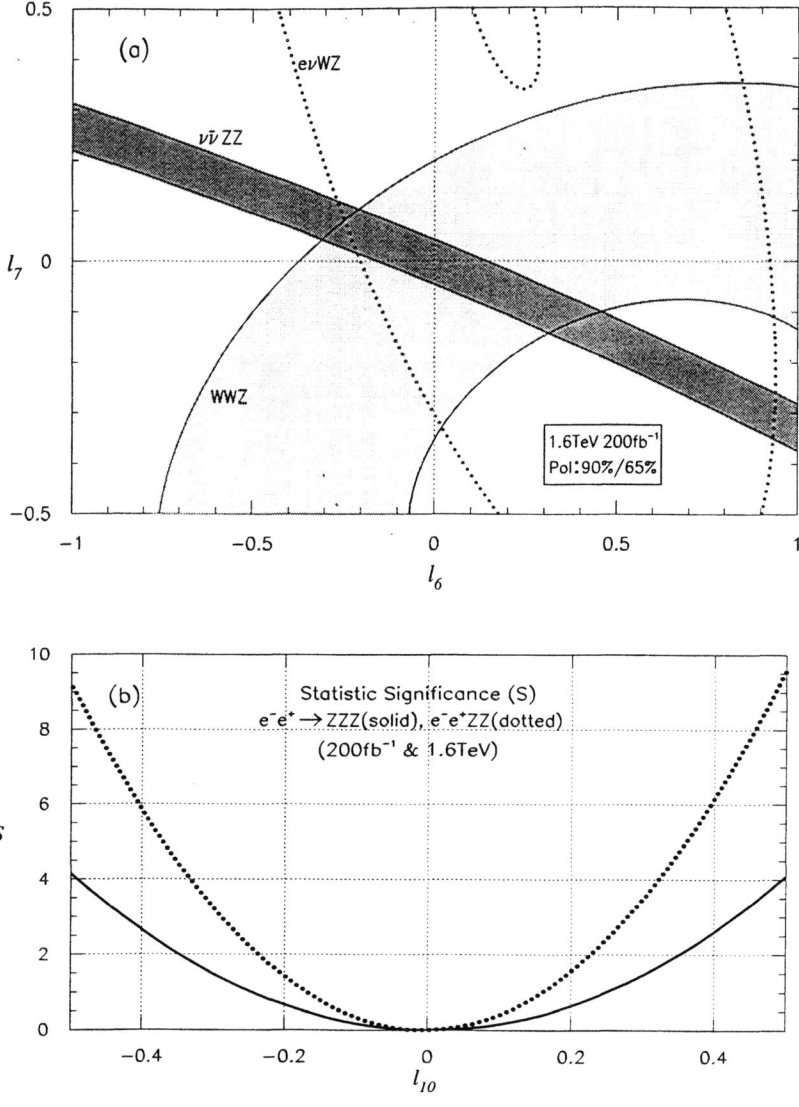

FIGURE 4. Interplay of the WW-fusion and WWZ/ZZZ-production for discriminating the $SU(2)_c$-breaking parameters ℓ_6-ℓ_7 and ℓ_{10} at $\sqrt{s} =1.6$ TeV with $\int \mathcal{L} =200$ fb^{-1}: (a). $\pm 1\sigma$ exclusion contours for $e^-e^+ \to \nu\bar{\nu}ZZ$, $e^+\nu W^-Z/e^-\bar{\nu}W^+Z$, and $e^-e^+ \to WWZ$ with polarizations (90% e^- and 65% e^+). (b). Statistic significance versus ℓ_{10} for $e^-e^+ \to ZZZ$, e^-e^+ZZ (with unpolarized $e^{\mp}$ beams).

Probing Anomalous Higgs Coupling through $\mu^+\mu^- \to H\gamma$

Ali Abbasabadi†, David Bowser-Chao§,¶,
Duane A. Dicus* and Wayne W. Repko‡

† *Department of Physical Sciences, Ferris State University, Big Rapids, Michigan 49307*
§ *Department of Physics, University of Illinois at Chicago, Chicago, Illinois 60607;*
* *Center for Particle Physics and Department of Physics, University of Texas, Austin, Texas 78712*
‡ *Department of Physics and Astronomy, Michigan State University, East Lansing, Michigan 48824*
¶ *Presented the talk.*

Abstract. The process $\mu^+\mu^- \to H\gamma$, though small compared to the resonance process, could be observable at proposed muon colliders, given expected integrated luminosities. The apparently leading diagrams occur at tree-level, and are proportional to the Higgs-muon coupling. We show, however, that the one-loop contribution is comparable to the tree-level amplitude. Furthermore, the one-loop diagrams, unlike those at tree-level, could be greatly enhanced by possible anomalous Higgs-top quark or Higgs-gauge boson couplings. For a 500 GeV unpolarized muon collider, the total cross section for $H\gamma$ associated production approaches 0.1 fb.

Recently, the possibility of using $\mu^+\mu^-$ colliders to investigate the properties of Higgs-bosons has received considerable attention [1,2]. There are significant advantages to studying Higgs-bosons with this type of collider, particularly if the mass is known from its discovery at, say, the LHC or NLC [2]. Under these circumstances, the width and branching ratios can be studied at the Higgs pole.

Resonance production, of course, is not the only channel for Higgs production at such colliders. Production in association with a photon has been considered in Ref. [3], where the leading tree-level diagrams (Fig. 1) involve radiation of the Higgs directly off the muon line, and thus are proportional the Higgs-muon coupling, which in turn (in the Standard Model) is proportional to the small but non-zero muon mass.

On the other hand, the analogous process $e^+e^- \to H\gamma$ has been studied [4], and found to have a cross-section comparable to the muon process, despite the relatively infinitesimal electron mass. In this case, it turns out that the one-loop diagrams are orders of magnitude larger than the tree-level diagrams, and in fact should not be considered radiative corrections to the latter — the former persist even in the limit of zero electron mass. This process can be generalized to the hadronic reaction $q\bar{q} \to H\gamma$ [5], and to the rare decays $H \to f\bar{f}\gamma$, where the fermion f may be massless [6]. Turning back to the muon reaction, we shall see below that the one-loop contribution is comparable to the tree-level amplitude. Furthermore, the one-loop diagrams, unlike those at tree-level, could be greatly enhanced by possible anomalous Higgs-top quark or Higgs-gauge boson couplings.

For both the tree-level and loop-level diagrams discussed in detail below, we label the momentum and helicity of the μ^- (μ^+) in the center of mass frame, respectively, by $p = (E, |\mathbf{p}|\hat{z})$ ($\bar{p} = (E, -|\mathbf{p}|\hat{z})$) and λ ($\bar{\lambda}$), the photon momentum and scattering angle by $k = (\omega, \omega\hat{k})$ and θ, and the photon polarization vector and helicity by ϵ and $\lambda_\gamma = \pm 1$. The tree level diagrams for $\mu^+\mu^- \to H\gamma$ are shown in Fig. 1. The amplitudes, labelled by muon and photon helicities, are found to be:

$$\mathcal{M}^{\text{tree}}_{\lambda\bar{\lambda}\lambda_\gamma} = -i\frac{egm_\mu}{\sqrt{2}\,m_W}\left(\frac{1}{2p\cdot k} + \frac{1}{2\bar{p}\cdot k}\right)$$
$$\cdot \begin{cases} \sin\theta\,[\lambda_\gamma(2|\mathbf{p}|^2 - E\omega) + |\mathbf{p}|\omega] & \lambda\bar{\lambda} = ++ \\ \sin\theta\,[\lambda_\gamma(2|\mathbf{p}|^2 - E\omega) - |\mathbf{p}|\omega] & \lambda\bar{\lambda} = -- \\ m_\mu\omega(1 + \lambda_\gamma\cos\theta) & \lambda\bar{\lambda} = +- \\ m_\mu\omega(1 - \lambda_\gamma\cos\theta) & \lambda\bar{\lambda} = -+ \end{cases}, \quad (1)$$

explicitly showing the extra suppression of the helicity flip amplitudes by a factor of m_μ relative to the non-flip amplitudes (where $\lambda = +$ denotes a μ^- helicity of $+1/2$.)

The one-loop amplitudes for $\mu^+\mu^- \to H\gamma$ receive contributions from pole diagrams involving virtual photon and Z exchange and from various box diagrams containing muons, gauge bosons and/or Goldstone bosons [4,8,7]. There are also double pole diagrams whose contribution vanishes. These are illustrated in Fig. 2. In the non-linear gauges we chose [4], the full amplitude consists of four separately gauge invariant terms: a photon pole, a Z pole, Z boxes and W boxes. These amplitudes take the form:

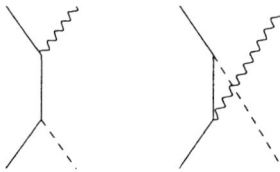

FIGURE 1. Tree level diagrams for $\mu^+\mu^- \to H\gamma$ are shown.

TABLE 1. The cross sections for the associated production of 100 GeV and 200 GeV Higgs bosons are shown together with the ratio of the signal to the square root of the background ($S/\sqrt{B}$) for several $\mu^+\mu^-$ collider energies. A luminosity of 100 fb^{-1} is assumed.

$\sqrt{s}$	$\sigma(m_H = 100$ GeV$)$	$S/\sqrt{B}$	$\sigma(m_H = 200$ GeV$)$	$S/\sqrt{B}$
500 GeV	6.78×10^{-2} fb	1.99	8.76×10^{-2} fb	3.06
1000 GeV	2.46×10^{-2} fb	1.50	3.87×10^{-2} fb	3.08
2000 GeV	8.76×10^{-3} fb	1.06	1.04×10^{-2} fb	1.72
4000 GeV	1.54×10^{-3} fb	0.36	2.17×10^{-3} fb	0.72

$$\mathcal{M}^{\gamma,Z}_{\text{pole}} = \mathcal{P}_{\gamma,Z}(s)\left(\delta_{\mu\nu}k\cdot(p+\bar{p}) - k_\mu(p+\bar{p})_\nu\right)\bar{v}(\bar{p})\gamma_\mu(v^{\mathcal{P}}_{\gamma,Z} + a^{\mathcal{P}}_{\gamma,Z}\gamma_5)u(p)\epsilon^*_\nu, \quad (2)$$

$$\mathcal{M}^{\gamma,Z}_{\text{box}} = [\mathcal{B}_{\gamma,Z}(s,t,u)(\delta_{\mu\nu}k\cdot p - k_\mu p_\nu) + p \leftrightarrow \bar{p}]\,\bar{v}(\bar{p})\gamma_\mu(v^{\mathcal{B}}_{\gamma,Z} + a^{\mathcal{B}}_{\gamma,Z}\gamma_5)u(p)\epsilon^*_\nu, \quad (3)$$

where $s = -(p+\bar{p})^2$, $t = -(p-k)^2$ and $u = -(\bar{p}-k)^2$. We have explicitly evaluated the form factors ($\mathcal{P}_\gamma$, etc.) in terms of the scalar functions defined in the appendices of our previous paper [4]; the results, though more complicated than those of the tree-level diagrams, are provided in Ref. [7] in closed form. In this case, as should be expected, it is the helicity *flip* contributions from the factors $\bar{v}(\bar{p})\gamma_\mu u(p)$ and $\bar{v}(\bar{p})\gamma_\mu\gamma_5 u(p)$ which survive in the $m_\mu \to 0$ limit, as can be seen, for example, by explicitly evaluating the spinor products above.

The differential cross section $d\sigma(\mu^+\mu^- \to H\gamma)/d\Omega_\gamma$ is given by

$$\frac{d\sigma(\mu^+\mu^- \to H\gamma)}{d\Omega_\gamma} = \frac{1}{256\pi^2}\frac{s - m_H^2}{\beta s^2}\sum_{\text{spin}}|\mathcal{M}^{\text{tree}} + \mathcal{M}^{\text{loop}}|^2, \quad (4)$$

with $\beta = \sqrt{1 - 4m_\mu^2/s}$. When integrating Eq. (4) to obtain the total cross section, the interference terms should be suppressed, since $\mathcal{M}^{\text{tree}}_{\pm\mp}$ and $\mathcal{M}^{\text{loop}}_{\pm\pm}$ both contain an additional factor of m_μ. This conclusion can only be invalid

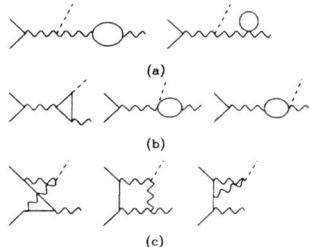

FIGURE 2. Typical diagrams for the double pole (a), single pole (b) and box (c) corrections are shown. An external solid line represents a muon, a wavy line a gauge boson, a dashed line a Higgs boson and an internal solid line a muon, gauge boson, Goldstone boson or ghost.

TABLE 2. Cross sections for the background process $\mu^+\mu^- \to \gamma b\bar{b}$ are given for several cuts on the $b\bar{b}$ invariant mass $m_{b\bar{b}}$. The last two columns are 5 GeV bins indicating, respectively, the background associated with a Higgs boson of mass 100 GeV or 200 GeV.

$\sqrt{s}$	45 GeV $< m_{b\bar{b}} < \sqrt{s}$	97.5 GeV $< m_{b\bar{b}} <$ 102.5 GeV	197.5 GeV $< m_{b\bar{b}} <$ 202.5 GeV
500 GeV	11.1 fb	1.16×10^{-1} fb	8.20×10^{-2} fb
1000 GeV	3.80 fb	2.69×10^{-2} fb	1.58×10^{-2} fb
2000 GeV	1.21 fb	6.83×10^{-3} fb	3.65×10^{-3} fb
4000 GeV	0.37 fb	1.81×10^{-3} fb	9.04×10^{-4} fb

if the angular integration of the muon propagator factors $(1 \pm \beta \cos\theta)^{-1}$ in Eqs. (1) produces inverse powers of m_μ. This is not the case. For the ++ or $--$ interference terms, the tree and loop amplitudes contain a factor of $\sin\theta$, which ensures that the angular integral is well behaved in the $\beta \to 1$ limit. The integral of $+-$ and $-+$ interference terms can produce a factor of β^{-1}, but this also is finite as $\beta \to 1$. As a consequence, we can simply add the tree [9] and one-loop cross sections to obtain $\sigma(\mu^+\mu^- \to H\gamma)$ for unpolarized muon beams. Working to leading order in m_μ, the polarized cross-sections are obtained simply by omitting the tree-level or loop-level contributions for helicity-flip or same-helicity beam polarization.

The result is illustrated in Fig. 3, where the tree, one-loop and total cross sections are plotted for several values of m_H as a function of the collider energy. For collider energies $\sqrt{s}$ above about 500 GeV, the one-loop contribution exceeds the tree contribution. Note that Fig. 3 should not be taken literally at $\sqrt{s} \approx m_H$. As $\omega \to 0$, the tree-level process must be considered together with the virtual QED correction to the resonance process $\mu^+\mu^- \to H$ to obtain an infrared-finite $O(\alpha)$ inclusive calculation [10]. Here, we are concerned with production of the Higgs with an observable, relatively hard photon. In Table I, the total cross section is shown as a function of m_H for several collider energies. At 500 GeV, luminosities of order 100 fb^{-1} are needed to probe this channel. To make this statement more precise, we investigated the principal background $\mu^+\mu^- \to b\bar{b}\gamma$ by adapting the amplitudes for $e\bar{e} \to \mu^+\mu^-\gamma$ [11]. In Table II, the background contributions are shown for several cuts on the $b\bar{b}$ invariant mass $m_{b\bar{b}}$. In addition to these invariant mass cuts, we require the transverse momenta of the b, $\bar{b}$ and γ to be greater than 15 GeV, their rapidities y to be less than 2.5 and the separation ΔR between the γ and the b and the γ and the $\bar{b}$ to be greater than 0.4. The background is compared to the signal in Table I for Higgs boson masses of 100 GeV and 200 GeV. This comparison shows that, while not a discovery mode for the Higgs boson, photon-Higgs associated production can be observed with signal to square root of background ratios $(S/\sqrt{B})$ greater than 2 when $m_H > 100$ GeV at a 500

GeV collider.

Finally, we comment on how the one-loop contribution makes it possible to use $\mu^+\mu^- \to H\gamma$ as a probe of the Higgs-boson coupling to W's, Z's and top quarks. To provide a sense of the sensitivity of the $H\gamma$ cross section to changes in Standard Model couplings, we have varied the $t\bar{t}H$ coupling by a factor λ [12]. The result is shown in Fig. 5, where the characteristic feature is the minimum in the cross section at the Standard Model value $\lambda = 1$. For $\lambda > 1$, the cross section rises significantly. At a 500 GeV collider, observation of $H\gamma$ production with a cross section of order 1 fb would indicate some type of anomalous coupling. Anomalous Higgs-gauge boson couplings have been studied [13] for the case of $e^+e^- \to H\gamma$ where enhancement of couplings such as HZZ could result in immense enhancements to the level of 10 fb. Thus, polarization of the muon collider *opposite* to that required for optimizing resonance Higgs production may be useful in highlighting the loop-level contributions to $\mu^+\mu^- \to H\gamma$, which if unexpectedly large could signal such anomalous couplings.

This research was supported in part by the U.S. Department of Energy

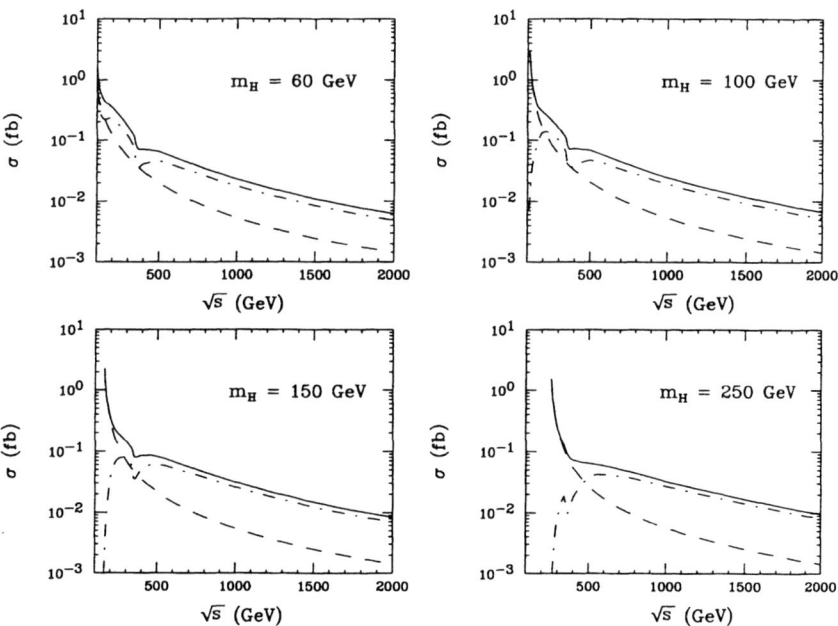

FIGURE 3. The cross section for $\mu^+\mu^- \to H\gamma$ resulting from the sum of the tree level and one-loop amplitudes is given for several values of m_H by the solid line. In each panel, the dashed line is the tree level contribution and the dot-dashed line is the one-loop contribution.

under Contract Nos. DE-FG013-93ER40757 and DE-FG02-84ER40173, and in part by the National Science Foundation under Grant No. PHY-93-07980.

REFERENCES

1. R. B. Palmer, A. Sessler and A. Tollestrup, *Progress on the design of a high luminosity $\mu^+\mu^-$ collider*, BNL-63245 (1996).
2. J. F. Gunion, *Muon Colliders: The Machine and The Physics*, UCD-97-17, hep-ph/9707379 (1997).
3. V. A. Litvin and F. F. Tikhonin, *Associated production of $H\gamma$ or HZ pairs at $\mu^+\mu^-$ collisions*, hep-ph/9704417 (1997).
4. A. Abbasabadi, D. Bowser-Chao, D. A. Dicus and W. W. Repko, Phys. Rev. D **52**, 3919 (1995).
5. A. Abbasabadi, D. Bowser-Chao, D. A. Dicus and W. W. Repko, hep-ph/9706335.
6. A. Abbasabadi, D. Bowser-Chao, D. A. Dicus and W. W. Repko, Phys. Rev. **D55**, 5647 (1997).
7. A. Abbasabadi, D. Bowser-Chao, D. A. Dicus and W. W. Repko, hep-ph/9708328, to appear in Phys. Rev. **D57**, (1998).

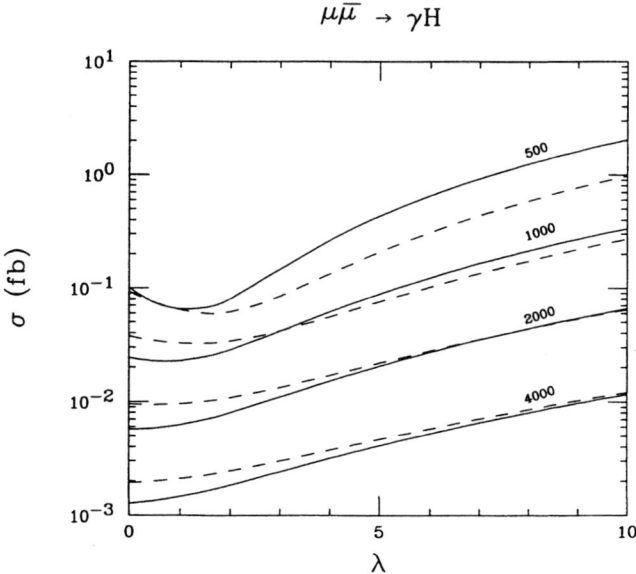

FIGURE 4. The cross section for $\mu^+\mu^- \to H\gamma$ obtained by scaling the Standard Model $t\bar{t}H$ coupling by a factor λ is shown for collider energies of 500 GeV, 1000 GeV, 2000 GeV and 4000 GeV. In each case, the solid line is $m_H = 60\,\text{GeV}$ and the dashed line is $m_H = 250\,\text{GeV}$.

8. A. Djouadi, V. Driesen, W. Hollick and J. Rosiek, University of Karlsruhe preprint KA-TP-21-96, hep-ph/9609420.
9. The $+-$ (and $-+$) amplitudes contribute at most a few percent to the tree level cross section for 500 GeV $\leq \sqrt{s} \leq$ 4 TeV and 60 GeV $\leq m_H \leq$ 300 GeV.
10. V. A. Litvin and F. F. Tikhonin, work in progress.
11. F. Berends *et al.*, Nucl. Phys. B **206**, 61 (1982); Z. Xu,D-H Zhang, and L. Chang, Nucl. Phys. B **291**, 392 (1987).
12. H. E . Haber, G. L. Kane and T. Sterling, Nuc. Phys. B **161**, 493 (1979).
13. G.J. Gounaris, F.M. Renard, and N.D. Vlachos, Nucl. Phys. **B459**, 51 (1996).

Strong Dynamics

Conveners: Estia Eichten, Fermilab
            Pushpa Bhat, Fermilab

Technicolor and the First Muon Collider [1]

Kenneth Lane

Department of Physics, Boston University, 590 Commonwealth Ave, Boston, MA 02215

Abstract.
The motivations for studying dynamical scenarios of electroweak and flavor symmetry breaking are reviewed and the latest ideas, especially topcolor-assisted technicolor, are summarized. Technicolor's observable low-energy signatures are discussed. The superb energy resolution of the First Muon Collider may make it possible to resolve the extraordinarily narrow technihadrons that occur in such models—π_T^0, ρ_T^0, ω_T—and produce them at very large rates compared to other colliders.

I OVERVIEW OF TECHNICOLOR

Technicolor—the strong interaction of fermions and gauge bosons at the scale $\Lambda_{TC} \sim 1\,{\rm TeV}$—describes the breakdown of electroweak symmetry to electromagnetism *without* elementary scalar bosons [1]. In its simplest form, technicolor is a scaled-up version of QCD, with massless technifermions whose chiral symmetry is spontaneously broken at Λ_{TC}. If left and right-handed technifermions are assigned to weak $SU(2)$ doublets and singlets, respectively, then $M_W = \cos\theta_W M_Z = \frac{1}{2} g F_\pi$, where $F_\pi = 246\,{\rm GeV}$ is the *technipion* decay constant, [2] analogous to $f_\pi = 93\,{\rm MeV}$ for the ordinary pion.

The principal signals in hadron and lepton collider experiments of "classical" technicolor were discussed long ago [2,3]. In the minimal technicolor model, with just one technifermion doublet, the only prominent collider signals are the enhancements in longitudinally-polarized weak boson production. These are the s-channel color-singlet technirho resonances near 1.5–2 TeV: $\rho_T^0 \to W_L^+ W_L^-$ and $\rho_T^\pm \to W_L^\pm Z_L^0$. The $\mathcal{O}(\alpha^2)$ cross sections of these processes are quite small

---

[1] Talk presented at the Workshop on Physics at the First Muon Collider and at the Front End of a Muon Collider
[2] The only technipions in minimal technicolor are the massless Goldstone bosons that become, via the Higgs mechanism, the longitudinal components $W_L^\pm$ and Z_L^0 of the weak gauge bosons.

at such masses. This and the difficulty of reconstructing weak-boson pairs with reasonable efficiency make observing these enhancements a challenge.

Nonminimal technicolor models are much more accessible because they have a rich spectrum of lower mass technirho vector mesons and technipion states into which they may decay. [3] If there are N_D doublets of technifermions, all transforming according to the same complex representation of the technicolor gauge group, there will be $4N_D^2 - 1$ technipions whose decay constant is

$$F_T = \frac{F_\pi}{\sqrt{N_D}}. \tag{1}$$

Three of these are the longitudinal weak bosons; the remaining $4N_D^2 - 4$ await discovery.

In the standard model and its extensions, the masses of quarks and leptons are produced by their Yukawa couplings to the Higgs bosons—couplings of arbitrary magnitude and phase that are put in by hand. This option is not available in technicolor because there are no elementary scalars. Instead, quark and lepton chiral symmetries must be broken explicitly *by gauge interactions alone*. The most economical way to do this is to employ extended technicolor, a gauge group containing flavor, color and technicolor as subgroups [4–6]. Quarks, leptons and technifermions are unified into a few large representations of ETC. The ETC gauge symmetry is broken at high energy to technicolor $\otimes$ color. Then quark and lepton hard masses arise from their coupling (with strength g_{ETC}) to technifermions via ETC gauge bosons of generic mass M_{ETC}:

$$m_q(M_{ETC}) \simeq m_\ell(M_{ETC}) \simeq \frac{g_{ETC}^2}{M_{ETC}^2} \langle \bar{T}T \rangle_{ETC}, \tag{2}$$

where $\langle \bar{T}T \rangle_{ETC}$ and $m_{q,\ell}(M_{ETC})$ are the technifermion condensate and quark and lepton masses renormalized at the scale M_{ETC}.

If technicolor is like QCD, with a running coupling α_{TC} rapidly becoming small above $\Lambda_{TC} \sim 1\,\text{TeV}$, then $\langle \bar{T}T \rangle_{ETC} \simeq \langle \bar{T}T \rangle_{TC} \simeq \Lambda_{TC}^3$. To obtain quark masses of a few GeV, $M_{ETC}/g_{ETC} \lesssim 30\,\text{TeV}$ is required. This is excluded: Extended technicolor boson exchanges also generate four-quark interactions which, typically, include $|\Delta S| = 2$ and $|\Delta B| = 2$ operators. For these not to be in conflict with $K^0\text{-}\bar{K}^0$ and $B_d^0\text{-}\bar{B}_d^0$ mixing parameters, M_{ETC}/g_{ETC} must exceed several hundred TeV [5]. This implies quark and lepton masses no larger than a few MeV, and technipion masses no more than a few GeV—a phenomenological disaster.

Because of this conflict between constraints on flavor-changing neutral currents and the magnitude of ETC-generated quark, lepton and technipion

---

[3] The technipions of nonminimal technicolor include the longitudinal weak bosons as well as additional Goldstone bosons associated with spontaneous technifermion chiral symmetry breaking. The latter must and do acquire mass—from the extended technicolor interactions discussed below.

masses, classical technicolor was superseded over a decade ago by "walking" technicolor [7]. Here, the strong technicolor coupling α_{TC} runs very slowly—walks—for a large range of momenta, possibly all the way up to the ETC scale of several hundred TeV. The slowly-running coupling enhances $\langle \bar{T}T \rangle_{ETC}/\langle \bar{T}T \rangle_{TC}$ by almost a factor of M_{ETC}/Λ_{TC}. This, in turn, allows quark and lepton masses as large as a few GeV and $M_{\pi_T} \gtrsim 100\,\text{GeV}$ to be generated from ETC interactions at $M_{ETC} = \mathcal{O}(100\,\text{TeV})$.

Walking technicolor requires a large number of technifermions in order that α_{TC} runs slowly. These fermions may belong to many copies of the fundamental representation of the technicolor gauge group, to a few higher dimensional representations, or to both. [4]

In many respects, walking technicolor models are very different from QCD with a few fundamental $SU(3)$ representations. One example of this is that integrals of weak-current spectral functions and their moments converge much more slowly than they do in QCD. Consequently, simple dominance of the spectral integrals by a few resonances cannot be correct. This and other calculational tools based on naive scaling from QCD and on large-N_{TC} arguments are suspect [10]. Thus, it is not yet possible to predict with confidence the influence of technicolor degrees of freedom on precisely-measured electroweak quantities—the S, T, U parameters to name the most discussed example [11].

The large mass of the top quark [12] motivated another major development in technicolor. Theorists have concluded that ETC models cannot explain the top quark's large mass without running afoul of experimental constraints from the ρ parameter and the $Z \to \bar{b}b$ decay rate [13]. This state of affairs has led to the proposal of "topcolor-assisted technicolor" (TC2) [14].

In TC2, as in top-condensate models of electroweak symmetry breaking [15], almost all of the top quark mass arises from a new strong "topcolor" interaction [16]. To maintain electroweak symmetry between (left-handed) top and bottom quarks and yet not generate $m_b \simeq m_t$, the topcolor gauge group under which (t,b) transform is usually taken to be a strongly-coupled $SU(3) \otimes U(1)$. The $U(1)$ provides the difference that causes only top quarks to condense. Then, in order that topcolor interactions be natural—i.e., that their energy scale not be far above m_t—without introducing large weak isospin violation, it is necessary that electroweak symmetry breaking remain due mostly to technicolor interactions [14].

Early steps in the development of the TC2 scenario have been taken in two recent papers [17]. The breaking of topcolor $SU(3) \otimes U(1)$ near the electroweak scale gives rise to a massive color octet of V_8 colorons and a color-singlet Z'.

---

[4] The last possibility inspired "multiscale technicolor" models containing both fundamental and higher representations, and having an unusual phenomenology [8]. In multiscale models, there typically are two widely separated scales of electroweak symmetry breaking, with the upper scale set by the weak decay constant, $F_\pi = 246\,\text{GeV}$. Multiscale models in which the entire top quark mass is generated by ETC interactions are excluded by such processes as $b \to s\gamma$ [9].

The $SU(3)$ may be broken by some of the same technifermion condensates that break electroweak $SU(2) \otimes U(1)$, so that the colorons (which are expected to be broad) have mass near 500 GeV. However, in order that the strong topcolor $U(1)$ interaction not contaminate the ordinary Z couplings to fermions, it and the weaker $U(1)$ acting on light fermions must be broken down to their diagonal subgroup, ordinary weak hypercharge, in the vicinity of 2 TeV. This suggests that the Z' mass is in the range 1–3 TeV, out of reach of all but the highest energy colliders. As I discussed in my talk at the FMC workshop, the Z' is so heavy that it may require a multi-TeV Big Muon Collider to find and study it. This subject deserves further study.

In TC2 models, ETC interactions are still needed to generate the light and bottom quark masses, contribute a few GeV to m_t, [5] and give mass to the technipions. The scale of ETC interactions still must be hundreds of TeV to suppress flavor-changing neutral currents and, so, the technicolor coupling still must walk.

Thus, even though the phenomenology of TC2 is still in its infancy, it is expected to share general features with multiscale technicolor: many technifermion doublets bound into many technihadron states, some at relatively low masses, some carrying ordinary color and some not. The lightest technihadrons may have masses in the range 100–300 GeV and should be accessible at the Tevatron collider in Run III if not Run II. All of them are easily produced and detected at the LHC at moderate luminosities. If technihadrons exist, they will be discovered at hadron colliders before the First Muon Collider (FMC) is built. As we shall see, this is a good thing for the FMC: Several of the lightest technihadrons are very narrow and can be produced in the s-channel of $\mu^+\mu^-$ annihilations. In the narrow-band FMC, it would be exceedingly difficult to find them by a standard scan procedure without a good idea of where to look.

II TECHNICOLOR AT THE FMC

A Technihadron Decay Rates

I assume that the technicolor gauge group is $SU(N_{TC})$ and take $N_{TC} = 4$ in calculations. Its gauge coupling must walk and I assume this is achieved by a large number of isodoublets of technifermions transforming according to the fundamental representation of $SU(N_{TC})$. I consider the phenomenology of only the lightest color-singlet technihadrons and assume that the constraint from the S-parameter on their spectrum still allows the lightest ones to be considered in isolation for a *limited* range of $\sqrt{s}$, the $\mu^+\mu^-$ center-of-mass energy, about their masses. These technihadrons carry isospin $I = 1$ and 0 and

---

[5] Massless Goldstone "top-pions" arise from top-quark condensation. This ETC contribution to m_t is needed to give them a mass in the range of 150–250 GeV.

consist of a single isotriplet and isosinglet of vectors, ρ_T^0, $\rho_T^\pm$ and ω_T, and pseudoscalars π_T^0, $\pi_T^\pm$, and $\pi_T^{0\prime}$. The latter are in addition to the longitudinal weak bosons, $W_L^\pm$ and Z_L^0—those linear combinations of technipions that couple to the electroweak gauge currents. I adopt TC2 as a guide for guessing phenomenological generalities. In TC2 there is no need for large technifermion isospin splitting associated with the top-bottom mass difference. This implies that the lightest ρ_T and ω_T are approximately degenerate. The lightest charged and neutral technipions also should have roughly the same mass, but there may be appreciable π_T^0–$\pi_T^{0\prime}$ mixing. If that happens, the lightest neutral technipions are really $\bar{U}U$ and $\bar{D}D$ bound states. Finally, for purposes of discussing signals at the FMC, we take the lightest technihadron masses to be

$$M_{\rho_T} \cong M_{\omega_T} \sim 200\,\text{GeV}; \qquad M_{\pi_T} \sim 100\,\text{GeV}. \tag{3}$$

The decays of technipions are induced mainly by ETC interactions which couple them to quarks and leptons. These couplings are Higgs-like, and so technipions are expected to decay into the heaviest fermion pairs allowed. Because only a few GeV of the top-quark's mass is generated by ETC, there is no great preference for π_T to decay to top quarks nor for top quarks to decay into them. Furthermore, the isosinglet component of neutral technipions may decay into a pair of gluons *if* its constituent technifermions are colored. Thus, the predominant decay modes of the light technipions are assumed to be

$$\begin{aligned}\pi_T^0 &\to \bar{b}b,\ \bar{c}c,\ \tau^+\tau^- \\ \pi_T^{0\prime} &\to gg,\ \bar{b}b,\ \bar{c}c,\ \tau^+\tau^- \\ \pi_T^\pm &\to c\bar{b},\ c\bar{s},\ \tau^+\nu_\tau\,.\end{aligned} \tag{4}$$

To estimate branching ratios we use the following decay rates (for later use in the technihadron production cross sections, we quote the energy-dependent width [3,18]): [6]

$$\Gamma(\pi_T \to \bar{f}'f) = \frac{1}{16\pi F_T^2}\, N_f\, p_f\, C_f^2 (m_f + m_{f'})^2$$

$$\Gamma(\pi_T^{0\prime} \to gg) = \frac{1}{128\pi^3 F_T^2}\, \alpha_S^2\, C_{\pi_T}\, N_{TC}^2\, s^{\frac{3}{2}}\,. \tag{5}$$

Here, C_f is an ETC-model dependent factor of order one *except* that TC2 suggests $|C_t| \lesssim m_b/m_t$; N_f is the number of colors of fermion f; p_f is the fermion momentum; α_S is the QCD coupling evaluated at M_{π_T}; and C_{π_T} is a Clebsch of order one. For $M_{\pi_T} = 110\,\text{GeV}$, $F_T = F_\pi/3 = 82\,\text{GeV}$, $m_b = 4.2\,\text{GeV}$, $N_{TC} = 4$, $\alpha_S = 0.1$, $C_b = 1$ for π_T^0 and $\pi_T^{0\prime}$, and $C_{\pi_T} = 4/3$:

$$\Gamma(\pi_T^0 \to \bar{b}b) = \Gamma(\pi_T^{0\prime} \to \bar{b}b) = 35\,\text{MeV}$$

$$\Gamma(\pi_T^{0\prime} \to gg) = 10\,\text{MeV}\,. \tag{6}$$

---

[6] The amplitude is taken to be $\mathcal{M}(\pi_T \to \bar{f}'(p_1)f(p_2)) = C_f(m_f + m_{f'})/F_T\, \bar{u}(p_2)\gamma_5 v(p_1)$.

If technicolor were like QCD, we would expect the main decay modes of the lightest technivector mesons to be $\rho_T^0 \to \pi_T^+ \pi_T^-$ and $\omega_T \to \pi_T^+ \pi_T^- \pi_T^0$ with the technihadrons all composed of the same technifermions. However, the large ratio $\langle \bar{T}T \rangle_{ETC}/\langle \bar{T}T \rangle_{TC}$ occurring in walking technicolor significantly enhances technipion masses compared to technivector masses. Thus, $\rho_T \to \pi_T \pi_T$ decay channels may well be closed. If this happens, then ρ_T^0 decays to $W_L^+ W_L^-$ or $W_L^\pm \pi_T^\mp$ and ω_T to $\gamma \pi_T^0$ or $Z^0 \pi_T^0$. [8,19,20,6]

We parameterize this for ρ_T decays by adopting a simple model of two isotriplets of technipions which are mixtures of $W_L^\pm$, Z_L^0 and mass-eigenstate technipions $\pi_T^\pm$, π_T^0. The lighter isotriplet ρ_T is assumed to decay dominantly into pairs of the mixed state of isotriplets $|\Pi_T\rangle = \sin\chi |W_L\rangle + \cos\chi |\pi_T\rangle$, where

$$\sin\chi = F_T/F_\pi. \tag{7}$$

Then, the energy-dependent decay rate for $\rho_T^0 \to \pi_A^+ \pi_B^-$ (where $\pi_{A,B}$ may be W_L, Z_L, or π_T) is given by

$$\Gamma(\rho_T^0 \to \pi_A^+ \pi_B^-) = \frac{2\alpha_{\rho_T} \mathcal{C}_{AB}^2}{3} \frac{p_{AB}^3}{s}, \tag{8}$$

where p_{AB} is the technipion momentum and α_{ρ_T} is obtained by *naive* scaling from the QCD coupling for $\rho \to \pi\pi$:

$$\alpha_{\rho_T} = 2.91 \left(\frac{3}{N_{TC}}\right). \tag{9}$$

The parameter $\mathcal{C}_{AB}^2$ is given by

$$\mathcal{C}_{AB}^2 = \begin{cases} \sin^4 \chi & \text{for } W_L^+ W_L^- \\ 2\sin^2 \chi \cos^2 \chi & \text{for } W_L^+ \pi_T^- + W_L^- \pi_T^+ \\ \cos^4 \chi & \text{for } \pi_T^+ \pi_T^- \end{cases} \tag{10}$$

Note that the ρ_T can be *very* narrow. For $\sqrt{s} = M_{\rho_T} = 210\,\text{GeV}$, $M_{\pi_T} = 110\,\text{GeV}$, and $\sin\chi = \frac{1}{3}$, we have $\sum_{AB} \Gamma(\rho_T^0 \to \pi_A^+ \pi_B^-) = 680\,\text{MeV}$, 80% of which is $W_L^\pm \pi_T^\mp$.

We shall also need the decay rates of the ρ_T to fermion-antifermion states. The energy-dependent widths are

$$\Gamma(\rho_T^0 \to \bar{f}_i f_i) = \frac{N_f \alpha^2}{3 \alpha_{\rho_T}} \frac{p_i (s + 2m_i^2)}{s} A_i^0(s). \tag{11}$$

Here, α is the fine-structure constant, p_i is the momentum and m_i the mass of fermion f_i, and the factors A_i^0 are given by

$$A_i^0(s) = |\mathcal{A}_{iL}(s)|^2 + |\mathcal{A}_{iR}(s)|^2,$$
$$\mathcal{A}_{i\lambda}(s) = Q_i + \frac{2\cos 2\theta_W}{\sin^2 2\theta_W} \zeta_{i\lambda} \left(\frac{s}{s - M_Z^2 + i\sqrt{s}\,\Gamma_Z}\right), \tag{12}$$
$$\zeta_{iL} = T_{3i} - Q_i \sin^2\theta_W, \qquad \zeta_{iR} = -Q_i \sin^2\theta_W.$$

For $M_{\rho_T} = 210\,\text{GeV}$ and other parameters as above, the $\bar{f}f$ partial decay widths are:

$$\begin{aligned}&\Gamma(\rho_T^0 \to \bar{u}_i u_i) = 5.8\,\text{MeV}\,, &&\Gamma(\rho_T^0 \to \bar{d}_i d_i) = 4.1\,\text{MeV}\\ &\Gamma(\rho_T^0 \to \bar{\nu}_i \nu_i) = 0.9\,\text{MeV}\,, &&\Gamma(\rho_T^0 \to \ell_i^+ \ell_i^-) = 2.6\,\text{MeV}\,.\end{aligned} \quad (13)$$

For the ω_T, phase space considerations suggest we consider only its $\gamma \pi_T^0$ and fermionic decay modes. The energy dependent widths are:

$$\Gamma(\omega_T \to \gamma \pi_T^0) = \frac{\alpha p^3}{3 M_T^2}\,,$$

$$\Gamma(\omega_T \to \bar{f}_i f_i) = \frac{N_f \alpha^2}{3 \alpha_{\rho_T}} \frac{p_i (s + 2 m_i^2)}{s} B_i^0(s)\,. \quad (14)$$

The mass parameter M_T in the $\omega_T \to \gamma \pi_T^0$ rate is unknown *a priori*; naive scaling from the QCD decay, $\omega \to \gamma \pi^0$, suggests it is several 100 GeV. The factor B_i^0 is given by

$$B_i^0(s) = |\mathcal{B}_{iL}(s)|^2 + |\mathcal{B}_{iR}(s)|^2\,,$$

$$\mathcal{B}_{i\lambda}(s) = \left[Q_i - \frac{4 \sin^2 \theta_W}{\sin^2 2\theta_W} \zeta_{i\lambda} \left(\frac{s}{s - M_Z^2 + i\sqrt{s}\,\Gamma_Z} \right) \right] (Q_U + Q_D)\,. \quad (15)$$

Here, Q_U and $Q_D = Q_U - 1$ are the electric charges of the ω_T's constituent technifermions. For $M_{\omega_T} = 210\,\text{GeV}$ and $M_{\pi_T} = 110\,\text{GeV}$, and choosing $M_T = 100\,\text{GeV}$ and $Q_U = Q_D + 1 = \frac{4}{3}$, the ω_T partial widths are:

$$\begin{aligned}&\Gamma(\omega_T \to \gamma \pi_T^0) = 115\,\text{MeV}\\ &\Gamma(\omega_T \to \bar{u}_i u_i) = 6.8\,\text{MeV}\,, &&\Gamma(\omega_T \to \bar{d}_i d_i) = 2.6\,\text{MeV}\\ &\Gamma(\omega_T \to \bar{\nu}_i \nu_i) = 1.7\,\text{MeV}\,, &&\Gamma(\omega_T \to \ell_i^+ \ell_i^-) = 5.9\,\text{MeV}\,.\end{aligned} \quad (16)$$

The beam momentum spread of the First Muon Collider has been quoted to be as narrow as $\sigma_p/p = 3 \times 10^{-5}$ at $\sqrt{s} = 100\,\text{GeV}$ and 10^{-3} at $\sqrt{s} = 200\,\text{GeV}$. These correspond to beam energy spreads of $\sigma_E = 5\,\text{MeV}$ at 100 GeV and 300 MeV at 200 GeV. The resolution at 100 GeV is less than the expected π_T^0, $\pi_T^{0\prime}$ widths. At 200 GeV it is sufficient to resolve the ρ_T^0, but not the ω_T, for the parameters we used. It is very desirable, therefore, that the 200 GeV FMC's energy spread be about factor of 10 smaller. Since each of these technihadrons can be produced as an s-channel resonance in $\mu^+ \mu^-$ annihilation, it would then be possible to sit on the peak at $\sqrt{s} = M$. As we see next, the peak cross sections are enormous, 2–3 orders of magnitude larger than can be achieved at a hadron collider and even at a linear $e^+ e^-$ collider because of the latter's inherent beam energy spread.

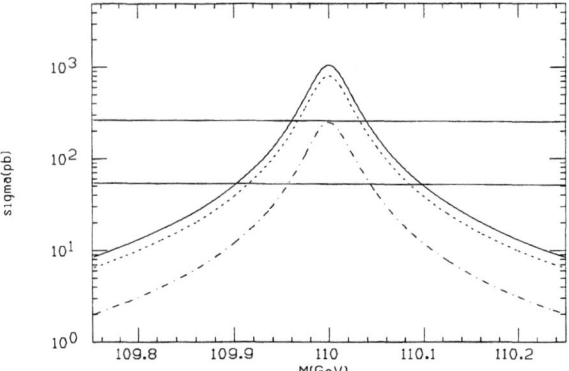

FIGURE 1. Theoretical (unsmeared) cross sections for $\mu^+\mu^- \to \pi_T^{0\prime} \to \bar{b}b$ (dashed), gg (dot-dashed) and total (solid) for $M_{\pi_T} = 110\,\text{GeV}$ and other parameters defined in the text. The solid horizontal lines are the backgrounds from γ, $Z^0 \to \bar{b}b$ (lower) and $Z^0 \to \bar{q}q$ (upper). Note the energy scale.

B Technihadron Production Rates

Like the standard Higgs boson, neutral technipions are expected to couple to $\mu^+\mu^-$ with a strength proportional to m_μ. Compared to the Higgs, however, this coupling is enhanced by a factor of $F_\pi/F_T = 1/\sin\chi$. This makes the resolution of the FMC well-matched to the π_T^0 width. Thus, the FMC is a technipion factory, overwhelming the rate at any other collider. Once a neutral technipion has been found in ρ_T or ω_T decays at a hadron collider, it should be relatively easy in the FMC to locate the precise position of the resonance and sit on it. The cross sections for $\bar{f}f$ and gg production are isotropic; near the resonance, they are given by

$$\frac{d\sigma(\mu^+\mu^- \to \pi_T^0 \text{ or } \pi_T^{0\prime} \to \bar{f}f)}{dz} = \frac{N_f}{2\pi}\left(\frac{C_\mu C_f m_\mu m_f}{F_T^2}\right)^2 \frac{s}{(s-M_{\pi_T}^2)^2 + s\,\Gamma_{\pi_T}^2}, \tag{17}$$

$$\frac{d\sigma(\mu^+\mu^- \to \pi_T^{0\prime} \to gg)}{dz} = \frac{C_{\pi_T}}{32\pi^3}\left(\frac{C_\mu m_\mu \alpha_S N_{TC}}{F_T^2}\right)^2 \frac{s^2}{(s-M_{\pi_T}^2)^2 + s\,\Gamma_{\pi_T}^2}.$$

Here, $z = \cos\theta$ where θ is the center-of-mass production angle.

The $\pi_T^{0\prime}$ production cross sections and the Z^0 backgrounds are shown in Fig. 1 for $M_{\pi_T} = 110\,\text{GeV}$ and other parameters as above ($C_\mu = C_f = 1$, $C_{\pi_T} = 4/3$, $F_T = 82\,\text{GeV}$, $\alpha_S = 0.1$, $N_{TC} = 4$). The peak signal rates approach 1 nb. The $\bar{b}b$ dijet rates are much larger than the $Z^0 \to \bar{b}b$ backgrounds, while the gg rate is comparable to $Z^0 \to \bar{q}q$. Details of these and the other calculations in this section, including the effects of the finite beam energy resolution, will appear in Ref. [21]. See Ref. [22] for another example of neutral scalars that may be produced in $\mu^+\mu^-$ annihilation.

The cross sections for technipion production via the decay of technirho and techniomega s-channel resonances are calculated using vector meson (γ, Z^0) dominance [3,8,19,20]. They are given by:

$$\frac{d\sigma(\mu^+\mu^- \to \rho_T^0 \to \pi_A\pi_B)}{dz} = \frac{\pi\alpha^2 p_{AB}^3}{s^{\frac{1}{2}}} \frac{A_\mu^0(s)\,C_{AB}^2\,(1-z^2)}{(s-M_{\rho_T}^2)^2 + s\Gamma_{\rho_T}^2},$$

$$\frac{d\sigma(\mu^+\mu^- \to \omega_T \to \gamma\pi_T^0)}{dz} = \frac{\pi\alpha^3 s^{\frac{1}{2}} p^3}{3\alpha_{\rho_T} M_T^2} \frac{B_\mu^0(s)\,(1+z^2)}{(s-M_{\omega_T}^2)^2 + s\Gamma_{\omega_T}^2}, \quad (18)$$

where A_μ^0 and B_μ^0 were defined in Eqs. 12 and 15, respectively. For $M_{\rho_T} = M_{\omega_T} = 210\,\text{GeV}$, $M_{\pi_T} = 110\,\text{GeV}$, and other parameters as above, the total peak cross sections are [21]:

$$\sum_{AB} \sigma(\mu^+\mu^- \to \rho_T^0 \to \pi_A\pi_B) = 1.1\,\text{nb}$$

$$\sigma(\mu^+\mu^- \to \omega_T \to \gamma\pi_T^0) = 8.9\,\text{nb}. \quad (19)$$

The technirho rate is 20% W^+W^- and 80% $W^\pm\pi_T^\mp$.

Finally, it is reasonable to expect a small nonzero isospin splitting between ρ_T^0 and ω_T. This would appear as a dramatic interference in the $\mu^+\mu^- \to \bar{f}f$ cross section *provided* the FMC energy resolution is good enough in the ρ_T–ω_T region. The cross section is most accurately calculated by using the full γ–Z^0–ρ_T–ω_T propagator matrix, $\Delta(s)$. With $\mathcal{M}_V^2 = M_V^2 - i\sqrt{s}\,\Gamma_V(s)$ for $V = Z^0, \rho_T, \omega_T$, this matrix is the inverse of

$$\Delta^{-1}(s) = \begin{pmatrix} s & 0 & -sf_{\gamma\rho_T} & -sf_{\gamma\omega_T} \\ 0 & s - \mathcal{M}_Z^2 & -sf_{Z\rho_T} & -sf_{Z\omega_T} \\ -sf_{\gamma\rho_T} & -sf_{Z\rho_T} & s - \mathcal{M}_{\rho_T}^2 & 0 \\ -sf_{\gamma\omega_T} & -sf_{Z\omega_T} & 0 & s - \mathcal{M}_{\omega_T}^2 \end{pmatrix}. \quad (20)$$

Here,

$$f_{\gamma\rho_T} = \sqrt{\frac{\alpha}{\alpha_{\rho_T}}}, \qquad f_{\gamma\omega_T} = \sqrt{\frac{\alpha}{\alpha_{\rho_T}}}(Q_U + Q_D)$$

$$f_{Z\rho_T} = \sqrt{\frac{\alpha}{\alpha_{\rho_T}}}\frac{\cos 2\theta_W}{\sin 2\theta_W}, \qquad f_{Z\omega_T} = -\sqrt{\frac{\alpha}{\alpha_{\rho_T}}}\frac{\sin^2\theta_W}{\sin 2\theta_W}(Q_U + Q_D). \quad (21)$$

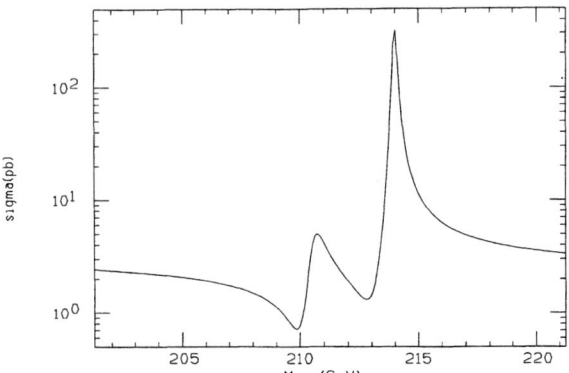

FIGURE 2. Theoretical (unsmeared) cross sections for $\mu^+\mu^- \to \rho_T^0, \omega_T \to e^+e^-$ for input masses $M_{\rho_T} = 210\,\text{GeV}$ and $M_{\omega_T} = 212.5\,\text{GeV}$ and other parameters as defined in the text.

Then, the cross section is given in terms of matrix elements of Δ by

$$\frac{d\sigma(\mu^+\mu^- \to \rho_T^0, \omega_T \to \bar{f}_i f_i)}{dz} = \frac{N_f \pi \alpha^2}{8s} \Big\{ \left(|\mathcal{D}_{iLL}|^2 + |\mathcal{D}_{iRR}|^2\right)(1+z)^2 \\ + \left(|\mathcal{D}_{iLR}|^2 + |\mathcal{D}_{iRL}|^2\right)(1-z)^2 \Big\}, \quad (22)$$

where

$$\mathcal{D}_{i\lambda\lambda'}(s) = s\bigg[Q_i Q_\mu \Delta_{\gamma\gamma}(s) + \frac{4}{\sin^2 2\theta_W}\zeta_{i\lambda}\zeta_{\mu\lambda'}\Delta_{ZZ}(s) \\ + \frac{2}{\sin 2\theta_W}\left(\zeta_{i\lambda}Q_\mu\Delta_{Z\gamma}(s) + Q_i\zeta_{\mu\lambda'}\Delta_{\gamma Z}(s)\right)\bigg]. \quad (23)$$

Figure 2 shows the theoretical ρ_T^0–ω_T interference effect in $\mu^+\mu^- \to e^+e^-$ for input masses $M_{\rho_T} = 210\,\text{GeV}$ and $M_{\omega_T} = 212.5\,\text{GeV}$ and other parameters as above. The propagator Δ shifts the nominal positions of the resonance peaks by $\mathcal{O}(\alpha/\alpha_{\rho_T})$. The theoretical peak cross sections are 5.0 pb at 210.7 GeV and 320 pb at 214.0 GeV. This demonstrates the importance of precise resolution in the 200 GeV FMC.

III CONCLUSIONS

Modern technicolor models predict narrow neutral technihadrons, π_T, ρ_T and ω_T. These states would appear as spectacular resonances in a $\mu^+\mu^-$ collider with $\sqrt{s} = 100\text{--}200\,\text{GeV}$ and energy resolution $\sigma_E/E \lesssim 10^{-4}$. This is a very strong physics motivation for building the First Muon Collider.

I thank other members of the First Muon Collider Workshop Strong Dynamics Subgroup for valuable interactions, especially Paul Mackenzie and Chris Hill for stressing the importance of a narrow π_T^0 at a muon collider, and Pushpa Bhat, Estia Eichten and John Womersley for considerable guidance. I am indebted to Torbjorn Sjostrand for first pointing out to me the likely importance of the ρ_T^0 and ω_T decays to fermions. This led to our consideration of ρ_T^0–ω_T interference, a phenomenon which may actually be observable for the first time in a muon collider.

REFERENCES

1. S. Weinberg, *Phys. Rev.* D **19**, 1277 (1979); L. Susskind, *Phys. Rev.* D **20**, 2619 (1979).
2. K. Lane, *The Scalar Sector of the Electroweak Interactions*, Proceedings of the 1982 DPF Summer Study on Elementary Particle Physics and Future Facilities, edited by R. Donaldson, R. Gustafson and F. Paige (Fermilab 1983), p. 222.
3. E. Eichten, I. Hinchliffe, K. Lane and C. Quigg, *Rev. Mod. Phys.* **56**, 579 (1984); *Phys. Rev.* D **34**, 1547 (1986).
4. S. Dimopoulos and L. Susskind, *Nucl. Phys.* B **155**, 237 (1979).
5. E. Eichten and K. Lane, *Phys. Lett.* B **90**, 125 (1980).
6. For a review of technicolor and its signatures up to 1996, see K. Lane, *Non-Supersymmetric Extensions of the Standard Model*, hep-ph/9610463, plenary talk at the 28th International Conference on High Energy Physics, edited by Z. Ajduk and A. K. Wroblewski, Vol. I, p. 367, Warsaw, July 25-31, 1996.
7. B. Holdom, *Phys. Rev.* D **24**, 1441 (1981); *Phys. Lett.* B **150**, 301 (1985); T. Appelquist, D. Karabali and L. C. R. Wijewardhana, *Phys. Rev. Lett.* **57**, 957 (1986); T. Appelquist and L. C. R. Wijewardhana, *Phys. Rev.* D **36**, 568 (1987); K. Yamawaki, M. Bando and K. Matumoto, *Phys. Rev. Lett.* **56**, 1335 (1986); T. Akiba and T. Yanagida, *Phys. Lett.* B **169**, 432 (1986).
8. K. Lane and E. Eichten, *Phys. Lett.* B **222**, 274 (1989); K. Lane and M. V. Ramana, *Phys. Rev.* D **44**, 2678 (1991).
9. B. Balaji, *Phys. Rev.* D **53**, 1699 (1996).
10. K. Lane, *Technicolor and Precision Tests of the Electroweak Interactions*, Proceedings of the 27th International Conference on High Energy Physics, edited by P. J. Bussey and I. G. Knowles, Vol. II, p. 543, Glasgow, June 20–27, 1994.
11. B. W. Lynn, M. E. Peskin and R. G. Stuart, in *Trieste Electroweak 1985*, 213 (1985); M. E. Peskin and T. Takeuchi, *Phys. Rev. Lett.* **65**, 964 (1990); A. Longhitano, *Phys. Rev.* D **22**, 1166 (1980); *Nucl. Phys.* B **188**, 118 (1981);

R. Renken and M. Peskin, *Nucl. Phys.* B **211**, 93 (1983); M. Golden and L. Randall, *Nucl. Phys.* B **361**, 3 (1990); B. Holdom and J. Terning, *Phys. Lett.* B **247**, 88 (1990); A. Dobado, D. Espriu and M J. Herrero, *Phys. Lett.* B **255**, 405 (1990); H. Georgi, *Nucl. Phys.* B **363**, 301 (1991).

12. F. Abe, et al., The CDF Collaboration, *Phys. Rev. Lett.* **73**, 225 (1994); *Phys. Rev.* D **50**, 2966 (1994); *Phys. Rev. Lett.* **74**, 2626 (1995); S. Abachi, et al., The DØ Collaboration, *Phys. Rev. Lett.* **74**, 2632 (1995).
13. R. S. Chivukula, S. B. Selipsky, and E. H. Simmons, *Phys. Rev. Lett.* **69**, 575 (1992); R. S. Chivukula, E. H. Simmons, and J. Terning, *Phys. Lett.* B **331**, 383 (1994), and references therein.
14. C. T. Hill, *Phys. Lett.* B **345**, 483 (1995).
15. Y. Nambu, in *New Theories in Physics*, Proceedings of the XI International Symposium on Elementary Particle Physics, Kazimierz, Poland, 1988, edited by Z. Adjuk, S. Pokorski and A. Trautmann (World Scientific, Singapore, 1989); Enrico Fermi Institute Report EFI 89-08 (unpublished); V. A. Miransky, M. Tanabashi and K. Yamawaki, *Phys. Lett.* B **221**, 171 (1989); *Mod. Phys. Lett.* **A4**, 1043 (1989); W. A. Bardeen, C. T. Hill and M. Lindner, *Phys. Rev.* D **D41**, 1647 (1990).
16. C. T. Hill, *Phys. Lett.* B **266**, 419 (1991); S. P. Martin, *Phys. Rev.* D **45**, 4283 (1992); *ibid* **D46**, 2197 (1992); *Nucl. Phys.* B **398**, 359 (1993); M. Lindner and D. Ross, *Nucl. Phys.* B **B370**, 30 (1992); R. Bönisch, *Phys. Lett.* B **268**, 394 (1991); C. T. Hill, D. Kennedy, T. Onogi, H. L. Yu, *Phys. Rev.* D **47**, 2940 (1993).
17. K. Lane and E. Eichten, *Phys. Lett.* B **352**, 382 (1995); K. Lane, *Phys. Rev.* D **54**, 2204 (1996).
18. J. Ellis, M. K. Gaillard, D. V. Nanopoulos and P. Sikivie, NPB **529**, 1981 (.)
19. E. Eichten and K. Lane, *Phys. Lett.* B **388**, 803 (1996).
20. E. Eichten, K. Lane and J. Womersley, PLB **405**, 305 (1997).
21. E. Eichten, K. Lane and J. Womersley, "Narrow Technihadron Production at the First Muon Collider", in preparation. Also see J. Womersley, "Technihadron Production at a Muon Collider", to appear in the FMC proceedings.
22. D. Bobrescu and C. T. Hill, FERMILAB-PUB-97-409-T, hep-ph/9712319 (Dec. 1997).

Topcolor and the First Muon Collider[1]

Christopher T. Hill

Fermi National Accelerator Laboratory
P.O. Box 500, Batavia, Illinois, 60510
and
The Department of Physics and Enrico Fermi Institute
The University of Chicago, Chicago, Illinois

Abstract. We describe a class of models of electroweak symmetry breaking that involve strong dynamics and top quark condensation. A new scheme based upon a seesaw mechanism appears particularly promising. Various implications for the first-stage muon collider are discussed.

TOPCOLOR I

The top quark mass may be large because it is a combination of a *dynamical condensate component*, $(1 - \epsilon)m_t$, generated by a new strong dynamics [1], together with a small *fundamental component*, ϵm_t, i.e, $\epsilon \ll 1$, generated by something else. The most obvious "handle" on the top quark for new dynamics is the color index. Invoking new dynamics involving the top quark color index leads directly to a class of Technicolor-like models incorporating "Topcolor". We expect in such schemes that the new strong dynamics occurs primarily in interactions that involve $\bar{t}t\bar{t}t$, $\bar{t}t\bar{b}b$, and $\bar{b}b\bar{b}b$.

In Topcolor I the dynamics at the ~ 1 TeV scale involves the following structure at the TeV scale (or a generalization thereof) [2]:

$$SU(3)_1 \times SU(3)_2 \times U(1)_{Y1} \times U(1)_{Y2} \times SU(2)_L \to SU(3)_{QCD} \times U(1)_{EM} \quad (1)$$

where $SU(3)_1 \times U(1)_{Y1}$ ($SU(3)_2 \times U(1)_{Y2}$) generally couples preferentially to the third (first and second) generations. The $U(1)_{Yi}$ are just strongly rescaled versions of electroweak $U(1)_Y$.

---

[1] Talk presented at the Workshop on Physics at the First Muon Collider and at the Front End of the Muon Collider

The fermions are then assigned $(SU(3)_1, SU(3)_2, Y_1, Y_2)$ quantum numbers in the following way:

$$(t,b)_L \sim (3,1,1/3,0) \qquad (t,b)_R \sim (3,1,(4/3,-2/3),0) \qquad (2)$$
$$(\nu_\tau, \tau)_L \sim (1,1,-1,0) \qquad \tau_R \sim (1,1,-2,0)$$

$$(u,d)_L, \ (c,s)_L \sim (1,3,0,1/3) \qquad (u,d)_R, \ (c,s)_R \sim (1,3,0,(4/3,-2/3))$$
$$(\nu,\ell)_L \ \ell = e, \mu \sim (1,1,0,-1) \qquad \ell_R \sim (1,1,0,-2)$$

Topcolor must be broken, which we describe by an (effective) scalar field:

$$\Phi \sim (3,\bar{3},y,-y) \qquad (3)$$

When Φ develops a VEV, it produces the simultaneous symmetry breaking

$$SU(3)_1 \times SU(3)_2 \to SU(3)_{QCD} \qquad \text{and} \qquad U(1)_{Y1} \times U(1)_{Y2} \to U(1)_Y \qquad (4)$$

$SU(3)_1 \times U(1)_{Y1}$ is assumed to be strong enough to form chiral condensates which will be "tilted" in the top quark direction by the $U(1)_{Y1}$ couplings. The theory is assumed to spontaneously break down to ordinary QCD $\times U(1)_Y$ at a scale of ~ 1 TeV, before it becomes confining. The isospin splitting that permits the formation of a $\langle \bar{t}t \rangle$ condensate but disables the $\langle \bar{b}b \rangle$ condensate is due to the $U(1)_{Yi}$ couplings. The b–quark mass in this scheme can arise from a combination of ETC effects and instantons in $SU(3)_1$. The θ–term in $SU(3)_1$ may manifest itself as the CP-violating phase in the CKM matrix. Above all, the new spectroscopy of such a system should begin to materialize indirectly in the third generation, perhaps at the Tevatron in top and bottom quark production, or possibly in a muon collider.

The symmetry breaking pattern outlined above will generically give rise to three (pseudo)–Nambu–Goldstone bosons $\tilde{\pi}^a$, or "top-pions", near the top mass scale. *This is the smoking gun of Topcolor.* [We were led to Topcolor by considering how strong dynamics might produce the analog of the decay $t \to H^+ + b$, considered to be a SUSY signature for a charged Higgs-boson H. This is an example of "SUSY-Technicolor/Topcolor duality".] If the Topcolor scale is of the order of 1 TeV, the top-pions will have a decay constant of $f_\pi \approx 50$ GeV, and a strong coupling given by a Goldberger–Treiman relation, $g_{tb\pi} \approx m_t/\sqrt{2}f_\pi \approx 2.5$, potentially observable in $\tilde{\pi}^+ \to t + \bar{b}$ if $m_{\tilde{\pi}} > m_t + m_b$.

We assume presently that ESB can be primarily driven by a Higgs sector or Technicolor, with gauge group G_{TC}[3] [4]. This gives the $\mathcal{O}(\epsilon)$ component of m_t. Technicolor can also provide condensates which generate the breaking of Topcolor to QCD and $U(1)_Y$.

The coupling constants (gauge fields) of $SU(3)_1 \times SU(3)_2$ are respectively h_1 and h_2 ($A^A_{1\mu}$ and $A^A_{2\mu}$) while for $U(1)_{Y1} \times U(1)_{Y2}$ they are respectively q_1

and q_2, $(B_{1\mu}, B_{2\mu})$. The $U(1)_{Yi}$ fermion couplings are then $q_i \frac{Y_i}{2}$, where Y_1, Y_2 are the charges of the fermions under $U(1)_{Y1}, U(1)_{Y2}$ respectively.

Topcolor I produces new gauge heavy bosons Z', and "colorons" B^A with couplings to fermions given by:

$$\mathcal{L}_{Z'} = g_1(Z' \cdot J_{Z'}) \qquad \mathcal{L}_B = g_3 \cot\theta (B^A \cdot J_B^A) \tag{5}$$

where the currents $J_{Z'}$ and J_B in general involve all three generations of fermions

$$J_{Z'} = -(J_{Z',1} + J_{Z',2})\tan\theta' + J_{Z',3}\cot\theta' \tag{6}$$
$$J_B = -(J_{B,1} + J_{B,2})\tan\theta + J_{B,3}\cot\theta$$

For example, for the third generation the currents read explicitly (in a weak eigenbasis):

$$J_{Z',3}^\mu = \frac{1}{6}\bar{t}_L\gamma^\mu t_L + \frac{1}{6}\bar{b}_L\gamma^\mu b_L + \frac{2}{3}\bar{t}_R\gamma^\mu t_R - \frac{1}{3}\bar{b}_R\gamma^\mu b_R \tag{7}$$
$$- \frac{1}{2}\bar{\nu}_{\tau L}\gamma^\mu \nu_{\tau L} - \frac{1}{2}\bar{\tau}_L\gamma^\mu \tau_L - \bar{\tau}_R\gamma^\mu \tau_R$$
$$J_{B,3}^{A,\mu} = \bar{t}\gamma^\mu \frac{\lambda^A}{2}t + \bar{b}\gamma^\mu \frac{\lambda^A}{2}b$$

where λ^A is a Gell-Mann matrix acting on color indices. We ultimately demand $\cot\theta \gg 1$ and $\cot\theta' \gg 1$ to select the top quark direction for condensation.

The attractive Topcolor interaction, for sufficiently large $\kappa = g_3^2 \cot^2\theta/4\pi$, would by itself trigger the formation of a low energy condensate, $\langle \bar{t}t + \bar{b}b \rangle$, which would break $SU(2)_L \times SU(2)_R \times U(1)_Y \to U(1) \times SU(2)_c$, where $SU(2)_c$ is a global custodial symmetry. On the other hand, the $U(1)_{Y1}$ force is attractive in the $\bar{t}t$ channel and repulsive in the $\bar{b}b$ channel. Thus, to make $\langle \bar{b}b \rangle = 0$ and $\langle \bar{t}t \rangle \neq 0$ we can have in concert critical and subcritical values of the combinations:

$$\kappa + \frac{2\kappa_1}{9N_c} > \kappa_{crit}; \qquad \kappa_{crit} > \kappa - \frac{\kappa_1}{9N_c}; \tag{8}$$

Here N_c is the number of colors and $\kappa_1 = g_1^2 \cot^2\theta'/4\pi$. (It should be mentioned that our analyses are performed in the context of a large-N_c approximation). This leads to "tilted" gap equations in which the top quark acquires a constituent mass, while the b quark remains massless. Given that both κ and κ_1 are large there is no particular fine-tuning occuring here, only "rough-tuning" of the desired tilted configuration. Of course, the NJL approximation is crude, but as long as the associated phase transitions of the real strongly coupled theory are approximately second order, analogous rough-tuning in the full theory is possible. The full phase diagram of the model is shown in Fig. 1. of [5].

TOPCOLOR II

If the above described "Topcolor I" is the analog of Weinberg's original version of the SM, incorporating standard fermions and the Z-boson, then Topcolor II is the analog of the original Georgi-Glashow model, which incorporated no new Z boson, but rather included additional fermions. [This is an example of "Weinberg—Georgi-Glashow" duality.] The strong $U(1)$ is present in the previous scheme to avoid a degenerate $\langle \bar{t}t \rangle$ with $\langle \bar{b}b \rangle$. However, we can give a model in which there is: (i) a Topcolor $SU(3)$ group but (ii) no strong $U(1)$ with (iii) an anomaly-free representation content. In fact the original model of [2] was of this form, introducing a new quark of charge $-1/3$. Let us consider a generalization of this scheme which consists of the gauge structure $SU(3)_Q \times SU(3)_1 \times SU(3)_2 \times U(1)_Y \times SU(2)_L$. We require an additional triplet of fermions fields (Q_R^a) transforming as $(3,3,1)$ and $Q_L^{\dot{a}}$ transforming as $(3,1,3)$ under the $SU(3)_Q \times SU(3)_1 \times SU(3)_2$.

The fermions are then assigned the following quantum numbers in $SU(2) \times SU(3)_Q \times SU(3)_1 \times SU(3)_2 \times U(1)_Y$:

$$
\begin{aligned}
(t,b)_L \ (c,s)_L &\sim (2,1,3,1) & Y &= 1/3 \\
(t)_R &\sim (1,1,3,1) & Y &= 4/3; \\
(Q)_R &\sim (1,3,3,1) & Y &= 0
\end{aligned}
\tag{9}
$$

$$
\begin{aligned}
(u,d)_L &\sim (2,1,1,3) & Y &= 1/3 \\
(u,d)_R \ (c,s)_R &\sim (1,1,1,3) & Y &= (4/3, -2/3) \\
(\nu,\ell)_L \ \ell = e,\mu,\tau &\sim (2,1,1,1) & Y &= -1; \\
(\ell)_R &\sim (1,1,1,1) & Y &= -2 \\
b_R &\sim (1,1,1,3) & Y &= 2/3; \\
(Q)_L &\sim (1,3,1,3) & Y &= 0;
\end{aligned}
$$

Thus, the Q fields are electrically neutral. One can verify that this assignment is anomaly free.

The $SU(3)_Q$ confines and forms a $\langle \bar{Q}Q \rangle$ condensate which acts like the Φ field and breaks the Topcolor group down to QCD dynamically. We assume that Q is then decoupled from the low energy spectrum by its large constituent mass. There is a lone $U(1)$ Nambu–Goldstone boson $\sim \bar{Q}\gamma^5 Q$ which acquires a large mass by $SU(3)_Q$ instantons.

TRIANGULAR TEXTURES

The texture of the fermion mass matrices will generally be controlled by the symmetry breaking pattern of a horizontal symmetry. In the present case we are specifying a residual Topcolor symmetry, presumably subsequent to some initial breaking at some scale Λ, large compared to Topcolor, e.g., the third generation fermions in Model I have different Topcolor assignments than do the second and first generation fermions. Thus the texture will depend in some way upon the breaking of Topcolor [5] [3].

Let us study a fundamental Higgs boson, which ultimately breaks $SU(2)_L \times U(1)_Y$, together with an effective field Φ breaking Topcolor as in eq.(4). We must now specify the full Topcolor charges of these fields. As an example, under $SU(3)_1 \times SU(3)_2 \times U(1)_{Y1} \times U(1)_{Y2} \times SU(2)_L$ let us choose:

$$\Phi \sim (3, \bar{3}, \frac{1}{3}, -\frac{1}{3}, 0) \qquad H \sim (1, 1, 0, -1, \frac{1}{2}) \qquad (10)$$

The effective couplings to fermions that generate mass terms in the up sector are of the form

$$\mathcal{L}_{M_U} = m_0 \bar{t}_L t_R + c_{33} \bar{T}_L t_R H \frac{\det \Phi^\dagger}{\Lambda^3} + c_{32} \bar{T}_L c_R H \frac{\Phi}{\Lambda} + c_{31} \bar{T}_L u_R H \frac{\Phi}{\Lambda}$$

$$+ c_{23} \bar{C}_L t_R H \Phi^\dagger \frac{\det \Phi^\dagger}{\Lambda^4} + c_{22} \bar{C}_L c_R H + c_{21} \bar{C}_L u_R H \qquad (11)$$

$$+ c_{13} \bar{F}_L t_R H \Phi^\dagger \frac{\det \Phi^\dagger}{\Lambda^4} + c_{12} \bar{F}_L c_R H + c_{11} \bar{F}_L u_R H + \text{h.c.}$$

Here $T = (t, b)$, $C = (c, s)$ and $F = (u, d)$. The mass m_0 is the dynamical condensate top mass. Furthermore $\det \Phi$ is defined by

$$\det \Phi \equiv \frac{1}{6} \epsilon_{ijk} \epsilon_{lmn} \Phi_{il} \Phi_{jm} \Phi_{kn} \qquad (12)$$

where in Φ_{rs} the first(second) index refers to $SU(3)_1$ ($SU(3)_2$). The matrix elements now require factors of Φ to connect the third with the first or second generation color indices. The down quark and lepton mass matrices are generated by couplings analogous to (11).

To see what kinds of textures can arise naturally, let us assume that the ratio Φ/Λ is small, $O(\epsilon)$. The field H acquires a VEV of v. Then the resulting mass matrix is approximately triangular:

$$\begin{pmatrix} c_{11} v & c_{12} v & \sim 0 \\ c_{21} v & c_{22} v & \sim 0 \\ c_{31} O(\epsilon) v & c_{32} O(\epsilon) v & \sim m_0 + O(\epsilon^3) v \end{pmatrix} \qquad (13)$$

where we have kept only terms of $\mathcal{O}(\epsilon)$ or larger.

This is a triangular matrix (up to the c_{12} term). When it is written in the form $U_L \mathcal{D} U_R^\dagger$ with U_L and U_R unitary and $\mathcal{D}$ positive diagonal, there automatically result restrictions on U_L and U_R. In the present case, the elements $U_L^{3,i}$ and $U_L^{i,3}$ are vanishing for $i \neq 3$, while the elements of U_R are not constrained by triangularity. Analogously, in the down quark sector $D_L^{i,3} = D_L^{3,i} = 0$ for $i \neq 3$ with D_R unrestricted. The situation is reversed when the opposite corner elements are small, which can be achieved by choosing $H \sim (1, 1, -1, 0, \frac{1}{2})$.

These restrictions on the quark mass rotation matrices have important phenomenological consequences. For instance, in the process $B^0 \to \overline{B^0}$ there are potentially large contributions from top-pion and coloron exchange. However, these contributions are proportional to the product $D_L^{3,2} D_R^{3,2}$. The same occurs in $D^0 - \bar{D}^0$ mixing, where the effect goes as products involving U_L and U_R off-diagonal elements. Therefore, triangularity can naturally select these products to be small.

The precise selection rules depend upon the particular symmetry breaking that occurs. This example is merely illustrative of the systematic effects that can occur in such schemes.

TOP-PIONS; INSTANTONS; THE B-QUARK MASS.

Since the top condensation is a spectator to the TC (or Higgs) driven ESB, there must occur a multiplet of top-pions. A chiral Lagrangian can be written:

$$L = i\overline{\psi}\slashed{\partial}\psi - m_t(\overline{\psi}_L \Sigma P \psi_R + h.c.) - \epsilon m_t \overline{\psi} P \psi, \qquad P = \begin{pmatrix} 1 & 0 \\ 0 & 0 \end{pmatrix} \qquad (14)$$

and $\psi = (t, b)$, and $\Sigma = \exp(i\tilde{\pi}^a \tau^a/\sqrt{2} f_\pi)$. With $\epsilon = 0$ this is invariant under $\psi_L \to e^{i\theta^a \tau^a/2}\psi_L$, $\tilde{\pi}^a \to \tilde{\pi}^a + \theta^a f_\pi/\sqrt{2}$. Hence, the relevant currents are left-handed, $j_\mu^a = \overline{\psi}_L \gamma_\mu \frac{\tau^a}{2} \psi_L$, and $<\tilde{\pi}^a|j_\mu^b|0> = \frac{f_\pi}{\sqrt{2}} p_\mu \delta^{ab}$. The Pagels-Stokar relation, eq.(1), then follows by demanding that the $\tilde{\pi}^a$ kinetic term is generated by integrating out the fermions. The top–pion decay constant estimated from eq.(1) using $\Lambda = M_B$ and $m_t = 175$ GeV is $f_\pi \approx 50$ GeV. The couplings of the top-pions take the form:

$$\frac{m_t}{\sqrt{2}f_\pi} \left[i\bar{t}\gamma^5 t\tilde{\pi}^0 + \frac{i}{\sqrt{2}}\bar{t}(1-\gamma^5)b\tilde{\pi}^+ + \frac{i}{\sqrt{2}}\bar{b}(1+\gamma^5)t\tilde{\pi}^- \right] \qquad (15)$$

and the coupling strength is governed by the relation $g_{bt\tilde{\pi}} \approx m_t/\sqrt{2}f_\pi$.

The small ETC mass component of the top quark implies that the masses of the top-pions will depend upon ϵ and Λ. Estimating the induced top-pion mass from the fermion loop yields:

$$m_{\tilde{\pi}}^2 = \frac{N\epsilon m_t^2 M_B^2}{8\pi^2 f_\pi^2} = \frac{\epsilon M_B^2}{\log(M_B/m_t)} \qquad (16)$$

where the Pagels-Stokar formula is used for f_π^2 (with $k = 0$) in the last expression. For $\epsilon = (0.03, 0.1)$, $M_B \approx (1.5, 1.0)$ TeV, and $m_t = 180$ GeV this predicts $m_{\tilde{\pi}} = (180, 240)$ GeV. The bare value of ϵ generated at the ETC scale Λ_{ETC}, however, is subject to very large radiative enhancements by Topcolor and $U(1)_{Y1}$ by factors of order $(\Lambda_{ETC}/M_B)^p \sim 10^1$, where the $p \sim O(1)$. Thus, we expect that even a bare value of $\epsilon_0 \sim 0.005$ can produce sizeable $m_{\tilde{\pi}} > m_t$. Note that $\tilde{\pi}$ will generally receive gauge contributions to it's mass; these are at most electroweak in strength, and therefore of order ~ 10 GeV.

Top-pions can be as light as ~ 150 GeV, in which case they would emerge as a detectable branching fraction of top decay [6]. However, there are dangerous effects in $Z \to b\bar{b}$ with low mass top pions and decay constamnts as small as ~ 60 GeV [8]. A more comfortable phenomenological range is slightly larger than our estimates, $m_{\tilde{\pi}} \gtrsim 300$ GeV and $f_\pi \gtrsim 100$ GeV.

The b quark receives mass contributions from ETC of $O(1)$ GeV, but also an induced mass from instantons in $SU(3)_1$. The instanton effective Lagrangian may be approximated by the 't Hooft flavor determinant (we place the cut-off at M_B):

$$L_{eff} = \frac{k}{M_B^2} e^{i\theta_1} \det(\bar{q}_L q_R) + h.c. = \frac{k}{M_B^2} e^{i\theta_1}[(\bar{b}_L b_R)(\bar{t}_L t_R) - (\bar{t}_L b_R)(\bar{b}_L t_R)] + h.c. \tag{17}$$

where θ_1 is the $SU(3)_1$ strong CP-violation phase. θ_1 cannot be eliminated because of the ETC contribution to the t and b masses. It can lead to induced scalar couplings of the neutral top-pion [5], and an induced CKM CP-phase, however, we will presently neglect the effects of θ_1.

We generally expect $k \sim 1$ to 10^{-1} as in QCD. Bosonizing in fermion bubble approximation $\bar{q}_L^i t_R \sim \frac{N}{8\pi^2} m_t M_B^2 \Sigma_1^i$, where $\Sigma_j^i = \exp(i\tilde{\pi}^a \tau^a/\sqrt{2} f_\pi)_j^i$ yields:

$$L_{eff} \to \frac{Nkm_t}{8\pi^2} e^{i\theta}[(\bar{b}_L b_R)\Sigma_1^1 + (\bar{t}_L b_R)\Sigma_1^2 + h.c.] \tag{18}$$

This implies an instanton induced b-quark mass:

$$m_b^* \approx \frac{3km_t}{8\pi^2} \sim 6.6\, k\, GeV \tag{19}$$

This is not an unreasonable estimate of the observed b quark mass, as we might have feared it would be too large.

TOP SEE-SAW

EWSB may occur via the condensation of the top quark in the presence of an extra vectorlike, weak-isoscalar quark [7]. The mass scale of the condensate is large, of order 0.6 TeV corresponding to the electroweak scale $f_\pi \approx 175$ GeV. The vectorlike iso-scalar then naturally admits a seesaw mechanism, yielding the physical top quark mass, which is then adjusted to the experimental value. The choice of a natural $\sim$TeV scale for the topcolor dynamics then determines the mass of the weak-isoscalar see-saw partner. The scheme is economical, requiring no additional weak-isodoublets, and therefore easily satisfies the constraints upon the S parameter using estimates made in the large-N approximation. The constraints on custodial symmetry violation, i.e., the value of the $\delta\rho$ or equivalently, T parameter, are easily satisfied, being principally the usual m_t contribution, plus corrections that are suppressed by the see-saw mechanism.

The dynamical fermion masses that are induced can be written as:

$$\mathcal{L} = -(\bar{t}_L , \bar{\chi}_L)\begin{pmatrix} 0 & m_{t\chi} \\ m_{\chi t} & m_{\chi\chi} \end{pmatrix}\begin{pmatrix} t_R \\ \chi_R \end{pmatrix} + \text{h.c.} \qquad (20)$$

Typically $\bar{\chi}_L\chi_R$ is the most attractive channel, and it is possible to arrange the $\langle\bar{\chi}_L\chi_R\rangle$ condensate to be significantly larger than the other ones, such that $m_{\chi\chi}^2 \gg m_{\chi t}^2 > m_{t\chi}^2$. As a result the physical top mass is suppressed by a seesaw mechanism:

$$m_t \approx \frac{m_{\chi t}m_{t\chi}}{m_{\chi\chi}}\left[1 + O\left(m_{\chi t,t\chi}^2/m_{\chi\chi}^2\right)\right] . \qquad (21)$$

The electroweak symmetry is broken by the $m_{t\chi}$ dynamical mass. Therefore, the electroweak scale is estimated to be given by

$$v^2 \approx \frac{3}{16\pi^2}m_{t\chi}^2 \ln\left(\frac{M}{m_{t\chi}}\right) . \qquad (22)$$

Thus, $v \approx 174$ GeV requires a dynamical mass $m_{t\chi} \sim 620$ GeV for $M \sim 5$ TeV (and $m_{t\chi} = 520$ GeV for $M \sim 10$ TeV). From eq. (21) follows then that a top mass of 175 GeV requires $m_{\chi t}/m_{\chi\chi} \approx 0.29$. The electroweak T parameter can be estimated in fermion-bubble large-N approximation as:

$$T \approx \frac{3m_t^2}{16\pi^2\alpha(M_Z^2)v^2}\frac{m_{t\chi}^2}{m_{\chi t}^2}\left[1 + O\left(m_{\chi t,t\chi}^2/m_{\chi\chi}^2\right)\right] , \qquad (23)$$

where α is the fine structure constant. Moreover, we obtain the usual Standard Model result for the S parameter. Requiring that our model does not exceed the 1σ upper bound on S and T, we obtain $m_{t\chi}/m_{\chi t} \leq 0.55$.

It should be emphasized that these results do not require excessive fine-tuning. The top-seesaw is therefore a plausible natural theory of dynamical EWSB with a minimal number of new degrees of freedom. This model also implies the existence of pseudo-Nambu-Goldstone bosons (pNGB's). A cursory discussion of that is given in ref.[7].

OBSERVABLES

There are several classes of possible experimental implications of the kinds of models we described above that may be relevant to the muon collider. We will describe them here briefly as lines to be developed further. These may be enumerated as follows:

1. $\mu\bar{\mu} \to Z'$; this is the province of high energy machine, since we expect $M_{Z'} \gtrsim 0.5$ TeV.

2. $\mu\bar{\mu} \to \pi_{top}$; the notion that the muon collider can see technipions, or other PNGB's, such as top-pions has emerged from discussions in this workshop, prompted by MacKenzie and myself. Lane has presented the multi-scale technicolor signal [4].

3. Effects in Z physics involving the third generation, such as $Z \to b\bar{b}$ [8].

4. Effects in top-quark pair production at threshold, e.g., see [11] for analogous case in e^+e^- and $p\bar{p}$ collider physics.

5. Induced GIM violation in low energy processes such as $K^+ \to \pi^+\nu\bar{\nu}$; we discuss this below as an example of a potential signature that can be enhanced by Topcolor wrt the Standard Model (this result was anticipated in ref[5] before the observation of the single event at Brookhaven E787).

6. Induced lepton family number violation, e.g. $\mu\bar{\mu} \to \tau\bar{\mu}$.

7. Flavor dependent production effects, e.g. anomalous $\mu\bar{\mu} \to b\bar{s}$, etc.

8. New physics in e.g. μp collisions, such as $d(u) + \bar{\mu} \to b(t) + \bar{\tau}$.

GIM and lepton family number violation arise because of the generational structure of topcolor. (It is actually more general than topcolor; the mere statement than the top mass is largely dynamical implies effects like this) In going to the mass eigenbasis, quark (and lepton) fields are rotated, e.g., by the matrices U_L, U_R (for the up-type left and right handed quarks) and D_L, D_R (for the down-type left and right handed quarks). For example, for the b-quark we make the replacement

$$b_L \to D_L^{bb} b_L + D_L^{bs} s_L + D_L^{bd} d_L \qquad (24)$$

and analogously for b_R. Thus there will be induced FCNC interactions. This provides constraints and opportunities. Thus, induced effects like $\mu\bar{\mu} \to b\bar{s}$ may be enhanced, and effects like $\mu\bar{\mu} \to \tau\bar{\mu}$ may occur. Since the muon is presumably closer in affiliation to the third generation than is the electron, such effects may show up in muon collider physics, but be inaccessible in electron linear colliders! Similarly, induced effects like $\mu\bar{\mu} \to b\bar{s}$ may be enhanced.

For the FMC, sensitive probes arise in e.g., K-physics. there is a Z' induced contact term at low energies of the form $b\bar{b}\nu_\tau\bar{\nu}_\tau$ (this assumes that the τ is associated with the third generation; nothing fundamentally compels this, but we shall assume it to be true in the following). The above mass rotation induces a $s\bar{d}\nu_\tau\bar{\nu}_\tau$ which contributes to $K^+ \to \pi^+\nu\bar{\nu}$. The ratio of the Topcolor amplitude to the SM is then

$$\frac{\mathcal{A}^{TC}}{\mathcal{A}^{SM}} = -\left(\frac{g_1 \cot\theta'}{M_{Z'}}\right)^2 \frac{\sqrt{2}\pi \sin^2\theta_W}{24\alpha\, G_F} \frac{\delta_{ds}}{\sum_j V_{js}^* V_{jd} D_j(x_j)} \sim -3 \times 10^9\, \delta_{ds}\, \frac{\kappa_1}{M_{Z'}^2} \tag{25}$$

where $\delta_{ds}^* = D_L^{bs} D_L^{bd*} - 2 D_R^{bs} D_R^{bd*}$. The form-factor $f_+(q^2)$ is experimentally well known. We expect, $|\delta_{ds}| \sim \lambda^{10}$ where λ is the Wolfenstein CKM parameter. For $M_{Z'} = 500$ GeV and $\kappa_1 = 1$ the ratio of amplitudes is about ~ 4.0, and the branching ratio is between 0.3 to $O(10)$, times the SM result, depending on the sign of the interference. The recent observation of one event by the Brookhaven E787 Collaboration [10] makes this an exciting channel in which to search for new physics. High sensitivity experiments are possible at the front-end muon collider with its copious K-meson yields.

REFERENCES

1. Y. Nambu, "BCS Mechansim, Quasi-Supersymmetry, and Fermion Mass Matrix," Talk presented at the Kasimirz Conference, EFI 88-39 (July 1988); "Quasi-Supersymmetry, Bootstrap Symmetry Breaking, and Fermion Masses," EFI 88-62 (August 1988); a version of this work appears in "*1988 International Workshop on New Trends in Strong Coupling Gauge Theories*," Nagoya, Japan, ed. Bando, Muta and Yamawaki; "Bootstrap Symmetry Breaking in Electroweak Unification," EFI Preprint, 89–08 (1989). V. A. Miransky, M. Tanabashi, K. Yamawaki, *Mod. Phys. Lett.* **A4**, 1043 (1989); *Phys. Lett.* **221B** 177 (1989); W. J. Marciano, *Phys. Rev. Lett.* **62**, 2793 (1989). W. A. Bardeen, C. T. Hill, M. Lindner *Phys. Rev.* **D41**, 1647 (1990).
2. C. T. Hill, *Phys. Lett.* **B266** 419 (1991); C. T. Hill, *Phys. Lett.*; **B345** 483, (1995); S. P. Martin, *Phys. Rev.* **D46**, 2197 (1992); *Phys. Rev.* **D45**, 4283 (1992), *Nucl. Phys.* **B398**, 359 (1993); M. Lindner and D. Ross, *Nucl. Phys.* **B 370**, 30 (1992); R. Bönisch, *Phys. Lett.* **B268** 394 (1991); C. T. Hill, D. Kennedy, T. Onogi, H. L. Yu, *Phys. Rev.* **D47** 2940 (1993); B. Pendleton, G.G. Ross, *Phys. Lett.* **98B** 291, (1981); C.T. Hill, *Phys. Rev.* **D24**, 691 (1981); C. T. Hill, C. N. Leung, S. Rao, *Nucl. Phys.* **B262**, 517 (1985).
3. K. Lane, E. Eichten, *Phys. Lett.* **B352**, 382, (1995); K. Lane, *Phys. Rev.* **D54** 2204, (1996).
4. See the talk by K. Lane, "Technicolor and the First Muon Collider" in these proceedings, or hep-ph/9801385.
5. G. Buchalla, G. Burdman, C. T. Hill, D. Kominis, *Phys. Rev.* **D53** 5185, (1996);
6. B. Balaji, *Phys. Lett.* **B393** 89 (1997).
7. "Electroweak Symmetry Breaking via a Top Seesaw," B. Dobrescu, C. T. Hill FERMILAB-PUB-97-409-T, hep-ph/9712319

8. G. Burdman, D. Kominis, *Phys. Lett.* **B403** 101 (1997). C. T. Hill, Xinmin Zhang, *Phys.Rev.* **D51** 3563, (1995).
9. R. S. Chivukula, B. Dobrescu, J. Terning, *Phys. Lett.* **B353**, 289 (1995); R. S. Chivukula, J. Terning, *Phys. Lett.* **B385** 209 (1996).
10. S. Adler *et al.* , *Phys. Rev. Lett.* **79** 2204 (1997).
11. M. Strassler and M. Peskin, *Phys. Rev.* **D43** 1500 (1991); see also C. T. Hill and S. Parke, *Phys. Rev.* **D49**, 4454 (1994).

Compositeness Test at the FMC with Bhabha Scattering

E. J. Eichten and S. Keller

*Fermi National Accelerator Laboratory, P. O. Box 500,
Batavia, IL 60510, U.S.A.*

Abstract. It is possible that quarks and/or leptons have substructure that will become manifest at high energies. Here we investigate the limits on the muon compositeness scale that could be obtained at the First Muon Collider using Bhabha scattering. We study this limit as a function of the collider energy and the angular cut imposed by the detector capability.

I INTRODUCTION

The presence of three generations of quarks and leptons, apparently identical except for mass, strongly suggests that they are composed of still more fundamental fermions. It is clear that, if substructure exist, the associated strong interaction energy scale Λ must be much greater than the quark and lepton masses. Long ago, 't Hooft figured out how interactions at high energy could produce essentially massless composite fermions: the answer lies in unbroken chiral symmetries of the underlying fermions and confinement of their new strong nonabelian gauge interactions [1]. There followed a great deal of theoretical effort to construct a realistic model of composite quarks and leptons (see, e.g., Ref. [2]) which, while leading to valuable insights on chiral gauge theories, fell short of its main goal.

It was pointed out that the existence of quark and lepton substructure will be signalled at energies well below Λ by the appearance of four-fermion "contact" interactions which differ from those arising in the standard model [3,4]. These interactions are induced by the exchange of bound states associated with the new gauge interactions. The main constraint on their form is that they must be $SU(3) \otimes SU(2) \otimes U(1)$ invariant because they are generated by forces operating at or above the electroweak scale. These contact interactions are suppressed by $1/\Lambda^2$, but the coupling parameter of the exchanges—analogous to the pion-nucleon and rho-pion couplings—is not small. Compared to the standard model, contact interaction amplitudes are of relative order $s/(\alpha\Lambda^2)$, where $\sqrt{s}$ is the center of mass energy of the process taking place and α the coupling constant of the standard model interaction. The appearance of $1/\alpha$ and the growth with s make contact-interaction effects the lowest-energy signal of quark and lepton substructure. They are sought in jet production at hadron and lepton colliders, Drell-Yan production of high invariant mass lepton pairs, Bhabha scattering, $e^+e^- \to \mu^+\mu^-$ and $\tau^+\tau^-$ [5], atomic parity violation [6], and polarized Möller scattering [7]. Hadron collider experiments can probe values of Λ from the 2–5 TeV range at the Tevatron to the 15–20 TeV range at the LHC (See Refs. [4,8]).

Here, we will study in some details one specific example for the First Muon Collider (FMC): the constraint that can be imposed on the scale of muon compositeness by measuring Bhabha scattering. The specific form for the muon contact interaction is presented in Section II. (All the results presented here are also applicable to electron compositeness at e^+e^- colliders with same energy and luminosity.)

CELLO at PETRA with a center of mass energy, $\sqrt{s}$, of 35 GeV and an integrated luminosity, $\mathcal{L}$, of 86 pb^{-1} was able to put a lower limit on the (electron) compositeness scale of the order of 2-4 TeV, depending on the specific model for compositeness [9]. This is about the same reach as the current Tevatron reach. This clearly show the potential for lepton colliders to probe compositeness; they have an enormous reach.

In section III, we study the reach versus the energy of the FMC, with the corresponding luminosity chosen for this workshop. We also study the effect of the angular cut on the reach. It is important to study that effect because large amount of radiation close to the beam will limit the capability of the detectors outside the central region.

II MUON COMPOSITENESS

We assume the muon has a substructure. For collider energy below the scale associated with this new structure, the effect can be parametrized by a four fermions interaction. Here, we use the flavor-diagonal, helicity-conserving contact interaction proposed by E. J. Eichten, K. Lane and M. E. Peskin [3]:

$$\mathcal{L} = \frac{g^2}{2\Lambda^2} \left[\eta_{LL}\ j_L j_L + \eta_{RR}\ j_R j_R + \eta_{LR}\ j_L j_R \right], \tag{1}$$

j_L and j_R are the left-handed and right-handed currents and Λ the compositeness scale. The coupling constant, $\alpha_{\text{new}} \equiv \frac{g^2}{4\pi}$, is assumed to be strong and set to one. By convention, the η have magnitude one.

The unpolarized cross section at lowest order, including the γ and Z exchange (s and t channel) and the contact interaction from Eq. 1 can be written in the following form, see Ref. [9]:

$$\frac{d\sigma}{d\Omega} = \frac{\alpha^2}{8s} \left[4B_1 + B_2(1 - \cos\theta)^2 + B_3(1 + \cos\theta)^2 \right] \tag{2}$$

where

$$B_1 = \left(\frac{s}{t}\right)^2 \left|1 + (g_V^2 - g_A^2)\,\xi + \frac{\eta_{RL} t}{\alpha \Lambda^2}\right|^2, \tag{3}$$

$$B_2 = \left|1 + (g_V^2 - g_A^2)\,\chi + \frac{\eta_{RL} s}{\alpha \Lambda^2}\right|^2, \tag{4}$$

$$B_3 = \frac{1}{2}\left|1 + \frac{s}{t} + (g_V + g_A)^2(\frac{s}{t}\xi + \chi) + \frac{2\eta_{LL} s}{\alpha \Lambda^2}\right|^2$$
$$+ \frac{1}{2}\left|1 + \frac{s}{t} + (g_V - g_A)^2(\frac{s}{t}\xi + \chi) + \frac{2\eta_{RR} s}{\alpha \Lambda^2}\right|^2, \tag{5}$$

$$\chi = \frac{G_F}{2\sqrt{2}} \frac{M_Z^2}{\pi \alpha} \frac{s}{s - M_Z^2 + iM_Z\Gamma}, \tag{6}$$

and

$$\xi = \frac{G_F}{2\sqrt{2}} \frac{M_Z^2}{\pi \alpha} \frac{t}{t - M_Z^2 + iM_Z\Gamma}. \tag{7}$$

α is the usual fine structure constant, θ the scattering angle between the incoming and outgoing muon, $t = -s/2(1 - \cos\theta)$, g_V and g_A the vector and axial vector coupling constant, M_Z and Γ the mass and width of the Z, and G_F the Fermi constant. We will consider four typical models: LL couplings ($\eta_{LL} = \pm 1$, $\eta_{RR} = \eta_{LR} = 0$), RR couplings ($\eta_{RR} = \pm 1$, $\eta_{LL} = \eta_{LR} = 0$), VV couplings ($\eta_{LL} = \eta_{RR} = \eta_{LR} = \pm 1$), and AA couplings ($\eta_{LL} = \eta_{RR} = -\eta_{LR} = \pm 1$). The positive and negative sign indicate the possible constructive or destructive interference between the electroweak (EW) and compositeness contributions.

III EXPERIMENTAL BOUNDS

The total cross section for the different energy and luminosity options considered at this workshop is presented in Table 1. The only detector effect included is

TABLE 1. Energy of the collider, luminosity, cross section (with $|\cos\theta| < 0.8$), and the expected number of events.

$\sqrt{s}$ (GeV)	$\mathcal{L}(fb^{-1})$	$\sigma(pb)$	N (10^3)
100	.6	125	75
200	1.	34	34
350	3.	11	33
500	7.	5	35

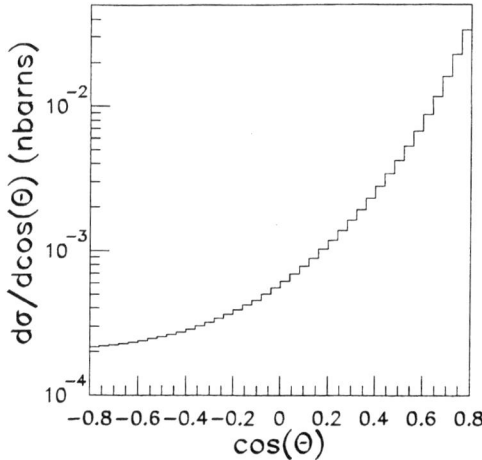

FIGURE 1. $\cos\theta$ distribution at 500 GeV.

an angular cut: $|\cos\theta| < 0.8$. No other detector effects were included in this analysis. As is well known, and can be seen in the set of equations presented earlier, the EW cross section decreases proportionally to s (except in the Z resonance region), whereas the interference term is independent of the energy and the pure compositeness term increases with s. This fact combined with the (almost) constant number of events expected as a function of the energy, see Table 1, clearly indicates that the best compositeness limit will come from the highest energy option.

In Fig. 1, the $\cos\theta$ distribution at $\sqrt{s} = 500$ GeV is presented. The typical t-channel, forward peaking is apparent. The $\cos\theta$ distributions at the other energies have the same shape and are therefore not shown.

To show the impact of the compositeness contribution we use the variable Δ, see Ref. [3]:

$$\Delta = \frac{\left(\frac{d\sigma}{d\cos\theta}\right)_{EW+\Lambda} - \left(\frac{d\sigma}{d\cos\theta}\right)_{EW}}{\left(\frac{d\sigma}{d\cos\theta}\right)_{EW}}, \tag{8}$$

the difference between the theory with and without the compositeness terms, divided by the EW contribution. The $\cos\theta$ distribution of Δ is presented in Fig. 2 for $\sqrt{s} = 100$ GeV. The Λ's were chosen such that the compositeness correction is of the order of 10% compared to the EW contribution. That requires that $\Lambda \simeq 30\sqrt{s}$

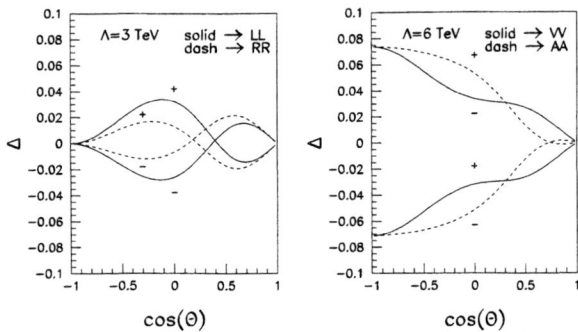

FIGURE 2. The variable Δ versus $\cos\theta$ at $\sqrt{s} = 100$ GeV for the four models, LL, RR, VV, and AA, for the two signs of the η's, indicate by $+$ and $-$ on the plot.

for the LL or RR ($\frac{s}{\alpha\Lambda^2} \sim .1$) couplings and $\Lambda \simeq 60\sqrt{s}$ for the VV or AA couplings ($\frac{s}{\alpha\Lambda^2} \sim .1/4$), there are four interference terms in these latter models, see section II. The results for the four models are shown in Fig. 2 for both sign of the $\eta's$. It is clear that one can get limits on the compositeness scale from the change of the shape of the distribution. Note that in the forward region, in term of sensitivity to the compositeness scale, the smaller change of the shape is compensated by the larger number of events. We also present the distribution at 200 and 500 GeV, in Fig. 3 and 4, respectively. The 300 GeV case is very similar to the 500 GeV case and is not shown.

The next step is to obtain a lower limit on Λ, assuming that the data follow

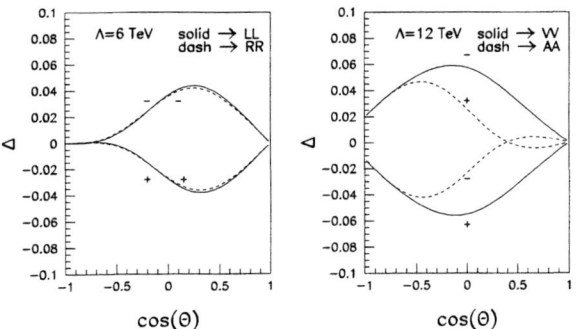

FIGURE 3. Same as Fig. 2 at 200 GeV, about the LEPII energy.

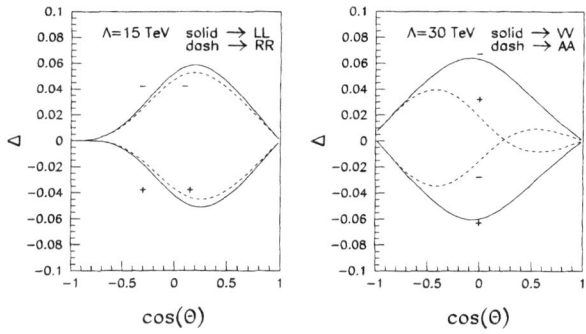

FIGURE 4. Same as Fig. 2 at 500 GeV.

TABLE 2. 95% CL limits (in TeV) for different energies (in GeV) of the muon collider, we used $|\cos\theta| < 0.8$. We also present the expected LEP limits for which we used $|\cos\theta| < 0.95$.

	LEP(91)	LEP(175)	100	200	350	500	4000
$\mathcal{L}(fb^{-1})$	.15	.1	.6	1.	3.	7.	450.
LL	4.0	5.8	4.8	10	20	29	243
RR	3.8	5.7	4.9	10	19	28	228
VV	6.9	12.	12	21	36	54	435
AA	3.8	7.2	12	13	21	32	263

the EW theory. Defining $x = 1/\Lambda^2$, on average (repeating the experiment many times) the central value of x resulting from a χ^2 fit will be zero because the data is assumed to follow the EW theory. For x small enough, the differential cross section is linear in x (the x^2 term is small). Within this approximation the χ^2 is quadratic in x, and the fit can be trivially done. The uncertainty on x, σ_x, is simply given by two times the inverse of the second derivative of χ^2 with respect to x (a constant). We used 20 intervals for the fit, such that the lowest number of events in one bin is still more than 100, which correspond to a maximum 10% statistical uncertainty in each bin. The 95% CL limit on Λ is then obtained from: $\Lambda^2 > 1/(1.64\sigma_x)$. The results are presented in Table 2, for different energies of the muon collider and $|\cos\theta| < 0.8$. As expected the highest energy machine put the strongest constraint on the compositeness scale. The limit that the 4 TeV machine will be able to put (with the luminosity scaled to maintain the number of events constant) is really impressive. The Λ limits are large enough such that the approximation used (x small) is valid. Because of the approximation used, central value of x equal to zero and the differential cross section linear in x, the limits are independent of the sign of the η. We have only included the statistical uncertainty in this analysis, and the limits presented here should be considered within that context. In particular, the

TABLE 3. 95% CL limits (in TeV) for different angular cuts at $\sqrt{s} = 500$ GeV, $\mathcal{L} = 7fb^{-1}$.

$|\cos\theta| <$	.6	.8	.9	.95
LL	26	29	31	32
RR	24	28	30	30
VV	50	54	56	57
AA	28	32	34	35

absolute normalization, which is used in this analysis, might be subject to large uncertainties. Our calculated limits for the CELLO case are compatible with their measurements (their central x value is of course non zero). In Table 2, we also have included the expected limit for LEP with the current integrated luminosity (per experiment) and its larger $\cos\theta$ coverage.

In Table 3 we explore the effect of the $\cos\theta$ cut on the 95% CL limit. Although any increase in coverage obviously increases the limit, the improvement between 0.8 and 0.95 is less than 10%. We therefore conclude that the coverage up to 0.8 is adequate for this measurement. It is not necessary to go down very close to the beam to do a very good measurement.

IV CONCLUSION

We investigated the limits on the muon compositeness scale that could be obtained at the First Muon Collider using the Bhabha scattering process. We considered four typical models for the four-fermion contact terms expected as a low-energy signal for compositeness: LL, RR, VV, and AA couplings.

As expected, the reach increases rapidly with energy. We find that the reach at the 500 GeV FMC is $\Lambda > 30 - 55$ TeV depending on the model. At a future 4 TeV muon collider the range extends to $\Lambda > 230 - 440$ TeV.

The likelihood of limited angular coverage in detectors (because of the unavoidable background of decaying muons) does not appear to poise a severe problem for the study of muon compositeness. We found that an angular coverage corresponding to $|\cos\theta| < 0.8$ is adequate to obtain 90% of the full reach in the compositeness scale Λ.

A number of detailed studies remain to be done. For example, it is clear that the polarization will help to differentiate between the four models considered here. Also we have only considered the statistical uncertainties, the systematic uncertainties in measurements of Bhabha scattering could be significant and need to be included in future more realistic studies.

REFERENCES

1. G. 't Hooft, in *Recent Developments in Gauge Theories*, edited by G. 't Hooft, et al. (Plenum, New York, 1980).
2. S. Dimopoulos, S. Raby and L. Susskind, Nucl. Phys. **B173**, 208 (1980); M. E. Peskin, Proceedings of the 1981 Symposium on Lepton and Photon Interactions at High Energy, edited by W. Pfiel, p. 880 (Bonn, 1981); I. Bars, Proceedings of the Rencontres de Moriond, *Quarks, Leptons and Supersymmetry*, edited by Tranh Than Van, p. 541 (1982).
3. E. J. Eichten, K. Lane and M. E. Peskin, Phys. Rev. Lett. **50**, 811 (1983)
4. E. J. Eichten, I. Hinchliffe, K. Lane and C. Quigg, Rev. Mod. Phys. **56**, 579 (1984).
5. For current collider limits on substructure, see F. Abe, et al., The CDF Collaboration, Phys. Rev. Lett. **77**, 438 (1996); and the Review of Particle Physics, Particle Data Group, *Phys. Rev.* **D54**, 1 (1996).
6. J. Rosner, Phys. Rev. **D53**, 2724 (1996), and references therein.
7. K. Kumar, E. Hughes, R. Holmes and P. Souder, "Precision Low Energy Weak Neutral Current Experiments", Princeton University (October 30, 1995), to appear in Modern Physics Letters A.
8. E. J. Eichten and K. Lane, "Electroweak and Flavor Dynamics at Hadron Colliders - II", Fermilab-CONF-96-298-T [hep-ph/9609298]; and the references therein.
9. CELLO collaboration, Z. Phys. **C51**, 143 (1991).

Constraints on Strong Dynamics from Rare B and K Decays

Gustavo Burdman

Department of Physics, University of Wisconsin, Madison WI 53706.

Abstract. We discuss the constraints from rare B and K decays on the Electroweak Symmetry Breaking (EWSB) sector, as well as on theories of fermion masses. We focus on models involving new strong dynamics and show that transitions involving Flavor Changing Neutral Currents (FCNC) play an important role in disentangling the physics in these scenarios. In a model-independent approach to the EWSB sector, the information from rare decays is complementary to precision electroweak observables in bounding the contributions to the effective lagrangian. We compare the pattern of deviations from the Standard Model (SM) that results from these sources, with the deviations associated with the mechanism for generating fermion masses.

INTRODUCTION

Two of the most intriguing questions in particle physics are the the EWSB mechanism and the origin of fermion masses. Although the SM remains a successful theory when compared with all the available data, it lacks predictability in the Higgs sector, which determines the masses of gauge bosons, as well as of fermions through *ad hoc* Yukawa couplings. This suggests the possibility that new physics beyond the SM might be associated with either of these questions. In general, the energy scales and dynamics behind the EWSB sector and the fermion masses may be unrelated. In order to avoid fine-tuning, the scale associated with EWSB cannot be much higher than a few TeV, whereas the scales where light fermion masses are generated could be much higher. If the mechanism responsible for the breaking of the electroweak symmetry involves some new strong dynamics, deviations from the SM might be observable in low energy signals even at energies much smaller than the scale of new physics Λ. Reaching this new frontier by direct observation of new physical states or even of tree-level effects in the couplings of SM particles, may require not only very large energies but also some previous knowledge of what (and what not) to expect. Thus, low energy measurements might be of paramount importance in planning experiments and search strategies at high energy machines. Among these low energy signals are electroweak measurements such as those at LEP and the Tevatron. On the other hand, processes involving Flavor Changing Neutral Currents (FCNC) can play a complementary role, since the fact that these processes are largely suppressed or forbidden in the SM may compensate the suppression by factors of m/Λ (with m the low energy scale, e.g. m_K, m_B, etc.). Here we address the potential of rare B and K decays as a complement to other low energy measurements in constraining models where strong dynamics is associated to either the EWSB sector and/or the origin of fermion masses. In the absence of a completely satisfactory theory of dynamical symmetry breaking and fermion masses, it is convenient to carry out a model-independent analysis that makes maximum use of the known properties of the electroweak interactions. This is the case with the EWSB sector, where an effective lagrangian approach allows us to parameterize the effects of the new strong dynamics in very much the same way chiral perturbation theory parameterizes low energy QCD. On the other hand, the effects from fermion mass generation can also be addressed by a general operator analysis. However, in addition, most theories predict the existence of relatively light states (scalars, pseudo-Goldstone bosons, etc.) which generally couple to mass in one way or another. To exemplify the effects of such states (which cannot be integrated out) we work with a particular set of models known as Topcolor-assisted Technicolor (TaTC). This provides a current example of how strong dynamics model building deals with the large top-quark mass and illustrates the distinct low energy phenomenology emerging from non-standard EWSB scenarios.

LOW ENERGY EFFECTS OF ELECTROWEAK SYMMETRY BREAKING

In the absence of a light Higgs boson the symmetry breaking sector is represented by a non-renormalizable effective lagrangian corresponding to the non-linear realization of the σ model. The essential feature is the spontaneous breaking of the global symmetry $SU(2)_L \times SU(2)_R \to SU(2)_V$. To leading order the interactions involving the Goldstone bosons associated with this mechanism and the gauge fields are described by the effective lagrangian [1]

$$\mathcal{L}_{LO} = -\frac{1}{4} B_{\mu\nu} B^{\mu\nu} - \frac{1}{2} \text{Tr}\left[W_{\mu\nu} W^{\mu\nu}\right] + \frac{v^2}{4} \text{Tr}\left[D_\mu U^\dagger D^\mu U\right], \tag{1}$$

where $B_{\mu\nu}$ and $W_{\mu\nu} = \partial_\mu W_\nu - \partial_\nu W_\mu + ig\left[W_\mu, W_\nu\right]$ are the the $U(1)_Y$ and $SU(2)_L$ field strengths respectively, the electroweak scale is $v \simeq 246$ GeV and the Goldstone bosons enter through the matrices $U(x) = e^{i\pi(x)^a \tau_a / v}$. The covariant derivative acting on $U(x)$ is given by $D_\mu U(x) = \partial_\mu U(x) +$

$igW_\mu(x)U(x) - \frac{i}{2}g'B_\mu(x)U(x)\tau_3$. To this order there are no free parameters once the gauge boson masses are fixed. The dependence on the dynamics underlying the strong symmetry breaking sector appears at next to leading order. To this order, a complete set of operators includes one operator of dimension two and nineteen operators of dimension four [1,2]. The effective lagrangian to next to leading order in the basis of Ref. [1] is given by

$$\mathcal{L}_{\text{eff.}} = \mathcal{L}_{LO} + \sum_{i=0}^{19} \alpha_i \mathcal{O}_i , \quad (2)$$

where $\mathcal{O}_0$ is a dimension two custodial-symmetry violating term absent in the heavy Higgs limit of the SM. If we restrict ourselves to CP invariant structures, there remain fifteen operators of dimension four. The coefficients of some of these operators are constrained by low energy observables. For instance precision electroweak observables constrain the coefficient of $\mathcal{O}_0$, which gives a contribution to the electroweak parameter T. The 3 σ limit requires

$$\alpha_0 < 6 \times 10^{-3}. \quad (3)$$

The combinations $(\alpha_1 + \alpha_8)$ and $(\alpha_1 + \alpha_{13})$ contribute to the electroweak parameters S and U. For instance, the constraint on S translates into

$$|\alpha_1 + \alpha_{13}| < 1.5 \times 10^{-2}. \quad (4)$$

In addition, the coefficients α_2, α_3, α_9 and α_{14} modify the triple gauge-boson couplings (TGC) and will be probed at LEPII and the Tevatron at the few percent level [3].

The remaining operators contribute to oblique corrections only to one loop and, in some cases, only starting at two loops. To the last group belong $\mathcal{O}_{11}$ and $\mathcal{O}_{12}$ given that their contributions to the gauge boson two-point functions only affect the longitudinal piece of the propagators. Of particular interest is the operator $\mathcal{O}_{11}$ defined by [1]

$$\mathcal{O}_{11} = \text{Tr}\left[(\mathcal{D}_\mu V^\mu)^2\right], \quad (5)$$

with $V_\mu = (D_\mu U)U^\dagger$ and the covariant derivative acting on V_μ defined by $\mathcal{D}_\mu V_\nu = \partial_\mu V_\nu + ig[W_\mu, V_\nu]$. The equations of motion for the $W_{\mu\nu}$ field strength imply [4]

$$\mathcal{D}_\mu V^\mu = \frac{2i}{v^2} \mathcal{D}_\mu J_w^\mu , \quad (6)$$

where the $SU(2)_L$ current is $J_w^\mu = \sum_\psi \left(\bar{\psi}_L \gamma^\mu \frac{\tau^a}{2} \psi_L\right)\tau^a$, ψ_L denote the left-handed fermion doublets. The dominant effect appears in the quark sector due to the presence of terms proportional to m_t. After the quark fields are rotated to the mass eigenstate basis, the operator $\mathcal{O}_{11}$ can be written as [5]

$$\mathcal{O}_{11} = \frac{m_t^2}{v^4} \left\{ (\bar{t}\gamma_5 t)^2 - 8 \sum_{i,j} V_{ti}^* V_{tj} (\bar{q}_{iL} t_R)(\bar{t}_R q_{jL}) \right\} + \cdots \qquad (7)$$

where $i, j = d, s, b$, the V_{ti} are Cabibbo-Kobayashi-Maskawa (CKM) matrix elements and the dots stand for terms suppressed by small fermion masses.

From the above discussion we see that the leading effects of the EWSB sector in FCNC processes are coming from the insertion of anomalous TGC vertices and four-fermion operators like (7). In the rest of this section, we review the status and future impact of these constraints on the symmetry breaking sector.

Four-fermion Operators

The effects of the four-fermion operators in (7) in rare B and K decays were considered in Ref. [6]. The loop insertion will result in contributions to several FCNC processes, that are controlled by both the coefficient α_{11} of the effective lagrangian (2) as well as by the high energy scale Λ. To one loop, only one parameter is needed, namely

$$y = \alpha_{11} \log \frac{\Lambda^2}{m_t^2} . \qquad (8)$$

This parameter also governs the contributions of (7) to other neutral processes, both flavor changing and flavor conserving. For instance, the $(\bar{b}_L t_R)(\bar{t}_R b_L)$ term in (7) gives a contribution to $Z \to b\bar{b}$, whereas the terms like $(\bar{b}_L t_R)(\bar{t}_R d_L)$ appear in $B^0 - \bar{B}^0$ mixing [5]. Thus the measurements of R_b and the rate of B mixing (together with all other CKM information) can be used to derive a bound on y. Although the bound carries some uncertainty mainly associated with CKM quantities like f_B and V_{ub}, we will take it to be, approximately [5,6]

$$|y| < 0.50 . \qquad (9)$$

Next, we use this as the allowed range for y in order to explore the possible impact of this physics in rare B and K decays. The one-loop insertion of the terms

$$\mathcal{O}_{11} = -\frac{8m_t^2}{v^4} \left\{ V_{ts}^* V_{tb} \bar{s}_L t_R \bar{t}_R b_L + V_{td}^* V_{tb} \bar{d}_L t_R \bar{t}_R b_L + V_{td}^* V_{ts} \bar{d}_L t_R \bar{t}_R s_L \right\} + \cdots \qquad (10)$$

induces new contributions to various FCNC vertices in B decays (the first two terms in (10)), as well as in K decays (third term in (10)).

First, let us consider $b \to q\gamma$ processes leading, for instance, to the inclusive $B \to X_s \gamma$, since this rate has been recently measured [7]. The one-loop insertion of the operator $\mathcal{O}_{11}$ does not give a contribution to these processes given

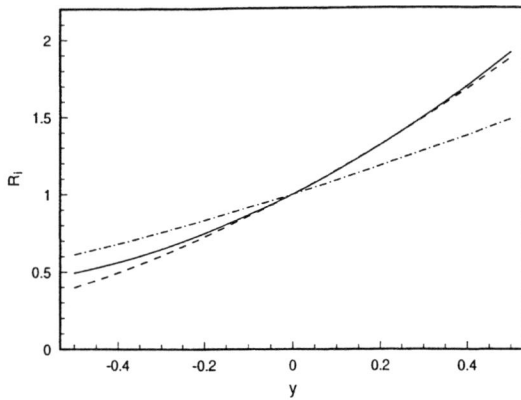

FIGURE 1. Ratio of the modified branching ratio to the standard model expectation as a function of $y = \alpha_{11} \log \frac{\Lambda^2}{m_t^2}$. The solid line corresponds to the ratio R_ℓ for $B \to X_{(s,d)}\ell^+\ell^-$ inclusive decays, the dashed line to R_ν for $B \to X_{(s,d)}\nu\bar{\nu}$ and the dot-dashed line to R_g for $b \to s\bar{s}s$ decays. From Ref. [6].

that it does not mix with the operator $\bar{s}_L \sigma_{\mu\nu} b_R$ responsible for the on-shell photon amplitude. Mixing only occurs at two loops, when QCD corrections are taken into account. As a result the effect, in all $b \to q\gamma$ transitions is expected to be only a few percent of the SM branching ratios [6].

On the other hand, the off-shell amplitudes for photons, Z's and gluons are non-zero at one loop. They generate contributions to processes such as $b \to q\ell^+\ell^-$, $b \to q\nu\bar{\nu}$, $b \to q\bar{q}'q'$; as well as to similar rare kaon decays like $s \to d\nu\bar{\nu}$, etc. In order to asses the potential effects we define

$$R_\ell \equiv \frac{Br(B \to X_{(s,d)}\ell^+\ell^-)}{Br(B \to X_{(s,d)}\ell^+\ell^-)_{SM}}, \qquad (11)$$

which is plotted in Fig. 1 as a function of the parameter y defined in (8), for the allowed range of y (9). Analogously, we can define the ratio R_ν, which tracks the effects in $B \to X_{s,d}\nu\bar{\nu}$ decays; whereas the contribution to gluon penguin processes such as $b \to s\bar{s}s$ is represented by the ratio R_g. As it is clear from Fig. 1, the effects of the operator $\mathcal{O}_{11}$ are very similar in all three types of B decays.

We see that, even with the R_b and $B^0 - \bar{B}^0$ mixing constraints, large deviations from the SM predictions for these modes are possible. The current experimental bounds on these processes are still not binding on y. However, sensitivity to SM branching ratios will be reached in the next round of experiments at the various B factories at Cornell, KEK, SLAC and Fermilab. The distinct feature of this effect is that no significant deviation is expected in $b \to s\gamma$, even when large deviations are observed in all the other modes.

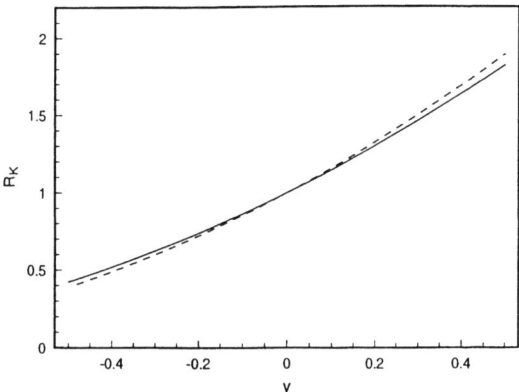

FIGURE 2. Ratio of the modified branching ratio to the standard model expectation for $K^+ \to \pi^+ \nu \bar{\nu}$ (solid line) and $K_L \to \pi^0 \nu \bar{\nu}$ (dashed line). From Ref. [6].

The effects are very similar in rare K decays such as $K^+ \to \pi^+ \nu \bar{\nu}$ and $K_L \to \pi^0 \nu \bar{\nu}$, etc. In Fig. 2 we plot R_K, a quantity analogous to R_ℓ in (11). Again, large effects of up to factors of 2 deviations, are allowed. The recently reported [8] observation of one event in $K^+ \to \pi^+ \nu \bar{\nu}$ roughly translates into $R_K \simeq (0.50 - 5.0)$, which is still not constraining.

Although in this model-independent approach we cannot, as a matter of principle, calculate the size of the coefficients α_i, we can use general arguments to estimate their approximate value. Using naive dimensional analysis [9] we have

$$\alpha_{11} \simeq \mathcal{O}(1) \times \frac{v^2}{\Lambda^2}, \qquad (12)$$

with the scale of new physics obeying $\Lambda \lesssim 4\pi v$. For instance, taking $\Lambda = 4\pi v$, one would obtain $y \simeq \mathcal{O}(1) \times 0.04$. On the other hand, if $\Lambda = 2\pi v$, one has $y \simeq \mathcal{O}(1) \times 0.12$. In any case, these are meant to be order of magnitude estimates. Therefore, the experimental relevance of the effect strongly depends on details of the dynamics we are not able to compute in a model-independent fashion.

Finally, we should note that rare B and K decays are the most sensitive signals for this effect. This is due to the fact that four-lepton operators are suppressed by the lepton masses, and that $\mathcal{O}_{11}$ does not mix quarks and leptons.

Triple Gauge-boson Couplings

Imposing C and P conservation, the most general form of the WWN ($N = \gamma, Z$) couplings can be written as [3]

$$\mathcal{L}_{WWN} = g_{WWN} \left\{ i\kappa_N W_\mu^\dagger W_\nu N^{\mu\nu} + ig_1^N \left(W_{\mu\nu}^\dagger W^\mu N^\nu - W_{\mu\nu} W^{\dagger\mu} N^\nu \right) \right. \tag{13}$$

$$\left. +i\frac{\lambda_N}{M_W^2} W_{\mu\nu}^\dagger W_\lambda^\nu N^{\nu\lambda} \right\} , \tag{14}$$

with the conventional choices being $g_{WW\gamma} = -e$ and $g_{WWZ} = -g\cos\theta$ [10]. In principle, there are six free parameters. Making contact with the electroweak lagrangian (2), these parameters can be expressed in terms of the next-to-leading order coefficients [4] $\alpha_1, \alpha_2, \alpha_3, \alpha_8, \alpha_9, \alpha_{13}$ and α_{14}. Conservation of the electromagnetic charge implies $g_1^\gamma = 1$. Furthermore, to this order in the energy expansion (2) $\lambda_N = 0$. Then we are left with κ_γ, κ_Z and g_1^Z. Finally, when considering rare B and K decays, we can neglect the contribution of κ_Z since it will be suppressed by powers of the small external momenta over m_Z. Thus, in this simplistic approach, there are only two parameters relevant at very low energies. The SM predicts $\kappa_\gamma = g_1^Z = 1$. The effects of anomalous TGC have been previously studied in the literature [11]. However, this hierarchical approach to the couplings has not been the one used in the various analyses and a more comprehensive study is needed. The experiments at LEP II and the next Tevatron run are going to be sensitive to deviations from the SM prediction at the $(5 - 10)\%$ level [10]. Effects of this size might be also observed in rare B and K decays. For instance, $\delta g_1^Z = g_1^Z - 1 = 0.10$ can produce enhancements in the branching ratios of $b \to s\ell^+\ell^-$ decay modes of up to $(60 - 70)\%$ [11]. In the near future, B factory experiments will have sensitivity to these processes at the SM level, turning these low energy measurements into an excellent complement of direct probes of the TGC.

FERMION MASSES AND ELECTROWEAK DYNAMICS

Up to now, we have only considered the effects of the dynamics associated with the EWSB. These are encoded in the effective lagrangian (2), which only involves the Goldstone boson and gauge fields. Additionally, it is possible that the new strong dynamics may also affect some or all fermions. We first comment on the effective lagrangian approach for non-SM couplings of fermions to gauge bosons, and then examine the effects of a prototypical class of theories (Topcolor) where the dynamical generation of fermion masses imply the existence of relatively light new states.

Anomalous Couplings of Fermions to Gauge Bosons

The effects of new dynamics on the couplings of fermions with the SM gauge bosons can be, in principle, also studied in an effective lagrangian approach. For instance, if in analogy with the situation in QCD, fermion masses are dynamically generated in association with EWSB, residual interactions of fermions with Goldstone bosons could be important [12] if the $m_f \simeq f_\pi \simeq v$. Thus residual, non-universal interactions of the third generation quarks with gauge bosons could carry interesting information about both the origin of the top quark mass and EWSB. In a very general parameterization, the anomalous couplings of third generation quarks can be written as

$$\Delta \mathcal{L} = -\frac{g}{\sqrt{2}} \{C_L \, (\bar{t}_L \gamma_\mu b_L) + C_R \, (\bar{t}_R \gamma_\mu b_R)\} W^{+\mu}$$
$$- \frac{g}{2 \, c\theta_W} \{N_L^t \, (\bar{t}_L \gamma_\mu t_L) + N_R^t \, (\bar{t}_R \gamma_\mu t_R)$$
$$+ N_L^b \, (\bar{b}_L \gamma_\mu b_L) + N_R^b \, (\bar{b}_R \gamma_\mu b_R)\} Z^\mu \,, \qquad (15)$$

where the parameters $C_{L,R}$, $N_{L,R}^{t,b}$ contain the residual, non-universal effects associated with the new dynamics, perhaps responsible for the large top quark mass. Then, if we assume that the new couplings are CP conserving, there are six new parameters. They are constrained at low energies by a variety of experimental information, mostly from electroweak precision measurements and the rate of $b \to s\gamma$. Several simplifications are usually made in order to reduce the number of free parameters. For instance, in most of the literature, it is assumed that $N_{L,R}^b = 0$ [14]. A stringent bound on the right-handed charged coupling is obtained from $b \to s\gamma$ [15]: $-0.05 < C_R < 0.01$. The bounds obtained on a particular coupling from electroweak observables such as S, T, U and R_b generally strongly depend on assumptions about the other couplings. For example, if $C_L = 0$, then the combination $(N_L^t - N_R^t)$ is strongly constrained since it contributes to T. On the other hand, if $C_L = N_L^t$, then $N_R^t < 0.02$ [12,16] since it is the only (linear) contribution to T. Thus, although in general most parameters are confined to a few percent, some of them are allowed to be as large as 0.30 under certain conditions. This "model-dependent" situation requires more experimental information. A global analysis of the effects of the couplings of eqn. (15) in rare B and K processes such as $b \to s\ell^+\ell^-$, $s \to d\nu\bar{\nu}$, etc. may help disentangle the various possible effects and perhaps will give constraints that may be of importance in interpreting data from higher energy experiments [17].

The effects of light states: the example of Topcolor

The description of the residual effects of strong dynamics at low energies on fermion couplings by using (15) corresponds to cases where the states associated with the new physics are heavy compared to the weak scale. Thus, integrating out the heavy states, leaves us with effective couplings which might be generated at tree level or through loops in the full theory. However, most theories in which electroweak symmetry and/or fermion masses have a dynamical origin also contain states with masses comparable to the weak scale. Such is the case, for instance, in Technicolor models where the breaking of large chiral symmetries imply the presence of pseudo-Goldstone bosons with masses of at most a few hundred GeV. It is also the case in Topcolor-assisted Technicolor (TaTC) models [18], where a top-condensation mechanism generated by the Topcolor interactions is responsible for the large dynamical top quark mass, whereas Technicolor breaks the electroweak symmetry giving (most of) the W and Z masses. The TaTC scenario is designed to relief the problems of Extended Technicolor (ETC) in generating a heavy top [19]. Although the new gauge bosons associated with the TaTC gauge group are heavier than 1 TeV, the presence of several scalar and pseudo-scalar states with masses in the few-hundred GeV range, forces us to take these into account directly in our calculations. From the point of view of their impact in low energy observables, the most important of these states are the top-pions $\vec{\pi}_t$, the triplet of Goldstone bosons associated with the breaking of the top chiral symmetry. Since top condensation does not fully break the electroweak symmetry ($f_{\pi_t} \simeq (60-70)$ GeV $< v$), after mixing with the techni-pions, there will be a triplet of physical top-pions in the spectrum, with a coupling to third generation quarks given by

$$i \frac{m_t}{\sqrt{2} f_{\pi_t}} \left\{ \bar{t}\gamma_5 t \pi^0 + \bar{t}_R b_L \pi^+ + \bar{b}_L t_R \pi^- \right\} . \qquad (16)$$

They acquire masses of a few hundred GeV due to explicit ETC quark mass terms. Additionally, in most models there are scalar and pseudo-scalar bound states due to the strong (although sub-critical) effective coupling of right-handed b-quarks. The closer the effective couplings are from criticality, the lighter these bound states tend to be. The spectrum and properties of these states, unlike those of top-pions, are not determined by model-independent features of the symmetry breaking pattern but depend on details of the model. Finally, in all TaTC models there will be pseudo-Goldstone bosons from the breaking of techni-fermion chiral symmetries. However, their couplings to third generation quarks are reduced with respect to (16) by m_{ETC}/m_t, where m_{ETC} is a small ETC mass of the order of 1 GeV[1]. The presence of the relatively light top-pions, as well as the additional bound states, imposes severe constraints on Topcolor models due to their potential loop effects in low energy observables, most notably R_b and rare B and K decays.

---

[1] Multi-scale Technicolor models such as the one in Ref. [20], in the absence of Topcolor, have un-suppressed top couplings to pseudo-scalars. This could lead to effects similar to those of top-pions.

Top-pion Effects in R_b : The one-loop contributions of top-pions to the $Z \to b\bar{b}$ process were studied in Ref. [21]. There it was shown that they shift R_b negatively by an amount controlled by m_{π_t} and f_{π_t}. For instance, for $f_{\pi_t} \simeq 60$ GeV the correction is about -1% for $m_{\pi_t} = 800$ GeV, and top-pions with masses in the expected $(100-300)$ GeV range give unacceptably large deviations. This value of the top-pion decay constant is obtained by using the Pagels-Stokar formula, which gives f_{π_t} a logarithmic dependence on the Topcolor energy scale, chosen here to be a few TeV. Potentially cancelling contributions by other states, such as the scalar and pseudo-scalar bound states, Topcolor vector and axial-vector mesons, etc., are either of the wrong sign or not large enough. Possible ways out of this constraint are: larger top-pion masses or larger values of f_{π_t}. The larger f_{π_t} is, the smaller the coupling, and the top-pions are more Goldstone-boson-like. For $f_{\pi_t} \simeq 120$ GeV, for instance, the shift of R_b is well within the experimentally allowed region even for $m_{\pi_t} \simeq (200-300)$ GeV. However, in order to obtain such an enhancement in the decay constant we must either assume large corrections to the Pagels-Stokar expression or introduce new and exotic fermion states.

Rare B and K Decays : The top-pions and other scalar states, give one-loop contributions to FCNC processes. These depend not only on f_{π_t} and m_{π_t} but typically also on one or more elements of the quark rotation matrices necessary to diagonalize the quark Yukawa couplings. The contributions of top-pions, as well as "b-pions" (scalar and pseudo-scalar bound states in models where b_R couples to the Topcolor interaction) to $b \to s\gamma$ depend on $D^{bs}_{L,R}$, the $b \to s$ element in the left or right down rotation matrix. Furthermore, the two contributions tend to cancel. Thus, the freedom in this model-dependent aspects of the prediction makes it possible to have quite low masses and still satisfy the bound from the experimental measurement of $B \to X_s\gamma$ [22]. The situation changes drastically in $b \to s\ell^+\ell^-$ processes, where the cancellations are much less efficient. Although experiments have not yet reached sensitivity to SM branching ratios [23], it will be soon achieved at both hadron and lepton B factories. As an example, we plot in Fig. 3 the Br($b \to s\ell^+\ell^-$) as a function of the top-pion mass with no other contributions, for $f_{\pi_t} = 70$ GeV. The $b \to s\gamma$ constraint is in this case (200 GeV $< m_{\pi_t} <$ 800 GeV). However, one can see that, even for heavier top-pions the effect can still be a $(30-60)\%$ enhancement over the SM prediction of 6×10^{-6}. On the other hand, in the presence of a 400 GeV charged b-pion the curve changes little, but the $b \to s\gamma$ bound is now $m_{\pi_t} < 600$ GeV. Finally, to compare the potential of these FCNC transitions with the R_b constraints, let us say that if we take $f_{\pi_t} \simeq 120$ GeV (which avoids conflict with R_b measurements), then the effect of a 400 GeV top-pion in $b \to s\ell^+\ell^-$ is still an enhancement of more than 50% with respect to SM expectations. Thus, the observation of these modes will further constrain Topcolor models beyond the R_b bounds. We expect similar effects due to top-pions and/or b-pions to be present in kaon processes such

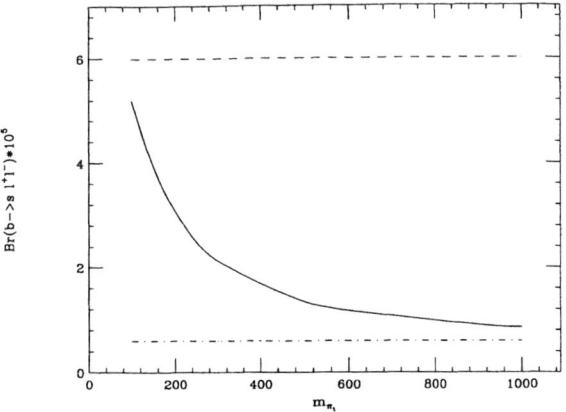

FIGURE 3. Br($b \to s\ell^+\ell^-$) vs. m_{π_t}, for $f_{\pi_t} = 70$ GeV. The dashed horizontal line is the current experimental limit for the inclusive rate [24], whereas the dot-dash line is the SM expectation.

as $K^+ \to \pi^+ \nu \bar{\nu}$.

CONCLUSIONS

We have seen that a complete, model-independent analysis of the effects of strong dynamics in rare B and K decays could shed light on the nature of the EWSB mechanism and the origin of fermion masses. The signals are also likely to be important in models where relatively light scalars couple strongly to mass, like in the case of TaTC. In most cases, the next round of experiments will have sensitivity to SM branching ratios. This will be the case, for instance, for the Tevatron experiments, as well the KEK and SLAC B factories in the $B \to X_{(s,d)}\ell^+\ell^-$ modes. It will also be the situation in the next generation of kaon experiments for $K^+ \to \pi^+\nu\bar{\nu}$ and $K_L \to \pi^0\nu\bar{\nu}$. The amount and variety of experimental information from these processes is such that suggests a parallel to the role of electroweak measurements at the Z pole as not only a constraint on new physics sources but also as guidance in the searches to be carried out at high energy machines such as the Tevatron in Run II, the LHC and eventually the NLC and/or the muon collider. It is possible to imagine a scenario where deviations from the SM in B and/or K decays point to a particular source, e.g. corrections to Goldstone boson propagators given by $\mathcal{O}_{11}$, anomalous TGC or anomalous couplings of third generation quarks to gauge bosons as in (15). The nature of the deviation might dictate the road to follow at high energies. As an example, if the source of an effect is in one the top quark couplings $N_{L,R}^t$, there would be a strong case for a lepton collider running at $t\bar{t}$ threshold. Other scenarios may not be so clear, and may require a comprehensive and careful analysis of all the data to come (including issues like hadronic uncertainties in B decays). This, however, constitutes a very well defined research program.

REFERENCES

1. A. Longhitano, *Phys. Rev.* D22, 1166 (1980), *Nucl. Phys.* B188, 118 (1981).
2. T. Appelquist and G. Wu, *Phys. Rev.* D48, 3235 (1993).
3. K. Hagiwara, K. Hikasa, R. D. Peccei and D. Zeppenfeld, *Nucl. Phys.* B282, 253 (1987); K. Hagiwara, S. Ishiara, R. Szalapski and D. Zeppenfeld, *Phys. Lett.* B283, 353 (1992) and *Phys. Rev.* D48, 2182 (1993).
4. F. Feruglio, *Int. J. Mod. Phys.* A8, 4937 (1993).
5. J. Bernabéu, D. Comelli, A. Pich and A. Santamaria, *Phys. Rev. Lett.* **78**, 2902 (1997).
6. G. Burdman, *Phys. Lett.* B409, 443 (1997).
7. R. Balest et al., the CLEO collaboration, *Phys. Rev. Lett.* 74, 2885 (1995).
8. S. Adler et al., the BNL 787 Collaboration, *Phys. Rev. Lett.* 79, 2204 (1997).
9. A. Manohar and H. Georgi, *Nucl. Phys.* B234, 189 (1984); H. Georgi, "*Weak Interactions and Modern Particle Theory*", Benjamin/Cummings, Menlo Park, California, 1984.
10. T. Barklow et al., SLAC-PUB-7366, In the Proceedings of "New Directions for High-Energy Physics (Snowmass 96), Snowmass, CO, 25 Jun - 12 Jul 1996.
11. S. P. Chia, *Phys. Lett.* B240, 465 (1990); K. A. Peterson, *Phys. Lett.* B282, 207 (1992); G. Baillie, Z. Phys. C61, 667 (1994).
12. R. D. Peccei and X. Zhang, *Nucl. Phys.* B337, 269 (1990); R. D. Peccei, S. Peris and X. Zhang, *Nucl. Phys.* B349, 305 (1991).
13. For a treatment of CP violating effects in B decays from these couplings see A. Abd El-Hady and G. Valencia, *Phys. Lett.* B414, 173 (1997).
14. E. Malkawi and C. P. Yuan, *Phys. Rev.* D50, 4462 (1994).
15. J. Hewett and T. Rizzo, *Phys. Rev.* D49, 319 (1994); K. Fujikawa and A. Yamada, *Phys. Rev.* D49, 5890 (1994);
16. B. Dobrescu and J. Terning, *Phys. Lett.* B416, 129 (1998).
17. G. Burdman, in preparation.
18. C. T. Hill, *Phys. Lett.* B345, 483 (1995).
19. E. Eichten and K. Lane, *Phys. Lett.* B352, 382 (1995).
20. E. Eichten and K. Lane, *Phys. Lett.* B222, 274 (1989); *ibid*, **388**, 803 (1996).
21. G. Burdman and D. Kominis, *Phys. Lett.* B403, 101 (1997).
22. G. Buchalla, G. Burdman, C. T. Hill and D. Kominis, *Phys. Rev.* D53, 5185 (1996).
23. C. Albajar et al., the UA1 collaboration, *Phys. Lett.* B262, 163 (1991); S. Glenn et al., the CLEO collaboration, CLNS-97/1514, hep-ex/9710003; B. Abbot et al., the D0 collaboration, hep-ex/9801027.

Technihadron Production at a Muon Collider

John Womersley

Fermi National Accelerator Laboratory[1]
Batavia, Illinois 60510

Abstract. Cross sections for relatively low-mass technihadron resonances at a $\mu^+\mu^-$ collider are presented. Such particles would give spectacular signals at the first muon collider. They could be studied in detail at this machine, making use of its good mass resolution and ability to reconstruct purely hadronic final states.

In low-scale models of walking technicolor [1], the lightest hadrons are technipions π_T. They are coupled (by extended technicolor interactions) to mass and hence are expected to decay into the heaviest fermions available:

$$\pi_T^0 \to t\bar{t},\ b\bar{b},\ \tau^+\tau^-$$
$$\pi_T^\pm \to t\bar{b},\ b\bar{c},\ \tau^\pm\nu_\tau$$

The technipion mass m_{π_T} is expected to be of the order of 100 GeV or more.

Pair production of these particles at a muon collider is greatly enhanced is there is an s-channel resonance such as a techni-rho or techni-omega:

$$\mu^+\mu^- \to \gamma^*/Z^* \to \rho_T^0$$
$$\mu^+\mu^- \to \gamma^*/Z^* \to \omega_T^0$$

---

[1] operated by the Universities Research Association for the U.S. Department of Energy

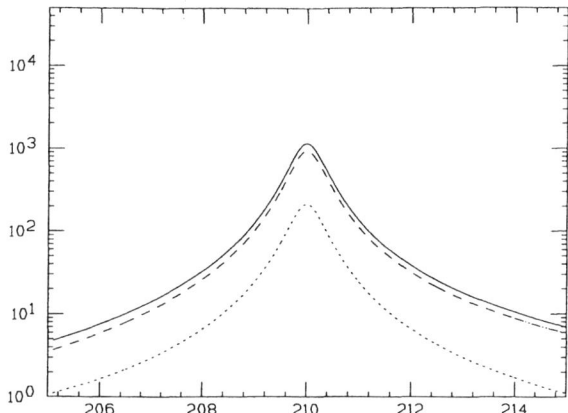

FIGURE 1. Cross section (pb) for technirho production at a muon collider as a function of $\sqrt{s}$ (GeV), for $m_{\rho_T} = 210$ GeV and $m_{\pi_T} = 110$ GeV. The solid curve is the total ρ_T cross section, the dashed curve that for $\rho_T \to W\pi_T$ and the dotted curve that for $\rho_T \to W^+W^-$.

We would expect the rho and omega to be (approximately) degenerate in mass, with masses $m_{\rho_T} \approx m_{\omega_T}$ of the order of 200 GeV or more.

The technirho would be seen in the detector in its decay modes:

- two technipions (including vector bosons) W^+W^-, $W^\pm \pi_T^\mp$, $\pi_T^\pm \pi_T^\mp$;
- fermion-antifermion pairs $q\bar{q}$, $\ell^+\ell^-$ and $\nu\bar{\nu}$.

The techniomega would be seen as:

- $\gamma \pi_T^0$, $Z\pi_T^0$;
- fermion-antifermion pairs $q\bar{q}$, $\ell^+\ell^-$ and $\nu\bar{\nu}$.

Figure 1 shows the cross section for technirho production at a muon collider as a function of $\sqrt{s}$ [2], for $m_{\rho_T} = 210$ GeV and $m_{\pi_T} = 110$ GeV. This choice of masses gives a very large peak cross section — 1 nb or 10^6 events per year at a luminosity of 10^{32} cm^{-2}s^{-1}. Figure 2 gives the corresponding cross section for techniomega production ($m_{\omega_T} = 210$ GeV). Here the peak cross section is even larger — 10 nb or 10^7 events per year at a luminosity of 10^{32} cm^{-2}s^{-1}. Note how extremely narrow the peak is: less than 1 GeV.

If the technirho mass is increased, the cross section will fall, as shown in Fig. 3. Here the masses simulated are $m_{\rho_T} = 400$ GeV and $m_{\pi_T} = 150$ GeV,

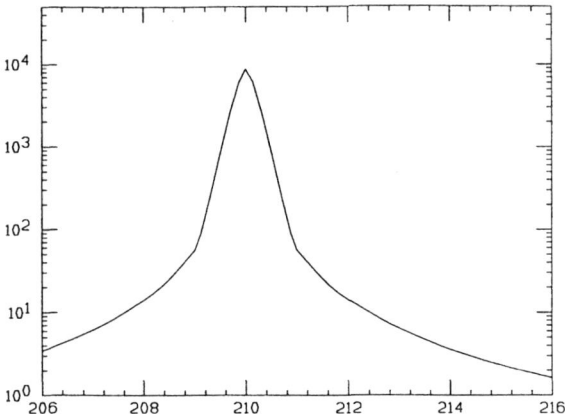

FIGURE 2. Cross section (pb) for techniomega production at a muon collider as a function of $\sqrt{s}$ (GeV), for $m_{\omega_T} = 210$ GeV and $m_{\pi_T} = 110$ GeV. The solid curve is the total ω_T cross section (the decay mode $\gamma \pi_T^0$ is dominant).

and so the decay mode $\rho_T \to \pi_T \pi_T$ is open; the peak cross section is much reduced and the resonance is much wider. Nonetheless, there are still 10^4 events per year on the peak at a luminosity of 10^{32} cm^{-2}s^{-1}.

The particular strengths of the muon collider in this context are:

- One can operate on the resonance and study all the decays and branching ratios of the technirhos, omegas and pions (the all-hadronic modes would, in contrast, be challenging at a hadron collider);
- The good mass resolution available would enable the widths of even the very narrow, low-mass resonances to be determined;
- One could study rho-omega mixing in detail, using the fermion-antifermion final state for example and scanning $\sqrt{s}$.

For this physics, the muon collider detector would need to be capable of identifying jets and leptons and of tagging b-jets; it would be interesting if charm jets could be distinguished from b's.

Clearly the first muon collider is capable of technihadron production with clear signals and some interesting studies can be carried out. It must be remembered, however, that one would not learn much about QCD from precision studies of the π, ρ and ω mesons. In the same way, precision studies of the π_T, ρ_T and ω_T will be of limited usefulness in eliciting the underlying dynamics of technifermions out of which they are made. If nature has chosen to operate

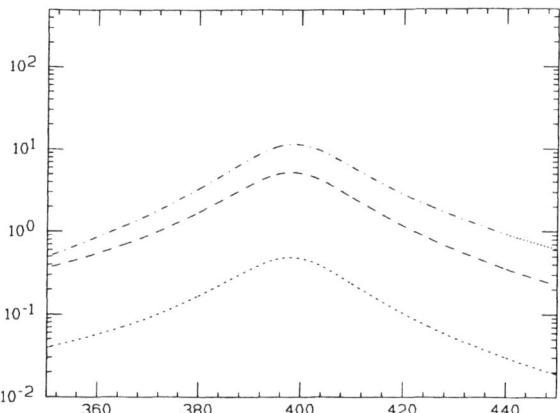

FIGURE 3. Cross section (pb) for technirho production at a muon collider as a function of $\sqrt{s}$ (GeV), for $m_{\rho_T} = 400$ GeV and $m_{\pi_T} = 150$ GeV. The dot-dash curve is the cross section for $\rho_T \to \pi_T \pi_T$, the dashed curve that for $\rho_T \to W \pi_T$ and the dotted curve that for $\rho_T \to W^+ W^-$.

in this way, we shall then need the maximum possible $\sqrt{s}$ muon collider as a follow-on machine in order to explore the full spectrum of technihadrons, and even to probe technifermion interactions through, for example, $W_L W_L$ production.

In conclusion, relatively low-mass technihadron resonances would give spectacular signals at the first muon collider. They could be studied in detail at such a machine, making use of its good mass resolution and ability to reconstruct purely hadronic final states.

REFERENCES

1. For an introduction to Technicolor, see the contribution by K. Lane to the proceedings of this workshop.
2. Cross sections were calculated using a Fortran program kindly provided by K. Lane and E. Eichten.

Testing Technicolor with Scalars at the First Muon Collider

Bogdan A. Dobrescu

Fermi National Accelerator Laboratory
P.O. Box 500, Batavia, Illinois, 60510, USA

Abstract. An interesting class of models of dynamical electroweak symmetry breaking allows only the third generation fermions to acquire dynamical masses, such that the masses of the first two generations should be given by coupling to a nonstandard "Higgs" doublet. The scalars in this case have large couplings to the second generation, so that they are copiously produced at a muon collider. We analyze the potential for discovery of the neutral scalars in the s-channel, and we show that at resonance there will be observed in excess of 10^5 events per year.

INTRODUCTION

Although the electroweak symmetry breaking and the quark and lepton spectrum may have a common origin, it is quite possible that they are generated by some new physics which manifests itself at low energy as distinct sectors. For example, the effective theory at a scale of order 1 TeV can include a sector that gives rise to the masses of the W, Z and third generation fermions, as well as a sector responsible for the masses of the lighter fermions.

This is the case when the electroweak symmetry is broken dynamically by some new strong gauge interactions, and only the third generation quarks and leptons couple to the fields charged under these interactions. In the context of technicolor models, this situation is discussed in [1,2]. The minimal mechanism for producing the masses of the first two generations requires a weak-doublet scalar transforming as the Higgs doublet, but having a positive squared-mass, as in the technicolor models with a scalar [3,4] or in bosonic technicolor models [5,6].

By coupling to the dynamical symmetry breaking sector, the scalar acquires a small vacuum expectation value (VEV). The existence of the scalars is not troublesome as long as the theory described here is the low energy manifestation of a TeV scale theory which includes scalar compositness or softly broken supersymmetry.

Because the VEV of the weak-doublet scalar is smaller than the electroweak scale, the Yukawa couplings of the "nonstandard-Higgs" bosons to the first and second generations must be larger than in the standard model. This is relevant for future collider searches. In particular, since these Yukawa couplings are proportional to the fermion mass, the s-channel production is very large at a muon collider.

Here we discuss briefly the discovery potential and resonant production of the "nonstandard-Higgs" bosons at a First Muon Collider, operating at a center of mass energy of up to 500 GeV. A comprehensive study of the collider phenomenology of the "nonstandard-Higgs" will be given elsewhere [7].

TECHNICOLOR WITH SCALARS

We assume that the electroweak symmetry breaking and the third generation fermion masses have a dynamical origin. To be specific, we consider a technicolor model constructed along the lines of ref. [2]. The second and first generation fermions acquire masses by coupling to a scalar, ϕ, which transforms under the gauge group like the standard model Higgs doublet.

In addition to the standard model fermions, consider one doublet of technifermions, P and N, and three scalars, ω, χ and ϕ, which transform under the $SU(4)_{TC} \times SU(3)_C \times SU(2)_W \times U(1)_Y$ gauge group as:

$$\Psi_R = \begin{pmatrix} P_R \\ N_R \end{pmatrix} : (4,1,2)_0 , \quad P_L : (4,1,1)_{+1} , \quad N_L : (4,1,1)_{-1} ,$$

$$\omega : (4,\bar{3},1)_{-\frac{1}{3}} , \quad \chi : (4,1,1)_{+1} , \quad \phi : (1,1,2)_{+1} . \tag{1}$$

The most general Yukawa interactions of ω and χ include only terms linear in the quark or lepton fields, such that there is a particular eigenstate basis in which only the third generation couples to the technicolored fields:

$$\mathcal{L}_Y^{\omega,\chi} = C_q \overline{\Psi}_R q_L^3 \omega + C_t \bar{t}_R P_L \omega^\dagger + C_b \bar{b}_R N_L \omega^\dagger + C_l \overline{\Psi}_R l_L^3 \chi + C_\tau \bar{\tau}_R N_L \chi^\dagger + \text{h.c.} \tag{2}$$

Using a phase redefinition on the third generation fields, $q_L^3, t_R, b_R, l_L^3, \tau_R$, we can choose the Yukawa coupling constants, $C_q, C_t, C_b, C_l, C_\tau$, to be positive.

We assume that the ω and χ techniscalars are sufficiently heavy to be integrated out, such that their effects in the low energy theory are given by four-fermion operators involving two technifermions and two fermions of the third generation. As in QCD, the $SU(4)_{TC}$ technicolor interactions trigger the formation of technifermion condensates,

$$\langle \overline{P}P \rangle \approx \langle \overline{N}N \rangle \approx 2\sqrt{3}\pi f^3 , \tag{3}$$

which breaks the electroweak symmetry at a scale f. This also results in masses for the t, b and τ [1,2]:

$$m_t \approx \frac{\sqrt{3}}{2} \frac{C_q C_t}{M_\omega^2} \pi f^3 ,$$

$$\frac{m_b}{m_t} = \frac{C_b}{C_t} ,$$

$$\frac{m_\tau}{m_t} = \frac{C_\tau C_l}{C_t C_q} \left(\frac{M_\omega}{M_\chi}\right)^2 . \qquad (4)$$

The ϕ has Yukawa couplings to the standard fermions (these are similar with the standard model couplings of the Higgs doublet, but with different coupling constants), and also to the technifermions:

$$\mathcal{L}_Y^\phi = \lambda_{jk}^e \bar{l}_L^j e_R^k \phi + \lambda_{jk}^u \bar{q}_L^j u_R^k i\tau^2 \phi^\dagger + \lambda_{jk}^d \bar{q}_L^j d_R^k \phi$$
$$+ \lambda_+ \overline{\Psi}_L P_R i\tau^2 \phi^\dagger + \lambda_- \overline{\Psi}_L N_R \phi + \text{h.c.} \qquad (5)$$

where $i, j = 1, 2, 3$ are generational indices, and the λ's are coupling constants. We consider the case where ϕ has a positive squared-mass, $M_\phi^2 > 0$, so in contrast to the standard model, the ϕ sector does not induces a VEV by itself. However, when the technifermions condense, the last two terms in the above Lagrangian give rise to tadpole terms, such that ϕ develops a VEV whose magnitude is [3,4]

$$\frac{f'}{\sqrt{2}} \approx (\lambda_+ + \lambda_-) \frac{2\sqrt{3}\pi f^3}{M_\phi^2} . \qquad (6)$$

Note that we neglected a possible quartic term in the ϕ potential. If the high energy theory that accounts for the existence of ϕ turns out to produce such a quartic term (and possibly higher dimensional terms), then its effects can be easily included. The observed W and Z masses require

$$f^2 + f'^2 = v^2 \approx (246 \text{ GeV})^2 . \qquad (7)$$

The first three terms of $\mathcal{L}_Y^\phi$ are responsible for the masses of the quarks and leptons of the first two generations, and for the CKM elements.

The effective theory below a scale of order 1 TeV where the technicolored fields are integrated out includes only the standard fermions and gauge bosons, an iso-triplet of Nambu-Goldstone bosons, π^a, $a = 1, 2, 3$, associated with the chiral symmetry breaking of the technifermions, and the components of ϕ [we assume $M_\phi \lesssim 1$ TeV, although eq. (6) does not exclude multi-TeV values]. ϕ decomposes into an iso-singlet σ, and an iso-triplet π'^a:

$$\phi = \frac{1}{\sqrt{2}} e^{-i\pi'^a \tau^a / f'} \begin{pmatrix} 0 \\ \sigma + f' \end{pmatrix} . \qquad (8)$$

The triplets π^a and π'^a mix and give rise to the longitudinal W and Z, and to a triplet of physical pseudo-scalars. However, the large top mass suggests that $f \approx v$, so that

$$f' \ll f, \qquad (9)$$

which implies that the mixing is small, and we will neglect it. Another consequence of inequality (9) is that

$$\lambda_+ + \lambda_- \ll \frac{1}{2\pi\sqrt{6}}\left(\frac{M_\phi}{f}\right)^2. \qquad (10)$$

In this situation, the neutral real scalars σ and π'^3, and the charged scalars, $\pi'^\pm = (\pi'^1 \mp i\pi'^2)/\sqrt{2}$, are almost degenerate, with a mass M_ϕ. The splittings in their masses are of order $(f'/f)^2$ and $\lambda_\pm^2(M_\phi/f)^2$.

SIGNALS AT THE FIRST MUON COLLIDER

If the model presented in the previous section is indeed the correct description of physics up to a TeV scale, then the only direct discovery accessible at a $\mu^+\mu^-$ collider with $\sqrt{s}$ below the first technihadron resonance will be the existence of the components of the ϕ doublet. Furthermore, for $M_\phi < \sqrt{s} < 2M_\phi$, only the neutral scalars, σ and π'^3, can be produced.

The couplings of σ and π'^3 to quarks and leptons are in general flavor non-diagonal. However, because the off-diagonal couplings are constrained by flavor-changing neutral current measurements, we are going to consider only the flavor-diagonal couplings. These are proportional with the corresponding fermion masses in the case of the first two generations. The couplings to the third generation are more arbitrary, because they are proportional only with the small contributions to the fermion mass, δm_f with $f = t, b, \tau$, from the ϕ VEV.

Since the bulk of electroweak symmetry breaking is provided by the technicolor sector, the couplings of the neutral scalars, σ and π'^3, to W^+W^- and ZZ are smaller by a factor of v/f' than the corresponding standard model couplings of the Higgs boson. On the contrary, the couplings of σ and π'^3, to the fermions of the second and first generations are enhanced by a factor of v/f' compared to the standard model Higgs boson. The couplings to the third generation are also enhanced by v/f', but they are suppressed by $m_f/\delta m_f$.

The total decay widths of the σ and π'^3 scalars are equal, and given by

$$\Gamma \approx \frac{M_\phi}{8\pi f'^2}\left[3m_c^2 + 3m_s^2 + m_\mu^2 + 3(\delta m_t)^2\left(1 - \frac{4m_t^2}{M_\phi^2}\right)^{1/2}\theta(M_\phi - 2m_t)\right.$$

$$\left. + 3(\delta m_b)^2 + (\delta m_\tau)^2\right] + \Gamma(W^+W^- + ZZ). \qquad (11)$$

We take the VEV of ϕ in the range

$$1 \text{ GeV} \lesssim f' \lesssim 10 \text{ GeV}, \tag{12}$$

where the lower bound is chosen to avoid Yukawa coupling constants larger than order one, and the upper bound is chosen to satisfy condition (9). In this case, the width for scalar decay into pairs of gauge bosons, $\Gamma(W^+W^- + ZZ)$, is at most a few percent of the width for $\sigma, \pi'^3 \to c\bar{c}$, and we neglect it. Generically, there is no reason to expect that δm_f is larger than the corresponding second generation mass. For simplicity we assume $(\delta m_f)^2 \ll m_c^2$, such that the width of the σ and π'^3 scalars is dominated only by the $c\bar{c}$ final state:

$$\Gamma \approx \frac{3m_c^2 M_\phi}{8\pi f'^2} \approx 13.2 \text{ GeV} \left(\frac{3 \text{ GeV}}{f'}\right)^2 \left(\frac{M_\phi}{500 \text{ GeV}}\right). \tag{13}$$

Given the enhanced couplings to the second generation, the s-channel production of the neutral scalars at a $\mu^+\mu^-$ collider is large. The natural spread in the muon collider beam energy, $\sigma_{\sqrt{s}}$, is much smaller than Γ, and can be ignored in computing the effective s-channel resonance cross section [8]:

$$\bar{\sigma}(\mu^+\mu^- \to \sigma, \pi'^3 \to X) \approx \frac{4\pi\Gamma^2}{(s - M_\phi^2)^2 + M_\phi^2\Gamma^2}$$

$$\times B(\sigma, \pi'^3 \to \mu^+\mu^-) B(\sigma, \pi'^3 \to X). \tag{14}$$

We are interested especially in the case where the final state is $X \equiv c\bar{c}$. The branching fractions for the decays into a pair of muons, respectively into c-quarks, are given by

$$B(\sigma, \pi'^3 \to \mu^+\mu^-) \approx \frac{m_\mu^2}{3m_c^2} \approx 0.2\%$$

$$B(\sigma, \pi'^3 \to c\bar{c}) \approx 1 - \frac{m_s^2}{3m_c^2} - \ldots \tag{15}$$

where the ellipsis stands mainly for the branching fractions into W^+W^-, ZZ, $\mu^+\mu^-$, etc. Therefore,

$$\bar{\sigma}(\mu^+\mu^- \to \sigma, \pi'^3 \to c\bar{c}) \approx \frac{8\pi\Gamma^2}{(s - M_\phi^2)^2 + M_\phi^2\Gamma^2} \left(\frac{m_\mu^2}{3m_c^2}\right). \tag{16}$$

The main background comes from $\mu^+\mu^- \to \gamma^*, Z^* \to c\bar{c}$, and amounts to

$$\sigma_B(\mu^+\mu^- \to c\bar{c}) \approx 0.7 \text{ pb} \frac{(500 \text{ GeV})^2}{s}. \tag{17}$$

We first study the discovery potential of a $\mu^+\mu^-$ collider operating at a maximum center of mass energy of 500 GeV. The beam energy can be reduced at the expense of luminosity. A decrease in the beam energy by a factor of two leads to a decrease in luminosity by a factor of ten [9]. The number of scan points can be optimized as follows. Consider that the adjacent scan points are separated by an energy difference $x\Gamma$, implying that

$$\overline{\sigma}(\mu^+\mu^- \to \sigma, \pi'^3 \to c\bar{c}) \geq \frac{8\pi}{(4x^2+1)M_\phi^2}\left(\frac{m_\mu^2}{3m_c^2}\right)$$

$$\approx \frac{80 \text{ pb}}{4x^2+1}\left(\frac{500 \text{ GeV}}{M_\phi}\right)^2. \quad (18)$$

To observe at a particular scan point with $\sqrt{s}$ a number of $c\bar{c}$ final state events which is 5σ over the background requires an integrated luminosity

$$L(s) \geq \frac{5\sigma_B(\mu^+\mu^- \to c\bar{c})}{r(c\bar{c})[\overline{\sigma}(\mu^+\mu^- \to \sigma, \pi'^3 \to c\bar{c})]^2}$$

$$\approx \frac{5.6 \times 10^{-4} \text{ pb}^{-1}}{(500 \text{ GeV})^2} \frac{(4x^2+1)^2 M_\phi^2}{r(c\bar{c})s}, \quad (19)$$

where $r(c\bar{c})$ is the efficiency for observing the $c\bar{c}$ final state, and is given basically by the square of the c-tagging efficiency. The integrated luminosity necessary for searching the scalar resonance over the whole range of beam energy is

$$L = \frac{1}{x\Gamma}\int_{\sqrt{s}_{min}}^{\sqrt{s}_{max}} d(\sqrt{s})L(s)$$

$$\approx \frac{5.6 \times 10^{-4} \text{ pb}^{-1}}{(500 \text{ GeV})^2} \frac{(4x^2+1)^2}{r(c\bar{c})x} \frac{M_\phi^4}{\Gamma}\left(\frac{1}{\sqrt{s}_{min}} - \frac{1}{\sqrt{s}_{max}}\right). \quad (20)$$

Clearly, x should be chosen to minimize L. At the minimum,

$$x = \frac{1}{2\sqrt{3}}, \quad (21)$$

which gives

$$L_{min} = \frac{0.14 \text{ fb}^{-1}}{r(c\bar{c})}\left(\frac{f'}{3 \text{ GeV}}\right)^2\left(\frac{M_\phi}{500 \text{ GeV}}\right)^3\left(\frac{500 \text{ GeV}}{\sqrt{s}_{min}} - \frac{500 \text{ GeV}}{\sqrt{s}_{max}}\right). \quad (22)$$

For $\sqrt{s}_{max} = 500$ GeV, $\sqrt{s}_{min} = 250$ GeV, and $r(c\bar{c}) \approx 10\%$,

$$L_{min} = 1.4 \text{ fb}^{-1}\left(\frac{f'}{3 \text{ GeV}}\right)^2\left(\frac{M_\phi}{500 \text{ GeV}}\right)^3. \quad (23)$$

The peak luminosity assumed at this workshop is 2×10^{34} cm^{-2}s^{-1} at $\sqrt{s} = 500$ GeV, which corresponds to roughly 2×10^{33} cm^{-2}s^{-1} at $\sqrt{s} = 250$ GeV. Note that the average luminosity is significantly lower, 7×10^{32} cm^{-2}s^{-1} at $\sqrt{s} = 500$ GeV, but given that the scalar resonance is broader than the spread in the beam energy, there is no need for a good energy resolution (the upper value given in [9], $\delta E/E = 0.12\%$, being sufficient), and therefore the peak luminosity can be used. This situation is in contrast with the search for a light standard model Higgs which is narrower than the spread in beam energy.

Eq. (23) shows that with a luminosity of order 100 fb^{-1}/year, the scalar resonance will be discovered in a short period of time, even for f' larger than 10 GeV.

Once the resonance is found, either at the muon collider by varying the beam energy, or at the LHC, the beam energy can be adjusted to the peak (even if this requires a significant reduction in the luminosity) and then the production cross section becomes very large:

$$\bar{\sigma}(\mu^+\mu^- \to \sigma, \pi'^3 \to c\bar{c}) \approx 80 \text{ pb} \left(\frac{500 \text{ GeV}}{M_\phi}\right)^2. \qquad (24)$$

With a luminosity of 2×10^{34} cm^{-2}s^{-1} (7×10^{32} cm^{-2}s^{-1}), and a c-tagging efficiency of $\sim 30\%$, there are going to be observed approximately 10^6 (5×10^4) events per year. This will make possible precision measurements of the couplings and masses of the neutral "nonstandard-Higgs" bosons, which in turn will open a window towards the TeV scale physics.

It is remarkable that the peak cross section is independent of the ϕ VEV. This is a consequence of the small couplings of the ϕ to pairs of gauge bosons.

CONCLUSIONS

The large resonance cross section computed in section 3 warrants the label "nonstandard-Higgs factory" for the muon collider. On the other hand, if the class of dynamical electroweak symmetry breaking models discussed here is correct, and the ϕ mass turns out to be significantly larger than 500 GeV, then no discovery will be made at the First Muon Collider, and there is need for a higher-energy muon collider. As discussed repeatedly at this workshop, the 4 TeV muon collider would also be a great tool for studying the strong dynamics sector. We emphasize that a "nonstandard-Higgs" with large couplings to the second generation might also be necessary in models where the strong dynamics is different than technicolor, for example in models that incorporate the top condensation seesaw mechanism [10], or in models with discrete horizontal symmetries [6].

Acknowledgements: I would like to thank the convenors of the Strong Dynamics working group, Estia Eichten and Pushpa Bhat, for creating a stimulating atmosphere. I am grateful to my collaborators, Tao Han and Hong-Jian He, for many useful comments.

REFERENCES

1. A. Kagan, *Proceedings of the 15th Johns Hopkins Workshop on Current Problems in Particle Theory*, G. Domokos and S. Kovesi-Domokos eds. (World Scientific, Singapore, 1992), p.217;
   B. A. Dobrescu, *Nucl. Phys.* **B449** (1995) 462, hep-ph/9504399.
2. A. Kagan, *Phys. Rev.* **D51** (1995) 6196, hep-ph/9409215;
   B. A. Dobrescu and J. Terning, *Phys. Lett.* **B 416** (1997) 129, hep-ph/9709297;
   B. A. Dobrescu and E. H. Simmons, work in progress.
3. E. H. Simmons, *Nucl. Phys.* **B312** (1989) 253;
   C. D. Carone and H. Georgi, *Phys. Rev.* **D49** (1993) 1427, hep-ph/9308205.
4. C. D. Carone and E. H. Simmons, *Nucl. Phys.* **B397** (1993) 591.
5. S. Samuel, *Nucl. Phys.* **B347** (1990) 625;
   M. Dine, A. Kagan and S. Samuel, *Phys. Lett.* **B243** (1990) 250;
   A. Kagan and S. Samuel, *Phys. Lett.* **B270** (1991) 37; *Phys. Lett.* **B252** (1990) 605.
6. A. Kagan and S. Samuel, *Int. J. Mod. Phys.* **A7** (1992) 1123.
7. B. A. Dobrescu, T. Han and H.-J. He, work in progress.
8. V. Barger, M.S. Berger, J.F. Gunion and T. Han, *Phys. Rev. Lett.* **75** (1995) 1462; *Phys. Rep.* **286** (1997) 1.
9. The $\mu^+\mu^-$ Collaboration, "$\mu^+\mu^-$ Collider: A Feasibility Study", BNL-52503, Fermi Lab-Conf.-96/092, LBNL-38946, July 1996; also in Proceedings of the Snowmass Workshop 96, to be published.
10. B. A. Dobrescu and C. T. Hill, FERMILAB-PUB-97-409-T, Dec 1997, hep-ph/9712319.

Search for Technicolor Particles in W+2jet with b-tag Channel at CDF *

Takanobu Handa[†], Kaori Maeshima[‡], Juan Valls[*], Rocio Vilar[‡]
(representing the CDF collaboration)

[†] *Hiroshima University* : *1-3-1 Kagamiyama Higashi-Hiroshima 739 Japan*
[‡] *Fermilab* : *PO Box 500 Batavia, IL 60510*
[*] *Rutgers University* : *Piscataway, New Jersey, 08854*

Abstract. We present a preliminary result from a search for walking technicolor particles using leptonically decayed W plus two jets with at least one b-tag using $109 \pm 7\,\mathrm{pb}^{-1}$ of data taken by the Collider Detector at Fermilab (CDF). We search for technipion mass peaks in the invariant mass distribution of the two jet system and technirho mass peaks in the invariant mass distribution of $W+2$jet system. We do not see any significant excess in our search sample, and set 95 % confidence level upper limits on the production cross section and exclude a region of the π_T mass v.s. ρ_T mass plane.

INTRODUCTION

A recent walking technicolor(TC) model expects color singlet technirho (ρ_T) production in high energy $p\bar{p}$ collisions [1,2]. At the Tevatron energy a W boson and a technipion (π_T) decay mode of a ρ_T has relatively large cross section times branching ratio when masses of the π_T and the ρ_T are around 90 GeV/c² and 180 GeV/c² respectively. At this mass combination, the cross section is approximately 12 pb ($\rho_T^{\pm} \to W^{\pm}\pi_T^0$) and 5 pb ($\rho_T^0 \to W^{\pm}\pi_T^{\mp}$). Here, we report a search for $\rho_T^{\pm} \to W^{\pm}\pi_T^0$ and $\rho_T^0 \to W^{\pm}\pi_T^{\mp}$ decay modes with leptonically (e or μ) decayed W's. The coupling of the π_T to fermions depends on the fermion masses and it is stronger for the larger fermion mass. Hence, the π_T would mostly decay to $b\bar{b}$ or $b\bar{c}$. The final states we search are: $\rho_T^{\pm} \to W\pi_T^0 \to e\nu b\bar{b}, \mu\nu b\bar{b}$ and $\rho_T^0 \to W\pi_T^{\mp} \to e\nu b\bar{c}, \mu\nu b\bar{c}$ as shown in Figure 1. For these final states, we use b-quark tagging to reduce the $W+2$jet backgrounds significantly. Following in this paper, we describe; event selection, signal Monte Carlo and efficiencies, background, mass distributions, topological and mass cuts, and finally our preliminary result.

---

*) Proceedings of 'Workshop on Physics at the First Muon Collider and at the Front End of a Muon Collider' Nov. 6-9, 1997, Fermilab.

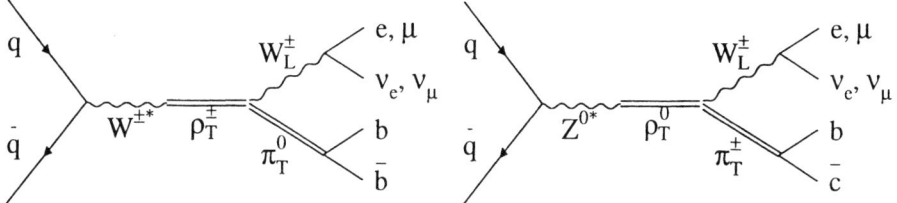

Figure 1 : Technirho production diagrams in $p\bar{p}$ collision decaying to W + Technipion.

I EVENT SELECTION

The W+2jet sample for this analysis is a subset of a sample of high-P_T inclusive lepton events which contain either an isolated electron with $E_T > 20$ GeV or an isolated muon with $P_T > 20$ GeV/c in the central region (pseudo-rapidity $|\eta| < 1.0$). Events that contain a second same flavor lepton of opposite charge are removed as Z boson candidates if the reconstructed ee or $\mu\mu$ invariant mass is between 75 and 105 GeV/c^2. An inclusive W boson sample is selected by further requiring missing transverse energy, $\not{E}_T > 20$ GeV and that the lepton be isolated from any jet activity. A W+2jet sample is selected by further requiring exactly two jets with $E_T > 15$ GeV. Jets are defined using a cone algorithm with $\Delta R = \sqrt{\Delta\phi^2 + \Delta\eta^2} = 0.4$. In order to separate TC from the large W+2jet background, we require that at least one of the jets be identified as a b jet candidate. Identification of b jets is done by reconstructing secondary vertices from b-quark decay using the Silicon Vertex Detector (SVX b-tagging). The SVX b-tagging algorithms are described in Ref. [3]. The number of selected events is 42, while the expected number of background events is 31.6 as shown in Table 1.

II SIGNAL MONTE CARLO AND EFFICIENCIES

Using PYTHIA 6.1 [4], we generate ρ_T^0 and $\rho_T^\pm$ at $\sqrt{s} = 1.8$ TeV $p\bar{p}$ collisions. The selected signatures are the followings: $q\bar{q} \to W^{*\pm} \to \rho_T^\pm \to W^\pm \pi_T^0$ ($\pi_T^0 \to b\bar{b}$:100 %) and $q\bar{q} \to Z^{0*}, \gamma^* \to \rho_T^0 \to W^\pm \pi_T^\mp$ ($\pi_T^\pm \to b\bar{c}, c\bar{b}$:95 % and $\pi_T^\pm \to c\bar{s}, s\bar{c}$:5 %) forcing W to decay either $e\nu$ or $\mu\nu$. Input parameters of the technicolor events in PYTHIA are all default values described in the TC model [2]. We generate 10k events for each π_T, ρ_T mass combination. These events are passed through the detector simulation. We choose about fifty mass combinations whose cross section is more than 5 pb. The acceptance and efficiencies are estimated using these Monte Carlo signal event samples. The result is shown in Figure 2.

W + 2jet at least 1 b-tag Background

CDF Preliminary

Source	Distribution	N_{event} (109 pb^{-1})
Mistag	DATA (W + 2jet)	5.1±2.0
Wbb Wcc	Herwig	9.4±2.5
Z+h.f.	VECBOS	1.4±0.5
Wc	Herwig	4.6±1.5
WW,WZ,Z$\tau\tau$	PYTHIA	1.5±0.5
non-W	DATA($\not{E}_T$,Iso method)	2.1±1.3
$t\bar{t}$	Herwig(σ=7.5 pb)	5.1±1.9
single top	Herwig(W* and Wg)	2.4±0.8
TOTAL		31.6±4.3

Table 1 : Expected number of background events in $W+2$jet with b-tag selection.

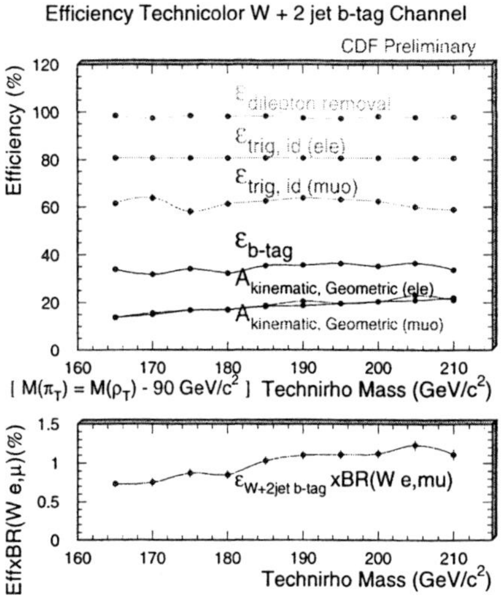

Figure 2 : Acceptances, efficiencies of each cut, and total efficiencies of $W+2$jet with b-tag selection including BR($W \to e\nu, \mu\nu$).

III DIJET MASS AND W+2JET MASS DISTRIBUTIONS

We reconstruct the invariant mass of the dijet system, M(jj) which would correspond to the technipion mass, and the invariant mass of the $W+2$jet system, M(Wjj) which would correspond to the technirho mass. A signal would appear as a narrow peak in the two mass distributions. To reconstruct M(Wjj), we need the P_z information of the neutrino. Since we only measure the missing E_T information, in order to obtain the P_z information we use the W mass constraint in a lepton-neutrino system and take the smaller P_z of the two solutions. If there is no solution for the P_z, we take the real part of the solution of the quadratic equation. Figure 3 and Figure 4 show the M(jj) and M(Wjj) distributions respectively. The upper plots show the technicolor Monte Carlo signal at M(π_T)=90 GeV/c^2 and M(ρ_T)=180 GeV/c^2 together with the background. The number of events for the signal and background are normalized to the expected number of events for 109 pb^{-1}. The bottom plots show the CDF data and the background.

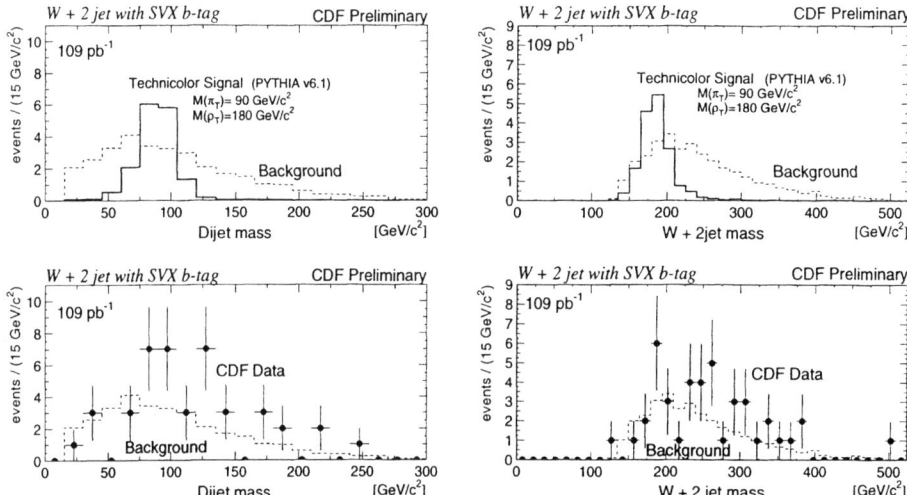

Figure 3: The invariant mass of the dijet system for the $W+2$jet sample with b-tag.

Figure 4: The invariant mass of the $W+2$jet system for the $W+2$jet sample with b-tag.

IV TOPOLOGY AND MASS WINDOW CUTS

The topology cuts are placed on the ϕ angle between two jets, $\Delta\phi(jj)$, and the P_T of the dijet system, $P_T(jj)$ [5]. Our TC signal search region has a characteristic which is $M(\pi_T) + M(W) \simeq M(\rho_T)$. In this case, technipions are produced nearly at rest, and consequently the $\Delta\phi(jj)$ is more back-to-back than the background and $P_T(jj)$ is smaller. Figure 5 shows the $\Delta\phi(jj)$ and $P_T(jj)$ distributions at $M(\pi_T) = 90\,\mathrm{GeV}/c^2$, $M(\rho_T) = 180\,\mathrm{GeV}/c^2$. In order to obtain optimum cut values, we apply the $\Delta\phi(jj)$ and the $P_T(jj)$ topology cuts simultaneously and maximize the $S/\sqrt{B}$ (signal over square-root of the background) values. We obtain different optimum topology cut values for each of the mass combinations. These cuts are very effective. The $S/\sqrt{B}$ is increased by typically 60%.

The final criteria to select a TC signal is to apply mass window cuts on the selected event sample. We require that $M(jj)$ and $M(Wjj)$ be within $\pm\,3\sigma$ of the mean values. The mean values and σ are obtained from the signal Monte Carlo for each mass combination considered.

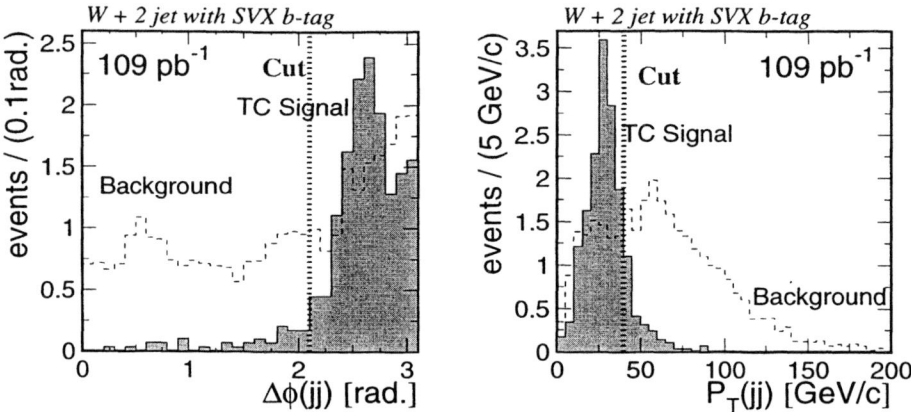

Figure 5 : The $\Delta\phi(jj)$ and $P_T(jj)$ distributions of the TC signal and the background for the $W+2$jet sample with b-tag. The dotted vertical lines show the optimum topological cut values for this particular mass combination.

V PRELIMINARY RESULTS

Table 2 summarizes our result. We do not see any significant excess in our search sample. As a consequence, we set 95 % C.L. upper limits on the production cross section taking into account a 26% systematic uncertainty on the signal efficiency, and we exclude some region in the $M(\pi_T)$ and $M(\rho_T)$ plane. Figure 6 shows the excluded region by this analysis as well as the expected theoretical cross section contours.

ACKNOWLEDGMENT

We would like to thank E.Eichten, K.Lane, J.Womersley, and T.Sjostrand for their theoretical and monte carlo work on which our analysis is based.

REFERENCES

1. Estia Eichten, Kenneth Lane 'Low-Scale Technicolor at the Tevatron'
   hep-ph/9607213, FERMILAB-PUB-96/075-T, Phys.Lett.B388:803-807,1996;

2. K. Lane and E. Eichten, 'Two Scale Technicolor' Phys. Lett. B222, 274 (1989);
   E. Eichten and K. Lane 'Electroweak and Flavor Dynamics at Hadron Colliders. 1.'
   FERMILAB-CONF-96/297-T (hep-ph/9609297)

→ Number of events & Cross Section Limits

CDF Preliminary

M_{π_T}, M_{ρ_T}	σ $\rho_T^\pm \to W^\pm + \pi_T^0$ $\rho_T^0 \to W^\pm + \pi_T^\mp$	$N_{Signal}^{allcuts}$ (109 pb^{-1})	$N_{B.G.}^{allcuts}$ (109 pb^{-1})	DATA	95% C.L. limits	
					$N_{95\%limit}$	$\sigma_{95\%limit}$
80,165	4.2 pb	2.7	4.1±0.6	5	8.1	12.6 pb
80,170	4.1 pb	2.9	5.4±0.7	5	7.2	10.4 pb
80,175	3.1 pb	2.1	7.6±1.0	9	10.4	15.6 pb
85,170	15.4 pb	11.1	3.8±0.5	5	8.3	11.5 pb
85,175	7.5 pb	4.9	5.5±0.8	5	7.2	11.0 pb
85,180	5.1 pb	3.7	6.5±0.9	7	8.8	12.1 pb
90,175	12.9 pb	10.3	4.2±0.6	5	8.0	10.0 pb
90,180	17.6 pb	13.3	5.7±0.8	5	7.1	9.4 pb
90,185	8.4 pb	6.9	7.2±1.0	7	8.4	10.3 pb
95,180	10.7 pb	10.8	5.6±0.8	6	8.3	8.3 pb
95,185	14.2 pb	13.7	6.4±0.9	6	7.8	8.1 pb
95,190	13.8 pb	12.1	7.5±1.0	8	9.3	10.6 pb
100,190	12.0 pb	12.1	6.5±0.9	6	7.7	7.7 pb
100,195	12.0 pb	10.9	7.1±1.0	8	9.6	10.5 pb
100,200	10.9 pb	10.0	9.7±1.3	14	15.1	16.4 pb
105,195	10.1 pb	10.4	6.6±0.9	6	7.6	7.3 pb
105,200	10.4 pb	10.7	7.4±1.0	8	9.3	9.1 pb
105,205	9.3 pb	8.8	8.8±1.2	12	13.3	14.1 pb
110,200	8.4 pb	9.1	7.5±1.0	8	9.2	8.4 pb
110,205	8.9 pb	9.5	8.2±1.1	10	11.2	10.4 pb
110,210	8.1 pb	8.6	9.8±1.3	13	13.6	12.8 pb
115,205	7.2 pb	8.9	8.2±1.1	8	8.8	7.2 pb
115,210	7.7 pb	8.5	8.4±1.2	10	11.0	9.9 pb
115,215	7.2 pb	7.6	9.0±1.2	10	10.5	9.9 pb

Table 2 : Expected number of signal and background events for 109 pb^{-1} after all cuts, number of remaining data events, and 95% C.L. upper limits of the TC signal.

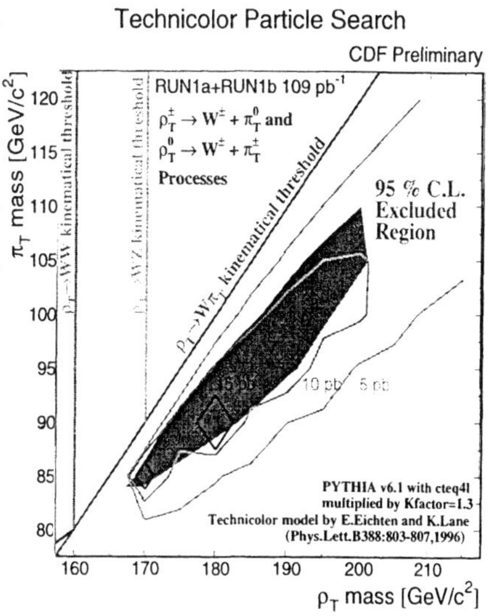

Technicolor Particle Search

CDF Preliminary

RUN1a+RUN1b 109 pb^{-1}

$\rho_T^\pm \to W^\pm + \pi_T^0$ and
$\rho_T^0 \to W^\pm + \pi_T^\mp$

Processes

95 % C.L. Excluded Region

PYTHIA v6.1 with cteq4l multiplied by Kfactor=1.3
Technicolor model by E.Eichten and K.Lane
(Phys.Lett.B388:803-807,1996)

ρ_T mass [GeV/c^2]

Figure 6 : 95% excluded region in the $M(\pi_T)$, $M(\rho_T)$ plane. Production cross section contour plot is shown on the same plane.

3. F. Abe et al., Phys. Rev. Lett. **74**, 2626 (1995)
4. Torbjorn Sjostrand : PYTHIA version 6.1
   'http://www.thep.lu.se/tf2/staff/torbjorn/Pythia.html'
5. Estia Eichten, Kenneth Lane, and John Womersley
   'Finding Low Scale Technicolor at Hadron Colliders' Phys.Lett.B405:305-311,1997

A Strong Electroweak Sector at Future $\mu^+\mu^-$ Colliders [1]

R. Casalbuoni[a,b,d], S. De Curtis[b], D. Dominici[a,b]

A. Deandrea[c], R. Gatto[d] and J. F. Gunion[e]

[a] *Dipartimento di Fisica Università di Firenze, I-50125 Firenze, Italia*
[b] *I.N.F.N., Sezione di Firenze, I-50125 Firenze, Italia*
[c] *Centre de Physique Théorique, CNRS, Luminy F-13288 Marseille, France*
[d] *Département de Physique Théorique, Université de Genève, CH-1211 Genève 4, Suisse*
[e] *Department of Physics, University of California, Davis, CA 95616, USA*

Abstract. We discuss the prospects for detecting at a muon collider the massive new vector resonances V and light pseudo-Nambu-Goldstone bosons P of a typical strongly interacting electroweak sector (as represented by the BESS model). Expected sensitivities to V's at a high energy collider are evaluated and the excellent prospects for discovering P's via scanning at a low energy collider are delineated.

INTRODUCTION

The exploration of technivector and technipion physics at muon colliders was the main focus of the strong dynamics working group at this meeting. Light states received particular attention [1]. In this contribution we consider some aspects of strong electroweak symmetry breaking at future $\mu^+\mu^-$ colliders. We will concentrate on the possibility of detecting new vector resonances and pseudo-Nambu-Goldstone bosons (PNGB's) originating from the strong interaction responsible for electroweak symmetry breaking.

This study will be performed within the framework of the BESS model [2] and its generalizations [3]. We recall the main features of this model. The BESS model is an effective lagrangian parameterization of the symmetry breaking mechanism, based on a symmetry $G = SU(2)_L \otimes SU(2)_R$ broken down to $SU(2)_{L+R}$. New vector particles are introduced as gauge bosons associated with a hidden $H' = SU(2)_V$. The symmetry group of the theory becomes $G' = G \otimes H'$. It breaks down spontaneously to $H_D = SU(2)$,

---

[1] Talk presented by D. Dominici.

which is the diagonal subgroup of G'. This gives rise to six Goldstone bosons. Three are absorbed by the new vector particles while the other three give mass to the standard model (SM) gauge bosons, after the gauging of the subgroup $SU(2)_L \otimes U(1)_Y \subset G$. The parameters of the BESS model are the mass of these new bosons M_V, their self coupling g'', and a third parameter b whose strength characterizes the direct couplings of the new vectors V to the fermions. However, due to the mixing of the V bosons with W and Z, the new particles are coupled to the fermions even when $b = 0$. The parameter g'' is expected to be large due to the fact that these new gauge bosons are thought of as bound states from a strongly interacting electroweak sector. By taking the formal $b \to 0$ and $g'' \to \infty$ limits, the new bosons decouple and the SM is recovered. By considering only the limit $M_V \to \infty$ they do not decouple.

The extension of the BESS model we will consider here is obtained by enlarging the original chiral symmetry $SU(2)_L \otimes SU(2)_R$ to the larger group $SU(8)_L \otimes SU(8)_R$ [3]. The main new feature of this extension is the presence of 60 PNGB's. Their masses come from the breaking of the chiral group provided by the $SU(3) \otimes SU(2)_L \otimes U(1)_Y$ gauge interactions and by Yukawa couplings [4]. We emphasize that most non-minimal models of a strongly interacting electroweak sector will contain PNGB's, although the number and their exact properties are model dependent.

BOUNDS ON THE PARAMETER SPACE FOR THE NEW VECTOR BOSONS

Bounds on the parameter space of the BESS model from existing data have already been studied (see for instance [5]). Future lepton colliders can improve these limits by testing virtual effects of the new vector particles, especially in the annihilation channels $l^+l^- \to W^+W^-$. In fact, the most relevant observable is the differential cross section $d\sigma(l^+l^- \to W^+_{L,T}W^-_{L,T})/d\cos\theta$, where θ is the overall center of mass scattering angle and the decays of the W^+ and W^- are used to reconstruct the final W polarizations. The analysis is performed by taking 19 bins in the angular region restricted by $|\cos\theta| \leq 0.95$. Since the new vectors strongly couple to the longitudinal W, the most relevant process is $l^+l^- \to W^+_L W^-_L$. We have studied the channel with one W decaying leptonically and the other one hadronically. In Fig. 1 we present the 90% C.L. contours in the plane $(b, g/g'')$ for $M_V = 1$ TeV. The solid (dashed) lines are the bounds from the combined differential $W_{L,T}W_{L,T}$ cross sections at a 500 GeV lepton collider with $L = 20\ fb^{-1}$, assuming effective branching ratios $B = 0.1$ ($B = 0.2$). [2] Statistical errors are taken into account and we have assumed a systematic error of 1.5%. The dot-dashed line is the bound

---

[2] The first B case is appropriate for an e^+e^- machine where the loss of luminosity from beamstrahlung is taken into account, and the second is appropriate for a muon collider or electron collider with a low beamstrahlung loss.

from the total cross section $pp \to W^\pm, V^\pm \to W^\pm Z \to \mu\nu\mu^+\mu^-$ at LHC if no deviation is observed with respect to the SM within the experimental errors. Statistical errors and a systematic error of 5% have been included. In all three cases, the regions within which one can exclude or detect the V at the 90% CL are the external ones. In conclusion, for models of new strong interacting vector resonances the measurement of $d\sigma(l^+l^- \to W^+_{L,T}W^-_{L,T})/d\cos\theta$ gives rather strong bounds, provided one is able to reconstruct the final W polarizations. These bounds become very stringent for increasing energy of the collider [5].

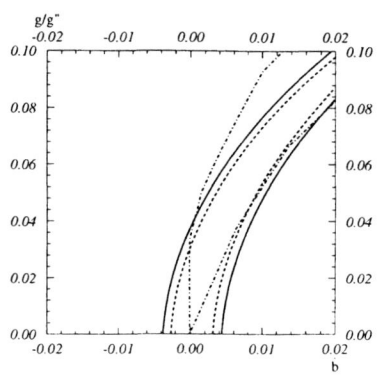
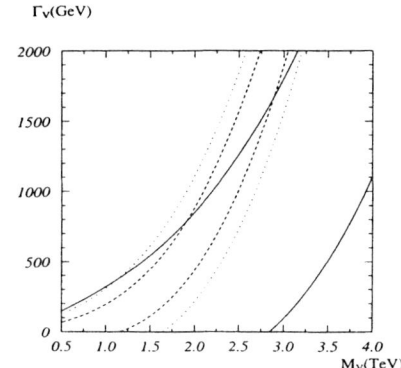

FIGURE 1. BESS model 90% C.L. contours in the $(b, g/g'')$ plane for $M_V = 1$ TeV. See text for details.

FIGURE 2. Partial wave unitarity bounds in the (M_V, Γ_V) plane for $\Lambda/M_V = 1.5$. See text for details.

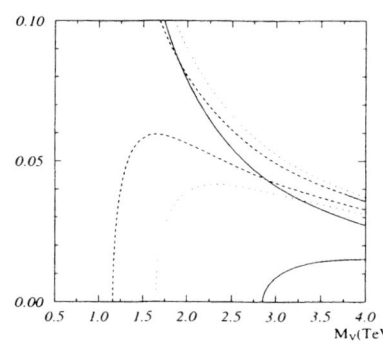
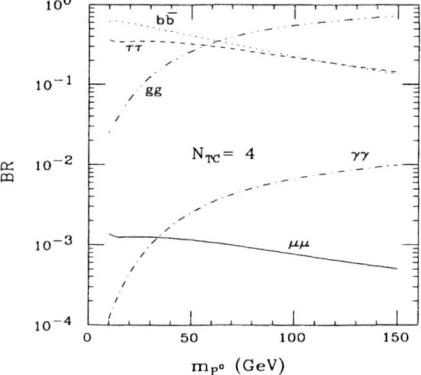

FIGURE 3. Partial wave unitarity bounds as in Fig. 2 but in the $(M_V, g/g'')$ plane.

FIGURE 4. Branching ratios for P^0 decay into $\mu^+\mu^-$, $\tau^+\tau^-$, $b\bar{b}$, $\gamma\gamma$, and gg.

Direct production of the new vector bosons can also be considered. A muon collider of $3-4$ TeV will enable us to completely explore the strongly interacting electroweak option [6]. Since one must rely on an effective non-renormalizable description, one has to take into account the partial wave unitarity limits from WW scattering. In fact when the mass of the new vectors is in the range $2-4$ TeV, these bounds come out to be quite restrictive. If we denote by $A(s,t,u)$ the amplitude for the scattering $W^+W^- \to ZZ$, one gets

$$A(s,t,u) = \left(1 - \frac{3}{4}\alpha\right)\frac{s}{v^2} + \frac{\alpha}{4}\frac{M_V^2}{v^2}\left(\frac{u-s}{t-M_V^2+iM_V\Gamma_V} + \frac{t-s}{u-M_V^2+iM_V\Gamma_V}\right)$$

where $\alpha = 192\pi v^2 \Gamma_V/M_V^3$ and $v = 246$ GeV. Projecting the components with definite isospin into the lower partial waves and requiring $a_{IJ} \leq 1$ up to energies, Λ, such that $\Lambda/M_V \leq 1.5$, we get the limitations in the plane (M_V, Γ_V) given in Fig. 2. There, the dashed, dotted, solid lines are the bounds from a_{00}, a_{20}, and a_{11}, respectively. The intersection of the three allowed regions gives a general upper bound of $M_V \sim 3$ TeV. The previous considerations can be applied also to the technirho case, which is obtained by taking $\alpha = 2$. In this model the unitarity bound turns out to be $M_{\rho_T} \leq 2$ TeV. In Fig. 3 we translate these limits into restrictions on the parameters of the BESS model, using the relation $\Gamma_V = M_V^5/(48\pi v^4 g''^2)$. We conclude that the unitarity bounds imply that one or more of the heavy vector resonances should be discovered at the LHC, NLC, a $\sqrt{s} \sim 500$ GeV muon collider, or, for certain, at a $3-4$ TeV muon collider, unless g'' is very large and b is very small so that they are largely decoupled.

PNGB PRODUCTION AT A MUON COLLIDER IN THE EXTENDED BESS MODEL

In this section, we consider s-channel production of the lightest neutral PNGB P^0 at a future $\mu^+\mu^-$ collider. Although we shall employ the specific P^0 properties as predicted by the extended BESS model with $SU(8) \otimes SU(8)$ symmetry [3], many of our results apply in general fashion to other models of a strongly interacting electroweak sector. General discussions of the production of isoscalar and isovector technipions appear in [1,7].

In the extended BESS model, the PNGB mass derives both from gauge contributions and from the effective low-energy Yukawa interactions between the PNGB's and ordinary fermions [4]. The lightest neutral PNGB's are the following combinations of the isosinglet and isotriplet components: $P^0 = (\tilde{\pi}_3 - \pi_D)/\sqrt{2}$, $P^{0\prime} = (\tilde{\pi}_3 + \pi_D)/\sqrt{2}$. The P^0 boson couples to the $T_3 = -1/2$ component of the fermion doublet while $P^{0\prime}$ couples to the $T_3 = 1/2$ component. It is the P^0 upon which we focus. The expressions for the P^0 and $P^{0\prime}$ masses are [4] $m_{P^0}^2 = \frac{2\Lambda^2}{\pi^2 v^2}m_b^2$, $m_{P^{0\prime}}^2 = \frac{2\Lambda^2}{\pi^2 v^2}m_t^2$, where Λ is an UV cut-off, situated in the TeV region. The first result above can be written as $m_{P^0} \sim 8$ GeV $\times \Lambda$(TeV). Thus, not only does the P^0 have the $\mu^+\mu^-$ coupling needed for s-channel production at a muon collider, but also, in this model, m_{P^0} should be relatively small, $\lesssim 80$ GeV for $\Lambda \lesssim 10$ TeV (as expected in the present model).

The P^0 Yukawa couplings to fermions are [4] $\mathcal{L}_Y = -i\lambda_b \bar{b}\gamma_5 b P^0$ $i\lambda_\tau \bar{\tau}\gamma_5 \tau P^0 - i\lambda_\mu \bar{\mu}\gamma_5 \mu P^0$, with $\lambda_b = \sqrt{\frac{2}{3}}\frac{m_b}{v}$, $\lambda_\tau = -\sqrt{6}\frac{m_\tau}{v}$, $\lambda_\mu = -\sqrt{6}\frac{m_\mu}{v}$. For the P^0, the $\gamma\gamma$ and gluon-gluon channels are also important; the corresponding couplings are generated by the ABJ anomaly. Clearly these couplings are model dependent and, as an example, we borrow them from technicolor theories [8]. We list here all the partial widths relevant for our analysis: $\Gamma(P^0 \to \bar{f}f) = C\frac{m_{P^0}}{8\pi}\lambda_f^2(1 - \frac{4m_f^2}{m_{P^0}^2})^{1/2}$; $\Gamma(P^0 \to gg) = \frac{\alpha_s^2}{48\pi^3 v^2}N_{TC}^2 m_{P^0}^3$; $\Gamma(P^0 \to \gamma\gamma) = \frac{2\alpha^2}{27\pi^3 v^2}N_{TC}^2 m_{P^0}^3$. Here, $C = 1(3)$ for leptons (down-type quarks) and N_{TC} is the number of technicolors. The corresponding branching ratios are shown in Fig. 4.

There are presently no definitive limits on the mass of the P^0. Potentially useful production modes arise through its ABJ anomaly coupling to pairs of electroweak gauge bosons [9,10]. At LEP the dominant production mode is $Z \to \gamma P^0$. The limit of [10], obtained by requiring a $Z \to \gamma\varphi$ decay width of $2\,10^{-6}$ GeV in order to make the φ visible in a sample of 10^7 Z bosons, can be rescaled to the case of the P^0. We find that for $N_{TC} \leq 9$ there is no limit on m_{P^0}, while, for instance, for $N_{TC} = 10$ $m_{P^0} \geq 12$ GeV is required.

Future limits from the Tevatron and LHC have also been considered [10]. In the single production mode, the best hope of finding the P^0 at hadron colliders is via the anomalous decay $P^0 \to \gamma\gamma$. The signal in this channel is similar to that of a Standard Model Higgs boson of the same mass, given the comparable branching ratio illustrated in Fig. 4. However, for the range of P^0 masses we are considering the signal will be hard to see since the $\gamma\gamma$ continuum background is very large at low mass. Another possibility would be to produce pairs of PNGB's, as for instance, in the resonant production $pp \to V^\pm \to P^\pm P^0 + X$ [11], where V is the vector resonance discussed in the Introduction. However, the discovery of the PNGB's via $\bar{t}b\bar{b}b$ or $\bar{t}bgg$ decays, needs a careful evaluation of backgrounds in the LHC environment. One could also consider the process $pp \to gg \to P^0 P^0$, mediated by the anomalous ggP^0 vertex, which could be detected by looking for equal mass pairs. Again, backgrounds will be large. Thus, as far as we know, reliable bounds will not be obtained at hadron colliders. Thus, it is clearly of great importance to find a means for discovering or eliminating a P^0 with mass between, roughly, 10 GeV and 100 GeV. For much of this mass range a muon collider would be the ideal probe. First, we note that the P^0 has a sizeable $\mu^+\mu^-$ coupling (see above). Second, the muon collider is unique in its ability to achieve the very narrow Gaussian spread, $\sigma_{\sqrt{s}}$, in $\sqrt{s}$ necessary to achieve a large P^0 cross section given the very narrow width of the P^0 (as plotted in Fig. 5). One can achieve $R = 0.003\%$ beam energy resolution with reasonable luminosity at the muon collider, leading to $\sigma_{\sqrt{s}} \sim 1$ MeV $\left(\frac{R}{0.003\%}\right)\left(\frac{\sqrt{s}}{50\text{ GeV}}\right)$; in addition, the beam energy can be very precisely tuned ($\Delta E_{\text{beam}} \sim 10^{-5} E_{\text{beam}}$ is 'easy'; 10^{-6} is achievable) as crucial for scanning for a very narrow resonance.

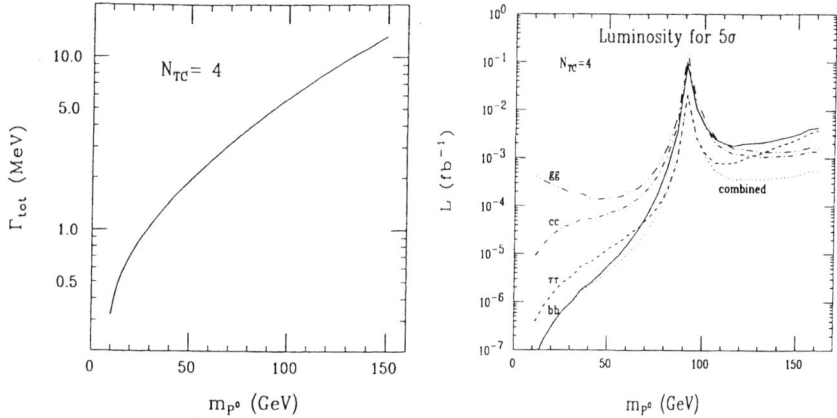

FIGURE 5. $\Gamma_{\text{tot}}(P^0)$ as a function of m_{P^0}. **FIGURE 6.** L for a 5σ P^0 signal. See text.

To quantitatively assess the ability of the muon collider to discover the P^0 we have proceeded as follows. We compute the P^0 cross section by integrating over the resonance using a $\sqrt{s}$ distribution given by a Gaussian of width $\sigma_{\sqrt{s}}$ (using $R = 0.003\%$) modified by bremsstrahlung photon emission. (Beamstrahlung is negligible at a muon collider.) See Ref. [12] for more details. We separate $\tau^+\tau^-$, $b\bar{b}$, $c\bar{c}$ and $q\bar{q}$, gg final states by using topological and τ tagging with efficiencies and mistagging probabilities as estimated by B. King [13]: $\epsilon_{bb} = 0.55$, $\epsilon_{cc} = 0.38$, $\epsilon_{bc} = 0.18$, $\epsilon_{cb} = 0.03$, $\epsilon_{qb} = \epsilon_{gb} = 0.03$, $\epsilon_{qc} = \epsilon_{gc} = 0.32$, $\epsilon_{\tau\tau} = 0.8$, $\epsilon_{\tau b} = \epsilon_{\tau c} = \epsilon_{\tau q} = 0$, where, for example, ϵ_{bb} (ϵ_{bc}) is the probability that a b-quark jet is tagged as a b (c). Only events in which the jets or τ's have $|\cos\theta| < 0.94$ (corresponding to a nose cone of $20°$) are considered. A jet final state is deemed to be: $b\bar{b}$ if one or more jets is tagged as a b; $c\bar{c}$ if no jet is tagged as a b, but one or more jets is tagged as a c; and $q\bar{q}, gg$ if neither jet is tagged as a b or a c. Background and signal events are analyzed in exactly the same manner. Note, in particular, that even though the P^0 does not decay to $c\bar{c}$, some of its $b\bar{b}$ and gg decays will be identified as $c\bar{c}$. In Fig. 6, we plot the integrated luminosity L required to achieve $S_i/\sqrt{B_i} = 5$ in a given channel, $i = b\bar{b}$, $\tau^+\tau^-$, $c\bar{c}$, or gg, (as defined after tagging), taking $\sqrt{s} = m_{P^0}$. Fig. 6 also shows the luminosity (labelled as 'combined') needed to achieve $\sum_k S_k/\sqrt{\sum_k B_k} = 5$, where the optimal choice of channels k is determined for each m_{P^0}. We observe that very modest L is needed unless $m_{P^0} \sim m_Z$. Of course, if we do not have any information regarding the P^0 mass, we must scan for the resonance. To estimate the luminosity required for scanning a given interval so as to either discover or eliminate the P^0, we have adopted the following approach. We imagine choosing $\sqrt{s}$ values separated by $2\sigma_{\sqrt{s}}$. We assume the worst case scenario in which the resonance sits midway between the two $\sqrt{s}$ values. The signal and (separately) background rates for these two

TABLE 1. Luminosity (in units of 0.01 fb^{-1}) required to scan from $M_{\min} + (m_Z - 90)$ to $M_{\min} + (m_Z - 90) + 5$ (GeV units) and either discover or eliminate the P^0 at the 3σ level. For scan details, see text.

$M_{\min}$	10	15	20	25	30	35	40	45	50	55
L	0.028	0.051	0.079	0.10	0.13	0.18	0.23	0.29	0.40	0.55
$M_{\min}$	60	65	70	75	80	85	90	95	100	105
L	0.77	1.2	2.2	5.3	17	166	274	52	23	15
$M_{\min}$	110	115	120	125	130	135	140	145	150	155
L	11	9.4	8.5	8.1	8.2	8.2	8.3	8.7	8.9	9.0

$\sqrt{s}$ values are summed together (for the optimal channel combination) and the net $N_{SD} \equiv (S_1 + S_2)/(B_1 + B_2)^{1/2}$ is computed. We require $N_{SD} = 3$ to claim a signal. The luminosity required for a successful scan of a given interval is computed assuming that the resonance lies between the *last* two scan points. This, in combination with the fact that $\sigma_{\sqrt{s}}$ for $R = 0.003\%$ is typically a factor of two smaller than $\Gamma_{\text{tot}}^{P^0}$ (implying that points further away than $\sigma_{\sqrt{s}}$ from the resonance could be usefully included in establishing a signal) will imply that the integrated luminosities given below are quite conservative. We give in Table 1 the integrated luminosity for a 3σ P^0 discovery after scanning the indicated 5 GeV intervals, assuming m_{P^0} lies within that interval. If the P^0 is as light as expected in the extended BESS model, then the prospects for discovery by scanning would be excellent. For example, a P^0 lying in the ~ 10 GeV to ~ 76 GeV mass interval can be either discovered or eliminated at the 3σ level with just 0.11 fb^{-1} of total luminosity, distributed in proportion to the (combined) luminosity plotted in Fig. 6. A P^0 with $m_{P^0} \sim m_Z$ would be much more difficult to discover unless its mass was approximately known. A 3σ scan of the mass interval from ~ 106 GeV to 161 GeV would require about 1 fb^{-1} of integrated luminosity.

CONCLUSION

We have demonstrated the very substantial suitability, and in many respects superiority, of a muon collider for exploring the full range of physics associated with a strongly-interacting electroweak sector as typified by the extended BESS model. Especially interesting is the potential for discovering any light pseudo-Nambu-Goldstone boson with lepton couplings by scanning. Such bosons are a general feature of models of a strongly interacting electroweak sector and may prove to be quite difficult to detect in any other way.

Acknowledgements: The research of RC, SDC, DD, AD and RG has been carried out within the Human Capital and Mobility Program:

"Tests of electroweak symmetry breaking and future European colliders", CHRXCT94/0579. JFG is supported by the U.S. Department of Energy under grant No. DE-FG03-91ER40674 and by the U.C. Davis Institute for High Energy Physics. JFG would like to thank V. Barger, M. Berger and T. Han for collaboration on Higgs discovery at a muon collider, as part of which project many of the techniques employed in the P^0 discussion were developed.

REFERENCES

1. K. Lane, these proceedings.
2. R. Casalbuoni, S. De Curtis, D. Dominici and R. Gatto, *Phys. Lett.* **B155**, (1985) 95; *Nucl. Phys.* **B282**, (1987) 235.
3. R. Casalbuoni, S. De Curtis, A. Deandrea, N. Di Bartolomeo, D. Dominici, F. Feruglio and R. Gatto, *Nucl. Phys.* **B409**, (1993) 257.
4. R. Casalbuoni, S. De Curtis, A. Deandrea, N. Di Bartolomeo, D. Dominici, F. Feruglio and R. Gatto, *Phys. Lett.* **B285**, (1992) 103.
5. R. Casalbuoni, A. Deandrea, S. De Curtis, D. Dominici and R. Gatto, contributed to Joint ECFA / DESY Study, Hamburg, 1996, hep-ph/9708287.
6. V. Barger, M. S. Berger, J. F. Gunion and T. Han, MADPH-96-949, hep-ph/9606417.
7. J. Womersley, these proceedings.
8. M. A. B. Bég, H. D. Politzer and P. Ramond, *Phys. Rev. Lett.* **43**, (1979) 170; S. Dimopoulos, *Nucl. Phys.*, **B168** (1980) 69; J. Ellis, M. K. Gaillard, D. V. Nanopoulos and P. Sikivie, *Nucl. Phys.*, **B182** (1981) 529; S. Dimopoulos, S. Raby and G. L. Kane, *Nucl. Phys.*, **B182** (1981) 77.
9. A. Manohar and L. Randall, *Phys. Lett.* **B246**, (1990) 537; L. Randall and E. H. Simmons, *Nucl. Phys.* **B380**, (1992) 3.
10. R. S. Chivukula, R. Rosenfeld, E. H. Simmons and J. Terning, BUHEP-95-07, Feb 1995, to be published in 'Electroweak Symmetry Breaking and Beyond the Standard Model, ed. by T. Barklow, et al. (World Scientific), hep-ph/9503202.
11. R. Casalbuoni, P. Chiappetta, S. De Curtis, A. Deandrea, D. Dominici, and R. Gatto, *Z. f. Physik* **C65**, (1995) 327.
12. V. Barger, M. Berger, J. Gunion and T. Han, *Phys. Rev. Lett.* **75** (1995) 1462; *Phys. Rep.* **286**, (1997) 1.
13. B. King, presentation at this conference.

T Tbar Factory

Conveners: Mike Berger, Indiana University
            Brian Winer, Ohio State University

Top Quark at the Upgraded Tevatron to Probe New Physics

Jin Min Yang [a,b]
and
A. Datta [b], M. Hosch [b], C. S. Li [c], R. J. Oakes [a],
K. Whisnant [b], Bing-Lin Young [b], X. Zhang [d]

[a] *Department of Physics, Northwestern University, Evanston, Il 60208*
[b] *Department of Physics, Iowa State University, Ames, IA 50011*
[c] *Department of Physics, Peking University, China*
[d] *IHEP, Academia Sinica, Beijing, China*

Abstract. This talk is a review of the recent studies on probing new physics through single top quark processes and probing exotic top quark decays at the upgraded Tevatron.

INTRODUCTION

The exceedingly heavy top quark is believed to be more sensitive to new physics than others. The upgraded Tevatron will provide a good opportunity to study top quark properties. Apart from the dominant top pair production which tests the top's QCD properties, the single top productions [1] are also interesting to study since they involve the electroweak interaction and can, therefore, be used to probe new physics in top's electroweak couplings. Analyses show that new physics effects that produce larger than 16% effect on the single top cross section should should be detectable at the upgraded Tevatron [2]. On the other hand, since top events will increase significantly at the upgraded Tevatron, it is interesting to search for exotic top decays predicted by new physics models.

This talk is a brief review of some recent studies on the ability of single top quark production at the upgraded Tevatron to probe new physics as well as the possibility of observing the exotic decay mode of the top quark predicted by Minimal Supersymmetric Model (MSSM). About the SUSY effects in top pair production, we refer to [3] and will not discuss them further.

MODEL-INDEPENDENT ANALYSIS FOR NEW PHYSICS IN SINGLE TOP PRODUCTION

Since no direct signal of new particles has been observed so far, it is very likely that the only observable effects of new physics at energies not too far above the SM energy scale could be in the form of new interactions affecting the couplings of the third-family quarks, and the untested sectors of the Higgs and gauge bosons. In this spirit, the new physics effects can be expressed as non-standard terms in an effective Lagrangian describing the interactions among third-family quarks, the Higgs and gauge bosons, which were enumerated in [4–6]. In the following we pick out those which affect single top production at the Tevatron.

Two typical operators which contribute to LEP I and LEP II observables, $\sigma_{t\bar{t}}$ and A^t_{FB} at NLC and $\sigma_{t\bar{b}}$ at the Tevatron are [5]:

$$O_{qW} = \left[\bar{q}_L\gamma^\mu\tau^I D^\nu q_L + \overline{D^\nu q_L}\gamma^\mu\tau^I q_L\right] W^I_{\mu\nu}, \quad (1)$$

$$O^{(3)}_{\Phi q} = i\left[\Phi^\dagger\tau^I D_\mu\Phi - (D_\mu\Phi)^\dagger\tau^I\Phi\right]\bar{q}_L\gamma^\mu\tau^I q_L. \quad (2)$$

Using 1σ bound of R_b we obtain the constraints [5]:

$$-0.0080 < \frac{4s_W c_W}{e}\frac{1}{\Lambda^2}\left[C_{qW}\frac{c_W m_Z^2}{4} - C^{(3)}_{\Phi q}\frac{vm_Z}{2}\right] < -0.0023 \quad (3)$$

Then the effects of $O^{(3)}_{\Phi q}$ are found to be negligibly small [5], while the effects of O_{qW} are

LEPII($e^+e^- \to b\bar{b}$) NLC ($e^+e^- \to t\bar{t}$) Tevatron($p\bar{p} \to t\bar{b} + X$)

$2.4\% < \frac{\Delta\sigma}{\sigma^0} < 8.4\%$ $8.6\% < \frac{\Delta\sigma}{\sigma^0} < 29.8\%$ $6.9\% < \frac{\Delta\sigma}{\sigma^0} < 24.0\%$ (4)

$0.3\% < \frac{\delta A_{FB}}{A^0_{FB}} < 1.0\%$ $16.3\% < \frac{\delta A_{FB}}{A^0_{FB}} < 56.8\%$

So the effects of O_{qW} may still be observable at LEP II, NLC and the upgraded Tevatron.

The following dimension-six CP-violating operators can give rise to transverse polarization asymmetry of top quark in single top production ($u + \bar{d} \to t + \bar{b}$, $\bar{u} + d \to \bar{t} + b$) at the Tevatron [6]:

$$\bar{O}_{qW} = i\left[\bar{q}_L\gamma^\mu\tau^I D^\nu q_L - \overline{D^\nu q_L}\gamma^\mu\tau^I q_L\right] W^I_{\mu\nu}, \quad (5)$$

$$\bar{O}_{tW\Phi} = i\left[(\bar{q}_L\sigma^{\mu\nu}\tau^I t_R)\tilde{\Phi} - (D^\mu\tilde{\Phi})^\dagger(\overline{D_\mu t_R q_L})\right] \quad (6)$$

Introducing the coordinate system in the top quark (or top antiquark) rest frame with the unit vectors $\vec{e}_z \propto -\vec{P}_{\bar{b}}$ and $\vec{e}_y \propto \vec{P}_u \times \vec{P}_{\bar{b}}$. Transverse polarization asymmetry is defined by [7] $A(\hat{y}) = \frac{1}{2}\left[\Pi(\hat{y}) - \bar{\Pi}(\hat{y})\right]$, where $\Pi(\hat{y})$ and

$\bar{\Pi}(\hat{y})$ are, respectively, the polarizations of the top quark and top antiquark in the direction $\hat{y}$. The polarizations are given by

$$\Pi(\hat{y}) = \frac{N_t(+\hat{y}) - N_t(-\hat{y})}{N_t(+\hat{y}) + N_t(-\hat{y})}, \quad \bar{\Pi}(\hat{y}) = \frac{N_{\bar{t}}(+\hat{y}) - N_{\bar{t}}(-\hat{y})}{N_{\bar{t}}(+\hat{y}) + N_{\bar{t}}(-\hat{y})}, \tag{7}$$

where $N_t(\pm\hat{y})$ [$N_{\bar{t}}(\pm\hat{y})$] is the number of $t(\bar{t})$ quarks polarized in the direction $\pm\hat{y}$.

Assuming $m_t = 175$ GeV, we obtain the asymmetry at hadron level as [6]

$$A(\hat{y}) = \begin{cases} -0.41 \frac{C_{qW} - 2C_{tW\Phi} - g_2 C_{Dt}/2}{(\Lambda/1 \text{ TeV})^2} & \text{at } \sqrt{s} = 2 \text{ TeV} \\ -0.84 \frac{C_{qW} - 2C_{tW\Phi} - g_2 C_{Dt}/2}{(\Lambda/1 \text{ TeV})^2} & \text{at } \sqrt{s} = 4 \text{ TeV} \end{cases} \tag{8}$$

Assume an observable level of ten percent of this asymmetry, the upgraded Tevatron will probe

$$\frac{C_{qW} - 2C_{tW\Phi} - g_2 C_{Dt}/2}{(\Lambda/1 \text{ TeV})^2} \text{ to } \begin{cases} 1/4 & \text{for } \sqrt{s} = 2 \text{ TeV} \\ 1/8 & \text{for } \sqrt{s} = 4 \text{ TeV} \end{cases} \tag{9}$$

This means that with a new physics scale at the order of 1 TeV, the further upgraded Tevatron can probe the coupling strength down to the level of 0.1.

PROBING SUSY IN SINGLE TOP PRODUCTION

In the R-parity Conserving MSSM, we found that within the allowed range of squark and gluino masses the supersymmetric QCD corrections can enhance the cross section by a few percent [8]. The Yukawa corrections [8] to single top quark production at the Tevatron can amount to more than a 15% reduction in the production cross section relative to the tree level result in the general two-Higgs-doublet model, and a 10% enhancement in the minimal supersymmetric model for the smallest allowed $\tan\beta$ ($\simeq 0.25$). The supersymmetric electroweak corrections [8] to the cross section are at most a few percent for $\tan\beta > 1$, but can exceed 10% for $\tan\beta < 1$. So the combined effects of SUSY QCD, SUSY EW, and the Yukawa couplings in the R-parity Conserving MSSM can exceed 10% for the smallest allowed $\tan\beta$ ($\simeq 0.25$) but are only a few percent for $\tan\beta > 1$.

In the R-parity violating MSSM, the processes induced by R-violating couplings are [9,10]

$$\lambda': \quad u\bar{d} \to \tilde{l} \to t + \bar{b}, \text{ (s - channel)} \tag{10}$$

$$\lambda'': \quad u\bar{d} \to t + \bar{b}, \text{ (t - channel)} \tag{11}$$

$$\lambda'': \quad cd \to \tilde{s} \to tb, \text{ (s - channel)} \tag{12}$$

$$\lambda'': \quad cs \to \tilde{d} \to tb, \text{ (s - channel)} \tag{13}$$

Their signature are an energetic charged lepton, missing E_T, and double b-quark jets. The backgrounds are (1) $q\bar{q}' \to W^* \to t\bar{b}$, (2) the quark-gluon process $qg \to q't\bar{b}$ with a W-boson as an intermediate state in either the t-channel or the s-channel of a subdiagram; (3) processes involving a b-quark in the initial state, $bq(\bar{q}) \to tq'(\bar{q}')$ and $gb \to tW$; (4) $Wb\bar{b}$; (5) Wjj; and (6) $t\bar{t} \to W^-W^+b\bar{b}$. For the upgraded Tevatron (LHC), the basic cuts are $p_T^l \geq 20$ GeV, $p_T^b \geq 20(35)$ GeV, $p_T^{\text{miss}} \geq 20(30)$ GeV, η_b, $\eta_l \leq 2.5(3)$ and ΔR_{jj}, $\Delta R_{jl} \geq 0.4$. Also we required reconstructed top quark mass $M(bW)$ to lie within the mass range $|M(bW) - m_t| < 30$ GeV, which can reduce the backgrounds $Wb\bar{b}$ and Wjj efficiently. The number of signal events required for discovery of a signal is $S \geq 5\sqrt{B}$.

For the s-channel process $u + \bar{d} \to \tilde{l} \to t + \bar{b}$ induced by λ', the histogram of the differential cross section versus the invariant mass of the $t\bar{b}$ system over the bin size of 10 GeV is shown in Fig.2 of Ref. [9]. The resonance behavior is already manifested. Because of their narrow widths, for each slepton the contributions of the λ'-couplings are negligible for a couple of bins away from the resonance. This will help to identify the signal of the slepton production. The value of $\lambda'_{111}\lambda'_{133} + \lambda'_{211}\lambda'_{233} + \lambda'_{311}\lambda'_{333}$ versus the slepton mass for $u\bar{d} \to \tilde{l} \to t\bar{b}$ to be observable under the criteria $S \geq 5\sqrt{B}$ is shown in Fig.4 of Ref. [10], which show that the LHC can do better than the upgraded Tevatron in further probing the couplings, especially for higher mass sleptons.

For the s-channel process $cd \to \tilde{s} \to tb$ induced by λ'' couplings, the value of $\lambda''_{212}\lambda''_{332}$ versus strange-squark mass for it to be observable under the criteria $S \geq 5\sqrt{B}$ is shown in Fig.1 of Ref. [10], which show that both the LHC and the upgraded Tevatron can efficiently probe the relevant couplings, and the LHC serves a more powerful probe than the upgraded Tevatron.

For the s-channel process $cs \to \tilde{d} \to tb$ induced by λ'', The value of $\lambda''_{212}\lambda''_{331}$ versus down-squark mass for it to be observable under the criteria $S \geq 5\sqrt{B}$ is shown in Fig.3 of Ref. [10], which show that this process cannot be probed as efficiently as $cd \to \tilde{s} \to tb$ because of the relative suppression of the strange quark structure function compared to the valence down quark.

SEARCHING FOR EXOTIC TOP DECAY MODES

The FCNC Decays in the SM, R-conserving MSSM [11] and R-violating MSSM [12] are given by

	SM	MSSM	$\not{R}$ MSSM	
$B(t \to cg)$	10^{-10}	10^{-6}	10^{-3}	
$B(t \to c\gamma)$	10^{-12}	10^{-8}	10^{-5}	(14)
$B(t \to cZ)$	10^{-12}	10^{-8}	10^{-4}	
$B(t \to ch)$	10^{-7}	10^{-5}		

The FCNC top decays in R-violating MSSM might be observable at the upgraded Tevatron since for integrated luminosity of 10 (100) fb^{-1}, the detection sensitivity is [13] $Br(t \to cg) \simeq 5 \times 10^{-3}(1 \times 10^{-3})$, $Br(t \to c\gamma) \simeq 4 \times 10^{-4}(8 \times 10^{-5})$ and $Br(t \to cZ) \simeq 4 \times 10^{-3}(6 \times 10^{-4})$.

Let us look at the top decay to light stop, $t \to \tilde{t}_1 \tilde{\chi}_1^0$, in the framework of R-conserving MSSM with the lightest neutralino being the LSP. the parameters involved in $\Gamma(t \to \tilde{t}_1 \tilde{\chi}_1^0)$ are $M_{\tilde{t}_1}, M_2, M_1, \mu, \tan\beta$. In the region of parameter space allowed by R_b data and the $ee\gamma\gamma + \not{E}_T$ event [14], we obtain [15] $0.07 \leq B(t \to \tilde{t}_1 \tilde{\chi}_1^0) \leq 0.50$. The dorminant decay of a light stop is $\tilde{t}_1 \to c\tilde{\chi}_1^0$. This will give a new final state in $t\bar{t}$ production: $t\bar{t} \to Wb\bar{c}\tilde{\chi}_1^0\tilde{\chi}_1^0$. Its signature is an energetic charged lepton, one b-quark jet, one light c-quark jet, plus missing E_T from the neutrino and the unobservable χ_1^0's. The potential SM backgrounds are (1) $bq(\bar{q}) \to tq'(\bar{q}')$, (2) $q\bar{q}' \to W^* \to t\bar{b}$; (3) $Wb\bar{b}$; (4) Wjj; (5) $t\bar{t} \to W^-W^+b\bar{b}$; (6) $gb \to tW$ and (7) $qg \to q't\bar{b}$. Besides the basic cuts we impose a cut on transverse mass $m_T = \sqrt{(P_T^l + P_T^{\text{miss}})^2 - (\vec{P}_T^l + \vec{P}_T^{\text{miss}})^2} > 90$ GeV. Then we found (1) this final state is unobservable at Run 1 with $\sqrt{s} = 1.8$ TeV and $L = 0.1$ fb^{-1}, (2) Run 2 with $\sqrt{s} = 2$ TeV and $L = 10$ fb^{-1} can either discover this final state or provide the additional strong constraint given approximately by $M_{\tilde{t}_1} - M_{\tilde{\chi}_1^0} < 6$ GeV.

If charged Higgs is light enough, $t \to H^+b$ is also possible; we refer to [16] for its phenomenological implications at Tevatron.

CONCLUSION

(1) Single top quark processes at upgraded Tevatron can be meaningfuly used to probe new physics; (2) The FCNC top decays $t \to cV$ and top decay to light stop $t \to \tilde{t}_1 \tilde{\chi}_1^0$ predicted by R-violating MSSM might be observable at upgraded Tevatron, else further constraints can be set on the relevant couplings.

REFERENCES

1. S. Willenbrock and D. Dicus, *Phys. Rev.* **D34**, 155 (1986); S. Dawson and S. Willenbrock, *Nucl. Phys.* **B284**, 449 (1987); C.-P. Yuan, *Phys. Rev.* **D41**, 42 (1990); F. Anselmo, B. van Eijk and G. Bordes, *Phys. Rev.* **D45**, 2312 (1992); R. K. Ellis and S. Parke, *Phys. Rev.* **D46**, 3785 (1992); D. Carlson and C.-P. Yuan, *Phys. Lett.* **B306**, 386 (1993); G. Bordes and B. van Eijk, *Nucl. Phys.* **B435**, 23 (1995); A. Heinson, A. Belyaev and E. Boos, hep-ph/9509274. S. Cortese and R. Petronzio, *Phys. Lett.* **B306**, 386 (1993). T. Stelzer and S. Willenbrock, *Phys. Lett.* **B357**, 125 (1995). M. Smith and S. Willenbrock, *Phys. Rev.* **D54**, 6696 (1996); S. Mrenna and C.-P. Yuan, hep-ph/9703224; T. Tait and C.-P. Yuan, hep-ph/9710372.
2. A. P. Heinson, hep-ex/9605010.

3. C. S. Li, B. Q. Hu, J. M. Yang and C. G. Hu, *Phys. Rev.* **D52**, 5014 (1995); J. M. Yang and C. S. Li, *Phys. Rev.* **D52**, 1541 (1995); J. Kim, J. L. Lopez, D. V. Nanopoulos and R. Rangarajan, *Phys. Rev.* **D54**, 4364 (1996); J. M. Yang and C. S. Li, *Phys. Rev.* **D54**, 4380 (1996); C. S. Li, H. Y. Zhou, Y. L. Zhu, J. M. Yang, *Phys. Lett.* **B379**, 135 (1996); S. Alam, K. Hagiwara and S. Matsumoto, *Phys. Rev.* **D55**, 1307 (1997); Z. Sullivan, *Phys. Rev.* **D56**, 451 (1997); W. Hollik, W. M. Mosle and D. Wackeroth, hep-ph/9706218.
4. G. J. Gounaris, D. T. Papadamou and F. M. Renard, hep-ph/9609437.
5. K. Whisnant, J. M. Yang, B.-L. Young and X. Zhang, *Phys. Rev.* **D56**, 467 (1997).
6. J. M. Yang and B.-L. Young, *Phys. Rev.* **D56**, 5907 (1997).
7. D. Atwood, S. B. Shalom, G. Eilam and A. Soni, *Phys. Rev.* **D54**, 5412 (1996).
8. C. S. Li, R. J. Oakes and J. M. Yang, *Phys. Rev.* **D55**, 1672 (1997); *Phys. Rev.* **D55**, 5780 (1997); hep-ph/9706412.
9. A. Datta, J. M. Yang, B.-L. Young and X. Zhang, *Phys. Rev.* **D56**, 3107 (1997);
10. R. J. Oakes, K. Whisnant, J. M. Yang, B.-L. Young and X. Zhang, hep-ph/9707477.
11. C. S. Li, R. J. Oakes and J. M. Yang, *Phys. Rev.* **D49**, 293 (1994); J. M. Yang and C. S. Li, *Phys. Rev.* **D49**, 3412 (1994); G. Couture, C. Hamzaoui and H. Konig, *Phys. Rev.* **D52**, 171(1995); J. L. Lopez, D. V. Nanopoulos and R. Rangarajan, hep-ph/9702350; G. Couture, M. Frank and H. Konig, hep-ph/9704305; G. M. de Divitiis, R. Petronzio and L. Silvestrini, hep-ph/9704244.
12. J. M. Yang, B.-L. Young and X. Zhang, hep-ph/9705341.
13. T. Han, R. D. Peccei, and X. Zhang, *Nucl. Phys.* **B454**, 527 (1995); T. Han, K. Whisnant, B.-L. Young, and X. Zhang, *Phys. Rev.* **D55**, 7241 (1997); *Phys. Lett.* **B385**, 311 (1996).
14. S. Ambrosanio, G. L. Kane, G. D. Kribs, S. P. Martin, and S. Mrenna, *Phys. Rev. Lett.* **76**, 3498 (1996); S. Dimopolous, M. Dine, S. Raby and S. Thomas, *Phys. Rev. Lett.* **76**, 3502 (1996); D. Garcia and J. Sola, *Phys. Lett.* **B357**, 349 (1995).
15. M. Hosch, R. J. Oakes, K. Whisnant, J. M. Yang, B.-L. Young and X. Zhang, hep-ph/9711234; S. Mrenna and C. P. Yuan, *Phys. Lett.* **B367**, 188 (1996).
16. J. Guasch and J. Sola, hep-ph/9707535.

Top Quark Physics at the Next Linear Collider

Rajendran Raja

Fermi National Accelerator Laboratory
P.O. Box 500
Batavia, IL 60510

Abstract. We report on the physics capabilities of a linear e^+e^- collider operating on or above the top quark threshold.

INTRODUCTION

The initial phase of the linear e^+e^- collider [1,2] is expected to operate at a maximum center of mass energy of 500 GeV, which will enable it to explore the threshold dependence of the $t\bar{t}$ cross section. Detailed exploration of the energy dependence of the $t\bar{t}$ cross section near threshold can lead to a determination of a number of important properties of the top quark, namely its mass m_t, its width Γ_t, the CKM matrix parameter V_{tb}, the Yukawa coupling of the top quark to the Higgs boson β_H, as well as the QCD coupling constant α_s. This is because the threshold shape depends on all these parameters. Measurements of the decay properties of the top quark will yield information on the form factors governing its decay. The ability to polarize the electron beam will aid significantly in suppressing WW backgrounds to the $t\bar{t}$ channel as well as help unravel the form factor information. In what follows, we will summarize this physics. A more detailed write-up can be found in Frey et al. [3].

TOP CROSS SECTION NEAR THRESHOLD

The standard model width of the top quark of mass 175 GeV/c² is ≈1.42 GeV. A $t\bar{t}$ (toponium) bound state will thus have at least twice the width of the top quark, since either of the top quarks may decay. The level spacing of toponia is $\approx \alpha_s^2 m_t$. For top quark masses greater than 150 GeV/c², the level spacing of toponia is comparable to the width of each state. This implies that the S,P and D states of toponia form an indistinguishable enhancement below the $t\bar{t}$ threshold. The shape of this curve has been calculated using the Bethe-Salpeter equation by Strassler and Peskin [4] and has been extensively studied by others [5-9]. The shape depends on m_t and α_s. The greater α_s, the stronger the binding between $t\bar{t}$ pairs in the bound state and lower the position of the 1S state below threshold. The top cross section curve also depends on the top quark width and on the Higgs boson mass via the Yukawa coupling of the top quark and the Higgs boson. Figure 1(a) shows the theoretical top quark production cross section for a top quark mass of 175 GeV/c^2, infinite Higgs mass and $\alpha_s(M_Z^2) = 0.120$. At high energy, the center of mass energy resolution of e^+e^- colliders suffers from two effects, namely initial state radiation (ISR) and beamstrahlung. The effective QED expansion parameter at high energies for real photon emission is $\beta \equiv \frac{2\alpha}{\pi}(ln(s/m_e^2) - 1) \approx 0.12$ (rather than α/π) for $\sqrt{s}$=500 GeV. For a muon collider of the same energy, β is 0.07, resulting in much smaller beam energy smearing due to ISR. Beamstrahlung, the emission of radiation by one beam due to the action on it of the other, is governed by the parameter $\Gamma \equiv \gamma(B/B_c)$, where B is the effective magnetic field strength of the beam ($\approx 6 \times 10^2$ T for NLC design) and $B_c \equiv m_e^2 c^3 / e\hbar \approx 4 \times 10^9$ T. This results in $\Gamma \approx 0.08$ at $\sqrt{s} = 500$ GeV for the SLAC NLC design. Figure 1(b) has ISR effects taken into account [10]. Figure 1(c) has ISR and beamstrahlung effects [11] taken into account. Figure 1(d) has in addition, single beam energy spreads appropriate to the SLAC NLC design applied. The energy spectrum of the beam may be characterized by a δ function piece at the nominal energy with a low energy tail due to the effects described above. The fraction of the luminosity in the δ function piece at $\sqrt{s}$=500 GeV for the SLAC X-band design is 43%. For a comparison of this curve with the corresponding one for the muon collider, please see the contribution by M. Berger to these proceedings [12].

At $\sqrt{(s)} = 500$ GeV, the lowest order total cross section for m_t=180 GeV/c^2 is 0.54 pb for unpolarized beams. One expects roughly 90% polarization for electrons at the NLC. The lowest order top cross section is 0.74pb(0.34pb) for fully left-hand (right-hand) polarized electron beam. For an integrated luminosity of 50fb^{-1}, achievable in one year's running, one expects roughly 25,000 $t\bar{t}$ events. The background to $t\bar{t}$ production is the production of W pairs. This process is dramatically reduced if the electron polarization is right handed.

Measurements of m_t and α_s at threshold

As α_s increases, the threshold bump of the theoretical curve moves lower. This can be compensated for by raising the top quark mass, resulting in a strong positive correlation between the fitted values of α_s and m_t. A number of studies have been performed to simulate measurements at the $t\bar{t}$ threshold.

A JLC study [9] assuming $m_t = 150$ GeV/c^2 scans the threshold with 11 scan points of 1 fb^{-1} integrated luminosity per point. Assuming that the width of the top quark is given by the standard model, a simultaneous fit to α_s and m_t yields errors of 0.005 and 200 MeV/c^2 respectively. If one assumes that α_s is known from other sources to much better than this accuracy, one can fit for m_t alone yielding 100 MeV/c^2 for the top quark mass. An update for m_t =170 GeV/c^2 [13], yields errors of 0.007 and 350 MeV/c^2 for the same 11 point scan. A similar 10 point scan, with 5 fb^{-1} per point and m_t=180 GeV/c^2 [14], yields errors of 0.0025 and 120 MeV/c^2 for α_s and m_t respectively.

Measurements of top quark width and V_{tb}

Another quantity that contains information on α_s and Γ_t is the momentum distribution of the top quark pairs. After they are formed and move apart, the $t\bar{t}$ pairs are slowed down by the QCD potential. They decay before they have a chance to hadronize. The greater the value of α_s, the greater will be the QCD potential and the more the $t\bar{t}$ pairs will be slowed down. Similarly, the greater the value of Γ_t, the faster will be the decay and the greater the average momentum of the $t\bar{t}$ pair. This effect has been extensively studied theoretically [6,8] and phenomenologically [9,15]. A 100 fb^{-1} study of top momentum distributions yields an error on $\alpha_s = 0.00024$ (0.2%) and of 7% in Γ_t respectively. Another source of information on Γ_t is the threshold scan itself. Figure 2 shows the variation of the threshold curve as a function of Γ_t. Fujii et al [13] find that fixing α_s and fitting for m_t and Γ_t simultaneously for an 11 point scan with 1 fb^{-1} per point yields an error of 100 MeV/c^2 for m_t and and a fractional error of 16 % for Γ_t. With a 50 fb^{-1} scan, the fractional error in Γ_t can perhaps be made as low as 5%.

In order to measure V_{tb}, one needs the partial width $\Gamma(t \to Wb)$. Knowing the total Γ_t, this can be determined by measuring the branching ratio $B(t \to Wb)$. No detailed Monte Carlos have been done to ascertain the precision to which this can be measured, but Frey et al [3] cite reasonable arguments to show that $B(t \to Wb)$ can be determined to a precision of 2.5% with 25 fb^{-1} of data, leading to a precision in V_{tb} of 2.8%.

Another quantity that is sensitive to the top quark width is the forward backward asymmetry of $t\bar{t}$ pairs [9,17]. This is caused by the interference between S , D states of toponium produced via the vector coupling of the γ/Z intermediate state and the P states produced by the axial vector coupling. The degree of forward backward asymmetry depends on the overlap of these states which in turn depends on Γ_t. This technique is less sensitive than the threshold scan (by a factor of 10 in luminosity), but provides a cross check of measurements made using other means.

Measurements of Higgs Yukawa coupling to the top quark

The $t\bar{t}$ pairs produced are subject not only to the QCD potential but also to a Yukawa potential V_H associated with Higgs exchange.

$$V_H = -\frac{\lambda_t^2}{4\pi}\frac{e^{-m_H r}}{r}$$

where m_H is the Higgs mass and the Yukawa coupling λ_t is given by

$$\lambda_t = \beta_H m_t / v_{Higgs}$$

v_{Higgs} is the Higgs vacuum expectation value. The dimensionless constant β_H is unity in the standard model but can assume other values in theories with more complicated Higgs structure. Figure 3 shows the variation of the standard model top cross section for various values of the Higgs mass [16]. If one assumes that the Higgs boson has been discovered by the time the NLC turns on, then one can use the NLC to measure deviations in β_H from unity. Figure 4 shows the variation of the top cross section as a function of β_H, with all the beam effects included. Fujii et al [13] apply their 11 point scan method to show that with 11 fb^{-1} of data it is possible to determine β_H to 25% for $m_t = 170$ GeV/c^2.

It should also be possible to measure the Higgs Yukawa coupling by observing the direct emission of a Higgs boson associated with $t\bar{t}$ production if the center of mass energy is sufficient to allow such "higgsstrahlung".

FORM FACTORS AND DECAY DISTRIBUTIONS

Another rich source of information on the properties of top quarks is their decay modes and distributions. The search for rare decay modes of the top quark is perhaps better done in hadron machines with much larger rates of $t\bar{t}$ production. However, the existence of polarization and an initial state of well defined energy provide the lepton colliders with perhaps a unique means to explore the decay dynamics of top quarks.

The top neutral-current coupling can be generalized to the following form for the γ/Z-t-$\bar{t}$ or vertex factor:

$$\mathcal{M}^{\mu(\gamma,Z)} = e\gamma^\mu \left[Q_V^{\gamma,Z} F_{1V}^{\gamma,Z} + Q_A^{\gamma,Z} F_{1A}^{\gamma,Z}\gamma_5\right] \\ + \frac{ie}{2m_t}\sigma^{\mu\nu}k_\nu \left[Q_V^{\gamma,Z} F_{2V}^{\gamma,Z} + Q_A^{\gamma,Z} F_{2A}^{\gamma,Z}\gamma_5\right],$$

which reduces to the SM tree level expression with the form factors $F_{1V}^\gamma = F_{1V}^Z = F_{1A}^Z = 1$ and the others zero. The quantities $Q_{A,V}^{\gamma,Z}$ are the SM coupling

Form Factor	SM Value (Lowest Order)	Limit 68% CL	Limit 90% CL
$F_{1R}^W(P=0)$	0	± 0.13	± 0.18
$F_{1R}^W(P=80\%)$	0	± 0.06	± 0.10
F_{1A}^Z	1	1 ± 0.08	1 ± 0.13
F_{1V}^Z	1	1 ± 0.10	1 ± 0.16
F_{2A}^γ	0	± 0.05	± 0.08
F_{2V}^γ	0	± 0.07	$^{+0.13}_{-0.11}$
F_{2A}^Z	0	± 0.09	± 0.15
F_{2V}^Z	0	± 0.07	± 0.10
$\Im(F_{2A}^Z)$	0	± 0.06	± 0.09

TABLE 1. The upper and lower limits of the couplings in their departures from the SM values are given at 68% and 90% CL. All couplings, each with real and imaginary parts, can be determined in this way. The right-handed charged-current coupling is shown both for unpolarized and 80% left-polarized electron beam, whereas the other results assume 80% left-polarized beam only. $\Im$ is the imaginary part, otherwise the results listed here are for the real parts.

constants: $Q_V^\gamma = Q_A^\gamma = \frac{2}{3}$, $Q_V^Z = (1 - \frac{8}{3}\sin^2\theta_W)/(4\sin\theta_W\cos\theta_W)$, and $Q_A^Z = -1/(4\sin\theta_W\cos\theta_W)$. The non-standard couplings $F_{2V}^{\gamma,Z}$ and $F_{2A}^{\gamma,Z}$ correspond to electroweak magnetic and electric dipole moments, respectively. For the top charged-current coupling we can write the W-t-b vertex factor as

$$\mathcal{M}^{\mu,W} = \frac{g}{\sqrt{2}}\gamma^\mu \left[P_L F_{1L}^W + P_R F_{1R}^W\right]$$
$$+ \frac{ig}{2\sqrt{2}\,m_t}\sigma^{\mu\nu}k_\nu \left[P_L F_{2L}^W + P_R F_{2R}^W\right],$$

where the quantities $P_{L,R}$ are the left-right projectors. In the SM we have $F_{1L}^W = 1$ and all others zero. The form factor F_{1R}^W represents a right-handed, or $V+A$, charged current component. Using an integrated luminosity of 10 fb^{-1}, and the $t\bar{t}$ decay mode final state of lepton + jets, and an electron polarization of $\pm 80\%$, it can be shown [18,19] that the form factors given above can be determined to a precision shown in table 1 at a center of mass energy of 500 GeV.

CONCLUSIONS

The e^+e^- linear colliders (and the muon collider) offer an excellent means of probing the physics of the $t\bar{t}$ threshold and extracting fundamental parameters such as m_t, Γ_t, β_H as well as the form factors governing top decay to precisions difficult to attain using hadron machines.

REFERENCES

1. C. Adolphsen, et al., "Zeroth-Order Design for the NLC", SLAC Report 474, May 1996 may be found at http://www.slac.stanford.edu/accel/nlc/zdr/
2. JLC-I, KEK report 92-16 can be found at http://www-jlc.kek.jp/JLC.proposal-e.html
3. Top quark Physics: Future measurements, R.Frey et al, Fermilab-CONF-97-085, Apr 1997, LANL e-print Archive hep-ph/9704243, Proceedings of the 1996 DPF/DPB Summer Study on New directions for High Energy Physics, Snowmass Colorado.
4. M. Strassler and M. Peskin, Phys. Rev. **D43**, 1500 (1991).
5. V. Fadin and V. Khoze, JETP Lett. **46**, 525 (1987) and Sov. J. Nucl. Phys. **48**, 309 (1988).
6. M. Jezabek, J. Kuhn, and T. Teubner, Z. Phys. **C56**, 653 (1992).
7. M. Jezabek and T. Teubner, Z. Phys. **C59**, 669 (1993).
8. Y. Sumino, K. Fujii, K. Hagiwara, H. Murayama, and C.-K. Ng, Phys. Rev. **D47**, 56 (1993).
9. K. Fujii, T. Matsui, and Y. Sumino, Phys. Rev. **D50**, 4341 (1994).
10. E.A. Kuraev and V.S. Fadin, Sov. J. Nucl. Phys. **41**, 466 (1985).
11. Pisin Chen, Phys. Rev. **D46**, 1186 (1992).
12. " The $t\bar{t}$ threshold at Muon Colliders", M.Berger, these proceedings.
13. K. Fujii, proceedings of the 1995 SLAC Summer Institute.
14. P. Comas, R. Miquel, M. Martinez, and S. Orteu, "Recent Studies on Top Quark Physics at NLC", proceedings of the Workshop on Physics and Experiments with Linear Colliders (LCWS95), 1995.
15. P. Igo-Kemenes, M. Martinez, R. Miquel, and S. Orteu, proceedings of the Workshop on Physics and Experiments with Linear Colliders (LCWS93), Waikoloa, Hawaii, USA, 1993.
16. R. Harlander, M. Jezabek, and J.H. Kuhn, Acta. Phys. Polon. **27**, 1781 (1996), hep-ph/9506292 (1995).
17. H. Murayama and Y. Sumino, Phys. Rev. **D47**, 82 (1993).
18. R. Frey, "Top Quark Physics at a Future e^+e^- Linear Collider: Experimental Aspects," proceedings of the Workshop on Physics and Experiments with Linear Colliders (LCWS95), Iwate, Japan, Sept., 1995; hep-ph/9606201 (1996).
19. M. Fero, Proceedings of the 1996 DPF/DPB Summer Study on New directions for High Energy Physics, Snowmass Colorado.

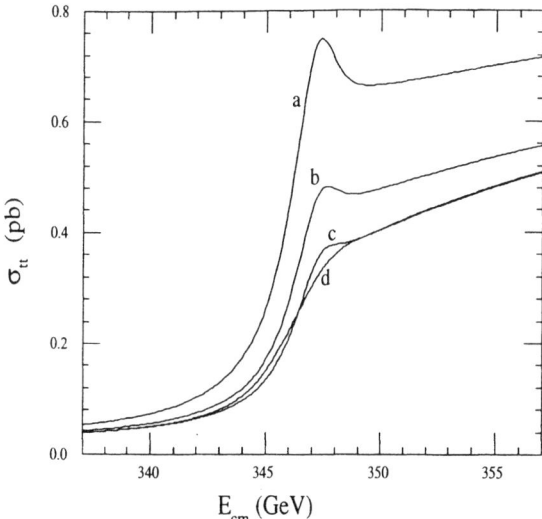

FIGURE 1. Production cross section for top-quark pairs near threshold for $m_t = 175$ GeV^2. The theoretical cross section is given by curve (a). The following energy redistribution effects have been applied to the theory for the remaining curves: (b) initial-state radiation (ISR); (c): ISR and beamstrahlung; (d): ISR, beamstrahlung, and single-beam energy spread.

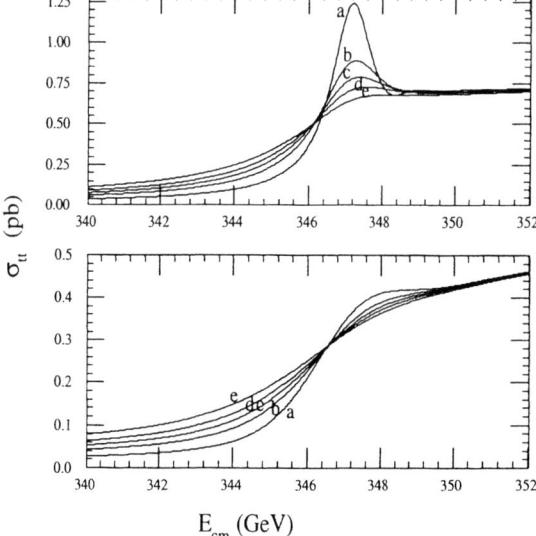

FIGURE 2. Threshold shape for various values of Γ_t. The upper plot is the theoretical prediction, while the lower plot includes all radiative and beam effects. The different curves correspond to Γ_t/Γ_t^{SM} = (a) 0.5, (b) 0.8, (c) 1.0, (d) 1.2, and (e) 1.5. We assumed $m_t = 175$ GeV, where the Standard Model width is $\Gamma_t^{SM} = 1.42$ GeV.

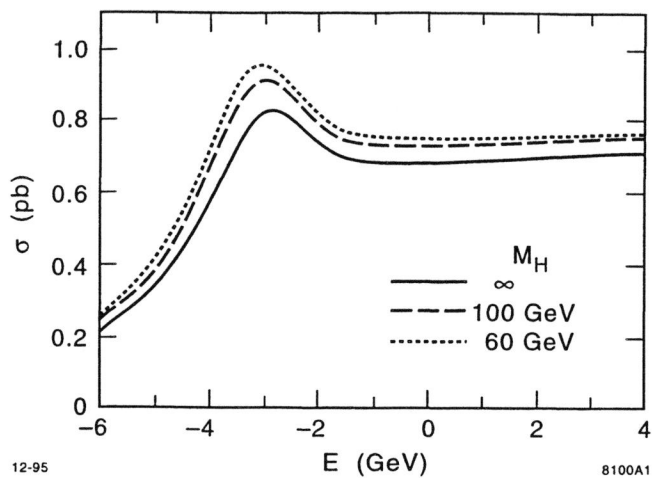

FIGURE 3. Cross section near threshold for different Higgs masses due to the Yukawa potential. $m_t = 180$ GeV/c^2 was assumed. The abscissa center-of-mass energy is relative to $2m_t$.

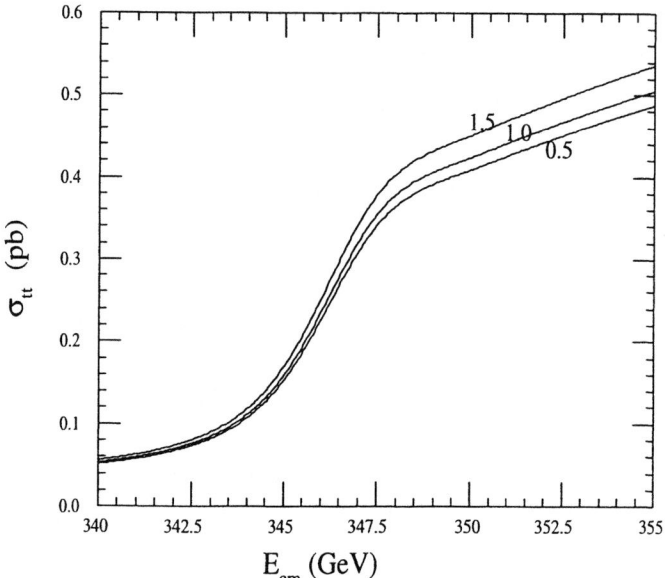

FIGURE 4. Threshold shape for various (real) values of the Yukawa coupling strength β_H. All radiative and beam effects are included, and $m_t = 175$ GeV, $m_H = 300$ GeV are used. The different curves corresond to $\beta_H = 1.5$, 1.0, and 0.5, as indicated.

The Top-Antitop Threshold at Muon Colliders

M. S. Berger

Indiana University
Bloomington Indiana 47405

Abstract. Muon colliders are expected to naturally have a small spread in beam energy making them an ideal place to study the excitation curve. We present the parameter determinations that are possible from measuring the total cross section near threshold at a $\mu^+\mu^-$ collider.

INTRODUCTION

Accurate measurements of particles masses, couplings and widths are possible by measuring production cross sections near threshold. The naturally small beam energy spread of a muon collider would provide an excellent opportunity to make these measurements. Pair production of W-bosons, $t\bar{t}$ production and the Bjorken process $\mu^+\mu^- \to ZH$ have been considered as possible places to study thresholds at a muon collider [1–3]. There is very rich physics associated with the $t\bar{t}$ threshold, including the determination of m_t, Γ_t ($|V_{tb}|$), α_s, and possibly m_h [4]. A precise value of the top-quark mass m_t could prove to be very valuable in theoretical studies.

TOP-QUARK MASS MEASUREMENT AT THE $\mu^+\mu^- \to t\bar{t}$ THRESHOLD

Fadin and Khoze first demonstrated that the top-quark threshold cross section is calculable since the large top-quark mass puts one in the perturbative regime of QCD, and the large top-quark width effectively screens nonperturbative effects in the final state [5]. Such studies have since been performed by several groups [6–13]. The phenomenological potential is given at small distance r by two-loop perturbative QCD and for large r by a fit to quarkonia spectra. In our analysis we make use of the Wisconsin potential [14] that interpolates these regimes.

The beam energy spread at a $\mu^+\mu^-$ collider is expected to naturally be small. The rms deviation σ in $\sqrt{s}$ is given by [15,16]

$$\sigma = (250 \text{ MeV}) \left(\frac{R}{0.1\%}\right) \left(\frac{\sqrt{s}}{350 \text{ GeV}}\right), \quad (1)$$

where R is the rms deviation of the Gaussian beam profile. With $R \lesssim 0.1\%$ the resolution σ is of the same order as the measurement one hopes to make in the top mass. For $t\bar{t}$ studies the exact shape of the beam is not important if $R \lesssim 0.1\%$. We take $R = 0.1\%$ here; the results are not improved significantly with better resolution[1].

Changing the value of the strong coupling constant $\alpha_s(M_Z)$ influences the threshold region. Large values lead to tighter binding and the peak shifts to lower values of $\sqrt{s}$. Weaker coupling also smooths out the threshold peak. These effects are illustrated in Fig. 1.

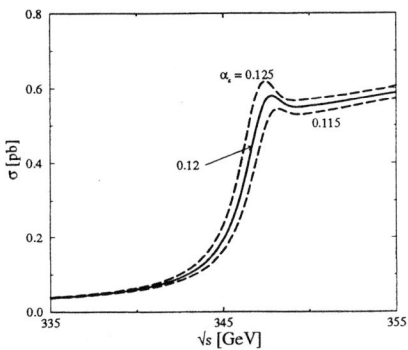

FIGURE 1. The cross section for $\mu^+\mu^- \to t\bar{t}$ production in the threshold region, for $m_t = 175$ GeV and $\alpha_s(M_Z) = 0.12$ (solid) and 0.115, 0.125 (dashes). Effects of ISR and beam smearing are included.

To assess the precision of parameter determinations from cross section measurements, we generate hypothetical sample data, shown in Fig. 2, assuming that 10 fb^{-1} integrated luminosity is used to measure the cross section at each energy in 1 GeV intervals. Since the top threshold curve depends on other quantities like $\alpha_s(M_Z)$, one must do a full scan to determine the shape of the curve and its overall normalization. To generate the ten data points in Fig. 2 we use nominal values of $m_t = 175$ GeV and $\alpha_s(M_Z) = 0.12$. Following Ref. [13] we assume a 29% detection efficiency for $W \to q\bar{q}$, including the decay branching fraction. The data points can then be fit to theoretical predictions for different values of m_t and $\alpha_s(M_Z)$; the likelihood fit that is obtained is shown as the $\Delta\chi^2$ contour plot in Fig. 3. The inner and

---

[1] The most recent TESLA design envisions a beam energy spread of $R = 0.2\%$ [17], and a high energy e^+e^- collider in the large VLHC tunnel would have a beam spread of $\sigma_E = 0.26$ GeV [18]

outer curves are the $\Delta\chi^2 = 1.0$ (68.3%) and 4.0 (95.4%) confidence levels respectively for the full 100 fb^{-1} integrated luminosity. Projecting the $\Delta\chi^2 = 1.0$ ellipse on the m_t axis, the top-quark mass can be determined to within $\Delta m_t \sim 70$ MeV, provided systematics are under control. (Systematic error issues will be discussed later.) A top-quark mass of 175 GeV can be measured to about 200 MeV at 90% confidence level with 10 fb^{-1} luminosity.

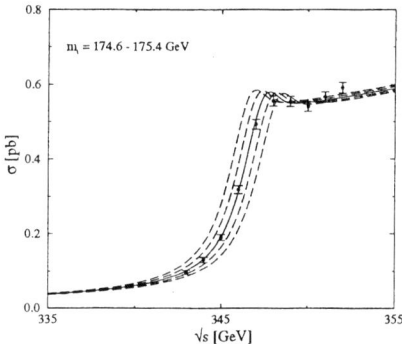

FIGURE 2. Sample data for $\mu^+\mu^- \to t\bar{t}$ obtained assuming a scan over the threshold region devoting 10 fb^{-1} luminosity to each data point. A detection efficiency of 29% has been assumed [13] in obtaining the error bars. The threshold curves correspond to shifts in m_t of 200 MeV increments. Effects of ISR and beam smearing have been included, and the strong coupling $\alpha_s(M_Z)$ is taken to be 0.12.

Since the exchange of a light Higgs boson can affect the threshold shape, a scan of the threshold cross section can in principle yield some information about the Higgs mass and its Yukawa coupling to the top quark. Figure 4 shows the dependence of the threshold curve on the Higgs mass, m_h. However, it may be difficult to disentangle such a Higgs effect from two-loop QCD effects, which are not yet fully calculated [19].

QCD measurements at future colliders and lattice calculations will presumably determine $\alpha_s(M_Z)$ to 1% accuracy (e.g. ± 0.001) [20] by the time muon colliders are constructed so the uncertainty in α_s will likely be similar to the precision obtainable at a $\mu^+\mu^-$ and/or e^+e^- collider with 100 fb^{-1} integrated luminosity. If the luminosity available for the threshold measurement is significantly less than 100 fb^{-1}, one can regard the value of $\alpha_s(M_Z)$ coming from other sources as an input, and thereby improve the top-quark mass determination.

There is some theoretical ambiguity in the mass definition of the top quark. The theoretical uncertainty on the quark pole mass due to QCD confinement effects is of order Λ_{QCD}, i.e., a few hundred MeV [21]. In the $\overline{MS}$ scheme of quark mass definition, the theoretical uncertainty is better controlled.

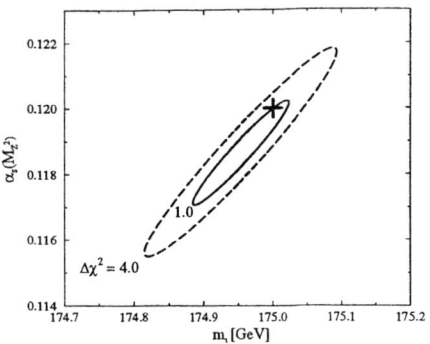

FIGURE 3. The $\Delta\chi^2 = 1.0$ and $\Delta\chi^2 = 4.0$ confidence limits for the sample data shown in Fig. 2. The "+" marks the input values from which the data were generated.

Systematic errors in experimental efficiencies are not a significant problem for the $t\bar{t}$ threshold determination of m_t. This can be seen from Fig. 2, which shows that a 200 MeV shift in m_t corresponds to nearly a 10% shift in the cross section on the steeply rising part of the threshold scan, whereas it results in almost no change in σ once $\sqrt{s}$ is above the peak by a few GeV. Not only will efficiencies be known to much better than 10%, but also systematic uncertainties will cancel to a high level of accuracy in the ratio of the cross section measured above the peak to measurements on the steeply-rising part of the threshold curve.

As Fig. 4 shows, it will be important to know the Higgs mass and the $ht\bar{t}$ coupling strength in order to eliminate this source of systematic uncertainty when extracting other quantities.

The measurements described in this section can be performed at either an e^+e^- or a $\mu^+\mu^-$ collider. The errors for m_t that we have found for the muon collider are smaller than those previously obtained in studies at the NLC electron collider primarily because the smearing of the threshold region by the energy spread of the beam is much less, and secondarily due to the fact that the reduced amount of initial state radiation makes the cross section somewhat larger.

CONCLUSION

With an integrated luminosity of 10 (100) fb^{-1}, the top-quark mass can be measured to 200 (70) MeV, using a 10-point scan over the threshold region, in 1 GeV intervals, to measure the shape predicted by the QCD potential. In the $t\bar{t}$ threshold study, differences of cross sections at energies below, at, and above the resonance peak, along with the location of the resonance peak, have different dependencies on the parameters m_t, α_s, m_h and $|V_{tb}|^2$ and should allow their determination.

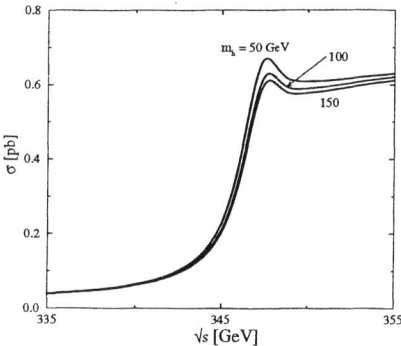

FIGURE 4. The dependence of the threshold region on the Higgs mass, for $m_h = 50, 100, 150$ GeV. Effects of ISR and beam smearing have been included, and we have assumed $m_t = 175$ GeV and $\alpha_s(M_Z) = 0.12$.

To utilize the highest precision measurements achievable at the statistical level, theoretical uncertainties and other systematics need to be under control. We are confident that uncertainty in α_s will not be a factor and we have noted that ratios of above-peak measurements to measurements on the steeply rising part of the threshold cross section will eliminate many experimental systematics related to uncertainties in efficiencies.

ACKNOWLEDGMENTS

I thank V. Barger, J. F. Gunion and T. Han for a pleasant collaboration on the issues reported here. This work was supported in part by the U.S. Department of Energy under Grant No. DE-FG02-91ER40661.

REFERENCES

1. V. Barger, M.S. Berger, J.F. Gunion and T. Han, Phys. Rev. **D56**, 1714 (1997).
2. M.S. Berger, talk presented at the *Workshop on Particle Theory and Phenomenology: Physics of the Top Quark*, Iowa State University, May 25–26, 1995, hep-ph/9508209.
3. V. Barger, M. S. Berger, J. F. Gunion and T. Han, Phys. Rev. Lett. **78**, 3991 (1997).
4. For a recent on the top-quark physics near the threshold, see *e.g.*, J.H. Kuhn, TTP-96-18, lectures delivered at SLAC Summer Institute, Stanford, July, 1995.
5. V.S. Fadin and V.A. Khoze, JETP Lett. **46**, 525 (1987); Sov. J. Nucl. Phys. **48**, 309 (1988).
6. J. Feigenbaum, Phys. Rev. **D43**, 264 (1991).

7. W. Kwong, Phys. Rev. **D43**, 1488 (1991).
8. M. Strassler and M. Peskin, Phys. Rev. **D43**, 1500 (1991).
9. M. Jezabek, J.H. Kuhn and T. Teubner, Z. Phys. **C56**, 653 (1992); M. Jezabek and T. Teubner, Z. Phys. **C59**, 669 (1993); M. Jezabek, talk presented at *DESY-Zeuthen Workshop on Elementary Particle Theory: "Physics at LEP200 and Beyond"*, Teupitz, Germany, April 1994 (hep-ph/9406411); M. Jezabek, Acta Phys. Pol. **B26**, 789 (1995); J.H. Kuhn, Acta Phys. Pol. **B26**, 711 (1995).
10. G. Bagliesi, et al., *Workshops on Future e^+e^- Linear Colliders*, Hamburg, Germany and Saariselka, Finland, Sep 2–3 and Sep 9–11, 1991, CERN-PPE/92-05.
11. Y. Sumino, K. Fujii, K. Hagiwara, H. Murayama and C.-K. Ng, Phys. Rev. **D47**, 56 (1993); H. Murayama and Y. Sumino, Phys. Rev. **D47**, 82 (1993); Y. Sumino, Acta Phys. Pol. **B25**, 1837 (1994).
12. P. Igo-Kemenes, M. Martinez, R. Miquel and S. Orteu, CERN-PPE/93-200, Contribution to *the Workshop on Physics with Linear e^+e^- Colliders at 500 GeV*.
13. K. Fujii, T. Matsui and Y. Sumino, Phys. Rev. **D50**, 4341 (1994).
14. K. Hagiwara, S. Jacobs, M. G. Olsson and K. J. Miller, Phys. Lett. **130B**, 209 (1983).
15. V. Barger, M.S. Berger, J.F. Gunion and T. Han, Phys. Rev. Lett. **75**, 1462 (1995).
16. V. Barger, M.S. Berger, J.F. Gunion and T. Han, Phys. Reports **286**, 1 (1997).
17. D. Miller, private communication.
18. J. Norem, private communication and http://www-ap.fnal.gov/VLHC/electrons/index.html.
19. A. H. Hoang, Phys. Rev. **D56**, 5851 (1997); Phys. Rev. **D56**, 7276 (1997); and these proceedings.
20. P. N. Burrows et al., SLAC-PUB-7371, to appear in *Proceedings of 1996 DPF/DPB Summer Study on New Directions for High-Energy Physics* (Snowmass 96), hep-ex/9612012.
21. M. C. Smith and S. Willenbrock, hep-ph/9612329.

Top Quark Pair Production at Threshold – Uncertainties and Relativistic Corrections

André H. Hoang

*Department of Physics, University of California, San Diego,
La Jolla, CA 92093-0319, USA*

Abstract. In this talk it is shown how nonrelativistic QCD (NRQCD) can be used to determine next-to-next-to-leading order relativistic and short-distance contributions to the total $t\bar{t}$ production cross section in the threshold regime at lepton colliders. A recipe for the calculation of all such contributions for the total photon mediated production cross section is presented and a review of the already known Abelian next-to-next-to-leading results is given.

INTRODUCTION

The production of $t\bar{t}$ pairs in the threshold region at future lepton colliders like the NLC (Next Linear Collider) or the FMC (First Muon Collider) offers a unique opportunity to carry out precision tests of QCD in a completely new environment. Due to the large top mass[1] ($M_t \approx 175$ GeV), which allows for the decay channel $t \to Wb$, hadronization effects can be neglected in a first approximation [1]. This makes the $t\bar{t}$ production cross section in the threshold regime (including various distributions) calculable from perturbative QCD (and electroweak interactions), which then allows for precise extractions of the top quark mass and the strong coupling once the cross section is measured. In fact, experimental simulations for the NLC and the FMC [2,3] have shown that experimental errors of around 100 MeV for the top quark mass and of around 0.002 for $\alpha_s(M_z)$ can be expected for a cross section measurement with a total integrated luminosity of $50 - 100 fb^{-1}$. In particular the prospect for the error in the top quark mass measurement beats any hadron collider experiment. However, the errors for M_t and α_s given above do not contain any theoretical uncertainties. At this point it is illustrative to recall that the

---
[1] Throughout this talk M_t is understood as the top quark pole mass.

standard present day formalism used for describing $t\bar{t}$ production at threshold consists of solving the nonrelativistic Schrödinger equation with a QCD potential which for small distances is given by perturbative QCD up to one loop [4,5] and for large and intermediate distances by fits to quarkonia spectra (and leptonic decay widths) [6,7]. The results are then modified by various $\mathcal{O}(\alpha_s)$ short-distance corrections which makes the results correct at the next-to-leading order (NLO) level, i.e., they properly include all $\mathcal{O}(\alpha_s)$ corrections.[2] NNLO ($\mathcal{O}(\alpha_s^2)$) corrections have never been taken into account so far. Their contributions, however, can be sizable. As an example, consider the $\mathcal{O}(\alpha_s^2)$ relativistic corrections to the total cross sections which can lead to a shift in the location of the 1S peak of order $M_t \alpha_s^4 \sim 150$ MeV and a corrections of order $\alpha_s^2 \sim 3\%$ in the size of the cross section (for $\alpha_s \sim \alpha_s(M_t \alpha_s) \sim 0.17$). Further, even the $\mathcal{O}(\alpha_s^2)$ short-distance corrections normalizing the total cross sections might be large if the huge size of the $\mathcal{O}(\alpha_s)$ corrections of order -20% is taken into account. From this point of view it is clear that the theoretical uncertainties in the present day analyses are certainly not negligible and that full control over all NNLO effects should be gained.

Unfortunately the formalism described above is constructed in a way that makes a systematic and rigorous implementation of all NNLO effects from first principles QCD conceptually difficult if not impossible – a consequence of the use of phenomenological information in the potential for large and intermediate distances. In principle, this formalism has to be considered as a (sophisticated) potential model approach which cannot be improved in a rigorous way at all. I therefore propose to rely on perturbative QCD only. This means that one applies the perturbative QCD potential for all distances. In fact, such a decision seems to be just natural if one takes into account that the $t\bar{t}$ system is almost insensitive to the large distance (i.e. non-perturbative) contributions in the QCD potential. This makes the framework in which effects beyond the NLO level shall be determined more transparent and still leaves the possibility to incorporate the non-perturbative effects later as a perturbation (see e.g. [8] for an approach of this sort).

In this talk I demonstrate how NRQCD [9] can be used to determine NNLO relativistic and short-distance corrections to the total $t\bar{t}$ production cross section. For simplicity only the production through a virtual photon is considered. The generalization for production through different currents is straightforward. I want to stress that I do not talk about the peculiar NNLO finite width effects coming from the off-shellness of the top quark, time dilatation effects and the interaction among the decay products. These effects have been addressed previously in a number of publications [10–12], but no rigorous and

---

[2] The solutions of the nonrelativistic Schrödinger equation with the one-loop corrected QCD potential contains, in the language of Feynman diagrams, the resummation of terms $\propto (\alpha_s/v)^n \times [1, \alpha_s]$, $n = 0, 1, \ldots, \infty$, (v being the top quark c.m. velocity) to all orders in α_s. Because in the threshold region $|v| \lesssim \alpha_s$ we count all terms $\propto \alpha_s/v$ of order one.

consistent description of them for the $t\bar{t}$ cross section has been found yet. To achieve full NNLO accuracy these finite width effects should eventually be included. For now they remain as an unsolved problem. As far as the NNLO relativistic corrections discussed in this talk are concerned I will use the naive replacement

$$E \equiv \sqrt{s} - 2M_t \longrightarrow \tilde{E} = E + i\Gamma_t \qquad (1)$$

in the spirit of [1] in order to examine their size and properties, where Γ_t represents a constant which is not necessarily the decay width of a free top quark.

THE $t\bar{t}$ CROSS SECTION IN NRQCD

NRQCD is an effective field theory of QCD designed to handle nonrelativistic systems of heavy quark–antiquark pairs to in principle arbitrary precision. It is based on the separation of long- and short-distance effects[3] by reformulating QCD in terms of a non-renormalizable Lagrangian containing all possible operators in accordance to the symmetries in the nonrelativistic limit. The NRQCD Lagrangian reads

$$\mathcal{L}_{\text{NRQCD}} = -\frac{1}{2}\text{Tr}\,G^{\mu\nu}G_{\mu\nu} + \sum_{q=u,d,s,c,b} \bar{q}\,i\slashed{D}\,q$$

$$+ \psi^\dagger \Bigg[iD_t + a_1\,\frac{\boldsymbol{D}^2}{2M_t} + a_2\,\frac{\boldsymbol{D}^4}{8M_t^3} + \frac{a_3\,g}{2M_t}\,\boldsymbol{\sigma}\cdot\boldsymbol{B}$$

$$+ \frac{a_4\,g}{8M_t^2}(\boldsymbol{D}\cdot\boldsymbol{E} - \boldsymbol{E}\cdot\boldsymbol{D}) + \frac{a_5\,g}{8M_t^2}\,i\boldsymbol{\sigma}\cdot(\boldsymbol{D}\times\boldsymbol{E} - \boldsymbol{E}\times\boldsymbol{D})\Bigg]\psi + \ldots \qquad (2)$$

The gluonic and light quark degrees of freedom are described by the conventional relativistic Lagrangian, whereas the top and antitop quarks are described by the Pauli spinors ψ and χ, respectively. For convenience all color indices are suppressed. The straightforward antitop bilinears are omitted and only those terms relevant for the NNLO cross section are displayed. D_t and $\boldsymbol{D}$ are the time and space components of the gauge covariant derivative D_μ and $E^i = G^{0i}$ and $B^i = \frac{1}{2}\epsilon^{ijk}G^{jk}$ the electric and magnetic components of the gluon field strength tensor (in Coulomb gauge). The short-distance coefficients $a_1,\ldots,a_5$ are normalized to one at the Born level.

To formulate the normalized total $t\bar{t}$ production cross section (via a virtual photon) $R = \sigma(^{e^+e^-}_{\mu^+\mu^-} \to \gamma^* \to t\bar{t})/\sigma_{pt}$ ($\sigma_{pt} = 4\pi\alpha^2/3s$) in the nonrelativistic region at NNLO in NRQCD we start from the fully covariant expression for the cross section

---

[3] In this context "long-distance" is not equivalent to "non-perturbative".

$$R(q^2) = \frac{4\pi Q_t^2}{s} \mathrm{Im}[\langle 0 | T \tilde{j}_\mu(q) \tilde{j}^\mu(-q) | 0 \rangle], \tag{3}$$

where $Q_t = 2/3$ is the electric charge of the top quark. We then expand the electromagnetic current (in momentum space) $\tilde{j}_\mu(\pm q) = (\bar{t}\gamma^\mu t)(\pm q)$ which produces/annihilates a $t\bar{t}$ pair with c.m. energy $\sqrt{q^2}$ in terms of $^3S_1$ NRQCD currents up to dimension eight $(i = 1, 2, 3)$[4]

$$\tilde{j}_i(q) = b_1 \left(\tilde{\psi}^\dagger \sigma_i \tilde{\chi}\right)(q) - \frac{b_2}{6M_t^2} \left(\tilde{\psi}^\dagger \sigma_i (-\tfrac{i}{2}\overleftrightarrow{\boldsymbol{D}})^2 \tilde{\chi}\right)(q) + \ldots, \tag{4}$$

where the constants b_1 and b_2 are short-distance coefficients normalized to one at the Born level. Inserting expansion (4) back into Eq. (3) leads to the NRQCD expression of the nonrelativistic cross section at the NNLO level

$$R_{\mathrm{NNLO}}^{\mathrm{thr}}(\tilde{E}) = \frac{\pi Q_t^2}{M_t^2} C_1(\mu_{\mathrm{hard}}, \mu_{\mathrm{fac}}) \mathrm{Im}\left[\mathcal{A}_1(\tilde{E}, \mu_{\mathrm{soft}}, \mu_{\mathrm{fac}})\right]$$

$$- \frac{4\pi Q_t^2}{3M_t^4} C_2(\mu_{\mathrm{hard}}, \mu_{\mathrm{fac}}) \mathrm{Im}\left[\mathcal{A}_2(\tilde{E}, \mu_{\mathrm{soft}}, \mu_{\mathrm{fac}})\right] + \ldots, \tag{5}$$

where

$$\mathcal{A}_1 = \langle 0 | (\tilde{\psi}^\dagger \vec{\sigma} \tilde{\chi})(\tilde{\chi}^\dagger \vec{\sigma} \tilde{\psi}) | 0 \rangle, \tag{6}$$

$$\mathcal{A}_2 = \tfrac{1}{2} \langle 0 | (\tilde{\psi}^\dagger \vec{\sigma} \tilde{\chi})(\tilde{\chi}^\dagger \vec{\sigma} (-\tfrac{i}{2}\overleftrightarrow{\boldsymbol{D}})^2 \tilde{\psi}) + \mathrm{h.c.} | 0 \rangle. \tag{7}$$

The cross section is expanded in terms of a sum of absorptive parts of nonrelativistic current-current correlators (containing long-distance physics) multiplied by short-distance coefficients C_i $(i = 1, 2, \ldots)$. In Eq. (5) I have also shown the dependences on the various renormalization scales: the soft scale μ_{soft} and the hard scale μ_{hard} are governing the perturbative expansions of the correlators and the short-distance coefficients[5], respectively, whereas the factorization scale μ_{fac} essentially represents the boundary between hard (i.e. of order M_t) and soft momenta. Because this boundary can in principle be chosen freely, both, correlators and the short-distance coefficients, in general depend on it (leading to new anomalous dimensions). Because the term in the second line in Eq. (5) is already of NNLO (i.e. suppressed by v^2) we can set $C_2 = 1$ and ignore the factorization scale dependence of the correlator $\mathcal{A}_2$. The calculation of all terms in expression (5) proceeds in two basic steps.

---

[4] Only the spatial components of the current contribute. The expansion of $\tilde{j}_\mu(-q)$ is obtained from Eq. (4) via charge conjugation symmetry.

[5] The scales μ_{soft} and μ_{hard} arise from the light degrees of freedom in $\mathcal{L}_{\mathrm{NRQCD}}$ and are already present in full QCD, whereas μ_{fac} is generated by "new" UV divergences in NRQCD diagrams. It is crucial to consider μ_{soft} and μ_{hard} as independent scales. Because both, nonrelativistic correlators and short-distance coefficients, are calculated perturbatively a residual dependence on μ_{soft} and μ_{hard} remains.

1. *Calculation of the nonrelativistic correlators.* – Determination of the correlators in Eq. (5) by taking into account the interactions up to NNLO displayed in $\mathcal{L}_{\text{NRQCD}}$.

2. *Matching calculation.* – Calculation of the constant C_1 up to $\mathcal{O}(\alpha_s^2)$ by matching expression (5) to the fully covariant cross section at the two-loop level in the (formal) limit $\alpha_s \ll v \ll 1$ where an expansion in (first) α_s and (then) v is feasible.

Calculation of the nonrelativistic correlators: In Coulomb gauge the gluon propagation is separated into a *longitudinal*, instantaneous (i.e. energy-independent) and a *transverse*, non-instantaneous (i.e. energy-dependent) propagation. The longitudinal exchange between the $t\bar{t}$ pair is described by an instantaneous potential. (The Coulomb potential is just the LO interaction caused by the longitudinal exchange.) The transverse exchange, however, leads to a temporally retarded interaction, closely related to Lamb-shift type effects known in QED. Fortunately, because the $t\bar{t}$ pair is produced/annihilated in a color singlet (S-wave) configuration, the energy dependence of the transverse gluon exchange leads only to NNNLO (i.e. $\mathcal{O}(\alpha_s^3)$ relative to the effects of the LO Coulomb exchange) contributions.[6] We therefore ignore the energy dependence of the transverse gluons which allows us to formulate all interactions contained in the NRQCD Lagrangian in terms of instantaneous potentials. In other words, as far as the nonrelativistic correlators at NNLO in Eq. (5) are concerned, NRQCD reduces to a two-body (top-antitop) Schrödinger theory. The potential in the resulting Schrödinger equation is determined by considering $t\bar{t} \to t\bar{t}$ one gluon exchange t-channel scattering amplitudes in NRQCD. To NNLO (i.e. including potentials suppressed by at most α_s^2, α_s/M_t or $1/M_t^2$ relative to the Coulomb potential) all potentials are already known and read ($a \equiv C_F \alpha_s(\mu_{\text{soft}})$, $C_F = 4/3$, $C_A = N_c = 3$, $r \equiv |\vec{r}|$)

$$V_{\text{BF}}(\vec{r}) = \frac{a\pi}{M_t^2}\left[1 + \frac{8}{3}\vec{S}_t\vec{S}_{\bar{t}}\right]\delta^{(3)}(\vec{r}) + \frac{a}{2M_t^2 r}\left[\vec{\nabla}^2 + \frac{1}{r^2}\vec{r}(\vec{r}\vec{\nabla})\vec{\nabla}\right]$$

$$- \frac{3a}{M_t^2 r^3}\left[\frac{1}{3}\vec{S}_t\vec{S}_{\bar{t}} - \frac{1}{r^2}(\vec{S}_t\vec{r})(\vec{S}_{\bar{t}}\vec{r})\right] + \frac{3a}{2M_t^2 r^3}\vec{L}(\vec{S}_t + \vec{S}_{\bar{t}}), \quad (8)$$

$$V_{\text{NA}}(\vec{r}) = -\frac{C_A}{C_F}\frac{a^2}{2M_t r^2}, \quad (9)$$

---

[6] This can be seen by either using formal counting rules (see e.g. [13,14]) or from positronium results where this phenomenon is well known. From the physical point of view the suppression comes from the fact that the transverse gluon is radiated after the $t\bar{t}$ pair is produced and absorbed before the $t\bar{t}$ pair is annihilated. This process is already suppressed by v^2 due to the dipole matrix element for transverse gluon radiation/absorption. If the gluon also carries energy, another (phase space) factor v arises because the gluon essentially becomes real. For the same reason no top quark self energy or crossed ladder diagrams have to be considered.

where $\vec{S}_t$ and $\vec{S}_{\bar{t}}$ are the top and antitop spin operators and $\vec{L}$ is the angular momentum operator. $V_{\rm BF}$ is the Breit-Fermi potential known from positronium and $V_{\rm NA}$ a purely non-Abelian potential generated through non-analytic terms in one-loop NRQCD (or QCD) diagrams containing the triple gluon vertex (see e.g. [15] for an older reference). The Coulomb potential $V_c(\vec{r}) = -a/r\,[1+$ corrections up to $\mathcal{O}(\alpha_s^2)]$ is not displayed due to lack of space. Its $\mathcal{O}(\alpha_s)$ and $\mathcal{O}(\alpha_s^2)$ corrections have been determined in [4,5] and [16], respectively.

The nonrelativistic correlators are directly related to the Green function of the Schrödinger equation

$$\left(-\frac{\vec{\nabla}^2}{M_t} - \frac{\vec{\nabla}^4}{4M_t^3} + V_c(\vec{r}) + V_{\rm BF}(\vec{r}) + V_{\rm NA}(\vec{r}) - \tilde{E}\right) G(\vec{r},\vec{r}',\tilde{E})$$
$$= \delta^{(3)}(\vec{r}-\vec{r}'), \qquad (10)$$

where $V_{\rm BF}$ is evaluated for the $^3S_1$ configuration only. The correlators read

$$\mathcal{A}_1 = 6\,N_c\left[\lim_{|\vec{r}|,|\vec{r}'|\to 0} G(\vec{r},\vec{r}',\tilde{E})\right], \qquad (11)$$

$$\mathcal{A}_2 = M_t\,\tilde{E}\,\mathcal{A}_1. \qquad (12)$$

Relation (11) can be easily inferred by taking into account that the Green function $G(\vec{r},\vec{r}',\tilde{E})$ describes the propagation of a top-antitop pair which is produced and annihilated at distances $|\vec{r}|$ and $|\vec{r}'|$, respectively. (A more formal derivation can be found e.g. in [6].) Relation (12), on the other hand, is obtained through the equation of motion (10). Because the exact solution of Eq. (10) seems to be an impossible task and we start with the well-known Coulomb Green function $G_c^{(0)}$ [17] ($V_c^{(0)}(\vec{r}) \equiv -a/r$),

$$\left(-\frac{\nabla^2}{M_t} - V_c^{(0)}(\vec{r}) - \tilde{E}\right) G_c(\vec{r},\vec{r}',\tilde{E}) = \delta^{(3)}(\vec{r}-\vec{r}'), \qquad (13)$$

and incorporate the higher order terms using Rayleigh-Schrödinger time-independent perturbation theory (TIPT). It should be noted that the limit $|\vec{r}|,|\vec{r}'| \to 0$ causes UV divergences which have to be regularized using either an explicit short-distance cutoff or dimensional regularization. The freedom in the choice of the regularization parameter is the origin of the factorization scale dependence mentioned before.

Matching calculation: The determination of the short-distance coefficient C_1 up to $\mathcal{O}(\alpha_s^2)$ can be carried out in two ways: One either calculates the constants b_1 in Eq. (4) through two-loop matching at the amplitude level for the electromagnetic vertex in full QCD and NRQCD or one matches expression (5) directly to the two-loop cross section calculated in full QCD. Both ways of matching must be carried out for stable top quarks ($\Gamma_t = 0$) and are performed in the (formal) limit $\alpha_s \ll v \ll 1$ where NRQCD and full QCD are applicable in the conventional multiloop approximation.[7] In our case one has to match at the two-loop level including terms up to NNLO in an expansion in v. Technically the second way of matching, called "direct matching" [19], is simpler because it allows for a sloppier treatment of the regularization procedure. (In fact, using the first way one has to be very careful to use exactly the same regularization for the matching calculation as in the calculation of the correlators. This is a quite tricky task if one wants to avoid solving the Schrödinger equation in D dimensions.) The disadvantage of direct matching, however, is that matching is carried out at the level of the final result which practically eliminates the possibility to use the calculated short-distance coefficient for any other process. Further, it requires that a multiloop expression for the cross section is at hand. For convenience, I use the "direct matching" method.

SOME EXPLICIT RESULTS

This talk would not be complete without making the somewhat general discussion before more explicit. In the following I will carry out the program described above for all Abelian contributions, i.e., for those effects which also exist in QED. I will not present any technical details and refer the interested reader to [20], where the calculations have actually been carried out.[8] I start with the well-known expression for the Coulomb Green function $G_c^{(0)}$ [17]

$$G_c^{(0)}(0,\vec{r},\tilde{E}) = -i\frac{M_t^2 \tilde{v}}{2\pi} e^{iM_t\tilde{v}r} \int_0^\infty dt\, e^{2iM_t\tilde{v}rt} \left(\frac{1+t}{t}\right)^{\frac{ia}{2\tilde{v}}}, \quad \tilde{v} \equiv \left(\frac{\tilde{E}}{M_t}\right)^{\frac{1}{2}}, \quad (14)$$

The NNLO corrections coming from the kinetic term $-\vec{\nabla}^4/4M_t^3$ and the Breit-Fermi potential $V_{\rm BF}$ are calculated via TIPT and read

$$\left[\delta G(0,0,\tilde{E})\right]_{\rm Abel}^{\rm NNLO} = \int d\vec{x}^3\, G_c^{(0)}(0,\vec{x},\tilde{E}) \left[\frac{\vec{\nabla}^4}{4M_t^3} - V_{\rm BF}(\vec{x})\right] G_c^{(0)}(\vec{x},0,\tilde{E}). \quad (15)$$

Abelian corrections coming from the one- and two-loop contributions in the Coulomb potential do not exist because we define $\alpha_s \equiv \alpha_{s,\overline{\rm MS}}^{(n_l=5)}$. As mentioned before, taking the limit $|\vec{r}|, |\vec{r}'| \to 0$ in Eq. (14) and the integral (15) leads to UV divergences which I regularize with the short-distance cutoff $\mu_{\rm fac}$. This leads to the following result for the correlator $\mathcal{A}_1$ at NNLO

$$\left[\mathcal{A}_1\right]_{\rm Abel} = \frac{3aM_t^2}{2\pi}\left\{i\tilde{v} - a\left[\ln(-i\frac{M_t\tilde{v}}{\mu_{\rm fac}}) + \gamma + \Psi\left(1 - i\frac{a}{2\tilde{v}}\right)\right]\right\}^2 \quad (16)$$

$$+\frac{9M_t^2}{2\pi}\left\{i\left(\tilde{v} + \frac{5}{8}\tilde{v}^3\right) - a\left(1 + 2\tilde{v}^2\right)\left[\ln(-i\frac{M_t\tilde{v}}{\mu_{\rm fac}}) + \gamma + \Psi\left(1 - i\frac{a(1+\frac{11}{8}\tilde{v}^2)}{2\tilde{v}}\right)\right]\right\},$$

---

[7] Other matching points like $v = 0$ are also possible, see e.g. [18]).
[8] In Ref. [20] the calculations have not been formulated in the framework of NRQCD. The results, of course, are not affected by this.

where γ is the Euler constant and Ψ the digamma function. All power divergences $\propto \mu_{\text{fac}}/M_t$ are freely dropped and μ_{fac} is defined in a way that expression (16) takes the simple form shown above. For the correlator $\mathcal{A}_2$ only the LO contribution in (16) is relevant and we arrive at

$$\mathcal{A}_2 = \tilde{v}^2 \frac{9 M_t^4}{2\pi} \left\{ i\tilde{v} - a \left[\ln(-i\frac{M_t \tilde{v}}{\mu_{\text{fac}}}) + \gamma + \Psi\left(1 - i\frac{a}{2\tilde{v}}\right) \right] \right\}. \quad (17)$$

There are no non-Abelian contributions to $\mathcal{A}_2$. The Abelian contributions to C_1 are calculated by expanding expression (5), expanding (first) for small α_s and (then) v, keeping terms up to order α_s^2 and NNLO in the v expansion,[9] and demanding equality to the corresponding two-loop cross section calculated in full QCD [20,21] for $\Gamma_t = 0$ ($v = (E/M_t)^{1/2}$),

$$\left[R^{\text{NNLO}}_{\text{2loop QCD}}\right]_{\text{Abel}} = N_c Q_t^2 \left\{ \left[\frac{3}{2}v - \frac{17}{16}v^3\right] + \frac{C_F \alpha_s}{\pi}\left[\frac{3\pi^2}{4} - 6v + \frac{\pi^2}{2}v^2\right] \right.$$

$$\left. + \alpha_s^2 \left[\frac{C_F^2 \pi^2}{8v} - 3C_F^2 + \left(\frac{49 C_F^2 \pi^2}{192} + \frac{3}{2} C_{\text{Abel}} - C_F^2 \ln v\right) v\right] \right\}, \quad (18)$$

where $C_{\text{Abel}} = C_F^2[\frac{1}{\pi^2}(\frac{39}{4} - \zeta_3) + \frac{4}{3}\ln 2 - \frac{35}{18}] + C_F T[\frac{4}{9}(\frac{11}{\pi^2} - 1)]$. The result for C_1 then reads

$$\left[C_1\right]_{\text{Abel}} = 1 - 4C_F \frac{\alpha_s(\mu_{\text{hard}})}{\pi} + \alpha_s^2(\mu_{\text{hard}}) \left[C_{\text{Abel}} + \frac{2}{3} C_F^2 \ln\left(\frac{M_t}{\mu_{\text{fac}}}\right)\right]. \quad (19)$$

Expression (19) does not contain any energy dependent (or even IR divergent) contributions because NRQCD and QCD have the same low energy behavior, i.e., all the energy dependence is contained in the correlators. Apart from the dependence on the hard scale μ_{hard}, which remains because $[C_1]_{\text{Abel}}$ represents a truncated perturbative series,[10] there is a dependence on the factorization scale μ_{fac}. As can be seen in Eq. (16), for $\Gamma_t = 0$ this dependence is cancelled by a corresponding μ_{fac}-dependent term in $[\mathcal{A}_1]_{\text{Abel}}$.[11] For $\Gamma_t \neq 0$ there is

---

[9] At this point we set $\mu_{\text{soft}} = \mu_{\text{hard}}$ because for $\alpha_s \ll v \ll 1$ a distinction between the soft and the hard scale is irrelevant.
[10] The dependence on μ_{hard} of the $\mathcal{O}(\alpha_s)$ term in Eq. (19) is not cancelled by terms in the $\mathcal{O}(\alpha_s^2)$ contributions because non-Abelian and massless quark corrections are not considered.
[11] The invariance under changing the factorization scale μ_{fac} can be used to resum renormalization group logarithms. However, I would like to warn the reader to blindly apply renormalization group methods in the belief this would represent a resummation of "leading logarithms". Although it is true that a naive resummation of logarithms is possible in this way, the resummed logarithms would certainly not be "leading". This is a consequence of the fact that $Q\bar{Q}$ systems in the threshold regime are multi-scale problems. At the NNLO level, where all interactions can be treated as instantaneous, only the relative momentum of the top quarks $\sim M_t \alpha_s$ and the top quark mass M_t are relevant scales. At NNNLO, however, also the energy of the top quarks $\sim M_t \alpha_s^2$ arises as a relevant scale and leads to a much more complicated structure of the anomalous dimensions. It is in fact not clear whether not even lower scales $M_t \alpha_s^n$ ($n > 2$) become relevant if effects beyond the NNLO level are considered.

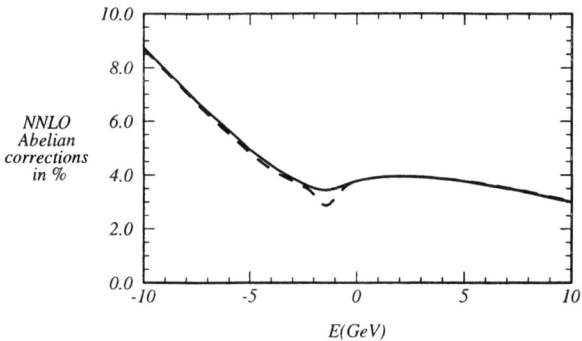

FIGURE 1. The NNLO Abelian corrections to the cross section in percent for $\Gamma_t = 1.56$ GeV (solid line) and 0.80 GeV (dashed line). See the text for more details.

a small contribution $\propto a \frac{\Gamma_t}{M_t} \ln \frac{M_t}{\mu_{\text{fac}}}$ which is not cancelled. This ambiguity arises from our ignorance of a consistent treatment of the NNLO finite width effect mentioned at the beginning. In Fig. 1 the relative size (in %) of the NNLO Abelian contributions $(R_{\text{NNLO}}^{\text{thr}} - R_{\text{NLO}}^{\text{thr}})/R_{\text{NLO}}^{\text{thr}}$ ($R_{\text{NLO}}^{\text{thr}}$ contains only the contribution from the Coulomb Green function, Eq. (14), in $\mathcal{A}_1$ and the terms in C_1 up to $\mathcal{O}(\alpha_s)$) is plotted in the energy range -10 GeV $< E < 10$ GeV for $M_t = 175$ GeV, $\alpha_s(M_z) = 0.118$ and $\Gamma_t = 1.56$ GeV (solid line)/0.80 GeV (dashed line). For the scales the choices $\mu_{\text{soft}} = 30$ GeV and $\mu_{\text{hard}} = \mu_{\text{fac}} = M_t$ have been made and two-loop running of the strong coupling has been used. It is evident that the corrections are indeed at the level of several percent and that they are fairly insensitive to the value of Γ_t indicating that the size of the Abelian NNLO contributions is not affected by the ignorance of a consistent treatment of the finite width effects. I also would like to note that the largest source of theoretical uncertainty in the cross section $R_{\text{NNLO}}^{\text{thr}}$ comes from the dependence on the soft scale μ_{soft} – clearly because it governs the perturbative series describing the long-distance (i.e. low energy) effects for which the convergence can be expected to be worse than for the short-distance coefficients. A thorough examination of this behavior, however, has to wait until all the NNLO relativistic corrections are calculated.

CONCLUSION

Due to the large top quark mass the $t\bar{t}$ system in the threshold regime offers a unique opportunity to study strong interactions in heavy-quark–antiquark pairs in the threshold regime using perturbative QCD. In this talk I have shown that NRQCD provides an ideal framework to determine the $t\bar{t}$ cross section at threshold at future lepton colliders. NRQCD, an effective field theory of QCD, allows for the calculation of the cross section (including various distributions) to in principle arbitrary precision by offering a systematic formalism which parameterizes all higher order effects from first principles QCD. In this respect NRQCD is superior to the present day potentialmodel-like approach used for analyses of $t\bar{t}$ production at threshold because NRQCD does not necessarily rely on any phenomenological input. I therefore propose that the potentialmodel-like approach should eventually be abandoned. In this talk I have given a detailed recipe how NNLO relativistic corrections to the total vector current induced cross section can be calculated using NRQCD, and I have presented explicit results for all Abelian NNLO contributions.

This work is supported in part by the U.S. Department of Energy under contract No. DOE DE-FG03-90ER40546.

REFERENCES

1. V. S. Fadin and V. A. Khoze, *JETP Lett.* **46** (1987) 525; *Sov. J. Nucl. Phys.* **48** (1988) 309; *Sov. J. Nucl. Phys.* **53** (1991) 692.
2. ECFA/DESY LC Physics Working Group, *Physics with e^+e^- linear colliders*, DESY 97-100 and hep-ph/9705442.
3. M. S. Berger, these proceedings and hep-ph/9712486.
4. W. Fischler, *Nucl. Phys.* **B 129** (1977) 157.
5. A. Billoire, *Phys. Lett.* **B 92** (1980) 343.
6. M. J. Strassler and M.E. Peskin, *Phys. Rev.* **D 43** (1991) 1500.
7. W. Kwong, *Phys. Rev.* **D 43** (1991) 1488;
   M. Jeżabek, J. H. Kühn and T. Teubner, *Z. Phys.* **C 56** (1992) 653.
8. V. S. Fadin and O. I. Yakovlev, *Sov. J. Nucl. Phys.* **53** (1991) 4.
9. W. E. Caswell and G. E. Lepage, *Phys. Lett.* **B 167** (1986) 437.
10. Y. Sumino, K. Fujii, K.Hagiwara, H. Murayama and C.-K. Ng, *Phys. Rev.* **D 47** (1993) 56.
11. M. Jeżabek and T. Teubner, *Z. Phys.* **C 59** (1993) 669.
12. W. Mödritsch and W. Kummer, *Nucl. Phys.* **B 430** (1994) 3.
13. P. Labelle, McGill Report No. McGill-96/33 and hep-ph/9608491.
14. B. Grinstein and I. Z. Rothstein, *Phys. Rev.* **D 57** (1998) 78.
15. S. N. Gupta, S. F. Radford and W. W. Repko, *Phys. Rev.* **D 26** (1982) 3305.
16. M. Peter, *Phys. Rev. Lett.* **78** (1997) 602, *Nucl. Phys.* **B 501** (1997) 471.
17. E. H. Wichmann and C. H. Woo, *J. Math. Phys.* **2** (1961) 178;
18. P. Labelle, Cornell Uni. Ph.D. thesis, UMI-94-16736-mc (microfiche), 1994.
19. A. H. Hoang, published in *Phys. Rev.* **D 57** (1998), hep-ph/9702331.
20. A. H. Hoang, *Phys. Rev.* **D 56** (1997) 5851.
21. A. H. Hoang, *Phys. Rev.* **D 56** (1997) 7276.

Flavor Changing Neutral Currents at $\mu^+\mu^-$ Colliders

Laura Reina[1]

Physics Department, University of Wisconsin
Madison, Wisconsin 53706

Abstract. We illustrate the possibility of observing signals from Flavor Changing Neutral Currents, originating from the scalar sector of a Two Higgs Doublet Model. In particular, we focus on the tree level process $\mu^+\mu^- \to \bar{t}c + \bar{c}t$, via scalar exchange in the s-channel, as a distinctive process for $\mu^+\mu^-$ colliders.

INTRODUCTION

The First Muon Collider (FMC) will reach a maximum center of mass energy of 500 GeV, exploring all the interesting intermediate regimes. In a second phase, the $\mu^+\mu^-$ collider should upgrade its energy to up to 4 TeV and therefore become a very high energy lepton collider, even compared to NLC.

A high energy lepton collider will be a very promising environment to look for *new physics* beyond the Standard Model (SM), taking advantage of the large enough energy which will become available at low background rates. As has become common knowledge by now, the great advantage of a $\mu^+\mu^-$ machine will be to allow the study of the s-channel production of scalar *Higgs like* particles, which should occur at a much more conspicuous rate with respect to an e^+e^- collider ($m_\mu \simeq 200\, m_e$).

A very important and direct application of this property will be the study of the Higgs sector of both the SM and its SuperSymmetric (SUSY) extensions, looking for positive or negative evidence of the scalar and pseudoscalar particles which are theoretically predicted in these models. Even if evidence is found at a different machine (for instance a hadron collider), a $\mu^+\mu^-$ machine will offer an optimal energy resolution to sit at resonance and study the properties of these particles, i.e. their masses and their couplings. This subject has been thoroughly covered in some plenary talks [1,2] and dedicated parallel sessions [3,4] during this workshop.

---

[1] Work done in collaboration with D. Atwood and A. Soni.

Along these lines, we want to discuss here the possibility of further studying the properties of the scalar sector of the SM as well as of many of its extensions, SUSY included, by looking for *anomalous* Flavor Changing Neutral Currents (FCNC) induced by scalar exchange. It is well known that any extended scalar sector containing identical replicas of the same representation of scalar fields (for instance: N identical doublets or N identical pairs of doublets, etc.) can induce FCNC at the tree level. This is due to the possibility of diagonalizing the mass matrix of the fermion fields without diagonalizing each single fermion-scalar coupling[2]. The interest in models containig many identical generations of scalars (sometimes even in a one-to-one correspondence with the generations of fermions) arises in many string-inspired Grand Unified Models (for instance E_6) and is therefore of a more general interest.

In order to be more predictive and to limit the number of parameters in our analysis, we focus on a minimal extension of the scalar sector of the SM, a Two Higgs Doublet Model (2HDM). This can be used as a simple model, which we are able to work out to the very last consequences in order to study the compatibility of its predictions with the existing experiments. On a more general ground, our analysis should provide useful hints for a diversity of extensions of the SM, which both theorist and experimentalist are encouraged to study.

Since FCNC are forbidden in the SM, their study can provide us with unambiguous evidence of *new physics*. However, severe constraints are imposed by the low energy physics of the K- and B-mesons, such that FCNC have practically to be avoided in this sector of the theory. This is naturally accomplished by the SM itself via the GIM mechanism, and has to be imposed *ad hoc* in any 2HDM, by introducing a discrete symmetry [5], which limits the possible Yukawa couplings between fermions and scalars.

Apart from the experimental constraints coming from K- and B-physics, there is no *a priori* theoretical reason not to have FCNC. Therefore, the assumption of this *ad hoc* discrete symmetry may be dropped in favor of a more natural one, which takes any Flavor Changing (FC) coupling to a scalar field to be proportional to the mass of the coupled fermions. The basic idea is that a natural hierarcy is provided by the observed fermion masses and this may be transfered to the couplings between fermions and scalar fields [6–9]. In this way, FCNC are naturally suppressed in the light sector of the theory, while dramatic effects may be seen in processes which involve the heavy quark fields of the third generation, i.e. the top quark.

We illustrate these ideas at work in the following sections, first presenting the model we refer to [10,11] and then focusing on some FC signals, namely $(\bar{t}c + \bar{c}t)$-production which, if possible at an e^+e^--collider [12], is even more enhanced and distinctive at a $\mu^+\mu^-$-collider [13].

---

[2] We assume Yukawa type couplings between fermions and scalars and fermion masses generated through spontaneous symmetry breaking.

THE MODEL

We explicitly consider in this context only the quark fields, assuming that the discussion of the quark and lepton sectors of the theory can proceed independently. Then, let us consider the quark Yukawa Lagrangian of a 2HDM, which we write as,

$$\mathcal{L}_Y^{(III)} = \eta_{ij}^U \bar{Q}_{i,L} \tilde{\phi}_1 U_{j,R} + \eta_{ij}^D \bar{Q}_{i,L} \phi_1 D_{j,R} + \xi_{ij}^U \bar{Q}_{i,L} \tilde{\phi}_2 U_{j,R} + \xi_{ij}^D \bar{Q}_{i,L} \phi_2 D_{j,R} + h.c. \quad (1)$$

where ϕ_i, for $i = 1, 2$, are the two scalar doublets of a 2HDM ($\tilde{\phi}_i = i\sigma^2 \phi_i$), while $\eta_{ij}^{U,D}$ and $\xi_{ij}^{U,D}$ are the non diagonal matrices of the Yukawa couplings. In order to prevent FCNC to arise at the tree level, the scalar potential and Yukawa Lagrangian need to be constrained by an *ad hoc* discrete symmetry [5],

$$\phi_1 \to -\phi_1 \quad \text{and} \quad \phi_2 \to \phi_2 \quad (2)$$
$$D_i \to -D_i \quad \text{and} \quad U_i \to \mp U_i \ .$$

Depending on whether the up-type and down-type quarks are coupled to the same or to two different scalar doublets respectively, one obtains the so called Model I and Model II 2HDM's. [14].

In contrast we want to consider the case in which no discrete symmetry is imposed and both up-type and down-type quarks then have FC couplings. For this type of 2HDM, that we call Model III, the Yukawa Lagrangian for the quark fields is as in Eq. (1) and no term can be dropped *a priori*, see also Refs. [10,12]. Since the two scalar doublet are completely independent, by a suitable rotation of the quark fields, we can chose the two scalar doublets in such a way that only the $\eta_{ij}^{U,D}$ couplings generate the fermion masses, i.e. such that

$$<\phi_1> = \begin{pmatrix} 0 \\ v/\sqrt{2} \end{pmatrix}, \quad <\phi_2> = 0 \ . \quad (3)$$

To the extent that the definition of the $\xi_{ij}^{U,D}$ couplings remains arbitrary, we will denote by $\xi_{ij}^{U,D}$ the new rotated couplings, such that the charged couplings look like $\xi^U \cdot V_{\text{CKM}}$ and $V_{\text{CKM}} \cdot \xi^D$. This form of the charged couplings is indeed peculiar to Model III compared to Models I and II and can have important phenomenological repercussions [16,11].

The scalar physical mass spectrum consists of two charged $\phi^\pm$ and three neutral spin 0 bosons, two scalars (H^0, h^0) and a pseudoscalar (A^0),

$$\begin{aligned} H^0 &= \sqrt{2}[(\text{Re}\,\phi_1^0 - v)\cos\alpha + \text{Re}\,\phi_2^0 \sin\alpha] \\ h^0 &= \sqrt{2}[-(\text{Re}\,\phi_1^0 - v)\sin\alpha + \text{Re}\,\phi_2^0 \cos\alpha] \\ A^0 &= \sqrt{2}(-\text{Im}\,\phi_2^0) \ , \end{aligned} \quad (4)$$

where α is a mixing phase which also determines the couplings between the neutral scalars and gauge bosons[3] ($W^{\pm}$, Z^0), i.e.

$\overline{H}^0, h^0 \quad\text{---}\!\!\!\sim\!\!\sim\!\!\sim Z^\mu \qquad\qquad \pm i \frac{g_W}{c_W} M_Z (\cos\alpha,\, \sin\alpha)\, g^{\mu\nu}$
$\qquad\qquad\quad\sim\!\!\sim\!\!\sim Z^\nu$

$\overline{H}^0, h^0 \quad\text{---}\!\!\!\sim\!\!\sim\!\!\sim W^\mu_+ \qquad\qquad \pm i g_W M_W (\cos\alpha,\, \sin\alpha)\, g^{\mu\nu}\quad .$
$\qquad\qquad\quad\sim\!\!\sim\!\!\sim W^\nu_-$

It is interesting to notice that for $\alpha = 0$: (1) H^0 corresponds exactly to the SM Higgs field, and $\phi^{\pm}$, h^0 and A^0 generate the new FC couplings; (2) h^0 does not couple to the gauge bosons, i.e. it behaves like the pseudoscalar field A^0.

Finally, we want to introduce some definite ansatz that will guide us in our phenomenological approach to the study of Model III. Because the Yukawa Lagrangian directly breaks the flavor symmetry among quarks, and this ultimately results into fermion mass generation, some major proposals exist in the literature [6–9] which suggest to take the new FC couplings to be proportional to the mass of the quarks involved in the coupling, i.e.

$$\xi_{ij} = \lambda_{ij} \frac{\sqrt{m_i m_j}}{v}\ , \qquad (5)$$

where for the sake of simplicity we take the λ_{ij} to be real (for more details see [12,11]). In this ansatz the residual degree of arbitrariness of the FC couplings is expressed through the λ_{ij} parameters, which need to be constrained by the available phenomenology. In particular we will see how $K^0 - \bar{K}^0$ and $B^0 - \bar{B}^0$ mixings (and to a less extent $D^0 - \bar{D}^0$ mixing) put severe constraints on the FC couplings involving the first family of quarks.

There is no doubt that the most interesting signals of these non-standard couplings are to come from the physics of the top quark, both production and decays. Therefore, we would like to single out the right processes and the right environment in which we could already have the possibility of testing the consequences of our assumptions.

---

[3] We remind that in a 2HDM the pseudoscalar field A^0 does not couple to the gauge bosons.

ANALYSIS OF THE CONSTRAINTS

The existence of FC couplings is very much constrained by the experimental results on $F^0 - \bar{F}^0$ flavor mixings (for $F = K, B$ and to a less extent D)

$$\Delta M_K \simeq 3.51 \cdot 10^{-15} \text{ GeV}$$
$$\Delta M_{B_d} \simeq 3.26 \cdot 10^{-13} \text{ GeV} \qquad (6)$$
$$\Delta M_D < 1.32 \cdot 10^{-13} \text{ GeV} ,$$

due to the presence of new tree level contributions to each of the previous mixings. We have analyzed the problem in detail in Ref. [11], taking into account both tree level and loop contributions. Indeed the two classes of contributions can affect different FC couplings, due to the peculiar structure of the charged scalar couplings (see previous section).

We find that, unless for scalar masses in the multi-TeV range, the tree level contributions need to be strongly suppressed, requiring that the corresponding FC couplings are much than one. Enforcing the ansatz made in Eq. (5), this amounts to demand that

$$\lambda_{ds}^D \ll 1 , \quad \lambda_{db}^D \ll 1 \text{ and } \lambda_{ud}^U \ll 1 . \qquad (7)$$

More generally, we can assume that the FC couplings involving the first generation are negligible. Particular 2HDM's have been proposed in the literature in which this pattern can be realized [15]. The remaining FC couplings, namely ξ_{ct}^U and ξ_{sb}^D are not so drastically affected by the $F^0 - \bar{F}^0$ mixing phenomenology. From the analysis of the loop contributions to the $F^0 - \bar{F}^0$ mixings (box and penguin diagrams involving the new scalar fields), we verify that many regions of the parameter space are compatible with the results in Eq. (6) [11]. Therefore we may want to look at other constraints in order to single out the most interesting scenarios.

Three are in particular the physical observables that impose strong bounds on the masses and couplings of Model III [16,11]

- the inclusive branching ratio for $B \to X_s \gamma$, which is measured to be [17]

$$BR(B \to X_s \gamma) = (2.32 \pm 0.51 \pm 0.29 \pm 0.32) \times 10^{-4} , \qquad (8)$$

- the ratio $R_b = \Gamma(Z \to b\bar{b})/\Gamma(Z \to \text{hadrons})$, whose present measurement [18] is such that $R_b^{\text{expt}} > R_b^{\text{SM}}$ ($\sim 1.4\sigma$),

$$R_b^{\text{expt}} = 0.2170 \pm 0.0009 \text{ while } R_b^{\text{SM}} = 0.2158 , \qquad (9)$$

- the corrections to the ρ parameter, which has become conventional to describe in terms of

$$\rho_0 = \frac{M_W^2}{\rho M_Z^2 \cos^2\theta_W} = 1 + \Delta\rho_0^{\text{NEW}} \qquad (10)$$

where ρ absorbs all the SM corrections to the gauge boson self energies and, in the presence of new physics, $\Delta\rho_0^{\text{NEW}}$ summarizes the deviation from the SM prediction (i.e. $\rho_0 = 1$). From the recent global fits of the electroweak data, which include the input for m_t from Ref. [19] and the new experimental results on R_b, ρ_0 turns out to be very close to unity. This imposes severe constraints on many extension of the SM, especially on the mass range of the new particles.

Since the experimental determination of R_b is not completely definite, we demand less strict agreement between R_b^{exp} and R_b^{SM}. In this case we find compatibility with the present experiments for

$$\lambda_{ct} \simeq O(1) \quad \text{and} \quad \lambda_{sb} \simeq O(1) \ . \qquad (11)$$

The value of the mixing angle α is not determinant, while the masses are mainly dictated by the fit to $Br(B \to X_s\gamma)$ and $\Delta\rho_0$ [16]. We are left with two possible scenarios,

$$M_H, M_h \leq M_c \leq M_A \quad \text{or} \quad M_A \leq M_c \leq M_H, M_h \ . \qquad (12)$$

We conclude that, given the existing constraints, Model III has the very interesting characteristic of providing sizable FC couplings for the top quark, in a way that will certainly be testable at the next generation of lepton and hadron colliders. We will discuss some of these phenomenological issues in the next section.

TOP-CHARM PRODUCTION: THE CASE OF A MUON COLLIDER

If we assume Eq. (11), ξ_{ct}^U becomes the most relevant FC coupling. The presence of a ξ_{ct}^U flavor changing coupling can be tested by looking at both top decays and top production (see Ref. [11] and references therein).

It is interesting that the first upper bounds on $t \to cV$ ($V = \gamma, Z^0$) are now coming from recent experimental analysis [20] (see Ref. [11] for the corresponding theoretical prediction in Model III). Encouraged by this progress, we want to concentrate here on top-charm production. The SM prediction for top-charm production is extremely suppressed and any signal would be a clear evidence of new physics with large FC couplings in the third family. The final state for this process has a unique kinematics, with a very massive jet recoiling against an almost massless one (very different from a bs production signal, for instance). This quite distinctive signature may allow to work even with relatively low statistics, as can be the case for a lepton collider. The much better statistics one could get at an hadron collider, would come at a cost of a much higher background (mostly, tree level SM background for a one-loop process). A nice analysis of the hadron collider case is presented in Ref. [21].

Here we want to focus on lepton colliders and in particular on the case of the FMC. In principle, the production of top-charm pairs arises both at the tree level, via the s-channel exchange of a scalar field with FC couplings, and at the one loop level, via corrections to the Ztc and γtc vertices.

The s-channel production is not relevant for an e^+e^- collider, because the coupling of the scalar fields to the electron is very suppressed ($m_\mu \simeq 200\, m_e$). An interesting proposal for a t-channel production via W^+W^--fusion at a $\sqrt{s}=$ 1 TeV e^+e^- collider has been pointed out in Ref. [22]. However, top-charm production at an e^+e^- collider remains mainly a loop effect and therefore (even at the energies of NLC) it is suppressed with respect to the corresponding production cross section at the FMC (see Ref. [12]) and in particular at the very high energy muon collider.

In fact, for a $\mu^+\mu^-$ collider, the s-channel top-charm production via a neutral scalar/pseudoscalar is a nice example of the kind of resonance production we briefly discussed in the Introduction. At resonance, the new scalars (let us denote them generically by $\mathcal{H}$) may be produced at an appreciable rate, and the effective cross section for any final state very much depends on the relation between the beam energy resolution and the width of the scalar particle ($\Gamma_\mathcal{H}$) produced in the s-channel [23].

Following [23], we define the effective cross section as the convolution of the Breit-Wigner σ_{tc}^{BW} cross section with a gaussian beam energy spread,

$$\sigma_{tc}^{eff} = \int d\sqrt{s'} \frac{\exp[-(\sqrt{s'}-\sqrt{s})^2/2\sigma^2]}{\sqrt{2\pi}\sigma} \sigma_{tc}^{BW}(s') , \qquad (13)$$

where the rsm of the gaussian distribution is defined in terms of the *resolution* parameter R as

$$\sigma = 7\,\text{MeV}\left(\frac{R}{0.01}\right)\left(\frac{\sqrt{s}}{100\,\text{GeV}}\right) . \qquad (14)$$

If $\sigma \gg \Gamma_H$ the effective cross section is suppressed as $\Gamma_\mathcal{H}/s$,

$$\sigma_{tc}^{eff} = \frac{\pi \Gamma_H}{2\sqrt{2\pi}\sigma} \sigma_{BW}(s=M_\mathcal{H}) , \qquad (15)$$

while the optimal case is reached if $\sigma \ll \Gamma_H$, when

$$\sigma_{tc}^{eff} = \sigma_{BW}(s=M_\mathcal{H}) , \qquad (16)$$

with a whole spectrum of possible intermediate cases. In our analysis we study

$$R_{tc} = \frac{\sigma_{tc}^{eff}}{\sigma_0} = R(\mathcal{H})\left(B(\mathcal{H} \to \bar{t}c) + B(\mathcal{H} \to \bar{c}t)\right) , \qquad (17)$$

where $\sigma_0 = \sigma(\mu^+\mu^- \to \gamma \to e^+e^-)$ and $R(\mathcal{H}) = \sigma_\mathcal{H}/\sigma_0$ for $\sigma_\mathcal{H}$ the total cross section for producing $\mathcal{H}$.

To be more explicit, let us consider the case of a scalar field $\mathcal{H} = h^0$. Then, according to what we discussed in a previous section,

$$C_{h^0 tc} = \frac{1}{\sqrt{2}}\left[\xi_{tc}P_R + \xi_{ct}^\dagger P_L\right]\cos\alpha \equiv \frac{g\sqrt{m_t m_c}}{2M_W}(\chi_R P_R + \chi_L P_L) . \qquad (18)$$

The total width Γ_{h^0} can be obtained from the literature (see for instance [14]), and varies with M_{h^0} because of the different decay channels that open up at higher M_{h^0}. On the other hand, the rate for top-charm production in Model III is explicitly given by,

$$\Gamma(h^0 \to t\bar{c}) = \frac{3g^2 m_t m_{\bar{c}} M_{h^0}}{32\pi M_W^2}\left(\frac{(M_{h^0}^2 - m_t^2)^2}{M_{h^0}^4}\right)\left(\frac{|\chi_R|^2 + |\chi_L|^2}{2}\right) . \qquad (19)$$

Therefore, our results depend on both M_{h^0} and Γ_{h^0}, and on how Γ_{h^0} compares to the resolution parameter R. We use the set of parameters posted for this workshop. In particular we consider the possibility to reach resolutions as accurate as $R \simeq 1\%$ and the availability of a average luminosity of $\mathcal{L}_{av} = 10^{32}$ for $\sqrt{s} = 200$ GeV up to $\mathcal{L}_{av} = 7 \times 10^{32}$ for $\sqrt{s} = 500$ GeV. For a given R, we vary M_{h^0} in the range,

$$100 \, \text{GeV} \leq M_{h^0} \leq 800 \, \text{GeV} , \qquad (20)$$

and for given M_{h^0} and R, we consider two different cases: $\alpha = 0$ and $\alpha = \pi/4$. In the first case Γ_{h^0} is smaller than in the second case, because the h^0 field does not couple to W^+W^- and Z^0Z^0. In both cases we assume all the χ_i couplings to be real and of $O(1)$. Our results are summarized in Fig. 1.

Note that for $\alpha = 0$, if M_{h^0} is below the $t\bar{t}$ threshold R_{tc} is about .2 − 3 and in fact tc makes up a large branching ratio. Above the $t\bar{t}$ threshold R_{tc} drops. In this *narrow width* case we clearly see how our result depends on R and how for smaller values of R the predictions get closer and closer to the pure Breit-Wigner case. For $\alpha = \pi/4$ the branching ratio is smaller due to the W^+W^- and Z^0Z^0 threshold at about the same mass as the tc threshold and so R_{tc} is around 10^{-3}. In this case the width is *broader* and there is almost no dependence on R. To be more specific, let us assume that $M_{h^0} = 300$ GeV, then $\sigma_0 \approx 1$ pb. For $\mathcal{L}_{av} = 10^{32}$ cm^{-2} s^{-1} and $R = 0.01$, $\alpha = 0$ will produce about 10^2 $(t\bar{c} + \bar{t}c)$ events and $\alpha = \pi/4$ will produce only a few events. Much higher statistics can be obtained improving on the average luminosity available, which we hope will remain one of the priorities in the study of the FMC. Given the distinctive nature of the final state and the

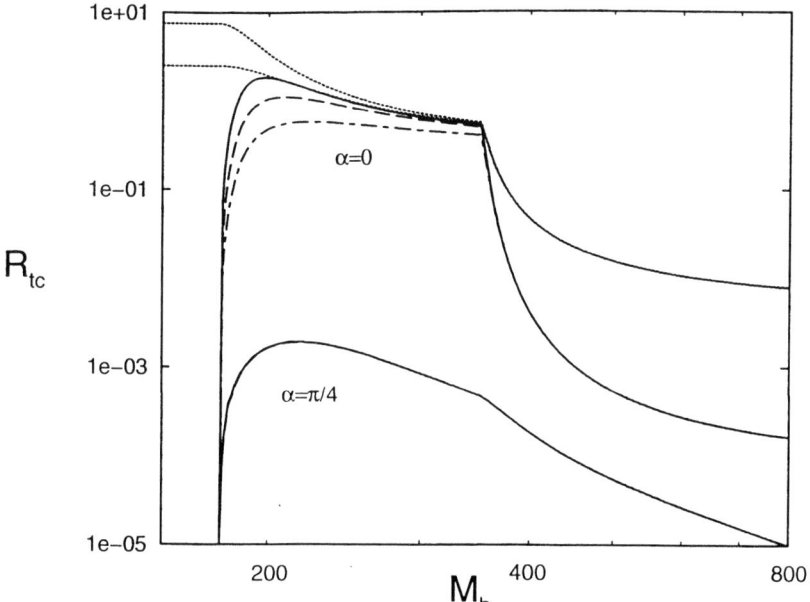

FIGURE 1. The value of $R(h^0)$ is shown as a function of M_{h^0} in a pure Breit-Wigner case (upper dotted line) and when the gaussian width distribution of the beam energy (for $R=0.01$) is taken into account (lower dotted line). The ratio R_{tc} is also shown for different values of the resolution parameter $R = 0.001$ (solid), 0.01 (dashed) and 0.03 (dot-dashed), when $\alpha=0$ (upper group of curves) and when $\alpha=\pi/4$ (lower group of curves).

lack of a Standard Model background, sufficient luminosity should allow the observation of such events.

If such events are observed, we will also have the possibility to extract the values of χ_L and χ_R, provided we determine the helicity of the produced top quark, expressed in termes of χ_L and χ_R as

$$\mathbf{H}_t = -\mathbf{H}_{\bar{t}} = \frac{|\chi_R|^2 - |\chi_L|^2}{|\chi_R|^2 + |\chi_L|^2} \ . \tag{21}$$

The helicity of the t quark cannot be determined directly, but has to be obtained from the decay distributions of the top [24], the number of events required to observe it with a significance of $3\,\sigma$ being

$$N_{3\sigma} = \frac{36}{\mathcal{E}_t^2 \mathbf{H}_t^2} \approx \frac{107}{\mathbf{H}_t^2} \ . \tag{22}$$

Thus at least 10^2 events are required to begin to measure the helicity of the top and hence the relative strengths of χ_L and χ_R. In the above numerical examples it is clear that for some combinations of parameters, particularly if the luminosity is higher than 10^{32} cm^{-2} s^{-1}, sufficient events to measure the helicity may be present.

ACKNOWLEGEMENT

This research was supported by the U.S. Department of Energy under Contract DE-FG02-95ER40896.

REFERENCES

1. V. Barger, this proceedings.
2. J. Gunion, this proceedings.
3. T. Han and M. Demarteau, this proceedings.
4. M. Carena and S. Protopopescu, this proceedings.
5. S.L. Glashow and S. Weinberg, *Phys. Rev.* **D15**, 1958 (1977).
6. T.P. Cheng and M. Sher, *Phys. Rev.* **D35**, 3484 (1987).
7. M. Sher and Y. Yuan, *Phys. Rev.* **D44**, 1461 (1991).
8. A. Antaramian, L.J. Hall, and A. Rasin, *Phys. Lett.* **B69**, 1871 (1992).
9. L.J. Hall and S. Weinberg, *Phys. Rev.* **D48**, R979 (1993).
10. M. Luke and M.J. Savage, *Phys. Lett.* **B307**, 387 (1993).
11. D. Atwood, L. Reina and A. Soni, *Phys. Rev.* **D55**, 3156 (1997).
12. D. Atwood, L. Reina and A. Soni, *Phys. Rev.* **D53**, 1199 (1996).
13. D. Atwood, L. Reina and A. Soni, *Phys. Rev. Lett.* **75**, 3800 (1995).
14. For a review see J. Gunion, H. Haber, G. Kane, and S. Dawson, *The Higgs Hunter's Guide*, Addison-Wesley, New York, 1990.
15. A. Das and C. Kao, Phys. Lett. **B372**, 106 (1996).
16. D. Atwood, L. Reina and A. Soni, *Phys. Rev.* **D54**, 3295 (1996).
17. R. Ammar *et al.*, CLEO Collaboration, *Phys. Rev. Lett.* **71**, 674 (1993); M.S. Alam *et al.*, CLEO Collaboration, *Phys. Rev. Lett.* **74**, 2885 (1995).
18. The LEP Electroweak Working group, LEPEWWG/97-02.
19. P. Giromini, Proccedings of the LP'97, Hamburg, 1997; A. Yagil, Proceedings of the International Europhysics Conference, Jerusalem, 1997.
20. F. Abe *et al.*, CDF Collaboration, FERMILAB-PUB-97-270; T. J. LeCompte, CDF Collaboration, Proceedings of the Symposium on FCNC, Santa Barbara, 1997.
21. T. Han, R.D. Peccei and X. Zhang, *Nucl. Phys.* **B454**, 527 (1995).
22. S. Bar-Shalom, G. Eilam, A. Soni, J. Wudka, *Phys. Rev. Lett.* **79**, 1217 (1997); hep-ph/9708358.
23. V. Barger, M.S. Berger, J.F. Gunion and T. Han, *Phys. Rev. Lett.* **75**, 1462 (1995); *Phys. Rept.* **286**, 1 (1997).
24. D. Atwood and A. Soni, *Phys. Rev.* **D52**, 6271 (1995).

Gluon Radiation in Top Production and Decay at Lepton Colliders

Lynne H. Orr*

Department of Physics and Astronomy
University of Rochester
Rochester, NY 14627-0171

Abstract. In this talk we discuss gluon radiation in top production and decay. After reviewing results for hadron colliders we consider soft gluon radiation at lepton colliders and present gluon distribuitons that are potentially sensitive to production-decay interference effects.

INTRODUCTION

Top quark events are often accompanied by gluons. This gluon radiation can occur in association with both the top production and top decay processes, and must be understood if we are to make sense of top physics. This is especially true for top momentum reconstruction, where a jet originating as a gluon may or may not be a top quark decay product. Top momentum reconstruction can play a crucial role not only in precision top mass measurements but also in identification of top events by using a mass cut. In addition, the pattern of the gluon radiation itself contains information about the top production and decay processes and their color structure, and can be sensitive to new physics.

In this talk we discuss results for gluon radiation in top events at hadron colliders, and then focus on lepton colliders, noting some similarities and differences along the way. Further details can be found in the references.

Review of Hadron Collider Results

In top production and decay at hadron colliders, strongly interacting particles in the intial, intermediate, and final states all give rise to gluon radiation. In reconstructing the top mass, it matters whether jets from these gluons are part of the top production or decay process. If the gluons arise in top production, they should be ignored in mass reconstruction, but if they are part of the top decay process then they must be included.[1]

In practice that seemingly straightforward rule is hard to follow for several reasons. First, gluon jets are not necessarily distinguishable from other jets from top decays. Second, even if the gluon jets can be identified one cannot easily distinguish between production- and decay-stage radiation. In a study of gluon radiation in top production and decay at the Tevatron [1] we found that for kinematic cuts meant to mimic typical detector capabilities, the amounts of production- and decay-stage radiation were comparable (ignoring radiation from W decay products). Furthermore, the two were not easily separated. Production-stage radiation is well spread out in rapidity, and although decay-stage radiation is more central, there is significant overlap in the relevant detector regions. Even proximity to one of the b-quark jets is not sufficient to uniquely identify a gluon as being associated with decay, for example. This underscores the importance of understanding gluon radiation in top processes.

At the LHC pp collider, the situation is slightly different. [2] Its higher energy and luminosity give a vast increase in the top production cross section, but at the same time the amount of gluon radiation is also increased, notably at the production stage. This is illustrated in Figure 1, which shows the ratio of cross sections for $t\bar{t}j$ and $t\bar{t}$ production at the Tevatron (solid line) and LHC (dashed line) as a function of the minimum transverse energy E_T of the gluon. The LHC is seen to have a vast increase in production-stage radiation over the Tevatron for any given E_T cut.

In contrast, the decay-stage radiation at the LHC looks similar to that at the Tevatron, because the kinematics of the produced top quarks are similar. The main difference is that the t's are more spread out in rapidity at the LHC, which gives rise to a decay-stage gluon distribution that is also more spread out.

At the LHC, then, the individual distributions of production- and decay-stage radiation are similar to those at the Tevatron, but production-stage radiation dominates by far. This can be a problem, for example, when those gluons overlap with jets from t decays, causing mismeasurement of parton energies that can feed into momentum reconstruction. Such effects cannot be avoided but they can be taken into account as long as they are incorporated into the relevant analyses.

GLUON RADIATION AT LEPTON COLLIDERS

At lepton colliders there is no gluon emission from the initial state. That means that, for purposes of studying gluon radiation in top production and

---

[1] Interference between gluons from production and those from decay is negligible provided their energies are large compared to the top width, a reasonable assumption for gluons that are to be detected as jets at hadron colliders. See the lepton collider section for further discussion.

decay, there is no difference between e^+e^- and $\mu^+\mu^-$ colliders. (There are important differences in, for example, precision studies at the $t\bar{t}$ threshold; see the talk by M. Berger. [3]) The absence of initial-state gluon radiation and the relatively simpler kinematics compared to hadron colliders allows for the possibility of more detailed study of radiation patterns at lepton colliders. The pattern of gluon radiation can give information about the dynamics of the underlying process, including interplay between top production and decay, and possible signals of new physics.

Soft Gluon Radiation

The study of soft gluon radiation patterns in high energy process, sometimes called "partonometry," is useful for a number of reasons. The infrared singu-

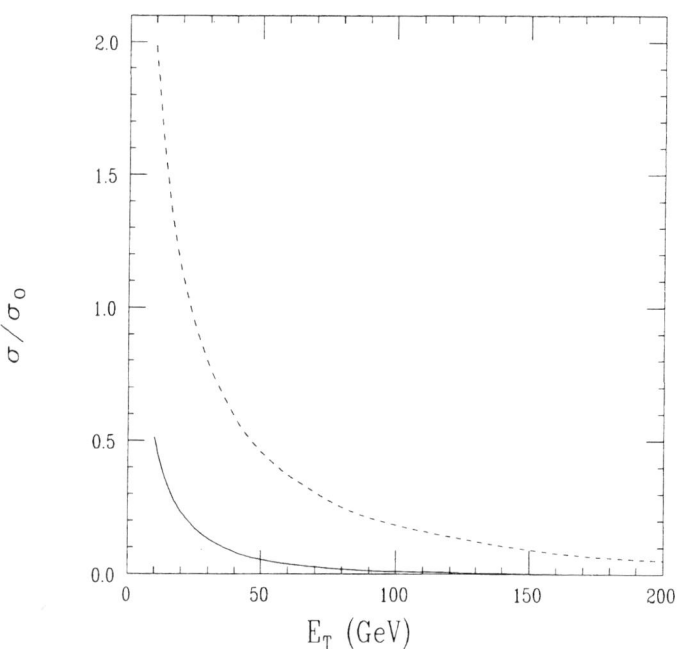

FIGURE 1. Ratio of $\sigma(t\bar{t}j)$ to $\sigma(t\bar{t})$ for $E_T^j > E_T$ at the Tevatron (solid line) and LHC (dashed line). [2]

larity in the cross section means that there is a high probability for emitting soft gluons, whose distribution determines the distributions of soft hadrons or minijets in such events. Studying these distributions then gives information about the underlying process. In doing calculations, the soft limit is useful because the cross section factorizes into the lowest order (differential) cross section and a piece due solely to the gluon radiation:

$$dN \equiv 1/\sigma_0 d\sigma_g = \frac{dE_g}{E_g} \frac{d\Omega}{4\pi} \frac{C_F \alpha_s}{\pi} \mathcal{R}, \tag{1}$$

where E_g and Ω denote the gluon energy and solid angle.

In $t\bar{t}$ production and decay at lepton colliders, although there is no initial state gluon radiation, there are still gluons emitted in both the top production and decay stages (respectively before and after the top quark goes on shell). It is straightforward to find $\mathcal{R}$ and decompose it into the contributions corresponding to radiation from the various stages; see [4].

One interesting result of such an analysis is that there can be interference between production- and decay-stage gluons when the gluon energy is roughly comparable to the top width Γ. This can be understood heuristically in terms of Breit-Wigner distributions for the square of the top momentum with and without the gluon, corresponding to radiation in the decay and production stages, respectively. The peak separation is roughly the gluon energy E_g, and since each distribution has width Γ, when $E_g \sim \Gamma$, the two distributions overlap, giving rise to non-negligible interference.[2] Because the presence of this interference for a given gluon energy depends on the value of Γ, it follows that gluon distributions can be sensitive to the top width.

In Ref. [4] the gluon distribution $\mathcal{R}$ was studied for specific kinematic configurations for a top mass of 140 GeV. Here we update those results for a 175 GeV top and for energies appropriate to muon collider studies. In Figure 2 we show the gluon distribution for center of mass energy 1 TeV for a configuration where the t quark decays to a backwards b and the $\bar{t}$ also decays to a backwards $\bar{b}$. The distribution is shown as a function of θ_g, the angle between the gluon and the top quark. For this configuration, the t and $\bar{b}$ quarks are located at $\theta_g = 0°$ and the $\bar{t}$ and b quarks are at $\theta_g = 180°$. The gluon energy is taken to be 5 GeV. The overall shape shows the "dead cone" behavior characteristic of radiation from heavy quarks — radiation along the quark direction is suppressed by the mass. The curves are for different values of the top width Γ, as labeled. We see that radiation in the SM case ($\Gamma = 1.5$ GeV) is suppressed compared to that for $\Gamma = 0$, due to destructive interference between the production and decay stages. As the width increases so does this

---

[2] We can now see why the production-decay interference is negligible at hadron colliders. We consider extra jets with minimum transverse energies of 20 and 40 GeV at the Tevatron and LHC, respectively; such energies are much larger than the Standard Model top width of about 1.5 GeV.

suppression, and in the limit $\Gamma \to \infty$, the gluon distribution is what it would be if the b quarks were produced directly: the contribution of radiation from the top quark disappears altogether.

Top quarks at high energies are not terribly likely to decay to backward b's, but interference effects remain if we look at a more likely configuration in Figure 3, where the b quarks come out at 90° to their parent top quarks. Here the t and $\bar{t}$ quarks are at 0° and 180°, respectively, and the b and $\bar{b}$ are at 90° and 270°. We see more interesting structure in this case, because the b directions do not coincide with those of the t's. Again we see destructive production-decay interference that increases with the top width, and in the $\Gamma \to \infty$ limit (dotted curve) there is no evidence of the top direction in the distribution.

The 1 TeV collision energy discussed above is certainly desirable from the

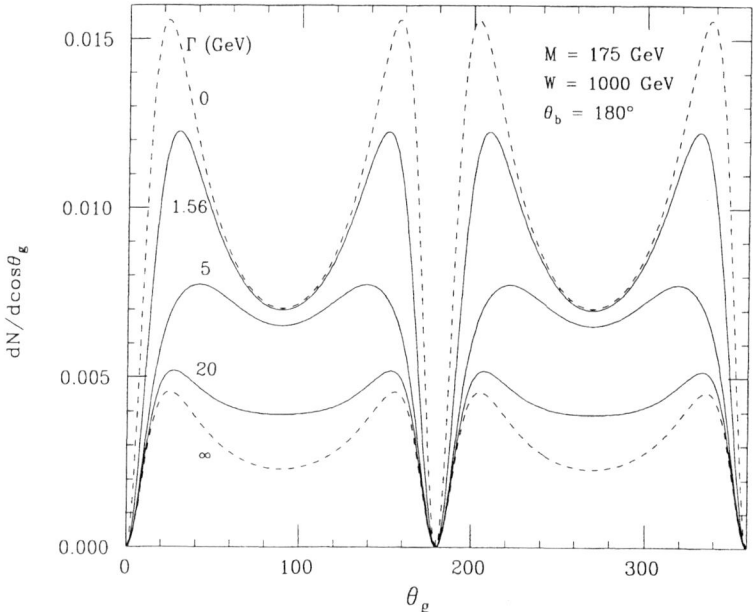

FIGURE 2. Soft gluon distribution in top production and decay at lepton colliders as described in the text, for t's decaying to backward b's and collision energy 1 TeV.

point of view of a variety of physics topics. However, as stated in the workshop guidelines, a first muon collider is more likely to operate at the lower energy of 500 GeV. How do top production-decay interference effects look in the gluon distribution at the lower collision energy? Not too promising, unfortunately, as can be seen in Figure 4, which is the same as Figure 2 except that the collision energy is 500 GeV. There is little sensitivity to the top width Γ. The reason is that the top quarks come out with lower energy and the radiation is dominated by that from the b quarks. Results for the configuration corresponding to Figure 3 show similar behavior.

At higher collision energies it appears that the distribution of soft gluon radiation may be sensitive to the top quark width, which suggests this as a method for measuring Γ. This will be challenging, to say the least. The results presented here are at the parton level only, and moreover they are for

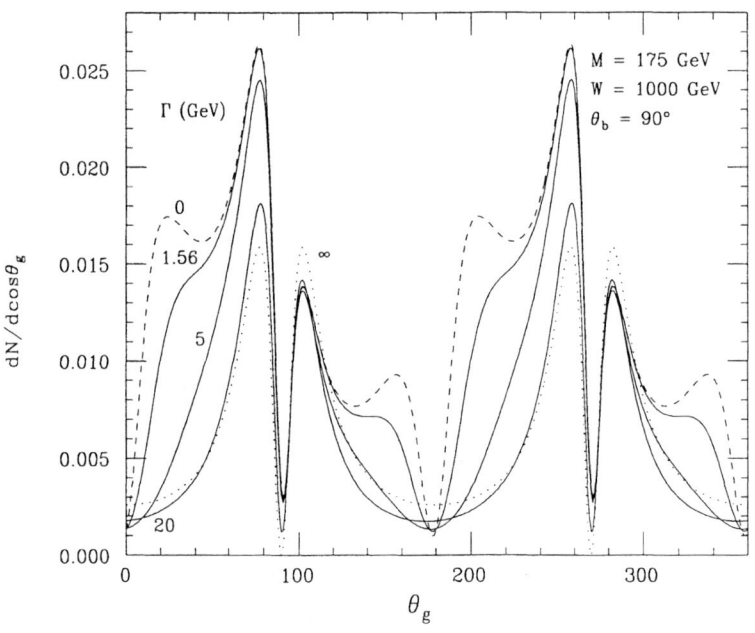

FIGURE 3. Soft gluon distribution in top production and decay at lepton colliders as described in the text, for t's decaying to b's at 90° and collision energy 1 TeV.

fixed kinematic configurations. A more realistic assessment requires a more detailed study of integrated cross sections. The result is most likely to be that the sensitivity leaves something to be desired. However, it will still be worth pursuing such measurements, because simply seeing the interference efects will be interesting. Moreover, the width that appears here is the *total* top width, independent of decay mode, and would serve as a consistency check to compare to other measurements. And it should be remembered that additional top decay modes, for example to supersymmetric particles, are likely to increase the total width, which would make the effects discussed here more easily observed.

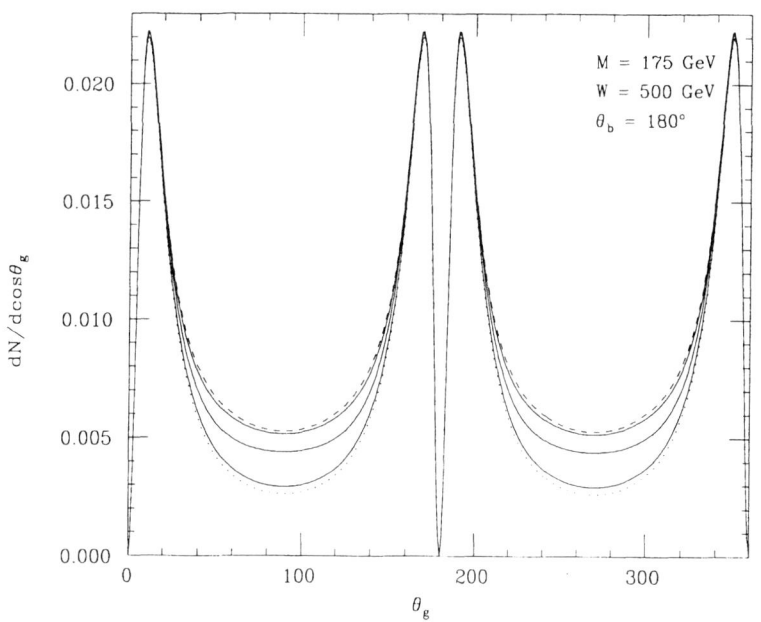

FIGURE 4. Same as Figure 2, except for collision energy 500 GeV.

CONCLUSION

In summary, we have discussed issues associated with gluon radiation in top production and decay at hadron and lepton colliders. We noted that for purposes of gluon radiation, there are no significant differences between muon and electron colliders, since there is no initial state gluon radiation. We presented some gluon distributions at lepton colliders that show interference effects between production- and decay-stage gluon radiation, and that are potentially sensistive to the value of the top quark width. Finally, we note that the clean environment of lepton colliders allows for detailsed experimental studies of QCD effects in top quark physics, both within the Standard Model and beyond.

REFERENCES

1. Orr, L.H., Stelzer, T., and Stirling, W.J., *Phys. Rev. D* **52**, 124 (1995).
2. Orr, L.H., Stelzer, T., and Stirling, W.J., *Phys. Rev. D* **56**, 446 (1997).
3. Berger, M., these proceedings.
4. Khoze, V.A., Orr, L.H., and Stirling, W.J., *Nucl. Phys.* **B378**, 413 (1992).

Top Quark Physics at a Polarized Muon Collider

Stephen Parke

Theoretical Physics Department
Fermi National Accelerator Laboratory
Batavia, IL 60510
USA
e-mail: parke@fnal.gov

Abstract. Top quark pair production is presented at a polarized Muon Collider above the threshold region. The off-diagonal spin basis is the natural basis for this discussion as the top quark pairs are produced in an essentially unique spin configuration for 100% polarization. Modest polarization, say 30%, can lead to 90% of all top quark pair events being in one spin configuration. This will lead to sensitive tests on anomalous top quark couplings.

Recently Parke and Shadmi [1] have shown that at a 100% polarized lepton collider that the top and anti-top quark pairs are produced in essentially a unique spin configuration. This spin basis has been called the "off-diagonal" basis and it interpolates between the beam direction at threshold and the top quark direction, i.e. helicity, far above threshold. The differential cross section using this basis is given by

$$\frac{d\sigma}{d\cos\theta^*}(\mu_L^- \mu_R^+ \to t_\uparrow \bar{t}_\uparrow \text{ or } t_\downarrow \bar{t}_\downarrow) = 0,$$

$$\frac{d\sigma}{d\cos\theta^*}(\mu_L^- \mu_R^+ \to t_\uparrow \bar{t}_\downarrow \text{ or } t_\downarrow \bar{t}_\uparrow) = \left(\frac{3\pi\alpha^2}{8s}\beta\right)$$

$$\left[f_{LL}(1+\beta\cos\theta^*) + f_{LR}(1-\beta\cos\theta^*) \right.$$

$$\left. \pm \sqrt{(f_{LL}(1+\beta\cos\theta^*) - f_{LR}(1-\beta\cos\theta^*))^2 + 4f_{LL}f_{LR}(1-\beta^2)} \right]^2 \quad (1)$$

where the f_{IJ}'s are the sum of the photon and Z-boson couplings corrected for the difference in the propagators. Details of this basis can be found in reference [1]. Figure 1 is the spin components for top quark pair production in the off-diagonal basis for both LR and RL incoming lepton helicities for a $\sqrt{s} = 400$ GeV collider. The sub-leading terms have been amplified by a factor of 100 so it is clear that the dominant configuration makes up more than 99% of the total cross section.

In the helicity basis the differential cross section is [2]

$$\frac{d\sigma}{d\cos\theta^*}(\mu_L^- \mu_R^+ \to t_L \bar{t}_L \text{ or } t_R \bar{t}_R) = \left(\frac{3\pi\alpha^2}{8s}\beta\right)(1-\beta^2)\sin^2\theta^* |f_{LL} + f_{LR}|^2,$$

$$\frac{d\sigma}{d\cos\theta^*}(e_L^- e_R^+ \to t_R \bar{t}_L \text{ or } t_L \bar{t}_R) = \left(\frac{3\pi\alpha^2}{8s}\beta\right)(1\mp\cos\theta^*)^2$$
$$\times |f_{LL}(1\mp\beta) + f_{LR}(1\pm\beta)|^2. \qquad (2)$$

Figure 2 is the corresponding plot for the helicity basis. Here the dominant spin configuration is less than 60% of the total.

In figure 3 I have plotted the dominant spin component's fraction of the total as function of the polarization of the beams,

$$\frac{(1+P^-)(1-P^+)\sigma_{LR\to UD} + (1-P^-)(1+P^+)\sigma_{RL\to UD}}{(1+P^-)(1-P^+)\sigma_{LR}^{tot} + (1-P^-)(1+P^+)\sigma_{RL}^{tot}} \qquad (3)$$

for the off-diagonal basis. Here $P \equiv (N_L - N_R)/(N_L + N_R)$ for the μ^+ and μ^- beams. Two different machines are included, the Muon Collider and the NLC. The Muon Collider is assumed to have equal but opposite polarization for the μ^+ and μ^- beams whereas the NLC is an electron-positron collider with only the electron beam polarized. From these curves a modest amount

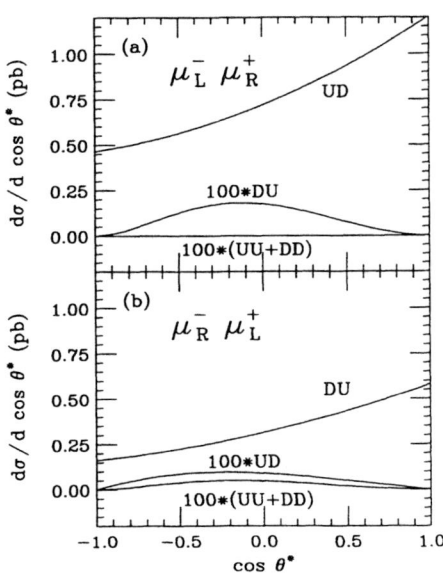

FIGURE 1. The spin configurations using the off-diagonal basis for both $\mu_L^+ \mu_R^-$ and $\mu_R^+ \mu_L^-$ for $\sqrt{s} = 400$ GeV. Note that the sub-leading configurations have been amplified by a factor of 100 in these figures.

of polarization, say 30%, at a Muon Collider can make the dominant spin configuration close to 90% of the total. Whereas at an electron-positron machine one requires 55% polarization to achieve the same goal.

Figure 4 is the corresponding plot for the helicity configuration. If one uses this spin basis the dominant spin configuration ranges from 41% to 52% of the total. Clearly polarization is not as important here without further cuts.

Since the top-quark pairs are produced in a nearly unique spin configuration, and the electroweak decay products of polarized top-quarks are strongly correlated to the spin axis, the top-quark events at $\mu^+\mu^-$ collider have a very distinctive topology. Deviations from this topology would signal anomalous couplings. In the Standard Model, the predominant decay mode of the top-quark is $t \to bW^+$, with the W^+ decaying either hadronically or leptonically. For definiteness we consider here the decay $t \to bW^+ \to be^+\nu$. The differential decay width of a polarized top-quark depends non-trivially on three angles. The first is the angle, χ_w^t, between the top-quark spin and the direction of motion of the W-boson in the top-quark rest-frame. Next is the angle between the direction of motion of the b-quark and the positron in the W-boson rest-frame. We call this angle $\pi - \chi_e^w$. Finally, in the top-quark rest-frame, we have the azimuthal angle, Φ, between the positron direction of motion and the top-quark spin around the direction of motion of the W-boson.

The differential polarized top-quark decay distribution in terms of these three angles is given by

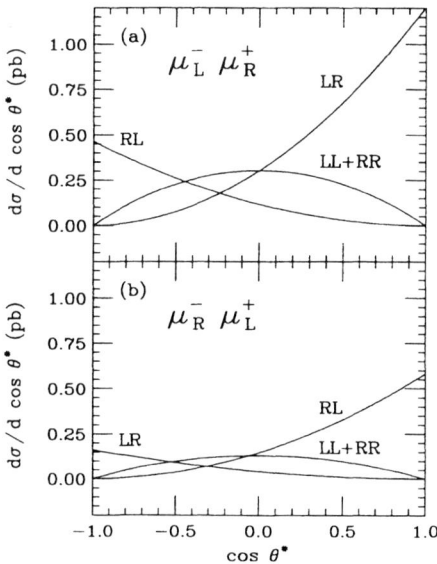

FIGURE 2. The spin configurations using the helicity basis for a $\sqrt{s} = 400\ GeV$ collider.

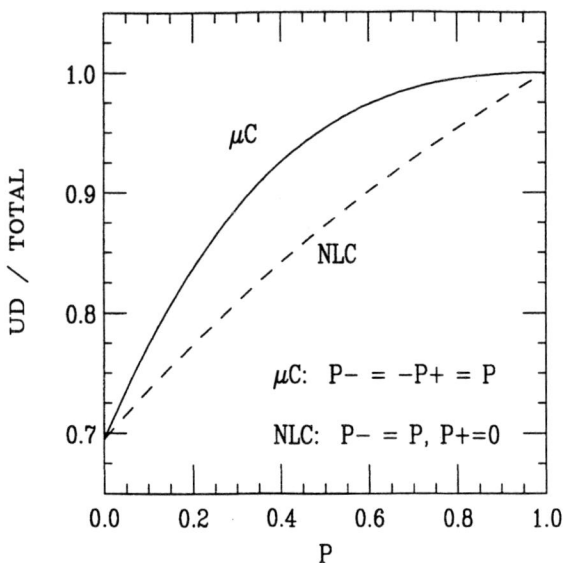

FIGURE 3. Fraction of the total cross section in the off-diagonal basis' Up-Down spin configuration as a function of the polarization. Both beams are assumed to be polarized for the Muon Collider (μC) but only one beam for the NLC.

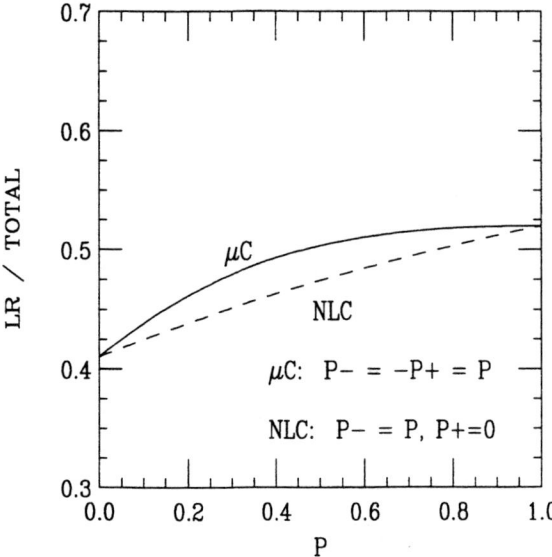

FIGURE 4. Same as Fig. 3 but the helicity basis is used.

$$\frac{1}{\Gamma_T}\frac{d^3\Gamma}{d\cos\chi_w^t\,d\cos\chi_e^w\,d\Phi}=\frac{3}{8\,(m_t^2+2m_W^2)}$$

$$\left[m_t^2(1+\cos\chi_w^t)\sin^2\chi_e^w+m_W^2(1-\cos\chi_w^t)(1-\cos\chi_e^w)^2\right.$$

$$\left.+2m_t m_W(1-\cos\chi_e^w)\sin\chi_e^w\sin\chi_w^t\cos\Phi\right], \quad (4)$$

where m_t is the top-quark mass, m_W is the W mass, and Γ_T is the total decay width (we neglect the b-quark mass). The first and second terms in (4) give the contributions of longitudinal and transverse W-bosons respectively. The interference term, given by the third term in (4), does not contribute to the total width, but its effects on the angular distribution of the top-quark decay products are sizable. Fig. 5 shows contour plots of the differential angular decay distribution in the $\chi_e^w - \chi_w^t$ plane after integrating over the azimuthal angle Φ. The peak at the center of the right hand side of this figure is due to the longitudinal W-bosons whereas the peak at the bottom left hand corner is caused by the transverse W-bosons.

There are also significant correlations of the angle between the top-quark spin and the momentum of the i-th decay product, χ_i^t, measured in the top-quark rest-frame, see figure 6. The differential decay rate of the top-quark is given by

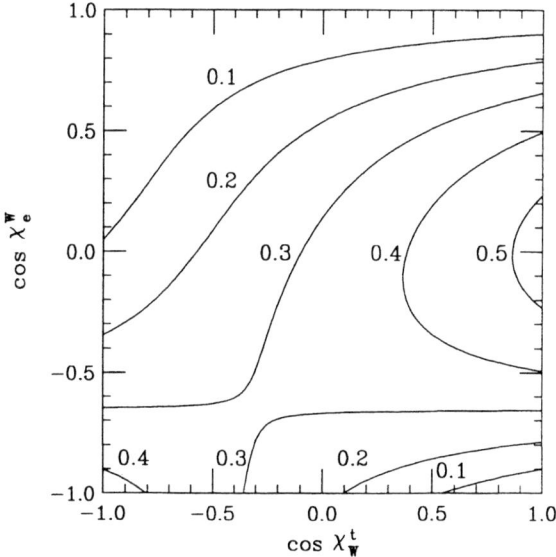

FIGURE 5. Correlations between the W-boson and the top spin direction in the top rest frame ($\cos\chi_W^t$) and the positron and the minus b-quark direction in the W-boson rest frame ($\cos\chi_e^W$).

$$\frac{1}{\Gamma_T} \frac{d\Gamma}{d\cos\chi_i^t} = \frac{1}{2}\left[1 + \alpha_i \cos\chi_i^t\right], \tag{5}$$

where $\alpha_b = -\alpha_W = -0.41$, $\alpha_\nu = -0.31$ and $\alpha_{e^+} = 1$, for $m_t = 175$ GeV, see ref. [3]. The interference between the longitudinal and transverse W-bosons is very importnat in determining these correlations. Note, the positron is more highly correlated with the spin of the top quark than its parent the W-boson!

Given that we know the spin configuration of the top quark pairs and the correlations of the top quark decay products there are many correlations studies that can be performed in top quark pair product at a lepton collider looking for anomalous couplings of the top quarks. These studies have been performed for the helicity basis [4] but need to be redone using the superior off-diagonal spin basis.

QCD effects modify this picture in only a minor fashion. The reason being that soft gluons cannot flip the spin of the heavy top quarks. Detailed studies of the effects of one loop calculations show that the dominant spin configuration is still more than 99% of the total even when QCD corrections are included [5].

In conclusion top quark pairs above the threshold region at Muon Colliders are a great place to search for anomalous couplings of the top quark. For $\sqrt{s} < 1\,TeV$ the off-diagonal basis is superior to the helicity basis in describing the events in the simplest possible terms. Polarization of the incoming

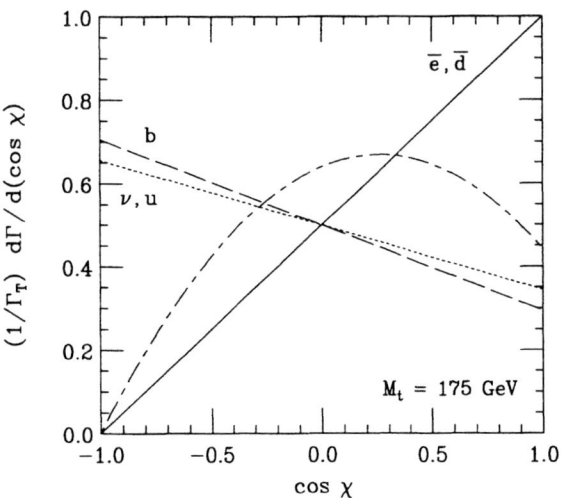

FIGURE 6. The straight lines are the correlations between the top quark decay products and the top quark spin direction in the rest frame of the top ($\cos\chi_i^t$). Whereas the curved line is the correlation between the b-quark and the positron or d-quark in the rest frame of the W-boson ($-\cos\chi_e^W$).

beams enhances this effect. Detailed studies of the one loop QCD corrections have been recently completed, showing no qualitative difference than the tree level analysis. An extensive analysis of the sensitivity to anomalous couplings of the top quark, using the off-diagonal basis, is now needed.

ACKNOWLEDGMENTS

Special thanks to the local organizers of this conference. Fermi National Accelerator Laboratory is operated by the Universities Research Association, Inc., under contract DE-AC02-76CHO3000 with the United States of America Department of Energy.

REFERENCES

1. S. Parke and Y. Shadmi, Phys. Lett. **B387**, 199, (1996); hep-ph/9606419.
2. M.E. Peskin and C.R. Schmidt, in *Physics and Experiments at Linear Colliders*, R. Orava, P. Eeorla and M. Nordberg, eds. (World Scientific, 1992).
3. M. Jeżabek and J.H. Kühn, Phys. Lett. **B329**, 317, (1994).
4. T.L. Barklow and C.R. Schmidt, in *DPF '94: The Albuquerque Meeting*, S. Seidel, ed. (World Scientific, Singapore, 1995);
   C.R. Schmidt Phys. Rev. **D54**, 3250, (1996).
5. J. Kodaira, T. Nasuno and S. Parke manuscript in preparation.
   M. Hori, Y. Kiyo, J. Kodaira, T. Nasuno and S. Parke hep-ph/9801370.

Accelerator

Conveners: Chuck Ankenbrandt, Fermilab
Bob Noble, Fermilab

Space-charge effects of the proposed high-intensity Fermilab booster

King-Yuen Ng and Zubao Qian

Fermi National Accelerator Laboratory,[1] P.O. Box 500, Batavia, IL 60510

Abstract. Space-charge effects on beam stabilities are studied for the proposed two-ring high-intensity Fermilab booster destined for the muon collider. This includes microwave instabilities and rf potential-well distortions. For the first ring, ferrite insertion is suggested to cancel the space-charge distortion of the rf wave form. To control the inductance of the ferrite during ramping and to minimize resistive loss, perpendicular biasing to saturation is proposed.

I INTRODUCTION

The proposed future Fermilab booster has been designed to be used as proton driver for the 50-50 GeV 5×10^{33} cm^{-2}s^{-1} luminosity muon collider also. To accomplish this, the booster should be able to deliver 2 bunches of protons, 5×10^{13} particles each, at 16 GeV with an rms length of ~1 ns and at a repetition rate of 15 Hz [1]. To lower the cost of the rf system, the booster is divided into 2 rings. The lower-energy ring accelerates protons from kinetic energy 1 to 4.5 GeV, has an rf harmonic of 2, a very large aperture, and a small circumference. The higher-energy ring accelerates protons of shorter bunch length up to 16 GeV, has a shorter rf wavelength, and a smaller aperture. A bunch rotation is performed to obtain a shorter bunch of rms length 1 ns just before extraction. Thus two bunch rotations can be performed with the two-ring system if necessary. Some designed parameters [2] for the two rings at injection are listed in Table 1. Laslett tune shifts at injection are given by [3]

$$\Delta\nu = -\frac{3hN_b r_p}{2\gamma^2 \beta \epsilon_{N95} B} = \begin{cases} -0.393 & \text{1st ring} \\ -0.388 & \text{2nd ring}, \end{cases} \quad (1.1)$$

where r_p is the classical proton radius, N_b the number per bunch, h the rf harmonic, ϵ_{N95} the 95% normalized emittance, and B the *bucket* bunching factor which assumes all buckets were filled. Experience at Fermilab [4], Brookhaven [5], and elsewhere tells us that the above tune depressions are practical when all stop bands are minimized. This has been the criterion with which the two rings are designed and the injection energy of the second ring is chosen [2].

Possible microwave instabilities are discussed in Sec. II. The distortion of the rf wave form by the space-charge force will be computed in Sec. III. To compensate this modification, we suggest in Sec. IV to use an insertion of a hollow ferrite cylinder in the beam pipe. The resistive loss in the ferrite is

---

[1] Operated by the Universities Research Association, Inc., under contract with the U.S. Department of Energy.

TABLE 1. Designed parameters for the 2 rings at injection.

	First Ring	Second Ring
Kinetic Energy (GeV)	1.0	4.5
Relativistic gamma γ	2.0658	5.7960
Relativistic beta β	0.8750	0.9850
Cycling rate (Hz)	15	15
Circumference, C (m)	180.649	474.203
Rf harmonic, h	2	21
Number of bunches M	2	2
Number per bunch, N_b	5.0×10^{13}	5.0×10^{13}
Bucket bunching factor, B	0.25	0.25
Transition γ_t	7	25
95% bunch area, A (eV-s)	1.0	1.0
95% normalized emittance, ϵ_{N95} (π-m)	200×10^{-6}	240×10^{-6}

computed and turns out to be large and position dependent along the bunch unless the bunch is very long. To control the inductance of the insertion and minimize loss, perpendicular biasing to saturation is proposed.

II MICROWAVE INSTABILITIES

The average beam currents at injection are, respectively, $I_{av} = eMN_b f_0 = 23.27$ and 9.977 Amp for the two rings, where $f_0 = \omega_0/(2\pi) = 1.452$ and 0.6227 MHz are the respective revolution frequencies, $M = 2$ is the number of bunches in each ring, and e is the proton charge. With a bucket bunching factor of $B = 0.25$, the peak currents become $I_{pk} = 93.06$ and 419.0 Amp for the two rings. For a parabolic bunch, the line distribution for particles at time τ ahead of the synchronous particle is

$$\lambda(\tau) = \frac{3eN_b}{4\hat{\tau}}\left(1 - \frac{\tau^2}{\hat{\tau}^2}\right). \quad (2.1)$$

The half bunch lengths are therefore $\hat{\tau} = 64.56$ and 14.34 ns for the two rings, or $\hat{\ell} = 16.94$ and 4.34 m. For a bunch area of 1 eV-s, the half momentum spreads are $\hat{\delta} = 3.322 \times 10^{-3}$ and 4.208×10^{-3}, respectively. We assume an average betatron amplitude of $\langle \beta \rangle = 25$ m and an average dispersion function of $\langle D \rangle = 1.8$ m. The average beam radius is then

$$a \approx \sqrt{\frac{\epsilon_{N95}\langle\beta\rangle}{\gamma\beta} + \left(\langle D\rangle\hat{\delta}\right)^2} = \begin{cases} 5.29 \text{ cm} & \text{1st ring} \\ 3.33 \text{ cm} & \text{2nd ring} \end{cases} \quad (2.2)$$

It is very probable that a beam pipe or vacuum chamber of radius $b = 8$ and 5 cm will be required, respectively, for the first ring and second ring, because the losses in rings with such high intensities should be minimized to less than 0.1%. The longitudinal space-charge impedances of the beam are then

$$\left.\frac{Z_0^\parallel}{n}\right|_{spch} = i\frac{Z_0}{2\gamma^2\beta}\left(1 + 2\ln\frac{b}{a}\right) = \begin{cases} 92.1 \text{ Ohms} & \text{1st ring} \\ 10.3 \text{ Ohms} & \text{2nd ring} \end{cases}, \quad (2.3)$$

where $Z_0 \approx 377$ Ohms is the free-space impedance. The limits of microwave instability driven by a broad-band impedance are given by the Boussard-modified Keil-Schnell criterion [6]:

$$\left|\frac{Z_0^\parallel}{n}\right| < F_\parallel \frac{E|\eta|}{e\beta^2 I_{\rm pk}} \left(\frac{\Delta E}{E}\right)^2_{\rm FWHM} = \begin{cases} 75.3 \text{ Ohms} & \text{1st ring} \\ 12.6 \text{ Ohms} & \text{2nd ring} \end{cases}, \qquad (2.4)$$

where the form factor $F_\parallel \approx 1$ for a parabolic bunch, η is the slippage factor, and the full-width-at-half-maximum energy spread is $(\Delta E/E)_{\rm FWHM} = \sqrt{2}(\widehat{\Delta E}/E)$. Different interpretation of these limits can lead to different stability results. Since both the low-energy and high-energy rings operate below transition, any realistic momentum distribution with a slope continuous at the edges of the distribution will enhance the space-charge side of the stability curve. Thus, the Boussard-modified Keil-Schnell limit is usually too pessimistic, and we think longitudinal microwave instability would not happen for the two rings near injection.

From injection to extraction, the slippage factor changes from -0.2139 to -0.0094 for the first ring, and from -0.0282 to -0.0015 for the second. Near extraction, the Keil-Schnell limits reduce to 1.04 and 0.21 Ohms, for the two rings, while the space charge impedances become $Z_0^\parallel/n = i16.0$ and $i1.2$ Ohms respectively. Thus flexible momentum-compaction lattices are highly recommended so that the γ_t's can be raised during rampings and lowered back to the design values when bunch rotation or extraction is performed.

A broad-band transverse impedance will also drive the transverse microwave instability. Here, again the most important coupling transverse impedance comes from the space-charge force, which gives the contribution

$$Z_1^\perp|_{\rm spch} = i\frac{RZ_0}{\beta^2\gamma^2}\left(\frac{1}{a^2} - \frac{1}{b^2}\right) = \begin{cases} 0.665 \text{ MOhms/m} & \text{1st ring} \\ 0.438 \text{ MOhms/m} & \text{2nd ring} \end{cases}, \qquad (2.5)$$

where R is the average radius of either ring. A similar Boussard-modified Keil-Schnell limit for transverse microwave instability driven by a broad-band impedance centered at the revolution harmonic n is

$$|Z_1^\perp| < F_\perp \frac{4\nu\beta}{eRI_{\rm pk}}(\Delta E)_{\rm FWHM}|(n-\nu)\eta + \nu\xi|, \qquad (2.6)$$

where the form factor $F_\perp \approx 1$ for a parabolic distribution, and ξ is the chromaticity. We use the cutoff harmonics of the beam pipes or vacuum chambers, $n_{\rm cutoff} = 2.405R/b = 987.8$ and 3685, as the central frequencies of the driving impedances for the two rings and $\nu \approx 2.2$ and ~ 12 as the betatron tunes. Then we obtain the limit $|Z_1^\perp| < 4.24$ and 4.87 MOhms/m, which are much larger than the respective space-charge values. However, near extraction, the stability limits are close to the space-charge impedances for both rings, due to the smaller slippage factors.

III POTENTIAL-WELL DISTORTION

Knowing the bunch length and the momentum spread, the synchrotron tune can be computed easily,

$$\nu_s = \frac{|\eta|\hat{\delta}}{\omega_0 \hat{\tau}} = \begin{cases} 0.001207 & \text{1st ring} \\ 0.002113 & \text{2nd ring} \end{cases} \quad (3.1)$$

Without consideration of the force due to space charge and any other impedance, the required rf voltage required to set up the bucket to fit the bunch is

$$V_{\rm rf} = \frac{2\pi\beta^2 E \nu_s^2}{|\eta| h} = \begin{cases} 31.73 \text{ kV} & \text{1st ring} \\ 250.13 \text{ kV} & \text{2nd ring} \end{cases} \quad (3.2)$$

The rf voltage seen by particle at a time advance τ from the bunch center is (head of bunch is $\tau = +\hat{\tau}$)

$$V_{\rm rf}\sin(-h\omega_0\tau) \approx -V_{\rm rf}\left(\frac{3\pi B}{2}\right)\left(\frac{\tau}{\hat{\tau}}\right) = \begin{cases} -37.38\left(\frac{\tau}{\hat{\tau}}\right) \text{ kV} & \text{1st ring} \\ -294.68\left(\frac{\tau}{\hat{\tau}}\right) \text{ kV} & \text{2nd ring,} \end{cases} \quad (3.3)$$

where $B = 0.25$ is the bucket bunching factor and the sinusoidal rf has been linearized, the parabolic longitudinal distribution has been assumed. The negative signs in Eq. (3.3) signify that the synchronous phase or stable fixed point is zero for operation below transition.

However, in our situation of low energy and high intensity, the space-charge force shaping the bunch distribution is large and dominates over that due to other coupling impedance. A particle at time advance τ from bunch center sees a longitudinal electric space-charge field

$$E_{z\,\rm spch} = -\frac{eZ_0}{4\pi\beta^2\gamma^2 c}\left(1 + 2\ln\frac{b}{a}\right)\frac{d\lambda}{d\tau}, \quad (3.4)$$

where $\lambda(\tau)$ is the line density of the bunch which, for a parabolic distribution, is given by Eq. (2.1). The voltage seen per turn is

$$V_{\rm spch} = E_{s\,\rm spch}C = \frac{3\pi I_b}{(\omega_0\hat{\tau})^2}\left|\frac{Z_0^\|}{n}\right|_{\rm spch}\left(\frac{\tau}{\hat{\tau}}\right), \quad (3.5)$$

which gives 29.1 kV at either end of the bunch for the first ring and 14.7 kV for the second ring. Here, $I_b = I_{\rm av}/M$ is the average current per bunch. Comparing Eqs. (3.3) and (3.5), we see that the space-charge distortion of the rf force will be small for the second ring but very large for the first ring. Thus, the rf voltage at injection must be increased to $V_{\rm rf} \approx 56.42$ kV for the first ring in order to cancel the effect of space charge. Another possibility to counteract this space-charge force is to compensate it by using a ferrite insert in the vacuum chamber, which we will study in the next section.

IV FERRITE COMPENSATION

A INDUCTANCE INSERTION

From Eq. (3.5), we see that the effect of the space-charge force on the rf potential can be minimized if the space-charge impedance is canceled by adding an inductance. This idea was first introduced by Neil and Briggs [7], with the wish to mitigate microwave instability, however. If a hollow cylinder of ferrite of length L, inner and outer radii b and d is encircling the beam, an inductive impedance

$$\left.\frac{Z_0^{\|}}{n}\right|_{\text{ind}} = -i\frac{Z_0\omega_0}{2\pi c}\mu L \ln\frac{d}{b}, \quad (4.1)$$

will be introduced, where μ is the relative permeability of the ferrite. For example, with $\mu = 1000$, $b = 8.0$ cm, and $d = 8.8$ cm, a length of $L = 52.96$ cm will be enough to cancel a space-charge impedance of $|Z_0^{\|}/n|_{\text{spch}} = 92.1$ Ohms for the first ring at injection.

B LOSSES

Unfortunately, ferrite of high permeability is often accompanied by high resistive losses. A conventional way to introduce loss is to replace the relative magnetic permeability by $\mu \to \mu' - i\mu''$. However, as is shown in Fig. 1, μ'' is highly frequency dependent, being nearly zero at low frequencies and reaching a maximum μ_R'' at some high frequency $\omega_R/(2\pi)$. It appears that the simplest representation of the ferrite impedance, which is proportional to $\omega\mu$, may be a broad-band resonance

$$\left.Z_0^{\|}(\omega)\right|_{\text{ferrite}} = \frac{R_s}{1 - iQ\left(\dfrac{\omega}{\omega_r} - \dfrac{\omega_r}{\omega}\right)}, \quad (4.2)$$

where Q is the quality factor. The three parameters are to be determined by

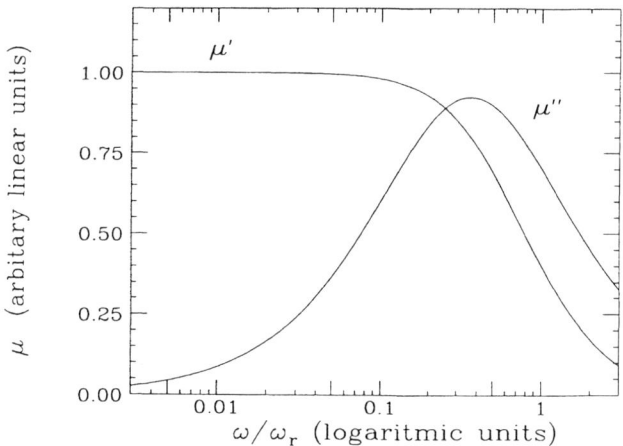

FIGURE 1. A typical plot of μ' and μ'' as functions of frequency.

three measured values of the ferrite, namely μ_R'', ω_R and μ_L', the latter being the value of μ' near zero frequency. Note that ω_R is the frequency at which Z/n attains the maximum and should therefore be slightly smaller than ω_r. However, we just approximate them to be equal here for simplicity. Equating Eq. (4.2) at low frequencies to the inductive part of the ferrite impedance in Eq. (4.1), the shunt impedance R_s is given by

$$R_s = \frac{Q\omega_r}{\omega_0} \left. \left| \frac{Z_0^{\|}}{n} \right| \right|_{\text{ind}}. \tag{4.3}$$

At resonance, the resistive part gives

$$\mathcal{R}e \left. Z_0^{\|}(\omega) \right|_{\text{ferrite}} = R_s = \frac{\omega_r \mu_R''}{\omega_0 \mu_L'} \left. \left| \frac{Z_0^{\|}}{n} \right| \right|_{\text{ind}}. \tag{4.4}$$

Comparing Eqs. (4.3) and (4.4), we obtain $\mu_R'' = Q\mu_L'$. Note that Q here relates the values of μ' and μ'' at *different* frequencies, and is not the usual industrial quoted Q which relates them at the *same* frequency.

We are now in a position to compute the loss. For a bunch distribution $\lambda(\tau)$, the energy *gained* per turn for a particle at time advance τ is

$$\Delta \mathcal{E}(\tau) = -e \int_\tau^\infty d\tau' \, \lambda(\tau') W_0'(\tau - \tau'), \tag{4.5}$$

where $W_0'(\tau)$ is the longitudinal wake function, which is the Fourier transform of the ferrite impedance of Eq. (4.2). For $\tau < 0$, it is given by

$$W_0'(\tau) = \frac{\omega_r R_s}{Q} e^{\alpha \tau} \left(\cos \bar{\omega}\tau + \frac{\alpha}{\bar{\omega}} \sin \bar{\omega}\tau \right), \tag{4.6}$$

where the shifted resonant frequency is $\bar{\omega} = \sqrt{\omega_r^2 - \alpha^2}$. Assuming $\omega_r/(2\pi) \approx 50$ MHz, and $\mu_R'' \approx \mu_L'$ (or $Q \approx 1$), the e-folding length of the wake is $\alpha^{-1} = 2Q/\omega_r = 6.4$ ns which is very much shorter than the length of the first-ring bunch at injection. We can therefore expand $\lambda(\tau')$ as a Taylor series about $\tau' = \tau$. Then, except for the very head of a bunch, the energy gained by a particle at τ becomes

$$\Delta \mathcal{E}(\tau) = e \sum_{n=0} \lambda^{(n)}(\tau) \frac{R_s}{Q \bar{\omega} \omega_r^{n-1}} \sin n\theta, \tag{4.7}$$

where $\lambda^{(n)}(\tau)$ is the n-th derivative of λ with respective to τ, $\sin \theta = \bar{\omega}/\omega_r$ and $\cos \theta = \alpha/\omega_r$. The approximation has been the upper limit of the integration of Eq. (4.5), which should be $\hat{\tau}$ for a finite bunch instead of ∞. For the parabolic distribution, there are only two terms. For particles that are not too close to the head of the bunch, the loss per turn is

$$\Delta \mathcal{E}(\tau) = \frac{e}{\omega_0} \left. \left| \frac{Z_0^{\|}}{n} \right| \right|_{\text{ind}} \left[\lambda'(\tau) + \frac{1}{Q\omega_r} \lambda''(\tau) \right], \tag{4.8}$$

where the Eq. (4.3) has been used. Substituting Eq. (2.1), we obtain

$$\Delta \mathcal{E}(\tau) = -\frac{3e\pi I_b}{(\omega_0 \hat{\tau})^2} \left. \frac{Z_0^{\|}}{n} \right|_{\text{ind}} \left(\frac{\tau}{\hat{\tau}}\right) - \frac{3e\pi I_b}{Q\omega_r \omega_0^2 \hat{\tau}^3} \left. \frac{Z_0^{\|}}{n} \right|_{\text{ind}}. \quad (4.9)$$

The first term is the contribution of the inductance of the ferrite, which cancels the space-charge voltage of Eq. (2.3) if $|Z_0^{\|}/n|_{\text{ind}}$ is chosen to equal to $|Z_0^{\|}/n|_{\text{spch}} = 92.1$ Ohms for the first ring. The second term gives the average loss of energy per particle per turn. When the space-charge force is canceled, this amount to 1.43 keV. The power loss is

$$P = \frac{3\pi I_b^2}{Q\omega_r \omega_0^2 \hat{\tau}^3} \left. \frac{Z_0^{\|}}{n} \right|_{\text{ind}}, \quad (4.10)$$

or 16.7 kw per bunch. The energy loss per particle is small at injection. We note that the loss is inversely proportional to the third power of the bunch length. As the protons are ramped to higher energies, the bunches becomes shorter. For example, when the bunches are prepared for extraction into the second ring, the half bunch length may become 14.34 ns or 4.5 times shorter. However, the longitudinal space-charge impedance will be 5.8 times smaller. Thus the energy loss per particle per turn will be increased by roughly 14 times to 20.2 keV. Usually $\mu_R'' < \mu_L'$ and ω_R may be smaller. Thus, the energy loss per particle can be tremendous. Also for a more accurate integration of Eq. (4.5) in the case of a short bunch, the amount of energy loss will depend on the position along the bunch, making compensation impossible. For this reason, it will be best to reduce the ferrite loss to a minimum, which we shall study next.

C PERPENDICULAR BIAS AT SATURATION

When the beam is ramped from the kinetic energy of 1 GeV at injection in the first ring to the kinetic energy of 4.5 GeV for extraction, the space charge impedance will be reduced by a factor of 8.862. We would like the inductance of the ferrite insertion to decrease by the same factor during the ramp. This can be accomplished by passing a dc bias field through the ferrite. To reduce loss, we suggest that the bias field should be perpendicular to the ac magnetic field generated by the beam particles. This dc biased field can be easily provided by placing a solenoid outside the ferrite cylinder.

One way is to set the dc biased field H_c in the beam- or z-direction so high that the magnetization $\vec{M}$ inside the ferrite is saturated and becomes M_s in the same direction. The ac field $\vec{H}_1$ from beam particles is in the transverse or x-y plane. This will produce an ac magnetization $\vec{M}_1$ which precesses about $\hat{z}H_c$ at the gyromagnetic circular frequency of $\omega_c = \gamma_g H_c$ where $\gamma_g = 2\pi \times 2.80$ MHz/Oersted. This precession creates an ac magnetization $\vec{M}_1$ in the transverse plane. Since the ferrite is at saturation, there will not be any hysteresis loss. Thus, we have

$$\vec{H} = \hat{z}H_c + \vec{H}_1, \qquad \vec{M} = \hat{z}M_s + \vec{M}_1. \quad (4.11)$$

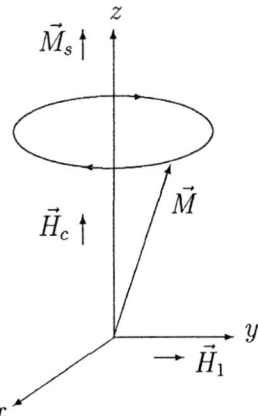

FIGURE 2. System with saturated perpendicular bias H_c in the z-direction. With the application of the ac field $\vec{H}_1$ in the y-direction, the magnetization $\vec{M}$ acquires an ac component in the x-y plane precessing about the z-axis.

The dc biased field is very much larger than the ac field from the beam, or $|\vec{H}_1| \ll H_c$. The system is represented schematically in Fig. 2. The equation of motion is then approximately

$$\frac{d\vec{M}}{dt} = \dot\gamma_g(\hat{z} M_s \times \vec{H}_1 + \vec{M} \times \hat{z} H_c). \tag{4.12}$$

Defining the reversible magnetic susceptibility tensor $\overleftrightarrow{\chi}_r$ as $\vec{M}_1 = \overleftrightarrow{\chi}_r \vec{H}_1$, the stationary solution of Eq. (4.12) is

$$\overleftrightarrow{\chi}_r = \begin{pmatrix} \chi & -j\kappa & 0 \\ j\kappa & \chi & 0 \\ 0 & 0 & 1 \end{pmatrix}, \tag{4.13}$$

where

$$\frac{\chi}{\mu_0} = \frac{\omega_c \omega_m}{\omega_c^2 - \omega^2}, \quad \frac{\kappa}{\mu_0} = \frac{\omega \omega_m}{\omega_c^2 - \omega^2}, \quad \omega_c = \gamma_g H_c, \quad \omega_m = \gamma_g \frac{M_s}{\mu_0}, \tag{4.14}$$

with μ_0 being the magnetic permeability of free space. There is a resonance at the gyromagnetic resonant frequency $\omega_c = \gamma_g H_c$, which is proportional to the dc H_c. This explains why we want H_c to be large so that the resonance effect can be avoided. Then, we obtain from Eqs. (4.14), $\kappa/\chi \approx \omega/\omega_c \ll 1$, implying that the off diagonal elements in $\overleftrightarrow{\chi}_r$ can be neglected.

The merit of this saturated biasing is the low loss, because the ferrite is saturated, there will not be hysteresis loss. The only loss is due to spin-wave propagation which is small. The disadvantage is that μ' is usually small at or above saturation. The loss is usually introduced by the factor α through $\omega_c \longrightarrow \omega_c - i\omega\alpha$ and write $\chi = \chi' - i\chi''$. In our application here, the ac field comes from the beam particles. So ω has the range of the bunch spectrum. For example, with the rms bunch length of 28.9 ns, the bunch frequency $\omega/(2\pi)$

has an rms value of ~ 5.5 MHz. It is more convenient to write χ' and χ'' as functions of ω instead with H_c held constant:

$$\chi' - i\chi'' = \frac{\left(\frac{\omega_m}{\omega_c}\right)\left[1-(1-\alpha^2)\left(\frac{\omega}{\omega_c}\right)^2\right] + i\alpha\left(\frac{\omega_m}{\omega_c}\right)\left(\frac{\omega}{\omega_c}\right)\left[1+(1+\alpha^2)\left(\frac{\omega}{\omega_c}\right)^2\right]}{\left[1-(1+\alpha^2)\left(\frac{\omega}{\omega_c}\right)^2\right]^2 + 4\alpha^2\left(\frac{\omega}{\omega_c}\right)^2}.$$

As an example, we choose Ferramic Q-1, which has a saturated flux density $B_s = 3300$ Gauss at $H_c = 25$ Oersted. Thus, the saturated magnetization is $M_s = 3275$ Gauss. At injection, we bias at $H_c = 25$ Oersted, which gives a resonant frequency of $\omega_c/(2\pi) = 70$ MHz, which is very much larger than the bunch spectrum spread. We see from Fig. 3 that up to 15 MHz, $\mu' \sim M_s/H_c = 131$. With the ferrite cylinder of inner and outer radii 8 and 10 cm, a length of $L = 2.96$ m is required to cancel $|Z_0^{\parallel}/n|_{\text{spch}} = 92.1$ Ohms. At extraction, μ' must be reduced to $131/8.862 = 14.78$. The biased field should therefore be raised to $H_c = M_s/\mu' = 221.6$ Oersted.

At low frequencies, the loss is $\mu'' \longrightarrow \alpha\omega\omega_m/\omega_c^2$. Taking a typical value of $\alpha = 0.05$, we find μ'' varies linearly from 0 and reaches 1.5 at 15 MHz when $H_c = 25$ Oersted at injection. This is illustrated in Fig. 3. At extraction, the loss is reduced by a factor of $8.862^2 = 74.30$ when $H_c = 221.6$ Oersted. Note that $\chi' - i\chi''$, when multiplied by the proper factor to become an impedance, is in the form of the resonant impedance given by Eq. (4.2) with $Q \approx 1/(2\alpha) \approx 10$ and $\omega_r/(2\pi) \approx \omega_c(2\pi) \approx 70$ MHz. Therefore, the loss per particle per turn at injection or extraction, according to Eq. (4.9), will be about 14 times less than the corresponding values quoted in Sec. IV.B.

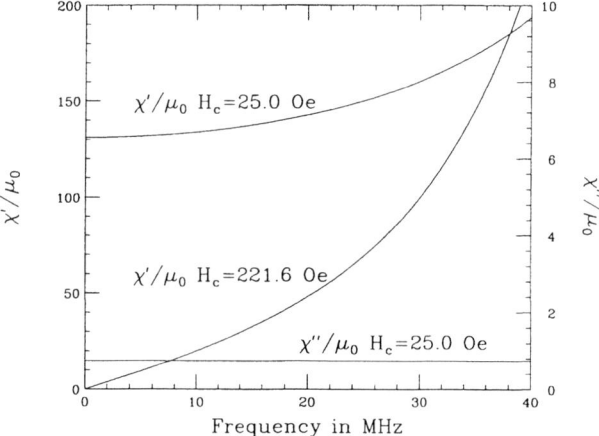

FIGURE 3. The real part of the ac susceptibility of the ferrite as functions of the beam frequency for the dc bias field 25.0 Oersted at injection and 221.6 Oersted at extraction. The imaginary part is shown only at 25.0 Oersted because it is too small to show at extraction.

V CONCLUSION

We have studied the single-bunch instabilities of the two rings of the proposed high-intensity Fermilab booster, and found that the bunches are stable against longitudinal and transverse microwave instabilities near injection, but can be unstable near extraction. The flexible momentum-compaction lattices are recommended, so that the γ_t's can be tuned larger during ramping to assure stability, and are brought down again only when the bunches are rotated at extraction.

For the first ring, the space-charge force will modify the rf waveform by very much, and a ferrite insertion is suggested so that the inductance can compensate the space-charge force. In order to control the ferrite induction during ramping and minimize resistive loss, perpendicular biasing at or above saturation is proposed. In this way, the gyromagnetic resonance arriving from large bias field will have a high frequency well above the frequency spread of the particle beam. Also the hysteresis loss in the ferrite can be avoided.

It is possible that the bunch areas in both rings will turn out to be 1.5 to 2 eV-s instead of 1 eV-s. However, the bunch lengths must be kept to the designed values in order that a 1 ns bunch can be delivered. In other words, the space-charge force will not be changed. If the bunch area is S times larger, the momentum spreads will be S times larger, which will certainly help in combating microwave instabilities. The rf voltages required without consideration of space charge will be S^2 times larger. This imply that the space-charge distortions of the rf potentials will be less severe. Nevertheless, the energy or power lost to the ferrite will remain unchanged.

REFERENCES

1. The $\mu^+\mu^-$ Collider Collaboration, $\mu^+\mu^-$ Collider: A Feasibility Study, June 18, 1996, BNL-52503, Fermilab-Conf.-96/092, LBNL-38946.
2. C. Ankenbrandt, private communication.
3. J. Laslett, Proc. 1963 Summer Study on Storage Rings, BNL-7534, p.324, 1963.
4. C. Ankenbrandt and S. Holmes, Proc. 1987 IEEE Part. Accel. Conf., Washington, DC, 1987, p.1066.
5. W.C. Weng, private communication.
6. E. Keil and W. Schnell, CERN Report TH-RF/69-48 (1969); V.K. Neil and A.M. Sessler, Rev. Sci. Instr. **36**, 429 (1965); D. Boussard, CERN Report Lab II/RF/Int./75-2 (1975).
7. V.K. Neil and R.J. Briggs, Plasma Physics **9**, 631 (1966).

High-intensity Muon Storage Rings for Neutrino Production: Lattice Design

C. Johnstone[1]

Fermi National Accelerator Laboratory[1] *P.O. Box 500, Batavia, Illinois 60510*

Abstract. Five energies, 250, 100, 50, 20, and 10 GeV, have been explored in the design of a muon storage ring for neutrino-beam production. The ring design incorporates exceptionally long straight sections with large beta functions in order to produce an intense, parallel neutrino beam *via* muon decay. To emphasize compactness and reduce the number of muon decays in the arcs, high-field superconducting dipoles are used in the arc design.

INTRODUCTION

The very intense muon sources being considered in current muon collider design studies are also an intense source of neutrinos due to muon decay. As first noted in [1], if the muons are stored in a ring with a long straight section, an intense neutrino beam is realizable. In this paper designs for such muon storage rings are discussed.

Since the source beam is retrieved and recirculated, muon storage rings are capable of generating very intense, uncontaminated neutrino beams in long field-free regions or straight sections. For acceleration and storage, muons are favored over pions because of their enhanced lifetime (factor of 100). The secondary neutrinos produced take on the characteristics of the parent muons, albeit with an increase in divergence given by the decay angle, p/m_μ. The emittance of the muon beam combined with the design optics therefore completely determine neutrino beam properties. The fact that bremsstrahlung may be ignored means that muons can be manipulated similar to protons of the same momentum, making lattice attributes and performance resemble proton machines. The rest of this paper is devoted to a discussion of simple lattices designed specifically to enhance neutrino production.

---

[1] Presented at the Workshop on Physics at the First Muon Collider and Front End of a Muon Collider, Fermilab, November 6-9, 1997. This work was performed at the Fermi National Accelerator Laboratory, which is operated by Universities Research Association, under contract DE-AC02-76CH03000 with the U.S. Department of Energy.

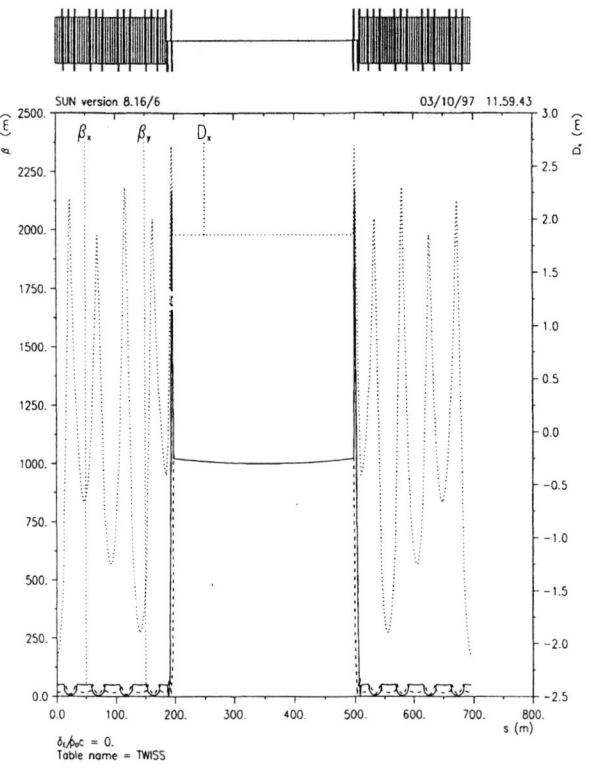

FIGURE 1. 250-GeV half-ring design.

RING DESIGN CRITERIA

Muon storage rings were designed for five possible energies: 250 100, 50, 20, and 10 GeV. As can be seen in Fig. 1 and Fig. 2, the basic ring design follows a simple racetrack layout. All rings contain superconducting dipoles with 8T poletip fields. For normal conducting dipoles, ring energies are reduced by a factor of 5 to 6; i.e. the energy of 10-GeV ring design becomes 1.5 GeV.

The use of superconducting dipoles is essential in a low-energy muon storage ring to minimize arc length and therefore the number of muon decays in the arcs. Respective arc lengths in the 250-GeV design and 10-GeV design are 400 m and 60 m. Initially the neutrino beam was directed vertically downward in both long straights at a 45° angle with arcs on opposite sides of the ring at different altitudes. Adding the needed 180° of vertical bend to align both

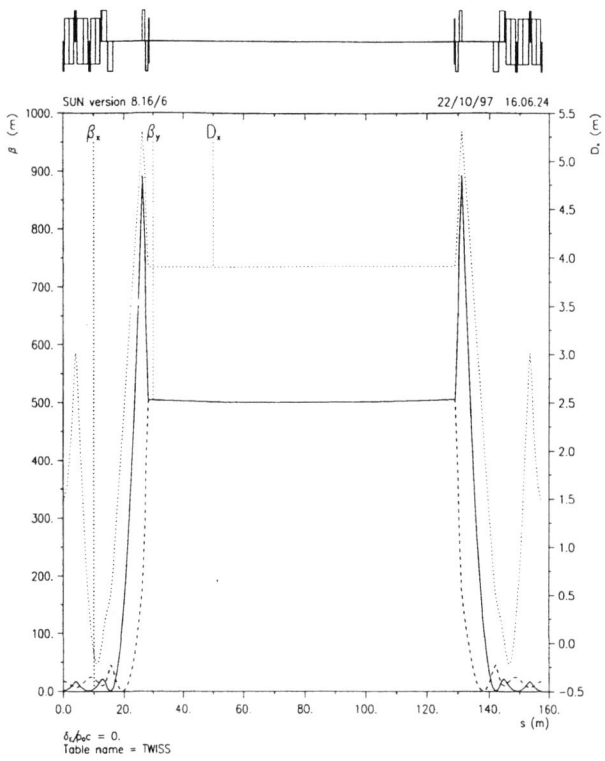

FIGURE 2. 10-GeV half-ring design.

arcs in a horizontal plane increased the circumference by more than 50% and compromised the compactness of the ring. A better approach is to tilt these tightly compacted rings vertically. Since the highest energy ring is 111 m in radius, a 51° pitch means the height changes by ±86 m at 250 GeV. For 10 GeV it is ±20 m. In either approach, one side of the ring produces a neutrino beam directed upwards so that a restricted radiation area will likely be needed where the upward neutrino beam emerges. The downward neutrino beam can be carefully aimed through the earth and onto a remote detector by measuring and controlling the trajectory of the parent muon beam [1].

The long production straights range in length from 300 m at 250 GeV decreasing to 100 m at 10 GeV and have a high beta consistent with the desired neutrino-beam properties; that is, the intrinisic divergence of the muon beam must be much less than the decay angle so that the decay kinematics dominate [1]. Initially, high-betas of 500 m to 1 km were chosen so that the muon beam emittance could be very large yet meet the divergence criteria in the long production straights. (Large muon emittances imply less cooling upstream of the storage ring.) With a beta of 1 km and a normalized rms emittance of $2000\pi\ mm-mr$, the divergence of a 250-GeV muon beam is only 0.03 mr, which is to be compared with the 0.4 mr decay angle. At 10 GeV, the decay angle is 10 mr. For the same normalized emittance, a high-beta of only 40 m gives a divergence which is much less then the decay angle in comparison–0.72 mr. (The decay angle scales as $1/\gamma$ while the divergence scales as the square root of $1/\gamma$.) The above parameters were kept consistent with 15-cm half-aperture quadrupoles which will be required to match the arcs into the production straights. In the following section the optical properties of the lattice will be described.

LATTICE OPTICS

As stated, the ring lattice is a simple racetrack with a doublet forming the high-betas in the long straights. Triplet quadrupole structures would remove the spike in the beta functions and will likely replace the simple doublets in future lattices. The highest energy lattices, the 250 and 100 GeV employ flexible momentum compaction (FMC) modules in the arcs and this parameter can be controlled. The lattice design reverts to a simple FODO structure at the other energies when the dipole becomes too strong for a short FMC module. Although the dipoles have 8T poletip fields nominally in all designs for compactness, the quadrupoles in the lower energy rings are normal conducting when they became too short.

The low beta values in the arcs (20-30 m) imply no significant natural chromaticity. The high-beta inserts, however, will require some local chromatic correction using sextupoles.

CONCLUSIONS

A muon storage ring (without a low-beta interaction region) is straightforward and similar in design to proton machines with the same momentum. The quoted normalized emittance acceptance appears drastically high for muons because of their small mass when compared with protons. Beam sizes are, however, comparable to proton beams. The lattices presented in this paper are designed for maximum compaction in the arcs and large transverse acceptance (the equivalent normalized rms acceptance for a proton beam would be $200\pi\ mm-mr$). The acceptance is ultimately constrained by the large apertures required for the high-beta insert quadrupoles. Unless significant field errors are present, the arcs will not require chromatic correction. Local chromatic correction of the high-beta insert will be required for good momentum acceptance ($\sim \pm 0.5\%$).

REFERENCES

1. S. Geer, Neutrino Beams from Muon Storage Rings: Characteristics and Physics Potential, Fermilab-PUB-97/389 (hep-ph/9712290), submitted to Phys. Rev. D.

Appendices

Group photograph showing a subset of the people attending the Conference.

LIST OF PARTICIPANTS

Albright, Carl H.	Northern Illinois University
Anderson, Greg	Northwestern University
Ankenbrandt, Chuck	Fermi National Accelerator Laboratory
Appel, Jeffrey A.	Fermi National Accelerator Laboratory
Arisaka, Katsushi	University of California, Los Angeles
Atac, Muzaffer	Fermi National Accelerator Laboratory
Bachman, Mark	University of California, Irvine
Balazs, Csaba	Michigan State University
Bardeen, William A.	Fermi National Accelerator Laboratory
Barger, Vernon	University of Wisconsin, Madison
Baur, Ulrich	State University of New York, Buffalo
Berger, Edmond	Argonne National Laboratory
Berger, Mike	Indiana University
Bernstein, Robert	Fermi National Accelerator Laboratory
Bhat, Pushpalatha	Fermi National Accelerator Laboratory
Blondel, Alain	CERN
Bolton, Tim	Kansas State University
Borzumati, Francesca	Universitat Zurich
Bowser-Chao, David	University of Illinois, Chicago
Brennan, Mike	Brookhaven National Laboratory
Buckley-Geer, Elizabeth	Fermi National Accelerator Laboratory
Burdman, Gustavo	University of Wisconsin, Madison
Burnstein, Ray	Illinois Institute of Technology
Caldwell, Allen	Columbia University
Campagnari, Claudio	University of California, Santa Barbara
Carabello, Steve	Purdue University
Carena, Marcela	Fermi National Accelerator Laboratory
Chakraborty, Dhiman	Fermi National Accelerator Laboratory
Cheng, Hsin-Chia	Fermi National Accelerator Laboratory
Cline, David	University of California, Los Angeles
Conrad, Janet	Fermi National Accelerator Laboratory
Cooper, Martin	Los Alamos National Laboratory

Cooper, William	Fermi National Accelerator Laboratory
Czarnecki, Andrzej	Brookhaven National Laboratory
de Barbaro, Lucyna	Fermi National Accelerator Laboratory
Demarteau, Marcel	Fermi National Accelerator Laboratory
Diehl, Thomas	Fermi National Accelerator Laboratory
Djilkibaev, Rashid	University of California, Irvine
Dobrescu, Bogdan	Fermi National Accelerator Laboratory
Dominici, Daniele	Universita' di Firenze
Drucker, Robert	Fermi National Accelerator Laboratory
Duchovni, Ehud	Weizmann Institute
Eichten, Estia	Fermi National Accelerator Laboratory
Ellis, Keith	Fermi National Accelerator Laboratory
Erler, Jens	University of California, Santa Cruz
Feng, Jonathan	Lawrence Berkeley National Laboratory
Fernow, Richard	Brookhaven National Laboratory
Fisher, Peter	Massachusetts Institute of Technology
Flattum, Eric	Fermi National Accelerator Laboratory
Flaugher, Brenna	Fermi National Accelerator Laboratory
Franklin, Gregg	Carnegie Mellon University
Fukui, Yasuo	Fermi National Accelerator Laboratory
Gallardo, Juan C.	Brookhaven National Laboratory
Gallas, Elizabeth	University of Texas, Arlington
Garren, Al	Lawrence Berkeley National Laboratory
Garvey, Gerald T.	Los Alamos National Laboratory
Geer, Steve	Fermi National Accelerator Laboratory
Goodman, Maury	Argonne National Laboratory
Gottschalk, Erik	Fermi National Accelerator Laboratory
Greening, Thomas	University of Wisconsin, Madison
Gunion, John	University of California, Davis
Gutierrez, Gaston	Fermi National Accelerator Laboratory
Han, Tao	University of Wisconsin
Hanson, Gail G.	Indiana University
Harris, Deborah	Fermi National Accelerator Laboratory
He, Hong-Jian	Michigan State University

Hill, Chris	Fermi National Accelerator Laboratory
Hinchliffe, Ian	Lawrence Berkeley National Laboratory
Hoang, Andre	University of Karlsruhe
Hsiung, Yee Bob	Fermi National Accelerator Laboratory
Hughes, Richard	The Ohio State University
Hungerford, Ed	University of Houston
Johnson, Rolland	
Johnstone, Carol	Fermi National Accelerator Laboratory
Kammel, Peter	Lawrence Berkeley National Laboratory
Kamyshkov, Yuri	Oak Ridge National Laboratory
Kao, Chung	University of Wisconsin, Madison
Kaplan, Daniel	Illinois Institute of Technology
Kawall, David	Yale University
Kayser, Boris	National Science Foundation
Keller, Stephane	Fermi National Accelerator Laboratory
Kelly, James	University of Wisconsin, Madison
King, Bruce	Brookhaven National Laboratory
Kirk, Harold	Brookhaven National Laboratory
Klasen, Michael	Argonne National Laboratory
Kobayashi, Takashi	KEK-IPNS, Tsukuba
Kobilarcik, Tom	Fermi National Accelerator Laboratory
Koltick, David	Purdue University
Kourbanis, Ioanis	Fermi National Accelerator Laboratory
Krakauer, Danny	Argonne National Laboratory
Krawczyk, Maria	Warsaw University
Kuno, Yoshitaka	KEK-IPNS
Landsberg, Greg	Fermi National Accelerator Laboratory
Lane, Kenneth	Boston University
Lebrun, Paul	Fermi National Accelerator Laboratory
Leeson, William	Argonne National Laboratory
Littenberg, Laurence	Brookhaven National Laboratory
Liu, Tingjun	University of California, Irvine
Lykken, Joseph	Fermi National Accelerator Laboratory
Ma, Hong	Brookhaven National Laboratory

Mackenzie, Paul	Fermi National Accelerator Laboratory
Magill, Stephen	Argonne National Laboratory
Malensek, Anthony	Fermi National Accelerator Laboratory
Marciano, William	Brookhaven National Laboratory
Marshall, Glen M.	TRIUMF
Matchev, Konstantin	Fermi National Accelerator Laboratory
McDonald, Kirk	Princeton University
McFarland, Kevin	Fermi National Accelerator Laboratory
Miller, David J.	University College London
Mo, Luke	Virginia Tech
Mohapatra, Rabi	University of Maryland
Mokhov, Nikolai	Fermi National Accelerator Laboratory
Molzon, William	University of California, Irvine
Moretti, Al	Fermi National Accelerator Laboratory
Morse, William	Brookhaven National Laboratory
Mrenna, Stephen	Argonne National Laoratory
Nachtman, Jane	University of California, Los Angeles
Naples, Donna	Kansas State University
Neuffer, David	Fermi National Accelerator Laboratory
Ng, King-Yuen	Fermi National Accelerator Laboratory
Noble, Robert	Fermi National Accelerator Laboratory
Norem, Jim	Fermi National Accelerator Laboratory
Norman, Douglas	Fermi National Accelerator Laboratory
Norton, Peter	Rutherford Appleton Laboratory
Olness, Fred	Fermi National Accelerator Laboratory
Orr, Lynne	University of Rochester
Paige, Frank	Brookhaven National Laboratory
Palmer, Robert	Brookhaven National Laboratory
Para, Adam	Fermi National Accelerator Laboratory
Parke, Stephen	Fermi National Accelerator Laboratory
Paschos, Emmanuel A.	University of Dortmund
Pauletta, Giovanni	Fermi National Accelerator Laboratory
Pawlik, Bogdan	Institute for Nuclear Physics
Pedrini, Daniele	Fermi National Accelerator Laboratory

Peoples, John	Fermi National Accelerator Laboratory
Peyaud, Bernard	DAPNIA, Saclay
Polonsky, Nir	Rutgers University
Popovic, Milorad	Fermi National Accelerator Laboratory
Porod, Werner	University of Vienna
Protopopescu, Serban	Brookhaven National Laboratory
Qian, Zubao	Fermi National Accelerator Laboratory
Quigg, Chris	Fermi National Accelerator Laboratory
Raby, Stuart	Ohio State University
Raja, Rajendran	Fermi National Accelerator Laboratory
Raychaudhuri, Sreerup	CERN
Reay, Neville W.	Kansas State University
Reina, Laura	University of Wisconsin, Madison
Ritchie, Jack	University of Texas, Austin
Ritz, Steven	Columbia University
Romosan, Alex	Lawrence Berkeley National Laboratory
Roser, Robert	Fermi National Accelerator Laboratory
Rossmanith, Robert	FZ Karlsruhe
Schellman, Heidi	Fermi National Accelerator Laboratory
Schnetzer, Steve	Rutgers University
Sculli, John	New York University
Shadmi, Yael	Fermi National Accelerator Laboratory
Shaevitz, Michael	Columbia University
Smith, Martin	University of Illinois
Spentzouris, Panagiotis	Fermi National Accelerator Laboratory
Stelzer, Tim	University of Illinois
Stern, Eric	Fermi National Accelerator Laboratory
Striganov, Sergei	Fermi National Accelerator Laboratory
Sullivan, Zack	University of Illinois
Summers, Don	University of Mississippi
Swallow, Earl	Elmhurst College
Tayloe, Rex	Los Alamos National Laboratory
Thomson, Gordon	Rutgers University
Tikhonin, Feodor	Institute for High Energy Physics

Tollestrup, Alvin	Fermi National Accelerator Laboratory
Tschirhart, Robert	Fermi National Accelerator Laboratory
Valencia, German	Iowa State University
Vejcik, Steve	Fermi National Accelerator Laboratory
Wagner, Carlos	CERN
Wagner, David	University of Colorado
Wan, Weishi	Fermi National Accelerator Laboratory
White, Christopher	Illinois Institute of Technology
White, D. Hywel	Los Alamos National Laboratory
White, Herman	Fermi National Accelerator Laboratory
White, James	Texas A&M University
Willenbrock, Scott	University of Illinois
Willocq, Stephane	Stanford Linear Accelerator Center
Winer, Brian	Ohio State University
Winstein, Bruce	The University of Chicago
Witten, Edward	Institute for Advanced Study
Womersley, John	Fermi National Accelerator Laboratory
Wu, Guo-Hong	Purdue University
Yagil, Avi	Fermi National Accelerator Laboratory
Yang, Jin Min	Northwestern University
Yu, Jaehoon	Fermi National Accelerator Laboratory

Author Index

A

Abbasabodi, A., 701
Ankenbrandt, C., 3

B

Bachman, M., 318, 460
Bailey, J. M., 427
Barger, V., 107
Bartl, A., 531
Baur, U., 675
Beer, G. A., 427
Berger, M. S., 227, 543, 663, 797
Beveridge, J. L., 427
Bhat, P. C., 208
Blondel, A., 597
Borzumati, F. M., 567
Boshier, M. G., 486
Bowser-Chao, D., 701
Burdman, G., 287, 742

C

Carena, M., 193
Casalbuoni, R., 772
Cheng, H.-C., 561
Cooper, M. D., 443
Czarnecki, A., 409

D

Datta, A., 783
Deandrea, A., 772
De Curtis, S., 772
Demarteau, M., 177
Dicus, D. A., 701
Djilkibaev, R. M., 481
Dobrescu, B. A., 758
Dominici, D., 772

E

Eberl, H., 531
Eichten, E. J., 208, 734
Erler, J., 669

F

Farrar, G. R., 567
Fisher, P., 139
Flattum, E., 525
Franklin, G. B., 82
Fujiwara, M. C., 427

G

Gatto, R., 772
Geer, S., 3, 384
Greening, T., 641
Gunion, J. F., 37, 772
Gutierrez, G., 121

H

Han, T., 177
Handa, T., 766
Harris, D. A., 376, 505
He, H.-J., 685
Hill, C. T., 723
Hoang, A. H., 803
Hosch, M., 783
Huber, T. M., 427
Hughes, V. W., 486
Hungerford, E. V., 437

J

Jacot-Guillarmod, R., 427
Johnstone, C., 851
Jungmann, K., 486

K

Kamal, B., 657
Kammel, P., 419, 427
Kao, C., 647
Kawall, D., 486
Kayser, B., 139
Keller, S., 734
Kelly, J. G., 555
Kim, S. K., 427
King, B. J., 334, 621
Klasen, M., 495
Knowles, P. E., 427
Kobayashi, T., 391
Kotwal, A. V., 398
Kraml, S., 531
Krawczyk, M., 635
Kuno, Y., 261
Kunselman, A. R., 427

L

Lane, K., 711
Langacker, P., 669
Li, C. S., 783
Littenberg, L., 121, 299
Liu, T., 475
Liu, W., 486

M

Ma, H., 308
Maeshima, K., 766
Maier, M., 427
Majerotto, W., 531
Marciano, W. J., 58, 409, 657
Marshall, G. M., 427
Mason, G. R., 427
McFarland, K. S., 376, 505
Melnikov, K., 409
Mohapatra, R. N., 358
Mokhov, N. V., 453
Molzon, W., 152
Mulhauser, F., 427

N

Nachtman, J., 519
Ng, K.-Y., 841

O

Oakes, R. J., 783
Olin, A., 427
Orr, L. H., 823

P

Paige, F. E., 91, 537
Palmer, R. B., 11
Parke, S., 831
Parsa, Z., 657
Paschos, E. A., 370
Petitjean, C., 427
Polonsky, N., 567
Porcelli, T. A., 427
Porod, W., 531
Protopopescu, S., 193

Q

Qian, Z., 841
Quigg, C., 242

R

Raby, S., 573
Raja, R., 583, 789
Reina, L., 813
Repko, W. W., 701
Ritz, S., 327

S

Schaller, L. A., 427
Schellman, H., 166
Sculli, J., 466
Seth, K. K., 274
Spentzouris, P., 66
Stefanski, R., 349
Stern, E. G., 511
Striganov, S. I., 453